Student Solutions Manual to Accompany

Organic Chemistry

Raber&Raber

Prepared by

Douglas J. Raber
University of South Florida

Nancy K. Raber
University of South Florida

with

James W. Leahy
University of South Florida

Ellen M. Salcines
University of South Florida

West Publishing Company
St. Paul New York Los Angeles San Francisco

WEST PUBLISHING CO.
50 W. Kellogg Boulevard
P.O. Box 64526
St. Paul, MN 55164-1003

Printed in the United States of America

ISBN 0-314-39742-6

to Wendy Elizabeth and Harriet Jessica

ACKNOWLEDGEMENT

This Solutions Manual is the result of true collaboration and teamwork among all four authors. It is almost completely computer generated: the text was produced using an IBM 3033 with Waterloo SCRIPT output to a XEROX 3700 laser printer, and chemical structures were generated using Macintosh computers with ChemDraw output to an Apple LaserWriter. Mass Spectra and other figures were generated using Macintosh computers with SuperPaint output to an Apple LaserWriter.

The assistance of the University Computer Services staff at the University of South Florida is gratefully acknowledged. We appreciate the help that has been provided by the editors at West, Pete Marshall and Maralene Bates.

We thank two of our colleagues who provided invaluable assistance in checking the accuracy of the answers: Terence C. Owen of the University of South Florida (who carefully reviewed both Stereochemistry and Carbohydrates) and Anne Richards of Tripos Associates, Inc. (who checked the entire manuscript).

All Mass Spectra are from the EPA/NIH Database, and we thank Sharon Lias (National Bureau of Standards) and William L. Budde (United States Environmental Protection Agency) for their assistance.

CONTENTS

Introduction

We have tried very hard to make this much more than just an "Answer Book". As with the text itself, our goal is to help students learn organic chemistry, and this includes learning how to learn. For this reason a large number of the answers presented in this manual include detailed explanations of the reasoning that can be used to reach the correct answer. In particular, we have always presented detailed discussions and analysis for road map problems, spectroscopic problems, and multistep synthesis problems.

Students should be very careful when using this manual to check whether or not one of their own answers is correct. Often there will be more than one way to correctly answer a question, particularly with synthetic schemes, but time and space have precluded us form putting down all the possibilities. If the answer that we present is not identical to that of the student, it does not mean that the student is wrong. Instead, it means that the students must sometimes take a few moments to verify that their reasoning is similar to (if not the same as) ours.

CHAPTER 1. ALKANES

ANSWERS TO EXERCISES

1.1 (a) (b)

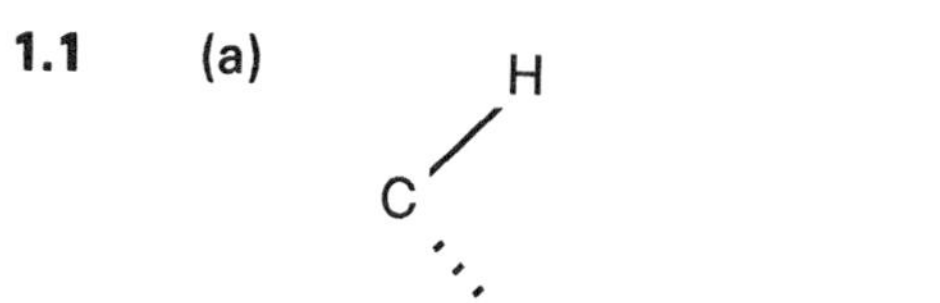

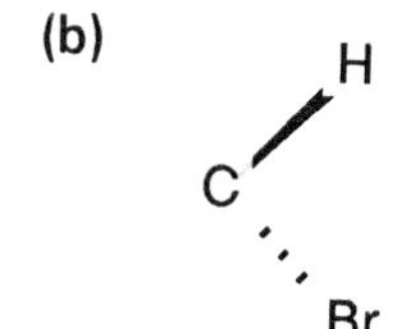

1.2 The second symmetry plane is the plane of the paper:

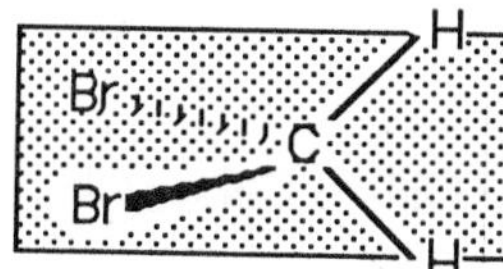

1.3

Cl
C
H
H
H

1.4

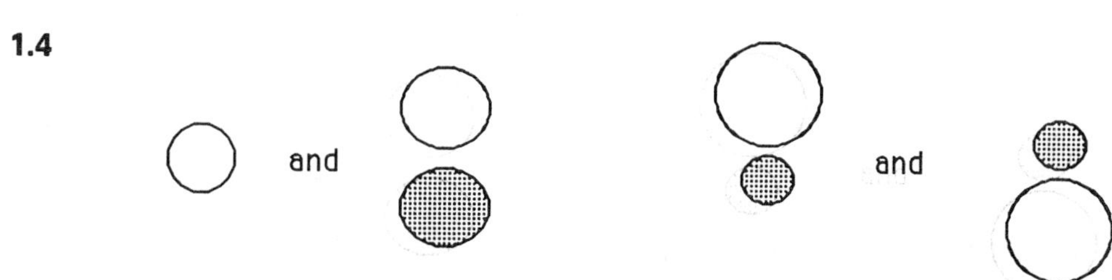

The first of these hybrid orbitals is identical to that shown in Figure 1.9; the second differs in phase but is otherwise the same.

1.5 $CH_3-CH_2-CH_2-CH_2-CH_2-CH_2-CH_2-CH_2-F$ 1-fluorooctane

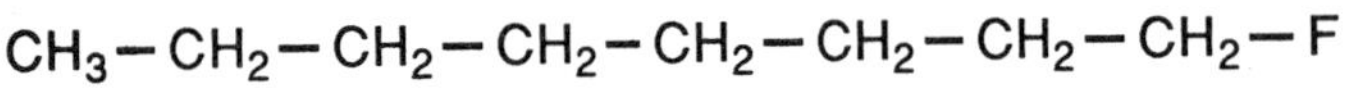

$CH_3-CH_2-CH_2-CH_2-CH_2-CH_2-CH(F)-CH_3$ 2-fluorooctane

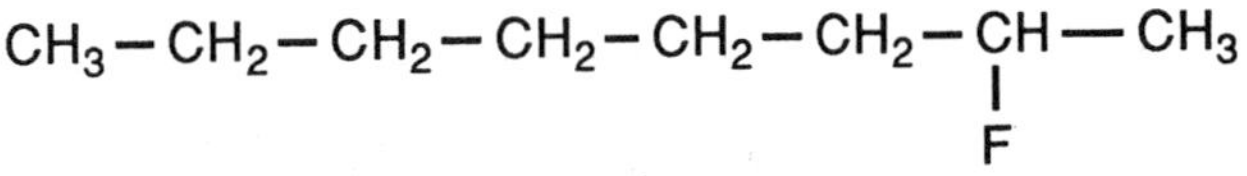

$CH_3-CH_2-CH_2-CH_2-CH_2-CH(F)-CH_2-CH_3$ 3-fluorooctane

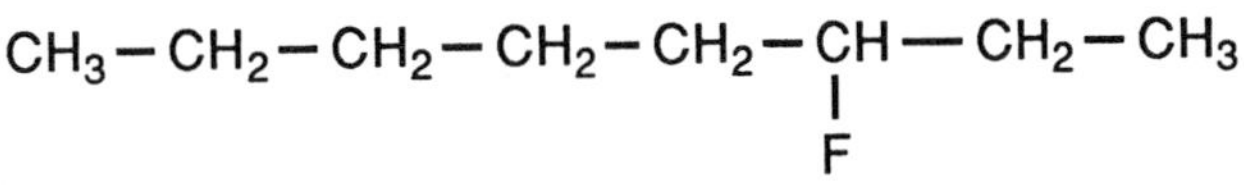

$CH_3-CH_2-CH_2-CH_2-CH(F)-CH_2-CH_2-CH_3$ 4-fluorooctane

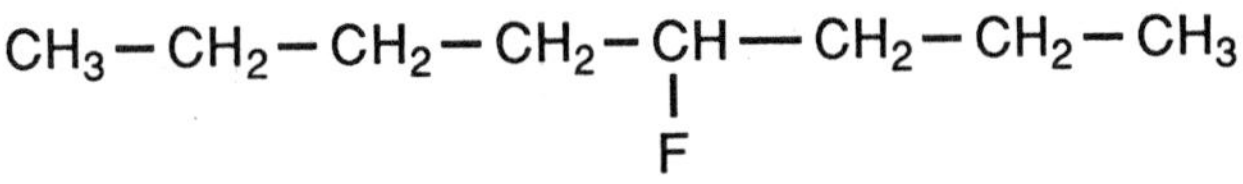

1.6 $CH_3-CH_2-CH_2-CH_2-CH_2-CH_2-CH_2-CH_2-CH_2-CH_2-I$ 1-iododecane

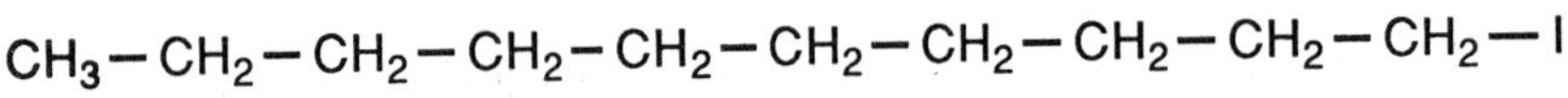

$CH_3-CH_2-CH_2-CH_2-CH_2-CH_2-CH_2-CH_2-CH(I)-CH_3$ 2-iododecane

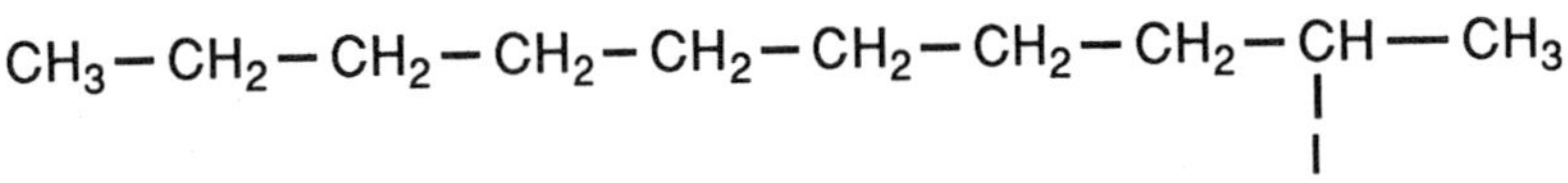

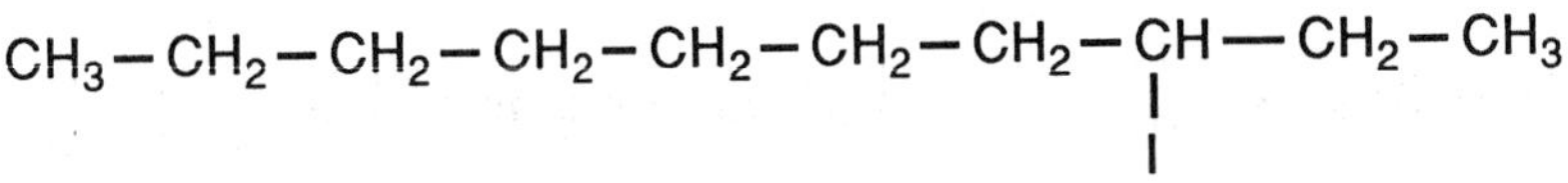

$CH_3-CH_2-CH_2-CH_2-CH_2-CH_2-CH_2-CH(I)-CH_2-CH_3$ 3-iododecane

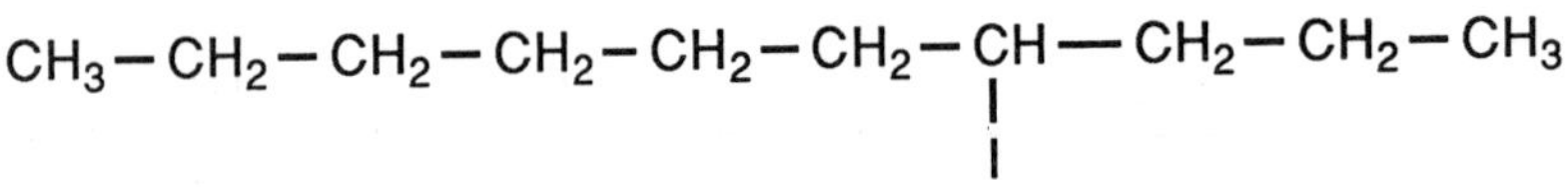

$CH_3-CH_2-CH_2-CH_2-CH_2-CH_2-CH(I)-CH_2-CH_2-CH_3$ 4-iododecane

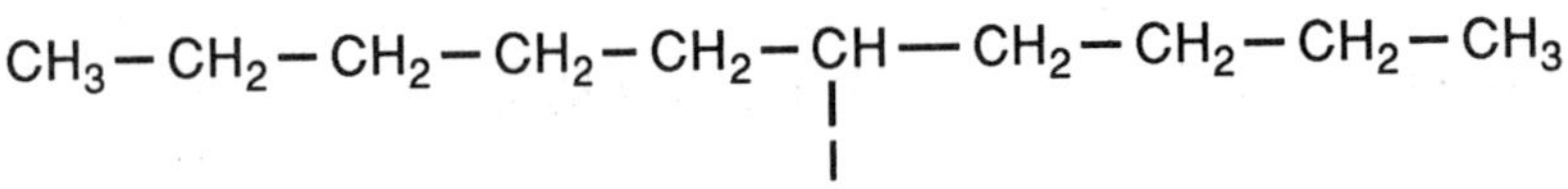

$CH_3-CH_2-CH_2-CH_2-CH_2-CH(I)-CH_2-CH_2-CH_2-CH_3$ 5-iododecane

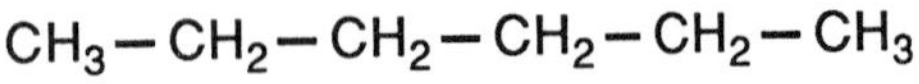

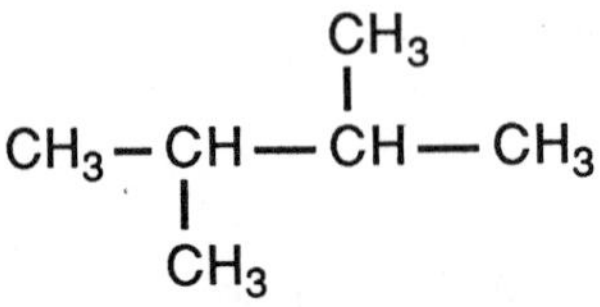

1.7

$CH_3-CH_2-CH_2-CH_2-CH_2-CH_3$

hexane

$CH_3-CH(CH_3)-CH(CH_3)-CH_3$

2,3-dimethylbutane

$CH_3-CH(CH_3)-CH_2-CH_2-CH_3$

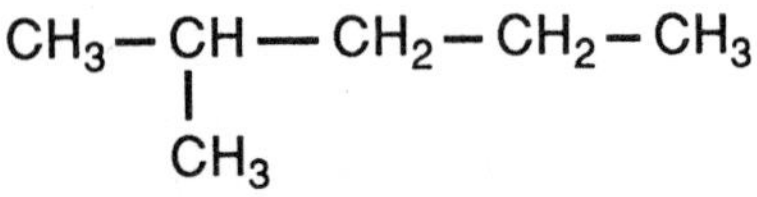

2-methlypentane

$$CH_3-CH_2-\underset{\underset{\displaystyle CH_3}{|}}{CH}-CH_2-CH_3$$

3-methylpentane

$$CH_3-\overset{\overset{\displaystyle CH_3}{|}}{\underset{\underset{\displaystyle CH_3}{|}}{C}}-CH_2-CH_3$$

2,2-dimethylbutane

1.8 $$Cl-CH_2-CH_2-\underset{\underset{\displaystyle CH_3}{|}}{CH}-CH_2-CH_2-CH_3$$

1-chloro-3-methylhexane

$$CH_3-\underset{\underset{\displaystyle Cl}{|}}{CH}-\underset{\underset{\displaystyle CH_3}{|}}{CH}-CH_2-CH_2-CH_3$$

2-chloro-3-methylhexane

$$CH_3-CH_2-\overset{\overset{\displaystyle Cl}{|}}{\underset{\underset{\displaystyle CH_3}{|}}{C}}-CH_2-CH_2-CH_3$$

3-chloro-3-methylhexane

$$CH_3-CH_2-\underset{\underset{\displaystyle CH_2Cl}{|}}{CH}-CH_2-CH_2-CH_3$$

3-(chloromethyl)hexane

$$CH_3-CH_2-\underset{\underset{\displaystyle CH_3}{|}}{CH}-\overset{\overset{\displaystyle Cl}{|}}{CH}-CH_2-CH_3$$

3-chloro-4-methylhexane

$$CH_3-CH_2-\underset{\underset{\displaystyle CH_3}{|}}{CH}-CH_2-\underset{\underset{\displaystyle Cl}{|}}{CH}-CH_3$$

2-chloro-4-methylhexane

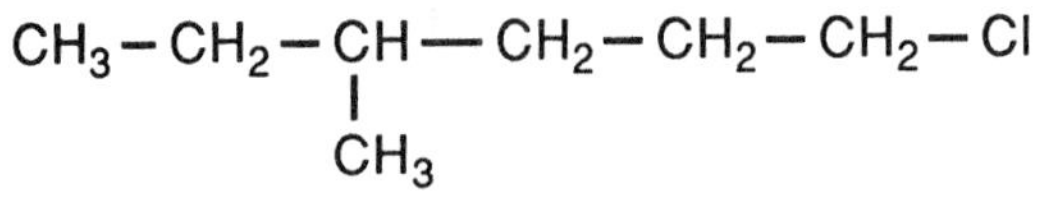

$$CH_3-CH_2-\underset{\displaystyle CH_3}{\underset{|}{CH}}-CH_2-CH_2-CH_2-Cl$$ 1-chloro-4-methylhexane

1.9 (a) (ii)

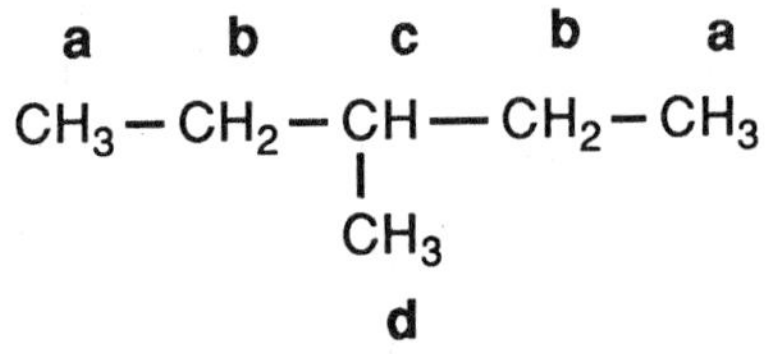

$$\overset{a}{CH_3}-\overset{b}{CH_2}-\underset{\underset{\displaystyle d}{\displaystyle CH_3}}{\underset{|}{\overset{c}{CH}}}-\overset{b}{CH_2}-\overset{a}{CH_3}$$

(iii)

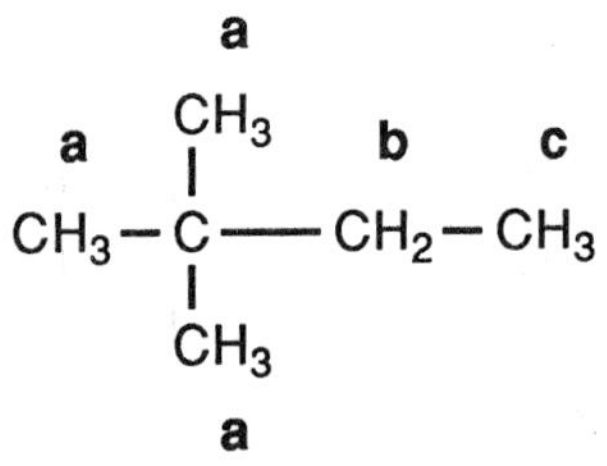

$$\overset{a}{CH_3}-\overset{\overset{\displaystyle a}{\displaystyle CH_3}}{\overset{|}{\underset{\underset{\displaystyle a}{\displaystyle CH_3}}{\underset{|}{C}}}}-\overset{b}{CH_2}-\overset{c}{CH_3}$$

(b) (i)

$$Cl-CH_2-\underset{\displaystyle CH_3}{\underset{|}{CH}}-CH_2-CH_2-CH_3$$

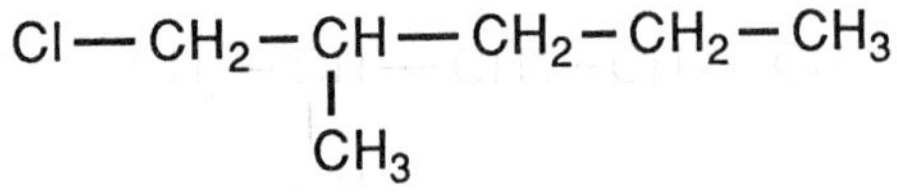

1-chloro-2-methylpentane

$$CH_3-\overset{\displaystyle Cl}{\overset{|}{\underset{\displaystyle CH_3}{\underset{|}{C}}}}-CH_2-CH_2-CH_3$$

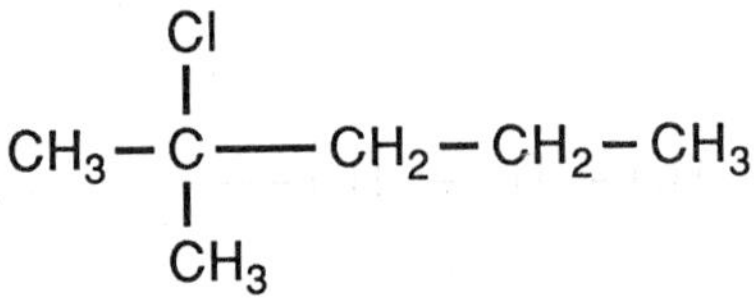

2-chloro-2-methylpentane

$$CH_3-\underset{\displaystyle CH_3}{\underset{|}{CH}}-\overset{\displaystyle Cl}{\overset{|}{CH}}-CH_2-CH_3$$

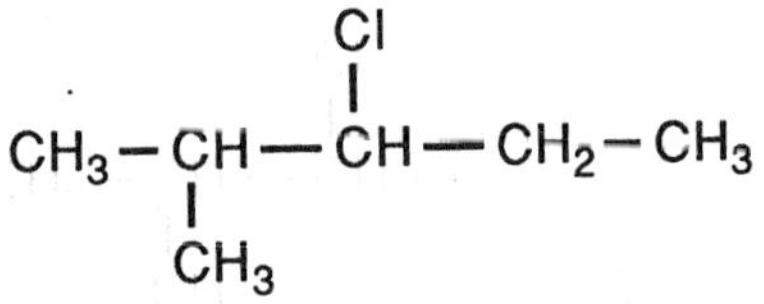

3-chloro-2-methylpentane

$$CH_3-\underset{\displaystyle CH_3}{\underset{|}{CH}}-CH_2-\overset{\displaystyle Cl}{\overset{|}{CH}}-CH_3$$

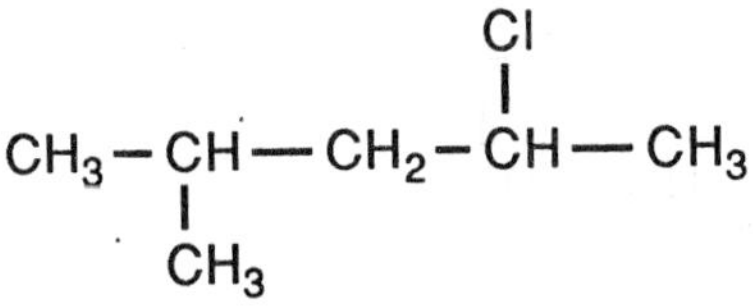

2-chloro-4-methylpentane

$$CH_3-\underset{\displaystyle CH_3}{\underset{|}{CH}}-CH_2-CH_2-CH_2-Cl$$

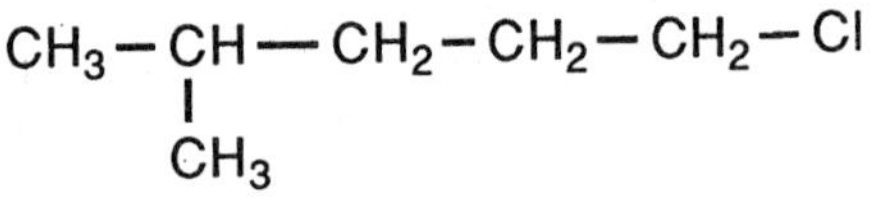

1-chloro-4-methylpentane

(ii)

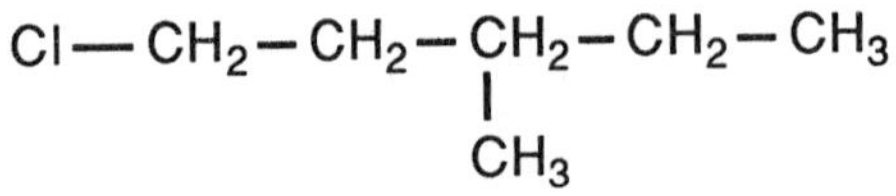

$$Cl-CH_2-CH_2-\underset{\displaystyle CH_3}{\underset{|}{CH_2}}-CH_2-CH_3$$

1-chloro-3-methylpentane

$$CH_3-\overset{\displaystyle Cl}{\overset{|}{CH}}-\underset{\displaystyle CH_3}{\underset{|}{CH}}-CH_2-CH_3$$

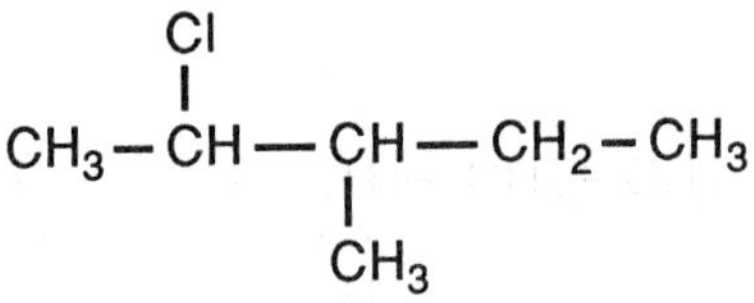

2-chloro-3-methylpentane

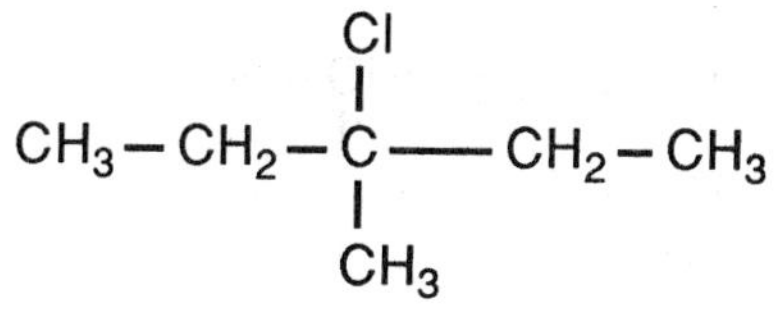

$$CH_3-CH_2-\underset{\underset{CH_3}{|}}{\overset{\overset{Cl}{|}}{C}}-CH_2-CH_3$$

3-chloro-3-methylpentane

(iii)

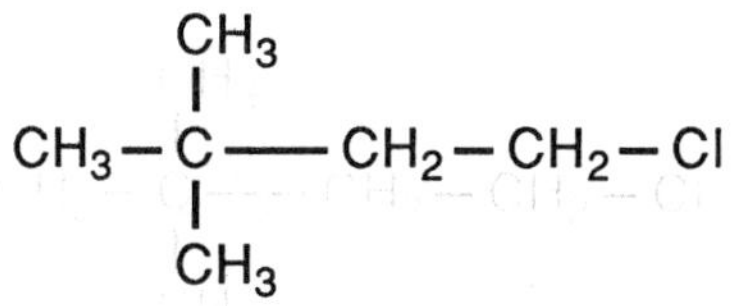

$$CH_3-\underset{\underset{CH_3}{|}}{\overset{\overset{CH_3}{|}}{C}}-CH_2-CH_2-Cl$$

1-chloro-3,3-dimethylbutane

$$CH_3-\underset{\underset{CH_3}{|}}{\overset{\overset{CH_3}{|}}{C}}-\overset{\overset{Cl}{|}}{CH}-CH_3$$

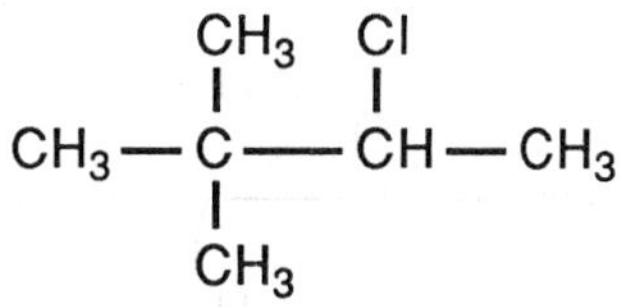

2-chloro-3,3-dimethylbutane

$$Cl-CH_2-\underset{\underset{CH_3}{|}}{\overset{\overset{CH_3}{|}}{C}}-CH_2-CH_3$$

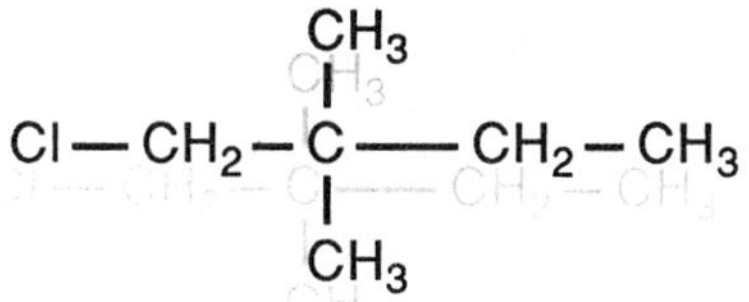

1-chloro-2,2-dimethylbutane

1.10 (a) Two possibilities are:

$$CH_3-CH_3 \quad \text{and} \quad CH_3-\underset{\underset{CH_3}{|}}{\overset{\overset{CH_3}{|}}{C}}-CH_3$$

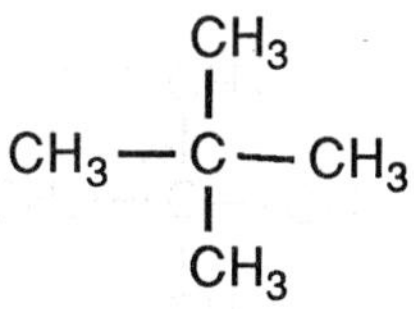

(b) Three examples are:

$$CH_3-CH_2-CH_3 \quad , \quad CH_3-CH_2-CH_2-CH_3 \quad , \quad CH_3-\overset{\overset{CH_3}{|}}{CH}-CH_3$$

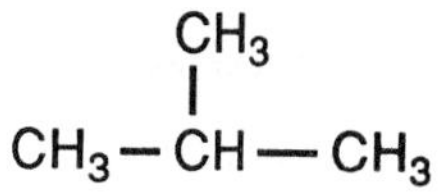

1.11 Carry out the reaction using a large excess of Cl_2 relative to the quantity of methane. When the reaction is complete, separate the CCl_4 from the unreacted, excess Cl_2.

1.12 (i)

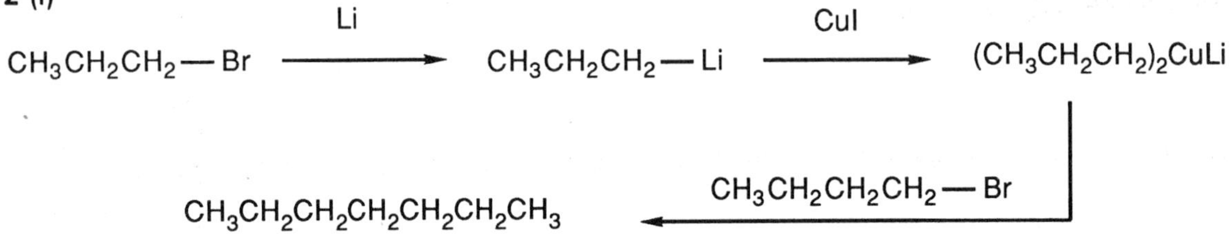

$$CH_3CH_2CH_2-Br \xrightarrow{Li} CH_3CH_2CH_2-Li \xrightarrow{CuI} (CH_3CH_2CH_2)_2CuLi$$

$$(CH_3CH_2CH_2)_2CuLi \xrightarrow{CH_3CH_2CH_2CH_2-Br} CH_3CH_2CH_2CH_2CH_2CH_2CH_3$$

(ii)

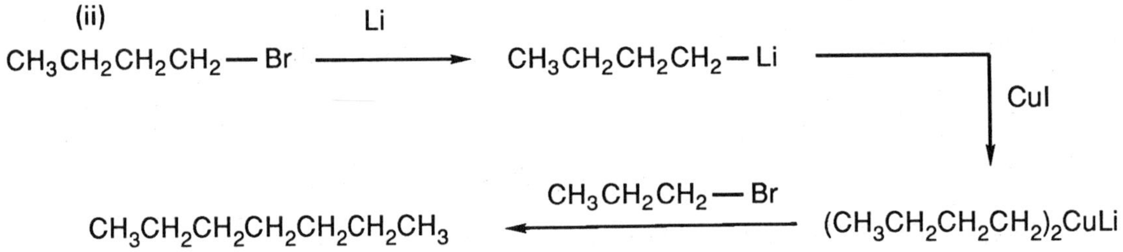

$$CH_3CH_2CH_2CH_2-Br \xrightarrow{Li} CH_3CH_2CH_2CH_2-Li \xrightarrow{CuI} (CH_3CH_2CH_2CH_2)_2CuLi$$

$$(CH_3CH_2CH_2CH_2)_2CuLi \xrightarrow{CH_3CH_2CH_2-Br} CH_3CH_2CH_2CH_2CH_2CH_2CH_3$$

1.13

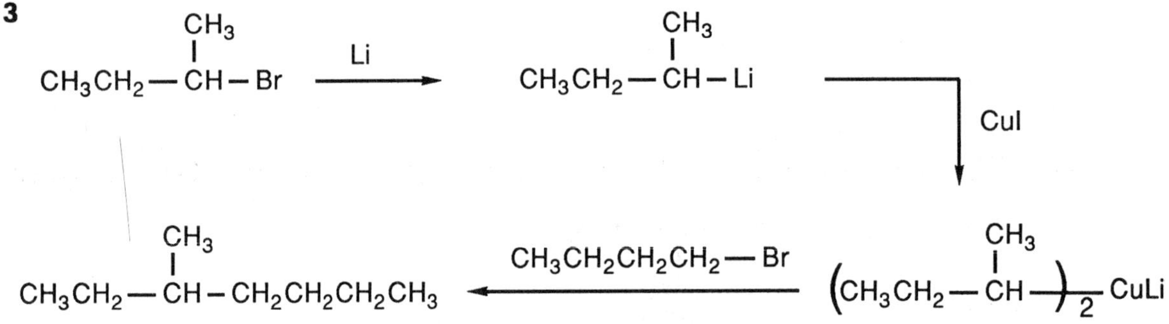

$$CH_3CH_2-CH(CH_3)-Br \xrightarrow{Li} CH_3CH_2-CH(CH_3)-Li \xrightarrow{CuI} (CH_3CH_2-CH(CH_3))_2CuLi$$

$$(CH_3CH_2-CH(CH_3))_2CuLi \xrightarrow{CH_3CH_2CH_2CH_2-Br} CH_3CH_2-CH(CH_3)-CH_2CH_2CH_2CH_3$$

The other possible sequence would require the reaction of an organocopper reagent (prepared from 1-bromobutane) with 2-bromobutane, a secondary alkyl halide. The use of a secondary alkyl halide in this reaction usually does not afford a good yield of product.

ANSWERS TO PROBLEMS

1.1 **(a)** $Br-CH_2-CH_2-Br$ **(b)** CCl_4 **(c)**

$$\begin{array}{ccccccccccccc} &&&&&&&&CH_3&&&&\\ &&&&&&&&|&&&&\\ Cl&-&CH_2&-&CH&-&CH_2&-&C&-&CH_2&-&CH_3\\ &&&&|&&&&|&&&&\\ &&&&CH_3&&&&CH_3&&&& \end{array}$$

(d)

$$\begin{array}{ccccccc} &&CH_3&&CH_3&&\\ &&|&&|&&\\ CH_3&-&C&-&C&-&CH_3\\ &&|&&|&&\\ &&CH_3&&CH_3&& \end{array}$$

(e)

$$\begin{array}{ccccccccc} Br&-&CH_2&-&CH_2&-&CH&-&CH_3\\ &&&&&&|&&\\ &&&&&&CH_3&& \end{array}$$

(f)

$$\begin{array}{ccccccccccccccccc} &&&&&&&&&&&&CH_3&&&&\\ &&&&&&&&&&&&|&&&&\\ CH_3&-&CH&-&CH_2&-&CH_2&-&CH&-&CH_2&-&CH&-&CH_2&-&CH_3\\ &&|&&&&&&|&&&&&&&&\\ &&CH_3&&&&&&CH&-&CH_2&-&CH_3&&&&\\ &&&&&&&&|&&&&&&&&\\ &&&&&&&&CH_3&&&&&&&& \end{array}$$

(g)

$$\begin{array}{ccccccccccccccc} &&&&CH_3&&&&&&&&&&\\ &&&&|&&&&&&&&&&\\ CH_3&-&CH_2&-&CH&-&CH_2&-&CH&-&CH_2&-&CH_2&-&CH_3\\ &&&&&&&&|&&&&&&\\ &&&&&&CH_3&-&C&-&CH_3&&&&\\ &&&&&&&&|&&&&&&\\ &&&&&&&&CH_3&&&&&& \end{array}$$

(h)

$$\begin{array}{ccccccc} &&CH_3&&&&\\ &&|&&&&\\ CH_3&-&CH&-&CH_2&-&Cl \end{array}$$

(i)

$$\begin{array}{ccccccc} &&CH_3&&&&\\ &&|&&&&\\ CH_3&-&C&-&CH_2&-&Br\\ &&|&&&&\\ &&CH_3&&&& \end{array}$$

(j)

$$\begin{array}{ccccccccccc} &&&&&&&&CH_3&&\\ &&&&&&&&|&&\\ I&-&CH_2&-&CH_2&-&CH&-&CH&-&CH_3\\ &&&&&&|&&&&\\ &&&&&&CH_2&-&CH_3&& \end{array}$$

1.2

(a) 2,2-dimethylpropane

(b) chloroform (trichloromethane)

(c) 2,2,6,6,7-pentamethyloctane

(d) 5-ethyl-2,6-dimethyloctane

(e) 4-isobutyl-2,7-dimethyloctane

(f) 2-bromopropane

(g) 2,4,4-tribromo-2,6-dimethyloctane

(h) iodomethane (methyl iodide)

(i) 1,1,1,2,2,2–hexafluoroethane

(j) 1,1,1–trichloro–7,7–dimethyl–5–propyloctane

1.3 (a)

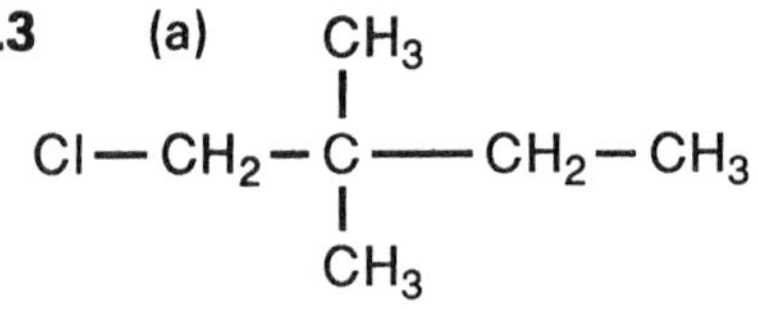

$$Cl-CH_2-\overset{CH_3}{\underset{CH_3}{C}}-CH_2-CH_3$$

1-chloro-2,2-dimethylbutane

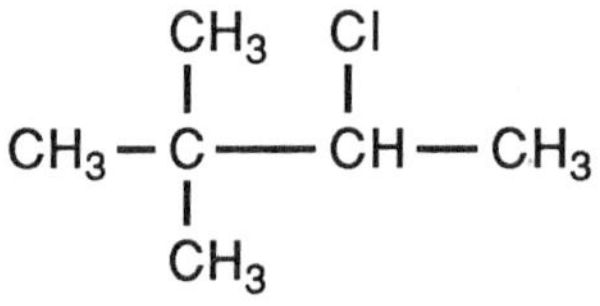

$$CH_3-\overset{CH_3}{\underset{CH_3}{C}}-\overset{Cl}{CH}-CH_3$$

2-chloro-3,3-dimethylbutane

$$CH_3-\overset{CH_3}{\underset{CH_3}{C}}-CH_2-CH_2-Cl$$

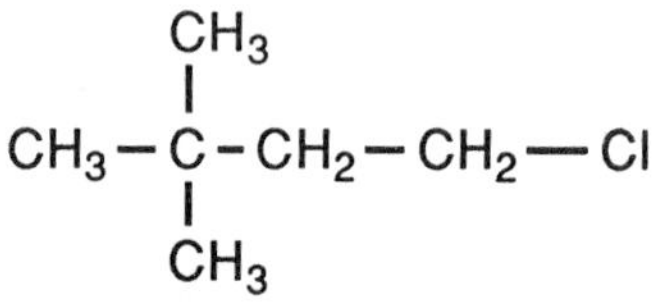

1-chloro-3,3-dimethylbutane

(b)

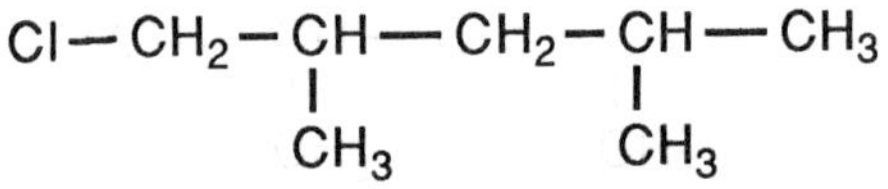

$$Cl-CH_2-\underset{CH_3}{CH}-CH_2-\underset{CH_3}{CH}-CH_3$$

1-chloro-2,4-dimethylpentane

$$CH_3-\overset{Cl}{\underset{CH_3}{C}}-CH_2-\underset{CH_3}{CH}-CH_3$$

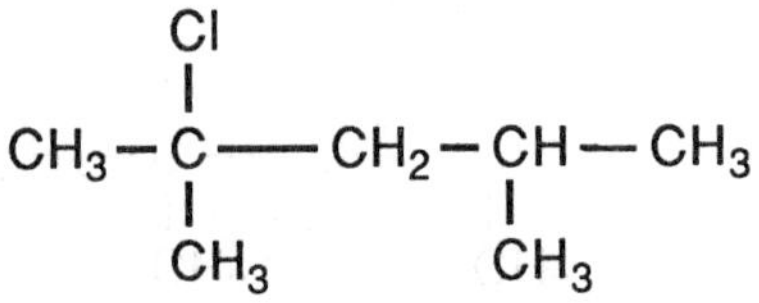

2-chloro-2,4-dimethylpentane

$$CH_3-\underset{CH_3}{CH}-\overset{Cl}{CH}-\underset{CH_3}{CH}-CH_3$$

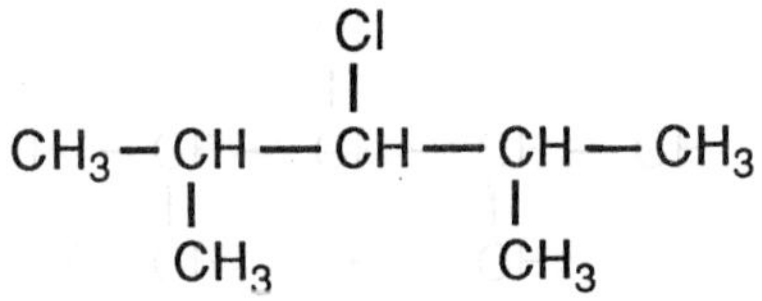

3-chloro-2,4-dimethylpentane

(c)

$$Cl-CH_2-CH_2-\underset{CH_3}{CH}-CH_2-CH_3$$

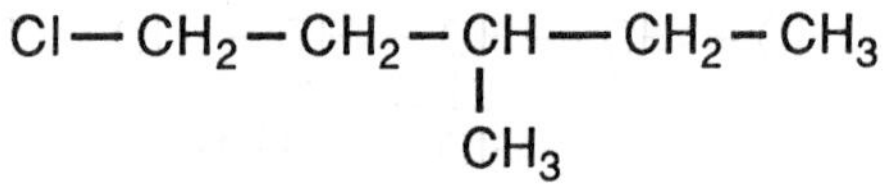

1-chloro-3-methylpentane

$$CH_3-\overset{Cl}{CH}-\underset{CH_3}{CH}-CH_2-CH_3$$

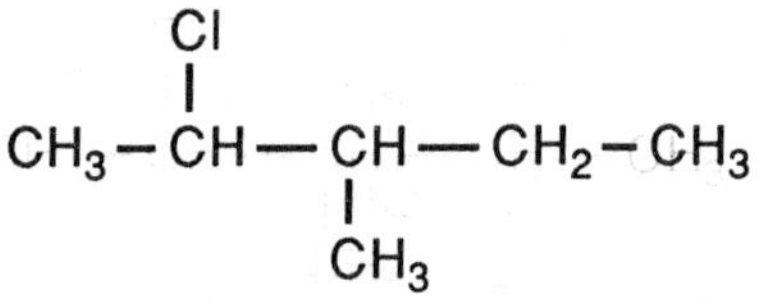

2-chloro-3-methylpentane

$$CH_3-CH_2-\overset{Cl}{\underset{CH_3}{C}}-CH_2-CH_3$$

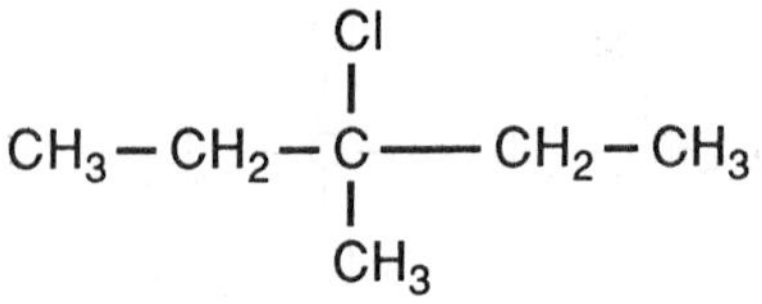

3-chloro-3-methylpentane

$$CH_3-CH_2-\underset{CH_2Cl}{CH}-CH_2-CH_3$$

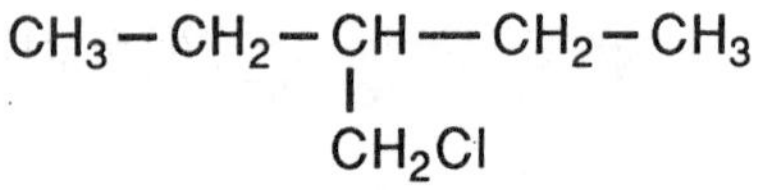

3-(chloromethyl)pentane

(d)

$$\begin{array}{c} \qquad\qquad\qquad CH_2-CH_3 \\ \qquad\qquad\qquad | \\ Cl-CH_2-CH_2-CH-CH_2-CH_3 \end{array}$$

1-chloro-3-ethylpentane

$$\begin{array}{c} \qquad\qquad CH_2-CH_3 \\ \qquad\qquad | \\ CH_3-CH-CH-CH_2-CH_3 \\ | \qquad\qquad\qquad \\ Cl \qquad\qquad\qquad \end{array}$$

2-chloro-3-ethylpentane

$$\begin{array}{c} CH_2-CH_3 \\ | \\ CH_3-CH_2-C-CH_2-CH_3 \\ | \\ Cl \end{array}$$

3-chloro-3-ethylpentane

(e)

$$\begin{array}{c} \quad CH_3 \qquad\qquad\qquad\qquad\qquad CH_3 \\ \quad | \qquad\qquad\qquad\qquad\qquad\quad | \\ Cl-CH_2-CH-CH-CH_2-CH_2-CH-CH_3 \\ | \\ CH_2-CH_3 \end{array}$$

1-chloro-3-ethyl-2,6-dimethylheptane

$$\begin{array}{c} CH_3 \quad Cl \qquad\qquad\qquad CH_3 \\ | \qquad | \qquad\qquad\qquad\quad | \\ CH_3-CH-C-CH_2-CH_2-CH-CH_3 \\ | \\ CH_2-CH_3 \end{array}$$

3-chloro-3-ethyl-2,6-dimethylheptane

$$\begin{array}{c} CH_3 \qquad\qquad\qquad\qquad\qquad CH_3 \\ | \qquad\qquad\qquad\qquad\qquad\quad | \\ CH_3-C-CH-CH_2-CH_2-CH-CH_3 \\ | \qquad | \\ Cl \quad CH_2-CH_3 \end{array}$$

2-chloro-3-ethyl-2,6-dimethylheptane

$$\begin{array}{c} CH_3 \qquad\qquad\qquad\qquad\qquad CH_3 \\ | \qquad\qquad\qquad\qquad\qquad\quad | \\ CH_3\text{-}CH-CH-CH_2-CH_2-CH-CH_3 \\ | \\ CH-CH_3 \\ | \\ Cl \end{array}$$

3-(1-chloroethyl)-2,6-dimethylheptane

$$\begin{array}{c} CH_3 \qquad\qquad\qquad\qquad\qquad CH_3 \\ | \qquad\qquad\qquad\qquad\qquad\quad | \\ CH_3\text{-}CH-CH-CH_2-CH_2-CH-CH_3 \\ | \\ CH_2-CH_2-Cl \end{array}$$

3-(2-chloroethyl)-2,6-dimethylheptane

CH_3 Cl CH_3
$CH_3-CH-CH-CH-CH_2-CH-CH_3$
CH_2-CH_3

4-chloro-3-ethyl-2,6-dimethylheptane

CH_3 Cl CH_3
$CH_3-CH-CH-CH_2-CH-CH-CH_3$
CH_2-CH_3

3-chloro-5-ethyl-2,6-dimethylheptane

CH_3 CH_3
$CH_3-CH-CH-CH_2-CH_2-C-CH_3$
CH_2-CH_3 Cl

2-chloro-5-ethyl-2,6-dimethylheptane

CH_3 CH_2-Cl
$CH_3-CH-CH-CH_2-CH_2-CH-CH_3$
CH_2-CH_3

1-chloro-5-ethyl-2,6-dimethylheptane

1.4 (a)

CH_3
$CH_3-CH-CH-Br$
Br

1,1-dibromo-2-methylpropane

CH_2-Br
$CH_3-CH-CH_2-Br$

1,3-dibromo-2-methylpropane

CH_3
CH_3-C-CH_2-Br
Br

1,2-dibromo-2-methylpropane

(b)

Br
$Br-CH_2-C-CH_2-CH-CH_3$
CH_3 CH_3

1,2-dibromo-2,4-dimethylpentane

Br Br
$CH_3-C-CH-CH-CH_3$
CH_3 CH_3

2,3-dibromo-2,4-dimethylpentane

Br Br
$CH_3-C-CH_2-C-CH_3$
CH_3 CH_3

2,4-dibromo-2,4-dimethylpentane

Br
$CH_3-C-CH_2-CH-CH_2-Br$
CH_3 CH_3

2,5-dibromo-2,4-dimethylpentane

(c)

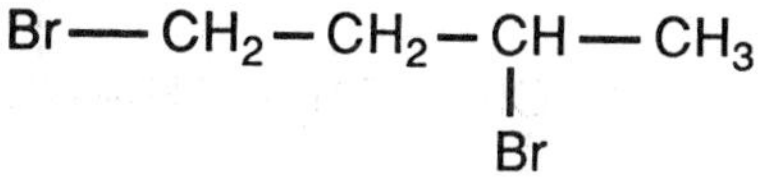

$Br{-}CH_2{-}CH_2{-}CH(Br){-}CH_3$

1,3-dibromobutane

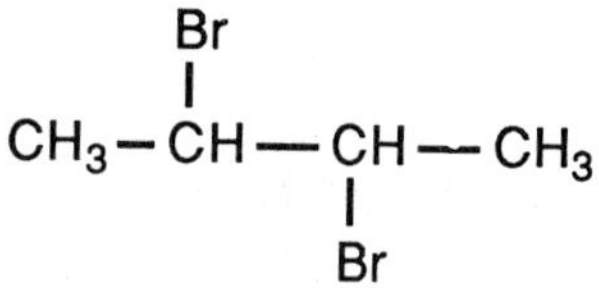

$CH_3{-}CH(Br){-}CH(Br){-}CH_3$

2,3-dibromobutane

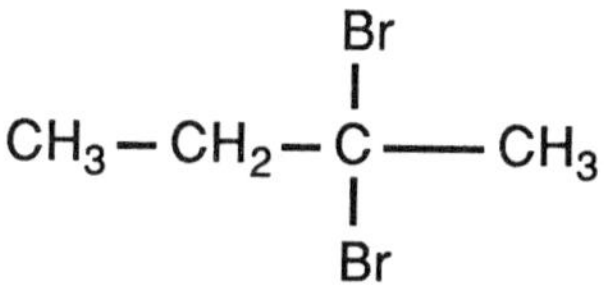

$CH_3{-}CH_2{-}C(Br)_2{-}CH_3$

2,2-dibromobutane

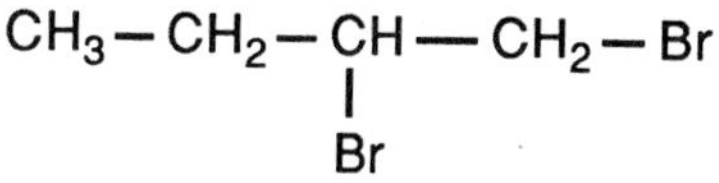

$CH_3{-}CH_2{-}CH(Br){-}CH_2{-}Br$

1,2-dibromobutane

(d

$Br{-}CH_2{-}CH_2{-}CH(CH_3){-}CH_2{-}Cl$

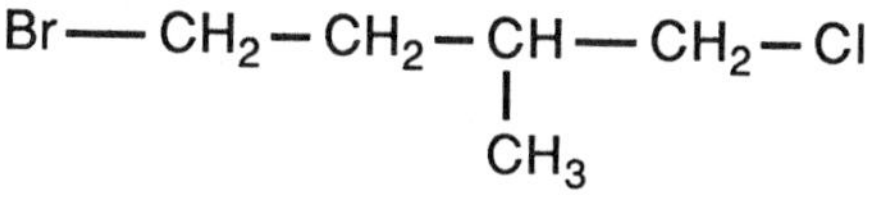

4-bromo-1-chloro-2-methylbutane

$CH_3{-}CH(Br){-}CH(CH_3){-}CH_2{-}Cl$

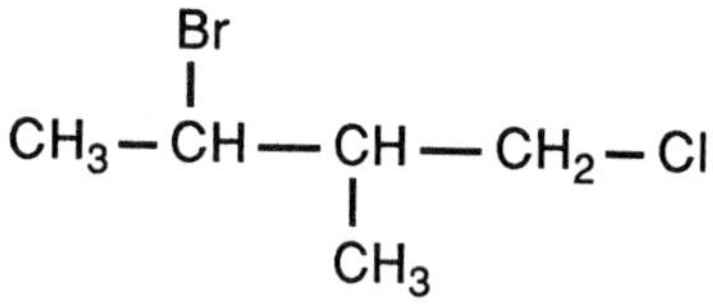

3-bromo-1-chloro-2-methylbutane

$CH_3{-}CH_2{-}C(Br)(CH_3){-}CH_2{-}Cl$

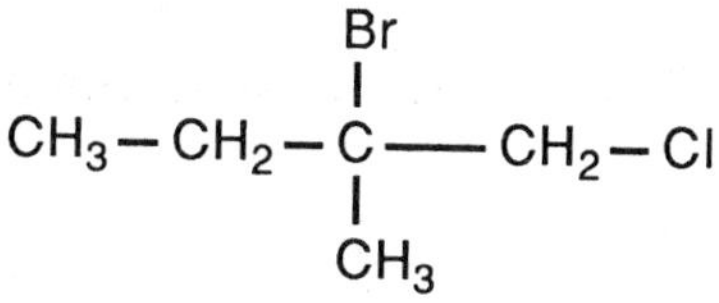

2-bromo-1-chloro-2-methylbutane

$CH_3{-}CH_2{-}CH(CH_2{-}Br){-}CH_2{-}Cl$

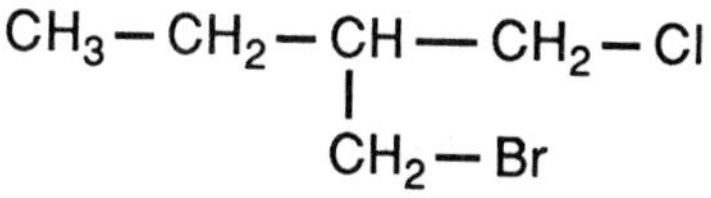

1-(bromomethyl)-3-chlorobutane

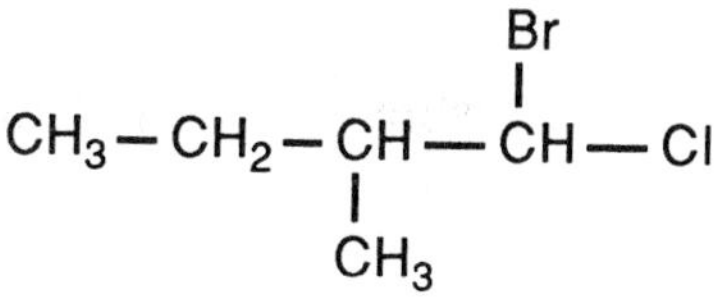

$CH_3{-}CH_2{-}CH(CH_3){-}CH(Br){-}Cl$

1-bromo-1-chloro-2-methylbutane

(e)

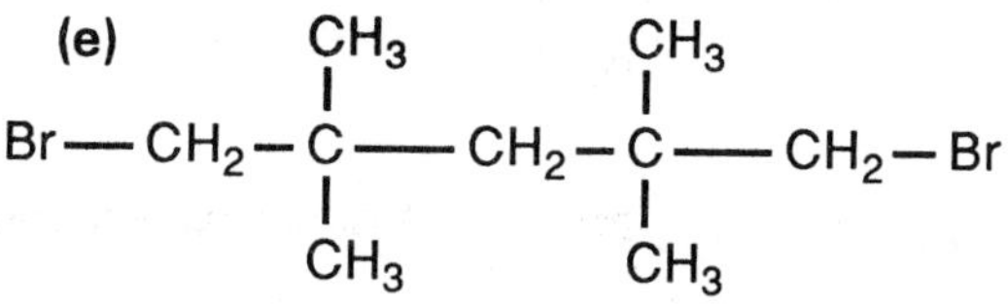

$Br{-}CH_2{-}C(CH_3)_2{-}CH_2{-}C(CH_3)_2{-}CH_2{-}Br$

1,5-dibromo-2,2,4,4-tetramethylpentane

$CH_3{-}C(CH_3)_2{-}CH(Br){-}C(CH_3)_2{-}CH_2{-}Br$

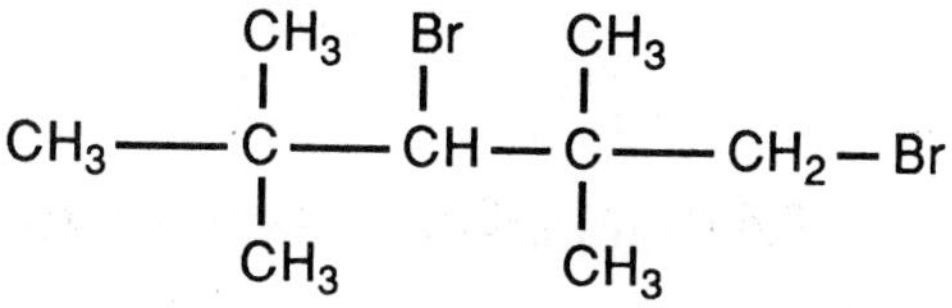

1,3-dibromo-2,2,4,4-tetramethylpentane

$$CH_3-\overset{\displaystyle CH_3}{\underset{\displaystyle CH_3}{C}}-CH_2-\overset{\displaystyle CH_2-Br}{\underset{\displaystyle CH_3}{C}}-CH_2-Br$$

1-bromo-2-(bromomethyl)-2,4,4-trimethylpentane

$$CH_3-\overset{\displaystyle CH_3}{\underset{\displaystyle CH_3}{C}}-CH_2-\overset{\displaystyle CH_3}{\underset{\displaystyle CH_3}{C}}-\overset{\displaystyle Br}{CH}-Br$$

1,1-dibromo-2,2,4,4-tetramethylpentane

1.5 **(a)** F; C; CH_3; CH_3; Cl

(b) CH_3; C; F; CH_3; Cl

(c) CH_3; C; F; Cl; CH_3

(d) Cl; F — C — CH_3; CH_3

1.6 (a) different

(b) constitutional isomers

(c) same

(d) different

(e) same

(f) different

(g) different

(h) different

(i) constitutional isomers

(j) different

1.7 (a) 20 (each carbon has 4 and each hydrogen has 1)

(b) 20 (the same as the total number of valence level atomic orbitals)

(c) 10 bonding; 10 antibonding

(d) 20 (each carbon has 4 and each hydrogen has 1)

(e) they are equal (Each bonding molecular orbital is doubly occupied).

(f) they are equal (Each doubly occupied molecular orbital corresponds to a bond).

1.8 The first step is formation of methyllithium:

$$CH_3{-}I \ + \ 2\ Li \longrightarrow CH_3{-}Li \ + \ LiI$$

This is followed by formation of lithium dimethylcuprate

$$2\ CH_3{-}Li \ + \ CuI \longrightarrow (CH_3)_2CuLi \ + \ LiI$$

1.9 (a) The target molecule is symmetrical, so a single alkyl halide can be used to generate the product. This is one of the few cases where the Wurtz reaction could be used successfully. Organocopper reagents offer a more general approach, and the synthesis is illustrated for 1-bromo-2-methylbutane as the starting material. (In this and subsequent examples, the iodo and chloro compounds could also be used).

$$CH_3{-}CH_2{-}CH(CH_3){-}CH_2Br \xrightarrow{Li} CH_3{-}CH_2{-}CH(CH_3){-}CH_2Li$$

$$CH_3{-}CH_2{-}CH(CH_3){-}CH_2Li \xrightarrow{CuI} (CH_3{-}CH_2{-}CH(CH_3){-})_2CuLi$$

$$(CH_3{-}CH_2{-}CH(CH_3){-})_2CuLi \xrightarrow{CH_3{-}CH_2{-}CH(CH_3){-}CH_2Br} CH_3{-}CH_2{-}CH(CH_3){-}CH_2{-}CH_2{-}CH(CH_3){-}CH_2{-}CH_3$$

(b) The target molecule contains 6 carbon atoms, but there is no way that it could be formed from two 3-carbon precursors. It is necessary to join C_2 and C_4 fragments using an organocopper reagent. The organocopper reagent must react with a primary alkyl halide, which corresponds to the C-2 fragment. The following sequence employs of bromoethane and 2-chlorobutane as starting materials.

$$CH_3-CH_2-\underset{}{\overset{CH_3}{\overset{|}{C}H}}-Cl \xrightarrow{Li} CH_3-CH_2-\overset{CH_3}{\overset{|}{C}H}-Li \xrightarrow{CuI} (CH_3-CH_2-\overset{CH_3}{\overset{|}{C}H}\!\!\rightarrow_2 CuLi$$

$$\downarrow CH_3-CH_2Br$$

$$CH_3-CH_2-\overset{CH_3}{\overset{|}{C}H}-CH_2-CH_3$$

(c) Working backwards, you find that a bond must be formed between a *tert*-butyl group and a propyl group. Therefore, a 1-halopropane (primary) must react with an organocopper reagent derived from a 2-halo-2-methylpropane:

$$CH_3-\underset{CH_3}{\underset{|}{\overset{CH_3}{\overset{|}{C}}}}-Br \xrightarrow{Li} CH_3-\underset{CH_3}{\underset{|}{\overset{CH_3}{\overset{|}{C}}}}-Br \xrightarrow{CuI} (CH_3-\underset{CH_3}{\underset{|}{\overset{CH_3}{\overset{|}{C}}}}\!\!\rightarrow_2 CuLi$$

$$\downarrow CH_3CH_2CH_2Br$$

$$CH_3-\underset{CH_3}{\underset{|}{\overset{CH_3}{\overset{|}{C}}}}-CH_2-CH_2-CH_3$$

1.10 (a) Li, then CuI

(b) Zn, HCl

(c) Zn, HCl

(d) The product consists of two fragments corresponding to the reactant. Treatment of the starting bromide with sodium metal would yield the product directly by the Wurtz reaction; alternatively an organocopper reagent could be used:

$$CH_3-\underset{\underset{CH_3}{|}}{CH}-CH_2Br \xrightarrow[2.\ CuI]{1.\ Li} (CH_3-\underset{\underset{CH_3}{|}}{CH}-CH_2)_2CuLi$$

$$\Big\downarrow \; CH_3-\underset{\underset{CH_3}{|}}{CH}-CH_2Br$$

$$CH_3-\underset{\underset{CH_3}{|}}{CH}-CH_2-CH_2-\underset{\underset{CH_3}{|}}{CH}-CH_3$$

(e) Cl_2, light

1.11 Compounds **E, F, G** and **H** must all have the same carbon skeleton; therefore compound **G** is 1-chloro-2,2-dimethylpropane and **E** contains a neopentyl group, $-CH_2-C(CH_3)_3$. Compound **D** must be a primary alkyl bromide, so **A** must contain the fragment, $-CH_2-CH_2-C(CH_3)_3$. Since **A** can be formed by the Wurtz reaction, it is symmetrical and must actually contain *two* of these groups at the ends of the carbon chain. This leaves only two carbons unaccounted for, and for a symmetrical alkane they must be $-CH_2-CH_2$. The structures of **A-G** must therefore be:

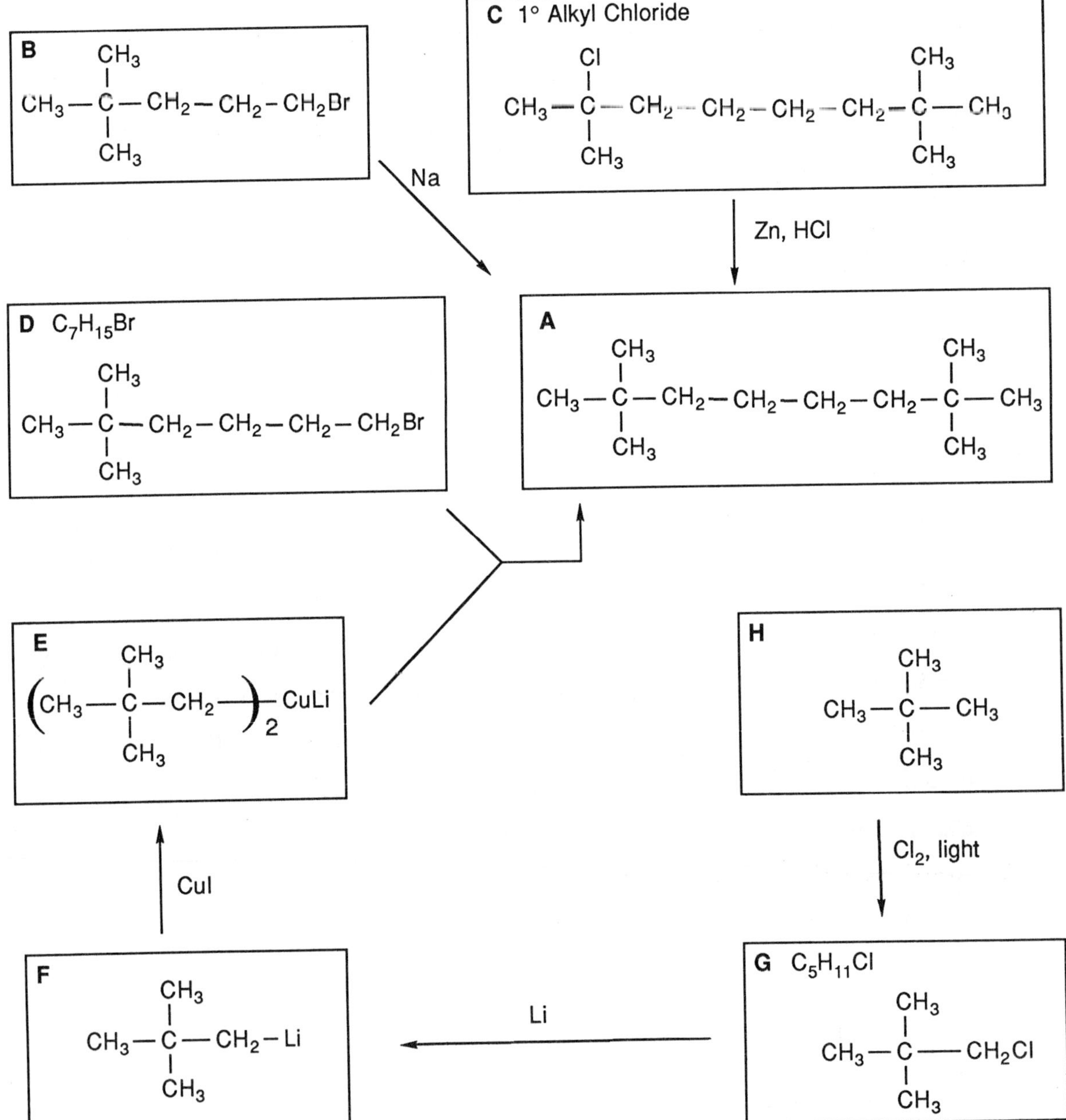
B
$CH_3-C(CH_3)_2-CH_2-CH_2-CH_2Br$
Na
C 1° Alkyl Chloride
$CH_3-CCl(CH_3)-CH_2-CH_2-CH_2-CH_2-C(CH_3)_2-CH_3$
Zn, HCl
D $C_7H_{15}Br$
$CH_3-C(CH_3)_2-CH_2-CH_2-CH_2-CH_2Br$
A
$CH_3-C(CH_3)_2-CH_2-CH_2-CH_2-CH_2-C(CH_3)_2-CH_3$
E
$(CH_3-C(CH_3)_2-CH_2)_2CuLi$
H
$CH_3-C(CH_3)_2-CH_3$
Cl_2, light
CuI
F
$CH_3-C(CH_3)_2-CH_2-Li$
Li
G $C_5H_{11}Cl$
$CH_3-C(CH_3)_2-CH_2Cl$

1.12 Only two primary C_4 alkyl groups exist, but there are five isomeric possibilities for the alkyl group of compounds **A–C**. This leads to 10 alternatives for **D**, and the key to the puzzle is the number of monochloro derivatives that could be formed. The possible structures for **D** are drawn below, and only one of these would afford just five monochloro derivatives.

Possibilities for **D**:

```
C-C-C-C-C-C-C-C-C            C-C-C-C-C-C-C-C
  |                            |         |
  C                            C         C

C-C-C-C-C-C-C-C      C-C-C-C-C-C-C      C-C-C-C-C-C-C
  |   |                |   |   |          | |
  C   C                C   C   C          C C-C

                        C-C-C              C-C-C
                        |                  |
C-C-C-C-C-C           C-C-C-C-C-C        C-C-C-C-C
  | |     |             |                  |   |
  C C-C   C             C                  C   C

C-C-C-C-C            C-C-C-C-C
  |                    |
  C-C-C-C-C            C-C-C-C
                           |
                           C
```

The correct structures of compounds **A-F** are shown as follows:

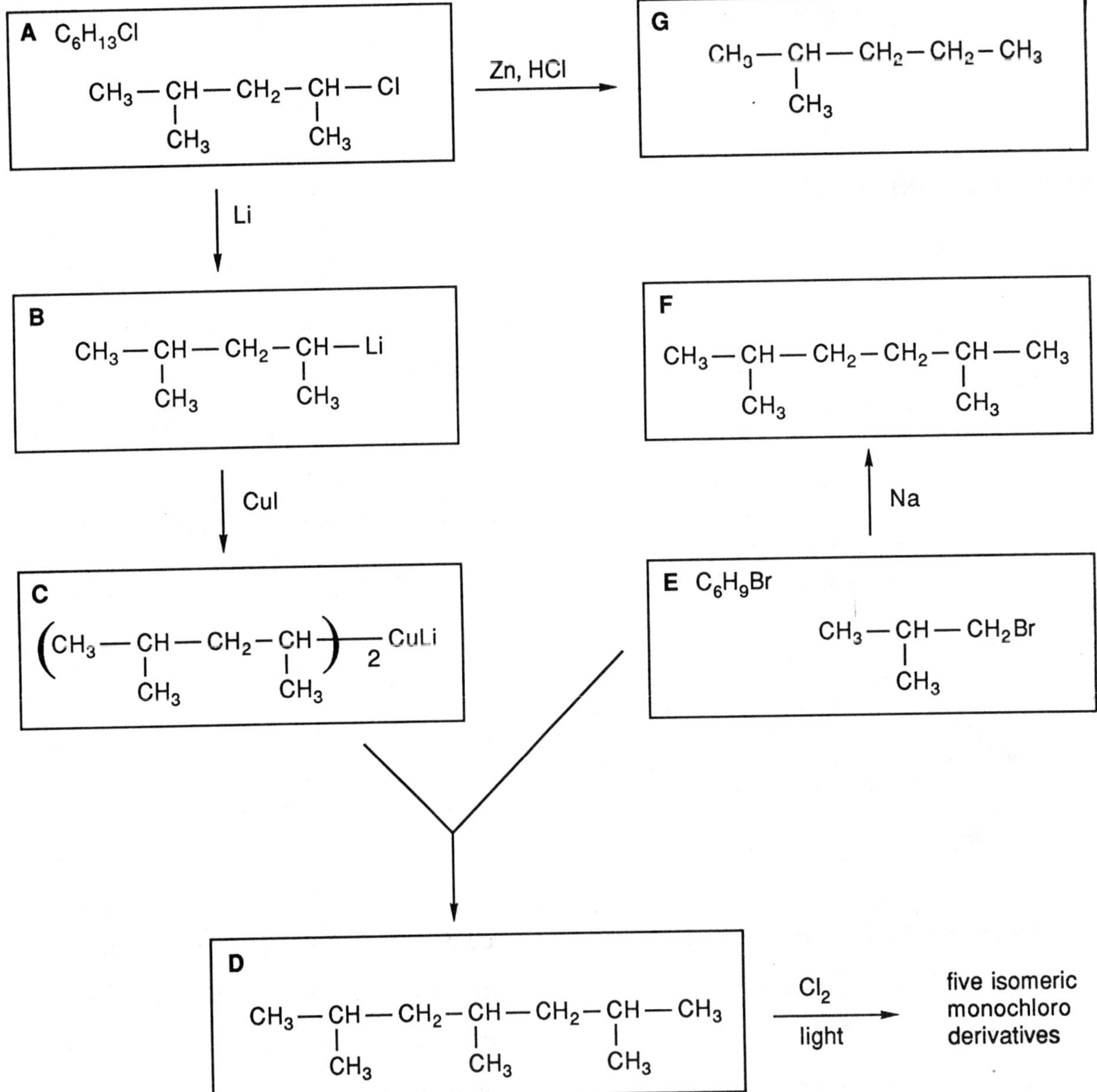

CHAPTER 2. STRUCTURE, STABILITY, AND REACTIVITY

ANSWERS TO EXERCISES

2.1

Energy

109°

Cl-C-Cl Bond Angle

The optimum bond angle (assumed to be 109° for convenience) is the minimum-energy structure. Any deviation from that geometry results in a structure of higher energy.

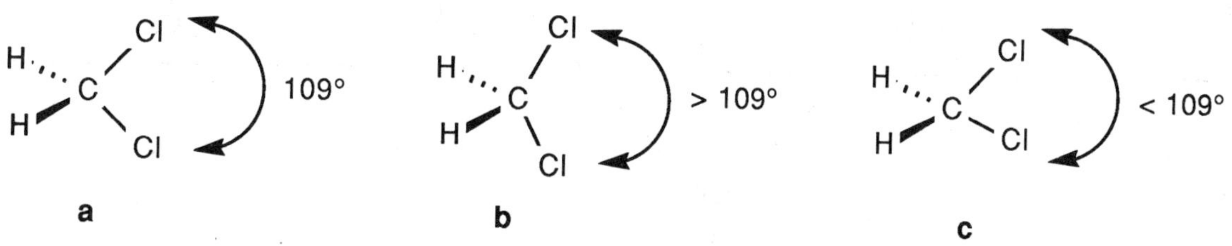

2.2 There are three minimum-energy staggered conformations (b), (d), (f), two of which (d) and (f) have the same energy. The staggered conformation at 180° together with the eclipsed structures at 120° and 240° on the curve all have severe methyl-methyl steric interactions, and the energies for these geometries are increased accordingly.

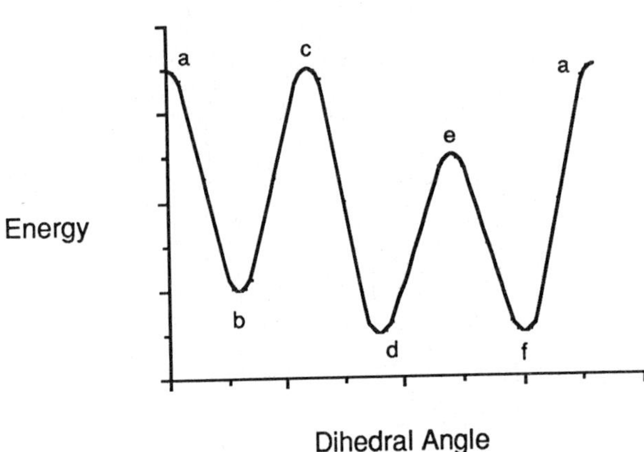

a b c d

e f

2.3 (a) Only a single staggered staggered conformation exists.

(b) There are three staggered conformations; two of them have the same energy. The third conformation (for which methyl-methyl interactions are minimized) is the most stable.

(c) Only a single staggered staggered conformation exists.

2.4

(a) initiation: $Cl-Cl \longrightarrow 2\ Cl^{\cdot}$

(b) propagation: $Cl^{\cdot} + CH_3-CH_3 \longrightarrow CH_3-CH_2^{\cdot} + H-Cl$

$CH_3-CH_2^{\cdot} + Cl-Cl \longrightarrow CH_3-CH_2-Cl + Cl^{\cdot}$

(c) termination: $2\ Cl^{\cdot} \longrightarrow Cl_2$

$2\ CH_3-CH_2^{\cdot} \longrightarrow CH_3-CH_2-CH_2-CH_2$

$CH_3-CH_2^{\cdot} + Cl^{\cdot} \longrightarrow CH_3-CH_2-Cl$

2.5 The energy differences in the reaction pathways for generating primary vs. secondary radicals should be similar for propane and butane, but the probability effects are different. Propane has six primary hydrogens and only two secondary hydrogens, while the corresponding values for butane are six and four. Therefore, to a first approximation you would expect the primary:secondary product ratio to be twice as high for propane as for butane, and this is indeed the case. For propane the ratio is 48:52 = 1.0, while for butane the ratio is 30:70 = 0.43.

2.6 The major product is that formed via a primary radical, and probability effects strongly favor that radical. The ratio of primary to tertiary hydrogens in 2-methylpropane is 9:1, which would be the expected product ratio if the energy effects were equal for the two pathways. That the observed product ratio is only 2.3:1 (i.e., 70:30) indicates that an energy effects favor the more stable tertiary radical by about a factor of four.

2.7 There are 12 equivalent primary hydrogens and 8 equivalent secondary hydrogens. The ratio is 3:2, which is the same as that for butane.

$$CH_3-CH_2-C(CH_2-CH_2^{\bullet})(CH_2-CH_3)-CH_2-CH_3 \qquad CH_3-CH_2-C(^{\bullet}CH-CH_3)(CH_2-CH_3)-CH_2-CH_3$$

The secondary radical is more stable, which favors formation of the secondary product. The probability effect favors the primary product, but it is only 3:2 so the secondary product should predominate (as it does in the case of butane).

2.8

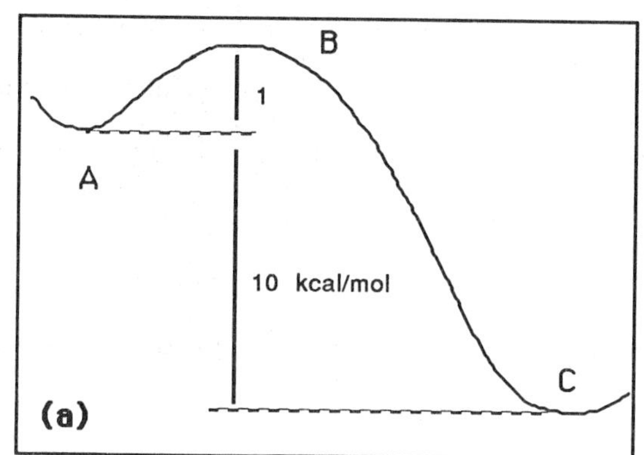

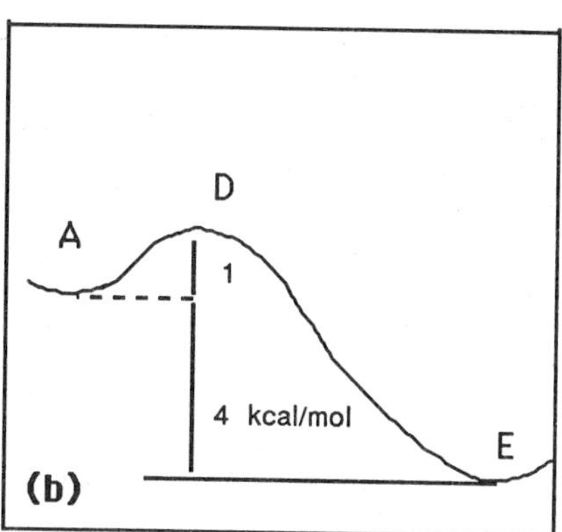

$$CH_3{-}\overset{\displaystyle CH_3}{\underset{\displaystyle CH_3}{C}}{-}H \;+\; Cl^{\cdot} \longrightarrow \left[CH_3{-}\overset{\displaystyle CH_3}{\underset{\displaystyle CH_3}{C}}\cdots H \cdots Cl \right]^{\cdot} \longrightarrow CH_2{-}\overset{\displaystyle CH_3}{\underset{\displaystyle CH_3}{C}}{}^{\cdot} \;+\; H{-}Cl$$

A — 2-methylpropane + $Cl^{\cdot}$

B

C — *tert*-butyl radical + HCl

↓

$$\left[CH_3{-}\overset{\displaystyle CH_2\cdots H\cdots Cl}{\underset{\displaystyle CH_3}{C}}{-}H \right]^{\cdot} \longrightarrow CH_3{-}\overset{\displaystyle CH_2{}^{\cdot}}{\underset{\displaystyle CH_3}{C}} \;+\; H{-}Cl$$

D

E — isobutyl radical + HCl

2.9

$$Cl_2 + CH_3\text{-}H \longrightarrow CH_3\text{-}Cl + HCl$$

Bonds broken:	Cl-Cl	+58
	CH_3-H	+104
Bonds formed:	CH_3-Cl	-84
	H-Cl	-103
	total	-25 kcal/mol

2.10 You would expect that a larger probability factor for the collision of a chlorine atom with methane. Statistically, methane has four hydrogens, while chloromethane has only three. From a different point of view, a chlorine atom could abstract a hydrogen from methane no matter what orientation the two species have at the time of collision. A collision between a chlorine atom and chloromethane, however, might not be productive if the chlorine atom approached the CH_3Cl molecule from the side of the chloro substituent.

ANSWERS TO PROBLEMS

2.1 (a) endothermic

(b) exothermic

(c) C

(d) transition state

(e) intermediate

(f) transition state

(g) transition state

(h) C

2.2 The relative amount of each product can be predicted by multiplying the number of each type of hydrogen by the reactivity of that type of hydrogen. There are four types of hydrogen in 2-chloro-2,4-dimethylpentane, leading to products a, b, c, and d. Using relative reactivities of 1, 4, and 5 for primary, secondary and tertiary hydrogens, respectively:

$$CH_3-\underset{}{CH}(CH_3)-CH_2-C(CH_3)(Cl)-CH_3 \longrightarrow Cl-CH_2-CH(CH_3)-CH_2-C(CH_3)_2-CH_3 \ (\textbf{A}) +$$

$$CH_3-CH(CH_3)-CH_2-C(CH_3)(Cl)-CH_2-Cl \ (\textbf{B}) + CH_3-CH(CH_3)-CH(Cl)-C(CH_3)(Cl)-CH_3 \ (\textbf{C}) +$$

$$CH_3-C(CH_3)(Cl)-CH_2-C(CH_3)(Cl)-CH_3 \ (\textbf{D})$$

Product A	6	1°H x 1	=	6
Product B	6	1°H x 1	=	6
Product C	2	2°H x 4	=	8
Product D	1	3°H x 5	=	5
				25

A 6/25 = 24%, B 6/25 = 24%, C 8/25 = 32%, D 5/25 = 20%

2.3 Using the data from Table 2.1 in kcal/mol:

A	B	C	D
+ 58	+ 58	+ 58	+ 58
+ 98	+ 98	+ 95	+ 92
- 81	- 81	- 80	- 80
-103	-103	-103	-103
- 28	- 28	- 30	- 33

2.4 The most stable conformation is that in which steric interactions between "large" substituents are minimized by avoiding *gauche* interactions between them. The least stable staggered conformations are those having *gauche* interactions between the largest groups. A methyl group (CH_3) is larger than a bromo (Br) group.

(a) CH_3, H, CH_3 / Br, H, CH_3

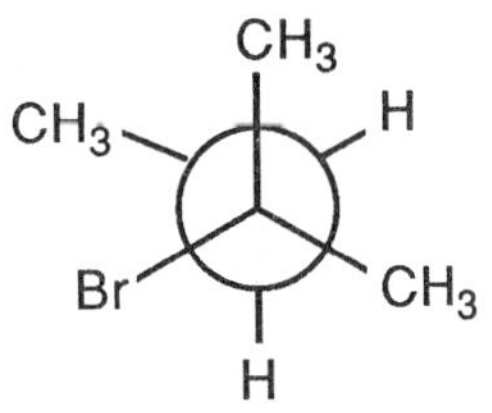

(b) H, BrH_2C, H / H, Br, H

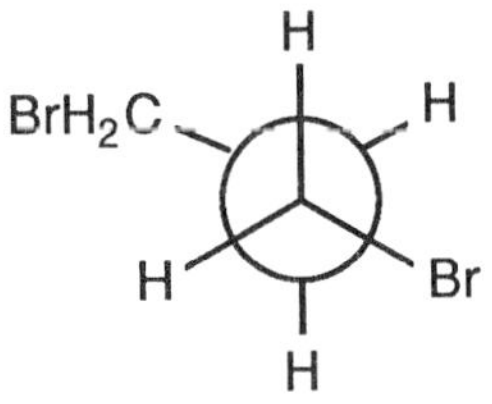

(c) CH_3, Br, H / H, CH_3, Br

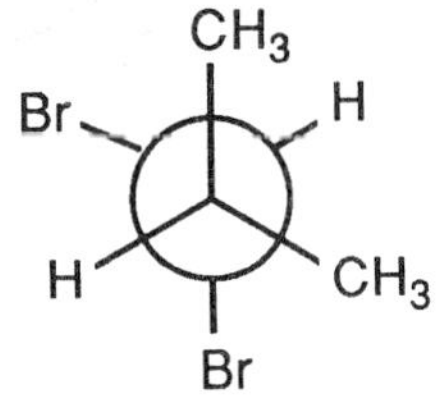

(d) CH_3, H, H / H, Br, CH_3

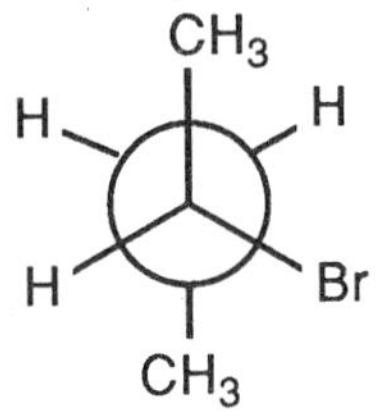

(e) CH_3–C(CH_3)–CH_3, H, H / H, CH_3, Cl

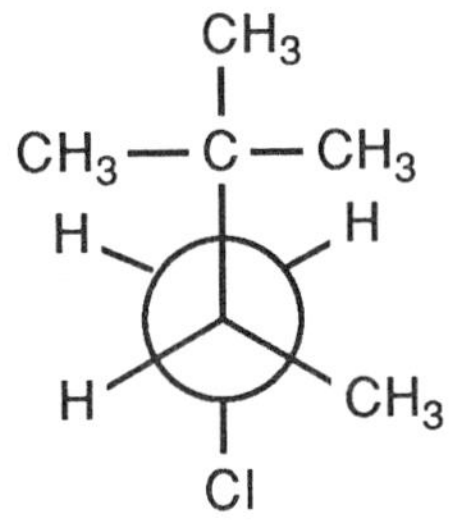

2.5 (a) CH_3, H, H_3C, H, Br, CH_3

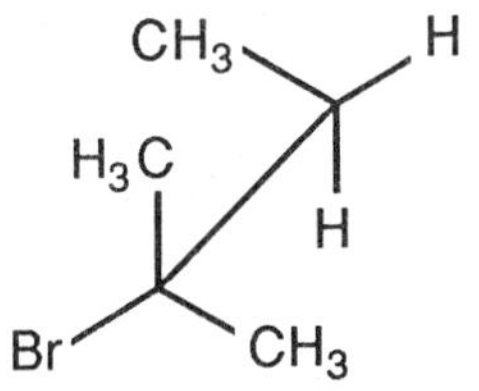

(b) BrH_2C, H, H, H, H, Br

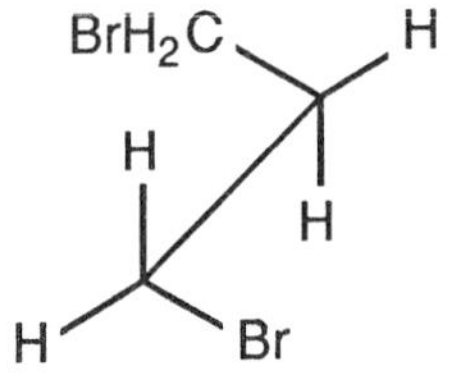

(c) Br, H, CH_3, Br, H, CH_3

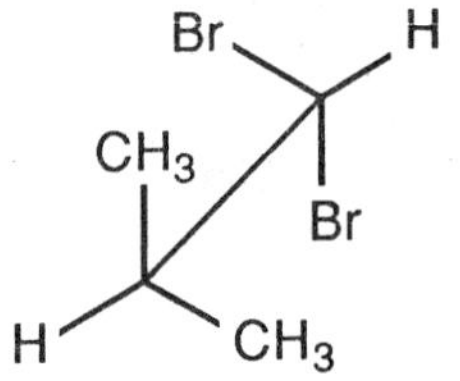

(d) H, H, H_3C, CH_3, H, Br

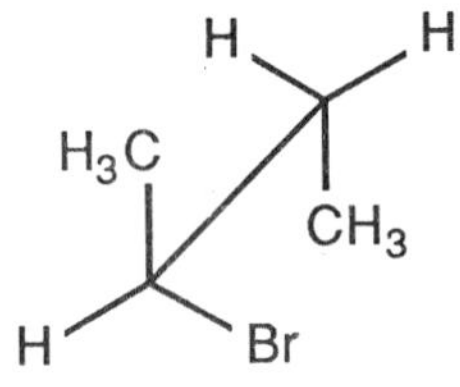

(e) H, H, $(H_3C)_3C$, Cl, H, CH_3

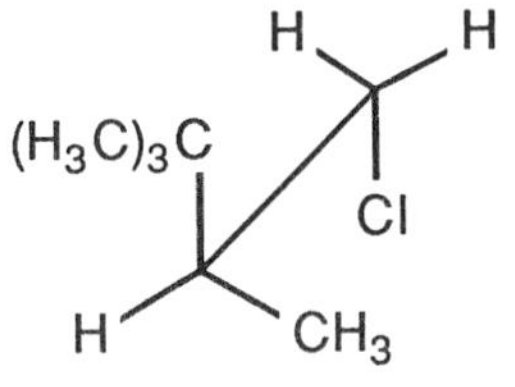

2.6 The least stable staggered conformations are those having *gauche* interactions between the largest groups. A methyl group (CH_3) is larger than a bromo (Br) group.

(a) CH_3, H, CH_3, Br, CH_3, H

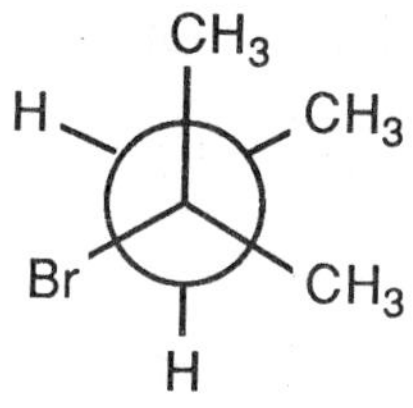

(b) Br, H, CH_2Br, H, H, H

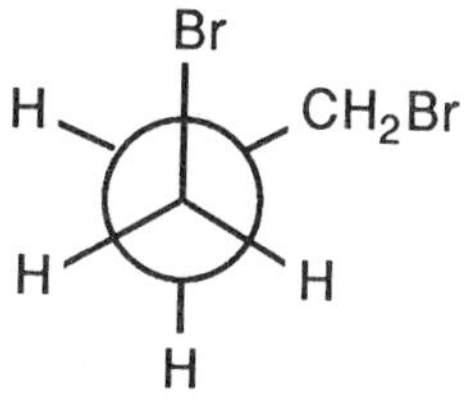

(c) Br, H, CH_3, H, Br, CH_3

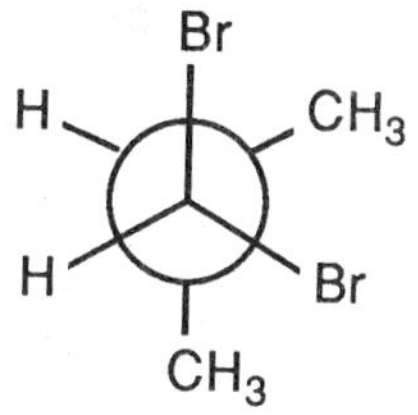

(d) CH_3, H, CH_3, H, Br, H

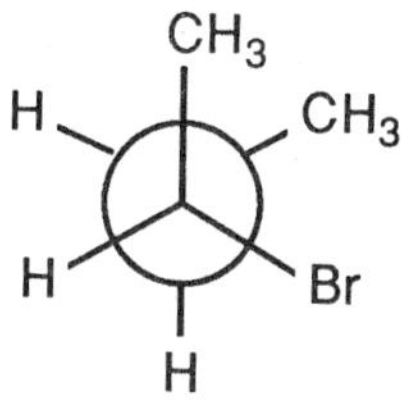

(e) CH_3—C(CH_3)—CH_3, H, Cl, H, CH_3, H

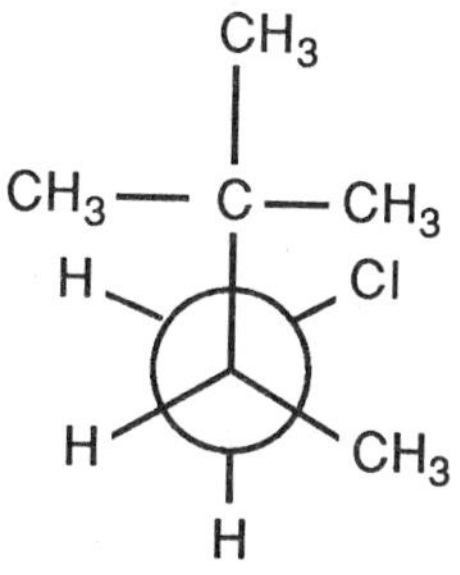

2.7

(a) H, CH_3, H_3C, H, Br, CH_3

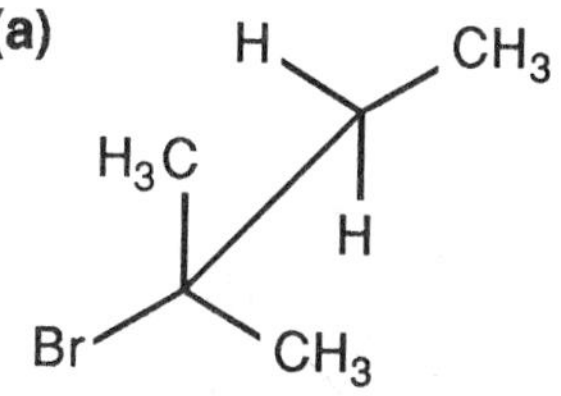

(b) H, CH_2Br, Br, H, H, H

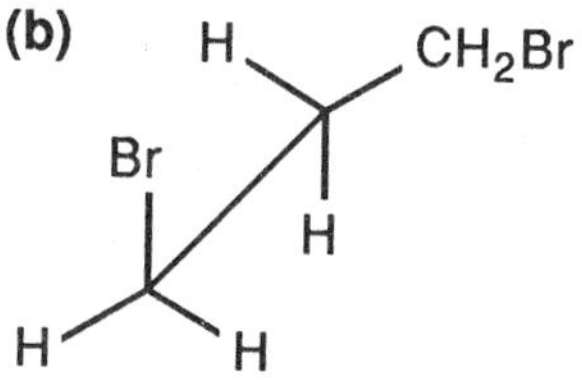

(c) H, CH_3, Br, CH_3, H, Br

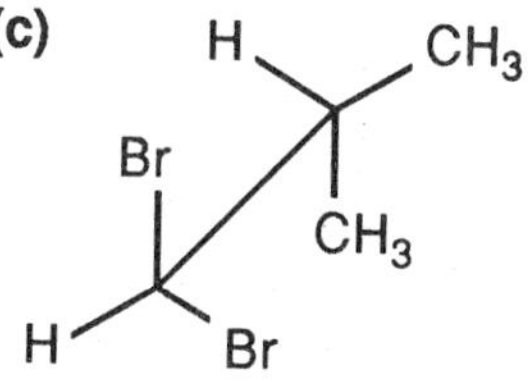

(d) H, CH_3, H_3C, H, H, Br

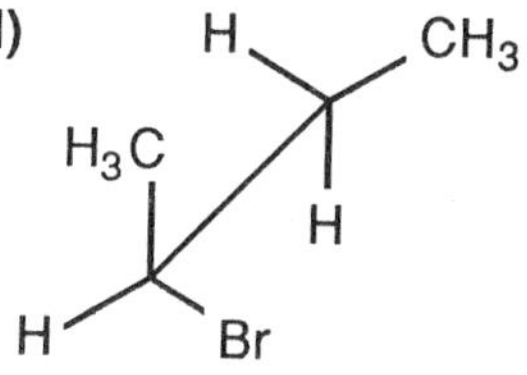

(e) H, Cl, $(CH_3)_3C$, H, H, CH_3

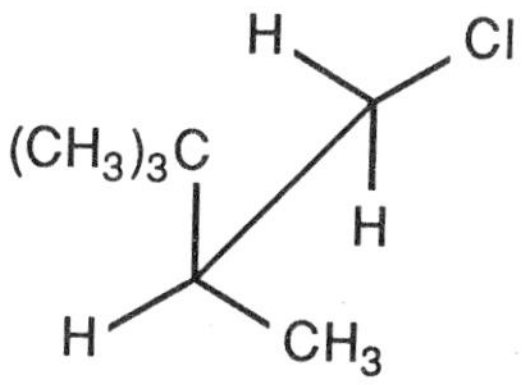

2.8 The least stable eclipsed conformations are those having eclipsing interactions between the largest groups. A methyl group (CH_3) is larger than a bromo (Br) group.

(a) (b) 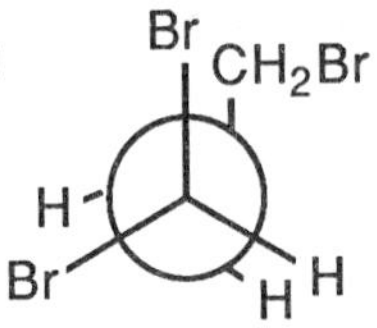 (c)

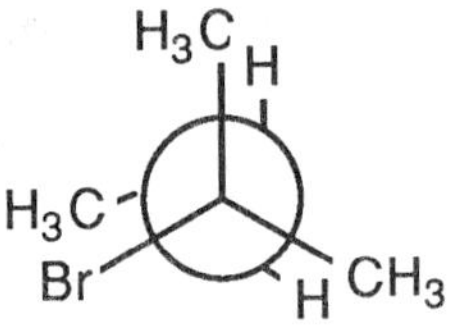

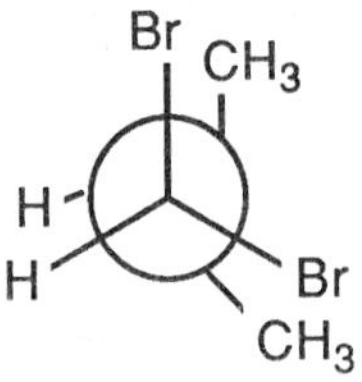

(d) 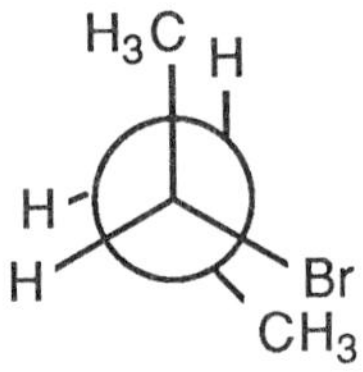 (e)

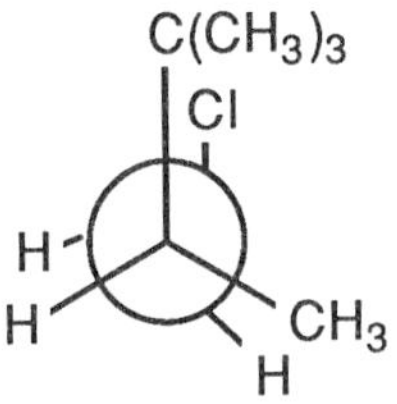

2.9

(a) (b) 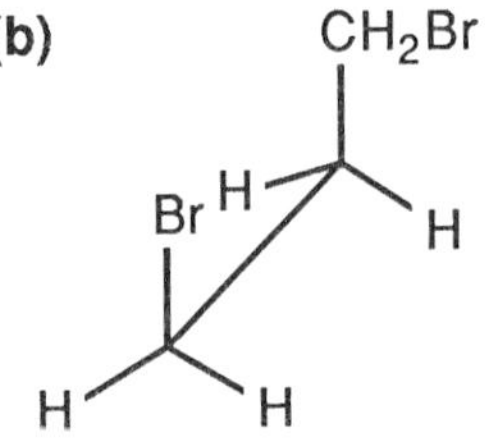 (c)

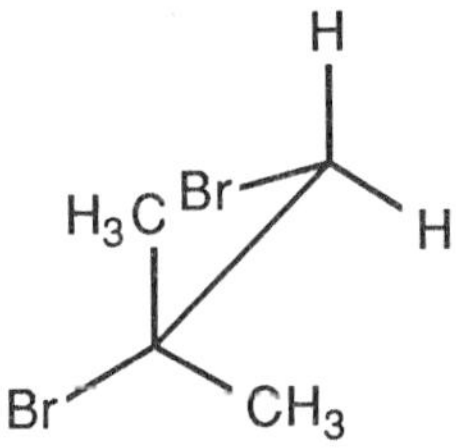

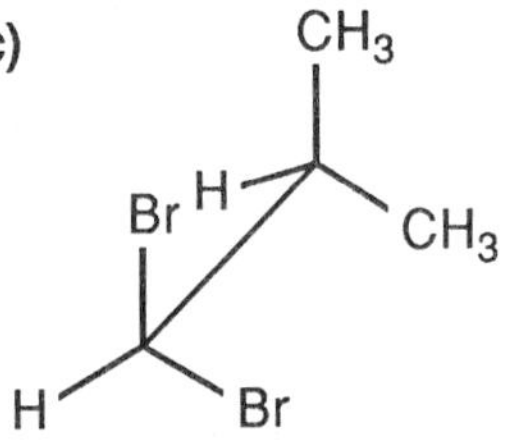

(d) 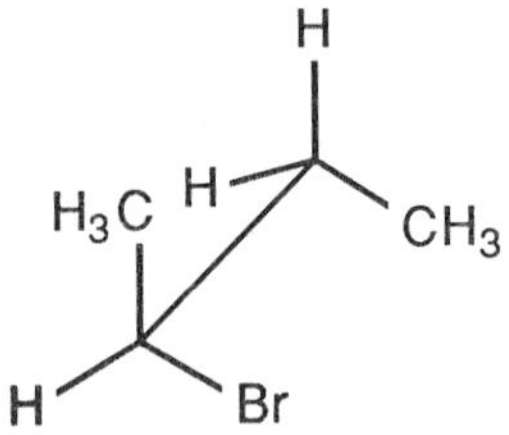 (e)

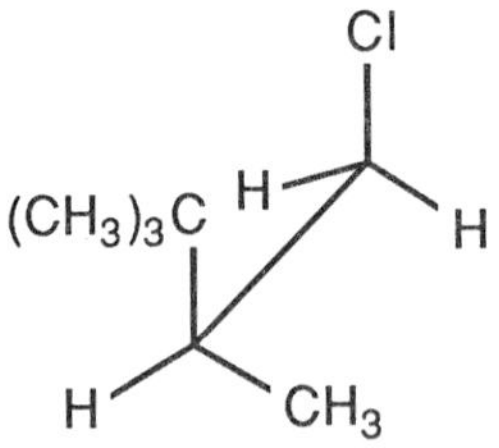

2.10 (a) 180°

(b) 60°

(c) 60°

(d) 0°

(e) 60°

(f) 120°

2.11 The reasonable choices are (d) and (h):

(a) A lower temperature would cause the reaction to proceed more slowly.

(b) Decreased concentrations would cause the reaction to proceed more slowly.

(c) The size of the flask is irrelevant; only the concentrations of the reactants are important.

(d) Increased concentrations would cause the reaction to proceed more rapidly, and this is the desired goal.

(e) This would not solve the problem. No matter how fast the students worked, the reaction would still be only 50% complete after a time that is twice as long as a laboratory class.

(f) These two changes would work against each other, so this is not a reasonable approach.

(g) The size of the flask is irrelevant; only the concentrations of the reactants are important.

(h) A higher temperature would cause the reaction to proceed more rapidly, and this is the desired goal.

(i) Who knows what will happen, but this certainly is not a reasonable approach.

2.12 In each case the ratio of the two conformations (i.e., the equilibrium constant) can be calculated from the expression:

$$\Delta G = -2.3RT\log K$$

where R is 2.0 cal/mol-deg and T is 300K. Since K is equal to [B]/[A] and the fraction of B is defined as [B]/([A] + [B]), algebraic manipulation gives

$$\text{fraction B} = K[A]/(K[A] + [A])$$

and this simplifies to

$$\text{fraction B} = K/(1+K)$$

(a)	0.54 A	0.46 B
(b)	0.70 A	0.30 B
(c)	0.84 A	0.16 B
(d)	0.97 A	0.03 B
(e)	0.9998 A	0.0002 B
(f)	≃1.0 A	10^{-7} B

2.13 Radicals (c) and (d) are both primary (and are in fact two drawings of the same radical); this is the least stable. Radical (a) is tertiary, while (b) is only secondary; therefore (a) is the most stable.

2.14 To answer this question, you must draw all possible C_5H_{12} isomers and evaluate the number of monochloro derivatives that could be formed from each. The unbranched isomer, pentane, would yield three isomers and 2,2-dimethylpropane would yield only one. Only 2-methylbutane would yield four isomeric monochloro derivatives.

$$CH_3—CH(CH_3)—CH_2—CH_3$$

2.15 As in the preceding problem, all possible C_6H_{14} isomers must be evaluated. The formation of only two monochloro derivatives indicates that the alkane has only two nonequivalent kinds of CH, and this should suggest to you that the alkane has symmetry. There are 5 possible isomers: hexane (3 dichloro products), 2-methylpentane (5 dichloro products), 3-methylpentane (4 dichloro products), 2,3-dimethylbutane (2 dichloro products), and 2,2-dimethylbutane (3 dichloro products). The correct answer is 2,3-dimethylbutane:

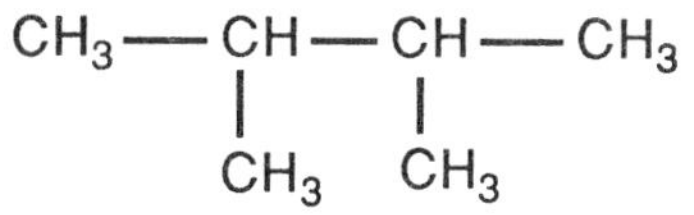

$$CH_3—\underset{\underset{CH_3}{|}}{CH}—\underset{\underset{CH_3}{|}}{CH}—CH_3$$

2.16 **Initiation:**

$$Br_2 \longrightarrow 2\ Br\cdot$$

Propagation:

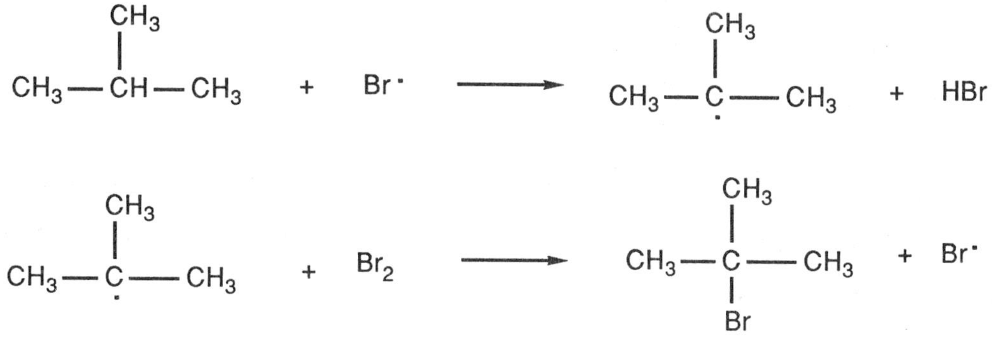

$$CH_3—\overset{\overset{CH_3}{|}}{CH}—CH_3 \quad + \quad Br\cdot \longrightarrow CH_3—\overset{\overset{CH_3}{|}}{\underset{\cdot}{C}}—CH_3 \quad + \quad HBr$$

$$CH_3—\overset{\overset{CH_3}{|}}{\underset{\cdot}{C}}—CH_3 \quad + \quad Br_2 \longrightarrow CH_3—\overset{\overset{CH_3}{|}}{\underset{\underset{Br}{|}}{C}}—CH_3 \quad + \quad Br\cdot$$

Termination:

$$Br\cdot \quad + \quad Br \longrightarrow Br_2$$

$$2\ CH_3—\overset{\overset{CH_3}{|}}{\underset{\cdot}{C}}—CH_3 \longrightarrow CH_3—\overset{\overset{CH_3}{|}}{C}—CH_3 \atop CH_3—\underset{\underset{CH_3}{|}}{C}—CH_3$$

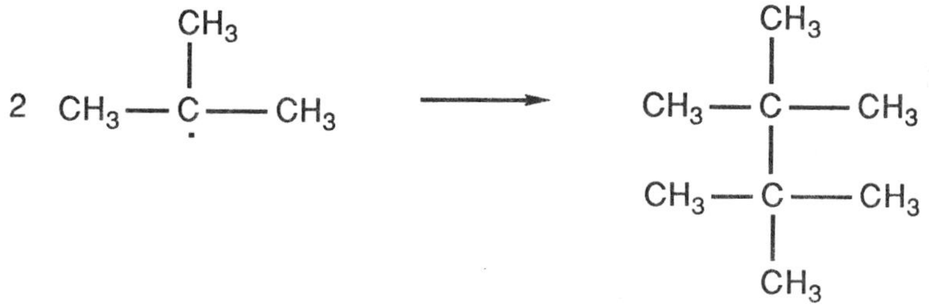

$$(CH_3)_3C\cdot + Br\cdot \longrightarrow (CH_3)_3C{-}Br$$

2.17 In the first step of the alternative mechanism a $Cl{-}CH_3$ bond is formed (−84 kcal/mol) and a $CH_3{-}H$ bond is broken (+104 kcal/mol) for a net change of +20 kcal/mol. In the second step a Cl−H bond is formed (−103) and a Cl−Cl bond is broken (+58) for a net change of −45 kcal/mol. The overall change for the two steps is −25 kcal/mol.

For the generally accepted mechanism the chlorine atom abstracts a hydrogen from methane (net change +1 kcal/mol) and in the second step of the chain reaction a methyl radical reacts with Cl_2 to form chloromethane and a chlorine atom (net change −26 kcal/mol). As you would expect (since overall energy changes are unrelated to the reaction mechanism), the overall change for the two steps is −25 kcal/mol.

The difference between the two mechanisms is found in the highly endothermic first step of the alternative mechanism. An activation energy of more than 20 kcal/mol is required for this pathway, so it would correspond to an extremely slow process in comparison to the accepted mechanism.

2.18 The ratio of 2-chloro-2-methylbutane to 1-chloro-2-methylbutane at 300K would be 7.5×10^{-7}.

2.19 To convert the tertiary halide, 2-chloro-2-methylbutane, to the primary isomer, 1-chloro-2-methylbutane, by a hypothetical process would require the following changes: Cleavage of a tertiary R–Cl (+80), formation of a tertiary R–H (−92), cleavage of a primary R–H (+98), and formation of a primary R–Cl (−81). The overall change is calculated as +5.0 kcal/mol, the amount by which the tertiary chloride is more stable. This is somewhat smaller than the observed value, but it still shows that the tertiary halide would be highly favored at equilibrium.

2.20 At room temperature (300K) a ratio of 97:3 would be expected, but much higher temperatures are normally used in industrial processes of this kind. If the equilibrium were established at 300°C (573K), the expected ratio would be only 85:15.

2.21 Five isomeric halides could be produced, and these are labeled as A–E in the following structures.

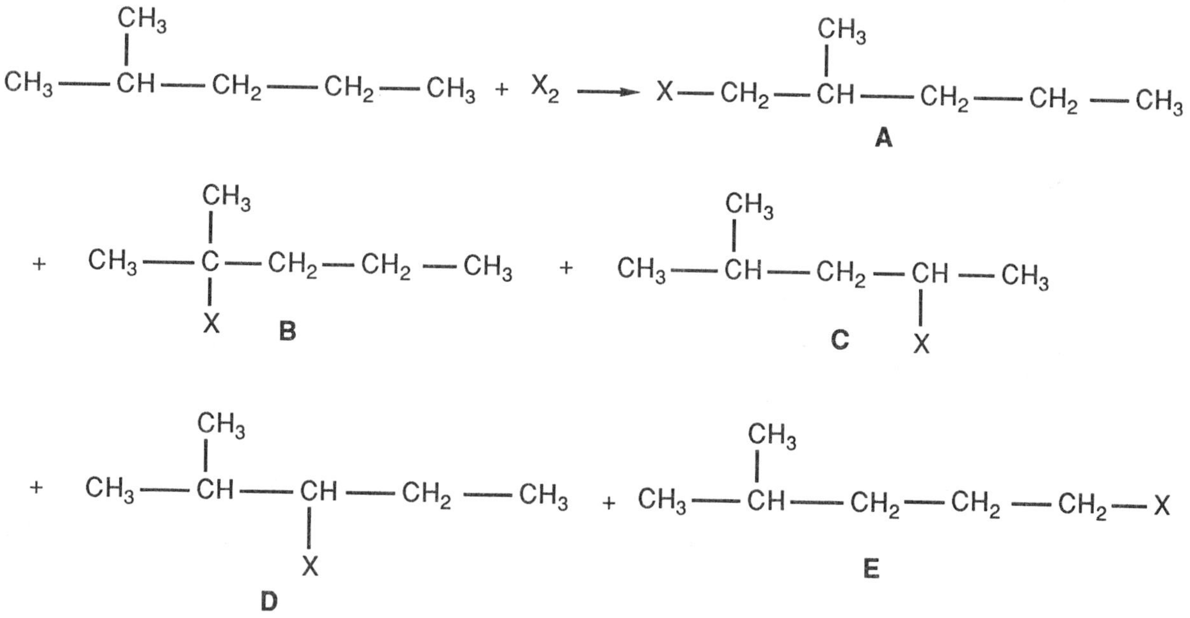

$$CH_3{-}CH(CH_3){-}CH_2{-}CH_2{-}CH_3 + X_2 \longrightarrow \underset{\textbf{A}}{X{-}CH_2{-}CH(CH_3){-}CH_2{-}CH_2{-}CH_3}$$

$$+\ \underset{\textbf{B}}{CH_3{-}C(CH_3)(X){-}CH_2{-}CH_2{-}CH_3} \ +\ \underset{\textbf{C}}{CH_3{-}CH(CH_3){-}CH_2{-}CH(X){-}CH_3}$$

$$+\ \underset{\textbf{D}}{CH_3{-}CH(CH_3){-}CH(X){-}CH_2{-}CH_3} \ +\ \underset{\textbf{E}}{CH_3{-}CH(CH_3){-}CH_2{-}CH_2{-}CH_2{-}X}$$

The selectivity calculations for chlorination are:

A (primary)	6H x 1	= 6	22%
B (tertiary)	1H x 4	= 4	15%
C (secondary)	2H x 3.5	= 7	26%
D (secondary)	2H x 3.5	= 7	26%
E (primary)	3H x 1	= 3	11%

2.22 Using the same structures as for the preceding problem, the selectivity calculations for bromination are:

A (primary)	6H x 1	= 6	0.3%
B (tertiary)	1H x 1600	= 1600	83%
C (secondary)	2H x 80	= 160	8%
D (secondary)	2H x 80	= 160	8%
E (primary)	3H x 1	= 3	0.2%

2.23 Compound A, which predominates at equilibrium, is more stable. The free energy difference for a 92:8 ratio at room temperature (300K) is 1.5 kcal/mol.

CHAPTER 3. CYCLOALKANES

ANSWERS TO EXERCISES

3.1 The reactant is symmetrical, so there are only two different kinds of cyclopropane C–C bonds. Cleavage of these with Br_2 would afford the following two possible products:

$$Br-C(CH_3)_2-CH_2CH_2Br \qquad BrCH_2-C(CH_3)_2-CH_2Br$$

In each case reaction would occur by reaction of a bromine atom with the cyclopropane ring to form a Br–C bond and generate a carbon free radical by rupture of the original C–C bond. The predominant product should be that which is formed via the most easily formed (i.e., most stable) free radical. This would be the first of the two products shown, because it would be formed via a tertiary radical:

$$\cdot C(CH_3)_2-CH_2-CH_2-Br$$

ANSWERS TO PROBLEMS

3.1 (a) 1-isopropyl-2-methylcyclohexane

(b) cyclodecane

(c) 1,2,3,4-tetramethylcyclobutane

(d) *trans*-1,4-dichlorocyclohexane

(e) *trans*-1-chloro-2-methylcyclopropane

(f) *trans*-1-*tert*-butyl-3-iodocyclohexane

(g) *cis*-1-ethyl-3-methylcyclopentane

(h) 1,1-dimethylcycloheptane

(i) *trans*-1-bromo-4-chlorocyclohexane

(j) *trans*-1,4-dimethylcyclohexane

3.2 (a) Br, CH_3 (b) $C(CH_3)_3$, H_3C, CH_3 (c) Br, Br

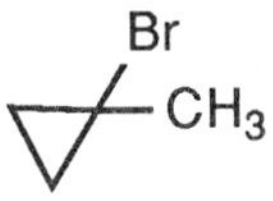

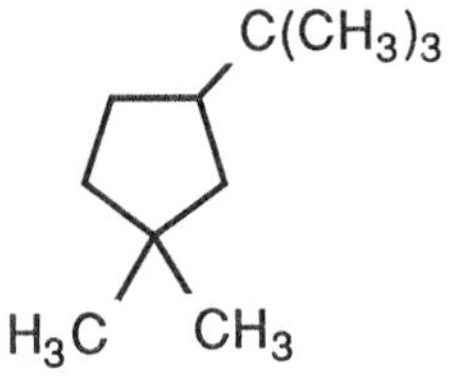

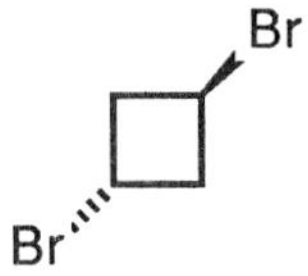

(d) Cl, $CH_2CH_2CH_3$ (e) F, F (f) Cl, Cl, Br

3.3 CH_2Cl, $CHCH_3$, CH_3

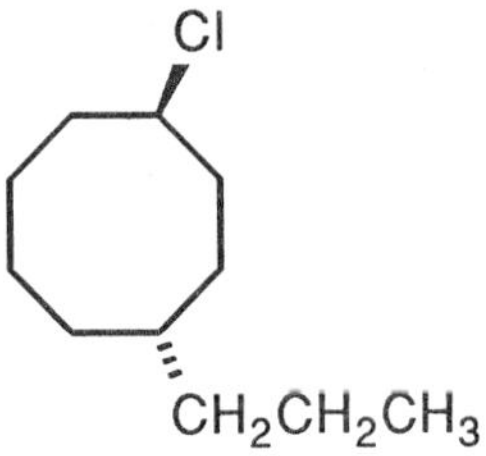

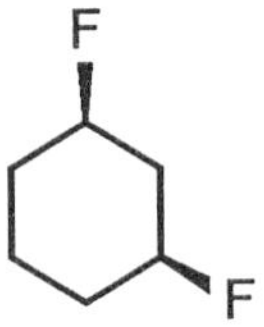

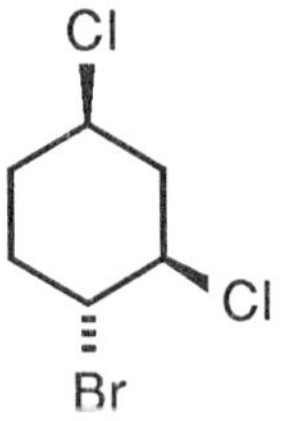

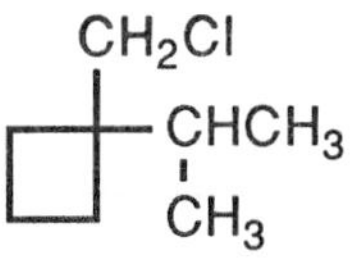

1-chloromethyl-
1-isopropylcyclobutane

CH_3, $CHCH_2Cl$, CH_3

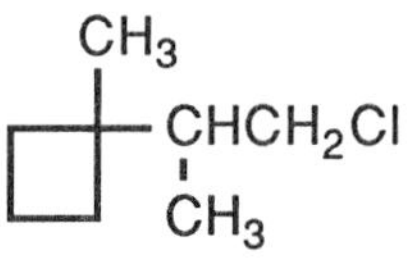

1-(2-chloro-1-methylethyl)-
1-methylcyclobutane

CH_3, $C(CH_3)_2$, Cl

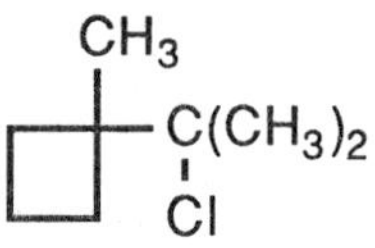

1-(1-chloro-1-methylethyl)-
1-methylcyclobutane

CH_3, $CHCH_3$, CH_3, Cl

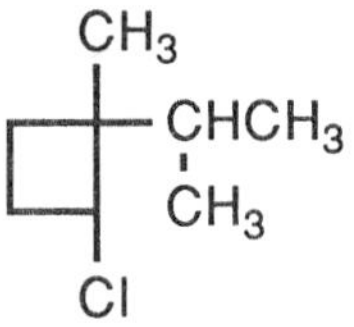

2-chloro-1-methyl-1-
isopropylcyclobutane

CH_3, $CHCH_3$, CH_3, Cl

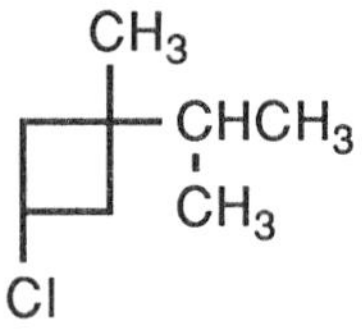

3-chloro-1-methyl-1-
isopropylcyclobutane

3.4

r-1-chloro-cis-2-isopropyl-2-methylcyclobutane

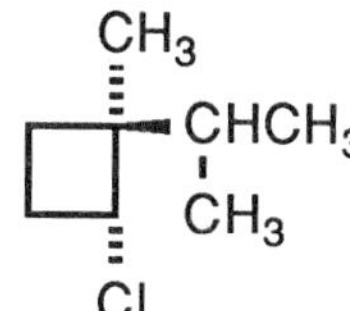

r-1-chloro-trans-2-isopropyl-2-methylcyclobutane

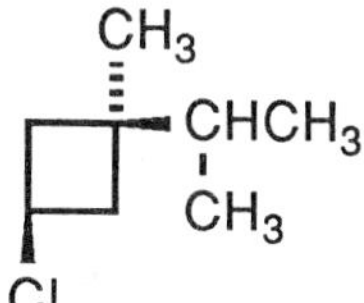

r-1-chloro-cis-3-isopropyl-3-methylcyclobutane

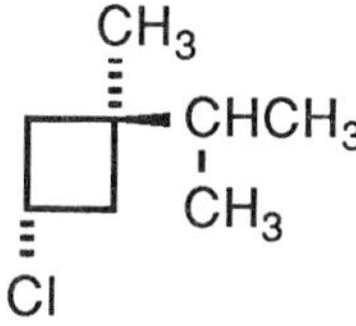

r-1-chloro-trans-3-isopropyl-3-methylcyclobutane

3.5

1,1-dichloromethylcyclopentane

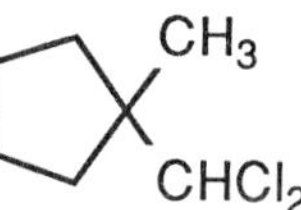

1-(dichloromethyl)-1-methylcyclopentane

2-chloro-1-(chloromethyl)-1-methylcyclopentane

3-chloro-1-(chloromethyl)-1-methylcyclopentane

3.6

r-1-chloro-cis-2-(chloromethyl)-2-methylcyclopentane

r-1-chloro-trans-2-(chloromethyl)-2-methylcyclopentane

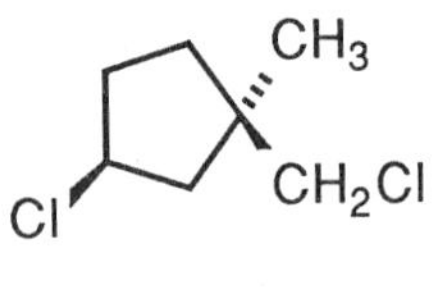

CH_3 CH_2Cl Cl

r-1-chloro-cis-3-(chloromethyl)-3-methylcyclopentane

CH_3 CH_2Cl Cl

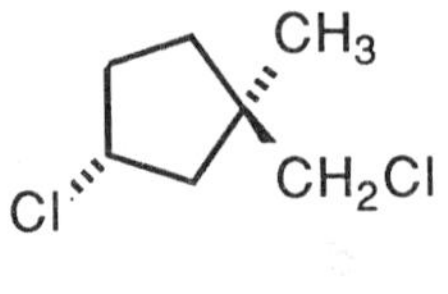

r-1-chloro-trans-3-(chloromethyl)-3-methylcyclopentane

3.7 (a) Stereoisomers

(b) Constitutional isomers

(c) Same (two drawings of the *cis* isomer)

(d) Same (two drawings of the *cis* isomer)

(e) Constitutional isomers (different carbon skeletons)

(f) Constitutional isomers (different carbon skeletons)

(g) Same (both are *cis*-1,4-diethylcyclodecane)

(h) Different (C_7 vs C_8)

(i) Same (both are *trans*1,4-dimethylcyclohexane)

(j) Different (C_6 vs C_5)

(k) Same (two conformations of *cis*-1-chloro-4-methylcyclohexane)

(l) Constitutional isomers (1,3- vs 1,4-dimethylcyclohexane)

3.8 **(a)** H, Cl, F, H

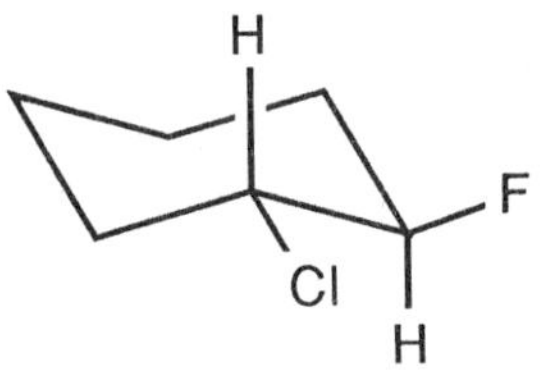

(b) H, Br, CH_3, H

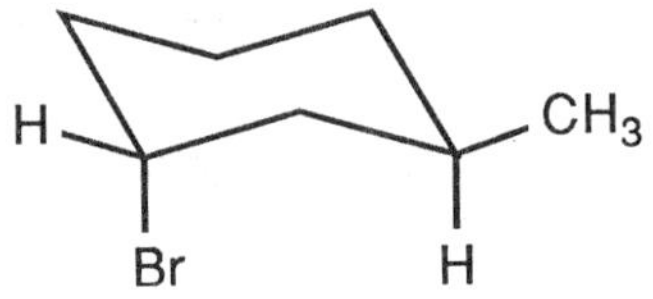

(c) H, CH_3CH_2, H, CH_3-CH_2

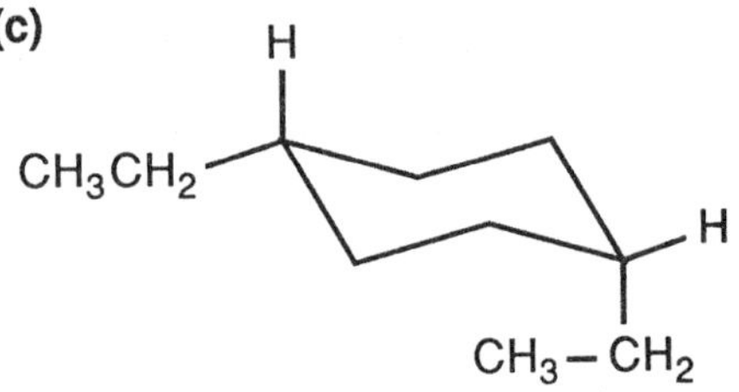

(d) $(CH_3)_3C$, H, CH_3, H

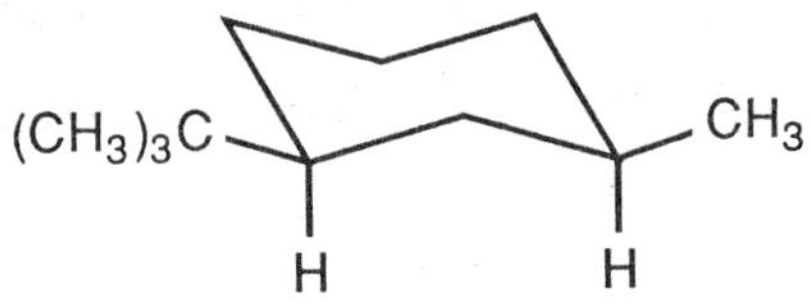

(e) H, CH_3-CH, CH_3, Br, H

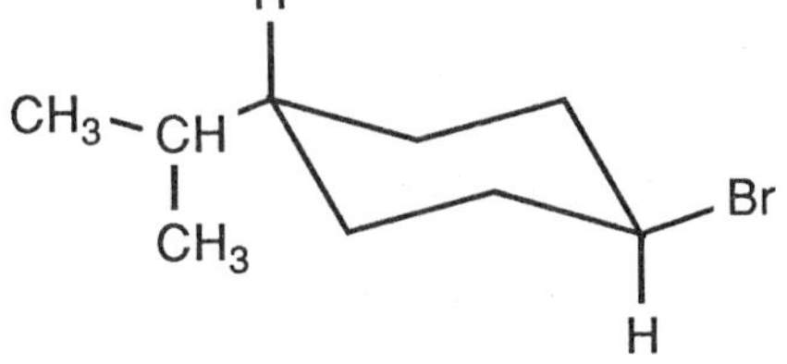

3.9 There is only a single way in which the target molecule can be dissected into two fragments, each containing six or fewer carbon atoms. One fragment must be a pentyl group and the other must be a 1-cyclobutylethyl group. The latter is secondary, so it must be employed as an organocuprate reagent in a reaction with a primary pentyl halide.

$$\text{C}_4\text{H}_7\text{–CH(CH}_3\text{)–Br} \xrightarrow[\text{2. CuI}]{\text{1. Li}} (\text{C}_4\text{H}_7\text{–CH(CH}_3\text{)–})_2\text{CuLi} \xrightarrow{CH_3CH_2CH_2CH_2CH_2Br} \text{C}_4\text{H}_7\text{–CH(CH}_3\text{)CH}_2\text{CH}_2\text{CH}_2\text{CH}_2\text{CH}_3$$

3.10 The directions for this problem require that a single cycloalkane be employed as the starting material, so a carbon–carbon bond cleavage must be involved. In other words high pressure hydrogenation must be used with either a cyclopropane or cyclobutane derivative. Several compounds could yield the target molecule by such hydrogenolysis: ethylcyclobutane, 1,2-dimethylcyclobutane, 1-ethyl-2-methylcyclopropane, and 1,2,3-trimethylcyclopropane. Only the last of the preceding hydrocarbons (because of its symmetry) would yield a single product upon hydrogenolysis, so this is the preferred method:

$$\text{1,2,3-trimethylcyclopropane} \xrightarrow[\text{high pressure}]{\text{Ni, }H_2} CH_3CH_2-\underset{\displaystyle CH_3}{CH}-CH_2CH_3$$

3.11 The reaction would be less exothermic for the compound that has a smaller potential energy (i.e., for the more stable isomer). This is the *cis* isomer for which both substituents can occupy equatorial positions.

HO, CH$_3$, H, H — *cis*   H, CH$_3$, OH, H — *trans*

3.12 The initiation step is light-induced cleavage of Br_2 into to bromine atoms. Propagation would then occur via the following steps:

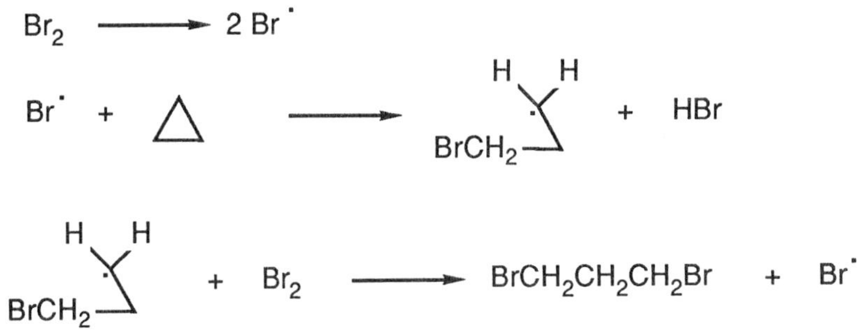

$$Br_2 \longrightarrow 2\ Br^{\cdot}$$

Br$^{\cdot}$ + △ ⟶ (H, H, BrCH$_2$–) + HBr

(H, H, BrCH$_2$–) + Br_2 ⟶ $BrCH_2CH_2CH_2Br$ + $Br^{\cdot}$

3.13 The two possible radical intermediates are:

(CH$_3$, CH$_3$, BrCH$_2$–) and (CH$_3$, CH$_3$, Br–, $^{\cdot}$CH$_2$–)

(i) (ii)

Radical (i) is more stable because it is tertiary. Radical (ii) would require more unfavorable interactions in its formation because the bromine atom would have to attack a hindered, quaternary carbon. Both factors indicate that radical (i) must be the actual intermediate.

3.14 Attack of a bromine atom on 1,1,2,2-tetramethylcyclopropane could occur at only two positions: at the secondary carbon C-3 to form intermediate (i), or at either of the quaternary carbons (C–1 or C–2) to form (ii) or (iii).

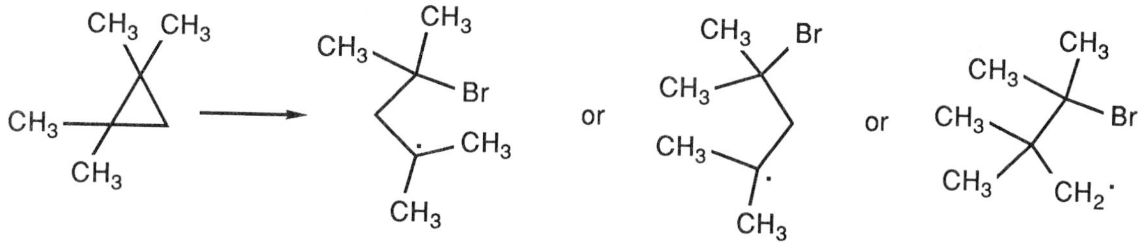

(i) or (ii) or (iii)

Formation of (iii), a primary radical, would not be expected. Moreover, steric hindrance would be expected to preclude attack at a quaternary carbon, so neither (ii) nor (iii) should be generated as intermediates. This means that (i) should be the intermediate in the reaction, and its subsequent reaction with Br_2 would yield 1,3–dibromo–2,2,3–trimethylbutane:

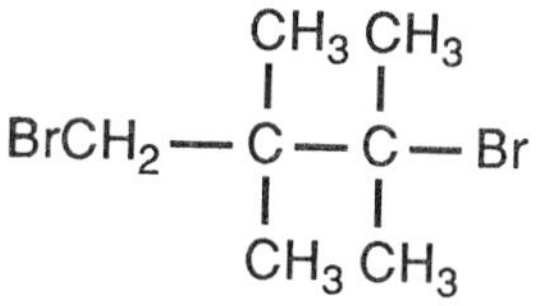

$$BrCH_2-C(CH_3)_2-C(CH_3)_2-Br$$

3.15

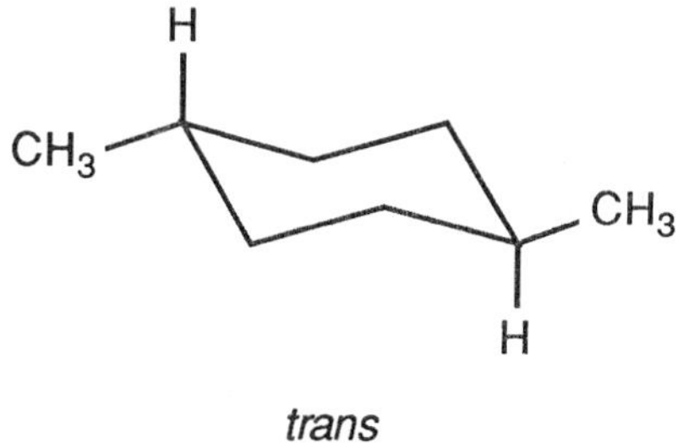

trans 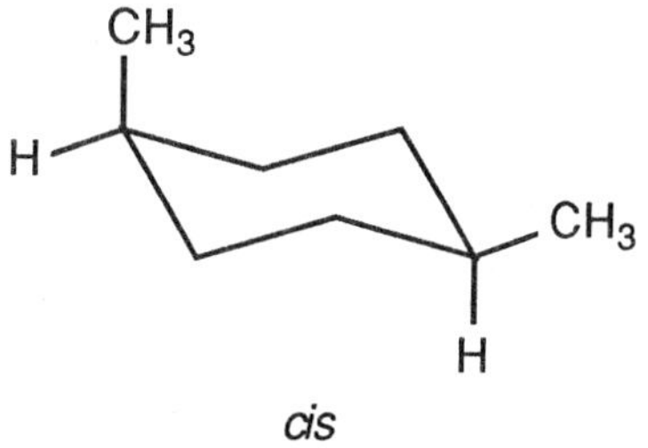 *cis*

The more stable isomer is the *trans* compound, for which both methyl groups can occupy equatorial positions. In the *cis* isomer one of the methyl groups must occupy an axial position. The energy difference of 1.9 kcal/mol between the two isomers is in very close agreement with the 1.7 kcal/mol attributed to an axial methyl group in Table 3.1.

3.16 One of the two substituents must be axial, either a methyl group (at a cost of 1.7

kcal/mol) or a bromine (at a cost of 0.4 kcal/mol).

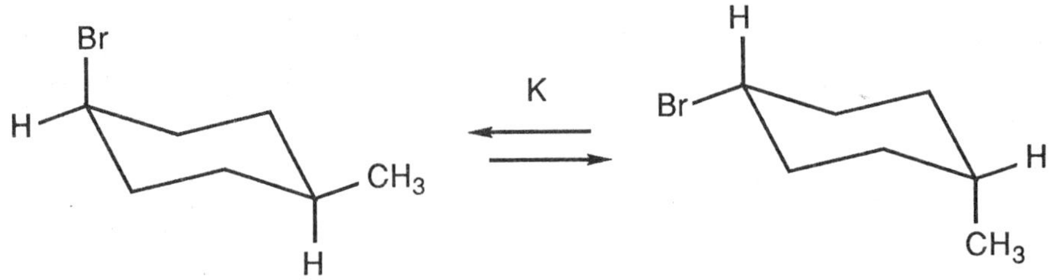

The first conformer, with an axial bromine, should be favored by 1.3 kcal/mol. From $\Delta G = -RT \log K$ the calculated equilibrium constant is 0.11, and the two conformations should be present in a distribution of about 90% to 10%.

3.17 From the sequence A → B → C → G, you know that compound G must have the molecular formula $C_{12}H_{22}$. It does not have a double bond, because no reaction occurs when it is subjected to mild conditions for hydrogenation. Vigorous hydrogenation yields H, which has the molecular formula $C_{12}H_{24}$, so this reactions has caused cleavage of a cyclopropane or cyclobutane ring. To deduce the nature of this ring, you can first determine where it is located. Compound A contains eight of the final 12 carbons of compound H, so it must contain the cyclohexane ring. Compound D is reduced by zinc to compound E, which must be C_4H_8. This molecular formula corresponds to a cycloalkane, so D is either cyclobutane or methylcyclopropane. Compound D must therefore be bromocyclobutane, (bromomethyl)cyclopropane, or 1-bromo-2-methylcyclopropane. If you draw the coupling product that would result from bromocyclobutane, you will find that subsequent hydrogenolysis could not produce the carbon skeleton of compound H. That means that D is a cyclopropane derivative. The preferred answer is (bromomethyl)cyclopropane for two reasons: First, it would lead cleanly to compound H by hydrogenolysis between the less hindered CH_2 groups (whereas a mixture might be expected from the sequence using 1-bromo-2-methylcyclopropane). Second, the reaction with an organocopper reagent proceeds better with a primary haloalkane. The full structures of A-G are shown:

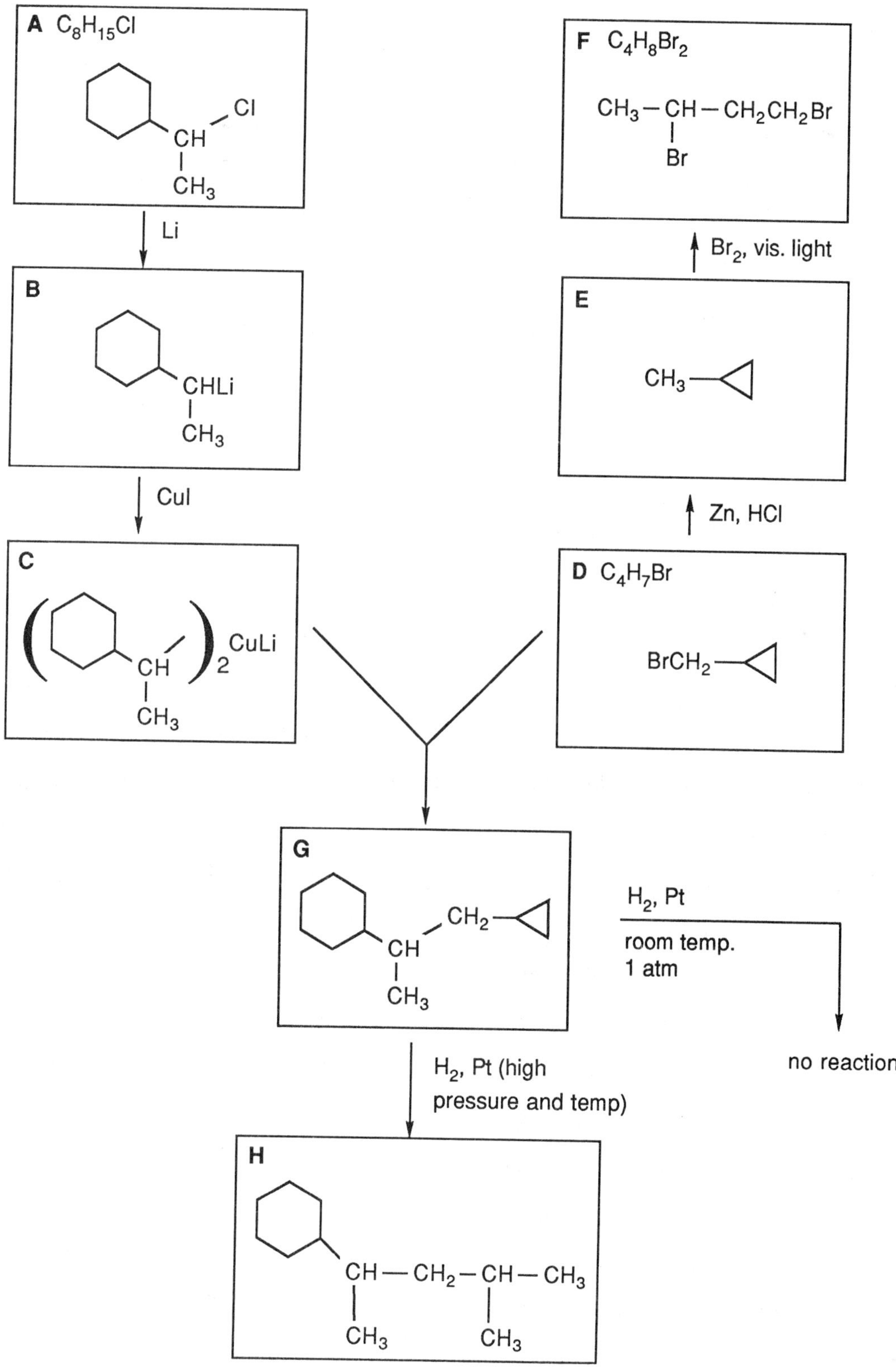
A $C_8H_{15}Cl$
Cl
CH
CH_3
Li
B
CHLi
CH_3
CuI
C
CH
CH_3
2
CuLi
F $C_4H_8Br_2$
$CH_3-CH-CH_2CH_2Br$
Br
Br_2, vis. light
E
CH_3
Zn, HCl
D C_4H_7Br
$BrCH_2$
G
CH_2
CH
CH_3
H_2, Pt
room temp.
1 atm
no reaction
H_2, Pt (high
pressure and temp)
H
$CH-CH_2-CH-CH_3$
CH_3
CH_3

CHAPTER 4. ACIDS, BASES AND INTERMOLECULAR FORCES

ANSWERS TO EXERCISES

4.1 **(a)**  H—Ö—H (two lone pairs on O) 8 protons, 8 electrons formal charge = 0

(b)  [H—Ö:]⁻ (three lone pairs on O) 8 protons, 9 electrons formal charge = -1

4.2 C_2H_5—O(+)—B(−)F₃ (O also bonded to a second C_2H_5; B bonded to three F)

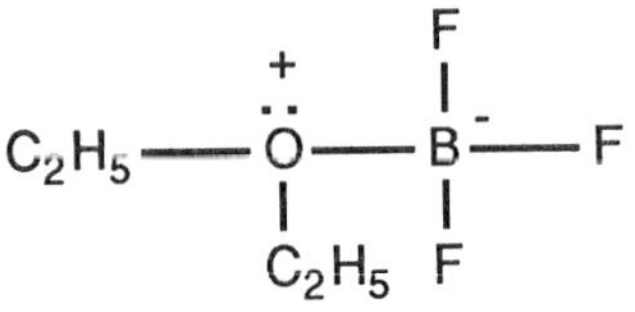

oxygen	boron
8 protons	5 protons
<u>7 electrons</u>	<u>6 electrons</u>
+1	-1

4.3

(a) 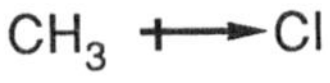 CH_3 ⟶ Cl (crossed arrow)

(b) 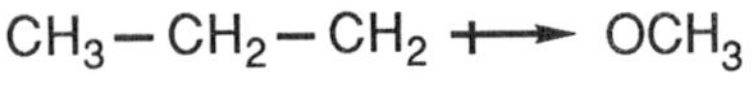 $CH_3-CH_2-CH_2$ ⟶ OCH_3 (crossed arrow)

(c) none

(d) 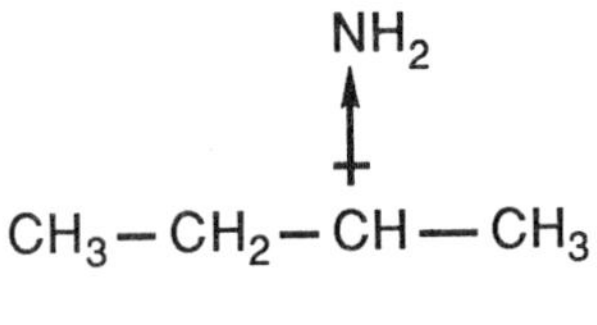 $CH_3-CH_2-CH(NH_2)-CH_3$, crossed arrow from CH to NH_2

(e) none

(f) 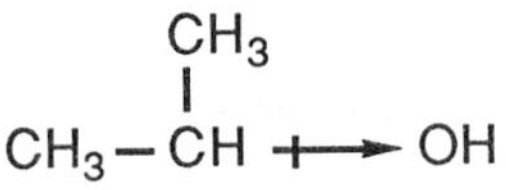 $CH_3-CH(CH_3)$ ⟶ OH (crossed arrow)

(g) 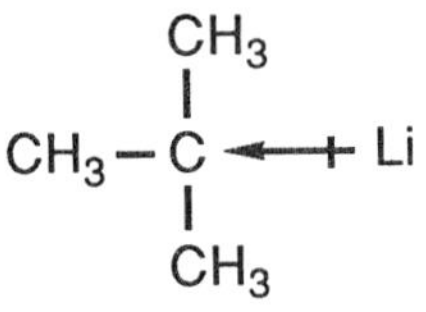 $CH_3-C(CH_3)_2$ ⟵ Li (crossed arrow)

(h)  CH_3-CH_2 ⟵ MgBr (crossed arrow)

(i) 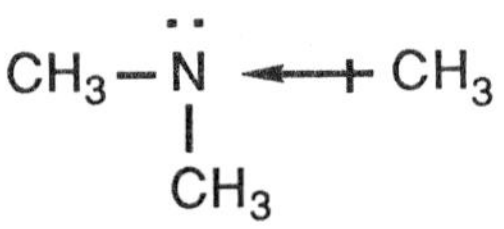 $CH_3-\ddot{N}(CH_3)$ ⟵ CH_3 (crossed arrow)

4.4 $(CH_3-CH_2)_3N > (CH_3-CH_2)_2O > CH_3-CH_2-F$

4.5 $CH_3-\overset{+}{F}H > CH_3-\overset{+}{O}H_2 > CH_3-\overset{+}{N}H_3$

4.6 $CH_3-CH_2-OH > CH_3-CH_2-NH_2 > CH_3-CH_2-CH_3$

4.7 $(CH_3)_3C^- > (CH_3)_2N^- > (CH_3)O^-$

4.8 **(a)** $CH_3-C(=O)-\overset{-}{C}H_2$ $\quad$ $CH_3-C(=O)-\overset{-}{N}H$

(b) $CH_3-C(=O)-\overset{-}{C}H_2 \longleftrightarrow CH_3-C(-O^-)=CH_2$

$CH_3-C(=O)-\overset{-}{N}H \longleftrightarrow CH_3-C(-O^-)=NH$

In each case the resonance form with negative charge on oxygen should contribute most to the overall structure.

4.9

$$CH_3-CH_2-\overset{-}{O}\overset{+}{Na} + CH_3-\underset{\underset{O}{\|}}{C}-CH_2-\underset{\underset{O}{\|}}{C}-CH_3 \longrightarrow CH_3-CH_2-OH + CH_3-\underset{\underset{O}{\|}}{C}-\overset{-}{C}H-\underset{\underset{O}{\|}}{C}-CH_3\ \overset{+}{Na}$$

$$\updownarrow$$

$$CH_3-\underset{\underset{O^-}{|}}{C}=CH-\underset{\underset{O}{\|}}{C}-CH_3\ \overset{+}{Na} \longleftrightarrow CH_3-\underset{\underset{O}{\|}}{C}-CH=\underset{\underset{O^-}{|}}{C}-CH_3\ \overset{+}{Na}$$

The two resonance forms with negative charge on oxygen contribute most two the overall structure of the ion; the resonance form with negative charge on carbon contributes much less.

4.10

A $$CH_3-\underset{\underset{O}{\|}}{C}-OH + \overset{+}{H} \longrightarrow CH_3-\underset{\underset{+OH}{\|}}{C}-OH \longleftrightarrow CH_3-\underset{\underset{OH}{|}}{C}=\overset{+}{O}H$$

B $$CH_3-\underset{\underset{O}{\|}}{C}-OH + \overset{+}{H} \longrightarrow CH_3-\underset{\underset{O}{\|}}{C}-\overset{+}{O}H_2$$

When the doubly bonded oxygen is protonated (**A**), you can draw two equivalent resonance forms that place positive charge on each of the two oxygen atoms. In contrast, no reasonable resonance forms can be drawn when the singly bonded oxygen is protonated (**B**).

4.11 (a) The hydroxyl-substituted carbocation is resonance stabilized.

$$H-\ddot{O}-\overset{+}{C}H_2 \longleftrightarrow H-\overset{+}{O}=CH_2$$

(b) The most important resonance form is an oxonium ion with both carbon and oxygen having an octet of electrons in its valence shell.

$$H-\overset{+}{O}=CH_2$$

(c) The oxygen atom bears most of the positive charge.

(d) If only inductive effects were operative, the electron withdrawing effect would destabilize $HO\text{-}CH_2^+$, and CH_3+ would be more stable.

$$CH_3^+$$

(e) Resonance effects appear to be more stable in this case, since the $HO\text{-}CH_3^+$ cation is more stable.

4.12 The material with the higher melting point (−50°) should be the more symmetrical of the two isomeric compounds.

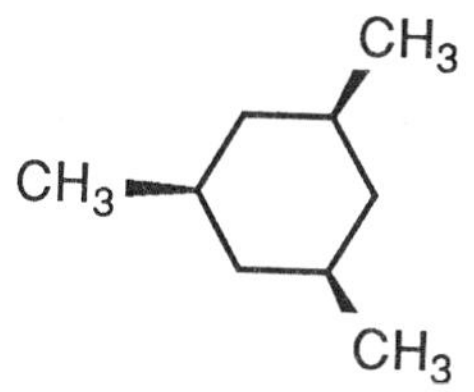

CH_3 CH_3 CH_3 (1,3,5-trimethylcyclohexane) — -50 °C

$CH_2CH_2CH_3$ (propylcyclohexane) — -95 °C

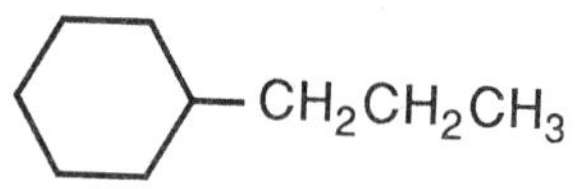

4.13 The two compounds have nearly identical molecular weights: hexane, MW 86 and $C_4H_{10}O_2$, MW 90. The major effect on boiling point should therefore arise from intermolecular forces. The dipolar attractions in the oxygen derivative should give it the higher boiling point. It differs by 14° from that of hexane (69°), so its boiling point must be 83°.

4.14 Diethylamine will undergo intermolecular hydrogen bonding, whereas triethylamine (which has no N-H group) cannot. The hydrogen bonding causes diethylamine to have a higher boiling point.

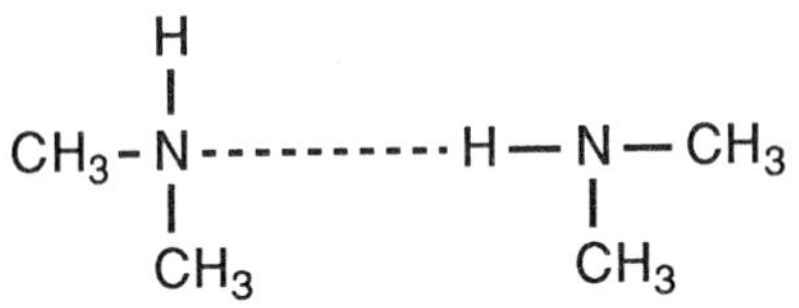

$$CH_3-\underset{\underset{CH_3}{|}}{\overset{\overset{H}{|}}{N}}\cdots\cdots\cdots H-\underset{\underset{CH_3}{|}}{N}-CH_3$$

ANSWERS TO PROBLEMS

4.1 **(a)** $CH_3-NH_2 \; + \; H^+ \longrightarrow CH_3-\overset{+}{N}H_3$ $\quad N = +1$

(b) $HO-\underset{\underset{O}{\|}}{S}-OH \longrightarrow H^+ \; + \; HO-\underset{\underset{O}{\|}}{S}-O^-$ $\quad O = -1$

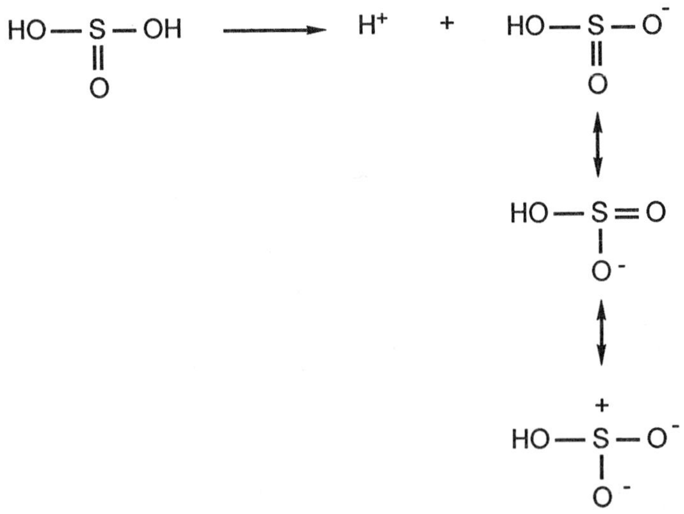

$\updownarrow$

$HO-\underset{\underset{O^-}{|}}{S}=O$ $\quad O = -1$

$\updownarrow$

$HO-\underset{\underset{O^-}{|}}{\overset{+}{S}}-O^-$ $\quad 2 \times O = -1$, $S = +1$

(c) $BF_3 \; + \; (CH_3)_3N \longrightarrow (CH_3)_3\overset{+}{N}\overset{-}{B}F_3$ $\quad N = +1$, $B = -1$

(d) $CH_3-CH_2-CH_2-OH \; + \; H^+ \longrightarrow CH_3-CH_2-CH_2-\overset{+}{O}H_2$ $\quad O = +1$

(e) $(CH_3)_2NH \longrightarrow H^+ \; + \; (CH_3)_2\overset{-}{N}$ $\quad N = -1$

(f) $CH_3-CH_2-\overset{\overset{\displaystyle O}{\|}}{C}-OH \longrightarrow H^+ + CH_3-CH_2-\overset{\overset{\displaystyle O}{\|}}{C}-O^-$

$\updownarrow$ O = -1

$CH_3-CH_2-\underset{\underset{\displaystyle O^-}{|}}{C}=O$

(g) $H-\overset{\overset{\displaystyle O}{\|}}{C}-NH_2 + CH_3O^-Na^+ \longrightarrow$

$CH_3OH + H-\overset{\overset{\displaystyle O^-\ Na^+}{|}}{C}=NH \longleftrightarrow H-\overset{\overset{\displaystyle O}{\|}}{C}-\overset{-}{N}H\ Na^+$

O = -1 N = -1

4.2

(a) $CH_3-CH_2-\overset{\overset{\displaystyle OH^*}{|}}{CH}-CH_3$

(b) $CH_3-CH_2-CH_2-\overset{*}{N}H-\overset{\overset{\displaystyle O}{\|}}{C}-CH_3$

(c) $H-\overset{\overset{\displaystyle O}{\|}}{C}-\overset{*}{C}H_2-CH_3$

(d) $CH_3-O-\overset{\overset{\displaystyle O}{\|}}{C}-\overset{*}{C}H_2-\overset{\overset{\displaystyle O}{\|}}{C}-CH_2-CH_3$

(e) $CH_3-\overset{\overset{\displaystyle O}{\|}}{C}-\underset{\underset{\displaystyle NH_2}{|}}{\overset{*}{C}H}-CH_3$

(f) $CH_3-\overset{\overset{\displaystyle OH^*}{|}}{CH}-CH_2-NH_2$

(g) $CH_3-\overset{\overset{\displaystyle OH}{|}}{CH}-\overset{\overset{\displaystyle O}{\|}}{C}-OH^*$

(h) $CH_3-\overset{\overset{\displaystyle OH^*}{|}}{CH}-\overset{\overset{\displaystyle O}{\|}}{C}-OCH_3$

4.3 (a)

$$CH_3-CH_2-CH_2-CH_2-OH$$

Alcohols ($K_a \approx 10^{-16}$) are more acidic than amines ($K_a \approx 10^{-34}$).

(b)

$$CH_3-CH_2-\overset{\displaystyle O}{\overset{\|}{C}}-CH_3$$

$$CH_3-\overset{-}{C}H-\overset{\displaystyle O}{\overset{\|}{C}}-CH_3 \longleftrightarrow CH_3-CH{=}\overset{\displaystyle O^-}{\overset{|}{C}}-CH_3$$

Only for the carbonyl compound can the anion be stabilized by resonance.

(c)

$$CH_3-O-\overset{\displaystyle O}{\overset{\|}{C}}-CH_2-\overset{\displaystyle O}{\overset{\|}{C}}-O-CH_3$$

$$CH_3-O-\overset{\displaystyle O}{\overset{\|}{C}}-\overset{-}{C}H-\overset{\displaystyle O}{\overset{\|}{C}}-O-CH_3 \longleftrightarrow CH_3-O-\overset{\displaystyle O}{\overset{\|}{C}}-CH{=}\overset{\displaystyle O^-}{\overset{|}{C}}-O-CH_3$$

$$\updownarrow$$

$$CH_3-O-\overset{\displaystyle ^-O}{\overset{|}{C}}{=}CH-\overset{\displaystyle O}{\overset{\|}{C}}-O-CH_3$$

For this compound the anion is resonance stabilized by two carbonyl groups.

(d)

$$CH_3-CH_2-\overset{\displaystyle O}{\overset{\|}{C}}-NH-CH_3$$

$$CH_3-CH_2-\overset{\displaystyle O}{\overset{\|}{C}}-\overset{-}{N}-CH_3 \longleftrightarrow CH_3-CH_2-\overset{\displaystyle ^-O}{\overset{|}{C}}{=}N-CH_3$$

A resonance-stabilized anion can be formed in both cases, but only for this compound is there an N-H. The most acidic hydrogens on the other compound are those of the CH_3 group.

(e) $CH_3-CH_2-\overset{\overset{O}{\|}}{C}-NH-CH_3$

$$CH_3-CH_2-\overset{\overset{O}{\|}}{C}-\overset{-}{N}-CH_3 \longleftrightarrow CH_3-CH_2-\overset{\overset{^{-}O}{|}}{C}=N-CH_3$$

The anion would result from ionization of an N-H group in both cases, but only for this compound is the resulting anion resonance stabilized.

(f)

$$CH_3-\overset{\overset{OH}{|}}{CH}-CH_2-CH_3$$

Alcohols ($K_a \simeq 10^{-17}$) are more acidic than ketones ($K_a \simeq 10^{-20}$). See Table 4.3.

(g)

$$CH_3-\underset{\underset{CH_3}{|}}{\overset{\overset{Cl}{|}}{C}}-\overset{\overset{O}{\|}}{C}-OH$$

The inductive effect of the chlorine will increase the acidity of this carboxylic acid.

(h)

$$CH_3-CH_2-CH_2-NH-CH_3$$

Amines have $K_a \simeq 10^{-34}$, whereas the other compound has only hydrogens that are bonded to carbon and therefore resembles an alkane ($K_a < 10^{-40}$). See Table 4.3.

4.4 (a) $CH_3-CH_2-CH_2-\overset{\longmapsto}{CH_2-OH}$ $\qquad$ $CH_3-CH_2-\overset{\longmapsto\ \longleftarrow\!\!+}{CH_2-NH-CH_3}$

(b)

$$CH_3-CH_2\overset{\uparrow}{\underset{+}{-}}\overset{\overset{O}{\|}}{C}-CH_3$$

(c)

$$CH_3-O\overset{\uparrow}{\underset{+}{-}}\overset{\overset{O}{\|}}{C}-CH_2\overset{\uparrow}{\underset{+}{-}}\overset{\overset{O}{\|}}{C}-O-CH_3 \qquad CH_3\overset{\uparrow}{\underset{+}{-}}\overset{\overset{O}{\|}}{C}-O-CH_2-O\overset{\uparrow}{\underset{+}{-}}\overset{\overset{O}{\|}}{C}-CH_3$$

(d) $CH_3-CH_2-\overset{O}{\overset{\|}{C}}-NH-CH_3$ $\quad$ $CH_3-CH_2-\overset{O}{\overset{\|}{C}}-\overset{CH_3}{\overset{|}{N}}-CH_3$

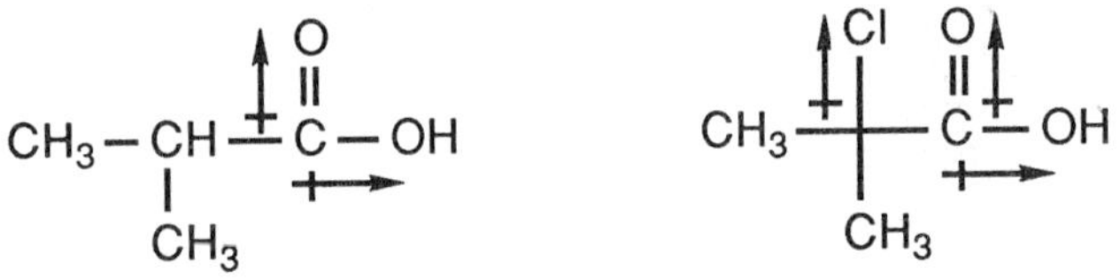

(e) $CH_3-CH_2-\overset{O}{\overset{\|}{C}}-NH-CH_3$ $\quad$ $CH_3-CH_2-CH_2-NH-CH_3$

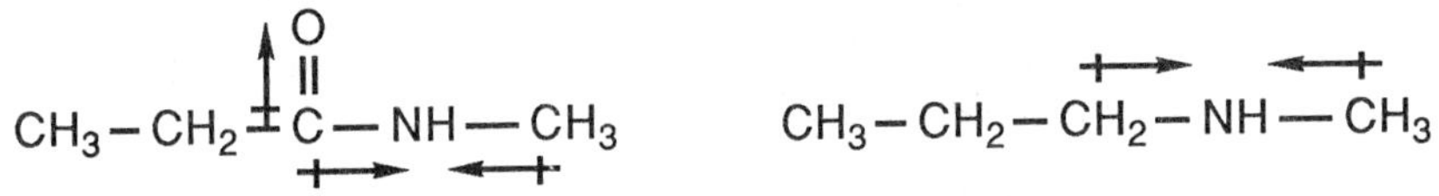

(f) $CH_3-\overset{O}{\overset{\|}{C}}-CH_2-CH_3$ $\quad$ $CH_3-\overset{OH}{\overset{|}{C}H}-CH_2-CH_3$

(g) $CH_3-\underset{CH_3}{\underset{|}{C}H}-\overset{O}{\overset{\|}{C}}-OH$ $\quad$ $CH_3-\overset{Cl}{\underset{CH_3}{C}}-\overset{O}{\overset{\|}{C}}-OH$

(h) $CH_3-CH_2-CH_2-NH-CH_3$ $\quad$ $CH_3-CH_2-CH_2-O-CH_3$

4.5 (a) $(CH_3CH_2CH_2)_3N$

The basicity of the other compound would be reduced by the inductive effect of the fluorine substituents.

(b) $CH_3CH_2O^-Na^+$

The other anion is resonance stabilized, which reduces its basicity.

(c) $CH_3CH_2NH_2$

Nitrogen is inherently more basic than oxygen (Table 4.1). In neither case would the protonated form be resonance stabilized.

(d) CH_3NHCH_3

Nitrogen is inherently more basic than oxygen (Table 4.1).

(e) $CF_3CH_2CH_2O^-K^+$

In the other compound the CF_3 group is separated from the negatively charged oxygen by one less CH_2 group, and it would better stabilize the anion through inductive effects. That compound would therefore be less basic than the one drawn here.

4.6 The boron in BF_3 has only 6 electrons in its valence shell. The three electron pairs are furthest apart when all three bonds lie in a single plane and are separated by 120°. In contrast the nitrogen of ammonia has 8 electrons in its valence shell, and electron pair repulsions are minimized with a tetrahedral geometry in which the bonds are separated by angles of 109°.

4.7 In each case the atom of the reactive center has 8 valence level electrons in the product, so a tetrahedral geometry is preferred.

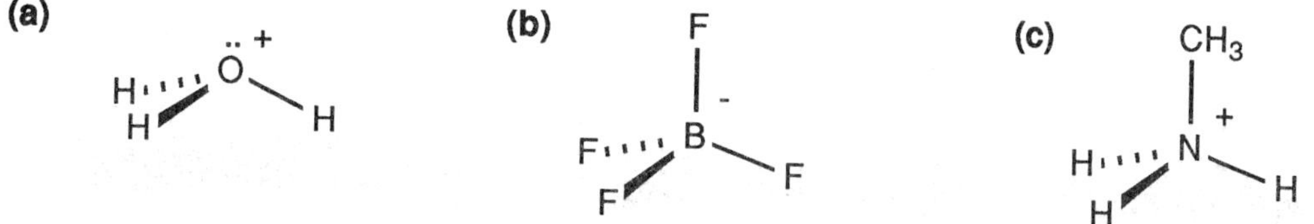

4.8 The methyl cation is electron deficient with only 6 electrons in the valence level of the central carbon, so a planar geometry is preferred. In contrast, the methyl anion has 8 valence electrons at the central carbon, so a tetrahedral geometry is preferred.

(a) (b)

4.9 (a) The compounds are isomers, and this one is more symmetrical.

$$CH_3-C(CH_3)_2-C(CH_3)_2-CH_3$$

(b) The compounds are isomers, and this one is more symmetrical.

$$CH_3-C(CH_3)_2-C(=O)-CH_3$$

(c) The compounds are isomers, and this one is more symmetrical.

[cyclohexane]

(d) The compounds have comparable molecular weights, but this one is a salt.

$$CH_3CH_2NH_3^{+}Cl^{-}$$

(e) The compounds are isomers, and this one is more symmetrical.

$$CH_3-C(CH_3)_2-C(=O)-OH$$

4.10 (a) The compounds are isomers, but this one can undergo intermolecular hydrogen bonding.

$$CH_3-CH(CH_3)-NH_2$$

(b) The compounds have similar structures, but this one has a higher molecular weight.

$$CH_3CH_2CH_2CH_2CH_2OH$$

(c) The compounds are isomers. The other can undergo intramolecular hydrogen bonding via a 6-membered ring. A less favorable 7-membered ring would be required for this compound, so intermolecular hydrogen bonding predominates and results in a higher boiling point.

$$HOCH_2CH_2CH_2CH_2OH$$

(d) The compounds are isomers, but this one can undergo intermolecular hydrogen bonding.

$$CH_3-CH_2-\overset{O}{\overset{\|}{C}}-NH_2$$

(e) The compounds are isomers, but this one can undergo intermolecular hydrogen bonding.

$$CH_3CH_2CH_2CH_2OH$$

4.11 (a) The oxygen atoms in this compound allow favorable hydrogen bonding with water, an interaction that is not possible with the other compound (a hydrocarbon).

(b) This compound is more polar, having the same polar group but fewer carbon atoms than the other compound.

$$CH_3-\overset{O}{\overset{\|}{C}}-OH$$

4.12 (a) Hexane is a nonpolar solvent, and CCl_4 would be more soluble because it is less polar than the other compound (an amide).

$$CCl_4$$

(b) Cyclohexane is less polar than the derivative that has six hydroxyl groups.

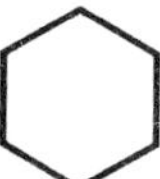

(c) This compound has chloro and amino substituents, but it is still less polar than the other compound (which is a salt).

$$ClCH_2CH_2NH_2$$

(d) This compound has a polar substituent (COOH), but it is still less polar than the other compound (which is a salt).

$$CH_3-CH_2-CH_2-\overset{\overset{O}{\|}}{C}-OH$$

4.13 (a) Water is a polar solvent, and the more polar amide would be more soluble. other compound (an amide).

$$CH_3-\overset{\overset{O}{\|}}{C}-NH_2$$

(b) The hydroxyl groups are polar and can undergo hydrogen bonding with the water.

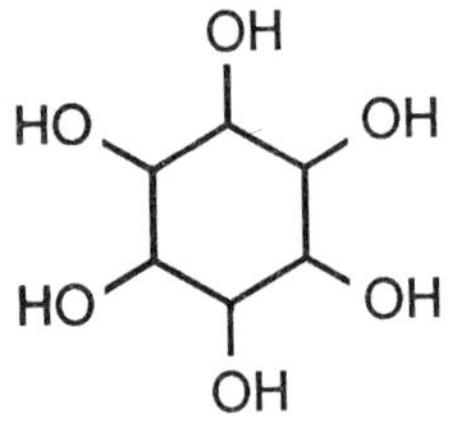

OH
HO OH
HO OH
OH

(c) The salt will be more soluble in water.

$$CH_3CH_2NH_3^+Cl^-$$

(d) The salt will be more soluble in water.

$$CH_3-CH_2-CH_2-\overset{\overset{O}{\|}}{C}-O^-\,K^+$$

4.14 Propionic acid itself exhibits a K_a of 1.3 x 10^{-5}, so these two derivatives have K_a's that are larger by factors of about 2500 and 500. Several clues are available from Figure 4.6 and from the discussion in Section 4.8. When a bromine is substituted on the carbon atom adjacent to the carbonyl group, the K_a of acetic acid increases by a factor of almost 80 (from 1.7 x 10^{-5} to 130 x 10^{-5}). The effect of fluorine substitution is comparable (a factor of about 100), but for a second fluorine on the same carbon K_a increases only by an additional factor of 30 for a total change (relative to acetic acid itself) of a factor of 3000. This is approximately the change observed for one of the unknown compounds, so it can be identified as 2,2-dibromopropionic acid. The data in Figure 4.6 show that the increase in acidity decreases dramatically when an additional CH_2 is interposed between the electron withdrawing group and the carboxyl group. 3,3,3-trifluoropropionic acid exhibits an increase in K_a of only 80 (relative to propionic acid) so it is not reasonable to suggest that to bromines on C-3 would increase K_a by a factor of 500. The only remaining alternative is 2,3-dibromopropionic acid, which would be consistent with a factor of 80 for the 2-bromo and a much smaller factor of 6 for the 3-bromo substituent.

4.15 Methane would produce an unsubstituted carbanion, but a cyano group can stabilize the negative charge by resonance.

$$CH_3-C\equiv N \longrightarrow H^+ + \overset{-}{C}H_2-C\equiv N \longleftrightarrow CH_2=C=N^-$$

The compound with two cyano groups can delocalize the negative charge onto both nitrogen atoms in addition to the central carbon atom.

$$N\equiv C-CH_2-C\equiv N \longrightarrow H^+ +$$

$$N\equiv C-\overset{-}{C}H-C\equiv N \longleftrightarrow N\equiv C-CH=C=\overset{-}{N} \longleftrightarrow \overset{-}{N}=C=CH-C\equiv N$$

4.16 The structure labelled **A** is more symmetrical (C_2 axis) than isomer **B.**

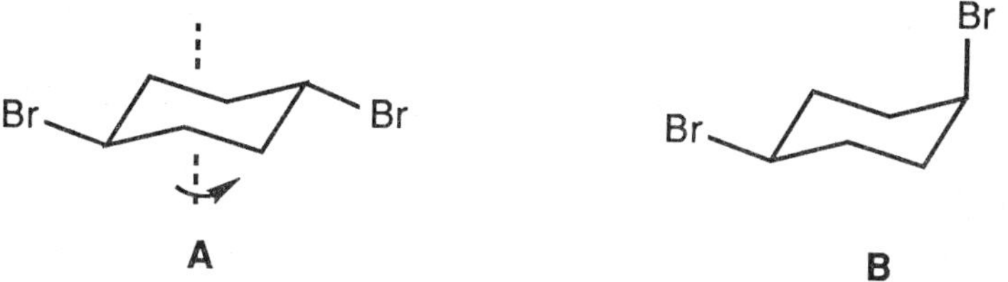

Therefore **A** is the higher melting isomer (mp 113°C), and **B** is the liquid.

4.17 Ammonium salts that can evaporate are those capable of dissociating into an amine (which is volatile) and a volatile acid such as HCl. Two such cases are illustrated in the following equations:

$$RNH_3^+Cl^- \rightleftharpoons RNH_2 + HCl$$

$$R_4N^+Cl^- \rightleftharpoons R_3N + HCl$$

The preceding equilibrium reactions proceed only to a very small extent, but this allows the volatile components to vaporize and recombine subsequently via the reverse reaction to form the salt.

4.18 (a) For ionization of the NH_2 group, a K_a of approximately 10^{-17} would be expected:

$$CH_3-\overset{\overset{\displaystyle O}{\|}}{C}-NH_2 \rightleftharpoons CH_3-\overset{\overset{\displaystyle O}{\|}}{C}-\overset{-}{N}H \quad + \quad H^+$$

ΔG at room temperature (300K) would be 22.1 kcal/mol. For ionization of the CH_3 group, a K_a of approximately 10^{-20} would be expected:

$$CH_3-\overset{\overset{\displaystyle O}{\|}}{C}-NH_2 \rightleftharpoons {}^{-}CH_2-\overset{\overset{\displaystyle O}{\|}}{C}-NH_2 \quad + \quad H^+$$

ΔG at room temperature (300K) would be 27.6 kcal/mol.

(b) The anion formed by deprotonation of the NH_2 group would be more stable by about 5.5 kcal/mol.

(c) The ratio of the NH anion to the CH_2 anion would be about 10^4:1.

CHAPTER 5. ALKENES

ANSWERS TO EXERCISES

5.1 Only (a) and (d) can exist as *cis-trans* isomers; (b) and (c) each have two identical substituents on one of the carbons of the double bond.

(a) CH_3, H, C=C, CH_2CH_3, H CH_3, H, C=C, H, CH_2CH_3

(d) Cl, H, C=C, CH_2Cl, H Cl, H, C=C, H, CH_2Cl

5.2 **(a)** 2,4-dimethyl-2-pentene

(b) (*E*)-5-bromo-3-ethyl-2-hexene

(c) (*Z*)-3-hexene

(d) 3-chloro-5-methylcyclopentene

5.3 **(a)** top to bottom: 4,5-dimethyl-1-hexene, 2,3-dimethyl-1-hexene, (*Z*)-4,5-dimethyl-2-hexene, (*E*)-4,5-dimethyl-2-hexene, (*E*)-2,3-dimethyl-3-hexene, 2,3-dimethyl-2-hexene.

(b) Catalytic hydrogenation of all six isomers would yield the same product,

2,3-dimethylhexane:

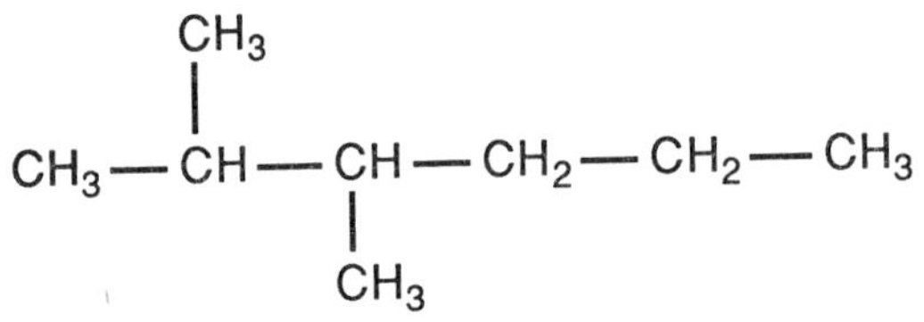

5.4 The two alkyl groups on this alkene are methyl and *tert*-butyl. When these two groups are *cis*, the compound is less stable than the other isomer by 4 kcal/mol. This is about four times large than the typical *Z*/*E* difference, suggesting that the steric bulk of a *tert*-butyl group is quite large.

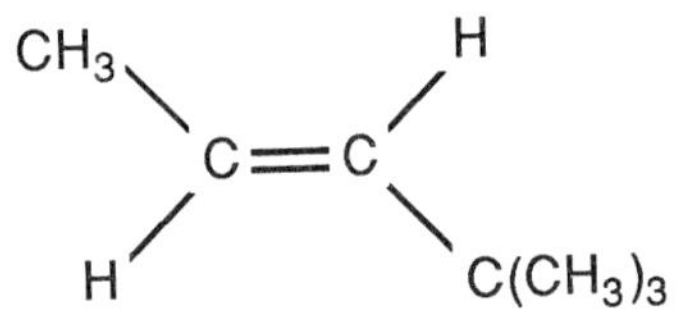

E

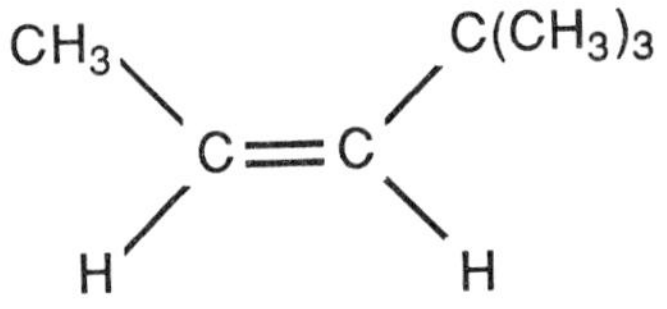

Z

5.5 (a)

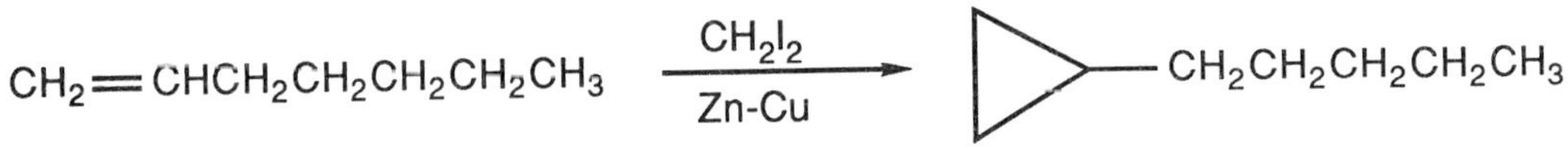

(b)

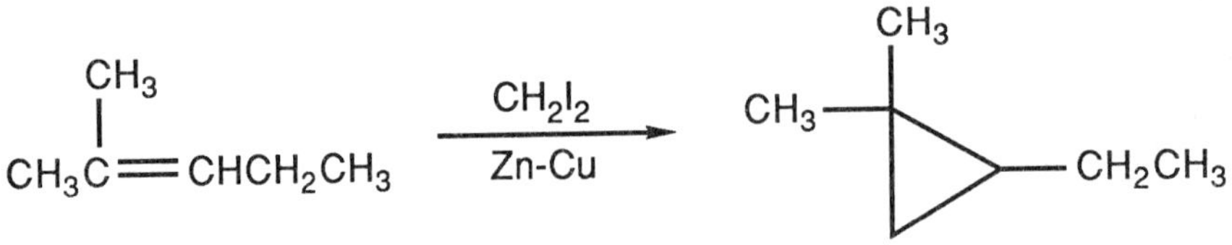

5.6 (a)

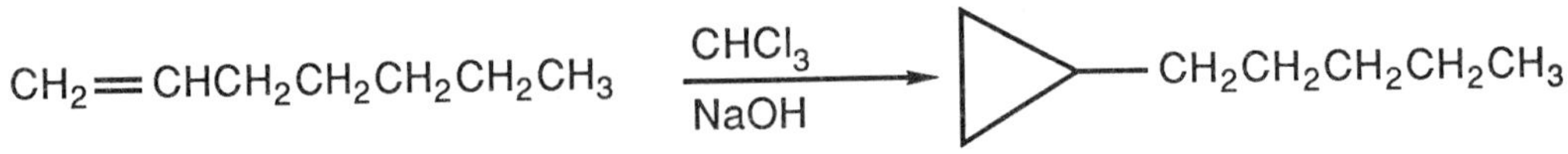

(b)

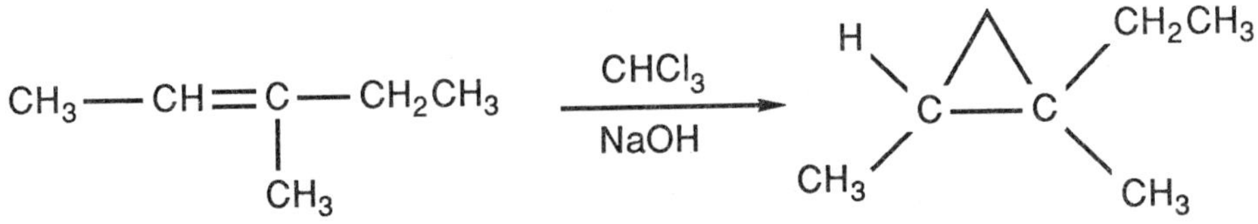

5.7

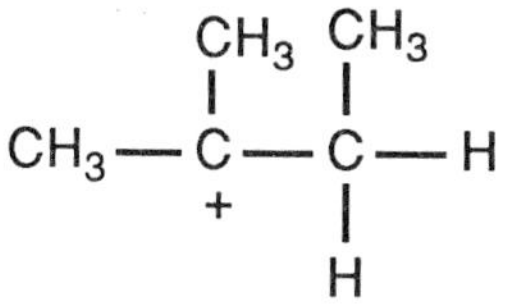

$$CH_3-\overset{+}{\underset{}{C}}(CH_3)-CH(CH_3)-H$$

3°
more stable

$$CH_3-CH(CH_3)-\overset{+}{C}(CH_3)-H$$

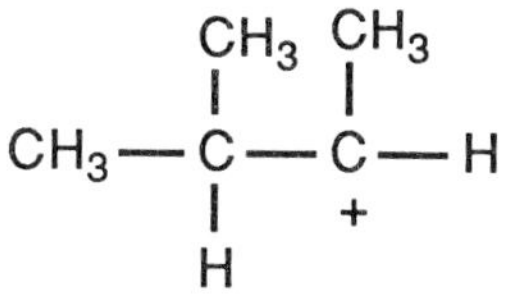

2°

The product is formed from the most stable carbocation.

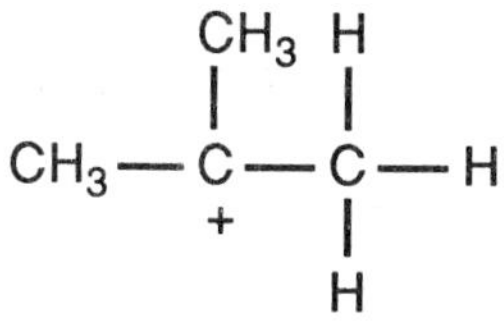

$$CH_3-\overset{+}{C}(CH_3)-CH_2-H$$

3°
more stable

$$CH_3-CH(CH_3)-\overset{+}{C}H-H$$

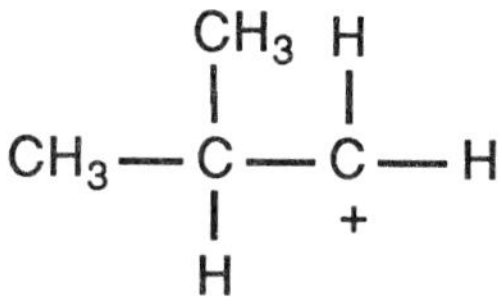

1°

The product is formed from the most stable carbocation.

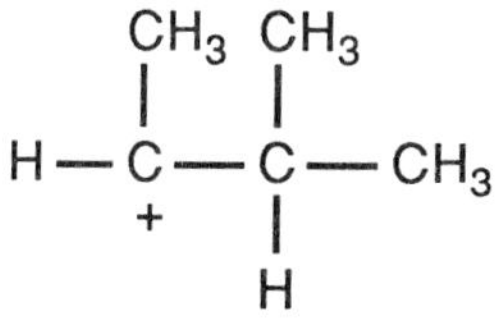

$$H-\overset{+}{C}(CH_3)-CH(CH_3)-CH_3$$

2°

$$H-CH(CH_3)-\overset{+}{C}(CH_3)-CH_3$$

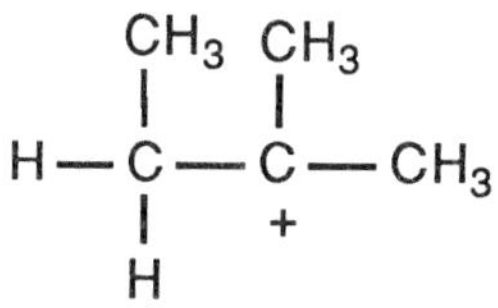

3°
more stable

The product is formed from the most stable carbocation.

5.8 **(a)** $CH_3-C(Br)(CH_3)-CH_2OH$ **(b)** $CH_3-C(Br)(CH_3)-\overset{+}{C}H_2$ **(c)** $CH_3-\overset{+}{C}(CH_3)-CH_2Br$ **(d) yes**

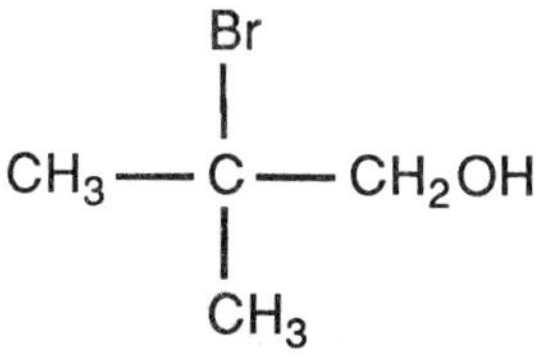

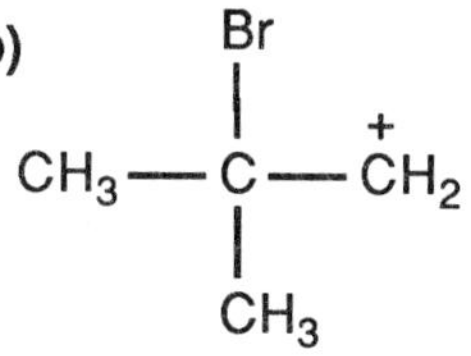

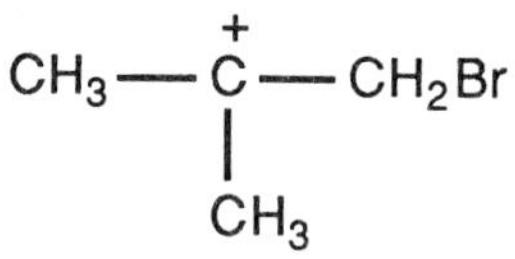

5.9 **(a)**

H, CH_3, C=C, H, H $\xrightarrow{Br—Br}$ H—C(H)(Br)—C(CH$_3$)(+)—H + Br^- → H—C(H)(Br)—C(CH$_3$)(H)—Br

(b)

H, CH_3, C=C, H, H $\xrightarrow{H—Cl}$ H—C(H)(H)—C(CH$_3$)(+)—H + Cl^- → H—C(H)(H)—C(CH$_3$)(Cl)—H

(c)

H, CH_3, C=C, H, H $\xrightarrow[H_2O]{Cl—Cl}$ H—C(H)(Cl)—C(CH$_3$)(+)—H + H—Ö—H → H—C(H)(Cl)—C(CH$_3$)(+O(H)—H)—H

loss of H^+

→ H—C(H)(Cl)—C(CH$_3$)(OH)—H

5.10 **(a)** hydride shift:

$$CH_3CH_2-\overset{\displaystyle CH_3}{\underset{+}{C}}-CH_3 \quad \text{(i)}$$

$$CH_3CH_2-\overset{\displaystyle CH_2CH_3}{\underset{\displaystyle CH_2CH_3}{C}}-CH_2-\underset{+}{CH}-CH_3 \quad \text{(ii)}$$

(b) methyl shift

$$CH_3CH_2-\overset{\displaystyle H}{\underset{+}{C}}-CH_2CH_3 \quad \text{(i)}$$

$$CH_3CH_2-\overset{\displaystyle CH_2CH_3}{\underset{\displaystyle CH_3CH_2}{C}}-\underset{\displaystyle CH_3}{CH}-\underset{+}{CH_2} \quad \text{(ii)}$$

(c) ethyl shift

$$CH_3-\overset{\displaystyle H}{\underset{+}{C}}-CH_2CH_2CH_3 \quad \text{(i)}$$

$$CH_3CH_2-\underset{\displaystyle CH_2CH_3}{\overset{+}{C}}-\overset{\displaystyle CH_2CH_3}{CH}-CH_2CH_3 \quad \text{(ii)}$$

For the first cation a hydride shift would be preferred because it would lead to a tertiary carbocation. For the second cation an ethyl shift would be preferred because it would lead to a tertiary carbocation.

5.11

(a) $$BrCH_2-\underset{\cdot}{CH}-CH_2-\overset{\displaystyle CH_3}{\underset{\displaystyle CH_3}{C}}-CH_3 \qquad \overset{\cdot}{C}H_2-\underset{\displaystyle Br}{CH}-CH_2-\overset{\displaystyle CH_3}{\underset{\displaystyle CH_3}{C}}-CH_3$$

(b) $$CH_3-\underset{+}{CH}-CH_2-\overset{\displaystyle CH_3}{\underset{\displaystyle CH_3}{C}}-CH_3 \qquad {+CH_2}-CH_2-CH_2-\overset{\displaystyle CH_3}{\underset{\displaystyle CH_3}{C}}-CH_3$$

In each case the secondary intermediate (radical or cation) would be more stable, and it would be this intermediate that would lead to product. In presence of peroxides this would be 1-bromo-4,4-dimethylpentane, and in the absence of peroxides the product would be 2-bromo-4,4-dimethylpentane.

5.12 **(a)** Acid catalyzed hydration of 3-methyl-1-butene would yield predominantly rearranged product:

$$CH_3-\underset{\displaystyle OH}{\overset{\displaystyle CH_3}{\underset{|}{\overset{|}{C}}}}-CH_2-CH_3$$

(b) Acid catalyzed hydration of 2,2,5,5-tetramethyl-3-hexene would also yield predominantly rearranged product:

$$CH_3-\underset{\displaystyle CH_3}{\overset{\displaystyle OH}{\underset{|}{\overset{|}{C}}}}-\underset{\displaystyle CH_3}{\underset{|}{CH}}-CH_2-\underset{\displaystyle CH_3}{\overset{\displaystyle CH_3}{\underset{|}{\overset{|}{C}}}}-CH_3$$

5.13 **(a)** $CH_3(CH_2)_6CH_2CH_2OH$ **(b)** $CH_3CH(CH_3)CH_2CH_2OH$ **(c)** $CH_3CH(OH)CH_2CH_2CH_3$ + $CH_3CH_2CH(OH)CH_2CH_3$

5.14 **(a)** Both reactions would yield the same two products:

$$CH_3CH_2CH(OH)CH_2CH_3 \quad + \quad CH_3CH(OH)CH_2CH_2CH_3$$

(b) *cis*: $CH_3CH_2(H)C{=}C(H)CH_2CH_3$ (both CH_3CH_2 groups on the same side; H and H on the same side) and *trans*: $CH_3CH_2(H)C{=}C(CH_2CH_3)H$ (CH_3CH_2 groups on opposite sides)

5.15 **(a)**

$$CH_3-\overset{\displaystyle CH_3}{\overset{|}{C}}{=}CH-CH_2CH_3 \xrightarrow[\text{2. } H_2O_2,\ OH^-]{\text{1. } BH_3} CH_3-\overset{\displaystyle CH_3}{\overset{|}{CH}}-\overset{\displaystyle OH}{\overset{|}{CH}}-CH_2CH_3$$

(b)

$$CH_3-\underset{\overset{|}{CH_3}}{\overset{}{CH}}-CH=CH_2 \xrightarrow[\text{2. } NaBH_4]{\text{1. } Hg(OCOCH_3)_2,\ CH_3OH} CH_3-\underset{\overset{|}{CH_3}}{CH}-\underset{\underset{OCH_3}{|}}{CH}-CH_3$$

5.16 $(CH_3)(H)C=C(CH_3)(CH_2CH_3)$

5.17 ionization:

$$CH_3-\overset{OH}{\overset{|}{CH}}-\underset{CH_3}{\overset{CH_3}{\overset{|}{\underset{|}{C}}}}-CH_3 \xrightarrow{H^+} CH_3-\overset{\overset{+}{O}H_2}{\overset{|}{CH}}-\underset{CH_3}{\overset{CH_3}{\overset{|}{\underset{|}{C}}}}-CH_3 \longrightarrow CH_3-\overset{+}{CH}-\underset{CH_3}{\overset{CH_3}{\overset{|}{\underset{|}{C}}}}-CH_3$$

rearrangement:

$$CH_3-\overset{+}{CH}-\underset{CH_3}{\overset{CH_3}{\overset{|}{\underset{|}{C}}}}-CH_3 \longrightarrow CH_3-\overset{CH_3}{\overset{|}{CH}}-\underset{CH_3}{\underset{|}{\overset{+}{C}}}-CH_3$$

loss of H^+:

$$CH_3-\overset{CH_3}{\overset{|}{CH}}-\underset{CH_3}{\underset{|}{\overset{+}{C}}}-CH_3 \longrightarrow CH_3-\underset{CH_3}{\underset{|}{CH}}-\underset{CH_3}{\underset{|}{C}}=CH_2 + (CH_3)_2C=C(CH_3)_2$$

These are the two alkenes formed as major products in the reaction with phosphoric acid. The minor product (2%) in that reaction is unrearranged.

5.18 $(CH_3)(H)C=C(CH_2CH_3)(H)$ + $(CH_3)(H)C=C(H)(CH_2CH_3)$ + $CH_2=CH\text{-}CH_2CH_2CH_3$

Z   *E*

major product

5.19

(a) $CH_3CH(Br)CH_3 + {}^-OtBu \rightarrow CH_3CH{=}CH_2 + t\text{-BuOH} + Br^-$

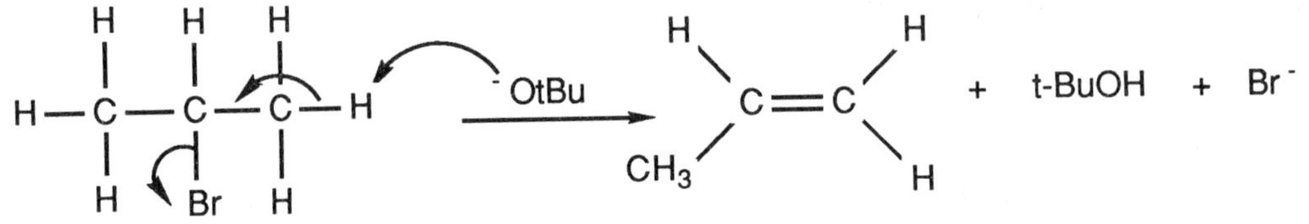

(b) $CH_3CH(Br)CH_3 + {}^-OH \rightarrow CH_3CH{=}CH_2 + H_2O + Br^-$

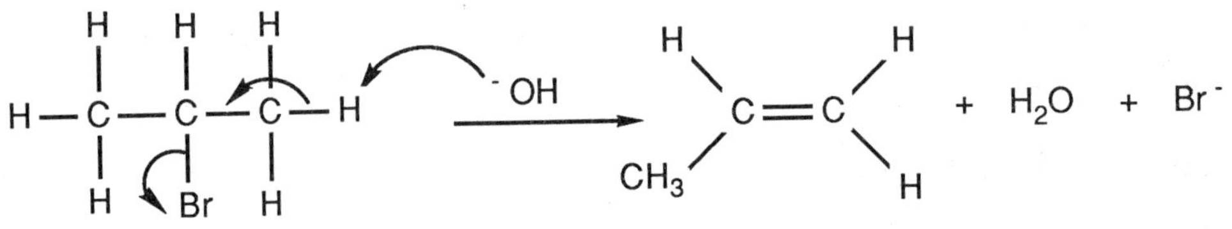

5.20 **(a)** ${}^-OH + CH_3CH_2CH_2CH_2CH(Br)C(CH_3)_2Br \rightarrow (CH_3CH_2CH_2CH_2)(Br)C{=}C(CH_3)_2 + H_2O + Br^-$

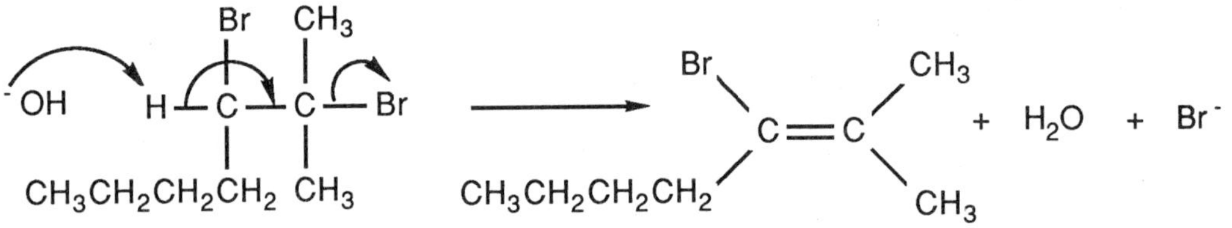

(b) The most stable alkene is that shown in (a). The other possible isomers all have disubstituted double bonds and should be of similar stabilities (although the second of the three with the alkyl substitutents *cis* should be least stable).

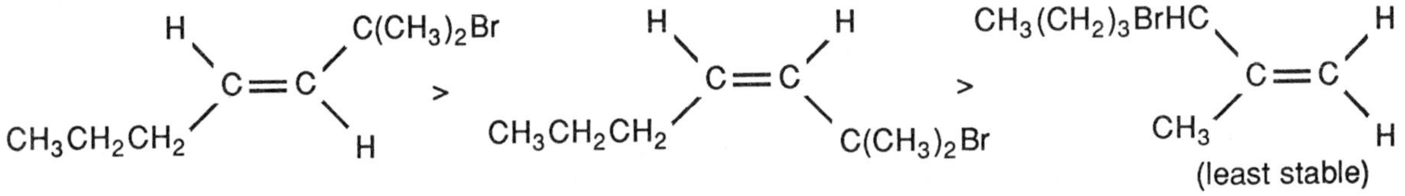

$(CH_3CH_2CH_2)HC{=}CH(C(CH_3)_2Br)$ (trans) > $(CH_3CH_2CH_2)HC{=}CH(C(CH_3)_2Br)$ (cis) > $CH_3(CH_2)_3BrHC(CH_3)C{=}CH_2$ (least stable)

5.21

$HBrC{=}C(CH_3)_2 \xrightarrow{\text{1. Li; 2. Cu}} (HC{=}C(CH_3)_2)_2LiCu \xrightarrow{CH_3CH_2CH_2CH_2Br} (CH_3CH_2CH_2CH_2)HC{=}C(CH_3)_2$

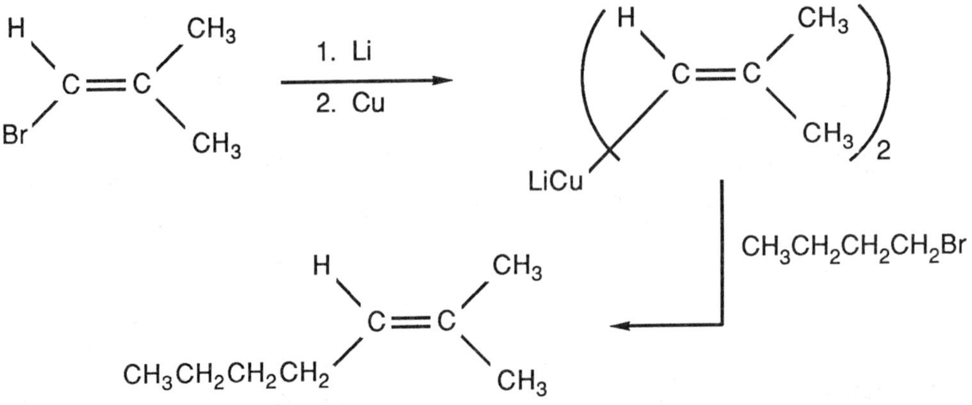

ANSWERS TO PROBLEMS

5.1 **(a)** 2-ethyl-1-butene **(b)** (*Z*)-3-heptene

(c) (*Z*)-3-methyl-3-heptene **(d)** vinylcyclohexane

(e) 3-methylcyclohexene **(f)** (*Z*)-2,6-dibromo-2-hexene

(g) (*E*)-2,5,6-trimethyl-3-octene **(h)** methylenecyclopentane

(i) (*Z*)-1-cyclopentyl-3-heptene **(j)** 6-chloro-2-methyl-2-hexene

5.2 **(a)** F, H; C=C; CH_3, $CHF-CH_3$

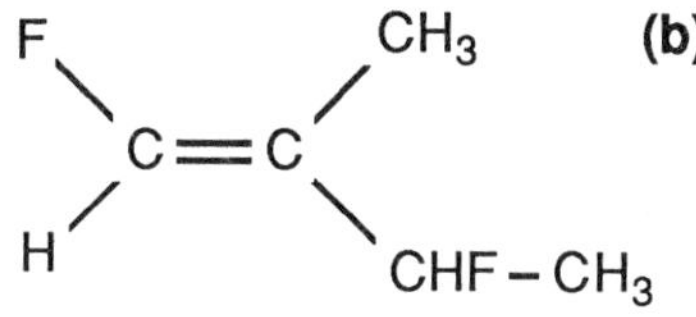

(b) $CH_2{=}CHCH_2CH_2CH_3$

(c) CH_3, CH_3

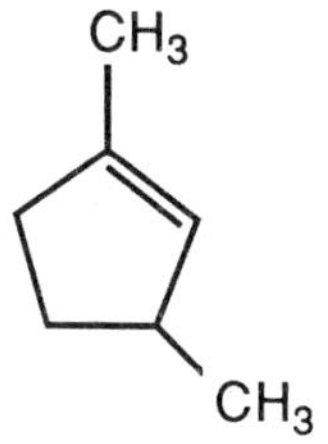

(d) CH_2

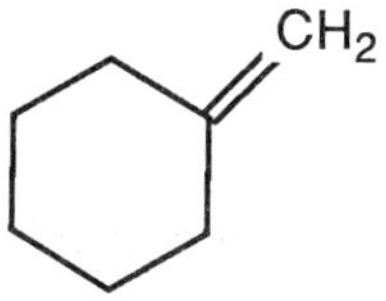

(e) CH_3, H; C=C; $CH_2CH_2CH_2Br$, $CH_2CH_2CH_3$

(f) Cl, Cl

(g) CH_3, CH_3; C=C; CH_3, CH_3

(h) CH_3CH_2, H; C=C; $C(CH_3)_3$, $CH_2CH_2CH_3$

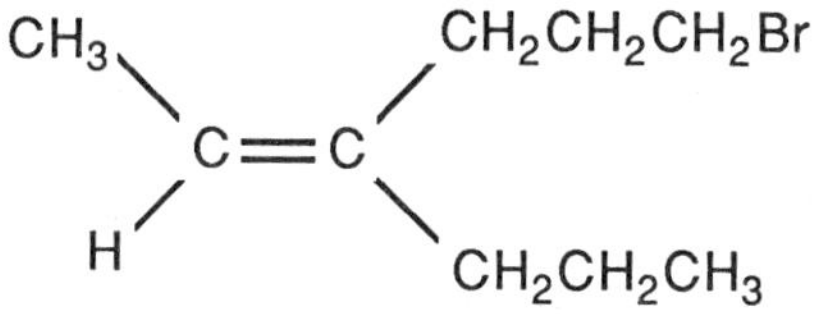

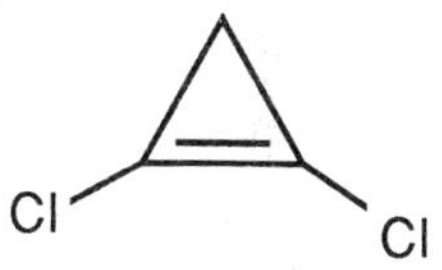

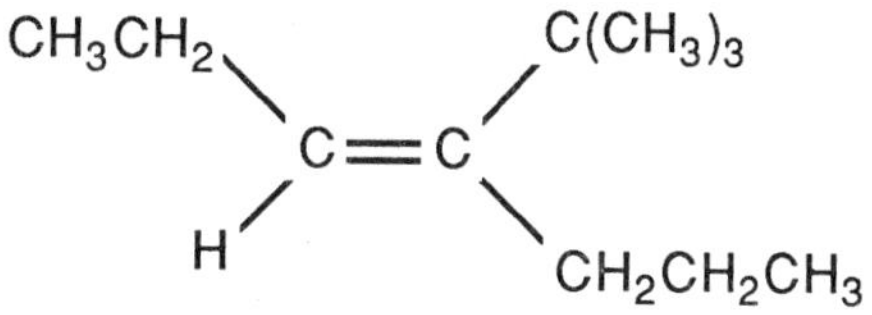

(i) $CH_3CH_2CH_2$, Cl; C=C; Cl, $CH_2CH_2CH_3$

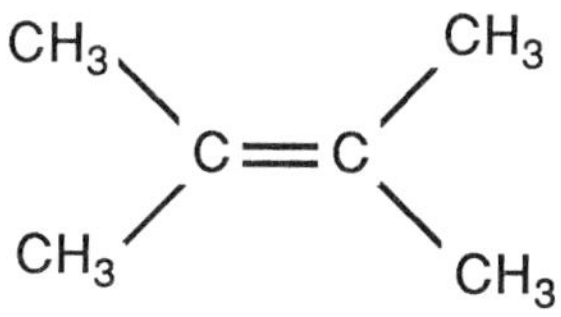

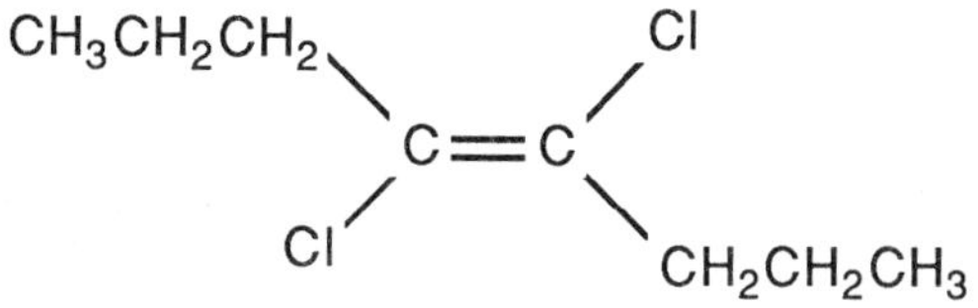

(j) H, Br; C=C; H, Br

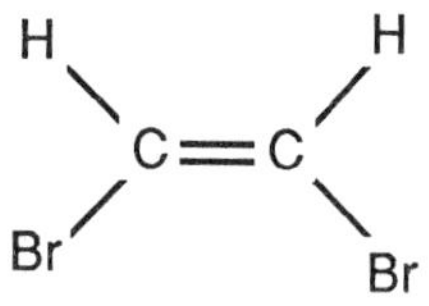

5.3 Further reaction of ethyl hydrogen sulfate with ethylene in conc. H_2SO_4 would yield diethyl sulfate:

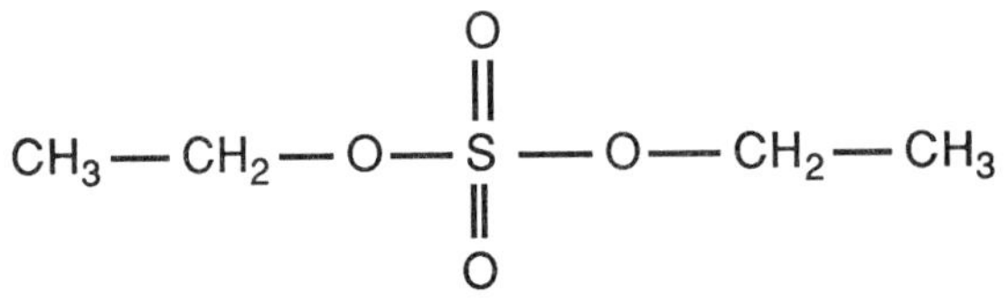

$$CH_3-CH_2-O-\overset{\overset{\displaystyle O}{\|}}{\underset{\underset{\displaystyle O}{\|}}{S}}-O-CH_2-CH_3$$

Hydrolysis of diethyl sulfate would yield H_2SO_4 and ethanol, CH_3CH_2OH.

5.4 **(a)** Same

(b) Same

(c) Constitutional isomers: both are dichloro derivatives of 2,3-dimethyl-2-butene, but the chlorines are on different relative positions for the two structures.

(d) Different: the first is C_5 but the second is only C_4.

(e) Same: both are 4-chlorocycloheptene.

(f) Stereoisomers: both are 3-chloro-2-pentene, but the first is *E* while the second is *Z*.

(g) Constitutional isomers: the first is 1-methylcyclopentene while the second is 4-methylcyclopentene.

(h) Stereoisomers: both are cyclodecene, but the first is *E* while the second is *Z*.

(i) Stereoisomers: both are derivatives of cyclohexane, but the substituents are *trans* in the first structure and *cis* in the second.

(j) Constitutional isomers: the first structure is 4-bromo-1-chloro-2-methyl-2-butene, while the second is 1-bromo-4-chloro-2-methyl-2-butene.

5.5 (a) cyclohexyl–$CH(CH_3)-CH_3$

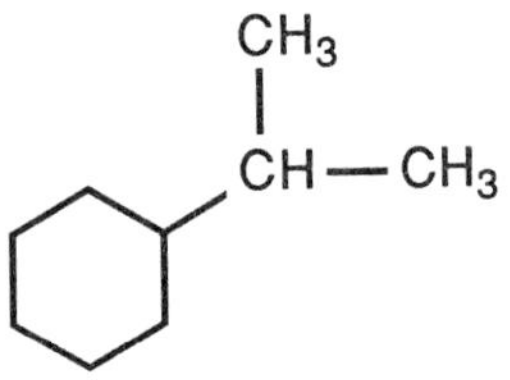

(b) 1-[$CH(CH_3)-CH_3$]-1-OH cyclohexane (ring carbon bearing $CH(CH_3)CH_3$ and OH)

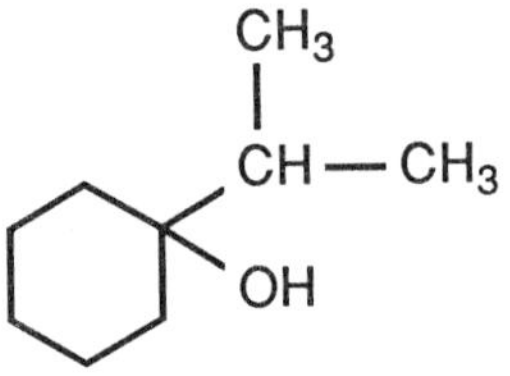

(c) 1-[$CH(CH_3)-CH_3$]-1-OCH_2CH_3 cyclohexane (ring carbon bearing $CH(CH_3)CH_3$ and OCH_2CH_3)

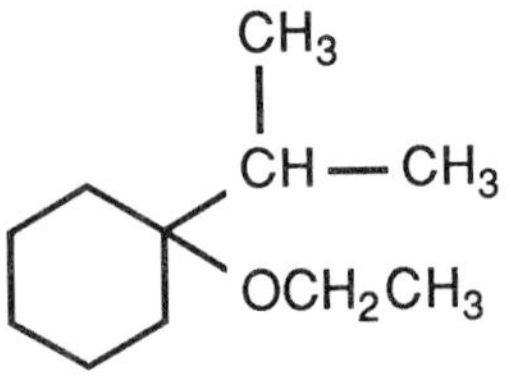

(d)

$$CH_3-\underset{\underset{\displaystyle CH_3}{|}}{\overset{\overset{\displaystyle Cl}{|}}{C}}-\overset{\overset{\displaystyle CH_3}{|}}{CH}-CH_2-C(CH_3)_3$$

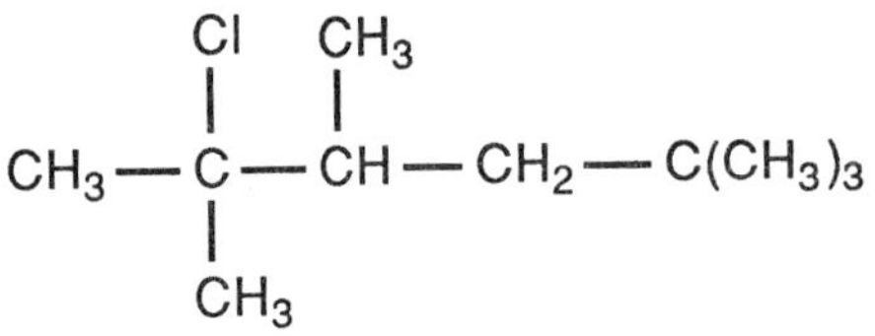

(e) 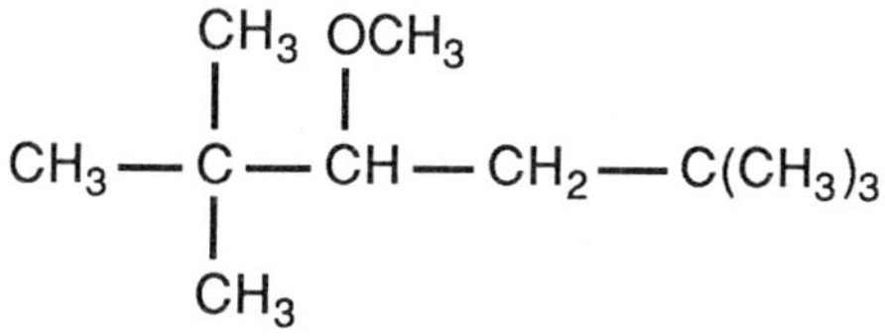 $CH_3-C(CH_3)_2-CH(OCH_3)-CH_2-C(CH_3)_3$

(f) 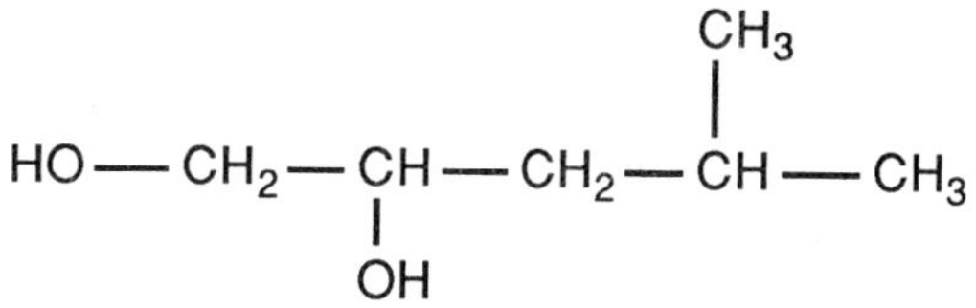 $HO-CH_2-CH(OH)-CH_2-CH(CH_3)-CH_3$

(g) 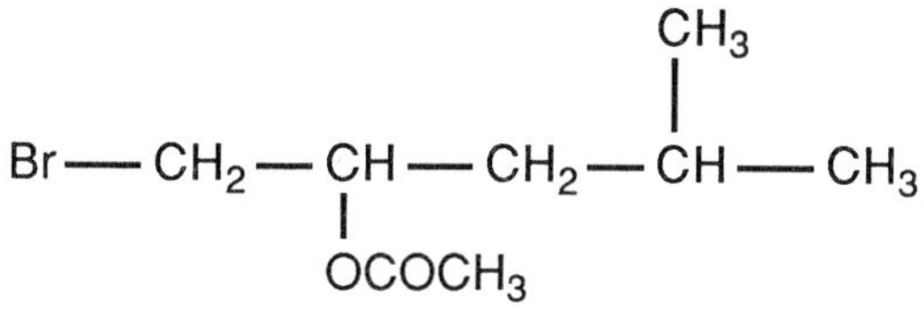 $Br-CH_2-CH(OCOCH_3)-CH_2-CH(CH_3)-CH_3$

(h) 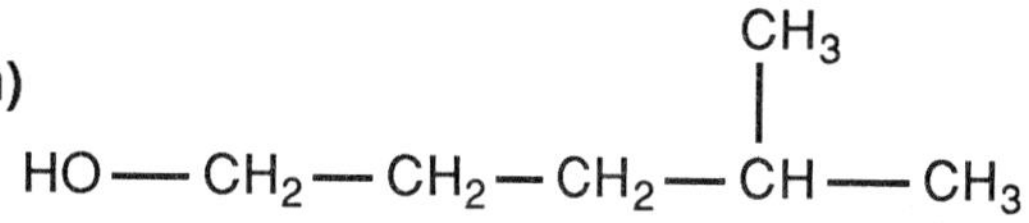 $HO-CH_2-CH_2-CH_2-CH(CH_3)-CH_3$

(i) 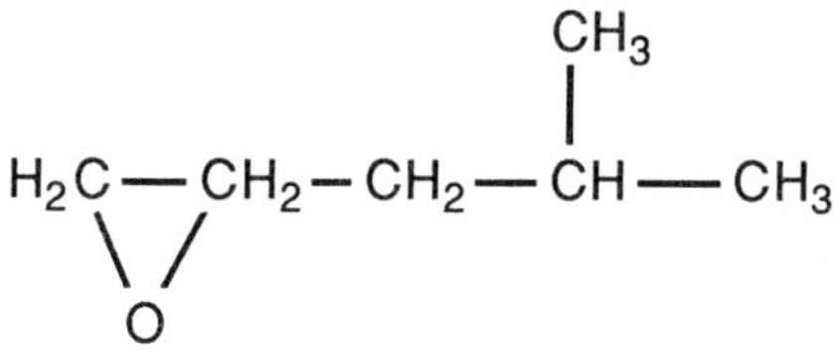 $H_2C(-O-)C-CH_2-CH_2-CH(CH_3)-CH_3$

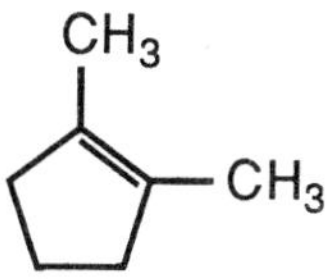

(j) CH_3, CH_3

(k) CH_3, Cl, CH_3O, CH_3

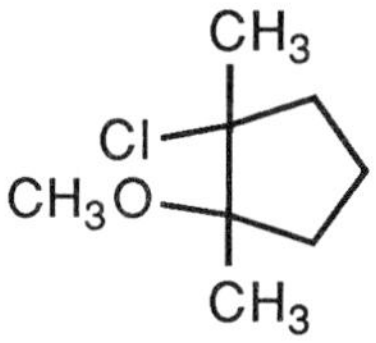

(l) CH_2Br, OH

(m) $H-C(=O)-H$ + cyclopentanone (O)

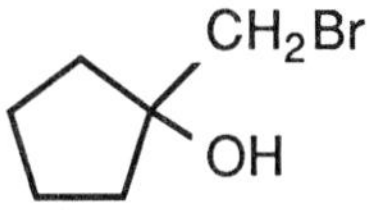

(n) CH_3, Br, CH_3

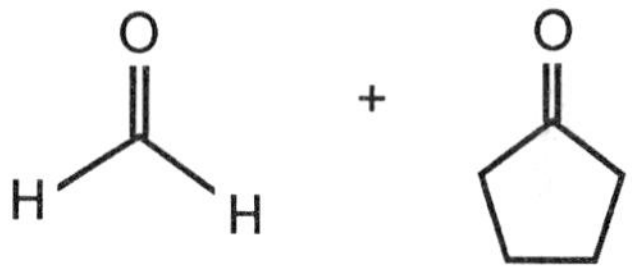

(o) CH_3, Cl, CH_3

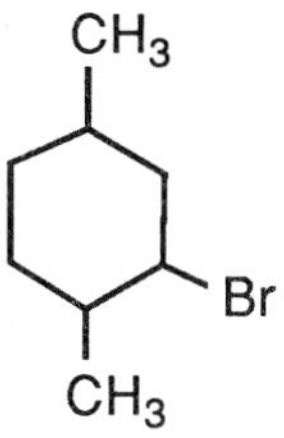

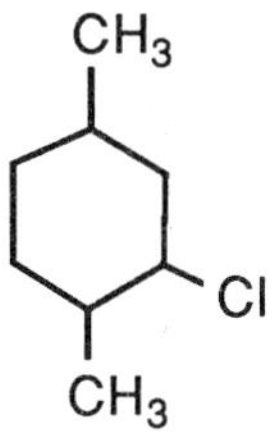

(p) 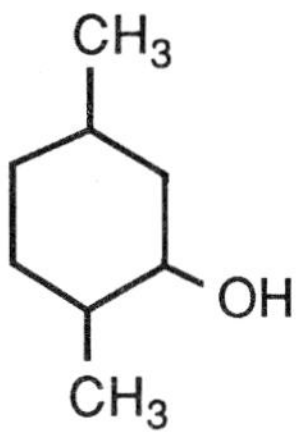 CH_3, OH, CH_3

(q) 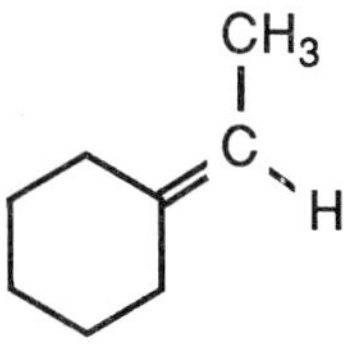 CH_3, C, H (cyclohexylidene)

(r) 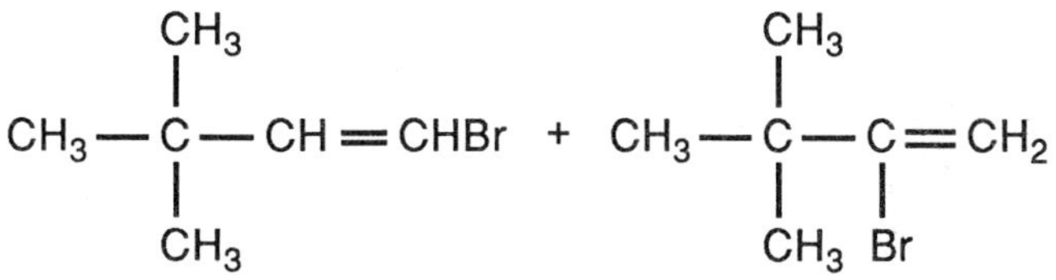 $CH_3-C(CH_3)_2-CH=CHBr$ + $CH_3-C(CH_3)_2-C(Br)=CH_2$

(s) 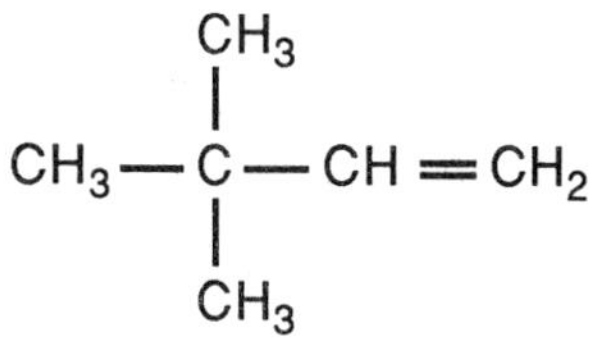 $CH_3-C(CH_3)_2-CH=CH_2$

(t) 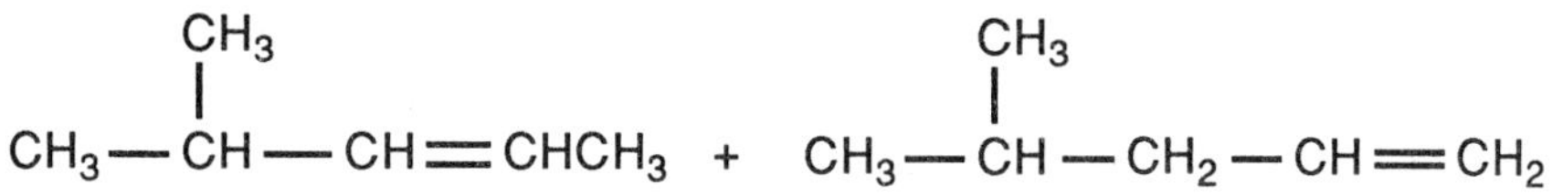 $CH_3-CH(CH_3)-CH=CHCH_3$ + $CH_3-CH(CH_3)-CH_2-CH=CH_2$

5.6 The hydrogenation experiment tells you that the carbon skeleton of the alkene is that of hexane, so the compound is one of the isomeric hexenes. Hydroboration and oxymercuration each afforded a single alcohol in high yield, and this could result from either of two possibilities: (1) the structure could be symmetrical; or (2) the two ends of the double bond differ in the degree of substitution. In the first case (i.e., 3-hexene) the symmetry would result in formation of the same alcohol by the two methods, but the problem indicates that two different alcohols are formed. Therefore situation (2) must describe the alkene, and its structure must be 1-hexene:

$$CH_3CH_2CH_2CH_2CH{=}CH_2$$

Hydroboration affords 1-hexanol and oxymercuration yields 2-hexanol.

5.7 (a)

Methylenecyclohexane ($=CH_2$) $\xrightarrow{\text{HBr/peroxides}}$ cyclohexyl–CH_2Br $\xrightarrow{(CH_3CH_2)_2CuLi}$ cyclohexyl–$CH_2CH_2CH_3$

(b)

$$CH_3-\underset{\underset{CH_3}{|}}{CH}-\overset{\overset{OH}{|}}{CH}-CH_3 \xrightarrow[H_3PO_4]{\text{conc. } H_2SO_4 \text{ or}} CH_3-\underset{\underset{CH_3}{|}}{C}{=}CH-CH_3$$

$$\downarrow Cl_2/CCl_4$$

$$CH_3-\underset{\underset{CH_3}{|}}{\overset{\overset{Cl}{|}}{C}}-\overset{\overset{Cl}{|}}{CH}-CH_3$$

(c)

$$CH_3-\underset{\underset{}{}}{\overset{CH_3}{\overset{|}{C}H}}-\underset{\underset{Br}{|}}{CH}-CH_2Br \xrightarrow{Zn} CH_3-\overset{CH_3}{\overset{|}{C}H}-CH=CH_2$$

$$\xrightarrow{\text{HBr/peroxides}} CH_3-\overset{CH_3}{\overset{|}{C}H}-CH_2-CH_2Br$$

(d)

$$(CH_3)(H)C=C(CH_3)_2 \xrightarrow[\text{2. } NaBH_4]{\text{1. } Hg(OCOCH_3)_2,\ CH_3OH} CH_3-CH_2-\underset{\underset{CH_3}{|}}{\overset{OCH_3}{\overset{|}{C}}}-CH_3$$

(e)

$$\text{cyclopentene} \xrightarrow{CH_3CO_3H} \text{cyclopentene oxide (epoxide, O)}$$

(f)

$$CH_3-CH_2-\overset{Br}{\overset{|}{C}H}-\underset{\underset{CH_3}{|}}{\overset{Br}{\overset{|}{C}}}-CH_3 \xrightarrow{Zn} (CH_3CH_2)(H)C=C(CH_3)_2$$

(g)

$$CH_3CH_2-\underset{\underset{H}{|}}{\overset{OH}{\overset{|}{C}}}-CH_3 \xrightarrow{\text{conc. } H_2SO_4} CH_3CH_2=CH-CH_3 \xrightarrow{H_2/Pd} CH_3CH_2CH_2CH_3$$

(h)

$$(CH_3)_2C=C(CH_2CH_3)(H) \xrightarrow[\text{2. } H_2O_2,\ NaOH]{\text{1. } BH_3} CH_3-\underset{\underset{H}{|}}{\overset{CH_3}{\overset{|}{C}}}-\underset{\underset{H}{|}}{\overset{OH}{\overset{|}{C}}}-CH_2CH_3$$

(i)

$(CH_3)_2C{=}C(Br)CH_3 \xrightarrow{(CH_3CH_2)_2CuLi} (CH_3)_2C{=}C(CH_2CH_3)CH_3 \xrightarrow{H_2/Pd} CH_3{-}CH(CH_3){-}CH(CH_3){-}CH_2CH_3$

(j) $CH_3CH_2CH_2CH_2CH_2Cl \xrightarrow[BuOH]{KOtBu} CH_3CH_2CH_2CH{=}CH_2 \xrightarrow{Cl_2,\ CCl_4} CH_3CH_2CH_2CH(Cl){-}CH_2Cl$

(k) 1-methylcyclohexanol (CH3, OH) $\xrightarrow{conc.H_2SO_4}$ 1-methylcyclohexene (CH3) $\xrightarrow[2.\ H_2O_2,\ NaOH]{1.\ BH_3}$ 2-methylcyclohexanol (CH3, OH)

(l) 1-methylcyclohexanol (CH3, OH) $\xrightarrow{conc.\ H_2SO_4}$ 1-methylcyclohexene (CH3) $\xrightarrow[CH_3OH;\ 2.\ NaBH_4]{1.\ Hg(OCOCH_3)_2,}$ 1-methoxy-1-methylcyclohexane (CH3, OCH3)

5.8 **(a)** Ordinarily, addition of bromine to an alkene affords a dibromo compound with the bromines on adjacent carbons. The 1,3-relationship of the bromines in the product of this reaction indicates that a rearrangement has occurred. The first bromine is introduced as "Br^+", so the cation that reacted with Br^- must have been:

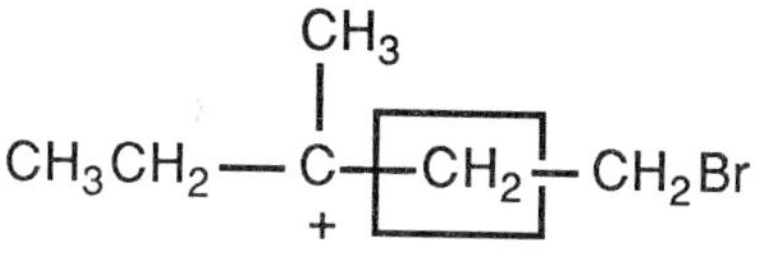

$$
\begin{array}{c}
CH_3 \\
| \\
CH_3CH_2-\overset{}{\underset{+}{C}}-\boxed{CH_2}-CH_2Br
\end{array}
$$

The original reaction with "Br^+" must have generated a cationic center on the carbon adjacent to the terminal CH_2Br group, and this carbon is highlighted in the preceding drawing. One of the two substituents on this carbon must have migrated, i.e., there was a *hydride* shift. The original alkene was therefore:

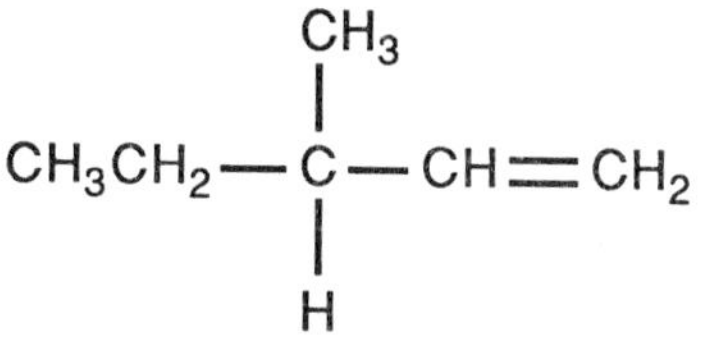

$$
\begin{array}{c}
CH_3 \\
| \\
CH_3CH_2-C-CH{=}CH_2 \\
| \\
H
\end{array}
$$

(b) This reaction involves overall addition of Br and OH to the double bond of an alkene. The carbons bearing those substitutents were originally the carbons of the double bond:

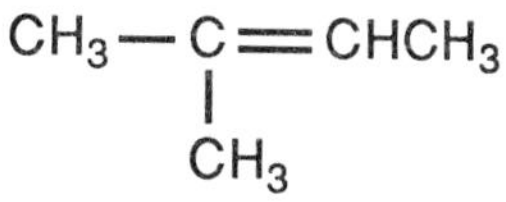

$$
\begin{array}{c}
CH_3-C{=}CHCH_3 \\
| \\
CH_3
\end{array}
$$

(c) The reactant differs from the product by a single molecule of HBr. The question therefore is, which bromine was introduced in the reaction? The addition of HBr in the presence of peroxides proceeds by initial reaction of a bromine atom to the less substituted end of a double bond, generating the more stable free radical. There is no way that this process could have introduced the tertiary bromine, because all the adjacent carbons are less substituted. Consequently it was the other bromine that was introduced, and the original alkene was:

$$
\begin{array}{c}
\quad CH_3 \qquad\qquad CH_3 \\
\quad | \qquad\qquad\quad | \\
CH_3-C-CH_2-C{=}CH_2 \\
| \qquad\qquad\qquad \\
Br \qquad\qquad\qquad
\end{array}
$$

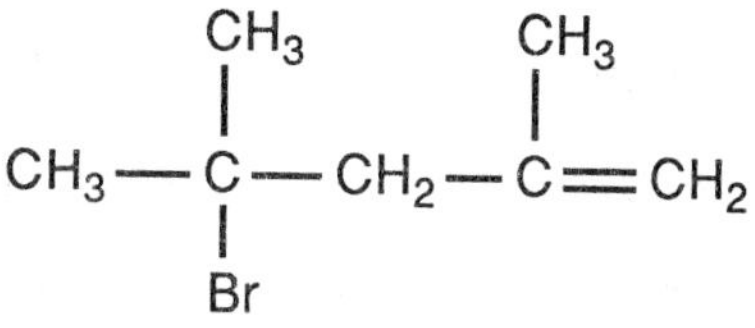

5.9 The ozonolysis reaction allows you to deduce the following partial structure for the alkene:

$$CH_3-CH=C$$

The alkene has a total of only 5 carbons, so the other carbon atom of the double bond either has two methyl substituents or one ethyl group and a hydrogen. In the latter case (i.e., 2-pentene) the two ends of the double bond would be comparably substituted and a mixture of products would be expected for both hydroboration and oxymercuration. The high yield of a single product in each case indicates that the one end of the alkene is more highly substituted, so the correct structure must be:

$$CH_3CH{=}C(CH_3)_2$$

5.10 The CF_3 group is strongly electron withdrawing (via inductive effects), and it destabilizes an adjacent cationic center. As a result addition of a proton (or other Lewis acid) to 3,3,3-trifluoro-1-propene occurs at C-2 to preferentially form positive charge at the (primary) carbon that is more distant from the CF_3 group.

$$CF_3-CH=CH_2 \longrightarrow CF_3-CH_2-CH_2^+$$

The effect of a CF_3 group is therefore opposite to that of a CH_3.

5.11 The iodo substituent is on C-2, so a hydrogen was introduced at either C-1 or C-3. In the latter case (2-hexene) both ends of the double bond have a single alkyl group, so a mixture of products would have been expected. The high yield of a single product requires the unsymmetrical double bond of 1-hexene:

$$CH_3CH_2CH_2CH_2CH=CH_2$$

5.12 The free radical addition of HBr is exothermic for both steps:

$$CH_3CH{=}CH_2 + Br\cdot \longrightarrow H{-}\dot{C}(CH_3){-}CH_2Br$$

π bond broken = 50
C-Br bond formed = -68
ΔH = -18 kcal/mol

$$H{-}\dot{C}(CH_3){-}CH_2Br \xrightarrow{H-Br} H{-}C(H)(CH_3){-}CH_2Br + Br\cdot$$

H-Br bond broken = 88
C-H bond formed = -95
ΔH = -7 kcal/mol

In contrast, the free radical addition of HCl would involve one step that is *endothermic*:

$$CH_3CH{=}CH_2 + Cl\cdot \longrightarrow H{-}\dot{C}(CH_3){-}CH_2Cl$$

π bond broken = 50
C-Cl bond formed = -80
ΔH = -30 kcal/mol

$$H{-}\dot{C}(CH_3){-}CH_2Cl \xrightarrow{H-Cl} H{-}C(H)(CH_3){-}CH_2Cl + Cl\cdot$$

H-Cl bond broken = 103
C-H bond formed = -95
ΔH = +8 kcal/mol

5.13 The addition of HCl involves a decrease in the oxidation state of C-1 (from -2 to -1) and a corresponding increase in the oxidation state of C-2 (from -1 to -2). There is no overall oxidation or reduction.

$$CH_3CH_2{-}\underset{2}{CH}{=}\underset{1}{CH_2} + HCl \longrightarrow CH_3CH_2\underset{2}{CH_2}\underset{1}{CH_2}Cl$$

5.14 The addition of Cl_2 involves a decrease in the oxidation state of both C-1 (from -2 to -1) and C-2 (from -1 to 0). The overall reaction is an oxidation.

$$CH_3CH_2{-}\underset{2}{CH}{=}\underset{1}{CH_2} + Br_2 \longrightarrow CH_3CH_2{-}\underset{2}{CH}(Br){-}\underset{1}{CH_2}Br$$

5.15 The reaction occurs with no change in the stereochemistry of the double bond:

$$\underset{\text{Cl}}{\overset{CH_3CH_2CH_2}{}}\!\!C{=}C\!\!\underset{\text{H}}{\overset{CH_3}{}} + (CH_3)_2CuLi \longrightarrow \underset{CH_3}{\overset{CH_3CH_2CH_2}{}}\!\!C{=}C\!\!\underset{\text{H}}{\overset{CH_3}{}}$$

Nevertheless, the starting material has the *E* configuration and the product has the *Z* configuration. This is simply a result of the rules of nomenclature: in the reactant the higher-priority groups at the ends of the double bond are Cl and methyl, but in the product they are propyl and methyl.

5.16 **(a)**

$$CH_3CH_2CH_2CH_2CH{=}CH_2 + H{-}Br \longrightarrow CH_3CH_2CH_2CH_2\overset{+}{C}H{-}CH_3 + Br^-$$

$$\downarrow$$

$$CH_3CH_2CH_2CH_2CHBrCH_3$$

(b)

$$ROOR \longrightarrow RO\cdot + \cdot OR$$

$$RO\cdot + HBr \longrightarrow Br\cdot + ROH$$

$$CH_3CH_2CH_2CH_2CH{=}CH_2 + \cdot Br \longrightarrow CH_3CH_2CH_2CH_2\dot{C}H{-}CH_2{-}Br$$

$$CH_3CH_2CH_2CH_2\dot{C}H{-}CH_2{-}Br \quad H{-}Br \longrightarrow CH_3CH_2CH_2CH_2{-}\overset{H}{\overset{|}{C}H}{-}CH_2Br + \cdot Br$$

5.17 **(a)**

$$CH_3CH_2CH_2-\underset{CH_3}{\overset{H}{C}}-CH{=}CH_2 + H-Br \longrightarrow CH_3CH_2CH_2-\underset{CH_3}{\overset{H}{C}}-\overset{+}{C}H-CH_3$$

$$\downarrow$$

$$CH_3CH_2CH_2-\underset{CH_3}{\overset{Br}{C}}-\underset{CH_3}{CH}-CH_3 \xleftarrow{Br^-} CH_3CH_2CH_2-\underset{CH_3}{\overset{+}{C}}-\underset{Br}{\overset{H}{C}}-CH_3$$

(b)

$$ROOR \longrightarrow RO\cdot + \cdot OR$$

$$RO\cdot + HBr \longrightarrow Br\cdot + ROH$$

$$CH_3CH_2CH_2-\underset{CH_3}{\overset{H}{C}}-CH{=}CH_2 + \cdot Br \longrightarrow CH_3CH_2CH_2-\underset{CH_3}{\overset{H}{C}}-\dot{C}H-CH_2Br$$

$$CH_3CH_2CH_2-\underset{CH_3}{\overset{H}{C}}-\dot{C}H-CH_2Br \quad H-Br \longrightarrow CH_3CH_2CH_2-\underset{CH_3}{\overset{H}{C}}-\overset{H}{C}H-CH_2Br$$

5.18 The initial protonation would yield a secondary cation:

$$CH_3CH_2CH_2-\underset{CH_3}{\overset{CH_3CH_2}{C}}-\underset{H}{C}{=}CH_2 \quad H^+ \longrightarrow CH_3CH_2CH_2-\underset{CH_3}{\overset{CH_3CH_2}{C}}-\underset{H}{\overset{+}{C}}-CH_3$$

This could rearrange via an ethyl shift:

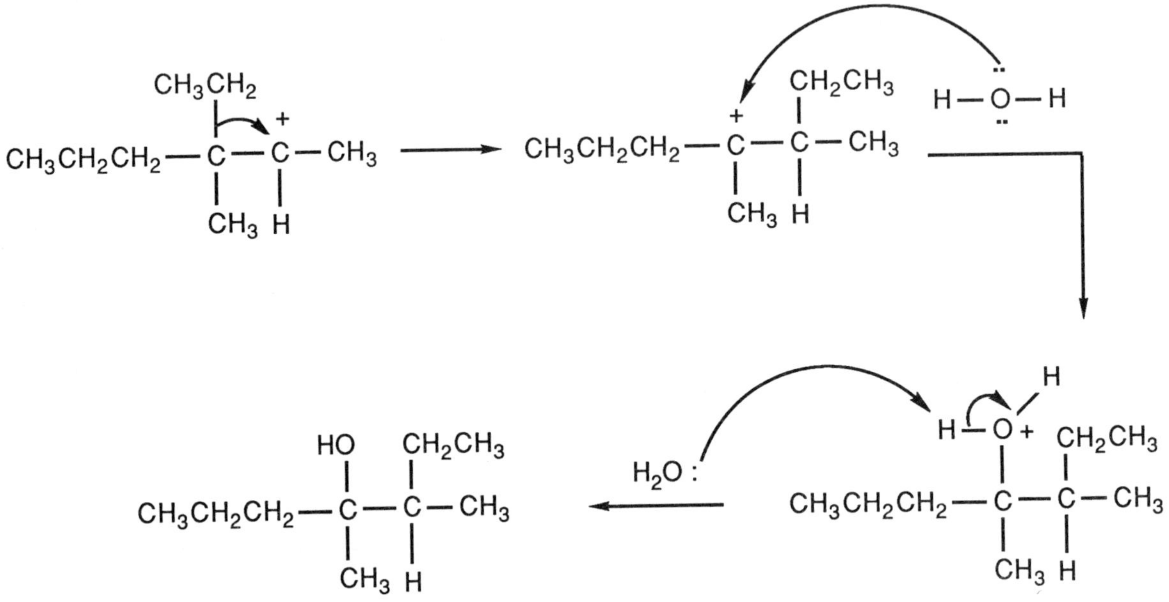

by a methyl shift:

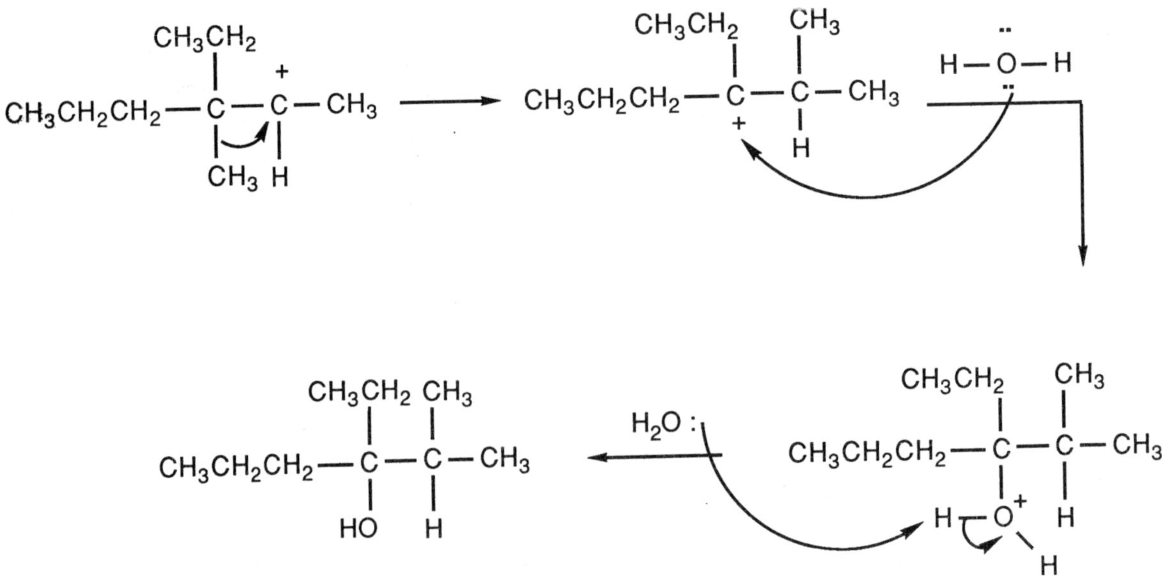

or by a propyl shift:

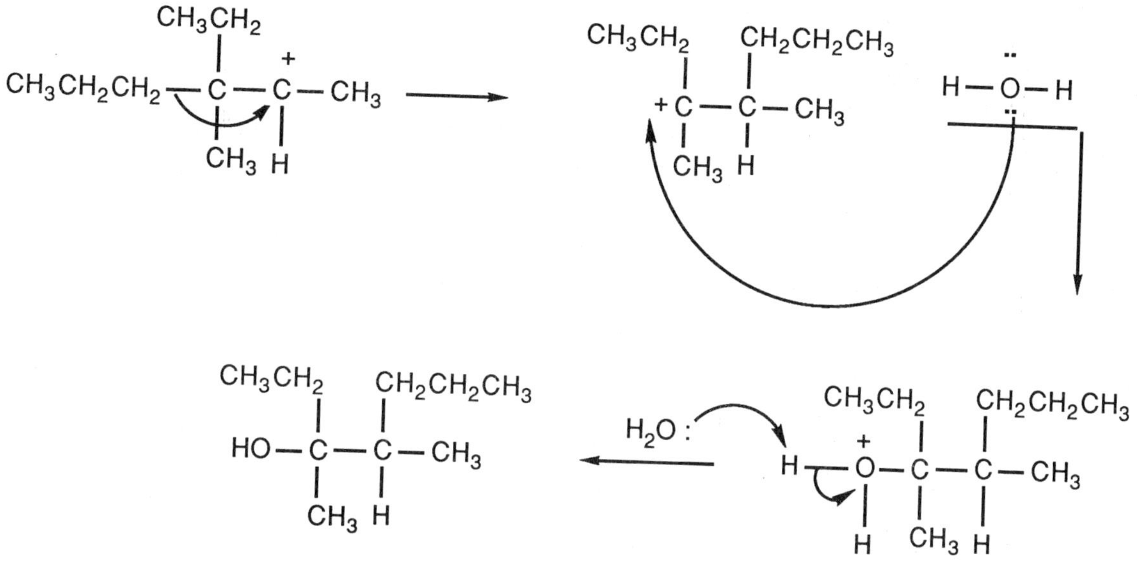

$CH_3CH_2CH_2-C(CH_2CH_3)(CH_3)-\overset{+}{C}H-CH_3 \longrightarrow \overset{+}{C}(CH_2CH_3)(CH_3)-C(CH_2CH_2CH_3)(H)-CH_3$

$H-\ddot{O}-H$

$H_2O: \quad H-\overset{+}{O}(H)-C(CH_2CH_3)(CH_3)-C(CH_2CH_2CH_3)(H)-CH_3 \longrightarrow HO-C(CH_2CH_3)(CH_3)-C(CH_2CH_2CH_3)(H)-CH_3$

5.19 $CH_2{=}CH_2$

5.20

(a) H^+, $:OH_2$, $:OH_2$; CH3 OH

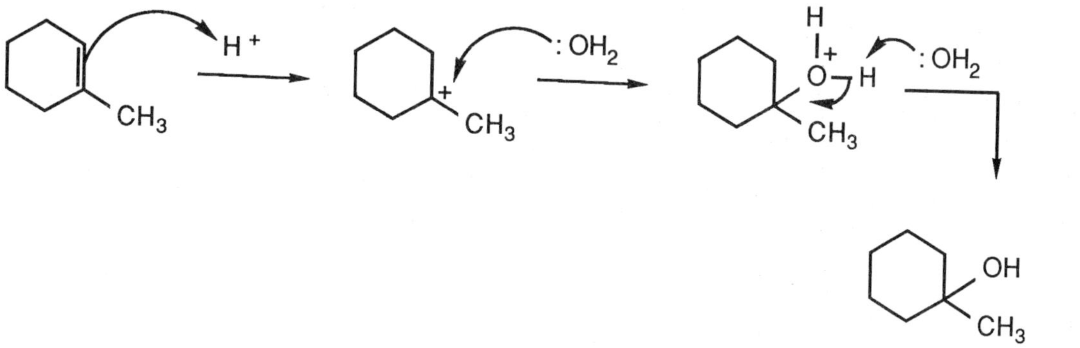

(b) $H-Br$, Br^-; Br CH3

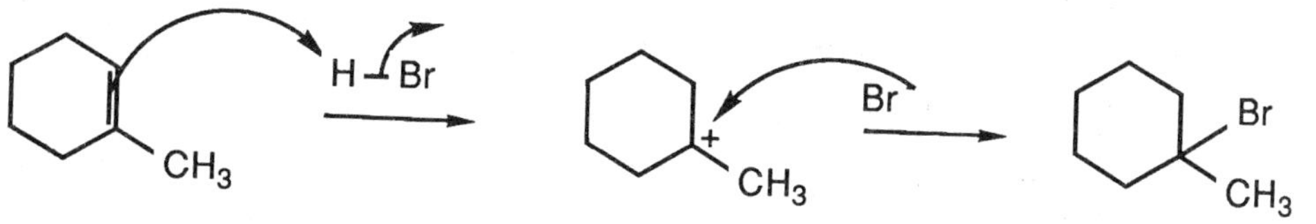

(c) ROOR ⟶ RO· + ·OR

RO· + HBr ⟶ Br· + ROH

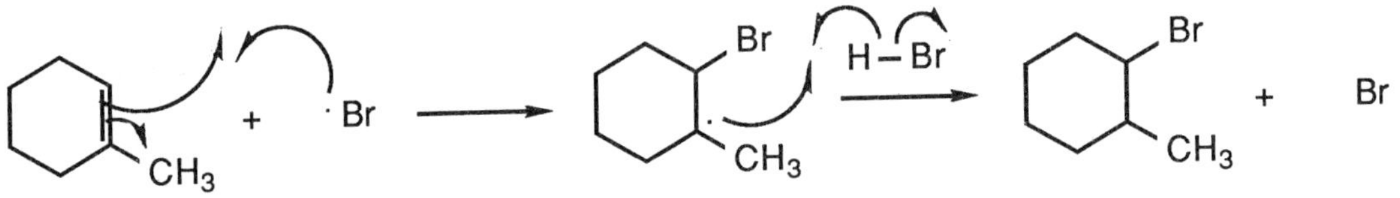

(d)

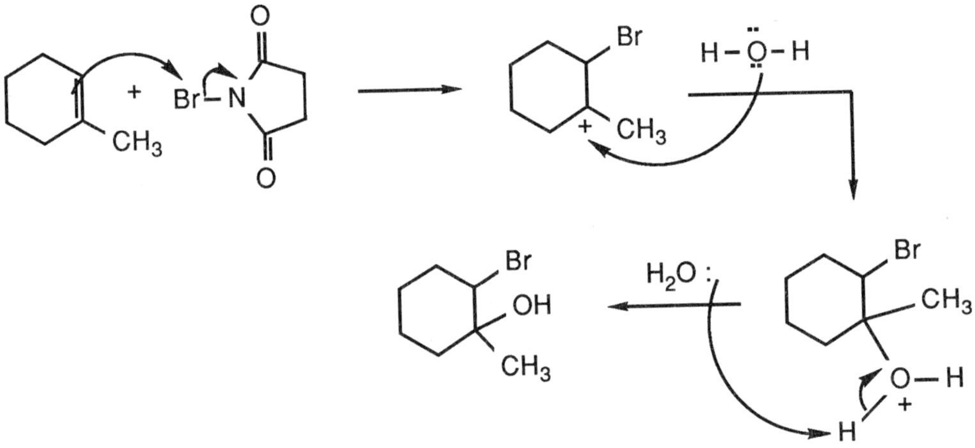

(e)

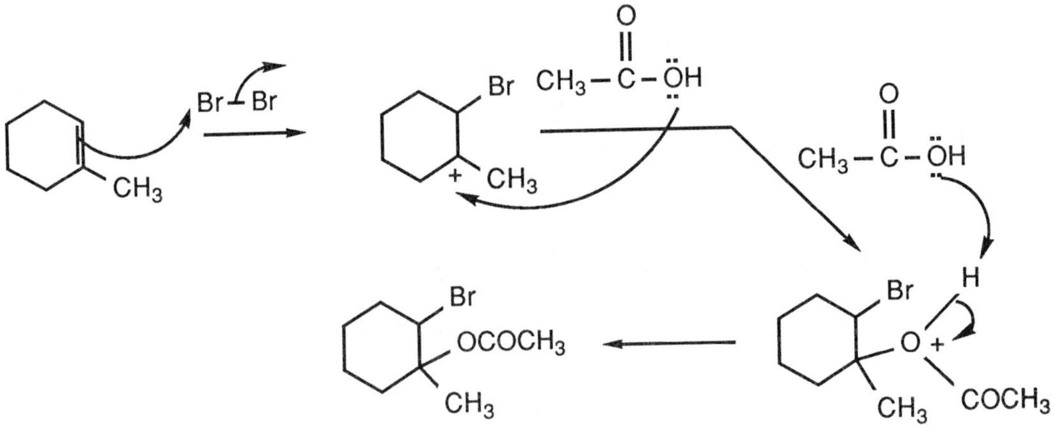

5.21

(a)

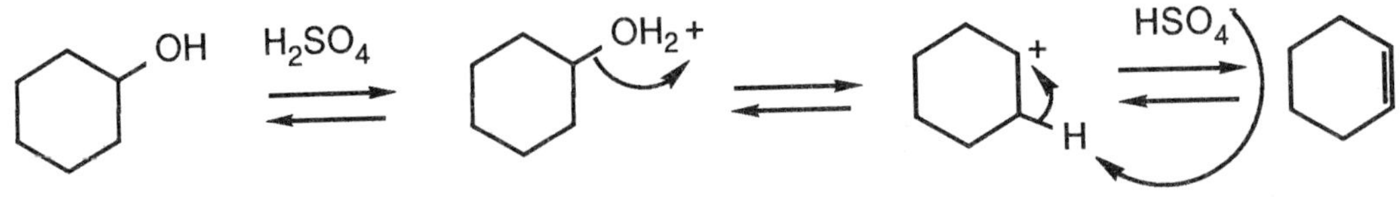

(b) [reaction scheme: trans-1,2-dibromocyclohexane + :Zn → cyclohexene]

$$\text{1,2-dibromocyclohexane} \xrightarrow{:Zn} \text{cyclohexene} + ZnBr_2$$

(c)

$$\text{bromocyclohexane} + {}^{-}OtBu \longrightarrow \text{cyclohexene} + tBuOH + KBr$$

5.22 **(a)** The *tert*-butyl group must be introduced as a 4-carbon unit here, and the key is to use lithium di-*tert*-butylcuprate (prepared from *tert*-butyl chloride:

$$\left(CH_3-\underset{CH_3}{\overset{CH_3}{C}}-\right)_2 CuLi + BrCH{=}C(CH_3)_2 \longrightarrow CH_3-\underset{CH_3}{\overset{CH_3}{C}}-CH{=}C(CH_3)_2$$

(b) The 5-carbon limit requires joining a cyclopentyl group to a vinyl group, and this can again be done using an organocopper reagent. (Bromocyclopentane can be used to prepare the necessary lithium dicyclopentylcuprate):

$$(C_5H_9)_2CuLi + BrCH{=}CH_2 \longrightarrow C_5H_9-CH{=}CH_2$$

(c) The target molecule contains six carbons, so the coupling of two 3-carbon compounds would be preferable. Propene would be the logical starting material for each of the 3-carbon compounds. Addition of HBr in the presence and absence of peroxides would yield 1- and 2-bromopropane, respectively. Conversion of the latter into the corresponding organocuprate could then be followed by a coupling reaction.

$$CH_3CH{=}CH_2 \xrightarrow{HBr} CH_3{-}\underset{}{\overset{CH_3}{\overset{|}{C}}}H{-}Br \xrightarrow[2.\ CuI]{1.\ Li} (CH_3{-}\overset{CH_3}{\overset{|}{C}}H)_2CuLi$$

$$CH_3CH{=}CH_2 \xrightarrow[peroxides]{HBr} CH_3CH_2CH_2Br$$

$$\downarrow$$

$$(CH_3)_2CHCH_2CH_2CH_3$$

(d) The target molecule contains a reactive functional group (the hydroxyl), which should be introduced last. Introduction of the OH could be accomplished by hydration of a double bond, and this suggests coupling of a C_3 alkenyl derivative with a C_5 alkyl compound. Either of these could be the organocopper reagent, but we have shown just one of the two possibilities:

$$(\text{cyclobutyl}{-}CH_2)_2CuLi \;+\; CH_3(Br)C{=}CH_2 \longrightarrow \text{cyclobutyl}{-}CH_2{-}\overset{CH_3}{\overset{|}{C}}{=}CH_2$$

(e) The target molecule again contains a hydroxyl group, and this suggests an organocopper reagent to couple an alkyl halide with an alkenyl halide (followed by addition of water to the double bond). The possible combinations are lithium divinylcuprate plus bromocyclohexane (a secondary but cyclic alkyl halide), lithium dicyclohexenylcuprate plus ethyl bromide, lithium dicyclohexylcuprate plus vinyl bromide, and lithium diethylcuprate plus 1-bromocyclohexene. We have shown the second of these alternatives:

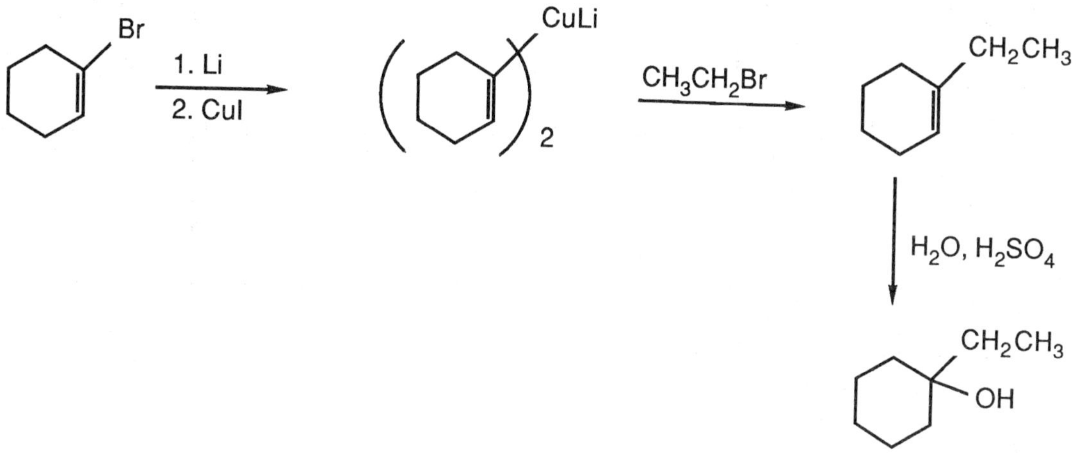

Br
1. Li
2. CuI
CuLi
2
CH_3CH_2Br
CH_2CH_3
H_2O, H_2SO_4
CH_2CH_3
OH

5.23 Both products have the same carbon skeleton, so there are no alkyl rearrangements in the HBr additions. The unsymmetrical alkene (A) (2,5-dimethyl-2-hexene) would yield the secondary bromide in reaction via the free radical mechanisms, and it would yield the tertiary bromide in an ionic addition. The symmetrical alkene, "diisopropylethylene", would also yield the secondary bromide in the addition of HBr in the presence of peroxides; this alkene would yield the tertiary bromide (via hydride shift to the corresponding tertiary cation) in an ionic addition. Two stereoisomers (B and C) are possible:

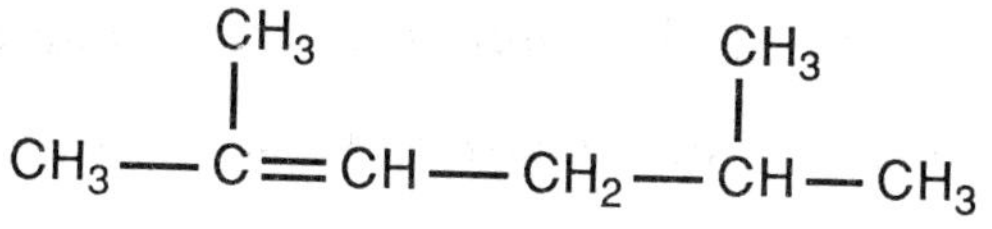

$CH_3—C(CH_3)=CH—CH_2—CH(CH_3)—CH_3$

A

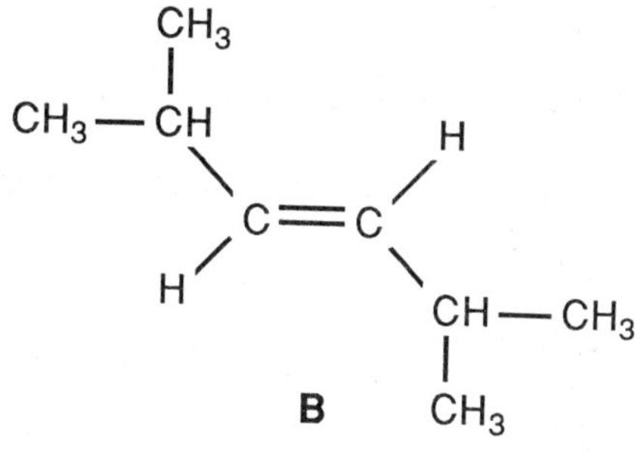

$(CH_3)_2CH$, H / C=C / H, $CH(CH_3)_2$

B

$(CH_3)_2CH$, $CH(CH_3)_2$ / C=C / H, H

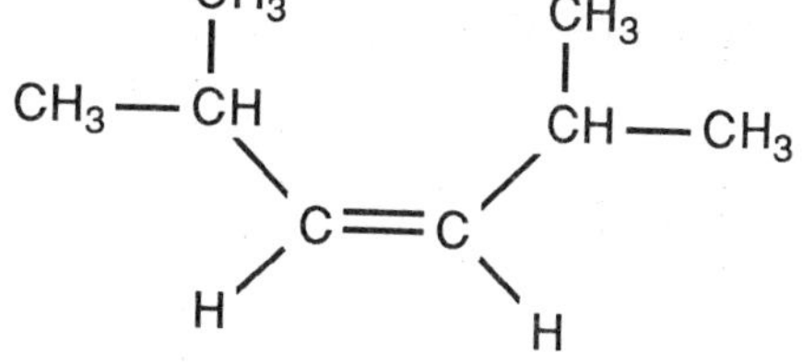

C

5.24

Ionic Addition

A:

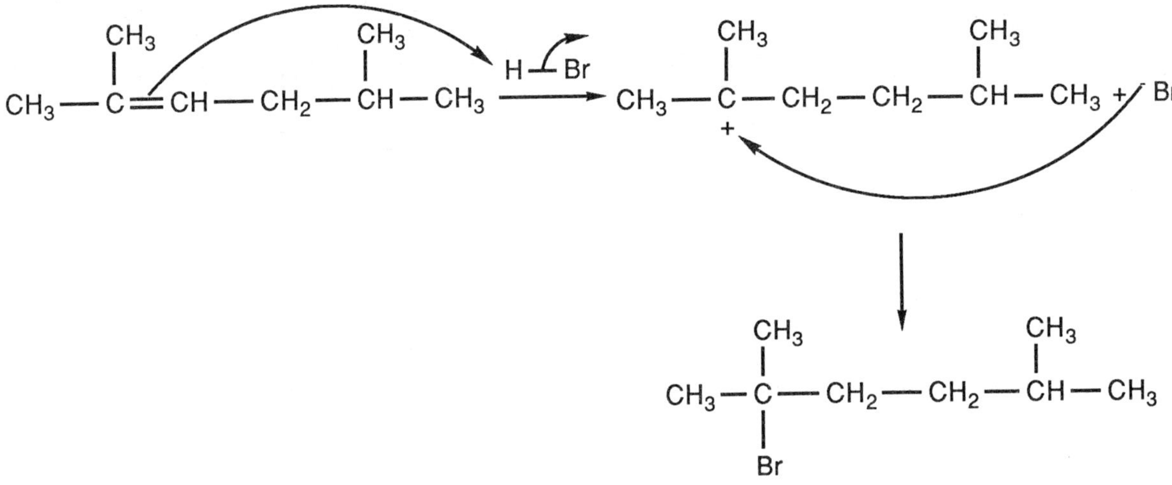

$$CH_3-\overset{\overset{CH_3}{|}}{C}=CH-CH_2-\overset{\overset{CH_3}{|}}{CH}-CH_3 \xrightarrow{H-Br} CH_3-\overset{\overset{CH_3}{|}}{\underset{+}{C}}-CH_2-CH_2-\overset{\overset{CH_3}{|}}{CH}-CH_3 + {}^{-}Br$$

$$\downarrow$$

$$CH_3-\overset{\overset{CH_3}{|}}{\underset{\underset{Br}{|}}{C}}-CH_2-CH_2-\overset{\overset{CH_3}{|}}{CH}-CH_3$$

B, C:

$$CH_3-\overset{\overset{CH_3}{|}}{CH}-CH=CH-\overset{\overset{CH_3}{|}}{CH}-CH_3 \xrightarrow{H-Br} CH_3-\overset{\overset{CH_3}{|}}{\underset{\underset{H}{|}}{C}}-\overset{+}{CH}-CH_2-\overset{\overset{CH_3}{|}}{CH}-CH_3$$

$$\downarrow$$

$$CH_3-\overset{\overset{CH_3}{|}}{\underset{\underset{Br}{|}}{C}}-CH_2-CH_2-\overset{\overset{CH_3}{|}}{CH}-CH_3 \longleftarrow CH_3-\overset{\overset{CH_3}{|}}{\underset{+}{C}}-CH_2-CH_2-\overset{\overset{CH_3}{|}}{CH}-CH_3 + {}^{-}Br$$

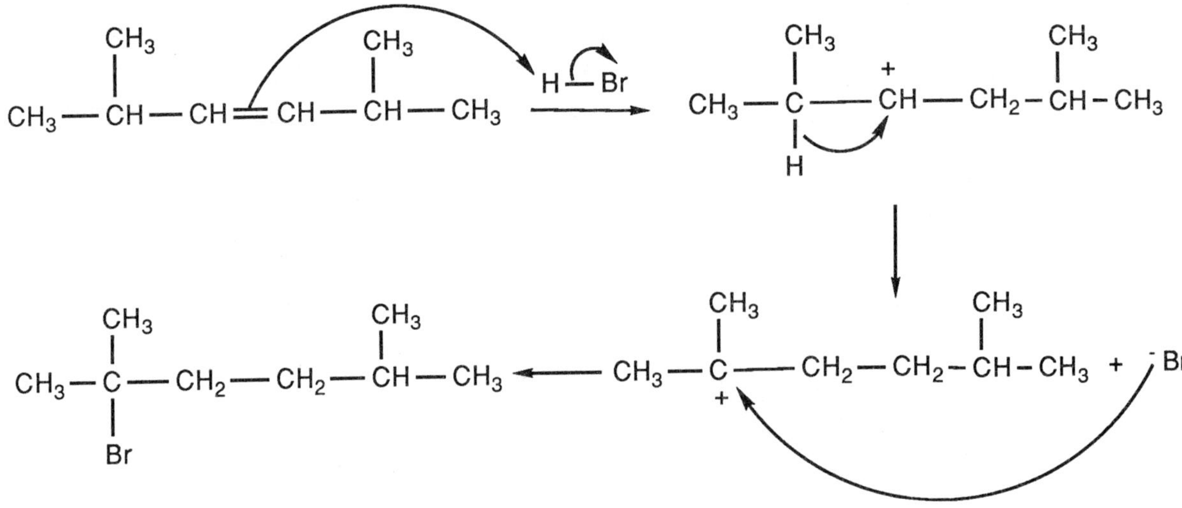

Polar Addition

$$ROOR \longrightarrow RO\cdot + \cdot OR$$

$$RO\cdot + HBr \longrightarrow Br\cdot + ROH$$

A:

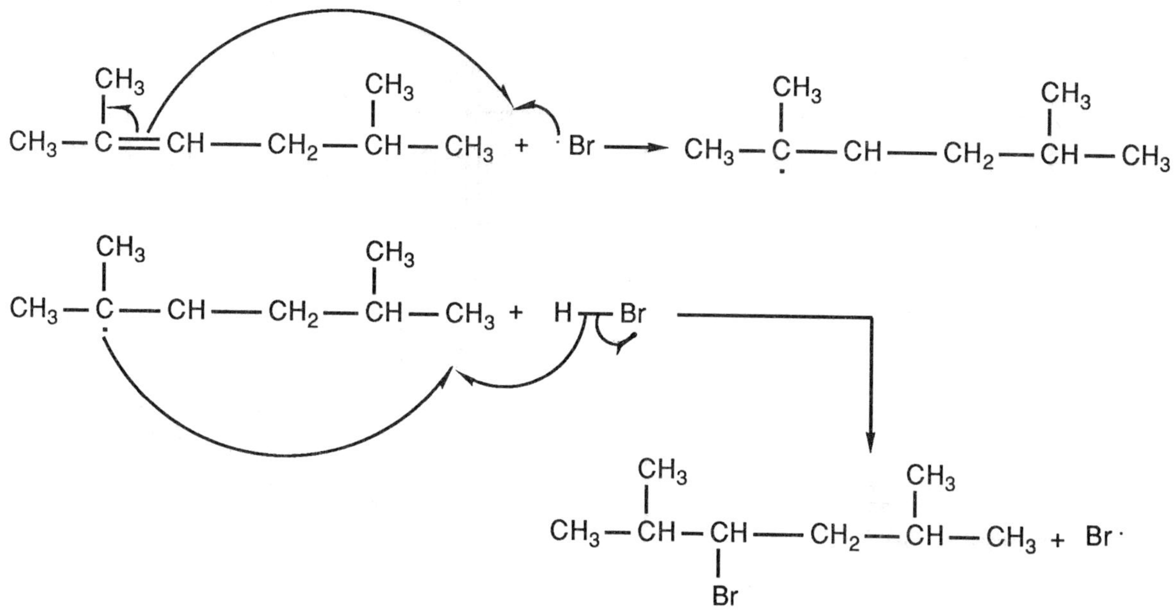

$$CH_3-\overset{CH_3}{\overset{|}{C}}=CH-CH_2-\overset{CH_3}{\overset{|}{CH}}-CH_3 + \cdot Br \longrightarrow CH_3-\overset{CH_3}{\overset{|}{\dot{C}}}-\underset{Br}{\underset{|}{CH}}-CH_2-\overset{CH_3}{\overset{|}{CH}}-CH_3$$

$$CH_3-\overset{CH_3}{\overset{|}{\dot{C}}}-\underset{Br}{\underset{|}{CH}}-CH_2-\overset{CH_3}{\overset{|}{CH}}-CH_3 + H-Br \longrightarrow CH_3-\overset{CH_3}{\overset{|}{CH}}-\underset{Br}{\underset{|}{CH}}-CH_2-\overset{CH_3}{\overset{|}{CH}}-CH_3 + Br\cdot$$

B, C:

$$CH_3-\overset{CH_3}{\overset{|}{CH}}-CH=CH-\overset{CH_3}{\overset{|}{CH}}-CH_3 + \cdot Br \longrightarrow CH_3-\overset{CH_3}{\overset{|}{CH}}-\dot{C}H-\underset{Br}{\underset{|}{CH}}-\overset{CH_3}{\overset{|}{CH}}-CH_3$$

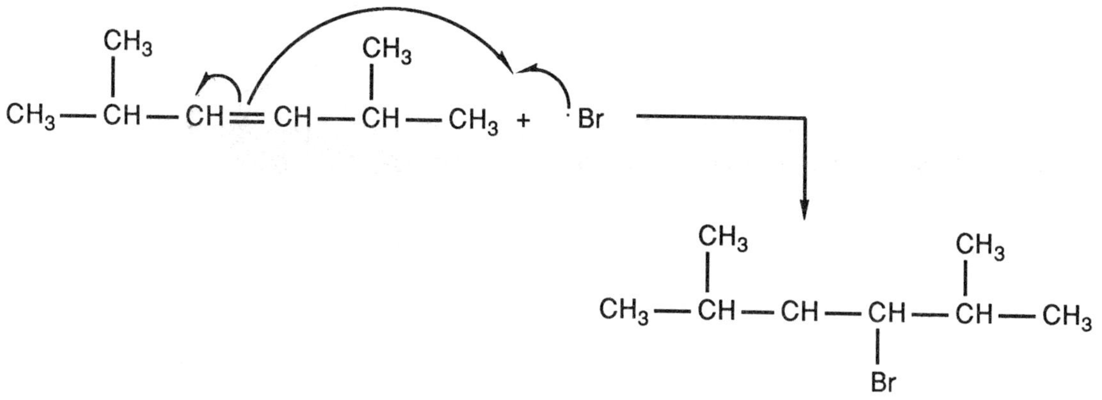

$$\mathrm{CH_3{-}\underset{\displaystyle}{\overset{\displaystyle CH_3}{\overset{|}{C}}H}{-}\dot{C}H{-}\underset{\underset{\displaystyle Br}{|}}{C}H{-}\overset{\displaystyle CH_3}{\overset{|}{C}}H{-}CH_3 \;+\; H{-}Br}$$

$$\longrightarrow\ \mathrm{CH_3{-}\overset{\displaystyle CH_3}{\overset{|}{C}}H{-}CH_2{-}\underset{\underset{\displaystyle Br}{|}}{C}H{-}\overset{\displaystyle CH_3}{\overset{|}{C}}H{-}CH_3 \;+\; Br\cdot}$$

5.25 You are given the structures of compounds E and F, and their different carbon skeletons tells you that a rearrangement must occur somewhere in the problem because you are also told that compounds A and C have the *same* carbon skeleton. The only place that a rearrangement is likely is in the acid mediated dehydration of A, so compounds A, C, D, and E all have the same skeleton. Two primary derivatives are possible, but only one of these would rearrange to the skeleton of B and F (via a 1,2-ethyl shift). The complete structures are:

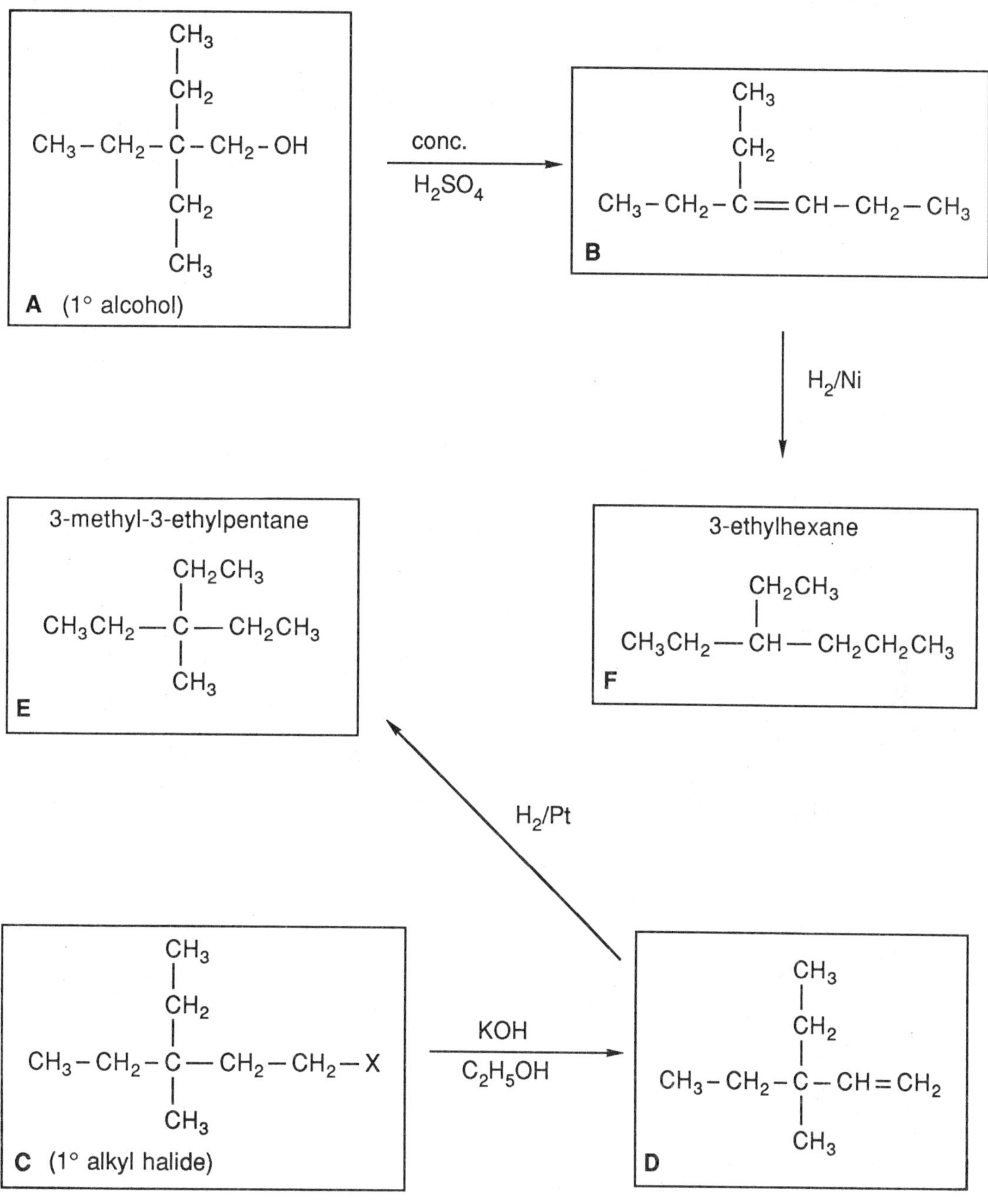

5.26 One key piece of structural information is provided by the ozonolysis reaction to form G and H. The intervening alkene must have been:

$(Et)_2C{=}C(Me)_2$

This could have been formed from either of two possible bromides, one of which is the correct structure for D:

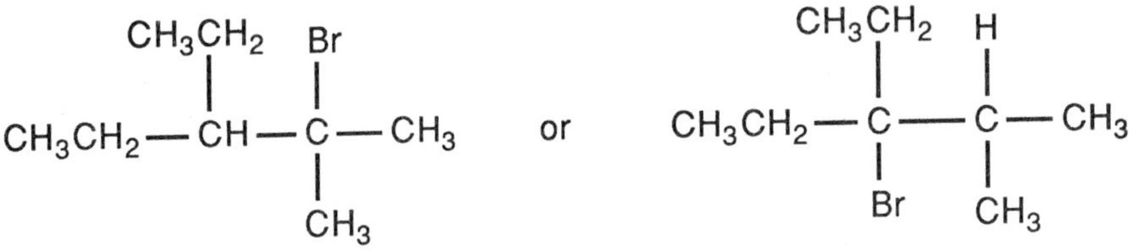

$$\begin{array}{ccccc} & CH_3CH_2 & & Br & \\ & | & & | & \\ CH_3CH_2 - & CH & - & C & - CH_3 \\ & & & | & \\ & & & CH_3 & \end{array} \quad \text{or} \quad \begin{array}{ccccc} & CH_3CH_2 & & H & \\ & | & & | & \\ CH_3CH_2 - & C & - & C & - CH_3 \\ & | & & | & \\ & Br & & CH_3 & \end{array}$$

A second important piece of evidence is provided by the different carbon skeletons of compound F and the alkene drawn above. This requires that a rearrangement (1,2-alkyl shift) has occurred somewhere in the scheme, presumably in the reaction of A with HBr. Either D or E (or possibly both) has been formed with rearrangement. The transformations leading to alkane C must occur without rearrangement, and this is different from alkane F, so compound E must be formed with rearrangement. Therefore you can be certain certain that E is also a tertiary bromide, so its structure must be:

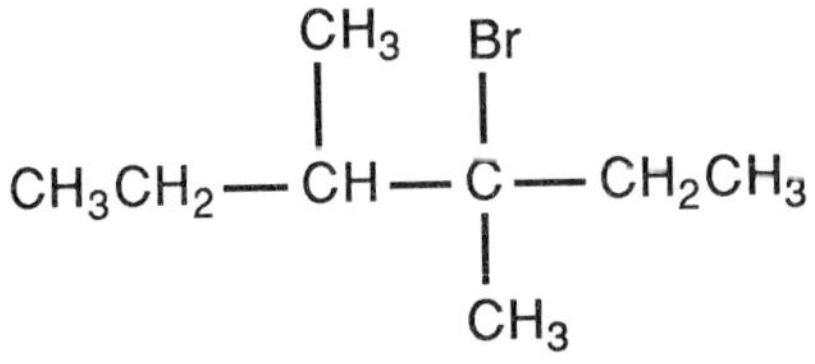

$$\begin{array}{ccccc} & CH_3 & & Br & \\ & | & & | & \\ CH_3CH_2 - & CH & - & C & - CH_2CH_3 \\ & & & | & \\ & & & CH_3 & \end{array}$$

Now you can draw the carbocation precursor for E as well as the two possibilities for D:

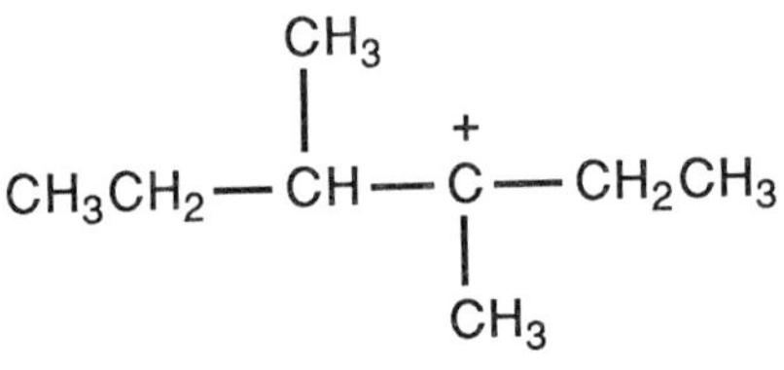

$$\begin{array}{ccccc} & CH_3 & & & \\ & | & & + & \\ CH_3CH_2 - & CH & - & C & - CH_2CH_3 \\ & & & | & \\ & & & CH_3 & \end{array}$$

E^+

$$\begin{array}{ccccc} & CH_3CH_2 & & & \\ & | & & + & \\ CH_3CH_2 - & CH & - & C & - CH_3 \\ & & & | & \\ & & & CH_3 & \end{array}$$

or

$$\begin{array}{ccccc} & CH_3CH_2 & & H & \\ & | & & | & \\ CH_3CH_2 - & C & - & C & - CH_3 \\ & + & & | & \\ & & & CH_3 & \end{array}$$

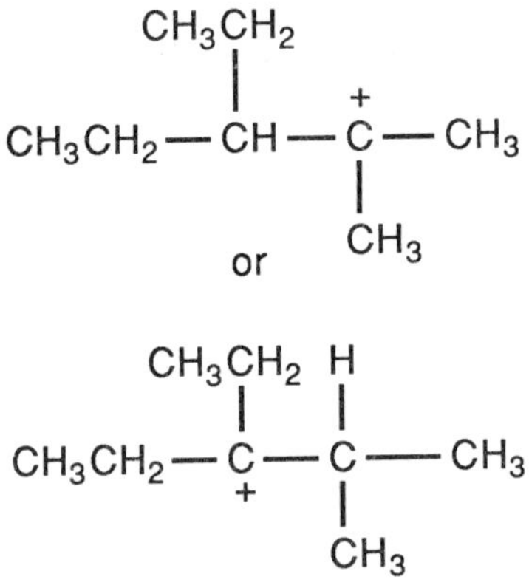

D^+

The question to be decided here, is what carbocation could yield both E^+ and one of the alternatives for D^+? Two cations are possible, and these are labeled A^+ to indicate their origin:

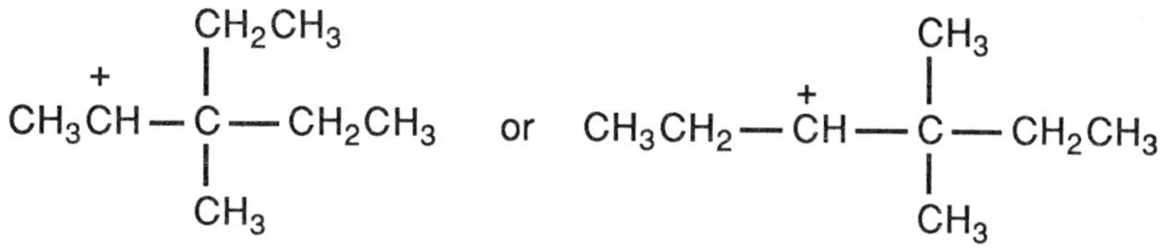

$$\begin{array}{ccccc} & & CH_2CH_3 & & \\ & \overset{+}{} & | & & \\ CH_3\overset{+}{C}H & - & C & - & CH_2CH_3 \\ & & | & & \\ & & CH_3 & & \end{array} \quad \text{or} \quad \begin{array}{ccccccc} & & & & CH_3 & & \\ & & & & | & & \\ CH_3CH_2 & - & \overset{+}{C}H & - & C & - & CH_2CH_3 \\ & & & & | & & \\ & & & & CH_3 & & \end{array}$$

For each of these possibilities rearrangement could occur by either a 1,2-ethyl shift or a 1,2-methyl shift to yield the D^+ and E^+ cations. A decision between the alternatives can be made by evaluating the possibilities for A itself:

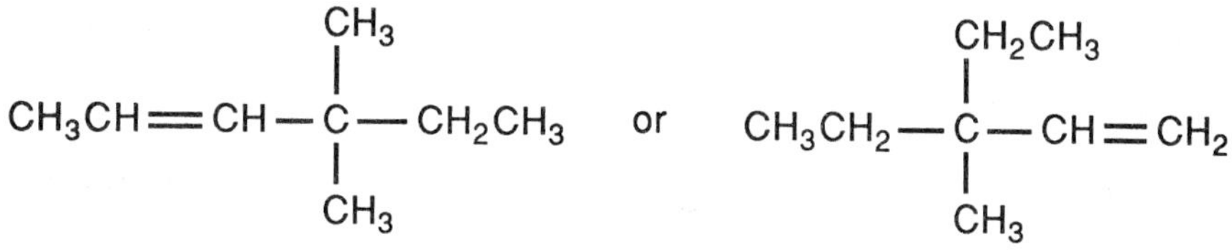

$$\begin{array}{ccccc} & & CH_3 & & \\ & & | & & \\ CH_3CH{=}CH & - & C & - & CH_2CH_3 \\ & & | & & \\ & & CH_3 & & \end{array} \quad \text{or} \quad \begin{array}{ccccc} & & CH_2CH_3 & & \\ & & | & & \\ CH_3CH_2 & - & C & - & CH{=}CH_2 \\ & & | & & \\ & & CH_3 & & \end{array}$$

One of these structures has a double bond with the partial structure:

$$CH_3\text{-}CH{=}CH\text{-}R$$

and the other has the partial structure:

$$CH_2{=}CH\text{-}R$$

The first of these possiblities could lead to addition of HBr with two orientations (both with and without peroxides), and this is not consistent with the products shown. The second possible partial structure would undoubtedly react at the terminal carbon with either H^+ or with a bromine atom, and it is the correct structure. This now allows the complete answer to be written:

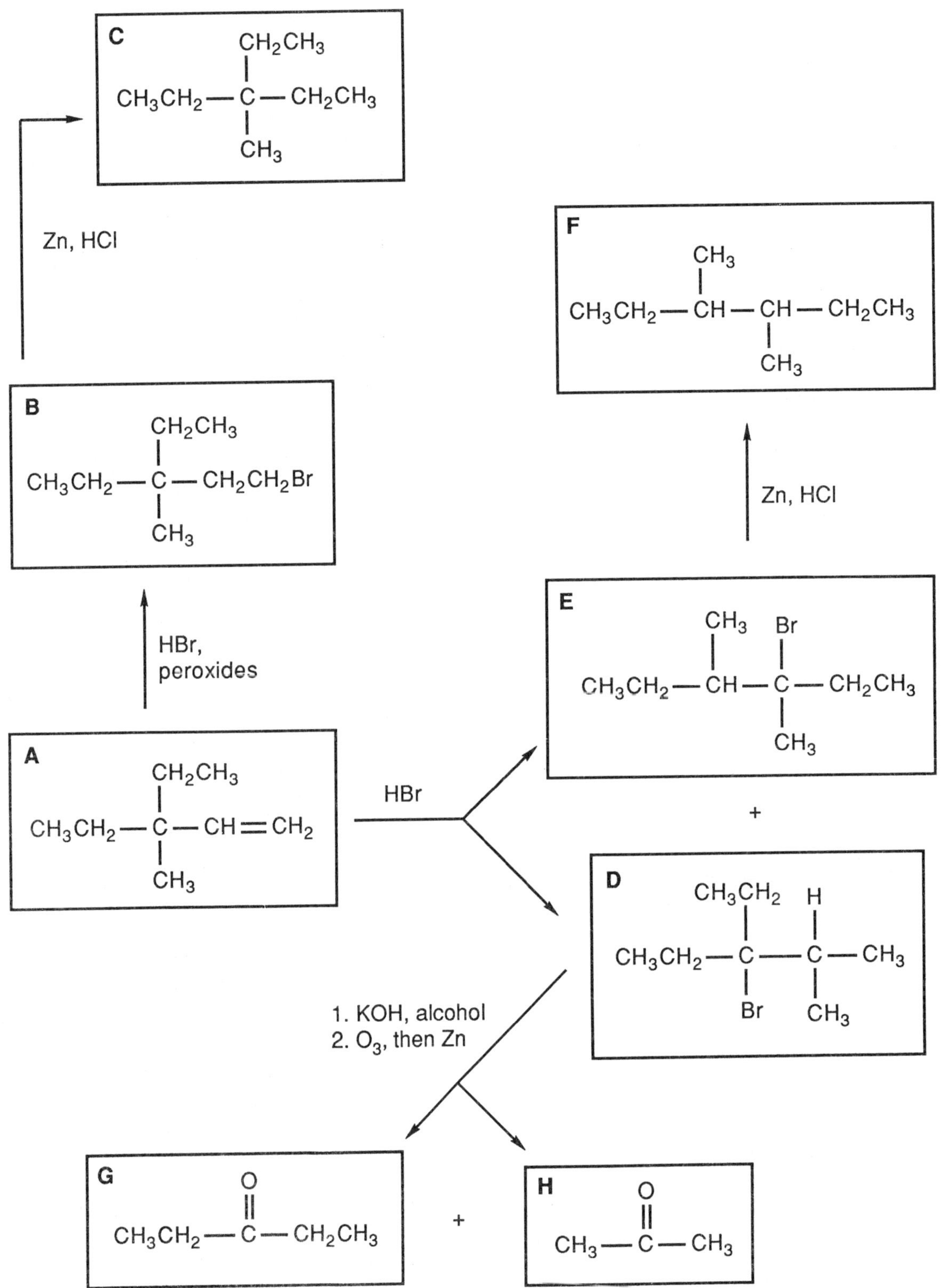
C
CH2CH3
CH3CH2—C—CH2CH3
CH3
Zn, HCl
F
CH3
CH3CH2—CH—CH—CH2CH3
CH3
B
CH2CH3
CH3CH2—C—CH2CH2Br
CH3
Zn, HCl
HBr,
peroxides
E
CH3 Br
CH3CH2—CH—C—CH2CH3
CH3
A
CH2CH3
CH3CH2—C—CH=CH2
CH3
HBr
+
D
CH3CH2 H
CH3CH2—C—C—CH3
Br CH3
1. KOH, alcohol
2. O3, then Zn
G
O
CH3CH2—C—CH2CH3
+
H
O
CH3—C—CH3

5.27 The only direct structural information provided in this problem is the structure of compound E. This was formed by the reaction of lithium dimethylcuprate with a *primary* alkyl halide. Only one primary alkyl halide would yield E in this reaction, and that gives you the structure for A:

$$CH_3-\underset{CH_3}{\overset{CH_3}{\overset{|}{\underset{|}{C}}}}-\overset{CH_3}{\overset{|}{CH}}-CH_2Br \xrightarrow{(CH_3)_2CuLi} CH_3-\underset{CH_3}{\overset{CH_3}{\overset{|}{\underset{|}{C}}}}-\overset{CH_3}{\overset{|}{CH}}-CH_2-\boxed{CH_3}$$

The reaction of A with KOH/alcohol is a straightforward elimination, and hydrogenation yields C. Finally, free radical bromination of C would take place at the only tertiary position to give D, and treatment of D with base could only give B again in an elimination reaction. The full answer follows:

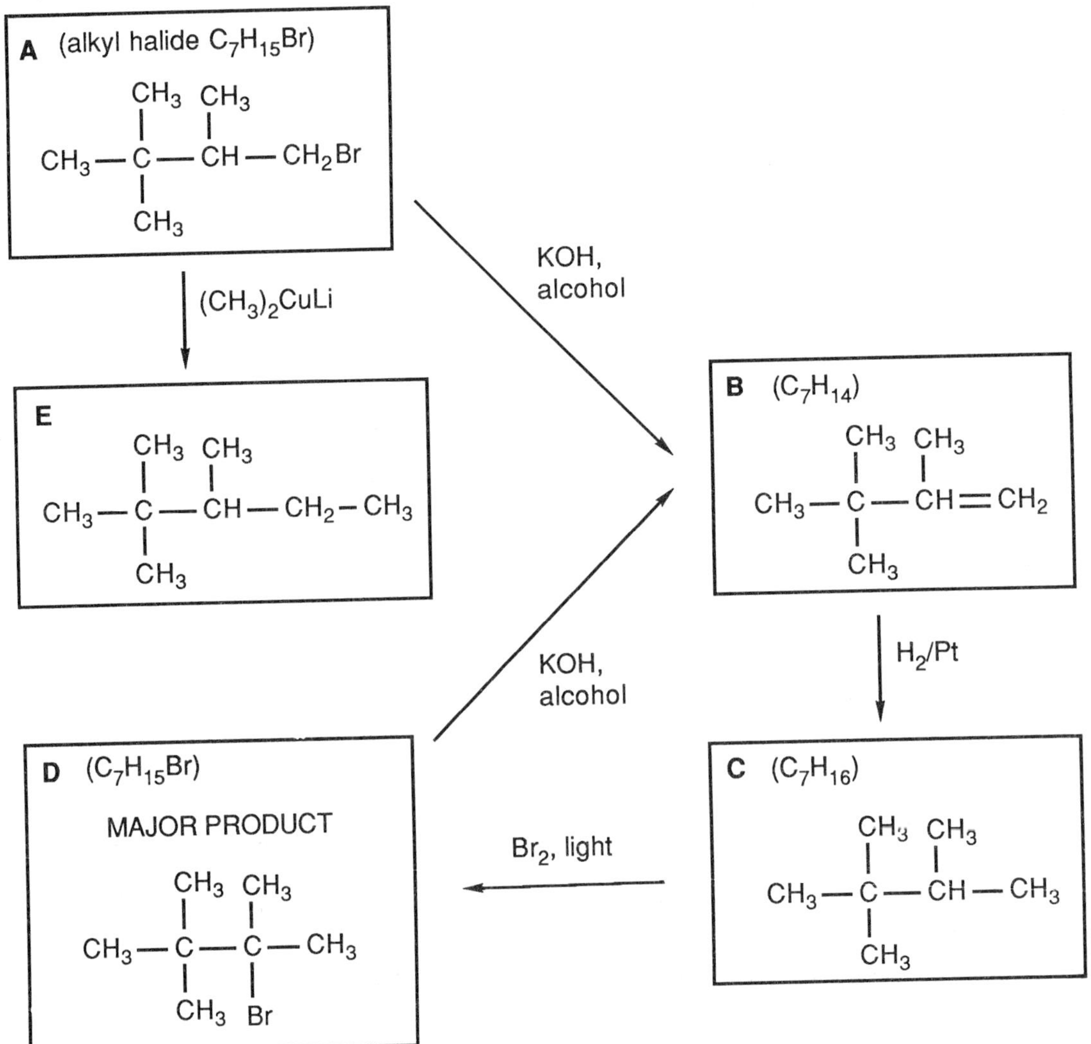

5.28 Dehydration should lead to only two isomeric alkenes, and the major product (B) is the more highly substituted alkene. The minor product is C. The alkyl halide (D) that would yield B as the sole product in elimination with base is 3-bromo-2,4-dimethylpentane, and formation of the terminal alkene C requires that E be the corresponding primary bromide, 1-bromo-2,4-dimethylpentane. Both of these halides would yield 2,4-dimethylpentane upon reduction, and the only remaining alkyl halide that would yield this alkane upon elimination followed by hydrogenation is the 2-bromo isomer. Therefore the structures of F and G are. The complete answer to the problem follows:

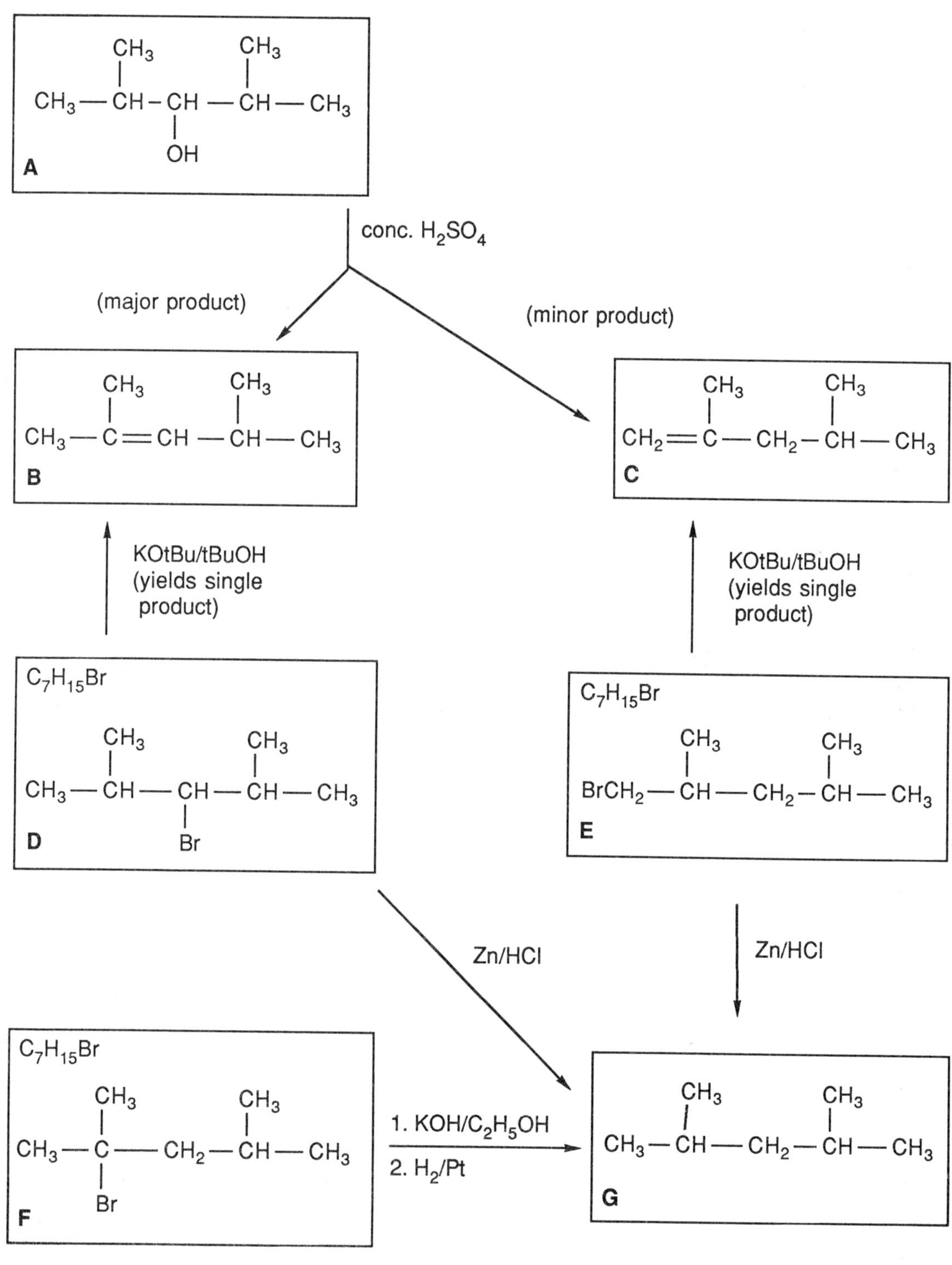

5.29 Neither base induced elimination nor catalytic hydrogenation leads to skeletal rearrangements, so B–D all have the same carbon skeleton as 2,3-dimethylpentane. The position of the double bond of A can be deduced from the formation of acetone in the ozonolysis reaction leading to B. Two different alkyl bromides would yield A upon elimination in base, and these must be C and D (or vice versa):

$$CH_3-\overset{\overset{CH_3}{|}}{\underset{\underset{Br}{|}}{C}}-\overset{\overset{CH_3}{|}}{CH}-CH_2CH_3 \quad \text{and} \quad CH_3-\overset{\overset{CH_3}{|}}{CH}-\overset{\overset{CH_3}{|}}{\underset{\underset{Br}{|}}{CH}}-CH_2CH_3$$

The formation of both C and D from a single alkene suggests 2,3-dimethylpentene, but that is compound A. Moreover, hydrogenation of E does not yield 2,3-dimethylpentene, so you know that addition of HBr to E must be accompanied by a skeletal rearrangement. Inspection of C and D tells you the structure of the two rearranged cations that finally reacted with Br^-:

$$\underset{\mathbf{C^+}}{CH_3-\overset{\overset{CH_3}{|}}{\underset{+}{C}}-\overset{\overset{CH_3}{|}}{CH}-CH_2CH_3} \qquad \underset{\mathbf{D^+}}{CH_3-\overset{\overset{CH_3}{|}}{CH}-\overset{\overset{CH_3}{|}}{\underset{+}{C}}-CH_2CH_3}$$

The cation labelled C^+ could have been formed by rearrangement of either an ethyl or a methyl group, but D^+ could only have been formed by a 1,2-methyl shift of the following cation, labelled E^+:

$$\underset{\mathbf{E^+}}{CH_3-\overset{+}{CH}-\overset{\overset{CH_3}{|}}{\underset{\underset{CH_3}{|}}{C}}-CH_2CH_3} \longrightarrow \underset{\mathbf{D^+}}{CH_3-\overset{\overset{CH_3}{|}}{CH}-\overset{\overset{CH_3}{|}}{\underset{+}{C}}-CH_2CH_3}$$

Notice that the other possible rearrangement of E^+, a 1,2-ethyl shift, would lead to the isomeric cation C^+. E^+ defines the carbon skeleton of compound E, and the location of the double bond is readily deduced from the formation of CH_2O as one of the ozonolysis products. The complete answer to the problem follows:

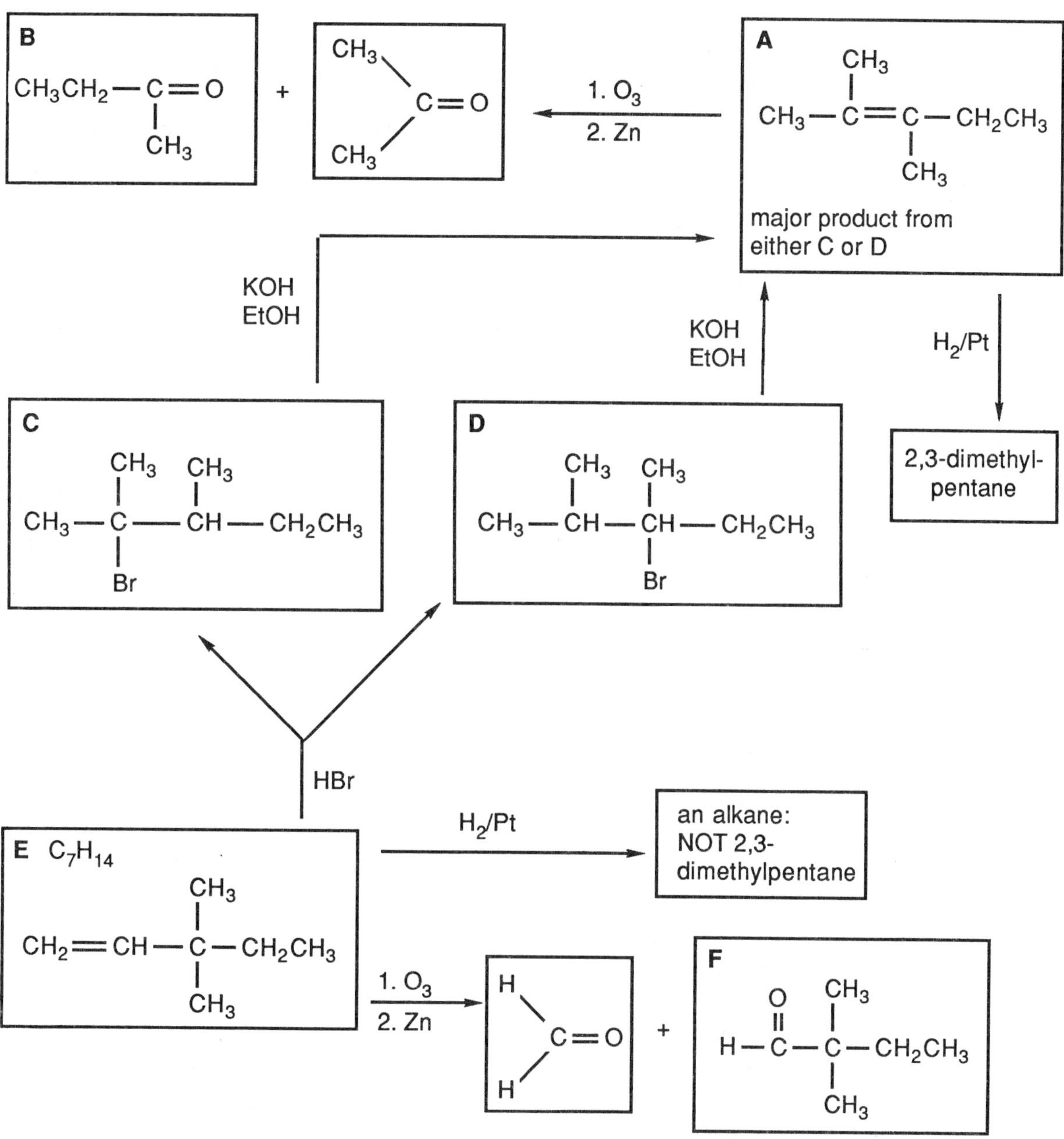

5.30

(a) $CH_3—CH(OH)—CH(OH)—CH_3$ **(b)** $CH_3CH_2CH_2CH_3$ **(c)** $CH_3CH_2—CH(OSO_3H)—CH_3$

(d) 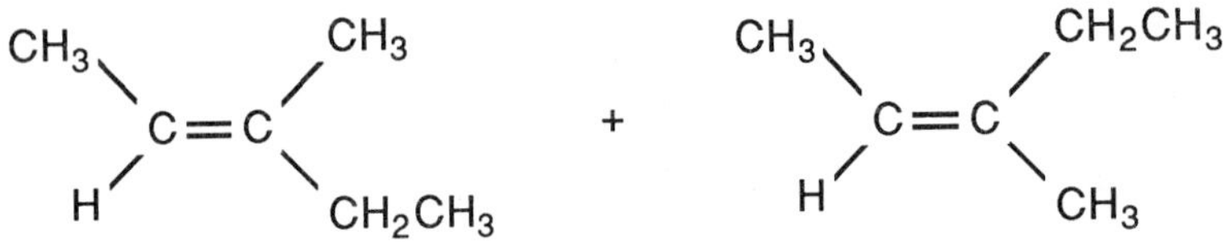 $CH_3(H)C{=}C(CH_3)(CH_2CH_3)$ + $CH_3(H)C{=}C(CH_2CH_3)(CH_3)$

(e) 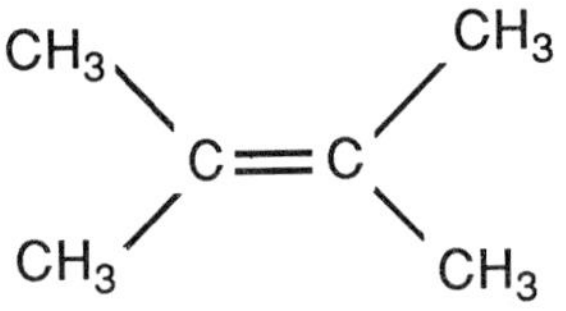 $(CH_3)_2C{=}C(CH_3)_2$

(f) 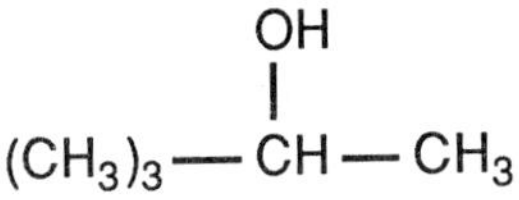 $(CH_3)_3{-}CH(OH){-}CH_3$

(g) 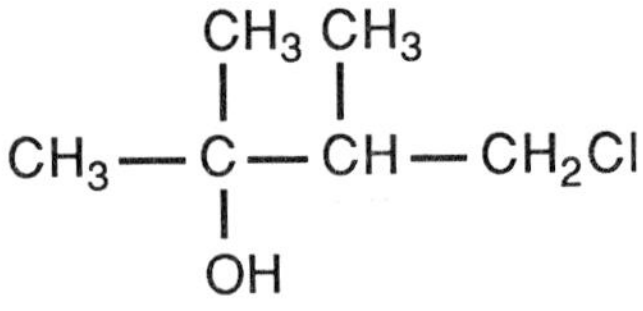 $CH_3{-}C(CH_3)(OH){-}CH(CH_3){-}CH_2Cl$

(h) $(CH_3)_3CCH_2CH_2Br$

(i) 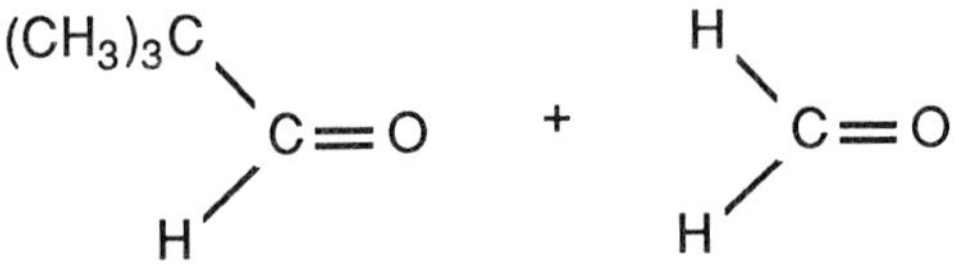 $(CH_3)_3C(H)C{=}O$ + $H(H)C{=}O$

(j) 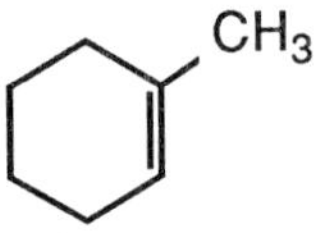 1-methylcyclohexene ring with CH_3

(k) 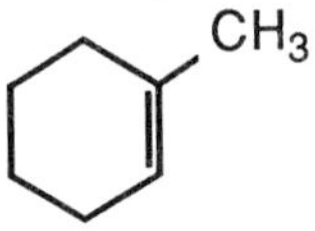 1-methylcyclohexene ring with CH_3

5.31

(a) 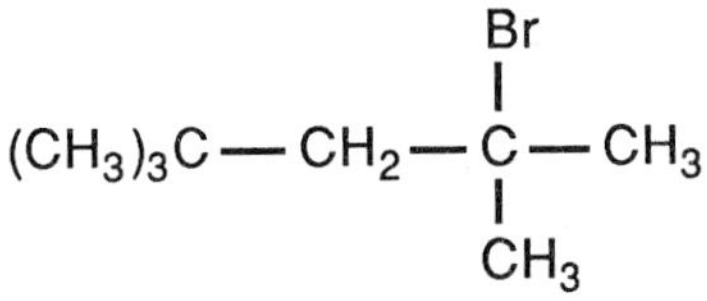 $(CH_3)_3C{-}CH_2{-}C(Br)(CH_3){-}CH_3$

(b) $(CH_3)_3C{-}CH(Br){-}CH(CH_3){-}CH_3$

(c) $(CH_3)_3C{-}CH_2{-}C(OH)(CH_3){-}CH_3$

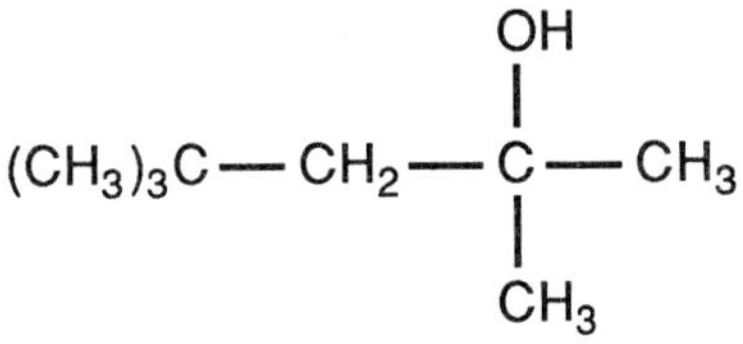

(d) 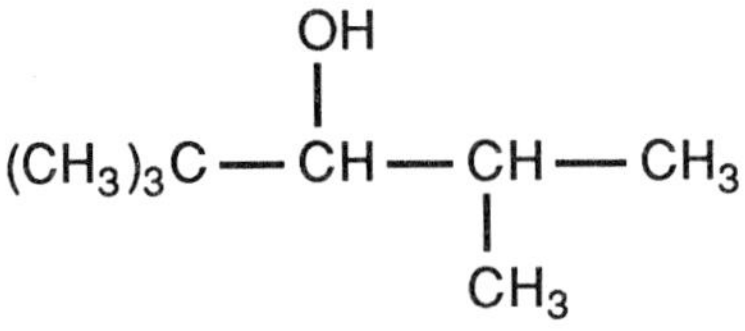 $(CH_3)_3C{-}CH(OH){-}CH(CH_3){-}CH_3$

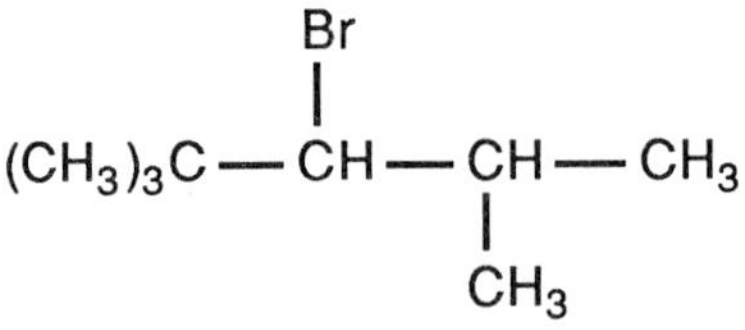

(e) $(CH_3)_3C-CH(Br)-C(Br)(CH_3)-CH_3$

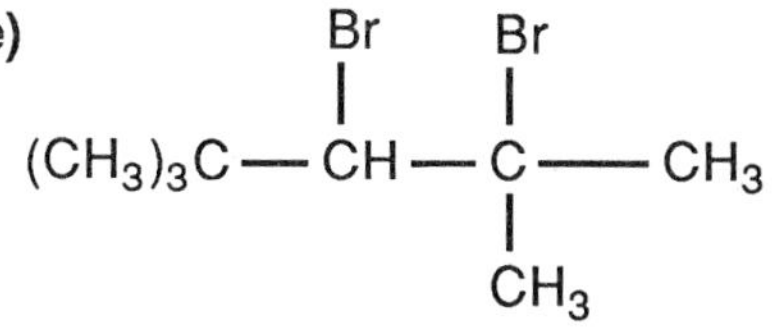

(f) 1-chloro-1-(CH(CH_3)-CH_3)cyclopentane + 1-Cl, 1-CH_3, 2-CH_3 cyclohexane

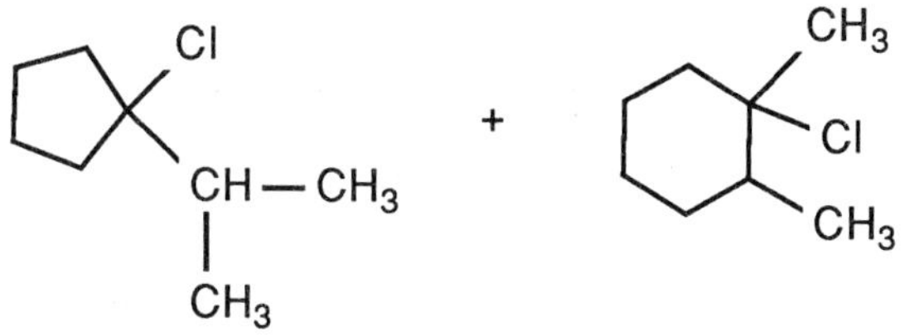

(g) $CH_3-C(CH_3)=CH-CH_2-CH_3$

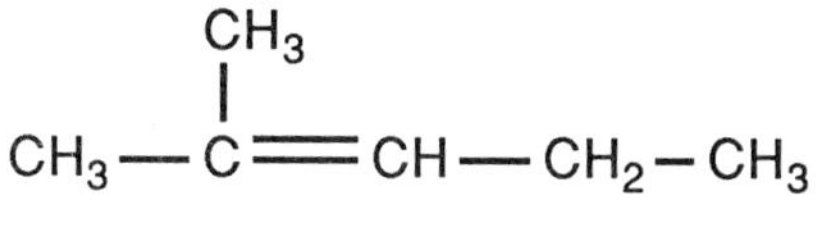

(h) cyclohexane with OH, OH

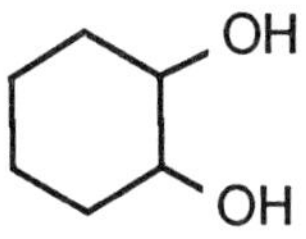

(i) cyclohexyl $O-C(=O)-CH_3$

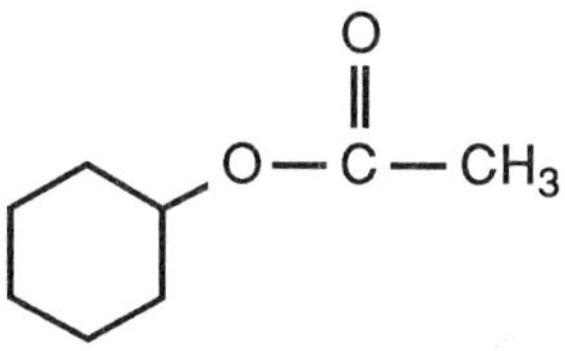

(j) cyclohexane with Cl, Cl

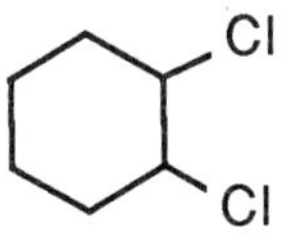

(k) cyclohexane with Cl

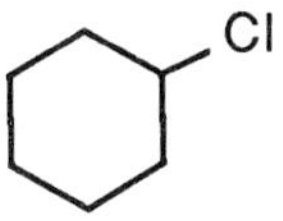

CHAPTER 6. STEREOCHEMISTRY

ANSWERS TO EXERCISES

6.1 **(a)** Constitutional isomers: the first structure is (*E*)-1-chloro-2-pentene and the second is (*E*)-1-chloro-3-pentene.

(b) Stereoisomers: the methyl groups are *cis* to each other in the first drawing but *trans* in the second drawing.

(c) Different, non-isomeric compounds: the first structure has five carbon atoms and the second has six.

(d) Same: both are drawings of (*E*)-2-pentene.

6.2 **(a)** Chiral (rotation in one direction tightens a screw, while rotation in the other direction loosens it); no plane of symmetry.

(b) Achiral; a cube has many planes of symmetry.

(c) Achiral; plane of symmetry. (This assumes that the tines of the fork are the same on the left and right.)

(d) Achiral; plane of symmetry. (This assumes that the you are not considering the writing on the label inside the shirt).

(e) Chiral; no plane of symmetry. For men's clothing the buttons are usually on the right and the buttonholes on the left; the opposite arrangement is typical for women's clothing.

(f) Achiral: two of the substituents on the central carbon are ethyl groups. A plane of symmetry lies perpendicular to the paper; it lies between the two ethyl groups and includes the central carbon together with its H and Cl substituents.

(g) Chiral; no plane of symmetry. The compound is 2-methyl-3-bromopentane, and

C-3 is a chiral center.

(h) Achiral: the compound is 2-chloro-2-methylbutane (*tert*-butyl chloride). There are three planes of symmetry, each defined by the chlorine together with C-2 and one of the methyl carbons.

(i) Chiral: the compound is 2,3-dimethylhexane, and C-3 is a chiral center (the four substituents are hydrogen, methyl, propyl, and isopropyl). No plane of symmetry.

(j) Achiral: this assumes that we are ignoring the numbers on the balls. (Depending on how the numbers were drawn, they might or might not have a symmetry plane passing through them.) The cue ball would have an infinite number of symmetry planes that bisect it.

6.3 Priorities are shown in parentheses after each substituent:

(a) *S*: Br(3), Cl(2), CH_3(1), H(0)

(b) *R*: Br(3), Cl(2), CH_3(1), H(0)

(c) *R*: OH(3), CH_2OH(2), CH_3(1), H(0)

(d) *S*: OH(3), CH_3CH_2(2), CH_3(1), H(0)

(e) *R*: Br(3), F(2), CF_3(1), H(0)

(f) *R*: OH(3), CH_2Br(2), CH_2F(1), CH_3(0)

(g) *S*: Cl(3), ring-$C(CH_3)_2$(2), ring-CH_2(1), H(0)

6.4 **(a)** Same: two drawings of 2-chloropropane. The structures are mirror images, but there is no chiral center.

(b) Enantiomers: the first structure is (*R*)-2-fluorobutane and the second has the *S* configuration.

(c) Enantiomers: the first structure is (*S*)-1-chloro-1-fluoropropane and the second has the *R* configuration.

(d) Same: two drawings of (*R*)-1,3-pentanediol in different conformations.

6.5 **(a)**

(b)

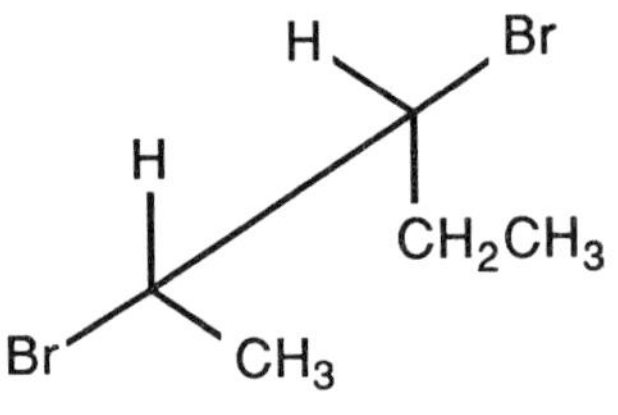

(c)

6.6 **(a)** (2*S*,3*R*)-2,3-dichloropentane

(b) (*R*)-1,3-dibromo-2,3-dimethylbutane

(c) (2*S*,3*S*)-2,3-dibromopentane

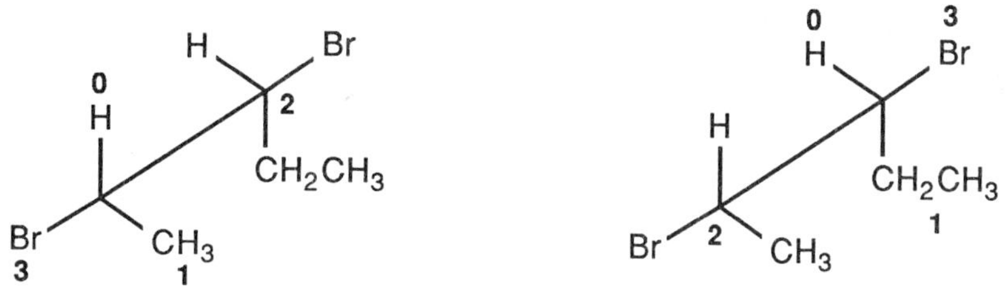

6.7 **(a)** Same compound: both are drawings of (*R*)-2-bromobutane.

(b) Diastereomers: with the carbon chain in the same orientation, the first structure has the hydroxyl groups on opposite sides but the second has both on the same side. The first structure is (2*R*,3*R*)-2,3-butanediol and the second is (2*R*,3*S*)-2,3-butanediol.

(c) Enantiomers: the first structure is (*S*)-2-bromobutane and the second is the *R* isomer.

(d) Diastereomers: the methyl substituents are *trans* in the first structure but *cis* in the second. (Each of the structures has a plane of symmetry, so that also tells you that they cannot be enantiomers).

6.8

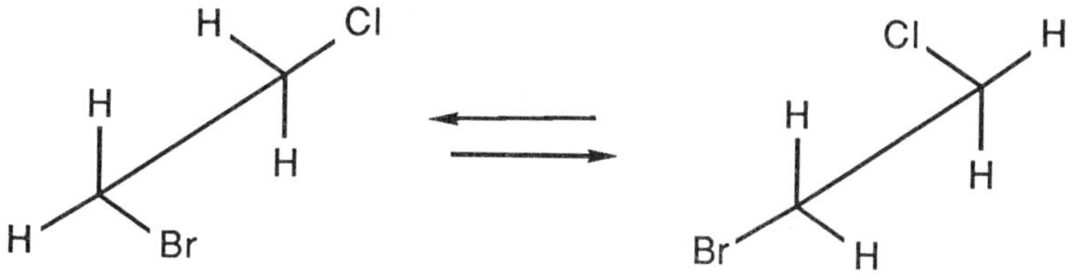

6.9 **(a)** The starting material can be drawn in the following three-dimensional representations.

(b)–(d) Six products with different physical properties would be formed as indicated in the following drawing:

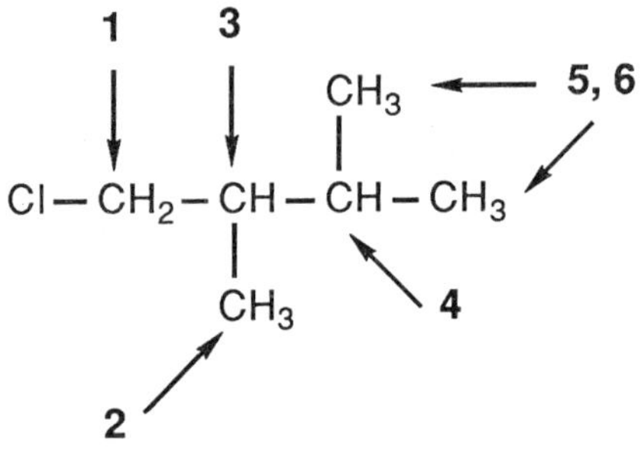

1 3 5, 6

$Cl-CH_2-CH(CH_3)-CH(CH_3)-CH_3$

2 4

The following drawings show the appropriate stereochemical information for each of the products:

1: $CH(CH_3)_2$, H, CH_3, S, $CH(Cl)_2$

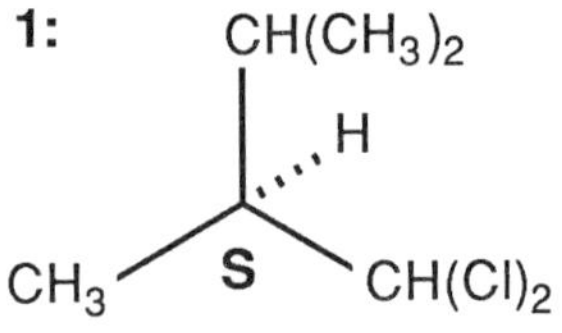

chiral
optically active

2: $CH(CH_3)_2$, H, ClH_2C, CH_2Cl

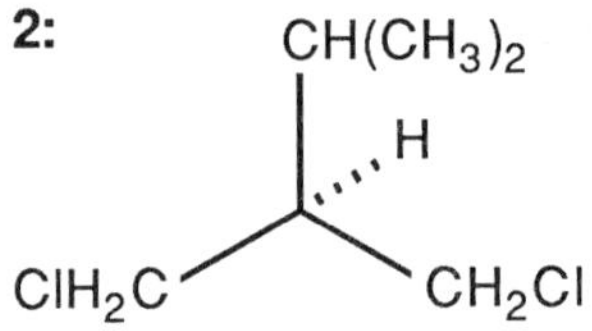

achiral
optically inactive

3: $CH(CH_3)_2$, CH_3, ClH_2C, R, Cl + $CH(CH_3)_2$, CH_3, Cl, S, CH_2Cl

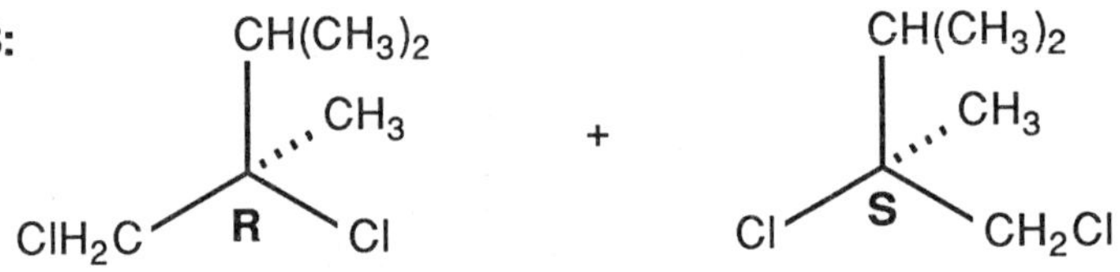

chiral chiral

optically inactive
(racemic)

4: $Cl-C(CH_3)_2$, H, CH_3, R, CH_2Cl

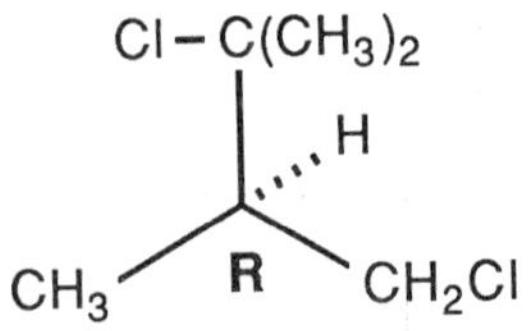

chiral
optically active

5: H_3C, R, CH_2Cl, H_3C, H, H, S, CH_2Cl

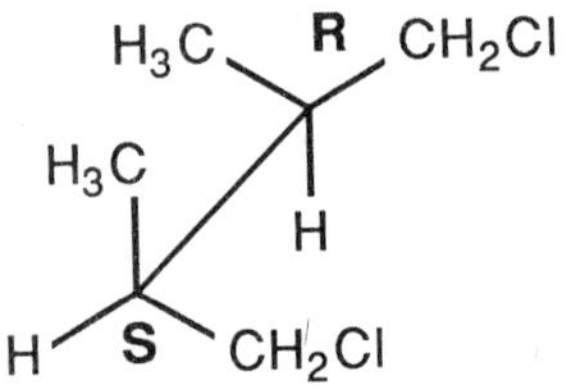

achiral (meso)
optically inactive

6: ClH_2C, S, CH_3, H_3C, H, H, S, CH_2Cl

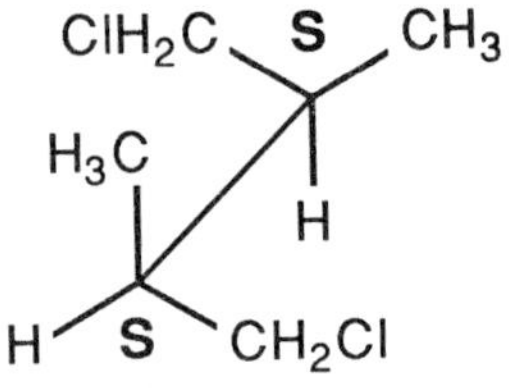

chiral
optically active

diastereomers

6.10 (a)

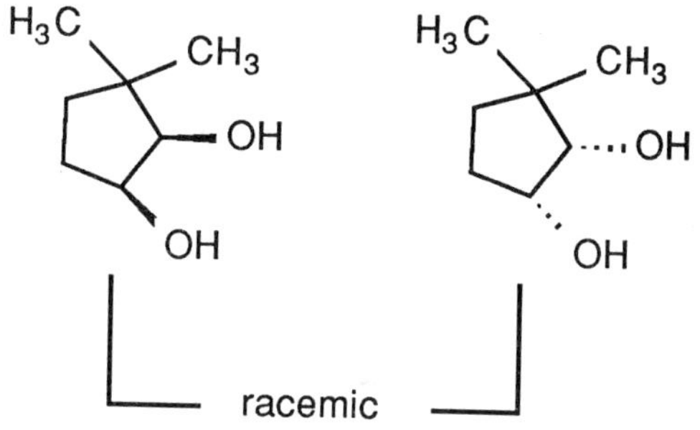

H_3C CH_3 OH OH — H_3C CH_3 OH OH

racemic

(b)

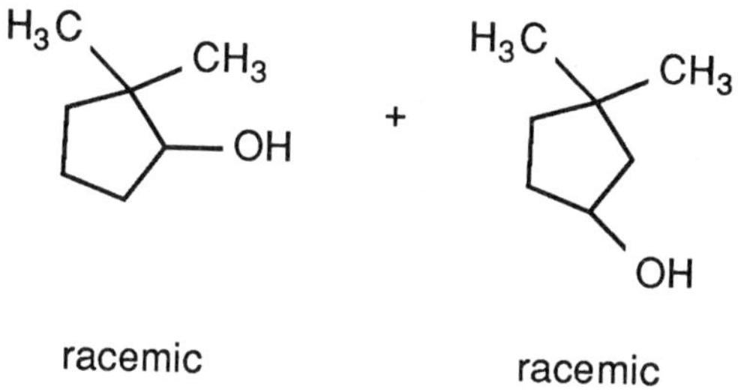

H_3C CH_3 OH + H_3C CH_3 OH

racemic racemic

(c)

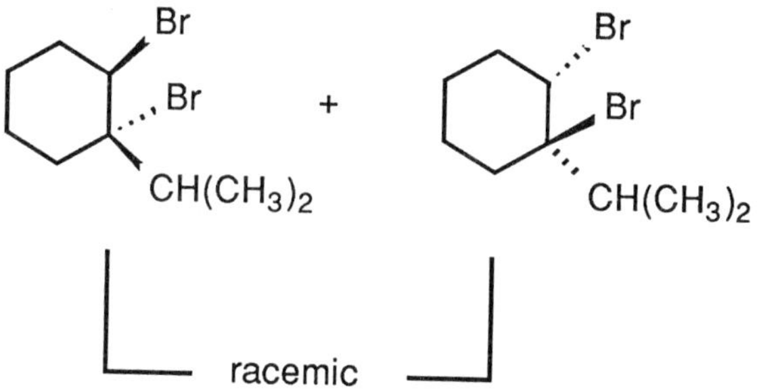

Br Br $CH(CH_3)_2$ + Br Br $CH(CH_3)_2$

racemic

ANSWERS TO PROBLEMS

6.1 (a) R

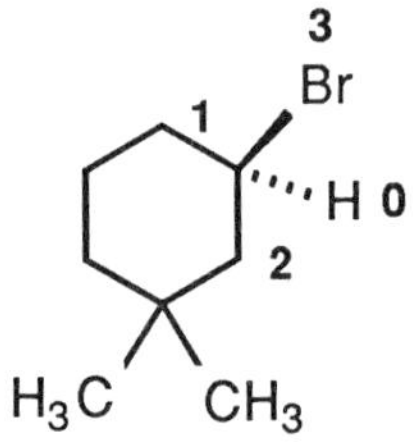

(b) S

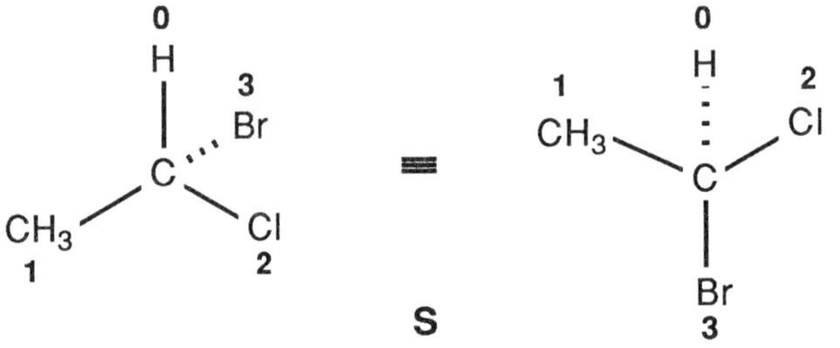

(c) R

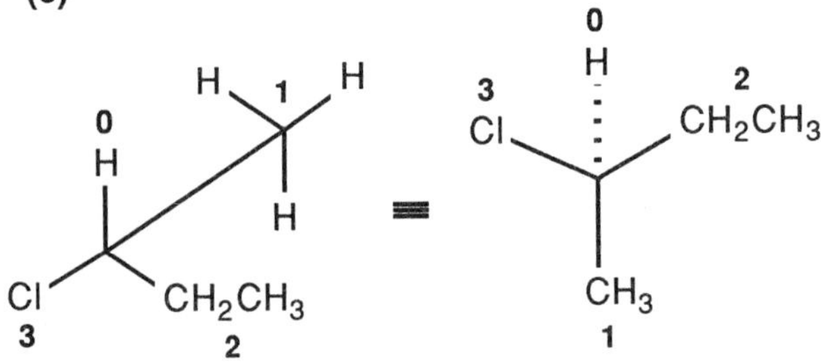

(d) S

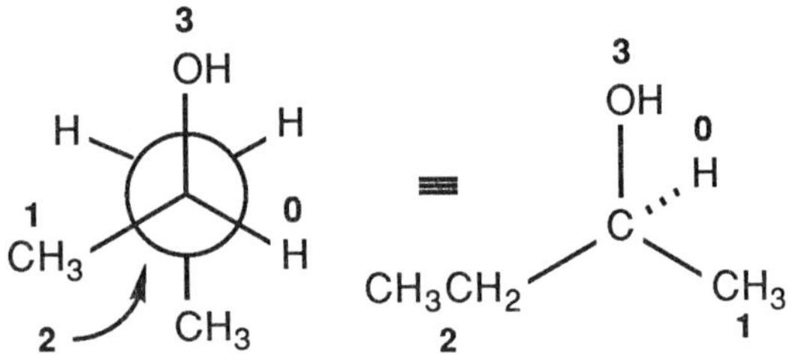

(e) R

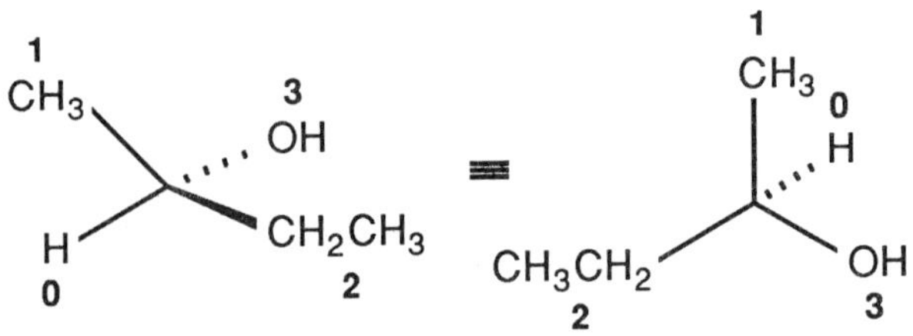

(f) S

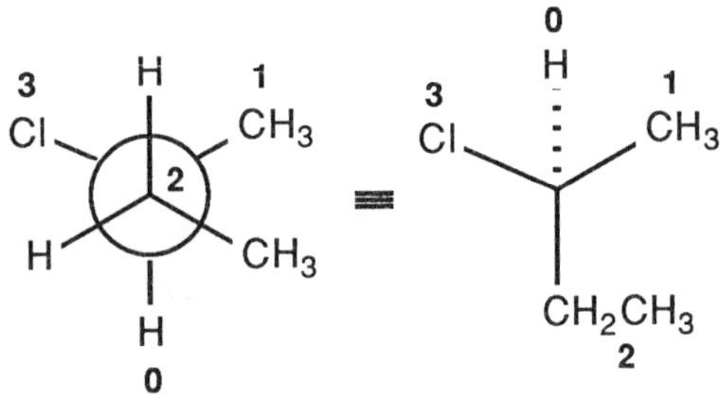

(g) S

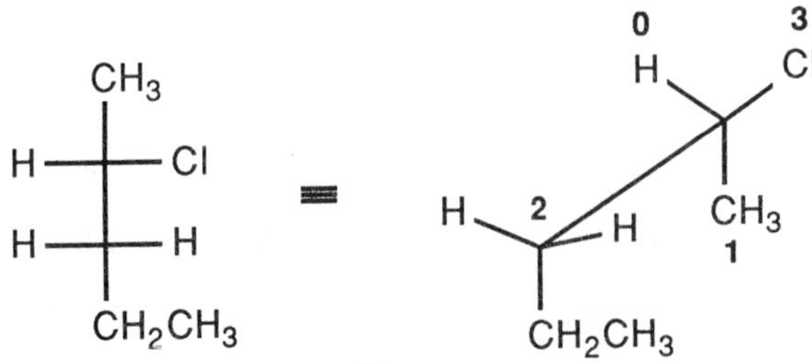

(h) R

(i)

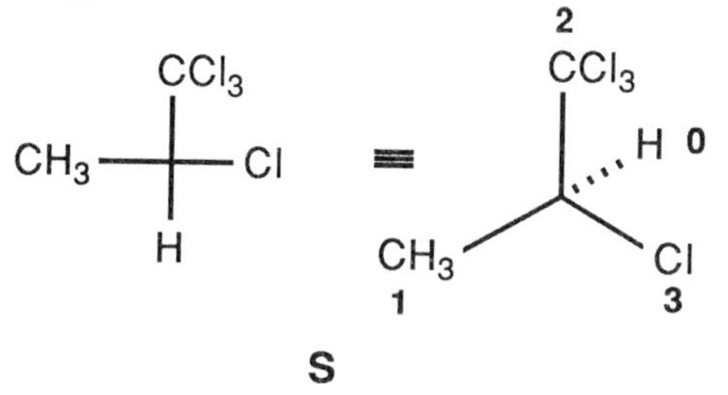

CH_3–C(CCl_3)(H)–Cl ≡ [wedge drawing: CCl_3 (2), H (0), CH_3 (1), Cl (3)]

S

(j)

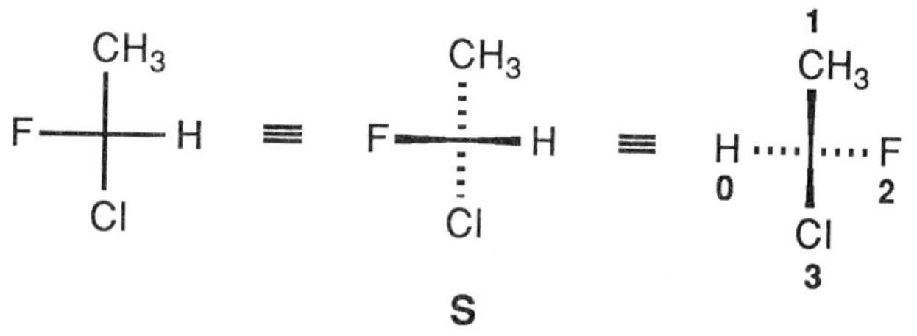

F–C(CH_3)(Cl)–H ≡ F–C(CH_3)(Cl)–H ≡ [CH_3 (1), H (0), F (2), Cl (3)]

S

(k)

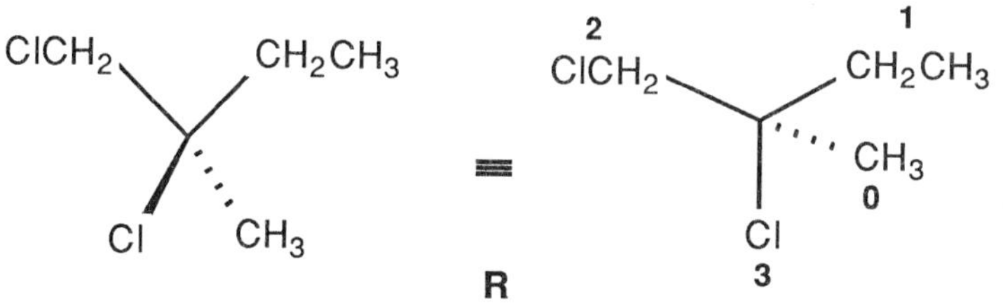

$ClCH_2$, CH_2CH_3, Cl, CH_3 ≡ [$ClCH_2$ (2), CH_2CH_3 (1), CH_3 (0), Cl (3)]

R

(l)

Cl **S** CH_3, H, H, HO **S** CH_3 ≡ [H (0), HO (3), CH_3 (1), $CHClCH_3$ (2)] ≡ [H (0), Cl (3), CH_3 (1), $CH(OH)CH_3$ (2)]

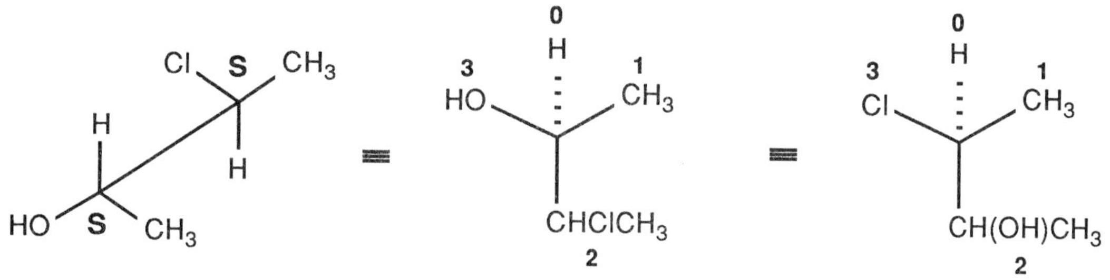

6.2 **(a)** Same: both are drawings of 3-chloro-3-methylpentane.

(b) Diastereomers: the first structure is (*E*)-cyclodecene (the two ring CH_2 groups on the double bond are *trans*) and the second structure is (*Z*)-cyclodecene (the two ring CH_2 groups on the double bond are *cis*.

(c) Diastereomers: the methyl and hydroxyl groups are *cis* in the first structure but *trans* in the second.

(d) Diastereomers: the first structure is (2*S*,3*R*)-2,3-dichloropentane and the second is (2*R*,3*R*)-2,3-dichloropentane. Alternatively, you can see that in the geometry of a Fischer projection (the second structure) having the methyl and ethyl groups eclipsed, the two chlorines are on opposite sides of the structure. In contrast, if you twist the back carbon of the first structure so that the chlorine is pointed up and is eclipsed with the chlorine on the front carbon, the methyl and ethyl will also be eclipsed.

(e) Constitutional isomers: the first compound is 1,4-cyclohexadiene and the second is 1,3-cyclohexadiene.

(f) Different compounds (not isomers): the first structure has six carbons but the second has only five.

(g) Same: both are drawings of (*R*)-3-chlorohexane.

(h) Enantiomers: each of the compounds has a C_2 axis, but rotation about that axis merely regenerates the same structure. The two drawings are presented as mirror images (with respect to an imaginary mirror-plane between them), and they are not superimposable. Alternatively, you could evaluate the configurations of the chiral centers; both are *R* in the first structure, and both are *S* in the second.

(i) Diastereomers: each structure is a 1,2,3-trisubstituted cyclopentane, but the first structure has the hydroxyls on C-1 and C-3 *cis* whereas the second structures has the same two groups *trans*. (The two structures differ at one but not all chiral centers.)

(j) Same: if you flip the first structure so that the hydroxyl-substituted carbon is in the back and the Br-substituted carbon is in the front, you will see that the two molecules differ only by a small change in conformation. Alternatively, you can evaluate each of the configurations to show that each is a drawing of (2*R*,3*R*)-3-bromo-2-pentanol.

(k) Diastereomers: the structure on the left is (2*R*,3*R*)-2,3-dibromopentane and that on the right is (2*R*,3*S*)-2,3-dibromopentane. (Alternatively, you can see that rotation about the central bond of the Newman projection to eclipse the alkyl substituents would also eclipse the bromines. In contrast, the alkyl substituents of the Fischer projection are already eclipsed but the bromines are not.)

(l) Enantiomers: the first structure is (*S*)-1-chloro-1-fluoropropane and the second is (*R*)-1-chloro-1-fluoropropane. If you superimpose the central carbon and any two substituents, the remaining two substituents will not match up.

(m) Same: both are drawings of (*R*)-1,3-pentanediol.

(n) Diastereomers: the sawhorse drawing has a C_2 axis but lacks reflection symmetry, so the first structure is chiral. In contrast, the Fischer projection of the second molecule clearly has a plane of symmetry; hence the two structures differ at some but not all chiral centers. Alternatively, you can see that rotation of the central bond of the sawhorse to eclipse the methyl groups would not eclipse the hydroxyls. Finally, by assigning configurations you can see that the first structure is (2*R*,3*R*)-2,3-butanediol whereas the second structure is (2*R*,3*S*)-2,3-butanediol.

(o) Same: if you flip the first drawing (i.e., rotate it about a horizontal axis by 180°) you will generate a structure identical to the second drawing. Alternatively, you can assign configurations and see that the double bond is *E* in each case and the carbon with the bromine is *S* in each case.

(p) Enantiomers: if you rotate the first drawing by 180°about an axis perpendicular to the page, you will generate the mirror image of the second structure. (Do not be confused by the fact that the CH_2 of the ethyl group is shown in three dimensions in the second structure.) Alternatively, you can assign configurations to show that the first drawing is (2*R*,3*R*)-1,2,3-hexanetriol whereas the second is (2*S*,3*S*)-1,2,3-hexanetriol.

(q) Constitutional isomers: the first drawing is 1-bromo-2,3-dichloro-2-methylpropane and the second is 1-bromo-3,3-dichloro-2-methylpropane.

(r) Different compounds (not isomers): the first structure has a total of eight carbons and the second has a total of nine.

(s) Diastereomers: the first drawing is (*E*,*E*)-2,4-hexadiene and the second is (*E*,*Z*)-2,4-hexadiene.

(t) Diastereomers: if you flip the first structure so that the hydroxyl-substituted carbon is in the front and the Cl-substituted carbon is in the back, you will see that the two molecules have the same configuration at the Cl-substituted carbon but differ in configuration at the hydroxyl-substituted carbon. Alternatively, complete assignment of configuration shows that the first structure is (2*S*,3*S*)-3-chloro-2-butanol and the second is (2*R*,3*S*)-3-chloro-2-butanol.

6.3 **(a)** The following drawings show the stereochemistry of the reactant and indicate the sites of attack that would yield the ten isomeric products that could be separated by distillation.

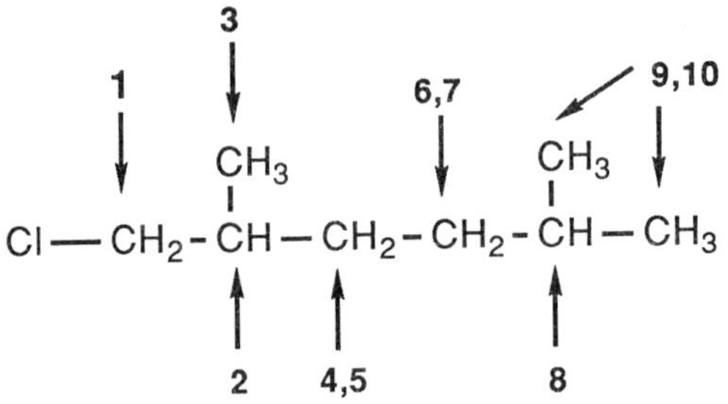

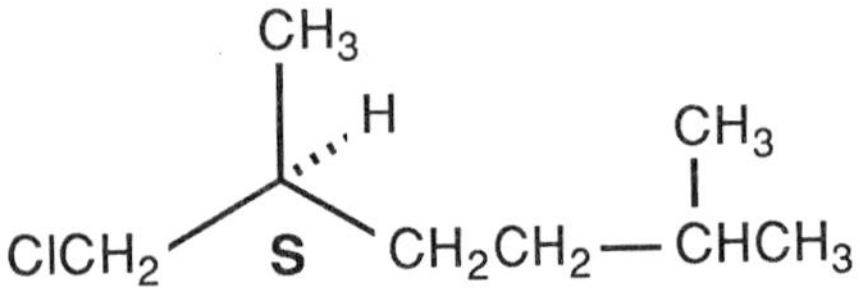

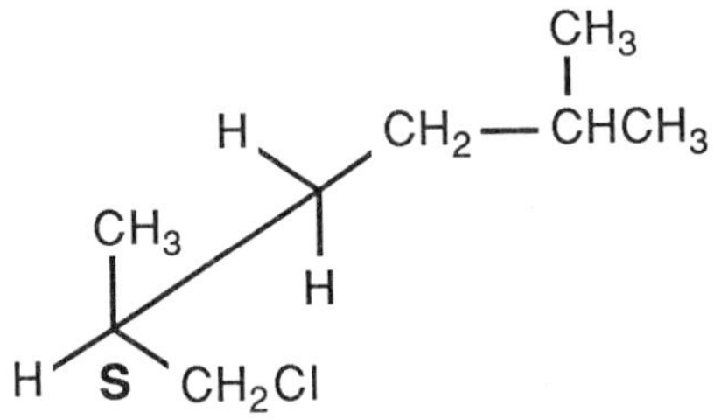

1 3 6,7 9,10

CH_3 CH_3

$Cl-CH_2-CH-CH_2-CH_2-CH-CH_3$

2 4,5 8

CH_3 H CH_3

$ClCH_2$ S $CH_2CH_2-CHCH_3$

CH_3

H CH_2-CHCH_3

CH_3 H

H S CH_2Cl

(b)-(e)

1: CH_3 H CH_3

Cl_2CH S $CH_2CH_2-CHCH_3$

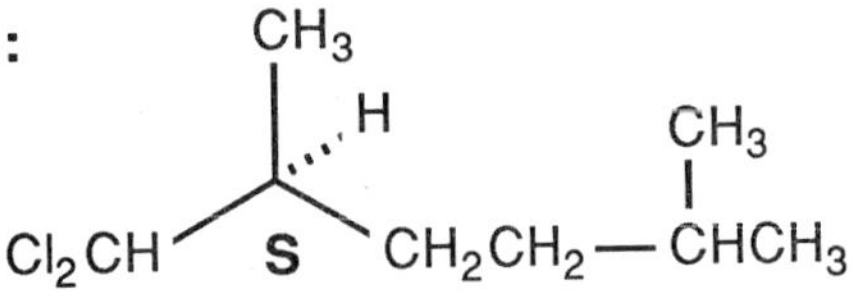

optically active; chiral

(S)-1,1-dichloro-2,5-dimethylhexane

2: CH_3 Cl CH_3

$ClCH_2$ R $CH_2CH_2-CHCH_3$

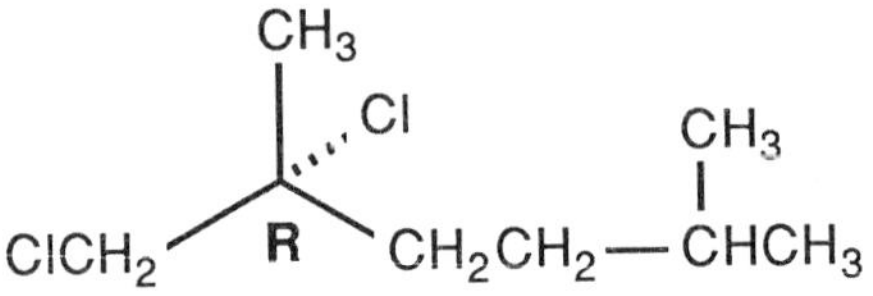

(R)-1,2-dichloro-2,5-dimethylhexane

+

Cl CH_3 CH_3

$ClCH_2$ S $CH_2CH_2-CHCH_3$

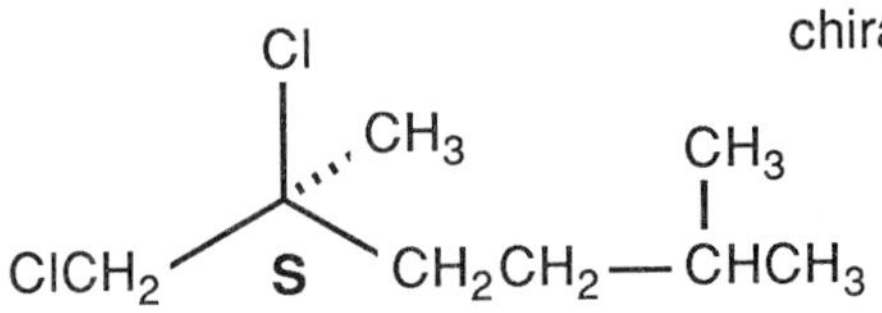

(S)-1,2-dichloro-2,5-dimethylhexane

optically inactive (racemic); chiral

3: CH_2Cl CH_3

$Cl-CH_2-CH-CH_2-CH_2-CH-CH_3$

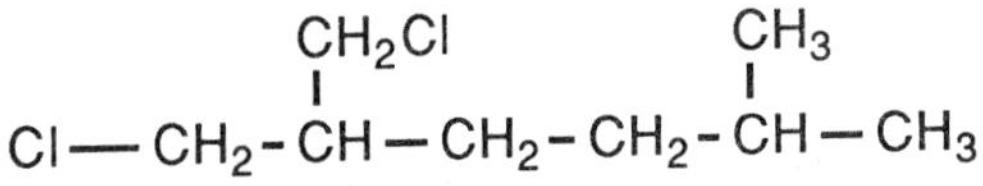

optically inactive; achiral

1-chloro-2-(chloromethyl)-5-methylhexane

4:

Structure with Cl, S, $CH_2—CHCH_3$ (with CH_3), CH_3, H, H, R, CH_2Cl

optically active; chiral

(2R,3S)-1,3-dichloro-2,5-dimethylhexane

5:

Structure with H, R, $CH_2—CHCH_3$ (with CH_3), CH_3, Cl, H, R, CH_2Cl

optically active; chiral

(2R,3R)-1,3-dichloro-2,5-dimethylhexane

4 and 5: diastereomers

6:

$ClCH_2$ S (CH_3, H) CH_2 R (Cl, H) $CH(CH_3)_2$

optically active; chiral

(2S,4R)-1,4-dichloro-2,5-dimethylhexane

7:

$ClCH_2$ S (CH_3, H) CH_2 S (H, Cl) $CH(CH_3)_2$

optically active; chiral

(2S,4S)-1,4-dichloro-2,5-dimethylhexane

6 and 7: diastereomers

8:

$ClCH_2$ S (CH_3, H) $CH_2CH_2-C(CH_3)_2$ (Cl)

optically active; chiral

(2S)-1,5-dichloro-2,5-dimethylhexane

9:

$ClCH_2$ S (CH_3, H) CH_2CH_2 R (CH_3, H) CH_2Cl

optically inactive; achiral

(2S,5R)-1,6-dichloro-2,5-dimethylhexane

10:

$ClCH_2$ S (CH_3, H) CH_2CH_2 S (CH_2Cl, H) CH_3

optically active; chiral

(2S,5S)-1,6-dichloro-2,5-dimethylhexane

9 and 10: diastereomers

6.4 **(a)** The following drawings show the stereochemistry of the reactant and indicate the sites of attack that would yield the five isomeric products that could be separated by distillation.

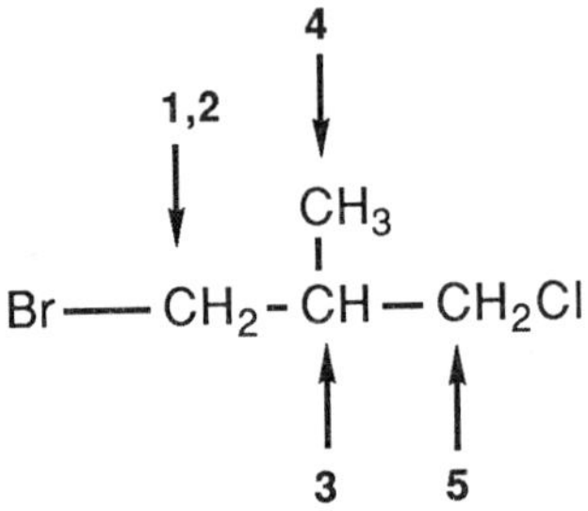

4

1,2

CH_3

Br—CH_2-CH—CH_2Cl

3 5

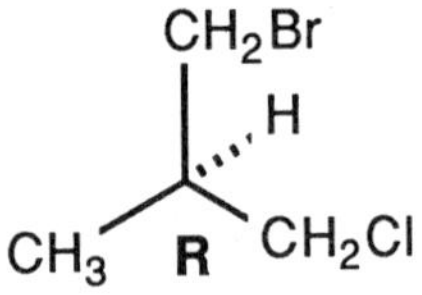

CH_2Br H CH_3 R CH_2Cl

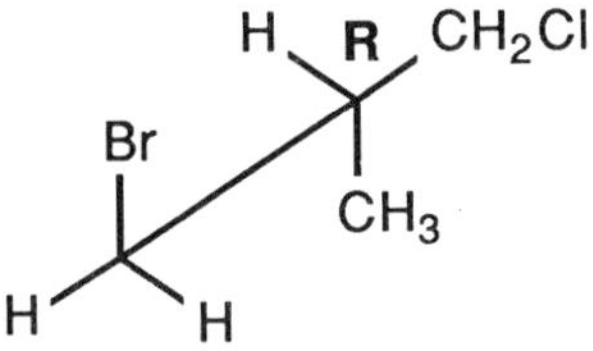

H R CH_2Cl Br CH_3 H H

(b)–(e)

1: H R CH_2Cl Br CH_3 H S Cl

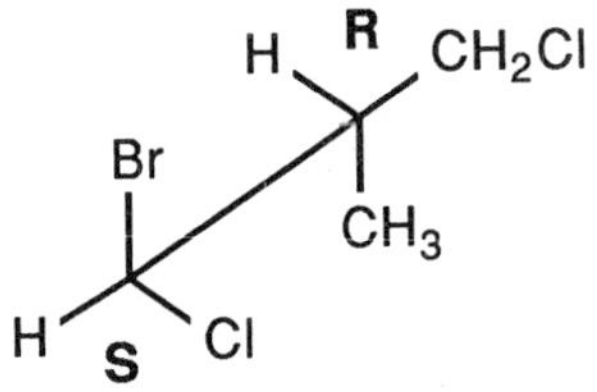

optically active; chiral

2: H R CH_2Cl Br CH_3 Cl R H

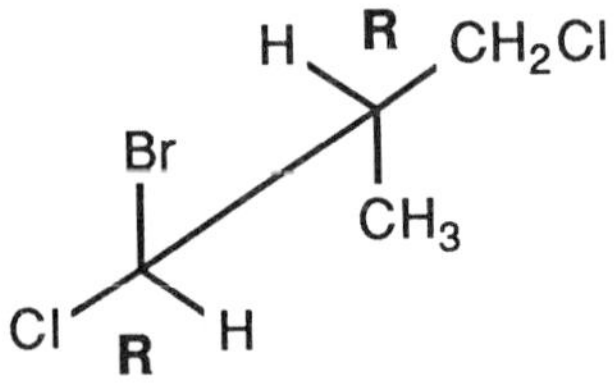

optically active; chiral

diastereomers

(1S,2R)-1-bromo-1,3-dichloro-2-methylpropane

(1R,2R)-1-bromo-1,3-dichloro-2-methylpropane

3: CH_2Br Cl CH_3 S CH_2Cl

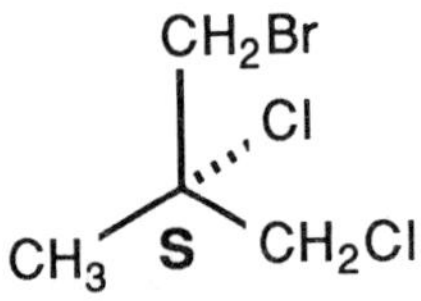

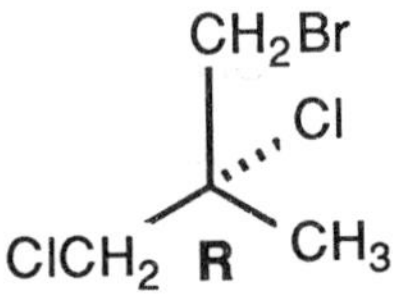

(S)-1-bromo-2,3-dichloro-2-methylpropane

chiral

CH_2Br Cl $ClCH_2$ R CH_3

(R)-1-bromo-2,3-dichloro-2-methylpropane

chiral

racemic

optically inactive

4:

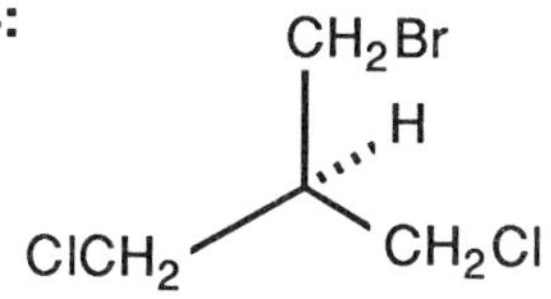

CH_2Br, H, $ClCH_2$, CH_2Cl

achiral
optically inactive
2-(bromomethyl)-1,3-dichloropropane

5:

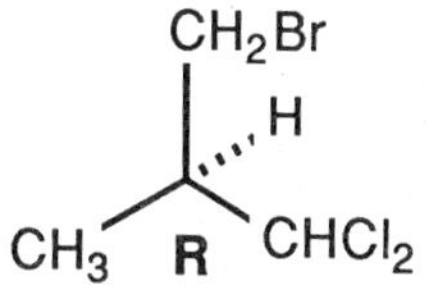

CH_2Br, H, CH_3 **R** $CHCl_2$

chiral
optically active
1-bromo-3,3-dichloro-2-methylpropane

6.5 **(a)** The following drawings show the stereochemistry of the reactant and indicate the sites of attack that would yield the twelve isomeric products that could be separated by distillation. (Five structures are used to show the stereochemistry of the reactant at different sites.)

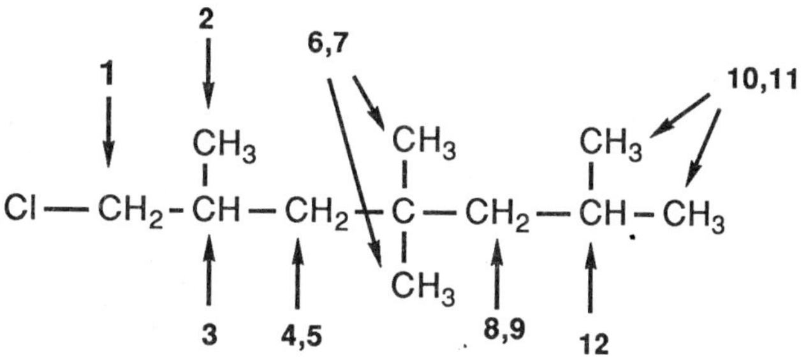

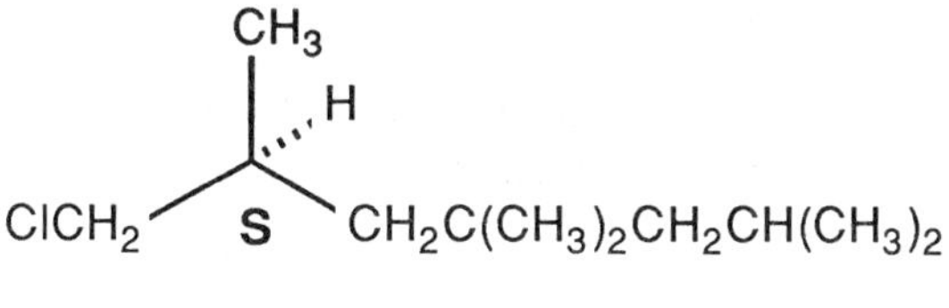

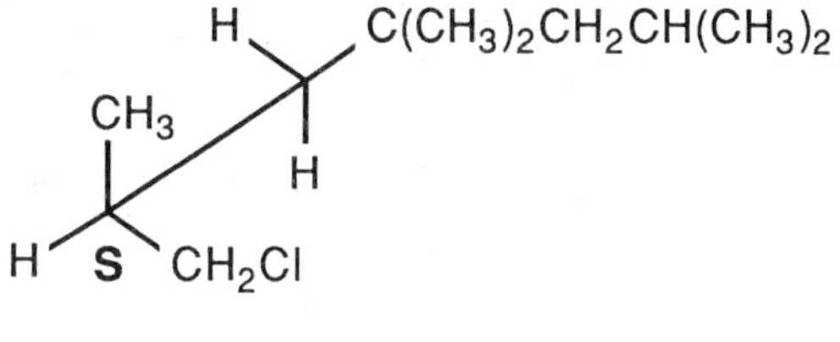

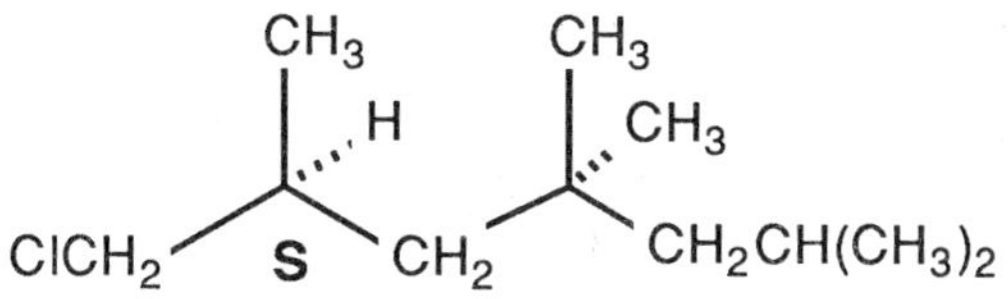

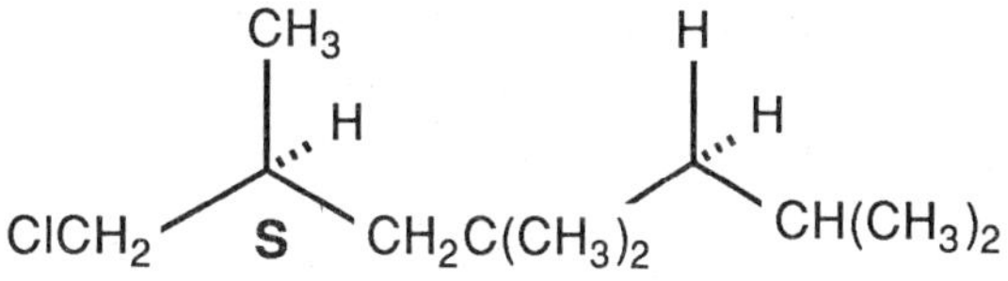

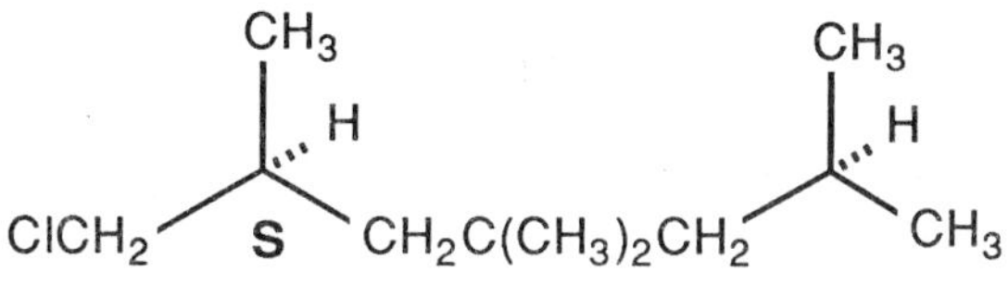

$Cl-CH_2-CH(CH_3)-CH_2-C(CH_3)_2-CH_2-CH(CH_3)-CH_3$

(sites of attack: 1, 2, 3, 4,5, 6,7, 8,9, 10,11, 12)

CH_3, H, $ClCH_2$ **S** $CH_2C(CH_3)_2CH_2CH(CH_3)_2$

H, $C(CH_3)_2CH_2CH(CH_3)_2$, CH_3, H, H **S** CH_2Cl

CH_3, H, CH_3, CH_3, $ClCH_2$ **S** CH_2, $CH_2CH(CH_3)_2$

CH_3, H, H, H, $ClCH_2$ **S** $CH_2C(CH_3)_2$, $CH(CH_3)_2$

CH_3, H, CH_3, H, $ClCH_2$ **S** $CH_2C(CH_3)_2CH_2$, CH_3

(b)–(e)

1: (S) — CH_3, H, Cl_2CH, $CH_2C(CH_3)_2CH_2CH(CH_3)_2$

optically active; chiral

(S)-1,1-dichloro-2,4,4,6-tetramethylheptane

2: CH_2Cl, H, $ClCH_2$, $CH_2C(CH_3)_2CH_2CH(CH_3)_2$

optically inactive; achiral

1-chloro-2-(chloromethyl)-4,4,6-trimethylheptane

3: (R) — CH_3, Cl, $ClCH_2$, $CH_2C(CH_3)_2CH_2CH(CH_3)_2$

(R)-1,2-dichloro-2,4,4,6-tetramethylheptane

chiral

(S) — Cl, CH_3, $ClCH_2$, $CH_2C(CH_3)_2CH_2CH(CH_3)_2$

(S)-1,2-dichloro-2,4,4,6-tetramethylheptane

chiral

racemic

optically inactive

4: H, S, $C(CH_3)_2CH_2CH(CH_3)_2$, CH_3, Cl, H, R, CH_2Cl

(2R,3S)-1,3-dichloro-2,4,4,6-tetramethylheptane

chiral, optically active

5: Cl, R, $C(CH_3)_2CH_2CH(CH_3)_2$, CH_3, H, H, R, CH_2Cl

(2R,3R)-1,3-dichloro-2,4,4,6-tetramethylheptane

chiral, optically active

diastereomers

6: CH_3, H, CH_2Cl, CH_3, $ClCH_2$, S, CH_2, S, $CH_2CH(CH_3)_2$

(2S,4S)-1-chloro-4-(chloromethyl)-2,4,6-trimethylheptane

chiral, optically active

7: CH_3, H, CH_3, CH_2Cl, $ClCH_2$, S, CH_2, R, $CH_2CH(CH_3)_2$

(2S,4R)-1-chloro-4-(chloromethyl)-2,4,6-trimethylheptane

chiral, optically active

diastereomers

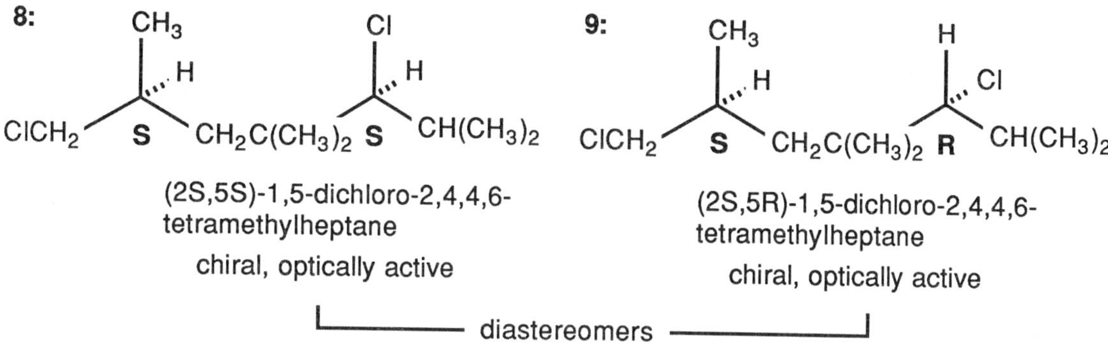

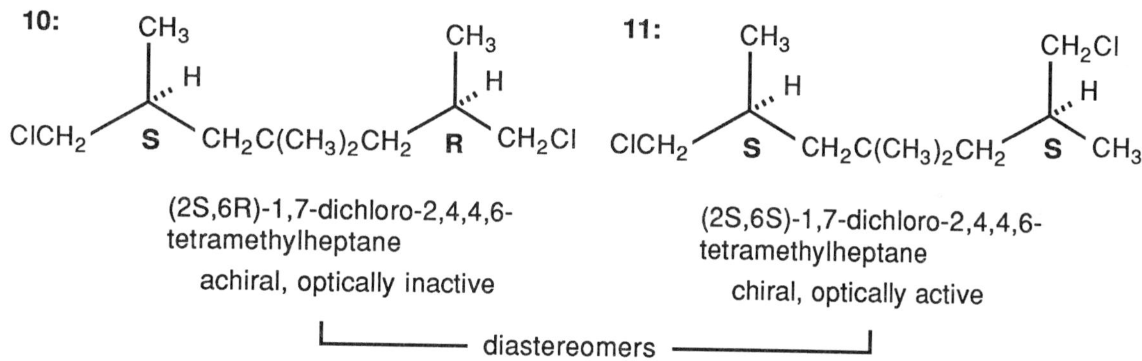

12:

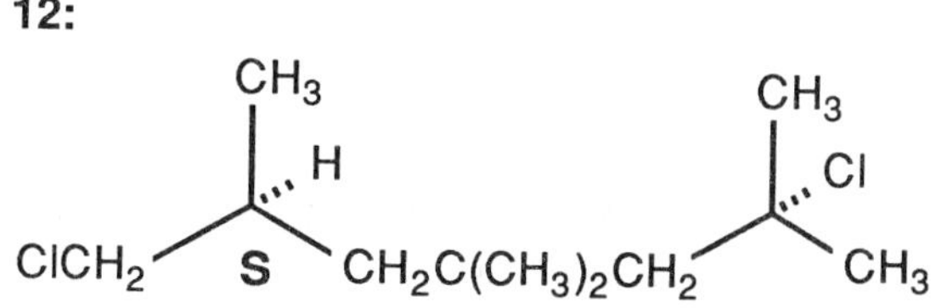

(2S)-1,6-dichloro-2,4,4,6-tetramethylheptane

chiral, optically active

6.6 If a saturated carbon is not chiral, at least two of its substituents must be identical. This means that unique priorities cannot be assigned. Even if you can assign priorities to two of the substituents, the remaining two substituents must have the same priority. As you can see in the example of 2-chloropropane, there is no difference in proceeding clockwise vs counterclockwise: in both cases the sequence is Cl–CH3–CH3.

3
Cl
0
H
CH3 CH3
2 2

6.7 The only chiral agent in Pasteur's resolution of sodium ammonium tartrate was Pasteur himself. It was his visual perception of the differences between the mirror image crystals that permitted him to separate them by hand.

6.8 **(a)** 100 x 3.12/5.20 = 60%

(b) The product must be predominantly the enantiomer of the material with a rotation of −5.20°, so it must have the *R* configuration. As shown by the following equation, this means that the reactant must have had the *S* configuration:

OH, H, $CH_2{=}CH$ **S** $CH_2CH_2CH_3$ → OH, H, CH_3CH_2 **R** $CH_2CH_2CH_3$

(c) (*S*)-1-hexen-3-ol

(d) The optical purity is 60%, which means that 60% is the *R* isomer and the remaining 40% (racemic) is a 50–50 mixture of *R* and *S*. The total percentage of the *R* isomer is therefore 80%.

6.9

(a) CH_3–CH–CH_3, D, D + enantiomer

(b) CH_3–CH–CH_3, OH (achiral)

(c) CH_3–CH–CH_3, OC_2H_5 (achiral)

(d) $CH_3-C(Cl)(CH_3)-CH(CH_3)-CH_2-C(CH_3)_3$

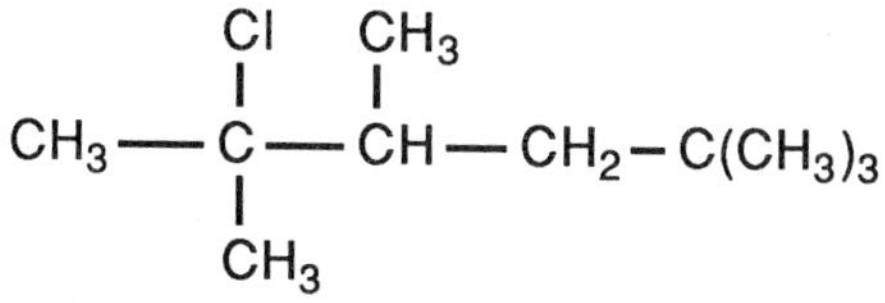

(racemic)

(e) $(CH_3)_3C$, H, D, OCH_3, H, $C(CH_3)_3$ + enantiomer

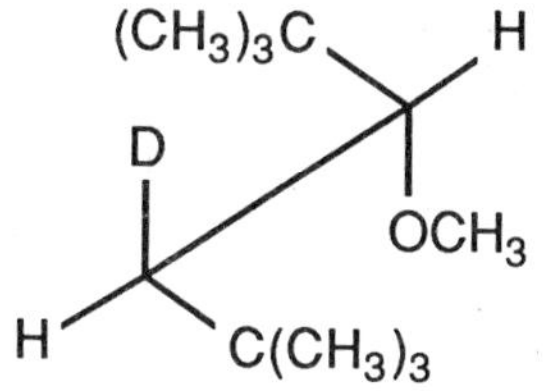

(f) OH, OH, CH_3 + enantiomer

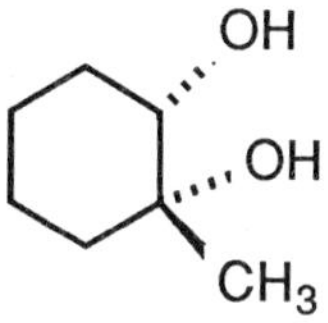

(g) Br, Br, CH_3 + enantiomer

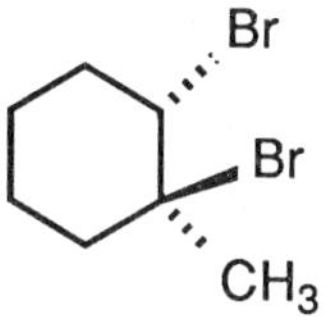

(h) OH, H, CH_3 + enantiomer

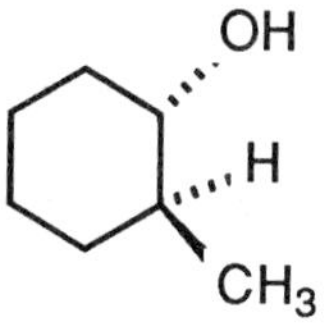

(i) H, O, CH_3 + enantiomer

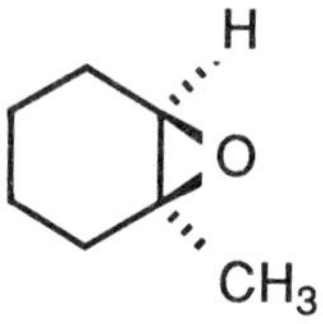

(j) O, $CH(CH_3)_2$ + enantiomer

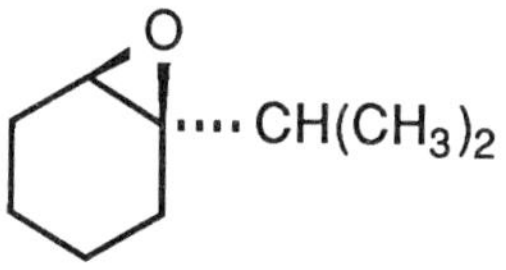

(k) CH_3, Cl, CH_3O, CH_3 + enantiomer

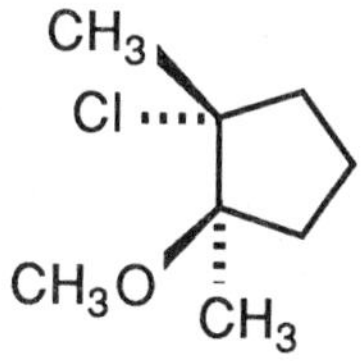

6.10

(a) CH_3CH_2, OH, OH, H, CH_3CH_2, H

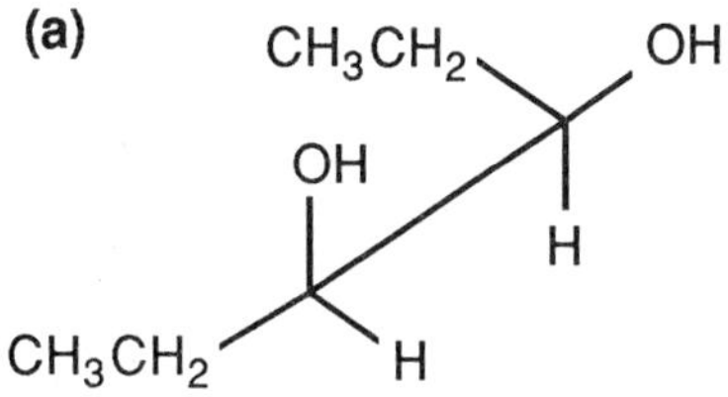

(achiral, meso)

(b) $CH_3CH_2-CH(Br)CH_2CH_2CH_3$

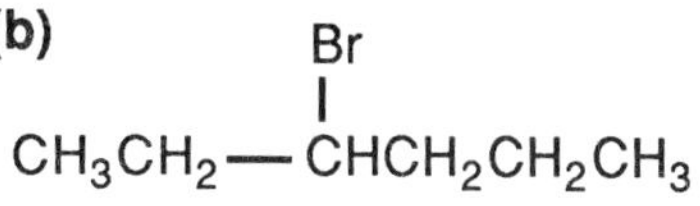

racemic

(c) $CH_3CH_2-CH(OH)CH_2CH_2CH_3$

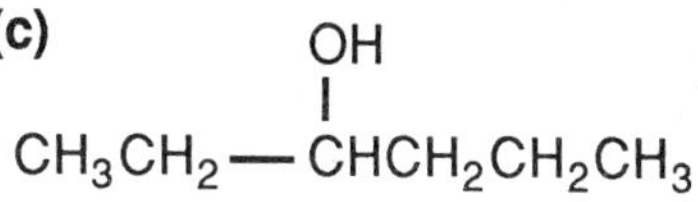

racemic

(d) 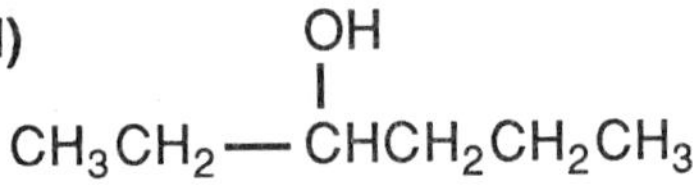 $CH_3CH_2-CH(OH)CH_2CH_2CH_3$

racemic

(e) 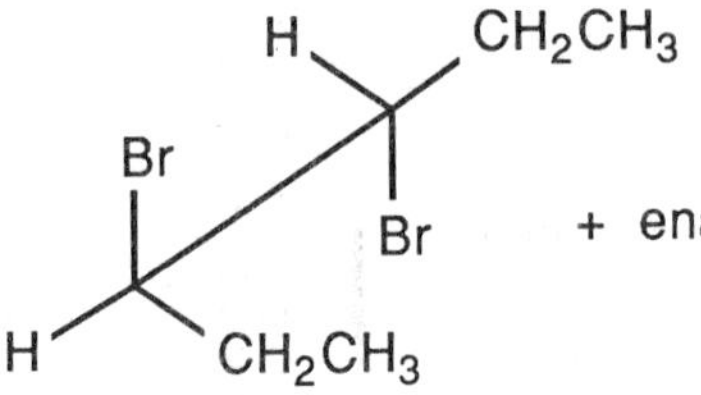 H, CH_2CH_3, Br, Br, H, CH_2CH_3 + enantiomer

(f) 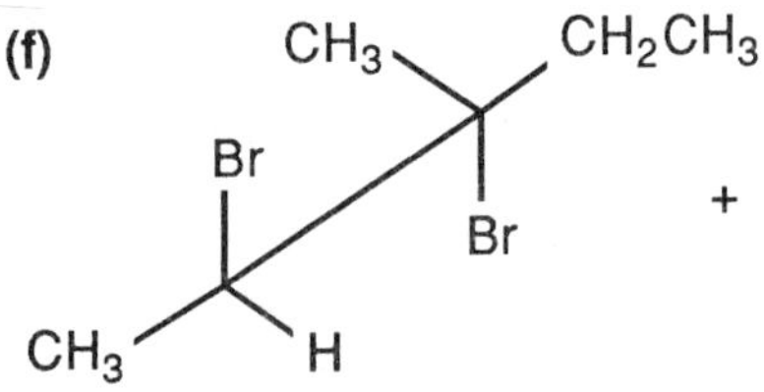 CH_3, CH_2CH_3, Br, Br, CH_3, H + enantiomer

(g) 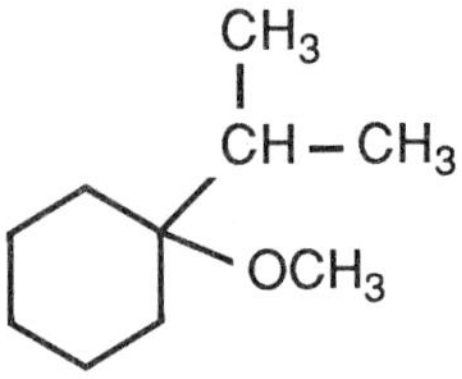 CH_3, CH−CH_3, OCH_3

(h) 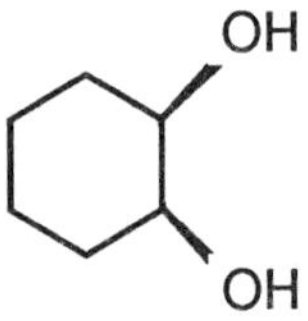 OH, OH

(achiral, meso)

(i) 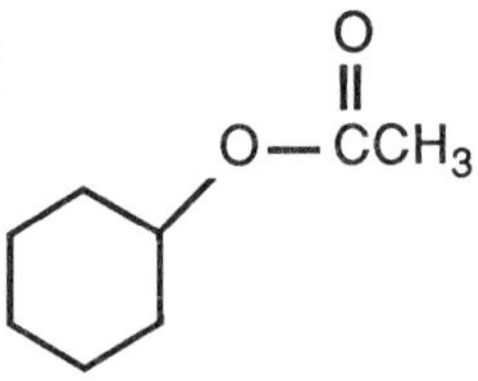 O−$\overset{O}{\overset{\|}{C}}CH_3$

(j) 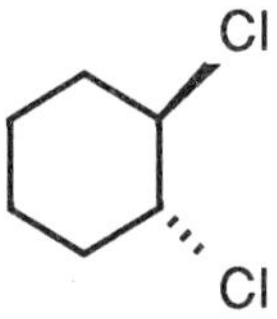 Cl, Cl

+ enantiomer

(k) 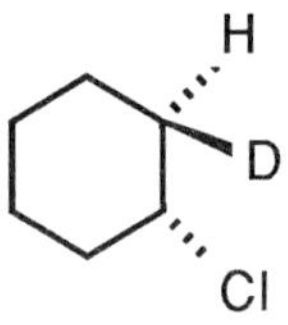 H, D, Cl

+ enantiomer

6.11 **(a)** From the molecular formula of the reactant you know that the reaction was elimination of HCl. A chlorine and hydrogen were originally bonded to the carbon atoms of the double bond in the product. The correct structure must be 3-chloro-2,4,6-trimethylheptane:

$$CH_3-\underset{}{\overset{CH_3}{\overset{|}{C}H}}-\underset{\underset{Cl}{|}}{CH}-\overset{CH_3}{\overset{|}{C}H}CH_2-\overset{CH_3}{\overset{|}{C}H}CH_3$$

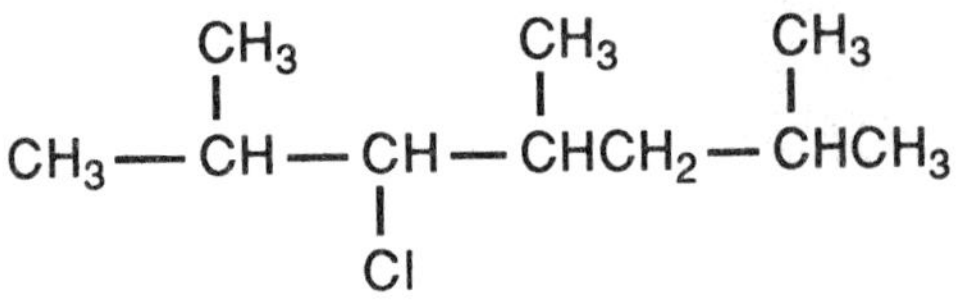

The other possibility (4-chloro-2,4,6-trimethylheptane) is achiral and could not be optically active.

(b) The product shown is racemic, and the reaction involves addition of Br_2 to an alkene. As shown in the following equation (formation of the first of the two product structures), *anti* addition of bromine requires the reactant to be

(Z)-4-methyl-2-pentene:

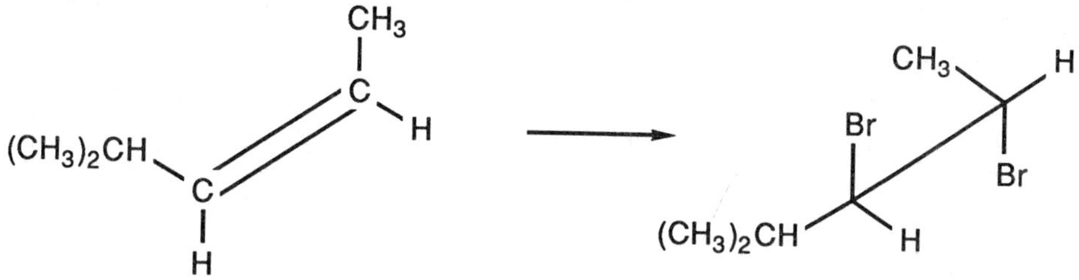

CH_3 C H $(CH_3)_2CH$ C H → CH_3 H Br Br $(CH_3)_2CH$ H

(c) The product is a meso compound, and it must have been formed by *anti* addition of Cl_2 to (E)-2-butene:

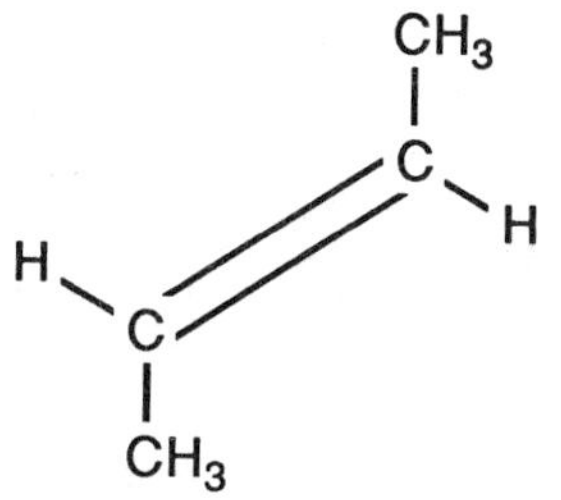

CH_3 C H H C CH_3

(d) Oxidation of an alkene to a diol with permanganate involves *syn* addition, so the racemic product must have been formed from (E)-3-hexene:

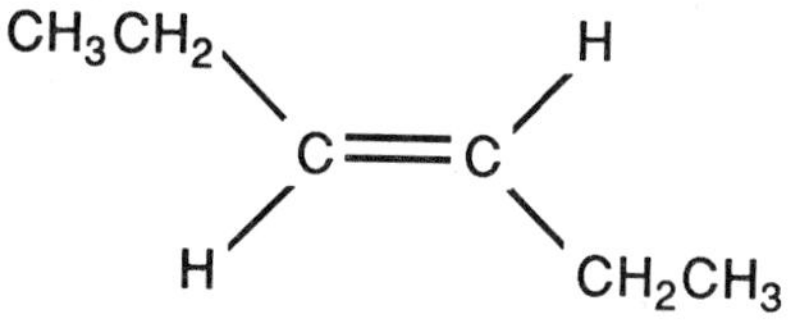

CH_3CH_2 H C=C H CH_2CH_3

(e) The racemic product has two chiral centers, so both must have been generated by addition to the original double bond. This allows you to locate the position of the double bond in the 5-membered ring. Notice that the hydrogen and the hydroxyl group introduced in the reaction are *cis* to each other.

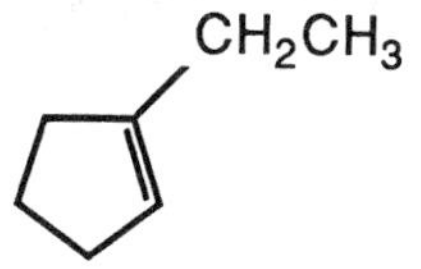

CH_2CH_3

(f) One or both of the chlorines must have been introduced via the addition of HCl to a double bond. But the product is optically active (only a single enantiomer is shown), so the chiral center must already have been present in the starting material. Therefore it is the other Cl that was introduced, and the reactant was

(*R*)-2-methyl-3-chloro-1-butene.

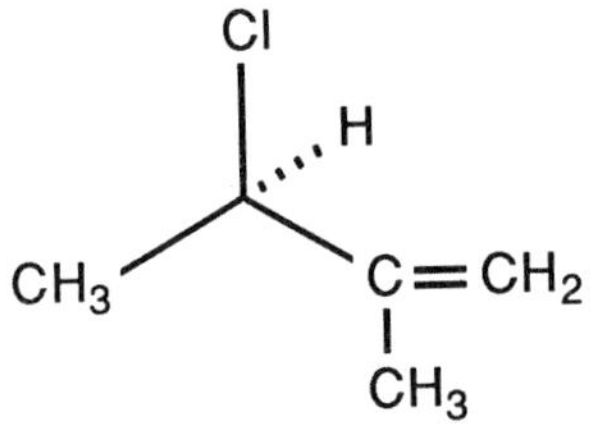

Cl

H

CH_3 C=CH_2

CH_3

6.12

(a) CH_3 CH_3

C=C

C_2H_5 H

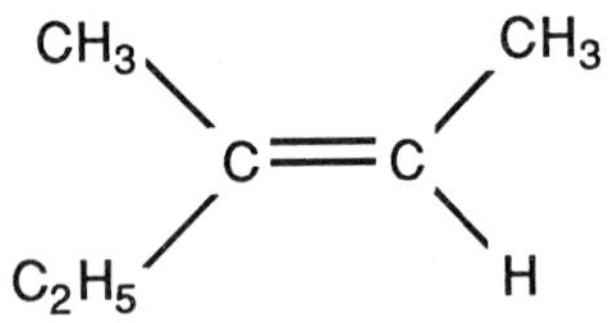

(b) CH_3 Br CH_2CH_3 OH + CH_3 OH CH_2CH_3 Br

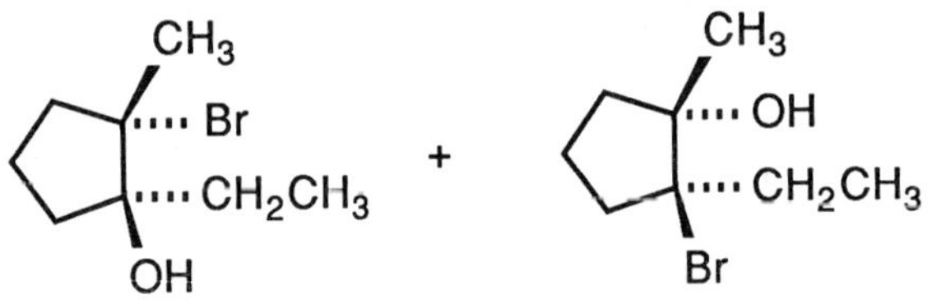

+ enantiomer + enantiomer

(c) CH_3

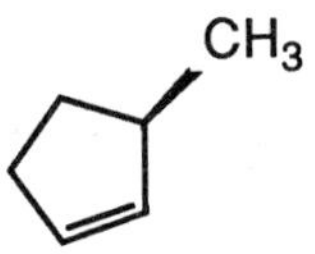

(d) CH_2CH_3 H Br

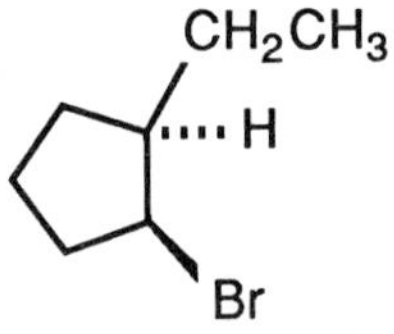

+ enantiomer

(e) CH_3 CH_3

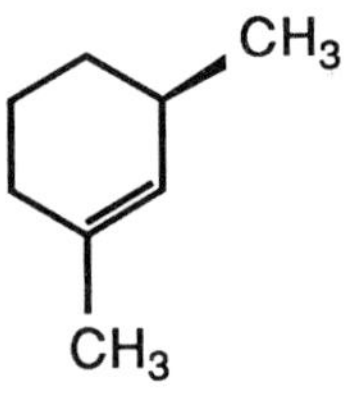

(f) O CH_3 CH_3 CH—CH_3 CH_3

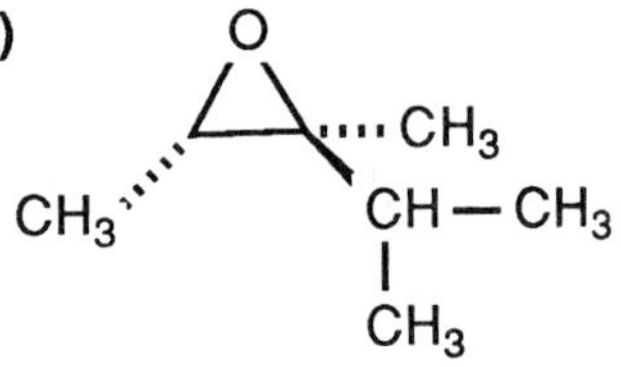

+ enantiomer

6.13 (a)

OH — conc. H_2SO_4 → (cyclohexene) — CH_3CO_3H → (epoxide, H, O, H) — H_2O, H_2SO_4 → OH, OH

(racemic)

(b)

H, C, CH_2CH_3, CH_3, C, H — cold, alkaline $KMnO_4$ →

H, CH_2CH_3, CH_3, H, OH, OH ≡ H, CH_2CH_3, OH, OH, H, CH_3 + enantiomer

(c)

H, H, C═C, CH_3, CH_2CH_3 — CH_3CO_3H → O, H, H, CH_3, CH_2CH_3 + enantiomer.

(d)

Br — KOtBu, tBuOH → (cyclohexene) — Br_2, H_2O → Br, OH + enantiomer

(e)

CH_3 C CH_2CH_3 H C CH_3

1. B_2H_6
2. H_2O_2, NaOH

H HO CH_3 CH_2CH_3 H CH_3 ≡ CH_3 H OH CH_2CH_3 H CH_3 + enantiomer

(f)

CH_3CH_2 H C=C H CH_2CH_3 —CH_3CO_3H→ O H CH_2CH_3 CH_3CH_2 H —H_2O, H_2SO_4→

CH_2CH_3 H CH_2CH_3 H OH OH ≡ H CH_2CH_3 OH OH CH_3CH_2 H

6.14 The key to this problem is the fact that all of the compounds except **B** are optically active. This means that at least one of the chiral centers of **G** and **H** was present in **A.** Only one of the two chiral centers has the same configuration in both **G** and **H.** Therefore that chiral center was already present in **A,** which must be (*R*)-3-bromo-3-methyl(methylenecyclohexane). When HBr adds to the double bond of **A** in the absence of peroxides, a new chiral center is formed with both possible configurations. When the new chiral center has the *S* configuration, **B** (a meso compound) results. The formation of both **B** and **C** from the reaction of **D** with HBr (all three compounds optically active) again requires that one chiral center (with the same configuration) be present in all three compounds. This must be the center with the *R* configuration, and several possibilities exist for the location of the double bond in **D**. It cannot be exocyclic to the six-membered ring, because that is compound **A.** Therefore **D** must be either 3-bromo-1,3-dimethylcyclohexene or 4-bromo-2,4-dimethylcyclohexene. If it were the latter, then one of the two products of Br_2 addition would be a meso compound and be optically inactive. Therefore the structure of **D** and the other compounds must be as shown:

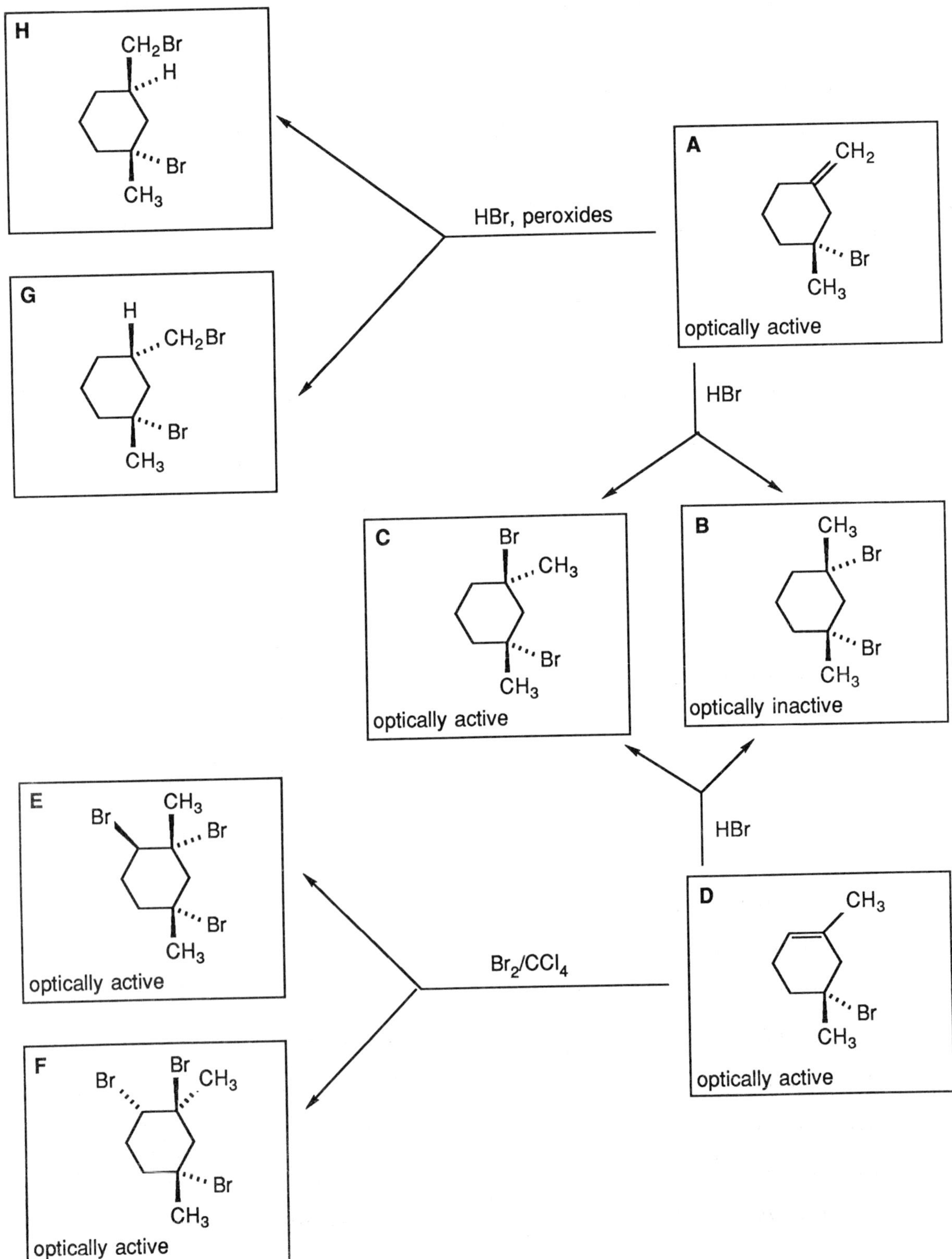
H
CH2Br
H
Br
CH3
G
H
CH2Br
Br
CH3
HBr, peroxides
A
CH2
Br
CH3
optically active
HBr
C
Br
CH3
Br
CH3
optically active
B
CH3
Br
Br
CH3
optically inactive
HBr
E
Br
CH3
Br
Br
CH3
optically active
Br2/CCl4
D
CH3
Br
CH3
optically active
F
Br
Br
CH3
Br
CH3
optically active

6.15 None of the reactions shown would cause any rearrangement in carbon skeleton, so all three compounds must have the carbon skeleton of the alkane, 2,2-dimethylpropane. With one CH_2Br and one CH_2Cl group the central carbon would not be chiral. Therefore both the Cl and Br must be on the same carbon, and that must be the chiral center. (Note that other products would also be formed in the halogenation reactions).

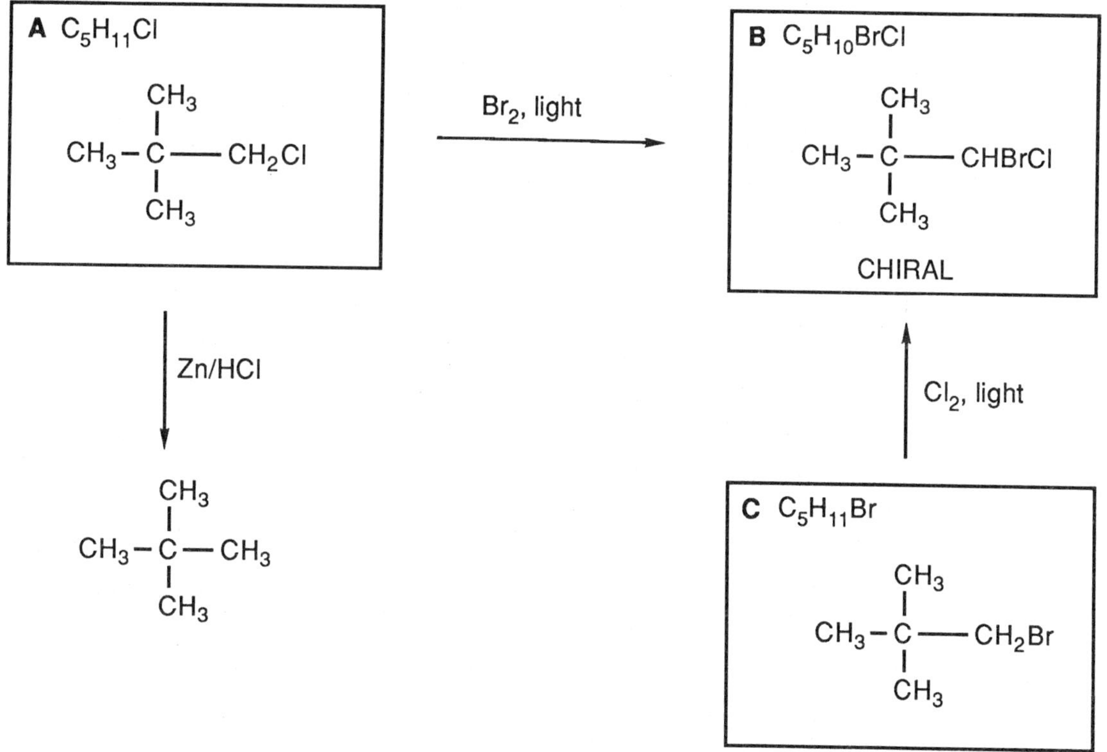

6.16 Initially you have little direct structural information, but the ozonolysis reaction tells you that compound **C** has a $C{=}CH_2$ group. The next major clue is provided by the formation of **B,** the only compound in the problem that is optically inactive. The addition of HBr in the presence of peroxides converts the $C{=}CH_2$ to a $CH{-}CH_2Br$ group, and this cannot change the configuration of the original chiral center. This can only mean that addition of HBr in the presence of peroxides has generated a second chiral center to produce a meso compound. This deduction is reinforced by the observation that *two* isomeric products are formed in this reaction. Now you can deduce that compound **B** contains *two* *CH–CH_2Br groups in which the starred carbons are chiral. Only two carbons remain unaccounted for, and these must be present as methyl substituents on the starred carbons (so that they will each have four different substituents). Assuming for convenience that the chiral center of **C** has the *R* configuration, the complete answer can be drawn in the following way:

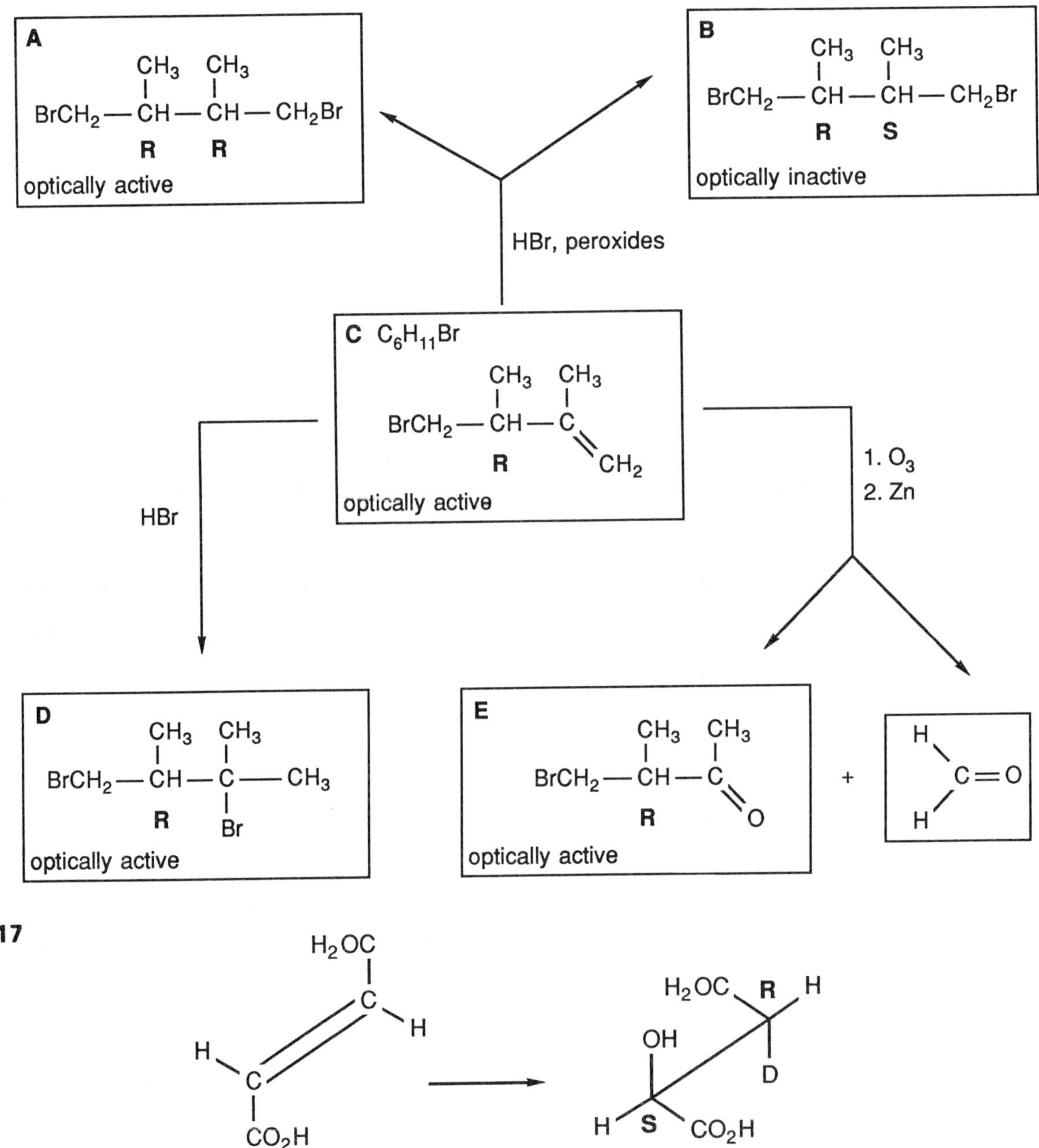

6.17

6.18 The two reactions indicate that compound **A** is a cyclopentane derivative with a single double bond. One end of the double bond is that which becomes the bromo-substituted carbon, and the other end must be one of the three adjacent carbons. The products of HBr addition are diastereomeric 1-bromo-1,2-dimethylcyclopentanes. The configuration at C-2 is *R* in both cases, so this chiral center must have been present in the starting material. If the second carbon of the double bond had been a CH_2 that became a CH_3 upon HBr addition, ozonolysis would have formed two products. Since all seven carbons of the starting material are present in **B**, the double bond of **A** must be in the ring. This requires the following structures for **A** and **B**:

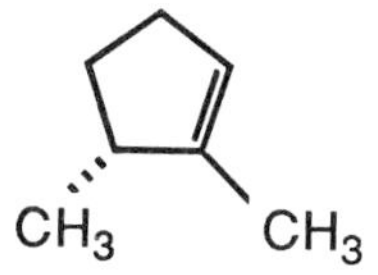

CH_3 CH_3 (1,5-dimethylcyclopentene)

A

$$\overset{O}{\overset{\|}{HC}}—CH_2CH_2—\underset{CH_3}{\underset{|}{CH}}—\overset{O}{\overset{\|}{C}}—CH_3$$

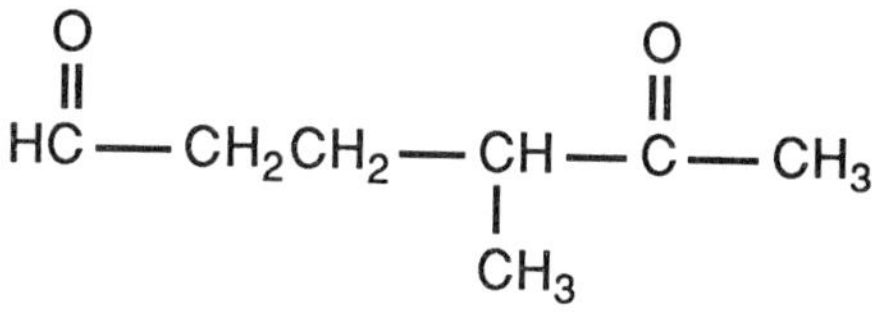

B

6.19 The key structural information is provided by the ozonolysis reaction, and this tells you that compound **B** must be 2-ethyl-1-pentene. Addition of HCl would generate the tertiary alkyl chloride, 3-chloro-3-methylhexane (**C**), and the isomeric, optically active chloroalkane **A** must be 3-(chloromethyl)hexane.

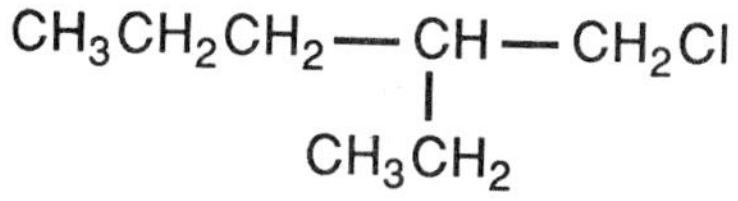

$$CH_3CH_2CH_2—\underset{CH_3CH_2}{\underset{|}{CH}}—CH_2Cl$$

A

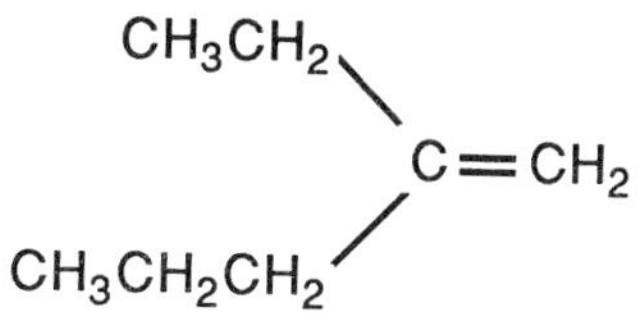

$$(CH_3CH_2)(CH_3CH_2CH_2)C{=}CH_2$$

B

$$CH_3CH_2CH_2—\underset{Cl}{\underset{|}{\overset{CH_3}{\overset{|}{C}}}}—CH_2CH_3$$

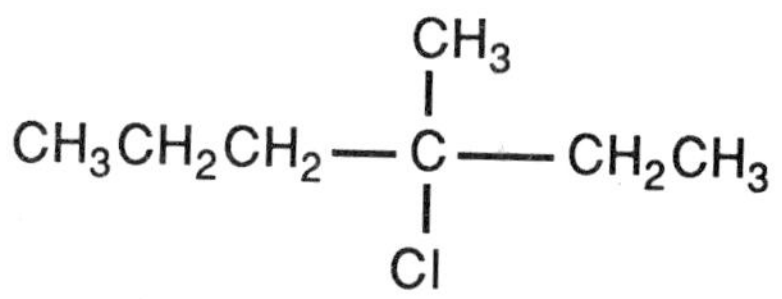

C

6.20

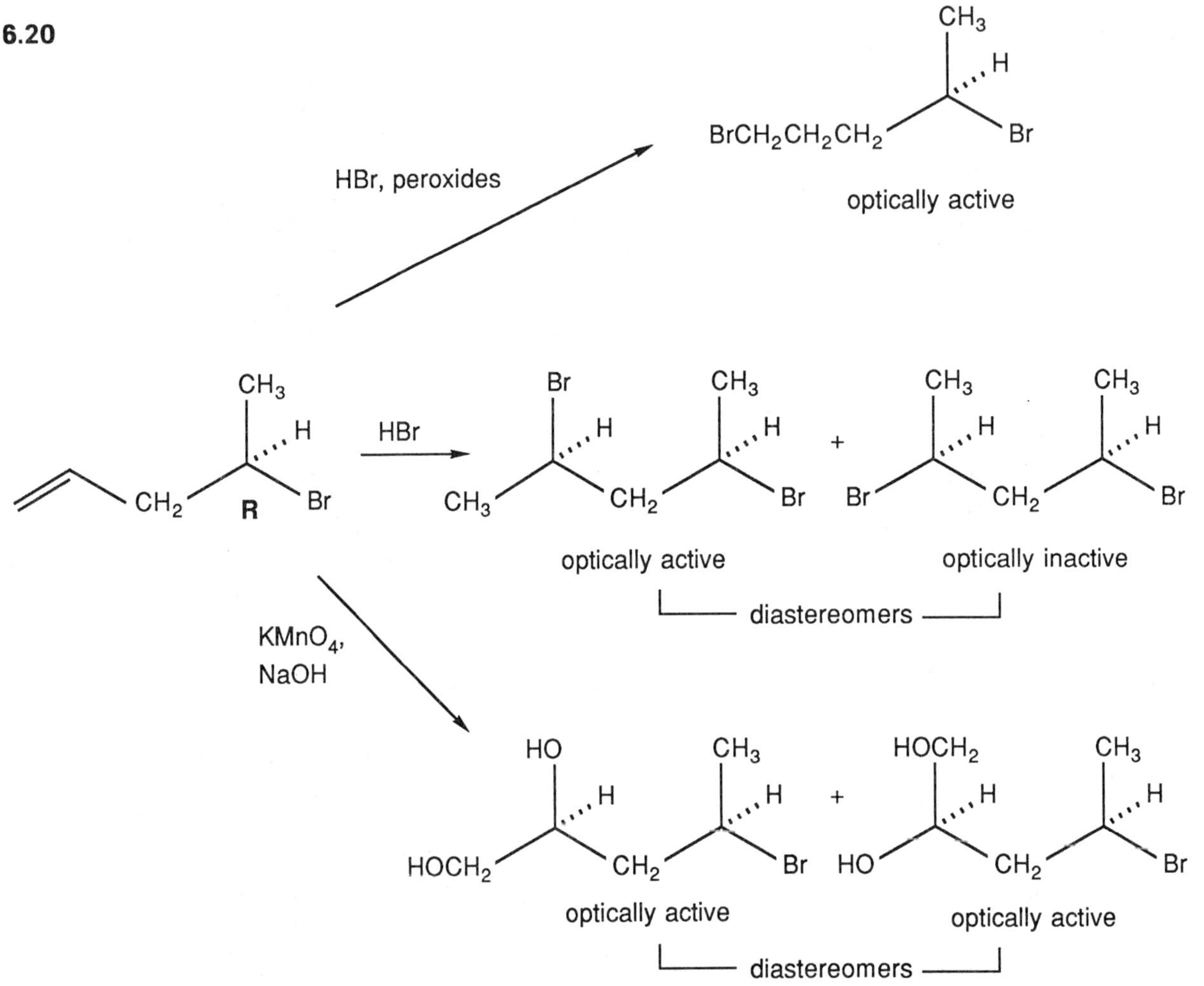

HBr, peroxides → $BrCH_2CH_2CH_2$–CH(CH_3)(H)Br — optically active

$CH_2{=}CHCH_2$–CH(CH_3)(H)Br (**R**) —HBr→ CH_3–CH(Br)(H)–CH_2–CH(CH_3)(H)Br (optically active) + Br–CH(CH_3)(H)–CH_2–CH(CH_3)(H)Br (optically inactive)

diastereomers

$KMnO_4$, NaOH → $HOCH_2$–CH(HO)(H)–CH_2–CH(CH_3)(H)Br (optically active) + HO–CH($HOCH_2$)(H)–CH_2–CH(CH_3)(H)Br (optically active)

diastereomers

6.21 As shown in the following scheme, *syn* hydroxylation ($KMnO_4$) of the (*Z*)-alkene affords the same product (meso) as *anti* hydroxylation of the (*E*)-alkene. Similarly, *anti* hydroxylation (CH_3CO_3H followed by acidic hydrolysis) of the (*Z*)-alkene affords the same product (racemic) as *syn* hydroxylation of the (*E*)-alkene:

Hydroxylation of the (Z)-alkene:

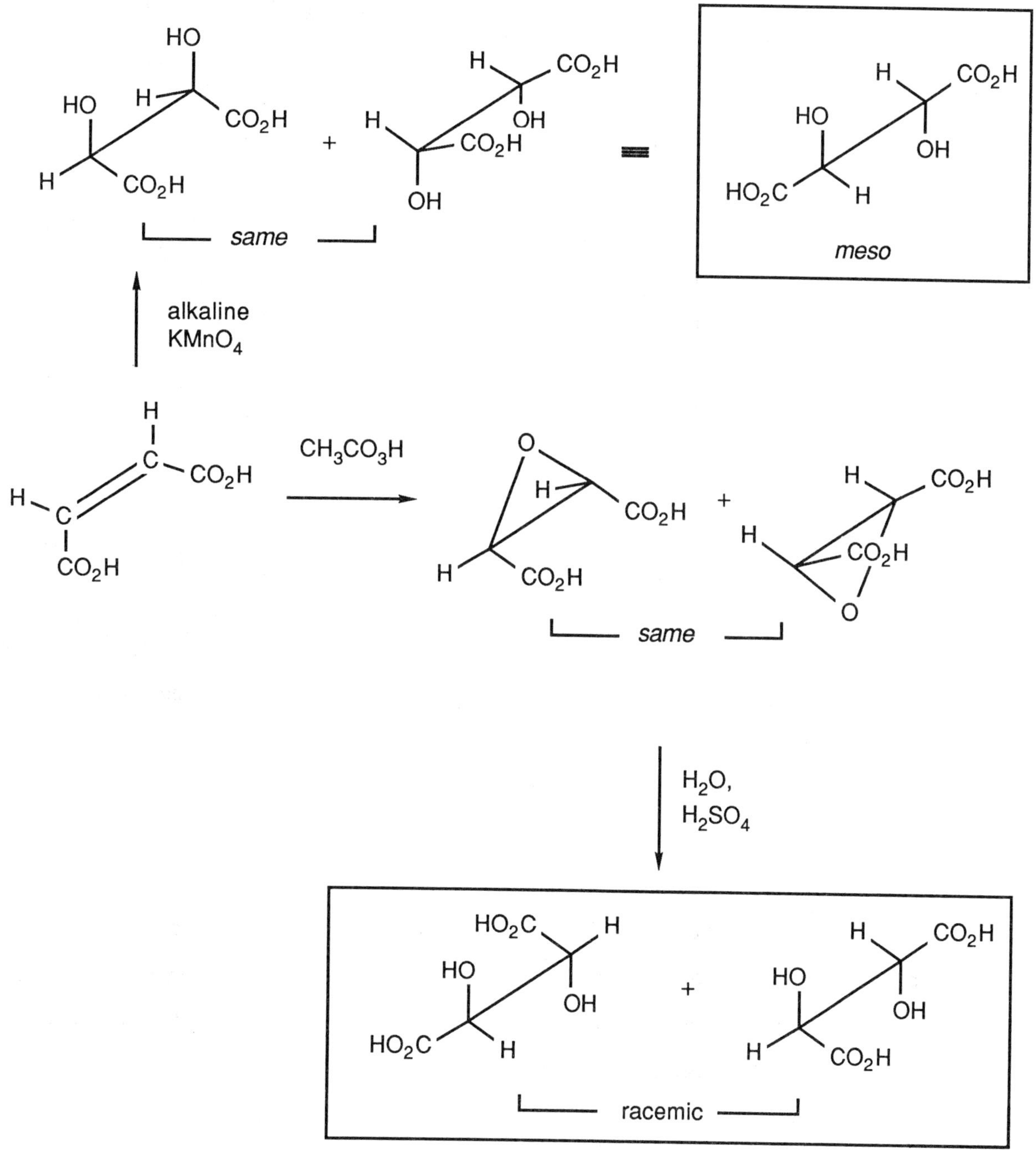

HO
HO H
CO$_2$H
H CO$_2$H
+
H CO$_2$H
H OH
CO$_2$H
OH
same
≡
H CO$_2$H
HO
OH
HO$_2$C H
meso

alkaline
$KMnO_4$

H
C CO$_2$H
H C
CO$_2$H

CH_3CO_3H

O
H
CO$_2$H
H CO$_2$H
+
H CO$_2$H
H
CO$_2$H
O
same

H_2O,
H_2SO_4

HO$_2$C H
HO
OH
HO$_2$C H
+
H CO$_2$H
HO
OH
H CO$_2$H
racemic

Hydroxylation of the (E)-alkene:

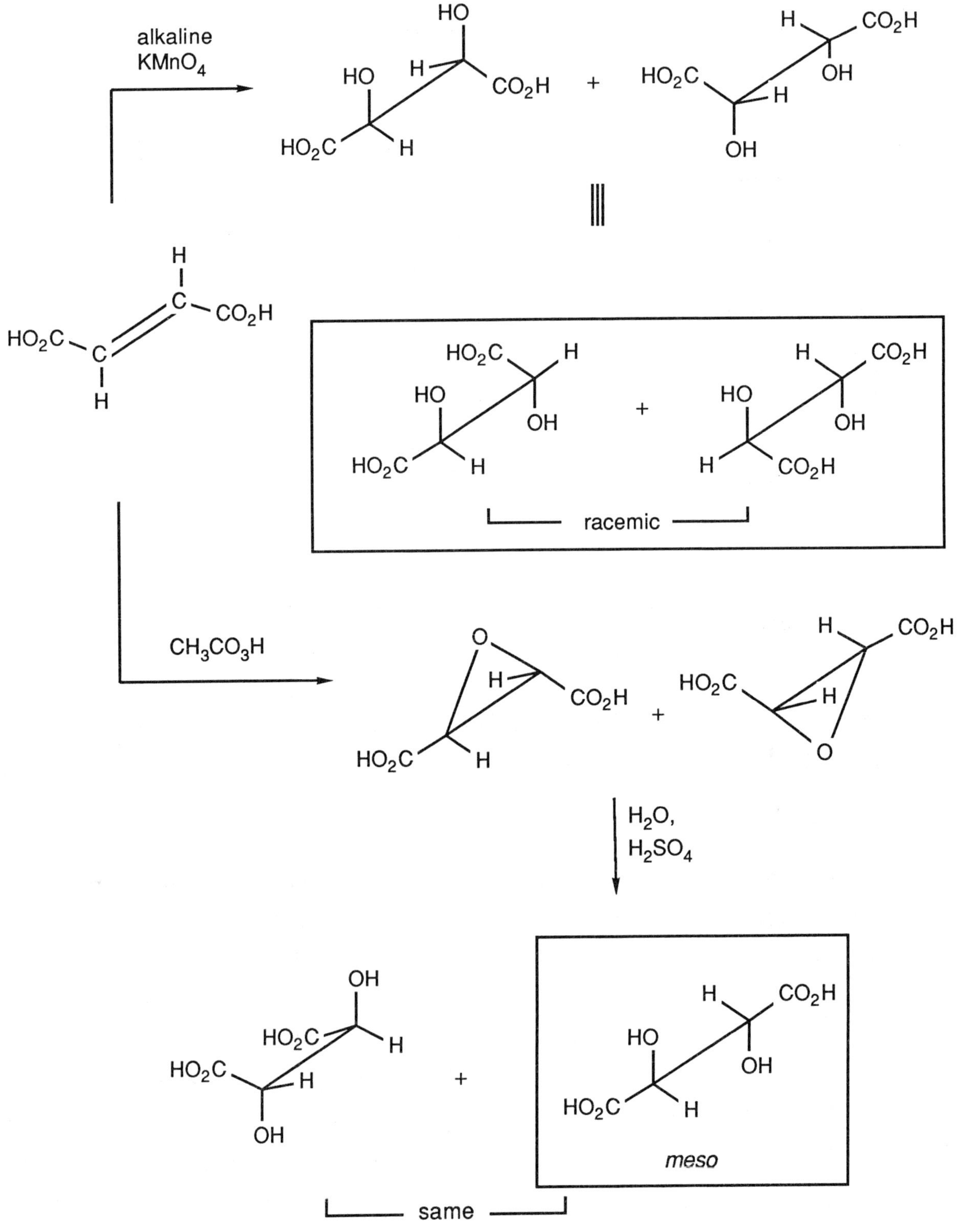

alkaline $KMnO_4$

HO H CO_2H + HO_2C H CO_2H HO HO_2C H OH H OH

racemic

CH_3CO_3H

H_2O, H_2SO_4

meso

same

6.22 There are at least two reasons why acid catalyzed dehydration is not stereospecific, and these can both be illustrated with the following deuterated cyclohexanol. First, the reaction is typically reversible, so the stereochemical relationships of the reactant can be interchanged via a nonproductive reaction:

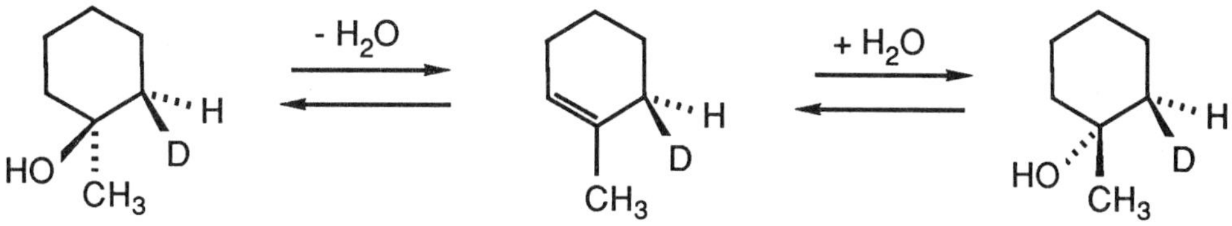

In the preceding equation the chirality of the methyl-substituted carbon has been lost in the intermediate alkene, and this allows the *trans* relationship between the hydroxyl and deuterium to be converted to *cis* so *anti* elimination would no longer proceed with loss of OH and D. A second reason for lack of stereospecificity involves carbocation formation:

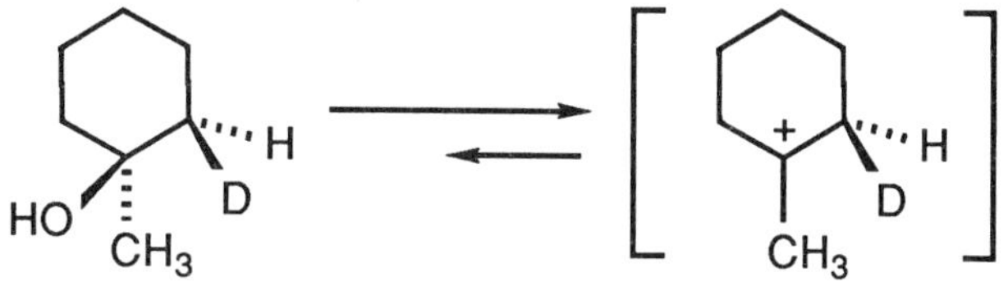

Once again, the chirality of the methyl-substituted carbon is destroyed in a planar carbocation, so there would be no stereochemical preference for loss of H vs D from the adjacent carbon atom.

6.23 **(a)** As shown in the following equations, the cyclic 6-membered transition state for ester pyrolysis can be achieved easily if the ester and hydrogen are gauche or eclipsed:

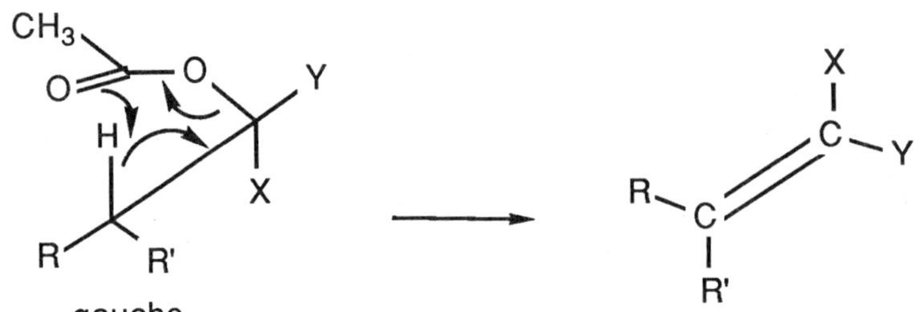

gauche

or

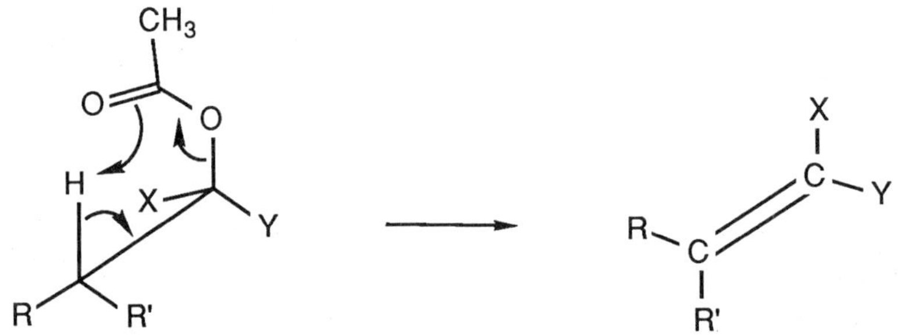

eclipsed

On the other hand, the following structures show that if the ester and H are *anti* to each other, they are too far apart to interact.

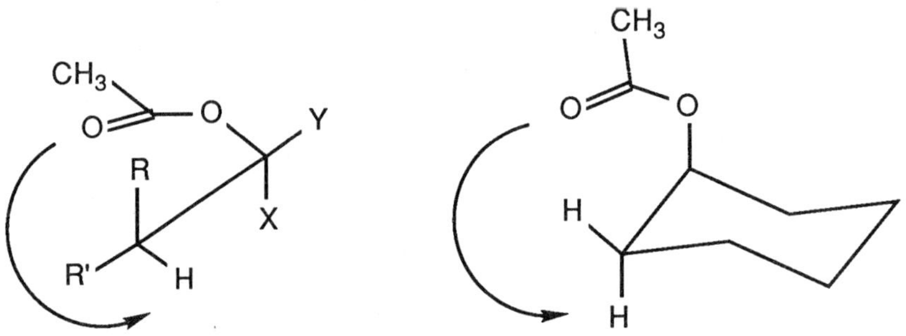

(b) For convenience we have shown the acetate in the equatorial position, and it has the same proximity to both the axial and equatorial hydrogens on the adjacent carbon. However, as shown in the following equations, cyclic elimination of the hydrogen *cis* to the acetate would yield (*Z*)-cyclohexene. In contrast, elimination of the hydrogen *trans* to the acetate would produce (*E*)-cyclohexene; a *trans* double bond in a six-membered ring would be highly unstable, so this process does not occur.

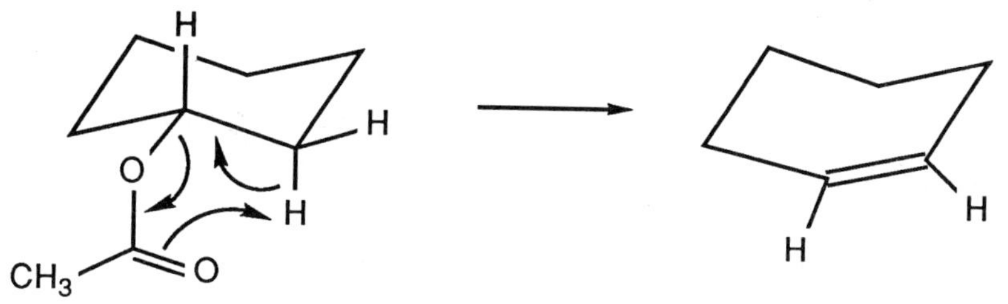

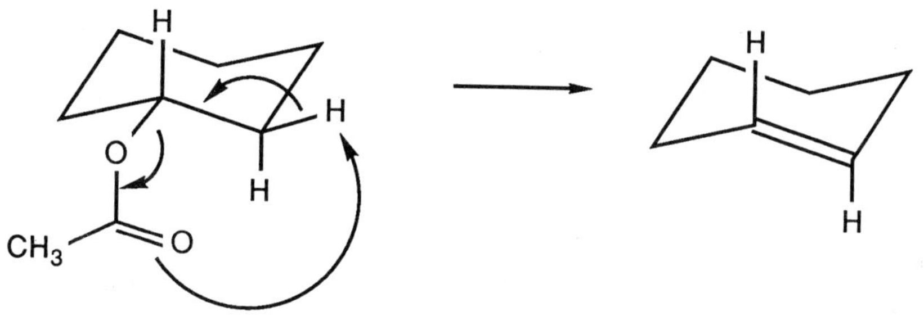

APPENDIX

MANIPULATION OF 3-DIMENSIONAL DRAWINGS

The study of stereochemistry requires the use of 2-dimensions (e.g., a printed page) to convey 3-dimensional information. We have introduced a variety of 3-dimensional representations for molecular structures, and it is frequently necessary to manipulate or reorient these drawings, much as you might pick up an actual molecular model and turn it around in your hands. The knowledge of several simple techniques together with a bit of practice should enable you to manipulate stereochemical drawings fairly easily.

The use of molecular models as a self-teaching device is almost certainly the most reliable way to develop proficiency in manipulating 3-dimensional drawings. We strongly urge you to use models *together with* these drawings as a means of developing the ability to utilize the drawings alone. A very small set of molecular models allows you to invent and check a nearly unlimited variety of problems using four different colors to indicate the four different substituents of a chiral center. First draw a set of structures such as the following (R = red, G = green, B = black, W = white):

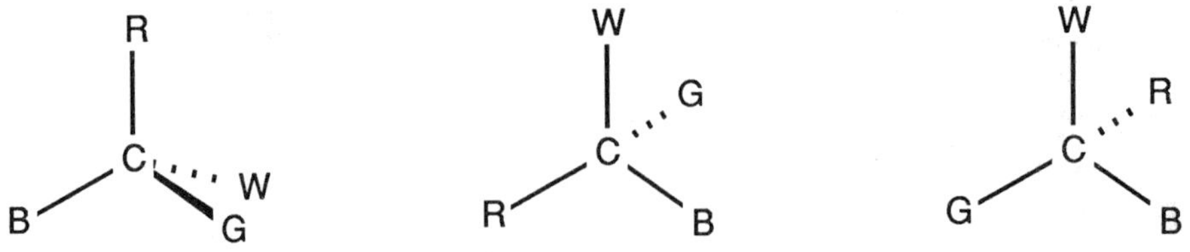

Then ask such questions as: Are all these the same? Or is one different than the other two? By designating a color as a particular substituent (perhaps white as methyl, green as chlorine, etc.) a very large number of structures can be evaluated quite easily, and you can very effectively teach yourself the techniques of drawing in 3-dimensions.

Reorientation of Drawings

In most cases the reorientation of a structure can be described in terms of three simple operations. None of these operations changes the conformation or configuration of the molecule in any way.

1. *Movement of the structural formula in the plane of the paper.** For example, we can reorient a drawing of (*R*)-2-chlorobutane in the following ways.

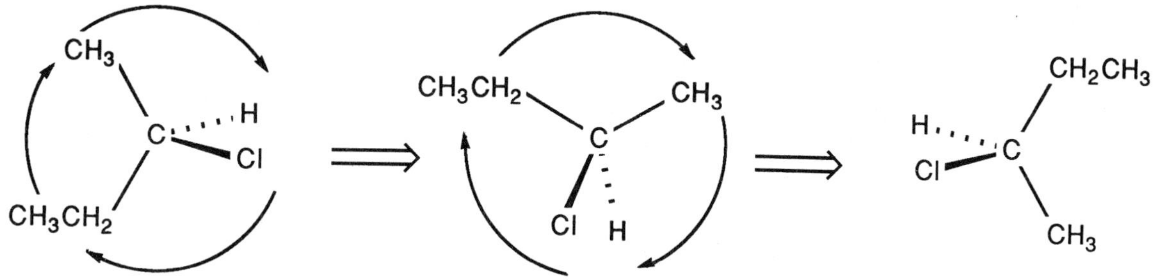

CH_3 CH_3CH_2 C H Cl ⟹ CH_3CH_2 CH_3 C Cl H ⟹ CH_2CH_3 H C Cl CH_3

In all three drawings the methyl and ethyl groups remain in the plane of the paper. These operations are analogous to reorienting some 3-dimensional object by sliding it around on a table.

2. *Rotation about an axis in the plane of the paper.* Using the plane of the paper as a reference, we can rotate the structure about an axis in this plane. Usually a horizonal or

---

*Note that this is not a legitimate operation with certain 3-dimensional representations such as sawhorse drawings and Fischer projections. The slope of the line connecting the "front" and "rear" carbon atoms of a sawhorse drawing has the very specific meaning that the atom which is lower on the page is closer to the viewer. Thus for (*R*)-2-chlorobutane it would not be proper to redraw structure (i) as structure (ii) where this line is horizontal.

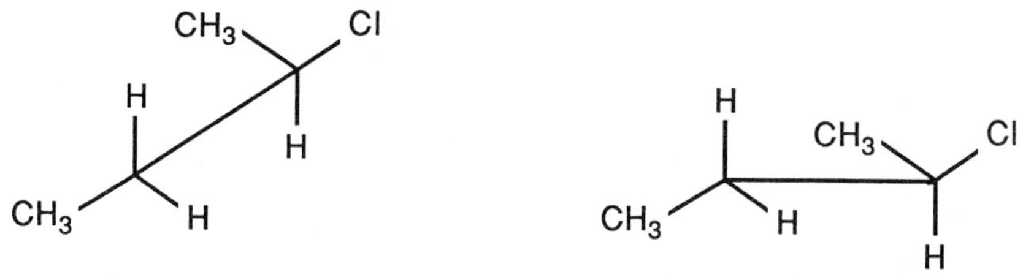

CH_3 Cl H H CH_3 H

(i)

H CH_3 Cl CH_3 H H

(ii) (improper drawing)

Reorientation of Fischer projections is discussed later in this Appendix.

vertical line is the easiest to use. For example, (*R*)-2-chlorobutane could be rotated about either a vertical or horizontal axis passing through the chiral center.

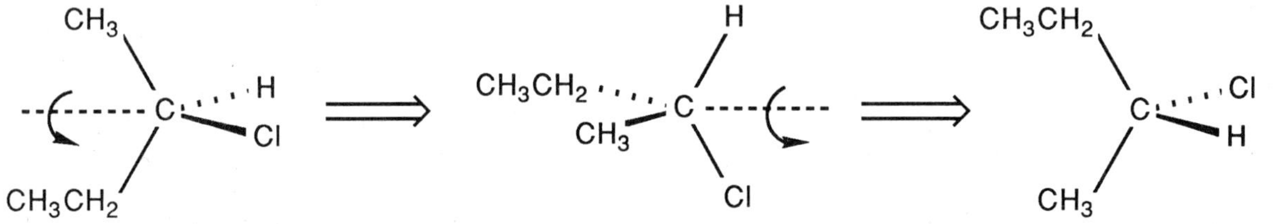

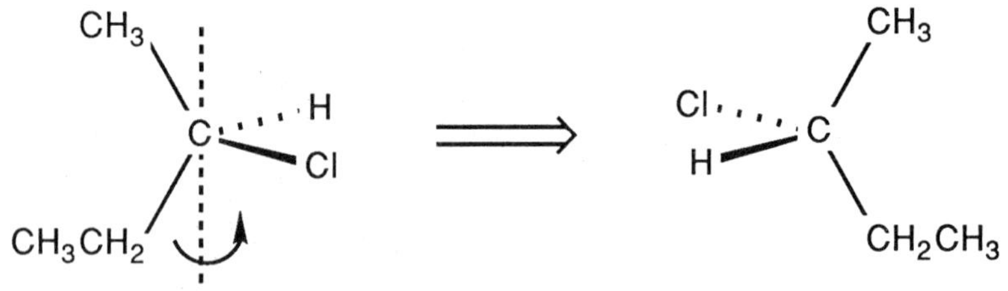

This type of rotation is a particularly useful way to reorient a sawhorse drawing by utilizing a horizontal axis through the midpoint of the central bond between the front and rear carbon atoms. The manipulation is somewhat analogous to the process of resting the back of your hand on a table (i.e., palm up) and then "flopping" your hand over so that your palm rests on the surface of the table.

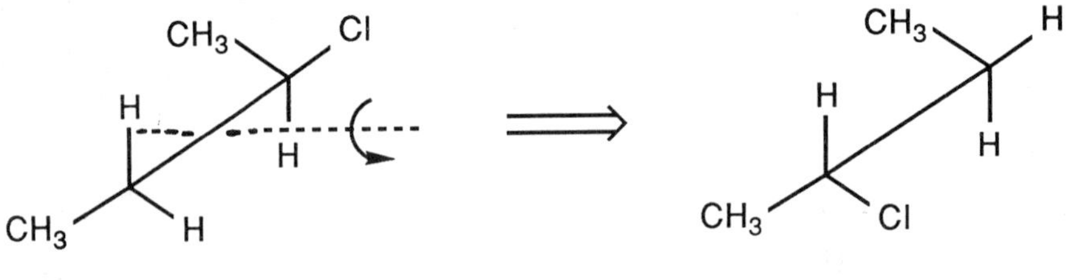

A vertical axis could also be used, although the substituents in the resulting drawing would not be drawn in the usual way.

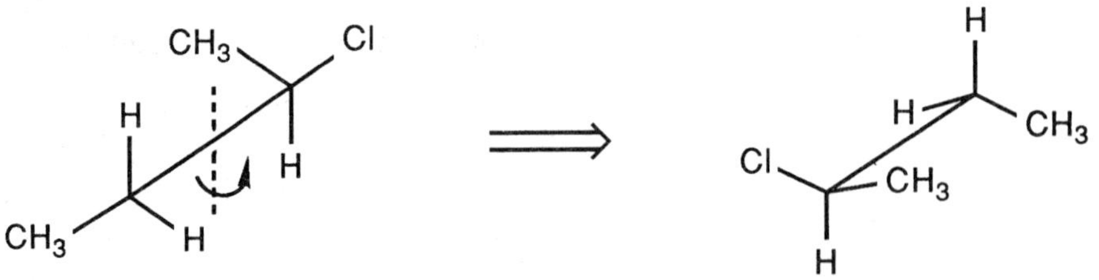

Note that these rotations of a sawhorse drawing must proceed a full 180° in order to regenerate the proper orientation of the central bond. Other kinds of 3-dimensional representations (for which a specific orientation is not required) may be rotated by any angle.

3. *Rotation of the entire molecule about an axis defined by one of the bonds.* This is easiest when the bond itself is horizontal, vertical or perpendicular to the plane of the paper. Such manipulations are analogous to picking up the molecule by one of its bonds and spinning the entire structure like a pinwheel. Again using (*R*)-2-chlorobutane:

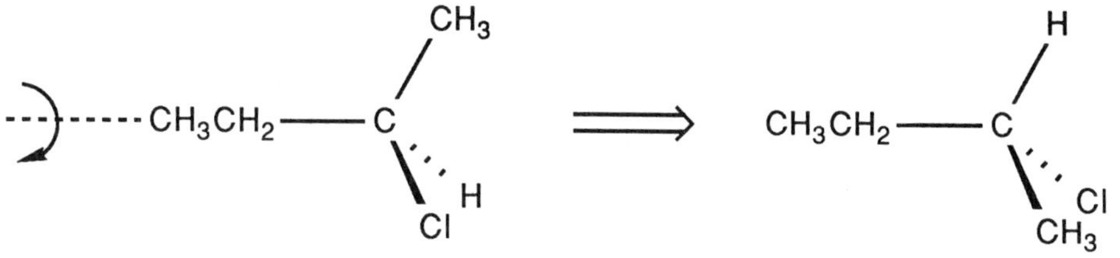

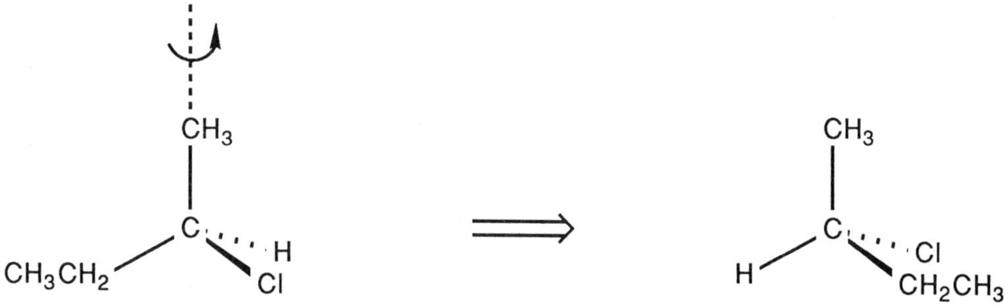

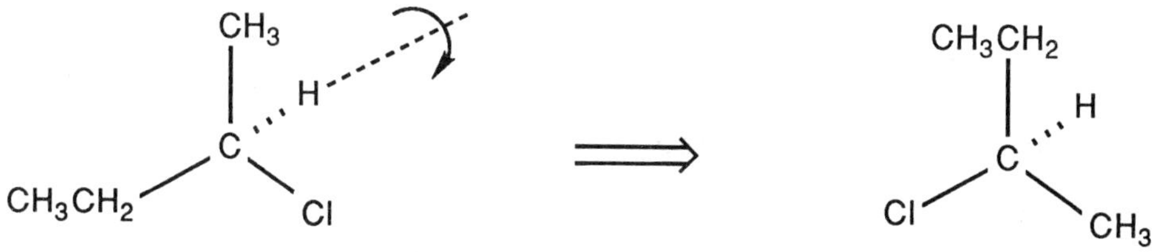

CH_3 CH_3CH_2—C, H, Cl ⟹ H CH_3CH_2—C, Cl, CH_3

CH_3 C, CH_3CH_2, H, Cl ⟹ CH_3 C, H, Cl, CH_2CH_3

CH_3 C, H, CH_3CH_2, Cl ⟹ CH_3CH_2 C, H, Cl, CH_3

You could also use bonds pointing in other directions but you might need more practice to do this easily.

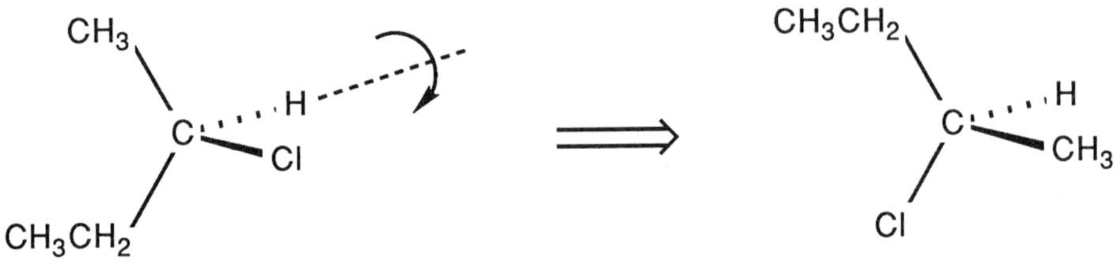

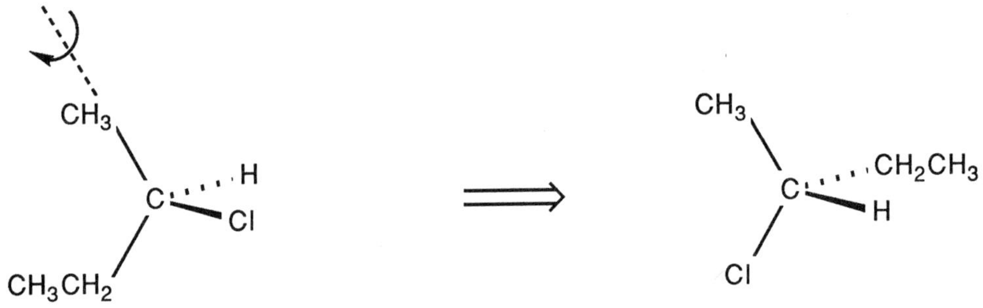

This is another useful technique for manipulating sawhorse projections.

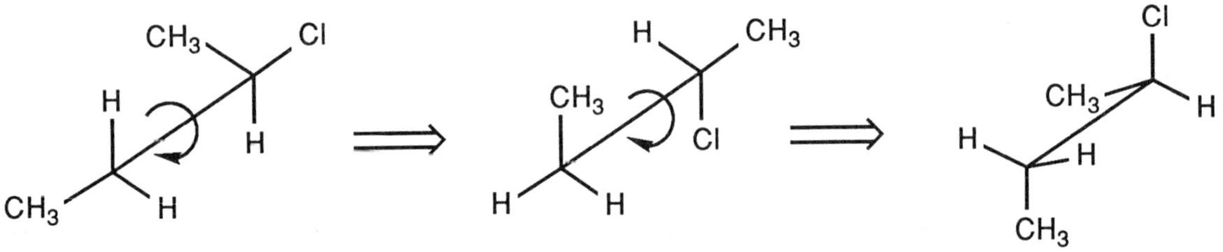

Rotation of Parts of Molecules About Single Bonds

This is not a reorientation but corresponds to drawing the molecule in a different conformation. As an example, we could rotate either the front or rear carbon atom of a sawhorse drawing about the central bond.

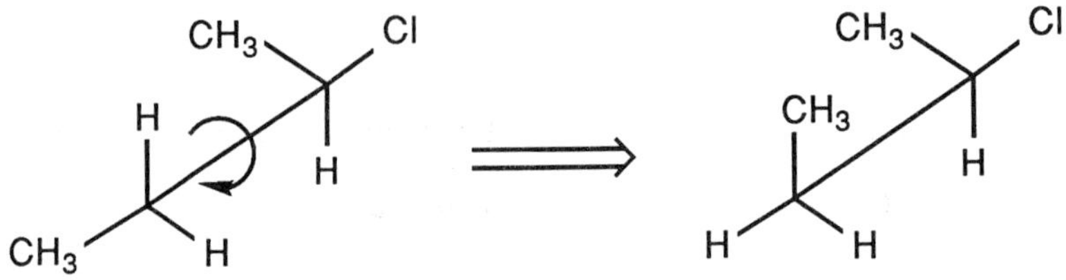

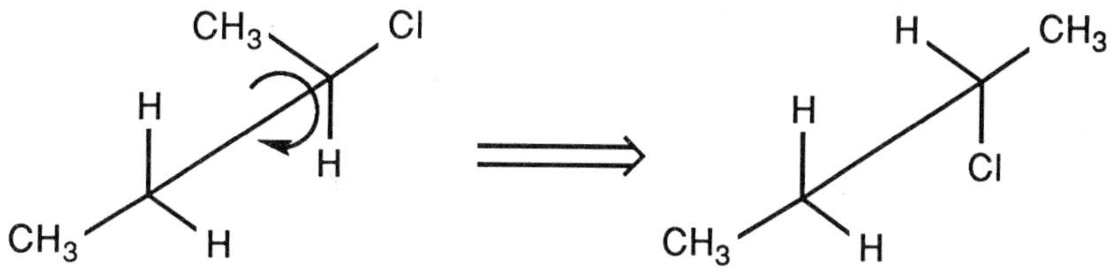

Example 6-A.1 Reorient one or both of the following drawings of (R)-2-chlorobutane to show that they are indeed the same.

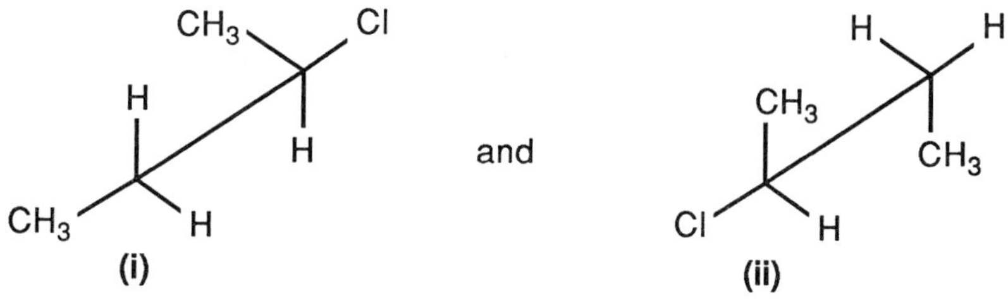

and

(i) (ii)

Answer. One approach to this problem would be to leave structure (i) unchanged and manipulate structure (ii) to put it in the same orientation as (i). This could be done in the following way.

First, we rotate the entire structure about a horizontal axis through the center of the molecule in order to place the two atoms of the central bond in the same relative positions, i.e., with C-2 at the back of the molecule as it is for (i).

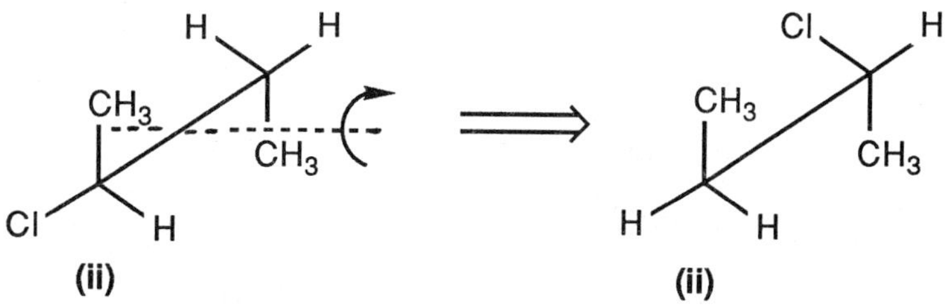

(ii) (ii)

Next, we rotate the entire molecule by 120° about the central bond in order to situate the hydrogen atom on C-2 directly below the carbon atom as it is in (i).

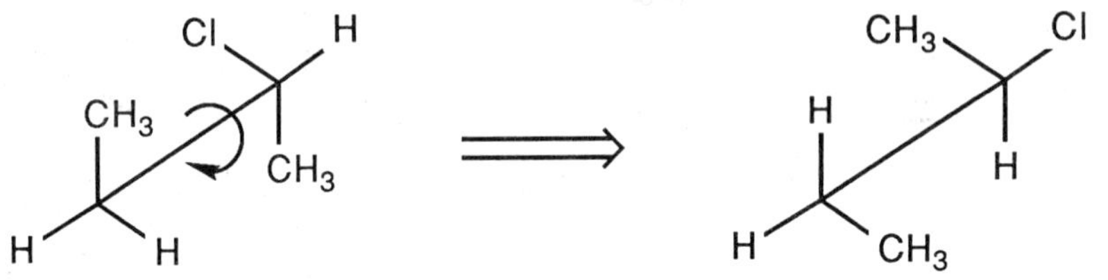

(ii) (ii)

The molecules are still in different conformations, and in the final step we rotate the front carbon of (ii) about the central bond.

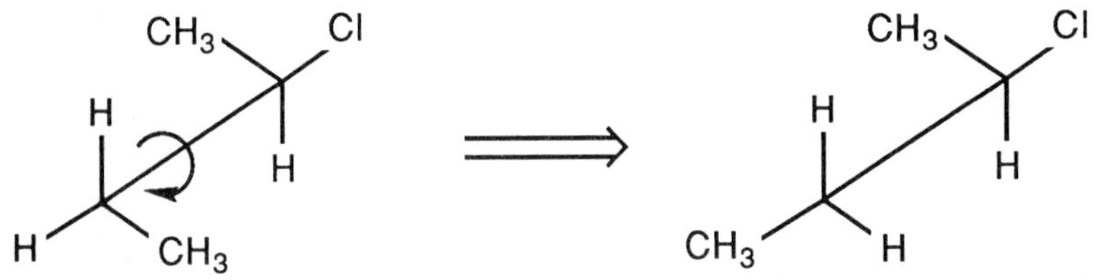

(ii) (ii) **identical to (i)**

Example 6-A.2 Reorient one or both of the following drawings of (*S*)-3-chlorohexane to show that they are indeed the same.

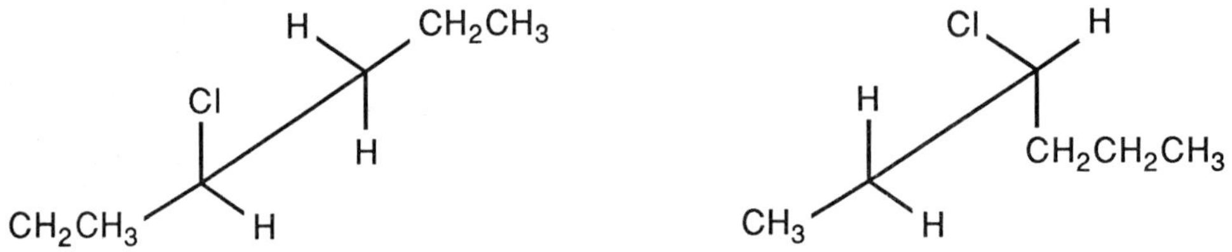

(i) (ii)

Answer. Note that the central bond of the drawings is between C-3 and C-4 in structure (i) but between C-2 and C-3 of structure (ii). If they are to be maintained as sawhorse drawings we must reconstruct one of them. We could do this in the following manner: (1) Rotate structure (i) about a horizontal axis through C-3 so that the bond between C-2 and C-3 is in the appropriate orientation for a sawhorse drawing. (2) Draw the "front" carbon atom in three dimensions. The conformation drawn was selected to show that (i) and (ii) are indeed the same.

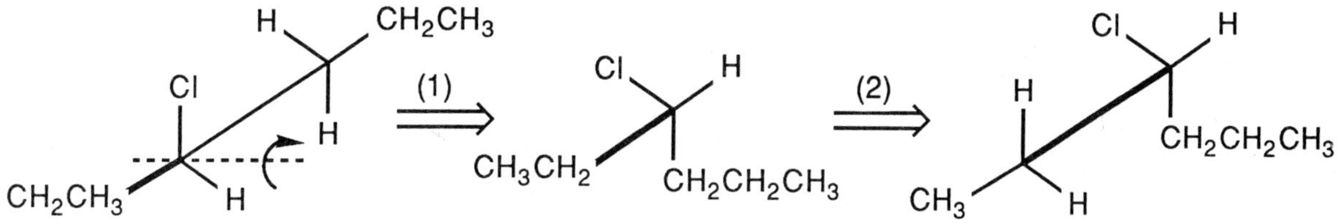

You can frequently answer problems such as this more easily by redrawing *both* structures with some other 3-dimensional representation. Only C-3 has four different substituents, so it is not necessary to show 3-dimensional structure at any of the other carbon atoms. Thus (i) and (ii) could be redrawn as:

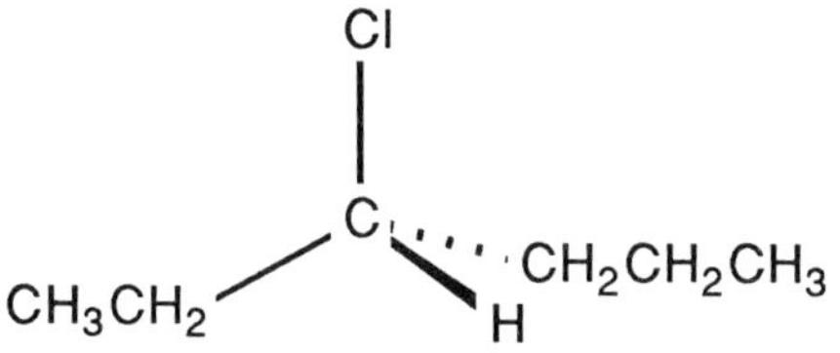

(i) and (ii)

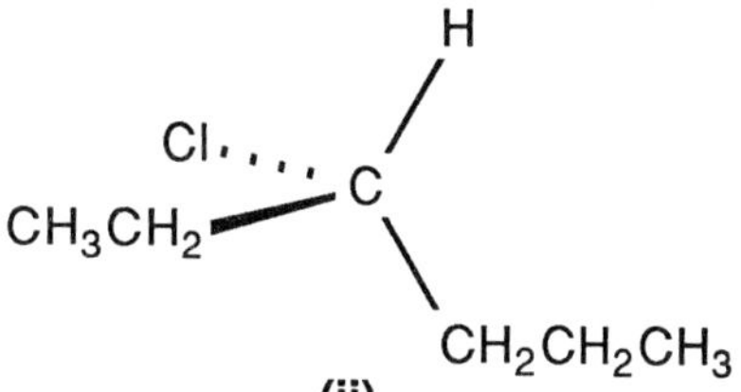

Problem 6-A.1 Reorient one or both of the following drawings in each pair of structures to show whether or not they are the identical compound.

(a)

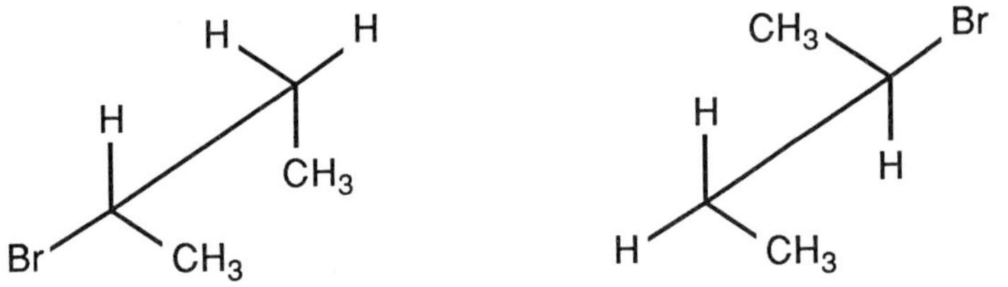

(b)

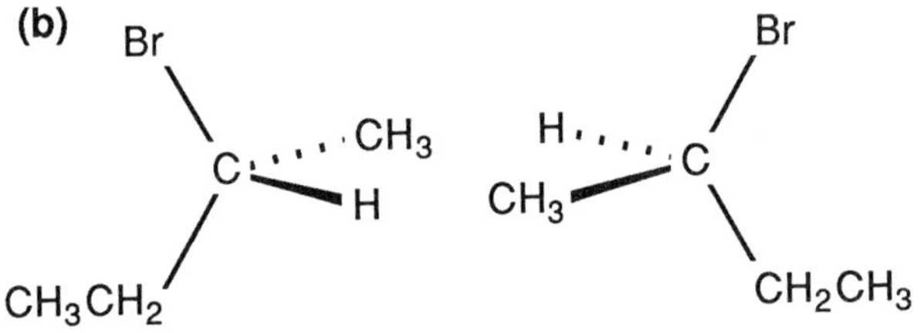

(c)

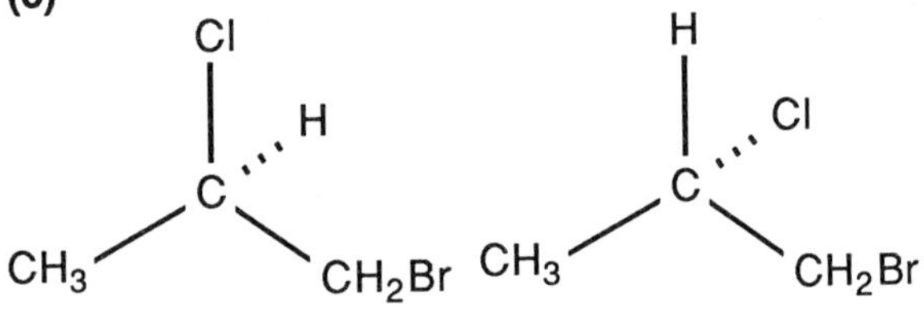

(d)

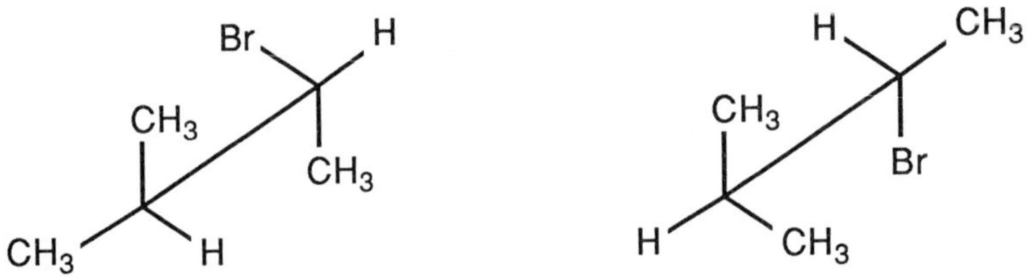

(e)

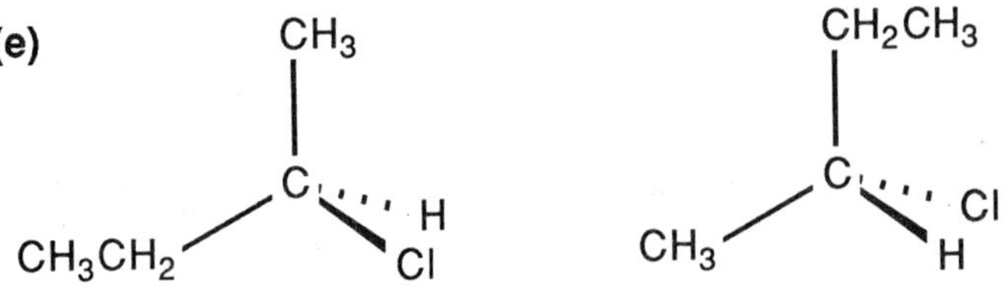

Fischer Projections

The only way that Fischer projections can be reoriented is by a rotation of 180° in the plane of the paper (Section 6.4).

CH_3 (top), H (left), Cl (right), CH_2OH (bottom) — same — CH_2OH (top), Cl (left), H (right), CH_3 (bottom)

Any other reorientation of the drawing would invalidate the convention that substituents on the vertical lines are situated behind the central carbon and those on the horizontal lines are in front of the central carbon. For example, turning a Fischer projection clockwise by 90° would interchange those substituents which were behind and in front of the chiral carbon atom.

(i) CH_2CH_3 (top), H (left), Cl (right), CH_3 (bottom) — not the same as — (ii) H (top), CH_3 (left), CH_2CH_3 (right), Cl (bottom)

We can see this by drawing (i) and (ii) with explicit stereochemical descriptions.

(i) CH_2CH_3 (top), H (left), Cl (right), CH_3 (bottom) — not the same as — (ii) H (top), CH_3 (left), CH_2CH_3 (right), Cl (bottom)

This comparison does afford an interesting result, however. 2-Chlorobutane contains a single chiral center, so the difference between (i) and (ii) must be in its configuration: in other words, (i) and (ii) are *enantiomers*. While there is only a single legitimate way in which we can *reorient* a Fischer projection, the structure can be *redrawn* in a variety of ways. For example, we can rotate *part of the molecule* about one of the bonds to the chiral center (as long as there is not a second chiral center which would be affected by the rotation). The following structure is rotated about the bond between the chiral center and the ethyl group.

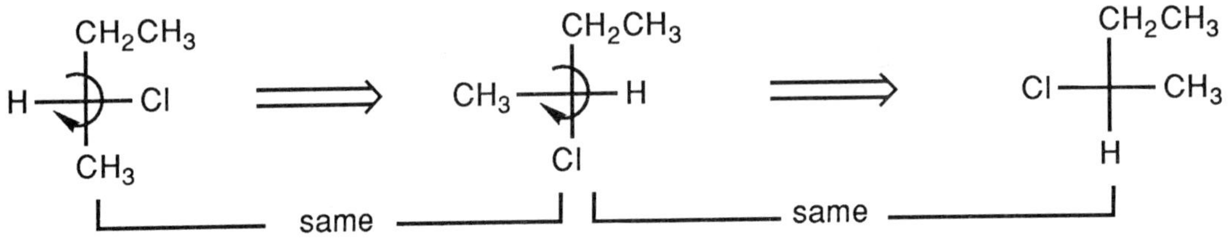

CH_2CH_3 / H—Cl / CH_3 ⟹ CH_2CH_3 / CH_3—H / Cl ⟹ CH_2CH_3 / Cl—CH_3 / H

same   same

One substituent (CH_3CH_2) remained in the original location, but the other three have all moved in this operation.

This type of manipulation of a Fischer projection can be quite useful for obtaining the orientation needed to assign *R/S* configuration. By placing the lowest priority group in either of the vertical positions (i.e., behind the chiral center) the other three groups are situated such that *R/S* can be assigned directly. For example, by rotation about the C–Cl bond, the hydrogen atom of 2-chlorobutane can be placed behind the chiral carbon atom:

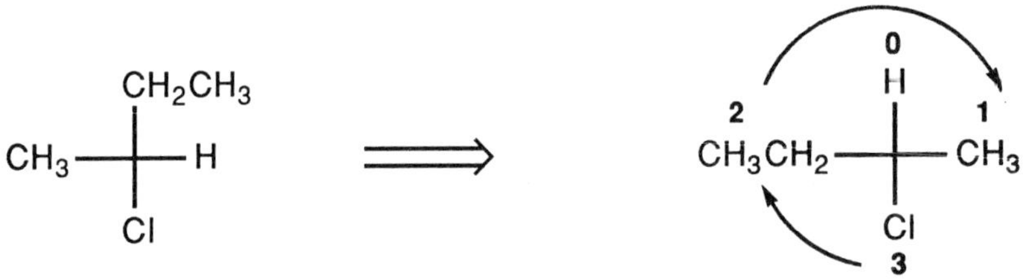

CH_2CH_3 / CH_3—H / Cl ⟹ 0 H / 2 CH_3CH_2—CH_3 1 / Cl 3

(R)-2-Chlorobutane

Interchange of Two Substituents on a Chiral Center

It is important to contrast the *rotation of three substituents* with the simple *interchange of any two substituents* on a chiral center. The interchange of two substituents on a chiral center will always result in a *change in configuration* (whether drawn with a Fischer projection or any other 3-dimensional representation). A second interchange will regenerate the original enantiomer (although it may not be an identical drawing).

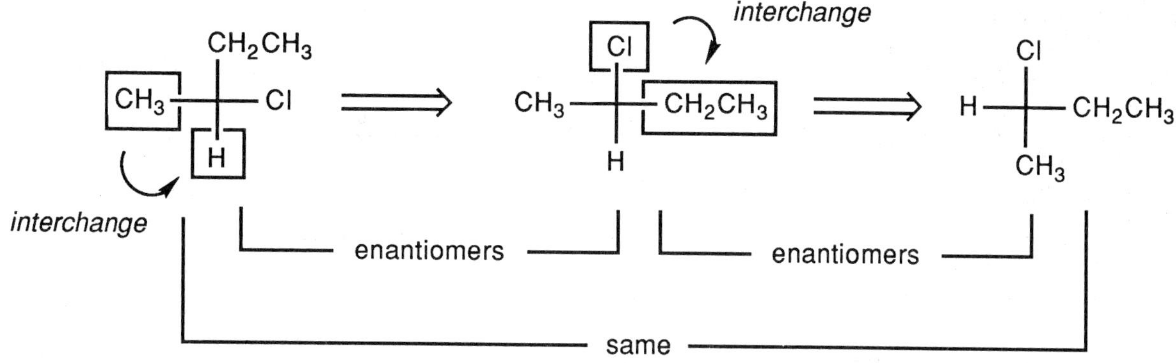

Problem A-6.2 Assign *R/S* configurations and redraw the structures of the preceding three compounds in order to demonstrate that the indicated relationships are correct.

The answers to Problems A-6.1 and A-6.2 are shown on the following page.

6–A.1 **(a)** enantiomers

(b) same

(c) enantiomers

(d) same

(e) same

6–A.2 left to right:

(a) *S*

(b) *R*

(c) *S*

CHAPTER 7. ALKYNES

ANSWERS TO EXERCISES

7.1 **(a)** $Cl-C\equiv C-Cl$ **(b)** $H_3C-C\equiv C-CH(CH_3)_2$

(c) $Br-CH_2-C\equiv C-CH_2-CH(CH_3)-CH_3$ **(d)** [cyclooctyne ring bearing Br and CH_3]

(e) $CH_3-CH_2-C\equiv C-CH(CH_2CH_3)-CH(CH_3)-CH_2F$

7.2

(a) (i) [2-pentanone skeletal structure] (ii) [pentanal skeletal structure, H, O]

(b) (i) [3-heptanone skeletal structure] + [4-heptanone skeletal structure] (ii) same as (i)

(c) (i) $C_6H_{11}-C(=O)-CH_3$ (ii) $C_6H_{11}-CH_2-C(=O)-H$

(d) (i) $C_6H_{11}-C(=O)-CH_2CH_3$ + $C_6H_{11}-CH_2-C(=O)-CH_3$ (ii) $C_6H_{11}-CH_2-C(=O)-CH_3$

(e) (i) cyclopentyl–C(=O)–CH_2–cyclopentyl (ii) cyclopentyl–C(=O)–CH_2–cyclopentyl

7.3 **(a)** H_2, Pd–$BaSO_4$ or hydroboration with disiamylborane or catecholborane followed by CH_3CO_2H.

(b) H_2, Pd–$BaSO_4$; hydroboration with disiamylborane or catecholborane followed by CH_3CO_2H; or Na, NH_3, NH_4SO_4.

(c) H_2 and Pd or Pt catalyst.

(d) Na or Li in NH_3 or another amine (e.g. $CH_3CH_2NH_2$) as solvent.

7.4 **(a)**

$CH_3CH_2CH_2C\equiv CH$ → (1. catecholborane (BH); 2. H_2O; 3. I_2, NaOH) → $(CH_3CH_2CH_2)(H)C{=}C(I)(H)$

(b)

$CH_3CH_2CH_2C\equiv CCH_2CH_2CH_3$ → (1. catecholborane (BH); 2. Br_2; 3. I_2, NaOH) → $(CH_3CH_2CH_2)(H)C{=}C(CH_2CH_2CH_3)(Br)$

The reaction could also be carried out by direct addition of (1 equivalent of) HBr to the alkyne, but that reaction is less reliable.

(c)

$CH_3CH_2CH_2C\equiv CCH_2CH_2CH_3$ → (Br_2 / CH_3CO_2H) → $(CH_3CH_2CH_2)(Br)C{=}C(CH_2CH_2CH_3)(Br)$

7.5 **(a)** $HC\equiv C-CH_3$ **(b)** $CH_3CH_2C\equiv CCH_2CH_3$

(c) NaBr + $CH_3C{\equiv}CH$ + $(CH_3)_2C{=}CH_2$

7.6 (a) from an alkene:

$$CH_3CH_2-C(CH_3)_2-CH{=}CH_2 \xrightarrow{Br_2} CH_3CH_2-C(CH_3)_2-CH(Br)-CH_2Br \xrightarrow{NaNH_2/NH_3} CH_3CH_2-C(CH_3)_2-C{\equiv}CH$$

from an aldehyde:

$$CH_3CH_2-C(CH_3)_2-CH_2-C(=O)-H \xrightarrow{PCl_5} CH_3CH_2-C(CH_3)_2-CH_2-CCl_2-H \xrightarrow{KOH/EtOH} CH_3CH_2-C(CH_3)_2-C{\equiv}CH$$

from a ketone:

$$CH_3CH_2-C(CH_3)_2-C(=O)-CH_3 \xrightarrow{PCl_5} CH_3CH_2-C(CH_3)_2-CCl_2-CH_3 \xrightarrow{KOH/EtOH} CH_3CH_2-C(CH_3)_2-C{\equiv}CH$$

(b) Three straightforward alternatives are possible; note that the choice of Br for the alkyl halide is arbitrary:

$$CH_3(CH_2)_4C{\equiv}CH \xrightarrow[\text{2. } CH_3Br]{\text{1. } NaNH_2} CH_3(CH_2)_4C{\equiv}CCH_3$$

$$HC{\equiv}CCH_3 \xrightarrow[\text{2. } CH_3(CH_2)_4Br]{\text{1. } NaNH_2} CH_3(CH_2)_4C{\equiv}CCH_3$$

$$HC{\equiv}CH \xrightarrow[\text{2. } CH_3(CH_2)_4Br]{\text{1. } NaNH_2} CH_3(CH_2)_4C{\equiv}CH \xrightarrow[\text{2. } CH_3Br]{\text{1. } NaNH_2} CH_3(CH_2)_4C{\equiv}CCH_3$$

7.7 Structures (a), (c), (d), and (f) are chiral. Structure (b) is a 1,3-diene (not an allene), and two of the substituents on one end of the allene in (e) are identical (the two carbons of the 6-membered ring).

7.8

$$CH_3CH_2CH{=}CH_2 + HBr \longrightarrow \boxed{CH_3CH_2{-}\underset{}{\overset{Br}{\overset{|}{C}}}H{-}CH_3}$$

$$\underset{H}{\overset{CH_3CH_2}{}}C{=}C\underset{H}{\overset{CH_3}{}} + ICH_2ZnI \longrightarrow \boxed{CH_3CH_2{-}\text{cyclopropane}{-}CH_3}$$

$$CH_3CH_2C{\equiv}CCH_3 \xrightarrow{H_2,\ Pd/BaSO_4} \boxed{\underset{CH_3CH_2}{\overset{H}{}}C{=}C\underset{CH_3}{\overset{H}{}}}$$

$$CH_3CH_2CH_2{-}\overset{O}{\overset{\|}{C}}{-}CH_3 + PCl_5 \longrightarrow \boxed{CH_3CH_2CH_2{-}\underset{Cl}{\underset{|}{\overset{Cl}{\overset{|}{C}}}}{-}CH_3}$$

$$CH_3CH_2CH_2C\equiv CH \xrightarrow[\text{2. } H_2O \quad \text{3. } I_2,\ NaOH]{\text{1. catecholborane (benzo-1,3,2-dioxaborole, O–BH–O)}} \boxed{\begin{matrix} H & & I \\ & C{=}C & \\ CH_3CH_2CH_2 & & H \end{matrix}}$$

7.9 **(a)** $HC\equiv CCH_2CH_3 \xrightarrow[\text{2. } CH_3CH(CH_3)CH_2Br]{\text{1. } NaNH_2} CH_3CH(CH_3)CH_2C\equiv CCH_2CH_3$

$$\xrightarrow{H_2/Pd\text{-}BaSO_4} \begin{matrix} CH_3CH(CH_3)CH_2 & & CH_2CH_3 \\ & C{=}C & \\ H & & H \end{matrix}$$

(b) $HC\equiv CH \xrightarrow[\text{2. } CH_3CH_2CH_2Br]{\text{1. } NaNH_2} HC\equiv CCH_2CH_2CH_3 \xrightarrow[\text{2. } CH_3CH_2CH_2Br]{\text{1. } NaNH_2} CH_3CH_2CH_2C\equiv CCH_2CH_2CH_3$

(c) $HC\equiv CH \xrightarrow[\text{2. } CH_3CH_2CH_2Br]{\text{1. } NaNH_2} CH_3CH_2CH_2C\equiv CH \xrightarrow[HgSO_4]{H_2O,\ H_2SO_4,} CH_3CH_2CH_2-\overset{\overset{\displaystyle O}{\|}}{C}-CH_3$

(d) 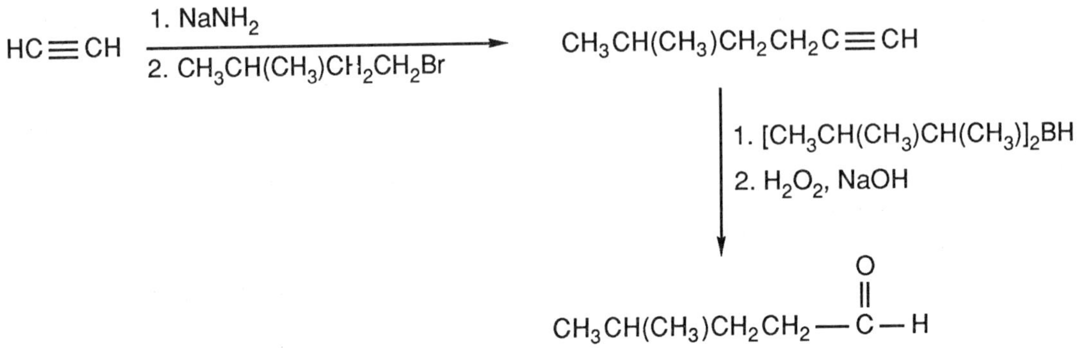 $HC{\equiv}CH \xrightarrow[\text{2. } CH_3CH(CH_3)CH_2CH_2Br]{\text{1. } NaNH_2} CH_3CH(CH_3)CH_2CH_2C{\equiv}CH$

$\downarrow$ 1. $[CH_3CH(CH_3)CH(CH_3)]_2BH$
2. H_2O_2, NaOH

$CH_3CH(CH_3)CH_2CH_2{-}\overset{\overset{\large O}{\|}}{C}{-}H$

ANSWERS TO PROBLEMS

7.1 (a) 6-(fluoromethyl)-2-hexyne

(b) 2-methyl-2,6-dibromo-3-hexyne

(c) 1-chloro-1,2-butadiene

(d) 1,2-dibromoacetylene

(e) 1-bromo-5-methyl-2-heptyne

7.2 (a)

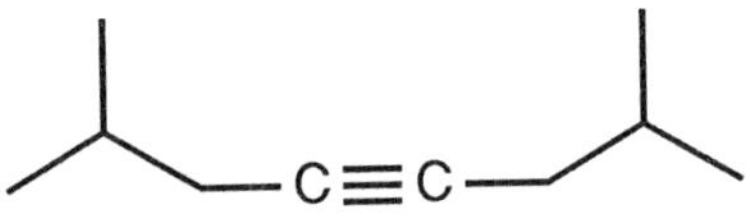

(b) $Cl_2CH{-}CH{=}C{=}CH{-}CH_2CH_3$

(c) 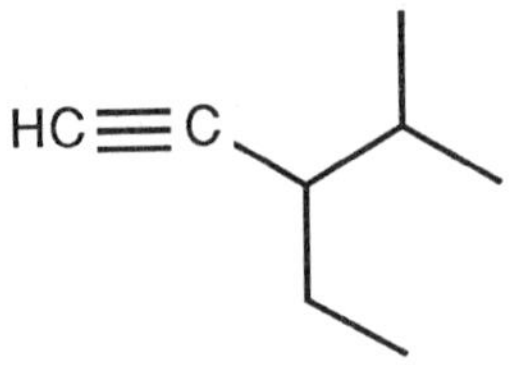 $HC{\equiv}C$

(d) 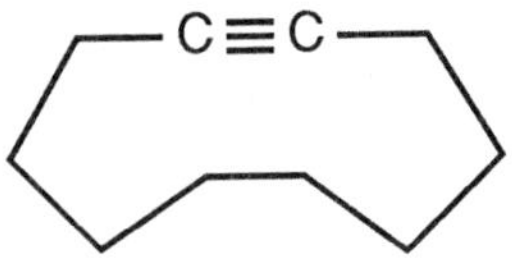 $C{\equiv}C$

(e) 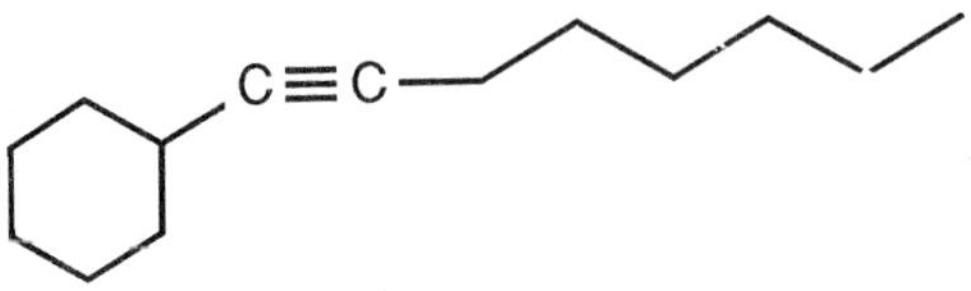 $C{\equiv}C$

7.3

(a) 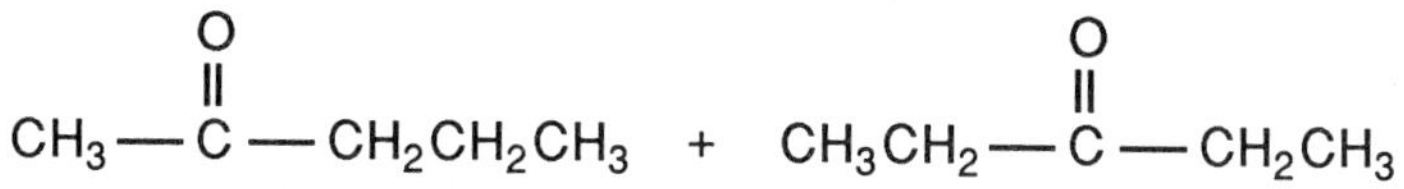 $CH_3-\overset{O}{\overset{\|}{C}}-CH_2CH_2CH_3$ + $CH_3CH_2-\overset{O}{\overset{\|}{C}}-CH_2CH_3$

(b) $CH_3CH_2CH_2CH_2CH_3$

(c) 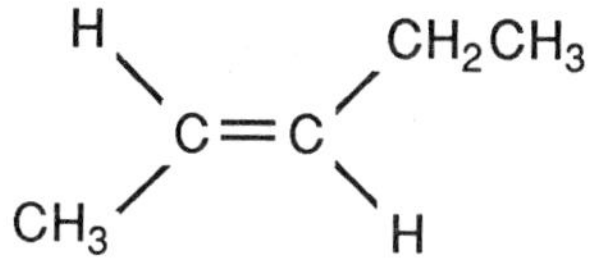 $\underset{CH_3}{\overset{H}{\;}}C=C\underset{H}{\overset{CH_2CH_3}{\;}}$

(d) $CH_3-\overset{O}{\overset{\|}{C}}-CH_2CH_2CH_3$ + $CH_3CH_2-\overset{O}{\overset{\|}{C}}-CH_2CH_3$

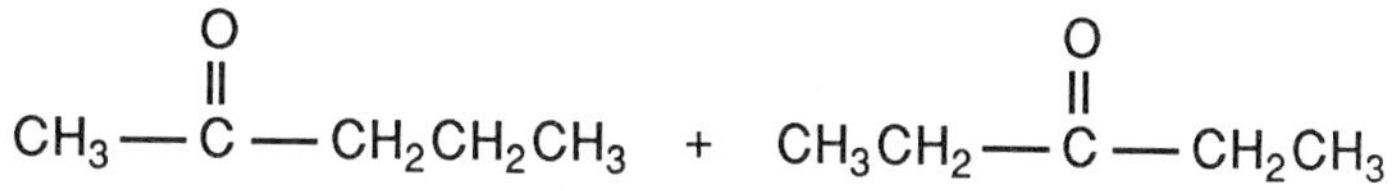

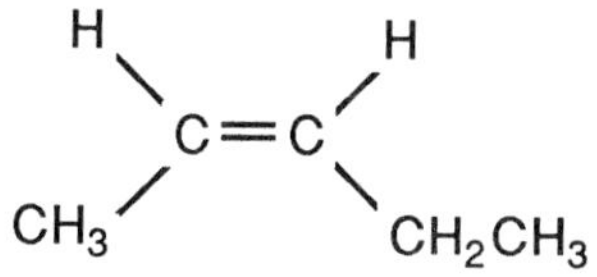

(e) $\underset{CH_3}{\overset{H}{\;}}C=C\underset{CH_2CH_3}{\overset{H}{\;}}$

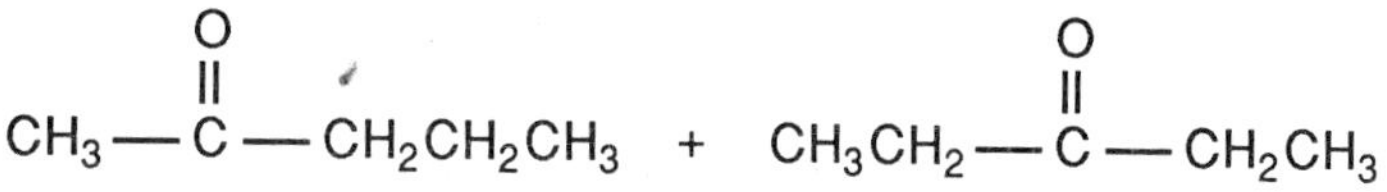

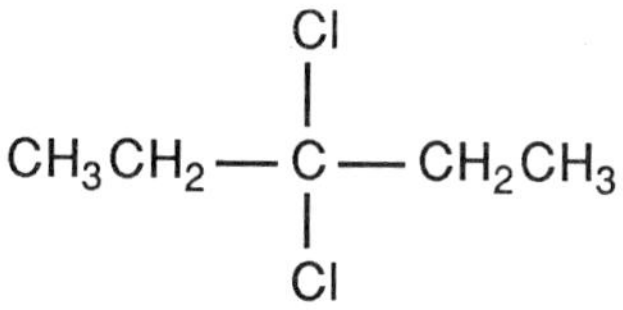

(f) $CH_3-\overset{O}{\overset{\|}{C}}-CH_2CH_2CH_3$ + $CH_3CH_2-\overset{O}{\overset{\|}{C}}-CH_2CH_3$

(g) $CH_3CH_2-\underset{Cl}{\overset{Cl}{C}}-CH_2CH_3$

(h) $\underset{CH_3}{\overset{H}{\;}}C=C\underset{Br}{\overset{CH_2CH_3}{\;}}$

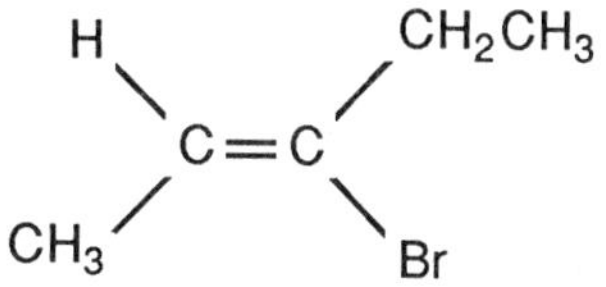

(i) 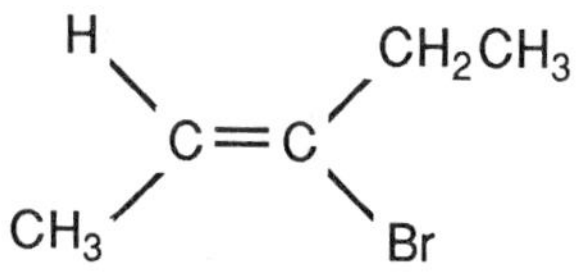 $\underset{CH_3}{\overset{H}{\;}}C=C\underset{Br}{\overset{CH_2CH_3}{\;}}$

(i) 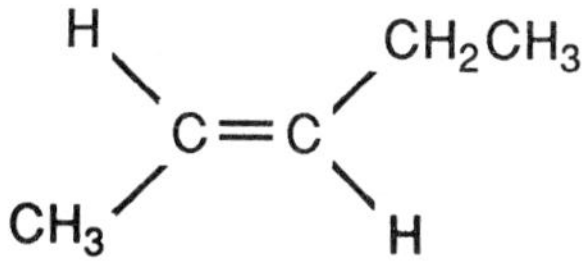 $\underset{CH_3}{\overset{H}{\;}}C=C\underset{H}{\overset{CH_2CH_3}{\;}}$

7.4

(a)  $H-\overset{O}{\overset{\|}{C}}-CH_2CH_2CH_2CH_3$

(b) $CH_3CH_2CH_2CH_2CH_3$

(c) $LiC\equiv CCH_2CH_2CH_3$

(d) 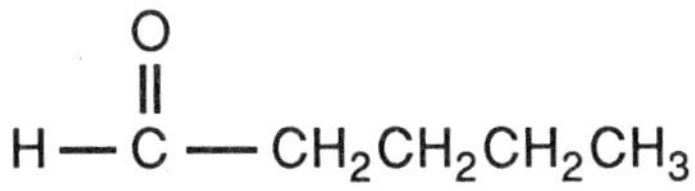 $H-\overset{O}{\overset{\|}{C}}-CH_2CH_2CH_2CH_3$

(e) $CH_2=CHCH_2CH_2CH_3$

(f) 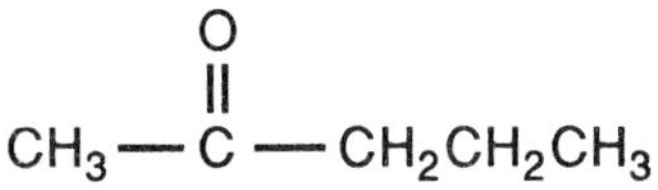 $CH_3-\overset{O}{\overset{\|}{C}}-CH_2CH_2CH_3$

(g) 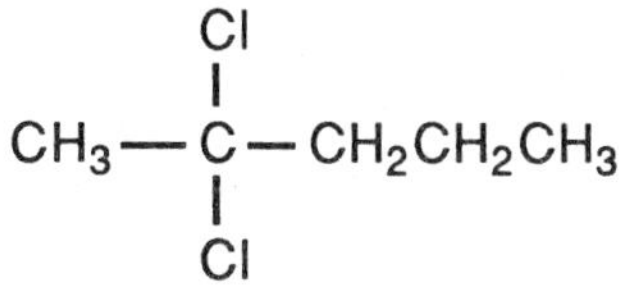 $CH_3-\underset{Cl}{\overset{Cl}{C}}-CH_2CH_2CH_3$

(h) 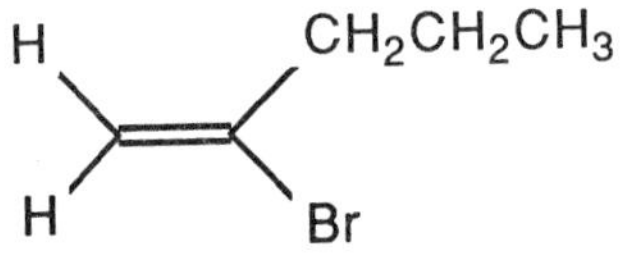 H, H on one carbon; $CH_2CH_2CH_3$, Br on the other

(i) 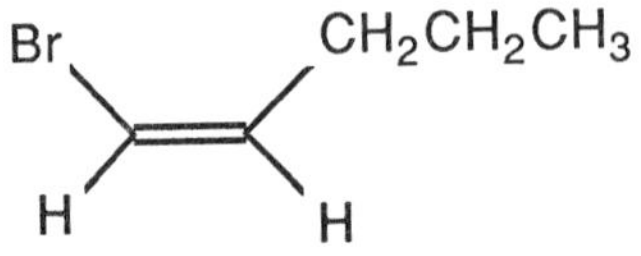 Br, H on one carbon; $CH_2CH_2CH_3$, H on the other (trans)

(j) $NaC \equiv CCH_2CH_2CH_3$

7.5 **(a)** Strong bases such as $NaNH_2$ in ammonia, KOH in alcohol, etc.

(b) Strong bases such as $NaNH_2$ in ammonia, KOH in alcohol, etc.

(c) $NaNH_2$ followed by CH_3CH_2Br.

(d) Strong bases such as $NaNH_2$ in ammonia, KOH in alcohol, etc.

7.6

(a) $CH_3CH_2CH(CH_3)CH_2-C(=O)-CH_3$

(b) $CH_3CH_2CH(CH_3)CH_2C \equiv CCu$

(c) $CH_3CH_2CH(CH_3)CH_2C \equiv CCH_3$

(d) $CH_2CH_2CH(CH_3)C \equiv CH$

(e) $CH_3CH_2-C(=O)-CH_2CH_2CH_3$

(f) 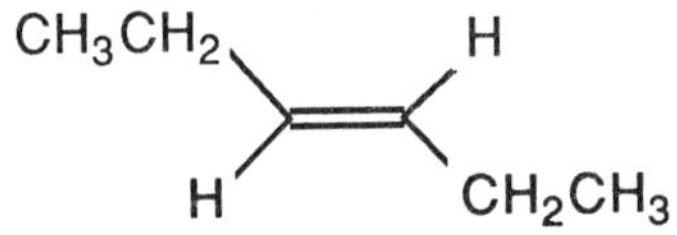 CH_3CH_2, H on one carbon; H, CH_2CH_3 on the other (trans)

7.7

(a) $CH_3CH_2CH(CH_3)CH_2-CCl_2-CH_3$

(b) 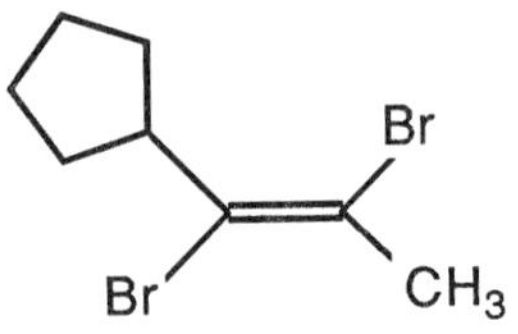 cyclopentyl, Br on one carbon; Br, CH_3 on the other

(c) 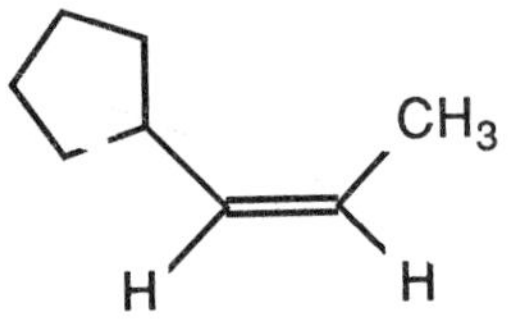 cyclopentyl, H on one carbon; CH_3, H on the other

(d) 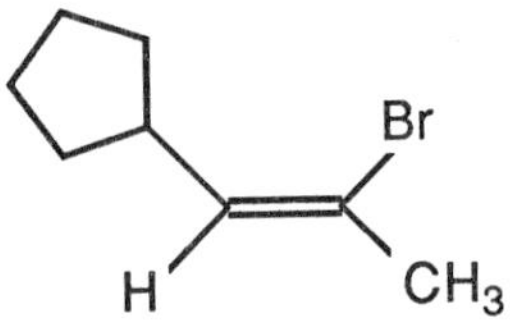 cyclopentyl, H on one carbon; Br, CH_3 on the other

(e) 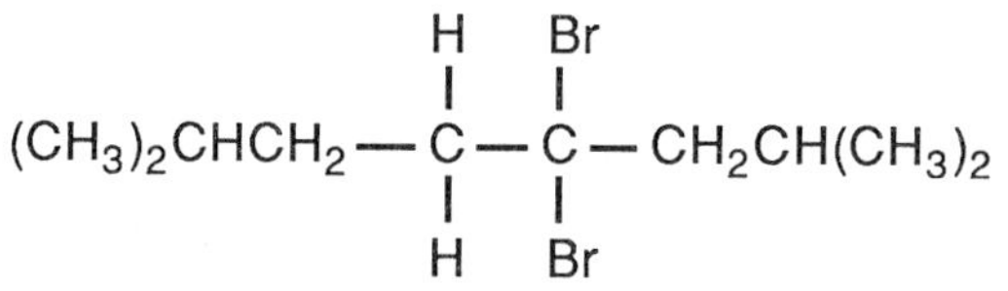 $(CH_3)_2CHCH_2-C(H)(H)-C(Br)(Br)-CH_2CH(CH_3)_2$

(f) 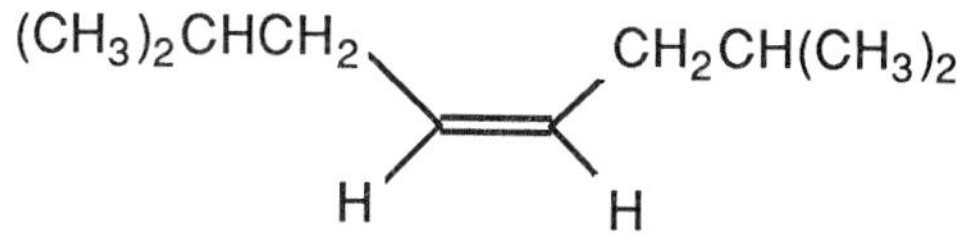 $(CH_3)_2CHCH_2CH=CHCH_2CH(CH_3)_2$ (cis; H, H on same side)

(g) 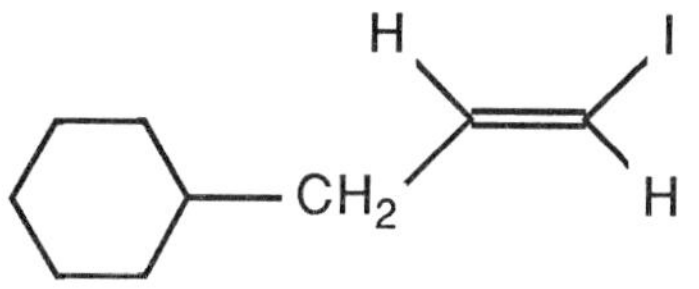 cyclohexyl$-CH_2CH=CHI$ (trans; H, 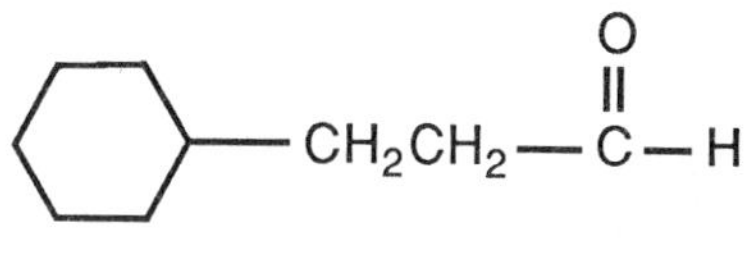 I)

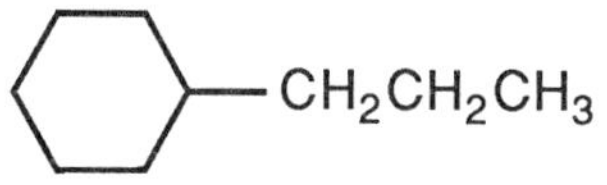

(h) cyclohexyl$-CH_2CH_2-C(=O)-H$

(i) cyclohexyl$-CH_2CH_2CH_3$

(j) 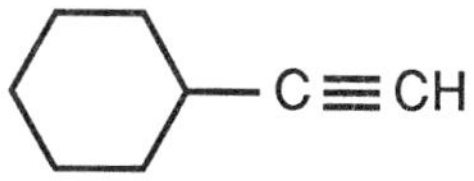 cyclohexyl$-C\equiv CH$

(k) $CH_3CH_2CH(CH_3)C\equiv CAg$

(l) $CH_3CH_2CH(CH_3)CH=CH_2$

(m) $CH_3CH_2C\equiv CCH(CH_3)_2$

7.8 **(a)** from a ketone:

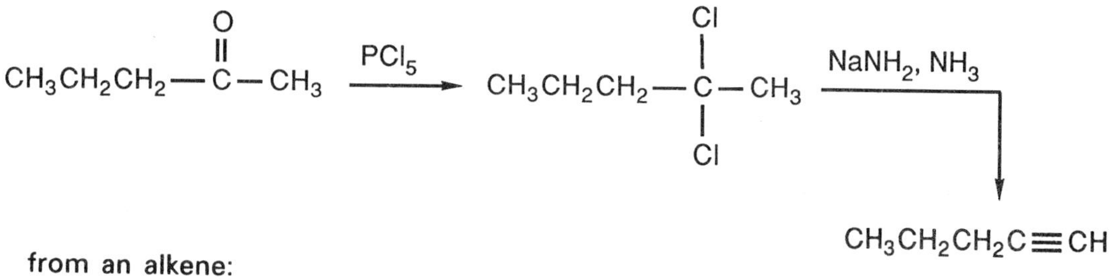

$$CH_3CH_2CH_2-C(=O)-CH_3 \xrightarrow{PCl_5} CH_3CH_2CH_2-CCl_2-CH_3 \xrightarrow{NaNH_2,\ NH_3} CH_3CH_2CH_2C\equiv CH$$

from an alkene:

$$CH_3CH_2CH_2CH=CH_2 \xrightarrow{Br_2} CH_3CH_2CH_2-CH(Br)-CH(Br)-H \xrightarrow{NaNH_2,\ NH_3} CH_3CH_2CH_2C\equiv CH$$

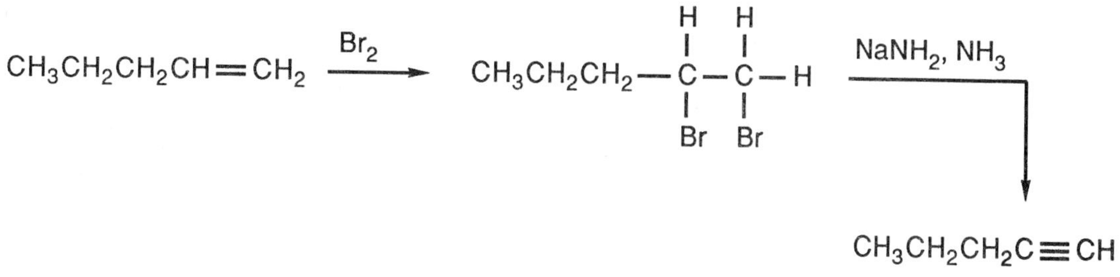

from an aldehyde:

$$CH_3CH_2CH_2CH_2{-}\overset{\overset{\displaystyle O}{\|}}{C}{-}H \xrightarrow{PCl_5} CH_3CH_2CH_2{-}\underset{H}{\overset{H}{C}}{-}\underset{Cl}{\overset{Cl}{C}}{-}H \xrightarrow{NaNH_2,\ NH_3} CH_3CH_2CH_2C{\equiv}CH$$

(b) Preparation from 3-heptanone would be unsatisfactory because elimination could occur in either of two directions. Therefore 3-heptene is the preferred starting material:

$$CH_3CH_2CH{=}CHCH_2CH_2CH_3 \xrightarrow{Br_2} CH_3CH_2\underset{Br}{\overset{H}{C}}{-}\underset{Br}{\overset{H}{C}}CH_2CH_2CH_3 \xrightarrow{NaNH_2,\ NH_3} CH_3CH_2C{\equiv}CCH_2CH_2CH_3$$

(c) Preparation from the alkene would be unsatisfactory, because a 1,2-alkyl shift at the adjacent tertiary center would be likely. The preferred method would involve reaction of a ketone:

$$\text{1-methylcyclohexyl}{-}\overset{}{\underset{\underset{\displaystyle O}{\|}}{C}}{-}CH_3 \xrightarrow{PCl_5} \text{1-methylcyclohexyl}{-}CCl_2CH_3 \xrightarrow{NaNH_2,\ NH_3} \text{1-methylcyclohexyl}{-}C{\equiv}CH$$

or of an aldehyde

$$\text{1-methylcyclohexyl}{-}CH_2CHO \xrightarrow{PCl_5} \text{1-methylcyclohexyl}{-}CH_2CCl_2H \xrightarrow{NaNH_2,\ NH_3} \text{1-methylcyclohexyl}{-}C{\equiv}CH$$

7.9 **(a)**

$$CH_3CH_2CH_2C{\equiv}CH \xrightarrow{NaNH_2} CH_3CH_2CH_2C{\equiv}CNa \xrightarrow{CH_3CH_2CH_2Br} CH_3CH_2CH_2C{\equiv}CCH_2CH_2CH_3$$

(b) Attempted alkylation of 1-pentyne with 2-bromopropane would lead mainly to elimination. The use of a primary alkyl halide is necessary:

$$(CH_3)_2CHC{\equiv}CH \xrightarrow{NaNH_2} (CH_3)_2CHC{\equiv}CNa \xrightarrow{CH_3CH_2CH_2Br} (CH_3)_2CHC{\equiv}CCH_2CH_2CH_3$$

(c)

$$HC{\equiv}CH \xrightarrow{NaNH_2} HC{\equiv}CNa \xrightarrow{CH_3CH_2Br} HC{\equiv}CCH_2CH_3$$

(d) Two alternatives are possible:

$$CH_3CH_2CH_2C{\equiv}CH \xrightarrow{NaNH_2} CH_3CH_2CH_2C{\equiv}CNa \xrightarrow{CH_3CH_2Br} CH_3CH_2CH_2C{\equiv}CCH_2CH_3$$

or

$$CH_3CH_2C{\equiv}CH \xrightarrow{NaNH_2} CH_3CH_2C{\equiv}CNa \xrightarrow{CH_3CH_2CH_2Br} CH_3CH_2C{\equiv}CCH_2CH_2CH_3$$

7.10 **(a)** $CH_3CH_2CH_2C{\equiv}CH \xrightarrow[\text{1 equiv.}]{HBr} (CH_3CH_2CH_2)(Br)C{=}CH_2$

(b) $CH_3CH_2CH_2C{\equiv}CH \xrightarrow[\text{2. } Br_2 \text{; 3. } NaOCH_3,\ CH_3OH]{\text{1. catecholborane (benzo-1,3,2-dioxaborole, BH)}}$ (CH$_3$CH$_2$CH$_2$)(H)C=C(H)(Br)

Disiamyl borane could have been used instead of catecholborane in the initial hydroboration step.

(c)

$$CH_3CH_2CH_2C{\equiv}CCH_2CH_2CH_3 \xrightarrow[\text{2. } Cl_2 \quad \text{3. } NaOCH_3,\ CH_3OH]{\text{1. catecholborane (BH)}} \text{(H)(}CH_3CH_2CH_2\text{)C=C(}CH_2CH_2CH_3\text{)(Cl)}$$

(d)

$$C_6H_{11}C{\equiv}C{-}CH_2CH_3 \xrightarrow[\text{2. } Br_2 \quad \text{3. } NaOCH_3,\ CH_3OH]{\text{1. catecholborane (BH)}} \text{(H)(}C_6H_{11}\text{)C=C(}CH_2CH_3\text{)(Br)}$$

Disiamyl borane could have been used instead of catecholborane in the initial hydroboration step.

(e)

$$C_5H_9{-}CH_2{-}C{\equiv}CH \xrightarrow[\text{2. } H_2O \quad \text{3. } I_2,\ NaOH]{\text{1. catecholborane (BH)}} \text{(H)(}C_5H_9CH_2\text{)C=C(I)(H)}$$

(f)

$$CH_3CH_2C{\equiv}CCH_2CH_3 \xrightarrow[\text{1 equiv.}]{HBr} \text{(}CH_3CH_2\text{)(H)C=C(Br)(}CH_2CH_3\text{)}$$

Alternatively, direct addition of HBr (1 equiv) could be employed, but addition of two molecules of HBr would occur as a side reaction.

(g)

$$CH_3CH_2C{\equiv}CCH_2CH_3 \xrightarrow[CCl_4]{Br_2} \text{(}CH_3CH_2\text{)(Br)C=C(Br)(}CH_2CH_3\text{)}$$

7.11 Complete structural information is given for **D**, but you can readily deduce that compound **A** is a terminal alkyne from its reaction with CuCl in aqueous ammonia. Alkylation of **A** with methyl iodide means that **B** must have the partial structure

$$-C{\equiv}C{-}CH_3$$

Compounds **B**, **C**, and **D** must all have the same carbon skeleton, but the sodium/ammonia reduction of **B** causes loss of optical activity. This means that two different substituents on the chiral center of **B** become equivalent as a result of the reduction and that compound **B** already had a propenyl group. The sodium/ammonia reduction then generates a second, equivalent *E*-propenyl group:

$$-C{\equiv}C-CH_3 \rightarrow -CH{=}CH-CH_3$$

This allows the complete structures to be drawn for **A-C:**

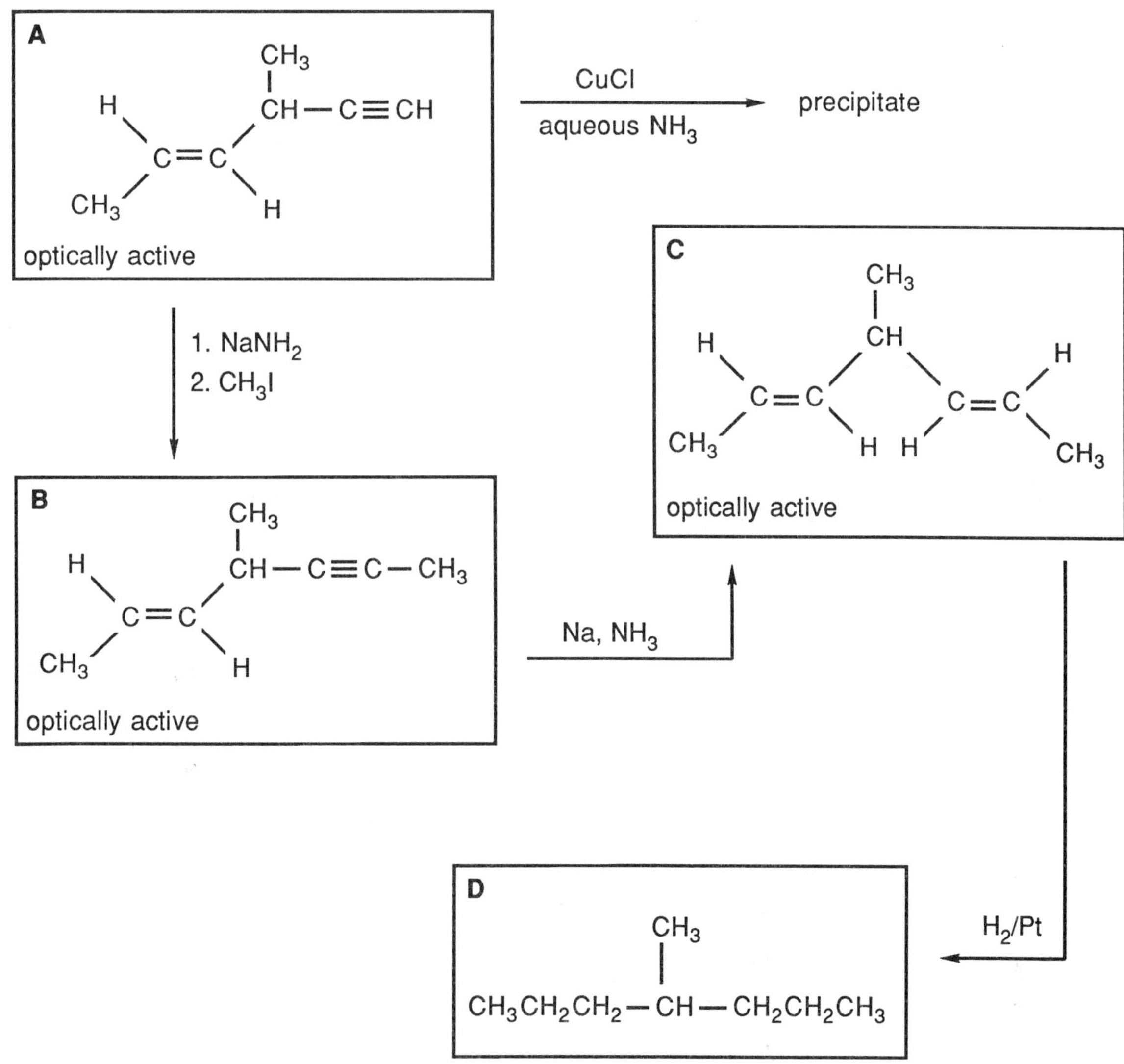

7.12 **(a)**

$$CH_3CH_2C\equiv CH \xrightarrow[\text{2. } H_2O_2\text{, NaOH}]{\text{1. catecholborane (BH)}} CH_3CH_2CH_2-\overset{O}{\overset{\|}{C}}-H$$

Alternatively, disiamylborane could be used in the hydroboration step.

(b)

$$CH_3CH{=}CH_2 \xrightarrow[\text{2. } NaNH_2\text{, } NH_3]{\text{1. } Br_2} CH_3C\equiv CH$$

(c)

$$(CH_3)(H)C{=}C(H)(CH_3) \xrightarrow[\text{2. } NaNH_2\text{, } NH_3]{\text{1. } Br_2} CH_3C\equiv CCH_3 \xrightarrow{H_2\text{, Pd-BaSO}_4} (CH_3)(H)C{=}C(CH_3)(H)$$

(d)

$$CH_3C\equiv CCH_3 \xrightarrow[\text{2. } Br_2 \text{; 3. } NaOCH_3\text{, } CH_3OH]{\text{1. catecholborane (BH)}} (CH_3)(H)C{=}C(Br)(CH_3)$$

Alternatively, direct addition of HBr could be use, but this reaction is unreliable and may not stop after monoaddition.

(e)

$$CH_3CH_2C\equiv CH \xrightarrow[\text{2. } H_2SO_4\text{, } HgSO_4]{\text{1. } H_2O} CH_3CH_2-\overset{O}{\overset{\|}{C}}-CH_3$$

(f)

$$CH_3C\equiv CCH_3 \xrightarrow[\text{2. } Br_2\text{, } CCl_4]{\text{1. } H_2\text{, Pd-BaSO}_4} CH_3CH(Br)CH(Br)CH_3$$

(g)

$$CH_3CH_2C{\equiv}CH \xrightarrow[\text{2. } H_2\text{, Pd-BaSO}_4 \text{ 3. Cold, alk. KMnO}_4]{\text{1. NaNH}_2\text{, NH}_3}$$ (HO, CH_2CH_3, HO, H, H, CH_2CH_3 diol)

Alternatively, reduction with Na or Li in ammonia to give the *E* alkene could be followed by epoxidation–hydrolysis to form the diol.

(h)

$$CH_3CH_2C{\equiv}CCH_2CH_3 \xrightarrow[\text{2. } H_2SO_4\text{, HgSO}_4]{\text{1. } H_2O} CH_3CH_2CH_2{-}\overset{O}{\overset{\|}{C}}{-}CH_2CH_3$$

(i)

$$CH_3CH_2C{\equiv}CH \xrightarrow[NH_4Cl]{Na,\ NH_3} CH_3CH_2C{=}CH_2 \xrightarrow[Zn(Cu)]{CH_2I_2}$$ CH_3CH_2-cyclopropane

(j)

Cyclohexyl–$\overset{O}{\overset{\|}{C}}$–$CH_2CH_3$ $\xrightarrow[\text{2. NaNH}_2\text{, NH}_3]{\text{1.PCl}_5}$ Cyclohexyl–$C{\equiv}CCH_3$

7.13 You know (Section 6.8) that free radical addition to an alkene proceeds by *anti* addition, and this is supported by the information that racemic 2,3-dibromobutane is formed from the addition of HBr/peroxides to (*Z*)-2-bromo2-butene:

(H, CH_3, C=C, CH_3, Br) ⟶ (H, CH_3, H, Br, CH_3, Br) + enantiomer

The Br and H are on opposite sides of the double bond in the *Z* isomer, so the original addition to the alkyne must also have proceeded via *anti* addition.

7.14 **(a)** diastereomers **(b)** same

(c) same (d) diastereomers

(e) diastereomers (f) diastereomers

(g) same (h) same

7.15 (a) Dissecting the target molecule into 5- and 3-carbon subunits does not help in this case; both fragments would be secondary alkyl groups, so joining them via an organocuprate coupling would not be expected to work well. The key functional group must be the cyclopropane ring, and a logical precursor is the alkene, which by reaction with $:CH_2$ would give the target molecule:

$$C_2H_5\text{-}CH{=}CH\text{-}CH(CH_3)_3$$

The preceding steps in the synthesis would then proceed via a typical alkyne sequence. Note that the alkene must have the *E* configuration.

$$HC{\equiv}CCH(CH_3)_2 \xrightarrow[\text{2. } CH_3CH_2Br]{\text{1. } NaNH_2} CH_3CH_2C{\equiv}CCH(CH_3)_2 \xrightarrow{Na,\ NH_3}$$

(E)-$CH_3CH_2(H)C{=}C(H)CH(CH_3)_2$ $\xrightarrow[\text{Zn-Cu}]{CH_2I_2}$ trans-1-ethyl-2-isopropylcyclopropane (CH_3CH_2, $CH(CH_3)_2$)

(b) Dissection of the target molecule into 4- and 3-carbon fragments affords one primary and one secondary alkyl group. When an alkyne is alkylated (with a primary halide) and then reduced, the net result is coupling of two *primary* alkyl groups. Consequently, an organocuprate reaction is necessary in this case. The secondary organocuprate must react with the primary alkyl halide:

$$CH_3CH_2CH(CH_3)Br \quad + \quad 2Li \longrightarrow CH_3CH_2CH(CH_3)Li \quad + \quad LiBr$$

$$2\ CH3CH2CH(CH3)Li \quad + \quad CuI \longrightarrow [CH_3CH_2CH(CH_3)]_2Li \quad + \quad LiI$$

$$[CH_3CH_2CH(CH_3)]_2Li \quad + \quad CH_3CH_2CH_2Br \longrightarrow CH_3CH_2CH(CH_3)CH_2CH_2CH_3 \quad + \quad CH_3CH_2CH(CH_3)Cu \quad + \quad LiBr$$

(c) This target molecule presents two problems: one of carbon-carbon bond formation and one of stereochemistry. In order to control the stereochemistry of the two adjacent chiral centers, you would want to introduce the hydroxyl groups into the corresponding alkene. The alkene could be prepared from the alkyne, which still would have four carbon atoms, so introduction of at least one methyl group would be necessary. In the following synthetic scheme the *Z* alkene undergoes *syn* hydroxylation with $KMnO_4$ (or OsO_4). Alternatively, the *E* alkene could be subjected to *anti* hydroxylation via epoxidation–hydrolysis.

$$HC{\equiv}CCH_3 \xrightarrow[\text{2. } CH_3Br]{\text{1. } NaNH_2} CH_3C{\equiv}CCH_3 \xrightarrow{H_2/Pd\text{-}BaSO_4} (Z)\text{-}CH_3CH{=}CHCH_3 \xrightarrow{KMnO_4} CH_3CH(OH)CH(OH)CH_3$$

(d) While the target molecule has only 9 carbons, the presence of the cyclopropyl ring makes it impossible to disect it into two fragments with 5 or fewer carbons. Assuming the cyclopropyl ring is formed by carbene addition to an alkene (which in turn is formed by reduction of the alkyne), you need only decide which alkyl group to introduce. As always it must be a primary alkyl group, so the isopropyl group must be present in your original starting material.

$$(CH_3)_2CHC{\equiv}CH \xrightarrow[\text{2. } CH_3CH_2CH_2Br]{\text{1. } NaNH_2} (CH_3)_2CHC{\equiv}CCH_2CH_2CH_3 \xrightarrow{H_2/Pd\text{-}BaSO_4} (Z)\text{-}(CH_3)_2CHCH{=}CHCH_2CH_2Cl \xrightarrow[Zn\text{-}Cu]{CH_2I_2} cis\text{-}1\text{-}(CH_3)_2CH\text{-}2\text{-}CH_2CH_2CH_3\text{-cyclopropane}$$

7.16 **(a)** The key functional group is the carbon–carbon double bond, which has three 3-carbon substituents. The overall synthetic problem can therefore be reduced to how these three propyl groups can be introduced onto an unsaturated 2-carbon starting material. Two of them could be introduced by alkylation of acetylene, and the final carbon–carbon bond be formed by an organocopper coupling reaction:

$$HC{\equiv}CH \xrightarrow[2.\ CH_3CH_2CH_2Br]{1.\ NaNH_2} HC{\equiv}CCH_2CH_2CH_3 \xrightarrow[2.\ CH_3CH_2CH_2Br]{1.\ NaNH_2} CH_3CH_2CH_2C{\equiv}CCH_2CH_2CH_3$$

$$CH_3CH_2CH_2C{\equiv}CCH_2CH_2CH_3 \xrightarrow[2.\ Br_2;\ 3.\ NaOCH_3,\ CH_3OH]{1.\ \text{catecholborane (C}_6\text{H}_4\text{O}_2\text{BH)}} (H)(CH_3CH_2CH_2)C{=}C(CH_2CH_2CH_3)(Br)$$

$$(H)(CH_3CH_2CH_2)C{=}C(CH_2CH_2CH_3)(Br) \xrightarrow[2.\ CuI;\ 3.\ CH_3CH_2CH_2Br]{1.\ Li} (H)(CH_3CH_2CH_2)C{=}C(CH_2CH_2CH_3)(CH_2CH_2CH_3)$$

(b) The CCl_2 group can be generated by addition of HCl an alkyne (or by PCl_5 reaction with a ketone that could be formed by hydration of an alkyne). The key is to avoid an alkyne that would yield two isomeric products, and a symmetrical alkyne will solve this problem. 4-Octyne can be prepared as shown in the preceding synthesis and then treated with HCl as follows:

$$CH_3CH_2CH_2C{\equiv}CCH_2CH_2CH_3 \xrightarrow{HCl} CH_3CH_2CH_2{-}CCl_2{-}CH_2CH_2CH_3$$

(c) The carbonyl group is the key functional group, and it is presumably generated by hydration of an alkyne. The obvious solution of alkylating propyne with a 4-carbon unit would not work, because the C_4 group is tertiary. Hence the carbon-carbon bond formation must occur by a method other than alkylation of a terminal alkyne. The use of lithium di-*tert*-butylcuprate and 3-bromo-1-propyne would afford the necessary intermediate:

$$(CH_3)_3C{-}Br \xrightarrow[2.\ CuI]{1.\ Li} ((CH_3)_3C)_2{-}CuLi \xrightarrow{BrCH_2C\equiv CH} (CH_3)_3CCH_2C\equiv CH$$

$$(CH_3)_3CCH_2C\equiv CH \xrightarrow{H_2O,\ H_2SO_4,\ HgSO_4} (CH_3)_3CCH_2{-}\overset{\overset{\displaystyle O}{\|}}{C}{-}CH_3$$

(d) The key functional group corresponds to the adjacent carbon atoms bearing chloro and methoxy groups. *Anti* addition of these substituents requires the *E* alkene. The alkene can be prepared by reduction of the corresponding alkene, and this in turn can be formed by alkylating acetylene twice with ethyl iodide:

$$HC\equiv CH \xrightarrow[2.\ CH_3CH_2I]{1.\ NaNH_2} HC\equiv CCH_2CH_3 \xrightarrow[2.\ CH_3CH_2I]{1.\ NaNH_2} CH_3CH_2C\equiv CCH_2CH_3$$

$$CH_3CH_2C\equiv CCH_2CH_3 \xrightarrow{Na,\ NH_3} (E)\text{-}CH_3CH_2CH{=}CHCH_2CH_3 \xrightarrow{CH_3OH,\ HCl} CH_3CH_2CH(OCH_3){-}CH(Cl)CH_2CH_3$$

7.17 **(a)** The double bond can be generated easily by reduction of the corresponding alkyne, and this in turn could be prepared by alkylation of acetylene with a 6-carbon alkyl halide:

$$HC\equiv CH \xrightarrow[2.\ \text{cyclopentyl-}CH_2CH_2Br]{1.\ NaNH_2} \text{cyclopentyl-}CH_2C\equiv CH \xrightarrow{H_2/Pd\text{-}BaSO_4} \text{cyclopentyl-}CH_2CH{=}CH_2$$

(b) The compound is symmetrical and can, moreover, be dissected into 5- and

3-carbon fragments. But you should remember that an alkane prepared by alkylation of a terminal alkyne will always have the subunit, $-CH_2CH_2CH_2-$. Therefore this compound cannot be prepared via alkylation of an alkyne, so an organocuprate must be employed. In order to have a primary alkyl halide in the reaction with the organocopper reagent, the organocuprate must be prepared from an isopropyl halide.

$$CH_3CH(CH_3)Br + 2Li \longrightarrow CH_3CH(CH_3)Li + LiBr$$

$$2\ CH3CH(CH3)Li + CuI \longrightarrow [CH_3CH(CH_3)]_2Li + LiI$$

$$[CH_3CH(CH_3)]_2Li + CH_3CH(CH_3)CH_2CH_2Br \longrightarrow CH_3CH(CH_3)CH_2CH_2CH(CH_3)_2 \quad CH_3CH(CH_3)Cu + LiBr$$

(Alternatively, isobutyl bromide could be used as a single starting material that would provide both reagents in the coupling reaction.)

(c) The key functional group corresponds to the adjacent carbon atoms with bromo substituents. Since the bromines must be introduced by an *anti* addition of bromine, the immediate precursor of the target molecule must be (*E*)-2-hexene. This could be prepared by metal/ammonia reduction of 2-hexyne, which in turn could be made from an allowed C_4 compound by alkylation of propyne:

$$HC{\equiv}CH \xrightarrow[2.\ CH_3Br]{1.\ NaNH_2} CH_3C{\equiv}CH \xrightarrow[2.\ CH_3CH_2CH_2Br]{1.\ NaNH_2} CH_3C{\equiv}CCH_2CH_2CH_3 \xrightarrow{Na,\ NH_3} (E)\text{-}CH_3CH_2CH_2CH{=}CHCH_3 \xrightarrow{Br_2}$$

[Structures: (E)-2-hexene with H, CH_3, $CH_3CH_2CH_2$, H on C=C; Br₂ addition gives the dibromide drawn in two equivalent sawhorse forms, each with Br, H, CH_3 on one carbon and Br, H, $CH_2CH_2CH_3$ on the other.]

(d) The target molecule contains 13 carbon atoms, so at least two carbon–carbon bonds must be generated in the synthesis. The key functional group is the carbon–carbon double bond, and one of the substituents is a primary alkyl group that could have been introduced via alkylation of a terminal alkyne. The other group (cyclopentyl) is secondary and could not be introduced by reaction of a sodium acetylide. It could, however, be introduced via an organocuprate reaction using an alkenyl halide. In the following sequence we have shown as the final step the coupling of lithium dicyclopentylcuprate with (*Z*)-1-bromo-3-cyclopentyl-1-propene. (It would also be possible to prepare the organocopper reagent from the alkenyl halide and allow it to react with bromocyclopentane. Recall from Section 5.14 that *cyclo*alkyl halides give satisfactory yields in this reaction.)

HC≡CH → (1. $NaNH_2$; 2. $BrCH_2$–cyclopentyl) → HC≡C—CH_2–cyclopentyl → (1. catecholborane (O, BH, O); 2. Br_2; 3. $NaOCH_3$, CH_3OH) → (H)(Br)C=C(H)(CH_2–cyclopentyl) → ((cyclopentyl)$_2$CuLi) → (H)(cyclopentyl)C=C(H)(CH_2–cyclopentyl)

7.18 Structural information is available in two places in this problem. The ozonolysis reaction shows that compound **A** has two double bonds, one of which must be in a ring (because only two products are formed). Reconstruction of the starting diene affords three possibilities for **A** (labeled **A-1**, **A-2**, and **A-3**):

A-1: cyclobutene–CH_2CH_2—CH=CH_2

A-2: cyclohexene–CH=CH_2

A-3: cyclopentene–CH_2—CH=CH_2

The other main piece of structural information comes from the alkylation of compound **C** with sodium acetylide. The product, **D,** must have the partial structure $-CH_2-C{\equiv}CH$, and compound **C** must be a primary alkyl bromide, $R-CH_2-Br$. The R group of **C** has only 5 carbons, so you can now rule out **A-2** (with a 6-membered ring) as a possibility for **A.** The ring size for **A** is therefore either 5 or 4, and the hydrogenation experiments allow you to choose between them. Reduction of **A** to **B** under ordinary conditions gives the same product as reduction of **D** at high temperature and pressure. This rules out the presence of a 4-membered ring (which would be cleaved under those conditions), and you can rule out **A-1.** Structure **A-3** is therefore the correct structure for **A,** and the R group of **C** is cyclopentyl. The structures of **A–D** are:

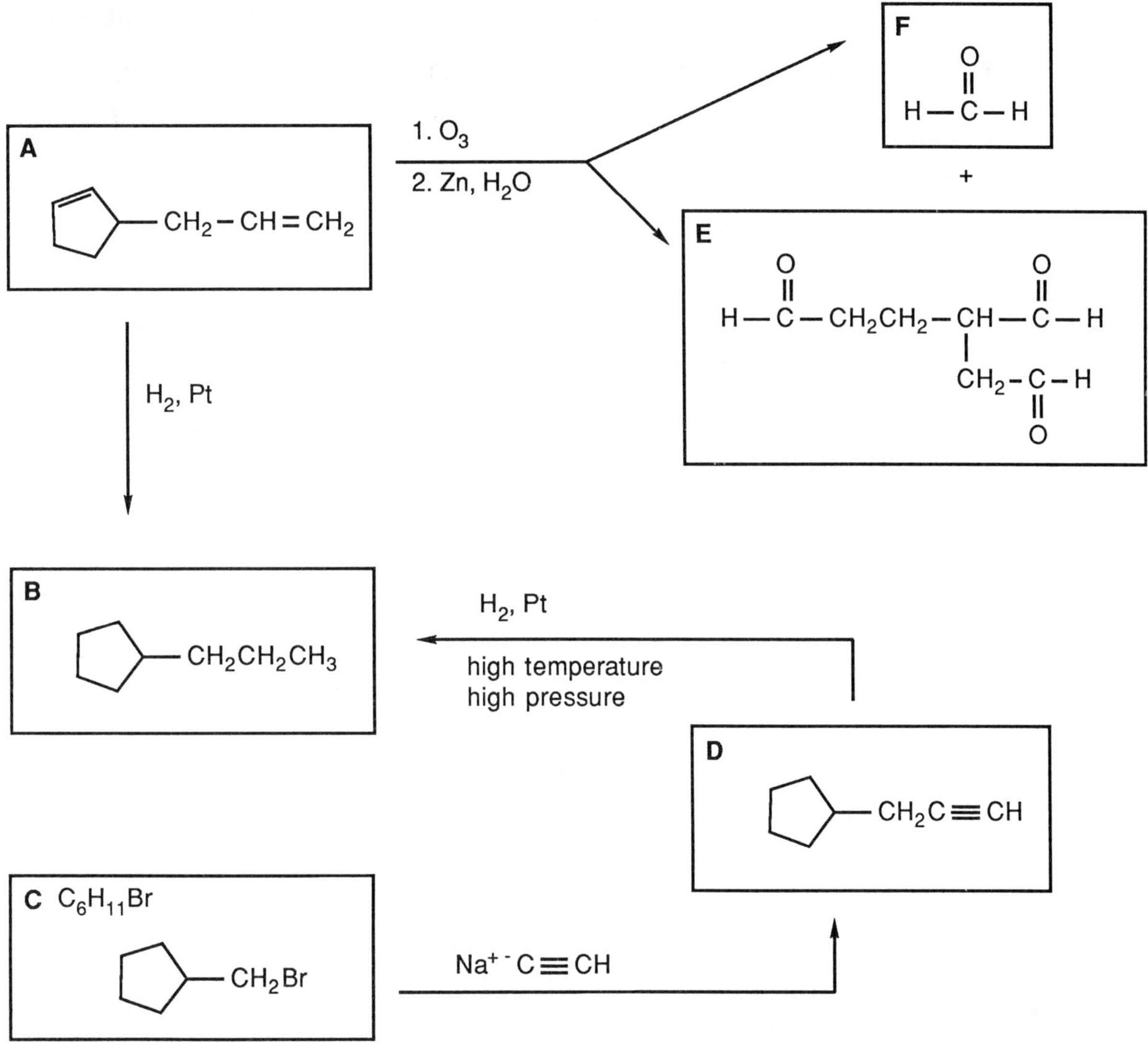

7.19 The structural information in this problem is scattered among the various reactions. Compound **A** has two degrees of unsaturation, and it must be a primary alkyl bromide because it can be alkylated by the sodium salt of propyne. Compound **E** has a single degree of unsaturation, and this must be a double bond because it reacts with diiodomethane and zinc-copper couple; this reaction affords a cyclopropane. The formation of **E** by hydroboration of **F** demonstrates that **F** is an alkyne (but not a terminal alkyne, because it does not yield a precipitate with

CuCl/ammonia). The hydroboration reaction tells you that the stereochemistry of the double bond in **E** has the *Z* configuration, and that the substituents on the cyclopropane ring of **D** must be *cis*.

The next level of structural deduction comes from the information on optical activity. Reduction of **B** to either **C** or **D** causes lack of optical activity, and this has major significance. Remember that there are only a few ways that optical activity can be destroyed: by breaking (and reforming) one of the bonds to the chiral center, by formation of a second (constitutionally equivalent) chiral center, or by causing two originally different substituents on a chiral center to become equivalent. No bonds would be broken in the reduction reactions, so simple racemization cannot be the explanation. The sodium/ammonia reduction converts a triple bond to a double bond (with the *E* configuration), and this provides the key: *compound* **B** must already have a chiral center with an *E* double *bond*! Generation of a second such group results in symmetry and the loss of optical activity. You will quickly see that this idea is also supported by the hydrogenation reaction. Both the double and triple bonds would be converted to equivalent $-CH_2CH_2-$ group, and this also results in the loss of optical activity.

The next step in deducing the structures comes from a consideration of the molecular formula and symmetry. You now know that **D** (which is a disubstituted cyclopropane) must be symmetrical, so the two R groups must be the same. One of the R groups is derived from the alkylation of a (primary) bromide with the sodium salt of propyne. This generates a

$$-CH_2-C\equiv C-CH_3$$

group, and its reduction must afford

$$-CH_2CH_2CH_2CH_2$$

The two R groups of **D** must therefore be butyl groups, and deduction of the remaining structures is straightforward. The complete answer is shown below:

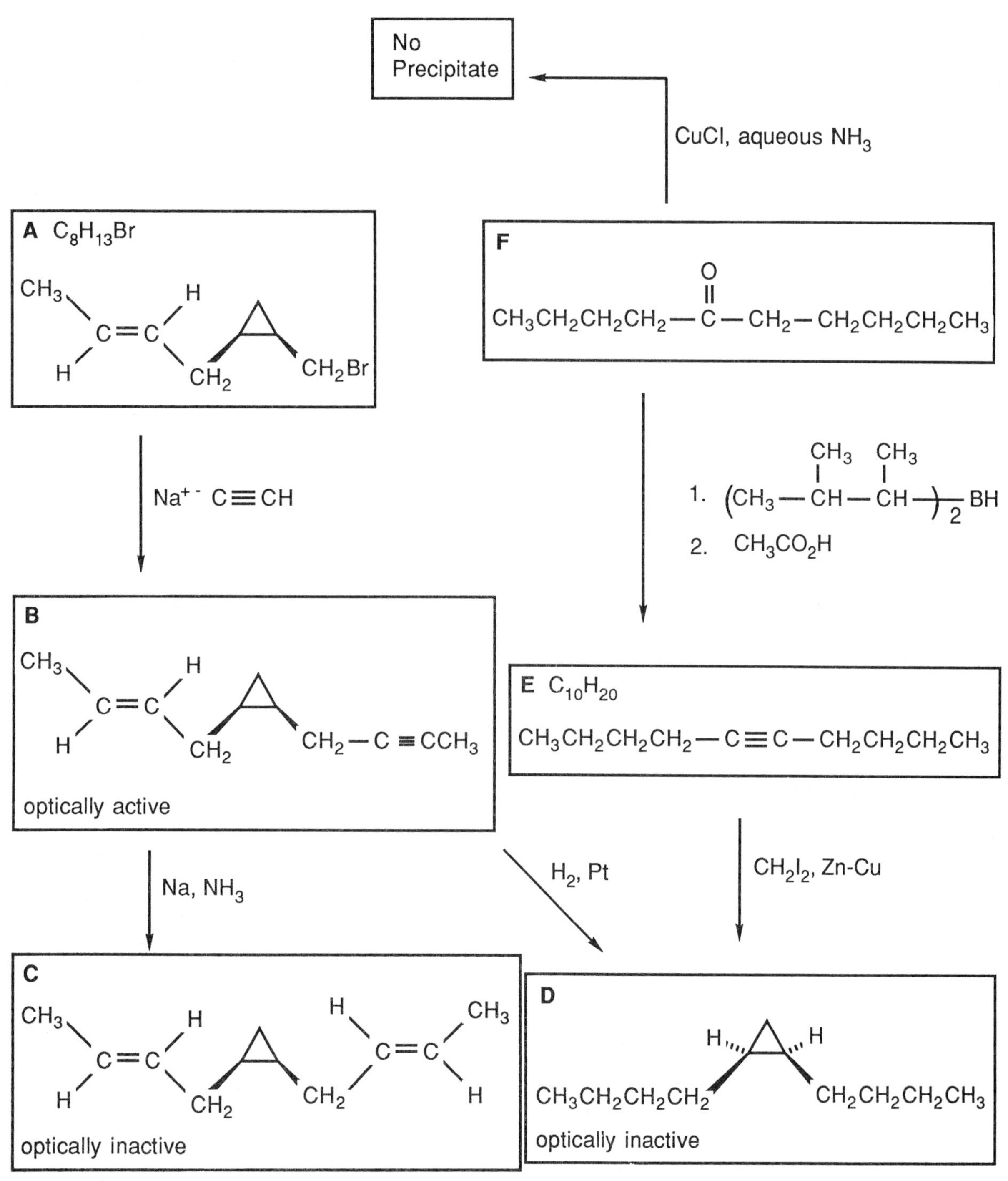

7.20 Compound **A** has three degrees of unsaturation, and the formation of a precipitate with alcoholic silver nitrate tells you that two of the correspond to a terminal alkyne subunit. The other must be either a ring or a double bond. Alkylation with **B**

followed by two reduction steps affords 2-methyl-7-ethylnonane, and this defines the carbon skeletons of all of the compounds. (Note that catalytic hydrogenation under mild conditions would not cleave a 3- or 4-membered ring, so these could not have been present.) Dissection of **E** into 7- and 5-carbon fragments could be done in two ways, but only one would correspond to the fragments needed for alkylation of a terminal alkyne with a primary alkyl halide. You know from its molecular formula that **B** is 1-bromo-3-methylbutane, which is achiral. Consequently, the optical activity of **C** and **D** must result from compound **A** (which must also be optically active, although this is not stated in the problem). The remaining degree of unsaturation must be a double bond as stated above, and this must be situated in such a way that the molecule has a chiral center. There is only one possibility:

$$\begin{array}{c} \quad\quad\quad CH_3CH_2 \\ \quad\quad\quad | \\ CH_2{=}CH{-}CH{-}C{\equiv}CH \end{array}$$

Compounds **C** and **D** retain their optical activity, but once reduction of the double bond occurs, the original chiral center has two identical substituents (ethyl groups) and **E** is optically inactive. The complete structures of **A-E** are shown below:

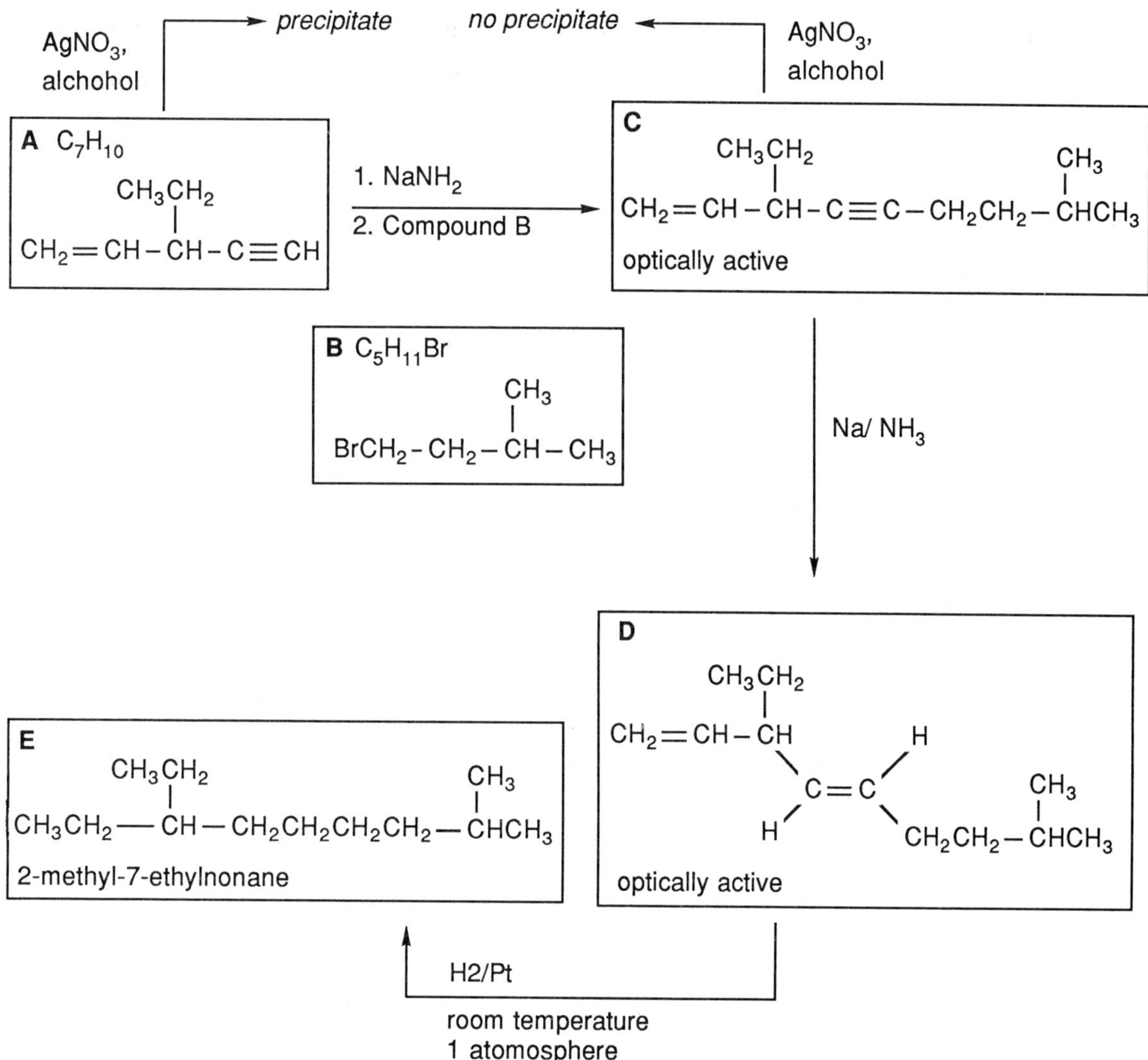

7.21 To answer this problem you must first recognize that the primary structural information is the molecular structure shown for **K**. Ozonolysis of **K** affords two products, one of which is acetone (and this is confirmed by the structure shown in the ozonolysis of **E**). The other product is a keto dialdehyde:

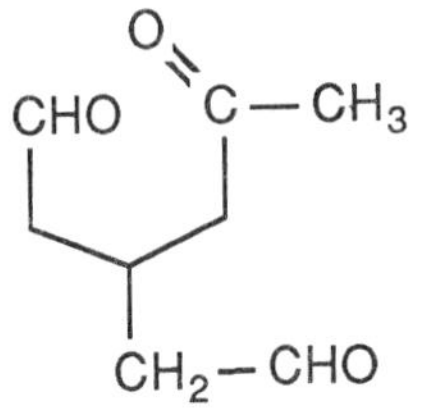

This compound is symmetrical (there are two $-CH_2-CHO$ groups), so "recombination" of either aldehyde group with the keto group would regenerate **K.** This allows you to deduce the structure of **A,** which also affords **B** and **C** upon ozonolysis; **A** must correspond to "recombination" of the two aldehyde groups:

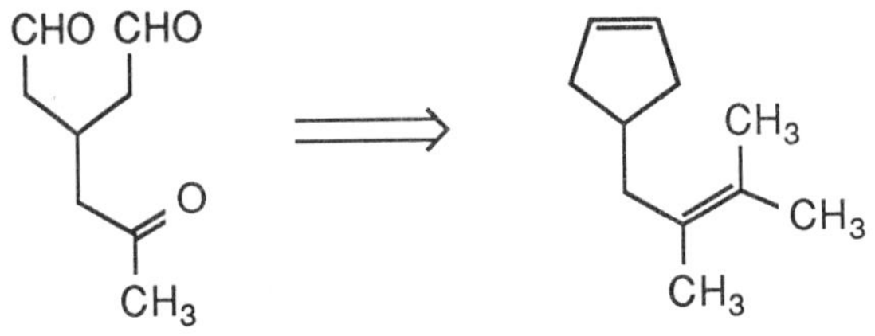

C

Hydrogenation of **A** would yield (2,3-dimethylbutyl)cyclopentane, and the carbon skeleton of **E** must be the same as that of **A.** The formation of acetone (**B**) in the ozonolysis of **E** means that both **E** and **A** share the feature of a $CH{=}C(CH_3)_2$ group. Only two products are formed in the ozonolysis of **E,** so if it has a second multiple bond it must be in the 5-membered ring.

In order to deduce the structure of **E,** you must utilize the information provided by the molecular formulas and reactions of compounds **H** and **I.** Compound **H** has two degrees of unsaturation, and after hydrogenation to compound **I,** only one remains; this must be the 5-membered ring. Now you know that **E** has only the single double bond discussed above and is an isomer of **H,** because **H** is converted to **E** by addition of HCl followed by elimination of HCl. Only two chlorides could give **E** upon treatment with KOH/ethanol, and these are compounds **F** and **G** (it does not matter which one you designate **F**). The reasonable assumption is that **F** and **G** result from addition of HCl to either end of a tetrasubstituted double bond:

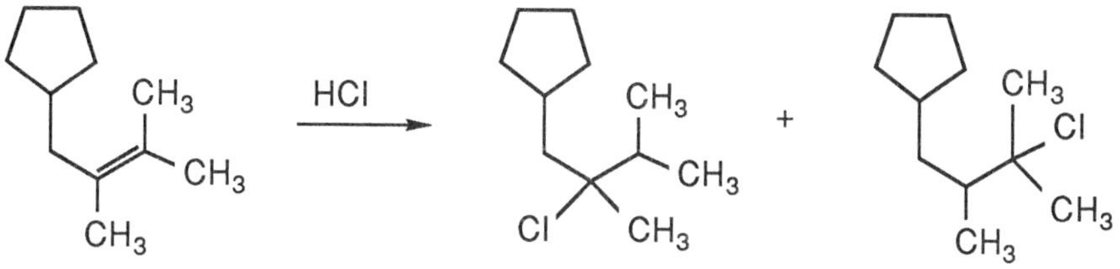

This reaction would certainly proceed as indicated but the alkene cannot be compound **H**, because that structure has already been assigned to **E**. The only possible explanation is that a skeletal rearrangement occurred during the addition of HCl to **H**; this is confirmed by the observation that hydrogenation of these two yields **I** and **D**, which are different compounds. To determine the nature of the rearrangement, you must evaluate the carbocations that lead to **F** and **G**. These are labeled (**F**$^+$) and (**G**$^+$) in the following drawings:

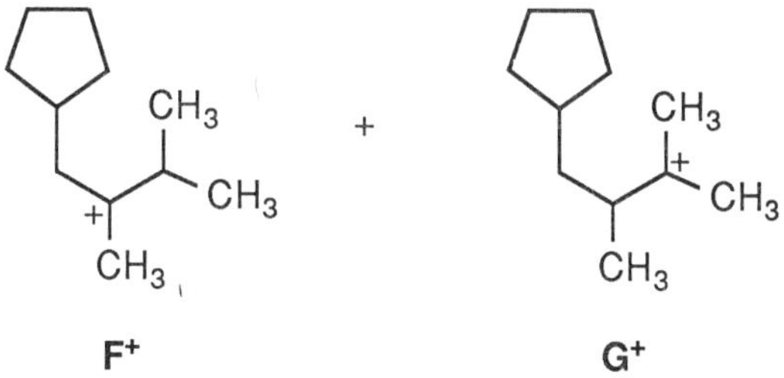

F$^+$ **G**$^+$

Remember that a carbocation rearrangement leads to a more stable carbocation in which the migrating alkyl group winds up on a carbon atom *adjacent* to the electron deficient center. For cation **F**$^+$ this could be either a methyl or a cyclopentyl group, corresponding to the two following cations as possible precursors:

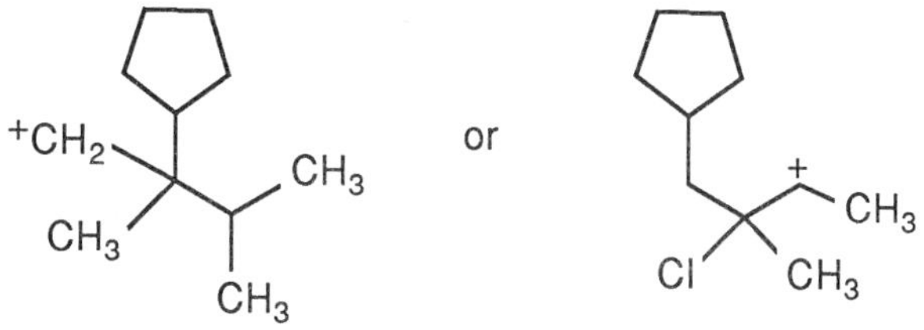

or

For cation **G**$^+$ this could be either a methyl or a C_5H_9-CH_2 group, corresponding to the two following cations as possible precursors:

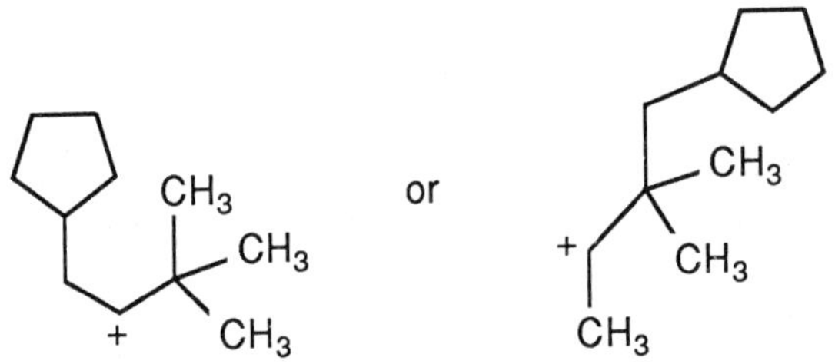

CH3 CH3 + CH3 or CH3 + CH3 CH3

Notice that one cation would be a common precursor to *both* $\mathbf{F}^+$ and $\mathbf{G}^+$, so this must be the actual intermediate. It in turn is clearly derived from protonation of a $-CH{=}CH_2$ group, so it is now possible to write the complete structure for **H:**

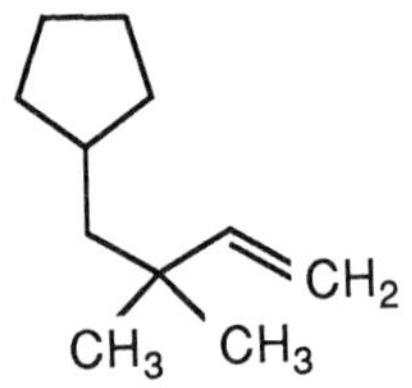

CH2 CH3 CH3

Finally, we have shown below the correct structures of **A–J**:

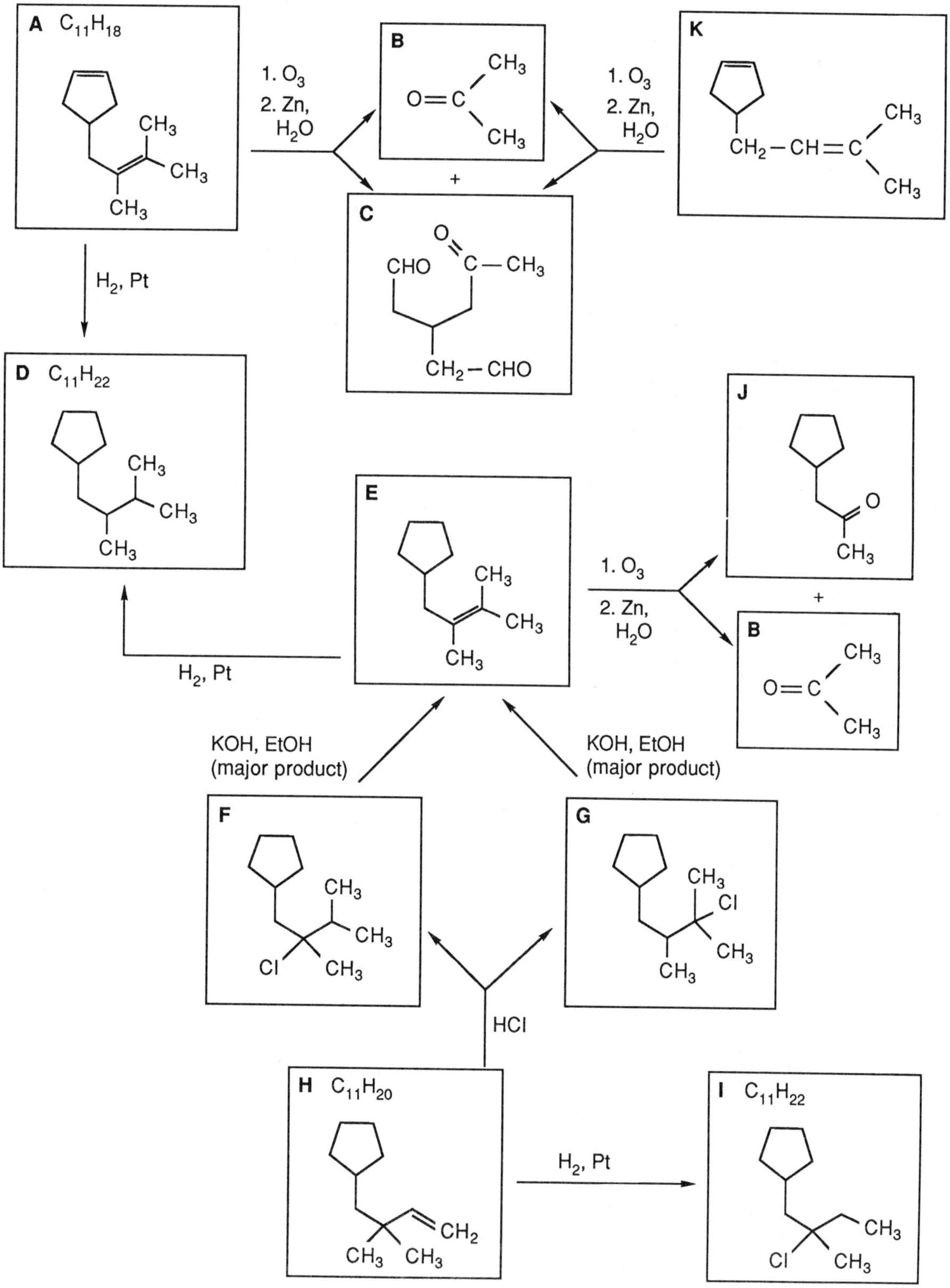
A $C_{11}H_{18}$
CH3
CH3
CH3
1. O3
2. Zn, H2O
B
O=C
CH3
CH3
+
C
CHO
O
C—CH3
CH2—CHO
1. O3
2. Zn, H2O
K
CH2—CH=C
CH3
CH3
H2, Pt
D $C_{11}H_{22}$
CH3
CH3
CH3
E
CH3
CH3
CH3
H2, Pt
J
O
CH3
1. O3
2. Zn, H2O
+
B
O=C
CH3
CH3
KOH, EtOH
(major product)
KOH, EtOH
(major product)
F
CH3
CH3
Cl
CH3
G
CH3
Cl
CH3
CH3
HCl
H $C_{11}H_{20}$
CH2
CH3
CH3
H2, Pt
I $C_{11}H_{22}$
CH3
Cl
CH3

7.22 From its reaction with sodium acetylide you know that **A** is a primary alkyl chloride. A second alkylation affords **C**, which must have the following structure:

$$C_5H_7\text{-}CH_2\text{-}C{\equiv}C\text{-}CH_2\text{-cyclohexyl}$$

Catalytic hydrogenation at high temperature affords **D**, which has two degrees of unsaturation. One of these is the cyclohexyl ring, and the other must be a 5-membered ring (because a 3- or 4-membered ring would have been cleaved by the hydrogenation reaction). Compound **D** is therefore:

$$\text{cyclopentyl-}CH_2\text{-}CH_2CH_2\text{-}CH_2\text{-cyclohexyl}$$

Compounds **A**, **B**, and **C** all have a second degree of unsaturation, which corresponds to a double bond in the 5-membered ring. The location of the double bond can be deduced from the optical activity of these compounds, since only one of the three possible locations would lead to a 5-membered ring that contains a chiral center.

Alternatively, you could deduce the complete structures of **C** and **E** from the ozonolysis information. From the construction of compound **C**, you can deduce that **F** must be:

$$\text{cyclohexyl-}CH_2\text{-}CHO$$

Reconstruction of the other fragment of **E** from the tricarbonyl compound **G** affords three possibilities: **E-1**, **E-2**, and **E-3**:

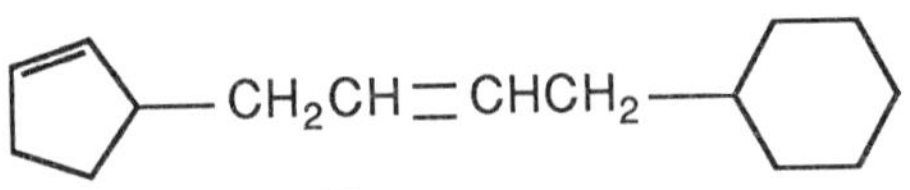

$CH_2CH{=}CHCH_2$ (cyclopentenyl … cyclohexyl)

E-1

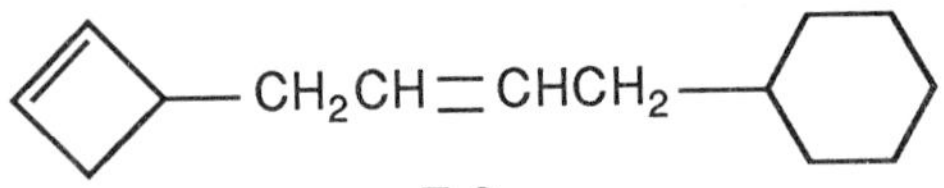

$CH_2CH{=}CHCH_2$ (cyclobutenyl … cyclohexyl)

E-2

$CH_2CH{=}CHCH_2$ (cyclohexenyl … cyclohexyl)

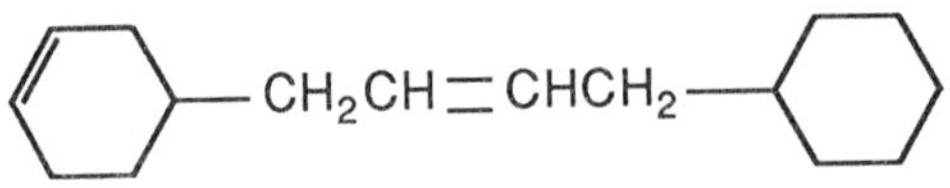

E-3

You can rule out **E-3** by referring back to the conversion of **A** to **B**, because the C_6H_9 group must be primary. The 4-membered ring in **E-2** allows you to discount it, so the correct structure of **E** must be **E-3.** This allows the complete structures of **A-F** to be drawn:

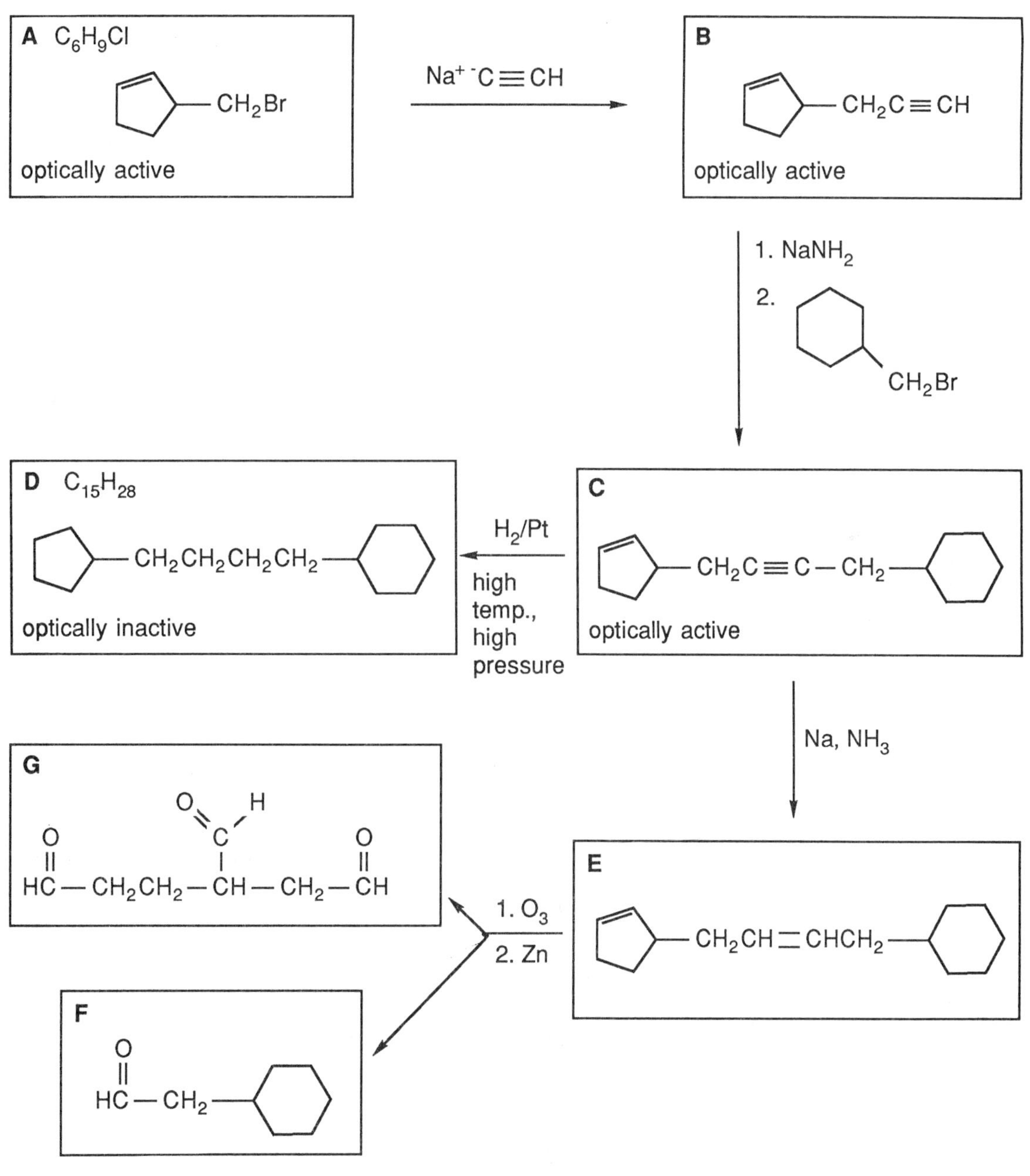

CHAPTER 8. COMPOUNDS WITH MORE THAN ONE REACTIVE CENTER

ANSWERS TO EXERCISES

8.1 (a) $CH_3-CH_2-\dot{C}(CH_3)-CH=CH_2 \longleftrightarrow CH_3-CH_2-C(CH_3)=CH-\dot{C}H_2$

(b) $CH_3-CH_2-C(CH_3)=CH-\dot{C}H_2 \longleftrightarrow CH_3-CH_2-\dot{C}(CH_3)-CH=CH_2$

(c) $\dot{C}H_2-C(CH_3)=CH-CH_2-CH_3 \longleftrightarrow CH_2=C(CH_3)-\dot{C}H-CH_2-CH_3$

(d) $CH_3-C(CH_3)=CH-\dot{C}H-CH_3 \longleftrightarrow CH_2-\dot{C}(CH_3)-CH=CH-CH_3$

8.2

(a) $CH_3-\bar{C}H-CH=CH_2$

(b) $CH_2=C(CH_3)-\overset{+}{C}H-CH_3$

(c) $CH_3-\overset{+}{C}(CH_3)-CH=CH_3$

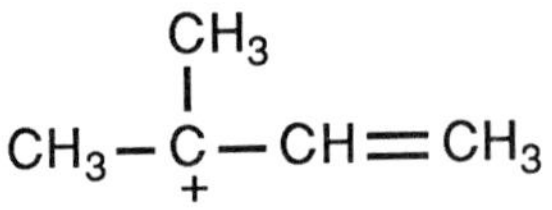

(d) $CH_3-\dot{C}(CH_3)-C(CH_3)=CH_2$

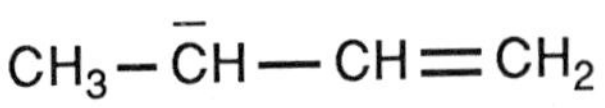

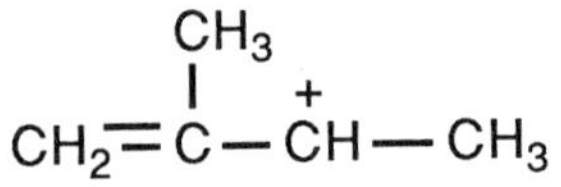

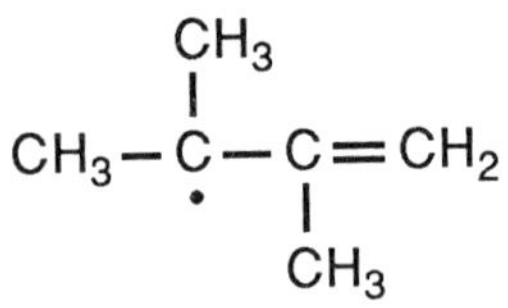

(e) 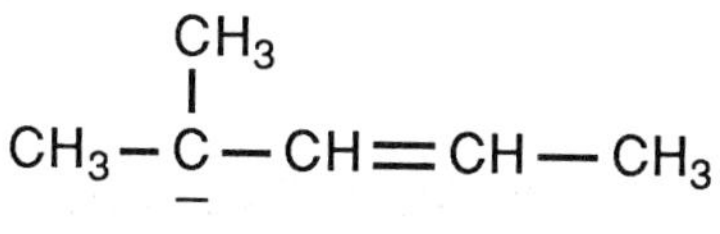 $CH_3-\overset{\overset{CH_3}{|}}{\underset{-}{C}}-CH{=}CH-CH_3$ (f) 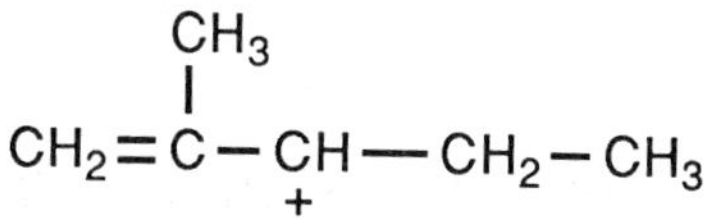 $CH_2{=}\overset{\overset{CH_3}{|}}{C}-\underset{+}{CH}-CH_2-CH_3$

8.3 From cyclohexene:

Br

Br

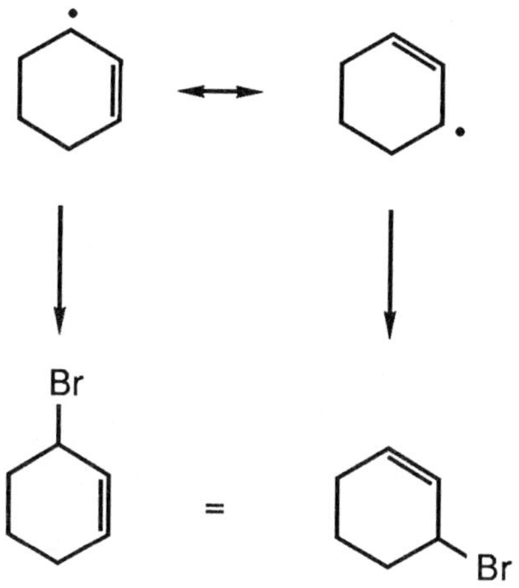

From bicyclic alkene:

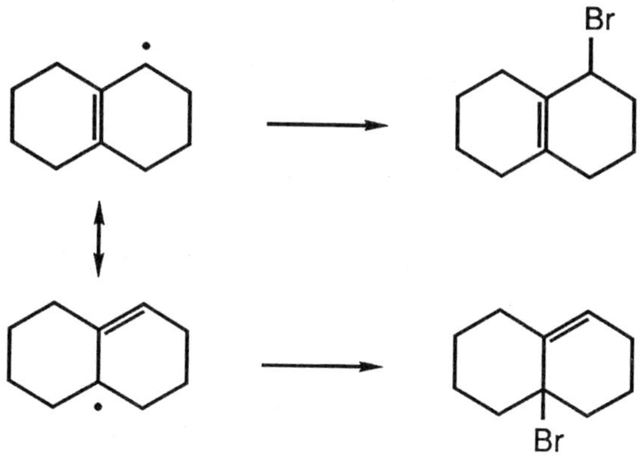

Br

Br

From 2-heptene:

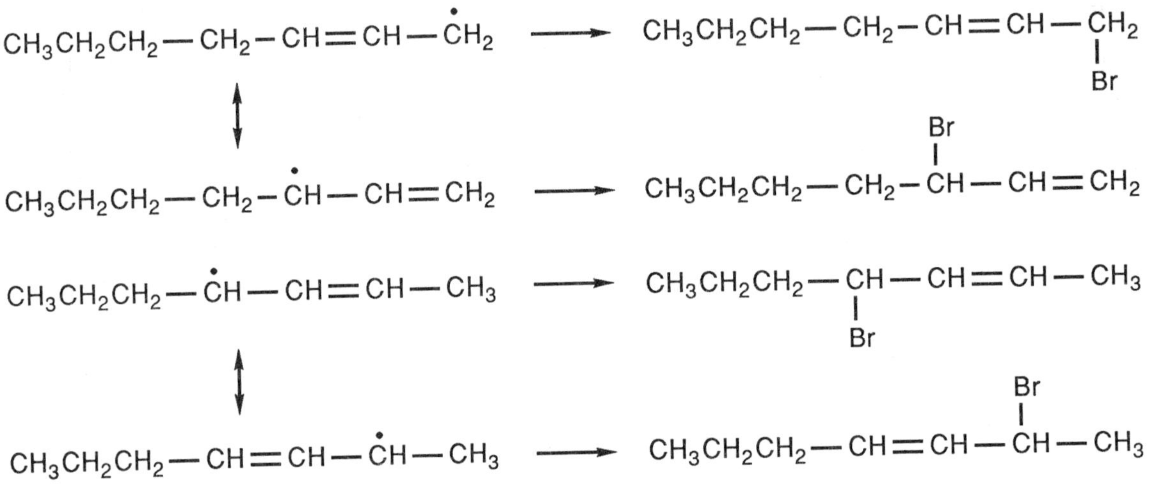

$$CH_3CH_2CH_2-CH_2-CH=CH-\dot{C}H_2 \longrightarrow CH_3CH_2CH_2-CH_2-CH=CH-\underset{\displaystyle Br}{\underset{|}{CH_2}}$$

$$\updownarrow$$

$$CH_3CH_2CH_2-CH_2-\dot{C}H-CH=CH_2 \longrightarrow CH_3CH_2CH_2-CH_2-\overset{\displaystyle Br}{\overset{|}{CH}}-CH=CH_2$$

$$CH_3CH_2CH_2-\dot{C}H-CH=CH-CH_3 \longrightarrow CH_3CH_2CH_2-\underset{\displaystyle Br}{\underset{|}{CH}}-CH=CH-CH_3$$

$$\updownarrow$$

$$CH_3CH_2CH_2-CH=CH-\dot{C}H-CH_3 \longrightarrow CH_3CH_2CH_2-CH=CH-\overset{\displaystyle Br}{\overset{|}{CH}}-CH_3$$

8.4 Part 1, bromination with NBS. Remember that you can predict the products on the basis of an intermediate allylic free radical.

(a)

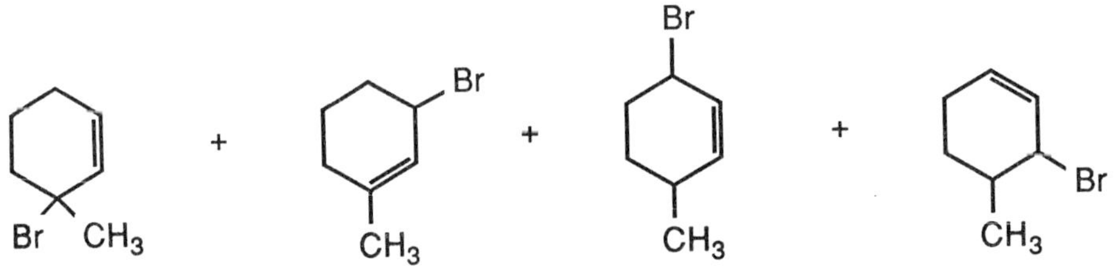

Br CH_3 + Br CH_3 + Br CH_3 + Br CH_3

(b)

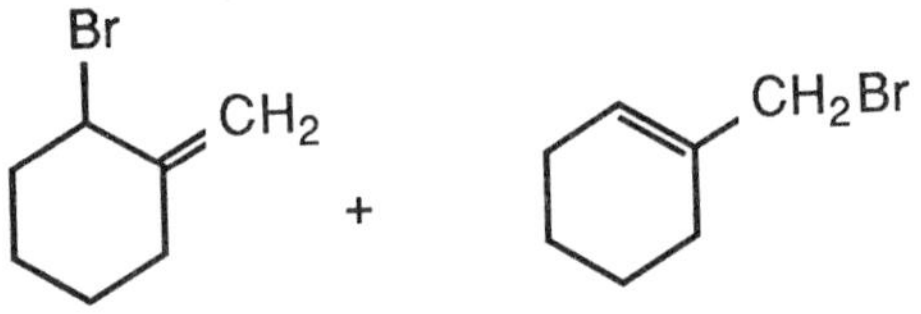

Br CH_2 + CH_2Br

(c)

$$CH_3CH_2-CH=CH-\underset{\displaystyle Br}{\underset{|}{C}}HCH_3 \quad + \quad CH_3CH_2-\underset{\displaystyle Br}{\underset{|}{CH}}-CH=CHCH_3$$

Part 2, oxidation with SeO_2. Remember that a free allylic radical is not formed and the double bond remains in its original position.

(a)

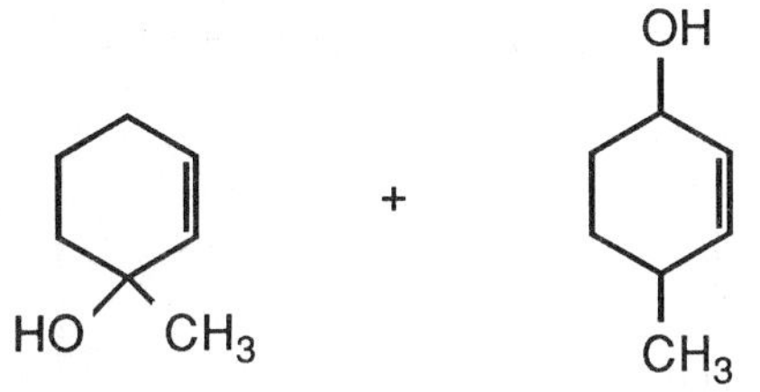

HO CH_3 + OH CH_3

(b)

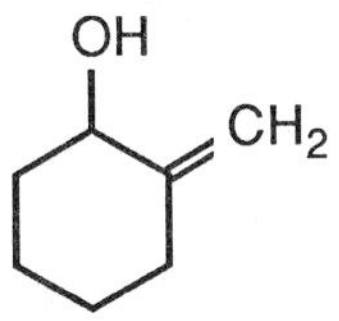

OH CH_2

(c) $CH_3CH_2{-}CH{=}CH{-}CH(OH)CH_3$

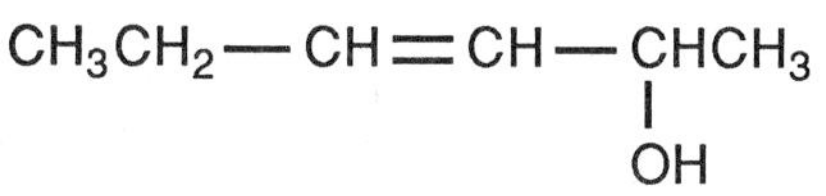

8.5 **(a)** Attack of Br^+ on 2,4-hexadiene:

$$CH_3{-}CH(Br){-}\overset{+}{C}H{-}CH{=}CH{-}CH_3 \longleftrightarrow CH_3{-}CH(Br){-}CH{=}CH{-}\overset{+}{C}H{-}CH_3$$

Attack of H^+ on 1,3-pentadiene:

$$CH_3{-}\overset{+}{C}H{-}CH{=}CH{-}CH_3 \longleftrightarrow CH_3{-}CH{=}CH{-}\overset{+}{C}H{-}CH_3$$

If protonation had instead occurred at the 4-position of 1,3-pentadiene, the resulting allylic cation would have had resonance contributors that were secondary and primary (rather than both secondary):

$$CH_2{=}CH{-}\overset{+}{C}H{-}CH_2{-}CH_3 \longleftrightarrow \overset{+}{C}H_2{-}CH{=}CH{-}CH_2{-}CH_3$$

(b) 1,2-addition:

$$CH_2{=}CH{-}CH{=}CH{-}CH_3 \xrightarrow{H^+} CH_3{-}\overset{+}{C}H{-}CH{=}CH{-}CH_3 \longrightarrow CH_3{-}\underset{}{\overset{Cl}{\overset{|}{C}H}}{-}CH{=}CH{-}CH_3$$

1,4-addition:

$$CH_2{=}CH{-}CH{=}CH{-}CH_3 \xrightarrow{H^+} CH_3{-}\overset{+}{C}H{-}CH{=}CH{-}CH_3 \longleftrightarrow CH_3{-}CH{=}CH{-}\overset{+}{C}H{-}CH_3 \longrightarrow CH_3{-}CH{=}CH{-}\overset{Cl}{\overset{|}{C}H}{-}CH_3$$

Both modes of addition yield 4-chloro-2-butene.

8.6 **(a)** $C_6H_5{-}\overset{OCH_3}{\overset{|}{C}H}{-}CH_2Br$ **(b)** $C_6H_5{-}CH{=}CH_2$

(c) [Stereostructures: OH, H, CH$_3$ / OH, H, C$_6$H$_5$] + [CH$_3$, H, OH / C$_6$H$_5$, H, OH]

(d) [Cyclopropane: C$_6$H$_5$, H, CH$_3$, H] + [Cyclopropane: H, C$_6$H$_5$, H, CH$_3$]

8.7 (a)

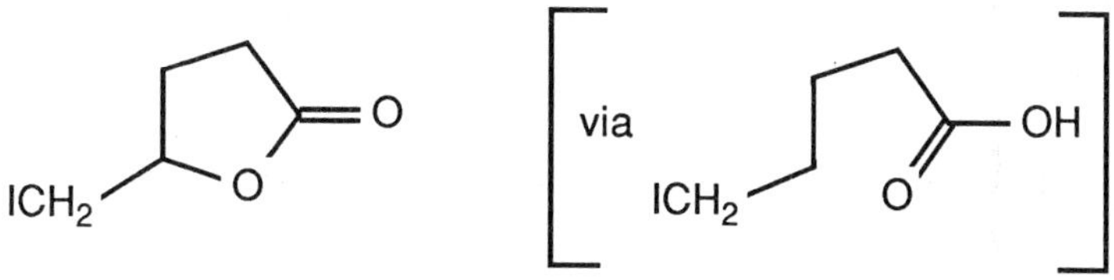

ICH_2–(γ-lactone, ring =O) [via $ICH_2CH_2CH_2CH_2C(=O)OH$]

(b)

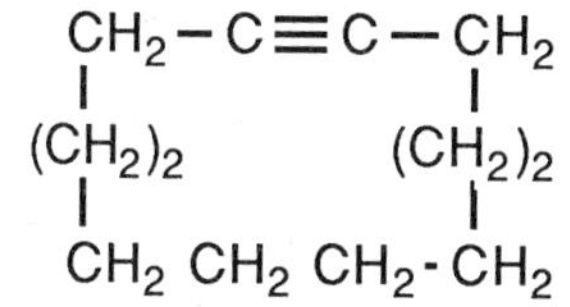

$CH_2-C\equiv C-CH_2$
$(CH_2)_2$ $(CH_2)_2$
CH_2 CH_2 CH_2-CH_2

(c)

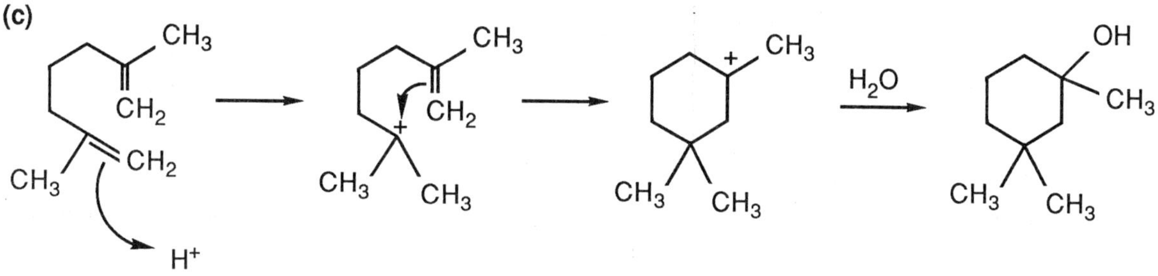

CH_3, CH_2, CH_3, CH_2, H^+ ⟶ CH_3, CH_2, +, CH_3, CH_3 ⟶ +, CH_3, CH_3, CH_3 —H_2O⟶ OH, CH_3, CH_3, CH_3

(d)

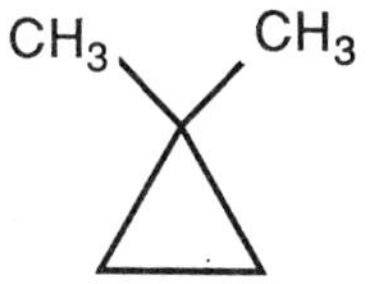

CH_3 CH_3 (1,1-dimethylcyclopropane)

ANSWERS TO PROBLEMS

8.1 (a)

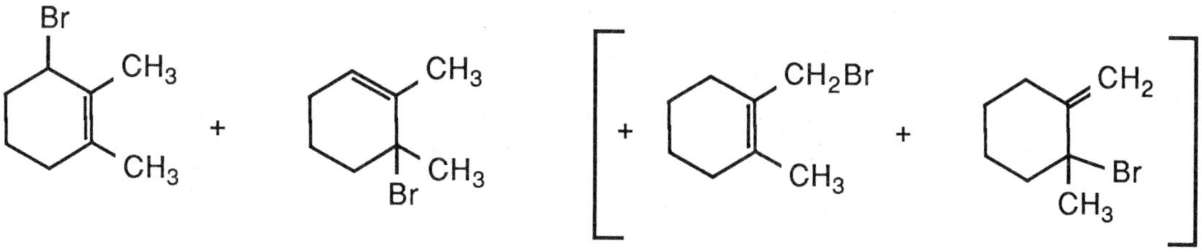

Br, CH_3, CH_3 + CH_3, CH_3, Br [+ CH_2Br, CH_3 + CH_2, Br, CH_3]

(b) 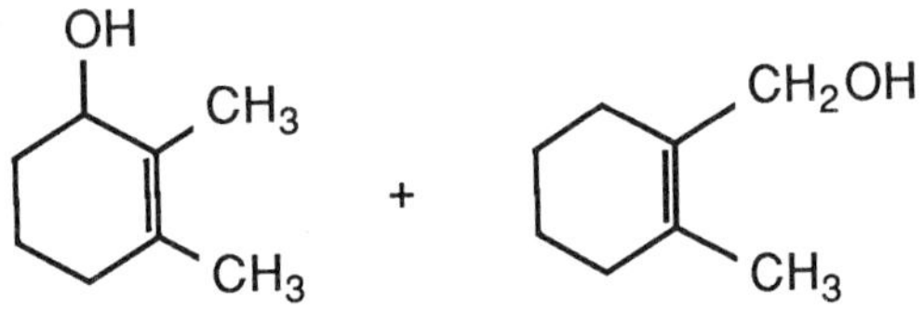 OH, CH3, CH3 + CH2OH, CH3

(c) 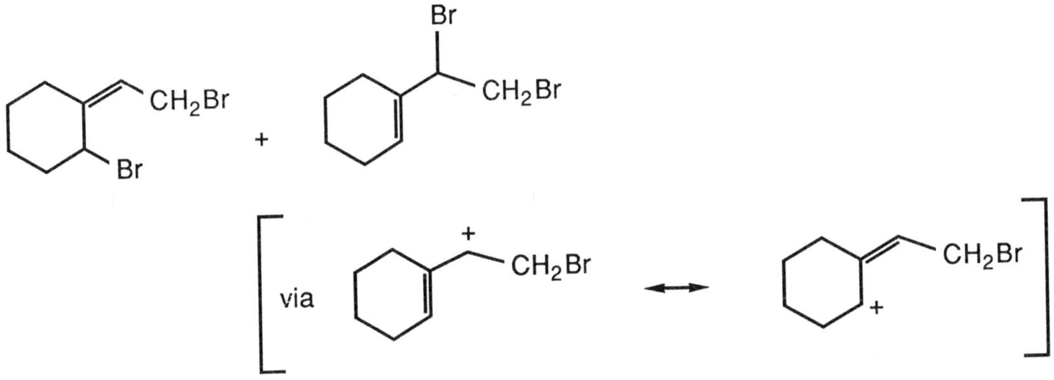 CH2Br, Br + Br, CH2Br

[via + CH2Br ⟷ CH2Br +]

(d) 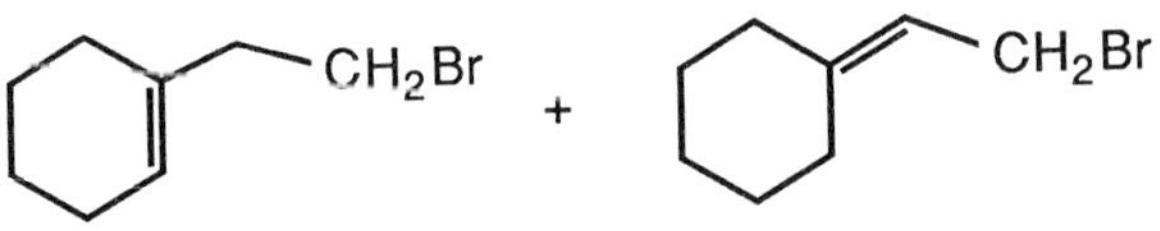 CH2Br + CH2Br

(e) 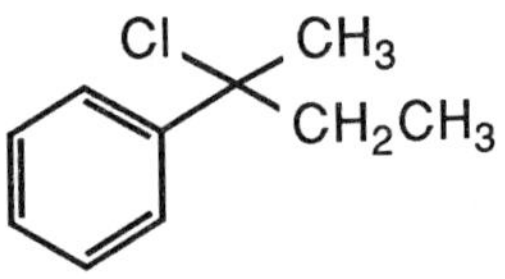 Cl, CH3, CH2CH3

(f) 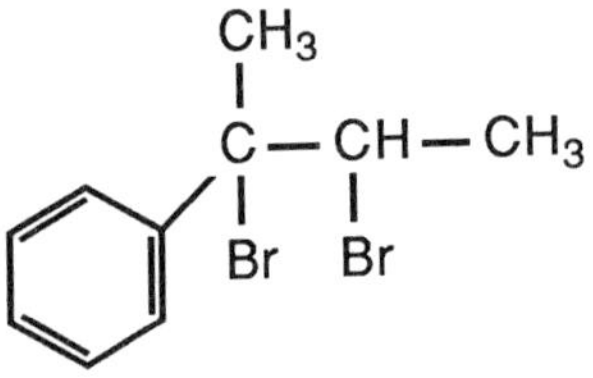 CH3, C—CH—CH3, Br Br

8.2 (a)  O

(b)

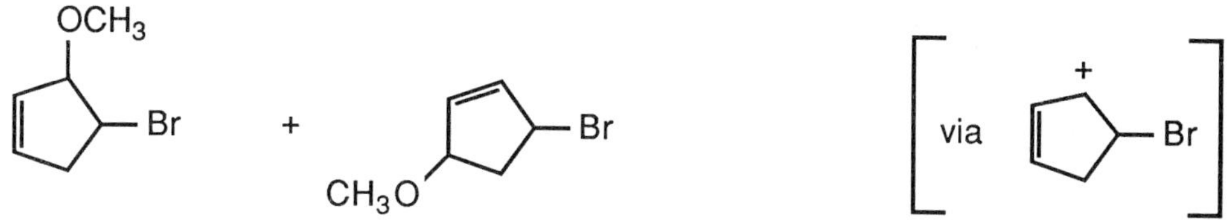

OCH3 — Br + Br — CH3O

[via + — Br]

(c) Br + Br

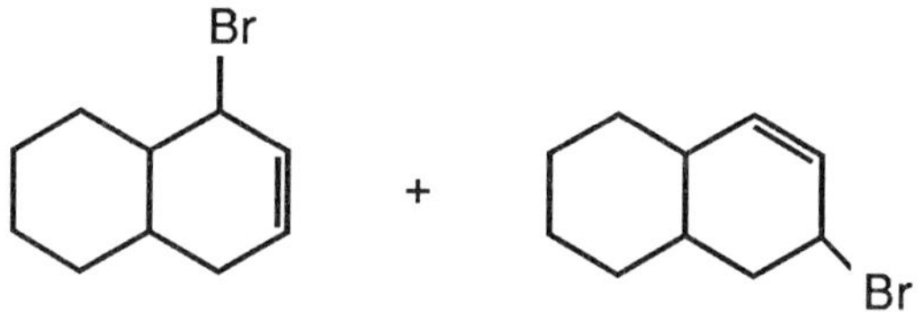

(d) OH

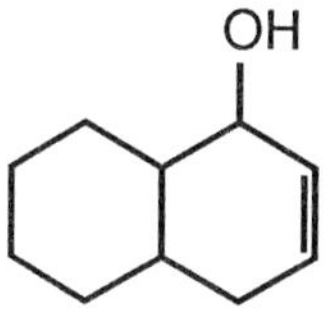

(e) O ‖ C — CH3

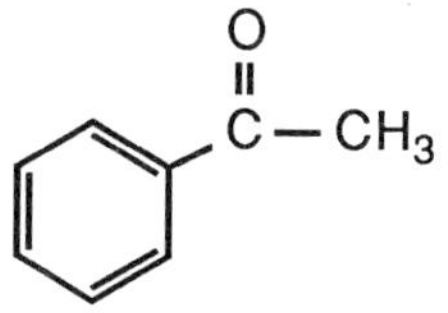

(f) CH3, CH3 — Br, CH2 + CH3, CH3 — CH2Br + CH3, BrCH2 — CH2

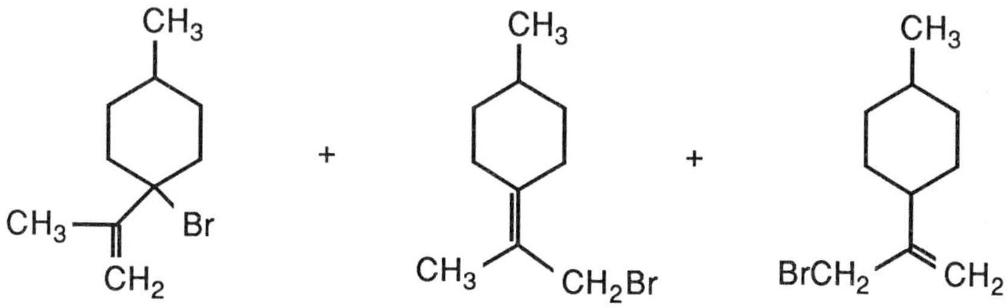

8.3 (a) CH_3

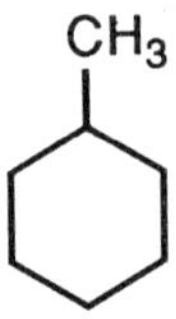

(b) Cl CH_3 + CH_3 Cl [via CH_3 + via]

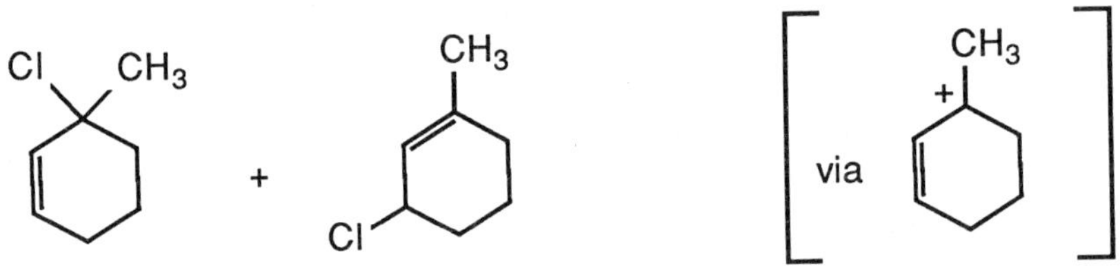

(c) CH_3 Cl Cl + Cl CH_3 Cl [via CH_3 Cl +]

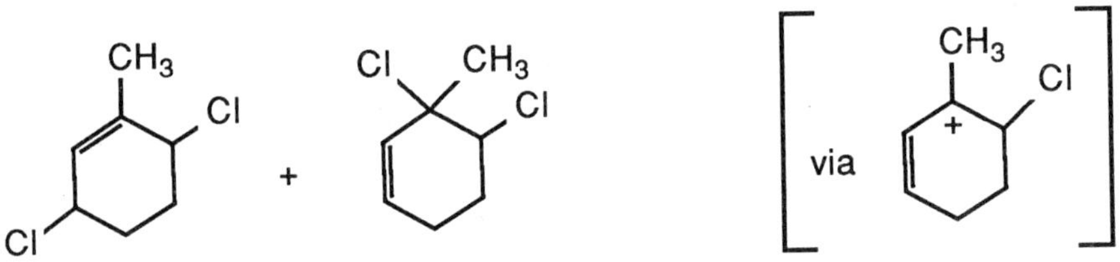

(d) OH CH — CH_2CH_3

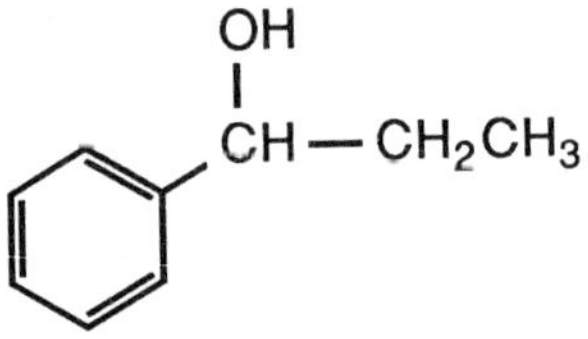

(e) O O I

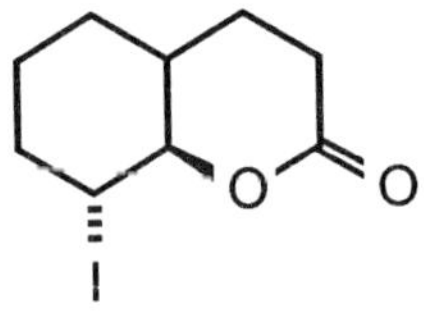

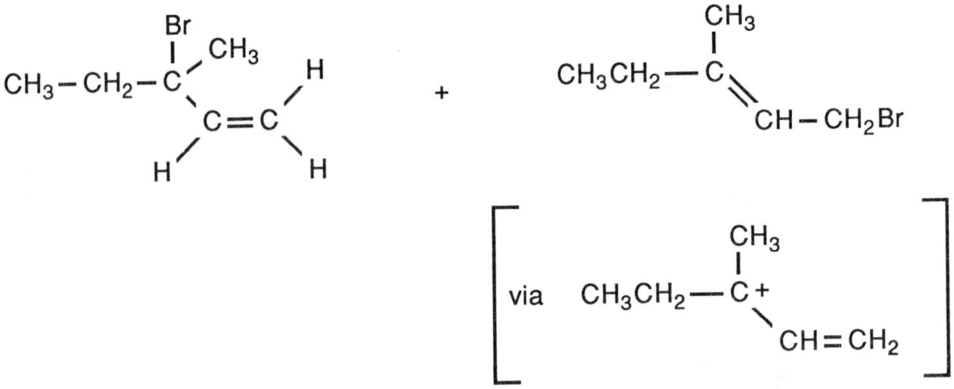

(f) Br CH_3 CH_3-CH_2-C C=C H H H + CH_3 CH_3CH_2—C CH — CH_2Br

[via CH_3 CH_3CH_2—C+ $CH=CH_2$]

8.4 **(a)** Reaction should occur at the less substituted of the two double bonds.

$$(CH_3)_2C{=}CH{-}CH_2{-}CH_2{-}CH_2{-}CH_2OH$$

(b) Reaction should occur at the more substituted of the two double bonds.

$$(CH_3)_2\overset{O}{C{-}CH}{-}CH_2{-}CH_2{-}CH{=}CH_2$$

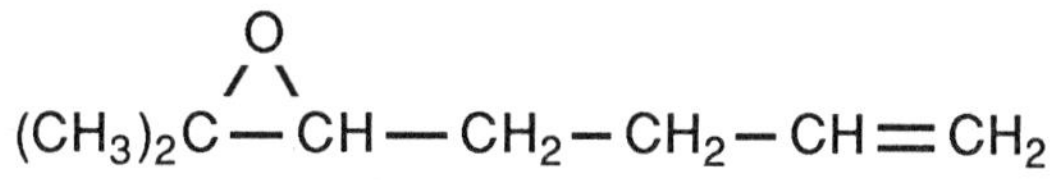

(c) Reaction should occur at both double bonds to generate a cyclic boron compound.

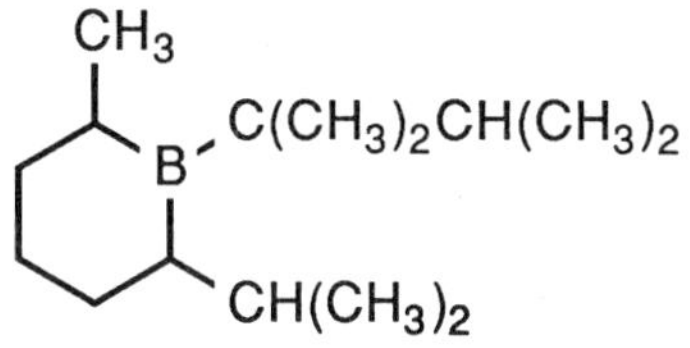

CH3, B, C(CH3)2CH(CH3)2, CH(CH3)2

(d) Reaction should occur at the less substituted of the two double bonds.

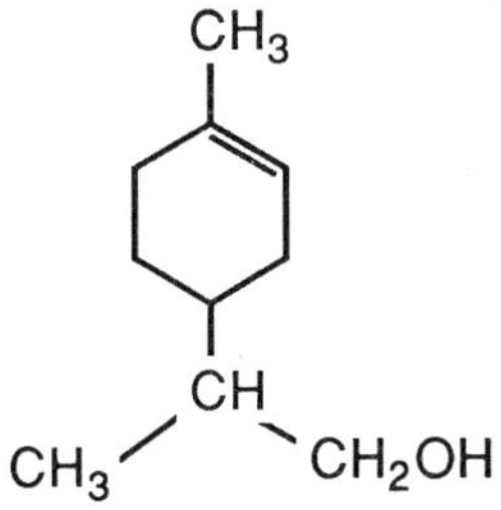

CH3, CH, CH3, CH2OH

(e)

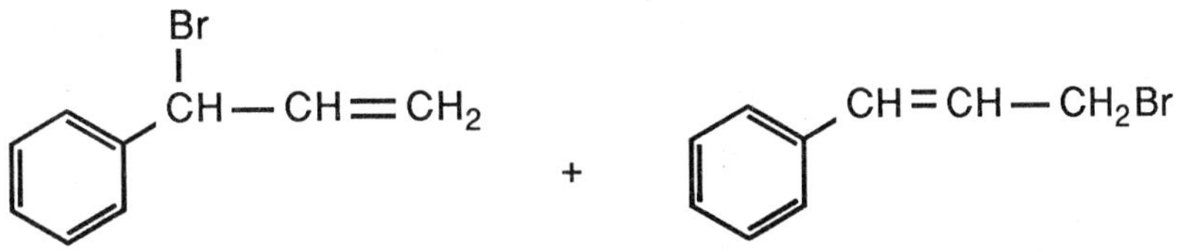

Br, CH—CH=CH2 + CH=CH—CH2Br

(f)

OH, CH—CH=CH2

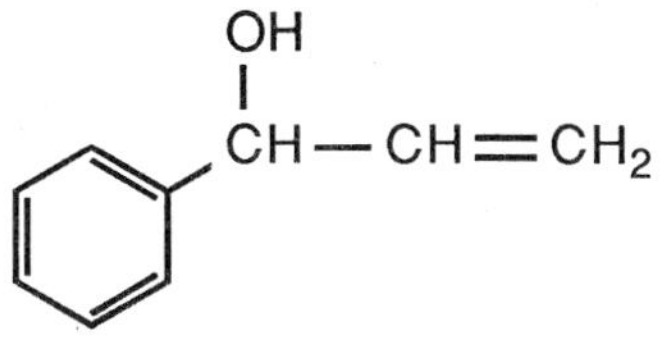

8.5 (a)

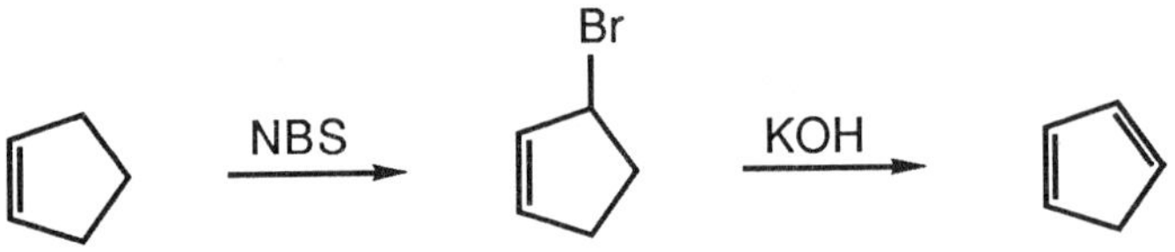

NBS → (Br) → KOH →

(b) NBS

(c) Na, NH_3

8.6 (a)

(b) $CH_2CH_2CO_2H$

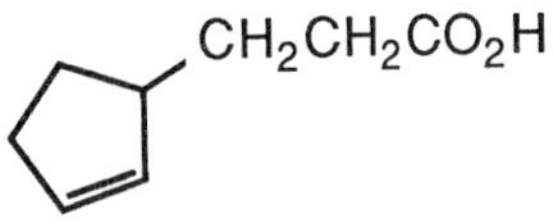

(c) The molecular formula fits either a cyclic alkene or an acyclic diene, so several alternatives are possible:

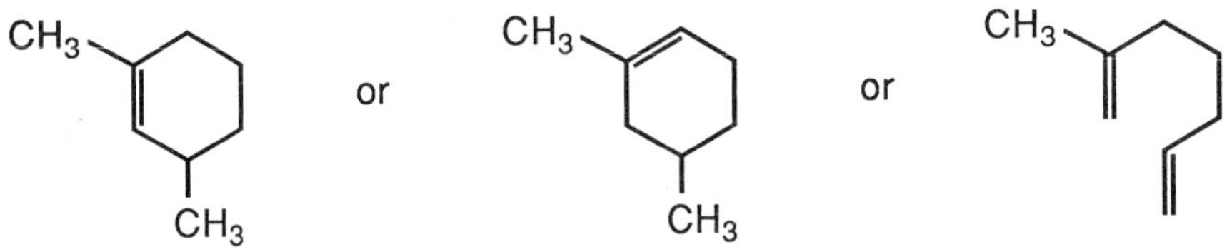

CH_3 CH_3 or CH_3 CH_3 or CH_3

(d) The reaction must have been addition of HBr to a conjugated diene. The double bond and the bromo-substituted carbon are three of the original four doubly-bonded carbons, and the fourth is now either the methyl-substituted carbon or the allylic CH_2. The latter (1-methyl-1,3-cyclohexadiene) would have protonated exclusively at the 4-position, so the reactant must have been:

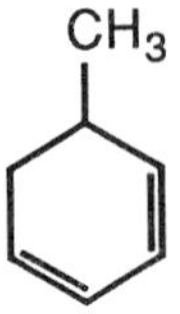

CH_3

8.7 **(a)**

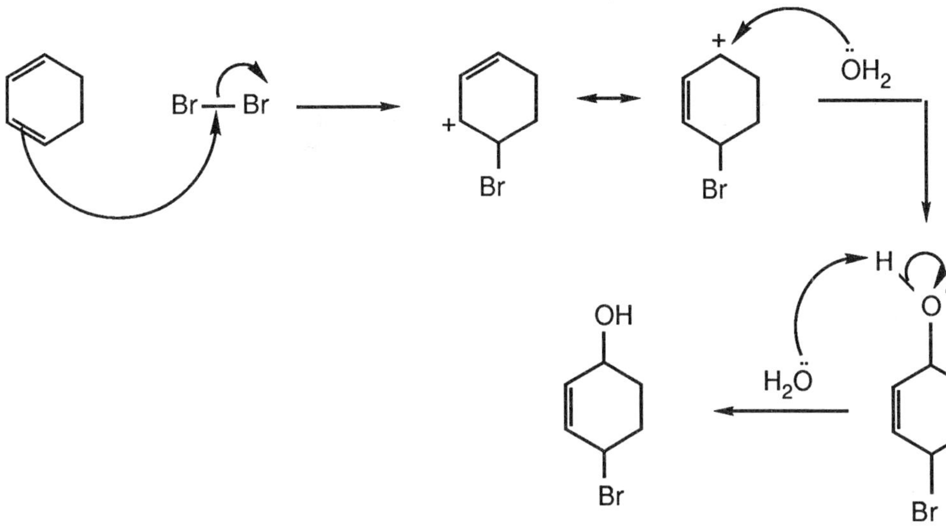

Br—Br, $\ddot{O}H_2$, Br, Br, $H_2\ddot{O}$, H, H, O, OH, Br, Br

(b)

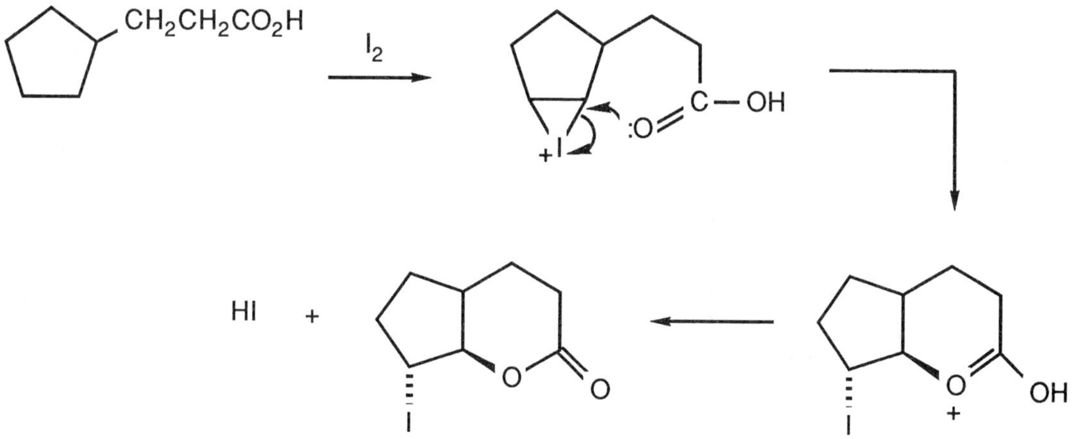

$CH_2CH_2CO_2H$ $\xrightarrow{I_2}$ C—OH, :O, +I

HI + O, O, I; O, OH, +, I

(c)

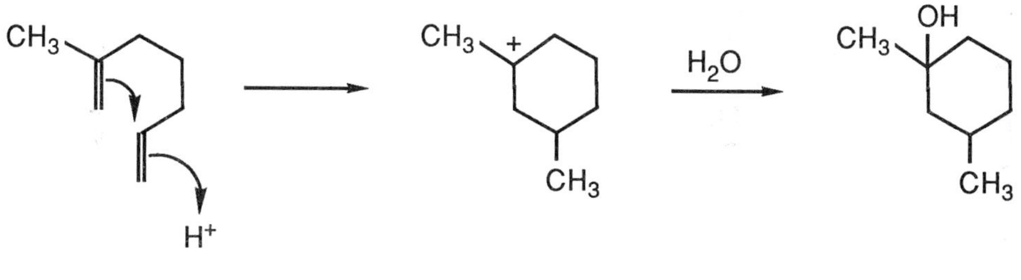

CH_3, H^+ ⟶ CH_3 +, CH_3 $\xrightarrow{H_2O}$ OH, CH_3, CH_3

(d)

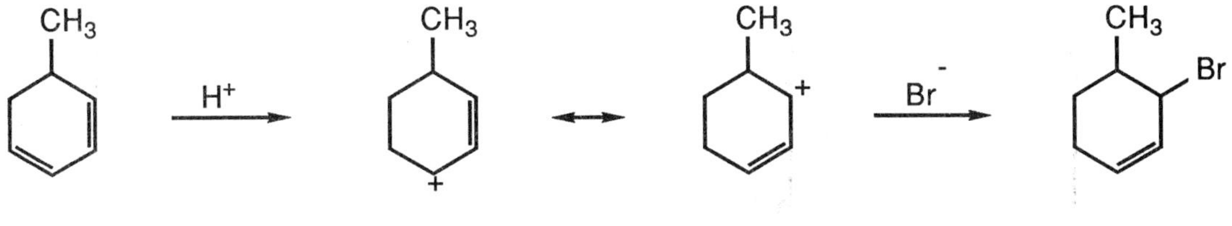

CH_3 $\xrightarrow{H^+}$ CH_3, + ⟷ CH_3, + $\xrightarrow{Br^-}$ CH_3, Br

8.8 $\xrightarrow{BH_3}$ BH + BH

8.9 Abstraction of a hydrogen from either allylic CH_2 group will generate the same radical, and only two of the various resonance forms are unique:

Reaction with bromine at each of the two sites having free radical character would yield the following two products:

Br + Br

8.10 The hindered disiamylborane will preferentially attack (at the less substituted end) of double bond that is less sterically crowded. Subsequent oxidation affords the primary alcohol.

CH_3 — CH(CH_3)CH_2OH

8.11 The initial product (C_5H_9Br) has one degree of unsaturation, either a ring or a double bond. Since the original bromo substituents had 1,3 rather than 1,2 relationships, the reaction with zinc must have produced a cyclopropane derivative. Catalytic hydrogenation should then cleave the cyclopropane between the least substituted carbon atoms.

$BrCH_2CH_2-CH(Br)-CH_2CH_2Br \longrightarrow$ CH_2CH_2Br $\longrightarrow (CH_3)_2CH-CH_2CH_2Br$

8.12 **(a)** Many different synthetic schemes are possible, but they are all limited by the presence of an isopropyl group combined with the 3-carbon requirement. (A secondary isopropyl group cannot be used to alkylate either an organocopper reagent or a sodium acetylide.) A difunctional compound would allow you to employ standard coupling reactions and introduce different substituents at the two ends of I-CH_2CH_2-Cl:

$$I-CH_2CH_2-Cl \xrightarrow{(CH_3CH_2)_2CuLi} CH_3CH_2CH_2CH_2-Cl \xrightarrow{[(CH_3)_2CH]_2CuLi} CH_3CH_2CH_2CH_2-\underset{}{\overset{CH_3}{\overset{|}{C}H}}-CH_3$$

(b) The same logic shown in part (a) will work again here:

$$I-CH_2CH_2-Cl \xrightarrow{[(CH_3)_2CH]_2CuLi} CH_3-\overset{CH_3}{\overset{|}{C}H}-CH_2CH_2Cl \xrightarrow{(CH_3CH_2-\overset{CH_3}{\overset{|}{C}H})_2CuLi} CH_3-\overset{CH_3}{\overset{|}{C}H}-CH_2CH_2-\underset{CH_2CH_3}{\underset{|}{C}H}-CH_3$$

(c) Once again, you could use a difunctional compound:

$$ClCH_2-\underset{CH_3}{\overset{CH_3}{\underset{|}{\overset{|}{C}}}}-CH_2I \xrightarrow{(CH_3CH_2-\overset{CH_3}{\overset{|}{CH}}\!\!\rightarrow_2 CuLi} ClCH_2-\underset{CH_3}{\overset{CH_3}{\underset{|}{\overset{|}{C}}}}-CH_2CH_2-\overset{CH_3}{\overset{|}{CH}}-CH_2CH_3$$

$$\big\downarrow (CH_3CH_2)_2CuLi$$

$$CH_3CH_2CH_2-\underset{CH_3}{\overset{CH_3}{\underset{|}{\overset{|}{C}}}}-CH_2CH_2-\overset{CH_3}{\overset{|}{CH}}-CH_2CH_3$$

8.13 **(a)** Same: both are 1,3-cyclopentadiene.

(b) Same: two conformations of 1,3-butadiene.

(c) Constitutional isomers: 1,3-cyclohexadiene and 1,4-cyclohexadiene.

(d) Constitutional isomers.

(e) Diastereomers: the first is (*Z,Z*)-1,6-cyclodecadiene and the second is (*E,E*)-1,6-cyclodecadiene.

(f) Constitutional isomers: the first is allylic and the second is not.

(g) Same: the two structures are resonance forms.

(h) Same: the two structures are resonance forms.

(i) Different: the two have different molecular formulas.

(j) Diastereomers: the substituents are *trans* in the first compound and *cis* in the second.

8.14 **(a)** The double bonds in the target molecule strongly suggest that you can join one C3 and two C4 fragments by using organocopper reagents and alkenyl halides or by alkylation of terminal alkynes. We will illustrate the latter in order to show how the stereochemistry can be controlled. The key is to use a 1,4-dihalobutane with *different* halogens as the starting material. This allow selective introduction of the unsaturated C4 and C3 groups. As shown in the following scheme it is essential to introduce the C4 group second because its *E*-stereochemistry requires a metal/ammonia reduction. This can be done with a carbon-carbon double bond elsewhere in the molecule, but a halogen substituent would be cleaved. The C-3 group is therefore introduced first, by displacement of the more reactive iodide of 1-chloro-4-iodobutane.

$$Cl-CH_2CH_2CH_2CH_2-I \xrightarrow{NaC\equiv CCH_3} ClCH_2CH_2CH_2CH_2-C\equiv CCH_3$$

$$\xrightarrow{H_2,\ Ba/PdSO_4} ClCH_2CH_2CH_2CH_2CH{=}CHCH_3 \ (\text{cis; } H, H \text{ on same side})$$

$$\xrightarrow{CH_3CH_2C\equiv CNa} CH_3CH_2C\equiv C-CH_2CH_2CH_2CH_2CH{=}CHCH_3 \ (\text{cis})$$

$$\xrightarrow{Na,\ NH_3} CH_3CH_2CH{=}CH-CH_2CH_2CH_2CH_2-CH{=}CHCH_3$$

(the new double bond *trans*: H and CH_3CH_2 on the same carbon, H and $CH_2CH_2CH_2CH_2$ on the other; the second double bond *cis*: H, H on the same side, CH_3 opposite the chain)

(b) The locations of the unsaturated linkages suggest that you could employ an unsymmetrical 1,6-dihalohexane as the starting unit for the synthesis. If terminal alkynes are used to introduce the other subunits, one of them must be reduced to a double bond. This requires a metal/ammonia reduction to obtain the desired *E*-stereochemistry, so it cannot be done in the presence of either a halogen or of the other triple bond of the target molecule. The only solution to this dilemma would be to add another bond-forming step to the sequence and carry out the reduction on a compound in which the second functional group is a *terminal* alkyne. (Recall that in the presence of $NaNH_2$ a terminal alkyne is converted to the corresponding salt, which is not reduced). The last steps of the synthesis would therefore be:

$$I-(CH_2)_6-Cl \xrightarrow{CH_3CH_2CH_2C\equiv CNa} CH_3CH_2CH_2C\equiv C(CH_2)_6Cl \xrightarrow{HC\equiv CNa} CH_3CH_2CH_2C\equiv C(CH_2)_6-C\equiv CH$$

$$CH_3CH_2CH_2C\equiv C(CH_2)_6-C\equiv CH \xrightarrow[\text{2. Na, } NH_3]{\text{1. } NaNH_2\text{, } NH_3} (H)(CH_3CH_2CH_2)C=C(H)((CH_2)_6-C\equiv CH)$$

$$\xrightarrow[\text{2. } CH_3CH_2CH_2Br]{\text{1. } NaNH_2} (H)(CH_3CH_2CH_2)C=C(H)((CH_2)_6-C\equiv CCH_2CH_2CH_3)$$

It would not be advantageous to introduce the terminal alkyne residue first, because this would produce a compound that is both a terminal alkyne and a primary halide (and it could undergo self-alkylation). Consequently the preferable synthesis would proceed by introducing the C_5 unit followed by the C_2 unit.

CHAPTER 9. SPECTROSCOPY AND MOLECULAR STRUCTURE

ANSWERS TO EXERCISES

9.1 Using the relationship, $E = hc/\lambda$(where hc has units of 2.86×10^{-5} kcal/mol), the values estimated from Figure 9.2 can be calculated more accurately.

(a) for λ = 100 m, E = 3×10^{-7} kcal/mol; Radio Frequency.

for λ = 0.1 m, E = 3×10^{-4} kcal/mol; Microwave.

for λ = 240×10^{-9} (240 nm), E = 119 kcal/mol; Ultraviolet.

(b) for E = 10^{7} kcal/mol, λ = 3×10^{-12} m; X-Ray.

for E = 50 kcal/mol, λ = 6×10^{-7} m; Visible.

for E = 10^{-5} kcal/mol, λ = 3 m; Radio Frequency.

9.2 **(a)** Four peaks; no two of the carbons are constitutionally equivalent.

(b) Five peaks: there are 7 carbon atoms in the molecule, but C-2 and C-6 or the ring are equivalent as are equivalent as are C-3 and C-5.

(c) Three peaks: there are 5 carbon atoms in the molecule, but two pairs of carbons are equivalent (C-2/C-5 and C-3/C-4).

(d) One peak; all six carbon atoms of the ring are equivalent.

(e) Three peaks; there are only three different kinds of carbon atom: the methyl carbons, the CH_2 carbons and the hydroxyl-substituted carbon.

9.3 The NMR data indicates that the reaction was successful. If no reaction had occurred, peaks corresponding to an alkene would have been expected in the region of 110-150 ppm. Alkyl carbons give peaks in the range of 10-40 ppm, and an oxygen substituent can cause an increase of 40 ppm in the chemical shift. This is consistent with the observed peaks (10–60 ppm) for the product.

9.4 **(a)**

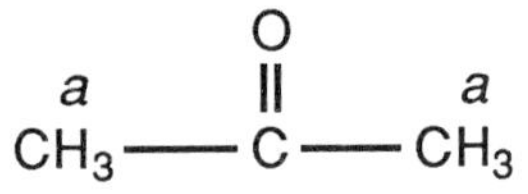

$$\overset{a}{CH_3}-\overset{\overset{O}{\|}}{C}-\overset{a}{CH_3}$$

1 signal

0.5 + 1.5	= 2.1 ppm	singlet

(b)

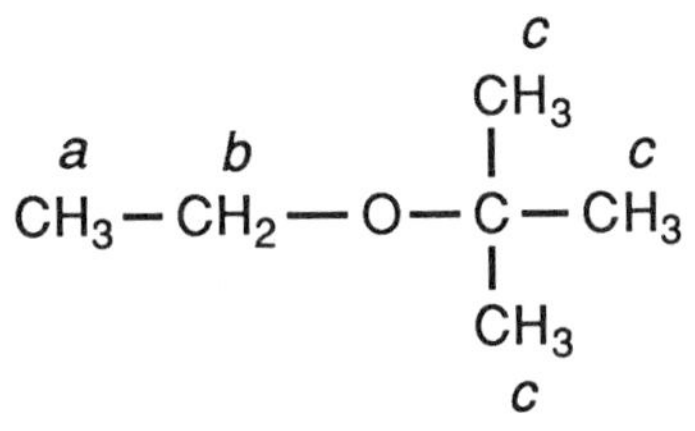

$$\overset{a}{CH_3}-\overset{b}{CH_2}-O-\underset{\underset{\underset{c}{CH_3}}{|}}{\overset{\overset{\overset{c}{CH_3}}{|}}{C}}-\overset{c}{CH_3}$$

3 signals (but *a* and *c* might overlap)

a	0.5 + 1.5	= 1.3 ppm	triplet	intensity = 3
b	0.5 + 0.4 + 2.6	= 3.5 ppm	quartet	intensity = 2
c	0.5 + 0.8	= 1.3 ppm	singlet	intensity = 9

(c)

$$\overset{a}{HO}-\overset{b}{CH_2}-\overset{c}{CH_2}-\overset{d}{CH_3}$$

4 signals

a	0.5–5 ppm (variable)		singlet	intensity = 1
b	0.5 + 0.4 + 2.7	= 3.6 ppm	triplet	intensity = 2
c	0.5 + 0.4 + 0.8	= 1.7 ppm	sextet	intensity = 2
d	0.5 + 0.4	= 0.9 ppm	triplet	intensity = 3

(d)

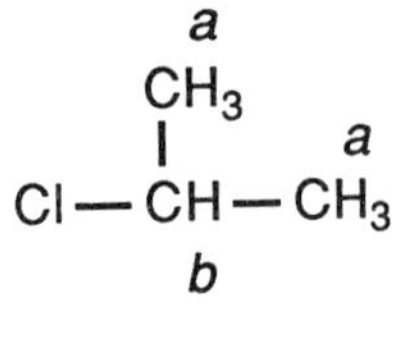

$$Cl-\underset{b}{\overset{\overset{\overset{a}{CH_3}}{|}}{CH}}-\overset{a}{CH_3}$$

2 signals

a	0.5 + 0.8	= 1.3 ppm	doublet	intensity = 6
b	0.5 + 2x0.4 + 2.6	= 3.6 ppm	septet	intensity = 1

(e)

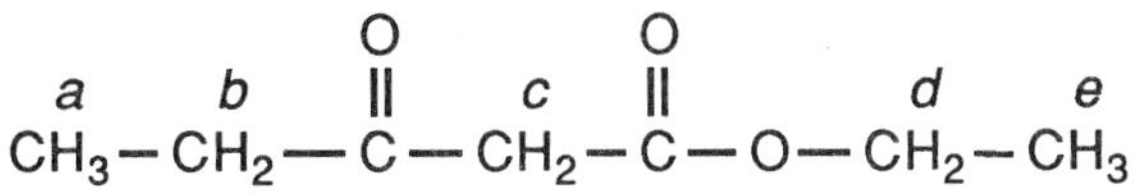

$$\overset{a}{CH_3}-\overset{b}{CH_2}-\overset{O}{\overset{\|}{C}}-\overset{c}{CH_2}-\overset{O}{\overset{\|}{C}}-O-\overset{d}{CH_2}-\overset{e}{CH_3}$$

5 signals

a	0.5 + 0.4	= 0.9 ppm	triplet	intensity = 3
b	0.5 + 0.4 + 1.6	= 2.5 ppm	quartet	intensity = 2
c	0.5 + 1.6 + 1.5	= 3.6 ppm	singlet	intensity = 2
d	0.5 + 0.4 + 3.2	= 4.1 ppm	quartet	intensity = 2
e	0.5 + 0.8	= 1.3 ppm	triplet	intensity = 3

9.5

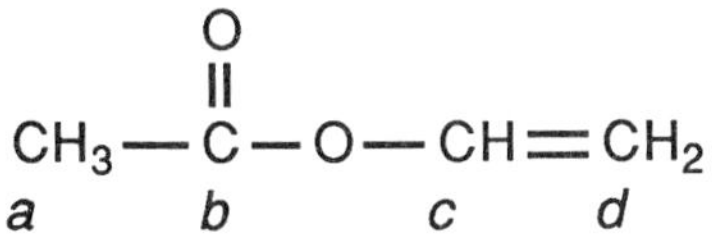

$$\underset{a}{CH_3}-\underset{b}{\overset{O}{\overset{\|}{C}}}-O-\underset{c}{CH}=\underset{d}{CH_2}$$

a	predicted: 10 ppm	observed: 20 ppm
b	predicted: 160–180 ppm	observed: 170 ppm
c	predicted: 150 ppm (110 + 40)	observed: 140 ppm
d	predicted: 110 ppm	observed: 100 ppm

9.6 **(a)** There are two alkyl substituents on the double bond, so the predicted λ_{max} is 235 nm (i.e., 215 + 2x10).

(b) There are four alkyl substituents on the double bonds, so the predicted λ_{max} is 235 nm (i.e., 215 + 4x5).

9.7

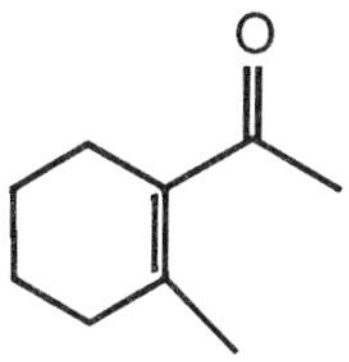

O

For structure (a) the double bond has three alkyl substituents, so so the predicted λ_{max} is 245 nm (i.e., 215 + 3x10). For structure (b) the double bond has only one alkyl substituent, so so the predicted λ_{max} is 225 nm (i.e., 215 + 1x10).

Clearly, structure (a) better matches the experimental λ_{max} of 245 nm.

9.8 The infrared spectrum shows a strong peak in the region of 3200–3400 cm^{-1} and this indicates the presence of an OH group. It is not, however, the OH of a carboxylic acid, which would give a much broader peak (typically 2500–3500 cm^{-1}). The correct structure must be (c).

$$CH_3-\underset{|}{\overset{OH}{CH}}-(CH_2)_6-CH_3$$

9.9 The molecular formula indicates two degrees of unsaturation. The strong peak at about 1750 cm^{-1} must be a carbonyl peak, so that accounts for one degree of unsaturation. The absence of any sharp peaks in the 1600 cm^{1} region indicates the absence of a double bond and suggests that the other degree of unsaturation is a ring. There molecule has only five carbon atoms, so the ring size must be five or smaller. The position of the carbonyl absorption falls in the range expected for a 5-membered ring cyclic ketone (1730–1755, Table 9.7), so the compound must be cyclopentanone:

9.10 **(a)** 86

(b) 116

(c) 106

(d) 214

9.11 **(a)** 100:6.6

(b) 100:7.7

(c) 100:5.5

(d) 100:4.4

9.12 **(a)** A ratio of 100:16.5 for the M to M+1 peaks indicates 15 carbon atoms (i.e., 15x1.1 = 16.5).

(b) A saturated, acyclic hydrocarbon with 15 carbons would have a molecular ion of 212 ($C_{15}H_{32}$). The observed molecular ion at 210 indicates a single degree of unsaturation, so the molecular formula of the unknown hydrocarbon must be $C_{15}H_{30}$.

ANSWERS TO PROBLEMS

9.1 **(a)** 3 peaks:

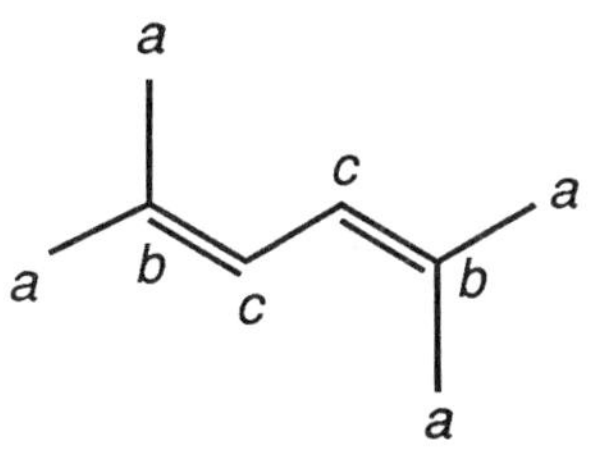

a, a, a, a, b, b, c, c

(b) 4 peaks:

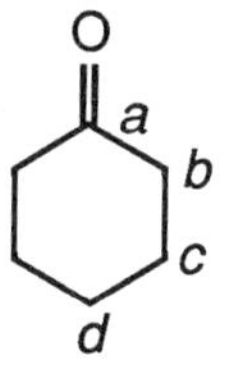

O, a, b, c, d

(c) 11 peaks:

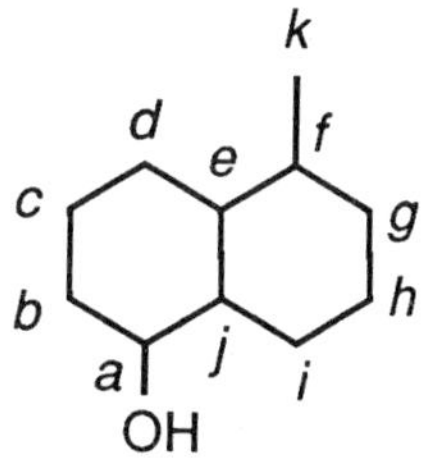

k, d, f, e, c, g, b, h, a, j, i, OH

(d) 6 peaks

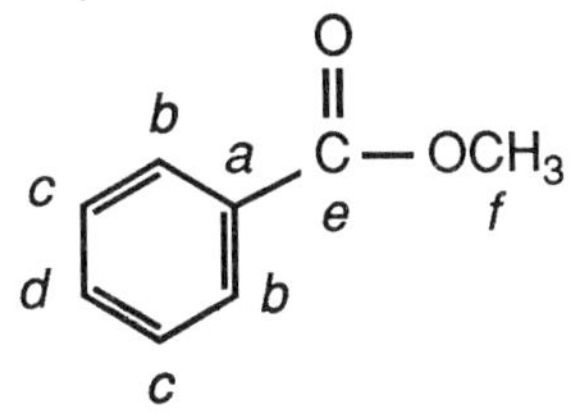

O, b, a, C—OCH$_3$, c, e, f, d, b, c

(e) 4 peaks:

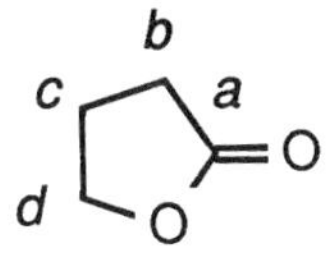

b, c, a, O, d, O

9.2 all values in ppm:

(a) *a* 10; *b* 130; *c* 120

(b) *a* 200–220; *b* 20; *c* 20; *d* 20

(c) *a* 60; *b* 20; *c* 20; *d* 20; *e* 30; *f* 30; *g* 20; *h* 20; *i* 20; *j* 30; *k* 10

(d) *a, b, c, d* 110–150; *e* 160–180; *f* 50

(e) *a* 160–180; *b* 20; *c* 20; *d* 50

9.3 The predicted chemical shifts are shown by each carbon atom:

(a)

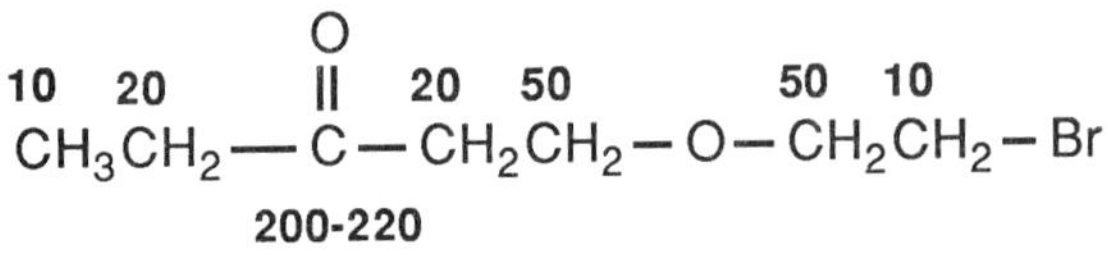

$$\overset{10}{CH_3}\overset{20}{CH_2}-\underset{200\text{-}220}{\overset{O}{\overset{\|}{C}}}-\overset{20}{CH_2}\overset{50}{CH_2}-O-\overset{50}{CH_2}\overset{10}{CH_2}-Br$$

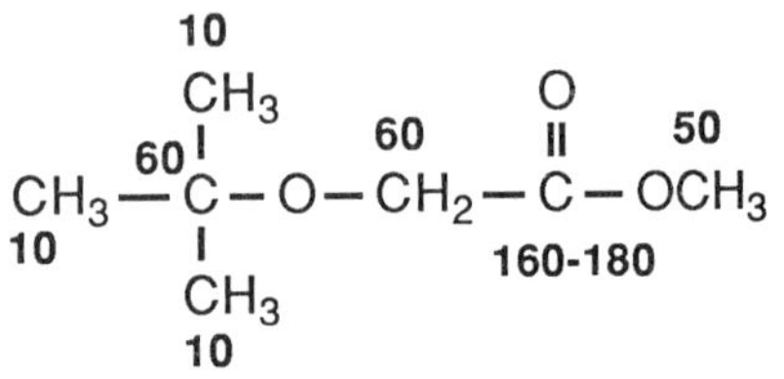

(b)

$$\underset{10}{CH_3}-\overset{60}{C}(\overset{10}{CH_3})(\underset{10}{CH_3})-O-\overset{60}{CH_2}-\underset{160\text{-}180}{\overset{O}{\overset{\|}{C}}}-\overset{50}{OCH_3}$$

(c)

$$\overset{10}{CH_3}-\underset{60}{C}(Cl)(Cl)-\overset{20}{CH_2}\overset{10}{CH_2}Br$$

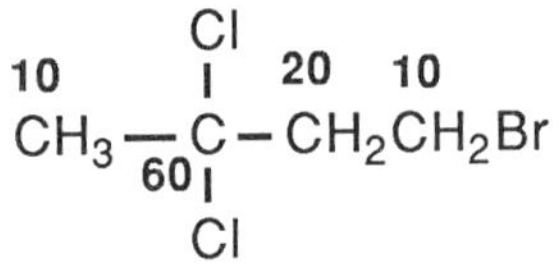

(d) Lactone ring (six-membered, ring O): C=O 160-180; CH_2 next to C=O 20; quaternary ring carbon 40 bearing two CH_3 groups (10, 10); ring CH_2 20; $O-CH_2$ 50.

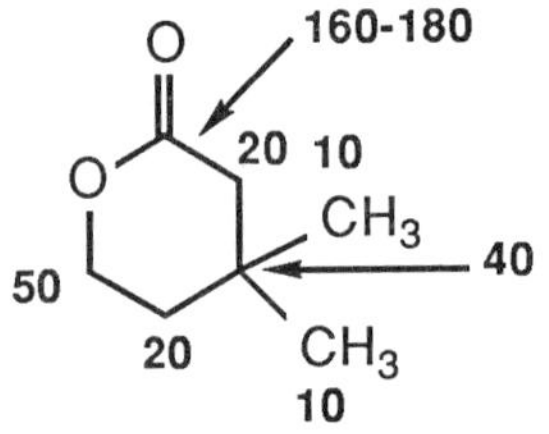

(e)

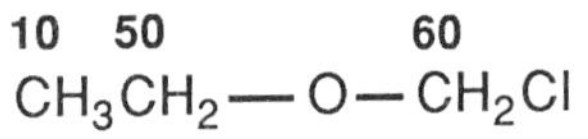

$$\overset{10}{CH_3}\overset{50}{CH_2}-O-\overset{60}{CH_2}Cl$$

(f) Cyclopentanone: C=O 200-220; ring CH_2 carbons 20, 20, 20; quaternary ring carbon 60 bearing CH_2OH (50) and CH_3 (10).

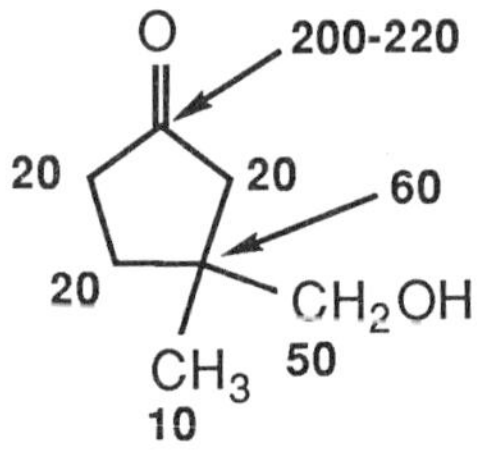

9.4 The ^{13}C NMR spectrum exhibits four peaks, and this allows you to decide which of the possible constitutional isomers of 1,2-dimethylcyclohexane is formed by catalytic hydrogenation:

1,1-dimethylcyclohexane:	5 nonequivalent carbons
1,2-dimethylcyclohexane:	4 nonequivalent carbons
1,3-dimethylcyclohexane:	5 nonequivalent carbons
1,4-dimethylcyclohexane:	3 nonequivalent carbons

Knowing that the reduction product is 1,2-dimethylcyclohexane, you must determine the position of the double bond in the reactant. Five constitutional isomers are possible with the necessary carbon skeleton:

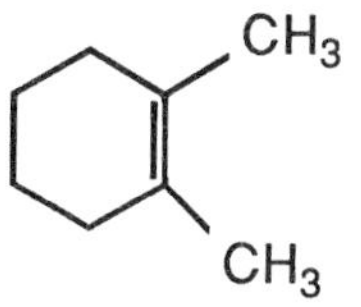

A 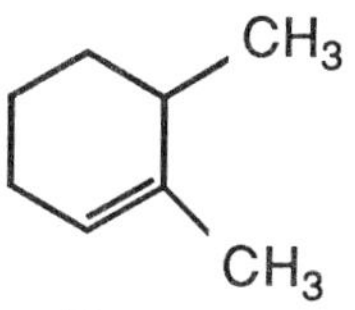 B 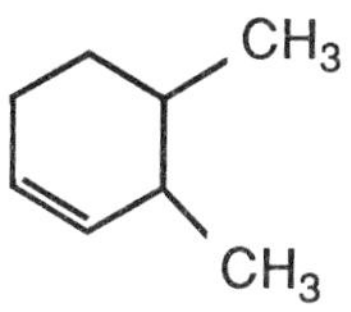 C 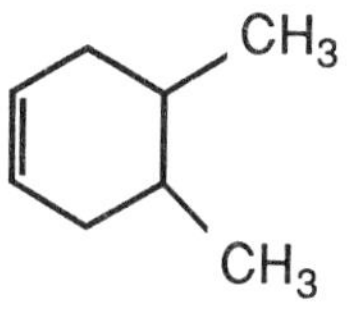 D 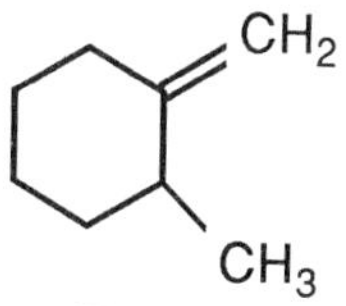 E

The ozonolysis reaction provides the key. Compound **A** would afford a diketone, whereas **B** would yield a keto aldehyde, and **E** would afford two monocarbonyl compounds. Compounds **B** and **C** would each yield a dialdehyde, but in the case of **D** the methyl groups would be equivalent. Therefore the unknown compound must be **C**, 3,4-dimethylcyclohexene.

9.5 Each compound has a single degree of unsaturation, so it has either a ring, a C=C double bond or a C=O double bond.

(a) All four carbons are different and the peak at 206 ppm must be a carbonyl carbon. This compound must be either 2-butanone or butyraldehyde. (The spectrum is actually that of 2-butanone.)

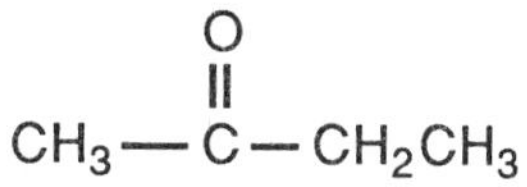

$$CH_3—\overset{\overset{\large O}{\|}}{C}—CH_2CH_3$$

(b) The absence of peaks with shifts of 150 ppm or greater indicates that there is a ring rather than a double bond. The peak at 68 ppm is characteristic of an oxygen-substituted carbon. There are only two types of carbon, and this suggests symmetry. Only tetrahydrofuran meets all these requirements:

(c) The peak at 205 ppm indicates a carbonyl carbon. There are only two other peaks, so two of the remaining three carbons are equivalent. These must correspond to an isopropyl group, so the compound is isobutyraldehyde:

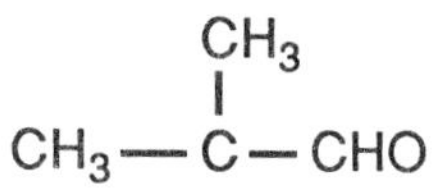

$$CH_3—\overset{\overset{\large CH_3}{|}}{C}—CHO$$

(d) All four carbons are different and the peak at 203 ppm must be a carbonyl carbon. This compound must be either 2-butanone or butyraldehyde. (The spectrum is actually that of butyraldehyde.)

$$CH_3CH_2CH_2\text{—}CHO$$

9.6 The compound has three degrees of unsaturation, and the presence of three signals integrating for 2H, 4H and 6H suggests that the compound is symmetrical. The triplet and quartet are the only signals that are split, and these show the characteristic pattern of an ethyl group. Integration indicates two ethyl groups, and they must be equivalent. We can therefore draw a partial structure with equivalent ethyl groups as the "ends" of the molecule:

$$CH_3CH_2\text{----------}CH_2CH_3$$

The chemical shift (4.3 ppm) of the CH_2 groups suggests that they are bonded to oxygen, so we can modify our partial structure to:

$$CH_3CH_2\text{-O------O-}CH_2CH_3$$

This leaves two hydrogens, two oxygens four carbons, and all three degrees of unsaturation still unaccounted for. The chemical shift of the singlet at 6.3 ppm is characteristic of vinyl hydrogen, so the partial structure can be further refined to:

$$CH_3CH_2\text{-O---CH=CH---O-}CH_2CH_3$$

Now we are missing two carbons, two oxygens, and two degrees of unsaturation, which corresponds to two carbonyl groups. The spectrum is that of maleic acid (although you cannot deduce the stereochemistry of the double bond):

$$CH_3CH_2\text{—O—}\overset{\overset{\displaystyle O}{\|}}{C}\text{—CH=CH—}\overset{\overset{\displaystyle O}{\|}}{C}\text{—O—}CH_2CH_3$$

9.7 The compound has 10 degrees of unsaturation, and the sharp peak at 7.3 ppm integrating for 10 H suggests the presence of two equivalent phenyl groups (each with four degrees of unsaturation) at the ends of the molecule:

C_6H_5----------------C_6H_5

The singlets integrating for 4H each at 2.7 and 5.1 ppm must be each correspond to two equivalent CH_2 groups. Since they are singlets, there can be no coupling, and the partial structure can be modified to:

C_6H_5---CH_2---CH_2----CH_2---CH_2---C_6H_5

This partial structure accounts for everything but two carbons, four oxygens, and two degrees of unsaturation: almost certainly, this corresponds to two carboxyl groups ($-CO_2-$). The phenyl groups appear as singlets, so they must be bonded to nonpolar CH_2 groups rather than to the carboxyl groups. The structure can now be refined to:

C_6H_5CH2---CH2----CH2---CH2C6H5

$-CO_2-$ $-CO_2-$

The compound must be symmetrical, and the benzylic CH_2 groups are (equivalent) singlets, so they cannot be bonded to the other CH_2 groups. Therefore the carboxyl groups must be interposed between the two types of CH_2 groups. The central CH_2 groups are bonded directly to each other but do not split each other because they are equivalent:

$C_6H_5CH_2$---CH_2CH_2---$CH_2C_6H_5$

$-CO_2-$ $-CO_2-$

The only remaining question is the relative orientation of the carboxyl group; in other words, two structures are possible:

A, $C_6H_5CH_2CO_2-CH_2CH_2-O_2CCH_2C_6H_5$

and

B, $C_6H_5CH_2O_2C-CH_2CH_2-CO_2CH_2C_6H_5$

Structure **A** should have peaks at 3.7 (0.5 + 1.7 for $-Ar$ + 1.5 for $-CO_2R$) and 4.4 ppm (0.5 + 3.2 for $-O_2CR$ + 0.8 for $-CH_2X$). Structure **B** should have peaks at 5.4 (0.5 + 1.7 for $-Ar$ + 3.2 for $-O_2C$) and 2.7 ppm (0.5 + 1.5 for $-CO_2R$ + 0.7 for $-CH_2Ar$, where CH_2Ar is the best approximation for CH_2-CO_2). The observed values of 2.7 and 5.1 ppm fit much better with **B,** which is the correct structure:

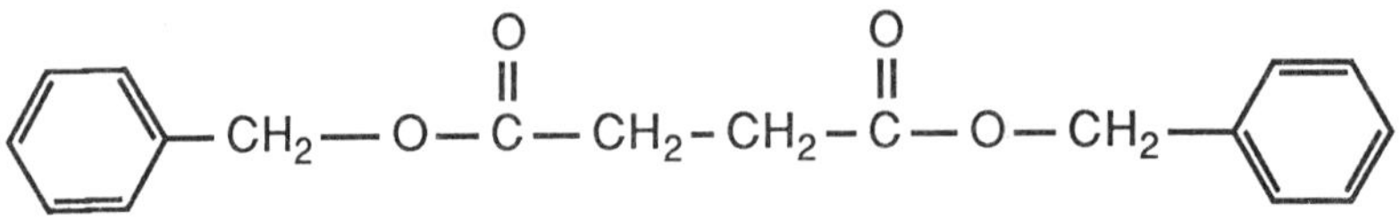

$$C_6H_5-CH_2-O-\overset{O}{\overset{\|}{C}}-CH_2-CH_2-\overset{O}{\overset{\|}{C}}-O-CH_2-C_6H_5$$

9.8 The compound has six degrees of unsaturation, and the presence of only two peaks in the NMR spectrum indicates a structure with substantial symmetry. The sharp peak at 8.1 ppm integrating for 4 hydrogens is strongly suggestive of a *para* disubstituted benzene derivative, and the singlet at 2.7 ppm presumably results from two equivalent methyl groups. We can propose a partial structure on the basis of this information:

$$CH_3\text{-----}C_6H_4\text{-----}CH_3$$

This accounts for all the hydrogens, leaving only two carbons, two oxygens, and two degrees of unsaturation: in other words two carbonyl groups. The compound has the following structure:

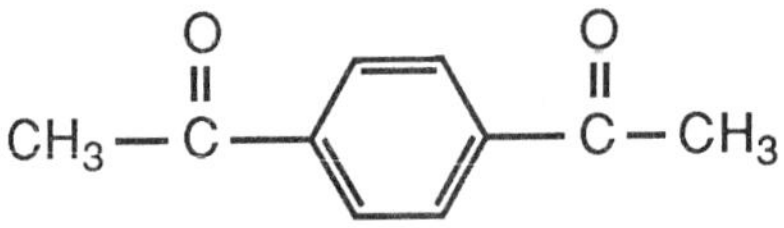

$$CH_3-\overset{O}{\overset{\|}{C}}-C_6H_4-\overset{O}{\overset{\|}{C}}-CH_3$$

9.9 The compound has a single degree of unsaturation. Despite the apparent complexity of the spectrum, you can decipher it readily if you utilize the information carefully. The complexity results largely from spin coupling, but this also gives you structural information. First, consider the degree of unsaturation; it could be a ring, a C=C double bond, or a C=O double bond. The absence of any peaks downfield of 5 ppm makes a C=C double bond unlikely, and the singlet integrating for 3 hydrogens is characteristic of a methyl group bonded to an unsaturated carbon atom. This provides the following as a likely partial structure:

$$CH_3\text{-}CO\text{----}$$

Next consider the multiplet at 4.9 ppm. Its chemical shift requires that this result from a CH that is bonded to oxygen, but even two alkyl groups and an OR substituent would lead to a predicted shift of only about 3.9 ppm. This suggests that the oxygen substituent on this CH is $-O_2CR$ (for which you would predict a shift of 4.5), and the partial structure can be revised:

$$CH_3\text{-}CO_2\text{-}CH\text{---}$$

Three carbons and 8 hydrogens remain unaccounted for, and spin coupling provides the structural key. The one-hydrogen multiplet at 4.9 ppm appears to be a sextet, suggesting that it is coupled to 5 neighboring hydrogens. This would have to be a CH_3 and a CH_2, and indeed there is a doublet integrating for 3 hydrogens at 1.2 ppm. This allows further revision of the partial structure:

$$CH_3\text{-}CO_2\text{-}CH(CH_3)\text{---}$$

Only two carbons and five hydrogens remain to be assigned, so the remaining structural fragment must be an ethyl group. The necessary peaks can be seen at 0.9 and 1.6 ppm for the CH_3 and CH_2 peaks, respectively. The former is a distorted triplet (not a clean triplet, because the CH_2 to which it is coupled has a similar chemical shift), and the CH_2 is a complex multiplet because it is also coupled to the -OCH hydrogen. The complete structure is:

$$CH_3\!-\!\overset{\overset{\displaystyle O}{\|}}{C}\!-\!O\!-\!\overset{\overset{\displaystyle CH_3}{|}}{CH}\!-\!CH_2CH_3$$

9.10 Spin coupling again provides the key to deducing the structure. The compound contains a single degree of unsaturation, and you might be suspicious about the possibility of a $-CO_2$ group. The triplet at 1.3 ppm is a methyl group bonded to a CH_2 group, and the latter would have to appear at least as a quartet. Only the peak at 4.2 ppm fits that description, and it is a quartet. Consequently, the molecule contains an isolated ethyl group:

$$CH_3CH_2\text{-----}$$

The remaining two peaks, at 2.9 and 3.6 ppm are both triplets, and each integrates for two hydrogens. Clearly, they must correspond to adjacent CH_2 groups:

$$CH_3CH_2\text{-----}CH_2CH_2\text{-----}$$

Two oxygens, one carbon, and a bromine remain to be placed in this partial structure. The bromine must be an "end", so we can refine the partial structure as follows:

$$CH_3CH_2\text{-----}CH_2CH_2\text{-----}Br$$

Assuming that the degree of unsaturation is a carbonyl group, we must consider the various possibilities for placing the C=O and the other oxygen. At least one of them must be located between the CH_2 groups that are not coupled to each other, and this leaves four possible structures:

A $CH_3CH_2C(O)CH_2CH_2\text{-}O\text{-}Br$ **B** $CH_3CH_2OCH_2CH_2\text{-}C(O)\text{-}Br$

C $CH_3CH_2CO_2CH_2CH_2\text{-}Br$ **D** $CH_3CH_2O_2CCH_2CH_2\text{-}Br$

Taking the CH_3 and three CH_2 groups in the sequence they are drawn above, we can predict the following chemical shifts for alternatives **A–D.** Note that the substituent effect for $-CH_2X$ or $-CH_2Ar$ (for $-CH_2\text{-}CO_2$) must be employed in most cases.

a	1.2	2.5	2.9	3.6
b	1.3	3.5	3.8	2.8
c	1.2	2.4	4.5	3.8
d	1.3	4.1	2.8	3.7

The observed values of 1.3, 4.2, 2.9 and 3.5 (or 3.5 and 2.9) fit best with alternative **D**, and this is indeed the correct structure:

$$CH_3CH_2\text{—}O\text{—}\overset{\overset{\displaystyle O}{\|}}{C}\text{—}CH_2CH_2Br$$

9.11 This spectrum appears to be fairly complex, but the complexity results from spin-coupling interactions and these actually allow you to interpret the spectrum rather easily. The compound has a single degree of unsaturation, which could be either a ring or a double bond. The absence of any signals with a chemical shift greater than 4.5 ppm makes a C=C double bond unlikely, whereas the presence of two oxygens in the molecular formula indicates that a C=O (or even CO_2) group is quite probable. The quartet at 4.15 ppm is so far downfield that it almost certainly must result from a CH_2-O- group, and the splitting pattern requires an adjacent methyl group (the triplet at 1.25 ppm):

$$-O-CH_2CH_3$$

Taking the hypothesis that the degree of unsaturation is a carbonyl group, three carbons and seven hydrogens must be accounted for, and this matches the three remaining signals in the spectrum (2H, 2H, and 3H). The signal at 2.3 ppm (a C_2 group) is a skewed triplet, so it must be adjacent to another CH_2 group -- the multiplet centered at 1.7 ppm. This leaves only the the CH_3 group, and it must be the skewed triplet at 1.0 ppm). Putting these three groups together, the fragment must be a propyl group:

$$CH_3CH_2CH_2-$$

The chemical shifts of all three signals differ only by a total of 1.7 ppm; the similarity in chemical shifts results in more complex spin-coupling interactions so that the triplets are highly skewed. The fragments can be joined to the carbonyl group to generate a proposed structure:

$$CH_3CH_2CH_2-\overset{\overset{\large O}{\|}}{C}-O-CH_2CH_3$$

The validity of the proposal is confirmed by comparing the observed and calculated shifts (from left to right in the preceding structure).

obs	1.0	1.7	2.3	4.15	1.25
calc	0.9	1.6	2.4	4.1	1.3

9.12 There are only three peaks in this spectrum and all are singlets. The molecular formula indicates five degrees of unsaturation, four of which presumably correspond to a phenyl group -- the 5H singlet at 7.2 ppm. The remaining two signals must result from isolated (i.e., uncoupled) CH_2 and CH_3 groups. We can therefore deduce the following partial structure:

$$C_6H_5\text{-----}CH_2\text{-----}CH_3$$

We have not yet accounted for one carbon, two oxygens and one degree of unsaturation. The degree of unsaturation is probably a carbonyl group, and we can propose several structures on this basis:

A C_6H_5-O-CH_2-C(O)CH_3 **B** C_6H_5-C(O)CH_2-O-CH_3

C C_6H_5-CH_2-O_2C-CH_3 **D** C_6H_5-CH_2-CO_2-CH_3

Structures **A** and **B** are unlikely because the presence of an electronegative substituent would lead to a multiplet for the phenyl group. The predicted chemical shifts for these possibilities are shown below:

A	5.4	2.1
B	5.1	3.1
C	5.4	2.0
D	3.7	3.7

Only for the last possibility do you predict two peaks with about the same chemical shift of 3.7 ppm, which also corresponds almost exactly with the observed peaks at 3.6 ppm. Hence the correct structure is **D:**

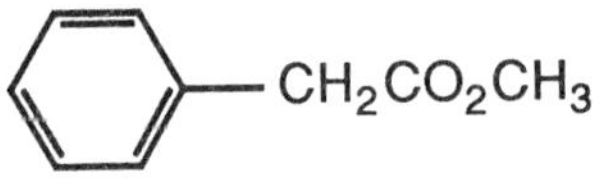

C_6H_5–$CH_2CO_2CH_3$

9.13 The compound has two degrees of unsaturation, and the peaks integrating for 2H in the region of 5.7-7.3 ppm suggest a disubstituted C=C double bond. The triplet at 1.3 ppm and the doublet at 4.2 ppm clearly indicate an *O*-ethyl group, so two partial structures are possible:

A CH_2=C----OCH_2CH_3 and

B -CH=CH----OCH_2CH_3

The remaining peak in the spectrum is a doublet at 1.9 ppm integrating for 3H, and this must be a CH_3 adjacent to a carbon with only single hydrogen. Therefore we know that **B** is the correct partial structure, and since a methyl group is an "end" it can be further refined to:

CH_3-CH=CH----OCH_2CH_3 Notice that you can now understand the coupling in the vinyl region: coupling between the vinyl hydrogens generates a doublet centered at about 5.8 ppm for the "isolated" CH, but the additional coupling to the methyl group results in an overlapping doublet of quartets centered at 7.0 ppm for the other vinyl CH. For the partial structure only one carbon, one oxygen and one degree of unsaturation remain unaccounted for -- a carbonyl group. The complete structure is therefore:

$$CH_3CH{=}CHCO_2CH_2CH_3$$

9.14 The compound has five degrees of unsaturation, and the peaks at 6.9-8.0 ppm suggest that four of the degrees of unsaturation correspond to a disubstituted aromatic ring. The symmetry of these signals further suggests a *para*-disubstituted benzene ring.

$$\text{---}C_6H_4\text{---}$$

The 2H triplet at 3.7 ppm indicates a CH_2 group with two adjacent hydrogens. These two adjacent hydrogens cannot be an isolated CH_2 group: that could explain the triplet at 3.1 ppm, but we would be unable to account for the splitting of the only remaining multiplet (2.2 ppm). The compound must therefore have three contiguous CH_2 groups.

$$\text{---}CH_2CH_2CH_2\text{---}$$

The CH_2 groups at the ends of this subunit must give rise to the triplets, and the central CH_2 must correspond to the quintet at 2.2 ppm. The two subunits can be combined in a partial structure

$$\text{---}C_6H_4\text{-----}CH_2CH_2CH_2\text{---}$$

This leaves one hydrogen (which must result in the singlet at 9.1 ppm), one carbon, one chlorine, two oxygens, and one degree of unsaturation to be accounted for. The following subunits are likely:

H- -O- C=O Cl-

Two of these are "ends" (H- and Cl-) but the hydrogen cannot be directly attached to either the aromatic ring (this would correspond to a phenyl group rather than a disubstituted benzene) or to the CH_2 group (which would make it a CH_3). Hence this

H must be bonded either to the carbonyl (to form an aldehyde, –CHO) or to the oxygen (to form a hydroxyl, –OH). Therefore, the fragments must be

–OH, –Cl, and C=O or –CHO, –Cl, and –O–

If there is a –CHO group, it must be bonded to the aromatic ring; otherwise it would be coupled to the CH_2 hydrogens. The other "end" of the molecule would be Cl, and the –O– could be at either end of the three-carbon chain:

A $OHC\text{-}C_6H_4\text{-}O\text{-}CH_2CH_2CH_2\text{-}Cl$ or **B** $OHC\text{-}C_6H_4\text{-}CH_2CH_2CH_2\text{-}O\text{-}Cl$

If the "ends" are –OH and –Cl, the C=O could occupy any of three locations, and six structures are possible

C $HO\text{-}C_6H_4\text{-}CO\text{-}CH_2CH_2CH_2\text{-}Cl$ or **D** $Cl\text{-}C_6H_4\text{-}CO\text{-}CH_2CH_2CH_2\text{-}OH$

E $HO_2CC_6H_4\text{-}CH_2CH_2CH_2\text{-}Cl$ or **F** $Cl\text{-}CO\text{-}C_6H_4\text{-}CH_2CH_2CH_2\text{-}OH$

G $HO\text{-}C_6H_4\text{-}CH_2CH_2CH_2\text{-}CO\text{-}Cl$ or **H** $Cl\text{-}C_6H_4\text{-}CH_2CH_2CH_2\text{-}CO_2H$ The peak at 9.1 ppm is reasonable for an aldehyde (9.4–11 ppm) or a phenol (4–8 ppm), but carboxylic acids (11.5–12.5 ppm) and alcohols (0.5–5 ppm) are much less likely. On this basis, we can narrow the possibilities to (a), (b), (c), and (g). Using substituent effects, the predicted chemical shifts of the three CH_2 groups in these isomers helps further narrow the possibilities. The center CH_2 (quintet) is listed last:

A	2.6/3.5	2.0
B	4.2/3.5	2.1
C	2.9/3.5	2.0
G	2.6/2.4	1.9
obs	3.1/3.7	2.2

The predictions for **C** are all within 0.2 ppm of the observed chemical shifts, whereas each of the other structures has a peak for which the discrepancy is at least 0.5 ppm. The correct structure is indeed **C:**

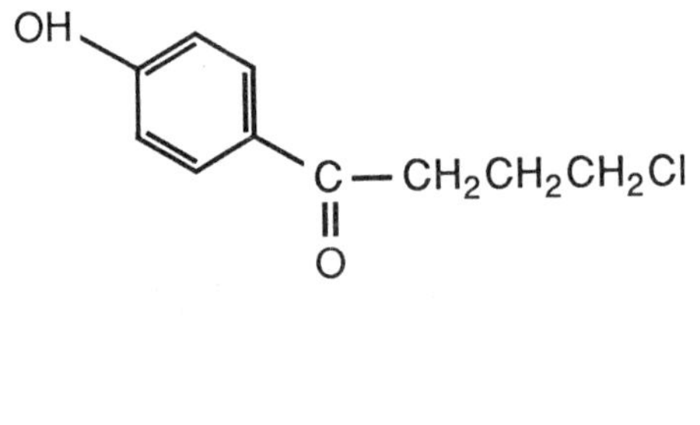

OH

C — $CH_2CH_2CH_2Cl$

O

9.15 The molecular formula shows that there are five degrees of unsaturation, and the peak integrating for 4H at about 7 ppm indicates an unsymmetrically substituted aromatic ring. The three remaining peaks are all singlets, so the two methyls and the CH_2 group are all isolated (in the sense that they are not bonded to a carbon that also has hydrogen substituents). The two methyl groups represent "ends" of the molecule, so we can draw the following partial structure:

$$CH_3{-}{-}{-}{-}C_6H_4{-}{-}{-}{-}CH_3$$

Inclusion of the CH_2 group gives:

$$CH_3{-}{-}{-}{-}C_6H_4{-}{-}{-}{-}CH_2{-}{-}{-}{-}CH_3$$

Only two oxygens, a carbon, and a degree of unsaturation remain unaccounted for, and these presumably correspond to –O– and C=O (perhaps combined as a $-CO_2-$ group). One of these insulating subunits must be situated between the CH_2 and CH_3 groups. The other subunit could be located at any of the three sites indicated by dashed lines in the partial structure. A total of six isomeric structures can be drawn:

A $CH_3{-}O{-}C_6H_4{-}CH_2{-}CO{-}CH_3$

B $CH_3{-}CO{-}C_6H_4{-}CH_2{-}O{-}CH_3$

C $CH_3{-}C_6H_4{-}O{-}CH_2{-}CO{-}CH_3$

D $CH_3{-}C_6H_4{-}CO{-}CH_2{-}O{-}CH_3$

E $CH_3{-}C_6H_4{-}CH_2{-}O_2C{-}CH_3$

F $CH_3{-}C_6H_4{-}CH_2{-}CO_2{-}CH_3$

Prediction of chemical shifts (CH_3, CH_3, CH_2) gives the following results

A	3.8/2.1	3.8
B	2.5/3.1	4.8
C	2.2/2.1	5.4
D	2.2/3.1	5.1
E	2.2/2.0	5.4
F	2.2/3.7	3.7
obs	2.1/3.8	3.7

Only for isomers **A** and **F** is the predicted chemical shift of the CH_2 group in reasonable agreement with experiment. In fact the overall agreement for both of these is good enough that you cannot distinguish between them on the basis of the information given. Similarly, you cannot determine whether the substitution pattern of the aromatic ring is *ortho* or *meta*. The actual structure is:

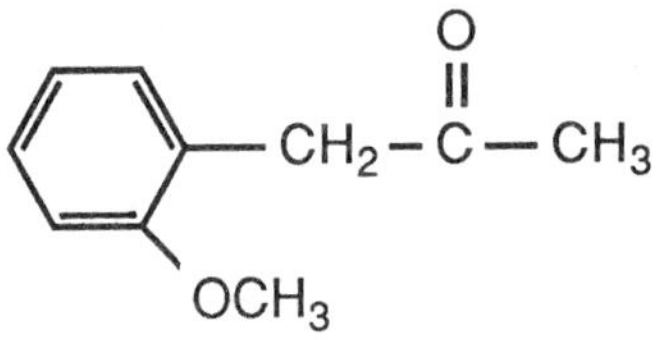

$$\text{C}_6\text{H}_4(\text{OCH}_3)\text{—CH}_2\text{—C(=O)—CH}_3$$

9.16 The molecular formula shows that there are six degrees of unsaturation. The proton spectrum indicates a phenyl group and two equivalent ethyl groups -- presumably *O*-ethyl groups (and likely even $-CO_2CH3$ groups) on the basis of chemical shifts. The ^{13}C spectrum show the four peaks expected for a phenyl group in the 127–138 ppm region (two of which have almost the same chemical shift of 128 ppm); remember that the two *ortho* and two *meta* carbons will be equivalent. Another peak at 169 ppm is typical for an ester (not ketone) carbonyl. This signal and that at 138 ppm for the *para* carbon of the phenyl group are quite weak because the carbons have no hydrogens. In the aliphatic region the ^{13}C spectrum shows four peaks, two corresponding to the (equivalent) ethyl groups and two additional peaks. This corresponds to the remaining peaks in the proton NMR, a CH_2 (doublet) and a CH (triplet). Clearly these two are coupled to each other; moreover they must lie on any symmetry element that makes the two ethyls equivalent:

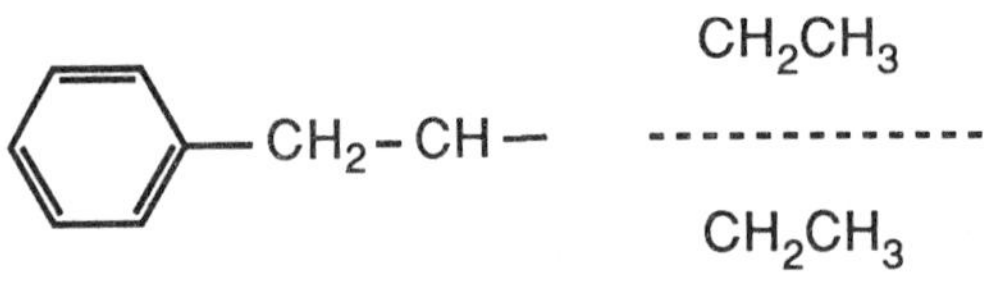

$$\text{C}_6\text{H}_5\text{—CH}_2\text{—CH—} \quad \begin{matrix} \text{CH}_2\text{CH}_3 \\ \text{-------------} \\ \text{CH}_2\text{CH}_3 \end{matrix}$$

This leaves two degrees of unsaturation, two carbons, and four oxygens to be accounted for -- again suggesting two $-CO_2-$ groups. There is only one way that these can be introduced symmetrically and have *O*-ethyl groups, and the complete structure is:

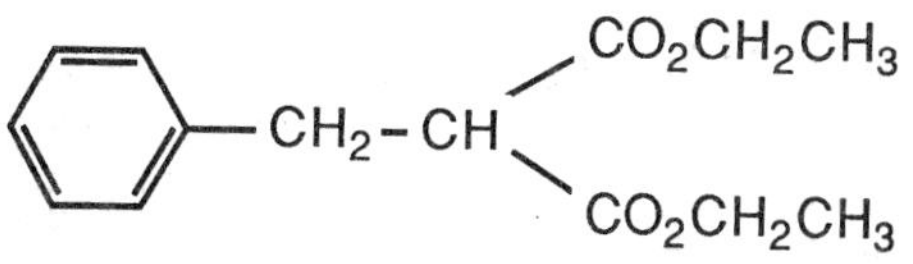

$$C_6H_5-CH_2-CH(CO_2CH_2CH_3)_2$$

9.17 Absorption in the ultraviolet region requires conjugated multiple bonds, and this structural feature is present in (c), (f), (g), and (h). Compounds (a) and (d) have no multiple bonds, and the double bonds of (b) and (e) are not conjugated.

9.18 For a compound to be colored there must be extended conjugation. While several of the compounds would seem close to meeting this requirement, all are actually colorless.

9.19 **(a)** 235 nm **(b)** 245 nm **(c)** 245 nm

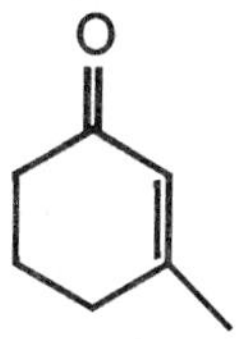

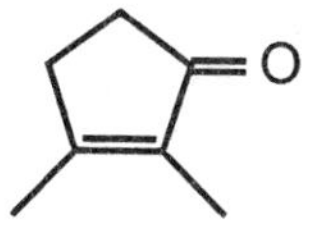

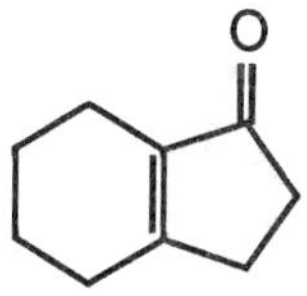

9.20 The predicted shifts are shown with each of the alternative structures:

(a) 230 nm **(b)** 235 nm **(c)** 225 nm

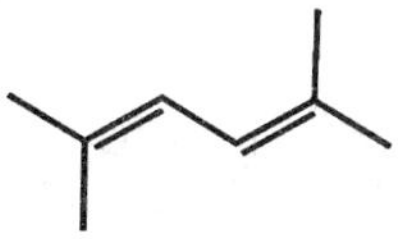

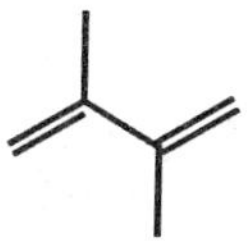

The best agreement is found for structure (a).

9.21 The predicted shifts are shown with each of the isomeric structures:

(a)

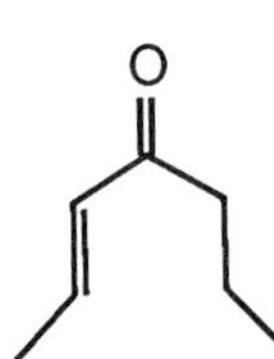

225 nm

(b)

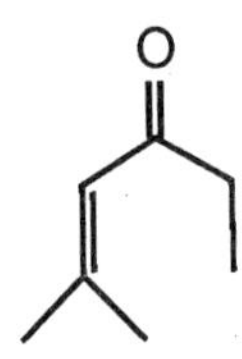

235 nm

(c)

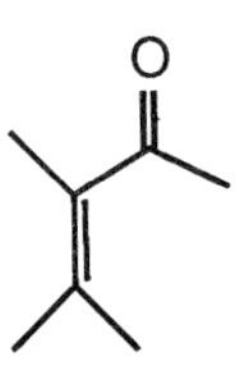

245 nm

The best agreement is found for structure (c).

9.22 The predicted shifts are shown with each of the isomeric structures:

(a)

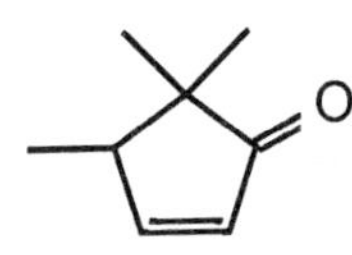

225 nm

(b)

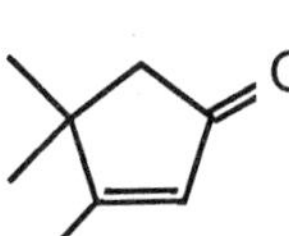

235 nm

(c)

245 nm

The best agreement is found for structure (a).

9.23 (a)

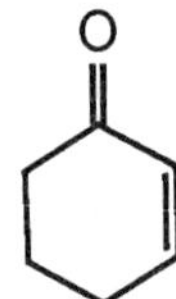

The substituent effect for α,β-unsaturation is $-40\ cm^{-1}$. The stretching frequency should be in the region of $1660–1685\ cm^{-1}$.

(b)

The substituent effect for a 5-membered ring is $+30\ cm^{-1}$. The stretching frequency should be in the region of $1730–1755\ cm^{-1}$.

(c)

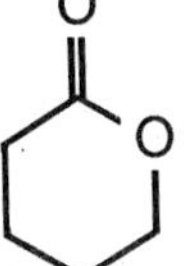

The substituent effect for an ester is +30 cm^{-1}. The stretching frequency should be in the region of 1730–1755 cm^{-1}.

(d)

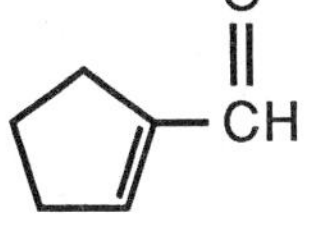

The substituent effects for α,β-unsaturation and for an aldehyde are −40 and +20 cm^{-1}, respectively. The stretching frequency should be in the region of 1680–1705 cm^{-1}.

(e)

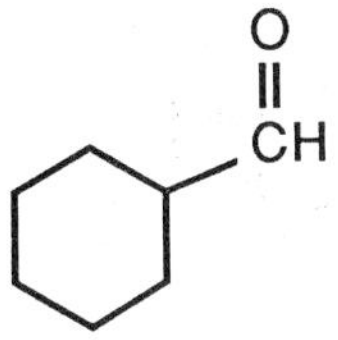

The substituent effect for an aldehyde is +20 cm^{-1}. The stretching frequency should be in the region of 1720–1745 cm^{-1}.

(f)

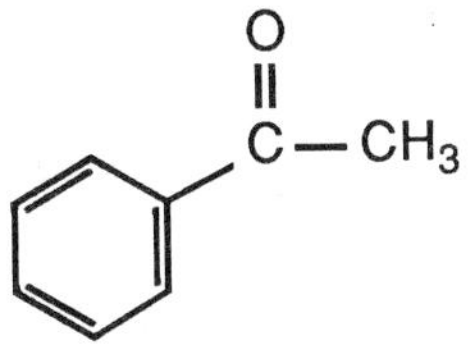

The substituent effect for an aromatic ketone is −25 cm^{-1}. The stretching frequency should be in the region of 1675–1700 cm^{-1}.

(g)

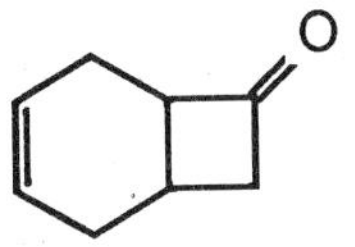

The substituent effect for a 4-membered ring is +60 cm^{-1}. The stretching frequency should be in the region of 1760–1785 cm^{-1}.

(h)

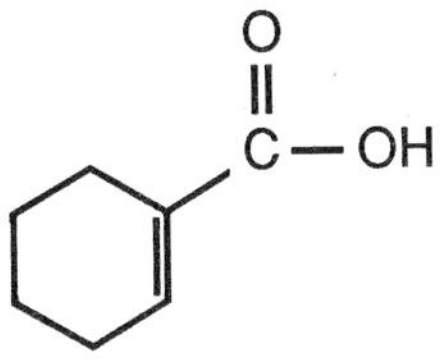

The substituent effect for α,β-unsaturation is −40 cm^{-1}; there is no substitutent effect for a carboxylic acid. The stretching frequency should be in the region of 1660–1685 cm^{-1}.

9.24 Hydration of 3-methyl-1-butyne via hydroboration-oxidation would be expected to yield the aldehyde, 3-methylbutanal. On the other hand, the isomeric ketone, 3-methyl-2-butanone, would be a reasonable by-product.

$$CH_3-\underset{CH_3}{CH}-\overset{O}{\overset{\|}{C}}-CH_3$$

The carbonyl peak at 1726 cm^{-1} is consistent with either and aldehyde or a ketone. The key is the peak at 2720 cm^{-1}, which is highly characteristic for an aldehyde. Hence the product of the reaction was indeed 3-methylbutanal.

$$CH_3-\underset{CH_3}{CH}-CH_2CHO$$

9.25 Hydrogenation of 1-nonyne could lead to the isolation of unreacted alkyne, of the desired alkene, or of the alkane if reduction proceeded too far. The peak near 1640 cm^{-1} clearly suggests that the compound is an alkene, and unreacted alkyne can be ruled out be the absence of any peaks between 2100 and 2250 cm^{-1}. The spectrum is that of 1-nonene.

$$CH_3CH_2CH_2CH_2CH_2CH_2CH_2CH=CH_2$$

9.26 The infrared spectrum of the product lacks any peaks corresponding to either a double bond (1600–1680 cm^{-1}) or a triple bond (2100 and 2250 cm^{-1}) Consequently, overreduction must have occurred and the compound is hexane:

$$CH_3CH_2CH_2CH_2CH_2CH_3$$

9.27 The molecular formula and structural information provided in the question actually give you enough information to answer the question, but first consider the infrared spectrum. You can see the series of low-intensity peaks from 1700 to 2000 cm^{-1} that result from the aromatic ring, but the strongest peak in the spectrum is located at 1688 cm^{-1}. This is undoubtedly a carbonyl peak and it is not an aldehyde, because the spectrum lacks the characteristic aldehyde peak at 2720 cm^{-1}. Therefore (since the compound has a single oxygen) it must be a ketone carbonyl, but it is outside the typical region of 1700–1725 cm^{-1} for "normal" ketones. This corresponds to a shift of between −12 and −37 cm^{-1}, and is typical for an *aromatic* ketone. Combining the subunit of an aromatic ketone with an ethyl group accounts for all nine carbons of the compound, so its structure must be:

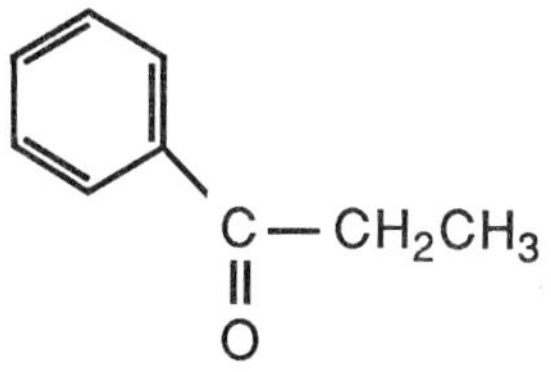

C_6H_5–C(=O)–CH_2CH_3

9.28 Compound **D** the structure of which is given, has the same carbon skeleton as compound **A.** Therefore you know that the ozonolysis reaction must cleave **A** into two 8-carbon fragments, i.e., cleavage occurs between the 6-membered ring and the 3-carbon chain. The infrared spectrum of compound B shows a carbonyl peak at 1718 cm^{-1}, which is characteristic of either an acyclic or 6-membered ring ketone. It must be the latter if cleavage affords two 8-carbon fragments, and the structures of compounds **A, B,** and **C** are shown below:

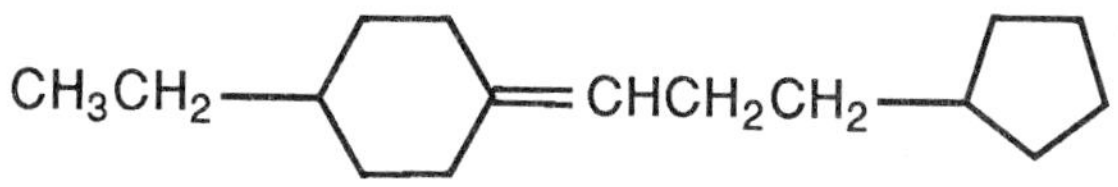

CH_3CH_2–(cyclohexylidene)=$CHCH_2CH_2$–(cyclopentyl)

A

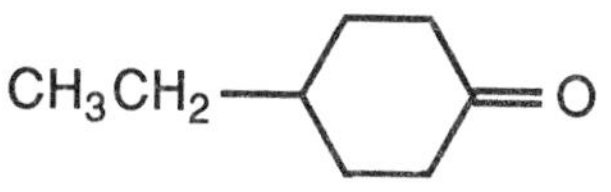

CH_3CH_2–(cyclohexanone)=O

B

HC(=O)–CH_2CH_2–(cyclopentyl)

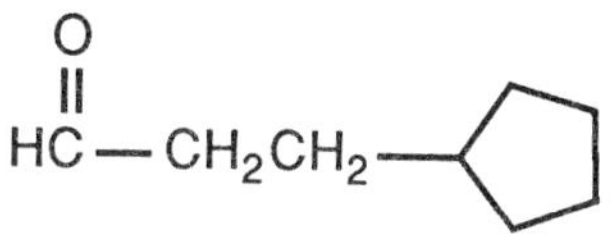

C

CHAPTER 10. AROMATIC COMPOUNDS

ANSWERS TO EXERCISES

10.1 **(a)** 1,2-dimethylbenzene (*o*-xylene)

(b) 2-methylbenzoic acid (*o*-toluic acid

(c) 2-methoxybenzoic acid (*o*-anisic acid)

(d) 2-methylaniline (*o*-toluidine)

(e) 2-methoxyaniline (*o*-anisidine)

(f) 2-methylphenol (*o*-cresol)

(g) 2-methoxybenzaldehyde (*o*-anisaldehyde)

(h) 2-hydroxybenzoic acid (salicylic acid)

(i) 1,2-benzenedicarboxylic acid (phthalic acid)

(j) 1,3-benzenedicarboxylic acid (isophthalic acid)

(k) 1,4-benzenedicarboxylic acid (terephthalic acid)

(l) 1,2-dihydroxybenzene (pyrocatechol)

(m) 1,3-dihydroxybenzene (resorcinol)

(n) 1,4-dihydroxybenzene (hydroquinone)

10.2 (a) CH_3 CH_3 (b) CO_2H CH_3 (c) CO_2H OCH_3

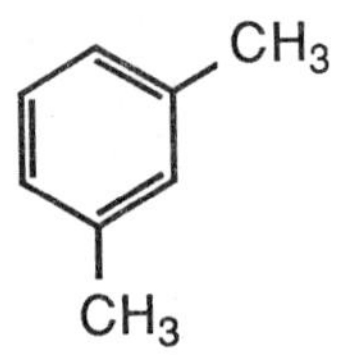

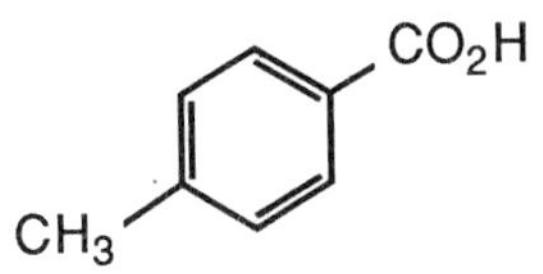

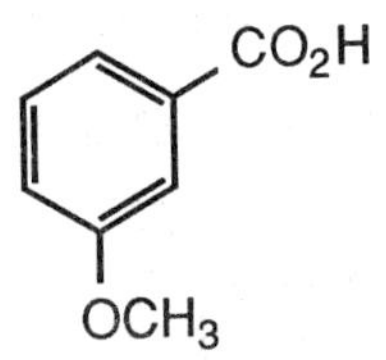

(d) NH_2, CH_3

(e) NH_2, CH_3O

(f) CH_3, HO

(g) CHO, OCH_3

10.3 **(a)** Six dichloro isomers would be possible:

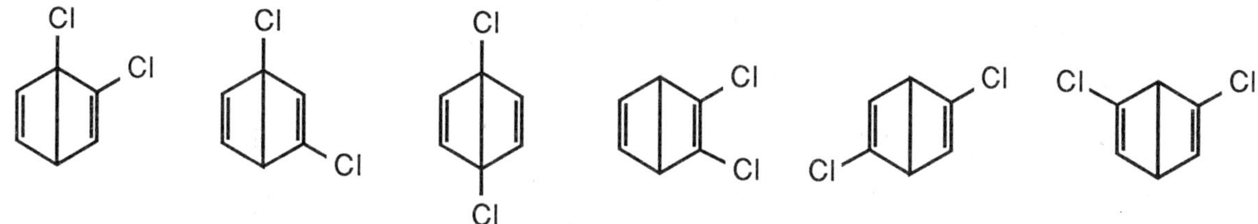

(b) This is clearly inconsistent with the experimental observation of only three isomeric dichlorobenzenes.

10.4 **(a)** Four dichloro isomers of prismane are possible. Three are constitutional isomers, while ii and iii are enantiomers.

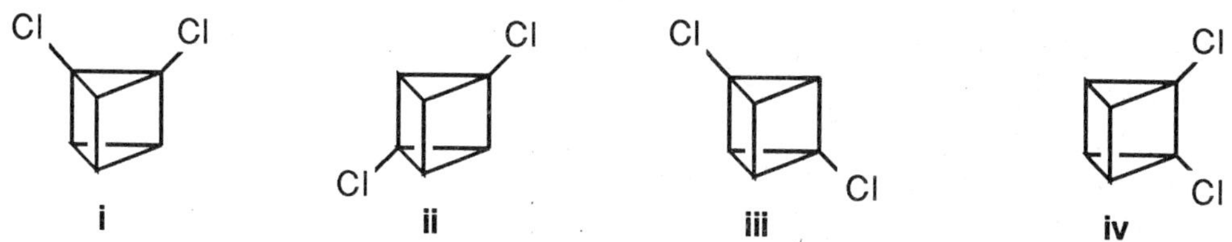

(b) Three constitutional isomers are possible. Two are chiral so enantiomers are possible. Structures i and v are enantiomers as are ii and iii.

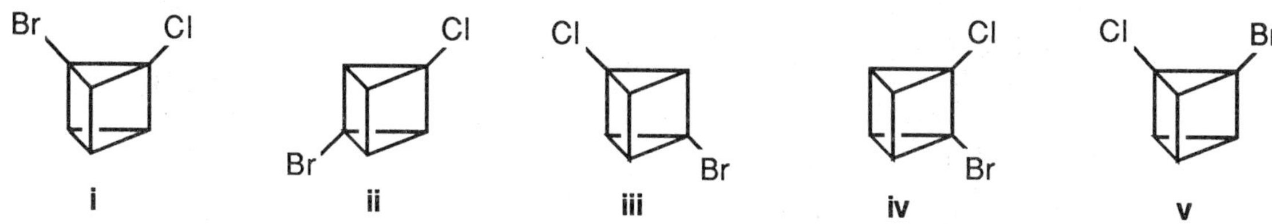

(c) There are four symmetry planes in unsubstituted prismane: one is the plane that lies parallel to (and is equidistant between) the three-membered rings; the other three are planes that include one of the bonds shared by two 4-membered rings and bisect both three-membered rings. When the two substituents are on different 3-membered rings but on the same 4-membered ring, they lie *in* a symmetry plane. Consequently structure iv is achiral in both parts (a) and (b). Structures ii and iii are chiral in both cases. The two chlorines in part (a) structure i are equivalent with respect to a symmetry plane, but this symmetry plane is not present for the corresponding structure in (b) where the substituents are different.

10.5 Eight isomers can be drawn:

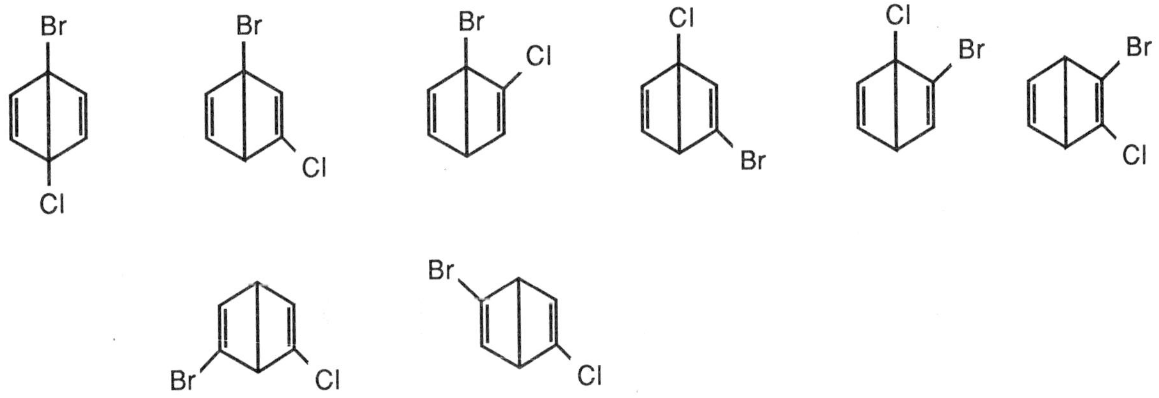

Br Cl Br Cl Br Cl Cl Br Cl Br Br Cl

Br Cl Br Cl

10.6

10.7 Elimination could occur to form 1-bromocyclobutene, which is incapable of losing another HBr to yield cyclobutadiene.

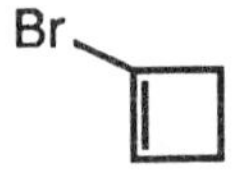

Br

Substitution (Chapter 12) could occur, resulting in replacement of –Br by –OH. Ring opening of the strained cyclobutene intermediates (Chapter 27) could also occur.

10.8 Addition occurs at one terminus of the conjugated π system to generate a resonance-stabilized carbocation. Subsequent reaction of the cation occurs at the other terminus of the π system:

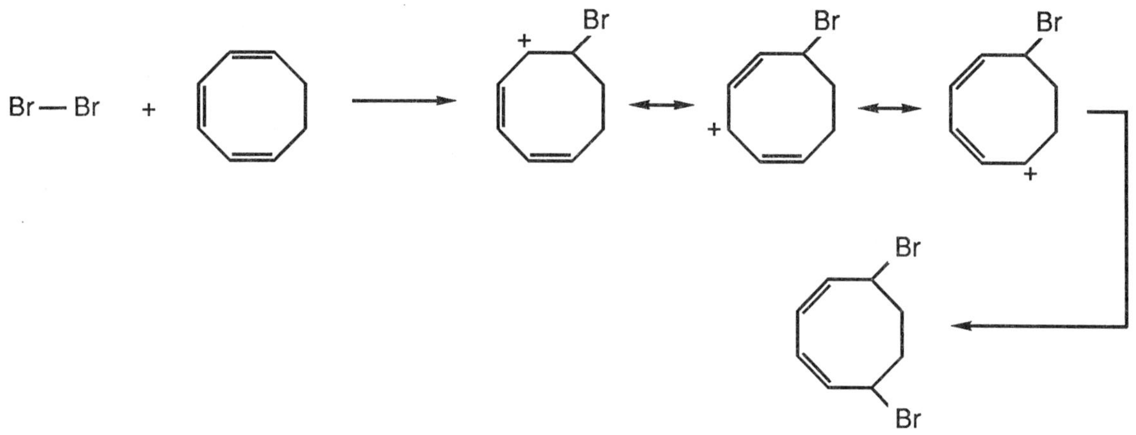

Br—Br + [cyclooctatetraene] → [Br-substituted cations, resonance structures] → [1,4-dibromocyclooctadiene]

10.9 Only four isomers would be possible if the compound were aromatic, corresponding to 1,2- 1,3- 1,4- and 1,5-substitution:

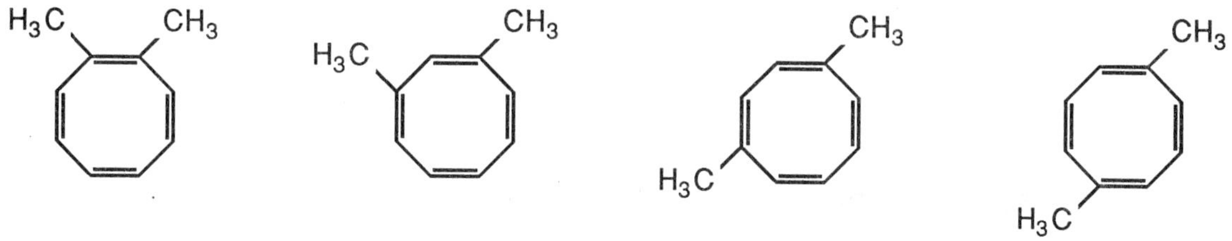

H_3C CH_3 H_3C CH_3 CH_3 H_3C CH_3 H_3C

10.10 The double bonds have fixed positions, so there are six different structures:

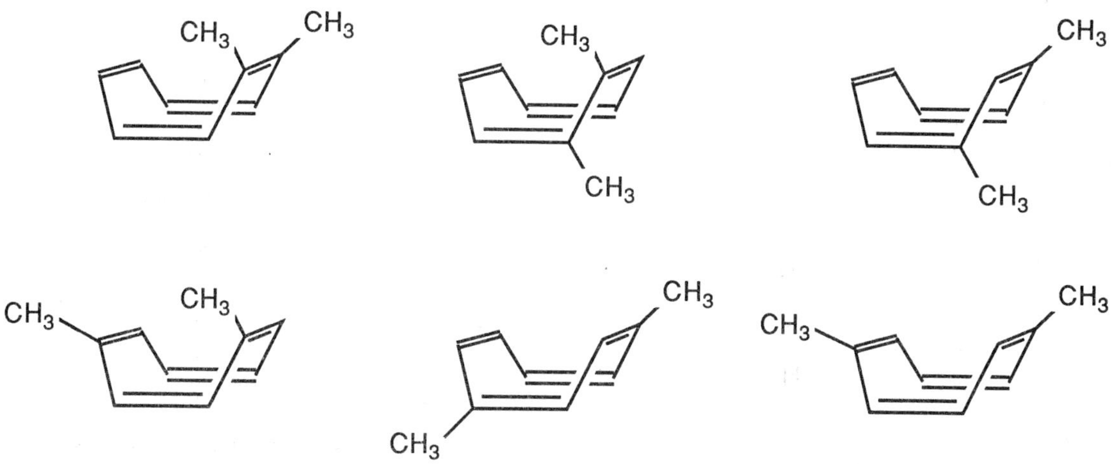

CH_3 CH_3 CH_3 CH_3 CH_3 CH_3

CH_3 CH_3 CH_3 CH_3 CH_3 CH_3

10.11 **(a)** All of the species shown are Huckel systems.

(b) Only i and iv should be aromatic; both are 6-π electron systems. Structure ii has eight π electrons, while iii and v both have four π electrons.

10.12 chlorination:

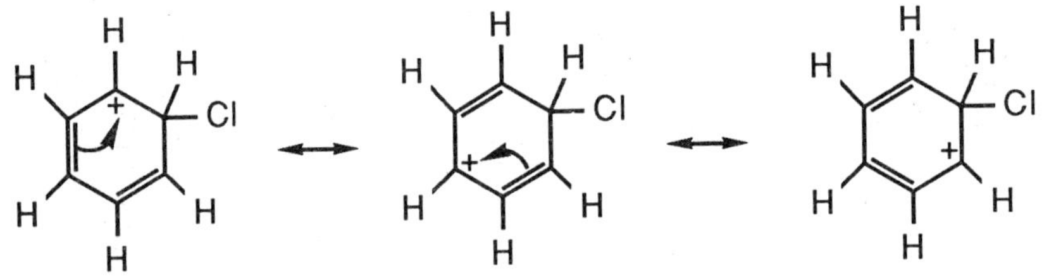

nitration:

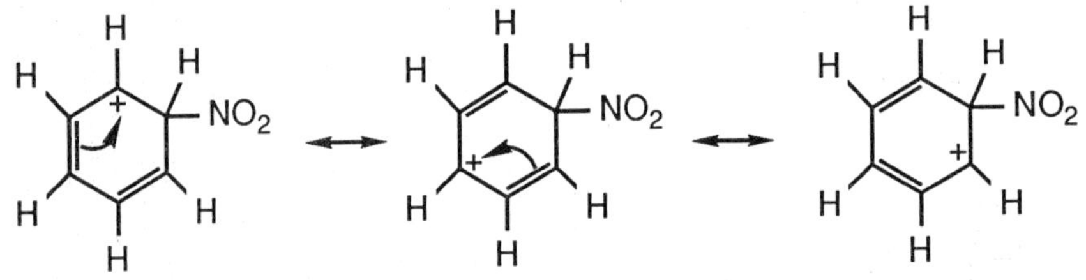

sulfonation:

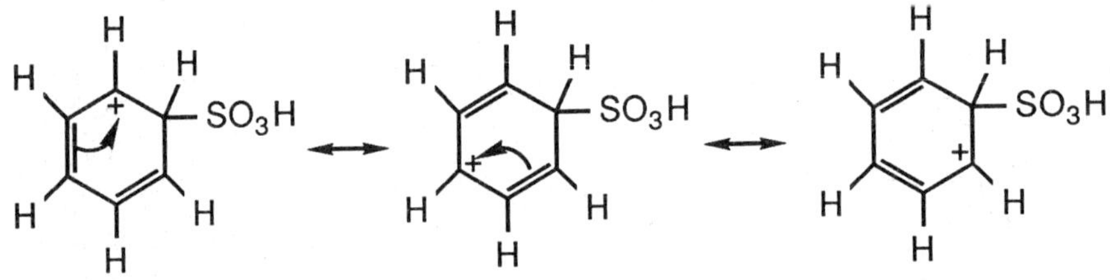

In each case positive charge resides on the three carbons that are *ortho* and *para* to the carbon atom that undergoes electrophilic attack.

10.13 (a)

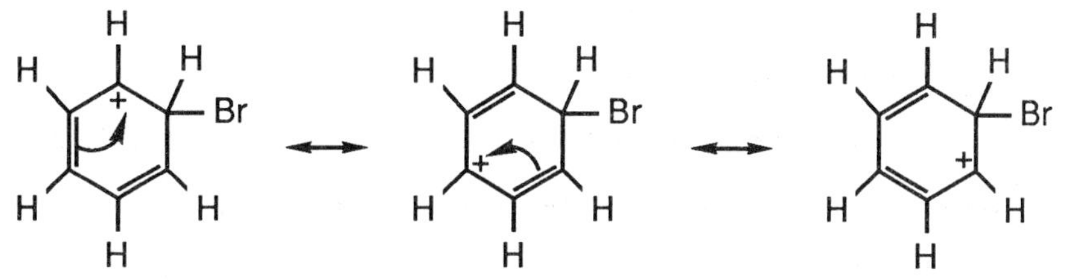

H, H, H, Br, H, H ⟷ H, H, H, Br, H, H ⟷ H, H, H, Br, H, H

(b)

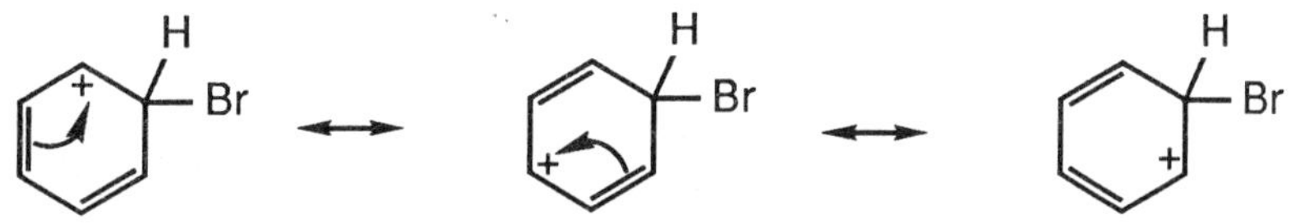

H, Br ⟷ H, Br ⟷ H, Br

10.14 *ortho*

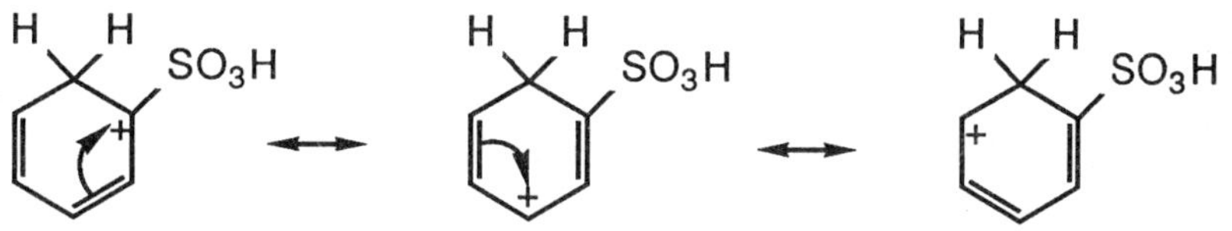

H H SO_3H ⟷ H H SO_3H ⟷ H H SO_3H

meta

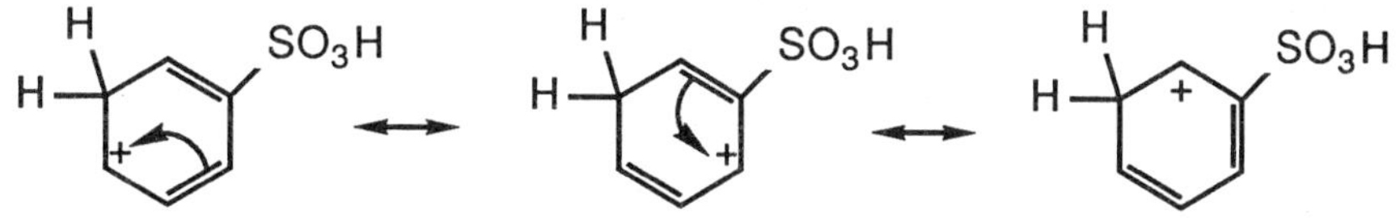

H H SO_3H ⟷ H H SO_3H ⟷ H H SO_3H

para

SO_3H H H ⟷ SO_3H H H ⟷ SO_3H H H

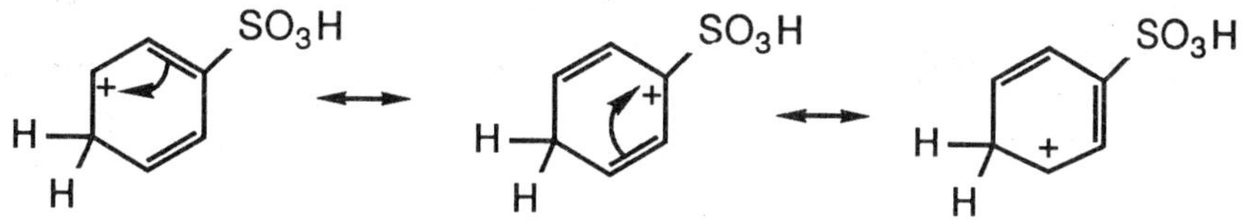

None of these intermediates could yield benzene by direct loss of the SO_3H group.

10.15 (a) The negative charge can reside on the phenolic oxygen, on the *ortho* or *meta* carbons, or on the nitro group. In all but the latter case two resonance forms for the nitro group can be drawn:

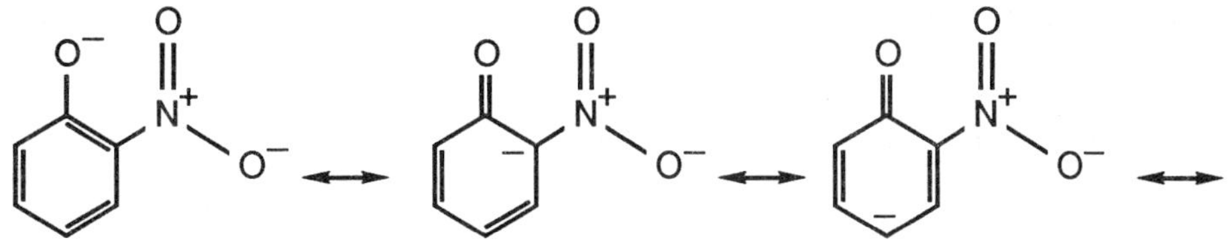

O^- O N^+ O^- ⟷ O O N^+ O^- ⟷ O O N^+ O^- ⟷

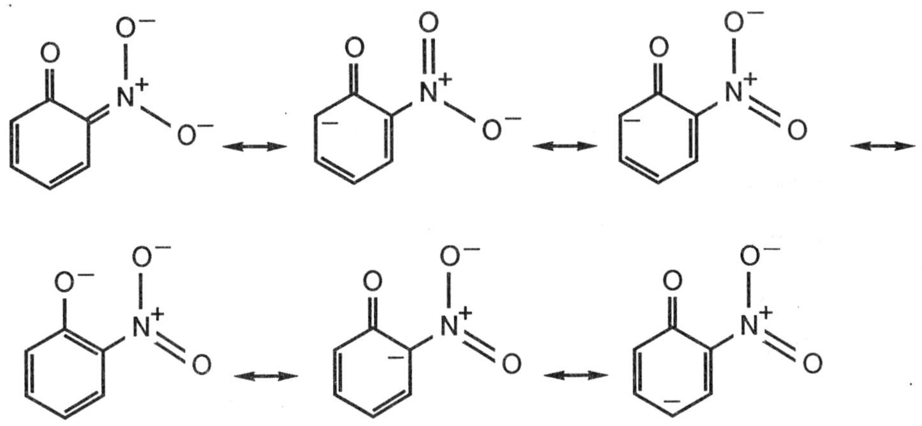

(b) The positive charge can reside on the carbon of the side chain, on the oxygen of the methoxy group, or on the *ortho* and *meta* carbons.

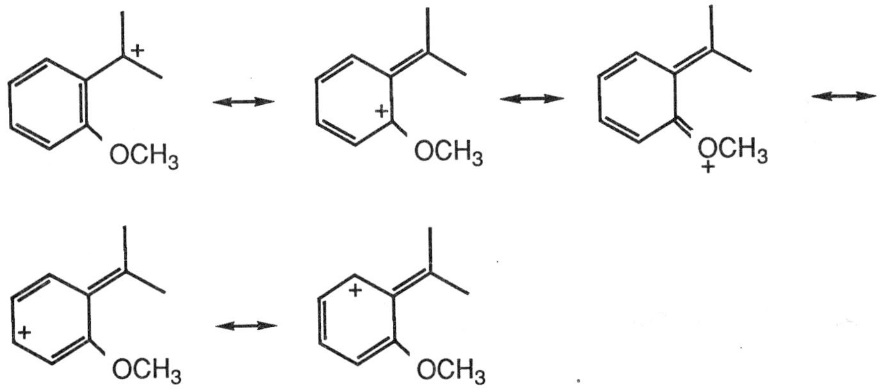

ANSWERS TO PROBLEMS

10.1 **(a)** 4-ethylphenol

(b) ethyl *m*-tolyl ether

(c) 3-bromo-5-iodo-*N*-methylaniline

(d) *o*-propylbenzoic acid

(e) 4-bromostyrene

(f) 2-chloro-3-fluorophenol

(g) 1-(*m*-iodophenyl)ethanol

(h) 2,6-dibromophenylacetic acid

(i) 2,6-dimethylbenzoic acid

(j) *o*-fluoropropiophenone

(k) *p*-nitroaniline

(l) 4-chloro-1,2-benzenedicarboxylic acid

10.2

(a) $C_6H_5CH_2OH$

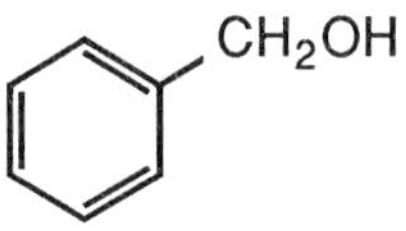

(b) $C_6H_5CH{=}CHOCCH_3$ (with C=O)

(c) $C_6H_5CH{=}CH$–cyclononyl

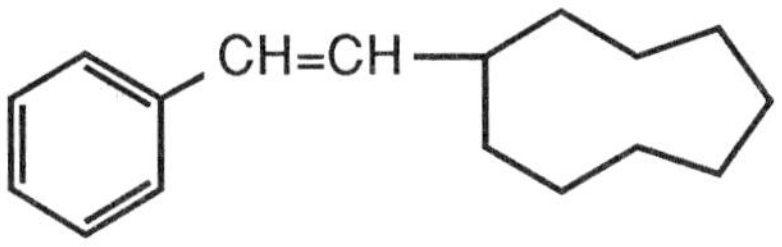

(d) $C_6H_5CH(OCH_3)CH_3$

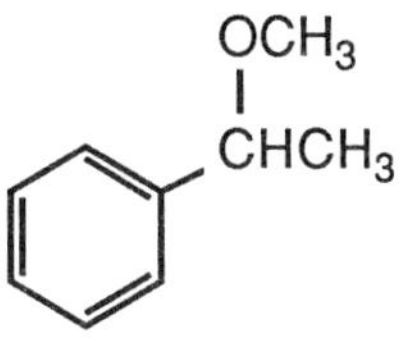

(e) $C_6H_5CH_2CH_2O{-}C({=}O){-}C_6H_5$

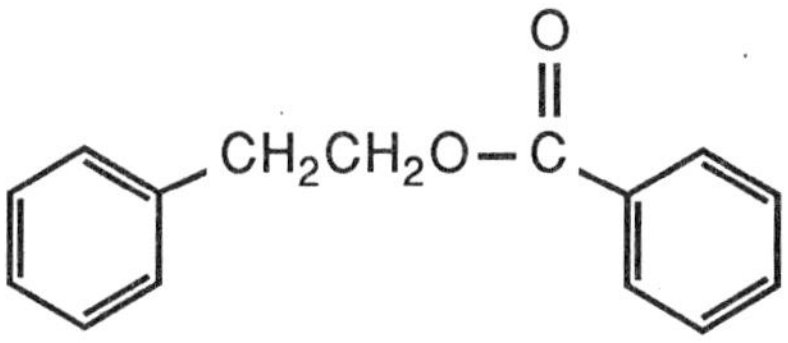

(f) $C_6H_5OCH_2(CH_2)_6CH_3$

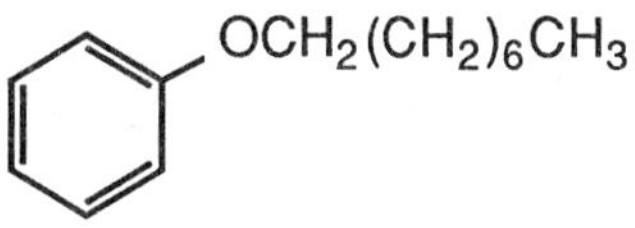

10.3

(a) $C_6H_5CH(CH_2Br)CH_2CO_2H$

(b) p-Br$C_6H_4CH(CH_3)CH_2CO_2H$

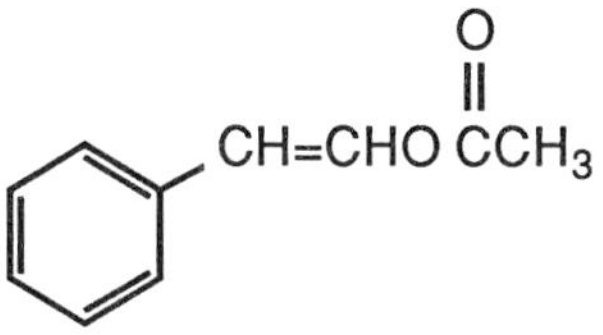

(c) $C_6H_5CH_2{-}C({=}O)OCH_2CH_3$

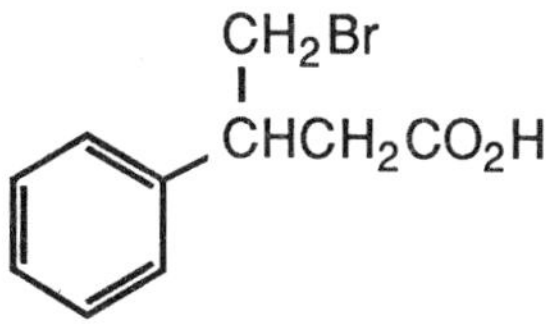

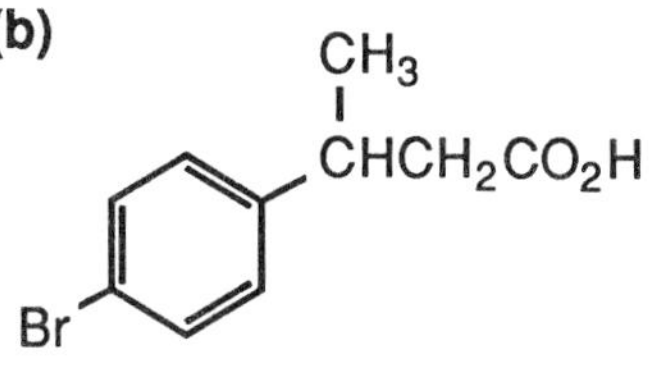

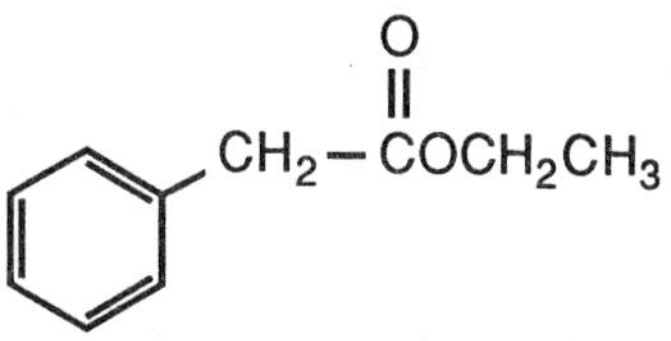

(d) CH_3CH_2 … $O-\overset{O}{\overset{\|}{C}}CH_3$

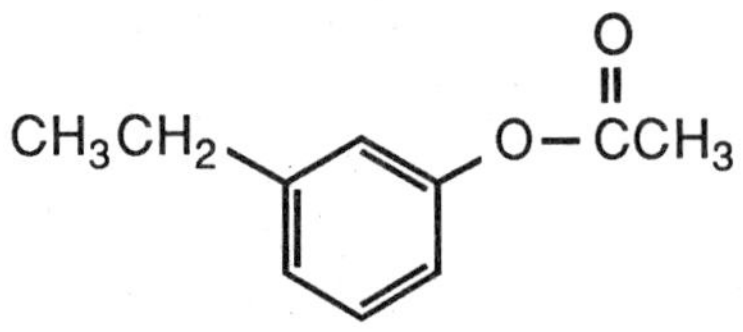

(e) NH_2, Cl, CH_2CH_3

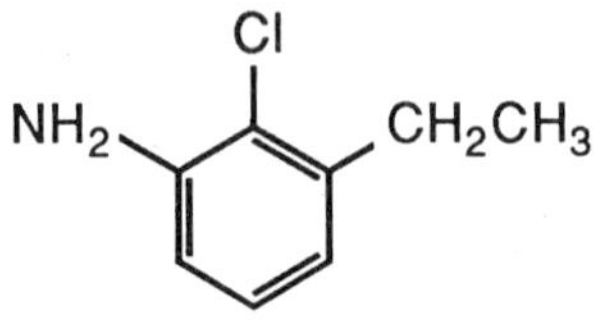

(f) NH_2, CH_2CH_2Cl

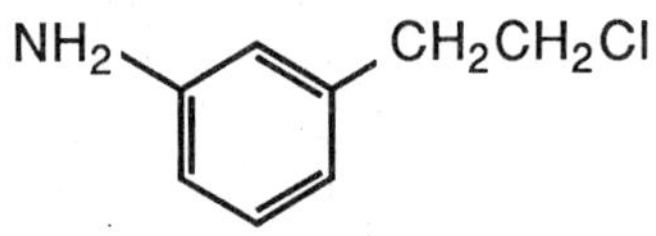

(g) NH_2, CH_2CH_3, CH_2CH_3

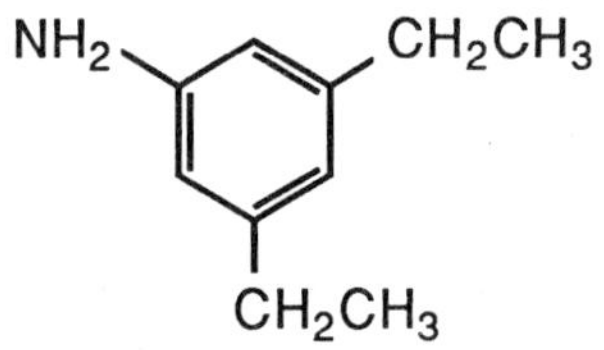

(h) $N(CH_2CH_3)_2$

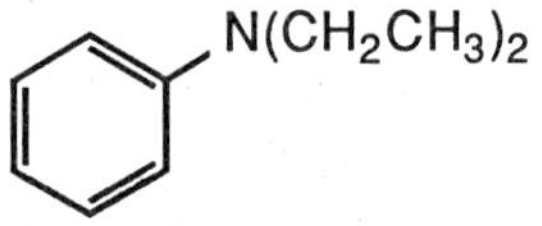

10.4

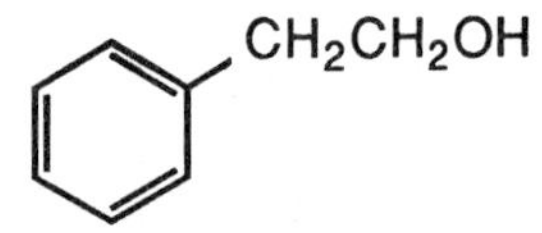

CH_2CH_2OH — ß-phenethyl alcohol

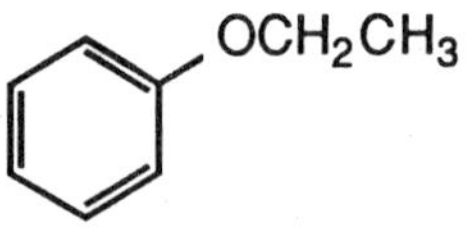

OCH_2CH_3 — ethyl phenyl ether

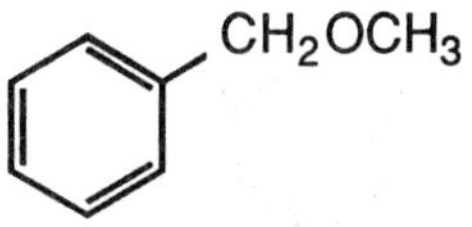

CH_2OCH_3 — benzyl methyl ether

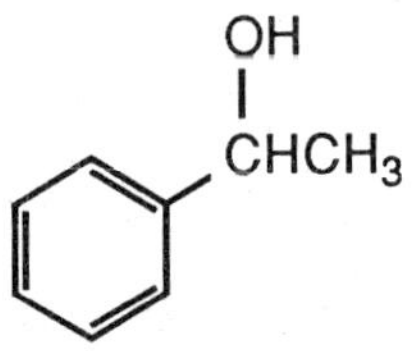

OH, $CHCH_3$ — α-phenethyl alcohol

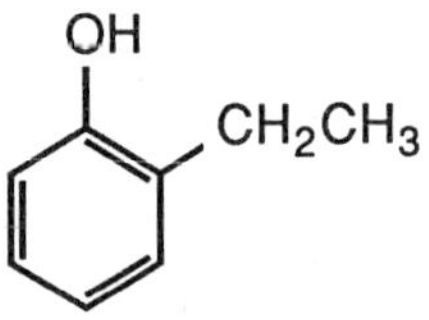

OH, CH_2CH_3 — o-ethylphenol

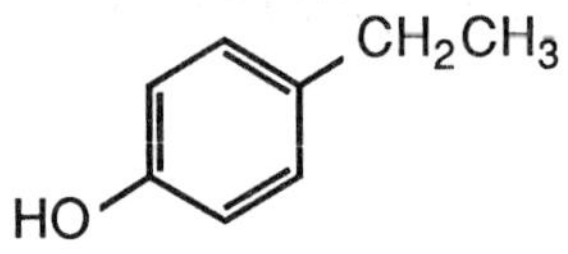

HO, CH_2CH_3 — p-ethylphenol

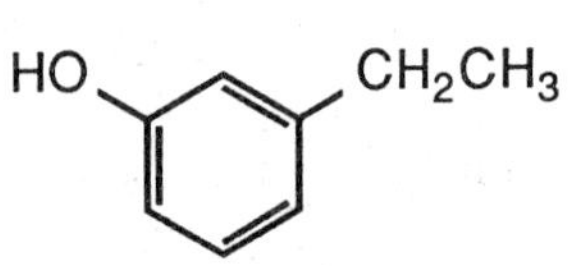

HO, CH_2CH_3 — m-ethylphenol

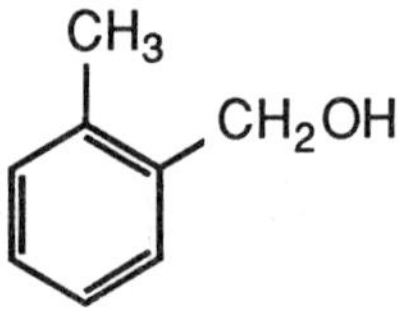

CH_3, CH_2OH — o-tolyl methanol

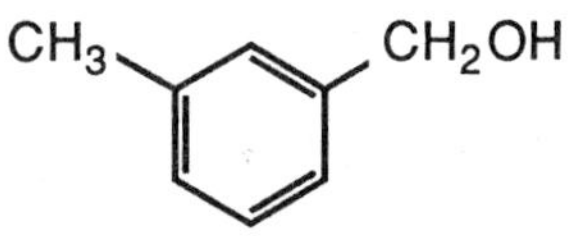

CH_3, CH_2OH — m-tolyl methanol

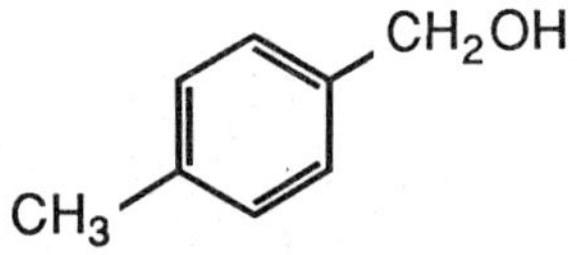

CH_2OH, CH_3 — p-tolyl methanol

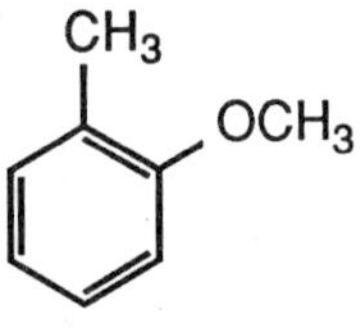

CH_3, OCH_3 — o-methylanisole

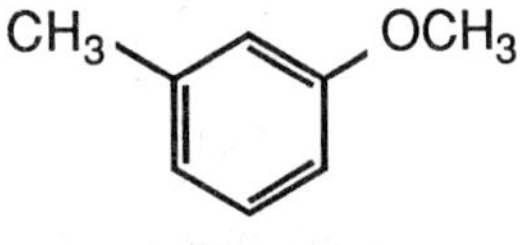

CH_3, OCH_3 — m-methylanisole

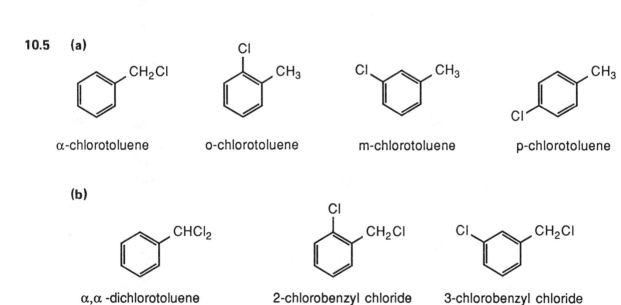

p-methylanisole

2,3-dimethylphenol

2,4-dimethylphenol

2,5-dimethylphenol

2,6-dimethylphenol

3,4-dimethylphenol

3,5-dimethylphenol

10.5 **(a)**

α-chlorotoluene

o-chlorotoluene

m-chlorotoluene

p-chlorotoluene

(b)

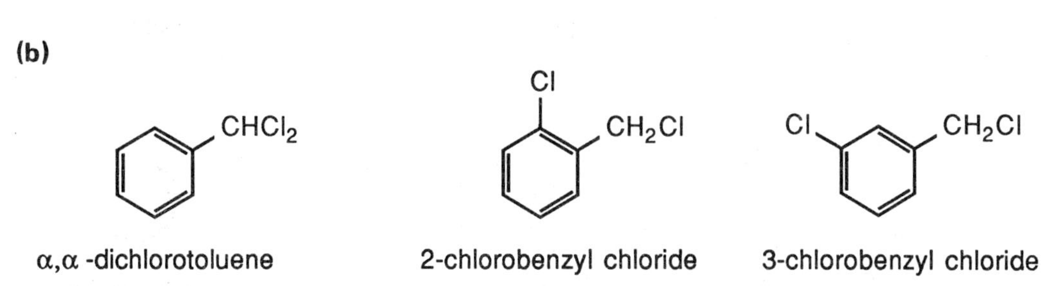

α,α -dichlorotoluene

2-chlorobenzyl chloride

3-chlorobenzyl chloride

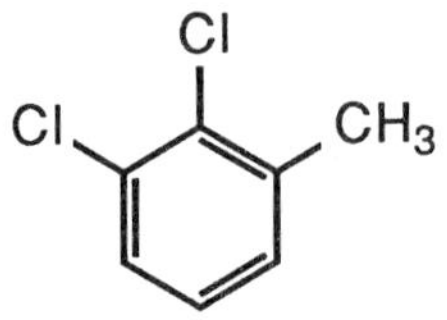

2,3-dichlorotoluene

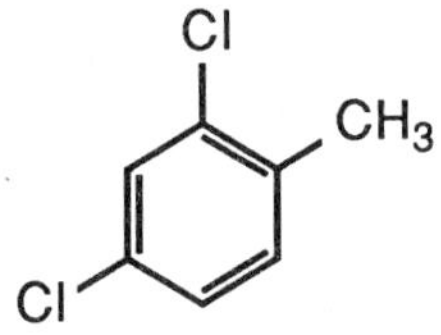

2,4-dichlorotoluene

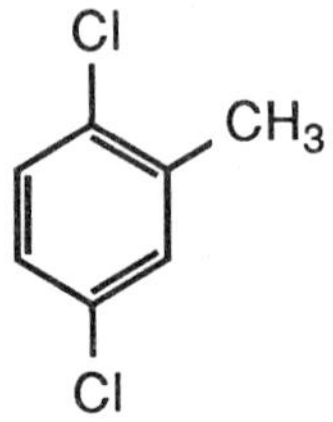

2,5-dichlorotoluene

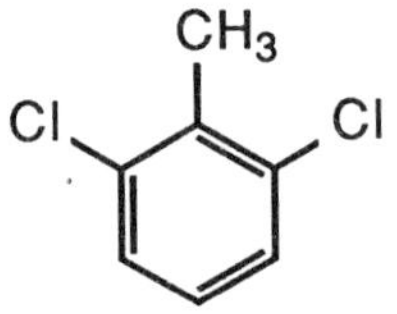

2,6-dichlorotoluene

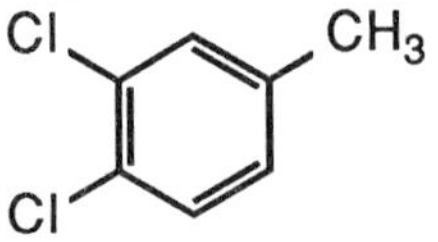

3,4-dichlorotoluene

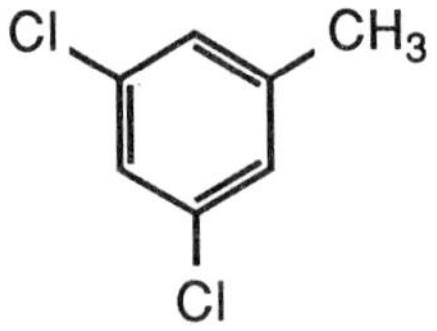

3,5-dichlorotoluene

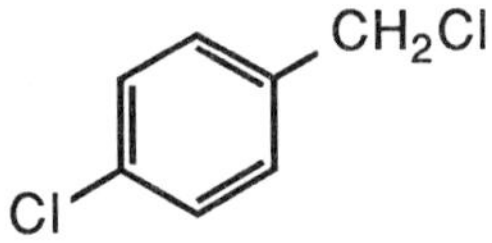

4-chlorobenzyl chloride

10.6 **(a)**

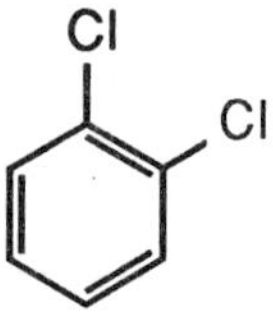

o-dichlorobenzene

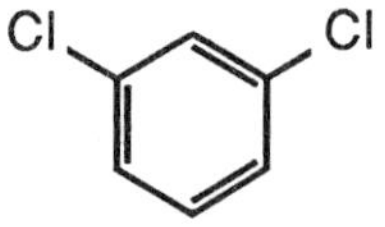

m-dichlorobenzene

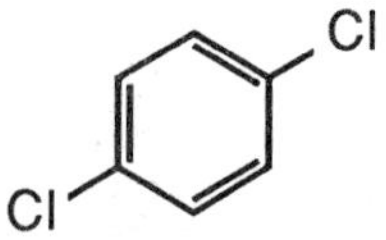

p-dichlorobenzene

(b)

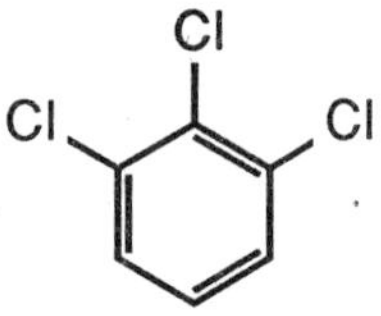

1,2,3-trichlorobenzene

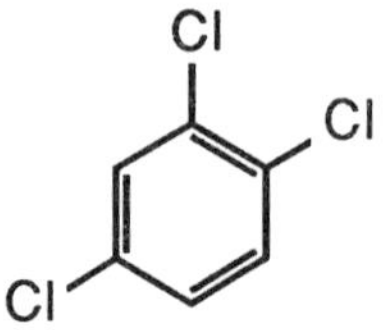

1,2,4-trichlorobenzene

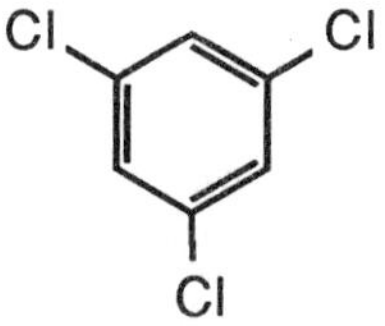

1,3,5-trichlorobenzene

10.7

(a) OH, O_2N, Cl

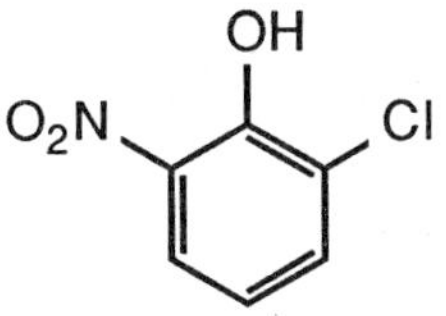

(b) OCH_3, $NHCH_3$

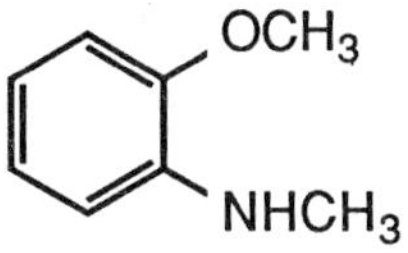

(c) O, $\overset{\|}{C}OCH_3$, $\overset{\|}{C}OCH_3$, O

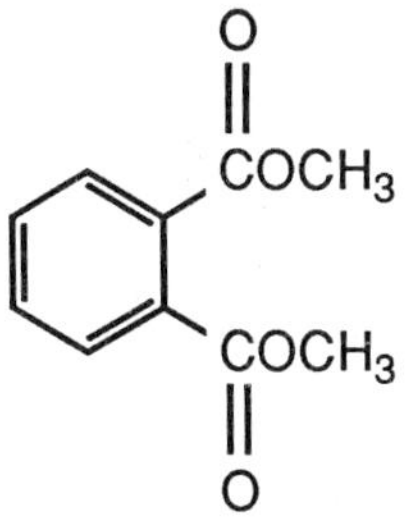

(d) O_2N, OH, NO_2

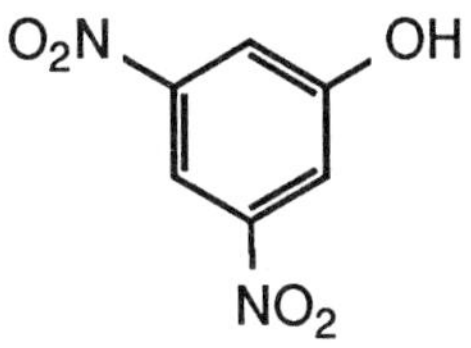

(e) Cl, NH_2

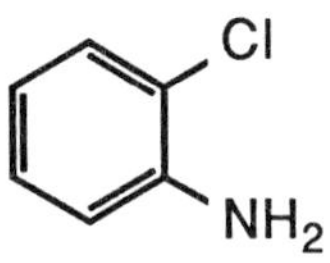

(f) CH_3, H_3C-HC, CO_2H

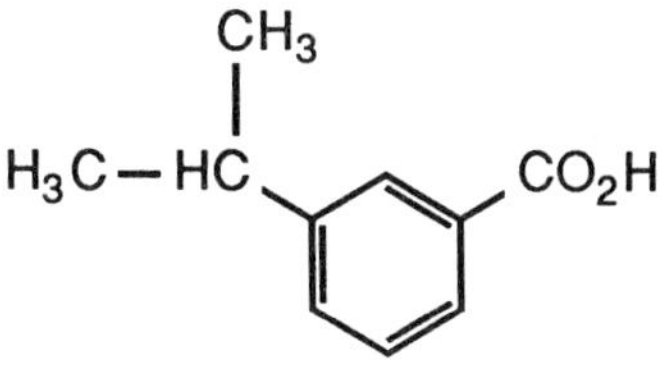

(g) O, $COCH_2CH_2CH_2CH_3$, OCH_3 (the ortho isomer is shown)

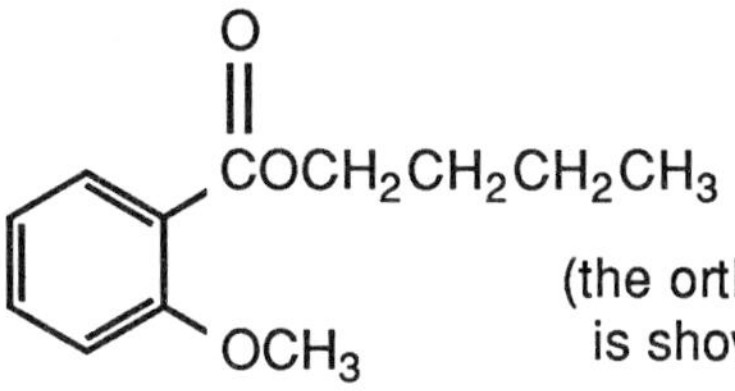

(h) H_2N, OH

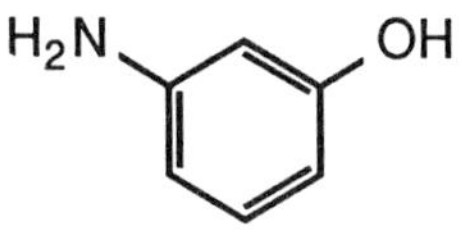

(i) CO_2H, Br

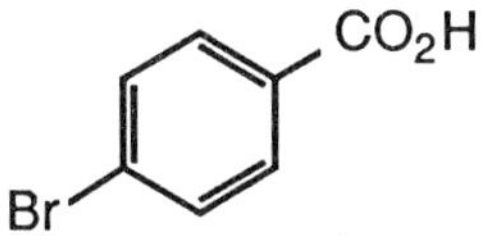

(j) Br, CH_3, Cl

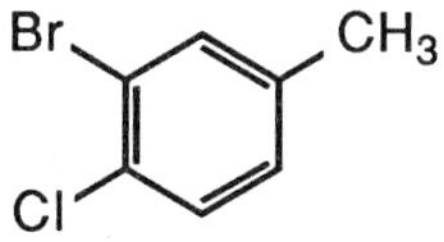

(k) CH_3, H_2N

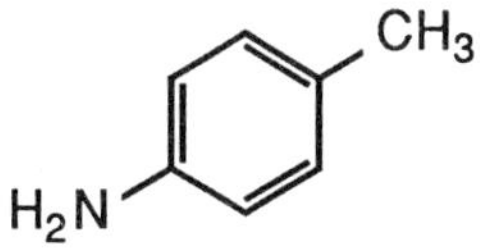

(l) CH_3, OH

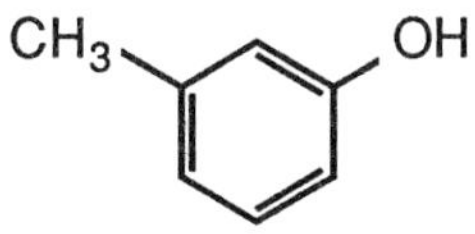

(m) CH_3, CH_3

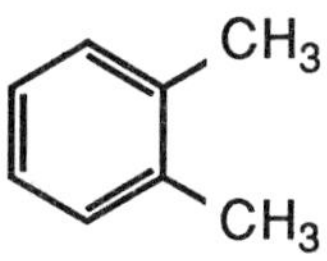

(n) OH

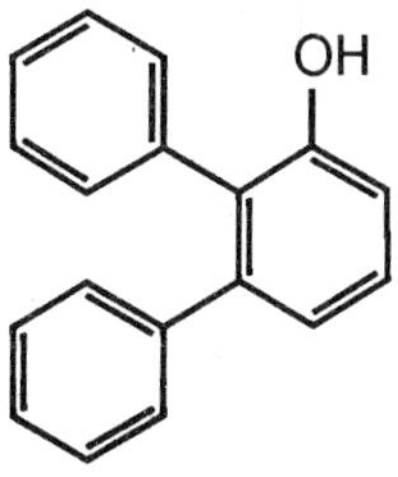

(o) CO_2H, CH_2

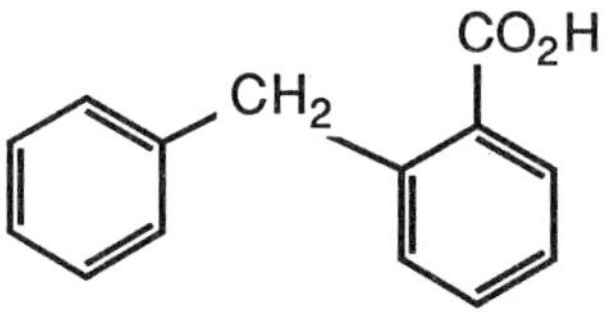

(p) O, C, CO_2H

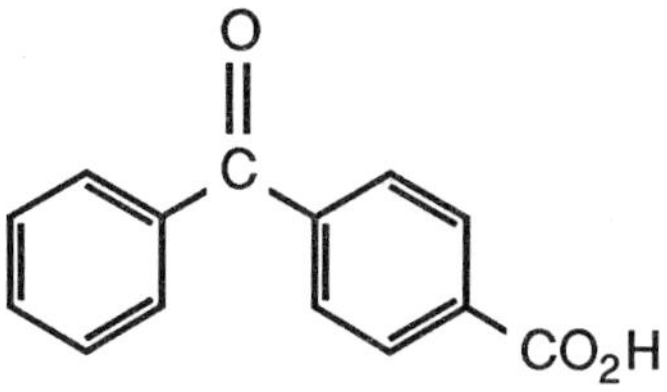

(q) CO_2H

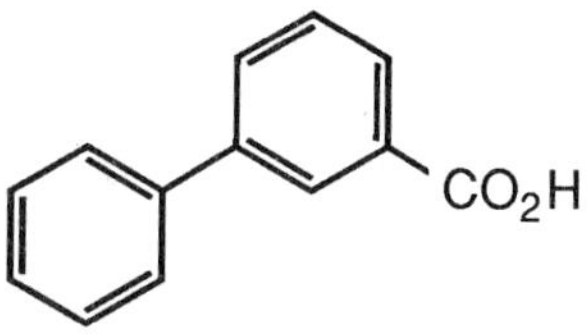

10.8 **(a)** *N*-acetyl-*p*-aminophenol

(b) ethyl *p*-aminobenzoate

(c) *p*-aminobenzoic acid

(d) *O*-acetylsalicylic acid

(e) 2,4,6-trinitrotoluene

(f) 2,4,6-trinitrophenol

10.9 The aromatic nucleus could correspond to a benzene ring, a pyridine ring, or a pyrrole ring. In the latter case a pyrrole ring has only three degrees of unsaturation, so the compound would also need another double bond as a substituent.

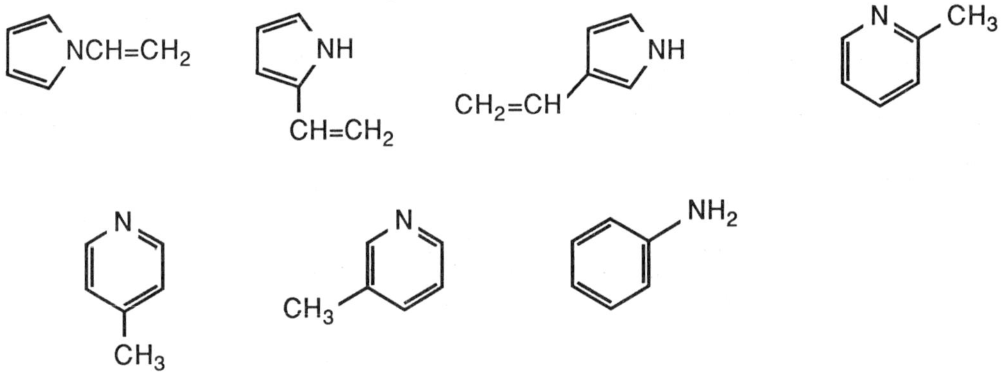

NCH=CH_2 NH CH=CH_2 CH_2=CH NH N CH_3

N CH_3 N CH_3 NH_2

10.10 Cyclononatriene would be more acidic than cycloheptatriene because the conjugate base of cyclononatriene would be a 10-π electron aromatic system. The conjugate base of cycloheptatriene, in contrast, would not be a $4n + 2$ system (having a total of 8 π electrons).

10.11

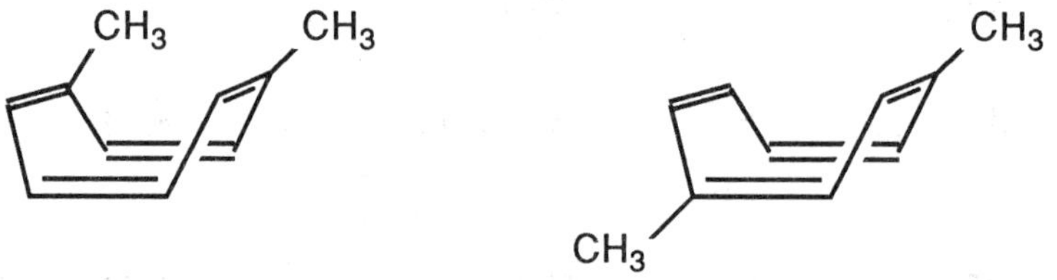

CH_3 CH_3 CH_3 CH_3

10.12 In a planar geometry there would be a severe steric interaction between the two internal hydrogens of the E double bonds:

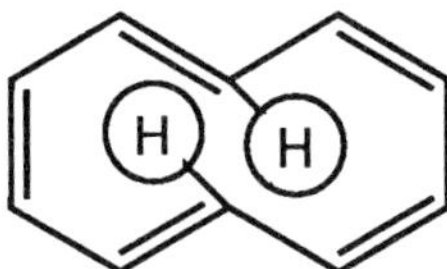

10.13 **(a)** The 7-chloro-1,3,5-cycloheptatriene would afford the aromatic cycloheptatrienyl cation, so it should ionize more readily.

(b) The cyclopropenyl cation is aromatic, whereas the cyclopentadienyl cation is not. The latter ion should therefore be more reactive, reacting rapidly with water to form 2,4-cyclopentadienol.

(c) The second compound (pyrrole) is a 6-π electron system, and protonation of the nitrogen could occur only with loss of aromaticity. The other compound can undergo protonation at the nonbonded pair of electrons that lies in the plane of the ring without loss of aromaticity.

(d) Cyclopentadienyl anion is a 6 π electron aromatic system, so it would be more stable.

(e) The cycloheptatrienyl cation is a planar, aromatic species. The nitrogen heterocycle is an 8-π electron system that is not aromatic and is not planar.

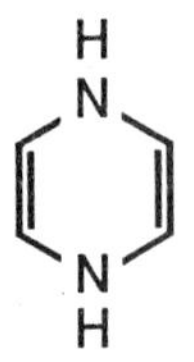

(f) The 5-membered ring cation would yield the aromatic compound, pyrrole, as its conjugate base and would therefore be more acidic:

In contrast, the 7-membered ring cation would yield an 8-π electron system as its conjugate base.

10.14 **(a)** This is not a Huckel system (the ring contains a saturated CH_2 group), so it is not aromatic.

(b) This is a Huckel system, but it has 8 π electrons and is not aromatic.

(c) This is not a Huckel system (the ring contains a saturated CH_2 group), so it is not aromatic.

(d) This is a Huckel system with 6 π electrons, so it is aromatic.

(e) This is not a Huckel system (the ring contains a saturated CH_2 group), so it is not aromatic.

(f) This is a Huckel system, but it has 8 π electrons and is not aromatic.

(g) This is not a Huckel system (the ring contains a saturated CH_2 group), so it is not aromatic.

(h) This is a Huckel system with 6 π electrons, so it is aromatic.

(i) This is a Huckel system with 6 π electrons, so it is aromatic.

(j) This is not a Huckel system (the ring contains a saturated CH_2 group), so it is not aromatic.

(k) This is not a Huckel system (the ring contains a saturated CH_2 group), so it is not aromatic.

10.15

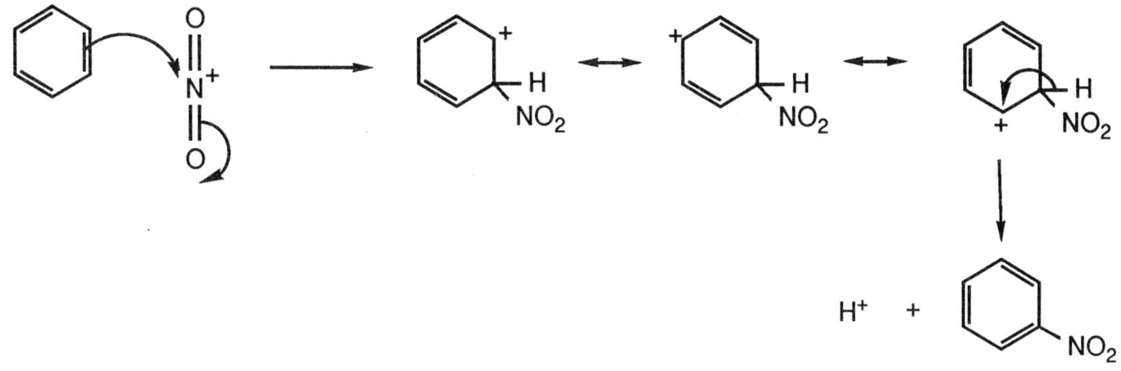

H

NO_2

H

NO_2

H

NO_2

H^+ + NO_2

10.16 While optimum π overlap of biphenyls would occur when the two rings were coplanar, this would introduce unfavorable steric interactions between the *ortho* hydrogens of the two rings. Therefore biphenyls adopt a twisted, chiral conformation. The two enantiomers of **B** are:

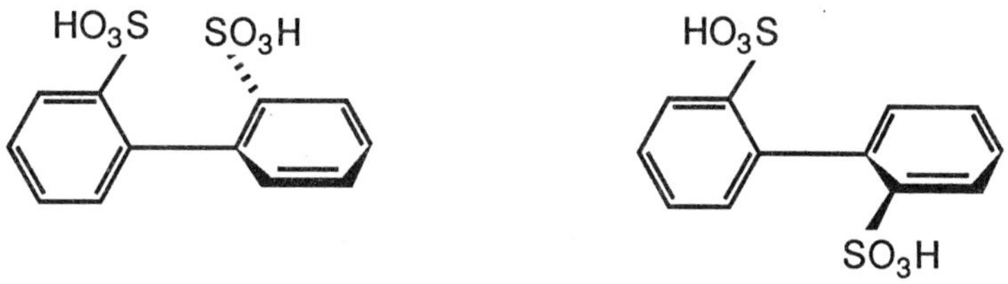

HO_3S SO_3H HO_3S

SO_3H

These enantiomers differ only by rotation about the bond between the aromatic rings, and this rotation is rapid for most biphenyls. When there are large *ortho* substituents, as in the present case, the rotation becomes slow and the enantiomeric forms can be isolated.

10.17 *ortho*: the methyl stabilizes the cation.

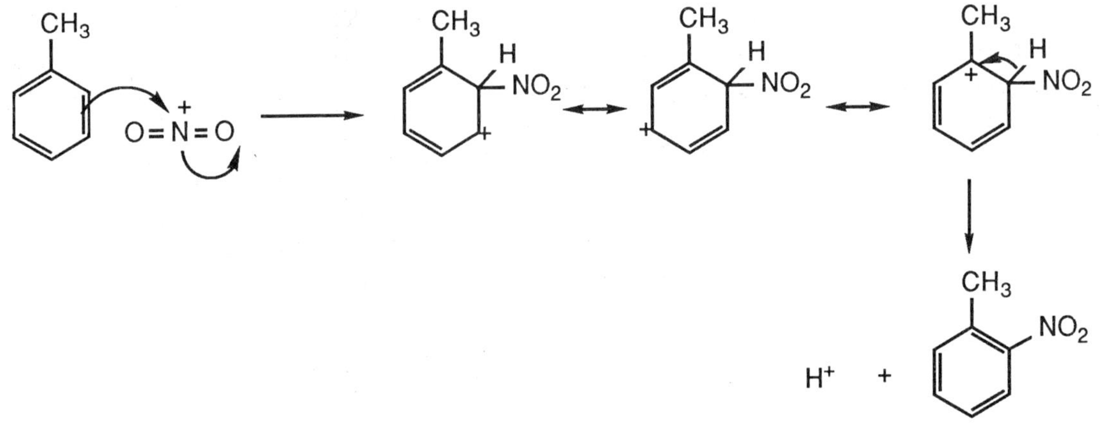

meta: the methyl does not stabilize the cation.

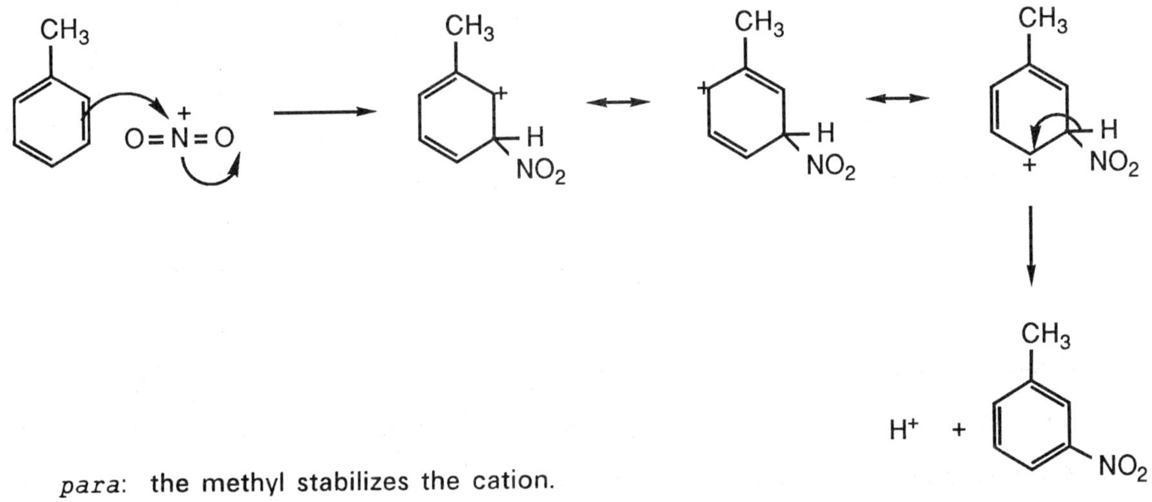

para: the methyl stabilizes the cation.

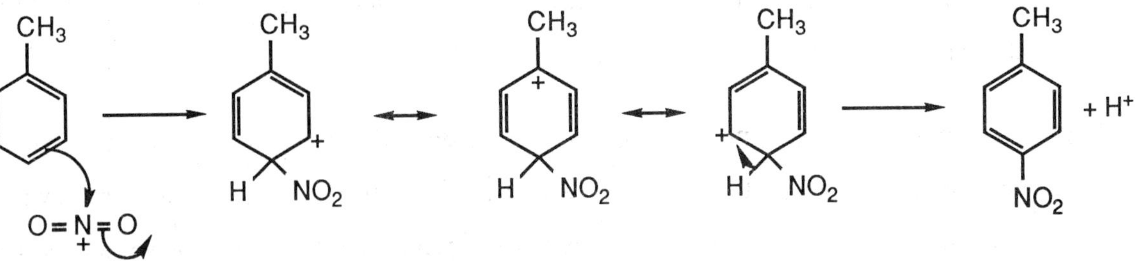

The expected products are those formed from the more stable intermediates: *ortho* and *para* substitution.

10.18

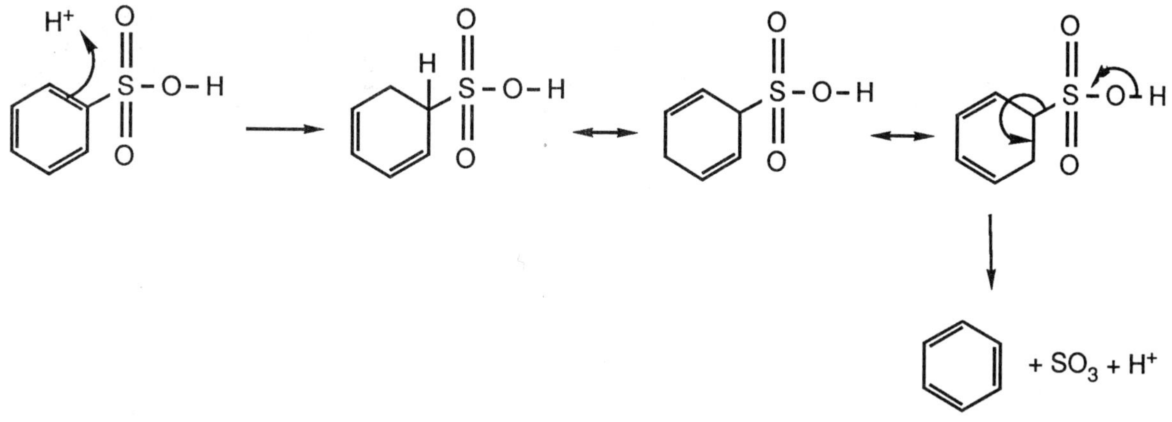

$$C_6H_5SO_3H + H^+ \rightarrow \cdots \rightarrow C_6H_6 + SO_3 + H^+$$

10.19 The ^{13}C spectrum shows only four peaks, all in the aromatic region of the spectrum (122–132 ppm). The simplest aromatic compound that contains bromine is bromobenzene, and this fits the spectrum well. (Note that one of the four peaks is less intense, corresponding to the carbon with no hydrogens.)

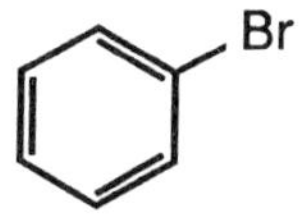

Br

Other compounds can be considered that would have only four signals in the NMR, but they would be less consistent with the spectrum. For example, 4,4′-dibromobiphenyl would have four peaks, but *two* of the carbons on each ring have no hydrogen and an intensity pattern different from that in the spectrum shown.

10.20 The influence of the nitro group on the acidity of the nitrobenzoic acids must be a combination of inductive (σ) and resonance (π) effects. While the inductive effects fall off with distance, even a *para* nitro group can effectively withdraw electron density from the carboxyl substituted carbon of the aromatic ring via resonance:

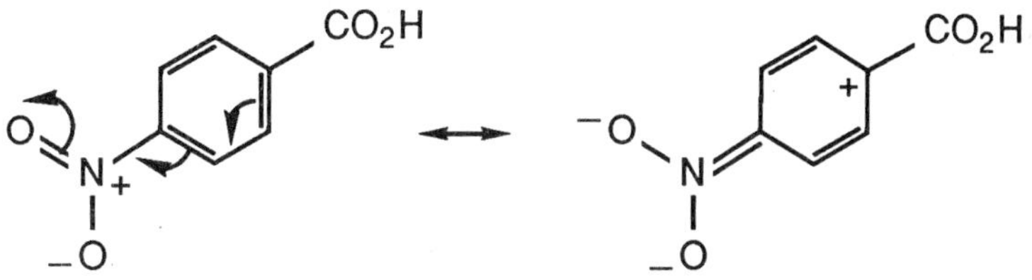

CO_2H $\longleftrightarrow$ CO_2H

10.21 This ion is a Huckel system. In the planar form there is a *p*-orbital on each carbon that is perpendicular to the plane of the ring. In addition to the eight π electrons corresponding to cyclooctatetraene, there are two additional electrons. Hence this is a 10 π electron, aromatic system.

CHAPTER 11. ALCOHOLS

ANSWERS TO EXERCISES

11.1 (a) 2-methyl-1-butanol

(b) 1-chloro-5,5-dimethyl-3-hexanol

(c) potassium 1-butanolate or potassium 1-butoxide

(d) 2-methyl-2-propanol (*tert*-butyl alcohol)

(e) *cis*-3-methylcyclohexanol

11.2 (a)

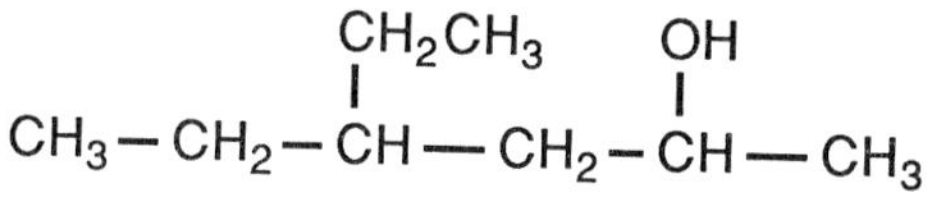

$$CH_3-CH_2-\underset{}{\overset{CH_2CH_3}{\overset{|}{C}H}}-CH_2-\overset{OH}{\overset{|}{C}H}-CH_3$$

(b) $Ca(OCH_2CH_3)_2$

(c)

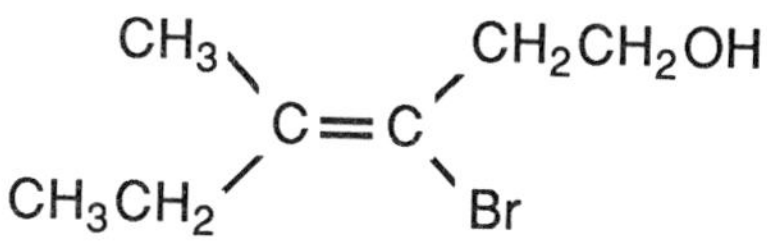

$$\begin{matrix} CH_3 & & CH_2CH_2OH \\ & C=C & \\ CH_3CH_2 & & Br \end{matrix}$$

(d) OH

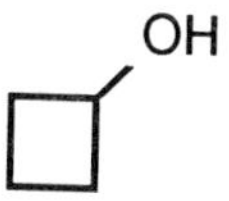

(e)

$$HC\equiv C-CH_2-CH_2-\overset{OH}{\overset{|}{C}H}-CH_2-CH_2-CH_3$$

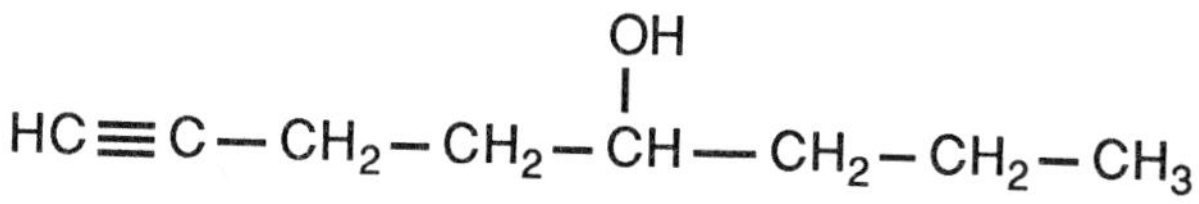

11.3 (a)

$$\text{C}_6\text{H}_{11}\text{CH}_2\text{OH} \xrightarrow{\text{KMnO}_4} \text{C}_6\text{H}_{11}\overset{\text{O}}{\overset{\|}{\text{C}}}\text{OH}$$

$$\text{C}_6\text{H}_{11}\text{CH}_2\text{OH} \xrightarrow{\text{CrO}_3\text{-pyridine}} \text{C}_6\text{H}_{11}\overset{\text{O}}{\overset{\|}{\text{C}}}\text{H}$$

(b) $(CH_3)_2CH-\overset{O}{\overset{\|}{C}}-CH_3$

11.4 (a) CrO_3, pyridine

(b) CrO_3

11.5 $C_6H_{11}-\overset{OH}{\overset{|}{CH}}-C_6H_{11}$

11.6 (a) $CH_3CH_2CH_2CH_2CH_2CH_2CH_2OH$

(b) $CH_3-CH_2-CH=CH-CH_2-CH_2-CH_2OH$

(c) $CH_3-\overset{OH}{\overset{|}{CH}}-CH_2-CH_2-CH_2OH$

(d) $CH_3-\overset{OH}{\overset{|}{CH}}-CH_2-CH_2-\overset{O}{\overset{\|}{C}}-OCH_2CH_3$

(e) $CH_3-\overset{OH}{\overset{|}{CH}}-CH_2-CH_2-CH_2OH$

11.7 (a) 1-ethylcyclopentanol (cyclopentane ring bearing CH_2CH_3 and OH on the same carbon)

(b) $C_6H_5-\overset{OH}{\overset{|}{\underset{CH_3}{\underset{|}{C}}}}-CH_3$

(c)

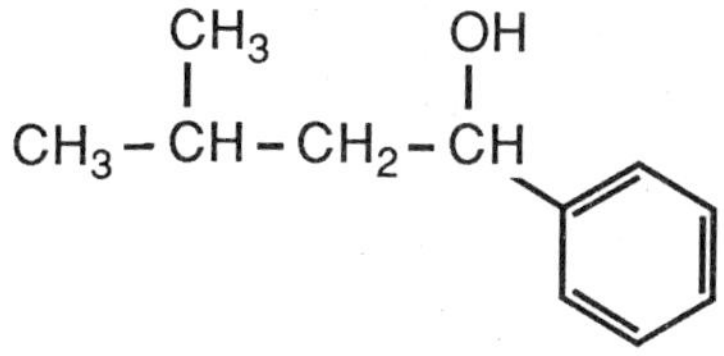

$$CH_3-\underset{}{\overset{CH_3}{\overset{|}{C}H}}-CH_2-\overset{OH}{\overset{|}{C}H}-C_6H_5$$

11.8

(a) 1. $CH_3-\overset{CH_3}{\overset{|}{C}}HMgBr$
2. H_2O

(b) 1. 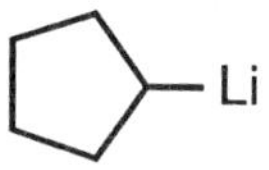 cyclopentyl—Li
2. H_2O

(c) 1. Mg
2. $H_2C=O$
3. H_2O

11.9

(a) $CH_3CH_2CH_2CH_2CH_2-\overset{OH}{\overset{|}{C}H}-CH_2CH_2CH_3$

(b) $CH_3CH_2CH_2CH_2OH$

(c) $CH_3CH_2CH_2-\overset{OH}{\overset{|}{C}H}-CH_2CH_2CH_3$

11.10

(a) $CH_3-\overset{CH_3}{\overset{|}{C}H}-\overset{OH}{\overset{|}{C}H}-CH_2-CH_3$

(b) $CH_3-\overset{CH_3}{\overset{|}{C}H}-CH_2-CH_2OH$

(c) CH_3CH_2Li

(d) $CH_3-CH-CH_2$ (epoxide, O bridging CH and CH_2)

11.11 When the oxygens are *ortho* or *para*, oxidation affords a diketone with two double bonds in the ring (first two structures below). But when the oxygens are *meta*, the corresponding diketone could not have two double bonds in the ring because two of the carbons are "isolated" (arrows in third structure):

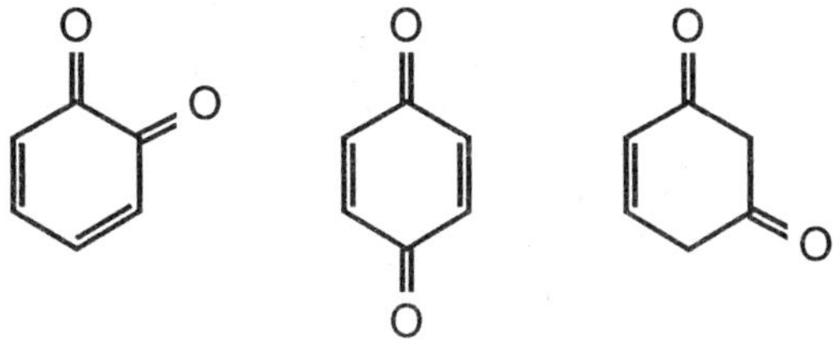

11.12 (a)

$$CH_3-\underset{\displaystyle CH_3}{\overset{|}{C}H}-CH_2CH_2OH \;+\; CH_3CH_2-\overset{\overset{\displaystyle O}{\|}}{C}-Cl \longrightarrow CH_3-\underset{\displaystyle CH_3}{CH}-CH_2CH_2-O-\overset{\overset{\displaystyle O}{\|}}{C}-CH_2CH_3$$

(b)

$$C_6H_5-CH_2OH \;+\; CH_3-\overset{\overset{\displaystyle O}{\|}}{C}-Cl \longrightarrow C_6H_5-CH_2O-O-\overset{\overset{\displaystyle O}{\|}}{C}-CH_3$$

(c)

$$C_5H_9-OH \;+\; CH_3-CH_2-\overset{\overset{\displaystyle O}{\|}}{C}-Cl \longrightarrow C_5H_9-O-\overset{\overset{\displaystyle O}{\|}}{C}-CH_2CH_3$$

(d)

$$CH_3CH_2OH \;+\; C_5H_9-\overset{\overset{\displaystyle O}{\|}}{C}-Cl \longrightarrow C_5H_9-\overset{\overset{\displaystyle O}{\|}}{C}-OCH_2CH_3$$

11.13 (a) (bicyclic hydrindane ring)–OTs (b) $CH_3-C(CH_3)_2-Cl$ (c) $(H)_2C=C(CH_2OSO_2CH_3)(CH_2CH_3)$

(d) PCl_5 (e) PBr_3

11.14 (a) Two different four-carbon aldehydes could be employed, $CH_3CH_2CH_2CHO$ or $CH_3CH(CH_3)CHO$. The use of the second alternative is shown in the following answer:

$$\mathrm{HC(=O)-CH(CH_3)-CH_3} \xrightarrow[\text{2. } H_2O]{\text{1. } CH_3CH_2CH_2MgCl} \mathrm{CH_3CH_2CH_2-CH(OH)-CH(CH_3)-CH_3} \xrightarrow[H_2SO_4]{CrO_3}$$

$$\mathrm{CH_3CH_2CH_2-C(=O)-CH(CH_3)-CH_3} \xrightarrow[\text{2. } H_2O]{\text{1. } CH_3CH_2CH_2MgCl} \mathrm{CH_3CH_2CH_2-C(OH)(CH_2CH_2CH_3)-CH(CH_3)-CH_3}$$

(b) The Grignard reagents are prepared as shown in Example 11.3.

$$\mathrm{HC(=O)-CH_3} \xrightarrow[\text{2. } H_2O]{\text{1. cyclopentyl-MgBr}} \mathrm{cyclopentyl-CH(OH)-CH_3} \xrightarrow[H_2SO_4]{CrO_3}$$

$$\mathrm{cyclopentyl-C(=O)-CH_3} \xrightarrow[\text{2. } H_2O]{\text{1. } CH_3CH_2CH_2CH_2CH_2MgCl} \mathrm{cyclopentyl-C(OH)(CH_3)-CH_2CH_2CH_2CH_2CH_3}$$

(c)

$$\mathrm{H_2C{=}O} \xrightarrow[\text{2. } H_2O]{\text{1. cyclopentyl-MgBr}} \mathrm{cyclopentyl-CH_2OH} \xrightarrow[\text{2. Mg}]{\text{1. } PBr_3} \mathrm{cyclopentyl-CH_2MgBr}$$

$$\mathrm{cyclopentyl-CH_2MgBr} \xrightarrow[\text{2. } H_2O]{\text{1. } H_2C{=}O} \mathrm{cyclopentyl-CH_2CH_2OH}$$

(d)

Cyclopentyl$-CH_2CH_2OH$ $\xrightarrow[2.\ Mg]{1.\ PBr_3}$ Cyclopentyl$-CH_2CH_2MgBr$ $\xrightarrow[2.\ H_3O^+]{1.\ \text{cyclohexanone}}$ 1-(2-cyclopentylethyl)cyclohexene

1-(2-cyclopentylethyl)cyclohexene $\xrightarrow[2.\ H_2O_2,\ OH^-]{1.\ BH_3}$ 2-(CH_2-CH_2-cyclopentyl)cyclohexanol (OH)

11.15 (a)

$$CH_3CH_2CH_2OH \xrightarrow[\text{pyridine}]{CrO_3} CH_3CH_2\overset{O}{\overset{\|}{C}}H$$

Cyclopentyl$-OH$ $\xrightarrow[2.\ Mg]{1.\ PBr_3}$ Cyclopentyl$-MgBr$ $\xrightarrow[2.\ H_2O]{1.\ CH_3CH_2\overset{O}{\overset{\|}{C}}H}$ Cyclopentyl$-\overset{OH}{\overset{|}{C}}HCH_2CH_3$

(b)

Cyclohexanone $\xrightarrow{NaBH_4}$ Cyclohexyl$-OH$ $\xrightarrow[2.\ Mg]{1.\ PBr_3}$ Cyclohexyl$-MgBr$ $\xrightarrow[2.\ H_2O]{1.\ CH_2\overset{O}{—}CH_2}$ Cyclohexyl$-CH_2CH_2OH$

Cyclohexyl$-CH_2CH_2OH$ $\xrightarrow[\text{pyridine}]{CrO_3}$ Cyclohexyl$-CH_2\overset{O}{\overset{\|}{C}}H$ $\xrightarrow[2.\ H_2O]{1.\ \text{Cyclohexyl}-MgBr}$ Cyclohexyl$-CH_2-\overset{OH}{\overset{|}{C}}H-$Cyclohexyl

(c) Both propyl groups can be introduced in a single step by the use of a propyl

Grignard reagent with ethyl formate. Conversion of the resulting alcohol into a Grignard reagent allows the hydroxyl group to be generated on the two-carbon side chain, so the elimination reaction places the double bond in the correct position:

$$HC(=O){-}OCH_2CH_3 \xrightarrow[2.\ H_2O]{1.\ CH_3CH_2CH_2MgBr} HO{-}CH(CH_2CH_2CH_3){-}CH_2CH_2CH_3$$

$$\xrightarrow[2.\ Mg]{1.\ PBr_3} BrMg{-}CH(CH_2CH_2CH_3){-}CH_2CH_2CH_3$$

$$\xrightarrow[2.\ H_2O]{1.\ CH_3CH{=}O} CH_3{-}CH(OH){-}CH(CH_2CH_2CH_3){-}CH_2CH_2CH_3$$

$$\xrightarrow[2.\ KOtBu/tBuOH]{1.\ PBr_5} CH_3{-}CH{=}C(CH_2CH_2CH_3){-}CH_2CH_2CH_3$$

ANSWERS TO PROBLEMS

11.1 **(a)** 2,5-dimethyl-3-hexanol

(b) 3-methyl-3,5-heptanediol

(c) 4-propyl-6-hepten-3-ol

(d) 1-bromo-5-chloro-4-methyl-2-heptanol

(e) 3-cyclobutyl-1-propanol

(f) 5-cyclopentyl-4-methyl-2-pentanol

(g) 3-ethyl-3-buten-2-ol

(h) 2-*tert*-butylcyclohexanol

11.2 (a) $CH_3-CH(OH)-CH(CH_3)-CH_2-CH_2-CH_3$

(b) cyclohexane ring: HO; CH_3, CH_3

(c) $CH_3-CH_2-CH_2-CH(CH_3)-CH_2-CH(CH(CH_3)-CH_3)-C(Cl)(Cl)-CH_2OH$

(d) $CH_3-CH_2-CH_2-CH(Li)-CH_3$

(e) $CH_3-CH_2-C(OLi)(CH_3)-CH_2-CH_3$

(f) $CH_3-CH_2-CH_2-CH=CH-C(CH_3)(CH_3)-CH(OH)-CH_2OH$

(g) cyclohexane ring: OH; $ClCH_2CH_2$

11.3 (a) $CH_3-CH_2-CH(CH_3)-CH_2Cl$

(b) $CH_3-CH_2-CH(CH_3)-CO_2H$

(c) $CH_3-CH_2-CH(CH_3)-CH_2ONa$

(d) $CH_3-CH(CH_3)-CH_2-CH(O-C(=O)-CH_2-CH_3)-CH(CH_3)-CH_3$

(e) $CH_3-CH(CH_3)-CH_2-CH(O-CH_3)-CH(CH_3)-CH_3$

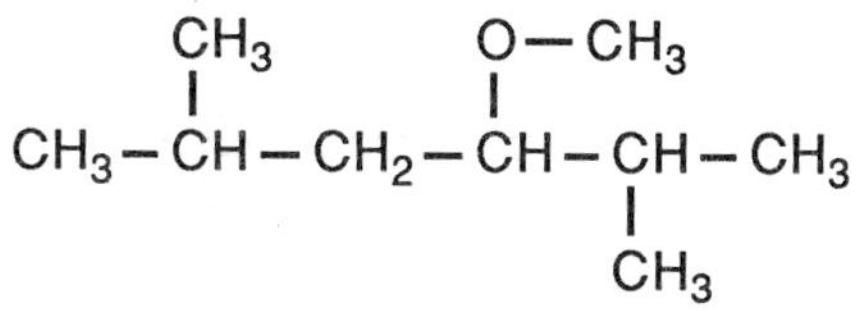

(f) $CH_3-CH(CH_3)-CH_2-CH(O-SO_2-CH_3)-CH(CH_3)-CH_3$

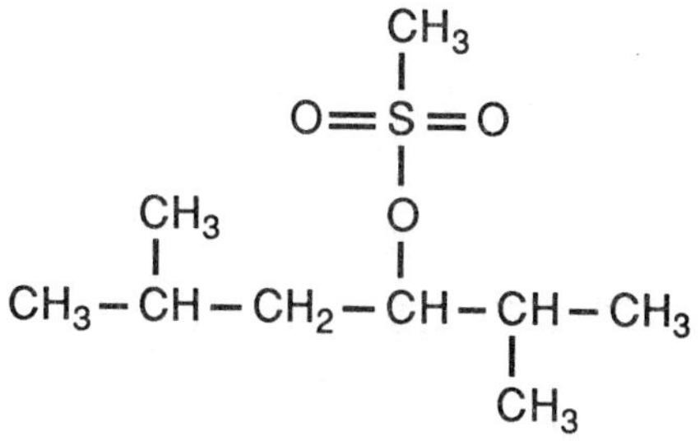

11.4

(a) $CH_3-CH(CH_3)-CH_2-CH_2OH$

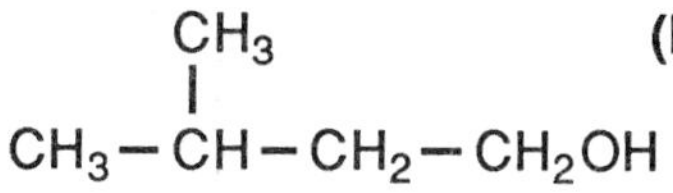

(b) $CH_3-CH(CH_3)-CH(OH)-CH_2OH$

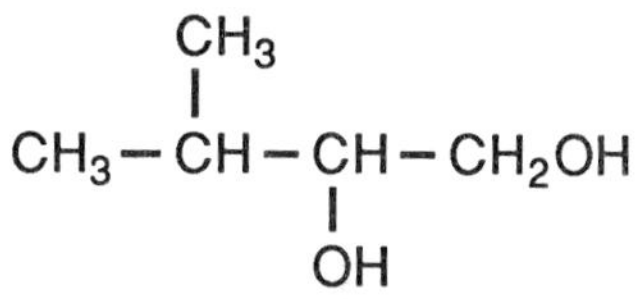

(c) $CH_3-C(CH_3)(OH)-CH=CH_2$

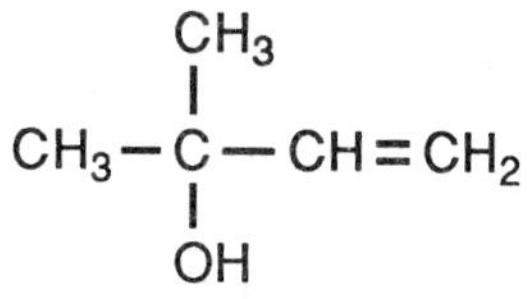

(d) $CH_3-CH_3-C(CH_3)(OH)H-CH(CH_3)-CH_3$

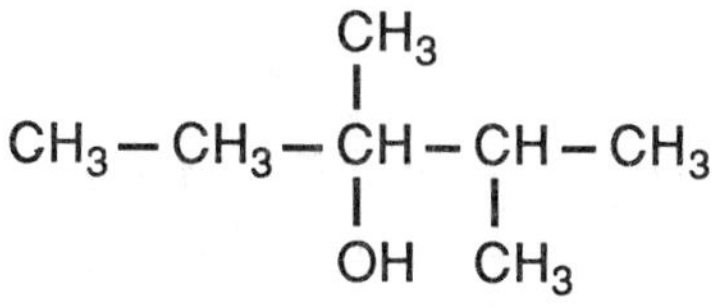

(e) $CH_3-CH_3-CH(OH)-CH(CH_3)-CH_3$

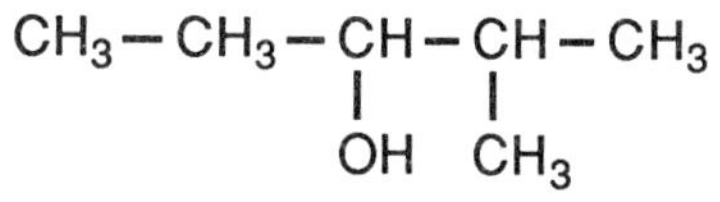

(f) $CH_3-CH_3-CH(OH)-CH(CH_3)-CH_3$

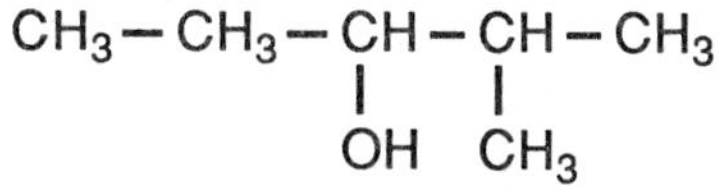

11.5

(a) cyclopentyl$-CH(CH_3)-CH_2OH$

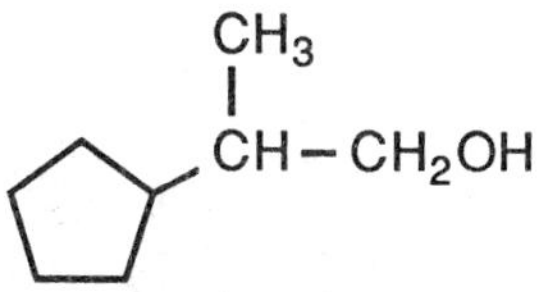

(b) cyclopentyl$-CH(CH_3)-CH_2OH$

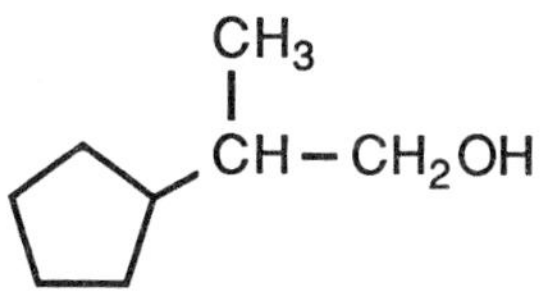

(c) $C(=O)-OCH_3$ on cyclohexane ring, OH at the 4-position

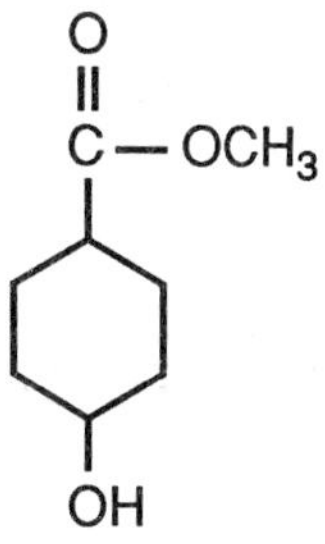

(d) CH_2OH on cyclohexane ring, OH at the 4-position

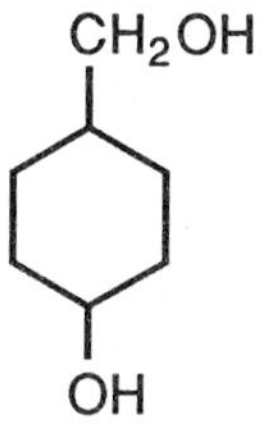

(e) CH_2OH on cyclohexene ring

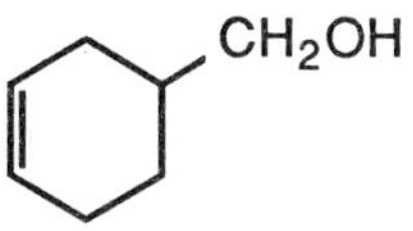

(f) CH_2OH on cyclohexane ring

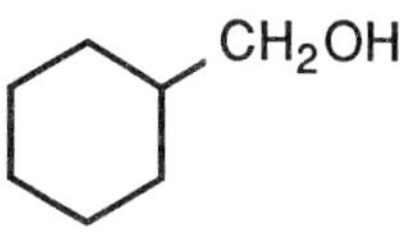

11.6 **(a)** 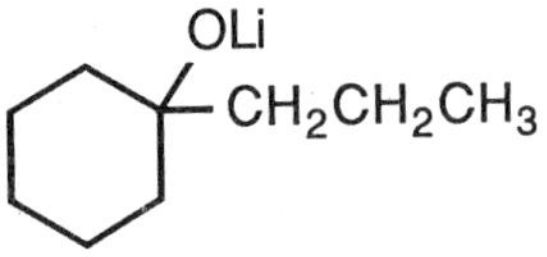 OLi, $CH_2CH_2CH_3$ **(b)** 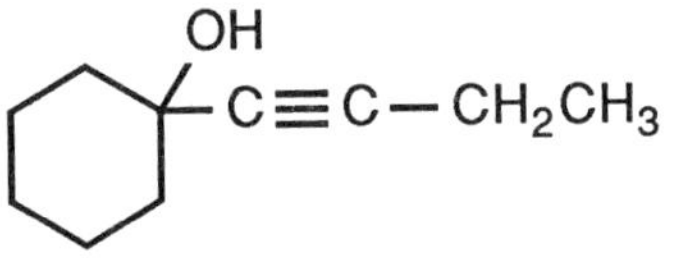 OH, $C\equiv C-CH_2CH_3$ **(c)** OH

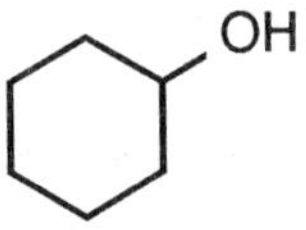

(d) 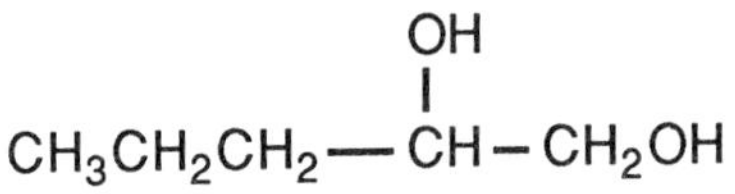 $CH_3CH_2CH_2-CH(OH)-CH_2OH$ **(e)** 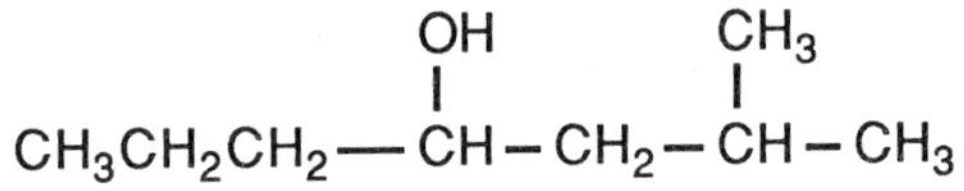 $CH_3CH_2CH_2-CH(OH)-CH_2-CH(CH_3)-CH_3$

(f) 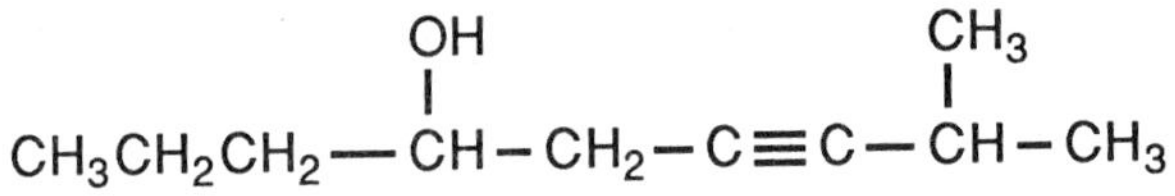 $CH_3CH_2CH_2-CH(OH)-CH_2-C\equiv C-CH(CH_3)-CH_3$

11.7 **(a)** 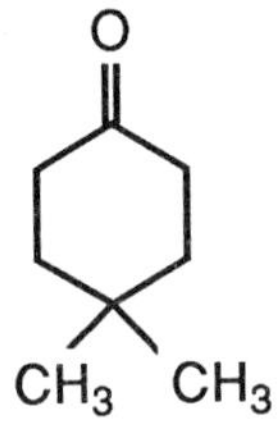 O, CH_3 CH_3 **(b)** 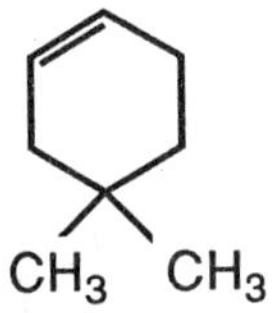 CH_3 CH_3 **(c)** 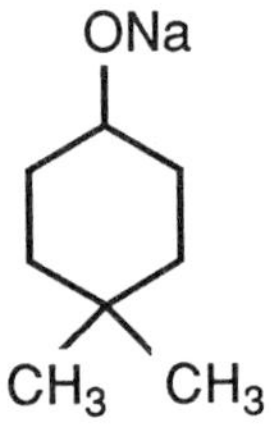 ONa, CH_3 CH_3

(d) 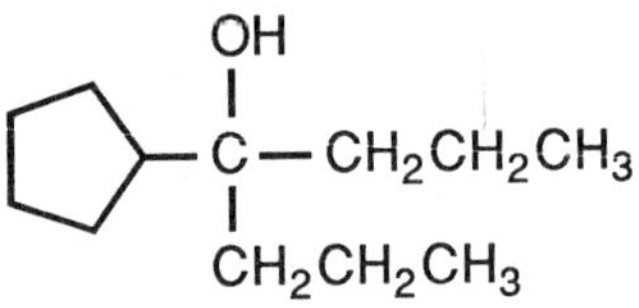 OH, $C-CH_2CH_2CH_3$, $CH_2CH_2CH_3$ **(e)** 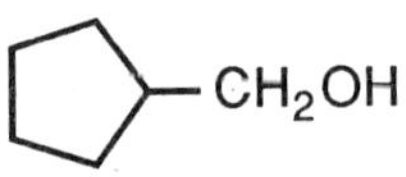 CH_2OH **(f)** 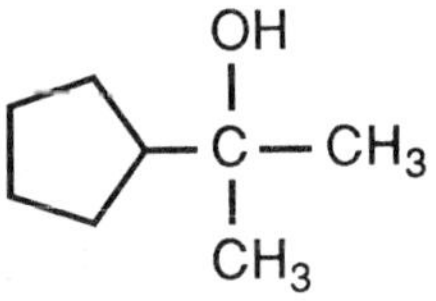 OH, $C-CH_3$, CH_3

11.8 **(a)** Either the Grignard reagent or the alkyllithium would work:

1. 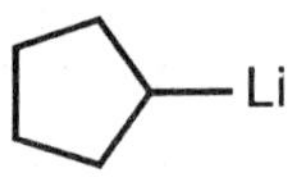 Li

2. H_2O

(b) $NaBH_4$ or $LiAlH_4$ followed by H_2O

(c) CrO_3-pyridine

(d) $LiAlH_4$ followed by H_2O

(e) H_2, Pt

11.9 **(a)**

$$\text{cyclopentanone} + \text{cyclopentyl-Li} \longrightarrow \text{1-cyclopentylcyclopentyl-}\bar{O}Li^+ \xrightarrow{H_2O} \text{1-cyclopentylcyclopentanol (OH)}$$

(b)

$$\text{cyclopentanone} + [H-BH_3]^- Na^+ \longrightarrow \text{cyclopentyl(H)-O-B} \xrightarrow[^-OH]{H_2O_2} \text{cyclopentyl(H)-OH}$$

(c) Not applicable -- the mechanism of CrO_3 has not been presented.

(d)

$$\text{cyclohexenyl-C(=O)-OCH}_3 \xrightarrow{[H-AlH_3]^-\ Li^+} \text{cyclohexenyl-C(O}^-\text{)(H)-OCH}_3 \longrightarrow \text{cyclohexenyl-CH=O}$$

$$\text{cyclohexenyl-CH=O} \xrightarrow{[H-AlH_3]^-\ Li^+} \text{cyclohexenyl-CH(O}^-\text{)-H} \xrightarrow{HO-H} \text{cyclohexenyl-CH}_2OH$$

(e)

catalyst (H H H) + cyclohexenyl-C(=O)-OCH$_3$ ⟶ catalyst (H H) bonded to ring carbon, H on adjacent carbon, ring-C(=O)-OCH$_3$ ⟶ catalyst (H); cyclohexane ring with H, H; C(=O)-OCH$_3$

11.10 **(a)** CH_3CH_2MgBr (or ethyllithium) followed by H_2O

(b) $LiAlH_4$ followed by H_2O

(c) $Hg(OAc)_2$ followed by $NaBH_4$ (or direct hydration by reaction with dilute H_2SO_4.

(d) BH_3 followed by H_2O_2, NaOH

(e) TsCl (or PBr_3, PCl_5, etc.) followed by KOtBu/tBuOH

(f) CrO_3 or $KMnO_4$

11.11 **(a)** SeO_2 followed by H_2, Pt (or HBr followed by KOtBu/tBuOH then BH_3 and finally H_2O_2, NaOH).

(b) $Hg(OAc)_2$ followed by $NaBH_4$ (or direct hydration by reaction with dilute H_2SO_4.

(c) 1. BH_3 followed by H_2O_2, NaOH to produce the primary alcohol.

2. CrO_3-pyridine to produce the aldehyde.

3. CH_3CH_2Li followed by H_2O.

(d) 1. BH_3 followed by H_2O_2, NaOH to produce the primary alcohol.

2. PBr_3 to yield the corresponding bromide.

3. Mg to form the Grignard reagent followed by reaction with CH_3CHO and then H_2O.

(e) 1. BH_3 followed by H_2O_2, NaOH to produce the primary alcohol.

2. PBr_3 to yield the corresponding bromide.

3. Mg to form the Grignard reagent followed by reaction with ethylene oxide and then H_2O.

11.12 **(b)**

H^+ CH$_2$ + CH$_3$ $\xrightarrow{H_2\ddot{O}}$ H +O H CH$_3$ $\xrightarrow{H_2\ddot{O}}$ OH CH$_3$

(e)

11.13 **(a)** $KMnO_4$ or CrO_3

(b) $SOCl_2$, PCl_3, PCl_5, etc.

(c) Several methods are possible: $SOCl_2$ followed by Zn/HCl; Conc. H_2SO&4 followed by H_2/Pt; or TsCl or $SOCl_2$ followed by reaction with KOtBu/tBuOH to form the alkene and then hydrogenation with H_2/Pt.

(d) CrO_3-pyridine.

(e) PBr_3 to generate the corresponding bromide (or TsCl to generate the tosylate) followed by elimination using KOtBu/tBuOH and finally allylic oxidation with SeO_2.

11.14 (c)

$$CH_3-CH_2-\underset{CH_3CH_2}{\underset{|}{CH}}-CH_2\ddot{O}H \xrightarrow{PCl_5} CH_3-CH_2-\underset{CH_3CH_2}{\underset{|}{CH}}-CH_2-PCl_4 + Cl^-$$

$$\longrightarrow CH_3-CH_2-\underset{CH_3CH_2}{\underset{|}{CH}}-CH_2-Cl \xrightarrow{Zn,\ HCl} CH_3-CH_2-\underset{CH_3CH_2}{\underset{|}{CH}}-CH_3$$

11.15 (a) CH_2—CH (epoxide, O bridging) -$\underset{CH_3}{\overset{CH_3}{C}}$—$CH_2CH_3$

(b) (five-membered lactone: ring C=O, ring O, carbon bearing CH_3, CH_3)

(c) cyclohexyl—$CH_2-\overset{O}{\overset{||}{CH}}$

(d) $CH_3-CH_2-\overset{CH_3}{\overset{|}{CH}}-CH_2-CH_2OH$

(e) $CH_3-CH_2-\overset{CH_3}{\overset{|}{CH}}-\overset{O}{\overset{||}{C}}-CH_3$

11.16 (a) cyclohexanol (OH) $\xrightarrow[\text{pyridine}]{CrO_3}$ cyclohexanone (O) $\xrightarrow[\text{2. } H_2O]{\text{1. } (CH_3)_2CuLi}$ 1-methylcyclohexanol (OH, CH_3)

(b) 1-methylcyclohexanol (OH, CH_3) $\xrightarrow{H_2SO_4}$ 1-methylcyclohexene (CH_3) $\xrightarrow[\text{2. } H_2O_2\text{, NaOH}]{\text{1. } BH_3}$ 2-methylcyclohexanol (OH, CH_3)

(c) 2-methylcyclohexanol (OH, CH_3) $\xrightarrow[\text{2. } CH_3I]{\text{1. } NaNH_2}$ 1-methoxy-2-methylcyclohexane (OCH_3, CH_3)

(d) OH, CH_3 — CrO_3 / pyridine → O, CH_3 — 1. CH_3CH_2Li 2. H_2O → OH, CH_2CH_3, CH_3

(e) OH, CH_3 — H_2SO_4 → CH_3 — O ‖ CH_3COOH → O, CH_3 — 1. CH_3CH_2Li 2. H_2O → CH_2CH_3, OH, CH_3

11.17 Structural information is provided in the two reactions of compound **C**, but the real key to solving this puzzle lies in understanding the $LiAlH_4$ reductions. The formation of *two* products in such a reduction is characteristic of an ester, so **A** and **D** must be isomeric esters. The oxidation of **C** to give an aldehyde tells you that it is the primary alcohol, cyclo-C_6H_{11}-CH_2OH, and the reaction with an acid chloride means that the structure of **D** is

$$(CH_3)_2CH\text{-}CO_2CH_2\text{-cyclo-}C_6H_{11}$$

Reduction of **D** yields (in addition to **C**) the primary alcohol, $(CH_3)_2CH$-CH_2OH. Finally, the only other ester that would yield both **B** and **C** is:

$$(CH_3)_2CH\text{-}CH_2O_2C\text{-cyclo-}C_6H_{11}$$

The complete answer is:

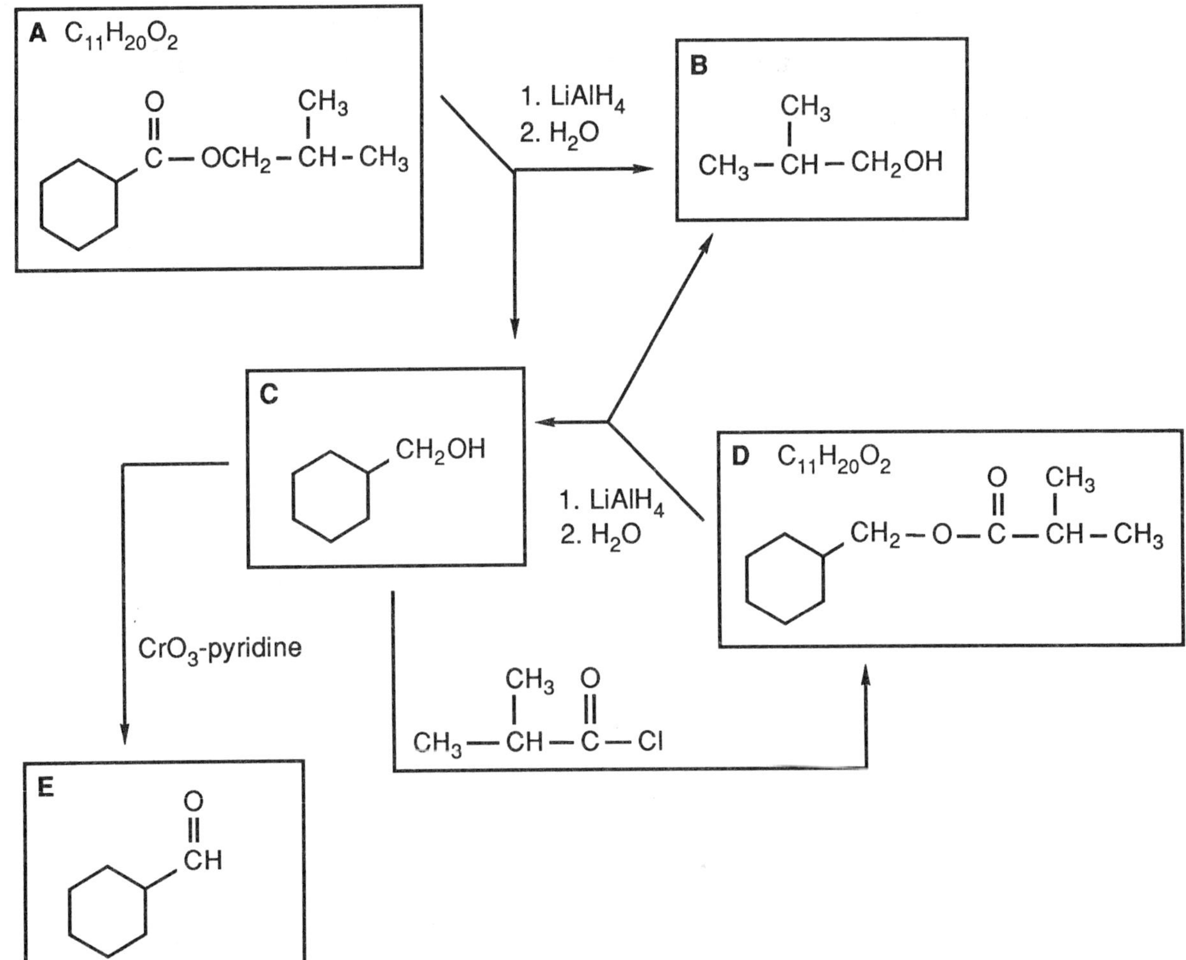

11.18 The only direct structural information is provided for compound **F.** The formation of **F** from **B, D,** and **E** by catalytic hydrogenation suggests that these three compounds are all alkenes, and this is borne out by the other reactions: formation from **A** by elimination reactions, and conversion to **C** by hydration reactions. The information on optical activity provides further structural clues, and **D** must be 3-methylcyclopentene. The reaction of **D** with dilute sulfuric acid could proceed with a hydride shift to generate a tertiary cation. This logic is confirmed by the formation of the same alcohol from both **B** and **E,** and the alcohol must be 1-methylcyclopentanol. The possible structures for **A** are limited by the fact that it is optically active and that it yields different products with *syn* and *anti*

eliminations; This requires that the methyl and hydroxyl groups be on adjacent carbons. The formation of **D** (3-methylcyclopentene) in the elimination of the tosylate requires that the hydroxyl and methyl groups of **A** be *trans*. This further explains the formation of optically inactive **B** via ester pyrolysis in the conversion of **A** to **B.** Finally, **E** must be methylenecyclopentane in order that it also yields 1-methylcyclopentanol in the oxymercuration reaction. The complete answer follows:

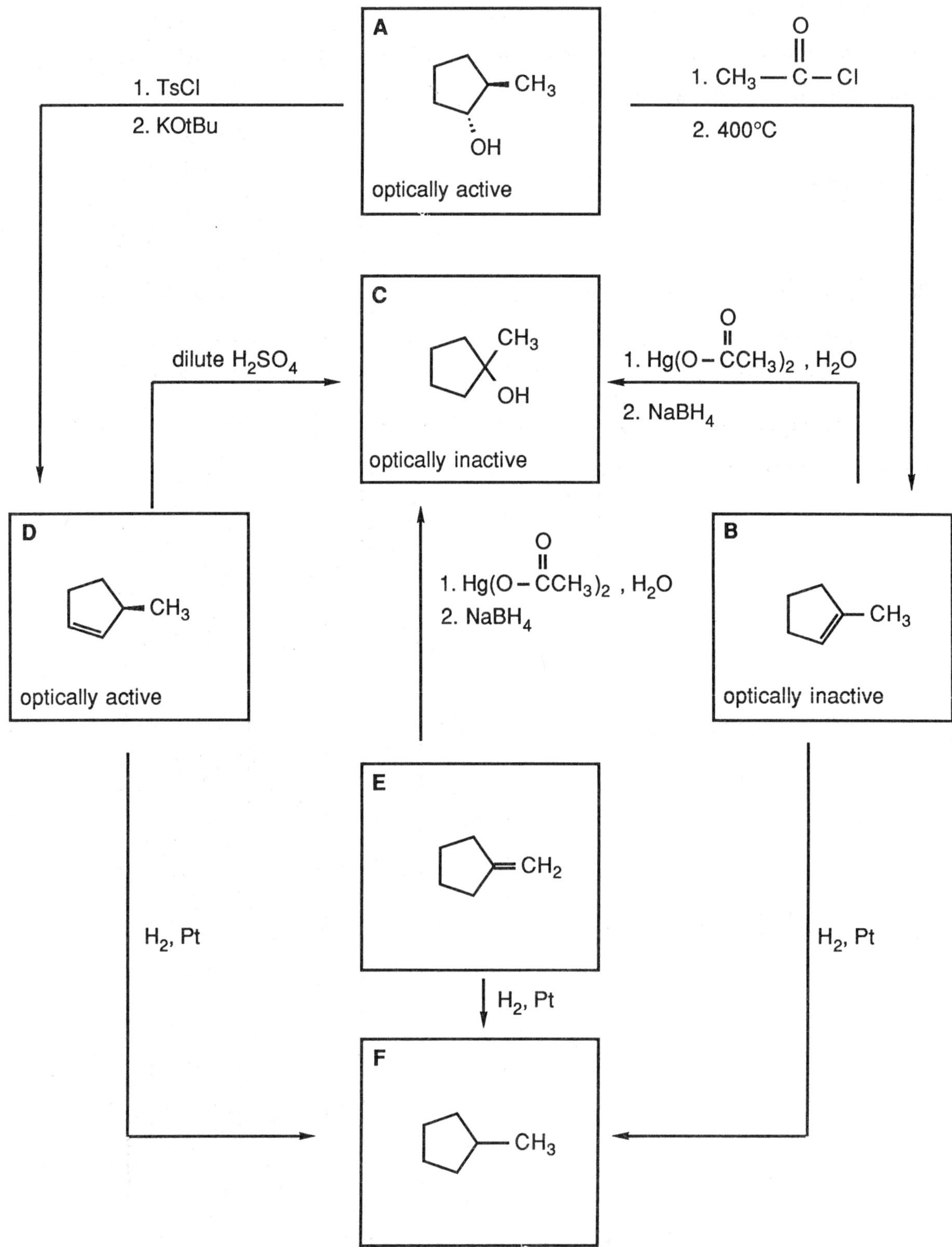
A
CH_3
OH
optically active
1. TsCl
2. KOtBu
1. $CH_3-\overset{O}{\overset{\|}{C}}-Cl$
2. 400°C
C
CH_3
OH
optically inactive
dilute H_2SO_4
1. $Hg(O-\overset{O}{\overset{\|}{C}}CH_3)_2$, H_2O
2. $NaBH_4$
D
CH_3
optically active
1. $Hg(O-\overset{O}{\overset{\|}{C}}CH_3)_2$, H_2O
2. $NaBH_4$
B
CH_3
optically inactive
E
CH_2
H_2, Pt
H_2, Pt
H_2, Pt
F
CH_3

11.19 Structural information is available for several of the compounds in this problem. The molecular formula of **B** tells you that only a single methyl group is introduced in the conversion of **D** to **F**, so the **B** → **D** → **F** sequence is:

CH_3, $-CH-CH_3$, OH ⟶ CH_3, $-C-CH_3$, $\| O$ ⟶ CH_3, CH_3, $-C-CH_3$, OH

Only a single possibility exists for the structure of alkene **A**, which must yield **B** upon oxymercuration and the isomeric alcohol **C** in acid-catalyzed hydration. (Note that the 1,2-methyl shift in the latter reaction is clearly defined by the structure of **G**). Finally, the structures of **D** and **E** can be deduced from the addition of HBr to **C** followed by base catalyzed elimination. The complete answer follows:

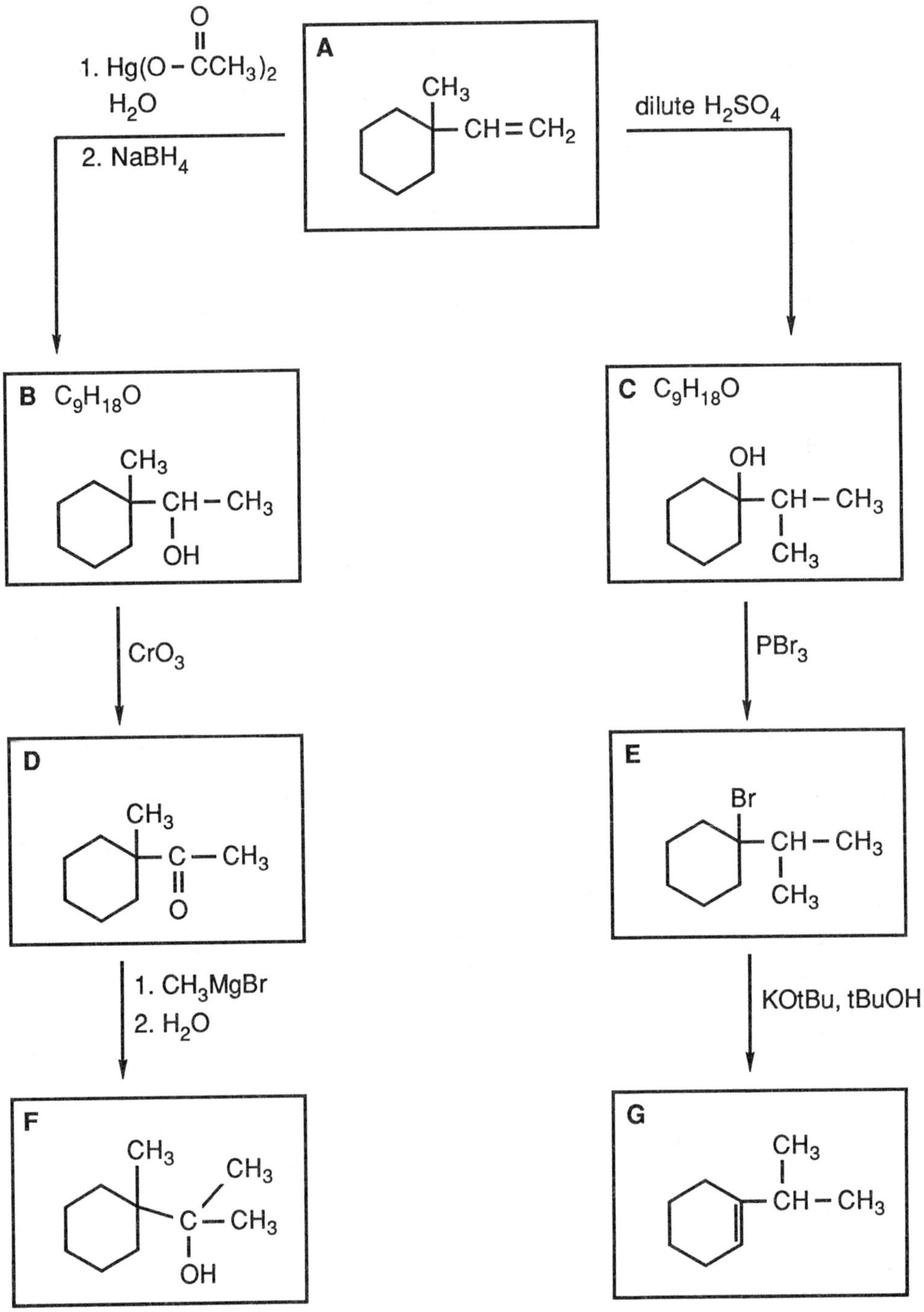

A (1-methyl-1-vinylcyclohexane: cyclohexane ring bearing CH_3 and $CH{=}CH_2$)

1. $Hg(O-\overset{O}{\overset{\|}{C}}CH_3)_2$, H_2O
2. $NaBH_4$

→ **B** $C_9H_{18}O$ (ring bearing CH_3 and $CH(OH)-CH_3$)

dilute H_2SO_4

→ **C** $C_9H_{18}O$ (ring bearing OH and $CH(CH_3)-CH_3$)

B → CrO_3 → **D** (ring bearing CH_3 and $C(=O)-CH_3$)

C → PBr_3 → **E** (ring bearing Br and $CH(CH_3)-CH_3$)

D → 1. CH_3MgBr 2. H_2O → **F** (ring bearing CH_3 and $C(OH)(CH_3)-CH_3$)

E → KOtBu, tBuOH → **G** (cyclohexene bearing $CH(CH_3)-CH_3$ on the double-bond carbon)

11.20 **(a)** The target molecule has a total of seven carbon atoms, so at least two carbon-carbon bonds must be formed. The C-OH is the key functional group, and the other three substituents on this carbon are a methyl, an ethyl, and a propyl group. Any one of these could be introduced by reaction of a Grignard (or alkyllithium) reagent with the corresponding ketone. The alternative shown here introduces the methyl group in this last step so that two 3-carbon subunits can be joined in the first step:

$$CH_3CH_2-\overset{O}{\overset{\|}{C}}H \xrightarrow[2.\ H_2O]{1.\ CH_3CH_2CH_2MgBr} CH_3CH_2CH_2-\overset{OH}{\overset{|}{C}}H-CH_2CH_3$$

$$CH_3CH_2CH_2-\overset{O}{\overset{\|}{C}}-CH_2CH_3 \xrightarrow[2.\ H_2O]{1.\ CH_3Li} CH_3CH_2CH_2-\underset{CH_3}{\underset{|}{\overset{OH}{\overset{|}{C}}}}-CH_2CH_3$$

(b) The key functional group in the target molecule is the chlorine, and it is presumably derived from the corresponding alcohol. The alcohol in turn is probably most easily generated via a Grignard or alkyllithium reaction. The target molecule can thus be considered in terms of two 3-carbon subunits and a single C_2 subunit. By using ethylene oxide as the C_2 subunit, the necessary functionality can be introduced at the carbon *adjacent* to the carbon to which a new bond is formed:

$$CH_3-\overset{Br}{\overset{|}{C}}H-CH_3 \xrightarrow[2.\ CH_2-CH_2\ (\text{epoxide, }O)\quad 3.\ H_2O]{1.\ Mg} HOCH_2-CH_2-\overset{CH_3}{\overset{|}{C}}H-CH_3 \xrightarrow[\text{pyridine}]{CrO_3}$$

$$HC(=O)-CH_2-\overset{CH_3}{\overset{|}{C}}H-CH_3 \xrightarrow[2.\ H_2O\quad 3.\ PCl_5]{1.\ CH_3CH_2CH_2Li} CH_3CH_2CH_2-\overset{Cl}{\overset{|}{C}}H-CH_2-CH-CH_3$$

(c) The target molecule has a total of eight carbon atoms, but hypothetical cleavage of one of the bonds to the key functional group (C-OH) would not generate two C_4 subunits. The target molecule could be prepared in a single step, however, via the reaction of an alkyllithium with an epoxide:

$$(CH_3)_3C\text{—}Li \xrightarrow[2.\ H_2O]{1.\ CH_2\text{—}C(O)(CH_3)\text{—}CH_3\ \text{(epoxide)}} CH_3\text{—}C(CH_3)_2\text{—}CH_2\text{—}C(CH_3)(OH)\text{—}CH_3$$

(d) The target molecule contains a total of eight carbon atoms, but the functionality is not situated in a way that would allow two four-carbon fragments to be joined. The terminal $-CH_2CHO$ group suggests the reaction of an organometallic reagent with ethylene oxide, and this could be preceded by the reaction of cyclobutyllithium (or the corresponding Grignard reagent) with acetaldehyde. One possible sequence follows:

$$\text{cyclobutyl—Li} \xrightarrow[2.\ H_2O]{1.\ CH_3\text{—}\overset{O}{\overset{\|}{C}}H} \text{cyclobutyl—}CH(CH_3)\text{—}OH \xrightarrow[2.\ CH_3CH_2CH_2CH_2Li]{1.\ PBr_5} \text{cyclobutyl—}CH(CH_3)\text{—}Li$$

$$\xrightarrow[\text{2. } H_2O\text{; 3. } CrO_3\text{-pyridine}]{1.\ CH_2\text{—}CH_2\ \text{(ethylene oxide, O)}} \text{cyclobutyl—}CH(CH_3)\text{—}CH_2\text{—}\overset{O}{\overset{\|}{C}}H$$

(e) The target molecule contains seven carbons, and you are only allowed a maximum of three carbons in your starting materials. The best approach to the last step would be to join C_3 and a C_4 subunits -- by reaction of acetone with *tert*-butyllithium (or the corresponding Grignard reagent). This in turn suggests the formation of *tert*-butyl alcohol from the reaction of acetone with methyllithium. The complete synthesis follows:

$$CH_3-\overset{O}{\overset{\|}{C}}-CH_3 \xrightarrow[\substack{3.\ PBr_5 \\ 4.\ CH_3Li}]{\substack{1.\ CH_3Li \\ 2.\ H_2O}} CH_3-\underset{CH_3}{\underset{|}{\overset{CH_3}{\overset{|}{C}}}}-Li \xrightarrow[2.\ H_2O]{1.\ CH_3-\overset{O}{\overset{\|}{C}}-CH_3} CH_3-\underset{OH}{\underset{|}{\overset{CH_3}{\overset{|}{C}}}}-\underset{CH_3}{\underset{|}{\overset{CH_3}{\overset{|}{C}}}}-CH_3$$

(f) The target molecule contains eight carbons, and it can easily be subdivided into two C_4 and one C_2 subunits:

$$\boxed{CH_3-\underset{CH_3}{\underset{|}{\overset{CH_3}{\overset{|}{C}}}}}-\boxed{CH_2-\overset{Br}{\overset{|}{CH}}}-\boxed{CH_2CH_2CH_2CH_3}$$

In order to form carbon-carbon bonds at *both* ends of the C_2 group, the best approaches would be to use either ethylene oxide or acetylene. Thus one approach would utilize the reaction of *tert*-butyllithium with ethylene oxide, oxidation of the resulting alcohol to the aldehyde, and finally, reaction of the aldehyde with butyllithium followed by conversion of -OH to -Br. Alternatively, the linear C_4 group could be introduced by alkylation of acetylene with 1-bromobutane, and the product could be converted to an epoxide after selective reduction to the alkene. The *tert*-butyl group could then be introduced using *tert*-butyllithium, and the complete synthesis follows:

$$CH_3CH_2CH_2CH_2Br \xrightarrow{HC\equiv CNa} CH_3CH_2CH_2CH_2C\equiv CH \xrightarrow[Pd\text{-}BaSO_4]{H_2} CH_3CH_2CH_2CH_2CH=CH_2$$

$$CH_3CH_2CH_2CH_2CH=CH_2 \xrightarrow{CH_3\overset{O}{\overset{\|}{C}}OOH} CH_3CH_2CH_2CH_2\overset{O}{CH-CH_2} \text{ (epoxide)}$$

$$\xrightarrow[2.\ PBr_5]{1.\ CH_3-\underset{CH_3}{\underset{|}{\overset{CH_3}{\overset{|}{C}}}}-Li} CH_3-\underset{CH_3}{\underset{|}{\overset{CH_3}{\overset{|}{C}}}}-CH_2-\overset{Br}{\overset{|}{CH}}-CH_2CH_2CH_2CH_3$$

11.21 **(a)** 2-Butanol can be used as the starting material for *both* halves of the final product:

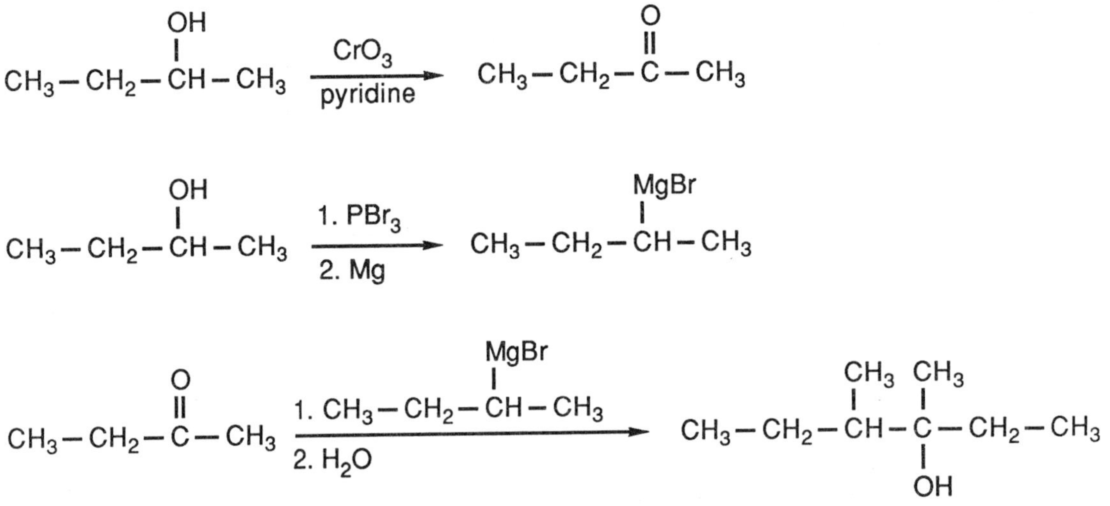

(b) Two six-carbon fragments can be used together with ethylene oxide in an efficient synthesis:

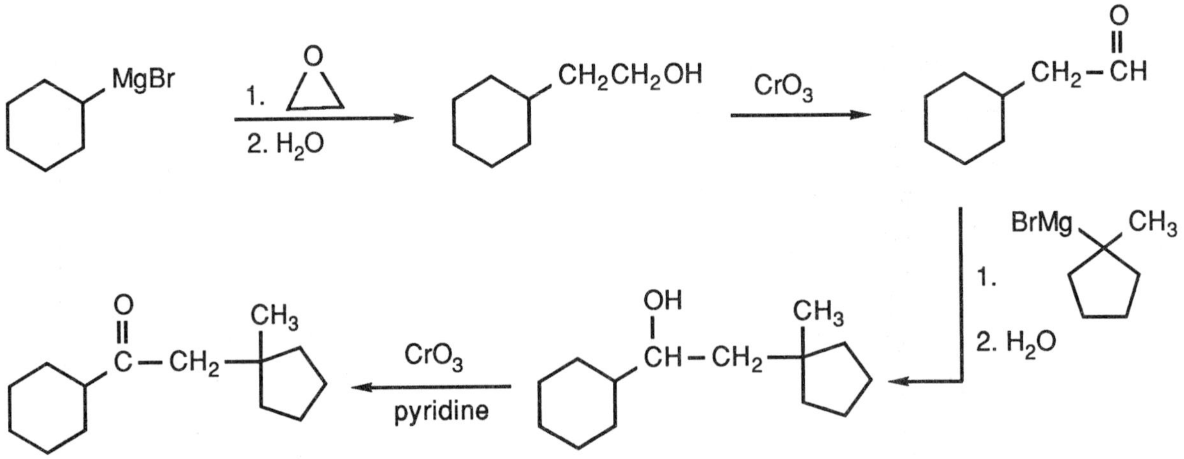

(c) Two groups are attached to the cyclohexane ring, and these must be introduced separately. The functionality is on the 3-carbon subunit, so it should be introduced last (via a Grignard or alkyllithium reaction). The following scheme illustrates one possible synthesis:

$$\text{cyclohexanone} \xrightarrow[\text{2. } H_2O]{\text{1. } CH_3MgBr} \text{1-methylcyclohexanol (HO, } CH_3) \xrightarrow[\text{3. } CH_3-\overset{O}{\overset{\|}{C}}-CH_3]{\text{1. PBr3; 2. Mg}} \text{C}_6\text{H}_{10}(CH_3)-C(CH_3)_2-OMgBr$$

$$\xrightarrow[\text{2. } CH_3-\overset{O}{\overset{\|}{C}}-Cl]{\text{1. } H_2O} \text{C}_6\text{H}_{10}(CH_3)-C(CH_3)_2-O-\overset{O}{\overset{\|}{C}}-CH_3$$

(d) The target molecule can not be prepared by alkylation of propyne with *tert*-butyl bromide, because elimination would occur exclusively. Instead, the *tert*-butyl group must be introduced using an organometallic derivative. One possibility would be to use lithium di-*tert*-butylcuprate with 1-bromopropene to prepare the alkene corresponding to the desired alkyne. (The double bond could then be converted to a triple bond as shown for the sequence that follows.) Alternatively, the reaction of *tert*-butyllithium with ethylene oxide would yield a primary alcohol that could be converted first to an alkene and then to the corresponding alkyne. Alkylation with methyl iodide would afford the desired product:

$$CH_3-C(CH_3)_2-Li \xrightarrow[\text{2. } H_2O]{\text{1. } CH_2-CH_2 \text{ (epoxide, O)}} CH_3-C(CH_3)_2-CH_2-CH_2OH \xrightarrow[\text{2. KOtBu}]{\text{1. } Cl-SO_2-C_6H_4-CH_3} CH_3-C(CH_3)_2-CH=CH_2$$

$$CH_3-C(CH_3)_2-CH=CH_2 \xrightarrow[\text{2. } NaNH_2]{\text{1. } Br_2} CH_3-C(CH_3)_2-C\equiv C^- \, Na^+ \xrightarrow{CH_3I} CH_3-C(CH_3)_2-C\equiv C-CH_3$$

(e) The two substituents are on adjacent carbons, and this suggests the use of an epoxide. Initial reaction with methyllithium (or the corresponding organocuprate) would yield an alcohol that could be converted first to a halide and then to a Grignard reagent. Reaction with formaldehyde would produce the desired carbon skeleton, and this compound could be oxidized to the target molecule:

Cyclohexene oxide → (1. CH_3Li; 2. H_2O) → 2-methylcyclohexanol (OH, CH_3) → (1. PBr_3; 2. Mg) → 2-methylcyclohexylmagnesium bromide (MgBr, CH_3) → (1. $H_2C{=}O$; 2. H_2O) → (CH_2OH, CH_3) → ($KMnO_4$) → (CO_2H, CH_3)

(f) Both substituents can be introduced onto the cyclopentane ring using organometallic reactions. The first of these reactions must leave the functionality as part of the cyclopentane ring, and the second must place the functional group on the side chain:

Cyclopentene → (1. BH_3; 2. H_2O_2, NaOH; 3. CrO_3) → cyclopentanone (O) → (1. CH_3MgBr; 2. H_2O) → (CH_3, OH) → (1. PBr_3; 2. Mg) → (CH_3, MgBr) → (1. $CH_3{-}\overset{O}{\overset{\|}{C}}H$; 2. H_2O) → (CH_3, $\overset{OH}{|}CH{-}CH_3$) → (CrO_3) → (CH_3, $\overset{O}{\overset{\|}{C}}{-}CH_3$)

11.22 The structures of the compounds in this problem must be deduced from a variety of small pieces of information. If the ester **A** (C_{16}) is written as $R{-}CO_2R'$, then the two alcohols that are formed by $LiAlH_4$ reduction must be RCH_2OH and HOR'. One of these is compound **C**, which is converted to Compound **H** (2,6-dimethylheptane) and therefore has the same carbon skeleton. **C** is also formed in the Grignard reaction of **A**, so it must be HOR′ and is not necessarily a primary alcohol. **C** is optically active, which allows only two possible isomers: 2,6-dimethyl-1-heptanol or 2,6-dimethyl-3-heptanol. The loss of optical activity upon oxidation precludes the first possibility, so **C** is 2,6-dimethyl-3-heptanol.

The reaction of **A** with methylmagnesium iodide will generate an alcohol with the structure $R-C(CH_3)_2OH$; replacement of the hydroxyl with a hydrogen generates Compound **H**, which can therefore be written as $R-CH(CH_3)_3$ Since **H** is 2,6-dimethylheptane, the R group of **A** must be 4-methylpentyl, and **A** is an ester of 5-methylhexanoic acid. The complete answer follows:

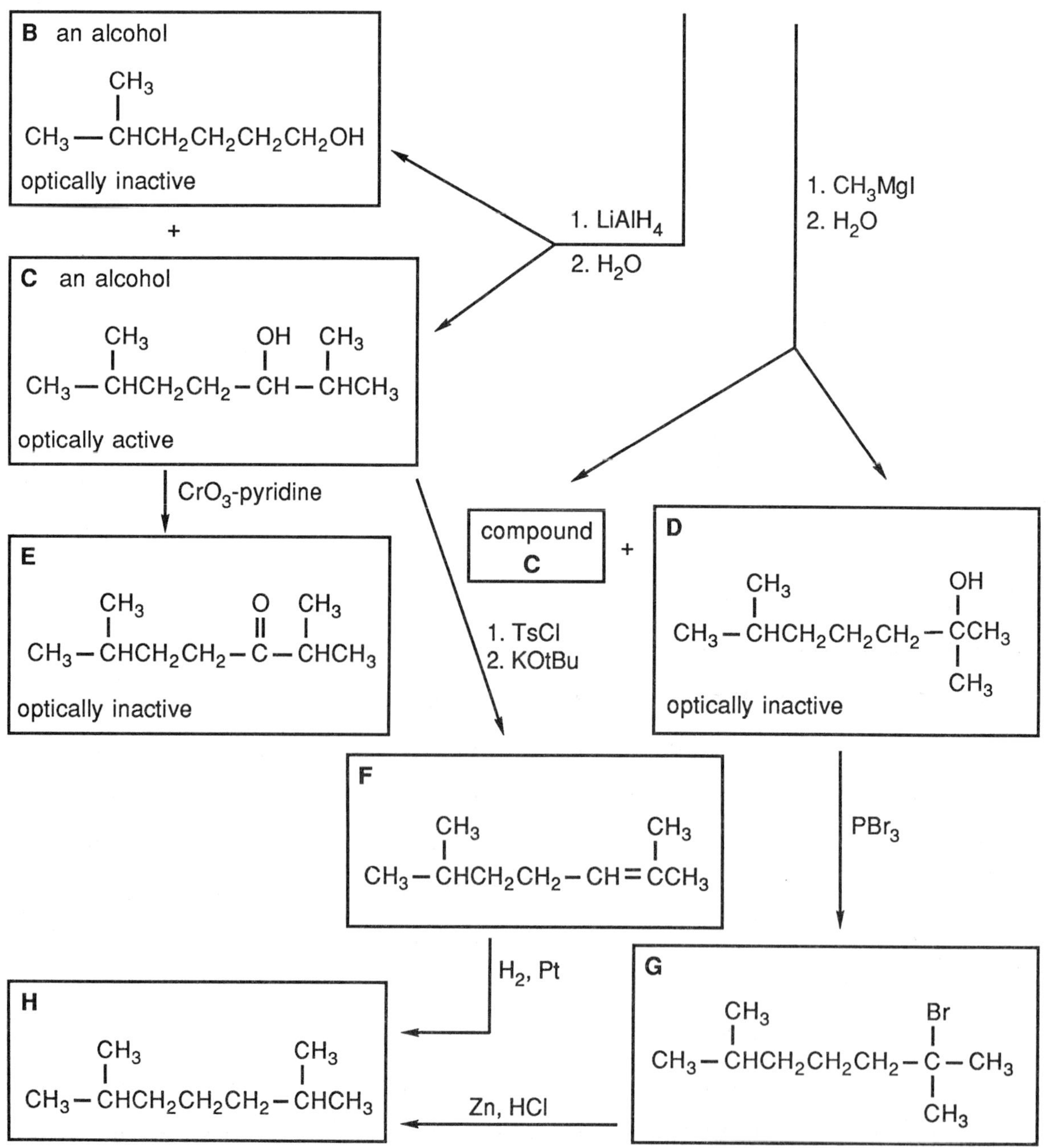
A an ester $C_{16}H_{32}O_2$
optically active
B an alcohol
optically inactive
+
C an alcohol
optically active
1. $LiAlH_4$
2. H_2O
1. CH_3MgI
2. H_2O
CrO_3-pyridine
E
optically inactive
1. TsCl
2. KOtBu
compound C
+
D
optically inactive
F
PBr$_3$
H_2, Pt
G
H
Zn, HCl

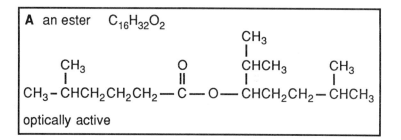

11.23 Despite the absence of any explicit notation of how many hydrogens correspond to each peak, you can deduce the correct structure of the alcohol. The key is the peak at 2.1 ppm, which is the only singlet in the spectrum and must therefore be the -OH signal. The triplet at 3.6 ppm is downfield and must correspond to the group bonded to the hydroxyl. It is approximately twice the size of the -OH signal, so it must be the CH_2 of a CHs'2.OH group. This leaves two signals, a doublet at 0.9 ppm and a multiplet at 1.5 ppm The doublet at 0.9 ppm corresponds to a methyl (or two methyls) split by a single hydrogen, so a partial structure can now be written:

$$(CH_3)_2CH\text{--------}CH_2OH$$

Since the CH_2 of the CH_2OH group is a triplet, it must be adjacent to a second CH_2 group. This signal from this CH_2 group must overlap with the tertiary CH to form the multiplet at 1.5 ppm, and allows the alcohol to be written as:

$$(CH_3)_2CHCH_2CH_2OH$$

The structures of the alcohol and aldehyde are:

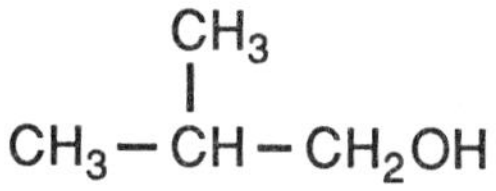

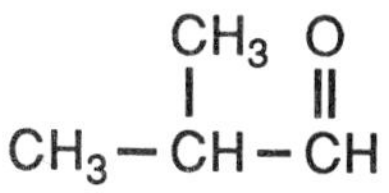

$$CH_3-\underset{}{\overset{\displaystyle CH_3}{\overset{|}{C}H}}-CH_2OH \qquad CH_3-\overset{\displaystyle CH_3}{\overset{|}{C}H}-\overset{\displaystyle O}{\overset{\|}{C}H}$$

11.24 As expected for an alcohol and the corresponding acetate, the two spectra are very similar. The hydrogens on the carbon bearing the OH group yield the multiplet at 3.8 ppm (shifted to 4.8 ppm in the ester), and the -OH peak must be the singlet at 1.9 ppm (replaced by a larger singlet for the methyl of the acetyl group in the ester). The remaining signals are very close together, but a triplet can be recognized at 0.9 ppm in both spectra. This corresponds to a methyl group adjacent to a CH_2, so there is an ethyl group in each structure. The doublet at about 1.2 ppm must correspond to a methyl bonded to a carbon with only a single hydrogen, so the following partial structure can be written:

$$CH_3CH_2\text{------}CHCH_3$$

Only the multiplets at about 1.5–1.6 and the downfield signal have not been accounted for, and these must be the CH_2 and CH of the preceding structure:

$CH_3CH_2\text{-}CH(OH)CH_3$

Indeed, the tertiary CH appears as a sextet in the spectrum of the ester, confirming the presence of five neighboring hydrogens. The complete structures of the alcohol and ester are:

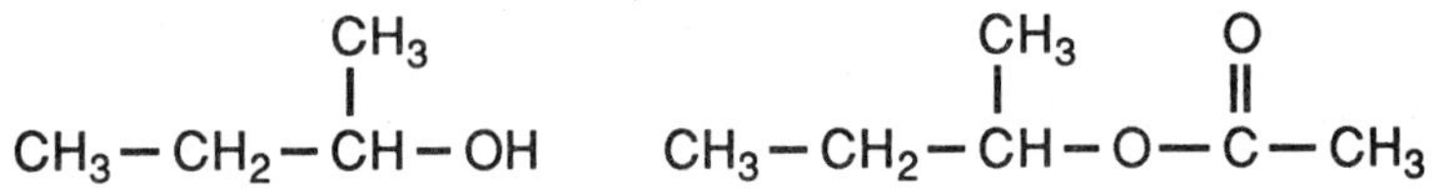

$$CH_3-CH_2-\overset{\displaystyle CH_3}{\overset{|}{CH}}-OH \qquad CH_3-CH_2-\overset{\displaystyle CH_3}{\overset{|}{CH}}-O-\overset{\displaystyle O}{\overset{\|}{C}}-CH_3$$

11.25 (a)  $\overset{-1}{CH_3-CH}=\overset{-2}{CH_2} \longrightarrow CH_3-\underset{\displaystyle Br}{\underset{|}{\overset{0}{CH}}}-\overset{-1}{CH_2Br}$ oxidation

(b)  $CH_3-CH_2-\overset{-1}{CH}=\overset{-1}{CH}-CH_3 \longrightarrow CH_3-CH_2-\overset{0}{CH}-\overset{0}{CH}-CH_3$ (epoxide, O bridging the two CH) oxidation

(c)  $\overset{-3}{CH_3}CH=CH_2 \longrightarrow HO\overset{-1}{CH_2}-CH=CH_2$ oxidation

(d)  cyclohexene (−1, −1) $\longrightarrow$ cyclohexane (−2, −2) reduction

(e) 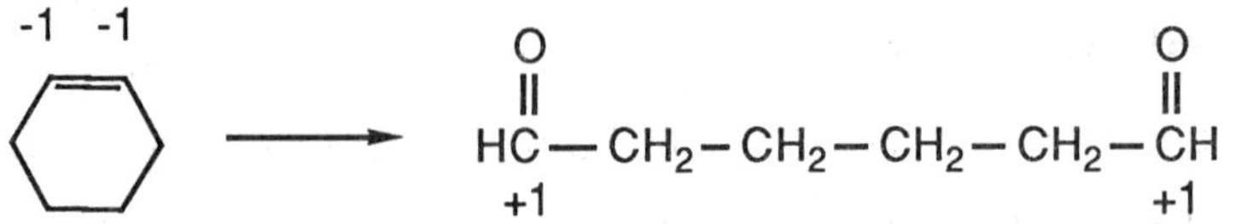 cyclohexene (−1, −1) $\longrightarrow \underset{+1}{\overset{\displaystyle O}{\overset{\|}{HC}}}-CH_2-CH_2-CH_2-CH_2-\underset{+1}{\overset{\displaystyle O}{\overset{\|}{CH}}}$ oxidation

(f)  $\overset{-3}{CH_3}-CH=CH_2 \longrightarrow Br\overset{-1}{CH_2}-CH=CH_2$ oxidation

(g)  $CH_3-\overset{-1}{CH}=\overset{-2}{CH_2} \longrightarrow CH_3-\underset{\displaystyle Br}{\underset{|}{\overset{0}{CH}}}-\overset{-3}{CH_3}$ no overall change in oxidation state

11.26 Aldehydes typically have lower boiling points than the corresponding primary alcohols. Consequently, when the aldehyde boils below 100°C it can be distilled from the reaction mixture before further oxidation occurs. The higher-boiling alcohol remains in the reaction mixture until oxidation occurs.

11.27 **(a)** In the reduction of pentanal you would look for disappearance of the carbonyl peak near 1700 cm^{-1}. An OH peak near 3300 cm^{-1} would appear in the product but the presence of water as an impurity could also produce such a peak and lead to an erroneous conclusion. In the NMR spectrum the aldehyde proton (a triplet near 9.7 ppm) would disappear and be replaced by a triplet integrating for two hydrogens near 3.6 ppm. In the ^{13}C NMR spectrum the peak in the 200–220 ppm region for the carbonyl carbon of the aldehyde would be replaced by a peak for the CH_2OH carbon in the vicinity of 50 ppm.

(b) Conversion of 2-pentanol to methyl 2-pentyl ether would result in the disappearance of the characteristic OH absorption near 3300 cm^{-1} infrared absorption. The major change in the NMR spectra would be the appearance of a new peak for the methyl group (a 3H singlet at about 3.1 ppm and a peak at near 40–50 ppm for proton and carbon spectra, respectively).

(c) Reduction of ethyl benzoate with $LiAlH_4$ would yield two alcohols (ethanol and benzyl alcohol). The carbonyl absorption near 1730 cm^{-1} would disappear and OH absorption near 3300 cm^{-1} would be observed. Interpretation of the NMR spectra would depend upon whether or not the crude product were purified, but normal aqueous workup would remove the ethanol. The resulting spectrum would therefore be just that of C_6H_5OH, and the characteristic ethyl pattern of the reactant would be replaced by a 2H singlet (about 4.9 ppm) in the proton NMR. The carbon spectrum would exhibit only one peak in the aliphatic region (the CH_2OH carbon in the vicinity of 50 ppm) instead of the original two (the ethyl group), and the peak from the carbonyl carbon (160–180 ppm) would be absent.

11.28 **(a)** In the oxidation of 1-octanol you would expect to see a carbonyl peak in the region of 1720–1745 cm^{-1} for the aldehyde but at 1700–1725 cm^{-1} for the acid. The aldehyde would show the weak but characteristic peak at 2720 cm^{-1}, whereas the acid would show the characteristic, very broad OH absorption from 2500–3500 cm^{-1}. The proton NMR spectra would be similar for the two compounds, the major difference being in the downfield (triplet) between 9.5 and 10 ppm for the aldehyde vs. a broad singlet between 11.5 and 12.5 ppm for the COOH proton. The absorptions for the carbonyl carbons would be quite different in the ^{13}C spectra: for a carboxylic acid the peak should be in the region of 160–180 ppm, whereas it should appear between 200 and 220 ppm for the aldehyde.

(b) The two possible isomeric alcohols that could be formed by addition of water to 4,4-dimethyl-2-hexene would afford extremely similar infrared spectra, and this would provide little information. Similarly, the ^{13}C NMR spectra would be quite similar, and you would need to employ selective decoupling techniques to distinguish between the compounds by this technique. Proton NMR spectra would provide a ready distinction between the two alternatives based on their structures in the vicinity of the hydroxyl group: 4,4-dimethyl-2-hexanol ($R-CH_2CH(OH)CH_3$) and 4,4-dimethyl-3-hexanol ($R-CH(OH)CH_2CH_3$). The second compound would exhibit a 1H triplet in the region of 4.0 ppm, whereas the first would show a complex 2H multiplet (5 neighboring hydrogens) in this region. Further information would be provided by the methyl groups of these two structures. In the case of 4,4-dimethyl-2-hexanol the methyl would appear as a clear doublet (it is coupled to a CH with quite different chemical shift), whereas the adjacent CH_2 of 4,4-dimethyl-3-hexanol would lead to a distorted triplet for the methyl group.

CHAPTER 12. NUCLEOPHILIC SUBSTITUTION

ANSWERS TO EXERCISES

12.1 (a) OH^-

(b) OH^-

(c) NH_2^-

12.2 (a) $CH_3—\overset{+}{O}H_2$ (b) $CH_3—\overset{+}{N}{\equiv}N$ (c) $CH_3—C(CH_3)_2—\overset{+}{N}(CH_3)_3$

$$\begin{array}{c} CH_3 \\ | \\ CH_3—C—\overset{+}{N}(CH_3)_3 \\ | \\ CH_3 \end{array}$$

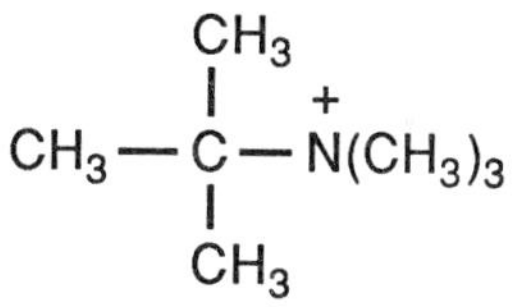

12.3 (a)

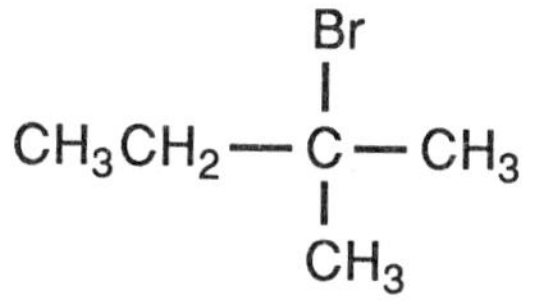

$$\begin{array}{c} Br \\ | \\ CH_3CH_2—C—CH_3 \\ | \\ CH_3 \end{array}$$

(b) $C_6H_5CH_2Cl$ (c)

$$\begin{array}{c} Br \\ | \\ CH_3—C—CH_2CH_2CH_3 \\ | \\ CH_3 \end{array}$$

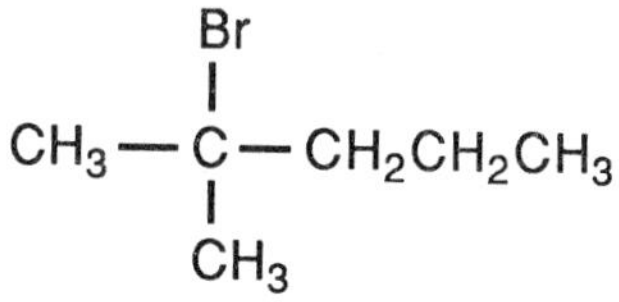

12.4 (a) A benzylic cation would be formed:

$$C_6H_5CH_2Cl$$

(b) A vinyl cation would be formed (whereas the other compound would react via direct displacement):

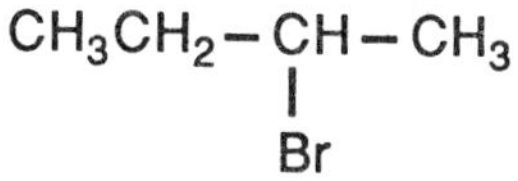

$$\begin{array}{c} CH_3CH_2-CH-CH_3 \\ | \\ Br \end{array}$$

(c) The cation formed would be stabilized by two phenyl groups:

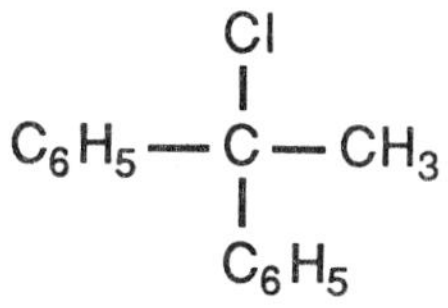

$$\begin{array}{c} Cl \\ | \\ C_6H_5—C—CH_3 \\ | \\ C_6H_5 \end{array}$$

12.5 For Exercise 12.3 (a) the first compound (2-bromo-2-methylbutane) is a tertiary alkyl halide that would react via a carbocation; the second compound (2-bromo-3-methylbutane) is a secondary halide that would react via direct displacement. For Exercise 12.3 (b) the first compound (benzyl chloride) could react by either mechanism depending upon the reaction conditions; the second compound (ethyl chloride) will always react via direct displacement. For Exercise 12.3 (c) the first compound (1-bromo-4-methyl-2-pentene) is an allylic derivative that could react via either mechanism depending on the conditions; the second compound (2-bromo-2-methylpentane) is a tertiary derivative that would always react via a carbocation.

For Exercise 12.4 (a) the first compound is again benzyl chloride, which could react via either mechanism; the second compound (1-chloro-2-phenylethane) is primary and would react via direct displacement. For Exercise 12.4 (b) the first compound (2-bromobutane) is secondary and would react via direct displacement; the second compound (2-bromo-2-butene) is a vinyl halide and would react via a carbocation mechanism. For Exercise 12.4 (c) the first compound (1-chloro-1,1-diphenylethane) is tertiary and would react via a (resonance stabilized) carbocation; the second compound (3-bromo-2,4-dimethylpentane) is a secondary halide that would react via direct displacement.

12.6 Using the expression for a bimolecular reaction step,

$$\text{rate} = k[(C_6H_5)_3C{-}F][NaOH]$$

the data in Table 12.1 afford calculated bimolecular rate constants of 0.34 and 0.15 $M^{-1}sec^{-1}$. Clearly, these are not constant. On the other hand, using the expression for a unimolecular reaction,

$$\text{rate} = k[(C_6H_5)_3C{-}F]$$

the data affords calculated unimolecular rate constants of 8.4×10^{-4} and 7.5×10^{-4} sec^{-1}. These values are nearly constant, in accord with a carbocation mechanism for the reaction.

Using the expression for a bimolecular reaction step,

$$\text{rate} = k[CH_3Br][\text{pyridine}]$$

the data in Table 12.2 afford calculated bimolecular rate constants of 4.7, 4.7, 4.7, and 4.8 $M^{-1}sec^{-1}$. These values are essentially constant, indicating that the reaction is indeed a direct displacement process.

12.7

(a) CH_3CH_2—C(H)(OCH_2CH_3)(CH_3)

(b) CH_3CH_2—C(H)(OCH_2CF_3)(CH_3)

(c) $CH_3CH_2CH_2$—C(CH_3)(CH_2CH_3)(OCH_2CH_3)

(d) $CH_3CH_2CH_2$—C(CH_3)(CH_2CH_3)(OCH_2CF_3)

(e) C_6H_5—C(CH_3)(OCH_2CH_3)(CH_2CH_3) + C_6H_5—C(CH_3)(CH_2CH_3)(OCH_2CH_3).

(f) C_6H_5—C(CH_3)(OCH_2CF_3)(CH_2CH_3) + C_6H_5—C(CH_3)(CH_2CH_3)(OCH_2CF_3)

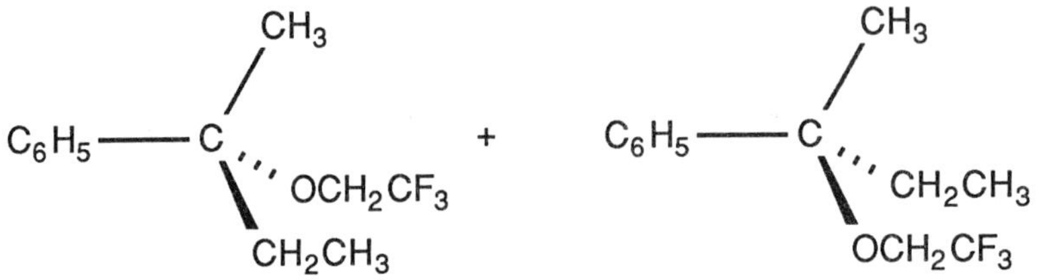

12.8

(a) CH_3CH_2—C(CH_3)(H)(OBs) $\xrightarrow{CF_3CH_2OH}$ CH_3CH_2—C(CH_3)(OCH_2CF_3)(H)

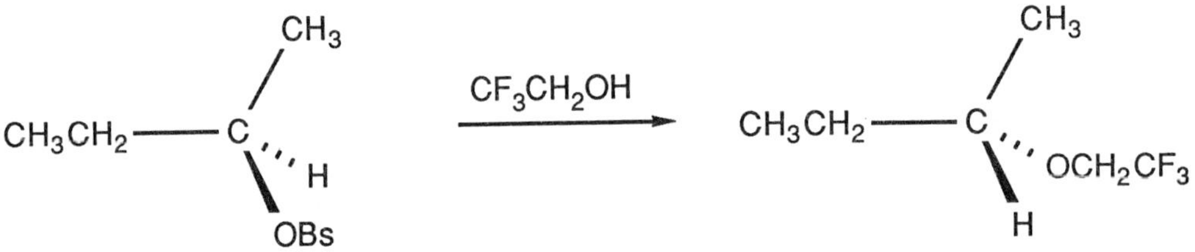

R S

(b) CH_3, Cl (cyclohexane) ⟶ CH_3, OCH_2CF_3 (cyclohexane)

(c) CH_3CH_2—C(CH_3)(Br)(C_6H_5) $\xrightarrow{CF_3CH_2OH}$ CH_3CH_2—C(CH_3)(OCH_2CF_3)(C_6H_5) + CH_3CH_2—C(CH_3)(C_6H_5)(OCH_2CF_3)

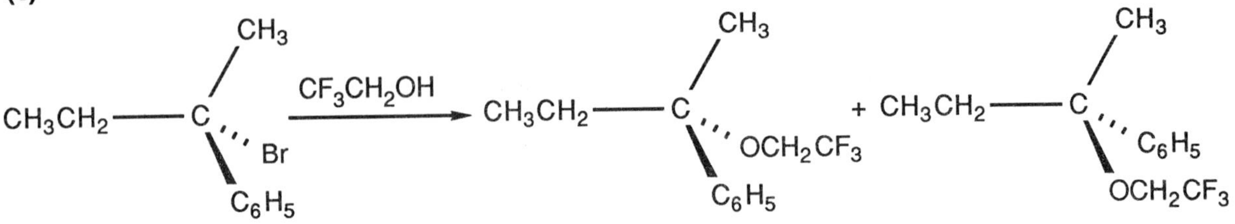

S S R

12.9 (a) CH_3CH_2—C(CH_3)(C_6H_5)(Cl) $\xrightarrow{CH_3CH_2OH}$ CH_3CH_2—C(CH_3)(OCH_2CH_3)(C_6H_5) + CH_3CH_2—C(CH_3)(C_6H_5)(OCH_2CH_3)

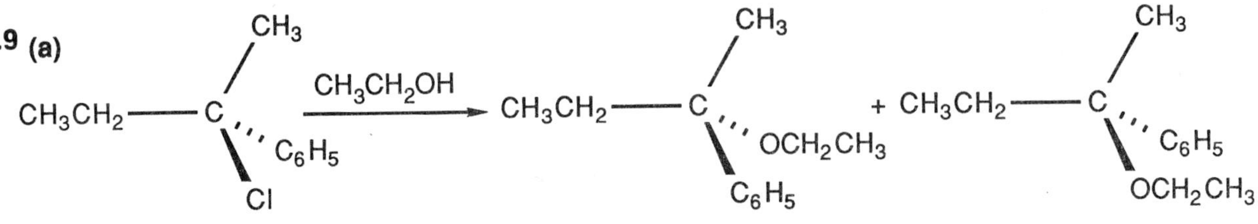

R S R

(b) 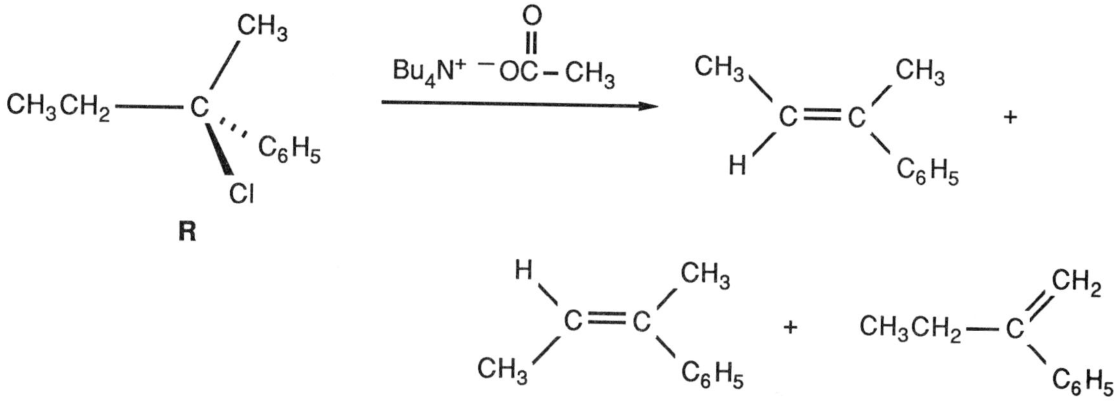 CH_3CH_2—C(CH_3)(C_6H_5)(Cl) (R) $\xrightarrow{Bu_4N^+ \ ^-OC(=O)-CH_3}$ (E/Z) $CH_3(H)C=C(CH_3)(C_6H_5)$ + $H(CH_3)C=C(CH_3)(C_6H_5)$ + CH_3CH_2—C(=CH_2)(C_6H_5)

12.10 (a) $CH_3-CH=CH-CH_3$ + $CH_3-CH_2-CH=CH_2$

(b) $CH_3-CH=CH-CH_3$ + $CH_3-CH_2-CH=CH_2$ + $CH_3-CH_2-CH(OCH_2CH_3)-CH_3$

(c) 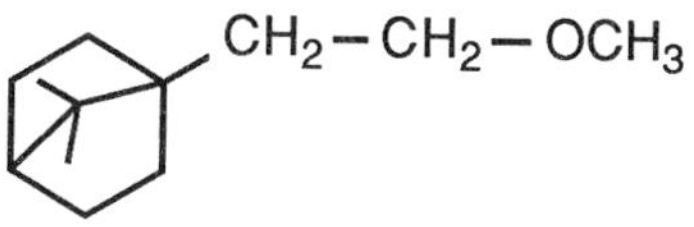 (bicyclic)—$CH_2-CH_2-OCH_3$

(d) 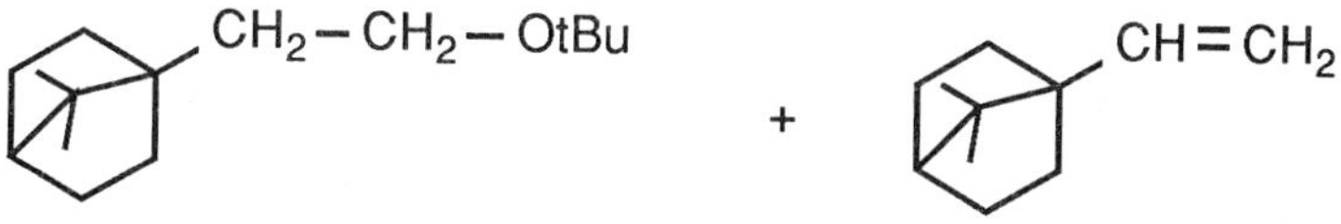 (bicyclic)—CH_2-CH_2-OtBu + (bicyclic)—$CH=CH_2$

(e) 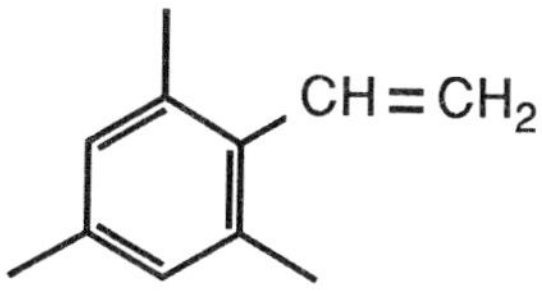 (2,4,6-trimethylphenyl)—$CH=CH_2$

(f) 2,4,6-trimethylphenyl–$CH_2-CH_2-OCH_3$ + 2,4,6-trimethylphenyl–$CH=CH_2$

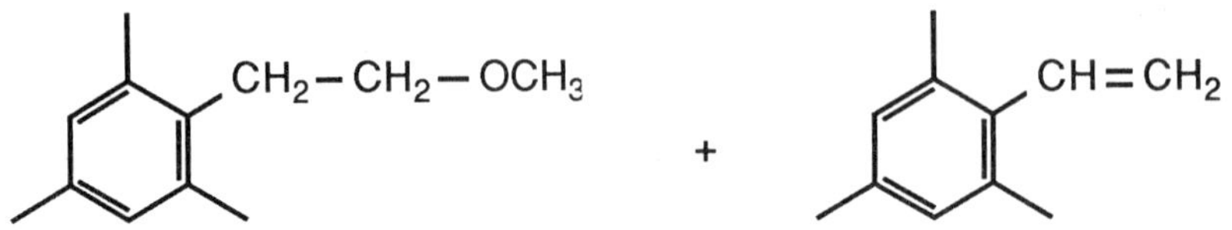

12.11 (a)

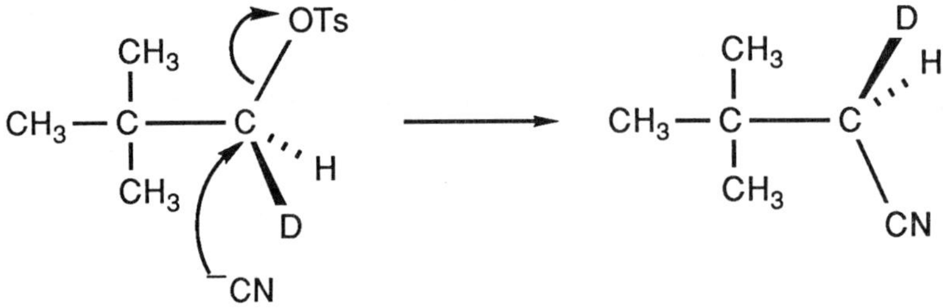

$$(CH_3)_3C-CHD-OTs \xrightarrow{^{-}CN} (CH_3)_3C-CHD-CN$$

(b)

Rearrangement:

$$CH_3-\underset{CH_3}{\overset{CH_3}{C}}-CH_2-Br \longrightarrow CH_3-\underset{CH_3}{\overset{+}{C}}-CH_2-CH_3$$

Elimination:

$$CH_3-\underset{CH_3}{\overset{+}{C}}-\overset{H}{CH}-CH_3 \longrightarrow CH_3-\underset{CH_3}{C}=CH-CH_3$$

Reaction with water:

$$CH_3-\underset{CH_3}{\overset{+}{C}}-CH_2CH_3 \xrightarrow{H_2O} CH_3-\underset{CH_3}{\overset{H-\overset{+}{O}H}{C}}-CH_2CH_3 \longrightarrow CH_3-\underset{CH_3}{\overset{OH}{C}}-CH_2CH_3$$

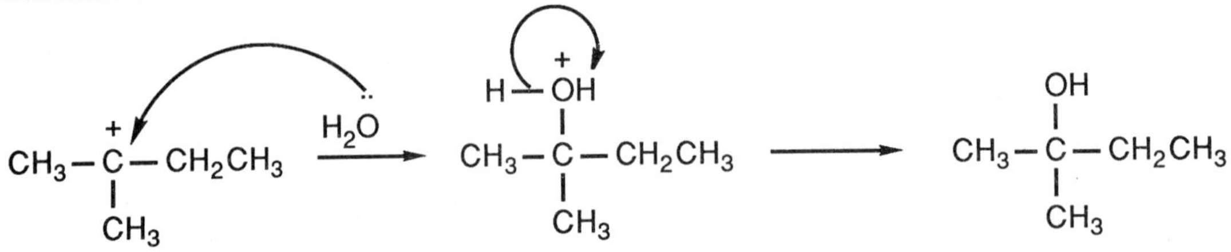

Reaction with ethanol:

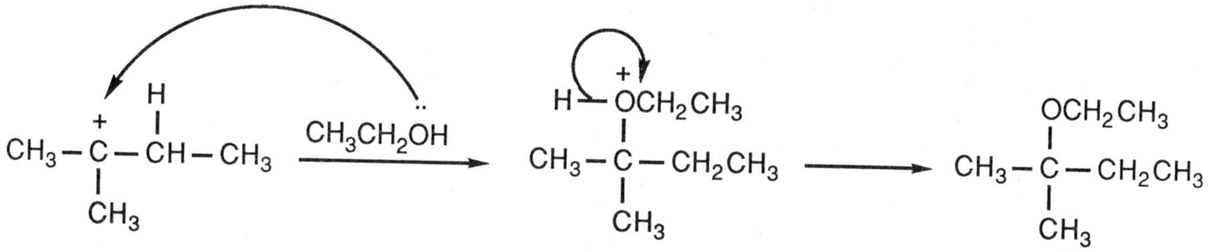

$$CH_3-\overset{+}{C}(CH_3)-CH(H)-CH_3 \xrightarrow{CH_3CH_2OH} CH_3-C(CH_3)(\overset{+}{O}(H)CH_2CH_3)-CH_2CH_3 \longrightarrow CH_3-C(CH_3)(OCH_2CH_3)-CH_2CH_3$$

(c)

$$CH_3-C(H)(CH_3)-CH(OTs)-CH_3 \longrightarrow CH_3-\overset{+}{C}(CH_3)-CH_2CH_3$$

$$CH_3-\overset{+}{C}(CH_3)-CH_2CH_3 \xrightarrow{CH_3-C(=O)-O-H} CH_3-C(CH_3)(O-C(=O)-CH_3)-CH_2CH_3$$

$$CH_3-\overset{+}{C}(CH_3)-CH_2CH_3 \xrightarrow{\text{loss of } H^+} CH_3-C(CH_3)=CHCH_3$$

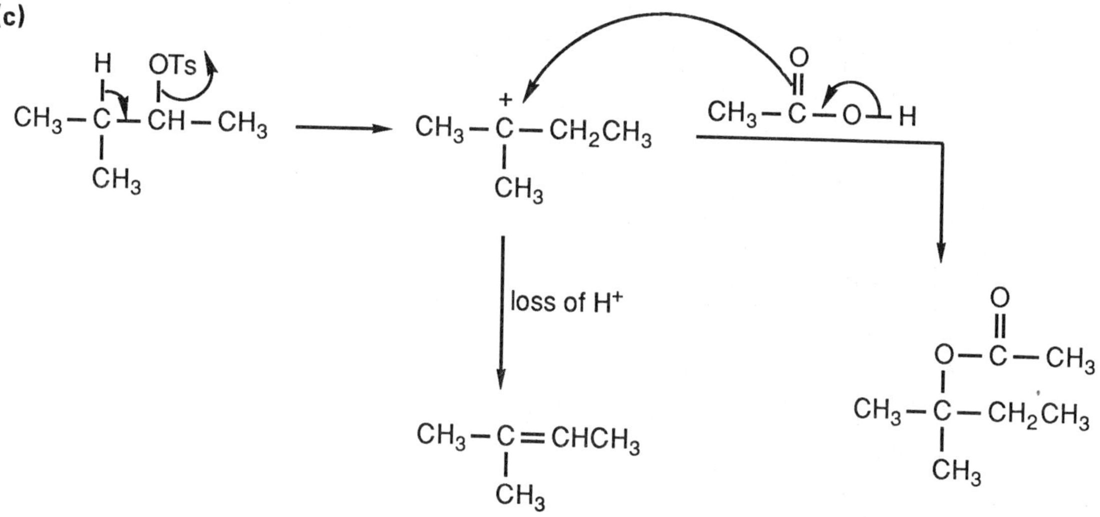

(d)

Rearranged:

$$\text{2-(OTs)cyclohex-3-ene (H shift)} \longrightarrow \text{cyclohexenyl cation} \xrightarrow{CH_3-C(=O)-O-H} \text{cyclohexenyl } O-C(=O)-CH_3$$

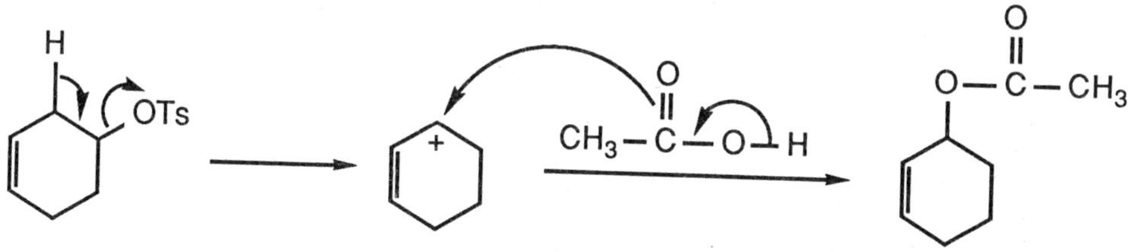

Unrearranged:

$$\text{cyclohexenyl-OTs} + CH_3-C(=O)-O-H \longrightarrow \text{cyclohexenyl } O-C(=O)-CH_3$$

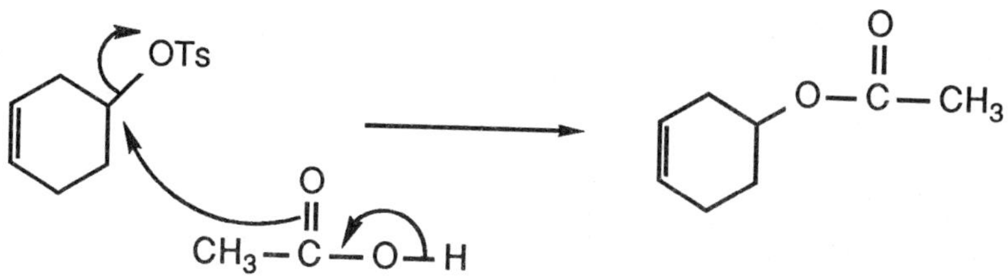

(e)

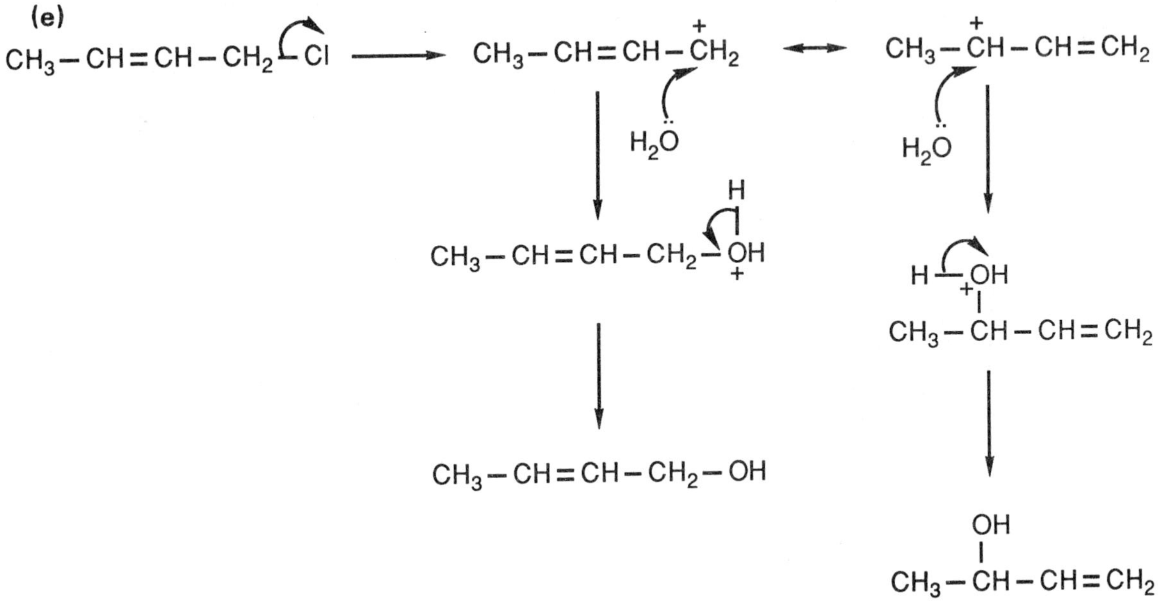

$CH_3-CH=CH-CH_2-Cl \longrightarrow CH_3-CH=CH-\overset{+}{C}H_2 \longleftrightarrow CH_3-\overset{+}{C}H-CH=CH_2$

$H_2\ddot{O}$ $H_2\ddot{O}$

$CH_3-CH=CH-CH_2-\overset{+}{O}H(H)$ $CH_3-CH(\overset{+}{O}H-H)-CH=CH_2$

$CH_3-CH=CH-CH_2-OH$ $CH_3-CH(OH)-CH=CH_2$

(f)

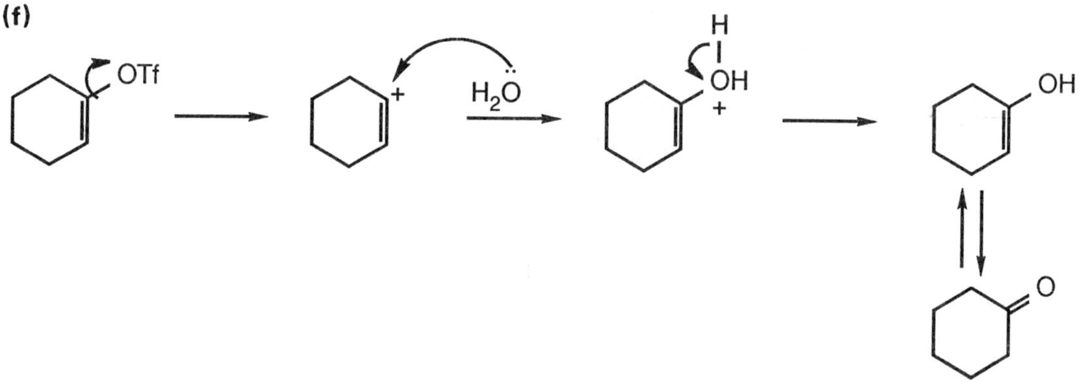

OTf $\longrightarrow$ + $\xrightarrow{H_2\ddot{O}}$ $\overset{+}{O}H(H)$ $\longrightarrow$ OH $\rightleftarrows$ O

(g)

$(C_6H_5)_2C=C(C_6H_5)OTs \longrightarrow (C_6H_5)_2C=\overset{+}{C}-C_6H_5 \xrightarrow{CH_3-C(=O)-O-H} (C_6H_5)_2C=C(C_6H_5)-O-C(=O)-CH_3$

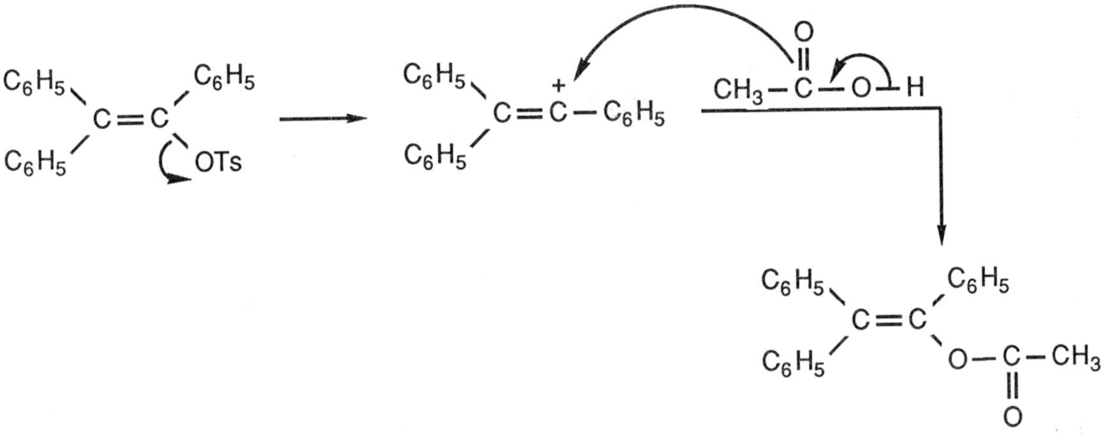

12.12 (a) $CH_3-C(OH)(CH_3)-CH(CH_3)-CH_3$

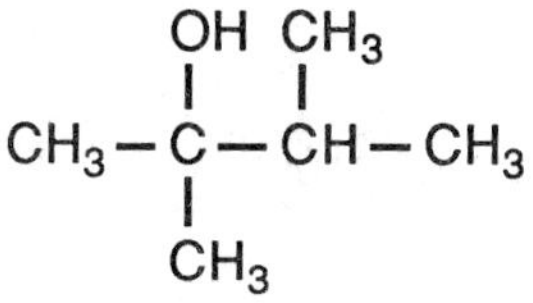

(b) 3-methylcyclohex-1-enyl acetate: ring with CH_3; $O-\overset{\overset{O}{\|}}{C}-CH_3$

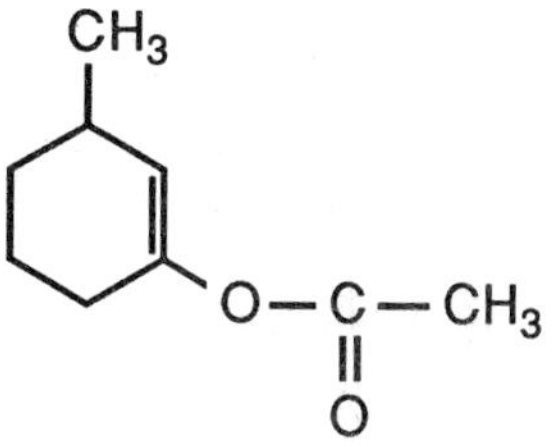

(c) ring with OCH_3 and CH_3 + ring with OCH_3 and CH_3

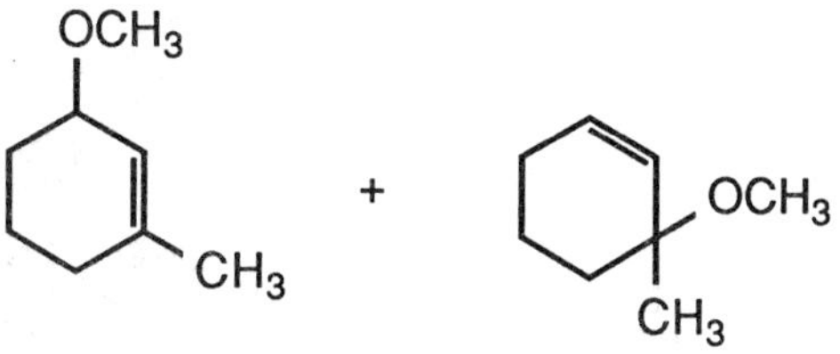

(d) $C_6H_5-\overset{\overset{O}{\|}}{C}-CH(C_6H_5)-C_6H_5$

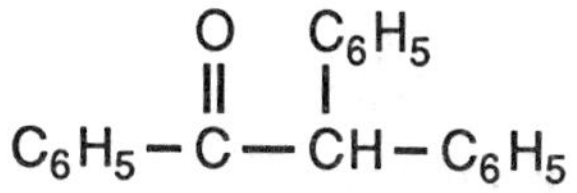

12.13

(a) $CH_3CH_2CH_2—C\equiv CH$ (b) $CH_3CH_2CH_2—O—\overset{\overset{O}{\|}}{C}—CH_3$ (c) $CH_3CH_2CH_2—C\equiv CH$

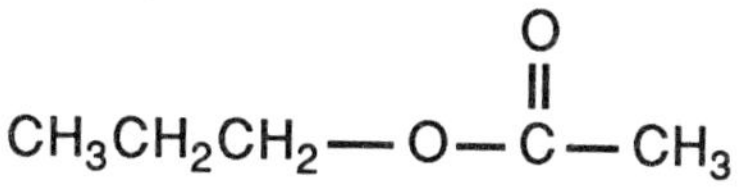

12.14

(a) $CH_3-CH=CH-CH_3$ (b) $CH_3-CH_2-CH(CH_3)-O-\overset{\overset{O}{\|}}{C}-CH_3$

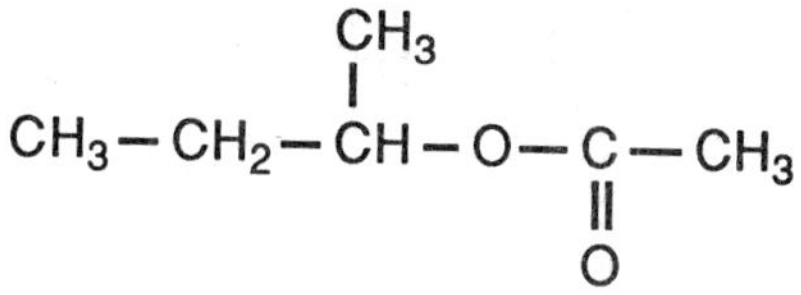

(c) $CH_3-CH_2-CH(CH_3)-C\equiv N$

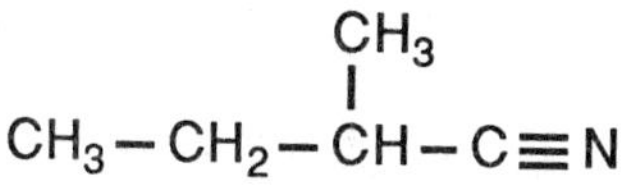

12.15 (a) tert-C_4H_9 cyclohexyl with $O-\overset{\overset{O}{\|}}{C}-CH_3$

(b) tert-C_4H_9 cyclohexyl with $C\equiv N$

(c) cyclohexane with OLi and $C\equiv C-CH_3$

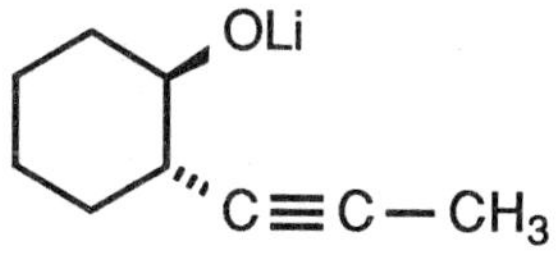

(d) cyclohexane with OH and D

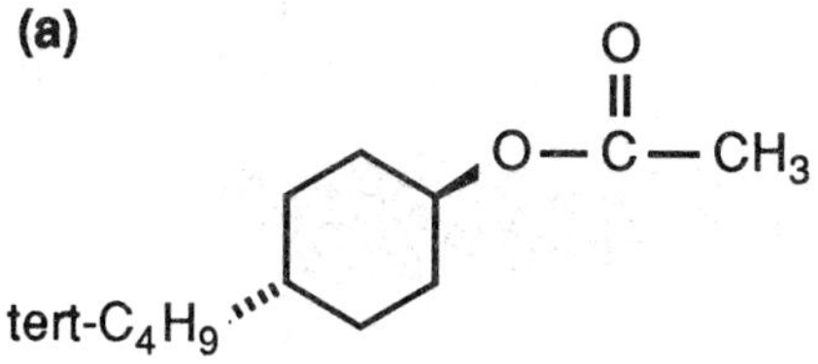

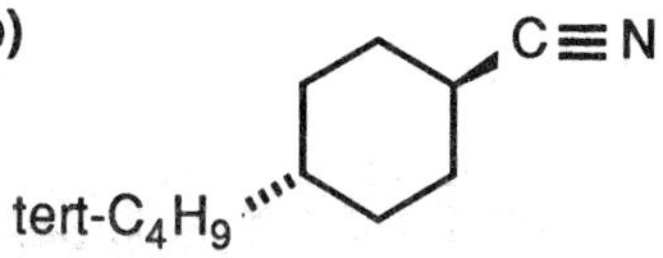

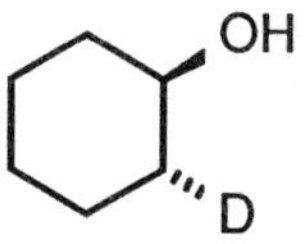

ANSWERS TO PROBLEMS

12.1 **(a)** CH_3O^-, which is a stronger base and stronger nucleophile than its conjugate acid, CH_3OH.

(b) CH_3NH^-, which is a stronger base and stronger nucleophile than a neutral amine (or, in this case, ammonia).

(c) NH_2^-, which is a stronger base and stronger nucleophile than neutral water.

(d) H_2O, which is more nucleophilic than neutral HI.

(d) Br^-, which is more nucleophilic than neutral CF_3CH_2OH.

12.2 **(a)** $C_6H_5CH_2OTs$, because OTs is a better leaving group than OH.

(b) $CH_3OSO_2CF_3$, because triflate is a better leaving group than brosylate.

(c) $CH_3CH_2OH_2^+$, because neutral water is a better leaving group than NH_2^-.

(c) $CH_3CH_2OH_2^+$, because neutral water is a better leaving group than NH_2^-.

(d) CH_3Br, because bromide is a better leaving group than hydroxide.

(e) $(CH_3)_2CHCH_2-S^+(CH_3)_2$, because neutral dimethylsulfide is a better leaving group than hydroxide ion.

(f) $CH_2=CH-CH_2-OBs$, because brosylate is a better leaving group than *p*-nitrobenzoate.

(g) $CH_3-OSO_2C_6H_5$, because benzenesulfonate is a better leaving group than chloride.

(h) $CH_3CH_2CH(C_6H_5)-Br$, because bromide is a better leaving group than *p*-nitrobenzoate.

12.3 The solvolysis of ethyl bromide in acetic acid is a direct displacement reaction of a primary halide in a moderately nucleophilic solvent.

(a) Neopentyl bromide would react much more slowly, because steric effects would retard the direct displacement reaction.

(b) Ethyl tosylate would react more rapidly because tosylate is a better leaving group than bromide.

(c) Solvolysis in water would be more rapid because water is more nucleophilic than acetic acid.

(d) Solvolysis in CF_3CO_2H would be slower because the electron withdrawing effect of the CF_3 group makes this solvent less nucleophilic.

(e) Addition of sodium acetate would increase the rate because acetate ion is a stronger nucleophile than neutral acetic acid.

12.4 For the initial reaction the product would be:

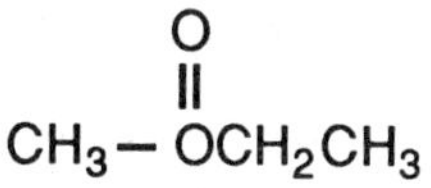

$$CH_3-\overset{\overset{\displaystyle O}{\|}}{C}-OCH_2CH_3$$

The products for the modified reactions are shown below:

(a) $CH_3-\overset{\overset{\displaystyle O}{\|}}{C}-O-\underset{\underset{\displaystyle CH_3}{|}}{\overset{\overset{\displaystyle CH_3}{|}}{C}}-CH_2CH_3$

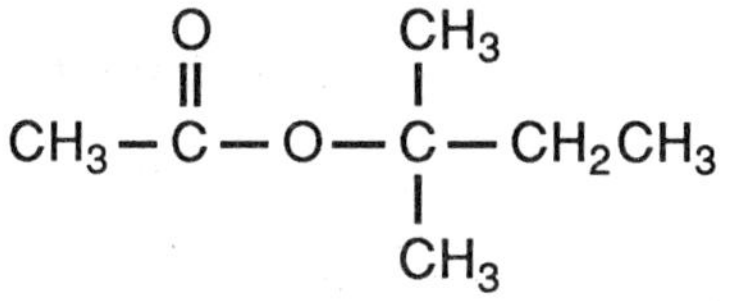

(b) $CH_3-\overset{\overset{\displaystyle O}{\|}}{C}-OCH_2CH_3$

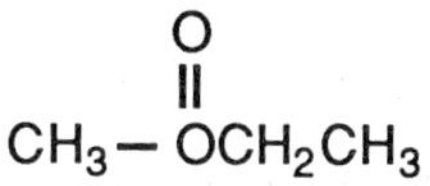

(c) CH_3-CH_2-OH

(d) $CF_3-\overset{\overset{\displaystyle O}{\|}}{C}-OCH_2CH_3$

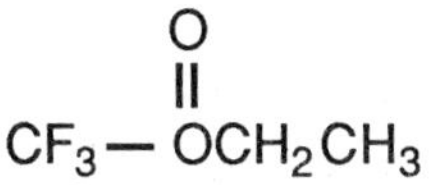

(e) $CH_3-\overset{\overset{\displaystyle O}{\|}}{C}-OCH_2CH_3$

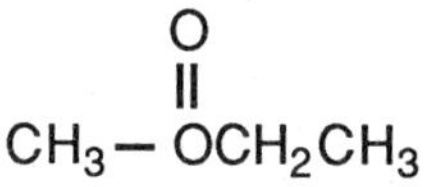

12.5 For the initial reaction the mechanism is:

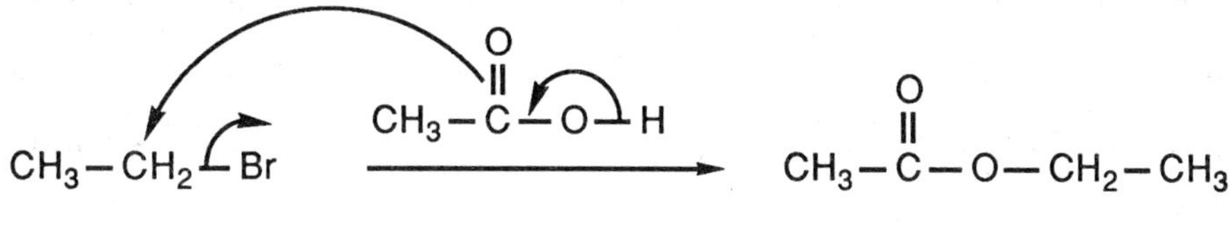

$$CH_3-CH_2-Br \xrightarrow{CH_3-\overset{\overset{\displaystyle O}{\|}}{C}-O-H} CH_3-\overset{\overset{\displaystyle O}{\|}}{C}-O-CH_2-CH_3$$

Each reaction except that in part (a) is a direct displacement, while that in (a) involves simultaneous ionization and rearrangement. The mechanisms are drawn in the following equations:

(a)

$$CH_3-\underset{CH_3}{\overset{CH_3}{C}}-CH_2-Br \longrightarrow CH_3-\underset{CH_3}{\overset{+}{C}}-CH_2-CH_3 \xrightarrow{CH_3-\overset{O}{\overset{\|}{C}}-O-H} CH_3-\overset{O}{\overset{\|}{C}}-O-\underset{CH_3}{\overset{CH_3}{C}}-CH_2-CH_3$$

(b)

$$CH_3-CH_2-OTs \xrightarrow{CH_3-\overset{O}{\overset{\|}{C}}-O-H} CH_3-\overset{O}{\overset{\|}{C}}-O-CH_2-CH_3$$

(c)

$$CH_3-CH_2-Br \xrightarrow{H_2\ddot{O}} CH_3-CH_2-\underset{+}{\overset{H}{O}}H \longrightarrow CH_3-CH_2-OH$$

(d)

$$CH_3-CH_2-Br \xrightarrow{CF_3-\overset{O}{\overset{\|}{C}}-O-H} CF_3-\overset{O}{\overset{\|}{C}}-O-CH_2-CH_3$$

(e)

$$CH_3-CH_2-Br \xrightarrow{CH_3-\overset{O^-}{\overset{|}{C}}=O} CH_3-\overset{O}{\overset{\|}{C}}-O-CH_2-CH_3$$

12.6 The solvolysis of 2-chloro-2-phenylpropane in 2,2,2-trifluoroethanol is a substitution reaction proceeding via a carbocation.

(a) Solvolysis in CF_3CO_2H would be faster because the CO_2H group results in an increase in solvent polarity (relative to the CH_2OH group of CF_3CH_2OH).

(b) The rate would decrease because replacement of trifluoroethanol as solvent with the less polar *tert*-butyl alcohol would retard carbocation formation.

(c) The rate would decrease because replacement of trifluoroethanol as solvent with the less polar ethanol would retard carbocation formation.

(d) The rate would decrease because replacement of the phenyl group with an ethyl group would result in a carbocation that is no longer resonance stabilized.

(e) The rate would decrease because *p*-nitrobenzoate is not as good a leaving group as chloride.

(e) The rate would increase because addition of a base would result in a second pathway (elimination) for reaction of the 2-phenyl-2-chloropropane.

12.7 The products for the original reaction would be:

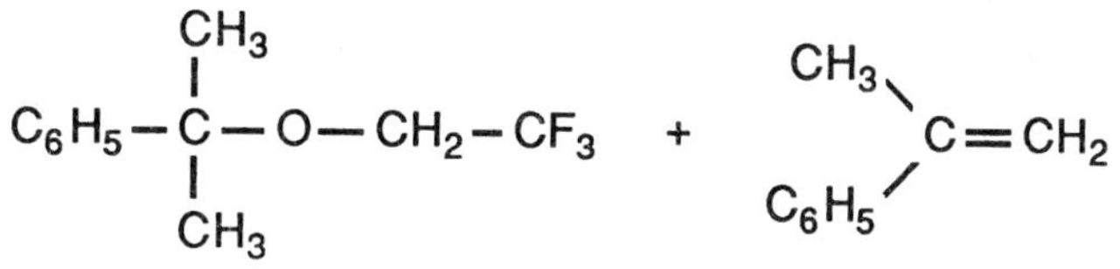

$$C_6H_5-C(CH_3)_2-O-CH_2-CF_3 \quad + \quad (CH_3)(C_6H_5)C{=}CH_2$$

The products for the modified reaction would be:

(a)

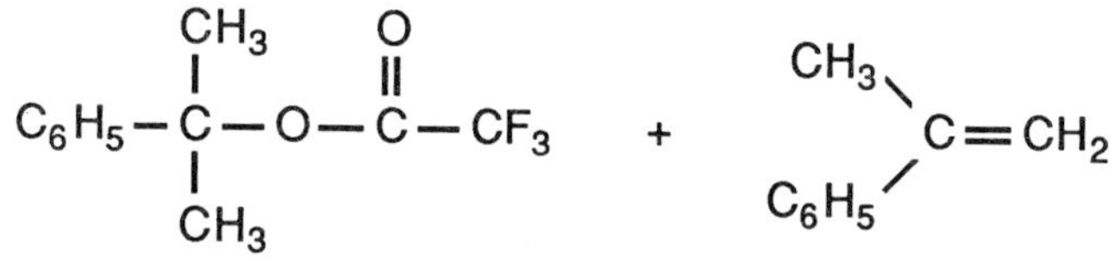

$$C_6H_5-C(CH_3)_2-O-C(=O)-CF_3 \quad + \quad (CH_3)(C_6H_5)C{=}CH_2$$

(b)

$$C_6H_5-C(CH_3)_2-O-C(CH_3)_2-CH_3 \quad + \quad (CH_3)(C_6H_5)C{=}CH_2$$

(c)

$$C_6H_5-C(CH_3)_2-O-CH_2-CH_3 \quad + \quad (CH_3)(C_6H_5)C{=}CH_2$$

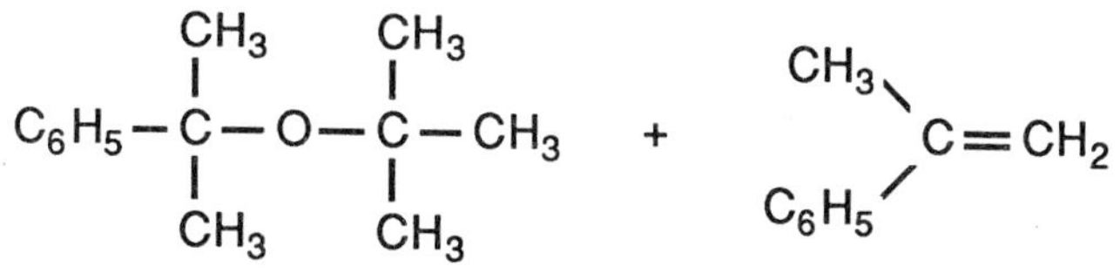

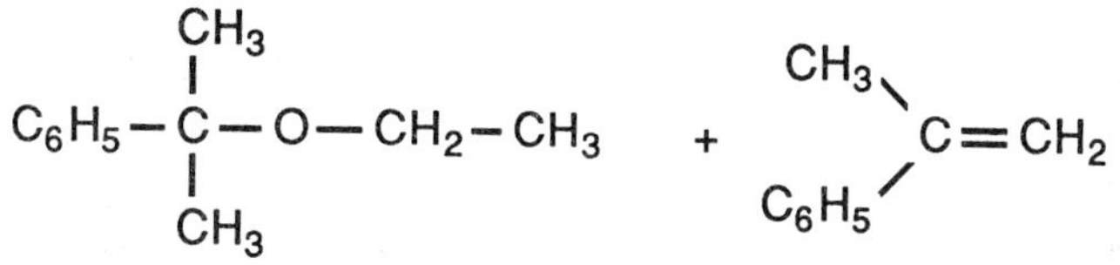

(d)

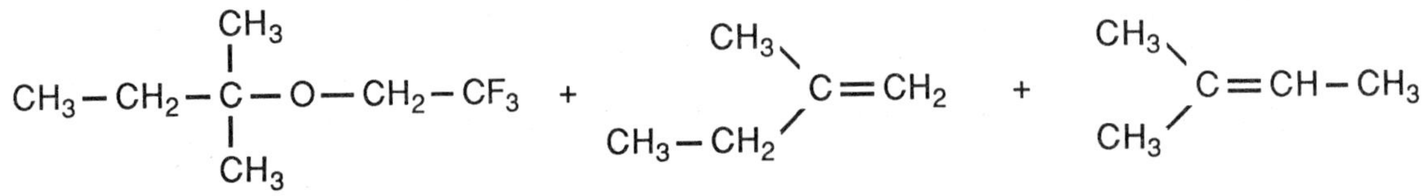

$$CH_3-CH_2-\underset{CH_3}{\overset{CH_3}{C}}-O-CH_2-CF_3 \quad + \quad \begin{matrix} CH_3 \\ CH_3-CH_2 \end{matrix}\!>\!C=CH_2 \quad + \quad \begin{matrix} CH_3 \\ CH_3 \end{matrix}\!>\!C=CH-CH_3$$

(e)

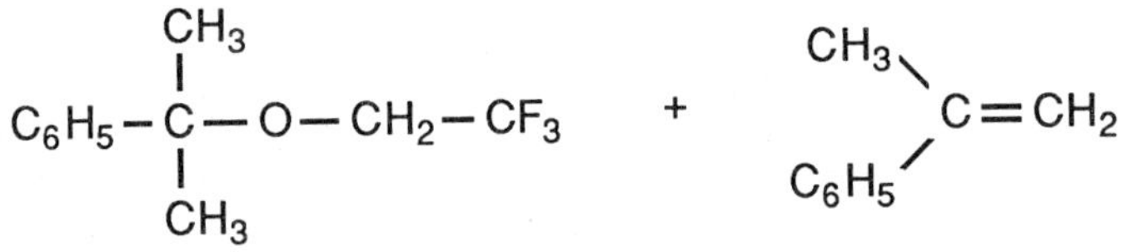

$$C_6H_5-\underset{CH_3}{\overset{CH_3}{C}}-O-CH_2-CF_3 \quad + \quad \begin{matrix} CH_3 \\ C_6H_5 \end{matrix}\!>\!C=CH_2$$

(f)

$$\begin{matrix} CH_3 \\ C_6H_5 \end{matrix}\!>\!C=CH_2$$

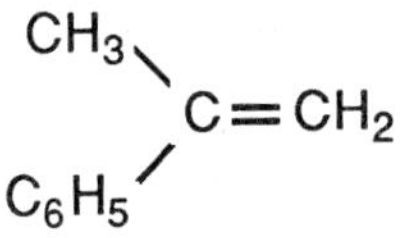

12.8 With the exception of the bimolecular elimination that contributes in part (f), all the reactions proceed via initial ionization to a carbocation:

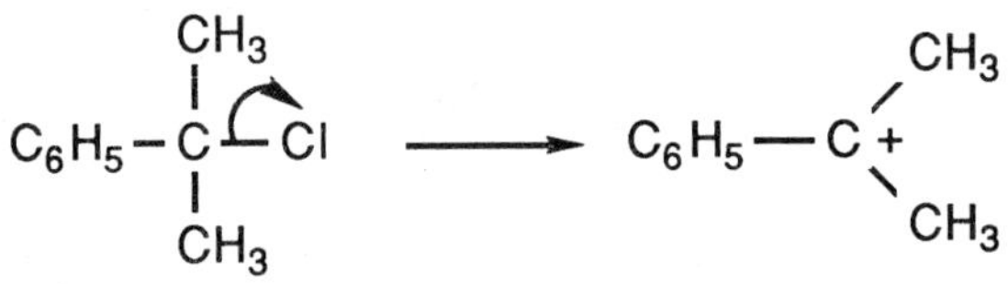

$$C_6H_5-\underset{CH_3}{\overset{CH_3}{C}}-Cl \longrightarrow C_6H_5-\overset{+}{C}\begin{matrix} CH_3 \\ CH_3 \end{matrix}$$

The elimination product is formed by solvent (shown here as SOH) acting as a base:

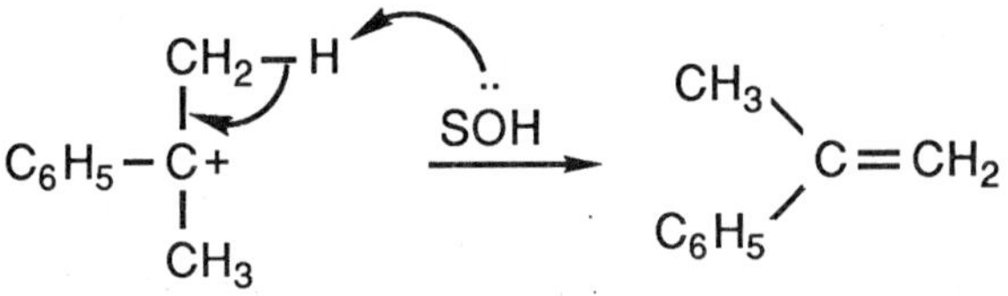

$$C_6H_5-\underset{CH_3}{\overset{CH_2-H}{C^+}} \xrightarrow{\ddot{S}OH} \begin{matrix} CH_3 \\ C_6H_5 \end{matrix}\!>\!C=CH_2$$

The substitution product for the original reaction occurs by reaction of the carbocation with solvent:

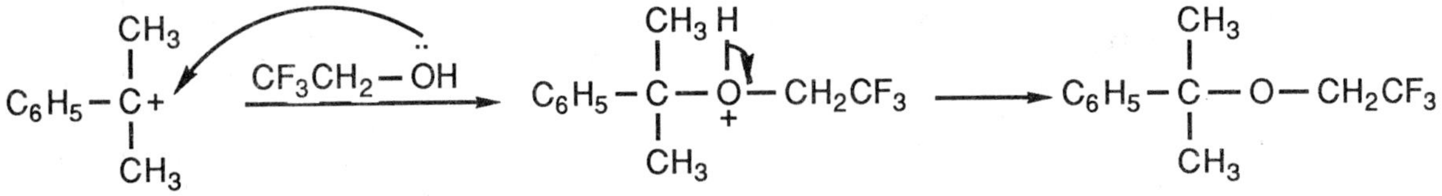

$$C_6H_5-\underset{CH_3}{\overset{CH_3}{C^+}} \xrightarrow{CF_3CH_2-\ddot{O}H} C_6H_5-\underset{CH_3}{\overset{CH_3}{C}}-\overset{H}{\underset{+}{O}}-CH_2CF_3 \longrightarrow C_6H_5-\underset{CH_3}{\overset{CH_3}{C}}-O-CH_2CF_3$$

(a)

$$C_6H_5-\overset{CH_3}{\underset{CH_3}{C}}{}^{+} \xrightarrow{CF_3-\overset{O}{\overset{\|}{C}}-OH} C_6H_5-\overset{CH_3}{\underset{CH_3}{C}}-O-\overset{\overset{H}{O^{+}}}{\overset{\|}{C}}-CF_3 \longrightarrow C_6H_5-\overset{CH_3}{\underset{CH_3}{C}}-O-\overset{O}{\overset{\|}{C}}-CF_3$$

(b)

$$C_6H_5-\overset{CH_3}{\underset{CH_3}{C}}{}^{+} \xrightarrow{HO-C(CH_3)_3} C_6H_5-\overset{CH_3}{\underset{CH_3}{C}}-\overset{H}{\underset{+}{O}}-\overset{CH_3}{\underset{CH_3}{C}}-CH_3 \longrightarrow C_6H_5-\overset{CH_3}{\underset{CH_3}{C}}-O-\overset{CH_3}{\underset{CH_3}{C}}-CH_3$$

(c)

$$C_6H_5-\overset{CH_3}{\underset{CH_3}{C}}{}^{+} \xrightarrow{CH_3CH_2-OH} C_6H_5-\overset{CH_3}{\underset{CH_3}{C}}-\overset{H}{\underset{+}{O}}-CH_2CH_3 \longrightarrow C_6H_5-\overset{CH_3}{\underset{CH_3}{C}}-O-CH_2CH_3$$

(d)

$$CH_3-CH_2-\overset{CH_3}{\underset{CH_3}{C}}-Cl \longrightarrow CH_3-CH_2-\overset{CH_3}{\underset{CH_3}{C}}{}^{+} \xrightarrow{CF_2CH_2-OH} CH_3-CH_2-\overset{CH_3}{\underset{CH_3}{C}}-\overset{H}{\underset{+}{O}}-CH_2CF_3 \longrightarrow CH_3-CH_2-\overset{CH_3}{\underset{CH_3}{C}}-O-CH_2CF_3$$

(e) The only difference from the original reaction is the ionization step:

$$C_6H_5-\overset{CH_3}{\underset{CH_3}{C}}-OPNB \longrightarrow C_6H_5-\overset{CH_3}{\underset{CH_3}{C}}{}^{+} \xrightarrow{CF_3CH_2OH} C_6H_5-\overset{CH_3}{\underset{CH_3}{C}}-OCH_2CF_3$$

(f) The major pathway would be bimolecular elimination:

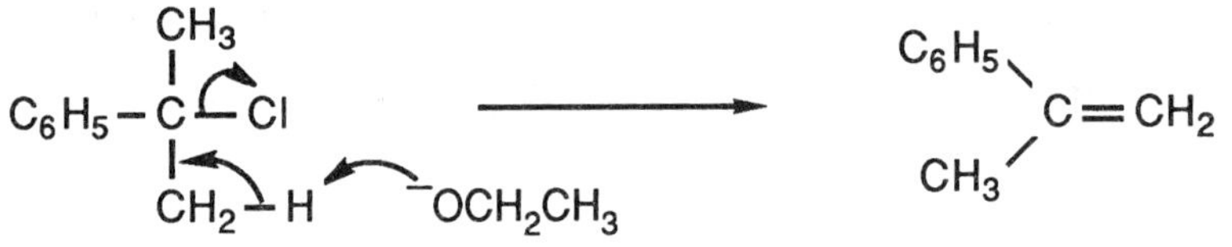

$$C_6H_5-\underset{CH_2-H}{\overset{CH_3}{C}}-Cl \quad {}^{-}OCH_2CH_3 \longrightarrow \begin{matrix} C_6H_5 \\ CH_3 \end{matrix}\!\!>C=CH_2$$

12.9 Trifluoroethanol is a polar solvent of very low nucleophilicity, so carbocation reactions are favored and direct displacement reactions will be very slow.

(a)

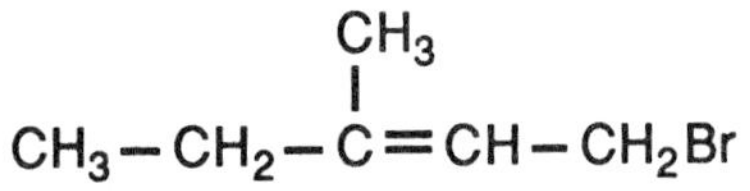

$$CH_3-CH_2-\overset{CH_3}{\overset{|}{C}}=CH-CH_2Br$$

This compound would yield the more stable allylic cation. The cation's resonance contributors are primary and tertiary,

$$CH_3-CH_2-\overset{CH_3}{\overset{|}{C}}=CH-\overset{+}{C}H_2 \longleftrightarrow CH_3-CH_2-\underset{+}{\overset{CH_3}{\overset{|}{C}}}-CH=CH_2$$

whereas those of the cation from the other bromide are primary and secondary.

(b) $C_6H_5CH_2OTs$ would react more rapidly because it would afford a resonance-stabilized cation.

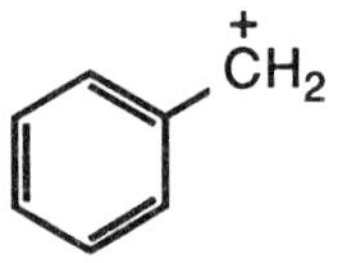

$$C_6H_5-\overset{+}{C}H_2$$

(c) $(CH_3)_3C-Br$ would react more rapidly because it is a tertiary bromide, whereas the other compound is only a secondary bromide.

(d) $(CH_3)_3C-CH_2CH_2OBs$ would react more rapidly than the isomer $CH_3CH_2C(CH_3)_2CH_2OBs$. Both compounds are primary and would react very slowly in this non-nucleophilic solvent, but the second compound is even less reactive because the reactive center is hindered by the adjacent quaternary carbon.

(e) $CH_3CH_2CH_2C(CH_3)_2Br$ is a tertiary bromide and would react more rapidly than the isomeric primary bromide.

(f) Both compounds are secondary brosylates that should react (slowly) by the direct displacement pathway. Cyclohexyl brosylate should react less slowly because it is less hindered:

12.10 The reaction rate of this compound increases by a factor of three when the concentration of RX is tripled (at the same concentration of NaN_3 On the other hand, doubling the concentration of azide ion (while maintaining constant RX concentration) has no significant effect on the rate.

(a) The data support a carbocation mechanism because the kinetics are first order in RX but zero order in azide ion.

(b) For [RX] = 0.80 M the rate should be five times faster than the value for [RX] = 0.16 M (at the same NaN_3 concentration). Therefore a rate of 1×10^4 sec^{-1} would be expected.

(c) Since the rate is essentially independent of azide ion concentration, a value comparable to the first entry in the Table would be expected: approximately 6×10^3 sec^{-1}

(d) From top to bottom in the Table, the calculated first order rate constants calculated from the equation

$$\text{rate} = k\,[\text{RX}]$$

are 1.2×10^4, 1.2×10^4, 1.3×10^4, and 1.3×10^4 sec^{-1}. These are essentially independent of the reaction conditions.

12.11 For a direct displacement reaction the rate equation would be:

$$\text{rate} = k\,[\text{RX}][\text{NaOH}]$$

The rate constant, k, is 0.033 M sec^{-1}, so you can calculate the rate for other concentrations of the reactants.

(a) 2.0×10^{-4} M sec^{-1}

(b) 9.9×10^{-4} M sec^{-1}

(c) 2.5×10^{-5} M sec^{-1}

(d) 2.5×10^{-4} M sec^{-1}

(e) 7.9×10^{-4} M sec^{-1}

12.12

(a) $CH_3-C(CH_3)_2-CH=CH_2$

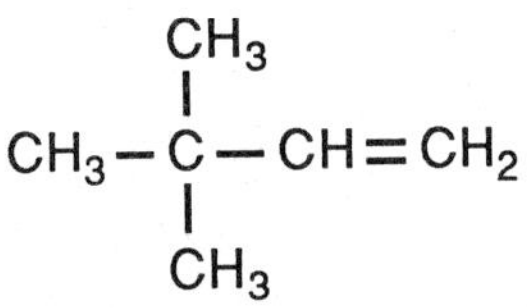

(b) $CF_3CH_2-C(CH_3)_2-CH(CH_3)-CH_3$ + $(CH_3)_2C=C(CH_3)_2$

(c) $O-C(=O)-CF_3$ (3-methylcyclopent-2-enyl, CH_3) + $O-C(=O)-CF_3$ (3-methylcyclopent-2-enyl, CH_3)

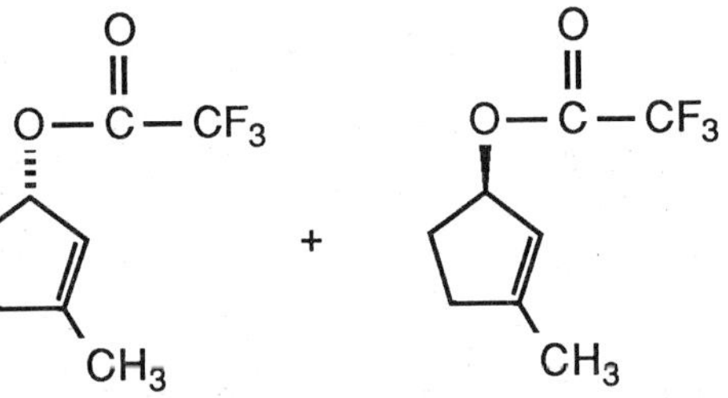

(d) $O-C(=O)-CH_3$ (3-methylcyclopent-2-enyl, CH_3)

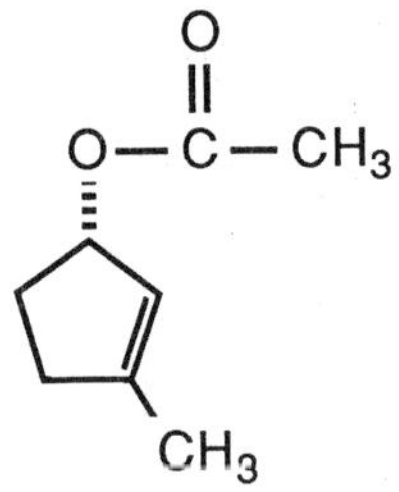

(e) $(CH_3)_3C$ — cyclopentane — OCH_3

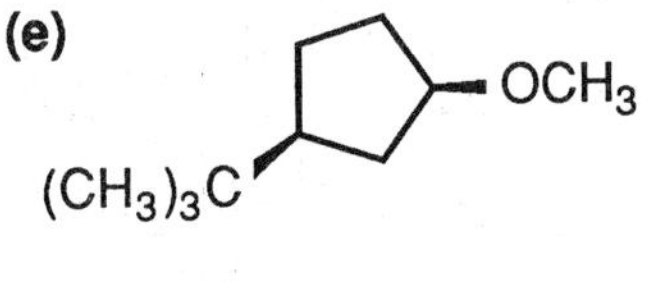

(f) $CH_2-CH_2-CH_2Br$ and OH on a benzene ring

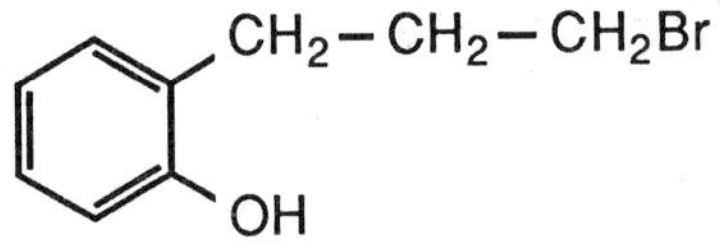

12.13

(a) C_6H_5–CH=CH–CH(CH_3)(H)–OCH_2CF_3 + C_6H_5–CH=CH–CH(CH_3)(H)–OCH_2CF_3 +

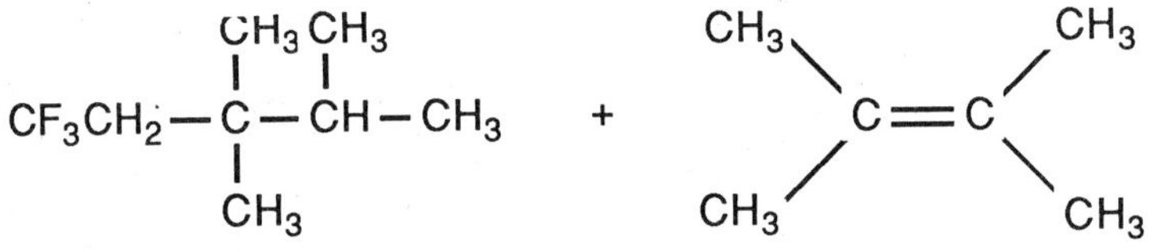

CH_3–CH=CH–CH(C_6H_5)(H)–OCH_2CF_3 + CH_3–CH=CH–CH(C_6H_5)(H)–OCH_2CF_3

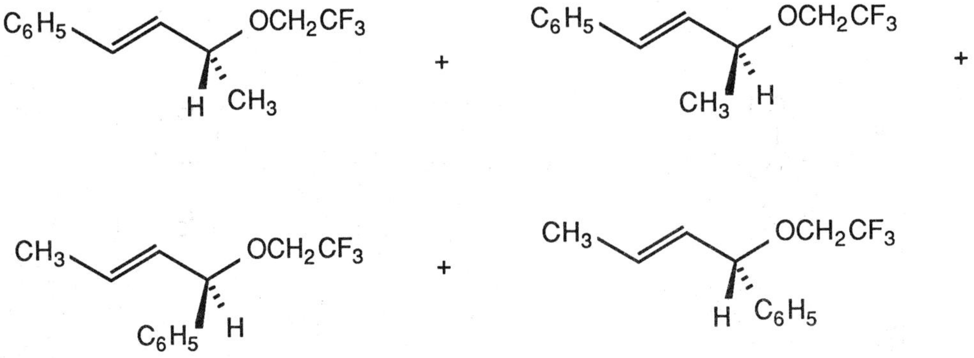

(b) $C_6H_5CH{=}CH{-}CH{=}CH_2$

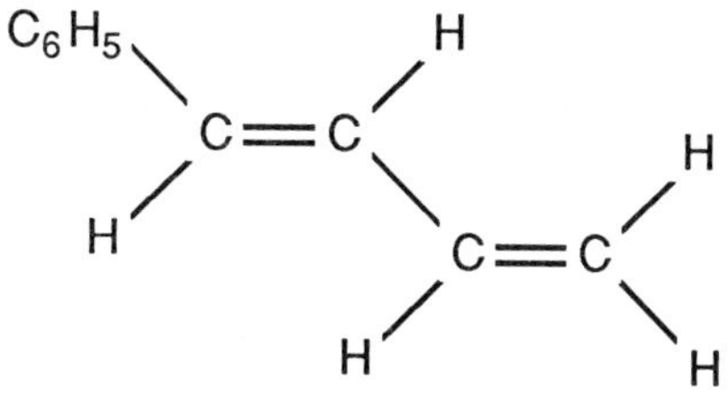

(c) CH_3O, CH_3 + CH_3

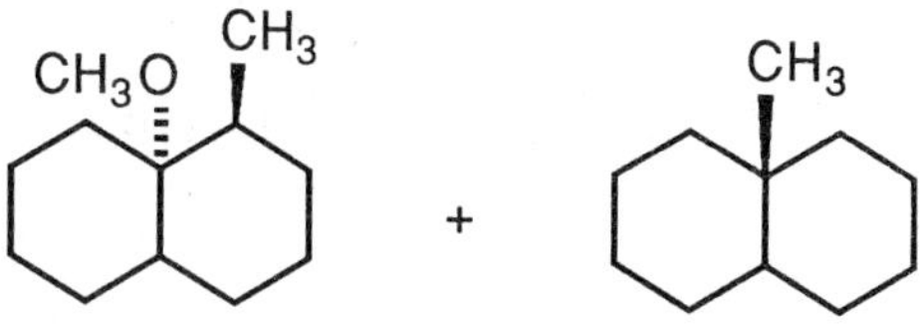

(d) 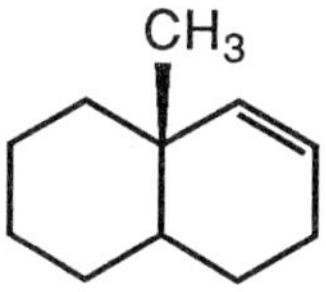 CH_3

(e)

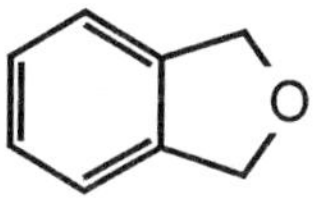

(f) 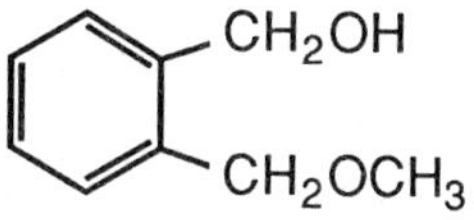 CH_2OH, CH_2OCH_3

12.14

(a)

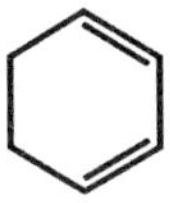

(b)  $O-CH_2-CF_3$ + $O-CH_2-CF_3$

(c) 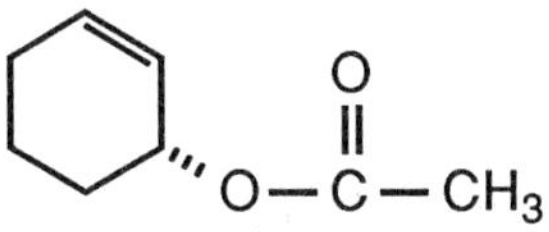 $O-\overset{O}{\overset{\|}{C}}-CH_3$

(d) 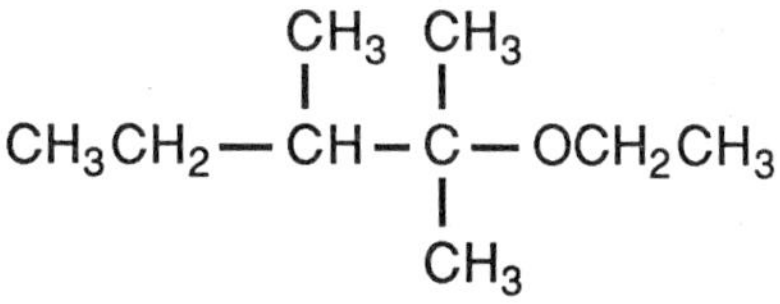 $CH_3CH_2-CH(CH_3)-C(CH_3)_2-OCH_2CH_3$

(e) 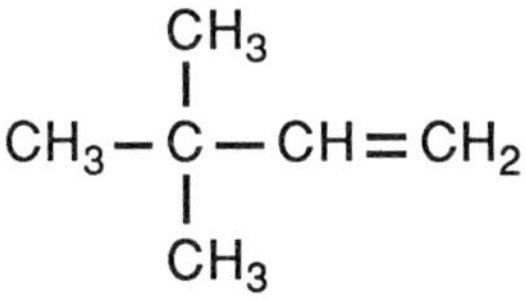 $CH_3-C(CH_3)_2-CH{=}CH_2$

(f) 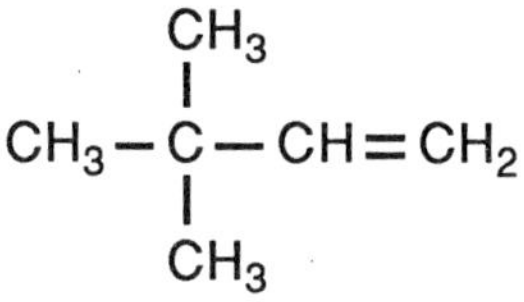 $CH_3-C(CH_3)_2-CH{=}CH_2$

12.15

(a) 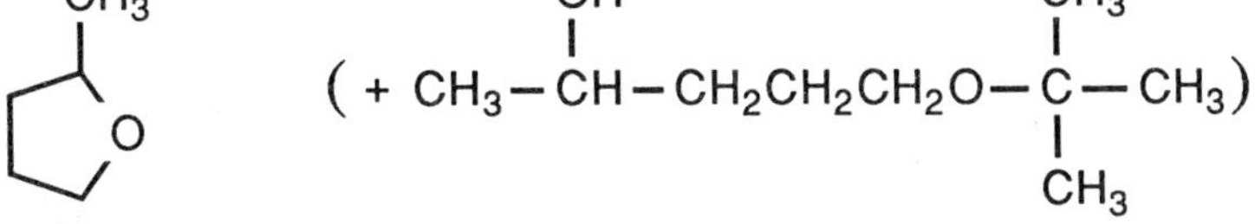 CH_3, O $(+\ CH_3-CH(OH)-CH_2CH_2CH_2O-C(CH_3)_2-CH_3)$

(b) 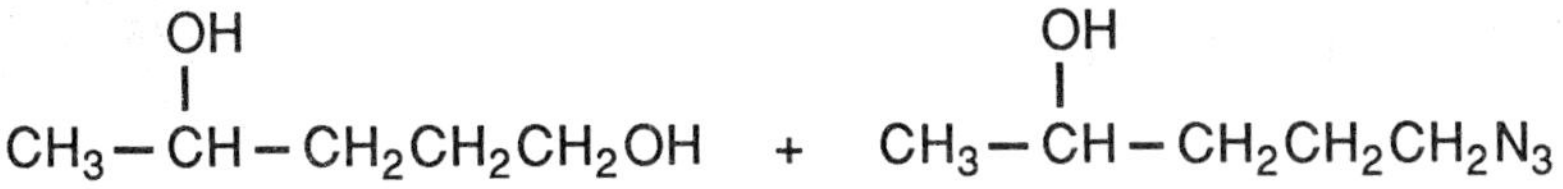 $CH_3-CH(OH)-CH_2CH_2CH_2OH$ + $CH_3-CH(OH)-CH_2CH_2CH_2N_3$

(c) 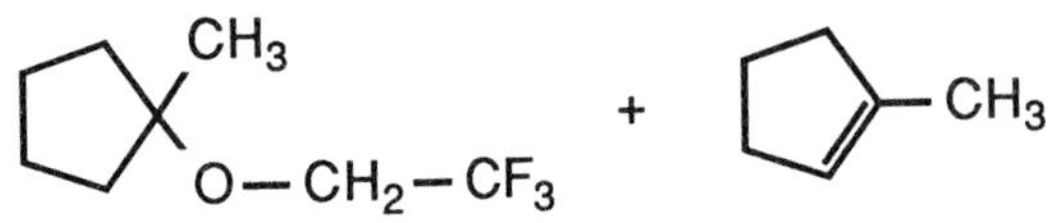 CH_3, $O-CH_2-CF_3$ + CH_3 (d) 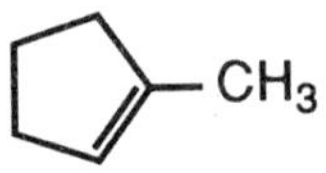 CH_3

(e) 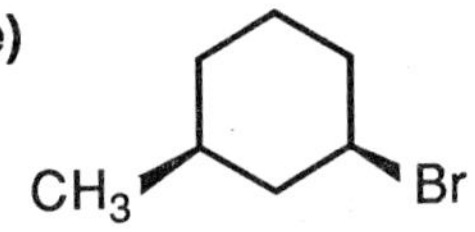 CH_3 Br (f) 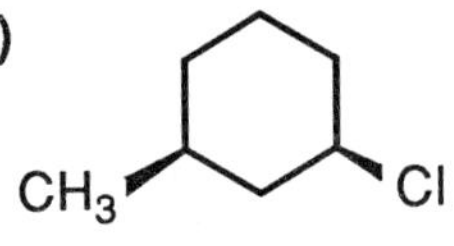 CH_3 Cl

12.16

(a) 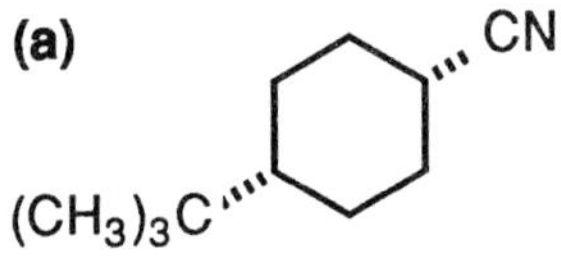 CN, $(CH_3)_3C$ (b) 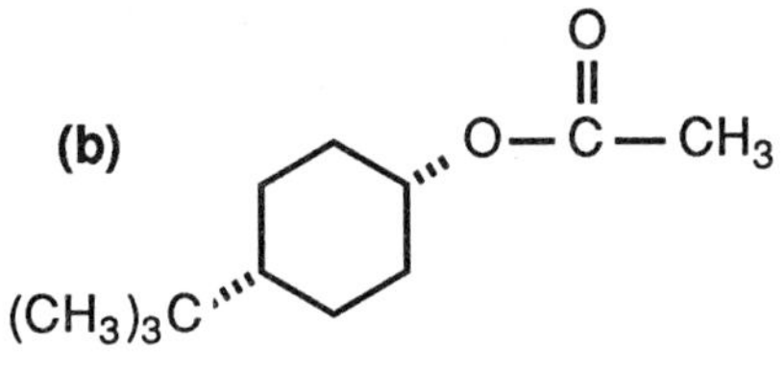 $O-C(=O)-CH_3$, $(CH_3)_3C$ (c) CF_3CH_2O, CH_3, CH_3

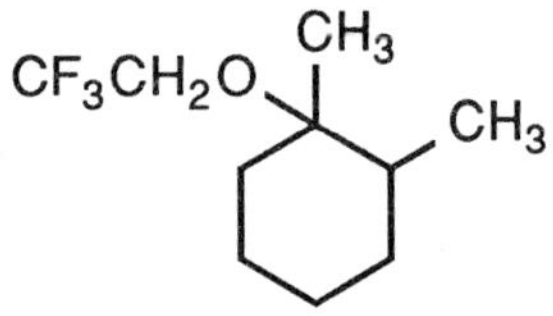

(d) 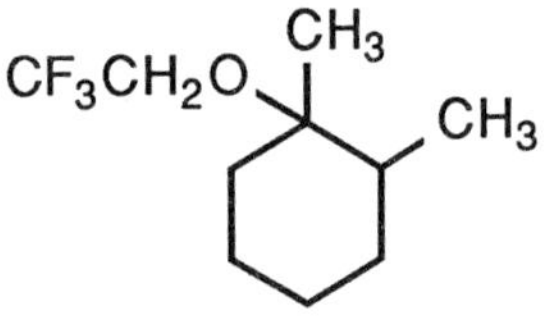 CF_3CH_2O, CH_3, CH_3 (e) 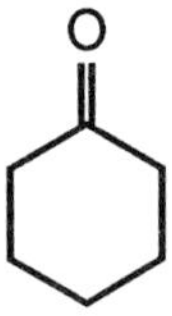 O (f) 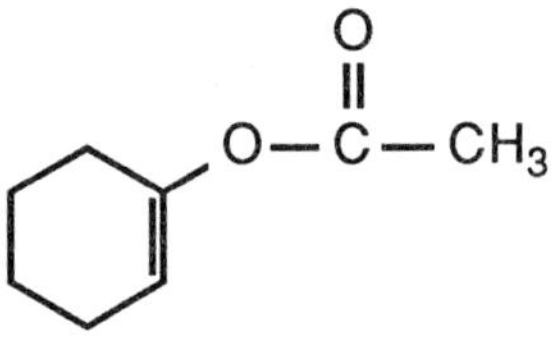 $O-C(=O)-CH_3$

12.17 (a) 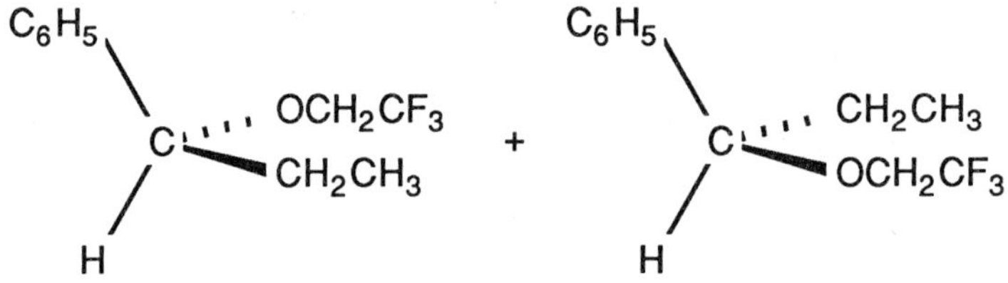 C_6H_5, C, OCH_2CF_3, CH_2CH_3, H + C_6H_5, C, CH_2CH_3, OCH_2CF_3, H (b) $C_6H_5-CH=CH-CH_3$

(c) 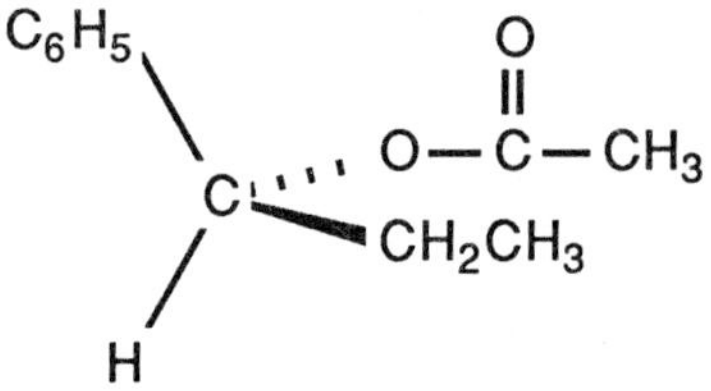 C_6H_5, C, $O-C(=O)-CH_3$, CH_2CH_3, H (d) 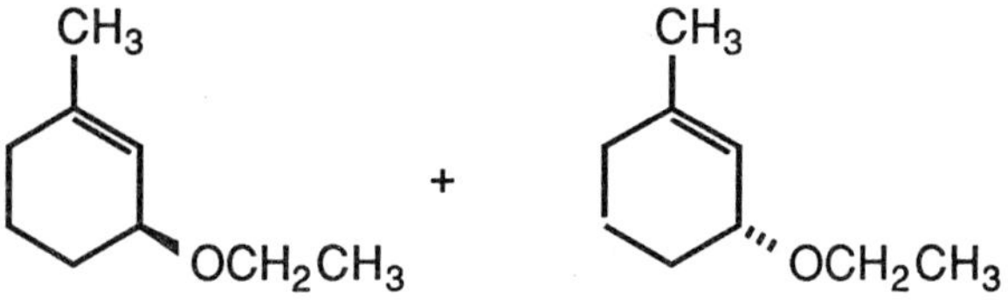 CH_3, OCH_2CH_3 + CH_3, OCH_2CH_3

(e) CH_3

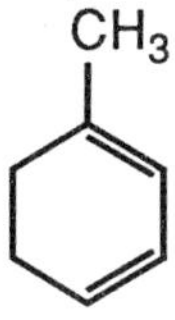

(f) CH_3 OH + CH_3 OH

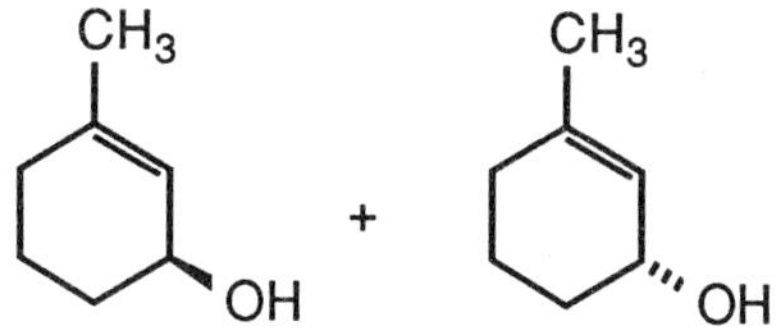

12.18 (a) CH_3 H H H Cl C_6H_5

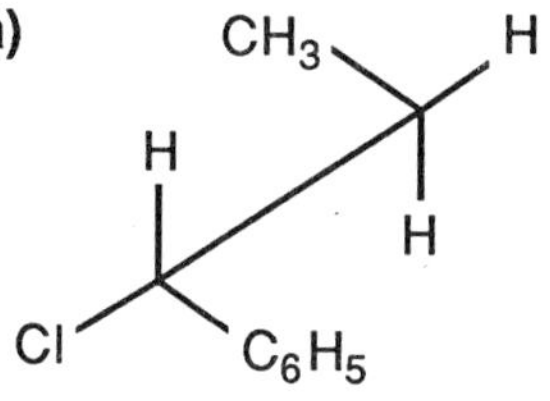

(b) CH_3 H H H C_6H_5 OCH_2CF_3 + CH_3 H H H CF_3CH_2O C_6H_5

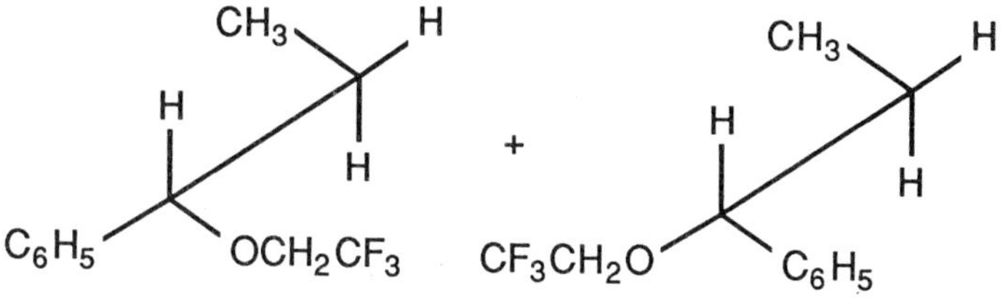

(c) H CH_3 C=C C_6H_5 H

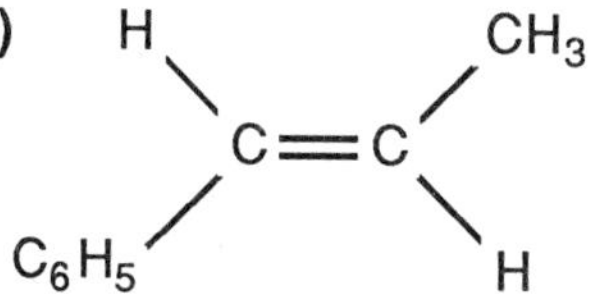

(d)

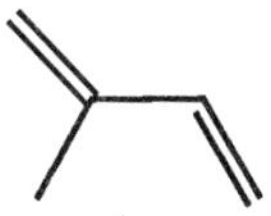

(e) CH_3 H C=C CH_3 CH_2O-CCH_3 O

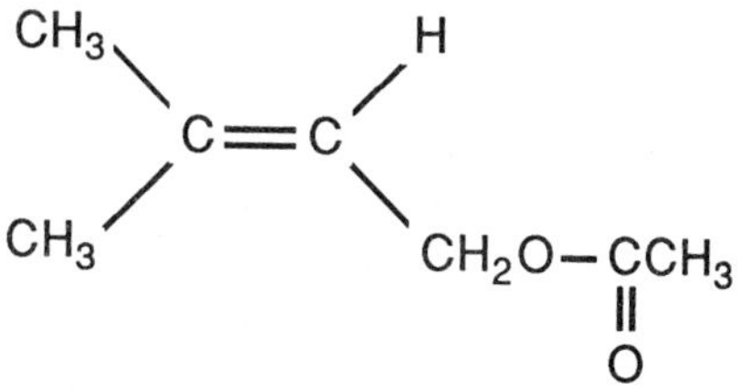

(f) CH_3 H C=C CH_3 $CH_2OCH_2CF_3$ + $O-CH_2CF_3$ $CH_3-C-CH=CH_2$ CH_3

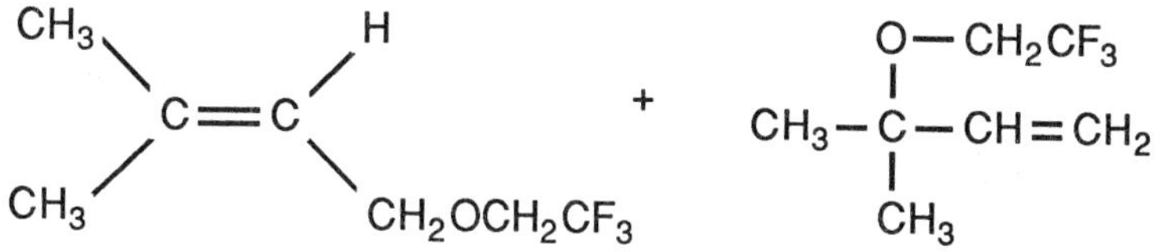

12.19 (a) CH_3 OCH_3

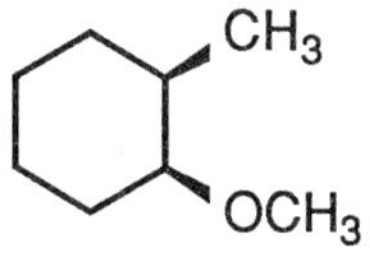

(b) CH_3

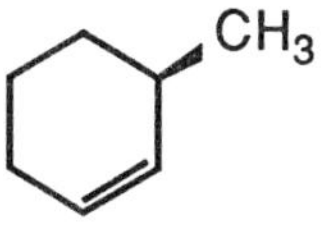

(c) CH_3 OCH_3 + OCH_3 CH_2-CH_3

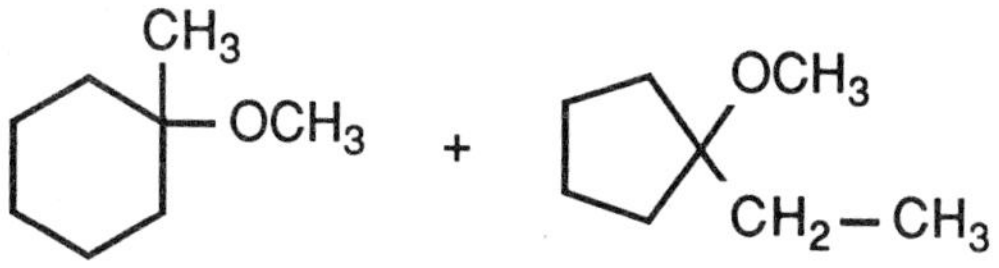

(d) H CH_3 CH_3 OH + H CH_3 HO CH_3

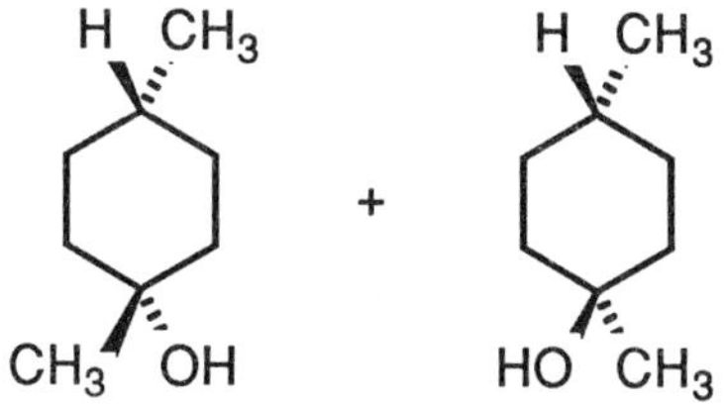

(e) CH_3

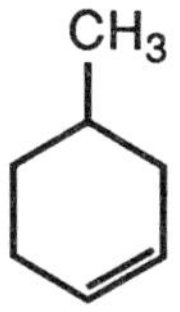

12.20

(a) OCH_2CF_3

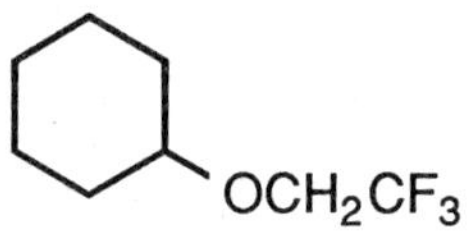

(b)

(c) $O-\overset{O}{\overset{\|}{C}}-CH_3$; CH_3

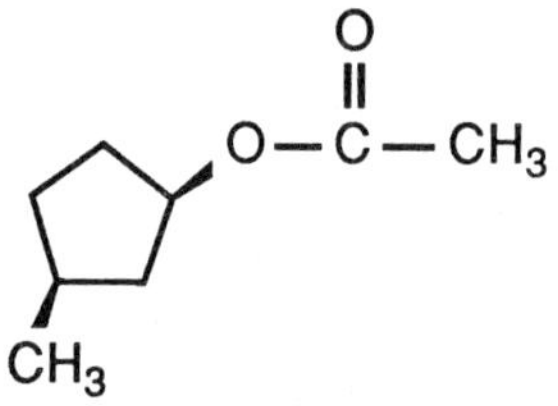

(d) CH_3 + CH_3 + OCH_3, CH_3

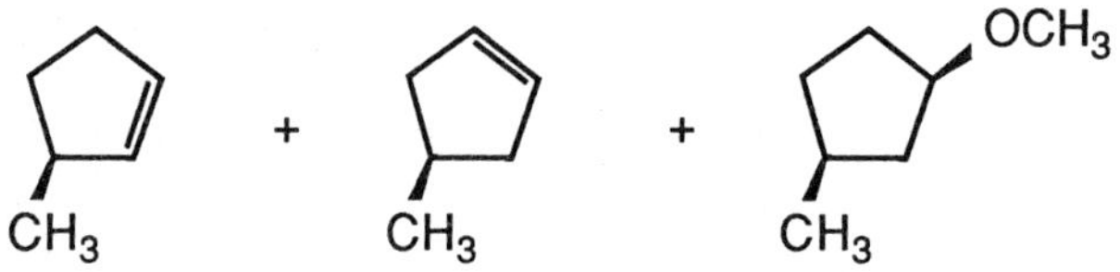

(e) $C_6H_5-CH_2-O-C(CH_3)_2-CH_3$

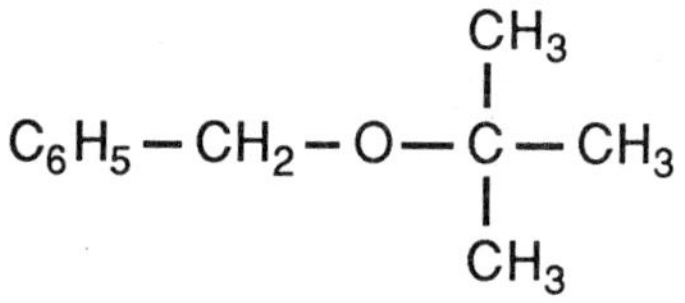

(f) $C_6H_5-CH_2-O-CH_2-CH_3$ + $C_6H_5-CH_2-OH$

12.21

(a) Br → $Bu_4\overset{+}{N}\ ^{-}OAc$ → $O-\overset{O}{\overset{\|}{C}}-CH_3$ → 1. $LiAlH_4$ 2. H_2O → OH → TsCl → OTs → H_2O acetone → OH

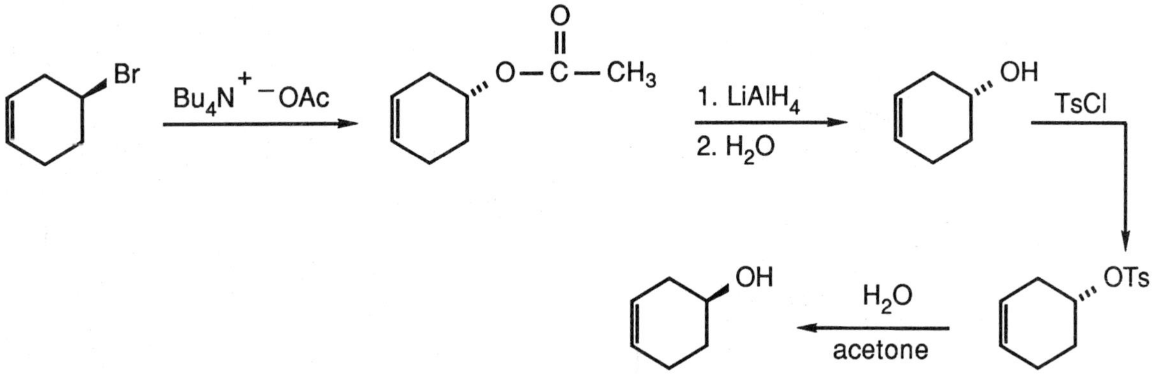

(b) H, CH_2CH_3, HO, H, CH_3, H → PBr_3 → H, CH_2CH_3, H, H, CH_3, Br → NaCN → H, CH_2CH_3, CN, H, CH_3, H

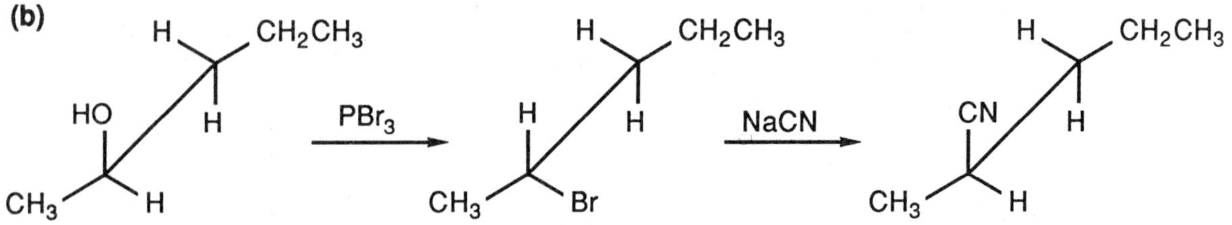

(c) D, CH_3-C, H, OH → PBr_3 → D, CH_3-C, Br, H → $Na^{+-}C\equiv CCH_3$ → D, CH_3-C, H, $C\equiv CCH_3$

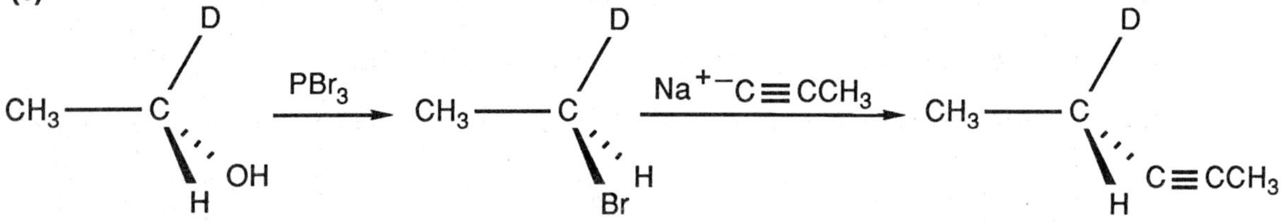

(d)

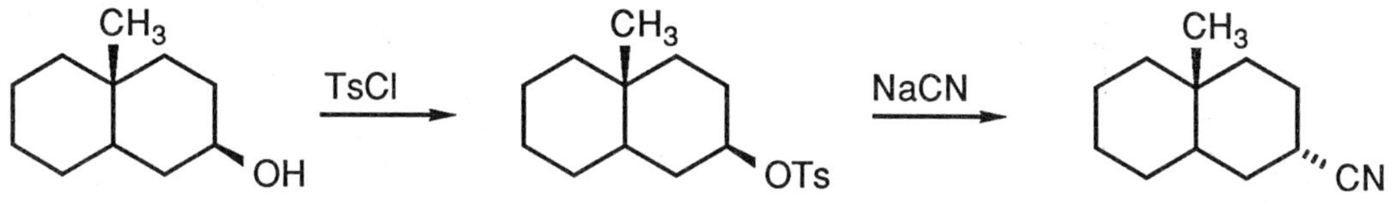

CH_3 ... OH —TsCl→ ... OTs —NaCN→ ... CN

(e)

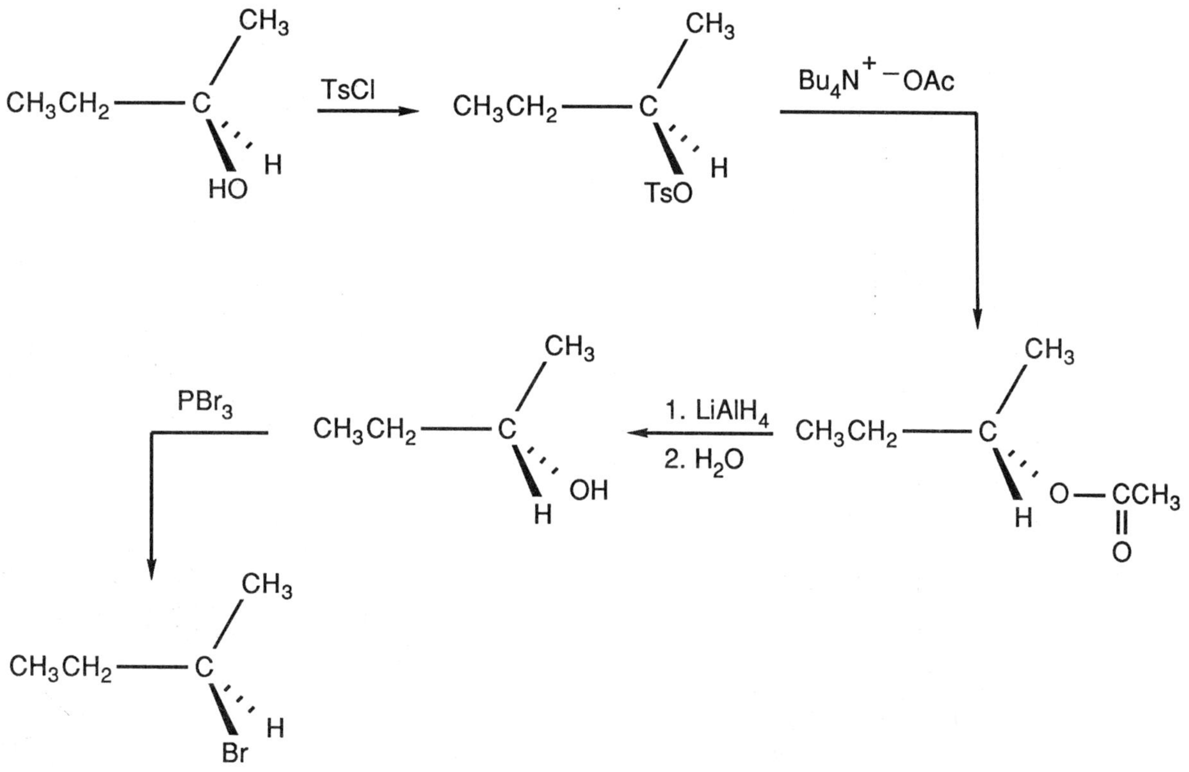

CH_3CH_2—C(CH_3)(H)(HO) —TsCl→ CH_3CH_2—C(CH_3)(H)(TsO) —$Bu_4N^+\ ^-OAc$→ CH_3CH_2—C(CH_3)(H)(O—$\underset{\|}{C}CH_3$, O)

—1. $LiAlH_4$ 2. H_2O→ CH_3CH_2—C(CH_3)(H)(OH) —PBr_3→ CH_3CH_2—C(CH_3)(H)(Br)

12.22 The complete structures are given for **G** and **H**, and they have the same carbon skeleton. This means that no rearrangements have occurred in any of the reactions. Acylation of **G** must afford (*R*)-1,3,3-trimethylpentyl acetate **(F)**, so **E** is (*S*)-1,3,3-trimethylpentyl acetate. This compound is formed by catalytic hydrogenation of a double bond in **D** (note that **A** and **B** both have one degree of unsaturation). There is only one possible location of a double bond in **D** that allows the chiral center to remain, so this compound is is (*S*)−1,3,3−trimethyl−4−pentenyl acetate. This is formed from the tosylate with inversion, and one further inversion of configuration occurs in the hydrolysis of the bromide **A**. The complete answer follows:

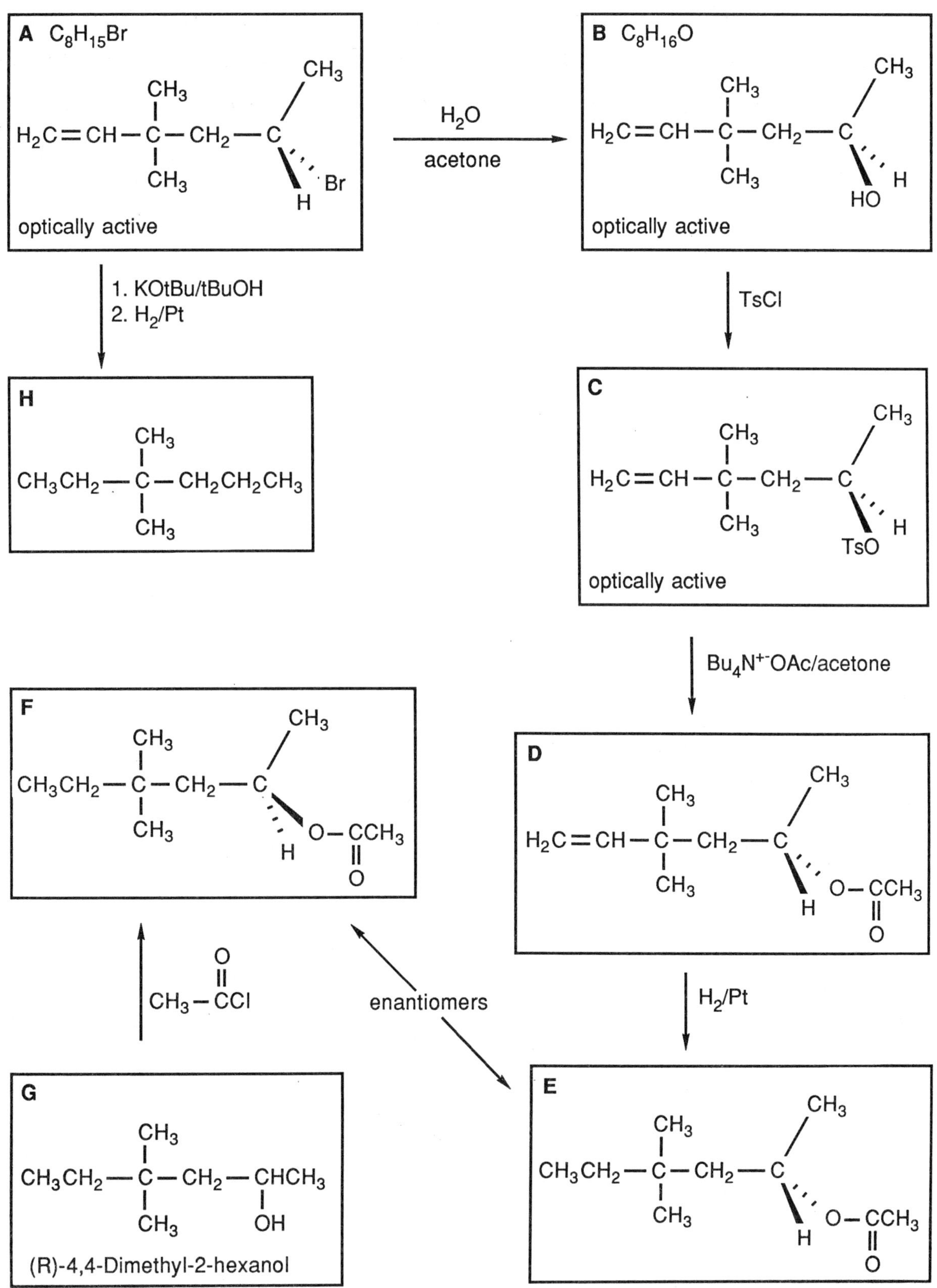
A $C_8H_{15}Br$
optically active
H_2O
acetone
B $C_8H_{16}O$
optically active
1. KOtBu/tBuOH
2. H_2/Pt
H
TsCl
C
optically active
$Bu_4N^{+-}OAc$/acetone
D
F
H_2/Pt
enantiomers
E
G
(R)-4,4-Dimethyl-2-hexanol

12.23

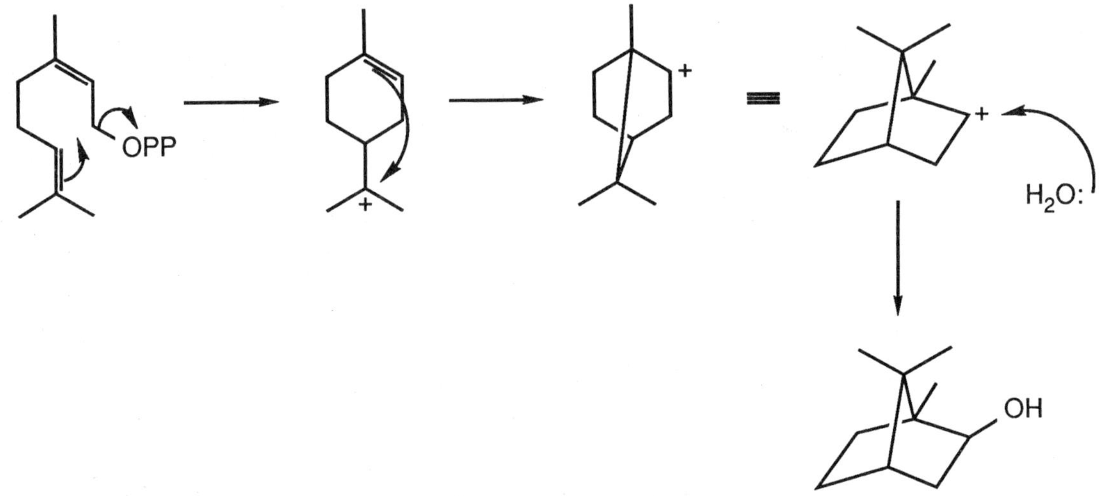

OPP

$H_2O:$

OH

12.24 In contrast to the case with 2,2-dimethyl-1-propyl tosylate, elimination could occur readily with 3,3-dimethyl-2-butyl tosylate. This elimination would occur much more rapidly than substitution.

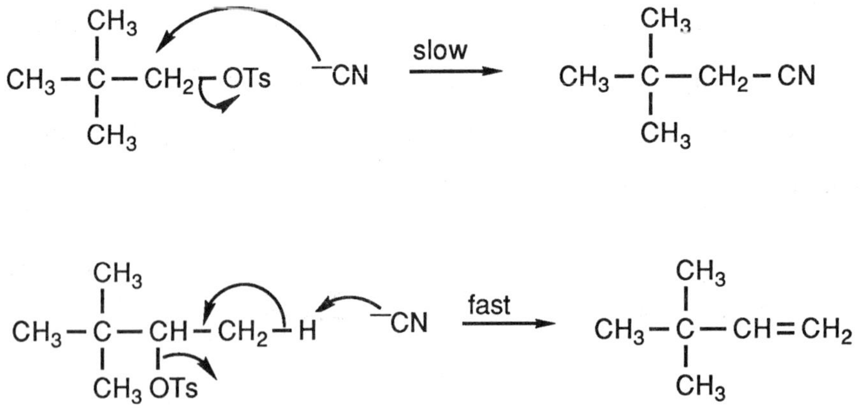

$$CH_3-C(CH_3)_2-CH_2-OTs \;\; {}^{-}CN \xrightarrow{\text{slow}} CH_3-C(CH_3)_2-CH_2-CN$$

$$CH_3-C(CH_3)_2-CH(OTs)-CH_2-H \;\; {}^{-}CN \xrightarrow{\text{fast}} CH_3-C(CH_3)_2-CH=CH_2$$

12.25 Substitution should still predominate, but the fraction of elimination should increase because there is less hindrance for removal of a proton in the absence of the *ortho* methyl groups:

$$CH_3C_6H_2(CH_3)_2\text{-}CH(H)\text{-}CH_2\text{-}OTs \xrightarrow{^-OCH_3} (CH_3)_3C_6H_2\text{-}CH{=}CH_2$$

$$C_6H_5\text{-}CH(H)\text{-}CH_2\text{-}OTs \xrightarrow{^-OCH_3} C_6H_5\text{-}CH{=}CH_2$$

12.26 In order for an E_1 elimination to occur the alkyl halide must meet two requirements: First, it must be able to form an appropriately stable carbocation. Second, it must have at least one hydrogen β to the halide. This second requirement is also the necessary structural feature for an E_2 reaction, and in the presence of strong base a bimolecular elimination almost always occurs more rapidly than ionization to a carbocation. For example:

$$\text{Base:}\ H\text{—}CH_2\text{—}CR_2\text{—}X \xrightarrow{\text{fast}} CH_2{=}CR_2$$

$$CH_3\text{—}CR_2\text{—}X \xrightarrow{\text{slow}} CH_3\text{—}C^+R_2$$

12.27 A difference in activation energies can be related to the ratio of rate constants:

$$\Delta E_a = -2.3RT \log(k_2/k_1)$$

In this specific case the equation can be written as:

$$E_a(B) - E_a(C) = -2.3RT \log (k_B/k_C)$$

Substitution of the energy differences (and using 300K as the temperature):

$$(13.2 - 14.0) \text{ kcal/mol} = (-2.3)(2\times10^{-3} \text{ kcal mol}^{-1}\text{deg}^{-1})(300 \text{ deg}) \log(k_B/k_C)$$

Rearrangement gives:

$$\log (k_B/k_C) = 0.8/(2.3 \times 0.6) = 0.58$$

and

$$k_B/k_C = 3.8$$

Therefore, the product ratio (B:C) would be 4:1.

CHAPTER 13. ALDEHYDES AND KETONES

ANSWERS TO EXERCISES

13.1 **(a)** 4-methylcyclohexanone

(b) 2-pentanone

(c) 4-oxopentanal

(d) 3-hydroxypentanal

(e) 2,4-hexanedione

13.2

(a) 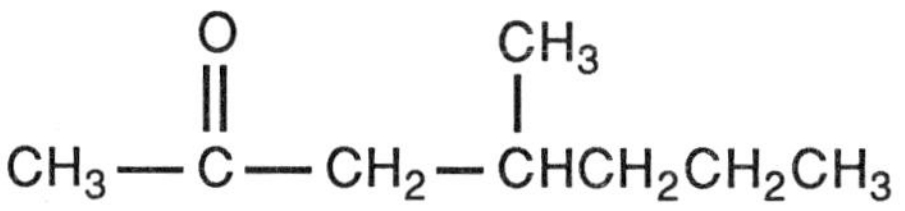

(b) 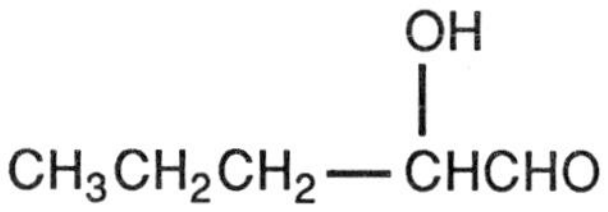

(c) 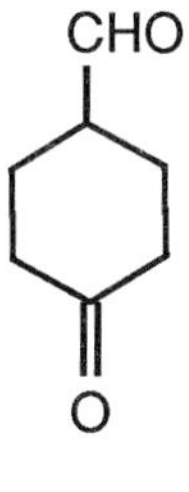

(d) 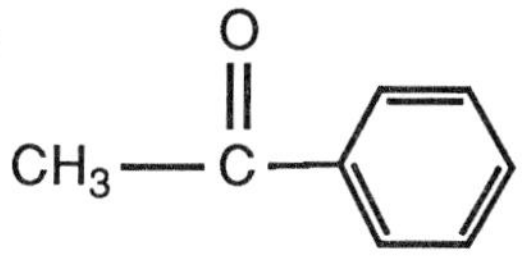

(e) 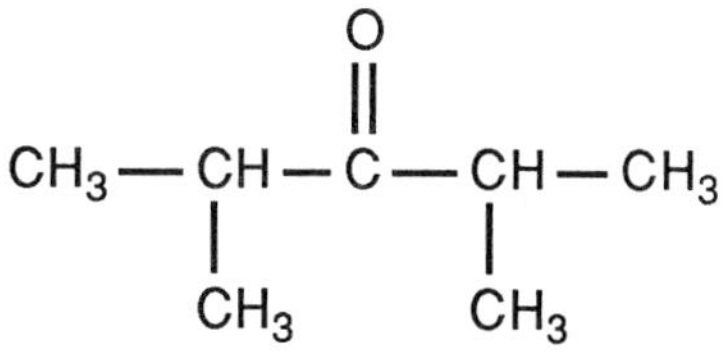

13.3

(a) $CH_3CH_2CH(CH_3)CHO$

(b) 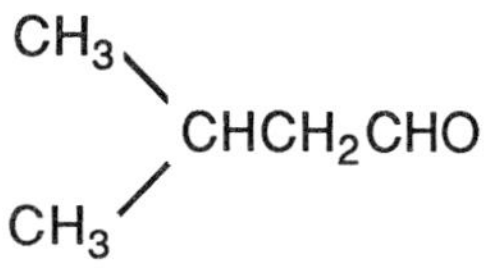

(c) CH_3CH_2CHO

(d) 2 CH_3CHO

(e) CH_3CH_2CHO + 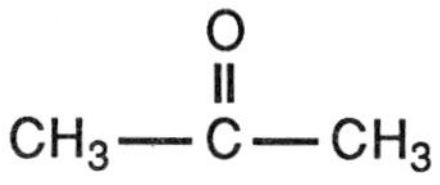

13.4 **(a)** OsO_4, $NaIO_4$ or O_2, then $(CH_3)_2S$

(b) CrO_3, pyridine

(c) $LiAl(OtBu)_3H$, then H_2O or H_2, Pd–$BaSO_4$

13.5 (a) $CH_3-CH(CH_3)-CH_2-C(=O)-CH_3$

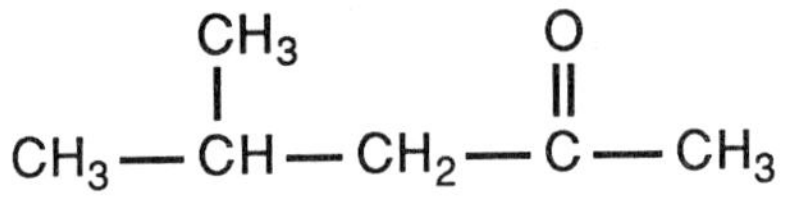

(b) $CH_3-CH(CH_3)-CH_2-C(OH)(CH_2CH_3)-CH_2CH_3$

(c) $CH_3CH_2CH_2-C(=O)-CH_3$

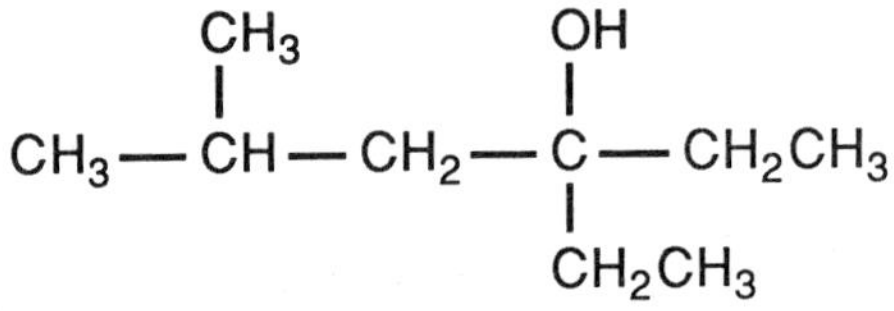

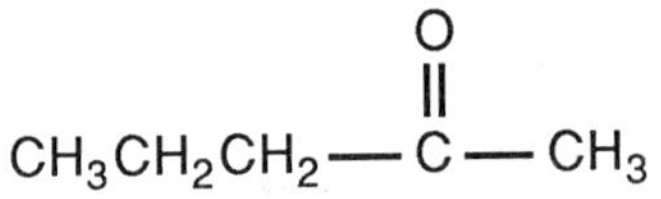

(d) $CH_3CH_2CH_2-C(=O)-CH_2CH_3$

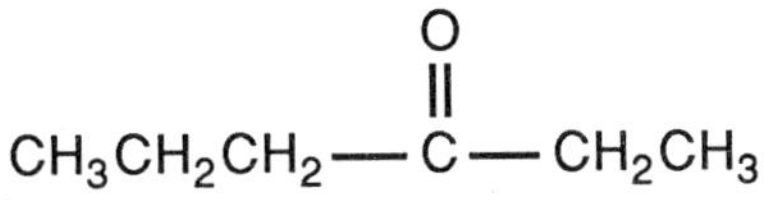

13.6 (a) $(CH_3CH_2CH_2)_2CuLi$ or $(CH_3CH_2CH_2)_2Cd$ (b) $CH_3CH_2CH_2Li$

13.7 (a) oxygen

(b) carbon

(c) oxygen

(d) carbon

(e) oxygen

(f) carbon

13.8 (a) $CH_3CH_2-CH(OH^*)-CH_2CH_3$

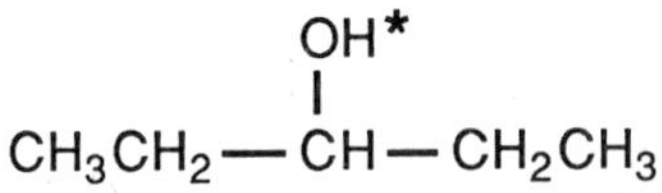

(b) $CH_3C^*H_2-C(=O)-C^*H_2CH_3$

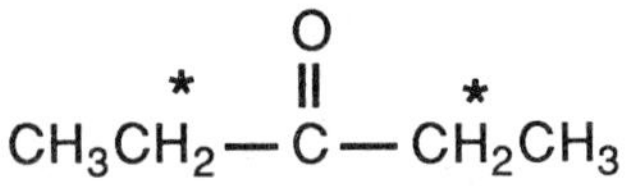

(c) $CH_3CH_2CH_2OH$ *

(d) $CH_3C^*H_2CHO$

13.9 (a) $CH_3CH_2-CH(OH)-C\equiv N$

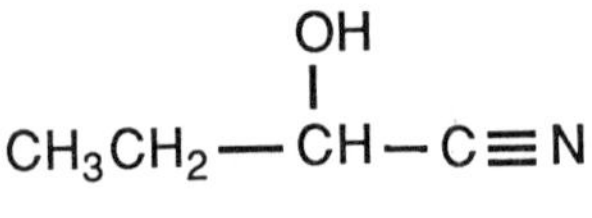

(b) $CH_3-C(OH)(OCH_3)-CH_2CH_3$

(c) $CH_3-C(OH)(SO_3^-\ Na^+)-CH_2CH_3$

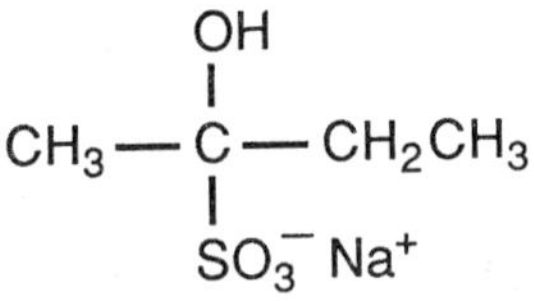

(d) cyclohexylidene $=NNHC_6H_5$

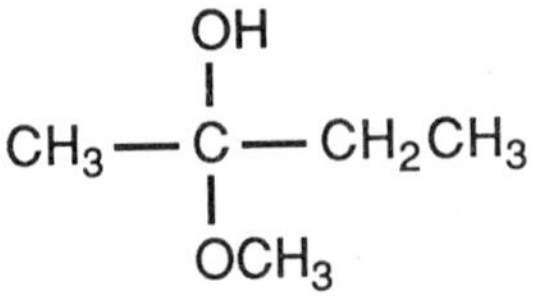

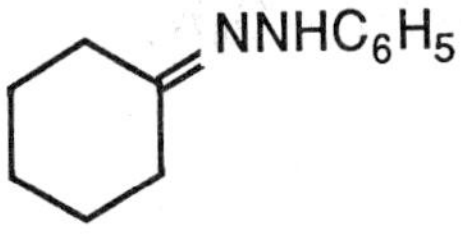

13.10

(a) 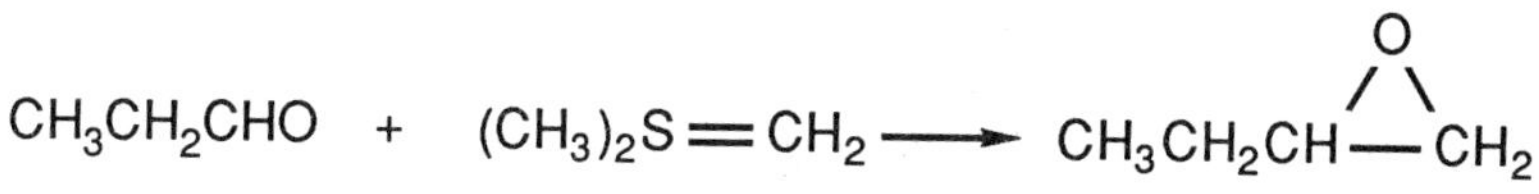 $CH_3CH_2CHO + (CH_3)_2S{=}CH_2 \longrightarrow CH_3CH_2CH\overset{O}{-}CH_2$ (epoxide)

(b) 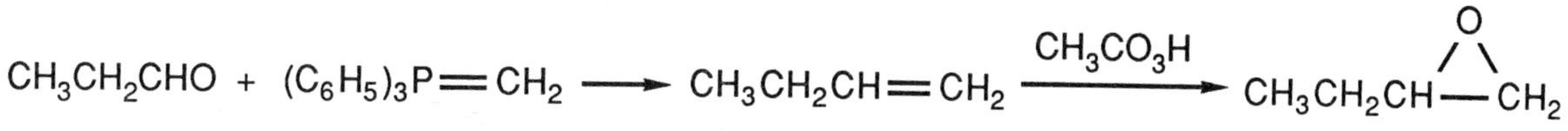 $CH_3CH_2CHO + (C_6H_5)_3P{=}CH_2 \longrightarrow CH_3CH_2CH{=}CH_2 \xrightarrow{CH_3CO_3H} CH_3CH_2CH\overset{O}{-}CH_2$ (epoxide)

13.11 (a) 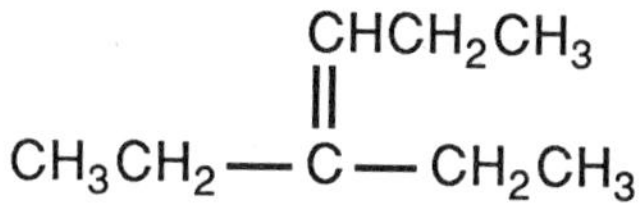 $CH_3CH_2-C(=CHCH_2CH_3)-CH_2CH_3$ (b) 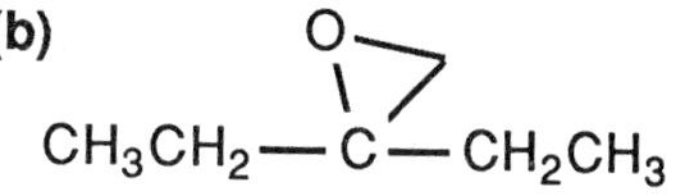 $CH_3CH_2-C(-O-)-CH_2CH_3$ (epoxide ring with O)

13.12

(a) 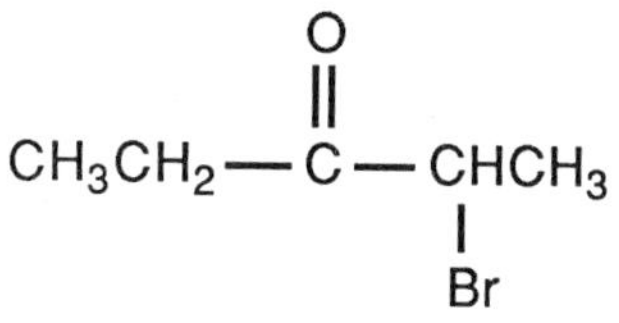 $CH_3CH_2-C(=O)-CHBrCH_3$ (b) $CH_3CHCl-C(=O)-C(CH_3)_2-CH_3$ (c) $C_6H_5-C(=O)-CO_2Na + CHI_3$

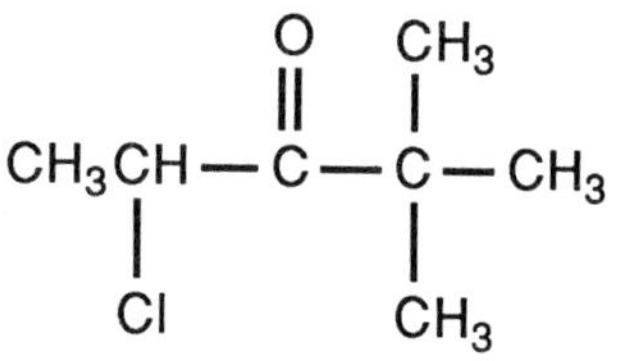

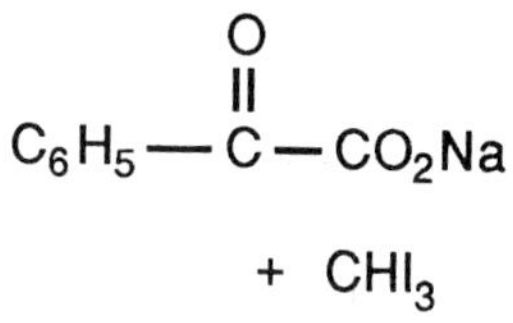

13.13

(a) 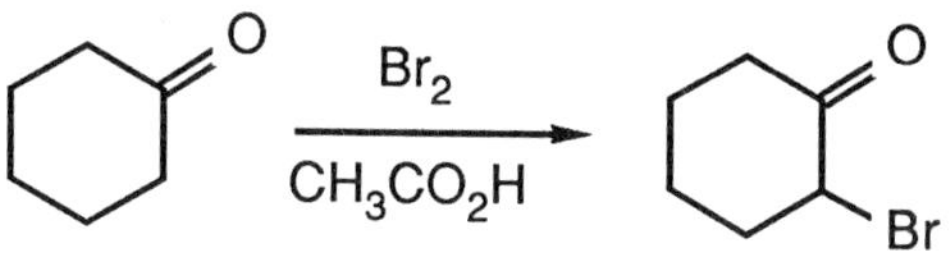 $\xrightarrow[CH_3CO_2H]{Br_2}$ Br

(b) 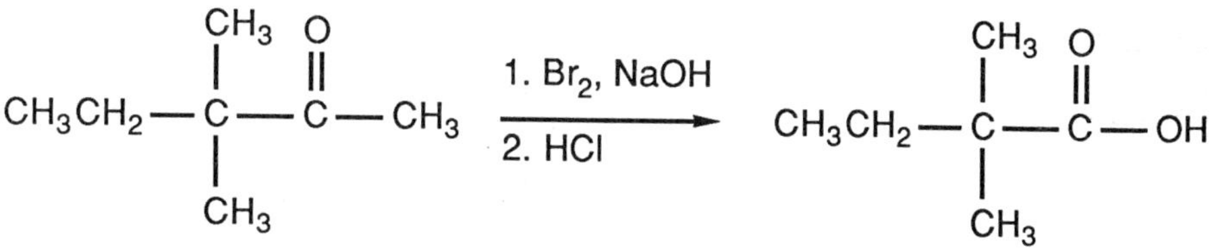 $CH_3CH_2-C(CH_3)_2-C(=O)-CH_3 \xrightarrow[2.\ HCl]{1.\ Br_2,\ NaOH} CH_3CH_2-C(CH_3)_2-C(=O)-OH$

13.14

(a) 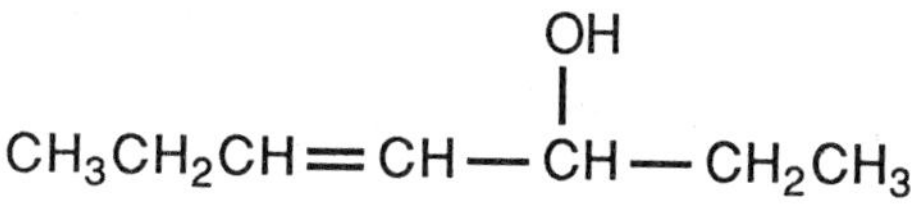 $CH_3CH_2CH{=}CH-CH(OH)-CH_2CH_3$ (b) 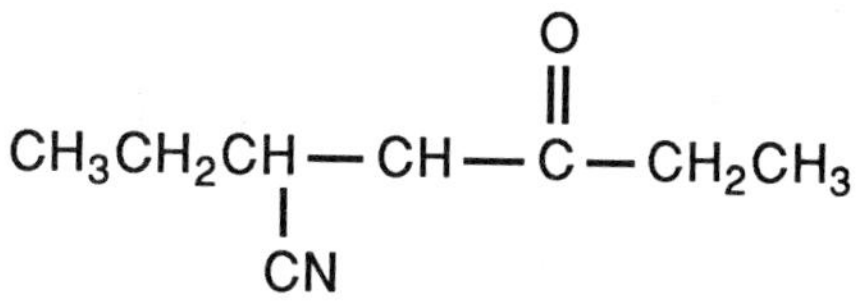 $CH_3CH_2CH(CN)-CH-C(=O)-CH_2CH_3$

(c) $CH_3CH_2CH{=}CH-C(OH)(CH_3)-CH_2CH_3$

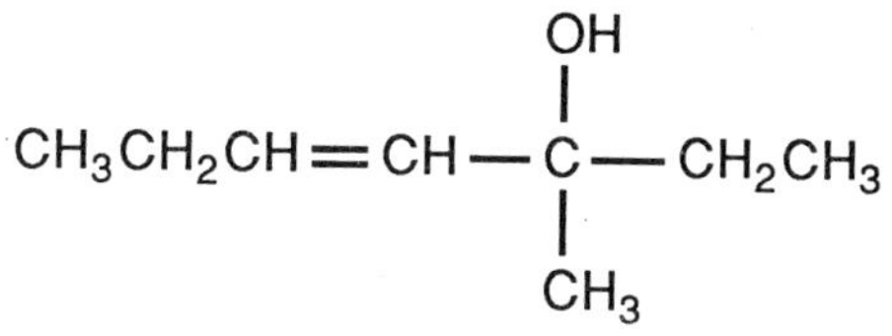

13.15 (a) 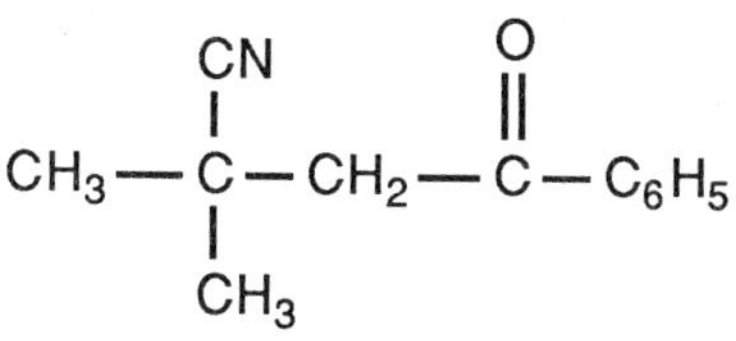 $CH_3-C(CN)(CH_3)-CH_2-C(=O)-C_6H_5$ (b) 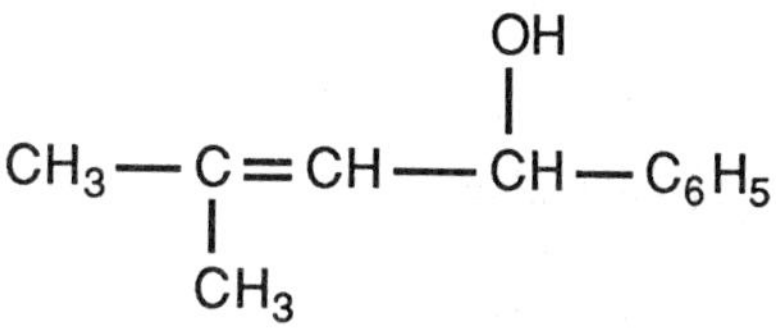 $CH_3-C(CH_3){=}CH-CH(OH)-C_6H_5$

(c) $CH_3-C(CH_3)_2-CH_2-C(=O)-C_6H_5$

13.16 (a) $CH_3CH_2-C(=O)-OCH_2CH_3$ (b) $CH_3-C(=O)-C(=O)-CH_2CH_3$ (c) $CH_3CH_2CH_2CH_2CH_3$

(d) $CH_3CH_2-CH(OH)-CH_2CH_3$

13.17 (a) $C_6H_5CH_2-C(=O)-OH$ (b) $C_6H_5CH_2CH_2OH$ (c) $C_6H_5CH_2CH_3$

(d) $C_6H_5CH_2-C(=O)-OH$

13.18 (a) i, ii, iii, vi, vii (b) iii (c) iii, vi (d) i, ii, vii (e) iii (f) iii (g) ii, iv, vii (h) i, ii, iii, vii

13.19 (a) [4-methylcyclohexanone] (b) [epoxy methylcyclohexanone] (c) [4-methylcyclohex-3-enol] (d) [spiro-epoxide of 4-methylcyclohex-3-ene]

13.20 (a) $CH_3CH{=}CHCH_2CH_2-CH(OH)-CH{=}CH_2$ (b) $CH_3CH{=}CHCH_2CH_2-C(=CH_2)-CH{=}CH_2$

(c) $CH_3CH{=}CHCH_2CH_2-C(OH)(CH_3)-CH{=}CH_2$ (d) $CH_3CH{=}CHCH_2CH_2-C(=O)-CH(O)CH_2$ (epoxide)

ANSWERS TO PROBLEMS

13.1 **(a)** 3-hexanone **(b)** 4-methyl-2-pentenal

(c) 3-propyl-2-cyclopenten-1-one **(d)** 7-methyl-6-octen-3-one

(e) 3,3-dimethylcycloheptanone **(f)** 3-methyl-4-phenylbutanal

(g) 3-methyl-2-propylpentanal **(h)** cyclopropyl ethyl ketone

(i) 4,5-dimethyl-2-hexanone **(j)** propiophenone

(k) phenyl cyclopentyl ketone **(l)** 2-propyl-1,3-cyclohexanedione

(m) 3-oxo-5-methylheptanal **(n)** 2,6-octanedione

(o) 5-cyclopentyl-2-hexanone **(p)** 4-bromo-1-phenyl-1,5-hexanedione

13.2

(a) $CH_3—CH(CH_3)—C(=O)—CH_2CH_2CH_2CH_3$

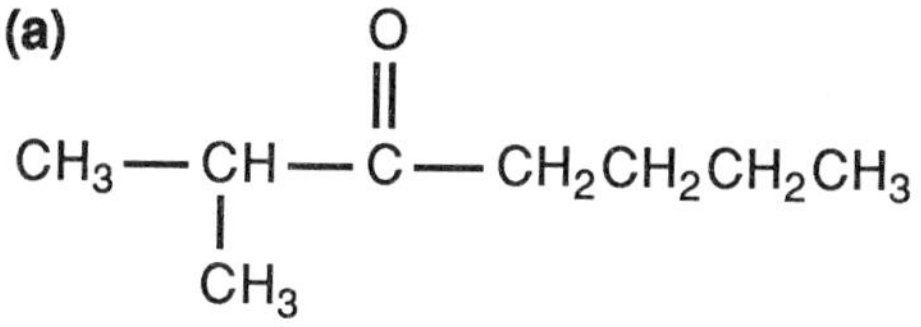

(b) O, Cl, CH_3

(c) $CH_3CH_2CH_2—C(CH_3)_2—CH_2CH_2CHO$

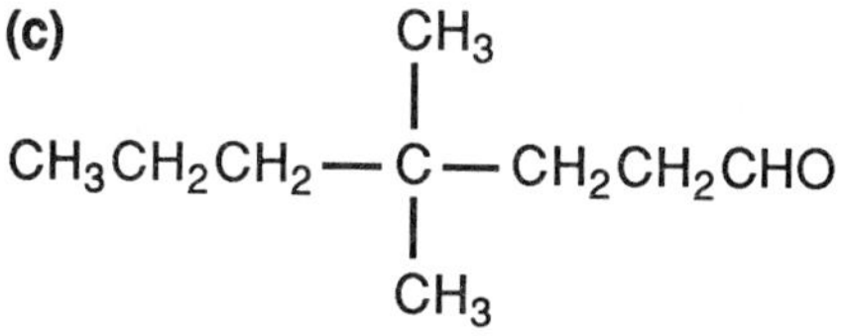

(d) $CH_3CH_2—C(=O)—C(=O)—CH_2CH_2CH_3$

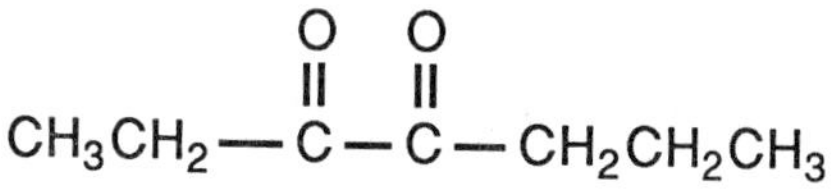

(e) $C_6H_5—C(=O)—CH_2CH_2CH_3$

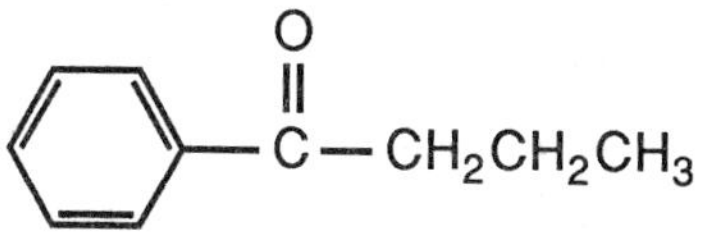

(f) $CH_3—C(\text{cyclobutyl})—C(=O)—CH_2—C(CH_3)_2—CH_2CH_2CH_3$

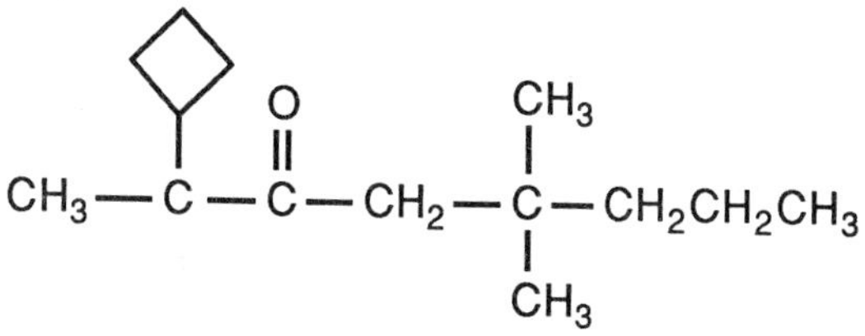

(g) O, CH_2CH_3, Cl, Cl

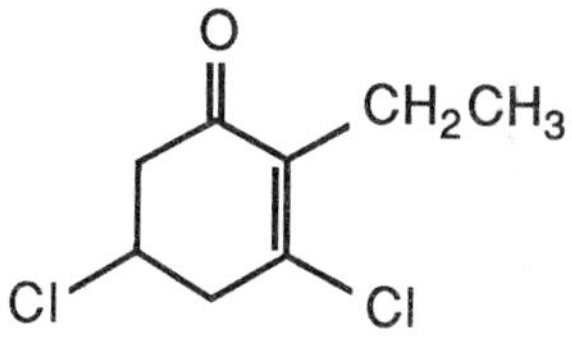

(h) $CH_3—C(=O)—C{\equiv}C—CH_2—CH(Br)—CH_2CH_3$

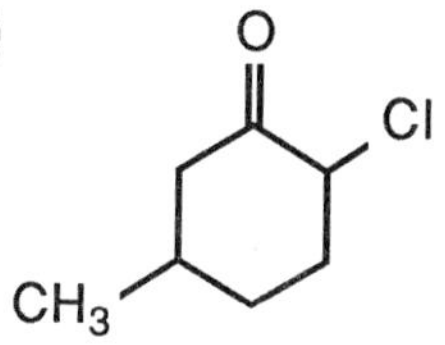

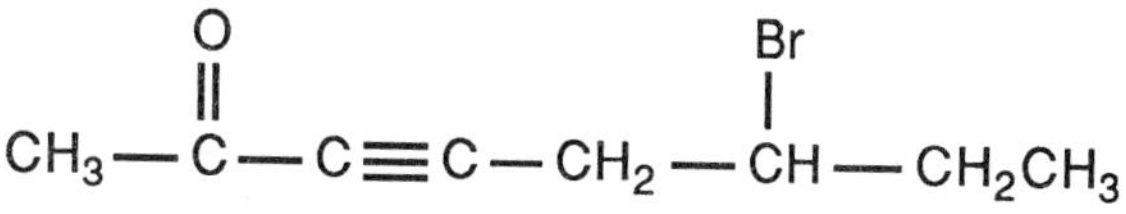

(i) 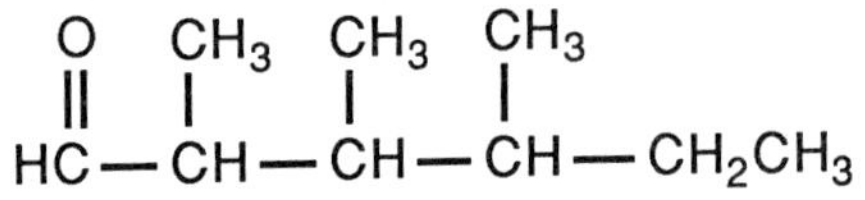 $HC(=O)-CH(CH_3)-CH(CH_3)-CH(CH_3)-CH_2CH_3$

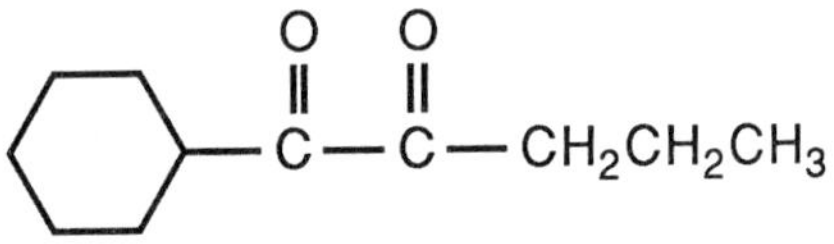

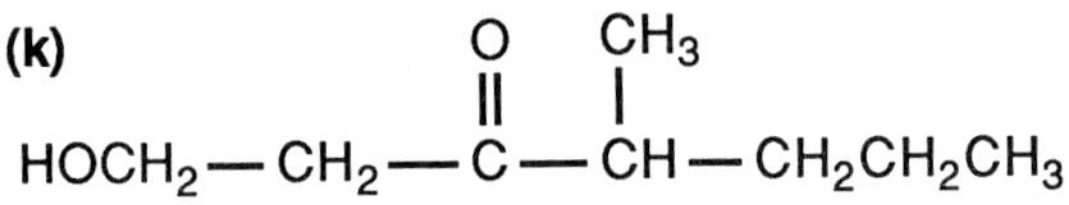

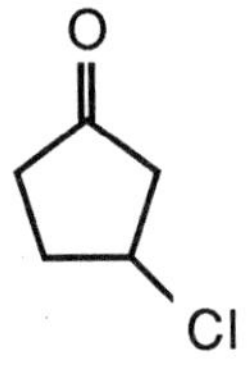

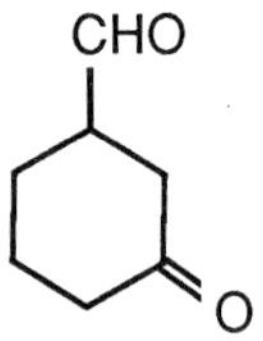

(j) $C_6H_{11}-C(=O)-C(=O)-CH_2CH_2CH_3$

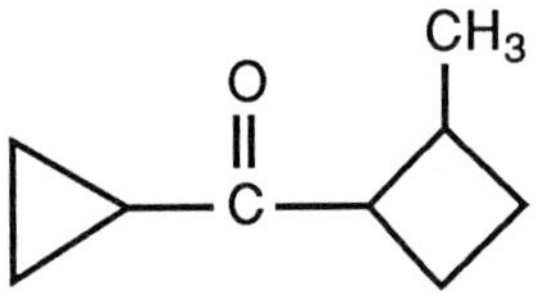

(k) $HOCH_2-CH_2-C(=O)-CH(CH_3)-CH_2CH_2CH_3$

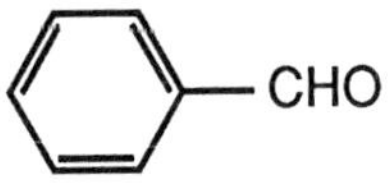

(l) O, Cl

(m) CHO, O

(n) $C_3H_5-C(=O)-C_4H_6(CH_3)$

(o) $C_6H_5-CH_2-C(=O)-CF_3$

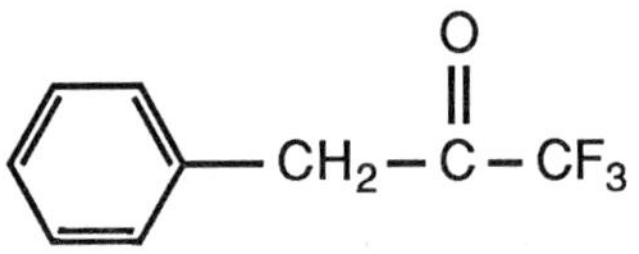

13.3

(a) 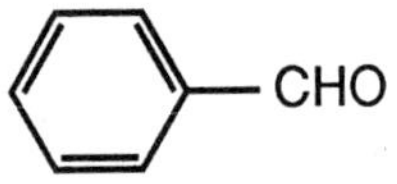 C_6H_5-CHO

(b) C_6H_5-CHO

(c) $C_6H_5-C(CH_3)_2-OH$

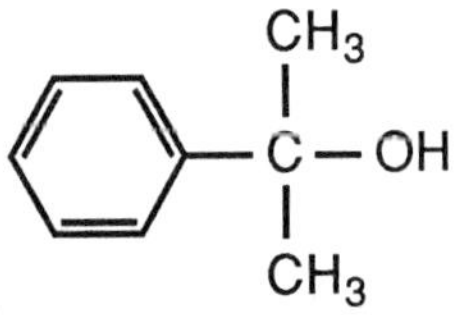

(d) 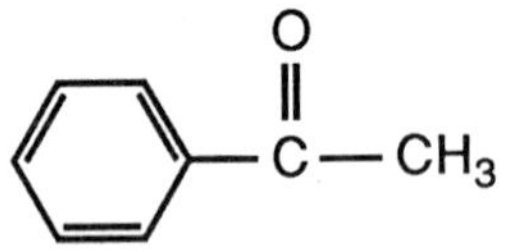 $C_6H_5-C(=O)-CH_3$

(e) 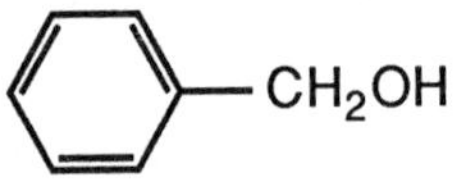 $C_6H_5-CH_2OH$

(f) 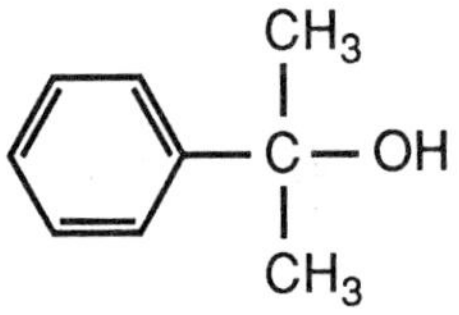 $C_6H_5-C(CH_3)_2-OH$

(g) 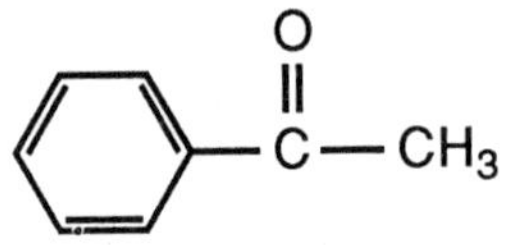 $C_6H_5-C(=O)-CH_3$

13.4

(a) $CH_3CH_2CH_2CHO$

(b) $CH_3CH_2CH_2CHO$

(c) 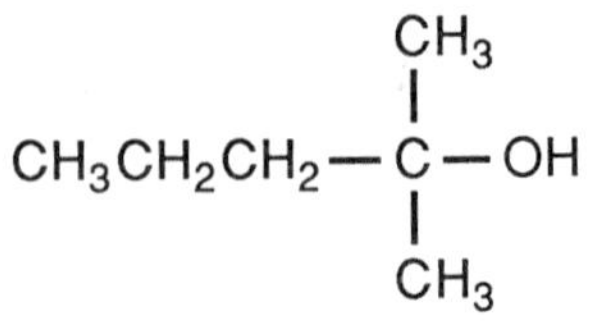 $CH_3CH_2CH_2-C(CH_3)_2-OH$

(d) $CH_3CH_2CH_2-C(=O)-CH_3$

(e) $CH_3CH_2CH_2CH_2OH$

(f) $CH_3CH_2CH_2-C(CH_3)_2-OH$

(g) $CH_3CH_2CH_2-C(=O)-CH_3$

13.5 (a) $CH_3CH_2-C(=O)-CH_3$

(b) 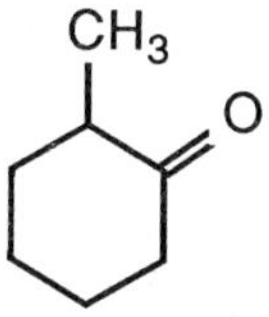 2-methylcyclohexanone (CH_3, O)

(c) $CH_3CH_2-CH(CH_3)-CH_2CO_2H$

(d) 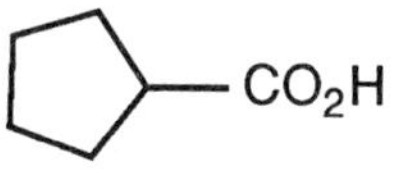 cyclopentyl—CO_2H

13.6 (a) $CH_3CH_2-C(=O)-CH_3$

(b) 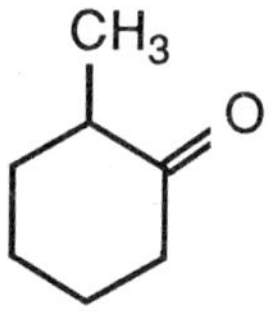 2-methylcyclohexanone (CH_3, O)

(c) $CH_3CH_2-CH(CH_3)-CH_2CHO$

(d) 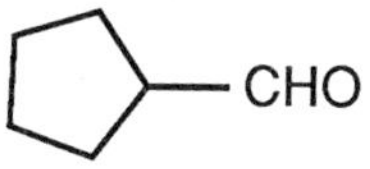 cyclopentyl—CHO

13.7

(a) CH_3CH_2CHO + $CH_3-C(=O)-CH_3$

(b) CH_3CH_2CHO + $CH_3-C(=O)-CH_3$

(c) $CH_3CH_2CO_2H$ + $CH_3-C(=O)-CH_3$

(d) CH_3CH_2CHO + $CH_3-C(=O)-CH_3$

13.8

(a)

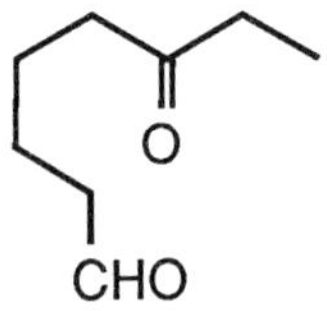
O
CHO

(b)

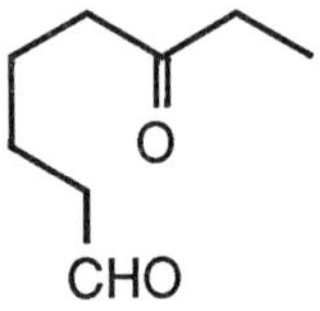
O
CHO

(c)

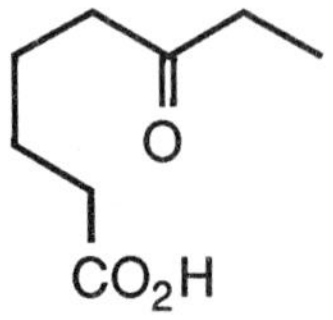
O
CO_2H

(d)

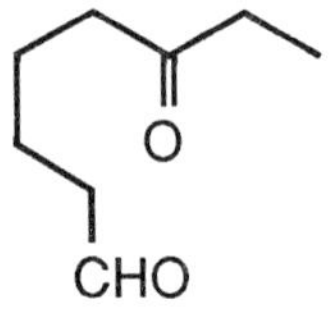
O
CHO

13.9 (a)

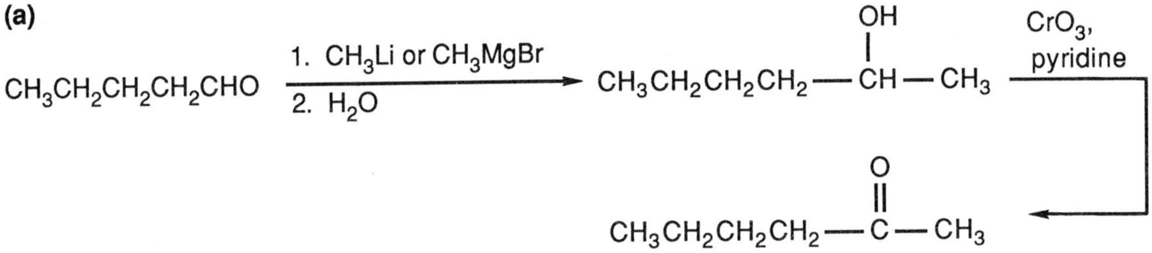
$CH_3CH_2CH_2CH_2CHO$
1. CH_3Li or CH_3MgBr
2. H_2O
OH
$CH_3CH_2CH_2CH_2—CH—CH_3$
CrO_3, pyridine
O
$CH_3CH_2CH_2CH_2—C—CH_3$

(b)

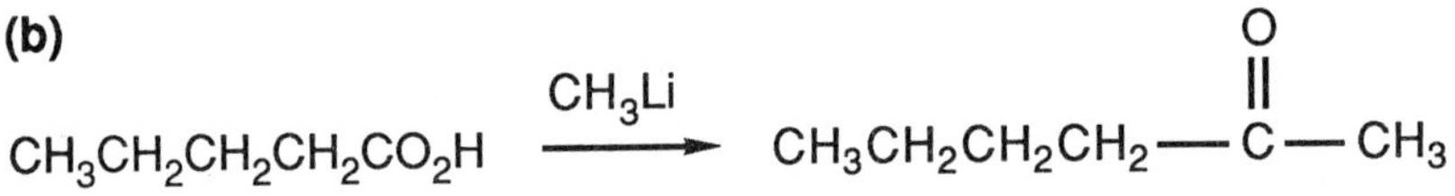
$CH_3CH_2CH_2CH_2CO_2H$
CH_3Li
O
$CH_3CH_2CH_2CH_2—C—CH_3$

(c)

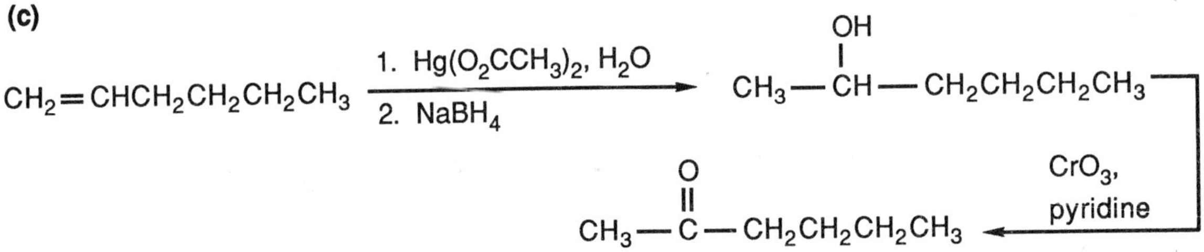
$CH_2=CHCH_2CH_2CH_2CH_3$
1. $Hg(O_2CCH_3)_2$, H_2O
2. $NaBH_4$
OH
$CH_3—CH—CH_2CH_2CH_2CH_3$
CrO_3, pyridine
O
$CH_3—C—CH_2CH_2CH_2CH_3$

(d)

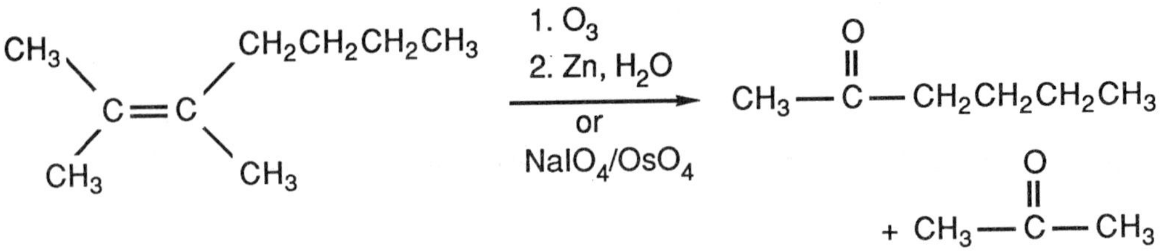
CH_3
$CH_2CH_2CH_2CH_3$
C=C
CH_3
CH_3
1. O_3
2. Zn, H_2O
or
$NaIO_4/OsO_4$
O
$CH_3—C—CH_2CH_2CH_2CH_3$
O
+ $CH_3—C—CH_3$

(e)

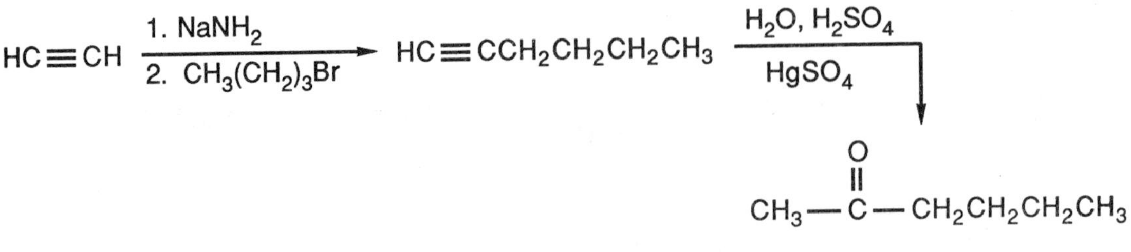
HC≡CH
1. $NaNH_2$
2. $CH_3(CH_2)_3Br$
$HC≡CCH_2CH_2CH_2CH_3$
H_2O, H_2SO_4
$HgSO_4$
O
$CH_3—C—CH_2CH_2CH_2CH_3$

(f)

$$CH_3-\overset{\overset{O}{\|}}{C}-Cl \xrightarrow[\text{[CH}_3\text{(CH}_2\text{)}_3\text{]}_2\text{CuLi}]{\text{[CH}_3\text{(CH}_2\text{)}_3\text{]}_2\text{Cd or}} CH_3-\overset{\overset{O}{\|}}{C}-CH_2CH_2CH_2CH_3$$

(g)

$$HC\equiv CCH_2CH_2CH_2CH_3 \xrightarrow[\text{HgSO}_4]{\text{H}_2\text{O, H}_2\text{SO}_4} CH_3-\overset{\overset{O}{\|}}{C}-CH_2CH_2CH_2CH_3$$

(h)

$$CH_3CH_2-\overset{\overset{OH}{|}}{CH}-\underset{\underset{CH_3}{|}}{\overset{\overset{OH}{|}}{CH}}-CH_2CH_2CH_2CH_3 \xrightarrow{\text{HIO}_4} CH_3-\overset{\overset{O}{\|}}{C}-CH_2CH_2CH_2CH_3$$

13.10

(a)

$$CH_2=CHCH_2CH_2CH_2CH_2CH_3 \xrightarrow[\text{NaIO}_4\text{/OsO}_4]{\text{1. O}_3 \quad \text{2. Zn, H}_2\text{O or}} CH_3CH_2CH_2CH_2CH_2CHO$$

(b)

$$CH_3CH_2CH_2CH_2CH_2CH_2OH \xrightarrow[\text{pyridine}]{\text{CrO}_3} CH_3CH_2CH_2CH_2CH_2CHO$$

(c)

$$CH_3CH_2CH_2CH_2CH_2CH_2Br \xrightarrow{\text{H}_2\text{O}} CH_3CH_2CH_2CH_2CH_2CH_2OH \xrightarrow[\text{pyridine}]{\text{CrO}_3} CH_3CH_2CH_2CH_2CH_2CHO$$

(d)

$$CH_3CH_2CH_2CH_2CH_2CO_2H \xrightarrow[\text{2. LiAl(OtBu)}_3\text{H}]{\text{1. PCl}_5} CH_3CH_2CH_2CH_2CH_2CHO$$

(e)

$$CH_3CH_2CH_2CH_2CH_2COCl \xrightarrow[H_2/Pd\text{-}BaSO_4]{LiAl(OtBu)_3H \text{ or}} CH_3CH_2CH_2CH_2CH_2CHO$$

(f)

$$CH_3-\underset{\displaystyle OH}{CH}-\underset{\displaystyle OH}{CH}-CH_2CH_2CH_2CH_2CH_3 \xrightarrow{HIO_4} CH_3CH_2CH_2CH_2CH_2CHO$$

(g)

$$HC\equiv CH \xrightarrow[2.\ CH_3(CH_2)_3Br]{1.\ NaNH_2} HC\equiv CCH_2CH_2CH_2CH_3 \xrightarrow[2.\ H_2O_2,\ NaOH]{1.\ [(CH_3)_2CHCHCH_3]_2BH} CH_3CH_2CH_2CH_2CH_2CHO$$

13.11 (a) $CH_3CH_2\overset{*}{C}H_2-\overset{O}{\overset{\|}{C}}-\overset{*}{C}H_2CH_2CH_3$

(b) $CH_3\overset{*}{C}H_2-\overset{O}{\overset{\|}{C}}-\overset{O}{\overset{\|}{C}}-\overset{*}{C}H_2CH_3$

(c) $CH_3CH_2\overset{*}{C}H_2-\overset{O}{\overset{\|}{C}}-H$

(d) $CH_3CH_2CH_2-\overset{O}{\overset{\|}{C}}-\overset{*}{O}H$

(e) $C_6H_5\overset{*}{C}H_2-\overset{O}{\overset{\|}{C}}-CH_2CH_3$

(f) $C_6H_5CH_2-\overset{O}{\overset{\|}{C}}-\overset{*}{C}H_2CHO$

(g) $(CH_3)_3C-\overset{O}{\overset{\|}{C}}-\overset{*}{C}H_3$

(h) $Br_2CH-\underset{\displaystyle CH_3}{\overset{\displaystyle CH_3}{C}}-\overset{O}{\overset{\|}{C}}-\overset{*}{C}H_3$

(i) $CH_3CH_2-\overset{O}{\overset{\|}{C}}-CH_2-\overset{O}{\overset{\|}{C}}-\overset{*}{O}H$

(j) $CH_3CH_2-\overset{\overset{*}{O}H}{\overset{|}{C}H}-CH_2-\overset{O}{\overset{\|}{C}}-H$

(k) $CH_3-\overset{Br}{\overset{|}{\overset{*}{C}H}}-\overset{O}{\overset{\|}{C}}-CH_2CH_2CH_3$

(l) $CH_3\overset{*}{C}H_2-\overset{O}{\overset{\|}{C}}-\underset{\displaystyle CH_3}{\overset{\displaystyle CH_3}{C}}-\overset{O}{\overset{\|}{C}}-\overset{*}{C}H_2CH_3$

13.12

(a) 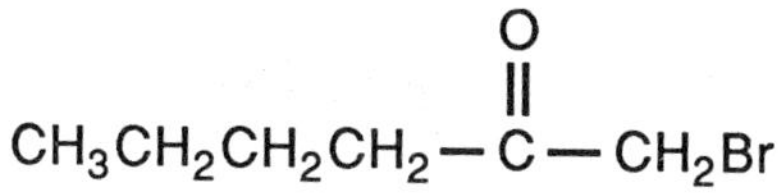 $CH_3CH_2CH_2CH_2-\overset{\overset{O}{\|}}{C}-CH_2Br$

(b) $CH_3CH_2CH_2CH_2O-\overset{\overset{O}{\|}}{C}-CH_3$

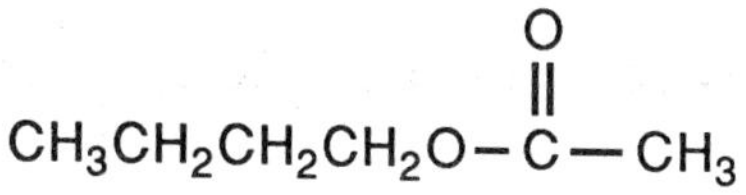

(c) $CH_3CH_2CH_2CH_2-\overset{\overset{OH}{|}}{CH}-CH_3$

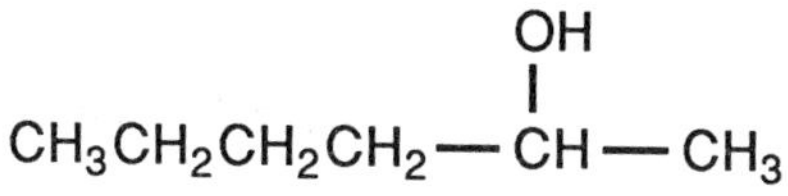

(d) 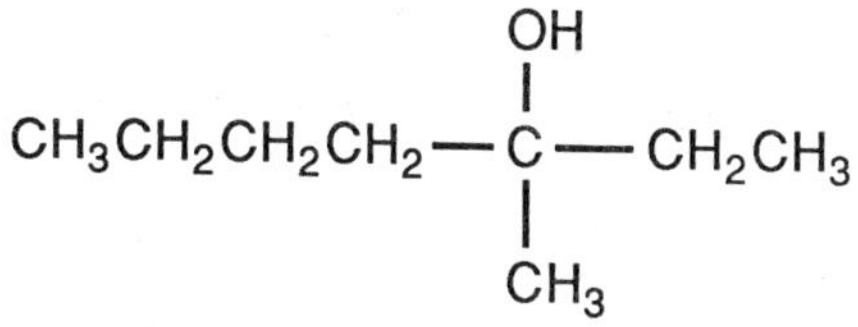 $CH_3CH_2CH_2CH_2-\underset{\underset{CH_3}{|}}{\overset{\overset{OH}{|}}{C}}-CH_2CH_3$

(e) 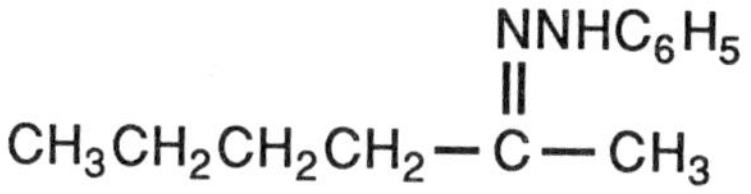 $CH_3CH_2CH_2CH_2-\overset{\overset{NNHC_6H_5}{\|}}{C}-CH_3$

(f) 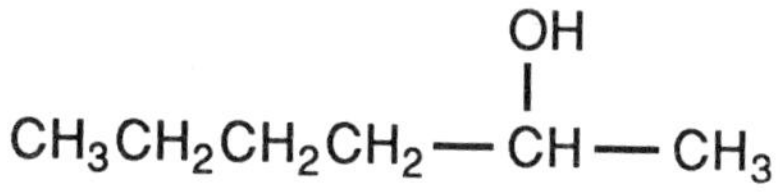 $CH_3CH_2CH_2CH_2-\overset{\overset{OH}{|}}{CH}-CH_3$

(g) $CH_3CH_2CH_2CH_2CH_2CH_3$

(h) 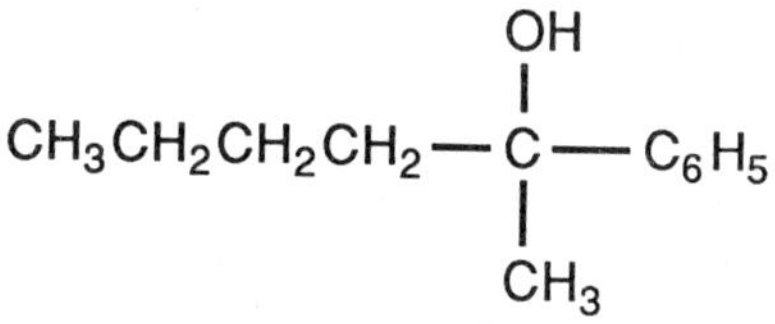 $CH_3CH_2CH_2CH_2-\underset{\underset{CH_3}{|}}{\overset{\overset{OH}{|}}{C}}-C_6H_5$

(i) 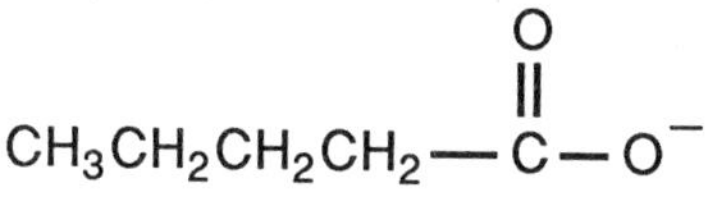 $CH_3CH_2CH_2CH_2-\overset{\overset{O}{\|}}{C}-O^- + CHBr_3$

(j) $CH_3CH_2CH_2CH_2CH_2CH_3$

(k) 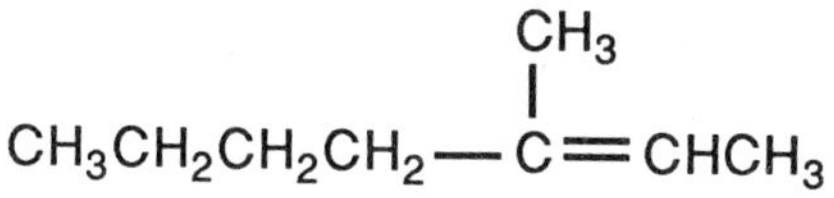 $CH_3CH_2CH_2CH_2-\overset{\overset{CH_3}{|}}{C}=CHCH_3$

(l) 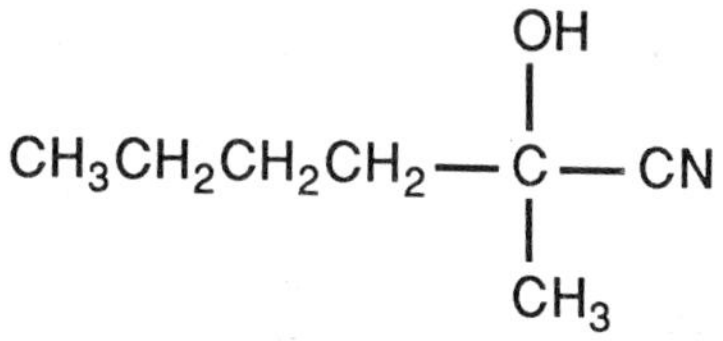 $CH_3CH_2CH_2CH_2-\underset{\underset{CH_3}{|}}{\overset{\overset{OH}{|}}{C}}-CN$

(m) 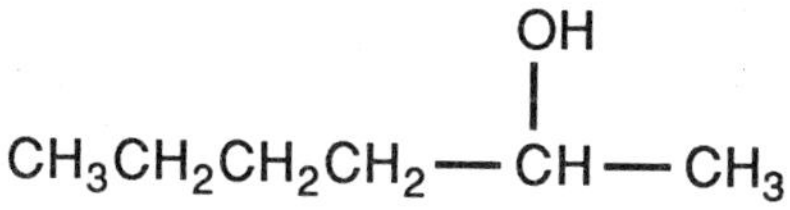 $CH_3CH_2CH_2CH_2-\overset{\overset{OH}{|}}{CH}-CH_3$

(n) 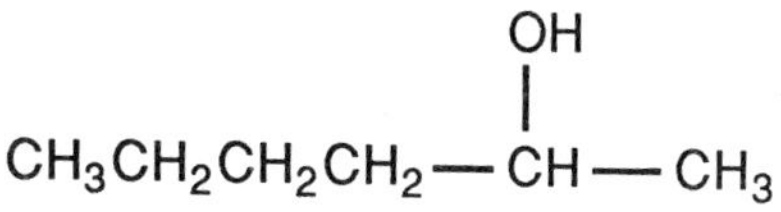 $CH_3CH_2CH_2CH_2-\overset{\overset{OH}{|}}{CH}-CH_3$

(o) $CH_3CH_2CH_2CH_2-\overset{\overset{CH_3}{|}}{C}-CH_2$ (epoxide: C and CH_2 bridged by O)

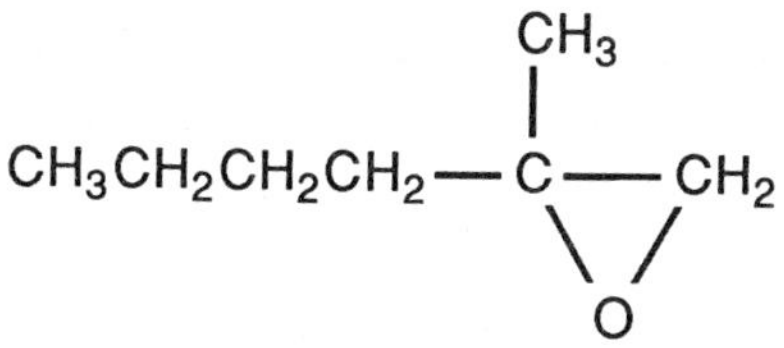

(p) 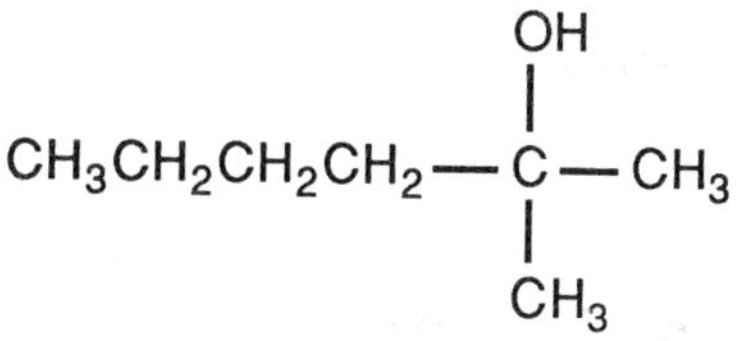 $CH_3CH_2CH_2CH_2-\underset{\underset{CH_3}{|}}{\overset{\overset{OH}{|}}{C}}-CH_3$

(q) 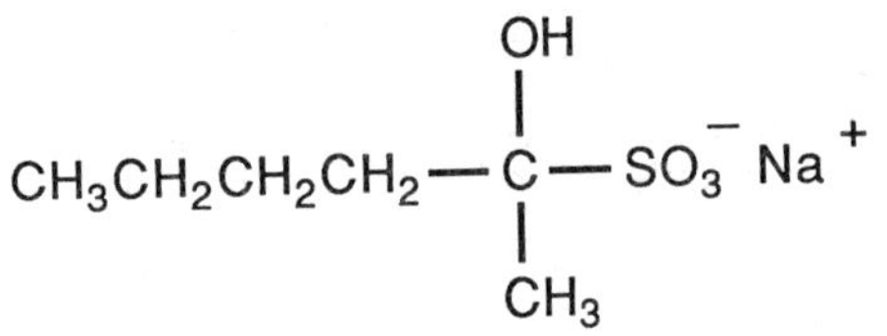 $CH_3CH_2CH_2CH_2-C(OH)(CH_3)-SO_3^- \ Na^+$

(r) 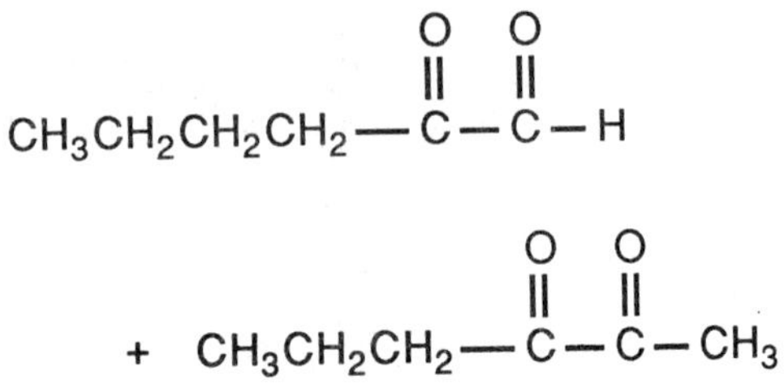 $CH_3CH_2CH_2CH_2-\overset{O}{\overset{\|}{C}}-\overset{O}{\overset{\|}{C}}-H$

$+ \ CH_3CH_2CH_2-\overset{O}{\overset{\|}{C}}-\overset{O}{\overset{\|}{C}}-CH_3$

13.13

(a) 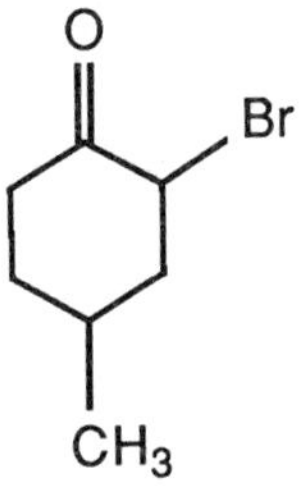 O, Br, CH_3

(b) 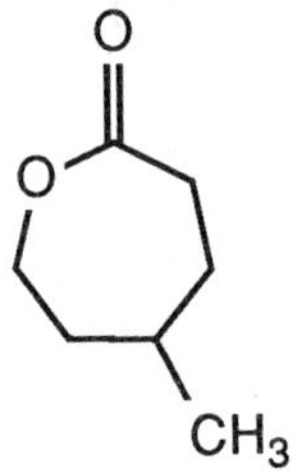 O, O, CH_3

(c)  OH, CH_3

(d) 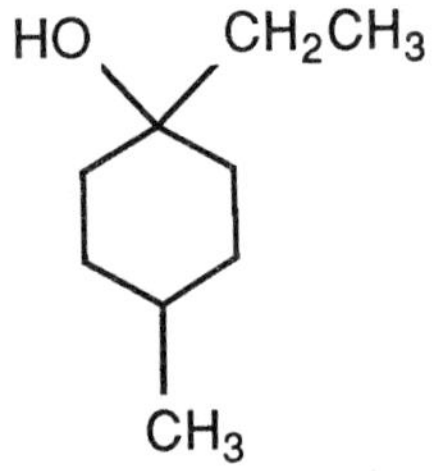 HO, CH_2CH_3, CH_3

(e) 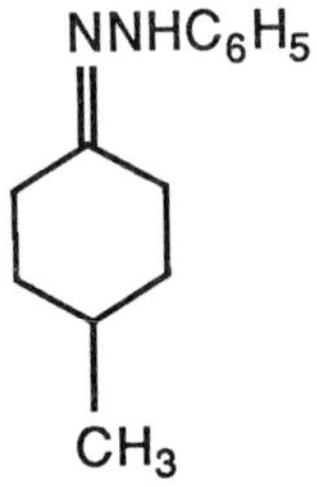 $NNHC_6H_5$, CH_3

(f)  OH, CH_3

(g) 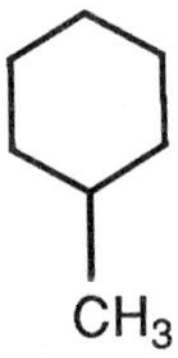 CH_3

(h) 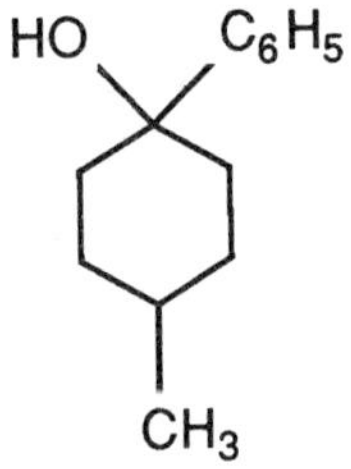 HO, C_6H_5, CH_3

(i) 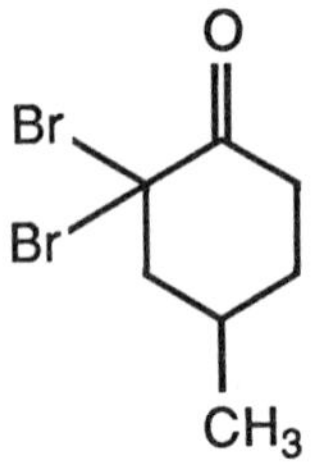 O, Br, Br, CH_3 + other alpha-bromo derivatives

(j) 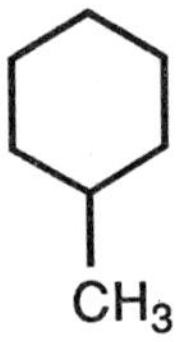 CH_3

(k) 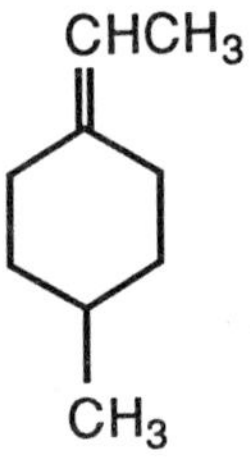 $CHCH_3$, CH_3

(l) 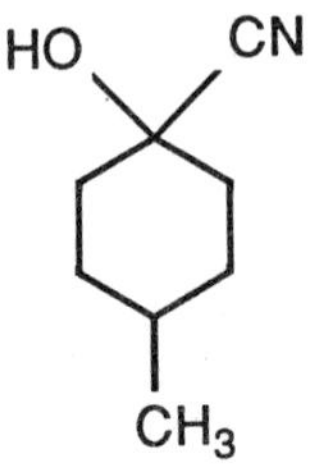 HO, CN, CH_3

(m)  HO, CH_3

(n)  HO, CH_3

(o) 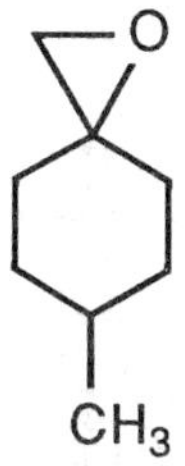 [structure: 1-oxaspiro compound — epoxide spiro-fused to cyclohexane bearing CH_3]

(p) 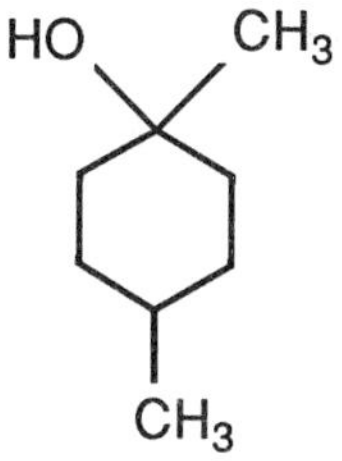 [structure: cyclohexane bearing HO and CH_3 on C-1, CH_3 on C-4]

(q) 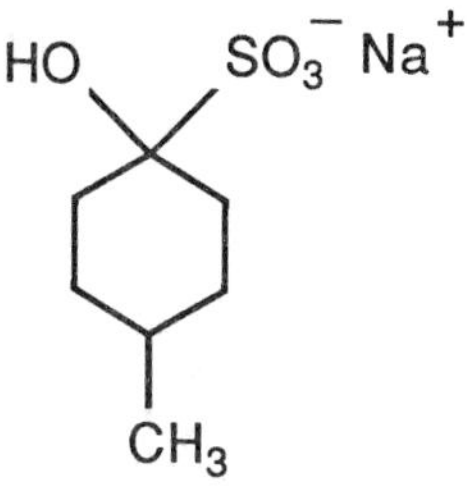 [structure: cyclohexane bearing HO and $SO_3^-\ Na^+$ on C-1, CH_3 on C-4]

(r) 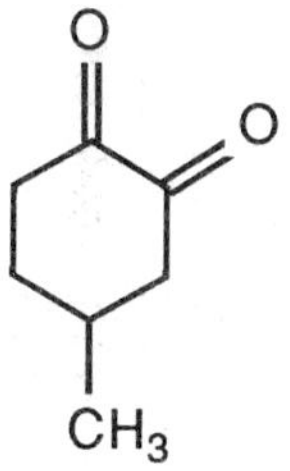 [structure: cyclohexane-1,2-dione (two C=O) with CH_3 on C-4]

13.14

(a) $CH_3CH_2CH_2CH_2CH_2CO_2H$

(b) $CH_3CH_2CH_2CH_2CH_2CO_2H$

(c) $CH_3CH_2CH_2CH_2CH_2CH_2OH$

(d) $CH_3CH_2CH_2CH_2CH_2-CH(OH)-CH_2CH_3$

(e) $CH_3CH_2CH_2CH_2CH_2-CH{=}NNHC_6H_5$

(f) $CH_3CH_2CH_2CH_2CH_2CH_2OH$

(g) $CH_3CH_2CH_2CH_2CH_2CH_3$

(h) $CH_3CH_2CH_2CH_2CH_2-CH(OH)-C_6H_5$

(i) $CH_3CH_2CH_2CH_2CH_2CO_2H$

(j) $CH_3CH_2CH_2CH_2CH_2CH_3$

(k) $CH_3CH_2CH_2CH_2CH_2CH{=}CHCH_3$

(l) $CH_3CH_2CH_2CH_2CH_2-CH(OH)-CN$

(m) $CH_3CH_2CH_2CH_2CH_2CH_2OH$ (n) $CH_3CH_2CH_2CH_2CH_2CH_2OH$

(o) $CH_3CH_2CH_2CH_2CH_2—CH—CH_2$ (epoxide, O bridging CH and CH_2) (p) $CH_3CH_2CH_2CH_2CH_2—CH(OH)—CH_3$

(q) $CH_3CH_2CH_2CH_2CH_2—CH(OH)—SO_3^-\,Na^+$ (r) $CH_3CH_2CH_2CH_2C(=O)—C(=O)—H$

13.15

(a) cyclopentyl—CO_2H (b) cyclopentyl—CO_2H (c) cyclopentyl—CH_2OH

(d) cyclopentyl—$CH(OH)—CH_2CH_3$ (e) cyclopentyl—$CH{=}NNHC_6H_5$ (f) cyclopentyl—CH_2OH

(g) cyclopentyl—CH_3 (h) cyclopentyl—$CH(OH)—C_6H_5$ (i) cyclopentyl—CO_2H

(j) cyclopentyl—CH_3 (k) cyclopentyl—$CH{=}CHCH_3$ (l) cyclopentyl—$CH(OH)—CN$

(m) cyclopentyl—CH_2OH (n) cyclopentyl—CH_2OH (o) cyclopentyl—$CH—CH_2$ (epoxide, O bridging CH and CH_2)

(p) cyclopentyl—$CH(OH)—CH_3$ (q) cyclopentyl—$CH(OH)—SO_3^-\,Na^+$

(r) Cannot yield dicarbonyl compound because the alpha carbon is tertiary.

13.16

(a) 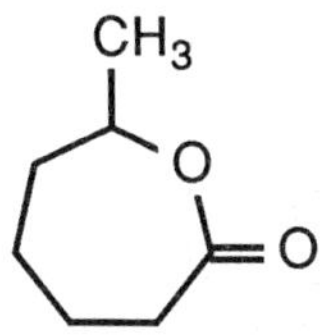 7-membered lactone ring (O and C=O in ring) with CH_3 on the carbon next to the ring O

(b) $CH_3CH_2CH_2CH_2O-\overset{O}{\overset{\|}{C}}-CH_3$

(c) $CH_3-\underset{CH_3}{\underset{|}{CH}}-O-\overset{O}{\overset{\|}{C}}-CH_2C_6H_5$

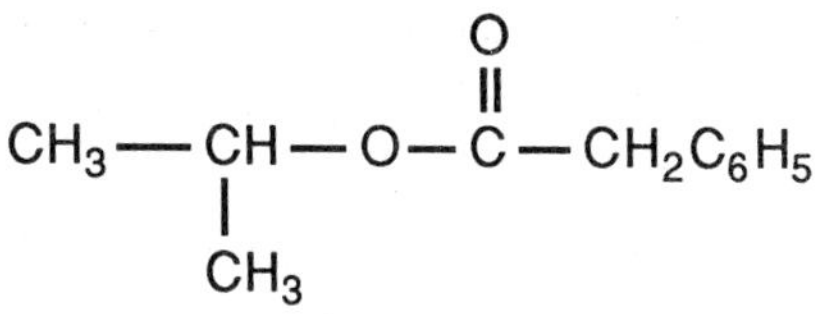

(d) 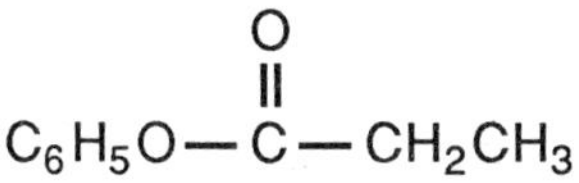 $C_6H_5O-\overset{O}{\overset{\|}{C}}-CH_2CH_3$

(e) 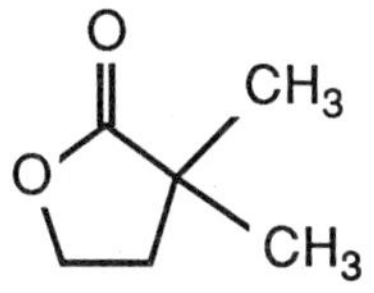 5-membered lactone ring (O, C=O) with two CH_3 groups on the carbon next to the C=O

(f) 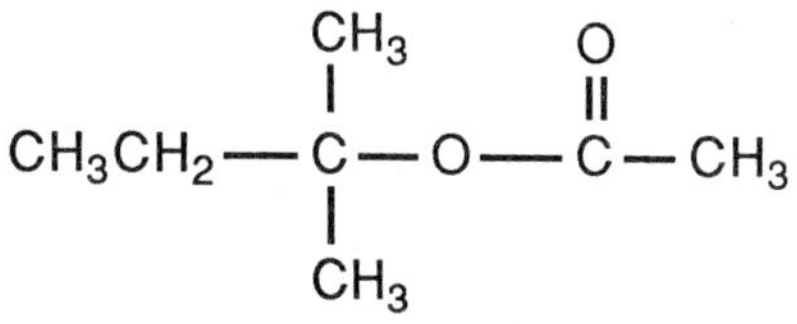 $CH_3CH_2-\overset{CH_3}{\overset{|}{\underset{CH_3}{\underset{|}{C}}}}-O-\overset{O}{\overset{\|}{C}}-CH_3$

13.17

(a) 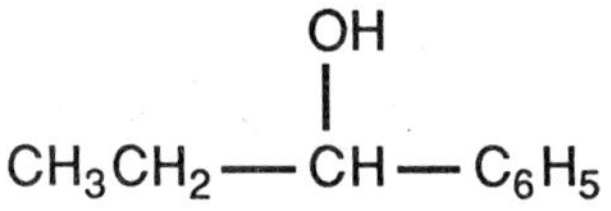 $CH_3CH_2-\overset{OH}{\overset{|}{CH}}-C_6H_5$

(b) $CH_3CH_2CH_2C_6H_5$

(c) 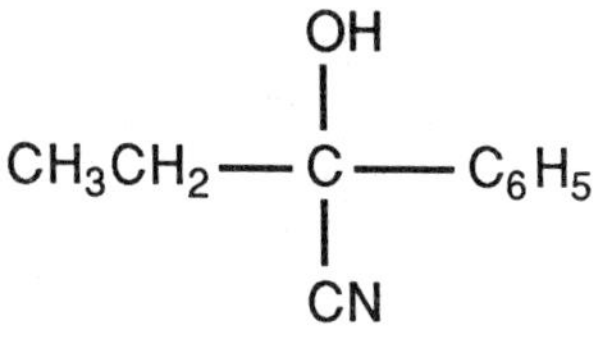 $CH_3CH_2-\overset{OH}{\overset{|}{\underset{CN}{\underset{|}{C}}}}-C_6H_5$

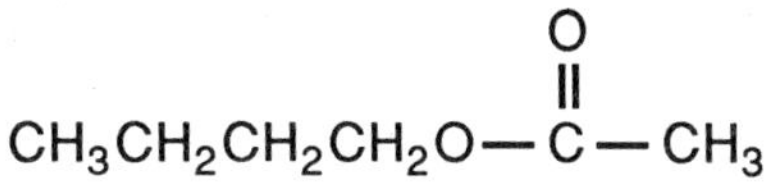

(d) 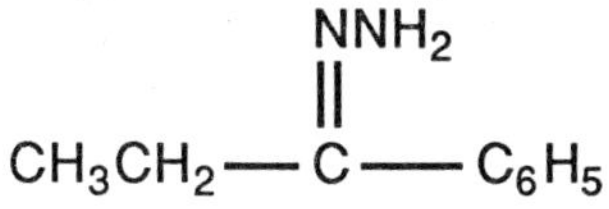 $CH_3CH_2-\overset{NNH_2}{\overset{\|}{C}}-C_6H_5$

(e) 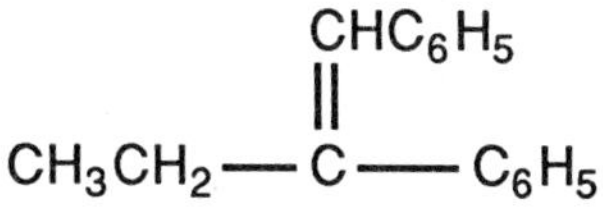 $CH_3CH_2-\overset{CHC_6H_5}{\overset{\|}{C}}-C_6H_5$

(f) 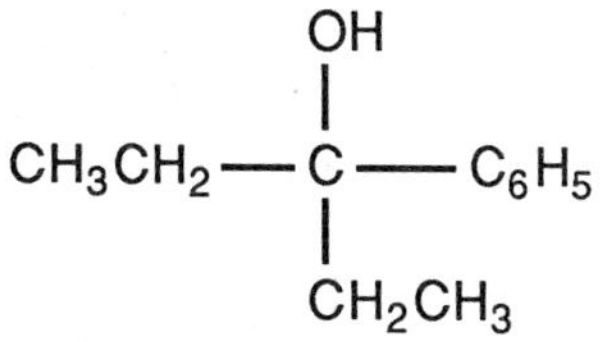 $CH_3CH_2-\overset{OH}{\overset{|}{\underset{CH_2CH_3}{\underset{|}{C}}}}-C_6H_5$

(g) 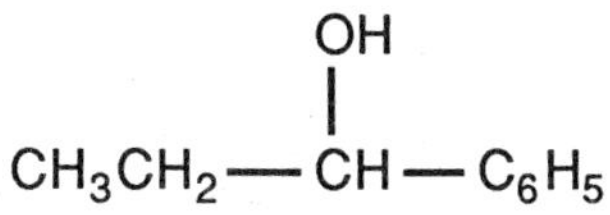 $CH_3CH_2-\overset{OH}{\overset{|}{CH}}-C_6H_5$

(h) 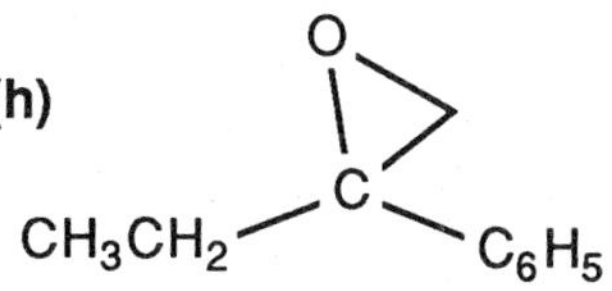 Epoxide: O and CH_2 ring with C bearing CH_3CH_2 and C_6H_5

(i) 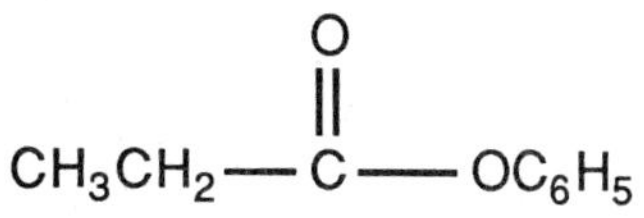 $CH_3CH_2-\overset{O}{\overset{\|}{C}}-OC_6H_5$

(j) 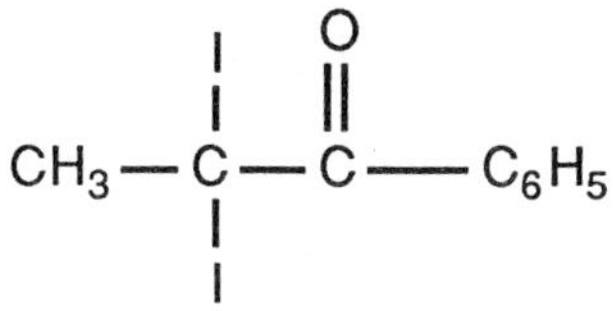 $CH_3-\overset{|}{\underset{|}{C}}-\overset{O}{\overset{\|}{C}}-C_6H_5$

13.18

(a)

$$CH_3CH_2-\overset{O}{\overset{\|}{C}}-C_6H_5 \xrightarrow{[H-BH_3]^-} CH_3CH_2-\overset{OH}{\overset{|}{C}H}-C_6H_5$$

(OH_2)

(b)

$$CH_3CH_2-\overset{O}{\overset{\|}{C}}-C_6H_5 \underset{}{\overset{:NH_2NH_2}{\rightleftharpoons}} CH_3CH_2-\overset{OH}{\underset{C_6H_5}{C}}-\overset{+}{N}H_2NH_2 \overset{\text{proton transfers}}{\rightleftharpoons}$$

(OH_2)

$$CH_3CH_2-\overset{OH}{\underset{C_6H_5}{C}}-NHNH_2 \rightleftharpoons CH_3CH_2-C(=\overset{+}{N}HNH_2)-C_6H_5 \overset{^-OH}{\rightleftharpoons} CH_3CH_2-C(=NNH)-C_6H_5 + OH_2$$

$$\rightleftharpoons CH_3CH_2-CH(-N=N-H)-C_6H_5 \overset{^-OH}{\rightleftharpoons} CH_3CH_2-CH(-N=N^-)-C_6H_5 \longrightarrow CH_3CH_2-CH_2-C_6H_5$$

(H_2O)

(c)

$$C_6H_5-\overset{O}{\overset{\|}{C}}-CH_2CH_3 \overset{H^+}{\rightleftharpoons} C_6H_5-\overset{\overset{+}{O}H}{\overset{\|}{C}}-CH_2CH_3 \overset{^-C\equiv N}{\rightleftharpoons} C_6H_5-\overset{OH}{\underset{CN}{C}}-CH_2CH_3$$

(d)

$$C_6H_5—\overset{O}{\overset{\|}{C}}—CH_2CH_3 \xrightarrow{H^+,\ :NH_2NH_2} C_6H_5—\overset{:\ddot{O}H}{\overset{|}{C}}—CH_2CH_3 \ (—NHNH_2) \rightleftharpoons C_6H_5—\overset{^+OH_2}{\overset{|}{C}}—CH_2CH_3 \ (—:NHNH_2)$$

$$\rightleftharpoons (C_6H_5)(CH_2CH_3)C{=}\overset{+}{N}HNH_2 \xrightleftharpoons{-H^+} (C_6H_5)(CH_2CH_3)C{=}NNH_2$$

(e)

$$C_6H_5—\overset{O}{\overset{\|}{C}}—CH_2CH_3 \xrightarrow{C_6H_5—\bar{C}H—\overset{+}{P}(C_6H_5)_3} C_6H_5—\overset{O^-}{\overset{|}{C}}(CH_2CH_3)—CH(\overset{+}{P}(C_6H_5)_3)—C_6H_5$$

$$\rightarrow (C_6H_5)(CH_3CH_2)C\langle CH, O \rangle P(C_6H_5)_3 \rightarrow C_6H_5—\overset{CH(C_6H_5)}{\overset{\|}{C}}—CH_2CH_3$$

(f)

$$C_6H_5—\overset{O}{\overset{\|}{C}}—CH_2CH_3 \xrightarrow{CH_3CH_2—Li} C_6H_5—\overset{O^-\ Li^+}{\overset{|}{C}}(CH_2CH_3)—CH_2CH_3 \xrightarrow{H_2O} C_6H_5—\overset{OH}{\overset{|}{C}}(CH_2CH_3)—CH_2CH_3$$

(g)

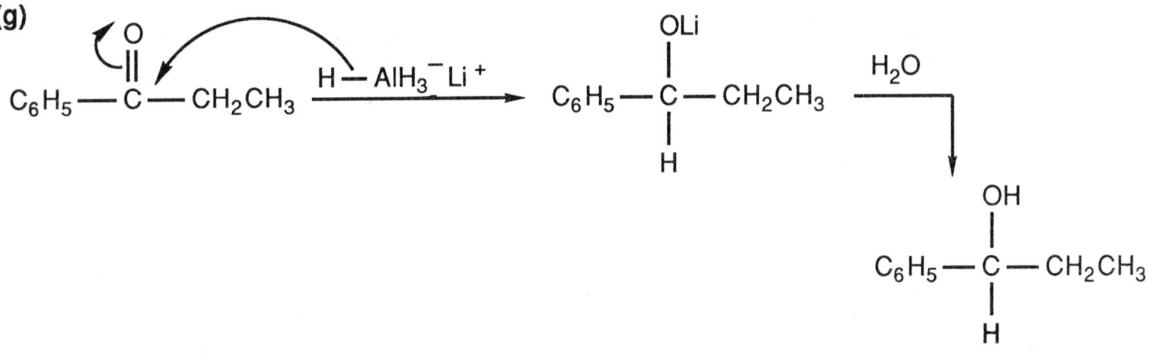

$C_6H_5—C(=O)—CH_2CH_3 \xrightarrow{H—AlH_3^- Li^+} C_6H_5—C(OLi)(H)—CH_2CH_3 \xrightarrow{H_2O} C_6H_5—C(OH)(H)—CH_2CH_3$

(h)

$C_6H_5—C(=O)—CH_2CH_3 \xrightarrow{^-CH_2—\overset{+}{S}(CH_3)_2} C_6H_5—C(O^-)(CH_2CH_3)—CH_2—\overset{+}{S}(CH_2)_2 \longrightarrow$ epoxide: C_6H_5, CH_2CH_3 on C, ring O—CH_2

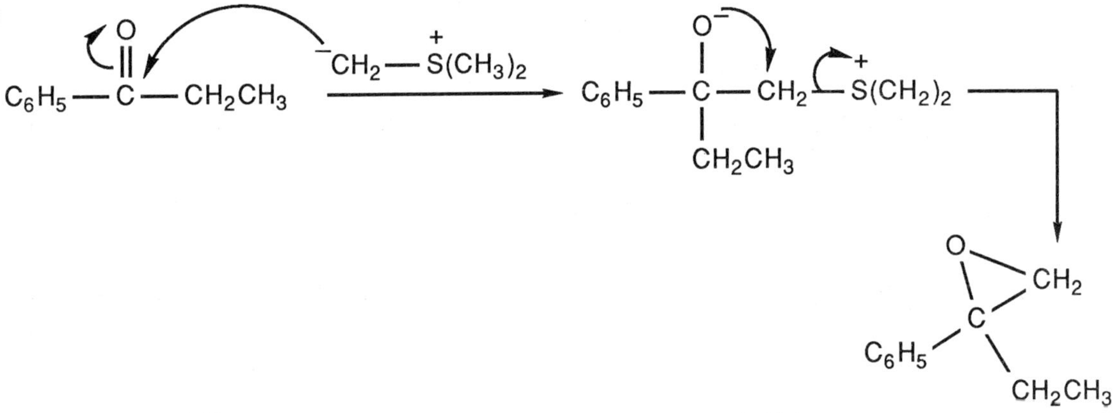

(i)

$C_6H_5—C(=\ddot{O}:)—CH_2CH_3 \overset{H^+}{\rightleftharpoons} C_6H_5—C(=\overset{+}{O}H)—CH_2CH_3 \xrightarrow{HOO—C(=O)—CH_3}$

$CH_2CH_3—C(OH)(C_6H_5)—\overset{+}{O}(H)—O—C(=O)—CH_3 \underset{\text{transfer}}{\overset{\text{proton}}{\rightleftharpoons}} CH_2CH_3—C(:OH)(C_6H_5)—O—O—C(=\overset{+}{O}H)—CH_3 \longrightarrow$

$CH_2CH_3—C(=\overset{+}{O}H)—OC_6H_5 \underset{\text{transfer}}{\overset{\text{proton}}{\rightleftharpoons}} CH_2CH_3—C(=O)—OC_6H_5$

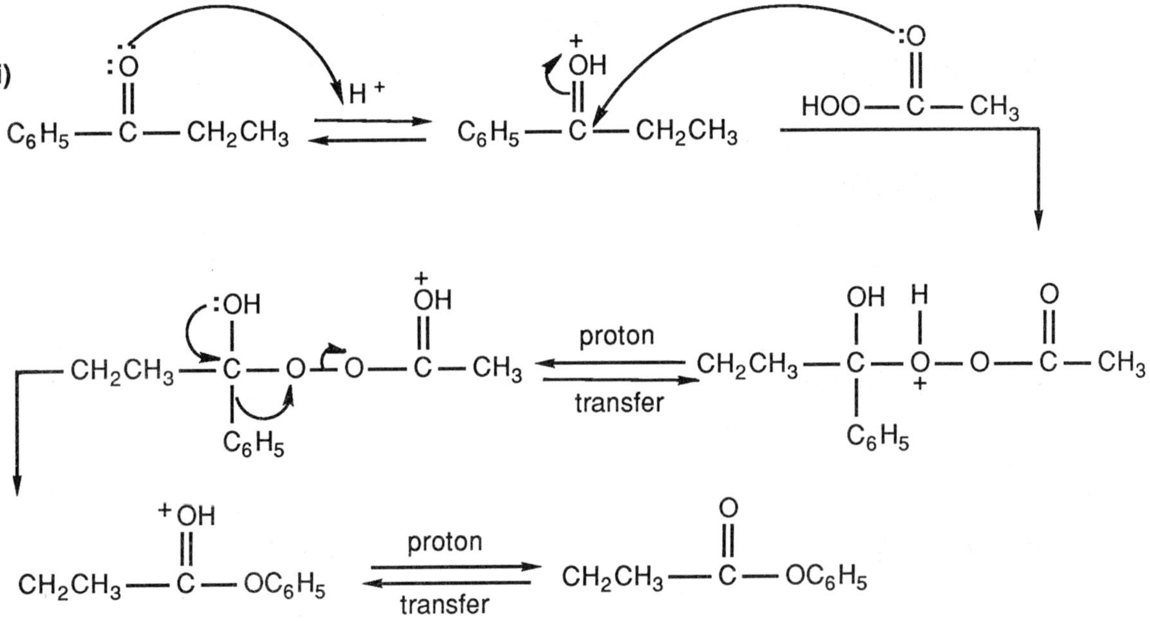

(j)

$$C_6H_5-\overset{O}{\overset{\|}{C}}-\underset{H}{\overset{H}{C}}-CH_3 \xrightarrow{^-OH} C_6H_5-\overset{O^-}{C}=C(H)(CH_3) \xrightarrow{I-I} C_6H_5-\overset{O}{\overset{\|}{C}}-\underset{I}{\overset{H}{C}}-CH_3$$

$$C_6H_5-\overset{O}{\overset{\|}{C}}-\underset{I}{\overset{H}{C}}-CH_3 \underset{}{\overset{^-OH}{\rightleftharpoons}} C_6H_5-\overset{O^-}{C}=C(I)(CH_3) \xrightarrow{I-I} C_6H_5-\overset{O}{\overset{\|}{C}}-\underset{I}{\overset{I}{C}}-CH_3$$

13.19

(a) $CH_3CH_2CH_2-\overset{OH}{CH}-CH_2CH_2CH_3$

(b) $CH_3CH_2CH_2CH_2CH_2CH_2CH_3$

(c) $CH_3CH_2CH_2-\underset{CN}{\overset{OH}{C}}-CH_2CH_2CH_3$

(d) $CH_3CH_2CH_2-\overset{NNH_2}{\overset{\|}{C}}-CH_2CH_2CH_3$

(e) $(CH_3CH_2CH_2)_2C=C(CH_3)_2$

(f) $CH_3CH_2CH_2-\underset{CH_2CH_3}{\overset{OH}{C}}-CH_2CH_2CH_3$

(g) $CH_3CH_2CH_2-\overset{OH}{CH}-CH_2CH_2CH_3$

(h) $CH_3CH_2CH_2-\overset{O-CH_2}{C}-CH_2CH_2CH_3$ (epoxide: O and CH_2 bridged to C)

(i) $CH_3CH_2CH_2O-\overset{O}{\overset{\|}{C}}-CH_2CH_2CH_3$

(j) $CH_3CH_2CH_2-\overset{O}{\overset{\|}{C}}-\overset{Br}{CH}CH_2CH_3$

(k) $CH_3CH_2CH_2-\overset{O}{\overset{\|}{C}}-\overset{O}{\overset{\|}{C}}-CH_2CH_3$

(l) $CH_3CH_2CH_2CH_2CH_2CH_2CH_3$

13.20 **(a)** CrO_3, pyridine

(b) 1. TsCl 2. KOtBu 3. OsO_4, HIO_4

(c) 1. PBr_3 2. Mg 3. CH_2O, then H_2O 4. CrO_3O, pyridine

(d) 1. TsCl 2. NaC≡CH 3. H_2SO_4, H_2O, $HgSO_4$

(e) 1. CrO_3, pyridine 2. C_2H_5Li, then H_2O 3. CrO_3

(f) 1. CrO_3, pyridine 2. $(C_6H_5)_3P{=}CH_2$

(g) 1. PBr_3 2. $(CH{=}CH{-})_2CuLi$

(h) 1. CrO_3, pyridine 2. $(CH_3)_2S{=}CH_2$

13.21

(a) cyclohexyl–CH_2–CH(O–H)–H $\xrightarrow{CrO_3,\ pyridine}$ cyclohexyl–CH_2–CH(=O)

We have not presented the detailed mechanism of this oxidation.

(b)

$C_6H_{11}CH_2CH_2\ddot{O}H$ + Ts—Cl → C_6H_{11}—CH_2—CH_2—OTs; $^-$OtBu

→ C_6H_{11}—CH=CH_2 + OsO_4 → osmate ester → (H_2O) → C_6H_{11}—CH(OH)—CH_2(OH)

↓ $O{=}I({=}O)_2{-}O^-$

→ periodate ester → C_6H_{11}—CHO

(c)

$C_6H_{11}CH_2CH_2\ddot{O}H$ + PBr_3 (Br—P(Br)—Br) → $C_6H_{11}CH_2CH_2$—$\overset{+}{O}(H)$—PBr_2; $^-$Br

→ $C_6H_{11}CH_2CH_2Br$ → (Mg) → $C_6H_{11}CH_2CH_2^-$ $MgBr^+$

1. H—C(=O)—H
2. H_2O

→ $C_6H_{11}CH_2CH_2CH_2OH$ → (CrO_3, pyridine) → $C_6H_{11}CH_2CH_2CHO$

(d)

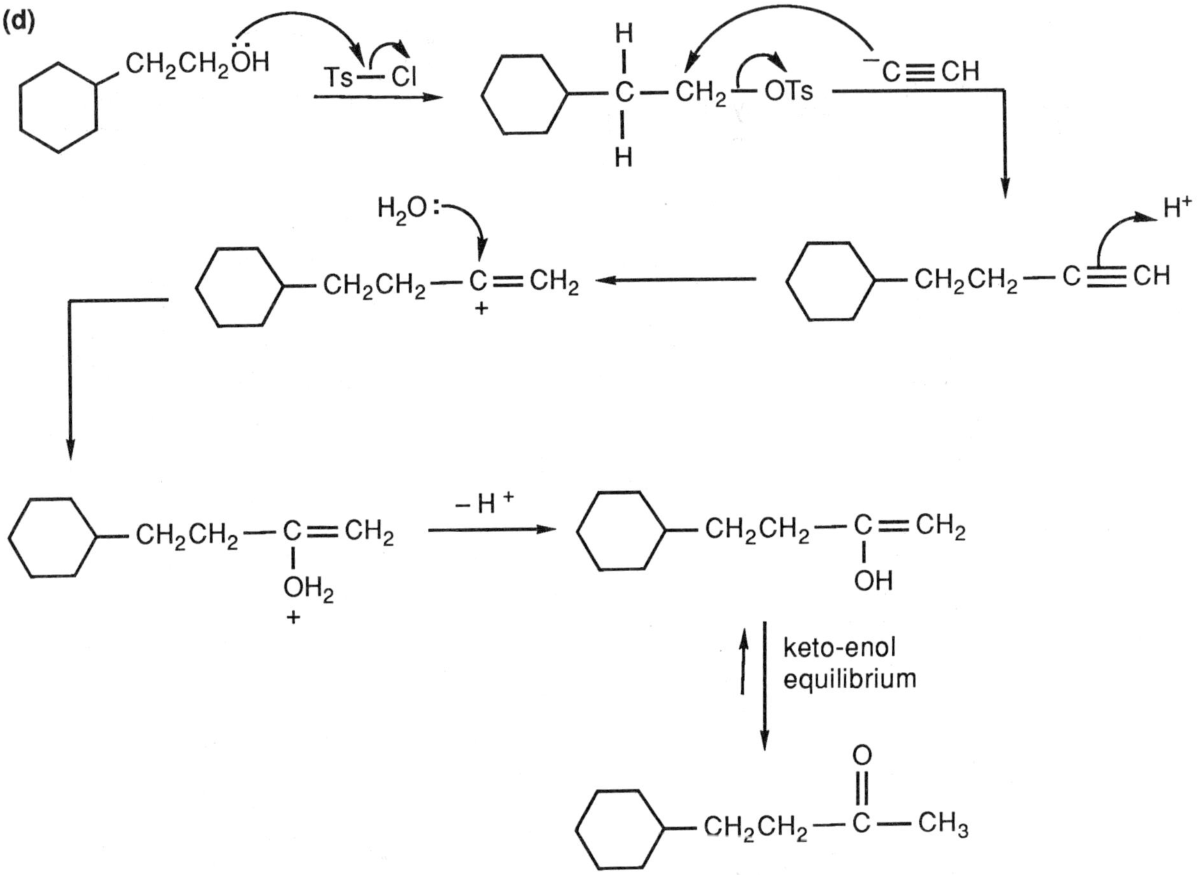

Cyclohexyl–$CH_2CH_2\ddot{O}H$ $\xrightarrow{Ts—Cl}$ Cyclohexyl–CH_2–CH_2–OTs $\xrightarrow{^-C\equiv CH}$ Cyclohexyl–$CH_2CH_2—C\equiv CH$ $\xrightarrow{H^+}$ Cyclohexyl–$CH_2CH_2—\overset{+}{C}=CH_2$ $\xrightarrow{H_2O:}$ Cyclohexyl–$CH_2CH_2—C(\overset{+}{O}H_2)=CH_2$ $\xrightarrow{-H^+}$ Cyclohexyl–$CH_2CH_2—C(OH)=CH_2$

keto-enol equilibrium

Cyclohexyl–$CH_2CH_2—C(=O)—CH_3$

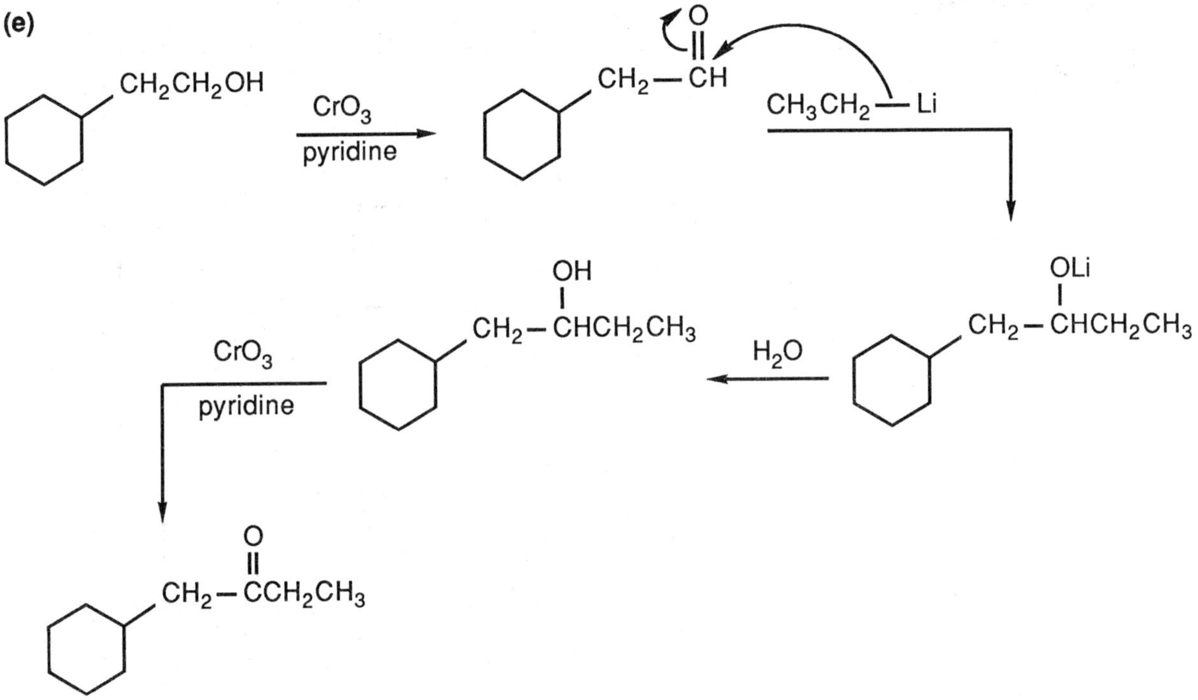

(e)

Cyclohexyl–CH_2CH_2OH $\xrightarrow[\text{pyridine}]{CrO_3}$ Cyclohexyl–$CH_2—CH(=O)$ $\xrightarrow{CH_3CH_2—Li}$ Cyclohexyl–$CH_2—CH(OLi)CH_2CH_3$ $\xrightarrow{H_2O}$ Cyclohexyl–$CH_2—CH(OH)CH_2CH_3$ $\xrightarrow[\text{pyridine}]{CrO_3}$ Cyclohexyl–$CH_2—C(=O)CH_2CH_3$

(f)

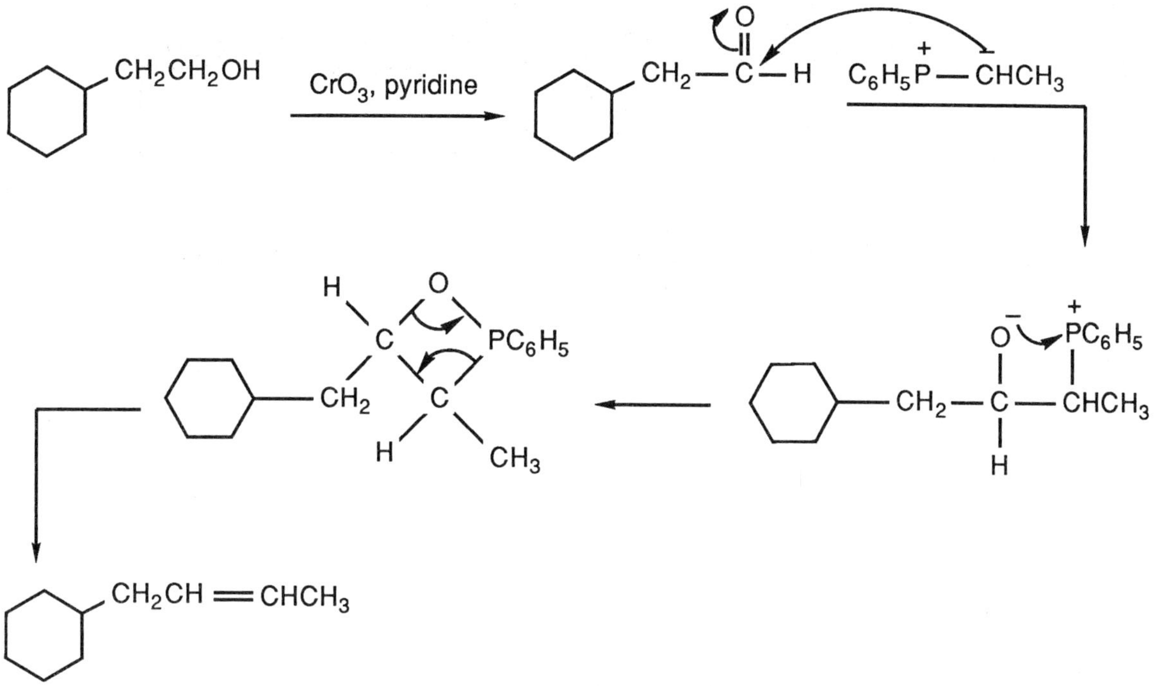

$C_6H_{11}CH_2CH_2OH \xrightarrow{CrO_3,\ pyridine} C_6H_{11}CH_2—C(=O)—H$ + $C_6H_5\overset{+}{P}—\overset{-}{C}HCH_3$

$\rightarrow C_6H_{11}CH_2—C(H)(\overset{-}{O})—CHCH_3(\overset{+}{P}C_6H_5)$

$\rightarrow$ H, C, O, PC_6H_5, $C_6H_{11}CH_2$, C, H, CH_3

$\rightarrow C_6H_{11}CH_2CH{=}CHCH_3$

(g)

$C_6H_{11}CH_2CH_2\ddot{O}H$ + $Br—P(Br)—Br \rightarrow C_6H_{11}CH_2CH_2—\overset{+}{O}(H)—Br_2$ + Br^-

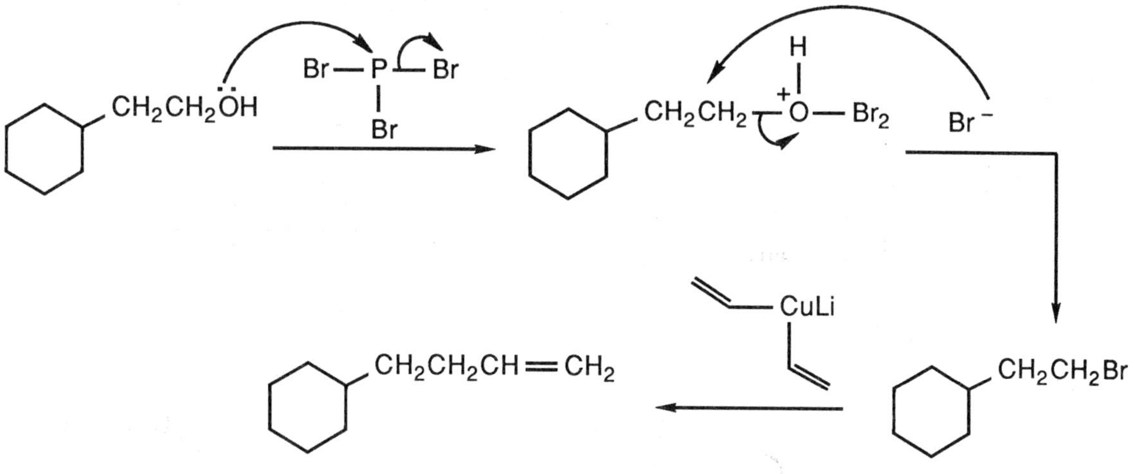

$\rightarrow C_6H_{11}CH_2CH_2Br$

CuLi

$\rightarrow C_6H_{11}CH_2CH_2CH{=}CH_2$

(h)

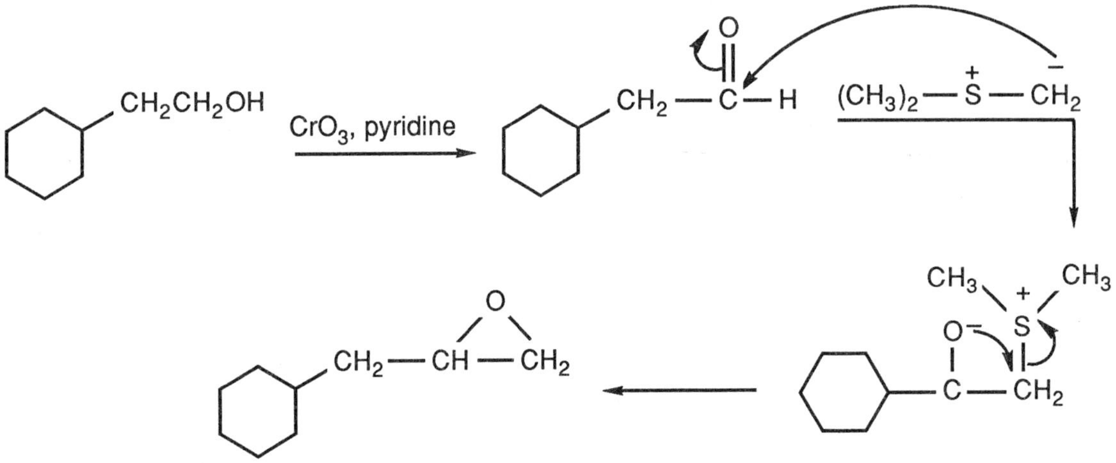

Cyclohexyl–CH_2CH_2OH $\xrightarrow{CrO_3,\ pyridine}$ Cyclohexyl–CH_2–C(=O)–H ; $(CH_3)_2\overset{+}{S}-\overset{-}{C}H_2$ → Cyclohexyl–CH_2–C(O^-)–CH_2–$\overset{+}{S}(CH_3)_2$ → Cyclohexyl–CH_2–CH–CH_2 (epoxide, O bridging)

13.22 (a) 1. dilute H_2SO_4 2. CrO_3

(b) 1. HBr 2. Mg 3. CH_2O 4. CrO_3

(c) 1. HBr 2. Mg 3.  ethylene oxide 4. CrO_3

(d) 1. HBr 2. Li, then CuI 3. CH_3CH_2COCl

(e) 1. HBr 2. Li, then CuI 3. 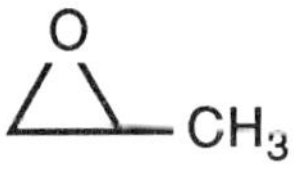 propylene oxide (CH_3-substituted epoxide) 4. CrO_3

13.23

(a)

Cyclopentene + H^+ → cyclopentyl cation + $:OH_2$ → cyclopentyl–$\overset{+}{O}H_2$ + $:OH_2$ ⇌ cyclopentanol $\xrightarrow{CrO_3,\ pyridine}$ cyclopentanone

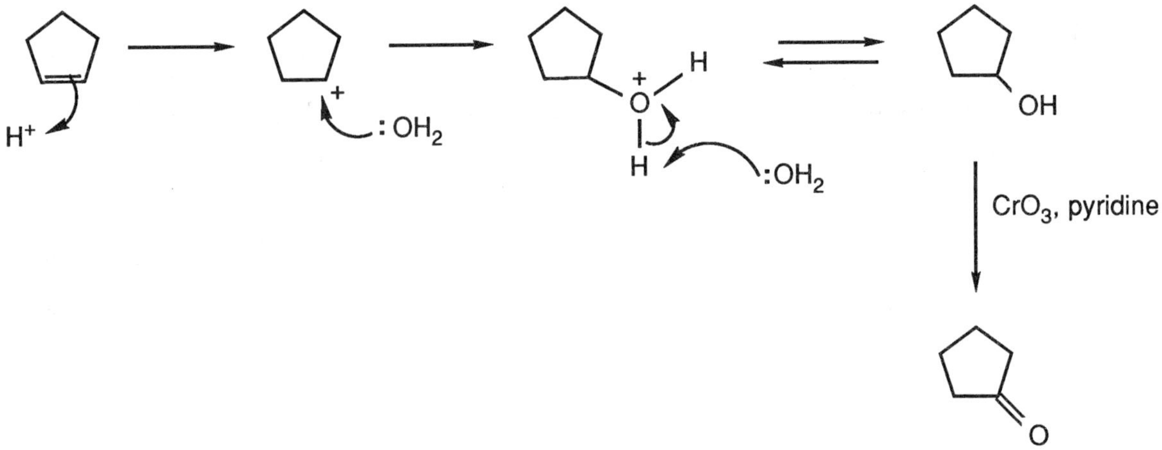

(b)

Cyclopentene + H—Br → cyclopentyl cation ($^+$) + Br^- → cyclopentyl–Br $\xrightarrow{Mg}$ cyclopentyl–MgBr

cyclopentyl–MgBr + H—C(=O)—H → cyclopentyl–CH_2OMgBr $\xrightarrow{H_2O,\ then\ CrO_3,\ pyridine}$ cyclopentyl–CHO

(c)

Cyclopentene $\xrightarrow[\text{as shown in (b)}]{1.\ HBr\quad 2.\ Mg}$ cyclopentyl–MgBr + ethylene oxide (O) → cyclopentyl–CH_2CH_2OMgBr $\xrightarrow{H_2O,\ then\ CrO_3,\ pyridine}$ cyclopentyl–CH_2CHO

(d)

Cyclopentene $\xrightarrow[\text{as in (b)}]{HBr}$ cyclopentyl–Br $\xrightarrow[2.\ CuI]{1.\ Li}$ (cyclopentyl)–Cu(Li)–(cyclopentyl)

+ CH_3CH_2—C(=O)—Cl → cyclopentyl–C(=O)—CH_2CH_3

(e)

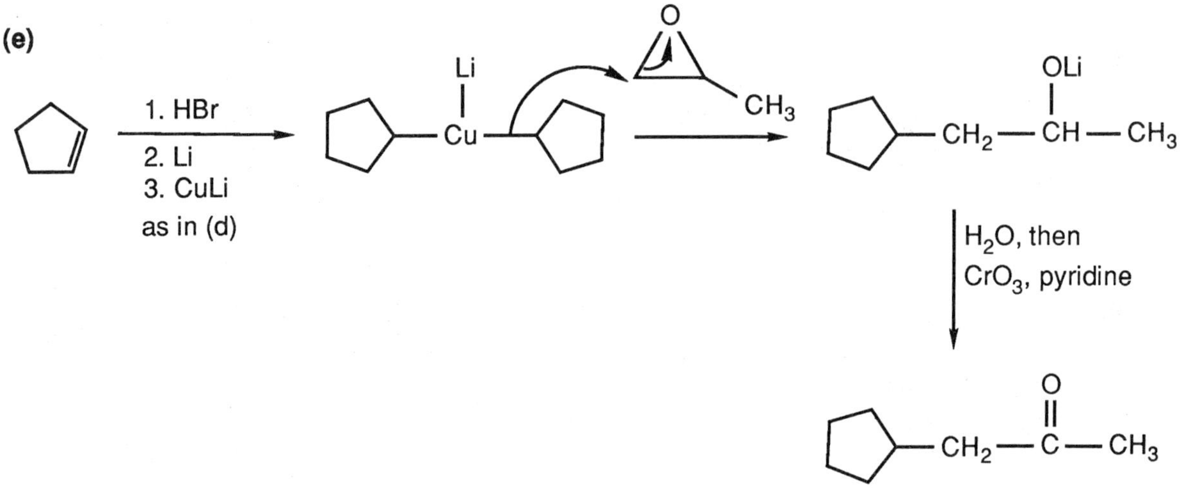

1. HBr
2. Li
3. CuLi
as in (d)

Li

Cu

O

CH_3

OLi

CH_2—CH—CH_3

H_2O, then
CrO_3, pyridine

O
||
CH_2—C—CH_3

13.24

(a)

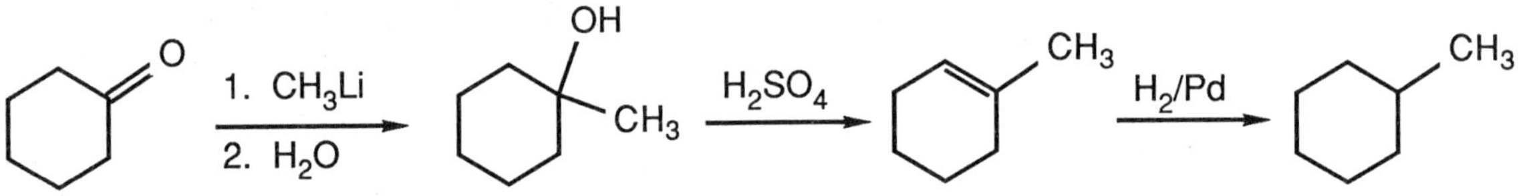

O

1. CH_3Li
2. H_2O

OH

CH_3

H_2SO_4

CH_3

H_2/Pd

CH_3

(b)

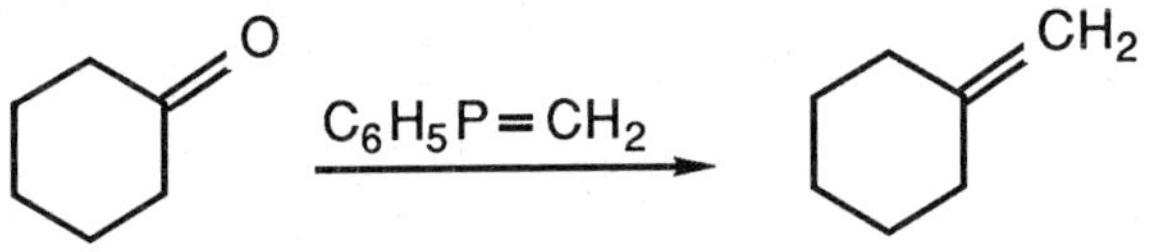

O

$C_6H_5P{=}CH_2$

CH_2

(c)

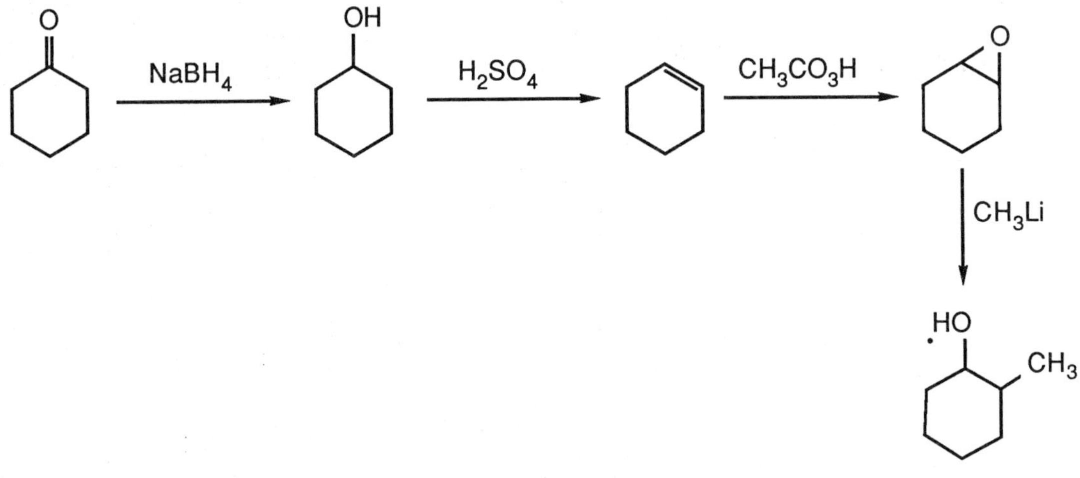

O

$NaBH_4$

OH

H_2SO_4

CH_3CO_3H

O

CH_3Li

HO

CH_3

(d) cyclohexanone $\xrightarrow{SeO_2}$ 1,2-cyclohexanedione

(e) cyclohexanone $\xrightarrow[2.\ H_2O]{1.\ LiAlH_4}$ cyclohexanol (OH) $\xrightarrow{PBr_3}$ bromocyclohexane (Br) $\xrightarrow[2.\ CuI]{1.\ Li}$ $(C_6H_{11})_2CuLi$ $\xrightarrow{Cl-\overset{O}{\overset{\|}{C}}-CH_3}$ cyclohexyl $\overset{O}{\overset{\|}{C}}CH_3$

(f) cyclohexanone $\xrightarrow[2.\ H_2O]{1.\ LiAlH_4}$ cyclohexanol (OH) $\xrightarrow{PBr_3}$ bromocyclohexane (Br) $\xrightarrow[2.\ CuI]{1.\ Li}$ $(C_6H_{11})_2CuLi$ $\xrightarrow[2.\ H_2O]{1.\ \text{ethylene oxide}}$ cyclohexyl CH_2CH_2OH $\xrightarrow[\text{pyridine}]{CrO_3,}$ cyclohexyl CH_2CHO

13.25

(a)

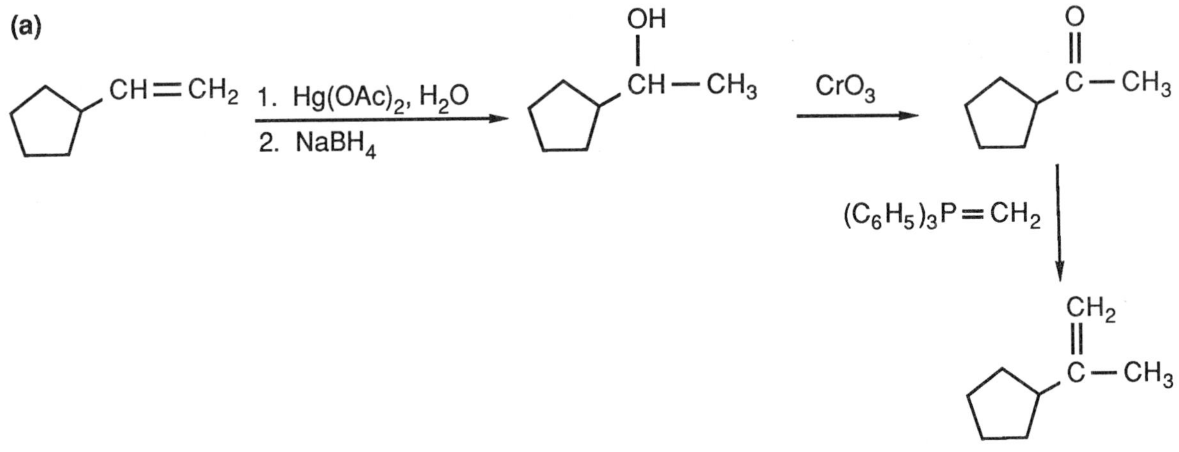

C_5H_9–CH=CH$_2$ $\xrightarrow[\text{2. NaBH}_4]{\text{1. Hg(OAc)}_2\text{, H}_2\text{O}}$ C_5H_9–CH(OH)–CH$_3$ $\xrightarrow{\text{CrO}_3}$ C_5H_9–C(=O)–CH$_3$ $\xrightarrow{(C_6H_5)_3P{=}CH_2}$ C_5H_9–C(=CH$_2$)–CH$_3$

(b)

C_5H_9–CH=CH$_2$ $\xrightarrow[\text{2. NaOH, H}_2\text{O}_2]{\text{1. BH}_3}$ C_5H_9–CH$_2$CH$_2$OH $\xrightarrow[\text{pyridine}]{\text{CrO}_3}$ C_5H_9–CH$_2$CHO $\xrightarrow{(C_6H_5)_3P{=}CH_2}$ C_5H_9–CH$_2$CH=CH$_2$

(c)

C_5H_9–CH=CH$_2$ $\xrightarrow{CH_3CO_3H}$ C_5H_9–CH–CH$_2$ (epoxide, bridged by O)

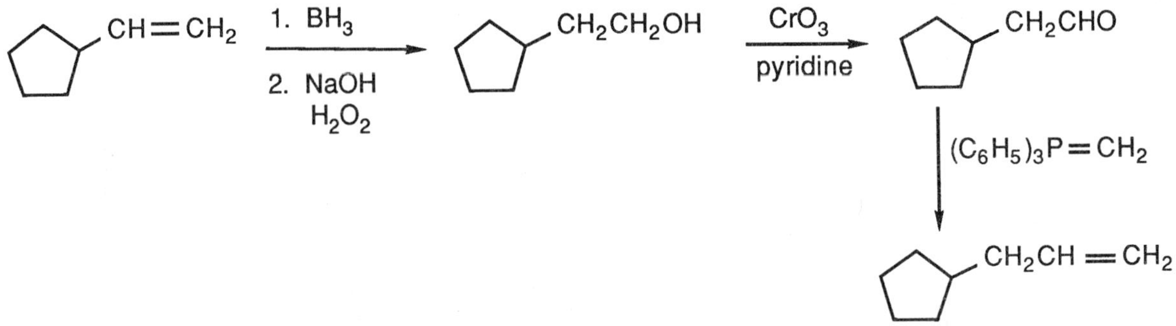

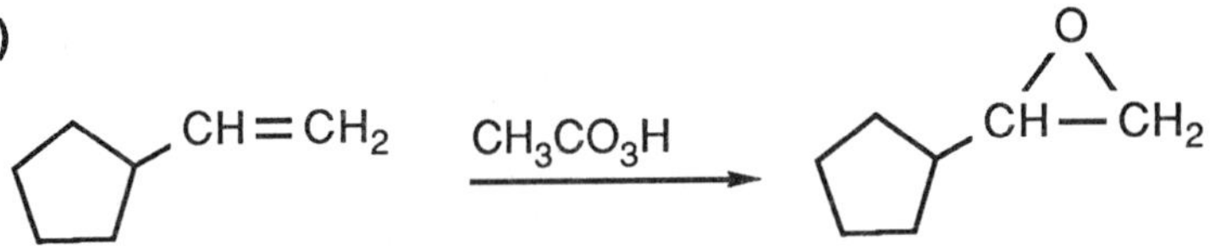

(d)

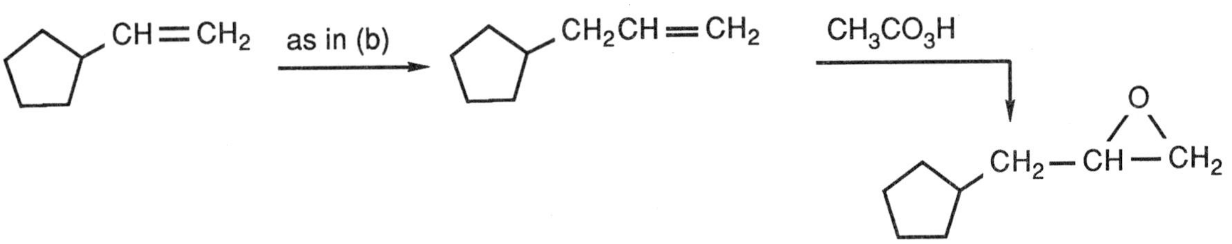

C_5H_9–CH=CH$_2$ $\xrightarrow{\text{as in (b)}}$ C_5H_9–CH$_2$CH=CH$_2$ $\xrightarrow{CH_3CO_3H}$ C_5H_9–CH$_2$–CH–CH$_2$ (epoxide, bridged by O)

(e)

C_5H_9–CH=CH$_2$ $\xrightarrow[\text{O}_3\text{, Zn}]{\text{OsO}_4\text{, HIO}_4\text{ or}}$ C_5H_9–CH=O

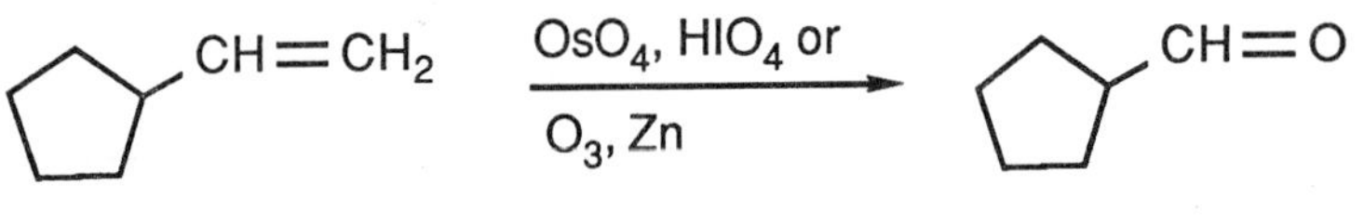

(f)

Cyclopentyl–$CH{=}CH_2$ $\xrightarrow{CH_3CO_3H}$ cyclopentyl–$CH{-}CH_2$ (epoxide, O bridging) $\xrightarrow[2.\ H_2O]{1.\ CH_3Li}$ cyclopentyl–$CH(OH){-}CH_2CH_3$

$\xrightarrow{CrO_3}$ cyclopentyl–$C(=O){-}CH_2CH_3$ $\xrightarrow{(CH_3)_2S{=}CH_2}$ cyclopentyl–$C(O{-}CH_2){-}CH_2CH_3$ (epoxide)

(g)

Cyclopentyl–$CH{=}CH_2$ $\xrightarrow[2.\ NaOH,\ H_2O_2]{1.\ BH_3}$ cyclopentyl–CH_2CH_2OH $\xrightarrow[pyridine]{CrO_3}$ cyclopentyl–CH_2CHO

(h)

Cyclopentyl–$CH{=}CH_2$ $\xrightarrow[H_2O_2]{HBr}$ cyclopentyl–CH_2CH_2Br $\xrightarrow[2.\ CH_2{=}O;\ 3.\ H_2O]{1.\ Mg}$ cyclopentyl–$CH_2CH_2CH_2OH$ $\xrightarrow[pyridine]{CrO_3}$ cyclopentyl–CH_2CH_2CHO

13.26 (a)

3-Methylcyclohex-2-enone $\xrightarrow{(C_6H_5)_3P{=}C(CH_3)_2}$ 1-methyl-3-(propan-2-ylidene)cyclohexene

(b) Li, NH_3 or H_2/Pd-C

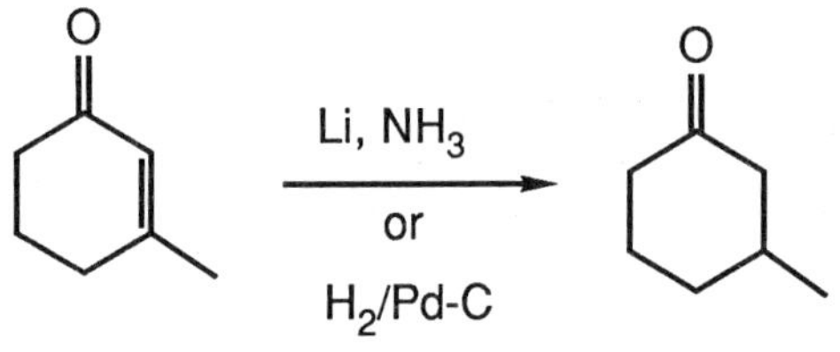

(c) H_2O_2 / NaOH

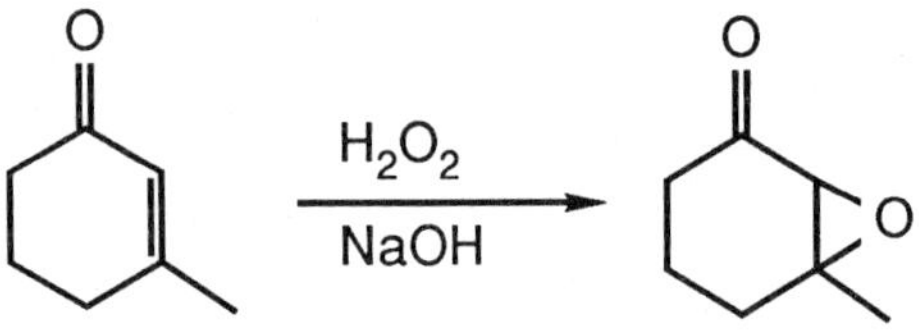

(d) $(CH_3)_2S=CH_2$

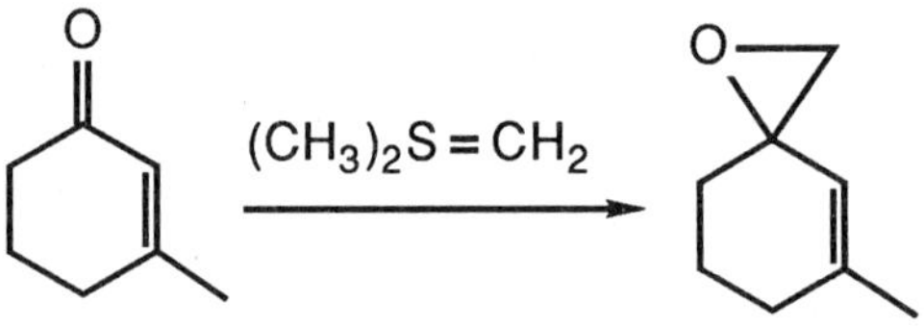

(e) HCN

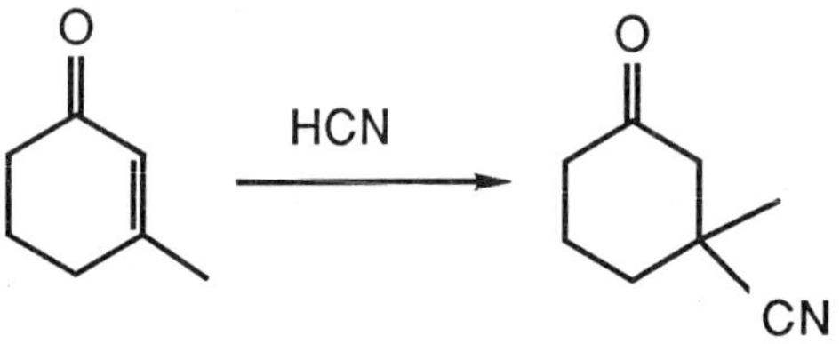

CN

(f) CH_3Li

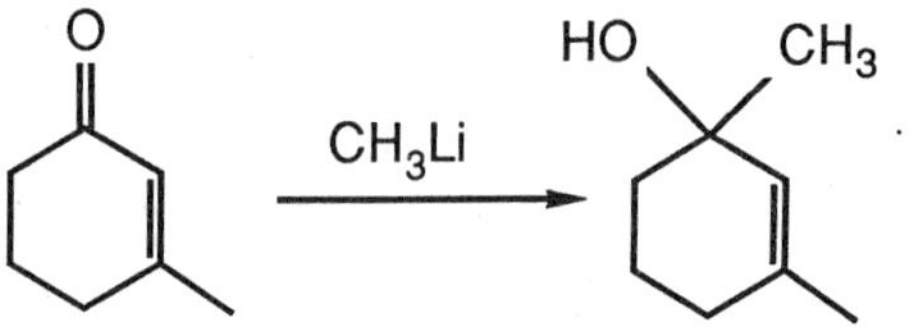

HO CH_3

(g) 1. $(CH_3)_2CuLi$ 2. H_2O

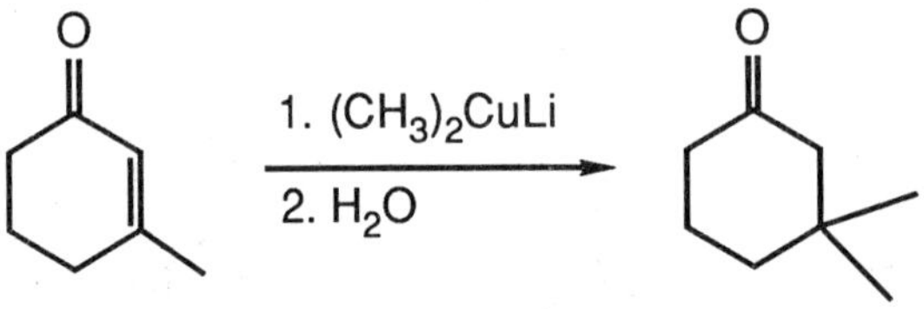

(h) 3-methylcyclohex-2-enone $\xrightarrow{\text{Li, } NH_3}$ 3-methylcyclohexanone $\xrightarrow{\text{Zn-Hg, HCl}}$ methylcyclohexane

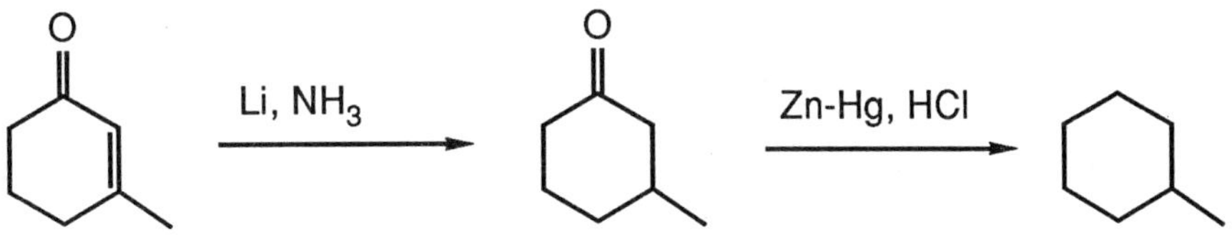

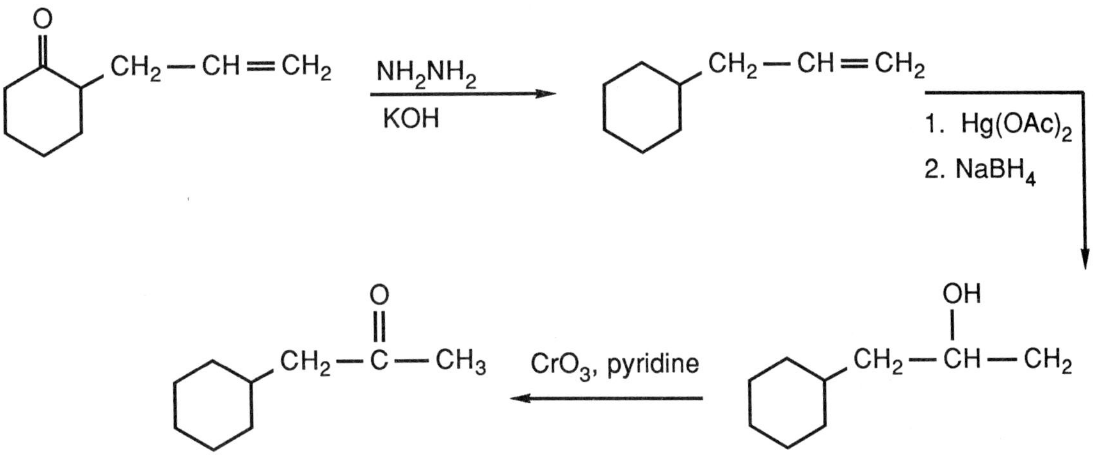

13.27

(a) 2-allylcyclohexanone ($CH_2-CH=CH_2$) $\xrightarrow[\text{KOH}]{NH_2NH_2}$ cyclohexyl–$CH_2-CH=CH_2$ $\xrightarrow{\text{1. } Hg(OAc)_2 \text{ 2. } NaBH_4}$ cyclohexyl–$CH_2-CH(OH)-CH_2$ $\xrightarrow{CrO_3\text{, pyridine}}$ cyclohexyl–$CH_2-C(=O)-CH_3$

(b) 2-allylcyclohexanone ($CH_2-CH=CH_2$) $\xrightarrow{H_2/Pd\text{-}C}$ 2-propylcyclohexanone ($CH_2CH_2CH_3$)

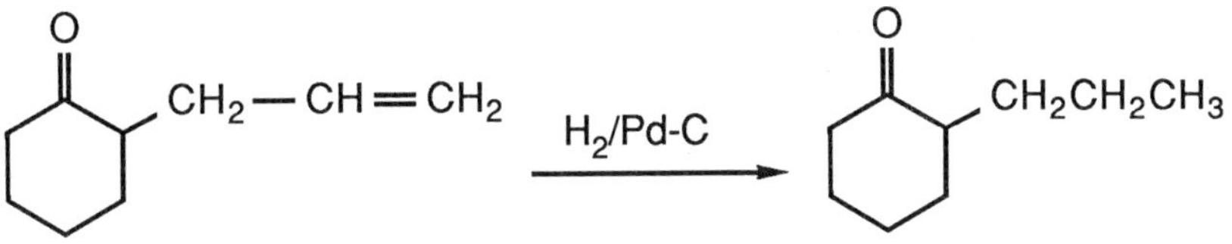

(c) 2-allylcyclohexanone ($CH_2-CH=CH_2$) $\xrightarrow{(C_6H_5)_3P=CH_2}$ 1-methylene-2-allylcyclohexane ($=CH_2$; $CH_2-CH=CH_2$)

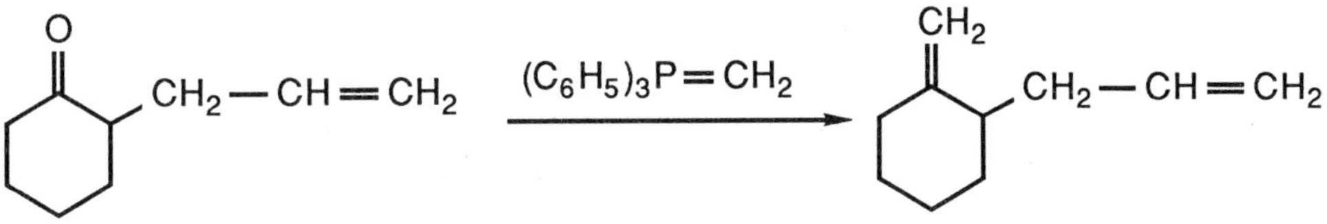

(d) 2-allylcyclohexanone ($CH_2-CH=CH_2$) $\xrightarrow{CH_3CO_3H}$ 2-(epoxypropyl)cyclohexanone ($CH_2-CH-CH_2$ with O bridging CH and CH_2)

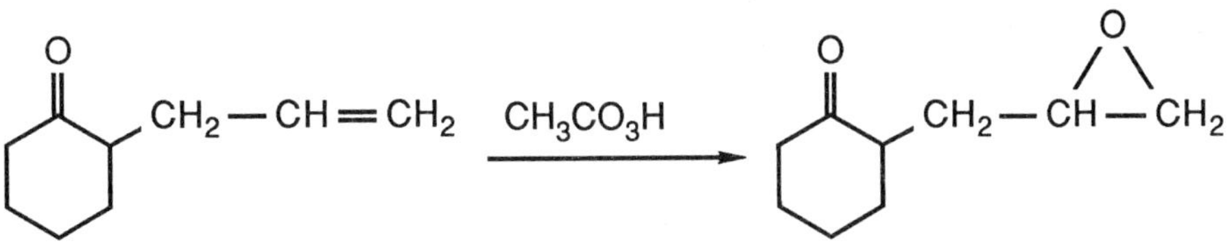

(e)

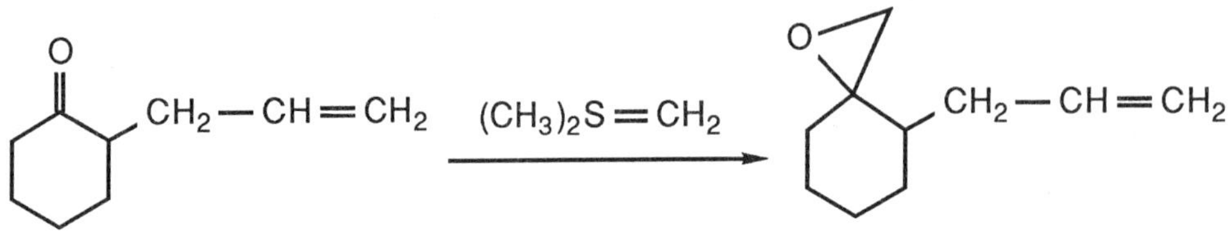

$$\text{2-allylcyclohexanone (cyclohexanone with } CH_2{-}CH{=}CH_2\text{)} \xrightarrow{(CH_3)_2S{=}CH_2} \text{spiro-epoxide (oxirane at C-1) with } CH_2{-}CH{=}CH_2$$

13.28 The major structural information provided in this problem is the complete structure of **G**, but equally important are classification of **A** as an ester together and the molecular formula of **D,** which has five degrees of unsaturation. The subsequent reactions of **D** (oxymercuration, oxidation, Baeyer–Villiger reaction) indicate that one of the degrees of unsaturation is a carbon–carbon double bond (with the remaining four corresponding to a phenyl group). The formation of **B** and **C** from both esters **A** and **F** requires that **B** and **C** are both *primary* alcohols. Compound **A** can be abbreviated as:

$$R{-}\overset{O}{\overset{\|}{C}}{-}OCH_2R'$$

and this allows us to draw **B, C** and **F**, respectively, as:

$$R{-}CH_2OH \qquad HOCH_2{-}R' \qquad R{-}CH_2{-}O{-}\overset{O}{\overset{\|}{C}}{-}R'$$

Identification of the R and R′ groups can be deduced from the reactions of **F.** Oxymercuration followed by chromic acid oxidation affords the ketone **E,** and this means that one of the CH_2 groups of **G** was the carbonyl carbon of **E** and one of the doubly bonded carbons in **D.** Only two isomeric alkenes are possible for **D:**

$$C_6H_5{-}\underset{}{\overset{CH_3}{\overset{|}{C}H}}{-}CH{=}CH{-}CH_3 \quad \text{or} \quad C_6H_5{-}\overset{CH_3}{\overset{|}{C}H}{-}CH_2{-}CH{=}CH_2$$

The formation of just **E** (rather than a mixture of isomeric products in the oxymercuration-oxidation sequence) allows you to deduce that **D** is 4-phenyl-1-pentene. The primary alkyl group migrates in preference to the methyl group in the Baeyer-Villiger reaction, so the complete answer can be written as follows:

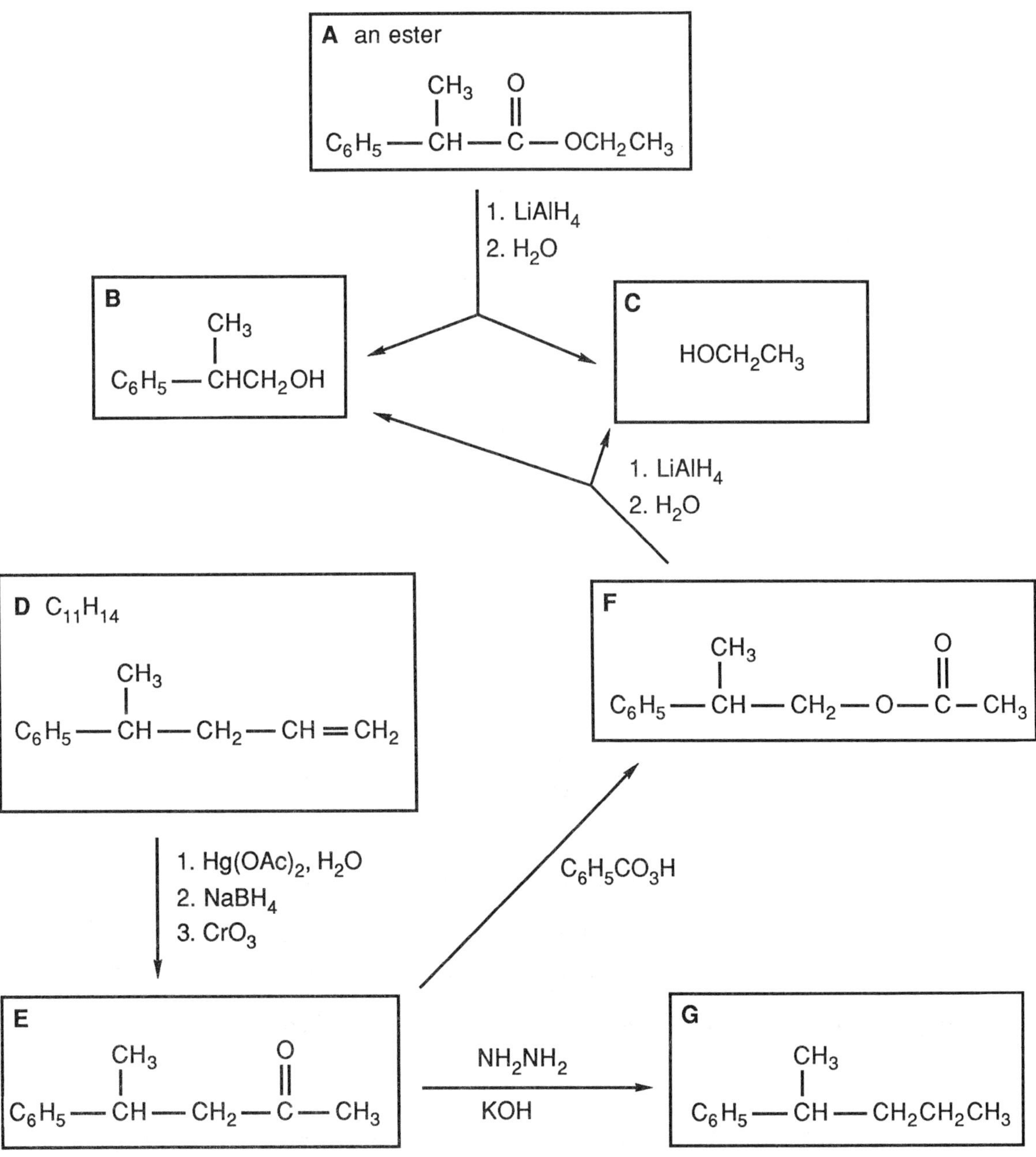

13.29 The reduction of ester **A** will give two alcohols, **B** and **C.** As least one of these compounds must be a primary alcohol, but the other could be methyl, primary, secondary, or tertiary. The problem states that **D** give a positive Tollens test, so is is an aldehyde derived from a primary alcohol. The formation of bromoform when **E** is treated with bromine in base, requires that **E** be a methyl ketone. On this basis you can conclude that the original ester, RCO_2R', affords RCH_2OH (1-pentanol) and a secondary alcohol (R'OH) that is oxidized to a methyl ketone. The molecular formula of **C** allows only two carbons in addition to the methyl group and the secondary carbon bearing the hydroxyl group. This compound must therefore be 2-butanol, and the complete answer follows:

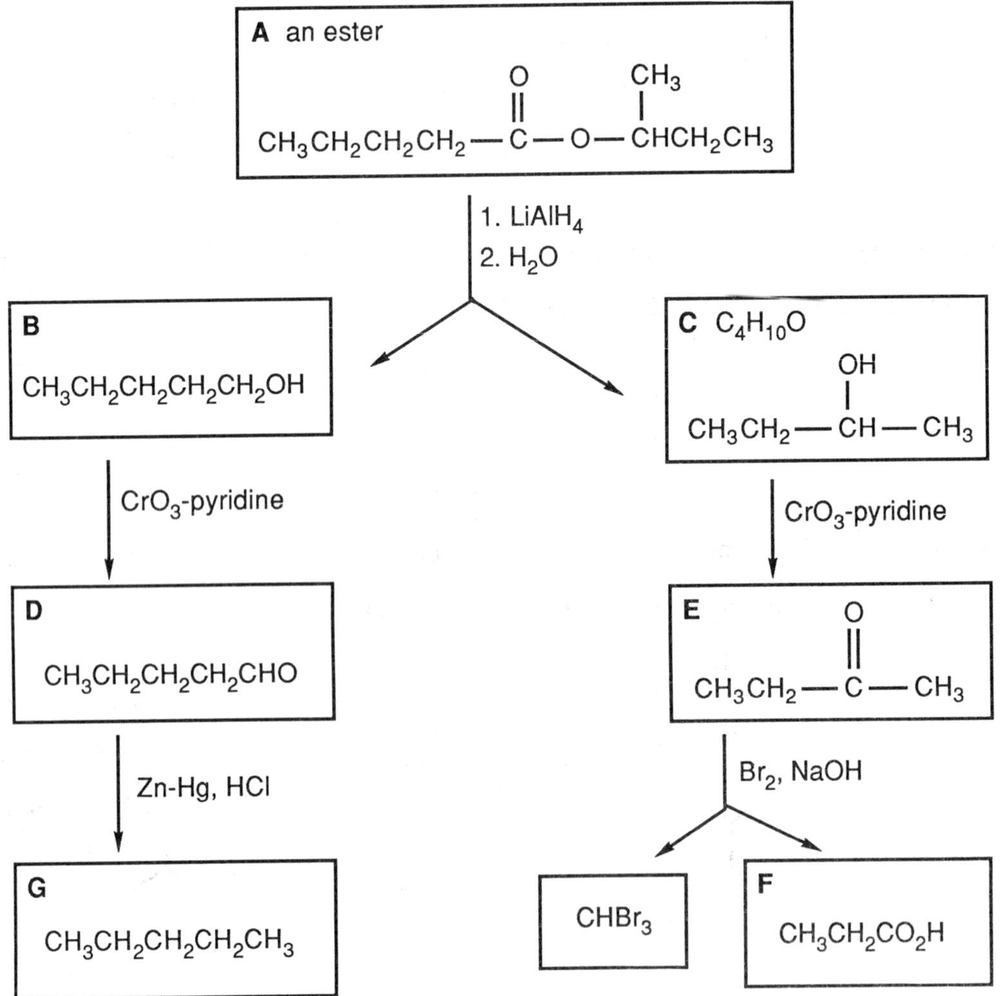

13.30 The deuterium exchange experiments clearly show that the X_3C^- anions are formed reversibly when the corresponding haloforms (HCX_3) are treated with aqueous base. The trend that exchange is faster when X is iodine or bromine than when it is chlorine is not in accord with inductive effects alone. The electronegativity is greatest for fluorine, decreasing for the other halogens as the atomic number increases. On that basis I_3C^- should be less stable (and less easily formed) than Br_3C^-. The experimental results therefore suggest that stabilization of these anions results in part (when X is Br or I) from delocalization of the lone pair of electrons into low-lying *d* orbitals on the halogens.

13.31 The unknown exhibits the peak between 9 and 10 ppm that is expected for an aldehyde, and this appears as a triplet. Clearly, this is coupled to the 2H signal at 3.7 ppm. The remaining 5H signal centered at 7.3 ppm is clearly a phenyl group. An additional oxygen atom cannot be excluded, so two reasonable structures are:

$$C_6H_5\text{-}CH_2\text{-}CHO \quad \text{and} \quad C_6H_5\text{-}OCH_2\text{-}CHO$$

Of these two possibilities, only the former gives a predicted chemical shift (3.8) in reasonable agreement with the observed spectrum:

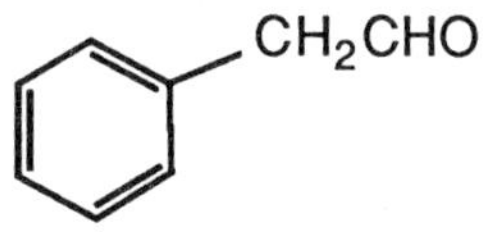

CH_2CHO

13.32 The close proximity of the secondary alcohol and aldehyde groups creates an optimum situation for intramolecular hemiacetal formation:

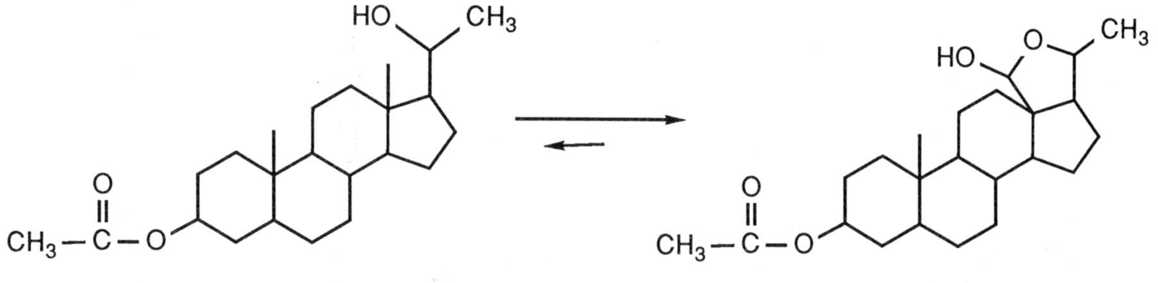

HO CH3

O ‖ CH_3-C-O

HO O CH3

O ‖ CH_3-C-O

13.33 The alkyne **A** gives negative tests for a terminal triple bond, so its structure can be abbreviated as R–C≡C–R′. Two products are formed in the hydration reaction, and these can be written as $RCOCH_2R'$ and RCH_2COR'. Compound **C** gives a positive iodoform test, and we can deduce that R′ is methyl and assign RCH_2COR' to **C**. Compound **B** therefore has the partial structure $RCOCH_2CH_3$. Since **A** has a total of 10 carbons, the unidentified R group must be a C_7 subunit. The NMR spectrum of **B** shows two singlets in addition to the expected ethyl pattern, and one of the singlets clearly corresponds to an aromatic ring. No integration is given for the signals, and two C_7 subunits are reasonable alternatives:

$$CH_3\text{-}C_6H_4\text{-} \quad \text{and} \quad CH_6H_5\text{-}CH_2\text{-}$$

The first of these is unacceptable because of two features of the NMR spectrum of **B**. First, the aromatic signal appears as a singlet, whereas a carbonyl group attached directly to the ring almost always results in a complex pattern. Second, the chemical shift of a CH_3 group should be about 2.2 ppm, whereas the remaining singlet in the spectrum appears at 3.6 ppm. The is consistent with the second alternative, however, and the correct structures of **A–C** follow:

$$C_6H_5\text{—}CH_2\text{—}C{\equiv}CCH_2CH_3$$

A

$$C_6H_5\text{—}CH_2\text{—}\overset{\overset{\large O}{\|}}{C}\text{—}CH_2CH_3$$

B

$$C_6H_5\text{—}CH_2CH_2\text{—}\overset{\overset{\large O}{\|}}{C}\text{—}CH_3$$

C

13.34 At nearly every stage of this reaction there is both strong base: CH_3Li or the nitrogen anion produced by addition of methyllithium to the nitrile. In addition the neutral nitrile (or subsequently the ketone) can serve as a proton source. Consequently, epimerization at the α-position can occur via enolate formation:

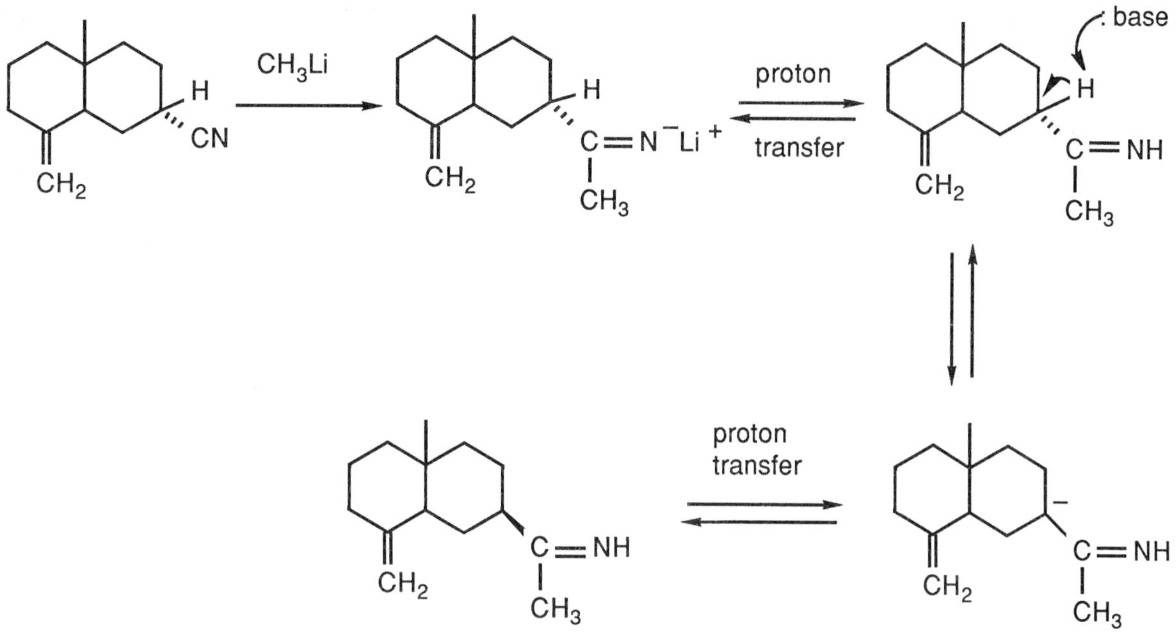

Alternatively, the isomerization could occur after the imine is hydrolyzed to the ketone:

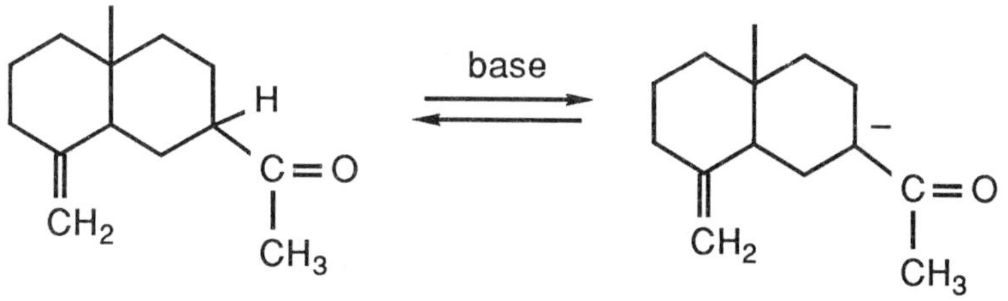

The driving force for the isomerization is formation of the product in which the substituent occupies the more stable *equatorial* position:

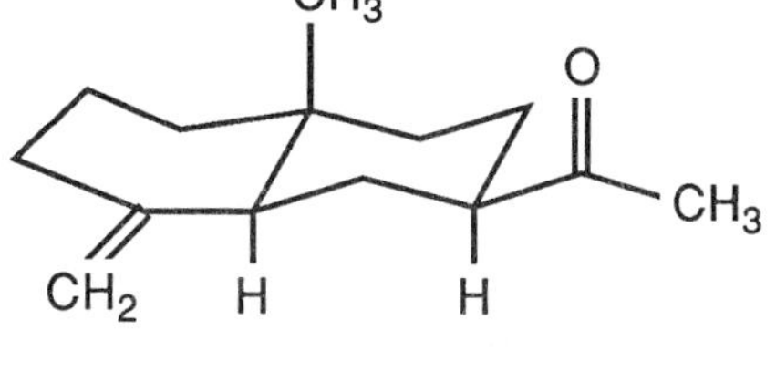

13.35 **(a)** The C=O peak in the region of 1675–1700 cm^{-1} would be absent in the product, and an OH absorption in the 3200–3400 cm^{-1} region would appear. The appearance of the CH_2 group in the NMR would change both in multiplicity (from coupling to the newly introduced CH) and in chemical shift (the –CH(OH)Ar would have a different influence than –COAr).

(b) The carbonyl absorption of the reactant would disappear in the infrared spectrum, and the NMR spectrum of the product would show an additional 2H signal in the aliphatic region.

(c) The carbonyl absorption of the reactant would disappear, and new infrared peaks would be seen in the 3200–3400 cm^{-1} region (OH) and in the triple bond region (2210–2260 cm^{-1}) for the cyano group. The NMR spectrum would again show changes in both position and splitting of the CH_2 group.

(d) The carbonyl absorption of the reactant would change because of differences between C=O and C=N, and new absorptions would be seen for NH in the 3300–3500 cm^{-1} region. The NMR spectrum would show a new absorption for the NH_2 group of the hydrazone.

(e) The carbonyl absorption of the reactant would disappear, and a new peak for the C=C double bond would be seen in the 1600–1680 cm^{-1} region. A new (and possibly overlapping) signal would be observed for the second phenyl group in the NMR spectrum.

(f) The C=O peak in the region of 1675–1700 cm^{-1} would be absent in the product, and an OH absorption in the 3200–3400 cm^{-1} region would appear. The chemical shift of the CH_2 group in the NMR would change, and both the triplet and quartet of the ethyl group would double in intensity.

(g) The changes would be the same as those indicated in (a).

(h) The carbonyl absorption of the reactant would disappear, and new peaks near 1250 cm^{-1} should appear for the C–O single bonds of the epoxide ring. The aliphatic region of the NMR spectrum should show new absorptions corresponding to the newly introduced CH_2 group.

(h) The carbonyl absorption of the reactant should be shifted as a result of the electronegative iodo substituents. The CH_2 group would disappear in the NMR spectrum, and the CH_3 group would appear as a singlet rather than as a triplet.

CHAPTER 14. CONDENSATION AND ALKYLATION REACTIONS OF ALDEHYDES AND KETONES

ANSWERS TO EXERCISES

14.1

(a) $CH_3-CH_2-CH=\underset{CH_3}{\underset{|}{C}}-\overset{O}{\overset{\|}{C}}H$

(b) O

(c) $CH_3-\overset{CH_3}{\overset{|}{C}}H-\underset{OH}{\underset{|}{C}}H-\underset{CH_3}{\underset{|}{\overset{CH_3}{\overset{|}{C}}}}-\overset{O}{\overset{\|}{C}}H$

14.2

(a) $CH_3-CH_2-\overset{O}{\overset{\|}{C}}-CH_2-CH_3$

(b) $C_6H_5-CH_2-\overset{O}{\overset{\|}{C}}H$

(c) O

14.3

$CH_3-\overset{O}{\overset{\|}{C}}-\underset{CH_3}{\underset{|}{C}}=\overset{CH_3}{\overset{|}{C}}-CH_2-CH_2-CH_3$

$CH_3-\overset{O}{\overset{\|}{C}}-\underset{CH_2CH_3}{\underset{|}{C}}=\overset{CH_3}{\overset{|}{C}}-CH_2-CH_3$

$CH_3-CH_2-\overset{O}{\overset{\|}{C}}-CH=\overset{CH_3}{\overset{|}{C}}-CH_2-CH_2-CH_3$

$CH_3-CH_2-CH_2-\overset{O}{\overset{\|}{C}}-CH=\overset{CH_3}{\overset{|}{C}}-CH_2-CH_3$

$CH_3-\overset{O}{\overset{\|}{C}}-\underset{CH_3}{\underset{|}{C}}=\overset{CH_3}{\overset{|}{C}}-CH_2-CH_3$

$CH_3-CH_2-\overset{O}{\overset{\|}{C}}-CH=\overset{CH_3}{\overset{|}{C}}-CH_2-CH_3$

$$CH_3-\overset{\overset{\displaystyle O}{\|}}{C}-\underset{\underset{\displaystyle CH_2-CH_3}{|}}{C}=\overset{\overset{\displaystyle CH_3}{|}}{C}-CH_2-CH_2-CH_3 \qquad CH_3-CH_2-CH_2-\overset{\overset{\displaystyle O}{\|}}{C}-CH=\overset{\overset{\displaystyle CH_3}{|}}{C}-CH_2-CH_2-CH_3$$

14.4

(a) $C_6H_5-CH=CH-\overset{\overset{\displaystyle O}{\|}}{C}-CH_3$ (b) $CH_3-\overset{\overset{\displaystyle CH_3}{|}}{CH}-CH=CH-\overset{\overset{\displaystyle O}{\|}}{C}-C_6H_5$

(c) $C_6H_5-CH=\underset{\underset{\displaystyle CH_2-CH_3}{|}}{C}-\overset{\overset{\displaystyle O}{\|}}{CH}$

14.5

(a) $C_6H_5-\overset{\overset{\displaystyle O}{\|}}{CH}$ and $CH_3-\overset{\overset{\displaystyle O}{\|}}{C}-CH_3$

(b) $CH_3O-C_6H_4-\overset{\overset{\displaystyle O}{\|}}{CH}$ (para) and $CH_3-\overset{\overset{\displaystyle O}{\|}}{C}-CH_2-CH_2-CH_3$

(c) cyclohexanone and $H\overset{\overset{\displaystyle O}{\|}}{C}-CH_3$ (d) $C_6H_5-\overset{\overset{\displaystyle O}{\|}}{CH}$ and $H\overset{\overset{\displaystyle O}{\|}}{C}-CH_2-CH_2-CH_3$

14.6 base catalysis (reaction via enolate formation at less substituted α carbon):

$$CH_3-CH_2-\overset{\overset{\displaystyle O}{\|}}{C}-CH=\overset{\overset{\displaystyle CH_3}{|}}{C}-CH_2-CH_3$$

acid catalysis (reaction via enolate formation at more substituted α carbon):

$$CH_3-\overset{\overset{\displaystyle O}{\|}}{C}-\overset{\overset{\displaystyle CH_3}{|}}{C}=\overset{\overset{\displaystyle CH_3}{|}}{C}-CH_2-CH_3$$

14.7

(a) 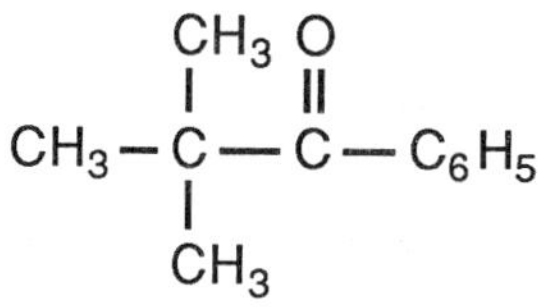 $CH_3-C(CH_3)_2-C(=O)-C_6H_5$ (b) $C_6H_5-CH(CH_2-CH=CH_2)-C(=O)-CH_2-CH_3$

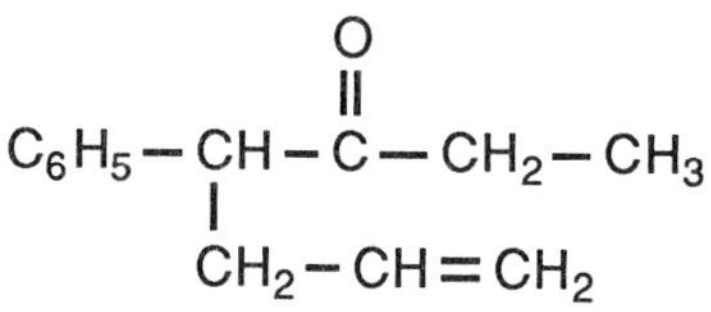

(c) $CH_3-CH_2-CH(CH_3)-C(=O)-CH(CH_2-CH_3)-CH_3$

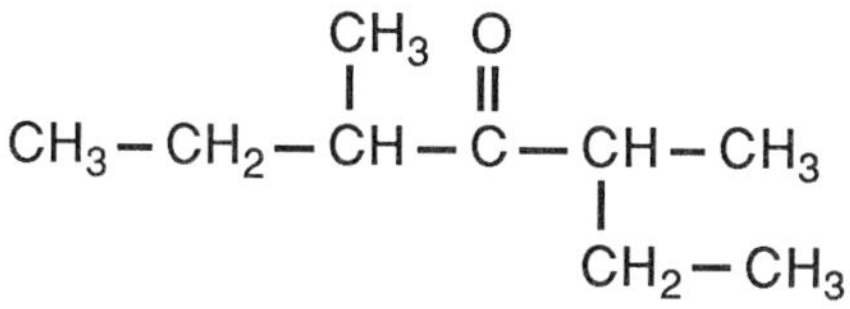

14.8

(a)

alkylation: $CH_3-CH_2-CH(CH_2-CH_3)-CH{=}O$ + dialkylation: $CH_3-CH_2-C(CH_2-CH_3)_2-CH{=}O$ + aldol: $CH_3-CH_2-CH_2-CH=C(CH_2-CH_3)-CH{=}O$

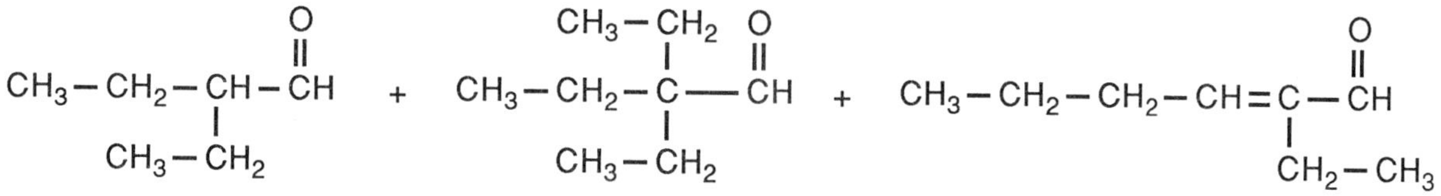

(b)

alkylation: $CH_3-CH(CH_3)-C(=O)-C_6H_5$ + dialkylation: $CH_3-C(CH_3)_2-C(=O)-C_6H_5$ + alkylation of base: $CH_3-O-CH_2-CH_3$

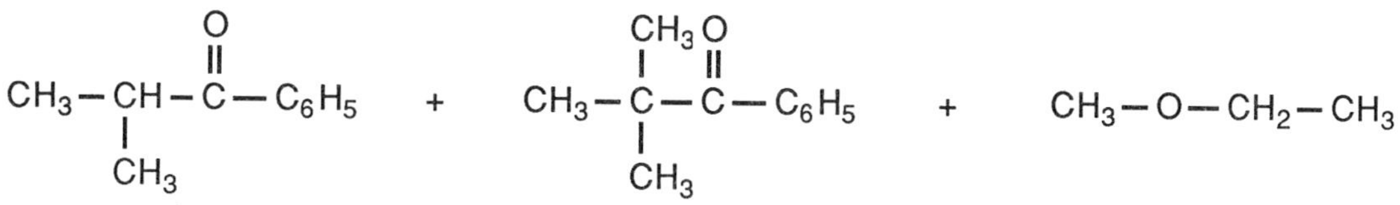

(c)

alkylation: $CH_3-CH(CH_3)-C(CH_3)_2-C(=O)-C_6H_5$ + elimination: $CH_2=CH-CH_3$

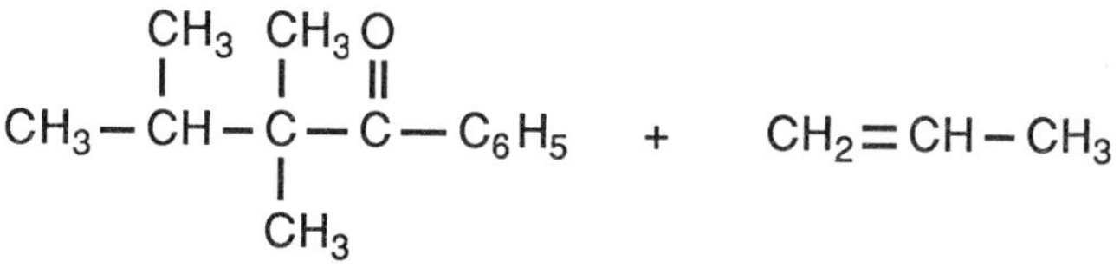

(d)

alkylation:

$$CH_3-CH(CH_2-C_6H_5)-C(=O)-CH_2-CH_3$$

+

polyalkylation:

$$CH_3-C(CH_2-C_6H_5)_2-C(=O)-CH_2-CH_3\text{, etc.}$$

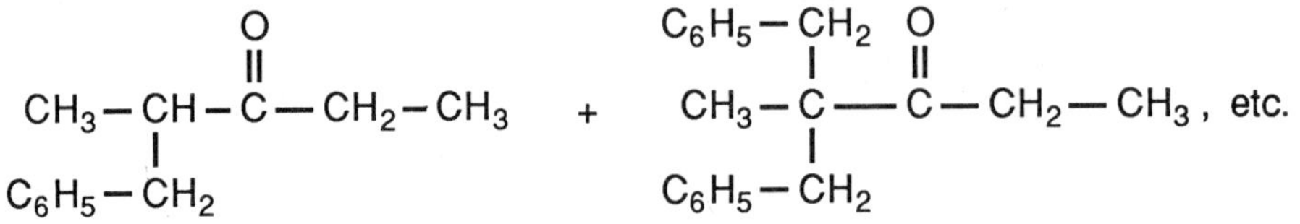

14.9 (a)

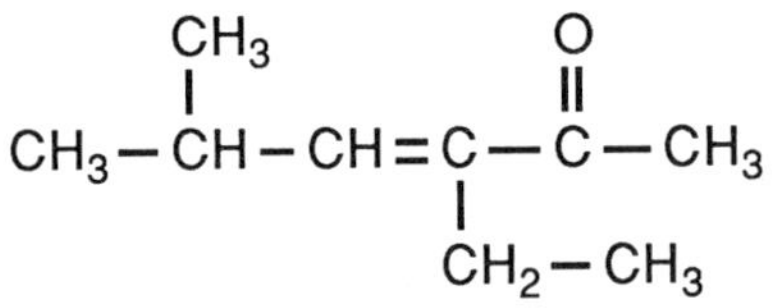

$$CH_3-CH(CH_3)-CH=C(CH_2-CH_3)-C(=O)-CH_3$$

(b)

$$CH_3-CH(CH_3)-CH_2-CH(CH_2-CH_3)-C(=O)-CH_3$$

(c) 6,6-dimethyl-2-cyclohexenone ring with CH_3, CH_3, O, and $CH_2-CH=CH_2$ substituent

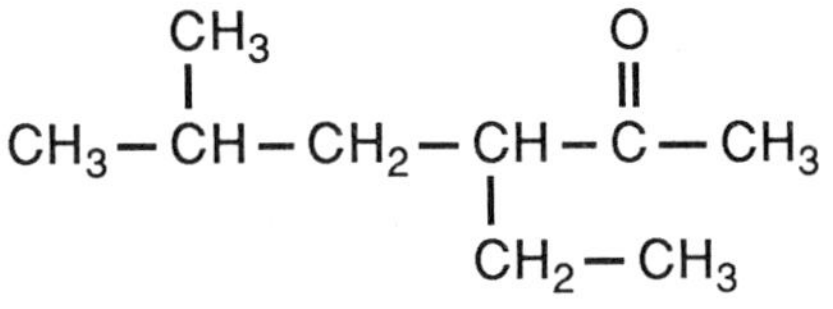

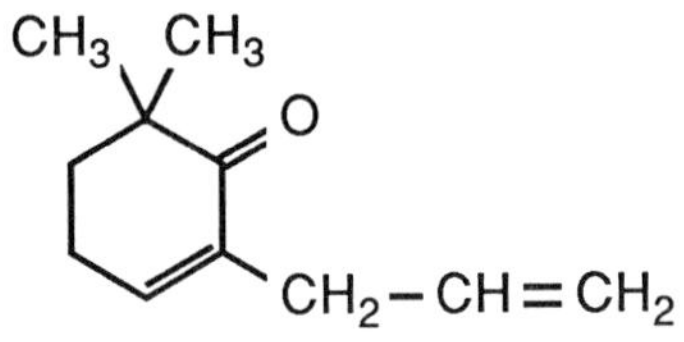

(d) 2,2-dimethylcyclohexanone ring with CH_3, CH_3, O, and $CH_2-CH=CH_2$ substituent

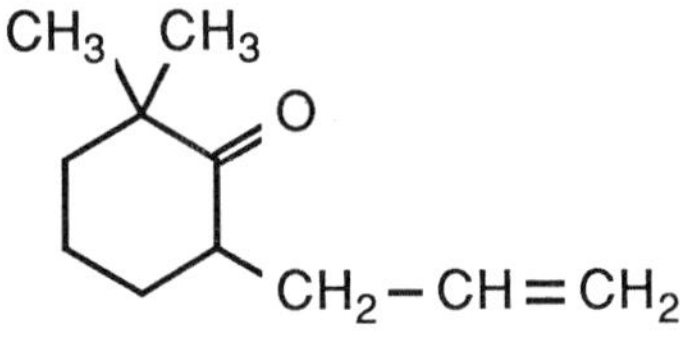

14.10 (a) 1. Li, NH_3 2. CH_3I

(b) 1. KOtBu 2. CH_3I

(c) 1. KOtBu (2 equiv) 2. CH_3I (2 equiv)

14.11

(a) cyclohexanone (O) substituted at C-3 with $C_6H_5-CH-C(=O)-C_6H_5$

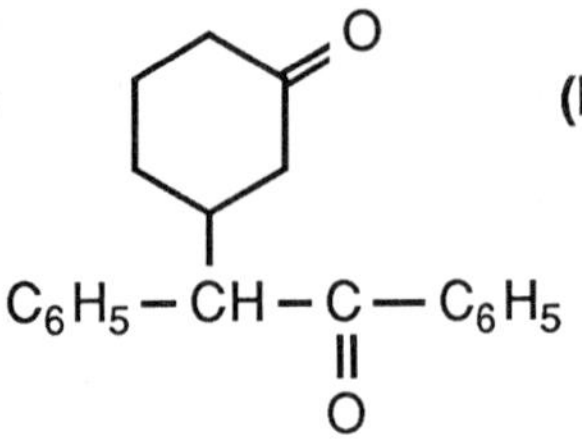

(b) cyclohexanone (O) substituted at C-2 with $CH(CH_3)-CH_2-C(=O)-C_6H_5$

(c) $CH_3-CH(\text{cyclopentan-3-one})-C(=O)-CH(CH_3)-CH_3$

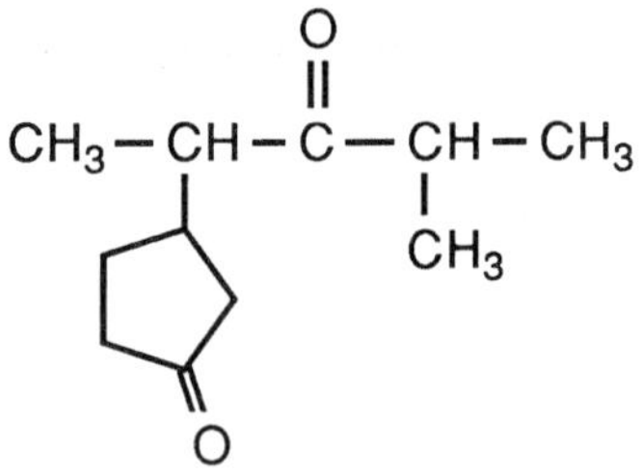

14.12

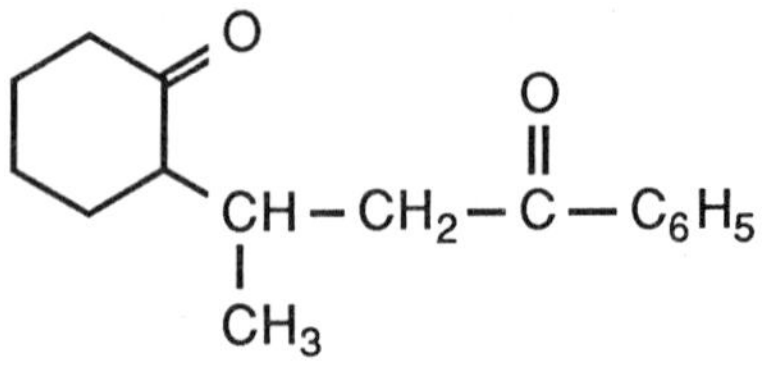

(a)

$$C_6H_5-C(=O)-CH=CH_2 \quad + \quad CH_3-CH(CH_3)-C(=O)-C_6H_5$$

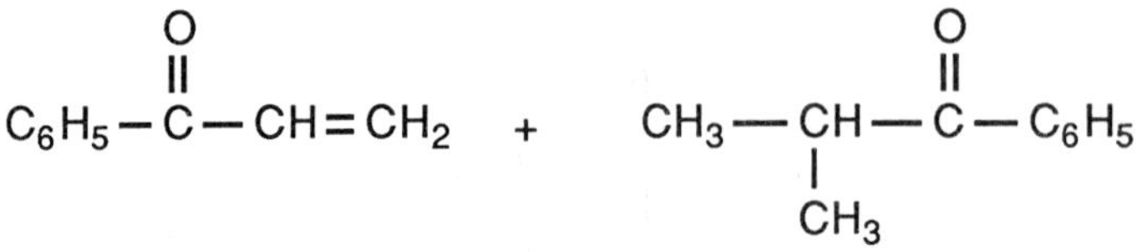

(b) 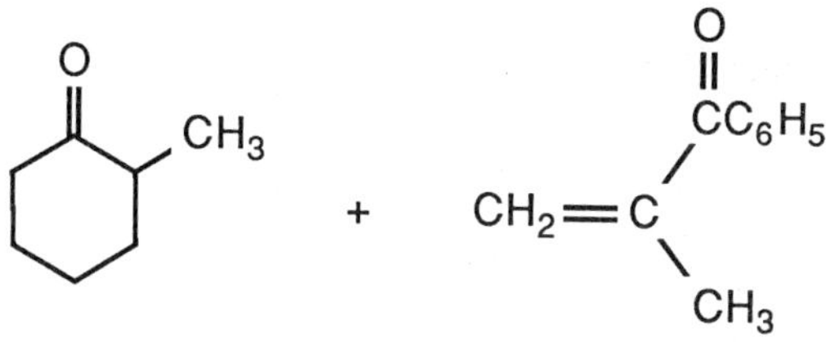 O, CH_3 + $CH_2{=}C(CH_3)\text{-}\overset{O}{\overset{\|}{C}}C_6H_5$

ANSWERS TO PROBLEMS

14.1 **(a)** alkylation

(b) aldol

(c) Michael

(d) aldol

(e) organolithium addition to ketone

(f) ylide addition to ketone

(g) alkylation

(h) Michael

(i) Conjugate addition of organocuprate (sometimes called "Michael-type" addition)

14.2

(a) 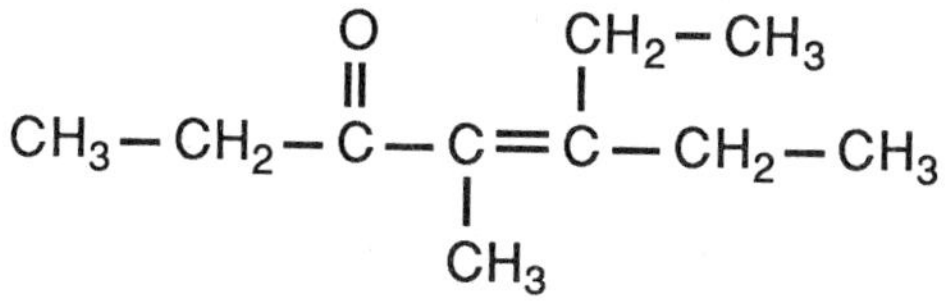 $CH_3-CH_2-\overset{O}{\overset{\|}{C}}-C(CH_3){=}C(CH_2-CH_3)-CH_2-CH_3$

(b) $CH_3-CH_2-CH_2-CH{=}C(CH_2-CH_3)-\overset{O}{\overset{\|}{C}}H$

(c) O

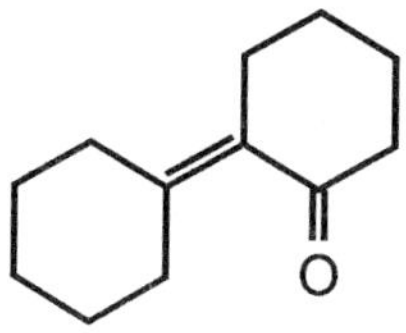

(d) 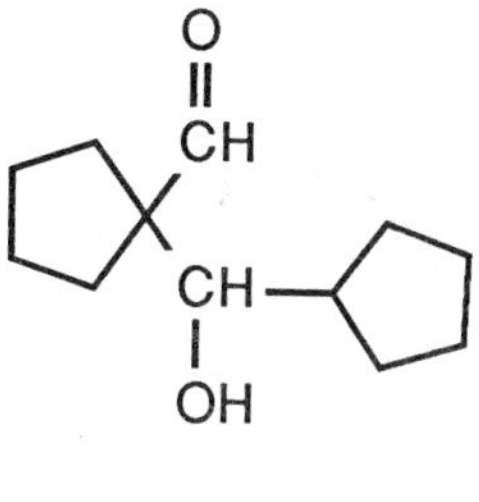 $\overset{O}{\overset{\|}{C}}H$, CH, OH

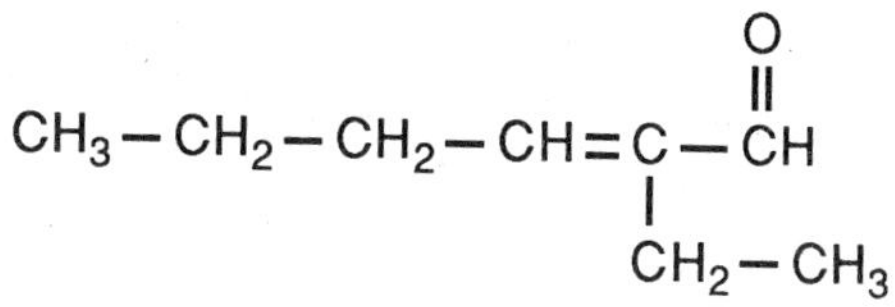

(e) 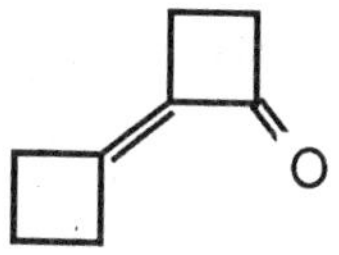 O

14.3

(a) p-$BrC_6H_4-CH{=}CH-\overset{O}{\overset{\|}{C}}-CH_3$

(b) p-$CH_3C_6H_4-CH{=}$(2-oxocyclopentylidene)

(c) 2-(propylidene)cyclohexanone: $CH-CH_2-CH_3$

(d) $C_6H_5-CH{=}CH-\overset{O}{\overset{\|}{C}}-CH_2-\underset{}{\overset{CH_3}{\overset{|}{C}H}}-CH_3$

(e) m-$ClC_6H_4-CH{=}$(2-methyl-6-oxocyclohexylidene); O, CH_3

(f) $C_6H_5-CH{=}\underset{\underset{CH_2-CH_2-CH_3}{|}}{C}-\overset{O}{\overset{\|}{C}}-CH_3$

(g) $C_6H_5-CH{=}\underset{\underset{CH_2-CH_3}{|}}{C}-\overset{O}{\overset{\|}{C}}H$

14.4 Four different enolates could be generated, corresponding to removal of a proton from the indicated sites:

$$CH_3CH_2CH_2-\boxed{CH_2}-\overset{O}{\overset{\|}{C}}-\boxed{CH_3} \quad + \quad CH_3CH_2-\boxed{CH_2}-\overset{O}{\overset{\|}{C}}-\boxed{CH_2}-CH_3$$

Each of these enolates could react with either of the two ketones and the following unsaturated ketones would result:

$$CH_3CH_2CH_2CH_2-\overset{O}{\overset{\|}{C}}-CH{=}\overset{CH_3}{\overset{|}{C}}-CH_2CH_2CH_2CH_3$$

$$CH_3CH_2CH_2CH_2-\overset{O}{\overset{\|}{C}}-CH{=}\overset{CH_2CH_3}{\overset{|}{C}}-CH_2CH_2CH_3$$

$$CH_3-\overset{O}{\overset{\|}{C}}-\underset{CH_2CH_2CH_3}{\underset{|}{C}}=\overset{CH_3}{\overset{|}{C}}-CH_2CH_2CH_2CH_3$$

$$CH_3-\overset{O}{\overset{\|}{C}}-\underset{CH_2CH_2CH_3}{\underset{|}{C}}=\overset{CH_2CH_3}{\overset{|}{C}}-CH_2CH_2CH_3$$

$$CH_3CH_2CH_2-\overset{O}{\overset{\|}{C}}-\underset{CH_3}{\underset{|}{C}}=\overset{CH_3}{\overset{|}{C}}-CH_2CH_2CH_2CH_3$$

$$CH_3CH_2CH_2-\overset{O}{\overset{\|}{C}}-\underset{CH_3}{\underset{|}{C}}=\overset{CH_2CH_3}{\overset{|}{C}}-CH_2CH_2CH_3$$

$$CH_3CH_2-\overset{O}{\overset{\|}{C}}-\underset{CH_2CH_3}{\underset{|}{C}}=\overset{CH_3}{\overset{|}{C}}-CH_2CH_2CH_2CH_3$$

$$CH_3CH_2-\overset{O}{\overset{\|}{C}}-\underset{CH_2CH_3}{\underset{|}{C}}=\overset{CH_2CH_3}{\overset{|}{C}}-CH_2CH_2CH_3$$

14.5

(a) Cyclopentanone (cyclopentane ring with C=O) $\quad C_6H_5-\overset{O}{\overset{\|}{C}}H$

(b) $C_6H_5-\overset{O}{\overset{\|}{C}}H \; + \; CH_3-\overset{O}{\overset{\|}{C}}-CH_2CH_2CH_3$

(c) $CH_3-\overset{O}{\overset{\|}{C}}-CH_3 \; + \; CH_3-\overset{O}{\overset{\|}{C}}H$

(d) $CH_3O-C_6H_4-\overset{O}{\overset{\|}{C}}H$ (para) $\; + \;$ $Cl-C_6H_4-\overset{O}{\overset{\|}{C}}-CH_3$ (para)

(e) $C_6H_5-\overset{O}{\overset{\|}{C}}-C_6H_5 \; + \; CH_3CH_2-\overset{O}{\overset{\|}{C}}-C_6H_5$

(f) $CH_3-\underset{CH_3}{\underset{|}{C}}H-CH_2-\overset{O}{\overset{\|}{C}}H \; + \; CH_3CH_2-\overset{O}{\overset{\|}{C}}-CH_2CH_3$

(g) $CH_3CH_2-\overset{O}{\overset{\|}{C}}H + C_6H_5CH_2-\overset{O}{\overset{\|}{C}}-CH_2C_6H_5$

(h) $C_6H_5-\overset{O}{\overset{\|}{C}}-CH_2CH_2CH_3$

14.6 (a) $C_6H_5-C(CH_3)_2-\overset{O}{\overset{\|}{C}}-C_6H_5$

(b) CH_3, CH_2CH_3, O

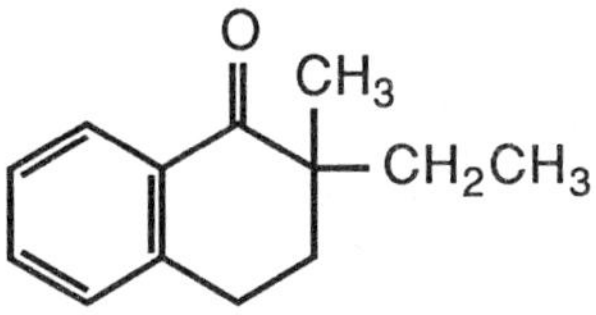

(c) CH_3, $CH_2-CH=CH_3$, O

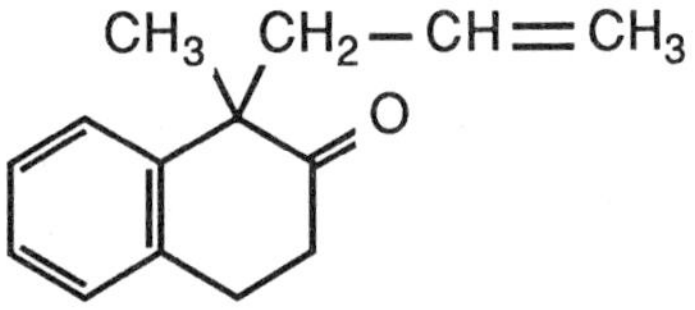

(d) $CH_3CH_2CH_2CH_2$, O + $CH_3CH_2CH_2CH_2$, $CH_3CH_2CH_2CH_2$, O

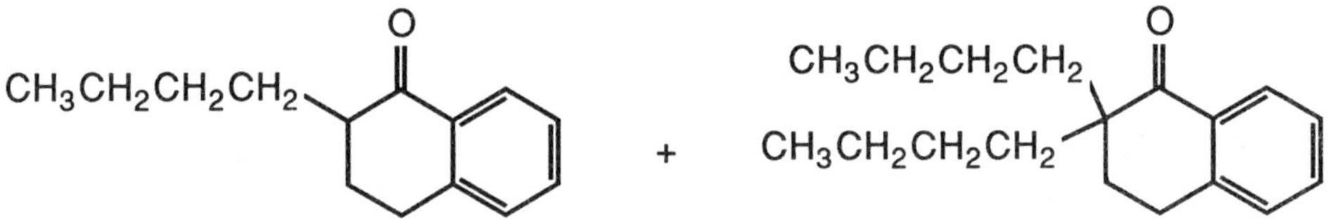

(e) O, $CH_2-CH=CH_2$, CH_3

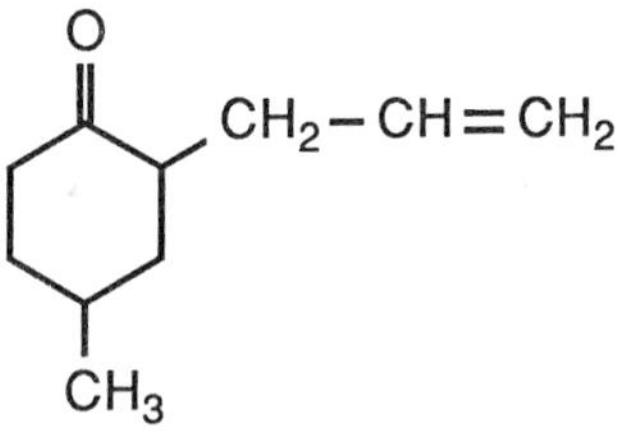

(f) CH_3, O, CH_3

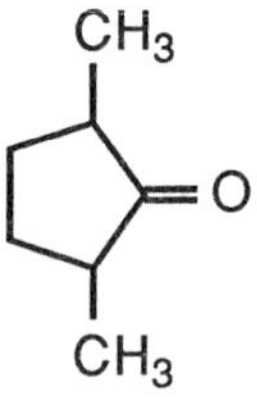

(g) CH_3, O, CH_2CH_3

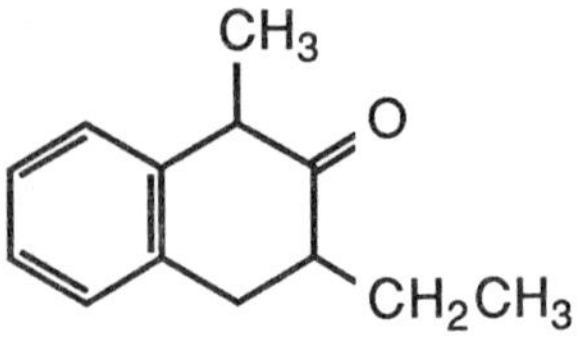

(h) O, CH_3, CH_3, CH_3 + H_3C, O, CH_3, CH_3, CH_3

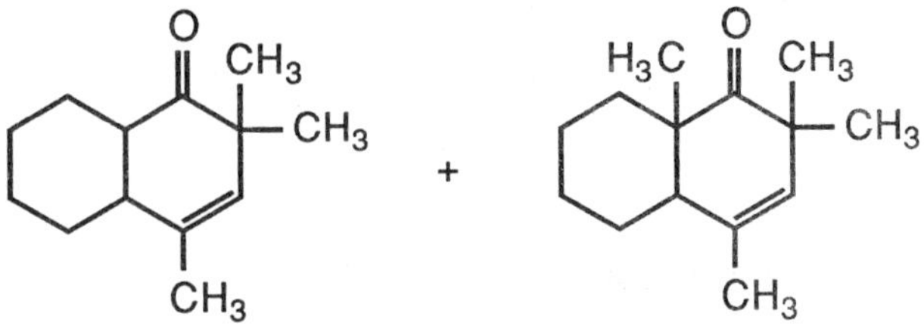

(i) CH_3, O, $CH_2CH_2CH_3$ + CH_3, O, $CH_3CH_2CH_2$, $CH_2CH_2CH_3$

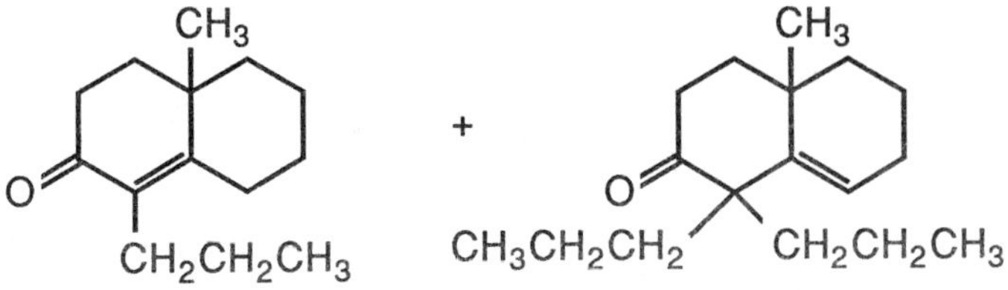

(j) O, $CH_2-CH=CH_2$

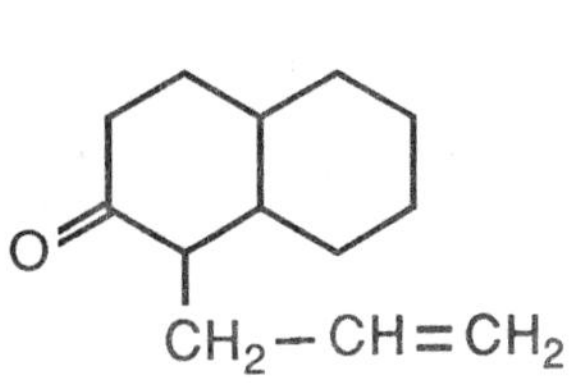

14.7 (a)

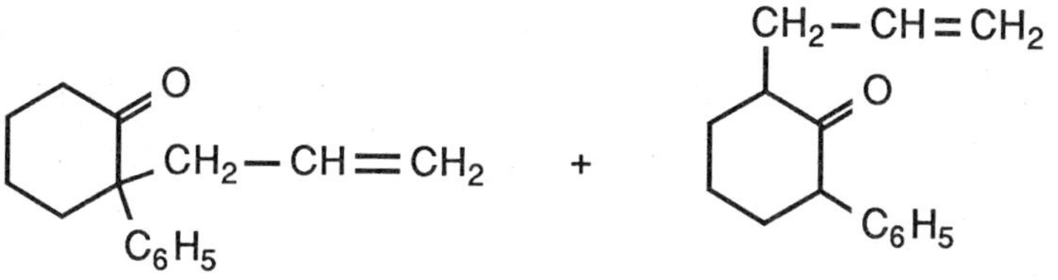

O
$CH_2-CH=CH_2$
C_6H_5

\+

$CH_2-CH=CH_2$
O
C_6H_5

\+ polyalkylation products

(b)

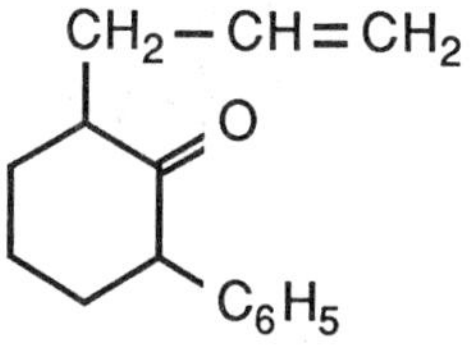

$CH_2-CH=CH_2$
O
C_6H_5

(c)

O
$CH_2-CH=CH_2$
C_6H_5

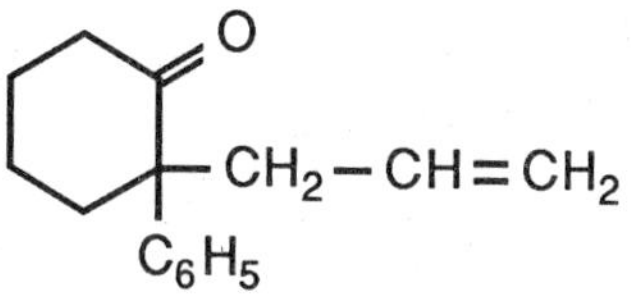

14.8 (a)

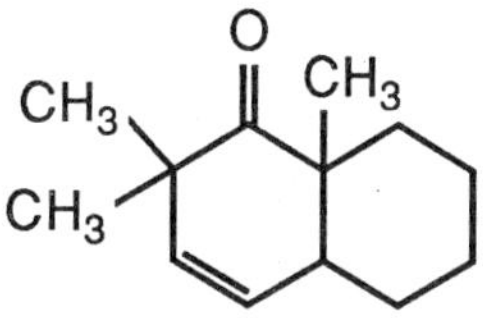

O
CH_3
CH_3
CH_3

(b)

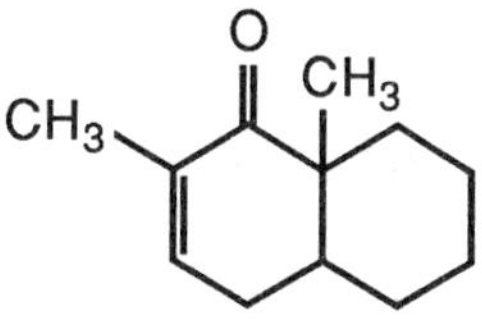

O
CH_3
CH_3

(c)

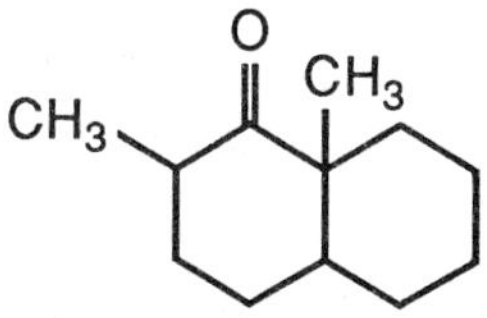

O
CH_3
CH_3

14.9

(a)

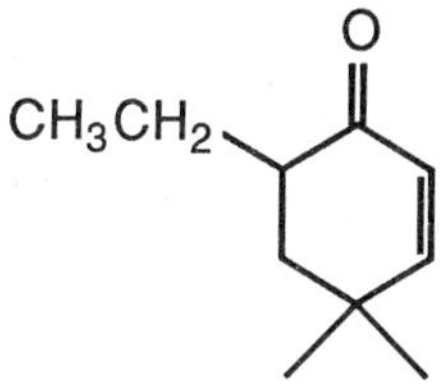

O
CH_3CH_2

(b)

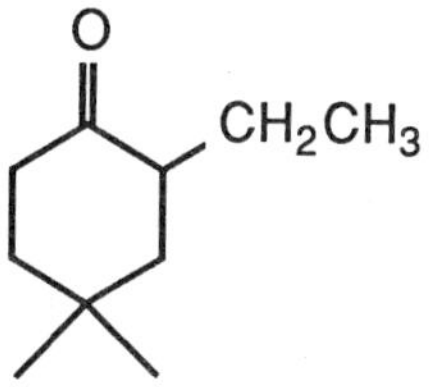

O
CH_2CH_3

(c)

O
CH_3CH_2

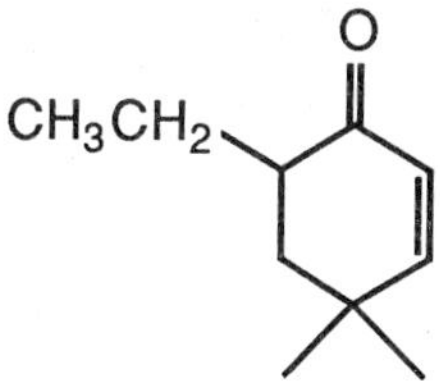

14.10 (a)

$$CH_3-\overset{\overset{\displaystyle O}{\|}}{C}-\underset{\underset{\displaystyle C_6H_5}{|}}{C}=CH_2 \quad + \quad CH_3-\overset{\overset{\displaystyle O}{\|}}{C}-CH_2CH_3$$

or

$$CH_3-\overset{\overset{\displaystyle O}{\|}}{C}-\underset{\underset{\displaystyle C_6H_5}{|}}{CH_2} \quad + \quad CH_2=CH-\overset{\overset{\displaystyle O}{\|}}{C}-CH_2CH_3$$

The second combination is better because one enolate (stabilized by the phenyl group) would be formed preferentially.

(b)

$$C_6H_5CH_2-\overset{\overset{\displaystyle O}{\|}}{C}-CH=CH_2 \quad + \quad C_6H_5CH_2-\overset{\overset{\displaystyle O}{\|}}{C}-C_6H_5$$

or

$$C_6H_5CH_2-\overset{\overset{O}{\|}}{C}-CH_3 \quad + \quad CH_2=\underset{\underset{C_6H_5}{|}}{C}-\overset{\overset{O}{\|}}{C}-C_6H_5$$

Only the first combination would work; only a single enolate is possible and it is also phenyl stabilized. The second combination would yield the wrong product via formation of a phenyl-substituted enolate.

(c)

$$CH_3-\underset{\underset{CH_3}{|}}{\overset{\overset{CH_3}{|}}{C}}-\overset{\overset{O}{\|}}{C}-CH_3 \quad + \quad CH_2=CH-\overset{\overset{O}{\|}}{CH}$$

or

$$CH_3-\underset{\underset{CH_3}{|}}{\overset{\overset{CH_3}{|}}{C}}-\overset{\overset{O}{\|}}{C}-CH=CH_2 \quad + \quad CH_3-\overset{\overset{O}{\|}}{CH}$$

Either combination should be acceptable. In each case only one enolate could form. Self condensation of acetaldehyde might be a problem for the second alternative.

14.11

(a) cyclohexanone bearing at C-2: $\underset{\underset{CH_3}{|}}{CH}-\overset{\overset{O}{\|}}{C}-C_6H_5$

(b) cyclohexanone bearing at C-2: $CH_2-CH_2-\overset{\overset{O}{\|}}{C}-C_6H_5$

(c)

$$C_6H_5-\underset{\underset{CH_3-\underset{\underset{O}{\|}}{C}-CH-C_6H_5}{|}}{CH}-CH_2-\overset{\overset{O}{\|}}{C}-C_6H_5$$

(d) cyclopentanone bearing at C-2: C_6H_5 and $CH_2-CH_2-\overset{\overset{O}{\|}}{CH}$

14.12

$$CH_3-\overset{O}{\overset{\|}{C}}-CH=\overset{CH_3}{\overset{|}{C}}-CH_3 \;\; {}^{-}OH \rightleftharpoons CH_3-\overset{O^-}{\overset{|}{C}}=CH-\overset{CH_3}{\overset{|}{\underset{OH}{\underset{|}{C}}}}-CH_3 \quad H_2O$$

$$\rightleftharpoons CH_3-\overset{O}{\overset{\|}{C}}-CH_2-\overset{CH_3}{\overset{|}{\underset{O-H}{\underset{|}{C}}}}-CH_3 \;\; {}^{-}OH$$

$$\rightleftharpoons CH_3-\overset{O}{\overset{\|}{C}}-CH_2-\overset{CH_3}{\overset{|}{\underset{O^-}{\underset{|}{C}}}}-CH_3$$

$$\rightleftharpoons CH_3-\overset{O}{\overset{\|}{C}}-CH_3 + CH_3-\overset{O}{\overset{|}{C}}=CH_2 \quad H_2O \rightleftharpoons CH_3-\overset{O}{\overset{\|}{C}}-CH_3$$

14.13

$$CH_3-\overset{\overset{\Large O:}{\|}}{C}-CH=\overset{\overset{\Large CH_3}{|}}{C}-CH_3 \;\; H^+ \rightleftharpoons CH_3-\overset{\overset{\Large \overset{+}{O}-H}{\|}}{C}-CH=\overset{\overset{\Large CH_3}{|}}{C}-CH_3 \;\; H_2O:$$

$$\rightleftharpoons$$

$$CH_3-\overset{\overset{\Large O-H}{|}}{C}=CH-\underset{\underset{\Large \overset{+}{O}H_2}{|}}{\overset{\overset{\Large CH_3}{|}}{C}}-CH_3 \;\; :OH_2$$

$$\rightleftharpoons$$

$$CH_3-\overset{\overset{\Large :O-H}{|}}{C}=CH-\underset{\underset{\Large OH}{|}}{\overset{\overset{\Large CH_3}{|}}{C}}-CH_3 \;\; H^+$$

$$\rightleftharpoons$$

$$CH_3-\overset{\overset{\Large \overset{+}{O}-H}{\|}}{C}-CH_2-\underset{\underset{\Large :OH}{|}}{\overset{\overset{\Large CH_3}{|}}{C}}-CH_3 \rightleftharpoons CH_3-\overset{\overset{\Large \overset{+}{O}-H}{\|}}{C}-CH_3 \;\; :OH_2 \;+\; CH_3-\overset{\overset{\Large :O-H}{|}}{C}=CH_2 \;\; H^+$$

$$\rightleftharpoons$$

$$CH_3-\overset{\overset{\Large O}{\|}}{C}-CH_3 \rightleftharpoons CH_3-\overset{\overset{\Large \overset{+}{O}-H}{\|}}{C}-CH_3 \;\; :OH_2$$

14.14

(a)

$$C_6H_5-CH=CH-\overset{O}{\overset{\|}{C}}-CH_3 \xrightarrow[2.\ H_2O]{1.\ Li,\ NH_3} C_6H_5-CH_2-CH_2-\overset{O}{\overset{\|}{C}}-CH_3$$

(b)

$$C_6H_5-CH=CH-\overset{O}{\overset{\|}{C}}-CH_3 \xrightarrow[2.\ H_2O]{1.\ Li,\ NH_3} C_6H_5-CH_2-CH_2-\overset{O}{\overset{\|}{C}}-CH_3 \xrightarrow[KOH]{NH_2NH_2} C_6H_5-CH_2-CH_2-CH_2-CH_3$$

(c)

$$C_6H_5-CH=CH-\overset{O}{\overset{\|}{C}}-CH_3 \xrightarrow{NaBH_4} C_6H_5-CH=CH-\overset{OH}{\overset{|}{C}H}-CH_3 \xrightarrow{PBr_3} C_6H_5-CH=CH-\overset{Br}{\overset{|}{C}H}-CH_3$$

$$C_6H_5-CH=CH-\overset{Br}{\overset{|}{C}H}-CH_3 \xrightarrow[tBuOH]{KOtBu} C_6H_5-CH=CH-CH=CH_2$$

(d)

$$C_6H_5-CH=CH-\overset{O}{\overset{\|}{C}}-CH_3 \xrightarrow{NaBH_4} C_6H_5-CH=CH-\overset{OH}{\overset{|}{C}H}-CH_3$$

(e)

$$C_6H_5-CH=CH-\overset{O}{\overset{\|}{C}}-CH_3 \xrightarrow[2.\ H_2O]{1.\ Li,\ NH_3} C_6H_5-CH_2-CH_2-\overset{O}{\overset{\|}{C}}-CH_3 \xrightarrow[KOH]{NH_2NH_2} C_6H_5-CH_2-CH_2-CH_2-CH_3$$

$$C_6H_5-CH_2-CH_2-CH_2-CH_3 \xrightarrow{NBS} C_6H_5-\overset{Br}{\overset{|}{C}H}-CH_2CH_2CH_3$$

(f)

$$C_6H_5-CH=CH-\overset{O}{\overset{\|}{C}}-CH_3 \xrightarrow[2.\ H_2O]{1.\ (CH_3)_2CuLi} C_6H_5-\overset{CH_3}{\overset{|}{C}H}-CH_2-\overset{O}{\overset{\|}{C}}-CH_3$$

(g)

$$C_6H_5-CH=CH-\overset{\overset{\displaystyle O}{\|}}{C}-CH_3 \xrightarrow[\text{2. } CH_3I]{\text{1. } LiN(iPr)_2} C_6H_5-CH=CH-\overset{\overset{\displaystyle O}{\|}}{C}-CH_2-CH_3$$

$$\xrightarrow[\text{2. } H_2O]{\text{1. Li, } NH_3} C_6H_5-CH_2-CH_2-\overset{\overset{\displaystyle O}{\|}}{C}-CH_2-CH_3$$

(h)

$$C_6H_5-CH=CH-\overset{\overset{\displaystyle O}{\|}}{C}-CH_3 \xrightarrow[\text{2. } H_2O]{\text{1. Li, } NH_3} C_6H_5-CH_2-CH_2-\overset{\overset{\displaystyle O}{\|}}{C}-CH_3 \xrightarrow[NaOH]{I_2} C_6H_5-CH_2-CH_2-\overset{\overset{\displaystyle O}{\|}}{C}-OH$$

(i)

$$C_6H_5-CH=CH-\overset{\overset{\displaystyle O}{\|}}{C}-CH_3 \xrightarrow[C_6H_5CH_2-\overset{\overset{\displaystyle O}{\|}}{C}-C_6H_5]{LiN(iPr)_2} \begin{array}{l} C_6H_5-CH-CH_2-\overset{\overset{\displaystyle O}{\|}}{C}-CH_3 \\ \qquad\;\; | \\ C_6H_5-CH-\underset{\underset{\displaystyle O}{\|}}{C}-C_6H_5 \end{array}$$

14.15 (a)

$$\text{2-cyclopentenone} \xrightarrow[\text{2. } H_2O]{\text{1. Li, } NH_3} \text{cyclopentanone}$$

(b)

$$\text{2-cyclopentenone} \xrightarrow{NaBH_4} \text{2-cyclopenten-1-ol}$$

(c) O= cyclopentenone $\xrightarrow[2.\ H_2O]{1.\ Li,\ NH_3}$ cyclopentanone (O) $\xrightarrow[KOH]{NH_2NH_2}$ cyclopentane

(d) O= cyclopentenone $\xrightarrow[2.\ H_2O]{1.\ (CH_3)_2CuLi}$ 3-methylcyclopentanone (O, CH_3)

(e) O= cyclopentenone $\xrightarrow[2.\ CH_3I]{1.\ Li,\ NH_3}$ 2-methylcyclopentanone (O, CH_3)

(f) O= cyclopentenone $\xrightarrow[2.\ H_2O]{1.\ CH_3Li}$ HO, CH_3 (1-methylcyclopent-2-en-1-ol)

(g) O= cyclopentenone $\xrightarrow[2.\ H_2O]{1.\ Li,\ NH_3}$ cyclopentanone (O) $\xrightarrow[2.\ H_2O]{1.\ CH_3MgBr}$ HO, CH_3 (1-methylcyclopentanol)

(h) O= cyclopentenone $\xrightarrow[2.\ H_2O]{1.\ Li,\ NH_3}$ cyclopentanone (O) $\xrightarrow{(C_6H_5)_3P{=}CHCH_3}$ ethylidenecyclopentane (CH_3)

(i) O= cyclopentenone $\xrightarrow[2.\ H_2O]{1.\ (CH_3)_2CuLi}$ 3-methylcyclopentanone (O, CH_3) $\xrightarrow{(C_6H_5)_3P{=}CH_2}$ 1-methylene-3-methylcyclopentane (CH_2, CH_3)

(j) 2-cyclopentenone $\xrightarrow[\text{2. } CH_3I]{\text{1. Li, } NH_3}$ 2-methylcyclopentanone $\xrightarrow{(C_6H_5)_3P{=}CH_2}$ 1-methylene-2-methylcyclopentane

(k) 2-cyclopentenone $\xrightarrow[\text{2. } CH_3I]{\text{1. LiN(iPr)}_2}$ 5-methyl-2-cyclopentenone

(l) 2-cyclopentenone $\xrightarrow[\text{2. } CH_3I]{\text{1. Li, } NH_3}$ 2-methylcyclopentanone $\xrightarrow[\text{2. } CH_3I]{\text{1. LiN(iPr)}_2}$ 2,5-dimethylcyclopentanone

(m) 2-cyclopentenone $\xrightarrow[\text{2. } H_2O]{\text{1. Li, } NH_3}$ cyclopentanone $\xrightarrow[\text{excess } CH_3I]{\text{KOtBu}}$ 2,2,5,5-tetramethylcyclopentanone

(n) 2-cyclopentenone $\xrightarrow{(C_6H_5)_3P{=}CH_2}$ 3-methylenecyclopentene $\xrightarrow{H_2,\ Pt}$ methylcyclopentane

14.16 Equilibrium will be established under more vigorous conditions, that is at higher temperature. The 25°C reaction therefore affords the product of kinetic control. The 80°C product (which has a conjugated double bond and is more stable) results from thermodynamic control:

$$CH_3-\overset{O}{\overset{\|}{C}}H + CH_3-\overset{CH_3}{\overset{|}{C}}H-\overset{O}{\overset{\|}{C}}H \rightleftharpoons CH_3-\overset{OH}{\overset{|}{C}}H-\underset{CH_3}{\underset{|}{\overset{CH_3}{\overset{|}{C}}}}-\overset{O}{\overset{\|}{C}}H \quad 25°C$$

$$CH_3-\overset{O}{\overset{\|}{C}}H + CH_3-\overset{CH_3}{\overset{|}{C}}H-\overset{O}{\overset{\|}{C}}H \rightleftharpoons CH_3-\overset{CH_3}{\overset{|}{C}}H-CH{=}CH-\overset{O}{\overset{\|}{C}}H \quad 80°C$$

14.17 Only one enolate could from from this unsymmetrical ketone:

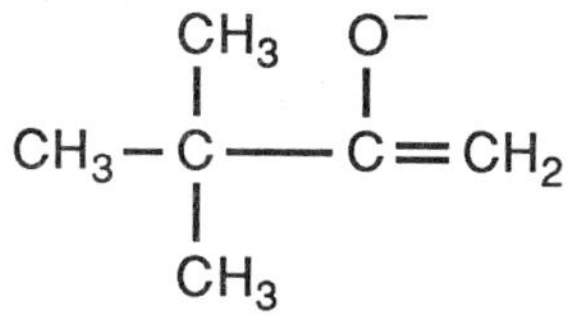

$$CH_3-C(CH_3)_2-C(O^-)=CH_2$$

Consequently, either potassium *tert*-butoxide or sodium amide would be satisfactory. Neither of these would cause major problems from reaction with methyl iodide: *tert*-butoxide ion is sterically hindered, and the amide ion is almost totally converted to ammonia in the acid-base equilibrium with the ketone. Sodium methoxide, on the other hand, would react as a nucleophile:

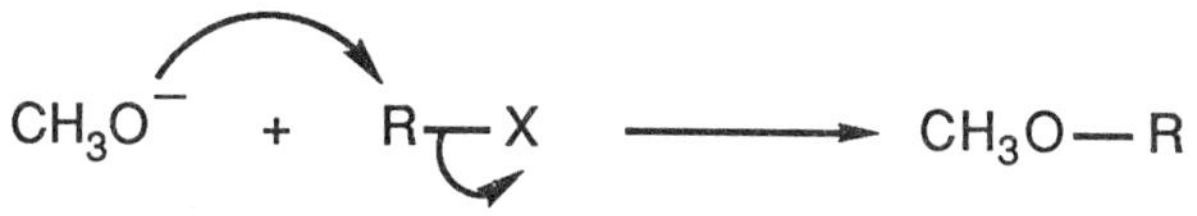

$$CH_3O^- + R-X \longrightarrow CH_3O-R$$

14.18 The interconversion is illustrated here for cyclohexenone:

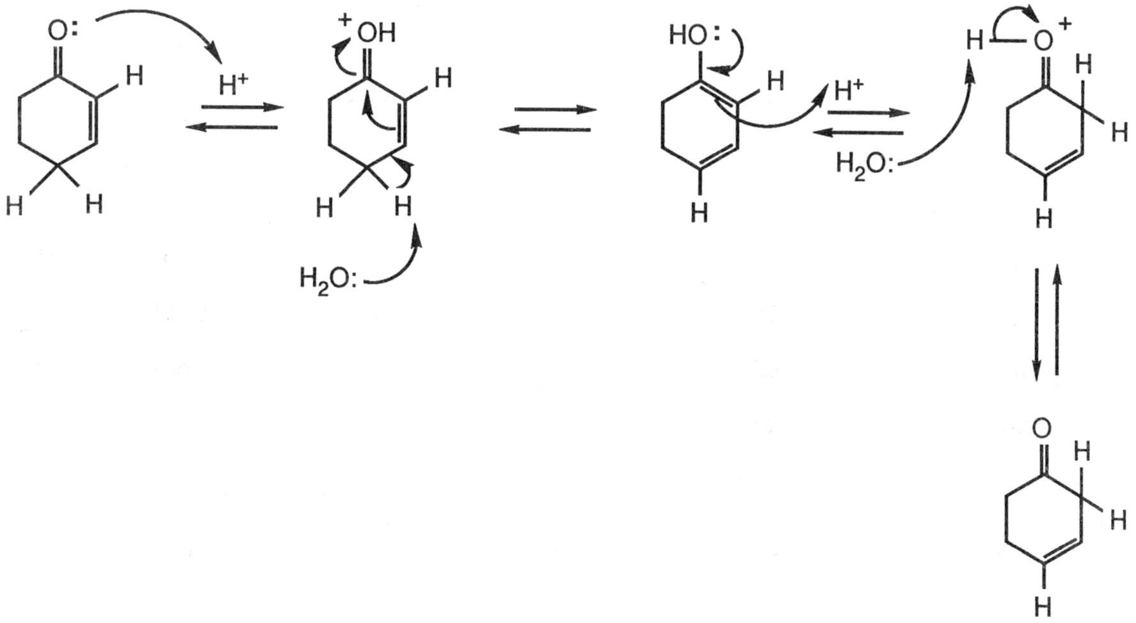

H⁺ ⇌ ⇌ H⁺ ⇌ ⇌

H_2O: H_2O:

14.19

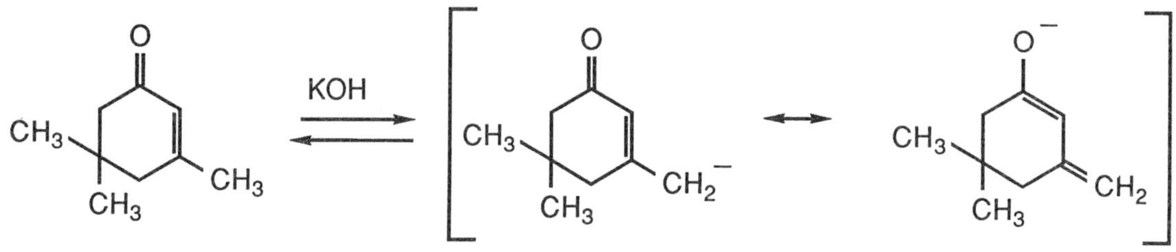

A

KOH

CH_3 CH_3 CH_3 CH_2^- CH_2 O O^-

14.20 (a)

CH_3-C-CH_3 (C=O:) ⇌ $CH_3-C(=\overset{+}{O}H)-CH_2-H$ ($:OH_2$) ⇌ $CH_3-C(:OH)=CH_2$ + $CH_3-C(=\overset{+}{O}H)-CH_3$

H^+ H_2O: H^+

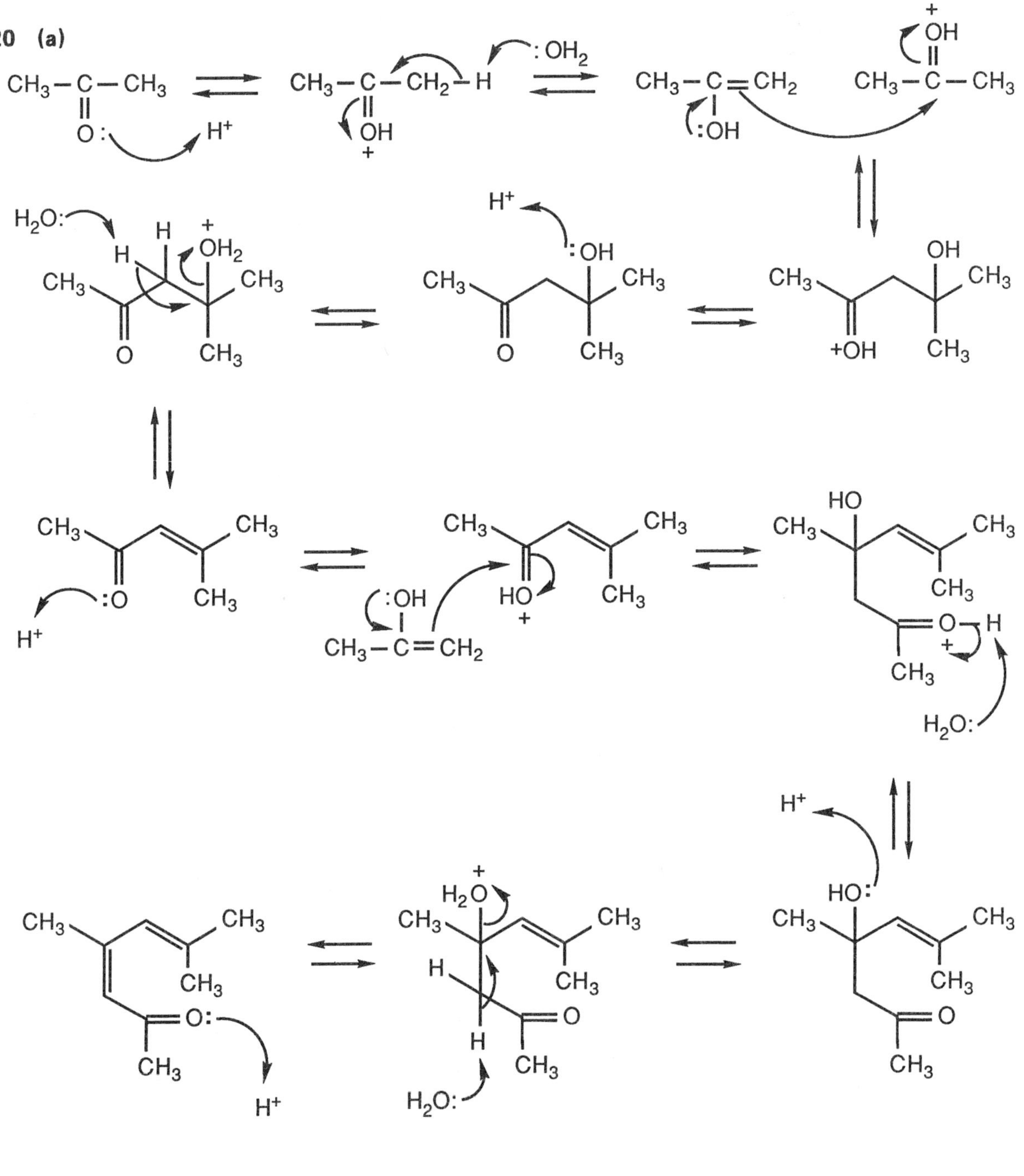

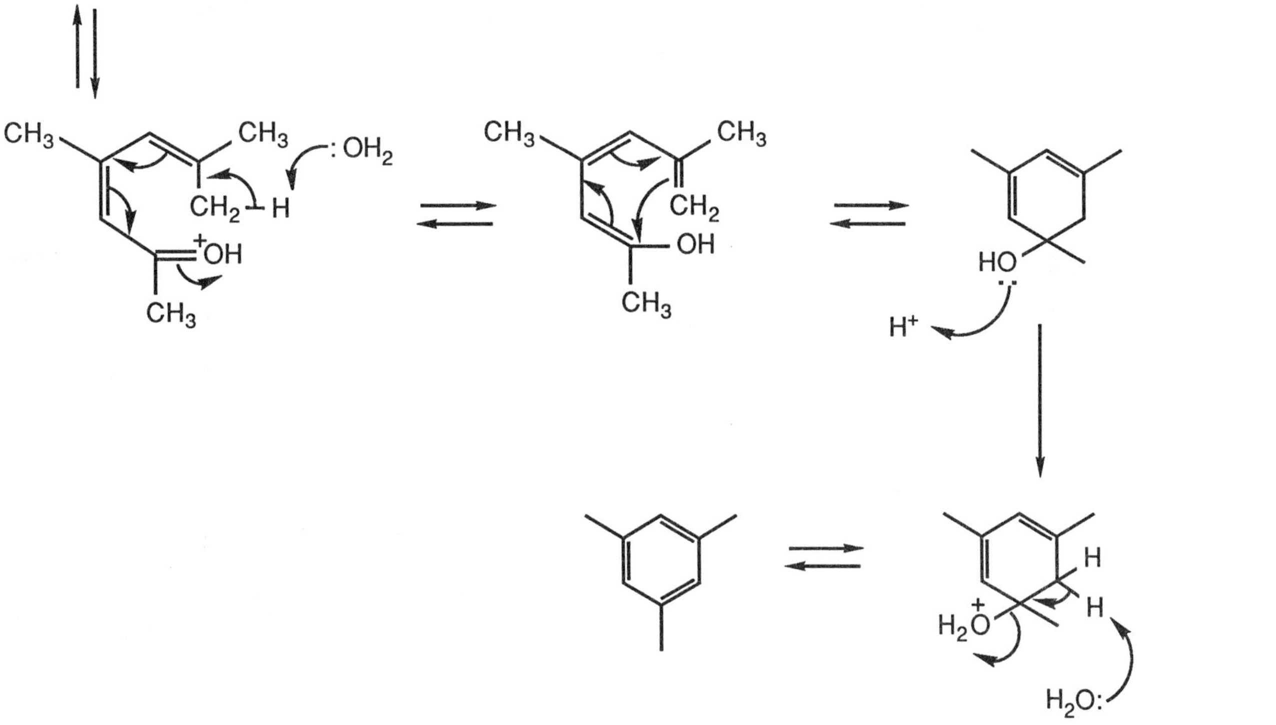
CH3
CH3
: OH2
CH2
H
+OH
CH3
CH3
CH3
CH2
OH
CH3
HO
H+
H
H
H2O+
H2O:

(b)

CH_3 CH_2 H ^-OH CH_3 CH_2 O H_2O CH_3-C-CH_3 ^-OH H H CH_3 CH_3 OH

CH_3 OH CH_3 H CH_3 O H CH_3 HO^- CH_3 CH_3 O CH_3 H_2O CH_3 CH_2

CH_3 CH_3 CH_3 O CH_3 ≡ CH_3 CH_3 CH_3 O CH_2 H ^-OH CH_3 CH_3 CH_3 O^-

CH_3 CH_3 CH_3 O H_2O CH_3 CH_3 CH_3 O^-

14.21

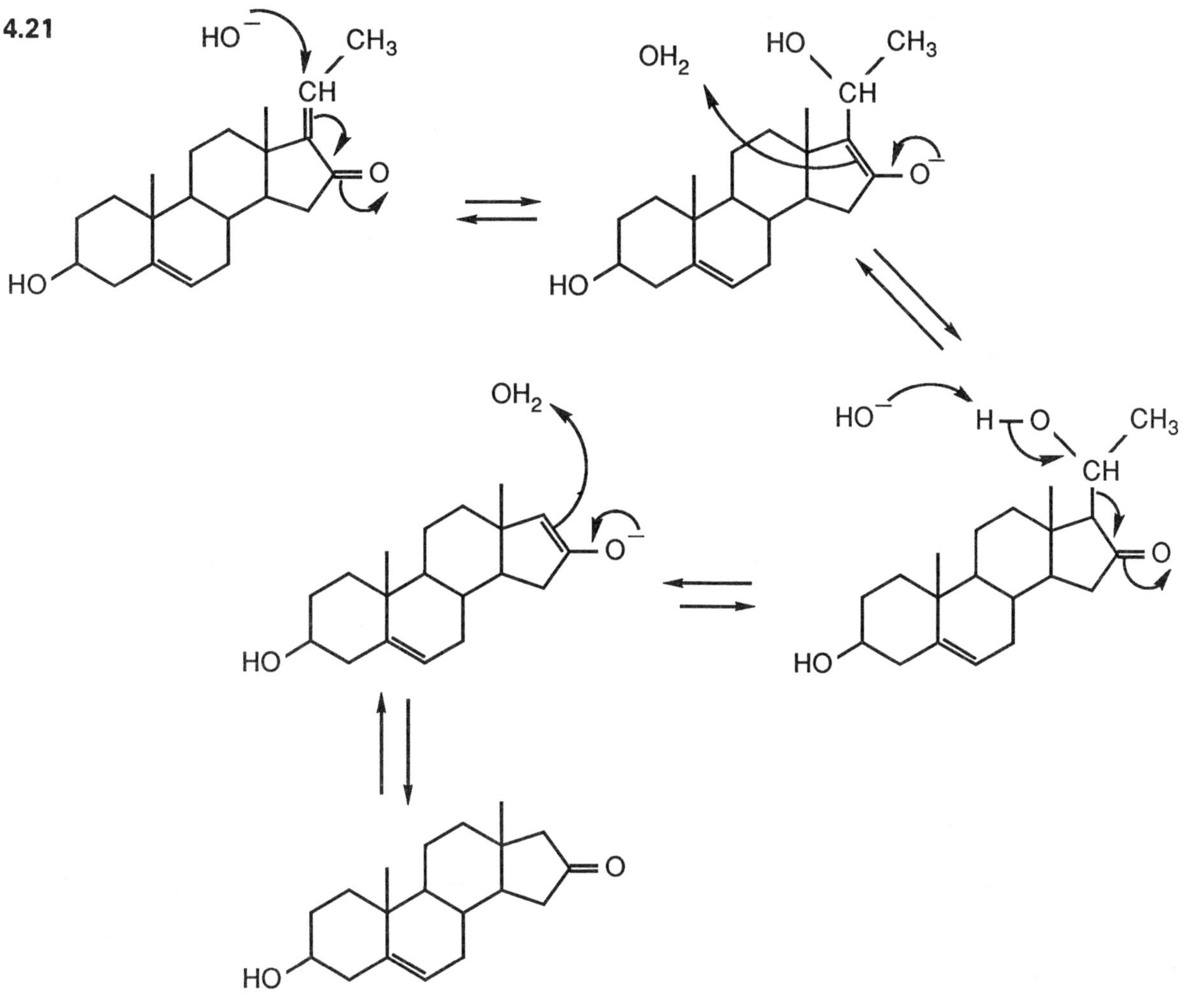

HO⁻ CH₃ CH O ⇌ OH₂ HO CH₃ CH O⁻ HO

HO⁻ H–O CH₃ CH O HO ⇌ OH₂ O⁻ HO ⇌ O HO

14.22 Under acidic conditions dehydration of the β-hydroxy ketone occurs readily:

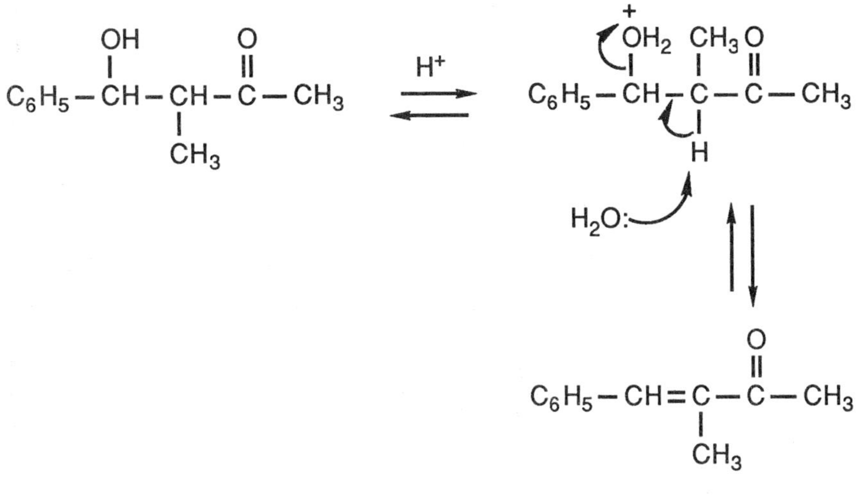

$$C_6H_5-CH(OH)-CH(CH_3)-C(=O)-CH_3 \underset{}{\overset{H^+}{\rightleftharpoons}} C_6H_5-CH(\overset{+}{O}H_2)-C(H)(CH_3)-C(=O)-CH_3$$

H_2O:

$$\rightleftharpoons C_6H_5-CH=C(CH_3)-C(=O)-CH_3$$

In base, however, a retroaldol condensation can occur to produce free benzaldehyde:

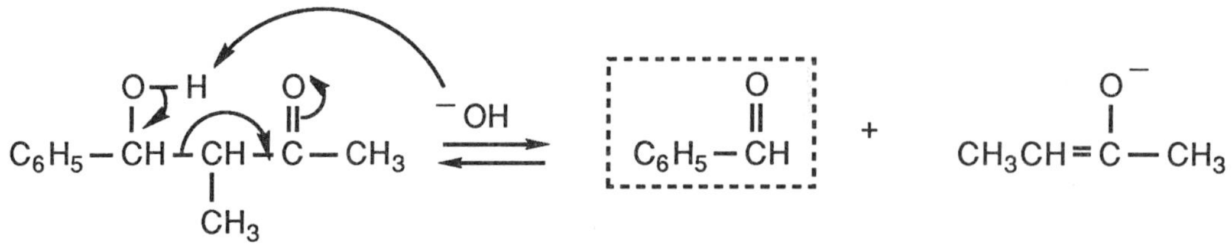

$$C_6H_5-CH(OH)-CH(CH_3)-C(=O)-CH_3 \xrightleftharpoons{^-OH} C_6H_5-CH(=O) + CH_3CH=C(O^-)-CH_3$$

The benzaldehyde can next react with the excess 2-butanone:

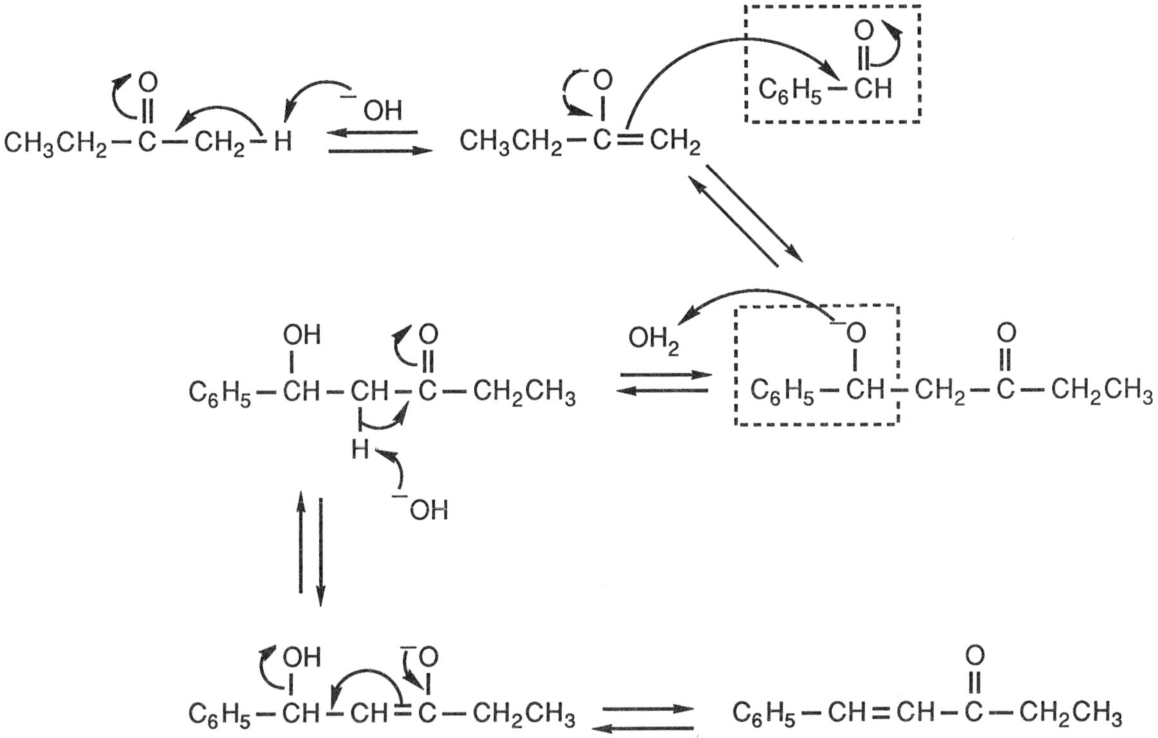

$$CH_3CH_2-C(=O)-CH_2-H \xrightleftharpoons{^-OH} CH_3CH_2-C(O^-)=CH_2 \;\; + \;\; C_6H_5-CH(=O)$$

$$\rightleftharpoons C_6H_5-CH(O^-)-CH_2-C(=O)-CH_2CH_3 \xrightleftharpoons{OH_2} C_6H_5-CH(OH)-CH_2-C(=O)-CH_2CH_3 \;\; (^-OH)$$

$$\rightleftharpoons C_6H_5-CH(OH)-CH=C(O^-)-CH_2CH_3 \rightleftharpoons C_6H_5-CH=CH-C(=O)-CH_2CH_3$$

14.23

(a)

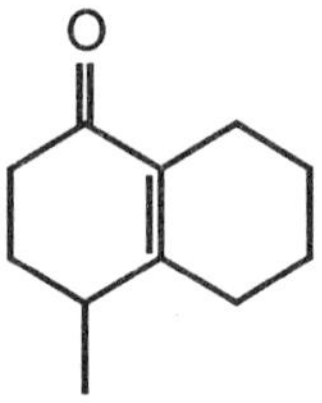

(b)

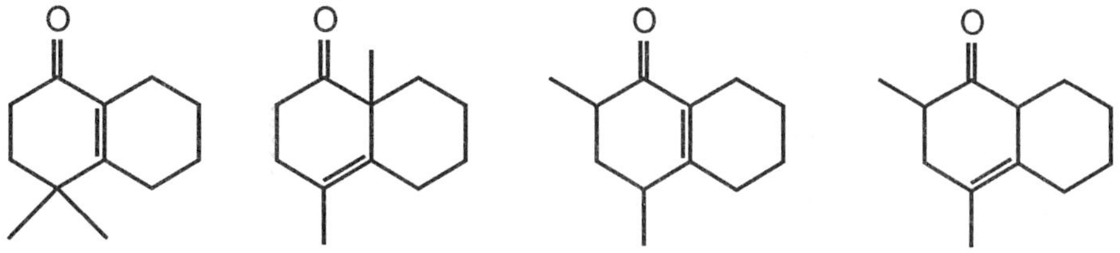

(c)

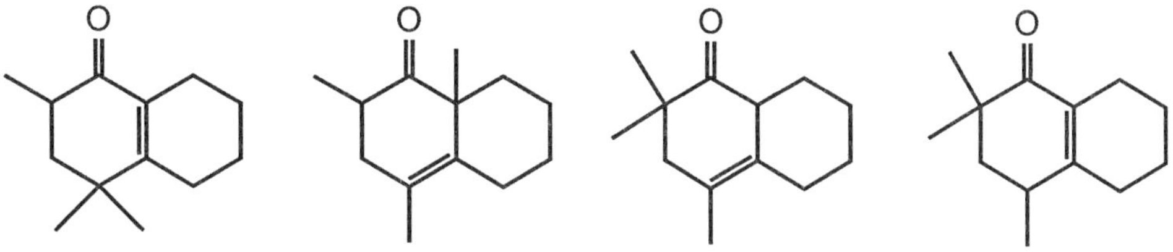

(d)

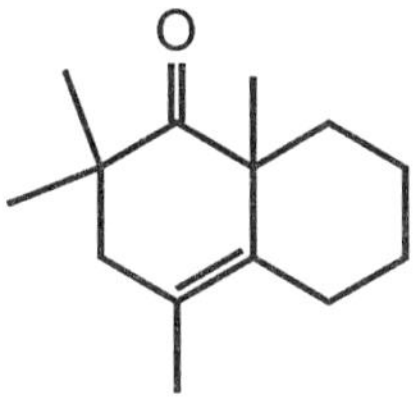

14.24

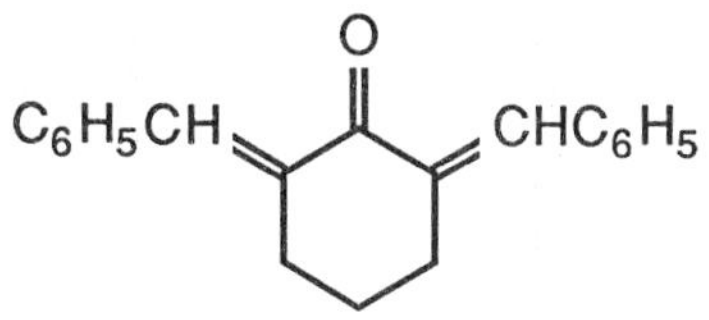

C_6H_5CH CHC_6H_5

14.25 In the reaction of 2-butanone with potassium *tert*-butoxide and methyl iodide you should expect alkylation to occur. This could occur at two different α carbons, and polyalkylation could occur as well. To interpret the NMR spectra of the products, you should first realize how the spectrum of the reactant would appear. 2-Butanone is methyl ethyl ketone, and the methyl group would afford a singlet near 2 ppm. The ethyl group would give rise to a Triplet near 2.5 ppm (CH_2) and a quartet near 1 ppm (CH_3). Compound **A** shows only the two peaks corresponding to an ethyl group, so alkylation must have occurred at the original isolated methyl group to produce a second CH_3CH_2:

$$CH_3-CH_2-\overset{\overset{\displaystyle O}{\|}}{C}-CH_2-CH_3$$

Compound **B** shows none of the peaks corresponding to 2-butanone, and only signals corresponding to isopropyl are seen. Both α positions must have been alkylated accordingly:

$$CH_3-\underset{\underset{\displaystyle CH_3}{|}}{CH}-\overset{\overset{\displaystyle O}{\|}}{C}-\underset{\underset{\displaystyle CH_3}{|}}{CH}-CH_3$$

Compound **C** shows a singlet at 2 ppm that would be expected for CH_3CO-, but a larger singlet appears at 1 ppm. This is the correct position for a methyl on a saturated carbon atom, but it is not split. This must be a consequence of three equivalent methyl groups:

$$CH_3-\overset{\overset{\displaystyle O}{\|}}{C}-\overset{\overset{\displaystyle CH_3}{|}}{\underset{\underset{\displaystyle CH_3}{|}}{C}}-CH_3$$

14.26 The expected products for treatment of acetone with base are those of aldol condensations. Condensation can occur at either (or both) of the two equivalent α positions. Compound **A** shows two singlets (one larger than the other) near 2 ppm, which correspond to methyl groups attached to a C=C or C=O bond. In addition there is a small peak at 6 ppm (the vinyl region). This compound is the result of a single aldol condensation followed by dehydration:

$$CH_3-\overset{\overset{\displaystyle CH_3}{|}}{C}=CH-\overset{\overset{\displaystyle O}{\|}}{C}-CH_3$$

The NMR spectrum for compound **B** is very similar to that for **A,** except that the methyl signals at 2 ppm are equal in intensity. This compound must have resulted from reaction at each of the α carbons:

$$CH_3-\overset{\overset{\displaystyle CH_3}{|}}{C}=CH-\overset{\overset{\displaystyle O}{\|}}{C}-CH=\overset{\overset{\displaystyle CH_3}{|}}{C}-CH_3$$

Why are there two different methyl peaks? The two methyl groups of a $C{=}C(CH_3)_2$ subgroup are *not* equivalent; one is *cis* to the C=O substituent and the other is *trans*. This in turn raises another question: why were there only two methyl signals for compound **A**? By coincidence one of the two methyl groups of the $C{=}C(CH_3)_2$ group, results in a peak at almost the identical position as that from the CH_3CO- group. The large singlet at 2.2 ppm for **A** therefore results from these two overlapping signals. The spectrum for compound **C** shows the peak at 2.2 ppm that corresponds to CH_3CO-, but there is no absorption for a $C{=}C(CH_3)_2$ group. The larger singlet at 1.2 ppm fits for two equivalent methyl groups, and the decreasing size of the peaks at 2.2, 2.6, and 3.7 ppm corresponds to integration for 3H, 2H, and 1H, respectively. All together these various peaks suggest subunits of CH_3CO-, $C(CH_3)_2$, $-CH_2-$, and $-OH$. This fits for the aldol product that has not undergone dehydration:

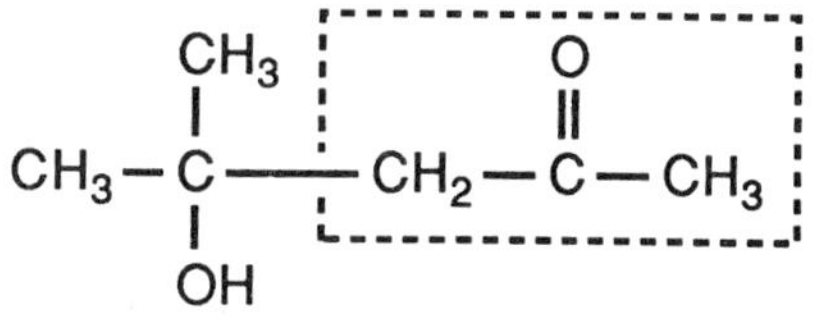

$$CH_3-C(CH_3)(OH)-CH_2-C(=O)-CH_3$$

14.27 The spectrum of this compound shows a 3H singlet at 2.1 ppm, corresponding to a methyl group bonded to C=C or C=O. There is a 1H singlet at 9.5 ppm corresponding to CHO, and there are six hydrogens in the vinyl-aromatic region between 7.2 and 7.5 ppm. Two possible sets of fragments can be considered, the first of which is:

$-CHO$, $-C_6H_4-$, $-CH{=}CH-$ (or $C{=}CH_2$), and CH_3

The only way that these could be combined so that both the aldehyde and methyl groups would not be coupled to other hydrogens are:

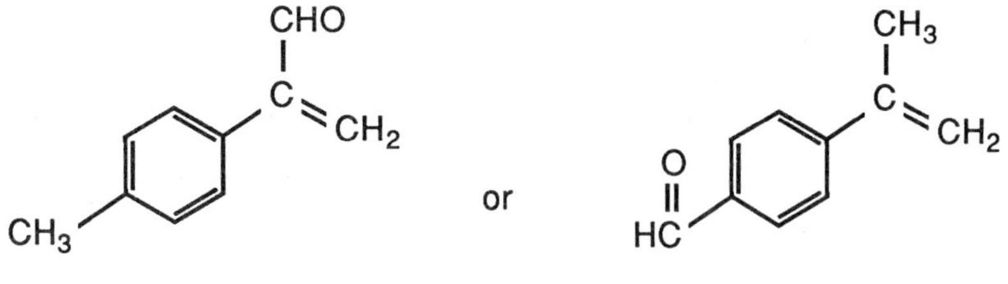

$$CH_3-C_6H_4-C(CHO){=}CH_2 \quad \text{or} \quad HC(=O)-C_6H_4-C(CH_3){=}CH_2$$

But neither of these corresponds to the product of an aldol condensation.

The other possible set of fragments involves a monosubstituted aromatic ring:

-CHO, $-C_6H_5$, C=CH, and CH_3

Three of these fragments are "ends", and the absence of coupling precludes attachment of either CH_3 or CHO to the CH end of the C=C group. Therefore the structure must be:

$$C_6H_5-CH=C(CH_3)-\overset{O}{\overset{\|}{C}}H$$

14.28 There are two types of methyl groups in the reactants for this problem, the CH_3 groups of acetone and the CH_3 group of 4-methylbenzaldehyde. Only one type of methyl appears in the product (the singlet at 2.4 ppm), so it seems that reaction has occurred at *both* α carbons of acetone. Dehydration would then result in the following compound, with all remaining signals appearing in the aromatic-vinyl region at 6.9-7.9 ppm.

$$CH_3-C_6H_4-CH=CH-\overset{O}{\overset{\|}{C}}-CH=CH-C_6H_4-CH_3$$

14.29 In contrast to many of the earlier spectroscopic problems that we have presented, infrared spectroscopy would be of little value in assigning structures to the products of these reactions. The major functionality (carbonyl groups and aromatic rings) is unchanged in the reactions, so only subtle changes in the infrared spectra would be expected. Of course, an infrared spectrum would still be very valuable in demonstrating that no major side reaction (such as reduction of a carbonyl group) had occurred. NMR spectroscopy would provide the most useful information. ^{13}C spectra would of course be highly informative, because the total number of signals would tell you the number of nonequivalent carbon atoms (and therefore how many new alkyl groups had been introduced). Nevertheless, within the context of this book the most reliable chemical shift dependency on structure involves proton NMR, so we will focus our answer on that method.

(a) Only one enolate ion is possible, so alkylation would convert a CH–CH_3 to a $C(CH_3)_2$ with two equivalent methyl groups:

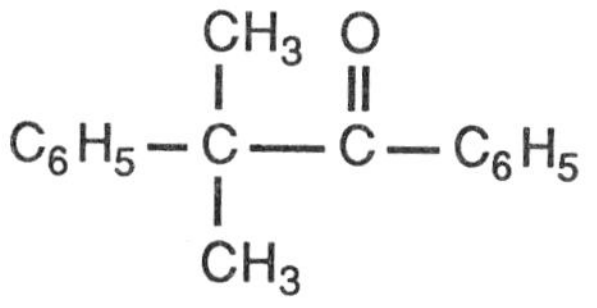

CH_3 O

$C_6H_5-C-C-C_6H_5$

CH_3

The original methyl doublet near 1 ppm would become a singlet with twice the intensity, and the 1H multiplet (in the 4–5 ppm region) would disappear.

(b) Again only one enolate ion is possible, so alkylation would convert a CH–CH_3 to a $C(CH_3)CH_2CH_3$:

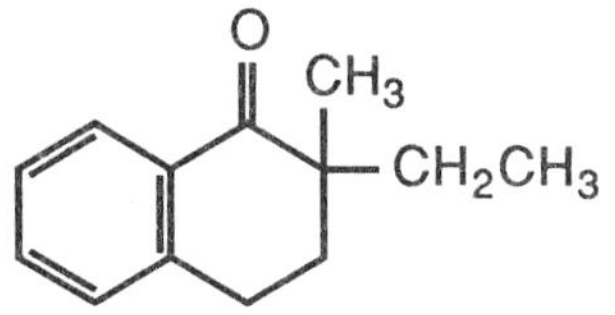

O CH_3

CH_2CH_3

The original methyl doublet near 1 ppm would become a singlet and an ethyl pattern (3H triplet near 1 ppm and 2H quartet near 1.7 ppm) would be observed. The original 1H multiplet in the 4–5 ppm region would disappear.

(c) Two enolates could be formed, but the expected product with sodium hydride (equilibrating conditions) would result from the more stable (conjugated) enolate.

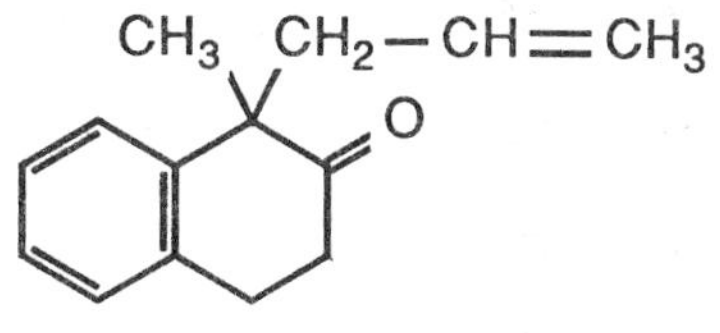

CH_3 $CH_2-CH=CH_3$

O

Again, the key features would be a change in the methyl signal from a double to a singlet together with appearance of the appropriate signals for the allyl group, $-CH_2-CH=CH_2$. If alkylation were to have occurred at the less substituted α carbon, this change in the methyl signal would not occur.

(d) Only one enolate ion is possible, but the aliphatic region of the spectra for both reactant and product would be complex:

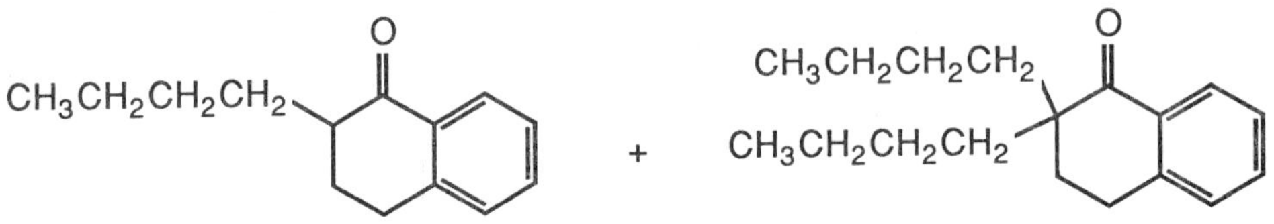

$CH_3CH_2CH_2CH_2$ + $CH_3CH_2CH_2CH_2$ $CH_3CH_2CH_2CH_2$

The most useful way to distinguish between mono- and disubstitution would be via integration of the methyl region (near 1 ppm) relative to the aromatic region.

(e) Because of symmetry only one enolate is possible.

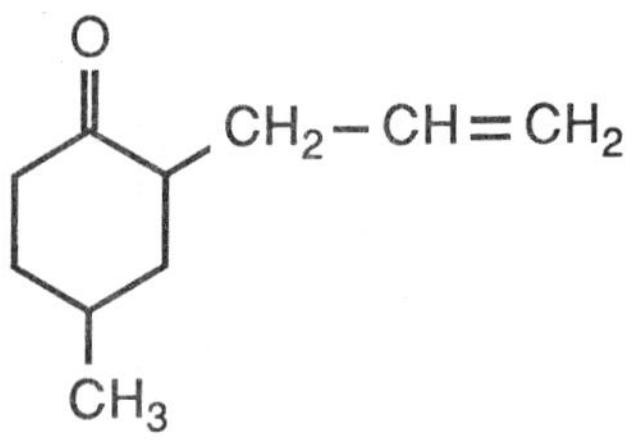

$CH_2-CH=CH_2$

CH_3

The potential problem of di- or polysubstitution could be analyzed by integration of the vinyl region relative to the doublet for the methyl group attached to the ring.

(f) Two enolates are possible, but the expected product is 2,5-dimethylcyclopentanone:

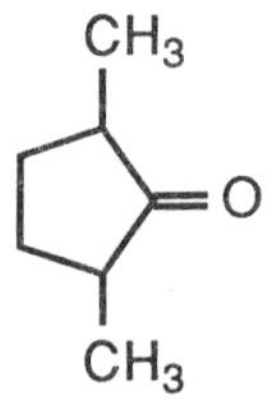

CH_3

O

CH_3

The two methyl groups are equivalent, and each should appear as a doublet; in other words, two peaks would be observed in the methyl region. The alternative product is (2,2-dimethylcyclopentanone). The two methyl groups are again equivalent, but in this compound they would not be coupled to an adjacent CH and would appear as a singlet.

(g) Two enolates are possible, but the indicated product would still exhibit a 3H doublet for the methyl group (whereas alkylation on the carbon bearing the methyl group would remove coupling to the methyl).

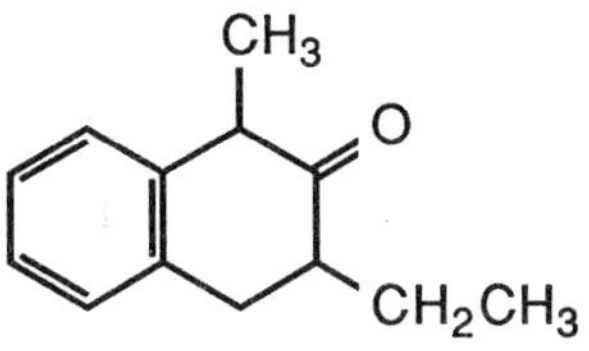

In addition the integration of the signals for the ethyl group would distinguish between mono- and polyalkylation.

(h) Both of the indicated products would differ from reactant by the change of the 3H doublet for the methyl on the bottom of the ring into a singlet.

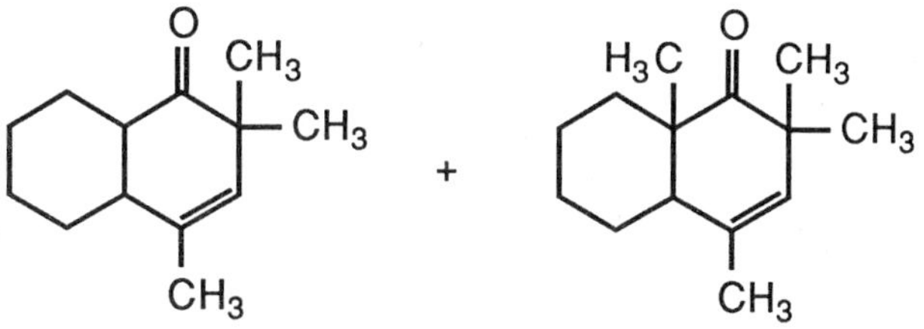

The two products have a total of three and four methyl groups, respectively, and all would appear in the NMR as singlet (although some of the signals might be very close together). Integration of the spectrum would indicate the total number of methyls if they were not sufficiently different in chemical shift.

(i) The expected product results from removal of a hydrogen in the ϕ position followed by alkylation at the α position and isomerization of the double bond so that it is conjugated with the ketone.

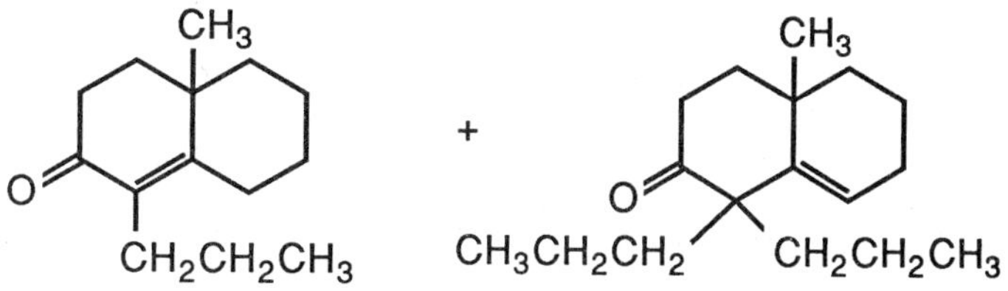

The original signal in the vinyl region would disappear, and a complex pattern for the propyl group would be found in the 1-2 ppm region.

(j) The reaction proceeds via reduction of the unsaturated ketone to give an α-substituted saturated ketone:

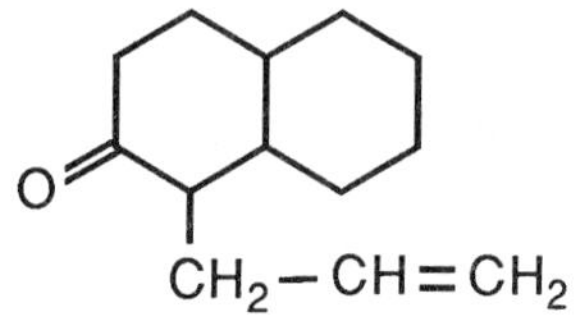

$CH_2-CH=CH_2$

The original vinyl hydrogen signal would no longer be present, but a complex 3H multiplet would be found in the vinyl region for the allyl group that is introduced. Other changes would be seen in the aliphatic region (in the vicinity of 1.5–2 ppm), but these would be difficult to interpret.

CHAPTER 15. ETHERS

ANSWERS TO EXERCISES

15.1 (a) 3-chloro-3-methyloxane

(b) 2-ethyl-3-methyloxirane (or 2,3-epoxypentane)

(c) 1,2-diethoxyethane

(d) 5-chloro-4-ethoxy-5-methyl-2-pentanone

15.2

(a) Cl; OCH_3

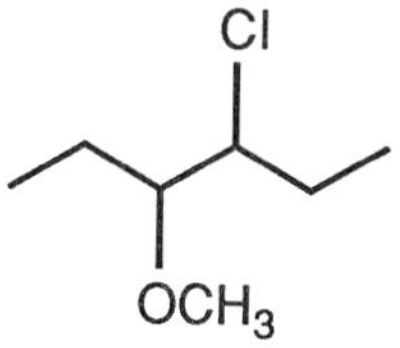

(b) CH_3; CH_3; O

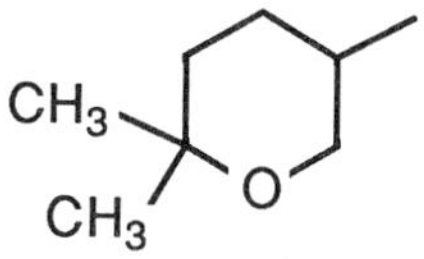

(c) CH_3; O

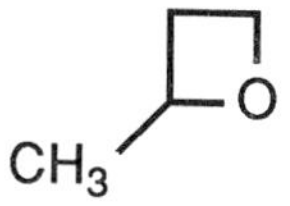

(d) CH_3; O

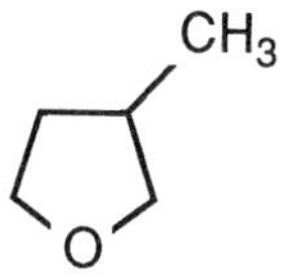

(e) O; O—CH; CH_3; CH_3

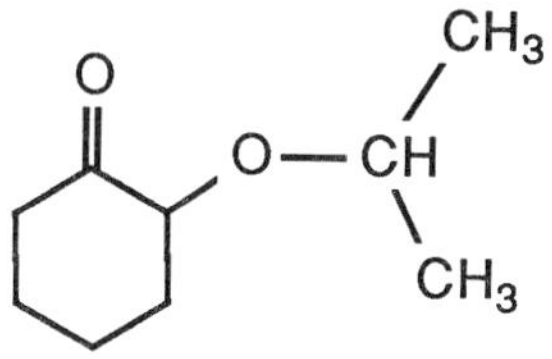

15.3

(a) OH; CH_3 + CH_3I

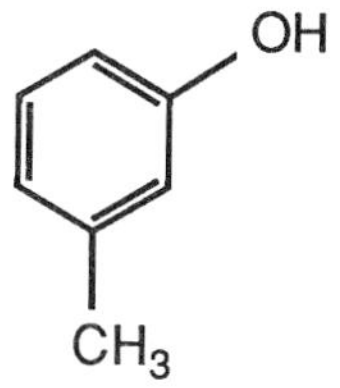

(b) CH_3—C(CH_3)(CH_3)—Br

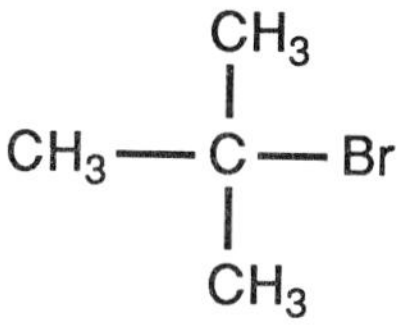

(c) $CH_3CH_2CH_2Br$ + CH_3—CH(CH_3)—Br

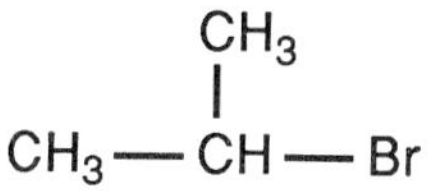

15.4

$$CH_3-\underset{\underset{CH_3}{|}}{C}\overset{O}{\diagup\diagdown}CH_2 \xrightarrow{NaBr} CH_3-\underset{\underset{CH_3}{|}}{\overset{\overset{OH}{|}}{C}}-CH_2Br$$

$$CH_3-\underset{\underset{CH_3}{|}}{C}\overset{O}{\diagup\diagdown}CH_2 \xrightarrow{HBr} \left[CH_3-\underset{\underset{CH_3}{|}}{\overset{+}{C}}-CH_2OH\right] \longrightarrow CH_3-\underset{\underset{CH_3}{|}}{\overset{\overset{Br}{|}}{C}}-CH_2OH$$

15.5 Acids such as HCl are unsatisfactory for the preparation of ethers because the anions (e.g., Cl^-) are relatively nucleophilic. The alcohols can therefore be converted to the corresponding chloride rather than to the desired ether.

15.6

$$CH_3CH_2CH_2-O-CH_2CH_2CH_3 \quad + \quad CH_3CH_2CH_2CH_2-O-CH_2CH_2CH_2CH_3 \quad +$$

$$CH_3CH_2CH_2-O-CH_2CH_2CH_2CH_3 \quad + \quad \text{rearranged ethers}$$

15.7

(a) $CH_3CH_2CH_2-O-CH_2CH_3$

(b)

$$CH_3-\underset{\underset{CH_3}{|}}{\overset{\overset{CH_3}{|}}{C}}-O-\underset{\underset{CH_3}{|}}{\overset{\overset{CH_3}{|}}{C}}-CH_3 \quad + \quad (CH_3)_2C{=}CH_2$$

(c)

$$CH_3CH_2-\underset{}{\overset{\overset{O-CH_2CH_2CH_3}{|}}{CH}}-CH_3$$

15.8 **(a)**

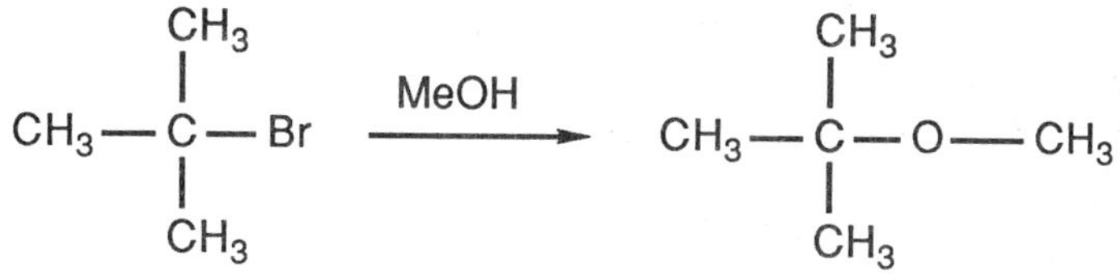

$$CH_3-\underset{CH_3}{\overset{CH_3}{\underset{|}{\overset{|}{C}}}}-Br \xrightarrow{MeOH} CH_3-\underset{CH_3}{\overset{CH_3}{\underset{|}{\overset{|}{C}}}}-O-CH_3$$

or

$$CH_3-\underset{CH_3}{\overset{CH_3}{\underset{|}{\overset{|}{C}}}}-OH \xrightarrow[2.\ CH_3I]{1.\ Na} CH_3-\underset{CH_3}{\overset{CH_3}{\underset{|}{\overset{|}{C}}}}-O-CH_3$$

or

$$(CH_3)_2C{=}CH_2 \xrightarrow[2.\ NaBH_4]{1.\ Hg(OAc)_2,\ CH_3OH} CH_3-\underset{CH_3}{\overset{CH_3}{\underset{|}{\overset{|}{C}}}}-O-CH_3$$

(b)

$$C_6H_5OH \xrightarrow[2.\ (CH_3)_2CHBr]{1.\ KOtBu} C_6H_5-O-\overset{CH_3}{\overset{|}{CH}}-CH_3$$

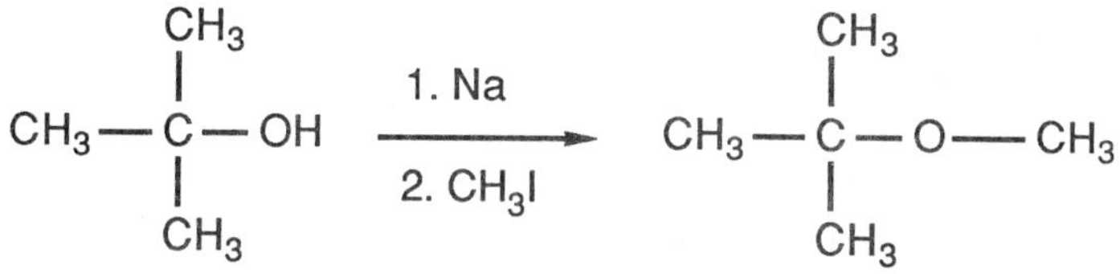

or

$$CH_3-CH{=}CH_2 \xrightarrow[\text{phenol}]{H_2SO_4} C_6H_5-O-\overset{CH_3}{\overset{|}{CH}}-CH_3$$

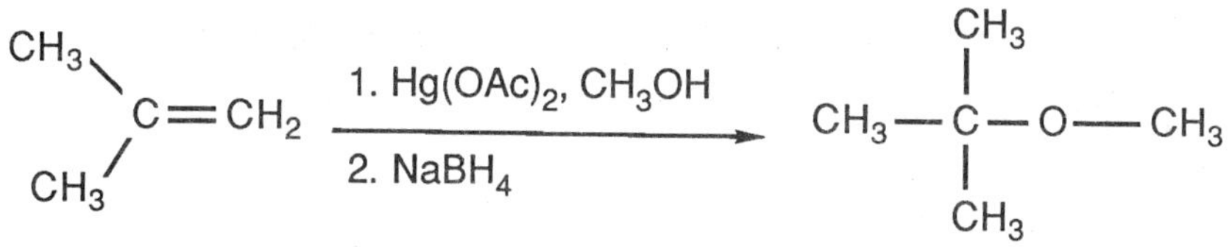

(c)

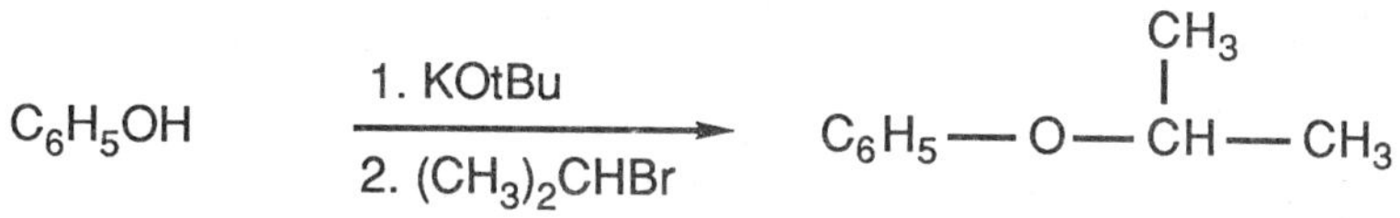

$$\text{cyclohexanol (cyclohexyl–OH)} \xrightarrow[2.\ C_2H_5Br]{1.\ Na} \text{cyclohexyl–}OCH_2CH_3$$

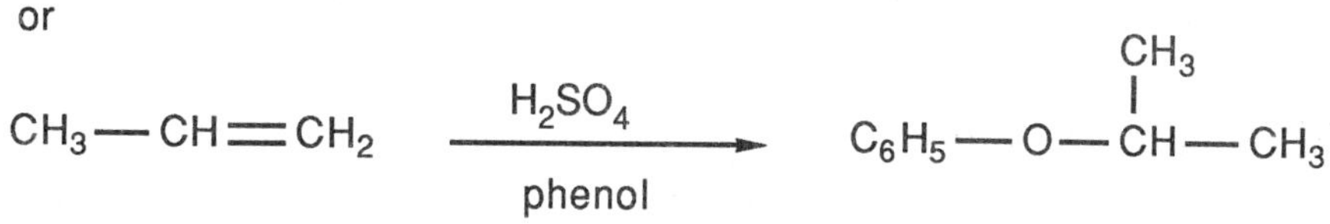

or

$$\text{cyclohexene} \xrightarrow[2.\ NaBH_4]{1.\ Hg(OAc)_2,\ C_2H_5OH} \text{cyclohexyl–}OCH_2CH_3$$

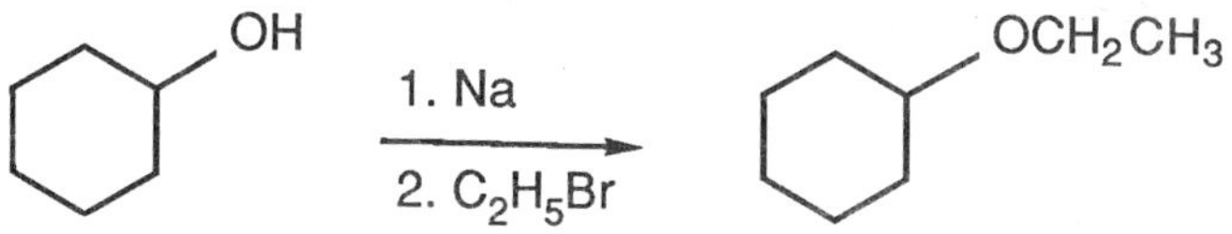

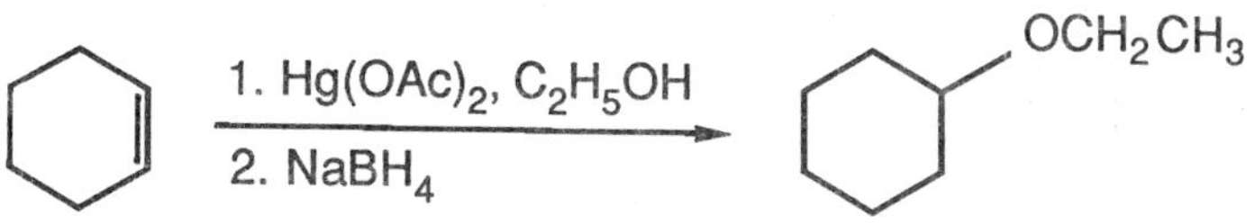

or

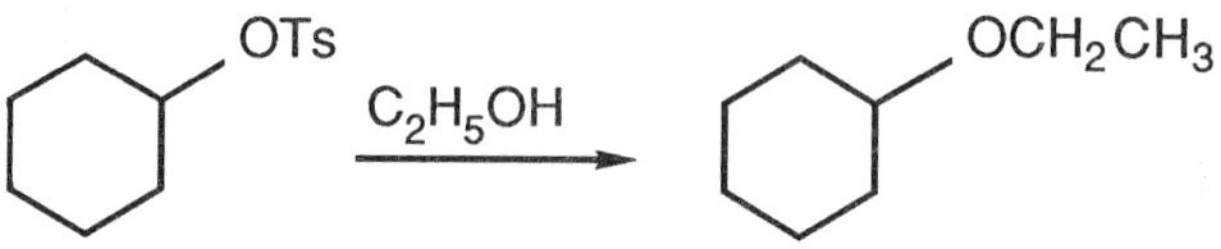

Cyclohexyl–OTs $\xrightarrow{C_2H_5OH}$ Cyclohexyl–OCH_2CH_3

(d) Cyclopentyl–Br $\xrightarrow{CH_3OH}$ CH_3O–cyclopentyl

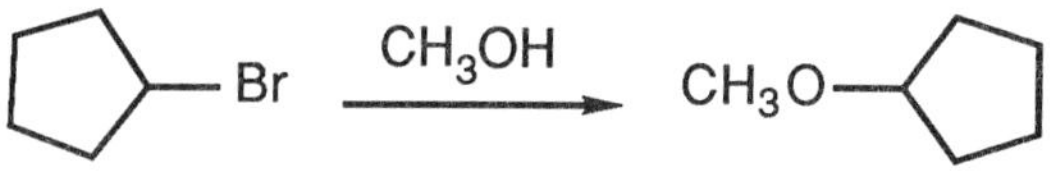

or

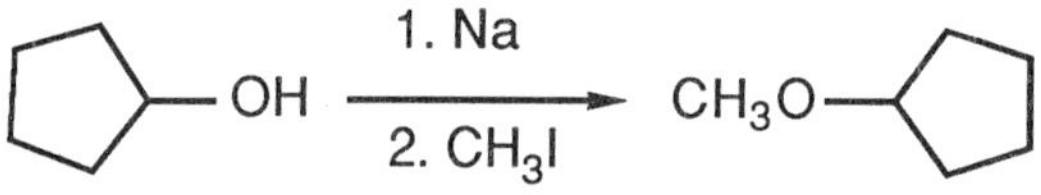

Cyclopentyl–OH $\xrightarrow[\text{2. } CH_3I]{\text{1. Na}}$ CH_3O–cyclopentyl

or

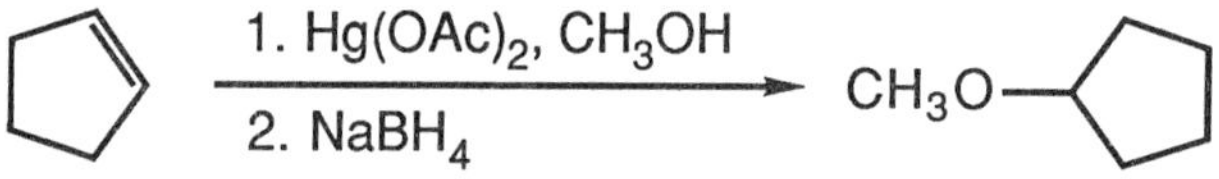

Cyclopentene $\xrightarrow[\text{2. } NaBH_4]{\text{1. } Hg(OAc)_2,\ CH_3OH}$ CH_3O–cyclopentyl

(e) $CH_3CH_2—CH(CH_3)—CH{=}CH_2 \xrightarrow{CF_3CO_3H} CH_3CH_2—CH(CH_3)—CH\overset{O}{—}CH_2$ (epoxide)

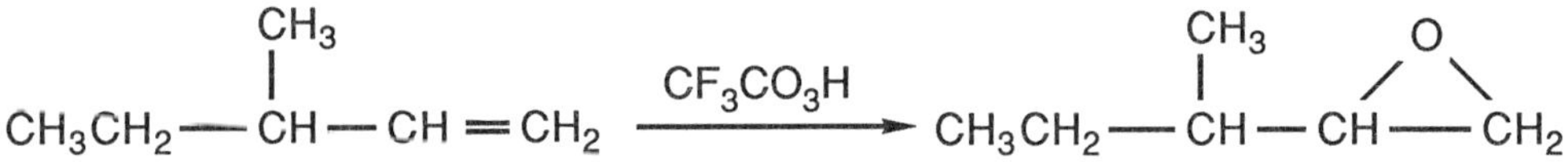

or

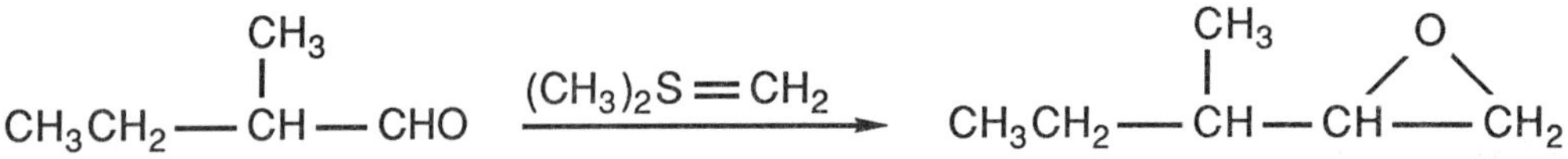

$CH_3CH_2—CH(CH_3)—CHO \xrightarrow{(CH_3)_2S{=}CH_2} CH_3CH_2—CH(CH_3)—CH\overset{O}{—}CH_2$ (epoxide)

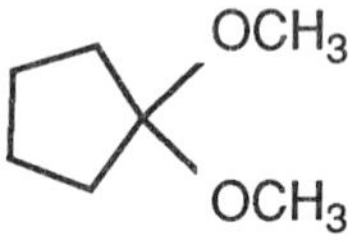

15.9 **(a)** 1,1-dimethoxycyclopentane (cyclopentane with OCH_3 and OCH_3 on the same carbon) **(b)** C_6H_5CHO **(c)** CH_3-substituted cyclohexane spiro-fused to a 1,3-dithiolane ring (S, S)

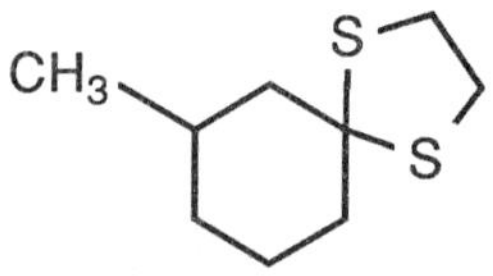

15.10 **(a)** $CH_3CH_2—\overset{O}{\overset{\|}{C}}—CH_3 \xrightarrow[C_6H_6]{HOCH_2CH_2OH,\ TsOH}$ CH_3CH_2–C(CH_3) in 1,3-dioxolane ring (O, O)

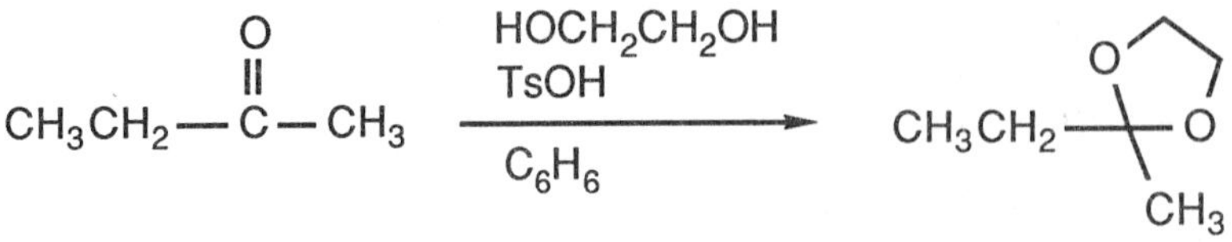

(b)

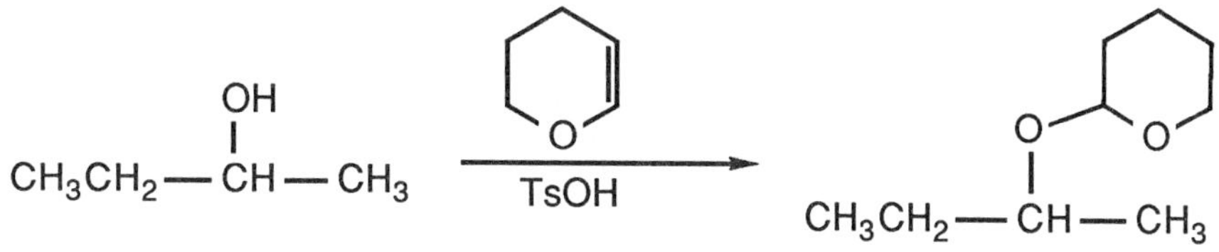

$$CH_3CH_2—CH(OH)—CH_3 \xrightarrow[\text{TsOH}]{\text{dihydropyran}} CH_3CH_2—CH(O\text{-THP})—CH_3$$

15.11 (a) HO–C_6H_{10}–CH_2OH (4-(hydroxymethyl)cyclohexanol)

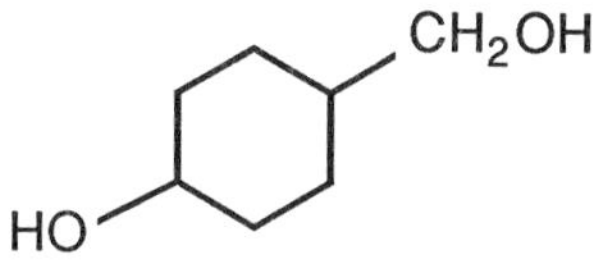

(b) ^-O–cyclohexenyl–$CH_2OSO_2CH_3$ $\xrightarrow[\text{CH}_3\text{SO}_3\text{CH}_2\text{-(4-oxocyclohexyl)}]{\text{excess base}}$ 2-oxocyclohexyl product bearing $CH_2OC_6H_5$ and a CH_2-(4-oxocyclohexyl) group

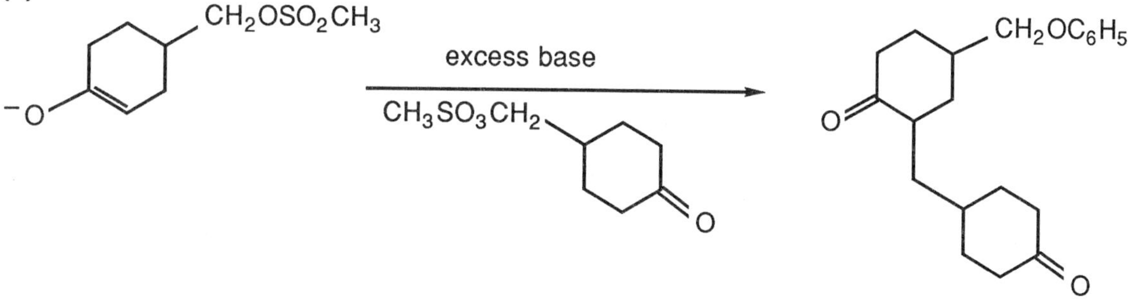

15.12 $Br(CH_2)_8$-O-$(CH_2)_8$-OH and products derived from this.

ANSWERS TO PROBLEMS

15.1 (a) 1-ethoxy-2-butanol

(b) 2-methoxybutane (or methyl *sec*-butyl ether)

(c) 1-chloro-2,3-epoxypentane or 2-(chloromethyl)-3-ethyloxirane

(d) 6-methoxy-5-methyl-2-hexanone

(e) 1,3,6-trimethoxyoctane

(f) 2,2-dimethyoxypropane

(g) 1-isopropoxy-2-methylbutane

15.2

(a) O

OCH$_3$

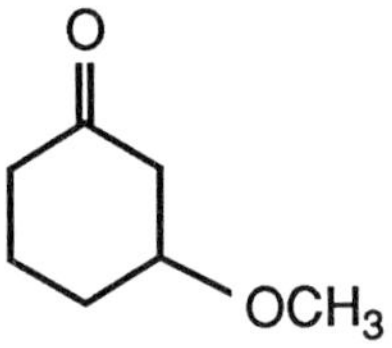

(b) Cl

$CH_3CH_2—CH—CH—CH_2CH_2CH_2CH_3$

$OCH_2CH_2CH_3$

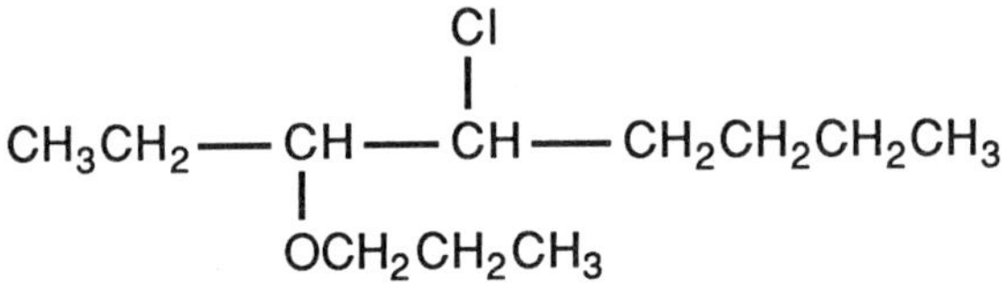

(c) O Br

CH_3 CH_3

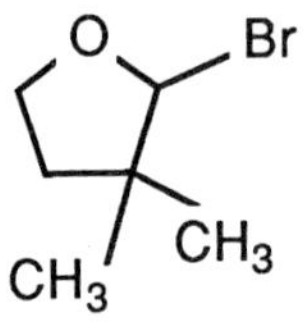

(d) O

$CH_3—CH—CH—CH—CH_2CH_3$

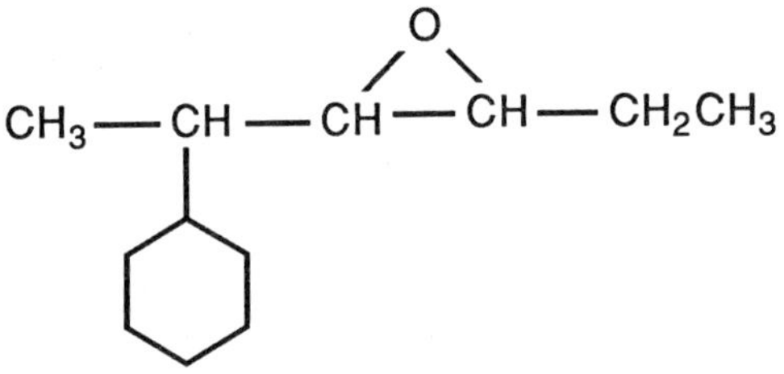

(e) OCH_3

$CH_3CH_2—C—CH_2CH_3$

OCH_3

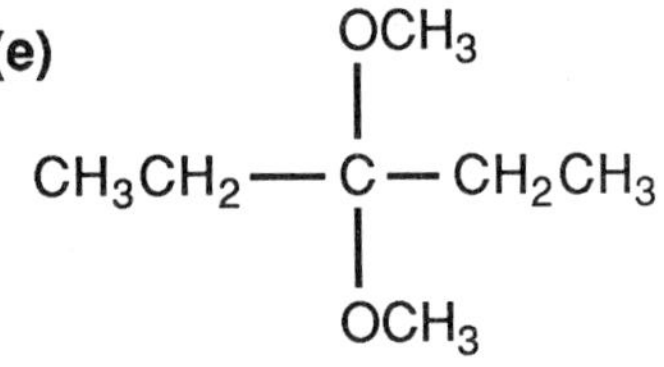

(f) OCH_3 CH_3 OH

$CH_3CH_2—CH—C—CH_2—CH_2$

CH_3

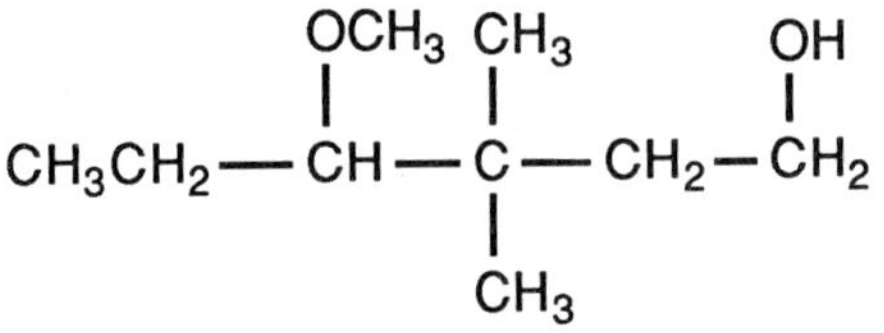

(g) $CH_3—CH—CH_2CH_2CH_2CH_2CH_2CH_3$

$OCH_2CH_2CH_2CH_2CH_2CH_3$

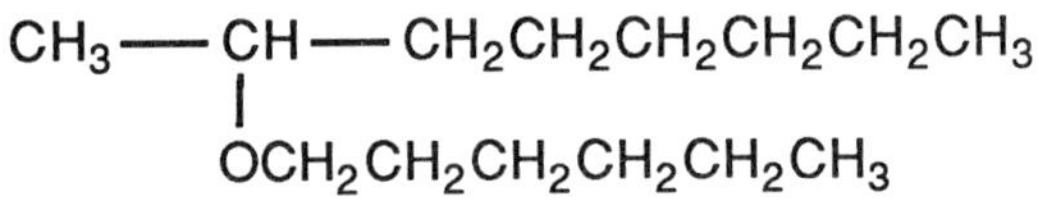

15.3

(a) $CH_3CH_2CH_2CH_2CH_2CH_2CH_2OCH_3$

(b) OH

$CH_3CH_2CH_2—CH—CH_2CH_2OCH_3$

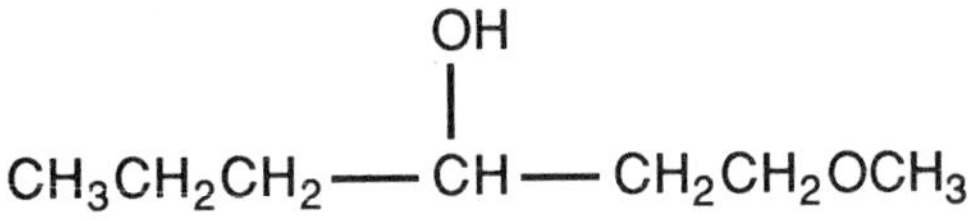

(c) $CH_3CH_2CH_2—CH—CH_2$

OH OH

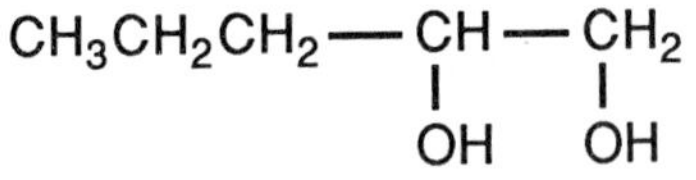

15.4 (a) $CH_3(CH_2)_4CH_2Br$ + CH_3Br

(b) $CH_3(CH_2)_4CH_2OH$ + $BrC(CH_3)_3$ (the tertiary center reacts more rapidly)

(c) $CH_3(CH_2)_4CH_2Br$ + C_6H_5OH

(d) $CH_3(CH_2)_4CH_2Br$

15.5

(a) Br

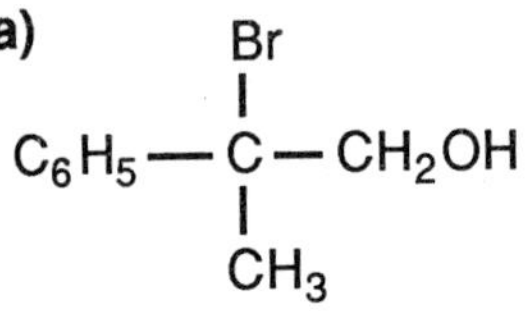

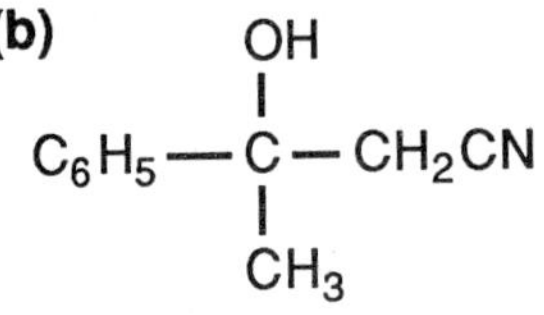

$C_6H_5—C—CH_2OH$

CH_3

(b) OH

$C_6H_5—C—CH_2CN$

CH_3

(c) OH Cl

$C_6H_5—C—CH_2OH$ + $C_6H_5—C—CH_2OH$

CH_3 CH_3

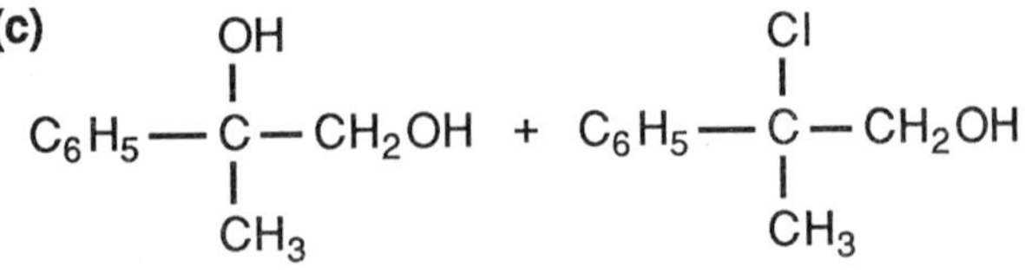

(d)

$$C_6H_5-\overset{\overset{\displaystyle OH}{|}}{\underset{\underset{\displaystyle CH_3}{|}}{C}}-CH_2Br$$

15.6 (a)

$$CH_3(CH_2)_4CH_2OH \xrightarrow[2.\ CH_3I]{1.\ Na} CH_3(CH_2)_4CH_2OCH_3$$

or

$$CH_3(CH_2)_4CH_2Br \xrightarrow{CH_3OH} CH_3(CH_2)_4CH_2OCH_3$$

(b)

$$C_6H_5CH_2OH \xrightarrow[2.\ BrCH_2CH_2CH_3]{1.\ NaNH_2} C_6H_5CH_2OCH_2CH_2CH_3$$

or

$$C_6H_5CH_2Br \xrightarrow{CH_3CH_2CH_2OH} C_6H_5CH_2OCH_2CH_2CH_3$$

or

$$CH_3CH_2CH_2OH \xrightarrow[2.\ C_6H_5CH_2Br]{1.\ Na} C_6H_5CH_2OCH_2CH_2CH_3$$

(c)

$$CH_3CH_2-\overset{\overset{\displaystyle Br}{|}}{C}-CH_3 \xrightarrow{CH_3CH_2OH} CH_3CH_2-\overset{\overset{\displaystyle OCH_2CH_3}{|}}{C}-CH_3$$

or

$$CH_3CH_2-\overset{\overset{\displaystyle OH}{|}}{C}-CH_3 \xrightarrow[2.\ CH_3CH_2I]{1.\ Na} CH_3CH_2-\overset{\overset{\displaystyle OCH_2CH_3}{|}}{C}-CH_3$$

or

$$CH_3CH{=}CHCH_3 \xrightarrow[2.\ NaBH_4]{1.\ Hg(OAc)_2,\ CH_3CH_2OH} CH_3CH_2-\overset{\overset{\displaystyle OCH_2CH_3}{|}}{C}-CH_3$$

(d) $C_6H_5OH \xrightarrow[\text{2. } CH_3CH_2CH_2Br]{\text{1. } NaNH_2} C_6H_5OCH_2CH_2CH_3$

(e) cyclohexanol (OH) $\xrightarrow[\text{2. } CH_3CH_2I]{\text{1. Na}}$ cyclohexyl OCH_2CH_3

or

bromocyclohexane (Br) $\xrightarrow{CH_3CH_2OH}$ cyclohexyl OCH_2CH_3

or

cyclohexene $\xrightarrow[\text{2. } NaBH_4]{\text{1. } Hg(OAc)_2,\ CH_3CH_2OH}$ cyclohexyl OCH_2CH_3

15.7

(a) 1-ethylcyclohexene (CH_2CH_3) $\xrightarrow[\text{2. NaOH, } H_2O_2]{\text{1. } BH_3}$ trans-2-ethylcyclohexanol (H, CH_2CH_3, OH) $\xrightarrow[\text{2. } CH_3CH_2I]{\text{1. Na}}$ trans-1-ethyl-2-ethoxycyclohexane (CH_2CH_3, OCH_2CH_3)

or

cyclohexene $\xrightarrow{C_6H_5CO_3H}$ cyclohexene oxide (O) $\xrightarrow{(CH_3CH_2)_2CuLI}$ trans-2-ethylcyclohexanol (CH_2CH_3, OH) $\xrightarrow[\text{2. } CH_3CH_2I]{\text{1. Na}}$ trans-1-ethyl-2-ethoxycyclohexane (CH_2CH_3, OCH_2CH_3)

(b) cis-$CH_3CH_2CH{=}CHCH_2C_6H_5$ $\xrightarrow{C_6H_5CO_3H}$ epoxide (O) with CH_3CH_2 and $CH_2C_6H_5$

(c)

Cyclopentene $\xrightarrow{C_6H_5CO_3H}$ cyclopentene oxide $\xrightarrow{(CH_3CH_2)_2CuLi}$ trans-2-methylcyclopentanol (CH$_3$, OH) $\xrightarrow[2.\ CH_3OH]{1.\ TsCl}$ cis-1-methoxy-2-methylcyclopentane (CH$_3$, OCH$_3$)

(d)

Cyclohexene $\xrightarrow[2.\ NaClO_3]{1.\ OsO_4}$ cis-1,2-cyclohexanediol (OH, OH) $\xrightarrow[2.\ CH_3I]{1.\ Na}$ cis-1,2-dimethoxycyclohexane (OCH$_3$, OCH$_3$)

(e)

Cyclohexene $\xrightarrow{CH_3CO_3H}$ cyclohexene oxide (O) $\xrightarrow{H_2O,\ H_2SO_4}$ trans-1,2-cyclohexanediol (OH, OH) $\xrightarrow[2.\ CH_3I]{1.\ NaNH_2}$ trans-2-methoxycyclohexanol (OH, OCH$_3$)

Cyclohexene oxide $\xrightarrow{CH_3OH,\ H_2SO_4}$ trans-2-methoxycyclohexanol (OH, OCH$_3$)

15.8

(a) $CH_3CH_2CHO \xrightarrow[H_2SO_4]{CH_3OH} CH_3CH_2—CH(OCH_3)—OCH_3$

(b) $CH_3CH_2—C(CH_3)_2—CH_2CH_2—C(=O)—CH_3 \xrightarrow[H_2SO_4]{CH_3OH} CH_3CH_2—C(CH_3)_2—CH_2CH_2—C(OCH_3)_2—CH_3$

(c) $C_6H_5CHO \xrightarrow[TsOH,\ C_6H_6]{HOCH_2CH_2OH}$ 1,3-dioxolane (O, O) with H and C_6H_5 on C-2

(d) CH_3COCH_3 $\xrightarrow[\text{TsOH, } C_6H_6]{HOCH_2CH_2OH}$ (cyclic ketal, CH_3, CH_3)

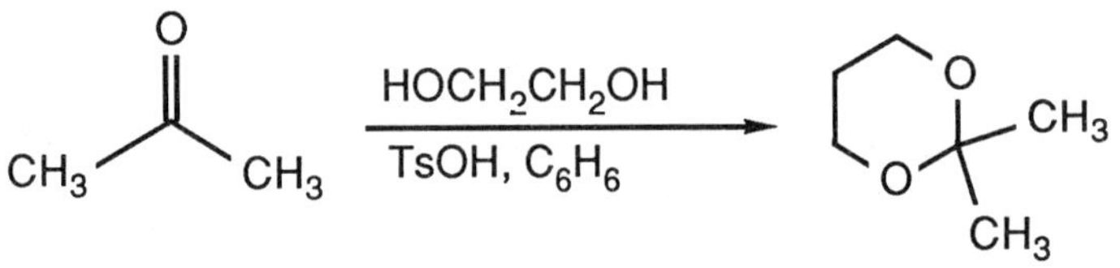

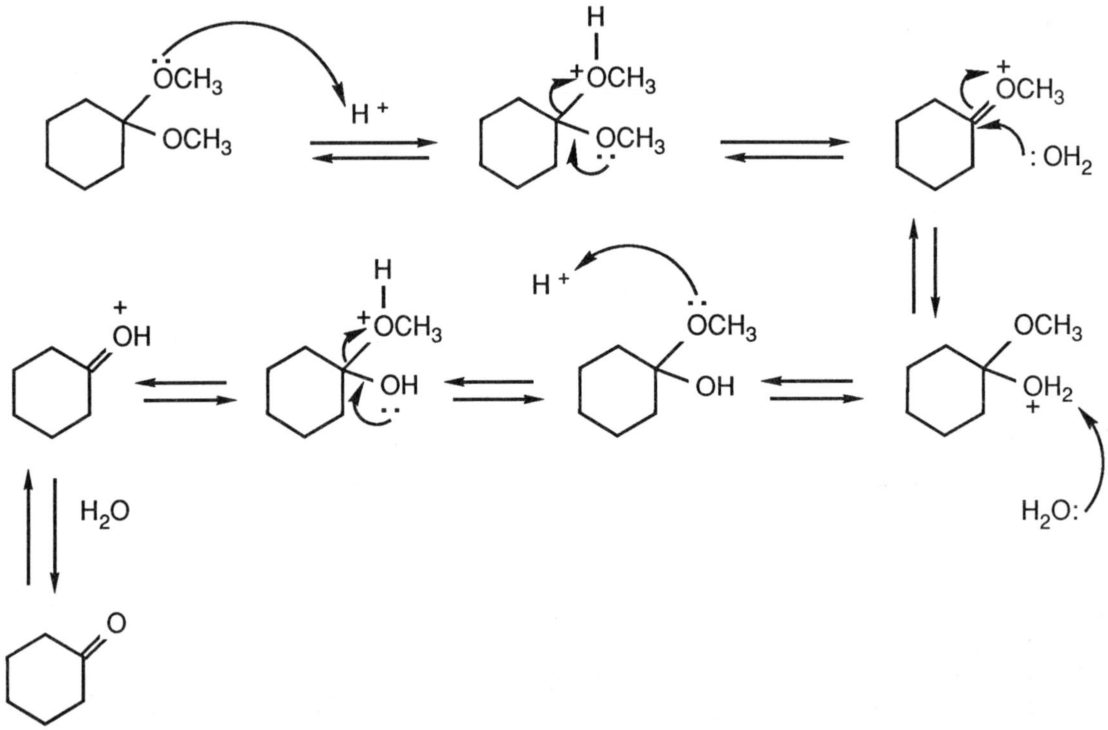

15.9

H⁺, OCH_3, OCH_3, H_2O, OH_2, OH

15.10

(a) cyclohexanone $\xrightarrow{NaBH_4}$ cyclohexanol (OH) $\xrightarrow[\text{2. } CH_3I]{\text{1. } NaNH_2}$ methoxycyclohexane (OCH_3)

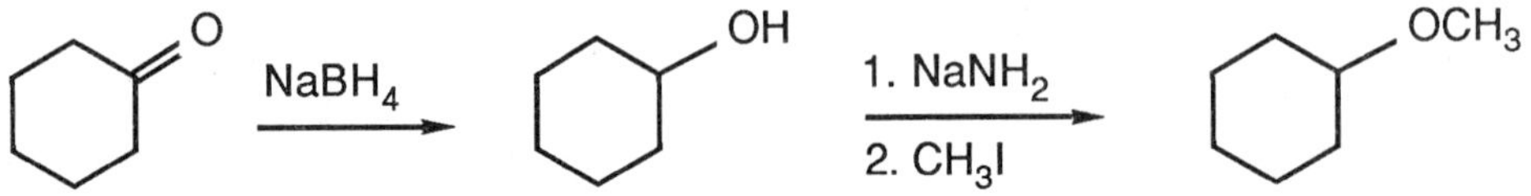

(b)

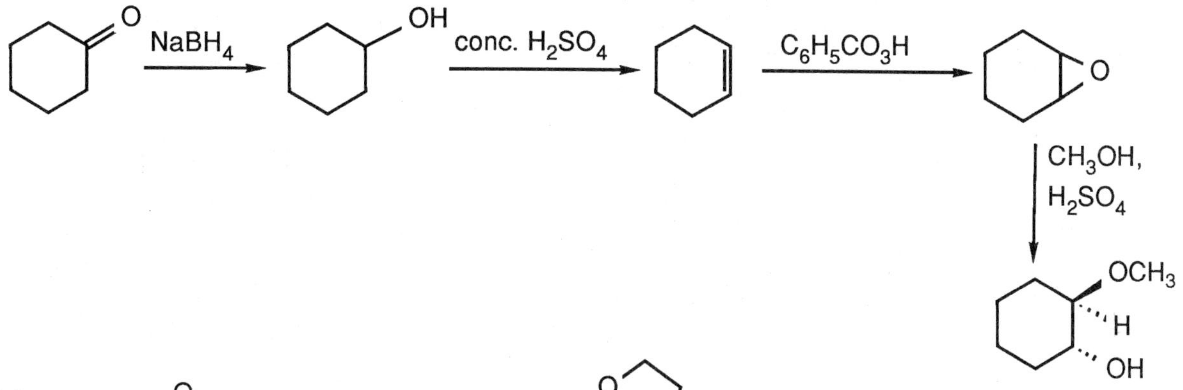

cyclohexanone $\xrightarrow{NaBH_4}$ cyclohexanol $\xrightarrow{\text{conc. } H_2SO_4}$ cyclohexene $\xrightarrow{C_6H_5CO_3H}$ cyclohexene oxide $\xrightarrow{CH_3OH,\ H_2SO_4}$ trans-2-methoxycyclohexanol (OCH$_3$, H, OH)

(c)

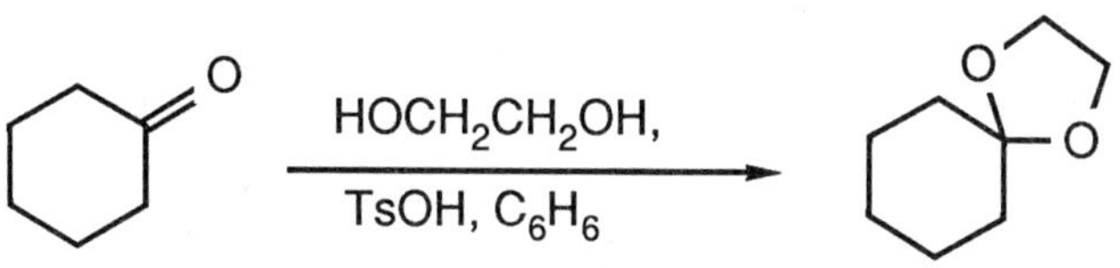

cyclohexanone $\xrightarrow[TsOH,\ C_6H_6]{HOCH_2CH_2OH,}$ cyclohexanone ethylene ketal (O, O)

(d)

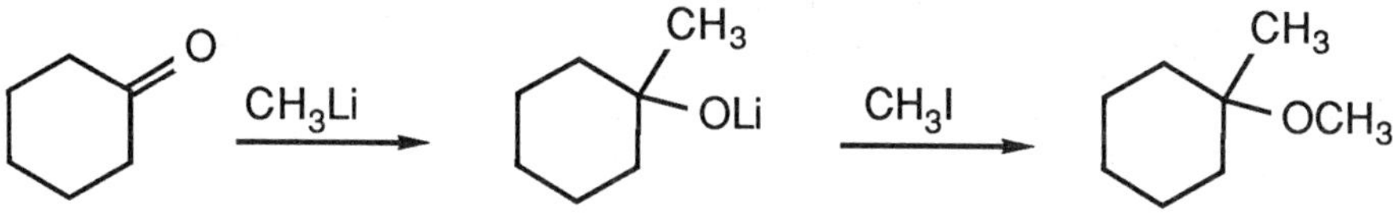

cyclohexanone $\xrightarrow{CH_3Li}$ 1-methylcyclohexyl OLi (CH$_3$, OLi) $\xrightarrow{CH_3I}$ 1-methoxy-1-methylcyclohexane (CH$_3$, OCH$_3$)

(e)

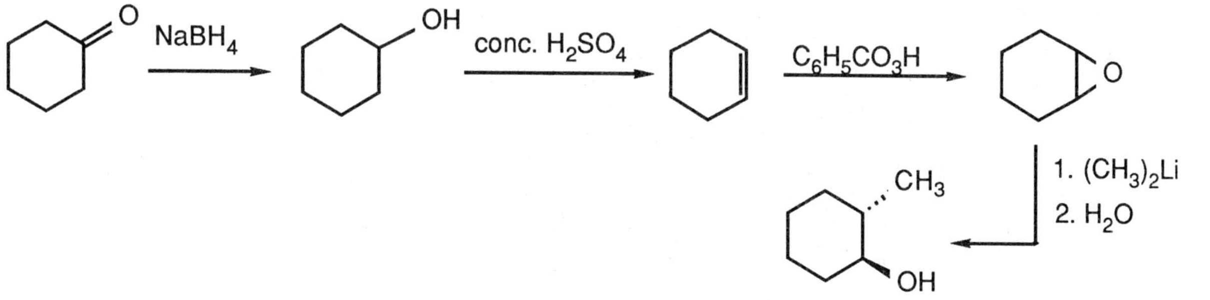

cyclohexanone $\xrightarrow{NaBH_4}$ cyclohexanol (OH) $\xrightarrow{\text{conc. } H_2SO_4}$ cyclohexene $\xrightarrow{C_6H_5CO_3H}$ cyclohexene oxide (O) $\xrightarrow{1.\ (CH_3)_2Li\quad 2.\ H_2O}$ trans-2-methylcyclohexanol (CH$_3$, OH)

(f)

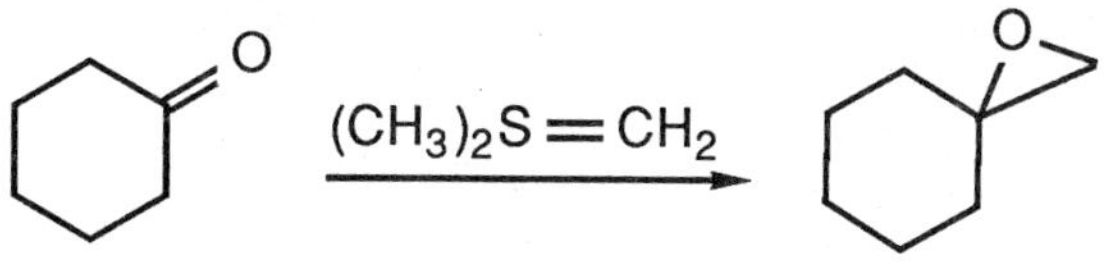

cyclohexanone $\xrightarrow{(CH_3)_2S{=}CH_2}$ spiro epoxide (O)

15.11

(a) $C_6H_5CHO \xrightarrow[\text{2. NaNH}_2\text{, CH}_3\text{CH}_2\text{I}]{\text{1. NaBH}_4} C_6H_5CH_2OCH_2CH_3$

(b) $C_6H_5CHO \xrightarrow{\text{NaBH}_4} C_6H_5CH_2OH \xrightarrow[\text{or H}_2\text{SO}_4]{\text{1. NaNH}_2\ \ \text{2. C}_6\text{H}_5\text{CH}_2\text{Br}} C_6H_5CH_2OCH_2C_6H_5$

(c) $C_6H_5CHO \xrightarrow[\text{HCl}]{\text{CH}_3\text{CH}_2\text{OH}} C_6H_5-CH(OCH_2CH_3)-OCH_2CH_3$

(d)

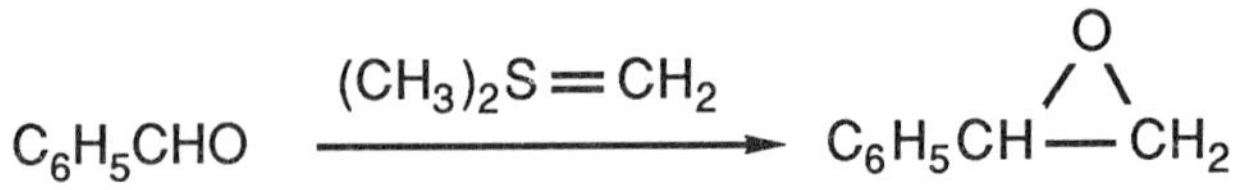

C6H5CHO
(CH3)2S=CH2
O
C6H5CH—CH2

(e)

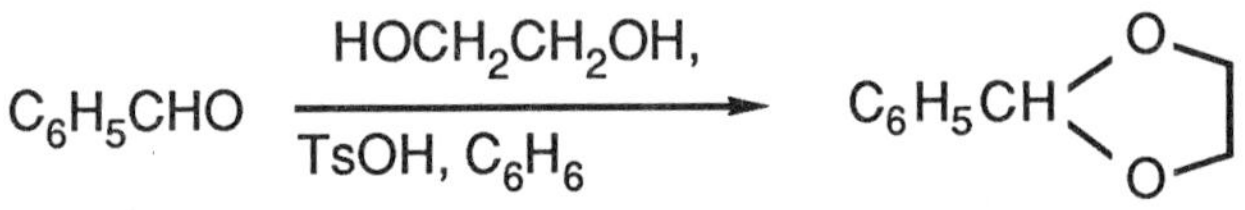

C6H5CHO
HOCH2CH2OH,
TsOH, C6H6
O
C6H5CH
O

(f)

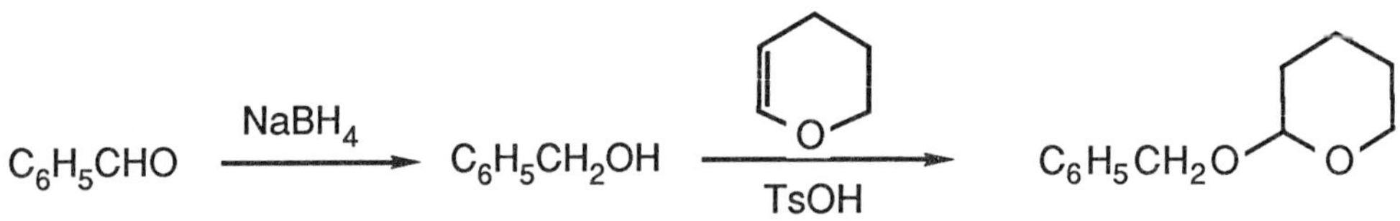

C6H5CHO
NaBH4
C6H5CH2OH
O
TsOH
C6H5CH2O
O

(g) $C_6H_5CHO \xrightarrow[\text{or Na-Hg, HCl}]{\text{NaNH}_2\text{, KOH}} C_6H_5CH_3$

(h)

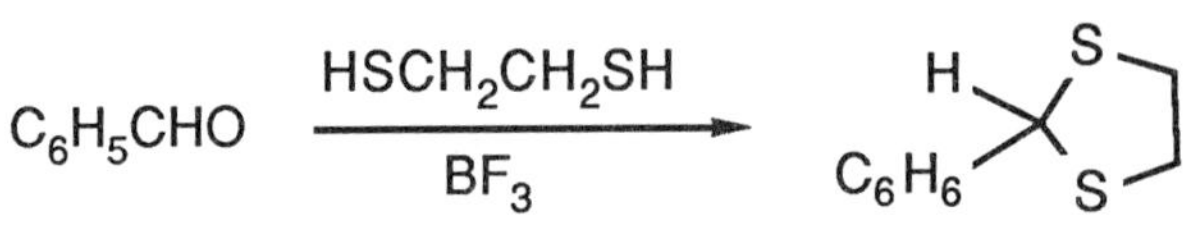

C6H5CHO
HSCH2CH2SH
BF3
H
S
C6H6
S

15.12

(a)

CHO — dihydropyran, TsOH → CHO (THP ether) — $KMnO_4$ → CO_2H (THP ether) — dilute HCl → CO_2H, OH

OH

(b)

CHO — $HOCH_2CH_2OH$, TsOH, C_6H_6 → (dioxolane), OH — CrO_3, pyridine → (dioxolane), O → CHO, O

OH

(c)

$CH{=}CH_2$ — $HOCH_2CH_2OH$, TsOH, C_6H_6 → $CH{=}CH_2$ (ketal) — 1. BH_3 2. H_2O_2, NaOH → CH_2CH_2OH (ketal) — 1. $NaNH_2$ 2. CH_3I → $CH_2CH_2OCH_3$ (ketal) — dilute HCl → $CH_2CH_2OCH_3$, O

O

(d) 3-oxocyclohexanecarboxylic acid $\xrightarrow[\text{2. HCl (neutralize)}]{\text{1. } NaNH_2\text{, KOH}}$ cyclohexanecarboxylic acid (CO_2H)

15.13 (a) cyclohexanol (OH) $\xrightarrow[\text{2. } CH_3I]{\text{1. } NaNH_2}$ methoxycyclohexane (OCH_3)

The conversion of OH to OCH_3 would result in the disappearance of the OH absorption in the 3200–3400 cm^{-1} region of the infrared spectrum. The OH signal in the NMR would also disappear, being replaced by a 3H singlet near 3.1 ppm.

(b) cyclohexene oxide $\xrightarrow[H_2SO_4]{CH_3OH,}$ 2-methoxycyclohexanol (OCH_3, H, OH)

The ring opening of the epoxide would result in the appearance of new peaks characteristic of both OH and OCH_3: an OH absorption in the 3200–3400 cm^{-1} region of the infrared spectrum, an OH signal in the NMR, and an OCH_3 NMR signal near 3.1 ppm.

(c) cyclohexanone (O) $\xrightarrow[\text{TsOH, } C_6H_6]{HOCH_2CH_2OH,}$ cyclohexanone ethylene ketal (O, O)

The C=O absorption near 1700 cm^{-1} in the infrared spectrum would disappear, and a peak integrating for 4H would arise in the NMR spectrum of the product near 3.9 ppm.

(d) cyclohexanone (O) $\xrightarrow[\text{2. } CH_3I]{\text{1. } CH_3Li}$ 1-methoxy-1-methylcyclohexane (CH_3, OCH_3)

The C=O absorption near 1700 cm^{-1} in the infrared spectrum would disappear, and a 3H signal for the OCH_3 group would be found near 3.1 ppm in the NMR spectrum.

(e)

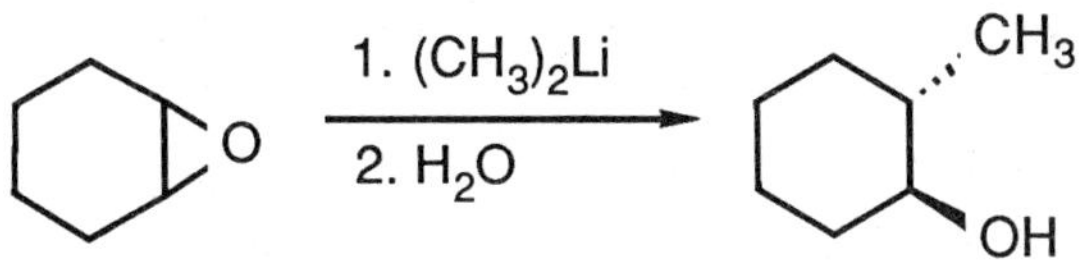

1. $(CH_3)_2Li$
2. H_2O

CH_3 OH

The ring opening of the epoxide would result in the appearance of new peaks characteristic of both OH and CH_3: an OH absorption in the 3200–3400 cm^{-1} region of the infrared spectrum, an OH signal in the NMR, and a doublet in the NMR spectrum integrating for 3H near 1 ppm.

(f)

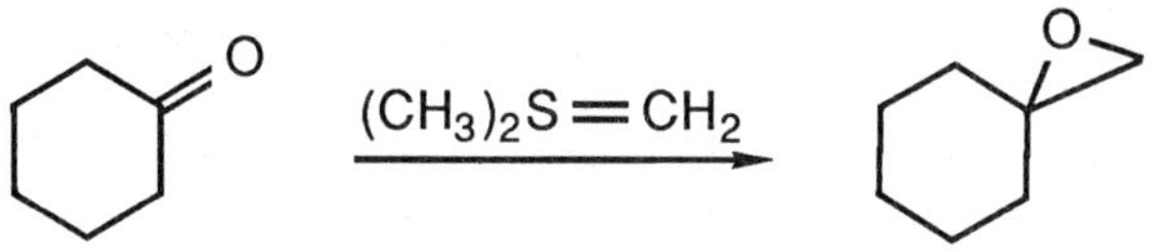

$(CH_3)_2S{=}CH_2$

The C=O absorption near 1700 cm^{-1} in the infrared spectrum would disappear, and a 2H signal for the CH_2 group of the epoxide would be found in the NMR spectrum.

CHAPTER 16. CARBOXYLIC ACIDS

ANSWERS TO EXERCISES

16.1 **(a)** 3-methylpentanoic acid

(b) 2-ethylpentanedioic acid (2-ethylglutaric acid)

(c) nonamide

(d) *N*-methylacetamide

(e) ethyl heptanoate

(f) 2-methylpentanoyl chloride

16.2 **(a)** ethyl 5-oxoheptanoate

(b) 3-hydroxy-4-methylpentanamide

(c) 4-oxopentanoyl chloride

(d) sodium 4-ethyl-6-heptenoate

16.3 **(a)** For two amides, the one with fewer carbons would be more soluble in water:

$$CH_3-CH_2-\overset{\overset{\displaystyle O}{\|}}{C}-NH_2$$

In the case of isomeric compounds, the more polar carboxylic would have higher water solubility than the ester.

(b) For a methyl ester and an anhydride of the same carboxylic acid the lower molecular weight of the ester would result in a lower boiling point:

$$CH_3(CH_2)_3\overset{\overset{\displaystyle O}{\|}}{C}-O-\overset{\overset{\displaystyle O}{\|}}{C}(CH_2)_3CH_3$$

Hydrogen bonding would result in a higher boiling point for a carboxylic acid than for an isomeric ester:

$$CH_3(CH_2)_3\overset{\overset{\displaystyle O}{\|}}{C}OH$$

(c) Hydrogen bonding and greater polarity would favor the crystalline state for a carboxylic acid (relative to the corresponding acid chloride) and would result in a higher melting point:

$$CH_3(CH_2)_3\overset{\overset{\displaystyle O}{\|}}{C}OH$$

Hydrogen bonding for the primary amide would favor the crystalline state and would result in a higher melting point:

$$CH_3-\underset{\underset{\displaystyle CH_3}{|}}{\overset{\overset{\displaystyle CH_3}{|}}{C}}-\overset{\overset{\displaystyle O}{\|}}{C}-NH_2$$

16.4 **(a)** (i) $CH_3-CH_2-\overset{\overset{\displaystyle ^{+}O-H}{\|}}{C}-Cl$ (ii) $CH_3-\bar{C}H-\overset{\overset{\displaystyle O}{\|}}{C}-Cl$

(b) (i) $CH_3-CH_2-\overset{\overset{\displaystyle ^{+}O-H}{\|}}{C}-NH_2$ (ii) $CH_3-CH_2-\overset{\overset{\displaystyle O}{\|}}{C}-\bar{N}H$

(c) (i) $CH_3CH_2-\overset{\overset{\displaystyle \overset{+}{O}-H}{\|}}{C}-OCH_2CH_3$ (ii) $CH_3-\bar{C}H-\overset{\overset{\displaystyle O-H}{\|}}{C}-OCH_2CH_3$

(d) (i) $CH_3CH_2-\overset{\overset{\displaystyle O}{\|}}{C}-O-\overset{\overset{\displaystyle ^{+}O-H}{\|}}{C}-CH_2CH_3$ (ii) $CH_3CH_2-\overset{\overset{\displaystyle O}{\|}}{C}-O-\overset{\overset{\displaystyle O}{\|}}{C}-\bar{C}H-CH_3$

(e) (i) $CH_3CH_2C{\equiv}\overset{+}{N}{-}H$ (ii) $CH_3\overset{-}{C}HC{\equiv}N$

(f) (i) $CH_3CH_2C(={}^{+}O{-}H){-}OH$ (ii) $CH_3CH_2C(=O){-}O^-$

16.5 The greater acidity of acetamide relative to acetonitrile results from the presence of hydrogens bonded to nitrogen:

$$CH_3C(=O){-}\ddot{N}H_2 \longrightarrow CH_3C(=O){-}\ddot{N}H^- + H^+$$

$$CH_3C{\equiv}N \longrightarrow {}^-{:}CH_2C{\equiv}N + H^+$$

N,N-Dimethylacetamide has no hydrogens bonded to nitrogen, and its acidity should be similar to that of acetonitrile.

$$CH_3C(=O){-}N(CH_3)_2 \longrightarrow {}^-{:}CH_2C(=O){-}N(CH_3)_2 + H^+$$

16.6

(a) $CH_3CH_2CH_2OH$ **(b)** $CH_3CH_2CH_2OH$ **(c)** $CH_3CH_2CH_2OH$

(d) $CH_3CH_2CH_2OH + CH_3OH$ **(e)** $CH_3CH_2CH_2NH_2$ **(f)** $CH_3CH_2CH_2NHCH_2CH_3$

(g) $CH_3CH_2CH_2N(CH_3)_2$ **(h)** $CH_3CH_2CH_2NH_2$

16.7

(a) $C_6H_5CH_2OC(=O){-}CH_3 + CO_2$ **(b)** $C_6H_5CH_2C(=O){-}OH + CO_2$

16.8 **(a)** $CH_3CH_2CH_2C(=O){-}CH_3$ **(b)** $CH_3CH_2CH_2C(OH)(CH_3){-}CH_3$ **(c)** $CH_3CH_2CH_2C(OH)(CH_3){-}CH_3$

(d) $CH_3CH_2CH_2C(=O)-CH_3$

16.9 (a) $CH_3CH_2CH_2C(=O)-OH$ (b) $CH_3CH_2CH_2C(OH)(CH_3)-CH_3$ (c) $CH_3CH_2CH_2C(OH)(CH_3)-CH_3$

(d) $CH_3CH_2CH_2C(=O)-CH_3$

16.10

(a) $CH_3CH_2CH_2C(=O)-CH_3$ (b) $CH_3CH_2CH_2C(=O)-O^-$ (c) $CH_3-CH(CH_3)-CH_2-C(=O)-OCH_3$

(d) no reaction

16.11 (a) $CH_3CH_2CO_2^-$ + CH_3CH_2OH (b) $CH_3CH_2C(=O)-N(Et)_2$ + CH_3CH_2OH

(c) $CH_3CH_2CO_2H$ + CH_3CH_2OH

16.12 (a) $CH_3CH_2CO_2^-$ (b) $CH_3CH_2C(=O)-N(Et)_2$ (c) $CH_3CH_2CO_2H$

16.13

(a) $CH_3CH_2CO_2^-$ (b) $CH_3CH_2C(=O)-N(Et)_2$ + $CH_3CH_2CO_2^-$ (c) $CH_3CH_2CO_2H$

16.14

(a)

$$CH_3CH_2C(=O)-OH \xrightarrow[\text{2. } NH_3]{\text{1. } PCl_5 \text{ (or } SOCl_2\text{, etc.)}} CH_3CH_2C(=O)-NH_2$$

(b)

$$CH_3CH_2C(=O)-OH \xrightarrow{PCl_5} CH_3CH_2C(=O)-Cl$$

(c)
$$CH_3CH_2\overset{O}{\overset{\|}{C}}-OH \xrightarrow[\text{or 1. }PCl_5\text{ 2. }CH_3OH]{CH_3OH,\ H_2SO_4} CH_3CH_2\overset{O}{\overset{\|}{C}}-OCH_3$$

(d)
$$CH_3CH_2\overset{O}{\overset{\|}{C}}-OH \xrightarrow[\text{or 1. }PCl_5\text{ 2. }(CH_3)_2CHOH]{(CH_3)_2CHOH,\ H_2SO_4} CH_3CH_2\overset{O}{\overset{\|}{C}}-OCH(CH_3)_2$$

(e)
$$CH_3CH_2\overset{O}{\overset{\|}{C}}-OH \xrightarrow[\text{2. }NH(CH_3)_2]{\text{1. }PCl_5} CH_3CH_2\overset{O}{\overset{\|}{C}}-N(CH_3)_2$$

(f)
$$CH_3CH_2\overset{O}{\overset{\|}{C}}-OH \xrightarrow[\text{2. }NH_3]{\text{1. }PCl_5} CH_3CH_2\overset{O}{\overset{\|}{C}}-NH_2 \xrightarrow{P_2O_5} CH_3CH_3C\equiv N$$

16.15 (a)
$$CH_3CH_2Br \xrightarrow[\text{2. }CO_2\text{ 3. HCl}]{\text{1. Mg}} CH_3CH_2\overset{O}{\overset{\|}{C}}-OH$$

$$CH_3CH_2Br \xrightarrow[\text{2. }H_2O,\ H_2SO_4]{\text{1. NaCN}} CH_3CH_2\overset{O}{\overset{\|}{C}}-OH$$

$$CH_3CH_2CH{=}CHCH_2CH_3 \xrightarrow{KMnO_4} CH_3CH_2\overset{O}{\overset{\|}{C}}-OH$$

(b)
$$(CH_3)_3CBr \xrightarrow[\text{2. }CO_2\text{ 3. HCl}]{\text{1. Mg}} (CH_3)_3CCO_2H$$

$$(CH_3)_3CBr \xrightarrow[\text{2. }H_2O]{\text{1. }H_2SO_4,\ CO} (CH_3)_3CCO_2H$$

$$(CH_3)_3C\overset{O}{\overset{\|}{C}}-CH_3 \xrightarrow[\text{2. HCl}]{\text{1. }Br_2,\ KOH} (CH_3)_3CCO_2H$$

16.16 (a) $CH_3CH_2-\underset{\displaystyle Br}{\underset{|}{CH}}-\overset{\displaystyle O}{\overset{\|}{C}}-Cl$ (b) $CH_3CH_2-\underset{\displaystyle Br}{\underset{|}{CH}}-\overset{\displaystyle O}{\overset{\|}{C}}-OH$ (c) $CH_3CH_2-\underset{\displaystyle Br}{\underset{|}{CH}}-\overset{\displaystyle O}{\overset{\|}{C}}-OCH_3$

16.17 (a) $HO(CH_2)_4CO_2^-$ (b) $HO(CH_2)_4CH_2OH$ (c) $HO(CH_2)_4\underset{\displaystyle CH_3}{\underset{|}{\overset{\displaystyle CH_3}{\overset{|}{C}}}}-OH$

16.18

(a) NH (b) $NH_2(CH_2)_4CO_2^-$

16.19 (b), (c), (d), (e), and (f): note that (f) would dissolve only after reacting with the aqueous base.

16.20 (c) and (e): all the other compound have too many carbon atoms.

16.21

$$C_6H_5-\underset{\displaystyle C_6H_5}{\underset{|}{CH}}-\overset{\displaystyle O}{\overset{\|}{C}}-NH_2$$

ANSWERS TO PROBLEMS

16.1 For R = H, CH_3, $CH(CH_3)_3$, $(CH_2)_4CH_3$, $CH_2C_6H_5$:

(a) formic acid, methyl formate, isopropyl formate, pentyl formate, phenyl formate, benzyl formate

(b) acetic acid, methyl acetate, isopropyl acetate, pentyl acetate, phenyl acetate, benzyl acetate

(c) isobutyric acid, methyl isobutyrate, isopropyl isobutyrate, pentyl isobutyrate, phenyl isobutyrate, benzyl isobutyrate

(d) heptanoic acid, methyl heptanoate, isopropyl heptanoate, pentyl heptanoate, phenyl heptanoate, benzyl heptanoate

(e) chloroacetic acid, methyl chloroacetate, isopropyl chloroacetate, pentyl chloroacetate, phenyl chloroacetate, benzyl chloroacetate

(f) cyclohexane carboxylic acid, methyl cyclohexanecarboxylate, isopropyl cyclohexanecarboxylate, pentyl cyclohexanecarboxylate, phenyl cyclohexanecarboxylate, benzyl cyclohexanecarboxylate,

16.2 For X = OH, Cl, NH_2, $NHCH_3$, $N(CH_3)_2$, CH_3-NH-CH_2CH_3:

(a) formic acid, formyl chloride, formamide, *N*-methylformamide, *N,N*-dimethylformamide, *N*-ethyl-*N*-methylformamide

(b) acetic acid, acetyl chloride, acetamide, *N*-methylacetamide, *N,N*-dimethylacetamide, *N*-ethyl-*N*-methylacetamide

(c) butyric acid, butyryl chloride, butyramide, *N*-methylbutyramide, *N,N*-dimethylbutyramide, *N*-ethyl-*N*-methylbutyramide

(d) benzoic acid, benzoyl chloride, benzamide, *N*-methylbenzamide, *N,N*-dimethylbenzamide, *N*-ethyl-*N*-methylbenzamide

(e) cyclohexanecarboxylic acid, cyclohexanecarbonyl chloride, cyclohexanecarboxamide, *N*-methylcyclohexanecarboxamide, *N,N*-dimethylcyclohexanecarboxamide, *N*-ethyl-*N*-methylcyclohexanecarboxamide

16.3 **(a)** $CH_3CH_2CH_2C(=O)-NHCH_3$

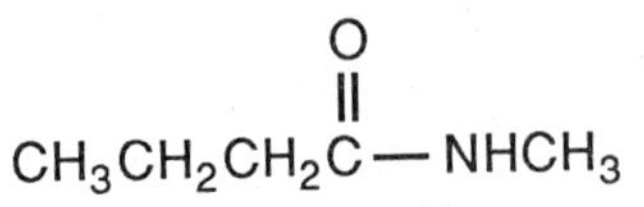

(b) HO (cyclohexane ring) CO_2H

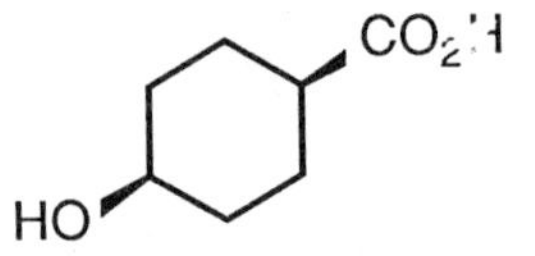

(c) $CH_3CH_2CH_2CH_2CH_2C(Cl)_2-CH_2-C(=O)-Cl$

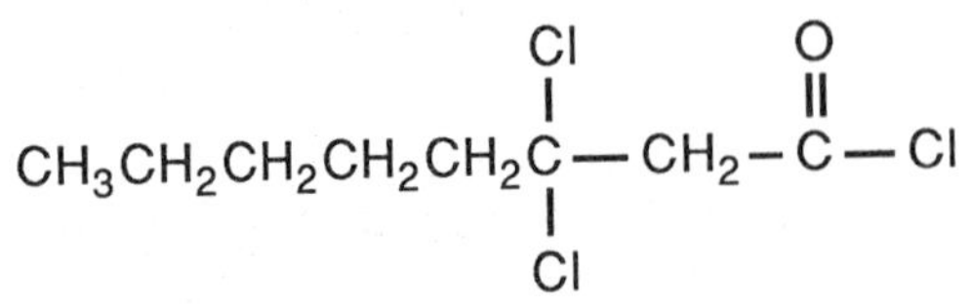

(d) $CH_3CH_2CH_2CH_2C(=O)-OCH_3$

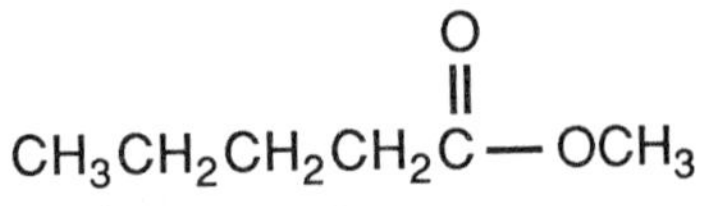

(e) 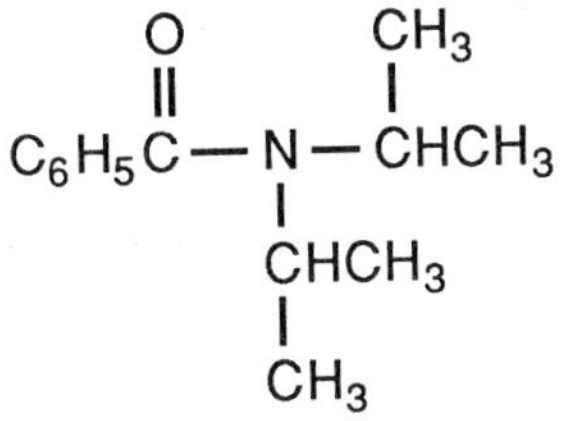

(f) 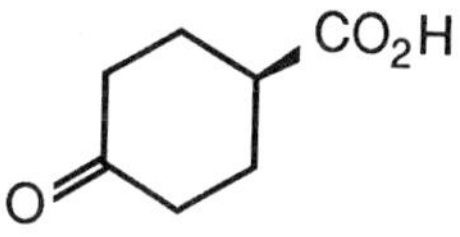

(g) 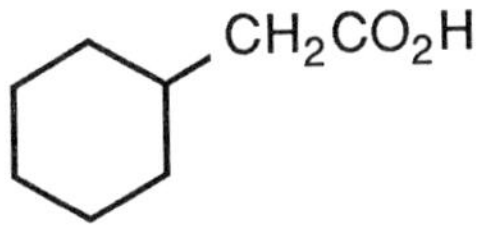

(h) 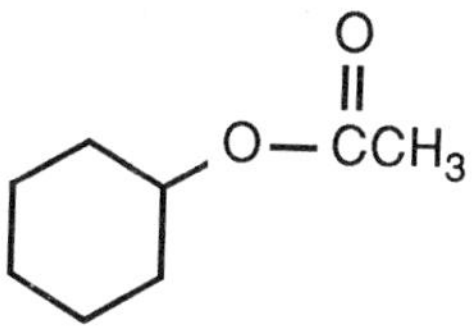

(i) 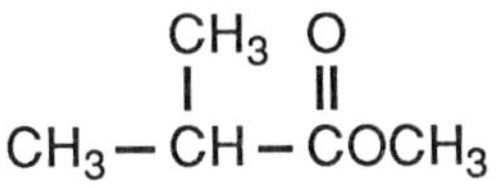

(j) 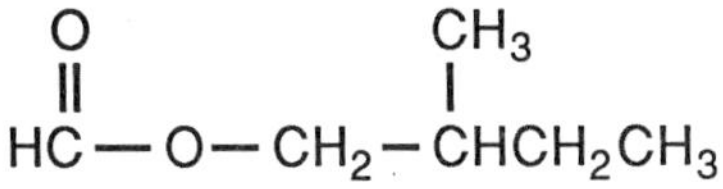

(k) 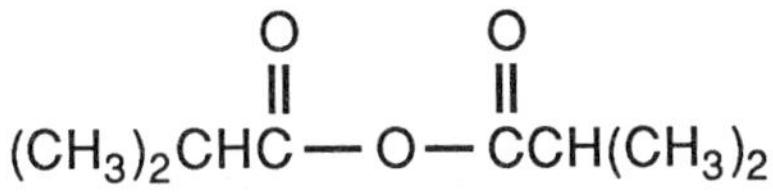

(l) $CH_3CH_2CH_2CH_2CO_2Na$

(m) 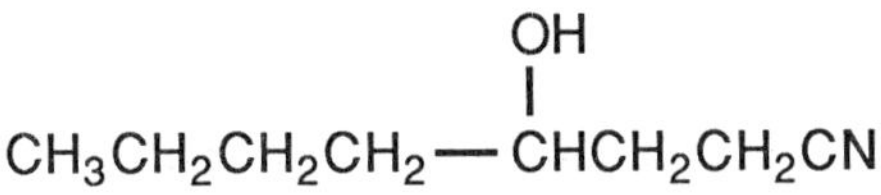

(n) 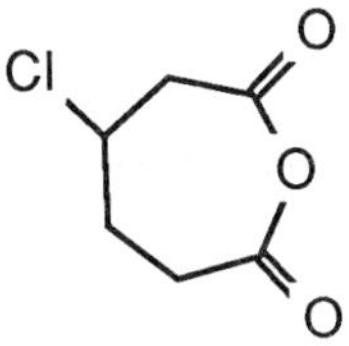

(o) $C_6H_5C\equiv N$

(p) 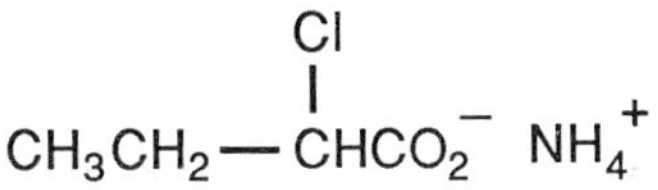

16.4 **(a)** The nitrile with fewer carbon atoms would be more soluble in water:

$$CH_3C\equiv N$$

(b) The diacid, with an additional polar group would be more water soluble:

$$HO_2C(CH_2)_4CO_2H$$

(c) The diacid, with the greater ability to undergo hydrogen bonding would have greater water solubility than the isomeric diester.

$$HO\overset{\overset{O}{\|}}{C}-\underset{\underset{CH_3}{|}}{CH}-CH_2-\underset{\underset{CH_3}{|}}{CH}-\overset{\overset{O}{\|}}{C}OH$$

16.5 **(a)** The nitrile with more carbon atoms would have a higher boiling point:

$$CH_3(CH_2)_4C\equiv N$$

(b) The diacid, with an additional polar (hydrogen bonding) group would be higher boiling:

$$HO_2C(CH_2)_4CO_2H$$

(c) The diacid, with the greater ability to undergo hydrogen bonding would have a higher boiling point than the isomeric diester.

$$HO\overset{\overset{O}{\|}}{C}-\underset{\underset{CH_3}{|}}{CH}-CH_2-\underset{\underset{CH_3}{|}}{CH}-\overset{\overset{O}{\|}}{C}OH$$

16.6

(a) $CH_3CH_2CO_2^- \quad + \quad CH_3OH$

(b) $CH_3CH_2CO_2^- \quad + \quad NH_3$

(c) $CH_3-\overset{\overset{CH_3}{|}}{C}HCO_2^-$

(d) $CH_3-\overset{\overset{CH_3}{|}}{C}HCH_2CO_2^-$

(e) $CH_3(CH_2)_4CO_2^- \quad + \quad NH_3$

(f) $C_6H_5CO_2^- \quad + \quad CH_3CH_2OH$

16.7 **(a)** $CH_3CH_2CO_2H \xrightarrow[H_2SO_4]{CH_3OH} CH_3CH_2CO_2CH_3$

(b) $CH_3CH_2CO_2H \xrightarrow[2.\ NH_3]{1.\ PCl_5} CH_3CH_2\overset{\overset{O}{\|}}{C}-NH_2$

(c) $CH_3-\overset{\overset{CH_3}{|}}{C}HCO_2H \xrightarrow{PCl_5} CH_3-\overset{\overset{CH_3}{|}}{C}HCOCl$

(e) $CH_3(CH_2)_4CO_2H \xrightarrow[2.\ NH_3]{1.\ PCl_5} CH_3(CH_2)_4\overset{\overset{O}{||}}{C}-NH_2 \xrightarrow{P_2O_5} CH_3(CH_2)_4CN$

(f) $C_6H_5CO_2H \xrightarrow[H_2SO_4]{CH_3CH_2OH} C_6H_5CO_2CH_2CH_3$

16.8

(a) $CH_3CH_2CH_2OH$ + CH_3OH

(b) $CH_3CH_2CH_2NH_2$

(c) $(CH_3)_2CHCH_2OH$

(d) $(CH_3)_2CHCH_2CH_2OH$

(e) $CH_3(CH_2)_4CH_2NH_2$

(f) $C_6H_5CH_2OH$ + CH_3CH_2OH

16.9

(a) $C_6H_5CO_2CH_2CH_3$ + CH_3OH

(b) $CH_3(CH_2)_4CO_2CH_2CH_3$

(c) $C_2H_5O_2C(CH_2)_3CO_2C_2H_5$

(d) $CH_3CH_2CO_2CH_2CH_3$ + C_6H_5OH

16.10

(a) $C_6H_5\overset{\overset{O}{||}}{C}-NHCH_2CH_2CH_2CH_3$ + CH_3OH

(b) $CH_3(CH_2)_4CO_2^-$ + $CH_3CH_2CH_2CH_2NH_3^+$

(c) $^-O_2C(CH_2)_3CO_2$ + $CH_3CH_2CH_2CH_2NH_3^+$

(d) $CH_3CH_2\overset{\overset{O}{||}}{C}-NHCH_2CH_2CH_2CH_3$ + C_6H_5OH

16.11 The products isolated after workup would be:

(a) $C_6H_5CH_2\underset{\underset{CH_3}{|}}{\overset{\overset{OH}{|}}{C}}-CH_3$

(b) $CH_3-\overset{\overset{CH_3}{|}}{C}H-CH_2-\overset{\overset{O}{||}}{C}-CH_3$

(c) 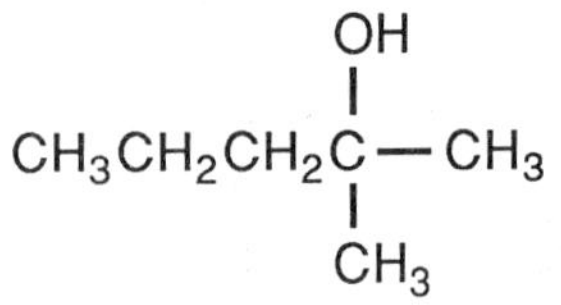
$$CH_3CH_2CH_2C(OH)(CH_3)-CH_3$$

(d) 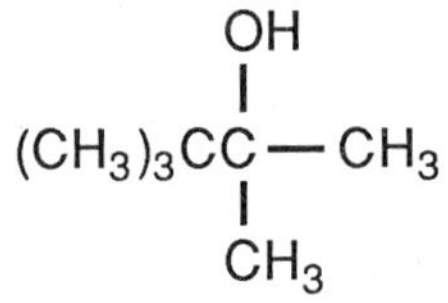
$$(CH_3)_3CC(OH)(CH_3)-CH_3$$

(e) 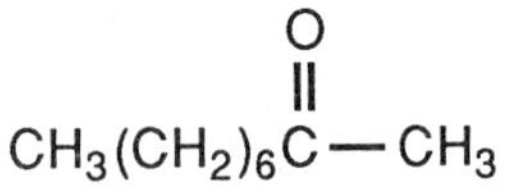
$$CH_3(CH_2)_6C(=O)-CH_3$$

16.12 The products isolated after workup would be:

(a) 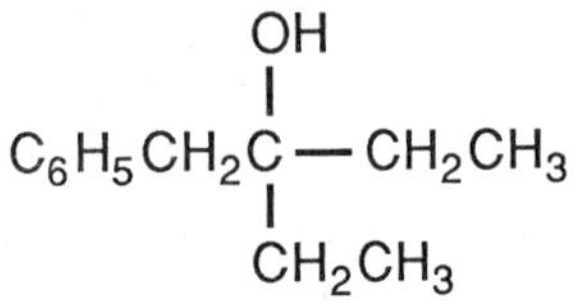
$$C_6H_5CH_2C(OH)(CH_2CH_3)-CH_2CH_3$$

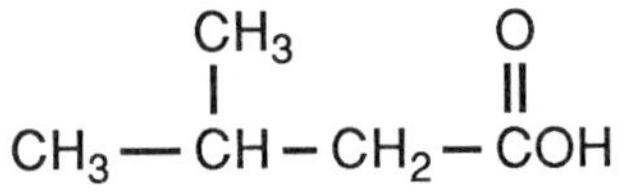

(b)
$$CH_3-CH(CH_3)-CH_2-C(=O)OH$$

(c)
$$CH_3CH_2CH_2C(OH)(CH_2CH_3)-CH_2CH_3$$

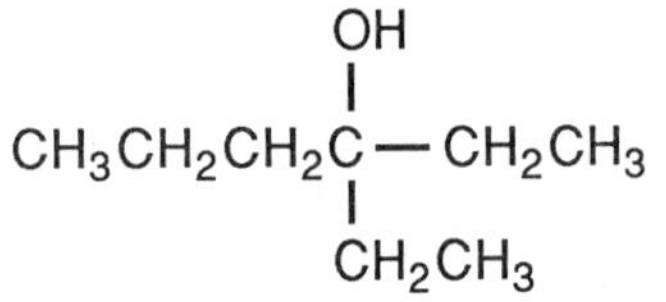

(d) 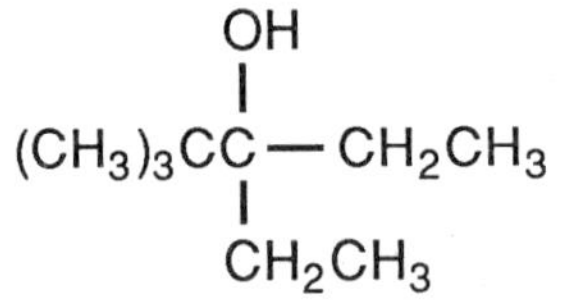
$$(CH_3)_3CC(OH)(CH_2CH_3)-CH_2CH_3$$

(e) 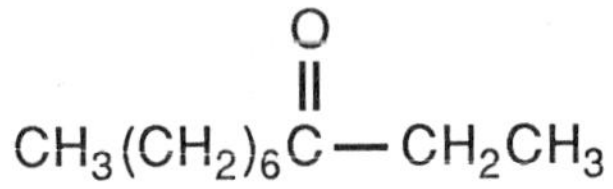
$$CH_3(CH_2)_6C(=O)-CH_2CH_3$$

16.13

(a) $C_6H_5CH_2CH_2OH$ + CH_3OH

(b) $(CH_3)_2CHCH_2CH_2OH$

(c) $CH_3CH_2CH_2CH_2OH$

(d) $(CH_3)_3CCH_2OH$ + CH_3CH_2OH

(e) $CH_3(CH_2)_6CH_2NH_2$

16.14

(a) $$C_6H_5CH_2CO_2H \xrightarrow[H_2SO_4]{CH_3OH} C_6H_5CH_2CO_2CH_3$$

(c) 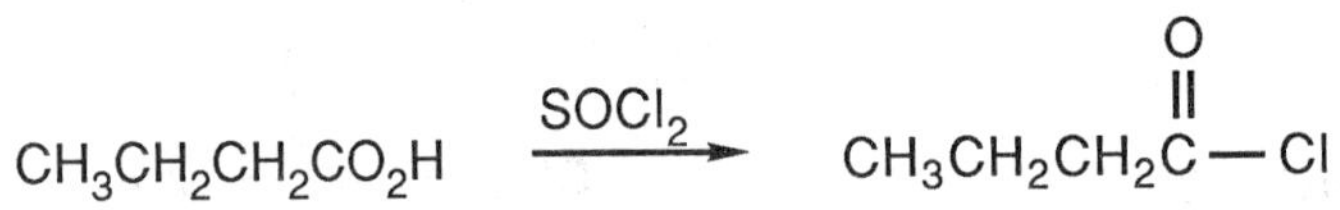 $$CH_3CH_2CH_2CO_2H \xrightarrow{SOCl_2} CH_3CH_2CH_2C(=O)-Cl$$

(d) $$(CH_3)_3CCO_2H \xrightarrow[H_2SO_4]{CH_3CH_2OH} (CH_3)_3CCO_2CH_2CH_3$$

(e) 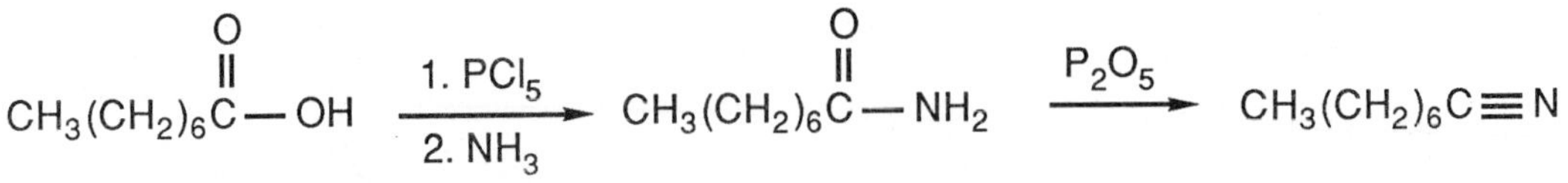 $$CH_3(CH_2)_6C(=O)-OH \xrightarrow[2.\ NH_3]{1.\ PCl_5} CH_3(CH_2)_6C(=O)-NH_2 \xrightarrow{P_2O_5} CH_3(CH_2)_6C\equiv N$$

16.15

(a) 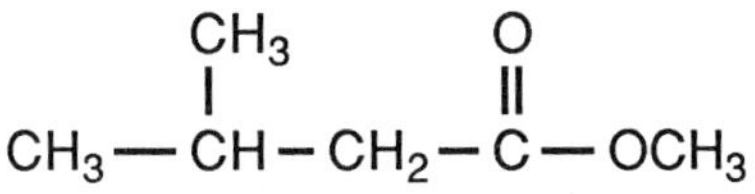 $CH_3-CH(CH_3)-CH_2-C(=O)-OCH_3$

(b) $CH_3-CH(CH_3)-CH_2-C(=O)-CH_3$

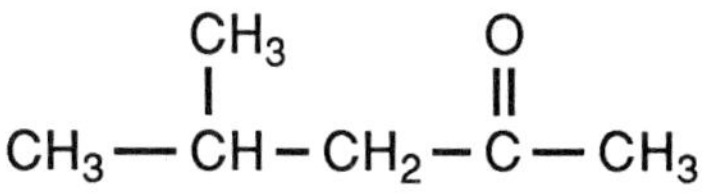

(c) $CH_3-CH(CH_3)-CH_2-C(=O)-OCH_2CH_3$

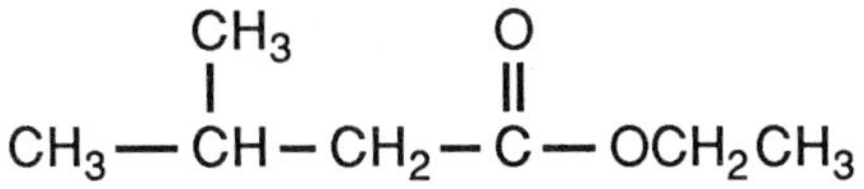

(d) 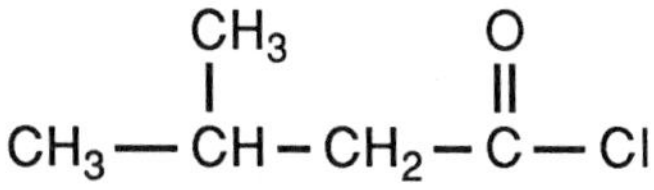 $CH_3-CH(CH_3)-CH_2-C(=O)-Cl$

(e) 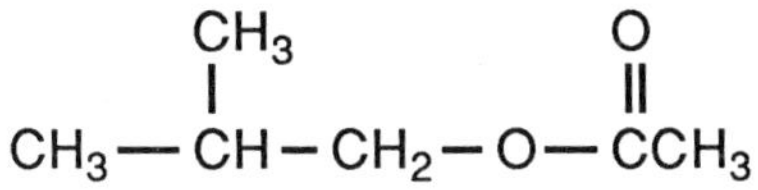 $CH_3-CH(CH_3)-CH_2-O-C(=O)CH_3$

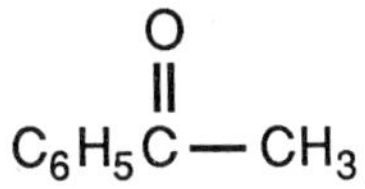

(f) 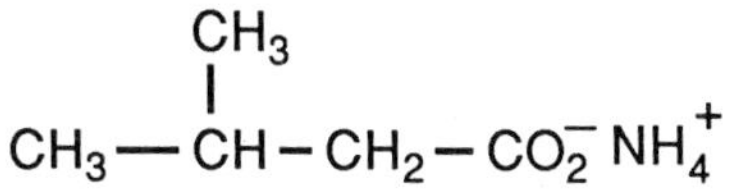 $CH_3-CH(CH_3)-CH_2-CO_2^-\ NH_4^+$

(g) 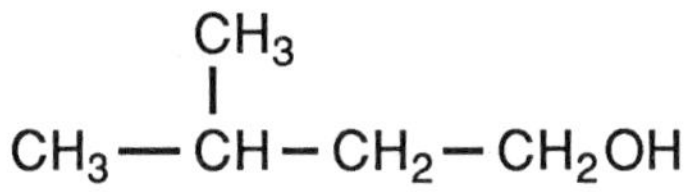 $CH_3-CH(CH_3)-CH_2-CH_2OH$

(h) 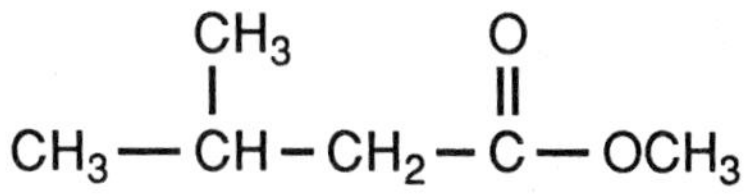 $CH_3-CH(CH_3)-CH_2-C(=O)-OCH_3$

16.16

(a) 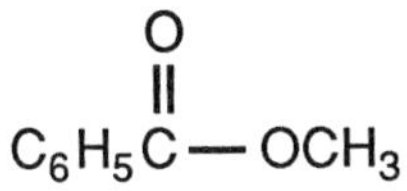 $C_6H_5C(=O)-OCH_3$

(b) $C_6H_5C(=O)-CH_3$

(c) $C_6H_5C(=O)-OCH_2CH_3$

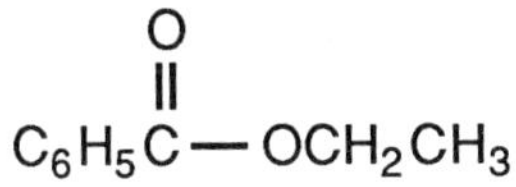

(d) 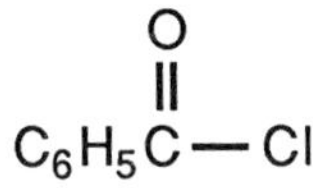 $C_6H_5C(=O)-Cl$

(e) 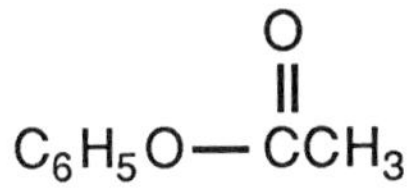 $C_6H_5O-C(=O)CH_3$

(f) $C_6H_5CO_2^-\ NH_4^+$

(g) $C_6H_5CH_2OH$

(h) 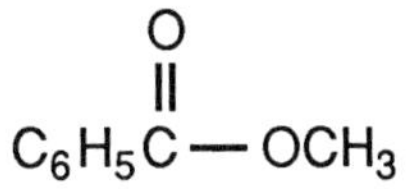 $C_6H_5C(=O)-OCH_3$

16.17

(a) $CH_3(CH_2)_4CH_2OH$

(b) 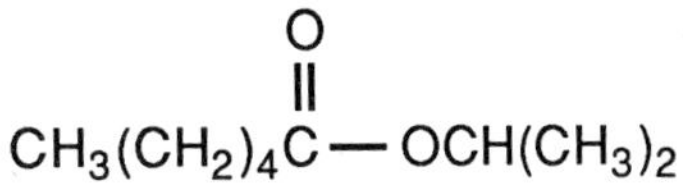 $CH_3(CH_2)_4C(=O)-OCH(CH_3)_2$

(c) 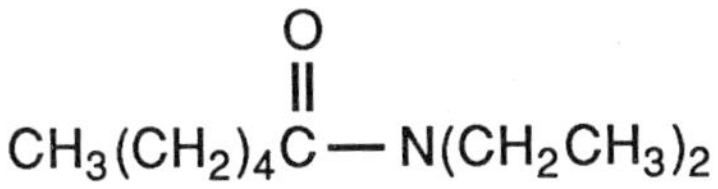 $CH_3(CH_2)_4C(=O)-N(CH_2CH_3)_2$

(d) 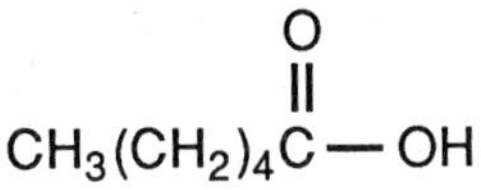 $CH_3(CH_2)_4C(=O)-OH$

(e) 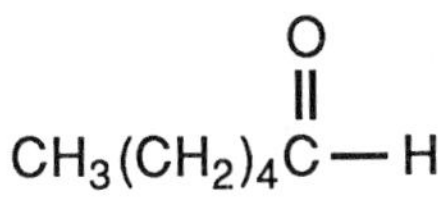 $CH_3(CH_2)_4C(=O)-H$

(f) 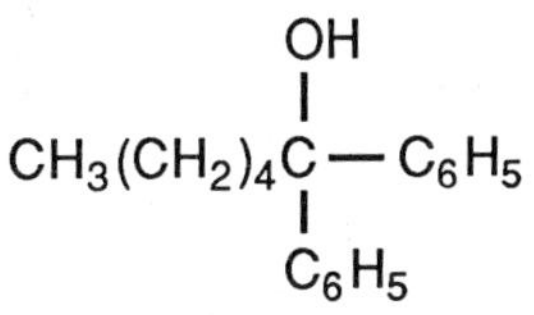 $CH_3(CH_2)_4C(OH)(C_6H_5)-C_6H_5$

(g) $CH_3(CH_2)_4C{\equiv}N$

(h) 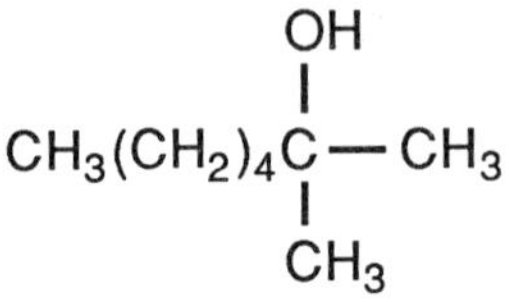 $CH_3(CH_2)_4C(OH)(CH_3)-CH_3$

(i) 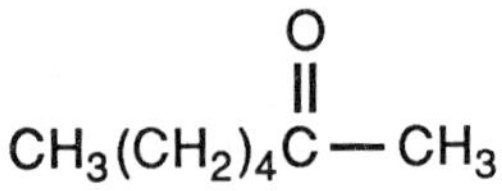 $CH_3(CH_2)_4C(=O)-CH_3$

16.18 (a) $C_6H_5CH_2OH$

(b) 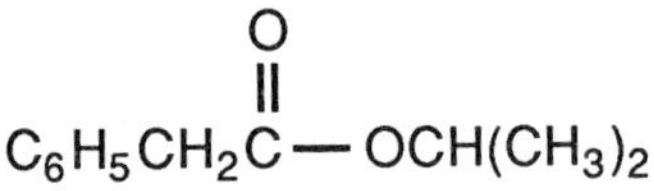 $C_6H_5CH_2C(=O)-OCH(CH_3)_2$

(c) $C_6H_5CH_2C(=O)-N(CH_2CH_3)_2$

(d) 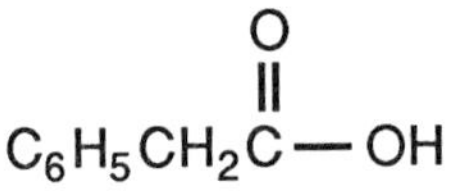 $C_6H_5CH_2C(=O)-OH$

(e) 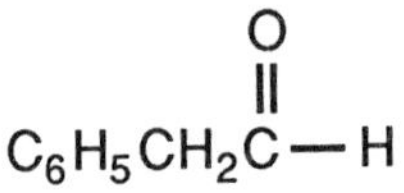 $C_6H_5CH_2C(=O)-H$

(f) 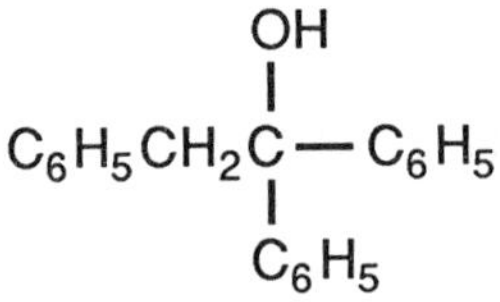 $C_6H_5CH_2C(OH)(C_6H_5)-C_6H_5$

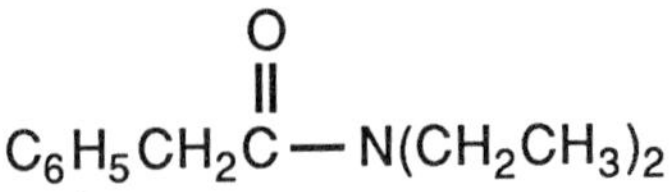

(g) $C_6H_5CH_2C{\equiv}N$

(h) 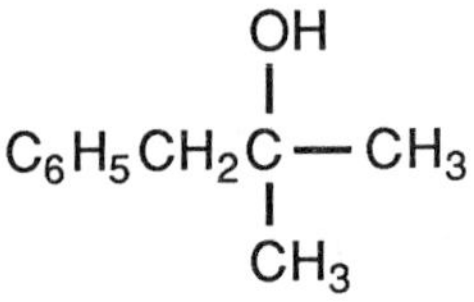 $C_6H_5CH_2C(OH)(CH_3)-CH_3$

(i) 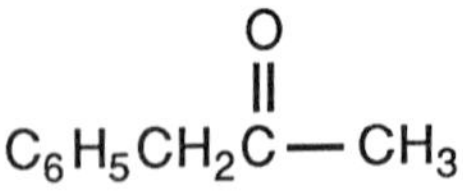 $C_6H_5CH_2C(=O)-CH_3$

16.19

(a) 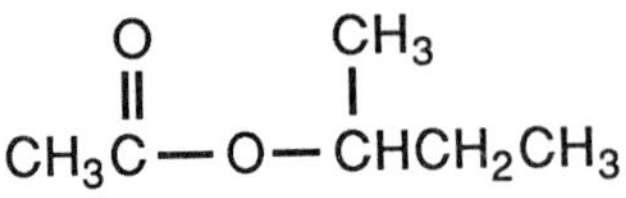 $CH_3C(=O)-O-CH(CH_3)CH_2CH_3$

(b) 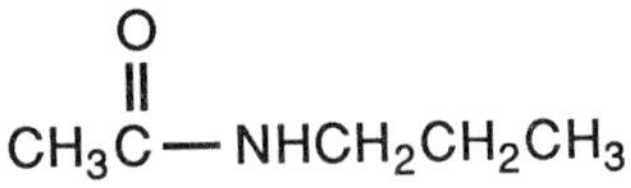 $CH_3C(=O)-NHCH_2CH_2CH_3$

(c) 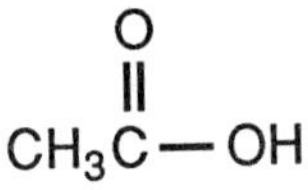 $CH_3C(=O)-OH$

(d) 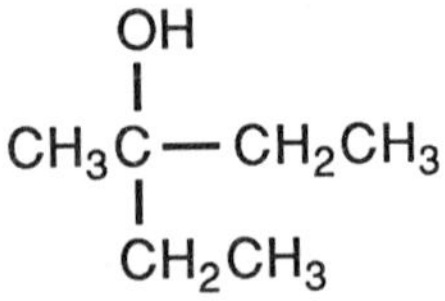 $CH_3C(OH)(CH_2CH_3)-CH_2CH_3$

(e) 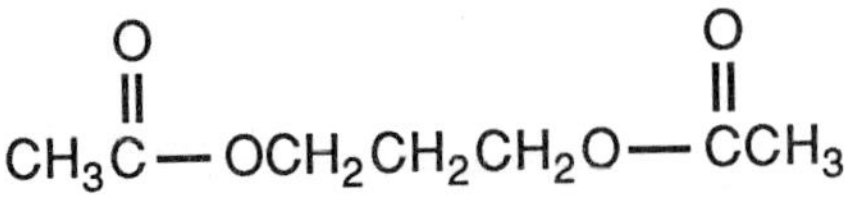 $CH_3C(=O)-OCH_2CH_2CH_2O-C(=O)CH_3$

(f) CH_3CH_2OH

16.20 (a) $C_6H_5C(=O)-O-CH(CH_3)CH_2CH_3$

(b) $C_6H_5C(=O)-NHCH_2CH_2CH_3$

(c) $C_6H_5C(=O)-OH$

(d) $C_6H_5C(OH)(CH_2CH_3)-CH_2CH_3$

(e) $C_6H_5C(=O)-OCH_2CH_2CH_2O-C(=O)C_6H_5$

(f) $C_6H_5CH_2OH$

16.21 (a) $CH_3CH_2C(=O)-OH \quad + \quad CH_3CH_2CH_2CH_2OH$

(b) $CH_3CH_2C(=O)-OCH_3 \quad + \quad CH_3CH_2CH_2CH_2OH$

(c) $CH_3CH_2CH_2OH \quad + \quad CH_3CH_2CH_2CH_2OH$

(d) $CH_3CH_2C(=O)-O^- \quad + \quad CH_3CH_2CH_2CH_2OH$

(e) $CH_3CH_2C(=O)-NHCH_2CH_2CH_2CH_2CH_3 \quad + \quad CH_3CH_2CH_2CH_2OH$

(f) $CH_3CH_2C(=O)-OCH_2CH_3 \quad + \quad CH_3CH_2CH_2CH_2OH$

(g) $CH_3CH_2C(OH)(C_6H_5)-C_6H_5 \quad + \quad CH_3CH_2CH_2CH_2OH$

16.22

(a) $CH_3C(=O)—OH$ + cyclohexanol (OH)

(b) $CH_3C(=O)—OCH_3$ + cyclohexanol (OH)

(c) CH_3CH_2OH + cyclohexanol (OH)

(d) $CH_3C(=O)—O^-$ + cyclohexanol (OH)

(e) $CH_3C(=O)—NHCH_2CH_2CH_2CH_2CH_3$ + cyclohexanol (OH)

(f) $CH_3C(=O)—OCH_2CH_3$ + cyclohexanol (OH)

(g) $CH_3C(OH)(C_6H_5)—C_6H_5$ + cyclohexanol (OH)

16.23

(a) $CH_3CH_2CH_2CH_2NH_2$

(b) $CH_3CH_2CH_2CO_2^-\ K^+$ + NH_3

(c) $CH_3CH_2CH_2C(=O)—NH^-\ {}^+MgBr$

(d) $CH_3CH_2CH_2C(=O)—OH$ + NH_4^+

16.24

(a) $CH_3CH_2CH_2NHCH_2CH_2CH_3$

(b) $CH_3CH_2CO_2^-\ K^+$ + $CH_3CH_2CH_2NH_2$

(c) $CH_3CH_2C(=O)—\bar{N}CH_2CH_2CH_3\ {}^+MgBr$

(d) $CH_3CH_2C(=O)—OH$ + $CH_3CH_2CH_2NH_3^+$

16.25

(a) $$CH_3(CH_2)_4CH_2OH \xrightarrow[\text{2. Li}]{\text{1. PBr}_3} CH_3(CH_2)_4CH_2Li \xrightarrow[\text{2. HCl}]{\text{1. CO}_2} CH_3(CH_2)_4CH_2CO_2H$$

or

$$CH_3(CH_2)_4CH_2OH \xrightarrow[\text{2. NaCN}]{\text{1. PBr}_3} CH_3(CH_2)_4CH_2CN \xrightarrow[\text{2. HCl}]{\text{1. KOH, H}_2\text{O}} CH_3(CH_2)_4CH_2CO_2H$$

(b) $CH_3CH_2C(CH_3)_2-OH \xrightarrow[\text{2. Li}]{\text{1. PBr}_3} CH_3CH_2C(CH_3)_2-Li \xrightarrow[\text{2. HCl}]{\text{1. CO}_2} CH_3CH_2C(CH_3)_2-CO_2H$

or

$CH_3CH_2C(CH_3)_2-OH \xrightarrow[\text{2. H}_2\text{O}]{\text{1. CO, H}_2\text{SO}_4} CH_3CH_2C(CH_3)_2-CO_2H$

(c) cyclohexyl–OH $\xrightarrow[\text{2. Li}]{\text{1. PBr}_3}$ cyclohexyl–Li $\xrightarrow[\text{2. HCl}]{\text{1. CO}_2}$ cyclohexyl–CO_2H

or

cyclohexyl–OH $\xrightarrow[\text{2. NaCN}]{\text{1. PBr}_3}$ cyclohexyl–CN $\xrightarrow[\text{2. HCl}]{\text{1. KOH, H}_2\text{O}}$ cyclohexyl–CO_2H

(d) 1-methylcyclohexyl (OH, CH_3) $\xrightarrow[\text{2. Li}]{\text{1. PBr}_3}$ 1-methylcyclohexyl (Li, CH_3) $\xrightarrow[\text{2. HCl}]{\text{1. CO}_2}$ 1-methylcyclohexyl (CO_2H, CH_3)

or

1-methylcyclohexyl (OH, CH_3) $\xrightarrow[\text{2. H}_2\text{O}]{\text{1. CO, H}_2\text{SO}_4}$ 1-methylcyclohexyl (CO_2H, CH_3)

(e) HO–cyclohexane–OH (1,4) $\xrightarrow[\text{2. NaCN}]{\text{1. PBr}_3}$ NC–cyclohexane–CN (1,4) $\xrightarrow[\text{2. HCl}]{\text{1. KOH, H}_2\text{O}}$ HO_2C–cyclohexane–CO_2H (1,4)

(f) $CH_3CH_2CH{=}CHCH_2OH \xrightarrow[\text{2. Li}]{\text{1. PBr}_3} CH_3CH_2CH{=}CHCH_2Li \xrightarrow[\text{2. HCl}]{\text{1. CO}_2} CH_3CH_2CH{=}CHCH_2CO_2H$

(g) $CH_3CH_2CH{=}CHCH_2OH \xrightarrow[\text{2. NaCN}]{\text{1. PBr}_3} CH_3CH_2CH{=}CHCH_2CN \xrightarrow[\text{2. HCl}]{\text{1. KOH, H}_2\text{O}} CH_3CH_2CH{=}CHCH_2CO_2H$

16.26 (a)

$$CH_3CH_2-\underset{}{\overset{CH_3}{\overset{|}{C}}}H-CH_2OH \xrightarrow{CrO_3} CH_3CH_2-\overset{CH_3}{\overset{|}{C}}H-CO_2H \xrightarrow[H_2SO_4]{CH_3OH} CH_3CH_2-\overset{CH_3}{\overset{|}{C}}H-CO_2CH_3$$

(b)

$$CH_3CH_2\overset{O}{\overset{\|}{C}}-CH_2CH_3 \xrightarrow{CH_3CO_3H} CH_3CH_2\overset{O}{\overset{\|}{C}}-OCH_2CH_3$$

(c)

$$CH_3\underset{CH_3}{\underset{|}{\overset{CH_3}{\overset{|}{C}}}}-\overset{O}{\overset{\|}{C}}CH_3 \xrightarrow[2.\ HCl]{1.\ NaOCl} CH_3\underset{CH_3}{\underset{|}{\overset{CH_3}{\overset{|}{C}}}}-\overset{O}{\overset{\|}{C}}OH \xrightarrow[H_2SO_4]{CH_3OH} CH_3\underset{CH_3}{\underset{|}{\overset{CH_3}{\overset{|}{C}}}}-\overset{O}{\overset{\|}{C}}OCH_3$$

(d)

$$CH_3\underset{CH_3}{\underset{|}{\overset{CH_3}{\overset{|}{C}}}}-\overset{O}{\overset{\|}{C}}CH_3 \xrightarrow{CH_3CO_3H} CH_3\underset{CH_3}{\underset{|}{\overset{CH_3}{\overset{|}{C}}}}-O-\overset{O}{\overset{\|}{C}}CH_3$$

(e)

$$CH_3CH_2CH_2CH_2CH_2\overset{O}{\overset{\|}{C}}-H \xrightarrow{CrO_3} CH_3CH_2CH_2CH_2CH_2\overset{O}{\overset{\|}{C}}-OH \xrightarrow[H_2SO_4]{CH_3OH} CH_3CH_2CH_2CH_2CH_2\overset{O}{\overset{\|}{C}}-OCH_3$$

16.27

(a) CO_2^- / CO_2^- (two carbons joined)

(b) CO_2CH_3 / CO_2^-

(c) CH_2OH / CH_2OH

(d) 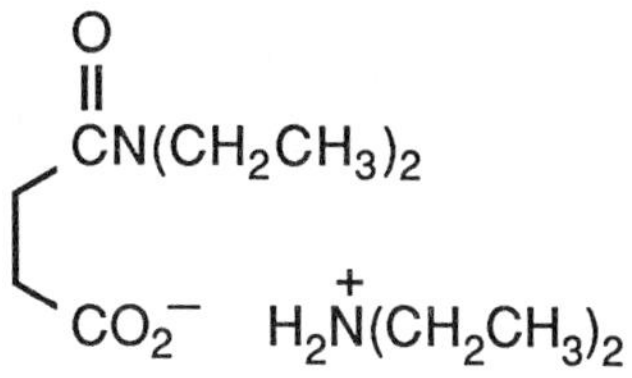 $\overset{O}{\overset{\|}{C}}N(CH_2CH_3)_2$ … CO_2^- $H_2\overset{+}{N}(CH_2CH_3)_2$

(e) 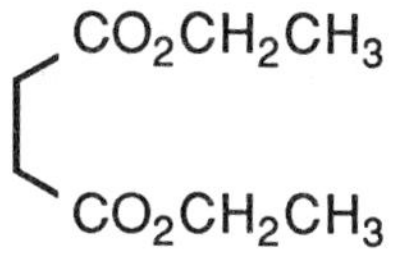 $CO_2CH_2CH_3$ … $CO_2CH_2CH_3$

(f) 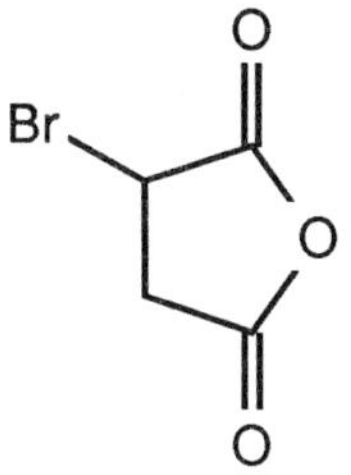 Br … O … O … O

16.28 (a) 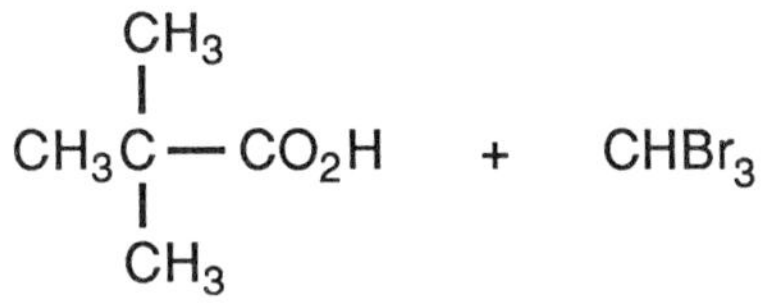 $CH_3C(CH_3)_2—CO_2H$ + $CHBr_3$

(b) 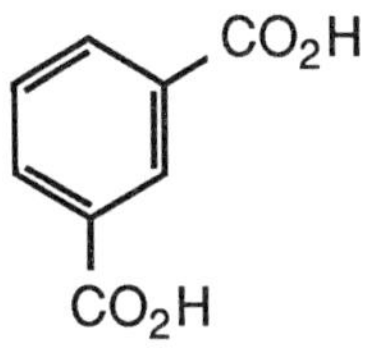 CO_2H … CO_2H

(c) 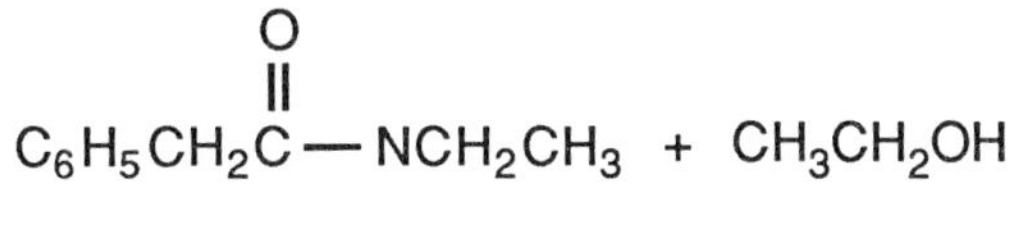 $C_6H_5CH_2\overset{O}{\overset{\|}{C}}—NCH_2CH_3$ + CH_3CH_2OH

(d) 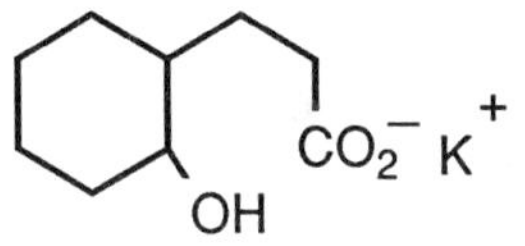 OH … CO_2^- $\overset{+}{K}$

(e) 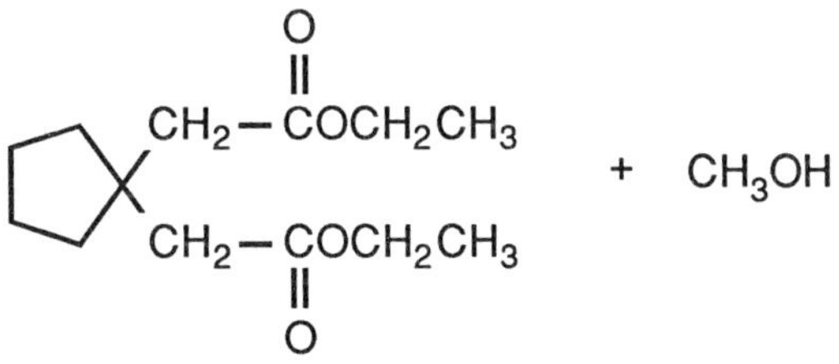 $CH_2—\overset{O}{\overset{\|}{C}}OCH_2CH_3$ … $CH_2—\underset{O}{\underset{\|}{C}}OCH_2CH_3$ + CH_3OH

(f) 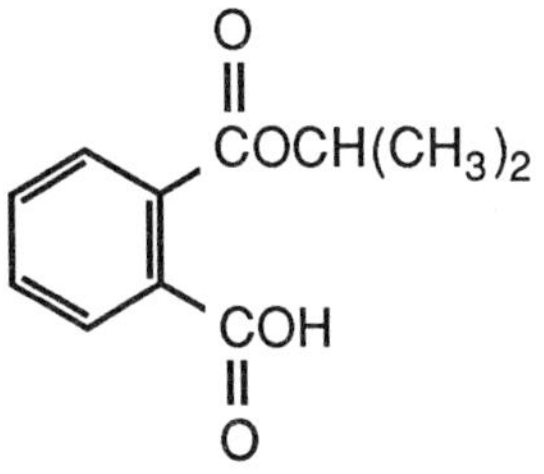 $\overset{O}{\overset{\|}{C}}OCH(CH_3)_2$ … $\underset{O}{\underset{\|}{C}}OH$

16.29

(a) 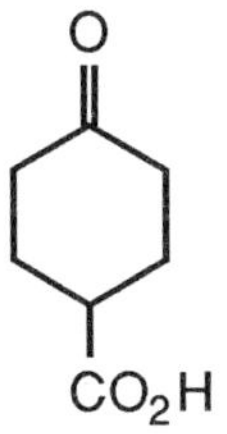 O … CO_2H

(b) 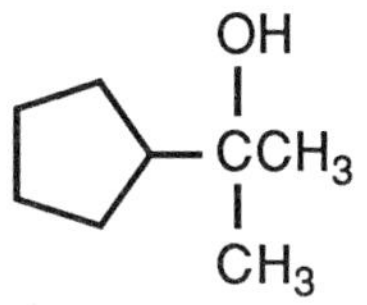 OH … CCH_3 … CH_3

(c) 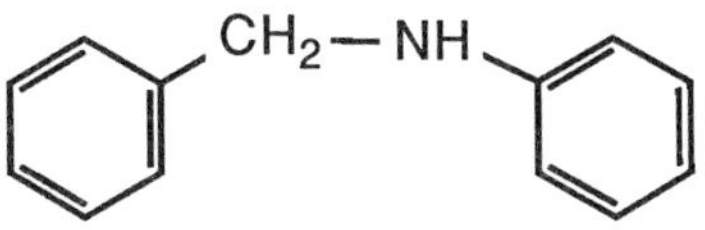 $CH_2—NH$

(d) 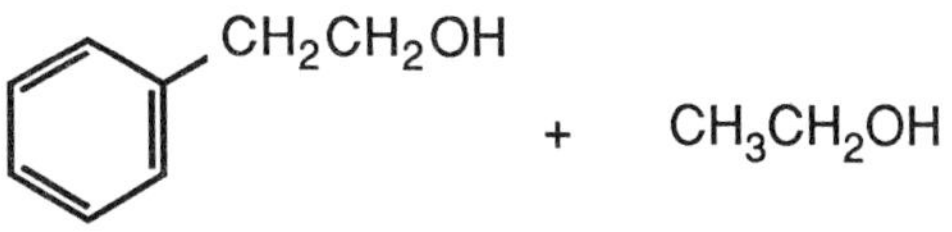 CH_2CH_2OH + CH_3CH_2OH

16.30 All of the structural information provided in this problem is indirect, and it is necessary to build up partial structures until the complete structure of at least one compound can be deduced. Ester **A** has a single degree of unsaturation, which must be the carbonyl group. Since both yield the same products (**B** and **C**), compounds **A** and **E** must be related as:

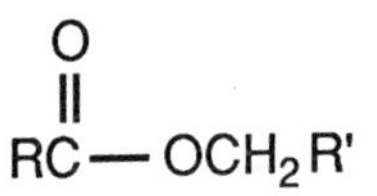

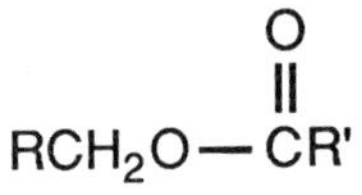

$$\underset{}{\overset{O}{\overset{\|}{RC}}}—OCH_2R' \qquad RCH_2O—\overset{O}{\overset{\|}{C}}R'$$

Therefore, **B** and **C** are both primary alcohols:

B: RCH_2OH **C:** $R'CH_2OH$.

Oxidation of **C** followed by reaction with $SOCl_2$ produces the acid chloride, **D** , which can be denoted as R'COCl. Reaction of this with **B** yields **E**, so **E** corresponds to:

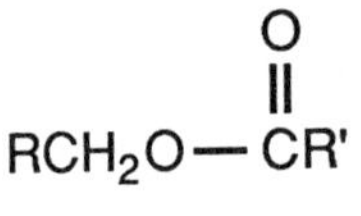

$$RCH_2O—\overset{O}{\overset{\|}{C}}R'$$

Reaction of **D** with lithium diisobutylcuprate affords a ketone with the partial structure **I**:

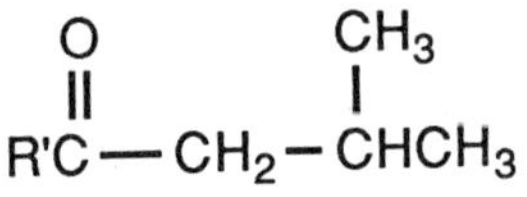

$$R'\overset{O}{\overset{\|}{C}}—CH_2—\overset{CH_3}{\overset{|}{C}}HCH_3$$

F is also an acid chloride, and it reacts with **C** to form **A**; therefore it corresponds to RCOCl. Reduction affords an aldehyde, so **G** is RCHO. Reaction with ethylmagnesium bromide yields **H**, and oxidation affords the ketone **I**:

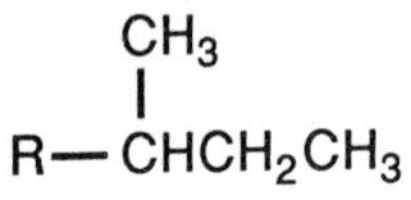

$$\underset{\mathbf{H}}{R—\overset{CH_3}{\overset{|}{C}}HCH_2CH_3} \qquad \underset{\mathbf{I}}{R\overset{O}{\overset{\|}{C}}—CH_2CH_3}$$

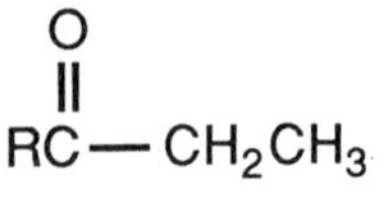

The two different designations for **I** permit the two alkyl groups R and R' to be identified as isobutyl and ethyl, respectively. The complete answer can now be shown as follows:

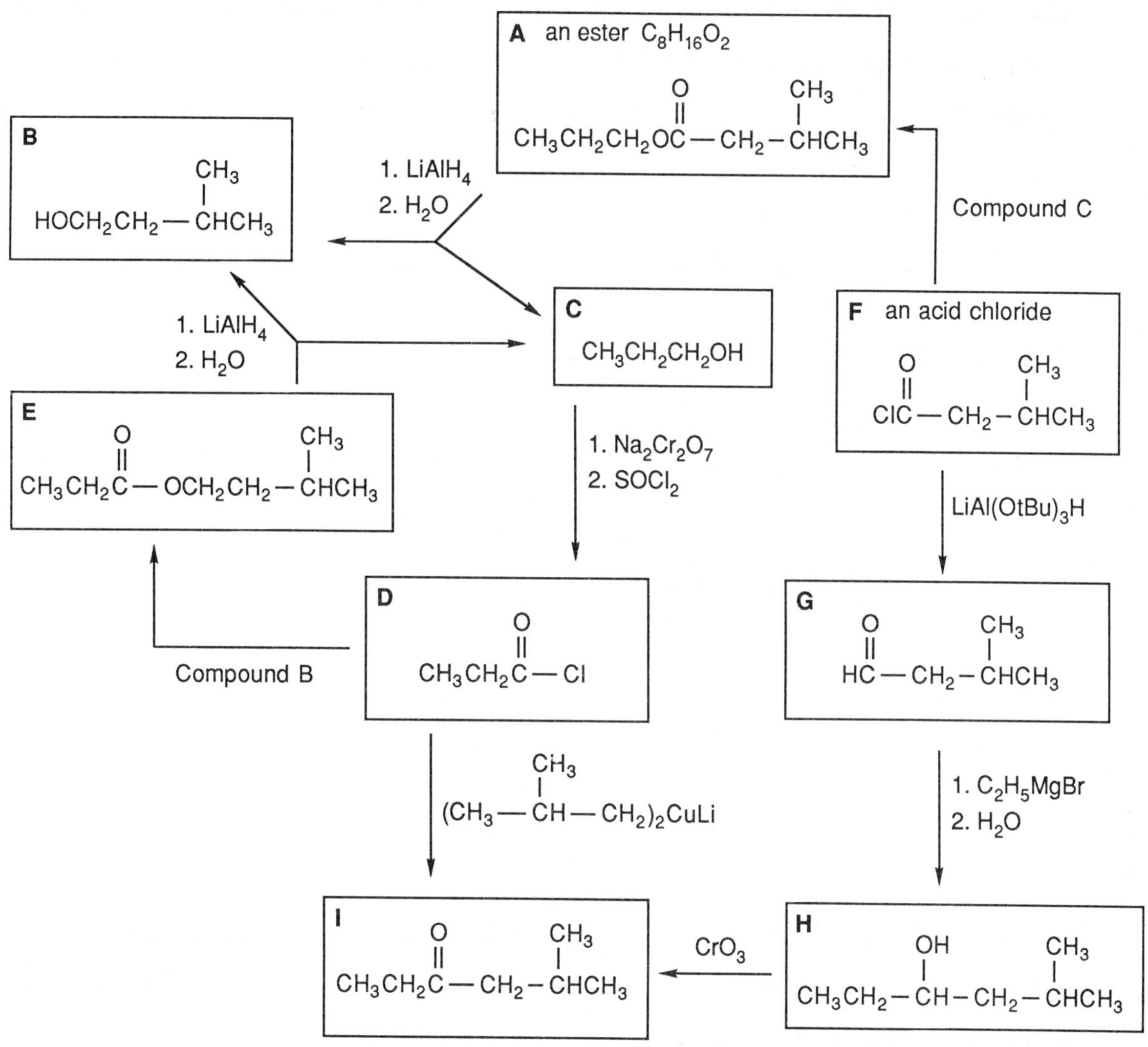

16.31 As in the preceding problem, the complete structure is not provided for any of the compounds, so you must build up a series of partial structures. The molecular formula for **G** indicates a single degree of unsaturation. This must correspond to the ester carbonyl, so there are no other double bonds or rings. The sequence to produce the acid chloride **C** from alcohol **A** allow us to write the following partial structures:

A: RCH_2OH **B**: RCO_2H **C**: $RCOCl$

The reaction of **C** with lithium dipropylcuprate yields **D**, which must have the following partial structure:

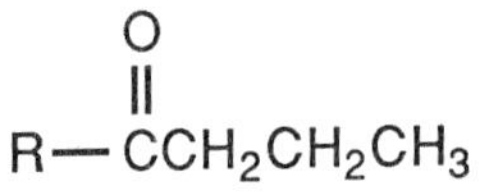

$$R-\overset{\overset{\displaystyle O}{\|}}{C}CH_2CH_2CH_3$$

Reaction of **D** with ethyllithium affords **E**:

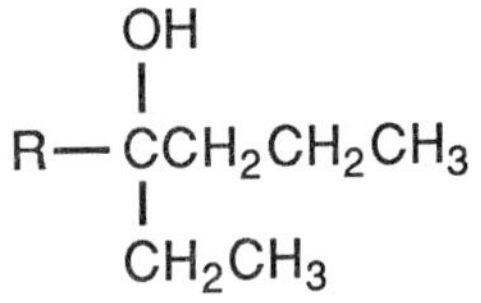

$$R-\underset{\underset{\displaystyle CH_2CH_3}{|}}{\overset{\overset{\displaystyle OH}{|}}{C}}CH_2CH_2CH_3$$

E is also formed by the reaction of ethylmagnesium bromide with **G**, so a second partial structure can be drawn:

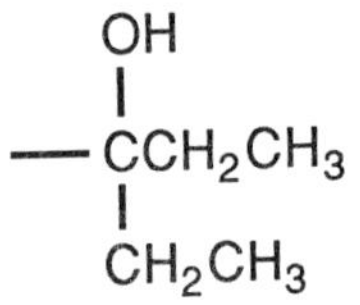

$$-\underset{\underset{\displaystyle CH_2CH_3}{|}}{\overset{\overset{\displaystyle OH}{|}}{C}}CH_2CH_3$$

This means that **E** has two ethyl groups and one propyl group, so its complete structure is:

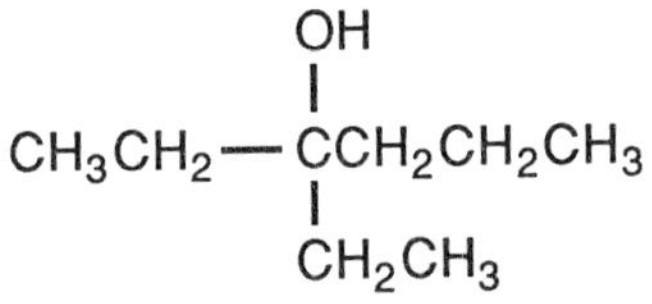

$$CH_3CH_2-\underset{\underset{\displaystyle CH_2CH_3}{|}}{\overset{\overset{\displaystyle OH}{|}}{C}}CH_2CH_2CH_3$$

The group we have labelled R must be ethyl, and the propyl group of **E** must have already been present in **G**. The molecular formula of **G** tells you that three carbons are unaccounted for, so the compound is either propyl butyrate or isopropyl butyrate. The choice between these two alternatives can be made on the basis of compounds **A** and **F**. Since R is ethyl, **F** is propyl alcohol, so **F** (which must be different) is isopropyl alcohol. The complete answer follows:

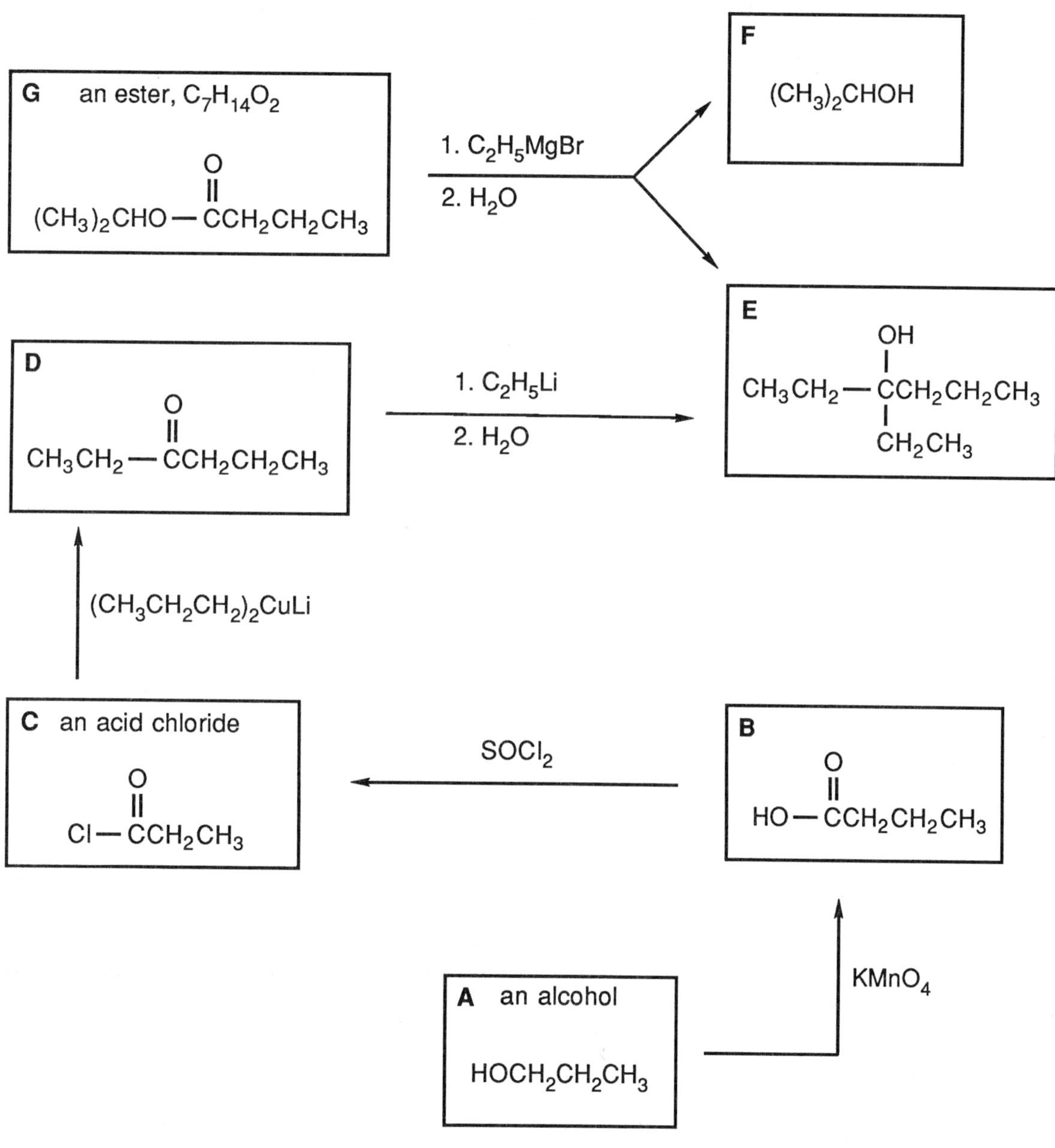

16.32

(a)

$$CH_3-\overset{CH_3}{\overset{|}{CH}}-CH_2-\overset{O}{\overset{\|}{C}}-OH \xrightarrow[H_2SO_4]{CH_3CH_2OH} CH_3-\overset{CH_3}{\overset{|}{CH}}-CH_2-\overset{O}{\overset{\|}{C}}-OCH_2CH_3$$

or

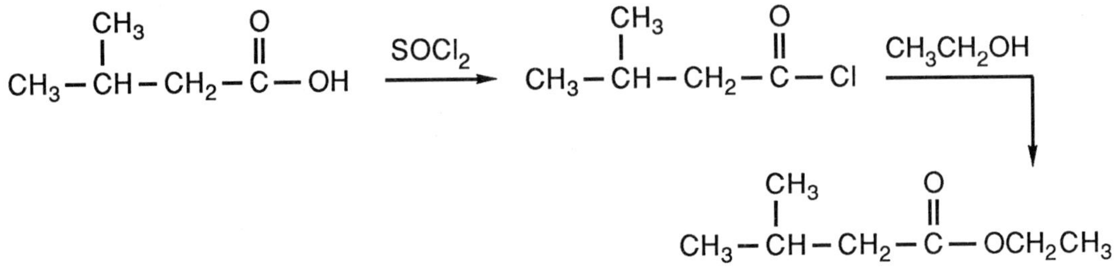

$$CH_3-\underset{}{\overset{CH_3}{\overset{|}{C}H}}-CH_2-\overset{O}{\overset{||}{C}}-OH \xrightarrow{SOCl_2} CH_3-\overset{CH_3}{\overset{|}{C}H}-CH_2-\overset{O}{\overset{||}{C}}-Cl \xrightarrow{CH_3CH_2OH} CH_3-\overset{CH_3}{\overset{|}{C}H}-CH_2-\overset{O}{\overset{||}{C}}-OCH_2CH_3$$

(b)

$$CH_3-\overset{CH_3}{\overset{|}{C}H}-CH_2-\overset{O}{\overset{||}{C}}-OH \xrightarrow[2.\ H_2O]{1.\ LiAlH_4} CH_3-\overset{CH_3}{\overset{|}{C}H}-CH_2-CH_2OH \xrightarrow[\text{or } Ac_2O]{CH_3-\overset{O}{\overset{||}{C}}Cl} CH_3-\overset{CH_3}{\overset{|}{C}H}-CH_2-O-\overset{O}{\overset{||}{C}}CH_3$$

(c)

2-bromotoluene (Br, CH_3) $\xrightarrow{Li}$ (Li, CH_3) $\xrightarrow{CO_2}$ (CO_2^-, CH_3) $\xrightarrow{HCl}$ (CO_2H, CH_3)

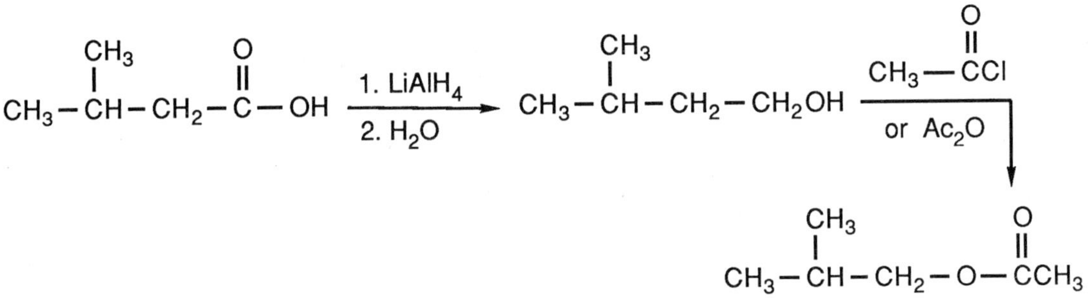

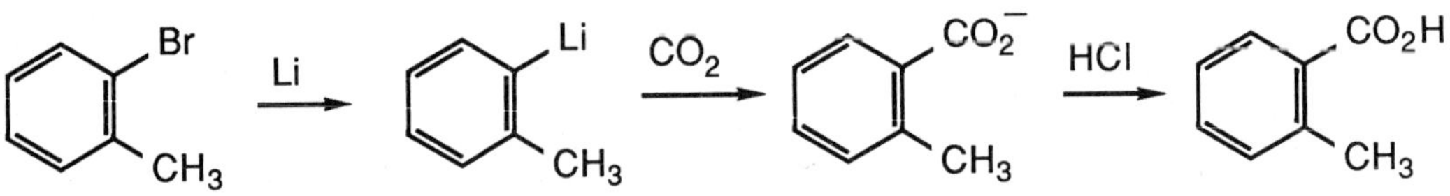

(d)

3-(CH_2Br)benzoic acid (CO_2H, CH_2Br) $\xrightarrow{KCN}$ (CO_2H, CH_2CN) $\xrightarrow[NaOH]{H_2O}$ (CO_2^-, $CH_2CO_2^-$) $\xrightarrow{HCl}$ (CO_2H, CH_2CO_2H)

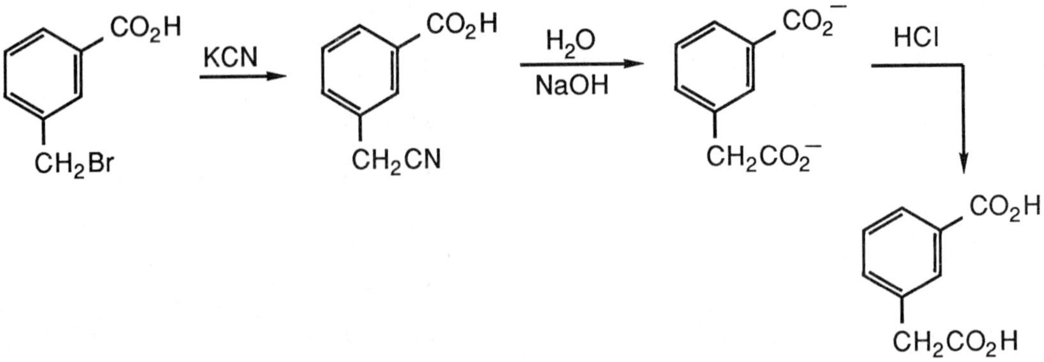

16.33

(a)

$$CH_3CH_2C(CH_3)_2-OH \xrightarrow{PBr_3} CH_3CH_2C(CH_3)_2-Br \xrightarrow{Li} CH_3CH_2C(CH_3)_2-Li \xrightarrow[2.\ H_2O]{1.\ CO_2} CH_3CH_2C(CH_3)_2-CO_2H$$

$$CH_3CH_2C(CH_3)_2-CO_2H \xrightarrow[2.\ NH_3]{1.\ SOCl_2} CH_3CH_2C(CH_3)_2-CONH_2 \xrightarrow{P_2O_5} CH_3CH_2C(CH_3)_2-CN$$

(b)

O, CO_2Et $\xrightarrow[TsOH\ /\ C_6H_6]{HOCH_2CH_2OH}$ O, O, CO_2Et

1. CH_3MgBr
2. H_2O

O, O, OH, CH_3, CH_3 $\xrightarrow{dil.\ H_2SO_4}$ O, OH, CH_3, CH_3

16.34 The ester has a total of 5 degrees of unsaturation, and all of these seem to be in the carboxylic acid portion: the carbonyl plus four for the aromatic ring that is indicated by the NMR spectrum. The aromatic ring is a sharp 4H singlet, suggesting a *p*-disubstituted benzene ring with no electronegative groups attached directly to the ring. The singlets at 2.3 and 3.5 ppm indicate isolated CH_2 and CH_3 groups, and the structure is readily deduced as CH_3-C_6H_4-CH_2-CO_2H.

The alcohol produced by hydrolysis of the ester must therefore have three carbons, so 1-propanol and 2-propanol are the only possibilities. The three multiplets (plus the 1H signal for hydroxyl) are consistent only with the former. The complete structures are:

$CH_3-C_6H_4-CH_2-C(=O)OCH_2CH_2CH_3$ (A)

$CH_3CH_2CH_2OH$ (B)

$CH_3-C_6H_4-CH_2CO_2H$ (C)

16.35 Each of the alcohols contains a phenyl group as indicated by the 5H signals at 7.2 ppm. The two triplets at 2.8 and 3.8 ppm for **B** indicate the partial structure $-CH_2-CH_2-$, and this leaves only C_6H_5- and $-OH$. The complete structure must be:

$$C_6H_5-CH_2CH_2OH$$

Alcohol **C** shows a 1H signal at 1.9 ppm (hydroxyl) and a 3H doublet at 1.5 ppm. The latter signal indicates a $CH-CH_3$ group, and the only possible structure for the alcohol is:

$$C_6H_5-CH(OH)CH_3$$

Reduction of the ester converts the carboxylic acid portion into a primary alcohol, so this corresponds to **B**, and the complete structures are:

$C_6H_5-CH_2-C(=O)-O-CH(CH_3)-C_6H_5$ (A)

$C_6H_5-CH_2CH_2OH$ (B)

$C_6H_5-CH(OH)CH_3$ (C)

16.36 In every case the characteristic, broad OH peak of the carboxylic acid (2500–3500 cm^{-1}) in the infrared spectrum will be absent in the product. Other changes are noted for the specific reactions.

(a)

$$CH_3-\overset{\overset{CH_3}{|}}{C}HCH_2-\overset{\overset{O}{\|}}{C}OH \longrightarrow CH_3-\overset{\overset{CH_3}{|}}{C}H-CH_2-\overset{\overset{O}{\|}}{C}-OCH_3$$

A new peak for the CH_3 group would be found in the NMR spectrum near 3.7 ppm.

(b)

$$CH_3-\overset{\overset{CH_3}{|}}{C}HCH_2-\overset{\overset{O}{\|}}{C}OH \longrightarrow CH_3-\overset{\overset{CH_3}{|}}{C}H-CH_2-\overset{\overset{O}{\|}}{C}-CH_3$$

A new peak for the CH_3 group would be found in the NMR spectrum near 2.1 ppm.

(c)

$$CH_3-\overset{\overset{CH_3}{|}}{C}HCH_2-\overset{\overset{O}{\|}}{C}OH \longrightarrow CH_3-\overset{\overset{CH_3}{|}}{C}H-CH_2-\overset{\overset{O}{\|}}{C}-OCH_2CH_3$$

A new peak for the C_2H_5 group would be found in the NMR spectrum near 4.1 ppm (3H quartet) and 1.3 ppm (3H triplet).

(d)

$$CH_3-\overset{\overset{CH_3}{|}}{C}HCH_2-\overset{\overset{O}{\|}}{C}OH \longrightarrow CH_3-\overset{\overset{CH_3}{|}}{C}H-CH_2-\overset{\overset{O}{\|}}{C}-Cl$$

There would be very few differences in the NMR spectrum, although the COOH signal (11.5–12.5 ppm) would be absent.

(e)

$$CH_3-\overset{\overset{CH_3}{|}}{C}HCH_2-\overset{\overset{O}{\|}}{C}OH \longrightarrow CH_3-\overset{\overset{CH_3}{|}}{C}H-CH_2-O-\overset{\overset{O}{\|}}{C}CH_3$$

A new peak for the CH_3 group would be found in the NMR spectrum near 1.9 ppm, and the CH_2 triplet near 2.0 ppm would be shifted to about 4.1 ppm.

(f)

$$CH_3-\overset{\overset{CH_3}{|}}{C}HCH_2-\overset{\overset{O}{\|}}{C}OH \longrightarrow CH_3-\overset{\overset{CH_3}{|}}{C}H-CH_2-CO_2^-\,NH_4^+$$

Both the infrared and NMR spectra would be different, but there are no key predictable changes that you would expect on the basis if the discussion in the text. The extremely broad OH peak in the infrared would be replaced by another broad peak for NH_4^+ in the 3000–3300 cm^{-1} region.

(g)

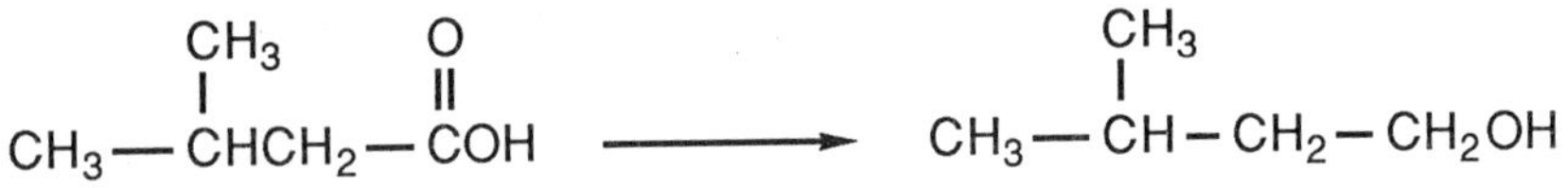

$$CH_3-\underset{}{\overset{CH_3}{\overset{|}{C}}}HCH_2-\overset{O}{\overset{\|}{C}}OH \longrightarrow CH_3-\overset{CH_3}{\overset{|}{C}}H-CH_2-CH_2OH$$

A new peak for the CH_2OH group would be found in the NMR spectrum near 3.6 ppm, and the alcohol would give rise to an OH signal in the infrared spectrum in the 3200–3400 cm^{-1} region.

(h)

$$CH_3-\overset{CH_3}{\overset{|}{C}}HCH_2-\overset{O}{\overset{\|}{C}}OH \longrightarrow CH_3-\overset{CH_3}{\overset{|}{C}}H-CH_2-\overset{O}{\overset{\|}{C}}-OCH_3$$

A new peak for the CH_3 group would be found in the NMR spectrum near 3.7 ppm.

16.37 We have not discussed the infrared spectra of acid chlorides, but they show a C=O peak near 1800 cm^{-1}, which will either be absent or substantially shifted for each of the products. Other changes are indicated for the specific examples.

(a)

$$CH_3(CH_2)_4-\overset{O}{\overset{\|}{C}}-Cl \longrightarrow CH_3(CH_2)_4CH_2OH$$

A new peak for the CH_2OH group would be found in the NMR spectrum near 3.6 ppm (2H triplet), and the alcohol would give rise to an OH signal in the infrared spectrum in the 3200–3400 cm^{-1} region.

(b)

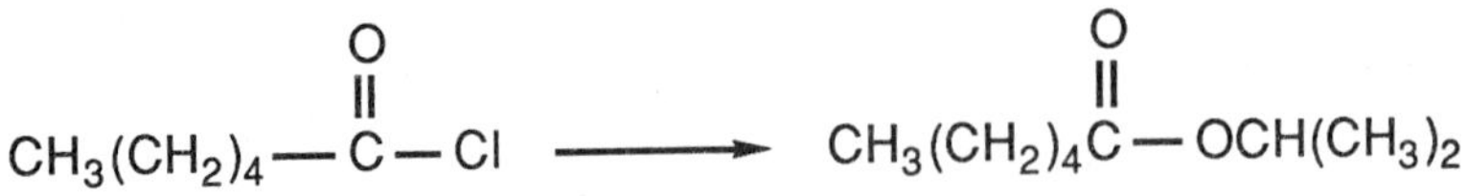

$$CH_3(CH_2)_4-\overset{O}{\overset{\|}{C}}-Cl \longrightarrow CH_3(CH_2)_4\overset{O}{\overset{\|}{C}}-OCH(CH_3)_2$$

The ester carbonyl would result in a peak in the 1730–1755 cm^{-1} region of the infrared spectrum. The isopropyl group would result in a doublet (6H) that would be expected to overlap other peaks in the region near 1.3 ppm of the NMR spectrum, and a septet (1H) of low intensity would arise in the 4.5–5 ppm region.

(c)

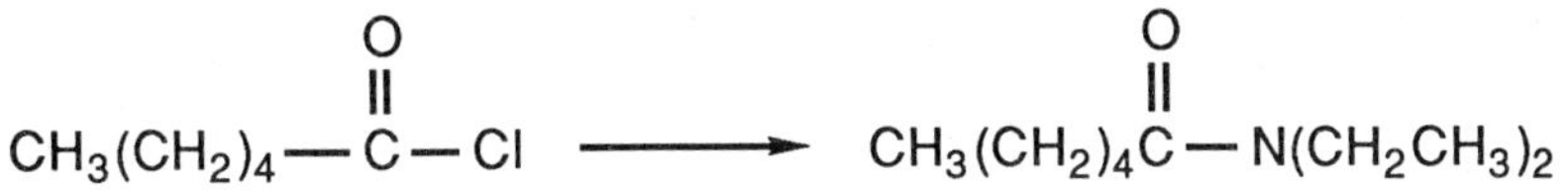

$$CH_3(CH_2)_4-\overset{\overset{\large O}{\|}}{C}-Cl \longrightarrow CH_3(CH_2)_4\overset{\overset{\large O}{\|}}{C}-N(CH_2CH_3)_2$$

The ester carbonyl would result in a peak in the 1730–1755 cm^{-1} region of the infrared spectrum. The ethoxy group would give rise to a characteristic ethyl pattern in the NMR spectrum (2H quartet near 4.1 ppm, 3H triplet near 1.1 ppm).

(d)

$$CH_3(CH_2)_4-\overset{\overset{\large O}{\|}}{C}-Cl \longrightarrow CH_3(CH_2)_4\overset{\overset{\large O}{\|}}{C}-OH$$

The acid carbonyl would result in a peak in the 1700–1725 cm^{-1} region of the infrared spectrum, and the COOH group would show the characteristic broad OH peak from 2500–3500 cm^{-1}. A new peak (1H) for the COOH group would be found in the 11.5–12.5 ppm region of the NMR spectrum.

(e)

$$CH_3(CH_2)_4-\overset{\overset{\large O}{\|}}{C}-Cl \longrightarrow CH_3(CH_2)_4\overset{\overset{\large O}{\|}}{C}-H$$

The aldehyde carbonyl would result in a peak in the 1720–1745 cm^{-1} region of the infrared spectrum, and the CHO group would show the characteristic aldehyde peak near 2720 cm^{-1}. A new peak (1H) for the CHO group would be found in the 9.5–9.8 ppm region of the NMR spectrum.

(f)

$$CH_3(CH_2)_4-\overset{\overset{\large O}{\|}}{C}-Cl \longrightarrow CH_3(CH_2)_4\underset{\underset{\large C_6H_5}{|}}{\overset{\overset{\large OH}{|}}{C}}-C_6H_5$$

There would be no carbonyl peak, but an OH peak in the 3200–3400 cm^{-1} region of the infrared spectrum would arise. The NMR spectrum would show the characteristic peak in near 6.9–7.2 ppm for the 10 hydrogens of the two equivalent phenyl groups.

(g)

$$CH_3(CH_2)_4{-}\overset{\overset{\displaystyle O}{\|}}{C}{-}Cl \longrightarrow CH_3(CH_2)_4C{\equiv}N$$

There would be no carbonyl peak, but characteristic peak in the 2210–2260 cm^{-1} region of the infrared spectrum would be observed for the carbon-nitrogen triple bond. There would be no qualitatively significant changes in the NMR spectrum.

(h)

$$CH_3(CH_2)_4{-}\overset{\overset{\displaystyle O}{\|}}{C}{-}Cl \longrightarrow CH_3(CH_2)_4\overset{\overset{\displaystyle OH}{|}}{\underset{\underset{\displaystyle CH_3}{|}}{C}}{-}CH_3$$

There would be no carbonyl peak, but the characteristic OH peak of an alcohol would be found in the 3200–3400 cm^{-1} region of the infrared spectrum. The NMR spectrum would show a 6H singlet near 1.1 ppm for the two equivalent methyl groups, and a 1H singlet for the OH group would be found in the 0.5–5 ppm region.

(i)

$$CH_3(CH_2)_4{-}\overset{\overset{\displaystyle O}{\|}}{C}{-}Cl \longrightarrow CH_3(CH_2)_4\overset{\overset{\displaystyle O}{\|}}{C}{-}CH_3$$

The carbonyl peak for the ketone would be found in the 1700–1725 cm^{-1} region of the infrared spectrum. A 3H singlet for the new methyl group would be observed near 2.1 ppm in the NMR spectrum.

CHAPTER 17. AMINES

ANSWERS TO EXERCISES

17.1 **(a)** 1-ethylpropylamine

(b) *N*-ethylisobutylamine (or *N*-ethyl-2-methylpropylamine)

(c) 5-methylamino-2-heptanone

(d) trimethylpropylammonium bromide

17.2 **(a)**

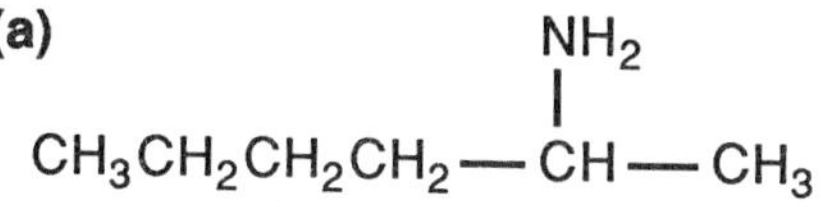

$$CH_3CH_2CH_2CH_2-\underset{}{\overset{NH_2}{\overset{|}{CH}}}-CH_3$$

(b)

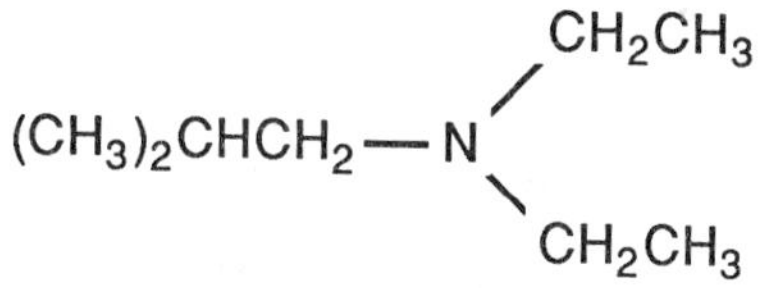

$$(CH_3)_2CHCH_2-N\begin{matrix} \diagup CH_2CH_3 \\ \diagdown CH_2CH_3 \end{matrix}$$

(c) $[(CH_3)_2CH]_3NH^+\ Br^-$

(d)

$$CH_3-\underset{NH_2}{\underset{|}{CH}}-\overset{O}{\overset{\|}{C}}-CH_2-\underset{OH}{\underset{|}{CH}}-CH_2CH_2CH_3$$

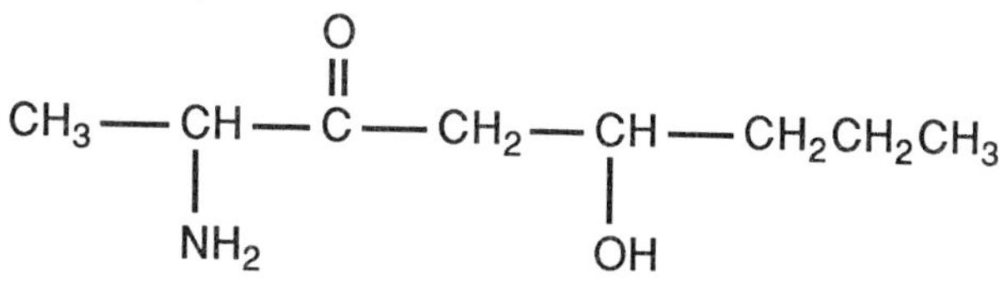

17.3 **(a)** The isomeric primary amine would have stronger hydrogen bonding interactions:

$$CH_3CH_2CH_2CH_2NH_2$$

(b) Alcohols form stronger hydrogen bonds than do amines:

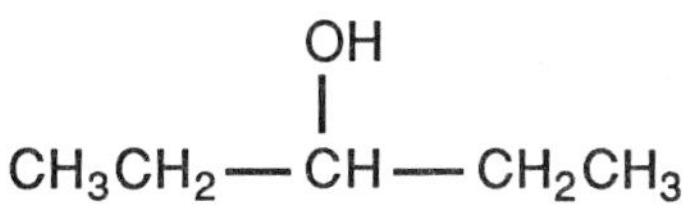

$$CH_3CH_2-\overset{OH}{\overset{|}{CH}}-CH_2CH_3$$

(c) The secondary amine (in contrast to the tertiary isomer) can form hydrogen bonds:

$$CH_3CH_2NHCH_3$$

17.4 **(a)** Both are tertiary amines; the compound with fewer carbon atoms would be more soluble in water:

$$(CH_3CH_2)_3N$$

(b) The presence of a second polar group (hydroxyl) would increase the water solubility:

$$CH_3CH_2CH_2CH_2NHCH_2CH_2CH_2CH_2OH$$

17.5

(a) $CH_3CH_2CH_2CH_2OH$

(b) 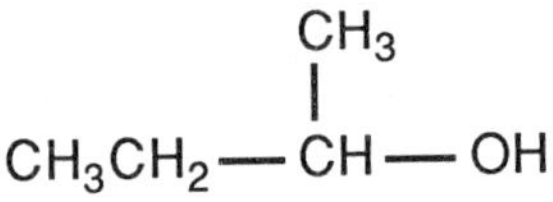 $CH_3CH_2—CH(CH_3)—OH$

(c) 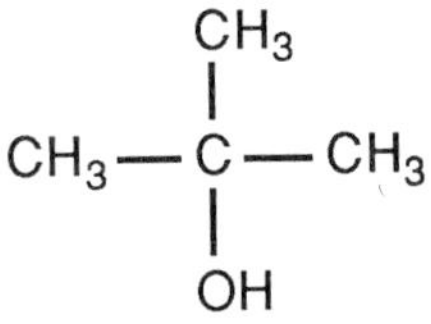 $CH_3—C(CH_3)(OH)—CH_3$

(d) 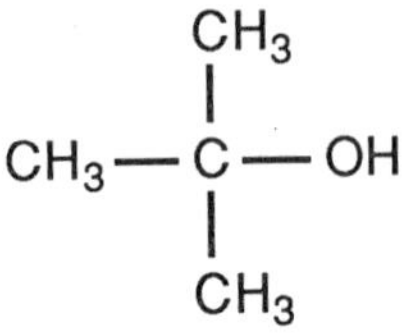 $CH_3—C(CH_3)_2—OH$

(e) 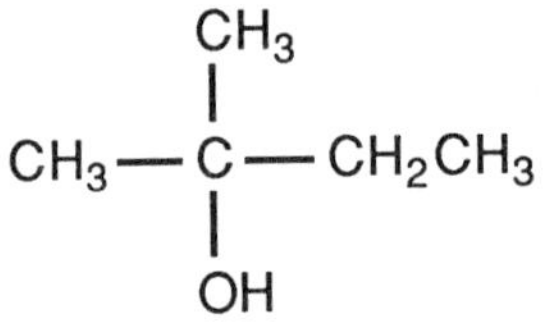 $CH_3—C(CH_3)(OH)—CH_2CH_3$

17.6 (a) $CH_3CH_2CH{=}CH_2$

(b) $(CH_3)_2CHCH{=}CH_2$ (major) + $(CH_3)_2C{=}CHCH_3$ (minor)

(c) $C_6H_5CH{=}CHCH_3$ + $C_6H_5CH_2CH{=}CH_2$

(d) 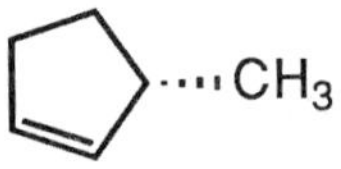 3-methylcyclopentene (····CH₃ hashed)

17.7

(a) $CH_3CH_2CH{=}CH_2$

(b) $(CH_3)_2CHCH{=}CH_2$ (minor) + $(CH_3)_2C{=}CHCH_3$ (major)

(c) $C_6H_5CH{=}CHCH_3$ + $C_6H_5CH_2CH{=}CH_2$

(d) 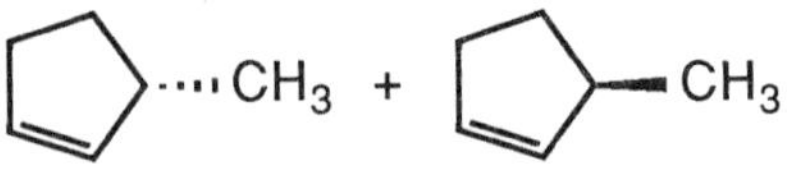 3-methylcyclopentene (····CH_3 hashed) + 3-methylcyclopentene (wedged CH_3)

17.8 **(a)**

$$C_6H_5-\overset{O}{\overset{\|}{C}}-NH_2 \xrightarrow[2.\ H_2O]{1.\ LiAlH_4} C_6H_5CH_2NH_2$$

$$C_6H_5CH_2NO_2 \xrightarrow[2.\ NaOH]{1.\ Fe,\ HCl} C_6H_5CH_2NH_2$$

$$C_6H_5CN \xrightarrow[2.\ H_2O]{1.\ LiAlH_4} C_6H_5CH_2NH_2$$

$$C_6H_5CH{=}NOH \xrightarrow[2.\ H_2O]{1.\ LiAlH_4} C_6H_5CH_2NH_2$$

(b)

$$CH_3-\overset{NOH}{\overset{\|}{C}}-CH_3 \xrightarrow[2.\ H_2O]{1.\ LiAlH_4} (CH_3)_2CH-NH_2$$

(c)

$$C_6H_5-NO_2 \xrightarrow[2.\ NaOH]{1.\ Fe,\ HCl} C_6H_5-NH_2$$

(d)

$$CH_3CH_2CH_2NH-\overset{O}{\overset{\|}{C}}-H \xrightarrow[2.\ H_2O]{1.\ LiAlH_4} CH_3CH_2CH_2NHCH_3$$

$$CH_3CH_2-\overset{O}{\overset{\|}{C}}-NHCH_3 \xrightarrow[2.\ H_2O]{1.\ LiAlH_4} CH_3CH_2CH_2NHCH_3$$

(e)

$$C_6H_{10}{=}NOH \text{ (cyclohexanone oxime)} \xrightarrow[2.\ H_2O]{1.\ LiAlH_4} C_6H_{11}NH_2 \text{ (cyclohexylamine)}$$

$$C_6H_{11}NO_2 \text{ (nitrocyclohexane)} \xrightarrow[2.\ NaOH]{1.\ Fe,\ HCl} C_6H_{11}NH_2 \text{ (cyclohexylamine)}$$

(f)

$$\text{C}_6\text{H}_{11}\text{-NH-}\overset{\overset{\displaystyle O}{\|}}{C}\text{-H} \xrightarrow[\text{2. } H_2O]{\text{1. LiAlH}_4} \text{C}_6\text{H}_{11}\text{-NHCH}_3$$

17.9 (a)

$$CH_3CH_2-\overset{\overset{\displaystyle O}{\|}}{C}-H \xrightarrow[\text{NaBH}_3\text{CN}]{\text{NH}_3\text{, HCl}} CH_3CH_2CH_2NH_2$$

$$CH_3CH_2CH_2Br \xrightarrow{NH_3} CH_3CH_2CH_2NH_2$$

$$CH_3CH{=}CH_2 \xrightarrow[\text{2. NH}_3\text{, aq. NaOCl}]{\text{1. BH}_3} CH_3CH_2CH_2NH_2$$

$$CH_3CH_2CH_2-\overset{\overset{\displaystyle O}{\|}}{C}-NH_2 \xrightarrow[\text{NaOH}]{\text{Br}_2} CH_3CH_2CH_2NH_2$$

(b)

$$CH_3CH_2-\overset{\overset{\displaystyle CH_3}{|}}{CH}-\overset{\overset{\displaystyle O}{\|}}{C}-H \xrightarrow[\text{NaBH}_3\text{CN}]{\text{NH}_3\text{, HCl}} CH_3CH_2-\overset{\overset{\displaystyle CH_3}{|}}{CH}-CH_2NH_2$$

$$CH_3CH_2-\overset{\overset{\displaystyle CH_3}{|}}{CH}-CH_2Br \xrightarrow{NH_3} CH_3CH_2-\overset{\overset{\displaystyle CH_3}{|}}{CH}-CH_2NH_2$$

$$(CH_3)(CH_3CH_2)C{=}CH_2 \xrightarrow[\text{2. NH}_3\text{, aq. NaOCl}]{\text{1. BH}_3} CH_3CH_2-\overset{\overset{\displaystyle CH_3}{|}}{CH}-CH_2NH_2$$

$$CH_3CH_2-\overset{\overset{\displaystyle CH_3}{|}}{CH}CH_2-\overset{\overset{\displaystyle O}{\|}}{C}-NH_2 \xrightarrow[\text{NaOH}]{\text{Br}_2} CH_3CH_2-\overset{\overset{\displaystyle CH_3}{|}}{CH}-CH_2NH_2$$

(c)

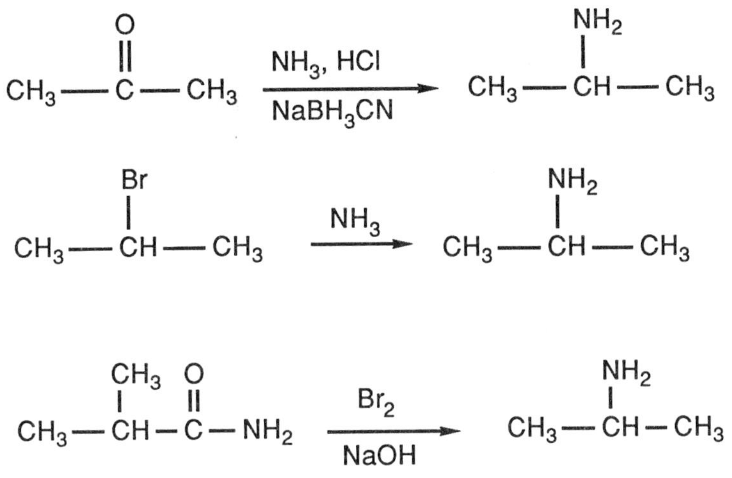

$$CH_3-\overset{\overset{O}{\|}}{C}-CH_3 \xrightarrow[NaBH_3CN]{NH_3,\ HCl} CH_3-\overset{\overset{NH_2}{|}}{CH}-CH_3$$

$$CH_3-\overset{\overset{Br}{|}}{CH}-CH_3 \xrightarrow{NH_3} CH_3-\overset{\overset{NH_2}{|}}{CH}-CH_3$$

$$CH_3-\overset{\overset{CH_3}{|}}{CH}-\overset{\overset{O}{\|}}{C}-NH_2 \xrightarrow[NaOH]{Br_2} CH_3-\overset{\overset{NH_2}{|}}{CH}-CH_3$$

(d)

$$CH_3CH_2-\underset{\underset{CH_3}{|}}{\overset{\overset{CH_3}{|}}{C}}-\overset{\overset{O}{\|}}{C}-NH_2 \xrightarrow[NaOH]{Br_2} CH_3CH_2-\underset{\underset{CH_3}{|}}{\overset{\overset{CH_3}{|}}{C}}-NH_2$$

$$CH_3CH_2-\underset{\underset{CH_3}{|}}{\overset{\overset{CH_3}{|}}{C}}-OH \xrightarrow[2.\ H_2O]{1.\ HCN,\ conc.\ H_2SO_4} CH_3CH_2-\underset{\underset{CH_3}{|}}{\overset{\overset{CH_3}{|}}{C}}-NH-\overset{\overset{O}{\|}}{C}-H$$

$$\downarrow KOH,\ H_2O$$

$$CH_3CH_2-\underset{\underset{CH_3}{|}}{\overset{\overset{CH_3}{|}}{C}}-NH_2$$

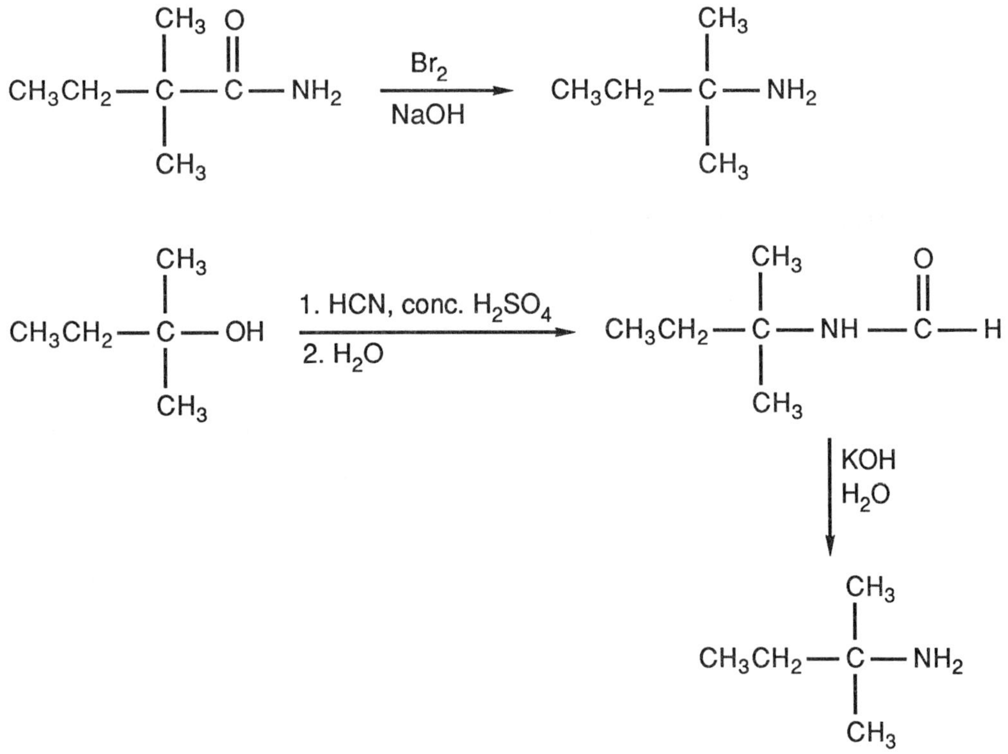

17.10

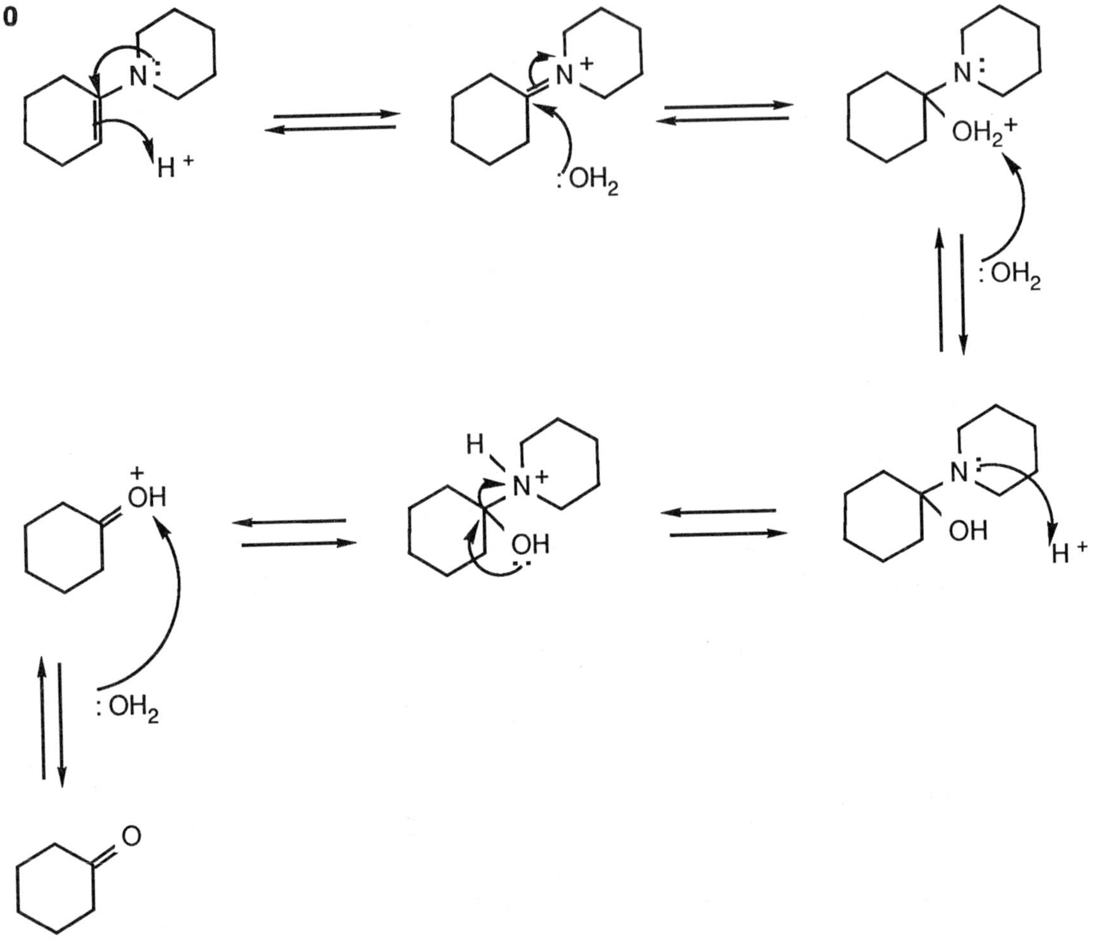

N: H^+ $\rightleftharpoons$ N^+ :OH_2 $\rightleftharpoons$ N: OH_2^+

:OH_2

+OH $\rightleftharpoons$ H N+ OH $\rightleftharpoons$ N: OH H^+

:OH_2

O

17.11 **(a)** CH_3 O $\xrightarrow[\text{TsOH}]{\text{NH}}$ CH_3 N $\xrightarrow{C_6H_5CH_2Br}$ CH_3 N+ $CH_2C_6H_5$

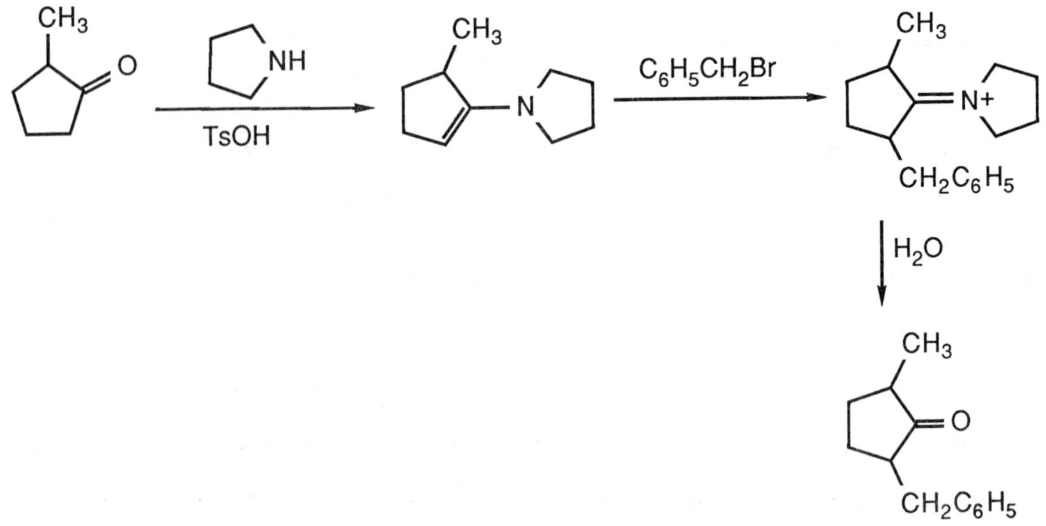

$\downarrow H_2O$

CH_3 O $CH_2C_6H_5$

(b)

4-Methylcyclohexanone $\xrightarrow{\text{morpholine (O, NH ring)}}$ 4-methyl-1-morpholinocyclohexene $\xrightarrow{CH_2{=}CHCO_2C_2H_5}$ iminium ion (N^+) bearing $CH_2CH_2CO_2C_2H_5$ $\xrightarrow{H_2O}$ 2-($CH_2CH_2CO_2C_2H_5$)-4-methylcyclohexanone (CH_3)

17.12 (a) $CH_3CH_2NH_2$ and (c) $CH_3-\overset{CH_3}{\underset{CH_3}{C}}-NH_2$

17.13 (a) $CH_3CH_2NH_2$ (b) $CH_3CH_2NHCH_2CH_3$ (c) $CH_3-\overset{CH_3}{\underset{CH_3}{C}}-NH_2$ (d) $C_6H_5NH_2$

17.14 Partition the reaction mixture between water and ether; the organic materials, including unreacted octanoic acid will dissolve in the ether layer. Then extract the ether layer with dilute aqueous sodium hydroxide (or sodium bicarbonate). The resulting acid-base reaction will convert the acid to sodium octanoate, which will dissolve in the aqueous layer while the ester will remain in the ether layer.

17.14 Isolation of the heptylamine could be accomplished by taking advantage of its acid-base properties. Acidification of the crude reaction mixture with aqueous HCl would convert the amine to the corresponding ammonium salt, and extraction with ether would remove other organic compounds from the aqueous layer. Neutralization of the aqueous layer with NaOH would convert the ammonium salt back to the amine, which could be extracted into ether. Evaporation of the ether would then afford the amine.

ANSWERS TO PROBLEMS

17.1 **(a)** triethylamine

(b) *N*-methylethylamine

(c) *tert*-butylamine

(d) *N*-ethyl-*sec*-butylamine (or *N*-ethyl-1-methylpropylamine)

(e) tetrapropylammonium bromide

(f) methyl 3-(methylamino)pentanoate

(g) 3-aminocyclohexanone

(h) 3-oxocyclohexyltrimethylammonium iodide

17.2 **(a)** $(CH_3)_2CHCH_2NH_3^+Cl^-$ **(b)** $NH(CH_2CH_3)_2$ **(c)** $NH_2CH_2CH_2CH_2CH_2CH_2CHO$

(d) $CH_3CH_2CH_2CH_2CH_2N(CH_2CH_2CH_2CH_3)_2$

(e)
```
                 CH3CH2CH2CH2
                           |
   CH3CH2CH2 — C — CH2NH2
                           |
                 CH2CH2CH2CH3
```

(f)
```
                      CH3
                       |
   CH3CH2CH2 — C — N(CH2CH2CH2CH3)2
                       |
                       H
```

(g)
```
          NH2            NH2
           |              |
   CH3 — C — CH2 — C — CH2CH3
           |              |
           H              H
```

(h) (cyclopentane ring bearing NH_2 groups at C-1 and C-3)

```
        NH2
        |
       / \
      |   |
       \_/
          \
           NH2
```

(i)
```
              (CH3)2N    O
                   |      ||
   CH3CH2 — CH — C — NH2
```

17.3 **(a)** Intermolecular hydrogen bonding would be present for the primary (but not for the tertiary) isomer:

$$\begin{array}{c} CH_3 \\ | \\ CH_3CH_2{-}C{-}NH_2 \\ | \\ CH_3 \end{array}$$

(b) The two compounds have almost the same molecular weight, but hydrogen bonding would result in higher boiling point for the amine:

$$CH_3(CH_2)_5CH_2NHCH_3$$

(c) The increased polarity and greater hydrogen bonding interactions of the difunctional compound would result in a higher boiling point

$$CH_3CH_2NHCH_2CH_2OH$$

(d) Hydrogen bonding is greater for the primary amine:

$$\begin{array}{c} NH_2 \\ | \\ CH_3CH_2CH_2{-}C{-}CH_3 \\ | \\ H \end{array}$$

17.4 **(a)** $CH_3CH_2CH{=}CHCH_3 \ + \ (CH_3)_3N$ **(b)** CH_3 $+ \ (CH_3)_3N$

(c) $CH_2{=}CHCH_2CH(CH_3)_2 \ + \ (CH_3)_3N$

(d)

$$\begin{array}{c} CH_3 \\ | \\ (CH_3)_2NCH_2{-}CH{-}CH_2CH_2CH_3 \end{array} \ + \ CH_2{=}CH_2$$

and $CH_3CH_2N(CH_3)_2 \ + \ CH_2{=}C(CH_3)CH_2CH_2CH_3$

$$CH_2{=}C\begin{array}{l} \diagup CH_3 \\ \diagdown CH_2CH_2CH_3 \end{array}$$

(e) $(CH_3)_3N$ + $(CH_3)_2C{=}CHC_6H_5$ + $CH_2{=}C(CH_2C_6H_5)(CH_3)$

(f) $(CH_3)_3N$ + 3-ethylcyclopentene (CH_2CH_3) + 1-ethylcyclopentene (CH_2CH_3)

(g) methylenecycloheptane (CH_2) + $(CH_3)_3N$ + 1-methylcycloheptene (CH_3)

(h) 3,3-dimethylcyclopentene (CH_3, CH_3) + $(CH_3)_3N$

17.5 (a) $CH_3CH_2CH{=}CHCH_3$ + $(CH_3)_2NOH$

(b) 1-methylcyclohexene (CH_3) + 3-methylcyclohexene (CH_3) + $(CH_3)_2NOH$

(c) $CH_3CH{=}CHCH(CH_3)_2$ + $CH_2{=}CHCH_2CH(CH_3)_2$ + $(CH_3)_2NOH$

(d) $CH_3{-}N(OH)CH_2{-}CH{-}CH_2CH_2CH_3$ $CH_2{=}CH_2$

and $CH_2{=}C(CH_3)(CH_2CH_3)$ + $CH_3{-}N(OH){-}CH_2CH_3$

(e) $(CH_3)_2C{=}CHC_6H_5$ + $CH_2{=}C(CH_2C_6H_5)(CH_3)$ + $(CH_3)_2NOH$

(f) [cyclopentene with CH_2CH_3] + $(CH_3)_2NOH$

(g) [1-methylcycloheptene, CH_3] + $(CH_3)_2NOH$ + [methylenecycloheptane, CH_2]

(h) [3,3-dimethylcyclopentene, CH_3, CH_3] + $(CH_3)_2NOH$

17.6 Only the secondary amine (c) would give a precipitate upon reaction with TsCl/NaOH:

$$CH_3-\underset{\displaystyle NHCH_3}{\underset{|}{CH}}-CH_2CH(CH_3)_2 \xrightarrow[\text{aq. NaOH}]{\text{TsCl}} CH_3-\underset{\displaystyle CH_3\diagup N \diagdown SO_2C_6H_4CH_3}{\underset{|}{CH}}-CH_2CH(CH_3)_2$$

The primary amines, (a), (b), (e), (f), (g), and (h), would all yield insoluble precipitates upon acidification of the alkaline solution:

$$CH_3CH_2-\overset{\displaystyle NH_2}{\overset{|}{CH}}-CH_2CH_3 \xrightarrow[\text{2. HCl}]{\text{1. TsCl/NaOH}} CH_3CH_2-\overset{\displaystyle NHSO_2C_6H_4CH_3}{\overset{|}{CH}}-CH_2CH_3$$

[cyclohexane with NH_2 and CH_3] $\xrightarrow[\text{2. HCl}]{\text{1. TsCl/NaOH}}$ [cyclohexane with $NHSO_2C_6H_4CH_3$ and CH_3]

$$CH_3-\underset{\displaystyle CH_3}{\overset{\displaystyle NH_2}{\overset{|}{\underset{|}{C}}}}-CH_2C_6H_5 \xrightarrow[\text{2. HCl}]{\text{1. TsCl/NaOH}} CH_3-\underset{\displaystyle CH_3}{\overset{\displaystyle NHSO_2C_6H_4CH_3}{\overset{|}{\underset{|}{C}}}}-CH_2C_6H_5$$

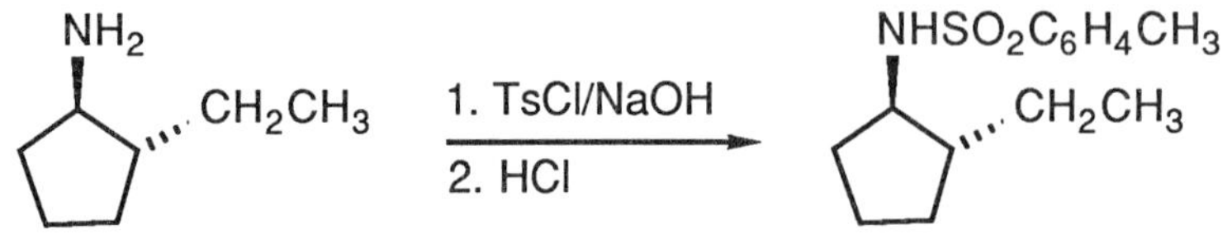

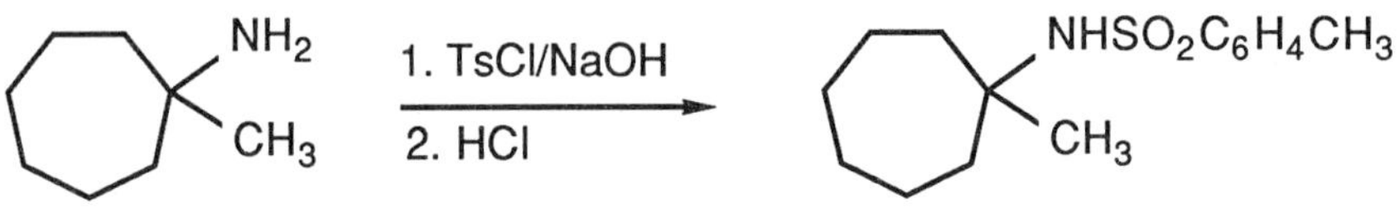

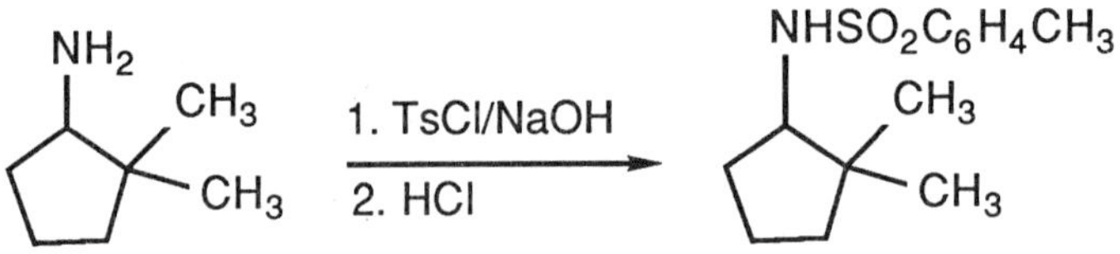

NH_2 / CH_2CH_3 cyclopentane $\xrightarrow[\text{2. HCl}]{\text{1. TsCl/NaOH}}$ $NHSO_2C_6H_4CH_3$ / CH_2CH_3

NH_2 / CH_3 cycloheptane $\xrightarrow[\text{2. HCl}]{\text{1. TsCl/NaOH}}$ $NHSO_2C_6H_4CH_3$ / CH_3

NH_2 / CH_3 / CH_3 cyclopentane $\xrightarrow[\text{2. HCl}]{\text{1. TsCl/NaOH}}$ $NHSO_2C_6H_4CH_3$ / CH_3 / CH_3

17.7 Evolution of nitrogen gas is characteristic of the reaction of nitrous acid with a primary amine. This would be expected for (a), (b), (e), (f), (g), and (h).

17.8 All of the amines except the tertiary amine (d) would be acylated by acetic anhydride. The following products would be formed:

(a) $CH_3CH_2—CH(NH—C(=O)—CH_3)—CH_2CH_3$

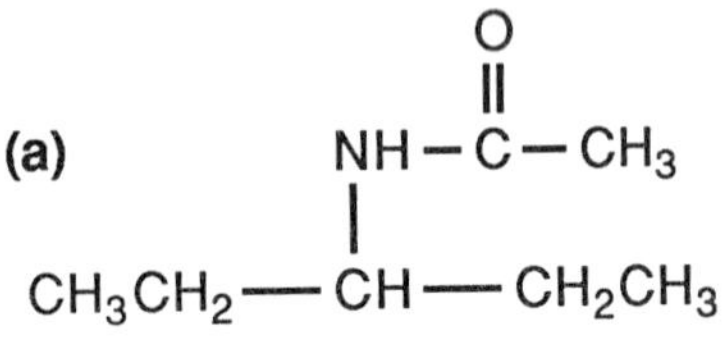

(b) CH_3 / $NH—C(=O)—CH_3$ (cyclohexane)

(c) $CH_3—CH(—CH_2CH(CH_3)_2)—N(CH_3)—C(=O)—CH_3$

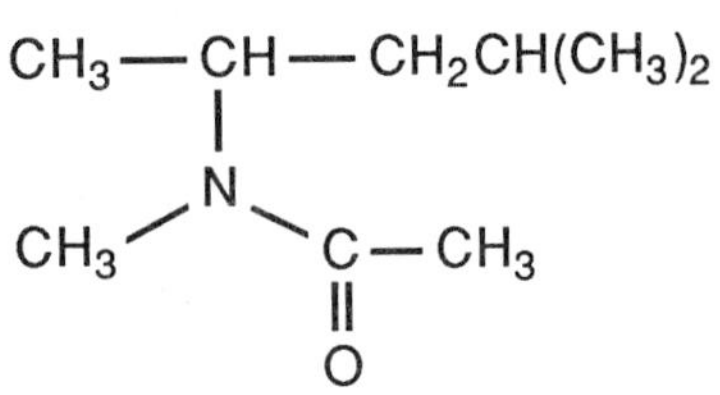

(e) $CH_3—C(CH_3)(NH—C(=O)—CH_3)—CH_2C_6H_5$

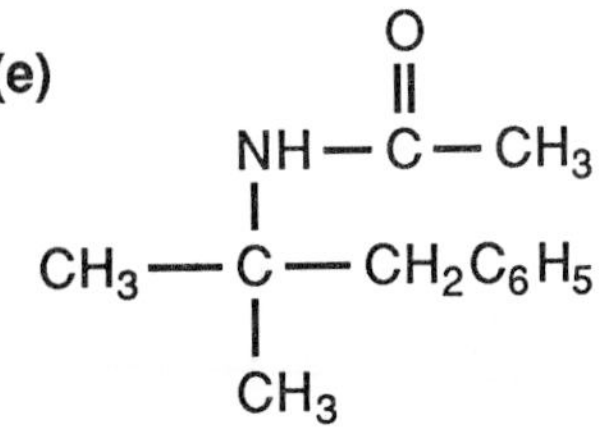

(f) CH_3 / $NH—C(=O)—CH_3$ (cyclopentane)

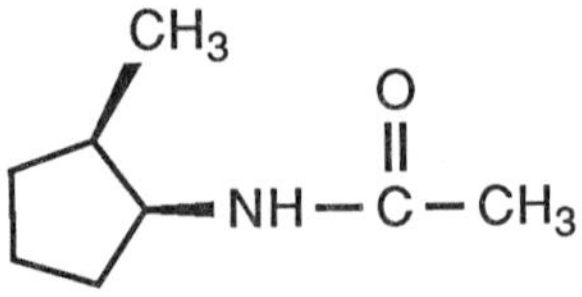

(g) CH_3 / $NH—C(=O)—CH_3$ (cycloheptane)

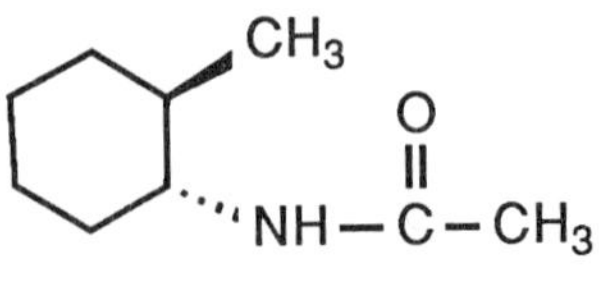

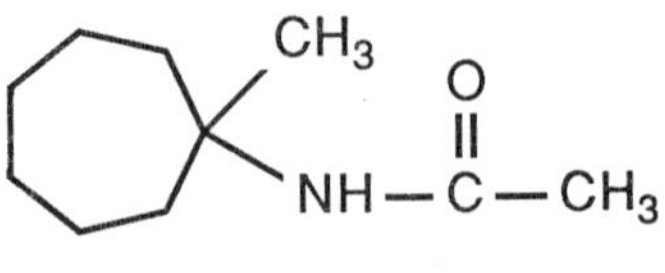

(h) NH—C(=O)—CH_3 ; CH_3 ; CH_3

17.9

(a) O ; NH_3, HCl ; $NaBH_3CN$; NH_2 ; or ; O ; NH_3 ; H_2/Ni ; NH_2

or

O ; NH_2OH ; NOH ; 1. $LiAlH_4$; 2. H_2O ; NH_2

(b) O ; 1. CH_3Li ; 2. H_2O ; CH_3 ; OH ; HCN ; H_2SO_4 ; CH_3 ; NH—CH(=O) ; H_2O ; NaOH ; CH_3 ; NH_2

(c) O ; C_6H_5MgBr ; H_2O ; C_6H_5 ; OH ; conc. H_2SO_4 ; C_6H_5 ; 1. BH_3 ; 2. NH_3, aq. NaOCl ; C_6H_5 ; NH_2

(d) cyclohexanone $\xrightarrow[\text{NaBH}_3\text{CN}]{(CH_3)_2NH,\ HCl}$ cyclohexyl–NH_2

or

cyclohexanone $\xrightarrow[\text{H}_2/\text{Ni}]{(CH_3)_2NH}$ cyclohexyl–NH_2

17.10

(a) cyclohexanone + H^+ ⇌ protonated cyclohexanone ($\overset{+}{O}H$) $\xrightleftharpoons{:NH_3}$ 1-hydroxy-1-($\overset{+}{N}H_3$)cyclohexane $\xrightleftharpoons[\text{transfer}]{\text{proton}}$ 1-($\overset{+}{O}H_2$)-1-(NH_2)cyclohexane

⇌ cyclohexyl iminium ($=\overset{+}{N}H_2$) $\xrightarrow{H-\overset{-}{B}H_2-CN}$ cyclohexyl–NH_2

(b)

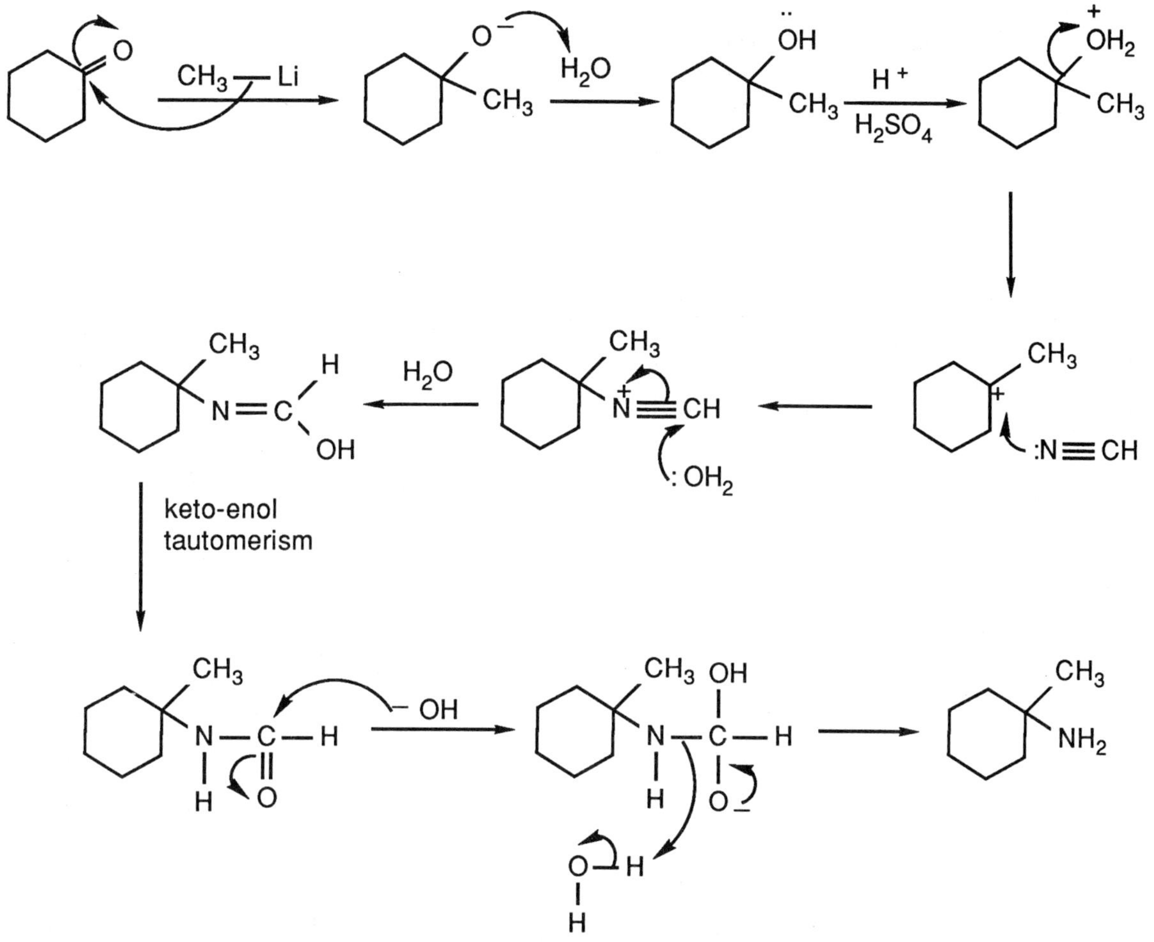

CH_3—Li

H_2O

H^+ H_2SO_4

CH_3 OH OH_2^+

H_2O

:N≡CH

$N=C$ H OH

$\overset{+}{N}$≡CH

: OH_2

keto-enol tautomerism

N—C—H

$^-$OH

NH_2

(c)

C_6H_5—MgBr

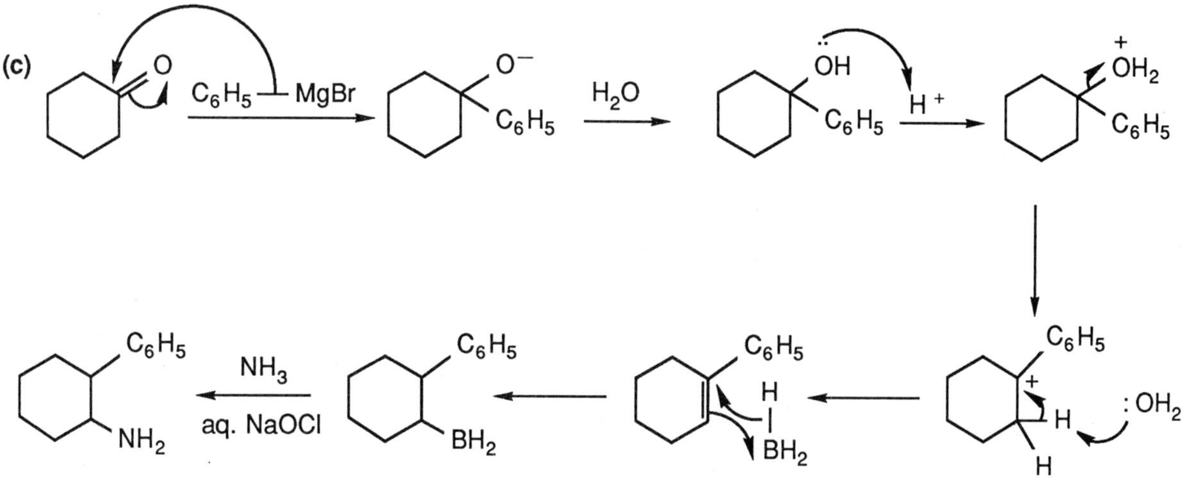

O^- C_6H_5

H_2O

H^+

$\overset{+}{O}H_2$

: OH_2

BH_2

NH_3 aq. NaOCl

NH_2

(d)

Cyclohexanone $\overset{H^+}{\rightleftharpoons}$ protonated cyclohexanone ($\overset{+}{O}H$) $\xrightleftharpoons{:NH(CH_3)_2}$ 1-hydroxy-1-($\overset{+}{N}H(CH_3)_2$)cyclohexane $\underset{\text{transfer}}{\overset{\text{proton}}{\rightleftharpoons}}$ 1-($\overset{+}{O}H_2$)-1-$NH(CH_3)_2$-cyclohexane $\rightleftharpoons$ cyclohexylidene=$\overset{+}{N}H(CH_3)_2$ $\xrightarrow{H_2\overset{-}{B}(H)-CN}$ cyclohexyl-$NH(CH_3)_2$

17.11

(a)

$$C_6H_5-\overset{O}{\overset{\|}{C}}-OH \xrightarrow{PCl_5} C_6H_5-\overset{O}{\overset{\|}{C}}-Cl \xrightarrow{NH_3} C_6H_5-\overset{O}{\overset{\|}{C}}-NH_2 \xrightarrow[2.\ H_2O]{1.\ LiAlH_4} C_6H_5CH_2NH_2$$

(b)

$$C_6H_5-\overset{O}{\overset{\|}{C}}-OH \xrightarrow[2.\ H_2O]{1.\ LiAlH_4} C_6H_5CH_2OH \xrightarrow{PCl_5} C_6H_5CH_2Cl \xrightarrow{NaCN} C_6H_5CH_2CN \xrightarrow[2.\ H_2O]{1.\ LiAlH_4} C_6H_5CH_2CH_2NH_2$$

(c)

$$C_6H_5-\overset{O}{\overset{\|}{C}}-OH \xrightarrow{PCl_5} C_6H_5-\overset{O}{\overset{\|}{C}}-Cl \xrightarrow{CH_3CH_2NH_2} C_6H_5-\overset{O}{\overset{\|}{C}}-NHCH_2CH_3 \xrightarrow[2.\ H_2O]{1.\ LiAlH_4} C_6H_5CH_2NHCH_2CH_3$$

(d)

$$C_6H_5-\overset{O}{\overset{\|}{C}}-OH \xrightarrow{PCl_5} C_6H_5-\overset{O}{\overset{\|}{C}}-Cl \xrightarrow{NH_3} C_6H_5-\overset{O}{\overset{\|}{C}}-NH_2 \xrightarrow{Br_2,\ KOH} C_6H_5NH_2$$

(e)

$$C_6H_5-\overset{O}{\overset{\|}{C}}-OH \xrightarrow{PCl_5} C_6H_5-\overset{O}{\overset{\|}{C}}-Cl \xrightarrow{(CH_3)_2CuLi} C_6H_5-\overset{O}{\overset{\|}{C}}-CH_3 \xrightarrow{NH_3} C_6H_5-\overset{\overset{+}{N}H}{\overset{\|}{C}}-CH_3 \xrightarrow[2.\ H_2O]{1.\ LiAlH_4} C_6H_5-\overset{NH_2}{\overset{|}{C}H}-CH_3$$

17.12

(a)

$$C_6H_5-\overset{O}{\overset{\|}{C}}-\ddot{O}H \xrightarrow{PCl_5} C_6H_5-\overset{O}{\overset{\|}{C}}-O-PCl_4 \xrightarrow{Cl^-} C_6H_5-\overset{O}{\overset{\|}{C}}-Cl \xrightarrow{:NH_3} C_6H_5-\overset{O}{\overset{\|}{C}}-NH_2$$

$$C_6H_5-\overset{O}{\overset{\|}{C}}-NH_2 \xrightarrow{LiAlH_3-H} C_6H_5-\overset{OAlH_3}{\underset{H}{\overset{|}{\underset{|}{C}}}}-\ddot{N}H_2 \longrightarrow C_6H_5-CH=\overset{+}{N}H_2 \xrightarrow{LiAlH_3-H} C_6H_5CH_2NH_2$$

(b)

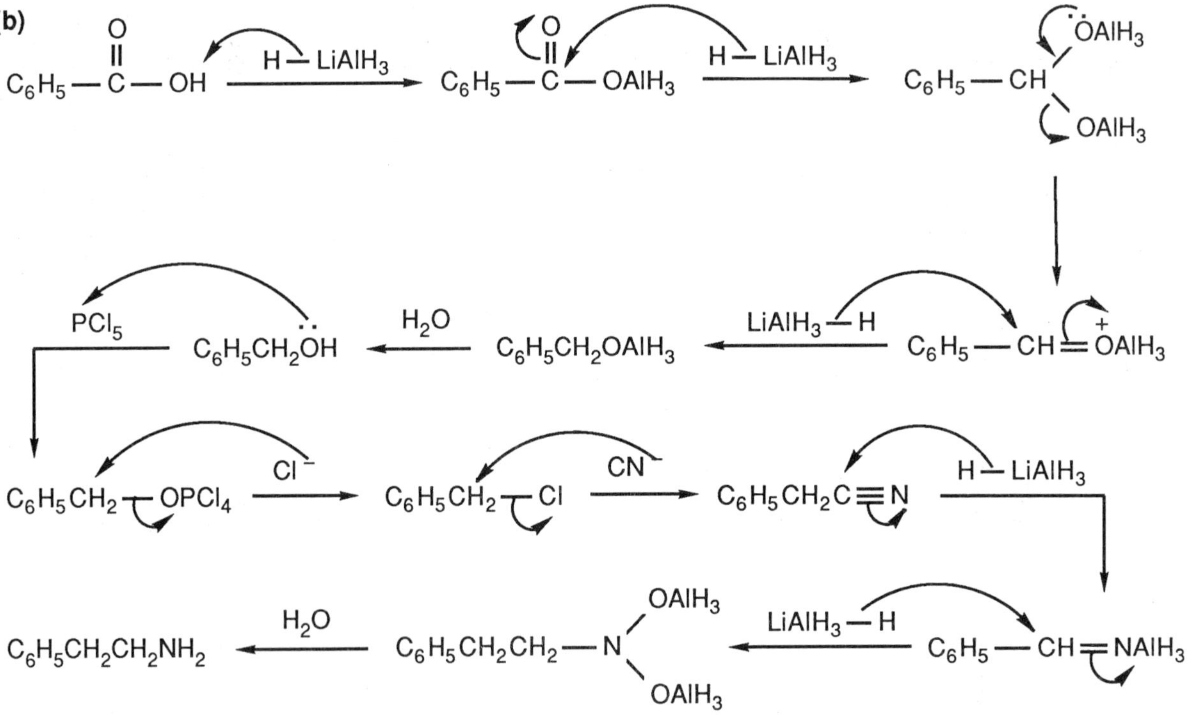

$$C_6H_5-\overset{O}{\overset{\|}{C}}-OH \xrightarrow{H-LiAlH_3} C_6H_5-\overset{O}{\overset{\|}{C}}-OAlH_3 \xrightarrow{H-LiAlH_3} C_6H_5-CH(\ddot{O}AlH_3)(OAlH_3)$$

$$\longrightarrow C_6H_5-CH=\overset{+}{O}AlH_3 \xrightarrow{LiAlH_3-H} C_6H_5CH_2OAlH_3 \xrightarrow{H_2O} C_6H_5CH_2\ddot{O}H \xrightarrow{PCl_5}$$

$$C_6H_5CH_2-OPCl_4 \xrightarrow{Cl^-} C_6H_5CH_2-Cl \xrightarrow{CN^-} C_6H_5CH_2C\equiv N \xrightarrow{H-LiAlH_3}$$

$$C_6H_5-CH=NAlH_3 \xrightarrow{LiAlH_3-H} C_6H_5CH_2CH_2-N(OAlH_3)_2 \xrightarrow{H_2O} C_6H_5CH_2CH_2NH_2$$

(c) prepared as in (a)

$$C_6H_5-\overset{O}{\overset{\|}{C}}-Cl + CH_3CH_2\ddot{N}H_2 \longrightarrow C_6H_5-\overset{O}{\overset{\|}{C}}-NHCH_2CH_3 \xrightarrow{LiAlH_3-H} C_6H_5-\overset{OAlH_3}{\underset{H}{C}}-\ddot{N}HCH_2CH_3$$

$$\longrightarrow C_6H_5-CH=\overset{+}{N}HCH_2CH_3 \xrightarrow{LiAlH_3-H} C_6H_5CH_2NHCH_2CH_3$$

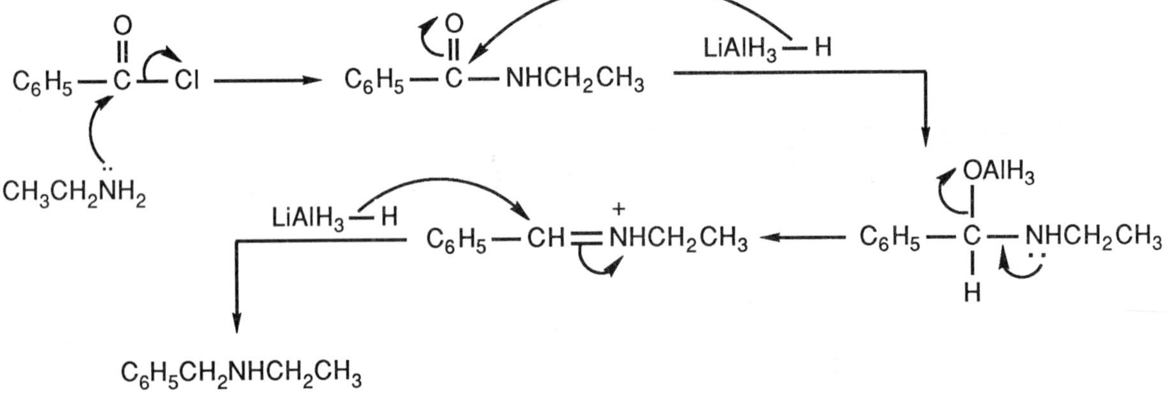

(d) prepared as in (a)

$$C_6H_5-\overset{O}{\overset{\|}{C}}-NH_2 \xrightarrow{^-OH} C_6H_5-\overset{O}{\overset{\|}{C}}-\overset{-}{N}H \xrightarrow{Br-Br} C_6H_5-\overset{O}{\overset{\|}{C}}-N(Br)H \xrightarrow{^-OH}$$

$$C_6H_5-\overset{O}{\overset{\|}{C}}-\overset{-}{N}-Br \longrightarrow O=C=N-C_6H_5 \xrightarrow{^-OH,\ H_2O} H-O-\overset{O}{\overset{\|}{C}}-NHC_6H_5 \xrightarrow{^-OH,\ H_2O} C_6H_5NH_2$$

(e)

$$C_6H_5-\overset{O}{\overset{\|}{C}}-Cl \xrightarrow{(CH_3)_2CuLi} C_6H_5-\underset{CH_3}{\overset{O^-}{\overset{|}{\underset{|}{C}}}}-Cl \longrightarrow C_6H_5-\overset{:O}{\overset{\|}{C}}-CH_3 \xrightarrow{H^+}$$

$$C_6H_5-\overset{\overset{+}{O}H}{\overset{\|}{C}}-CH_3 \xrightarrow{NH_3:} C_6H_5-\underset{CH_3}{\overset{OH}{\overset{|}{\underset{|}{C}}}}-\overset{+}{N}H_3 \xrightarrow{\text{proton transfer}} C_6H_5-\underset{CH_3}{\overset{\overset{+}{O}H_2}{\overset{|}{\underset{|}{C}}}}-\ddot{N}H_2$$

$$\longrightarrow C_6H_5-\overset{\overset{+}{N}H_2}{\overset{\|}{C}}-CH_3 \xrightarrow{H-LiAlH_3} C_6H_5-\overset{NH_2}{\overset{|}{C}H}-CH_3$$

17.13

(a) $CH_3CH_2CH_2CH_2CH_2OH \xrightarrow{PBr_3} CH_3CH_2CH_2CH_2CH_2Br \xrightarrow{NH_3} CH_3CH_2CH_2CH_2CH_2NH_2$

(b) $CH_3CH_2CH_2CH_2CH_2OH \xrightarrow{CrO_3} CH_3CH_2CH_2CH_2CO_2H \xrightarrow{PCl_5} CH_3CH_2CH_2CH_2COCl \xrightarrow{NH_3} CH_3CH_2CH_2CH_2CONH_2 \xrightarrow{Br_2,\ KOH} CH_3CH_2CH_2CH_2NH_2$

(c) $CH_3CH_2CH_2CH_2CH_2OH \xrightarrow{PBr_3} CH_3CH_2CH_2CH_2CH_2Br \xrightarrow{NaCN} CH_3CH_2CH_2CH_2CH_2CN \xrightarrow{1.\ LiAlH_4;\ 2.\ H_2O} CH_3CH_2CH_2CH_2CH_2CH_2NH_2$

17.14 **(a)** For NaOH the equilibrium strongly favors protonation of OH^- and generation of the free amine; reaction with $C_2H_5NH_2$ or $(C_2H_5)_3N$ will also deprotonate the ammonium salt; The use of excess reagent will shift the equilibrium. Only CH_3CO_2Na is not a strong enough base to deprotonate significant quantities of an ammonium salt.

(b) Side reactions would be expected for NaOH (hydrolysis) and for $C_2H_5NH_2$ (acylation).

(c) $(C_2H_5)_3N$ is most satisfactory for general use.

17.15 The two S=O double bonds allow greater delocalization of the negative charge in the conjugate base, thereby favoring its formation:

$$C_6H_5-\overset{O}{\overset{\|}{C}}-\overset{-}{N}CH_3 \longleftrightarrow C_6H_5-\overset{O^-}{\overset{|}{C}}=NCH_3$$

vs.

$$C_6H_5-\underset{O}{\underset{\|}{\overset{O}{\overset{\|}{S}}}}-\overset{-}{N}CH_3 \longleftrightarrow C_6H_5-\underset{O}{\underset{\|}{\overset{O^-}{\overset{|}{S}}}}=NCH_3 \longleftrightarrow C_6H_5-\underset{O^-}{\underset{|}{\overset{O}{\overset{\|}{S}}}}=NCH_3$$

17.16

(a)

$$CH_3CH_2CH_2-\overset{O}{\overset{\|}{C}}-OH \xrightarrow{PCl_5} CH_3CH_2CH_2-\overset{O}{\overset{\|}{C}}-Cl \xrightarrow{NH_3} CH_3CH_2CH_2-\overset{O}{\overset{\|}{C}}-NH_2$$

$$\xrightarrow{1.\ LiAlH_4;\ 2.\ H_2O} CH_3CH_2CH_2CH_2NH_2$$

$$CH_3CH_2CH_2CH_2NH_2 \xrightarrow{CH_3CH_2CH_2-\overset{O}{\overset{\|}{C}}-Cl} CH_3CH_2CH_2-\overset{O}{\overset{\|}{C}}-NH-CH_2CH_2CH_2CH_3$$

$$\xrightarrow{1.\ LiAlH_4;\ 2.\ H_2O} CH_3CH_2CH_2CH_2NHCH_2CH_2CH_2CH_3$$

(b)

$$CH_3CH_2CH_2-\overset{O}{\overset{\|}{C}}-OH \xrightarrow{PCl_5} CH_3CH_2CH_2-\overset{O}{\overset{\|}{C}}-Cl \xrightarrow{NH_3} CH_3CH_2CH_2-\overset{O}{\overset{\|}{C}}-NH_2$$

$$\xrightarrow{1.\ LiAlH_4;\ 2.\ H_2O} CH_3CH_2CH_2CH_2NH_2$$

$$CH_3CH_2CH_2CH_2NH_2 \xrightarrow[HCO_2H]{CH_2O} CH_3CH_2CH_2CH_2N(CH_3)_2$$

(c)

$$CH_3CH_2CH_2-\overset{O}{\overset{\|}{C}}-OH \xrightarrow{PCl_5} CH_3CH_2CH_2-\overset{O}{\overset{\|}{C}}-Cl \xrightarrow{NH_3} CH_3CH_2CH_2-\overset{O}{\overset{\|}{C}}-NH_2$$

$$CH_3CH_2CH_2-\overset{O}{\overset{\|}{C}}-NH_2 \xrightarrow{Br_2,\ KOH} CH_3CH_2CH_2NH_2$$

$$CH_3CH_2CH_2NH_2 \xrightarrow{CH_3CH_2CH_2-\overset{O}{\overset{\|}{C}}-Cl} CH_3CH_2CH_2-\overset{O}{\overset{\|}{C}}-NH-CH_2CH_2CH_3$$

$$CH_3CH_2CH_2-\overset{O}{\overset{\|}{C}}-NH-CH_2CH_2CH_3 \xrightarrow[2.\ H_2O]{1.\ LiAlH_4} CH_3CH_2CH_2CH_2NHCH_2CH_2CH_3$$

(d)

$$CH_3CH_2CH_2-\overset{O}{\overset{\|}{C}}-OH \xrightarrow{PCl_5} CH_3CH_2CH_2-\overset{O}{\overset{\|}{C}}-Cl \xrightarrow{NH_3} CH_3CH_2CH_2-\overset{O}{\overset{\|}{C}}-NH_2$$

$$CH_3CH_2CH_2-\overset{O}{\overset{\|}{C}}-NH_2 \xrightarrow[2.\ H_2O]{1.\ LiAlH_4} CH_3CH_2CH_2CH_2NH_2$$

$$CH_3CH_2CH_2CH_2NH_2 \xrightarrow{\text{excess } CH_3I} CH_3CH_2CH_2CH_2\overset{+}{N}(CH_3)_3\ I^-$$

17.17 In each case the removal of an allylic hydrogen occurs with the consequent formation of a conjugated double bond.

17.18 Prior to any isomerization, 5-(*N,N*-dimethylamino)-1-pentene could only yield 1,4-pentadiene:

$(CH_3)_2N$

Similarly, in the absence of any isomerization 1,3-pentadiene would arise from 4-(*N,N*-dimethylamino)-1-pentene (or from the 3-substituted isomer, which is not shown):

$$(CH_3)_2N\text{-CH(CH}_3\text{)-CH}_2\text{-CH=CH}_2 \longrightarrow CH_3\text{-CH=CH-CH=CH}_2$$

Assuming that no 4-membered rings were involved, the identification 4-(*N,N*-dimethylamino)-1-pentene would have been interpreted as requiring a 5-membered ring for piperidine:

$$\text{2-methylpyrrolidine (N-H)} \longrightarrow \text{N,N-dimethyl-2-methylpyrrolidinium (}N^+\text{, two }CH_3\text{)} \longrightarrow CH_3CH(N(CH_3)_2)CH_2CH{=}CH_2 \equiv (CH_3)_2N\text{-}CH(CH_3)CH_2CH{=}CH_2$$

17.19

(a) $C_6H_5CH_2NCH_2CH_2C_6H_5 \xrightarrow{\text{excess } CH_3I} C_6H_5CH_2-\overset{+}{N}(CH_3)_2-CH_2CH_2C_6H_5\ I^- \xrightarrow[H_2O]{Ag_2O}$

$C_6H_5CH_2-\overset{+}{N}(CH_3)_2-CH_2CH_2C_6H_5\ OH^- \xrightarrow{\text{pyrolysis}} C_6H_5CH_2NH_2 + CH_2{=}CHC_6H_5$

(b) $CH_3-CH(CH_3)-CH_2CH_2OH \xrightarrow{CrO_3} CH_3-CH(CH_3)-CH_2-\overset{O}{\overset{\|}{C}}-OH \xrightarrow[2.\ NH_3]{1.\ PCl_5}$

$CH_3-CH(CH_3)-CH_2-\overset{O}{\overset{\|}{C}}-NH_2 \xrightarrow[KOH]{Br_2} CH_3-CH(CH_3)-CH_2OH$

(c)

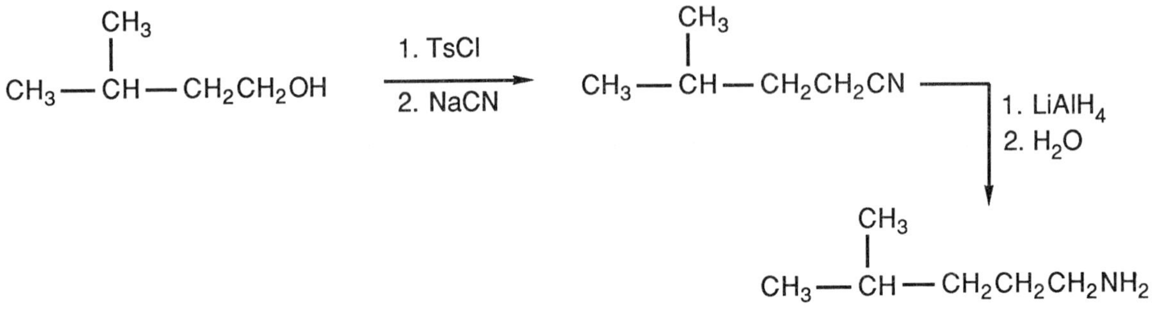

$$CH_3-\underset{\displaystyle CH_3}{\overset{|}{CH}}-CH_2CH_2OH \xrightarrow[\text{2. NaCN}]{\text{1. TsCl}} CH_3-\underset{}{\overset{CH_3}{\overset{|}{CH}}}-CH_2CH_2CN \xrightarrow[\text{2. } H_2O]{\text{1. } LiAlH_4} CH_3-\overset{CH_3}{\overset{|}{CH}}-CH_2CH_2CH_2NH_2$$

(d)

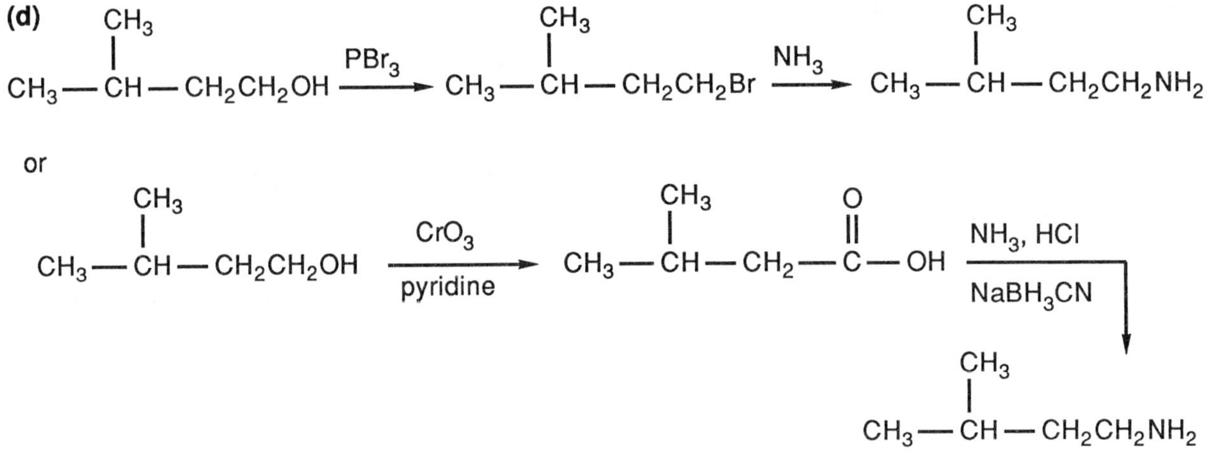

$$CH_3-\overset{CH_3}{\overset{|}{CH}}-CH_2CH_2OH \xrightarrow{PBr_3} CH_3-\overset{CH_3}{\overset{|}{CH}}-CH_2CH_2Br \xrightarrow{NH_3} CH_3-\overset{CH_3}{\overset{|}{CH}}-CH_2CH_2NH_2$$

or

$$CH_3-\overset{CH_3}{\overset{|}{CH}}-CH_2CH_2OH \xrightarrow[\text{pyridine}]{CrO_3} CH_3-\overset{CH_3}{\overset{|}{CH}}-CH_2-\overset{O}{\overset{\|}{C}}-OH \xrightarrow[NaBH_3CN]{NH_3,\ HCl} CH_3-\overset{CH_3}{\overset{|}{CH}}-CH_2CH_2NH_2$$

(e)

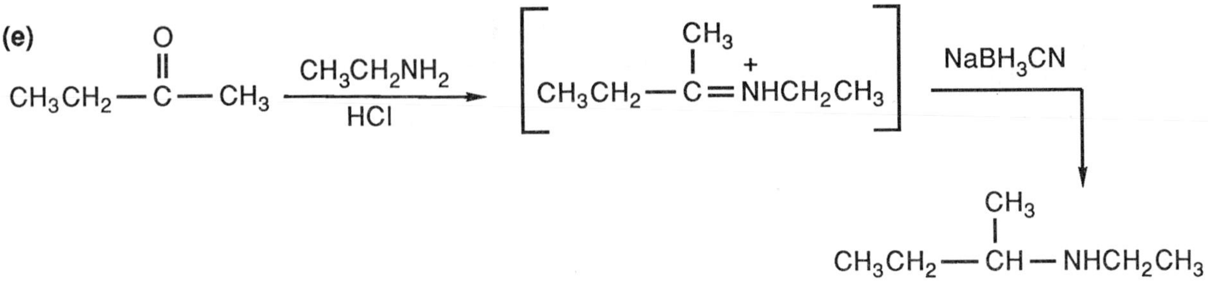

$$CH_3CH_2-\overset{O}{\overset{\|}{C}}-CH_3 \xrightarrow[HCl]{CH_3CH_2NH_2} \left[CH_3CH_2-\overset{CH_3}{\overset{|}{C}}=\overset{+}{N}HCH_2CH_3\right] \xrightarrow{NaBH_3CN} CH_3CH_2-\overset{CH_3}{\overset{|}{CH}}-NHCH_2CH_3$$

or

$$CH_3CH_2-\overset{\overset{\large O}{\|}}{C}-CH_3 \xrightarrow{CH_3CH_2NH_2} CH_3CH_2-\overset{\overset{\large NCH_2CH_3}{\|}}{C}-CH_3 \xrightarrow{H_2,\ Ni} CH_3CH_2-\overset{\overset{\large CH_3}{|}}{CH}-NHCH_2CH_3$$

17.20

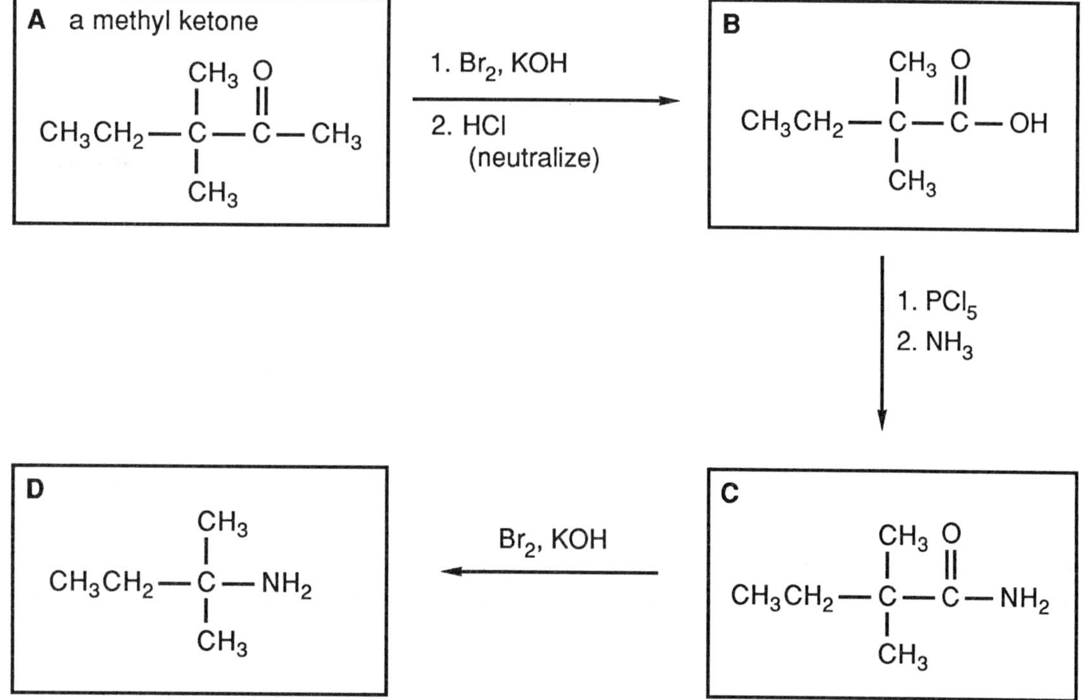

17.21 Conversion of **A** to **B**:

$$CH_3CH_2-C(CH_3)_2-C(=O)-CH_3 \xrightarrow{^-OH} CH_3CH_2-C(CH_3)_2-C(=O)-\bar{C}H_2 \xrightarrow{Br-Br} CH_3CH_2-C(CH_3)_2-C(=O)-CH_2Br$$

$$CH_3CH_2-C(CH_3)_2-C(=O)-CH_2Br \xrightarrow{OH^-} CH_3CH_2-C(CH_3)_2-C(=O)-\bar{C}HBr \xrightarrow{Br-Br} CH_3CH_2-C(CH_3)_2-C(=O)-CHBr_2$$

$$CH_3CH_2-C(CH_3)_2-C(=O)-CHBr_2 \xrightarrow{OH^-} CH_3CH_2-C(CH_3)_2-C(=O)-\bar{C}Br_2 \xrightarrow{Br-Br} CH_3CH_2-C(CH_3)_2-C(=O)-CBr_3$$

$$CH_3CH_2-C(CH_3)_2-C(=O)-CBr_3 \xrightarrow{OH^-} CH_3CH_2-C(CH_3)_2-C(O^-)(OH)-CBr_3 \longrightarrow CH_3CH_2-C(CH_3)_2-C(=O)-OH$$

$$CH_3CH_2-C(CH_3)_2-C(=O)-OH \xrightarrow{OH^-} CH_3CH_2-C(CH_3)_2-C(=O)-O^- \xrightarrow{HCl} CH_3CH_2-C(CH_3)_2-C(=O)-OH$$

Conversion of **B** to **C**:

$$CH_3CH_2-C(CH_3)_2-C(=O)-\ddot{O}H \xrightarrow{PCl_5} CH_3CH_2-C(CH_3)_2-C(=O)-O-PCl_4 \xrightarrow{Cl^-} CH_3CH_2-C(CH_3)_2-C(=O)-Cl$$

$$CH_3CH_2-C(CH_3)_2-C(=O)-Cl \xrightarrow{:NH_3} CH_3CH_2-C(CH_3)_2-C(=O)-NH_2$$

Conversion of **C** to **D**:

$$CH_3CH_2-C(CH_3)_2-C(=O)-NH_2 \;\xrightarrow{\;^-OH\;}\; CH_3CH_2-C(CH_3)_2-C(=O)-\bar{N}H \;\xrightarrow{\;Br-Br\;}\; CH_3CH_2-C(CH_3)_2-C(=O)-NHBr$$

$$CH_3CH_2-C(CH_3)_2-C(=O)-NHBr \;\xrightarrow{\;^-OH\;}\; CH_3CH_2-C(CH_3)_2-C(=O)-\bar{N}-Br \;\longrightarrow\; CH_3CH_2-C(CH_3)_2-N{=}C{=}O$$

$$CH_3CH_2-C(CH_3)_2-N{=}C{=}O \;\xrightarrow{\;^-OH,\ OH_2\;}\; CH_3CH_2-C(CH_3)_2-NH-C(=O)-O-H \;\xrightarrow{\;^-OH,\ OH_2\;}\; CH_3CH_2-C(CH_3)_2-NH_2$$

17.22 The problem states that **A** is an amine, and the Hinsberg test (formation of the sulfonamide **C**, which is soluble in neither acid nor base) indicates that the **A** is a secondary amine. The sequence of reactions converting **A** to **F** introduces a benzyl group, thereby forming a tertiary amine. Quaternization with methyl iodide follow by elimination yields **F** and **G**. Working backwards from **F** and **G** you can deduce the structure of **E**. Since **A** is a secondary amine, the cyclohexyl group and one of the methyl must have been present in that compound. The complete answer follows:

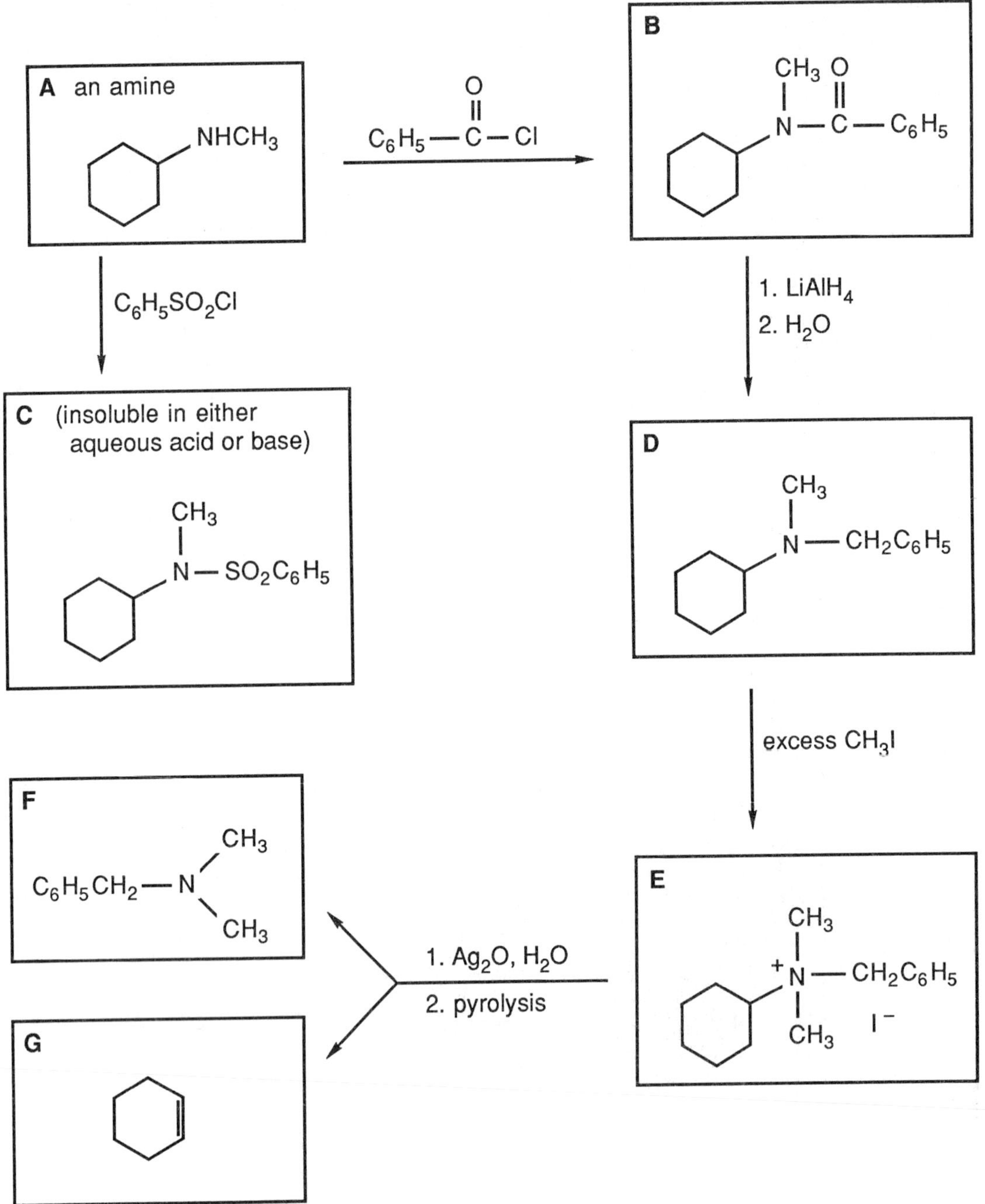

17.23 Denoting **A** as R-COCl, **B** must have the structure $R\text{-}NH_2$. Reaction of this with **A** followed by reduction would afford $R\text{-}CH_2\text{-}NH\text{-}R$. The other reaction of **A** (R-COCl) with $CH_3CH_2CH_2NH_2$ followed by reduction would yield $R\text{-}CH_2\text{-}NH\text{-}CH_2CH_2CH_3$, which is also **C.** Therefore, you can conclude that R is $CH_2CH_2CH_3$, and the complete answer follows:

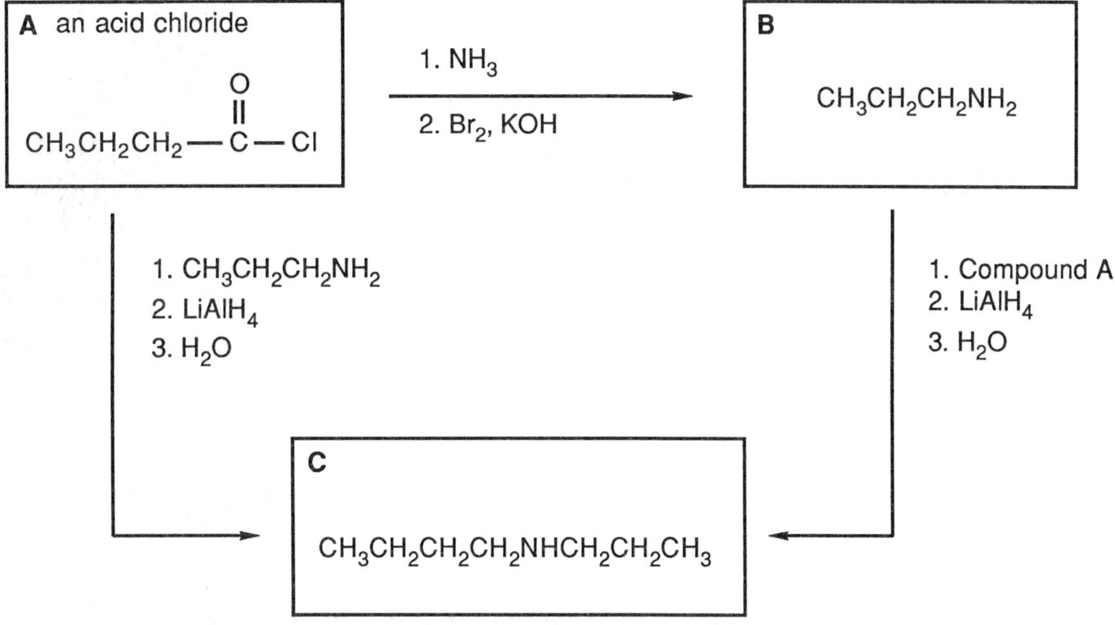

17.24

A $CH_3CH_2CH_2{-}\overset{O}{\overset{\|}{C}}{-}Cl$ $\xrightarrow{NH_3}$ $CH_3CH_2CH_2{-}\overset{O}{\overset{\|}{C}}{-}NH_2$ $\xrightarrow{Br_2,\ KOH}$ **B** $CH_3CH_2CH_2NH_2$

$\downarrow$ $CH_3CH_2CH_2{-}\overset{O}{\overset{\|}{C}}{-}Cl$

$CH_3CH_2CH_2CH_2{-}\overset{O}{\overset{\|}{C}}{-}NHCH_2CH_2CH_3$ $\xrightarrow[2.\ H_2O]{1.\ LiAlH_4}$ **C** $CH_3CH_2CH_2CH_2NHCH_2CH_2CH_3$

A

$$CH_3CH_2CH_2—\overset{\overset{\large O}{\|}}{C}—Cl \xrightarrow{CH_3CH_2CH_2NH_2} CH_3CH_2CH_2—\overset{\overset{\large O}{\|}}{C}—NHCH_2CH_2CH_3$$

1. $LiAlH_4$
2. H_2O

C

$CH_3CH_2CH_2CH_2NHCH_2CH_2CH_3$

17.25 The reaction of **A** with benzenesulfonyl chloride indicates that the amine is primary because the resulting sulfonamide is soluble in aqueous base. Denoting **A** as R-NH_2 the sequence from **A** to **C** produces R-NH-$CH_2C_6H_5$. Quaternization with methyl iodide followed bv elimination yields **G**, which still contains the benzyl group, and **E**, which must correspond to an alkene having the structure R (with loss of a hydrogen atom). Oxidative cleavage of **E** followed by treatment with base yields **H**, an α,β-unsaturated aldehyde that must have resulted from an aldol condensation. Working backward from **H**, you can deduce the structure of **F** as 3,3-dimethylhexanedial, so **E** must be 4,4-dimethylcyclohexene. Now there are two possibilities for **C**: the benzylamino group might have either a 1,4 or a 1,3 relationship with the dimethyl group on the 6-membered ring. But only the former would yield **E** as a single product. The complete answer follows:

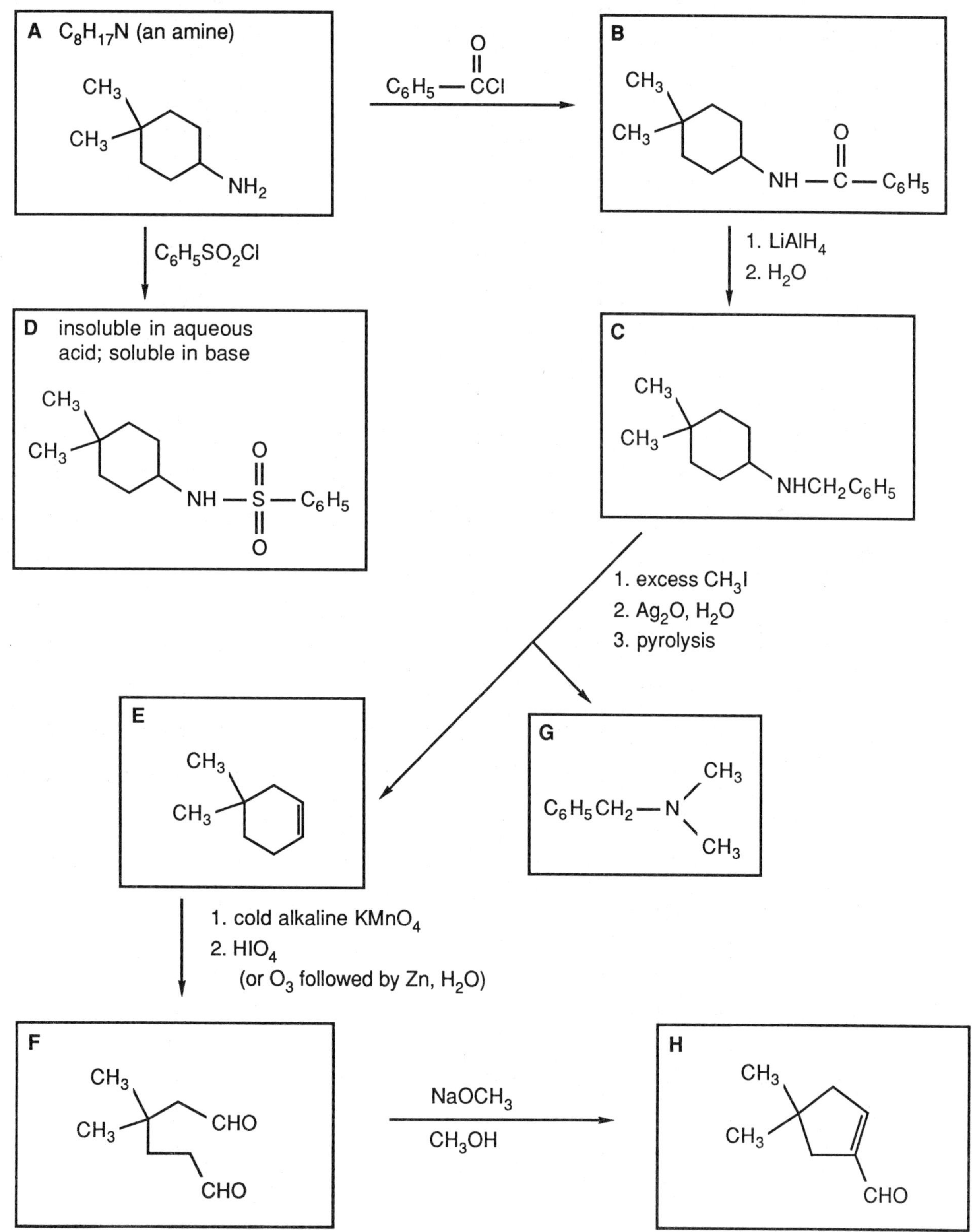

A $C_8H_{17}N$ (an amine)
CH_3
CH_3
NH_2
O
$C_6H_5—CCl$
B
CH_3
CH_3
O
$NH—C—C_6H_5$
$C_6H_5SO_2Cl$
1. $LiAlH_4$
2. H_2O
D insoluble in aqueous acid; soluble in base
CH_3
CH_3
O
$NH—S—C_6H_5$
O
C
CH_3
CH_3
$NHCH_2C_6H_5$
1. excess CH_3I
2. Ag_2O, H_2O
3. pyrolysis
E
CH_3
CH_3
G
CH_3
$C_6H_5CH_2—N$
CH_3
1. cold alkaline $KMnO_4$
2. HIO_4
(or O_3 followed by Zn, H_2O)
F
CH_3
CH_3
CHO
CHO
$NaOCH_3$
CH_3OH
H
CH_3
CH_3
CHO

17.26 The solubility properties of **A** tell you that it is an amine, and the solubility properties of the sulfonamide **B** tell you that the amine is secondary. Quaternization with ethyl iodide followed by elimination produces **C**, and its NMR spectrum shows a signal in the aromatic region together with a distinctive ethyl pattern. **C** must be a tertiary amine, and closer inspection of the NMR spectrum will show you that the relative sizes of the ethyl and aromatic signals must correspond to *one* phenyl and *two* ethyl groups. One of the original substituents in **A** must therefore be phenyl, and consideration of the molecular formula tells you that the other is an ethyl group. The complete structures are:

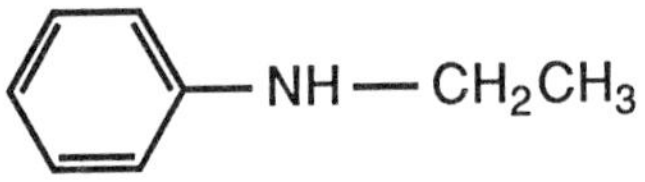

$C_6H_5-NH-CH_2CH_3$ **A**

$C_6H_5-N(CH_2CH_3)-SO_2-C_6H_5$ **B**

$C_6H_5-N(CH_2CH_3)-CH_2CH_3$ **C**

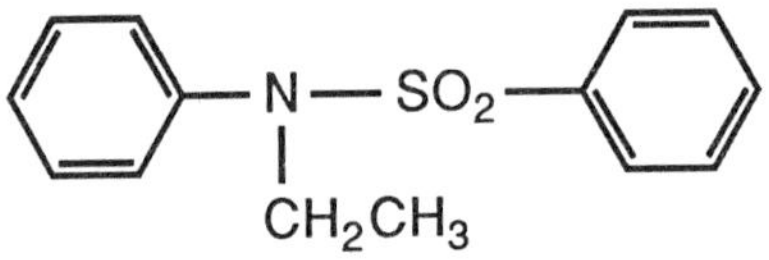

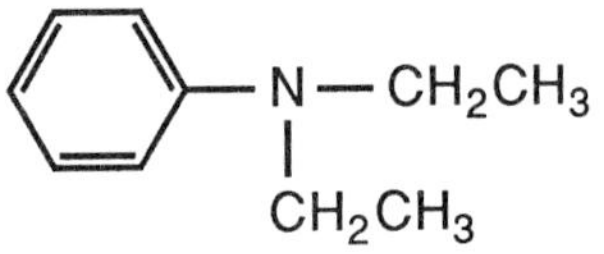

17.27 The pK_a values of the conjugate acids tell you that the anilinium ion is the weakest acid, so aniline is the strongest base with the sequence: aniline > 3-nitroaniline > 4-nitroaniline. The decreased basicity of 3-nitroaniline can be attributed to the electron withdrawing inductive effect of the nitro group. The larger decrease in basicity of 4-nitroaniline results from the direct resonance interaction between the amino and nitro groups:

$O_2N^{+}-C_6H_4-\ddot{N}H_2 \longleftrightarrow (^{-}O)_2{}^{+}N{=}C_6H_4{=}\overset{+}{N}H_2$

CHAPTER 18. AROMATIC SUBSTITUTION

ANSWERS TO EXERCISES

18.1 **(a)**

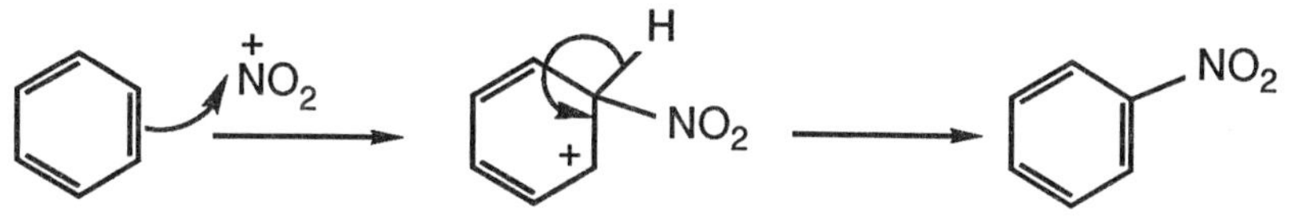

(b)

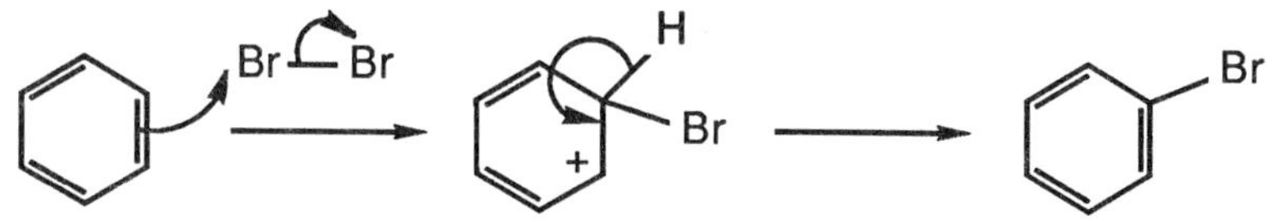

(c)

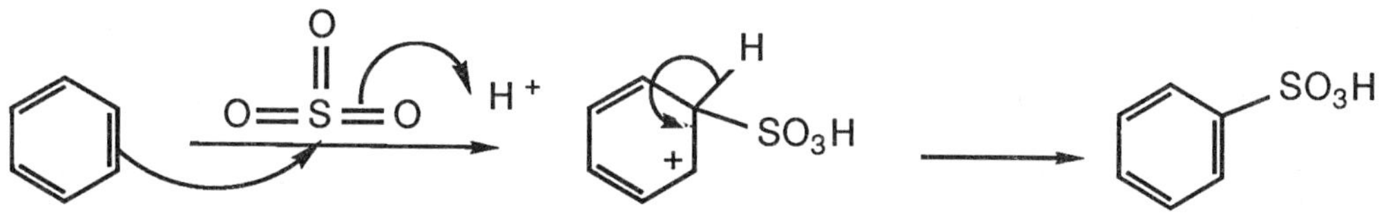

18.2 **(a), (b)**

$-OH$: activating and *o,p*-directing

$-SCH_3$: activating and *o,p*-directing

$-O_2CCH_3$: activating and *o,p*-directing

$-NH_2$: activating and *o,p*-directing

$-NH_3^+$: deactivating and *m*-directing

$-Br$: deactivating and *o,p*-directing

$-CO_2CH_3$: deactivating and *m*-directing

$-CH_2CH_2CH_2CF_3$: activating and *o,p*-directing (the inductive effect of the fluorines is minimal with three intervening CH_2 groups).

$-CD_3$: activating and *o,p*-directing

(c) There are many examples of groups that are *activating and o,p-directing* (these are typically oxygen or nitrogen substituents in which the heteroatom has nonbonding electrons; vinyl and aryl groups behave similarly). Groups that are *deactivating and m-directing* are also common and are characterized by that have positive charge on the atom attached to the aromatic ring (either a formal positive charge or a partial charge as in the carbon of a C=O group). The halogens are *deactivating and o,p-directing* a relatively uncommon situation that results from the electron withdrawing effect that is partially compensated by a weak electron donating resonance effect. There are no examples of substituents that are *activating and m-directing*

18.3 **(a)** CH_3, Br (para) + CH_3, Br (ortho)

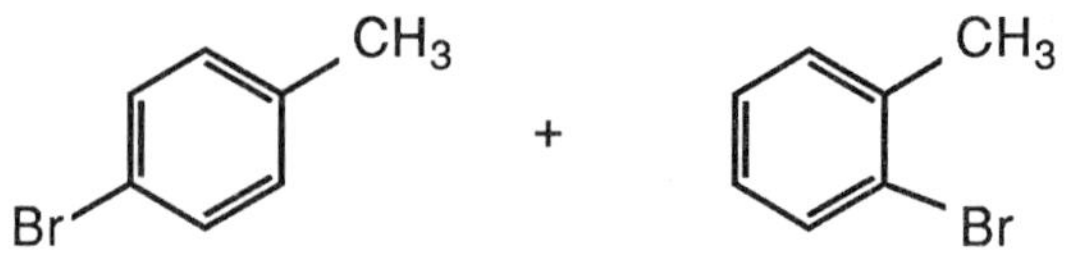

(b) OCH_3, HSO_3 (para) + OCH_3, SO_3H (ortho)

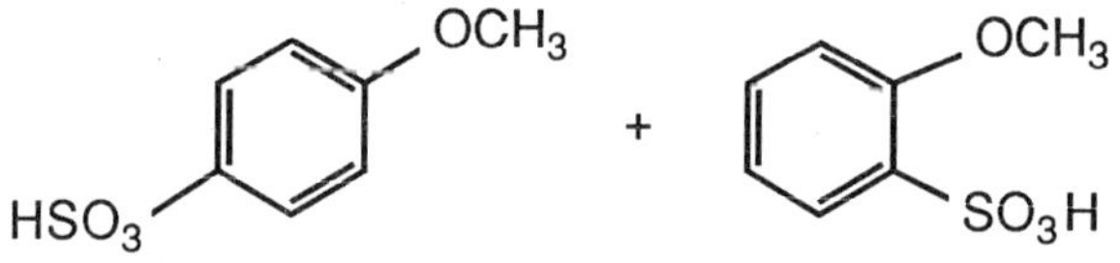

(c) $\overset{+}{N}(CH_3)_3$, NO_2 (meta)

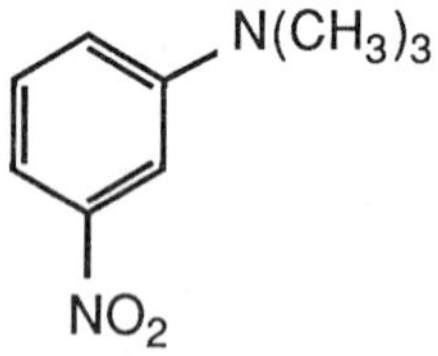

18.4 **(a)** The most unfavorable resonance contributors are highlighted within a box; reaction at the *meta* position is therefore preferred:

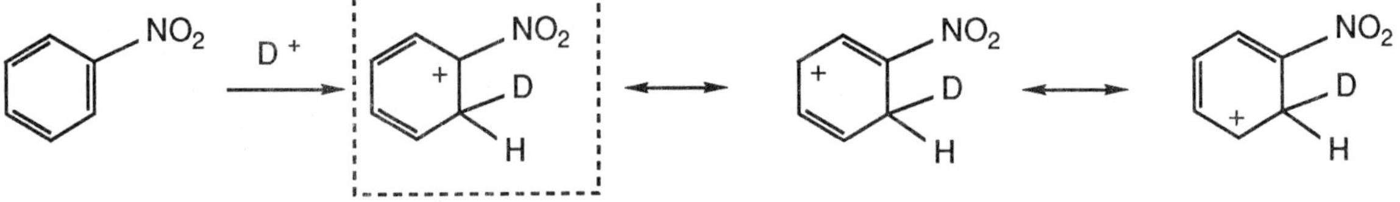

NO_2 —D^+→ [NO_2, D, H, +] ⟷ NO_2, D, H, + ⟷ NO_2, D, H, +

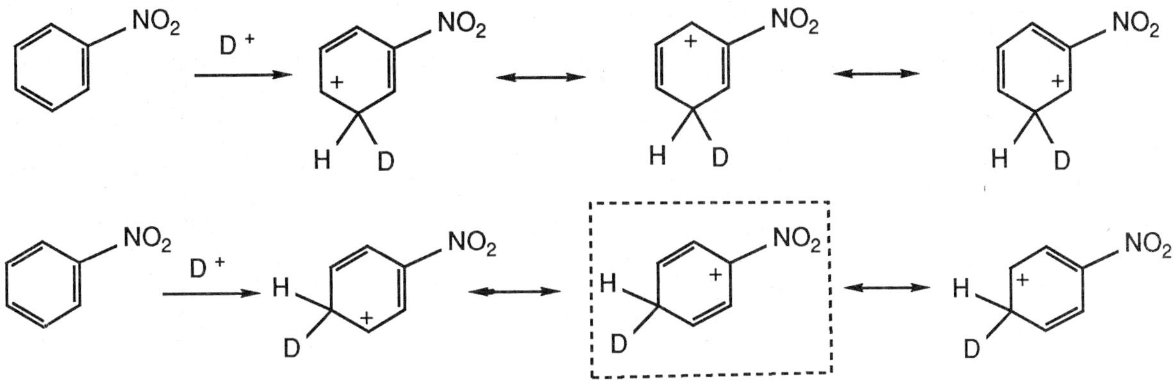

(b) The most favorable resonance contributors are highlighted within a box; reaction at the *ortho* and *para* positions are therefore preferred:

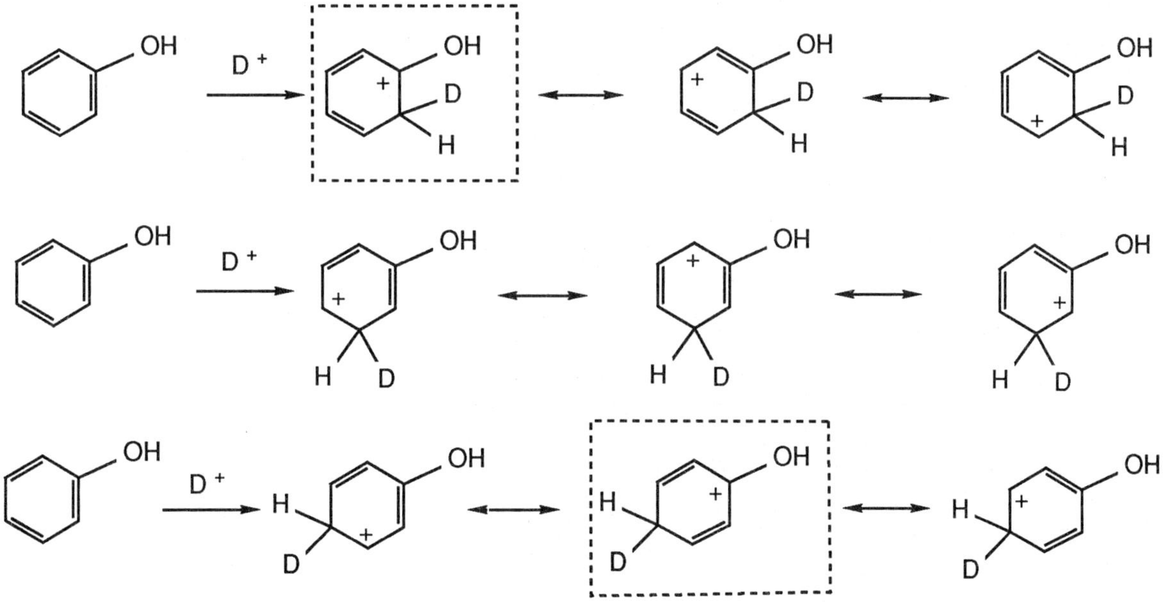

18.5 The rate effects estimated from Figure 18.1 are written next to each substituent in the reactant. The product is governed by the most activating substituent (and little or no 1,2,3-substitution is expected):

(a) 10^{-5} CO_2CH_3 (para: 10^{3} CH_3) → product with CO_2CH_3, NO_2, CH_3

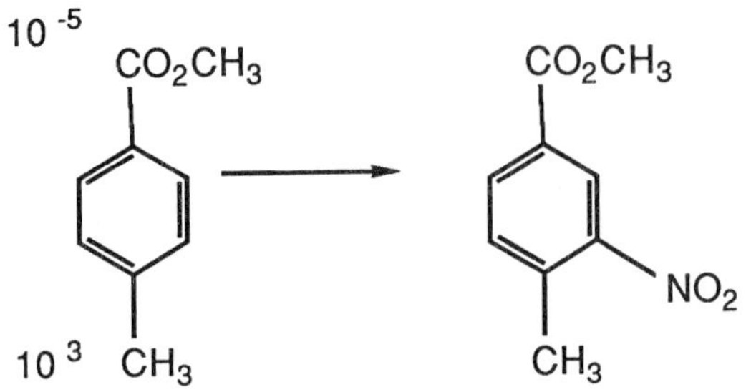

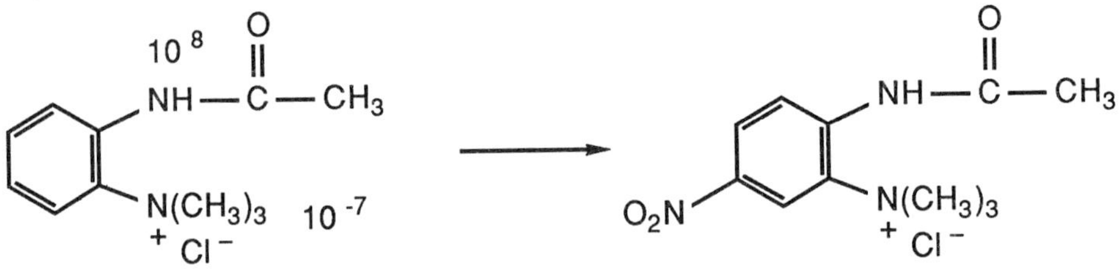

(b) 10^{8} $NH—C(=O)—CH_3$, $N(CH_3)_3^{+}\ Cl^{-}$ 10^{-7} → O_2N product with $NH—C(=O)—CH_3$, $N(CH_3)_3^{+}\ Cl^{-}$

(c) 10^{12} OH (para: 10^{-1} Cl) → product with OH, NO_2, Cl

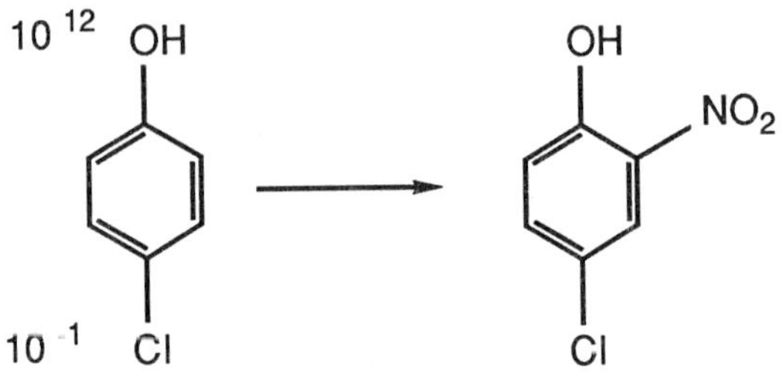

(d) 10^{-1} Cl, 10^{-10} NO_2 → product with Cl, NO_2, NO_2

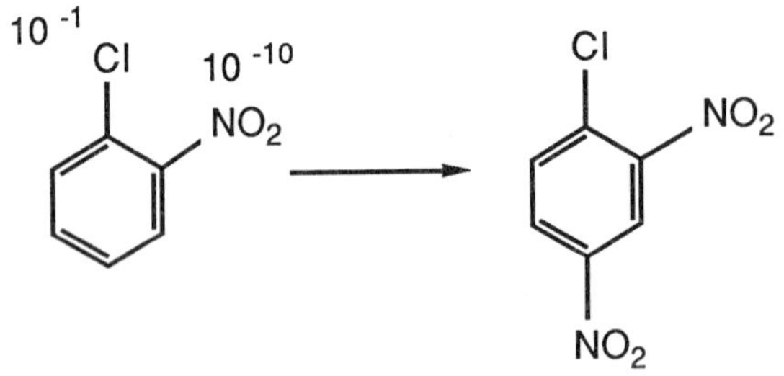

18.6 (a) CH_3, CH_2CH_3 (ortho) + CH_3, CH_2CH_3 (para)

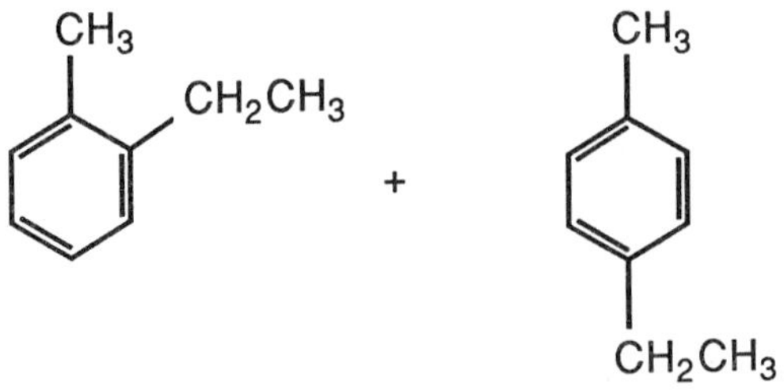

(b) Polysubstituted products such as diethyltoluene:

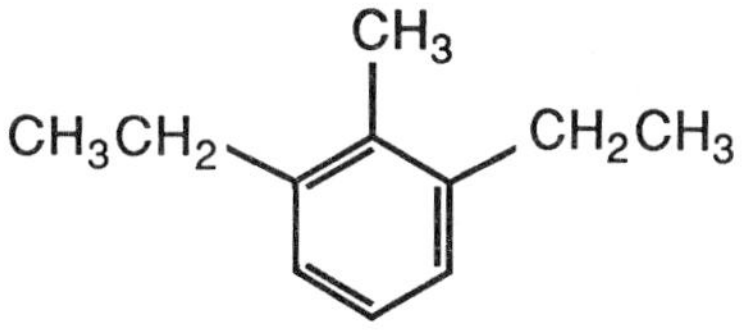

(c) Rearrangement to give an equilibrium mixture:

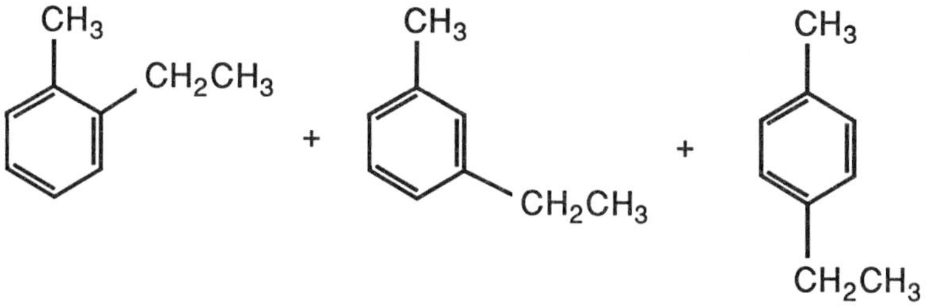

18.7

(a)

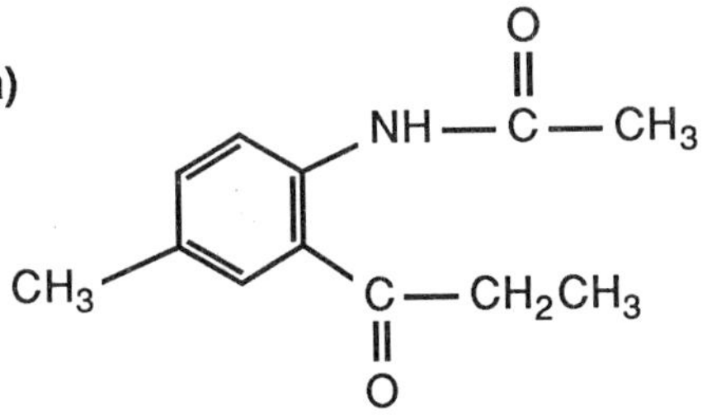

(b)

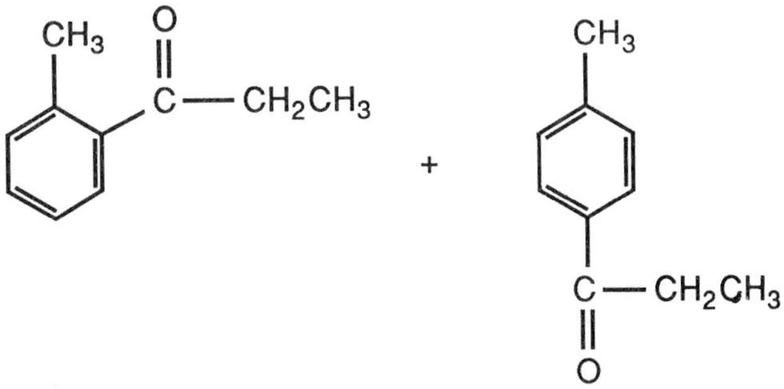

(c)

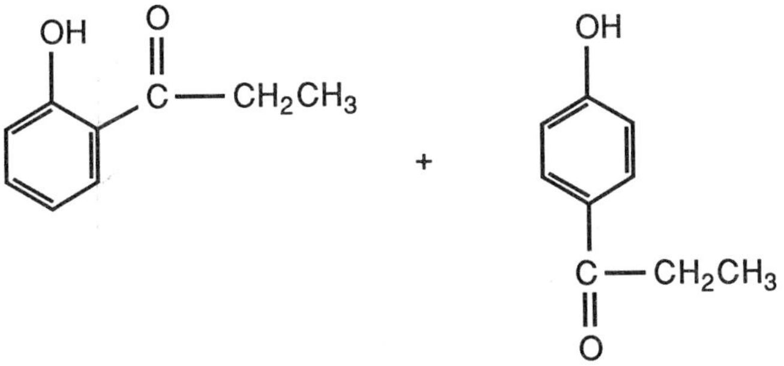

(d) CH_3 / CH_3 / $C(=O)-CH_2CH_3$

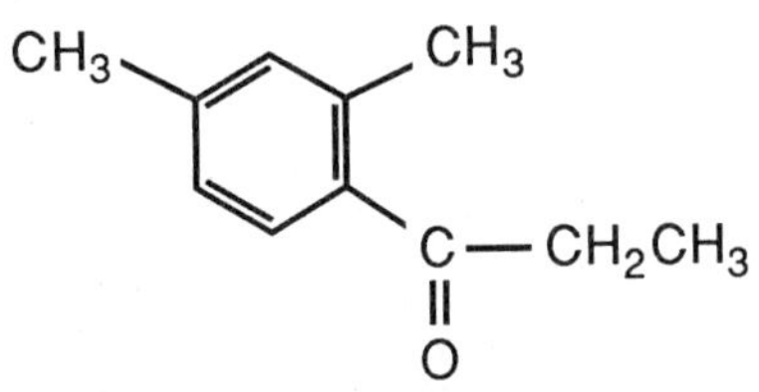

18.8 (a) The effects estimated from Figure 18.1 are written next to each substituent, and the overall predicted rate effect is enclosed in a box by each structure. Only those structures with an overall rate effect larger than 10^0 should be more reactive than benzene:

(i) OCH_3 10^9; NO_2 10^{-10} — 10^{-1}

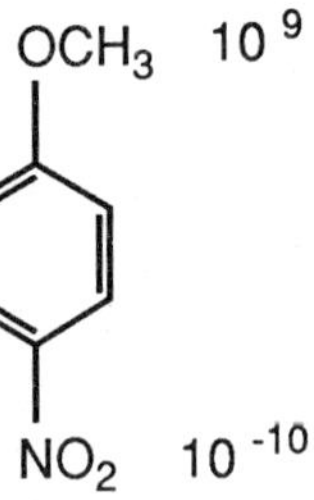

(ii) 10^3 CH_3; $C(=O)-CH_3$ 10^{-7}; CH_3 10^3 — 10^{-1}

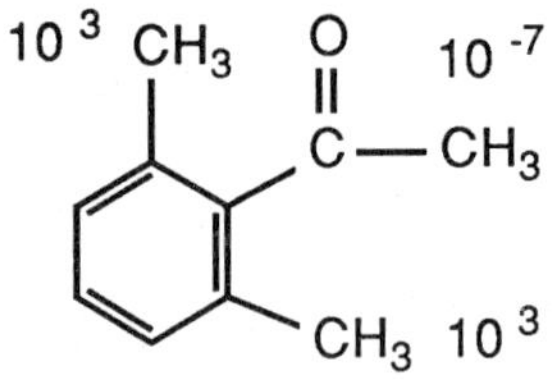

(iii) 10^{-5} CO_2H; CO_2H 10^{-5} — 10^{-10}

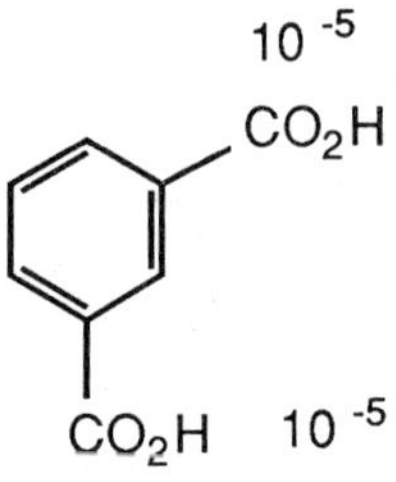

(iv) 10^{-7} NH_2; CH_3 10^{-3} — 10^{-10}

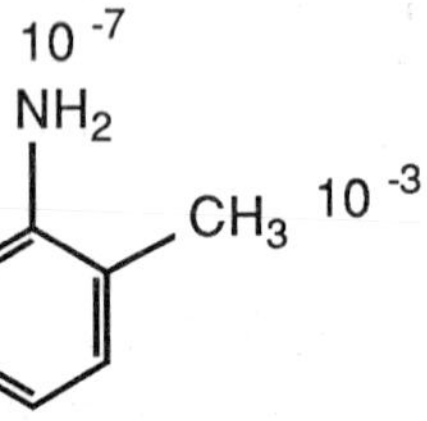

(v) 10^{-2} Br; CH_3 10^3 — 10^1

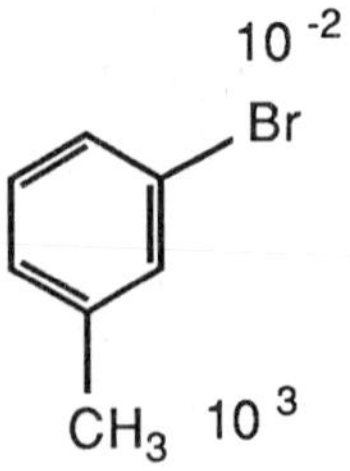

(vi) 10^{-5} CO_2H; CH_3O 10^9 — 10^4

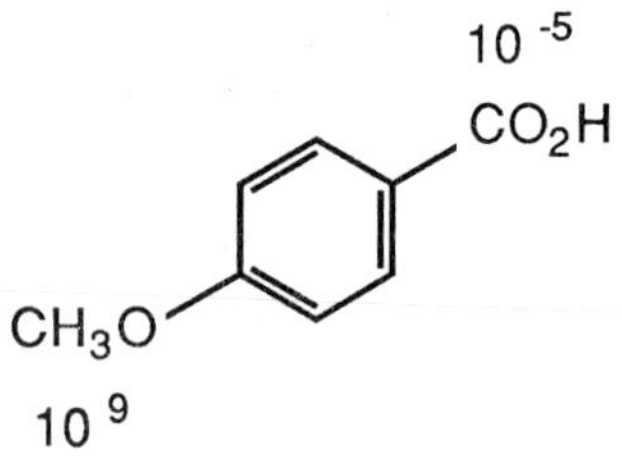

(vii)

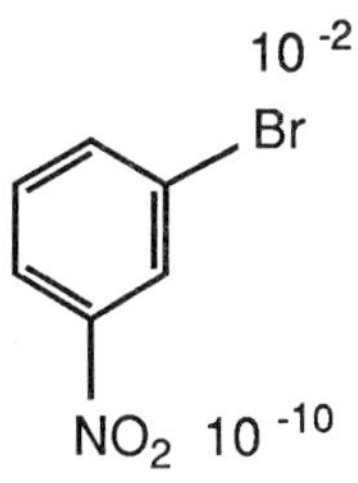

10^{-2} Br

NO_2 10^{-10}

10^{-12}

(b) Each of the structures has at least one deactivating substituent, and any that is less reactive than the halobenzenes (i.e., with an estimated rate effect of 10^{-3} or smaller) should undergo Friedel-Crafts acylation: (i) and (ii) might be borderline, but (v) and (vi) should react.

18.9

(a) C_6H_5–Cl $\xrightarrow{Mg}$ C_6H_5–MgCl $\xrightarrow[2.\ H_2SO_4]{1.\ CO_2}$ C_6H_5–CO_2H

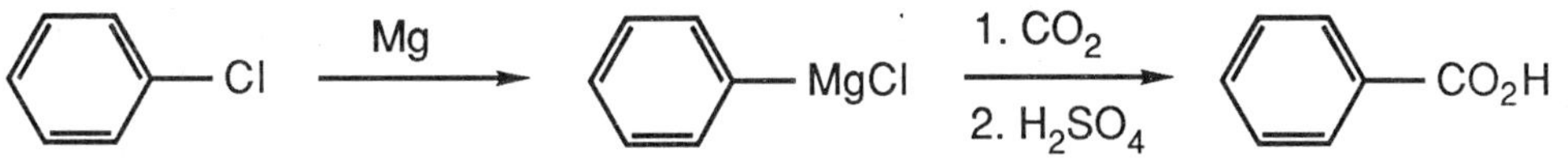

(b) C_6H_5–Cl $\xrightarrow[2.\ CuI]{1.\ Li}$ (C_6H_5)–CuLi $\xrightarrow{CH_3-C(=O)-Cl}$ C_6H_5–C(=O)–CH_3

or

C_6H_5–Cl $\xrightarrow{Mg}$ C_6H_5–MgCl $\xrightarrow[2.\ H_2O]{1.\ CH_3CHO}$ C_6H_5–CH(OH)–CH_3 $\xrightarrow{CrO_3}$ C_6H_5–C(=O)–CH_3

(c) C_6H_5–Cl $\xrightarrow{Li}$ C_6H_5–Li $\xrightarrow[2.\ H_2O]{1.\ \text{ethylene oxide}}$ C_6H_5–CH_2CH_2OH

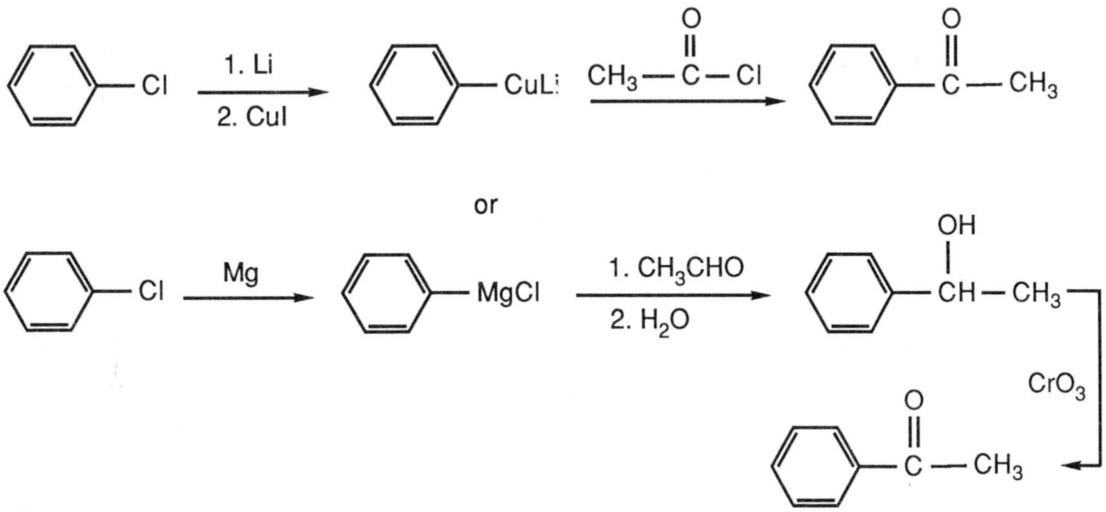

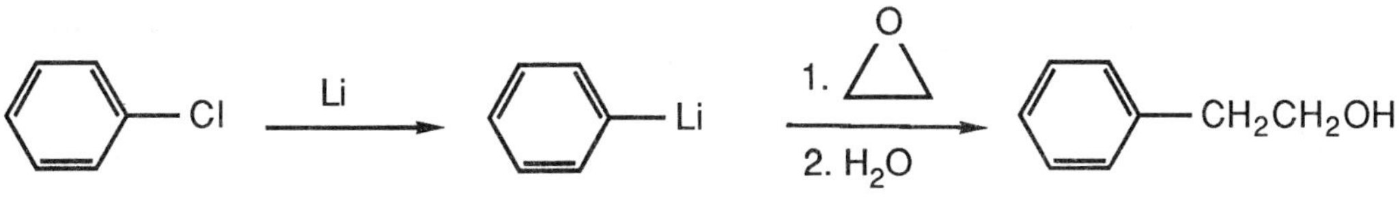

(d)

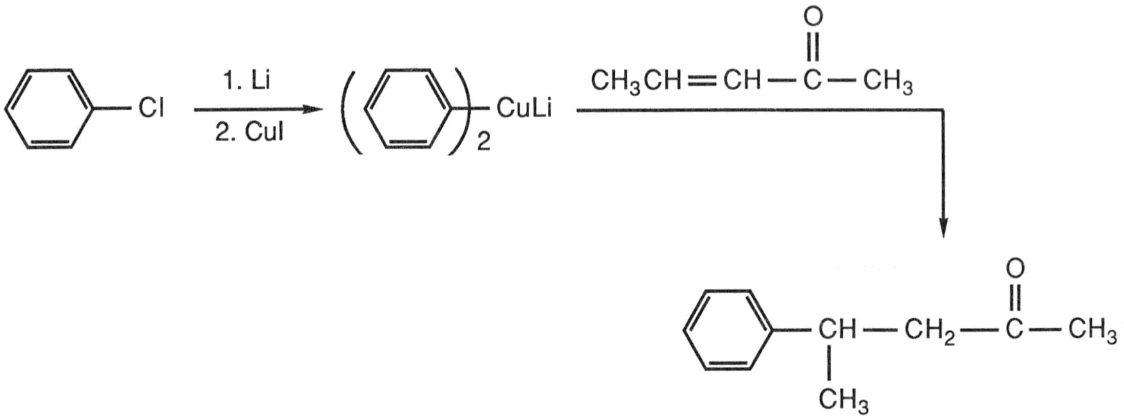

$C_6H_5Cl \xrightarrow[\text{2. CuI}]{\text{1. Li}} (C_6H_5)_2CuLi \xrightarrow{CH_3CH{=}CH{-}C(=O){-}CH_3} C_6H_5{-}CH(CH_3){-}CH_2{-}C(=O){-}CH_3$

(e)

$C_6H_5{-}SO_3H \xrightarrow[H_2O]{H_2SO_4} C_6H_6$

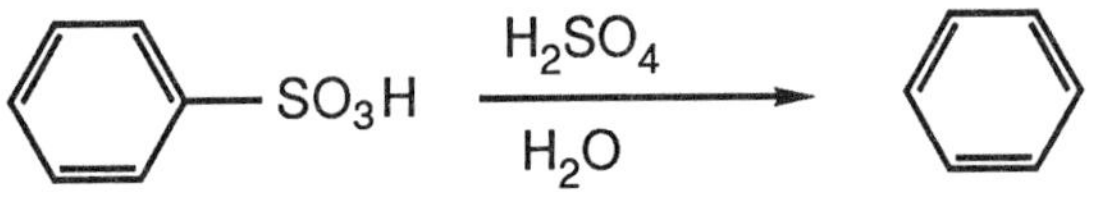

(f)

$C_6H_5{-}CH_3 \xrightarrow{KMnO_4} C_6H_5{-}CO_2H$

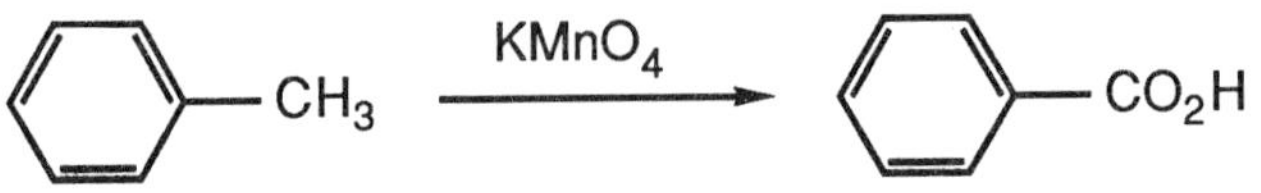

(g)

$C_6H_5{-}CH_3 \xrightarrow{NBS} C_6H_5{-}CH_2Br$

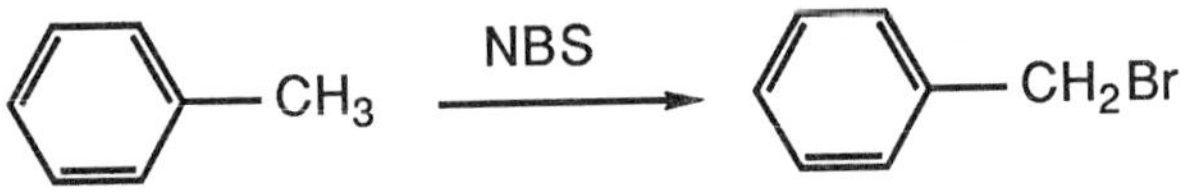

(h)

$C_6H_5{-}OH \xrightarrow[\text{2. } CH_3I]{\text{1. KOtBu}} C_6H_5{-}OCH_3$

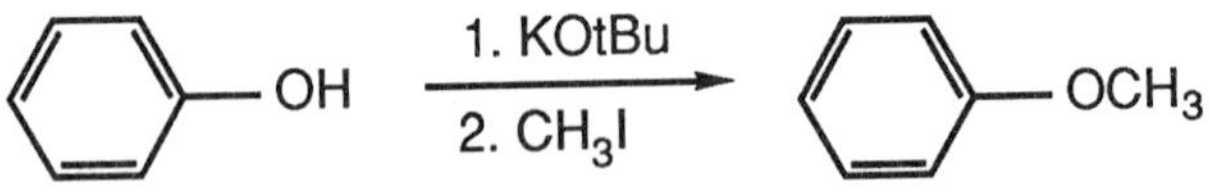

(i)

$C_6H_5{-}OH \xrightarrow{CH_3{-}C(=O){-}Cl} C_6H_5{-}O{-}C(=O){-}CH_3$

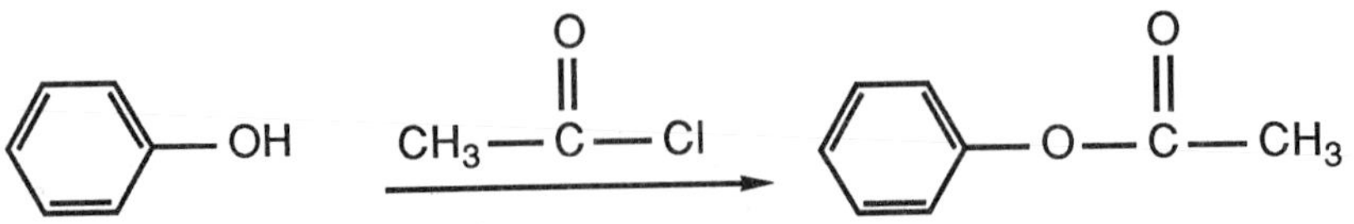

(j)

$C_6H_5{-}NO_2 \xrightarrow[\text{2. NaOH}]{\text{1. Fe, HCl}} C_6H_5{-}NH_2$

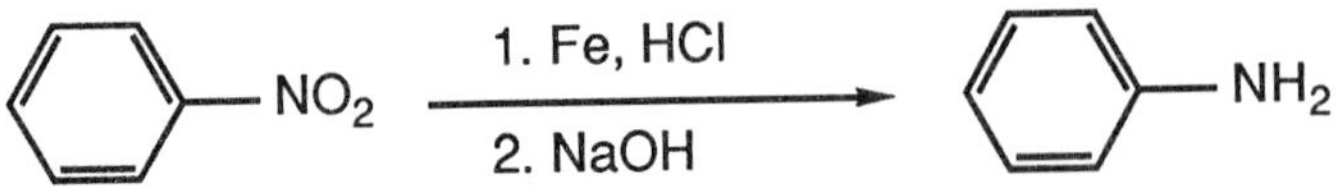

or

$C_6H_5{-}NO_2 \xrightarrow{H_2/Pt} C_6H_5{-}NH_2$

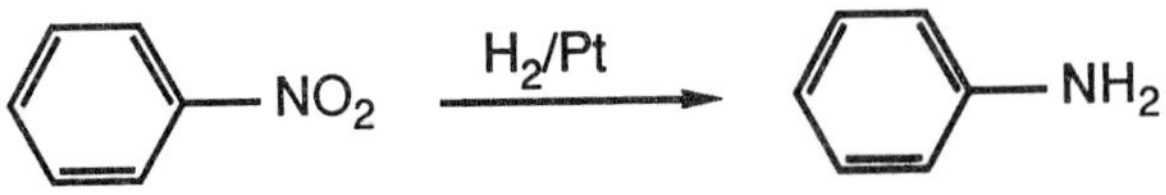

(k)

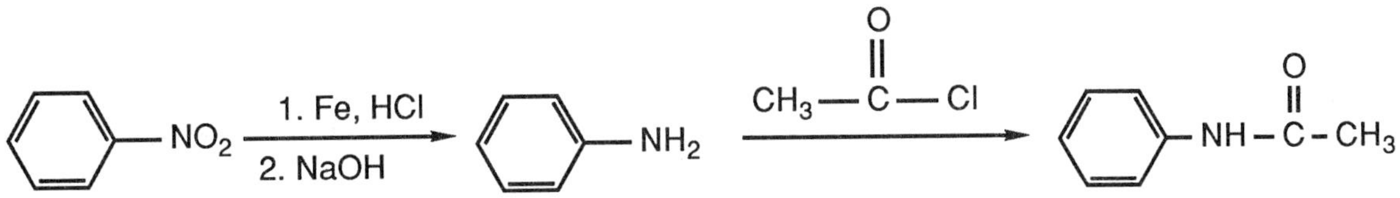

$C_6H_5NO_2 \xrightarrow[\text{2. NaOH}]{\text{1. Fe, HCl}} C_6H_5NH_2 \xrightarrow{CH_3-C(=O)-Cl} C_6H_5NH-C(=O)-CH_3$

(l)

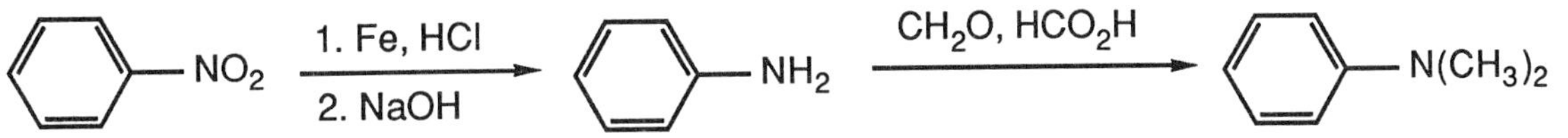

$C_6H_5NO_2 \xrightarrow[\text{2. NaOH}]{\text{1. Fe, HCl}} C_6H_5NH_2 \xrightarrow{CH_2O,\ HCO_2H} C_6H_5N(CH_3)_2$

or

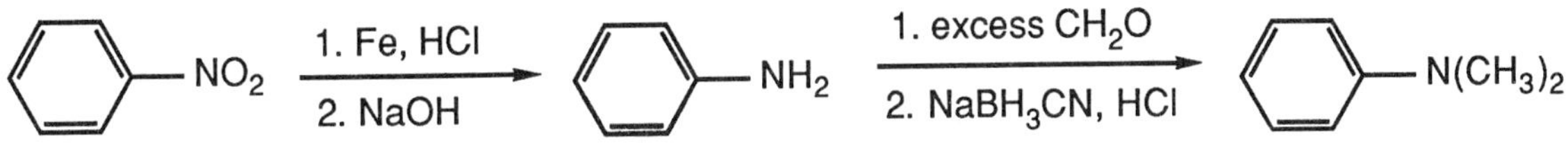

$C_6H_5NO_2 \xrightarrow[\text{2. NaOH}]{\text{1. Fe, HCl}} C_6H_5NH_2 \xrightarrow[\text{2. NaBH}_3\text{CN, HCl}]{\text{1. excess CH}_2\text{O}} C_6H_5N(CH_3)_2$

18.10

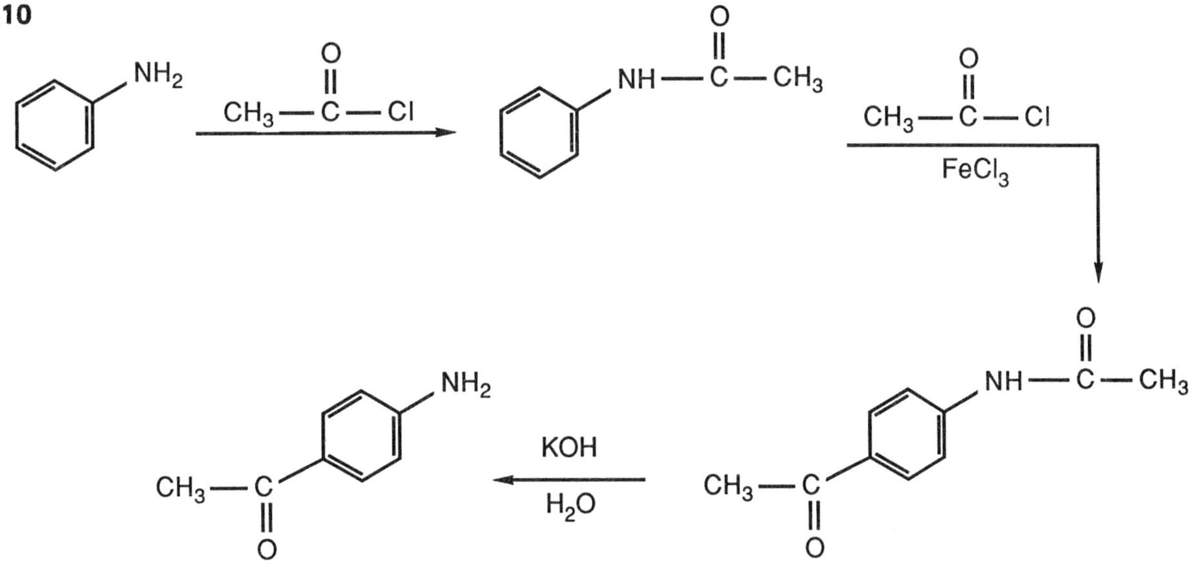

$C_6H_5NH_2 \xrightarrow{CH_3-C(=O)-Cl} C_6H_5NH-C(=O)-CH_3 \xrightarrow[FeCl_3]{CH_3-C(=O)-Cl} p\text{-}CH_3-C(=O)-C_6H_4-NH-C(=O)-CH_3 \xrightarrow[H_2O]{KOH} p\text{-}CH_3-C(=O)-C_6H_4-NH_2$

18.11

(a)

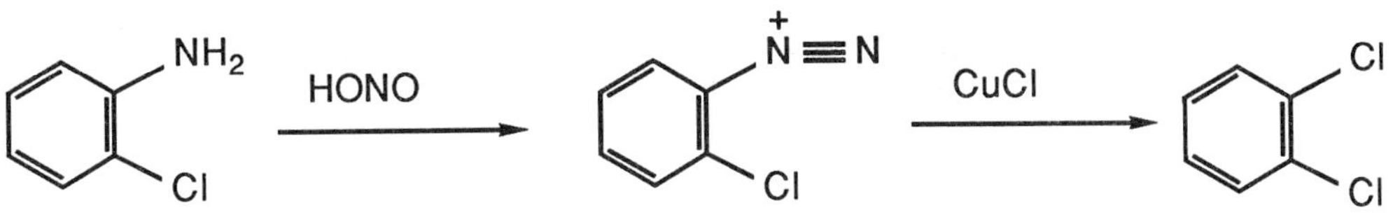

$o\text{-}ClC_6H_4NH_2 \xrightarrow{HONO} o\text{-}ClC_6H_4\overset{+}{N}\equiv N \xrightarrow{CuCl} o\text{-}ClC_6H_4Cl$

(b)

$o\text{-}ClC_6H_4NH_2 \xrightarrow{HONO} o\text{-}ClC_6H_4\overset{+}{N}\equiv N \xrightarrow{CuNO_2} o\text{-}ClC_6H_4NO_2$

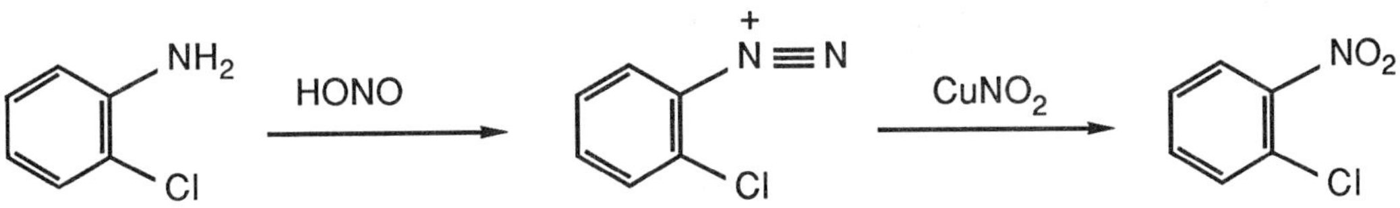

(c) 2-chloroaniline (NH$_2$, Cl) $\xrightarrow{\text{HONO}}$ 2-chlorobenzenediazonium ($\overset{+}{N}\equiv N$, Cl) $\xrightarrow[\text{or } NaBH_4]{H_3PO_2}$ chlorobenzene (Cl)

(d) 2-chloroaniline (NH$_2$, Cl) $\xrightarrow[\text{2. CuCN}]{\text{1. HONO}}$ 2-chlorobenzonitrile (CN, Cl) $\xrightarrow[H_2O]{H_2SO_4}$ 2-chlorobenzoic acid (CO_2H, Cl)

18.12

(a) CH_3–C$_6$H$_4$–NH$_2$ $\xrightarrow{\text{HONO}}$ CH_3–C$_6$H$_4$–$\overset{+}{N}\equiv N$ $\xrightarrow[H_2SO_4]{H_2O}$ CH_3–C$_6$H$_4$–OH

(b) CH_3–C$_6$H$_4$–NH$_2$ $\xrightarrow{\text{HONO}}$ CH_3–C$_6$H$_4$–$\overset{+}{N}\equiv N$ $\xrightarrow{\text{CuCN}}$ CH_3–C$_6$H$_4$–CN

(c) CH_3–C$_6$H$_4$–NH$_2$ $\xrightarrow{\text{HONO}}$ CH_3–C$_6$H$_4$–$\overset{+}{N}\equiv N$ $\xrightarrow{\text{CuBr}}$ CH_3–C$_6$H$_4$–Br

(d) CH_3–C$_6$H$_4$–NH$_2$ $\xrightarrow{\text{HONO}}$ CH_3–C$_6$H$_4$–$\overset{+}{N}\equiv N$ $\xrightarrow{HBF_4}$ CH_3–C$_6$H$_4$–F

(e) CH_3–C$_6$H$_4$–NH$_2$ $\xrightarrow{\text{HONO}}$ CH_3–C$_6$H$_4$–$\overset{+}{N}\equiv N$ $\xrightarrow[NaBH_4]{H_3PO_2 \text{ or}}$ CH_3–C$_6$H$_5$

18.13

Toluene (benzene–CH_3) $\xrightarrow[H_2SO_4]{HNO_3}$ o-nitrotoluene (CH_3, NO_2) $\xrightarrow[2.\ CH_3—\overset{O}{\overset{\|}{C}}—Cl]{1.\ Sn,\ HCl}$ (CH_3, $NH—\underset{\underset{O}{\|}}{C}—CH_3$)

$\downarrow$ HNO_3, H_2SO_4

(CH_3, $NH—\underset{\underset{O}{\|}}{C}—CH_3$, NO_2) $\xrightarrow[H_2O]{KOH}$ (CH_3, NH_2, NO_2) $\xrightarrow{HONO}$ (CH_3, $\overset{+}{N}\equiv N$, NO_2)

$\xrightarrow{NaBH_4}$ (CH_3, NO_2)

18.14

(a)

(NH_2) $\xrightarrow{Ac_2O}$ ($NH—\overset{O}{\overset{\|}{C}}—CH_3$) $\xrightarrow[H_2SO_4]{HNO_3}$ (O_2N, $NH—\overset{O}{\overset{\|}{C}}—CH_3$)

$\downarrow$ KOH, H_2O

(O_2N, NH_2) $\xrightarrow{HONO}$ (O_2N, $\overset{+}{N}\equiv N$) $\xrightarrow{CuCN}$ (O_2N, CN)

$\downarrow$ H_2O, H_2SO_4

(O_2N, CO_2H)

(b)

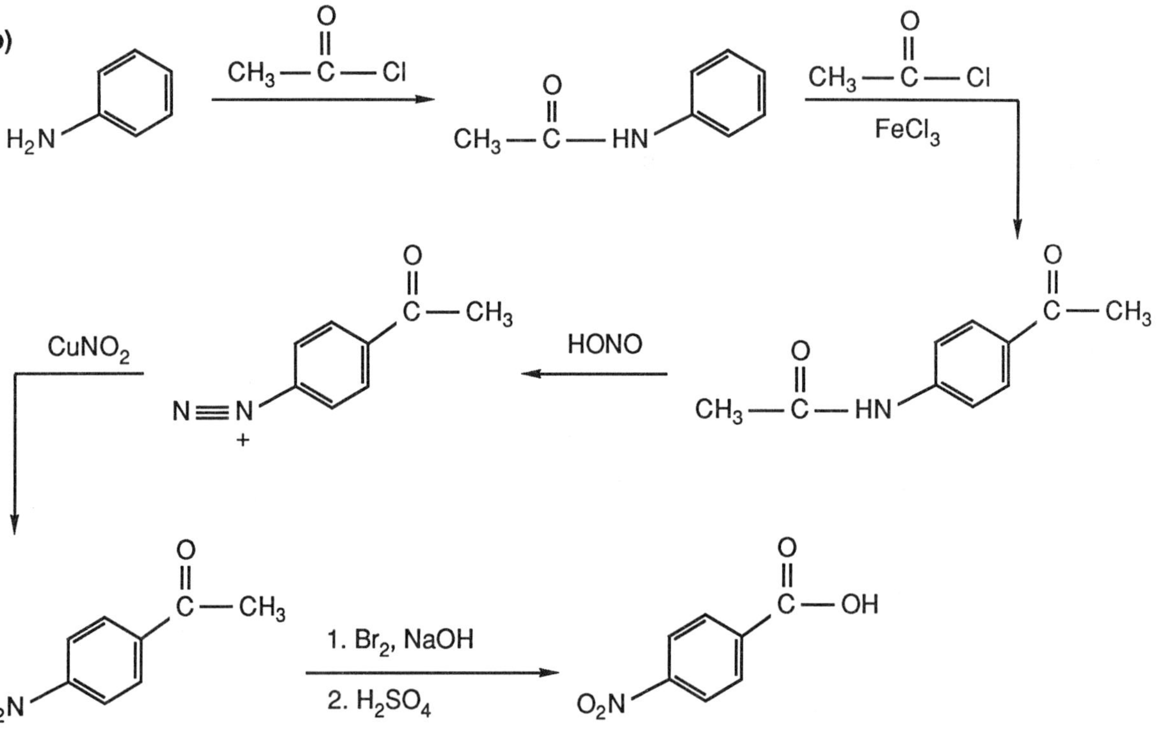

$H_2N-C_6H_5$ $\xrightarrow{CH_3-C(=O)-Cl}$ $CH_3-C(=O)-HN-C_6H_5$ $\xrightarrow[FeCl_3]{CH_3-C(=O)-Cl}$ $CH_3-C(=O)-HN-C_6H_4-C(=O)-CH_3$ $\xrightarrow{HONO}$ $N\equiv N^{+}-C_6H_4-C(=O)-CH_3$ $\xrightarrow{CuNO_2}$ $O_2N-C_6H_4-C(=O)-CH_3$ $\xrightarrow[2.\ H_2SO_4]{1.\ Br_2,\ NaOH}$ $O_2N-C_6H_4-C(=O)-OH$

18.15

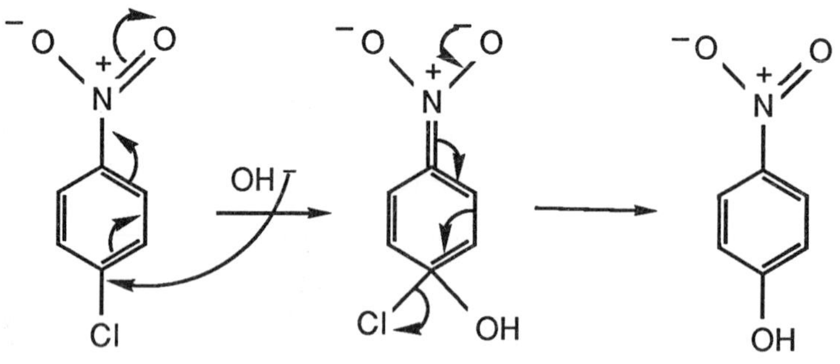

$^{-}O-N^{+}(=O)-C_6H_4-Cl$ $\xrightarrow{OH^-}$ $^{-}O-N^{+}(-O^{-})=C_6H_4(Cl)(OH)$ $\longrightarrow$ $^{-}O-N^{+}(=O)-C_6H_4-OH$

18.16

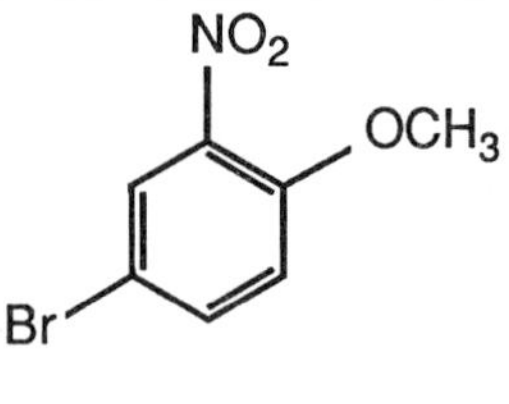

NO_2, OCH_3, Br

18.17

$C_6H_4(OCH_3)(Cl)$ $\xrightarrow{NaNH_2}$ OCH_3 (benzyne) $\xrightarrow{NH_3}$ $C_6H_3(OCH_3)(NH_2)(H)$ + $C_6H_3(OCH_3)(H)(NH_2)$

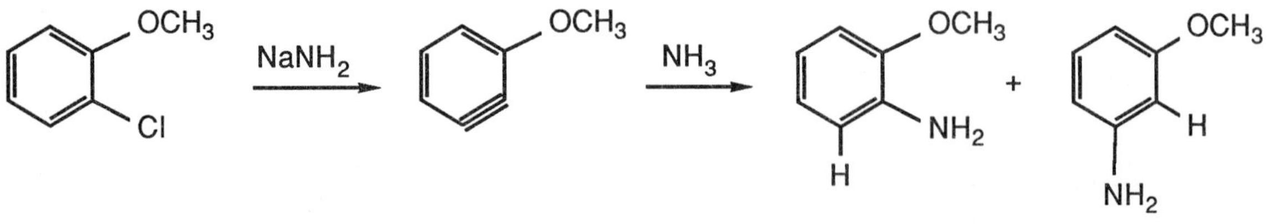

ANSWERS TO PROBLEMS

18.1 **(a)** $-CH_2CH_2CH_3$: more reactive than benzene; *o,p*-directing

(b) $-OC_6H_5$: more reactive than benzene; *o,p*-directing

(c) $-CF_3$: less reactive than benzene; *m*-directing

(d) $-COCH_3$: less reactive than benzene; *m*-directing

(e) $-NHCHO$: more reactive than benzene; *o,p*-directing

(f) $-O_2CCH_3$: more reactive than benzene; *o,p*-directing

(g) $-CO_2CH_3$: less reactive than benzene; *m*-directing

18.2 Using the substituent effects in Figure 18.1:

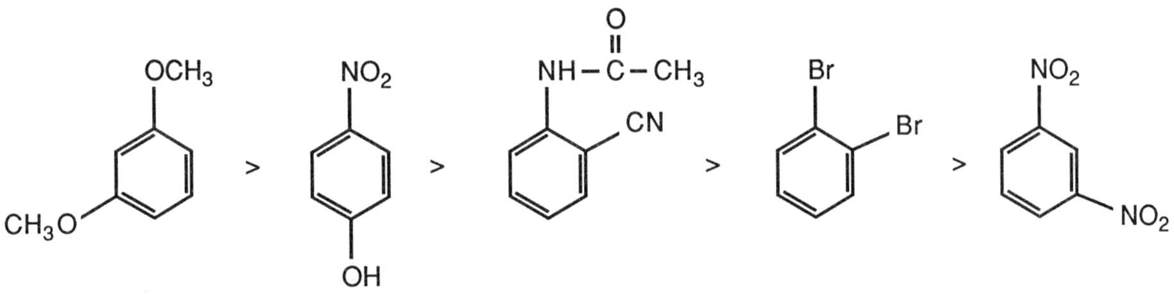

(c) 1,3-dimethoxybenzene (OCH_3, CH_3O) > (a) p-nitrophenol (NO_2, OH) > (e) o-($NH-C(=O)-CH_3$)benzonitrile (CN) > (b) o-dibromobenzene (Br, Br) > (d) m-dinitrobenzene (NO_2, NO_2)

18.3

(a) NO_2, NO_2, OH

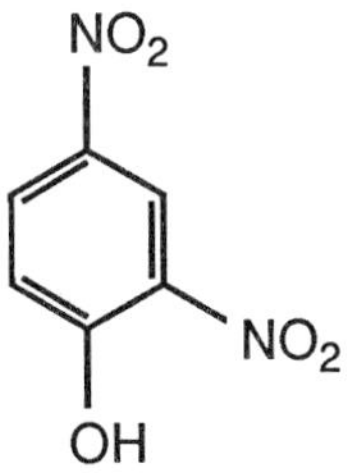

(b) Br, Br, NO_2

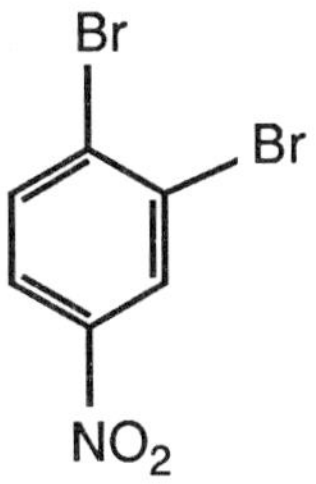

(c) OCH_3, CH_3O, NO_2

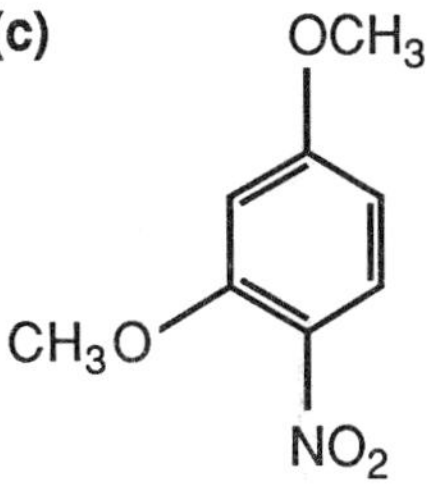

(d) NO_2, O_2N, NO_2

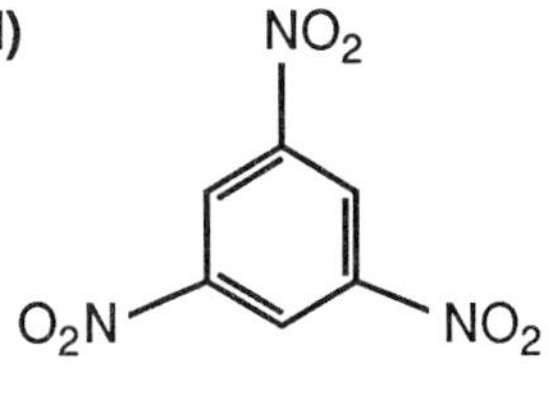

(e)

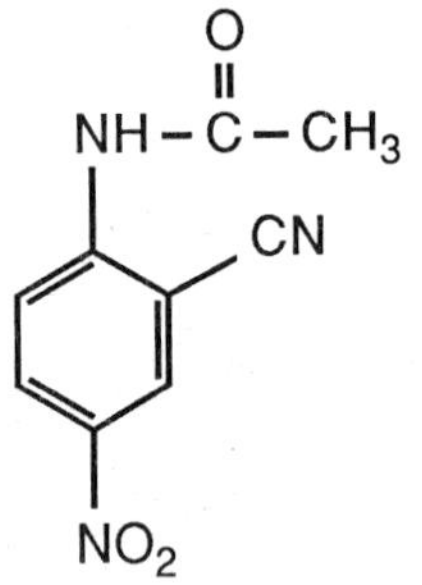

$$\text{NH}-\overset{\overset{\displaystyle O}{\|}}{C}-CH_3$$

CN

NO_2

18.4

NO_2^+

H NO_2 ⟷ H NO_2 ⟷ H NO_2

NO_2

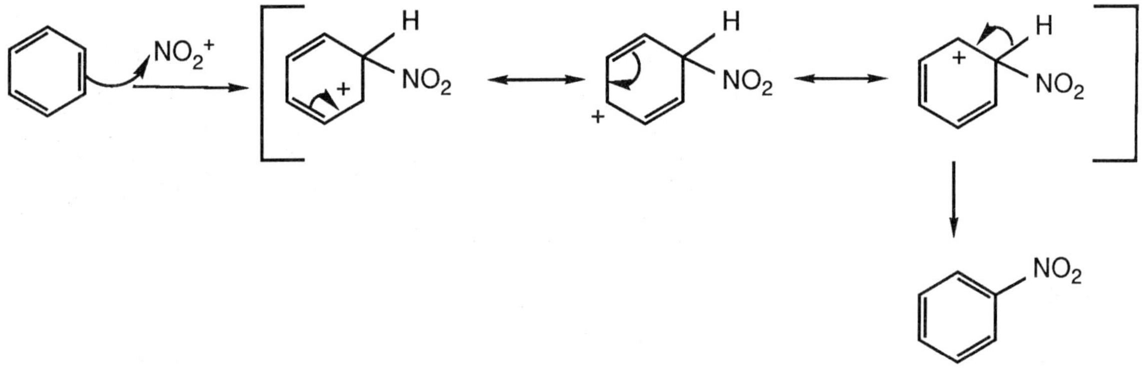

18.5

$$CH_3CH_2-\overset{\overset{\displaystyle O}{\|}}{C}-Cl \quad + \quad AlCl_3 \rightleftharpoons CH_3CH_2-\overset{\overset{\displaystyle O}{\|}}{C}-\overset{+}{Cl}\,\overset{-}{AlCl_3}$$

$$CH_3CH_2-\overset{\overset{\displaystyle O}{\|}}{C}-\overset{+}{Cl}\,\overset{-}{AlCl_3} \rightleftharpoons CH_3CH_2-\overset{+}{C}=\ddot{O}: \longleftrightarrow CH_3CH_2-C\equiv\overset{+}{O}: \quad + \quad AlCl_4$$

Ortho attack:

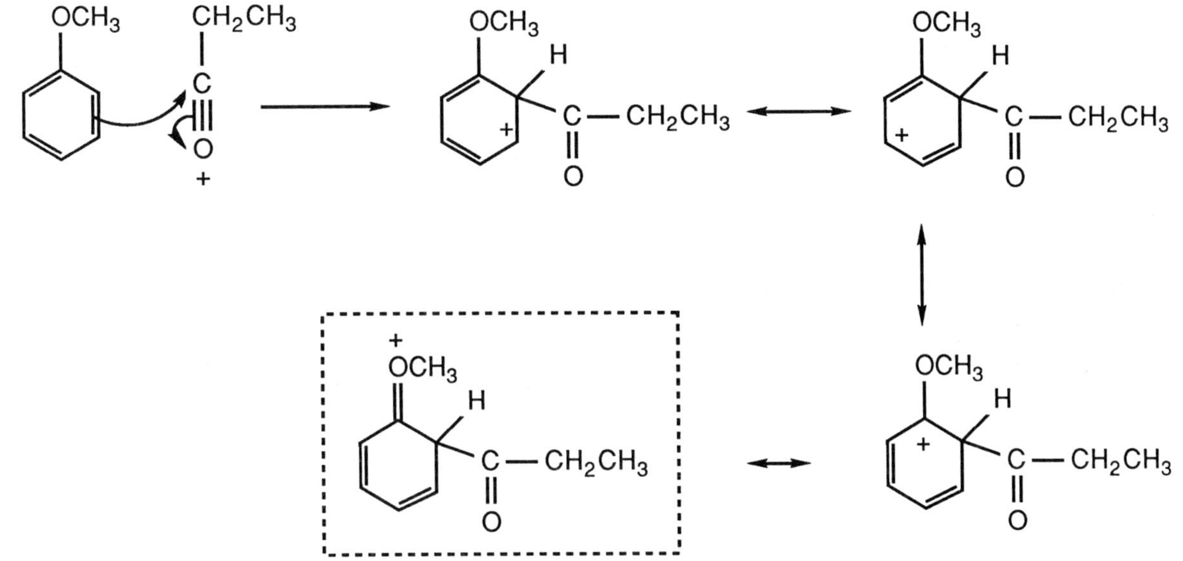

Meta attack:

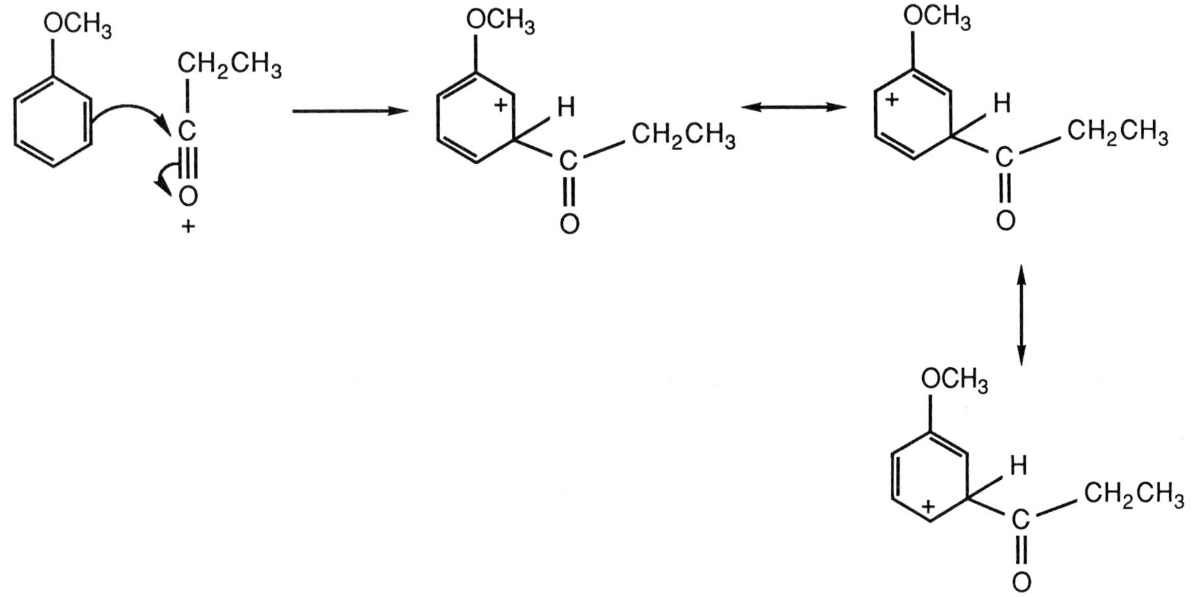

Para attack:

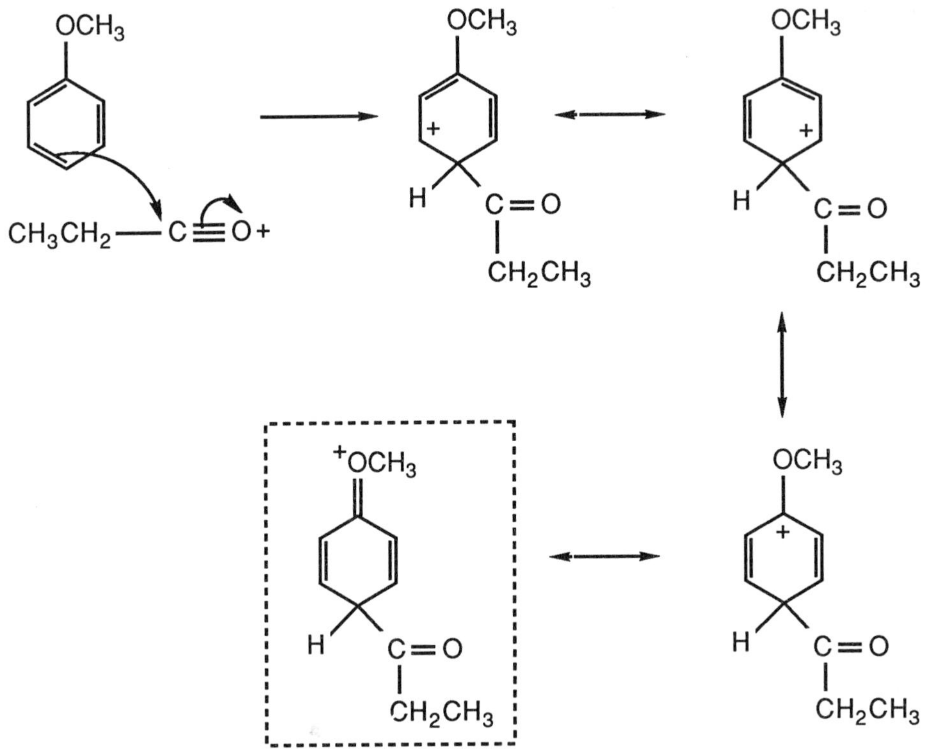

OCH$_3$ … $CH_3CH_2-C\equiv O^+$ → OCH$_3$, H, C=O, CH$_2$CH$_3$ ↔ OCH$_3$, H, C=O, CH$_2$CH$_3$ ↔ OCH$_3$, H, C=O, CH$_2$CH$_3$ ↔ $^+$OCH$_3$, H, C=O, CH$_2$CH$_3$

Product formation would result from *ortho* and *para* attack, which have favorable resonance contributors:

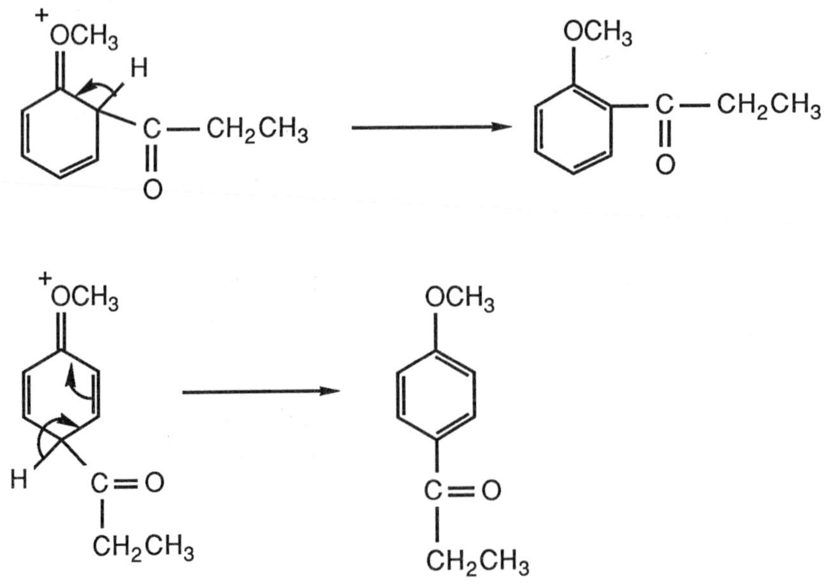

$^+$OCH$_3$, H, C—CH$_2$CH$_3$, O → OCH$_3$, C—CH$_2$CH$_3$, O

$^+$OCH$_3$, H, C=O, CH$_2$CH$_3$ → OCH$_3$, C=O, CH$_2$CH$_3$

18.6

$$Br-Br + FeBr_3 \rightleftharpoons Br-\overset{+}{Br}-\overset{-}{Fe}Br_3$$

Ortho attack:

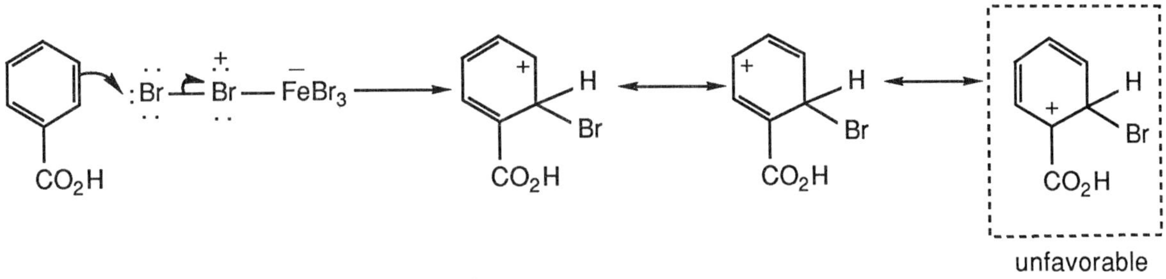

CO₂H

unfavorable

Meta attack:

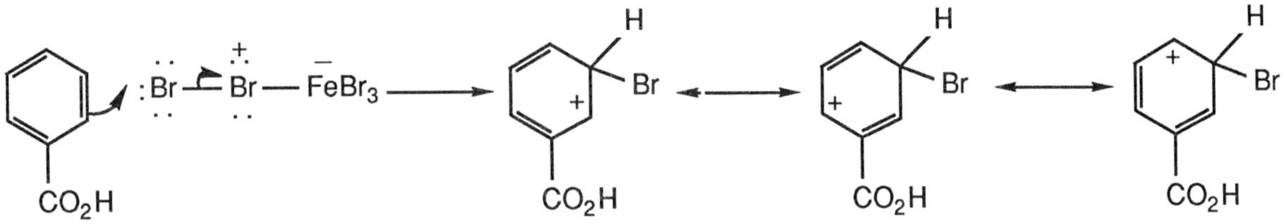

Para attack:

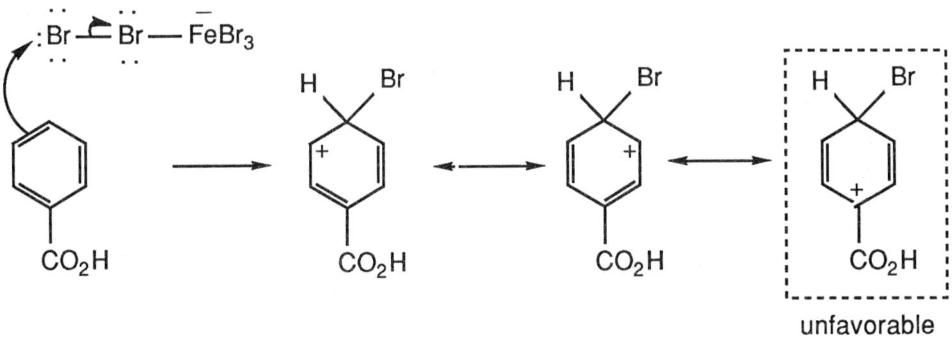

unfavorable

Product formation would result from *meta* attack, for which there is no unfavorable resonance contributor:

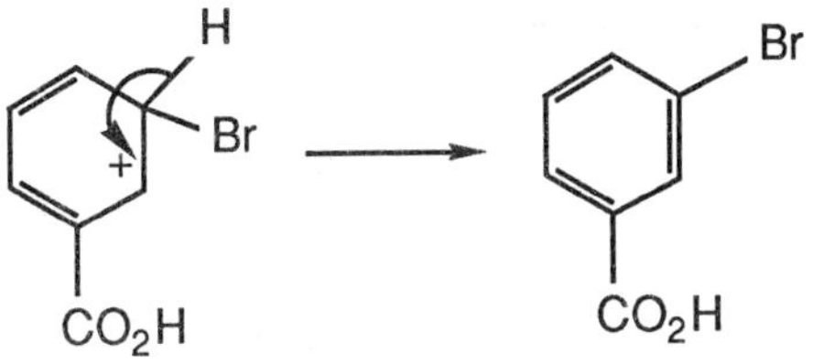

18.7 There are two possible sites of attack; an oxonium ion is possible only for attack *ortho* to the methoxy group:

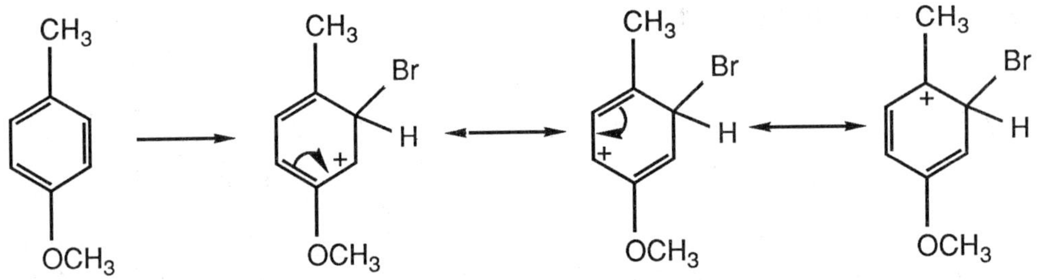

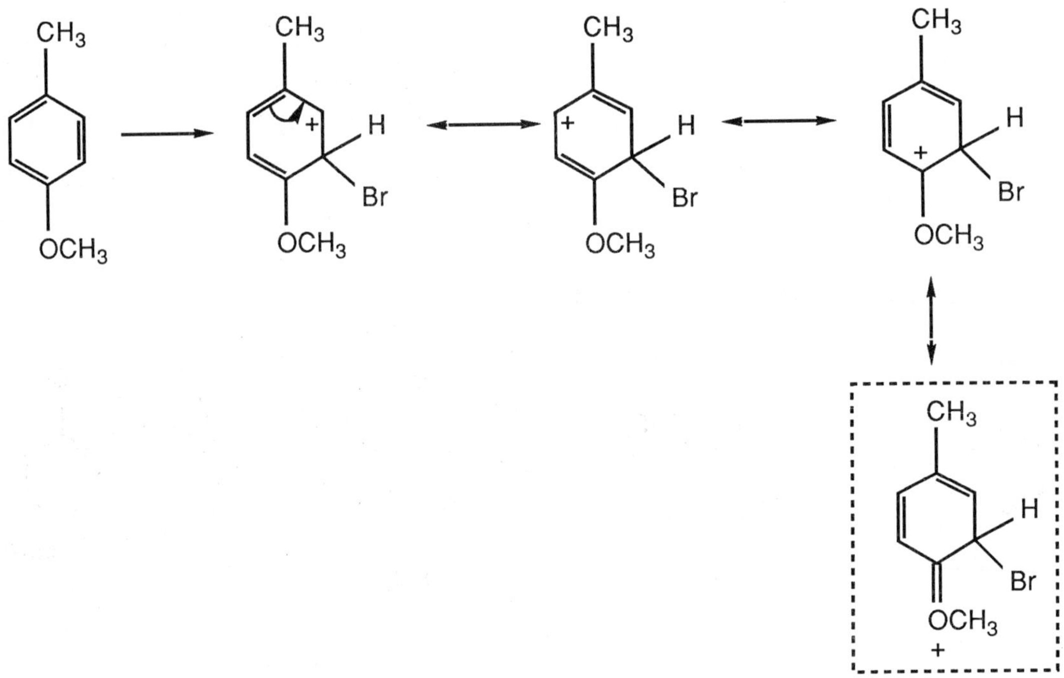

CH_3 OCH_3 → CH_3 Br H + OCH_3 ↔ CH_3 Br H + OCH_3 ↔ CH_3 Br + H OCH_3

CH_3 OCH_3 → CH_3 + H Br OCH_3 ↔ CH_3 + H Br OCH_3 ↔ CH_3 H + Br OCH_3

↕

CH_3 H Br OCH_3 +

oxonium ion

18.8 Fluorene is merely a CH_2-bridged biphenyl, and to a first approximation it should behave similarly to biphenyl. Attack *para* to the other phenyl group affords a phenyl-stabilized intermediate cation:

Br + H ↔ + Br H

However, the unfavorable steric interactions between *ortho* substituents cause biphenyl to adopt a twisted conformation, which prevents resonance stabilization of the cation by the other phenyl group:

H Br ⟷ H Br

18.9 Nitration at the position *ortho* to the isopropyl group would result in unfavorable steric interactions:

HNO_3 / H_2SO_4 → H NO_2

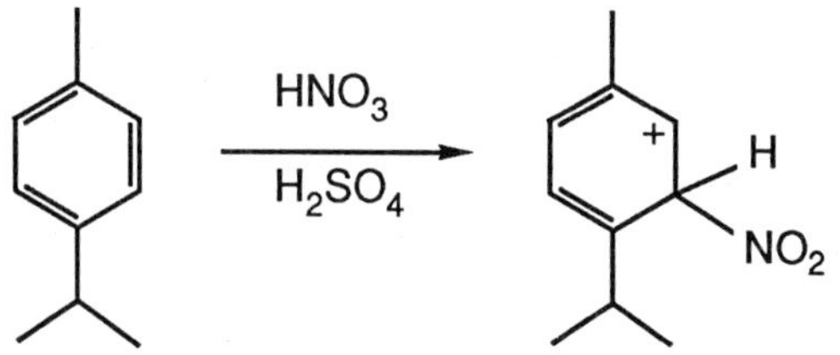

18.10 The same result would be expected:

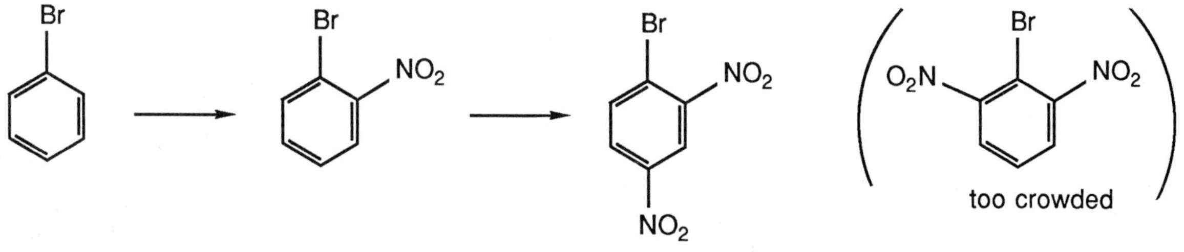

Br → Br NO_2 → Br NO_2 NO_2 (O_2N Br NO_2 too crowded)

18.11

NH_2 —HONO→ $\overset{+}{N}\equiv N$ —(OH)→ N=N— OH

18.12

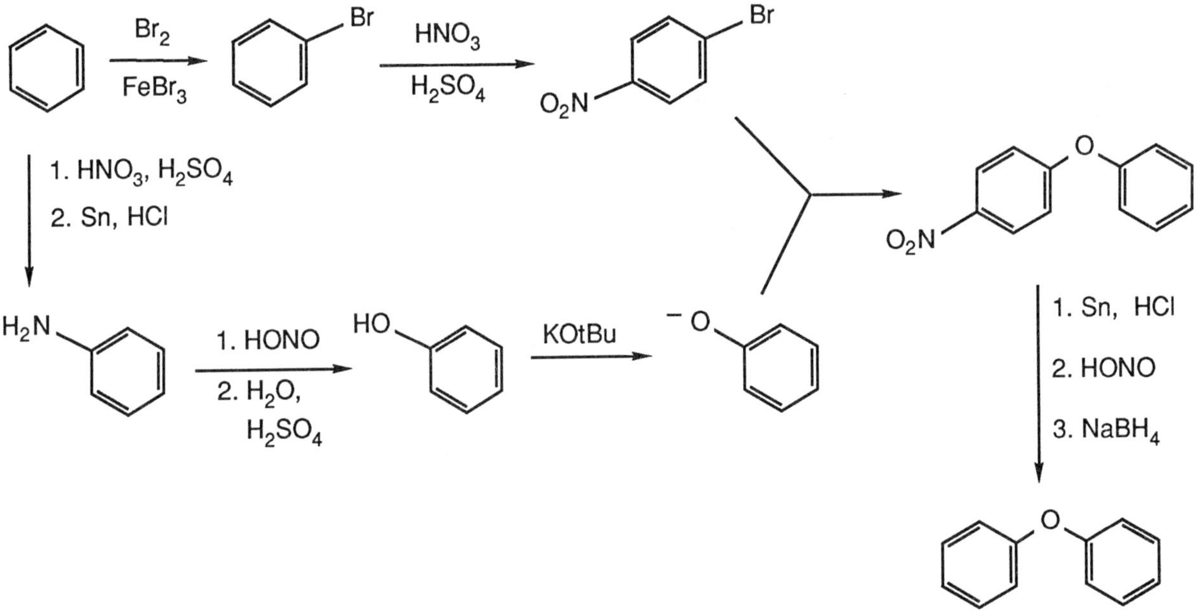

Br_2 / $FeBr_3$; HNO_3 / H_2SO_4; Br; O_2N

1. HNO_3, H_2SO_4
2. Sn, HCl

H_2N

1. HONO
2. H_2O, H_2SO_4

HO; KOtBu; ^-O

O; O_2N

1. Sn, HCl
2. HONO
3. $NaBH_4$

O

18.13

HNO_3 / H_2SO_4; NO_2; Fe / HCl; NH_2; $CH_3-C(=O)-Cl$

$NH-C(=O)-CH_3$; HNO_3 / H_2SO_4; $NH-C(=O)-CH_3$; O_2N; NO_2

H_2O / NaOH

NH_2; O_2N; NO_2; 1. HONO 2. HBF_4; F; O_2N; NO_2

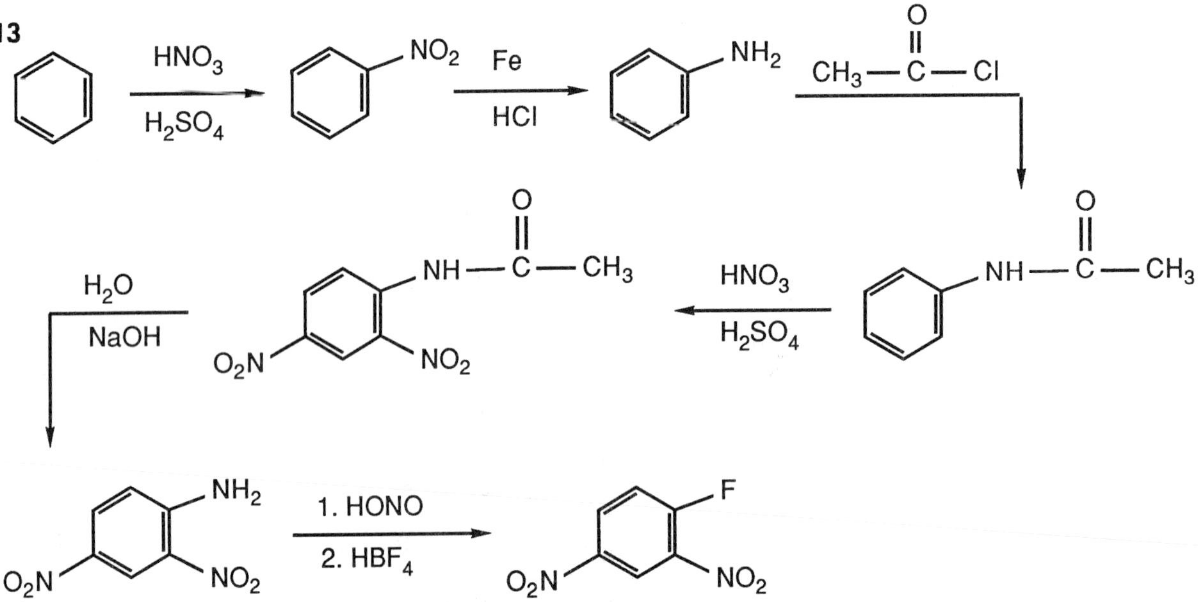

18.14 Friedel–Crafts acylation occurred in the presence of PCl_5 as a Lewis acid catalyst to give the cyclic product:

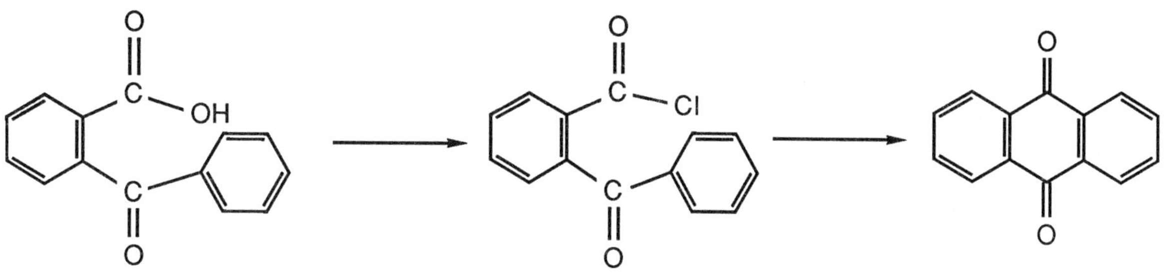

18.15 Acylation necessarily occurs in two stages. The first acylation deactivates the ring having the acetyl group, so reaction occurs at the *para* position of the other ring:

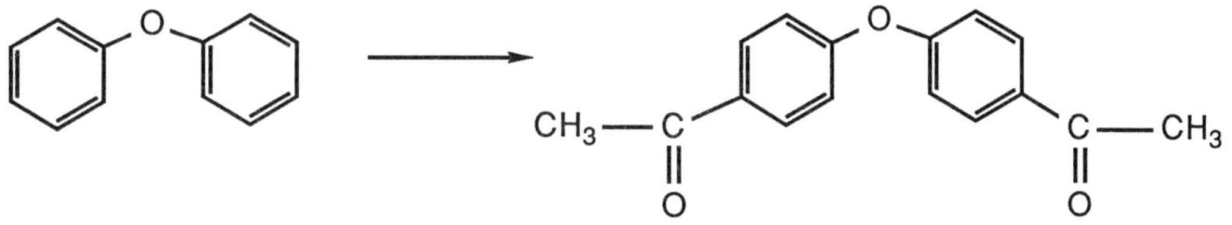

18.16 The ring positions are similar with regard to electronic effects, but reaction will occur at the less hindered site:

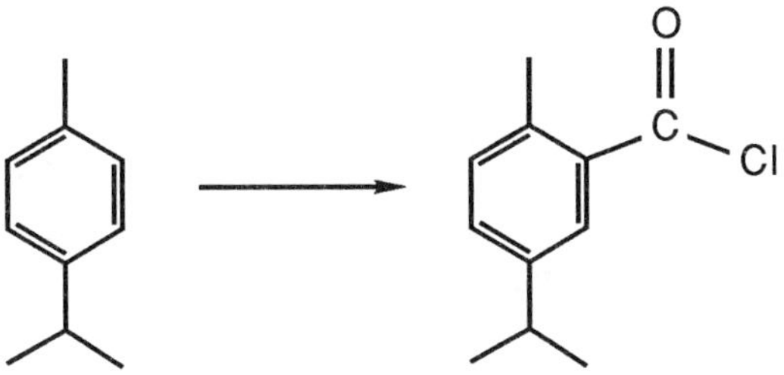

18.17 A methoxy group is strongly activating, but when it is *ortho* to the other phenyl group electronic interaction between the two aromatic rings results in increased steric interaction between the groups. As a result reaction occurs on the ring bearing the methoxy:

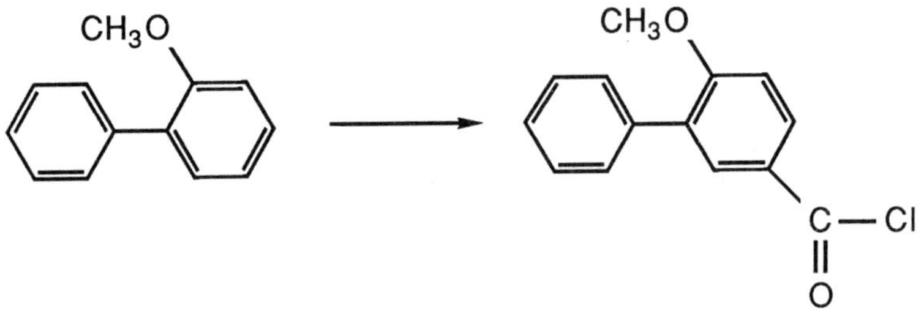

CH_3O CH_3O C—Cl ‖ O

via

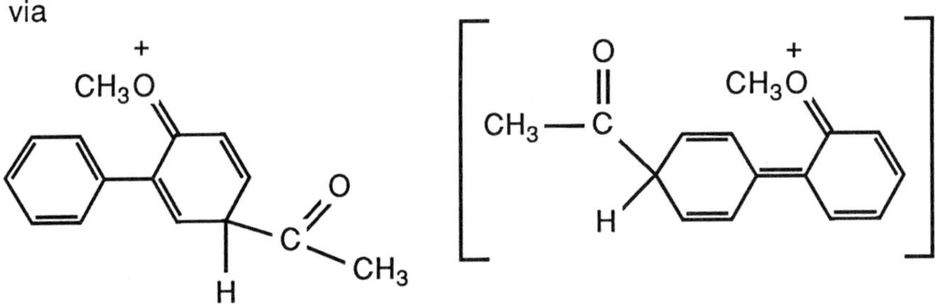

\+ CH_3O O C CH_3 H

O ‖ CH_3—C H \+ CH_3O

reaction of the p-position of the other ring would generate unfavorable steric interactions

When the methoxy and phenyl groups have a 1,4-relationship, electronic interaction between the two rings does not cause unfavorable steric interactions with the methoxy. In this case reaction takes place at the *p*-position of the unsubstituted ring:

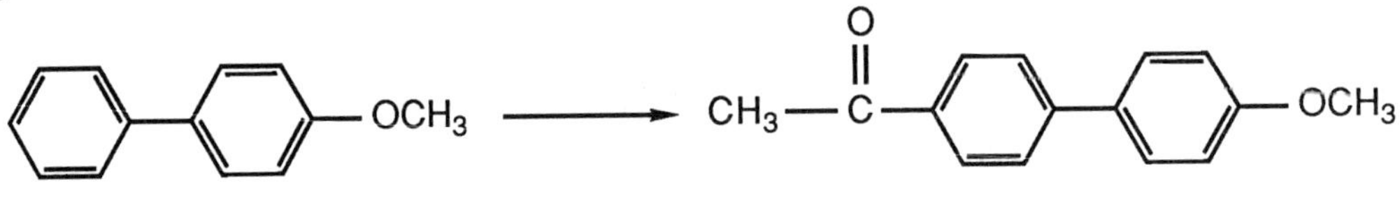

OCH_3 ⟶ CH_3—C(=O)— OCH_3

via

CH_3—C(=O) H \+ OCH_3

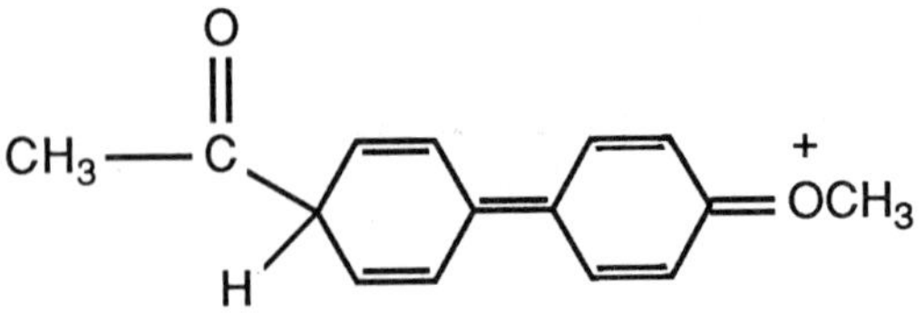

18.18 **(a)** 2,4-Dinitrobenzonitrile is strongly deactivated toward electrophilic substitution by a species such as NO_2^+:

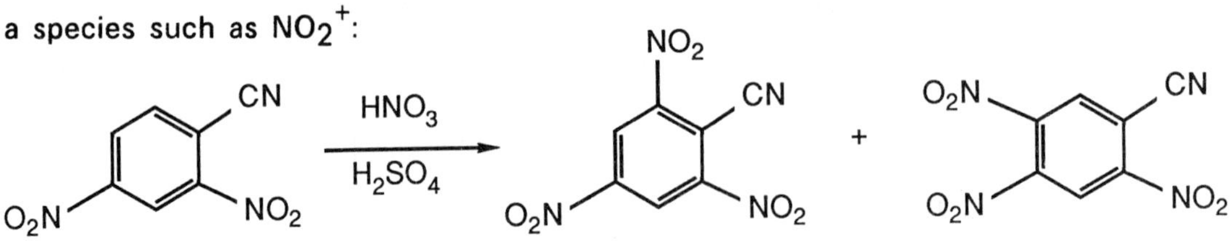

$$\xrightarrow[H_2SO_4]{HNO_3}$$

CN O_2N NO_2 ⟶ NO_2 CN O_2N NO_2 + O_2N CN O_2N NO_2

The reaction with methanolic KOH, however, is *nucleophilic* aromatic substitution.

For this reaction the nitro groups are activating:

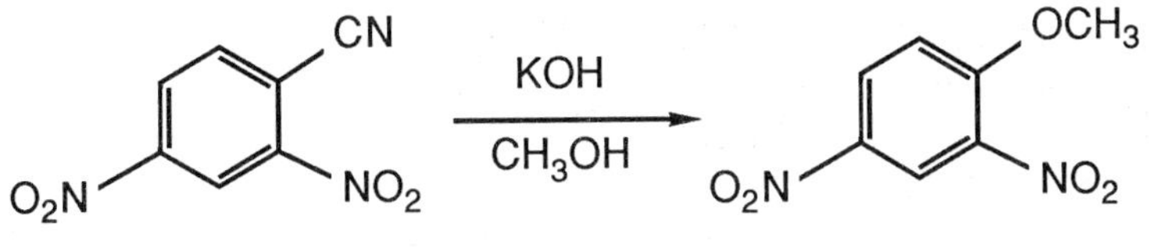

(b) The cyano group is highly deactivating (whereas chloro is only slightly deactivating) in electrophilic substitution:

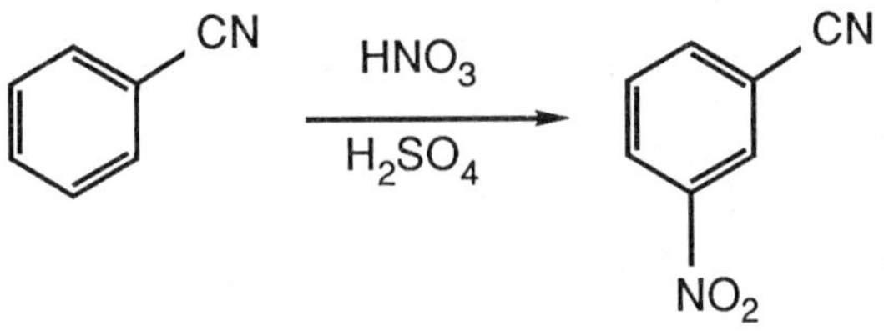

For the reactions with methanolic KOH, we are considering two different reactions. Chlorobenzene would react extremely slowly in a nucleophilic aromatic substitution reaction. Benzonitrile, however, would react smoothly in a *hydrolysis* reaction:

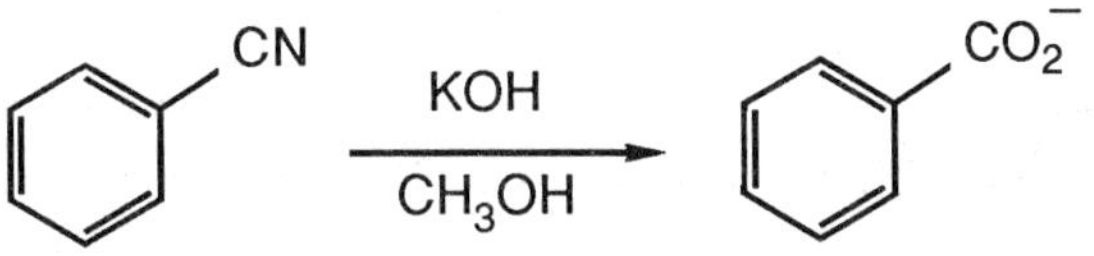

18.19 The desired pathway is nitrile hydrolysis, but the *para* relationship of the cyano and bromo groups activates the compound toward an aromatic nucleophilic substitution reaction:

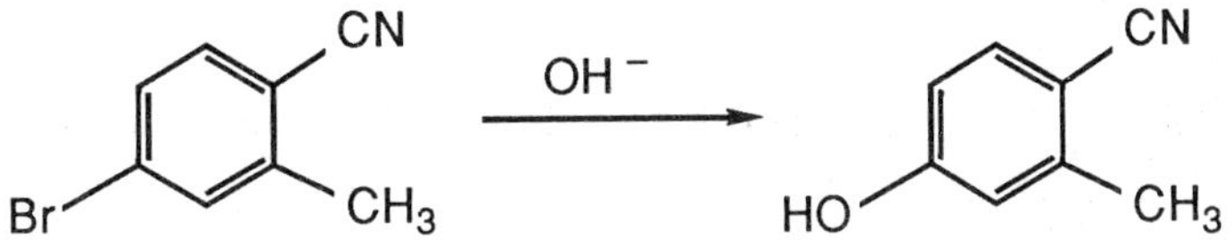

18.20

(a)

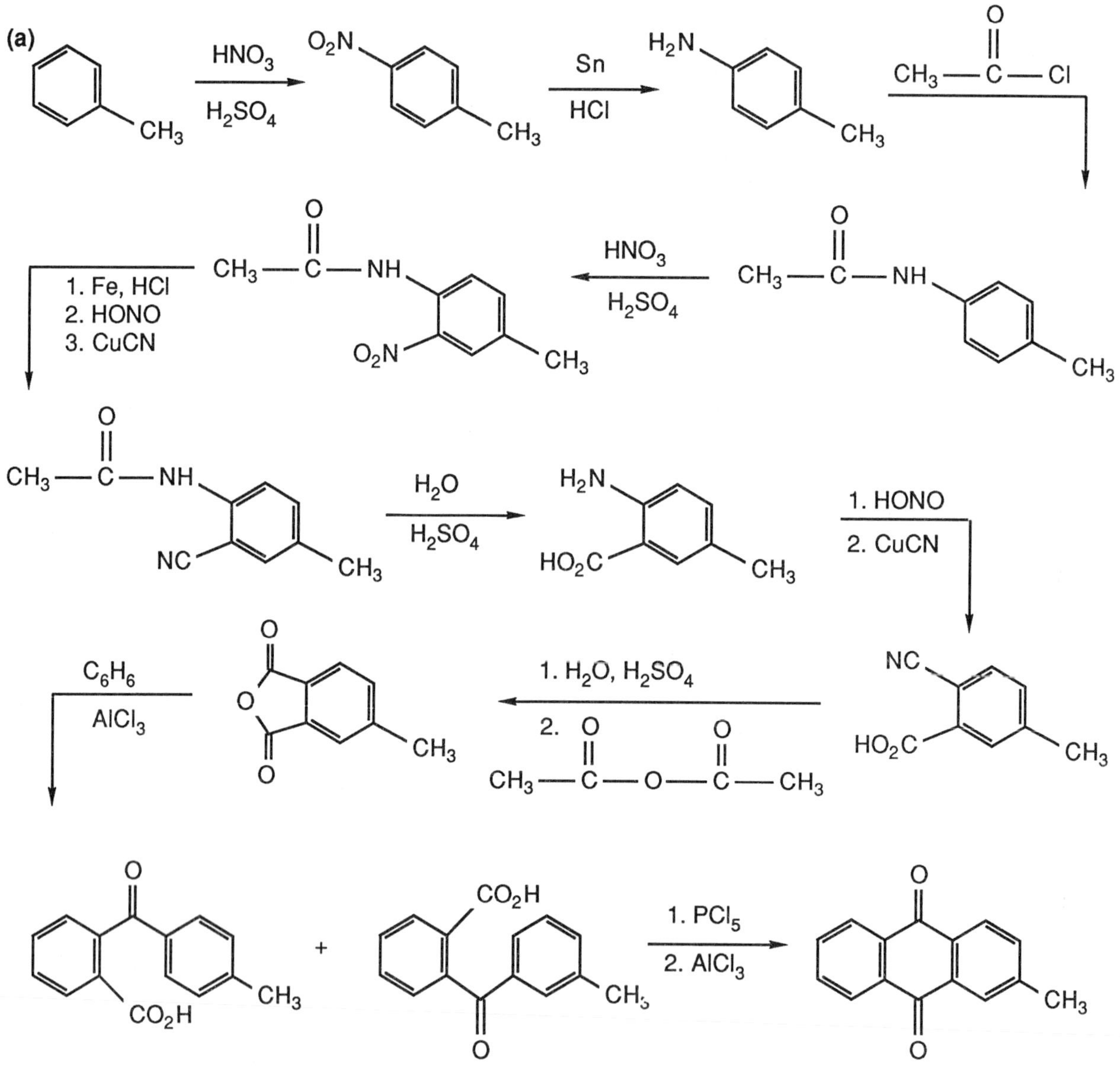

CH_3 (toluene) $\xrightarrow[H_2SO_4]{HNO_3}$ O_2N, CH_3 $\xrightarrow[HCl]{Sn}$ H_2N, CH_3 $\xrightarrow{CH_3-C(=O)-Cl}$ $CH_3-C(=O)-NH$, CH_3 $\xrightarrow[H_2SO_4]{HNO_3}$ $CH_3-C(=O)-NH$, O_2N, CH_3

1. Fe, HCl
2. HONO
3. CuCN

$\rightarrow$ $CH_3-C(=O)-NH$, NC, CH_3 $\xrightarrow[H_2SO_4]{H_2O}$ H_2N, HO_2C, CH_3

1. HONO
2. CuCN

$\rightarrow$ NC, HO_2C, CH_3

1. H_2O, H_2SO_4
2. $CH_3-C(=O)-O-C(=O)-CH_3$

$\rightarrow$ O, O, O, CH_3 (anhydride) $\xrightarrow[AlCl_3]{C_6H_6}$

O, CO_2H, CH_3 + CO_2H, O, CH_3

1. PCl_5
2. $AlCl_3$

$\rightarrow$ O, O, CH_3

(b)

CH_3

HNO_3

H_2SO_4

CH_3

NO_2

Sn

HCl

CH_3

NH_2

O

$CH_3—C—Cl$

CH_3

$HN—C—CH_3$

O

Br_2

$FeBr_3$

CH_3

Br

$HN—C—CH_3$

O

KOH

H_2O

CH_3

Br

NH_2

1. HONO
2. $NaBH_4$

CH_3

Br

(c)

HNO_3

H_2SO_4

NO_2

Sn, HCl

NH_2

Br_2

Br

Br

Br

NH_2

1. HONO
2. $NaBH_4$

Br

Br

Br

(d)

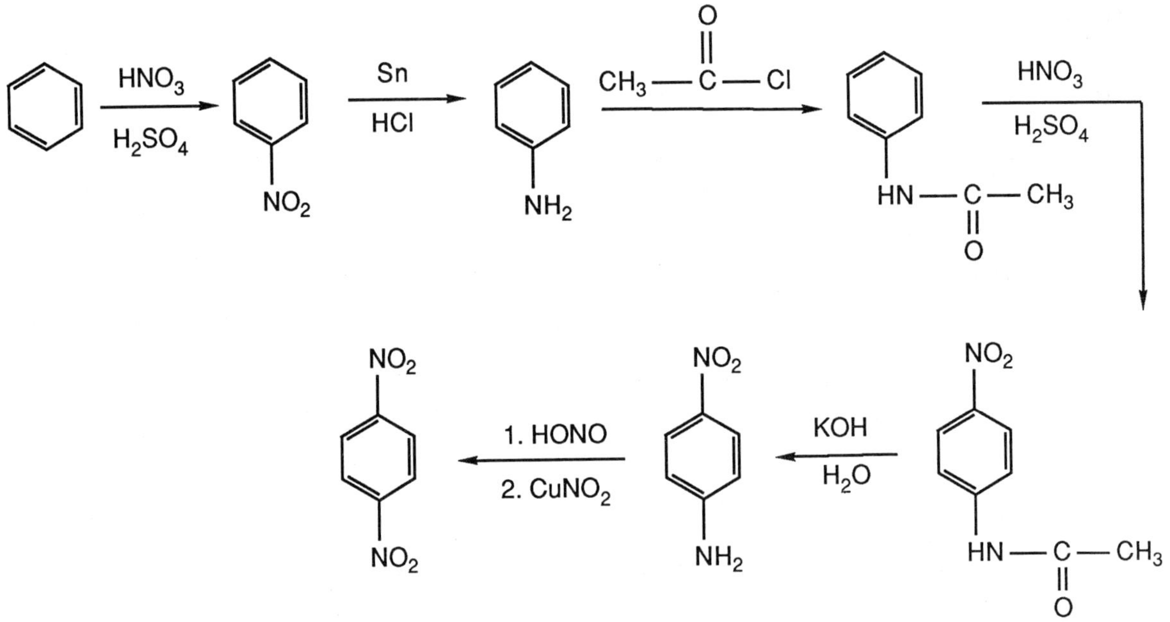

Benzene $\xrightarrow[H_2SO_4]{HNO_3}$ nitrobenzene (NO_2) $\xrightarrow[HCl]{Sn}$ aniline (NH_2) $\xrightarrow{CH_3-\overset{O}{\overset{\|}{C}}-Cl}$ acetanilide ($HN-\overset{}{C}(=O)-CH_3$) $\xrightarrow[H_2SO_4]{HNO_3}$

p-nitroacetanilide (NO_2; $HN-C(=O)-CH_3$) $\xrightarrow[H_2O]{KOH}$ p-nitroaniline (NO_2; NH_2) $\xrightarrow[2.\ CuNO_2]{1.\ HONO}$ p-dinitrobenzene (NO_2; NO_2)

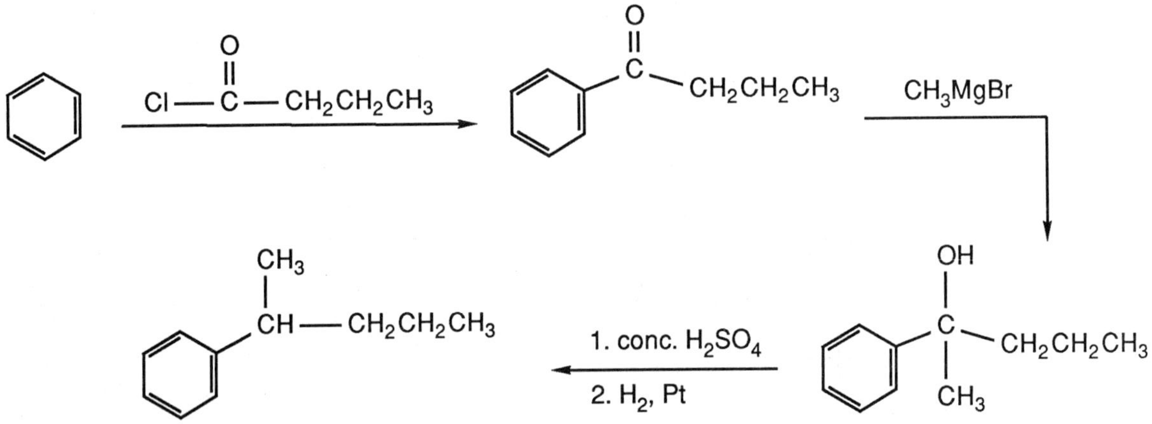

(e)

Benzene $\xrightarrow{Cl-\overset{O}{\overset{\|}{C}}-CH_2CH_2CH_3}$ $C_6H_5-\overset{O}{\overset{\|}{C}}-CH_2CH_2CH_3$ $\xrightarrow{CH_3MgBr}$ $C_6H_5-C(OH)(CH_3)-CH_2CH_2CH_3$ $\xrightarrow[2.\ H_2,\ Pt]{1.\ conc.\ H_2SO_4}$ $C_6H_5-CH(CH_3)-CH_2CH_2CH_3$

(f)

Toluene (CH_3) $\xrightarrow[H_2SO_4]{HNO_3}$ p-nitrotoluene (CH_3; NO_2) $\xrightarrow[HCl]{Sn}$ p-toluidine (CH_3; NH_2) $\xrightarrow{CH_2O,\ HCO_2H}$ N,N-dimethyl-p-toluidine (CH_3; $N(CH_3)_2$)

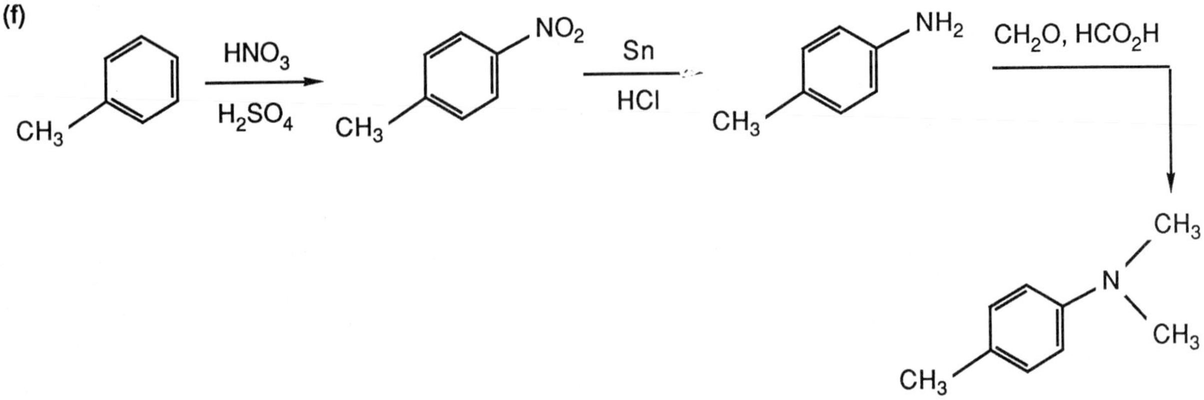

18.21

(a)

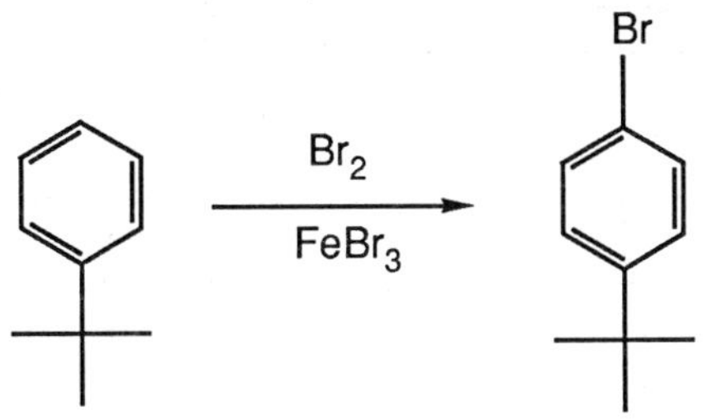

Br$_2$, FeBr$_3$

Br

(b)

HNO_3, H_2SO_4

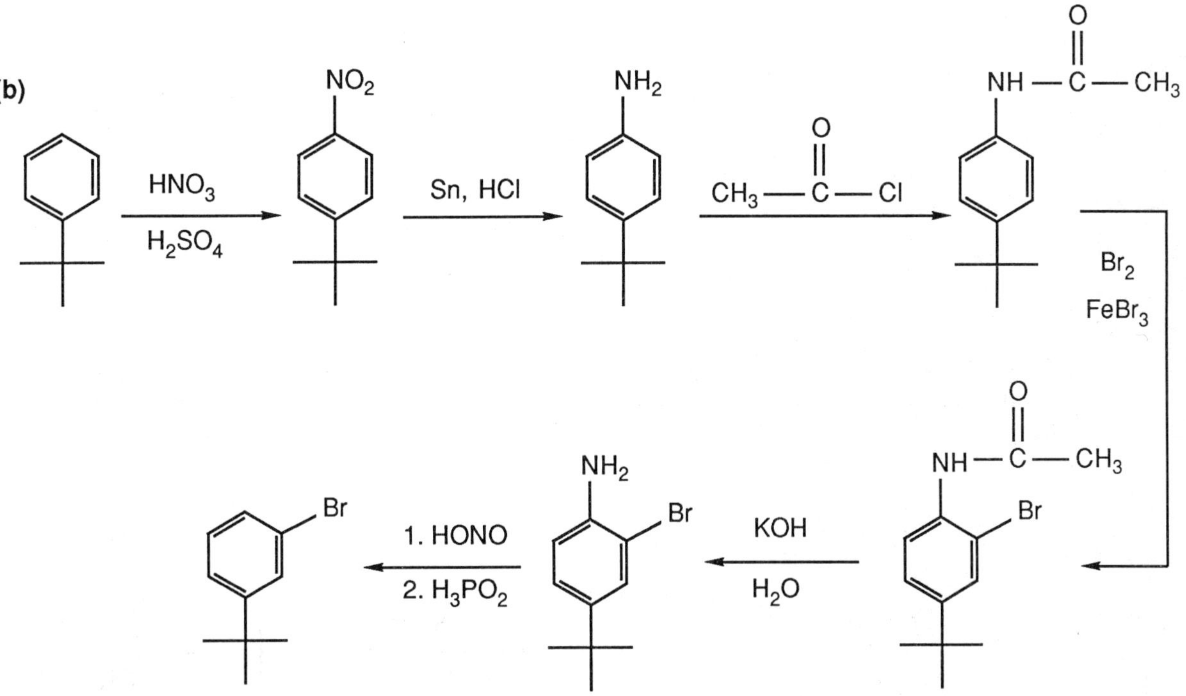

NO_2

Sn, HCl

NH_2

$CH_3-\overset{O}{\overset{\|}{C}}-Cl$

$NH-\overset{O}{\overset{\|}{C}}-CH_3$

Br$_2$, FeBr$_3$

$NH-\overset{O}{\overset{\|}{C}}-CH_3$ Br

KOH, H_2O

NH_2 Br

1. HONO
2. H_3PO_2

Br

(c)

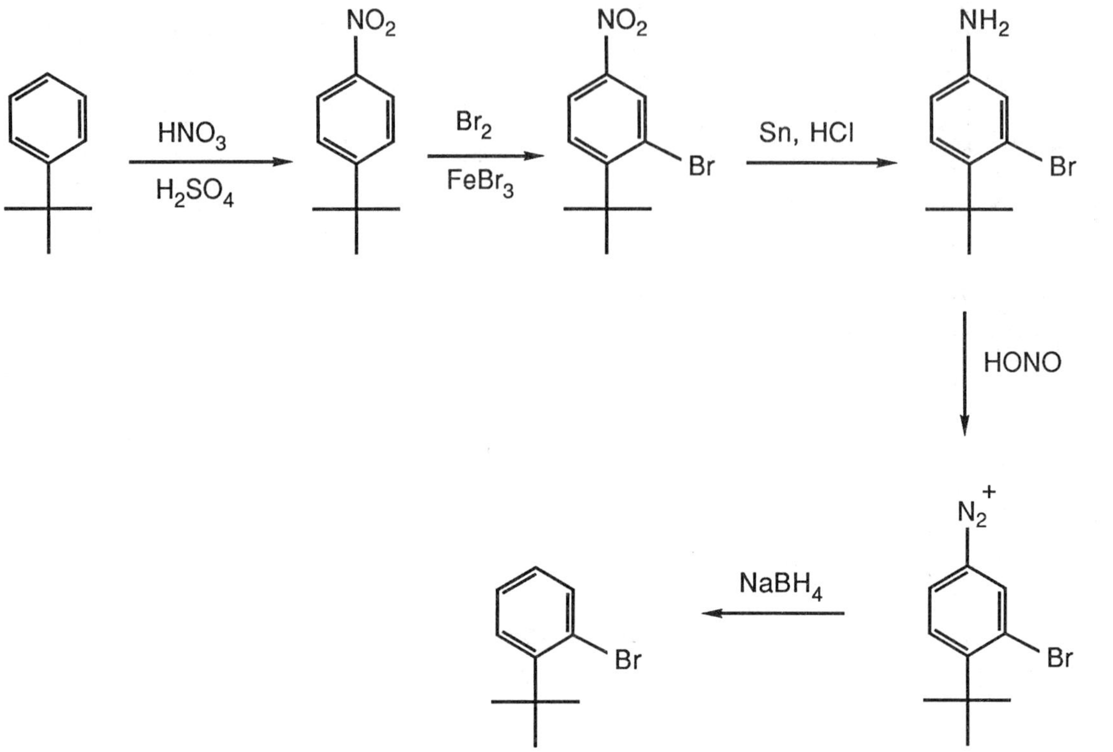

tert-butylbenzene → (HNO_3, H_2SO_4) → 4-nitro-tert-butylbenzene (NO_2) → (Br_2, $FeBr_3$) → (NO_2, Br) → (Sn, HCl) → (NH_2, Br) → (HONO) → (N_2^+, Br) → ($NaBH_4$) → (Br)

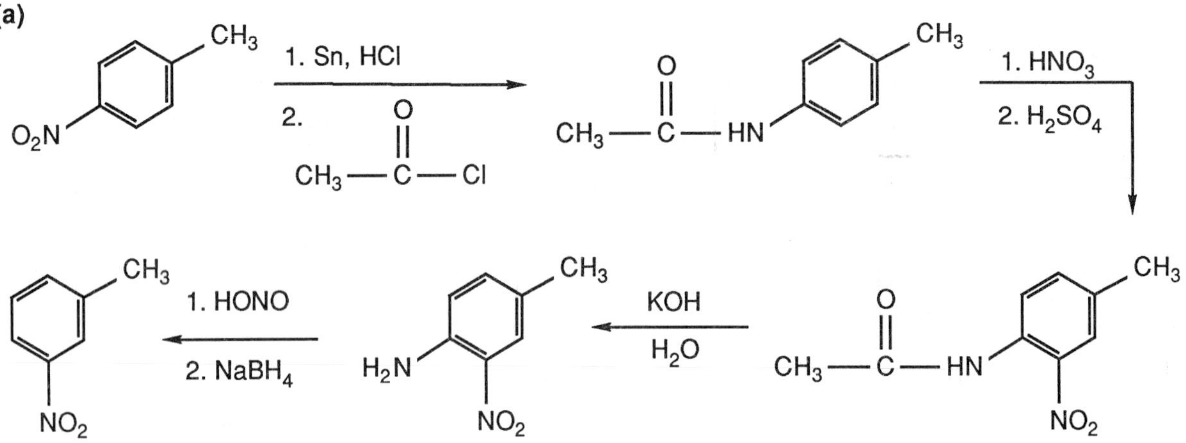

18.22

(a)

O_2N–C₆H₄–CH_3 → (1. Sn, HCl; 2. CH_3—C(=O)—Cl) → CH_3—C(=O)—HN–C₆H₄–CH_3 → (1. HNO_3; 2. H_2SO_4) → CH_3—C(=O)—HN–C₆H₃(NO_2)–CH_3 → (KOH, H_2O) → H_2N–C₆H₃(NO_2)–CH_3 → (1. HONO; 2. $NaBH_4$) → CH_3–C₆H₄–NO_2

(b)

O_2N–C₆H₄–CH_3 → ($KMnO_4$) → O_2N–C₆H₄–CO_2H

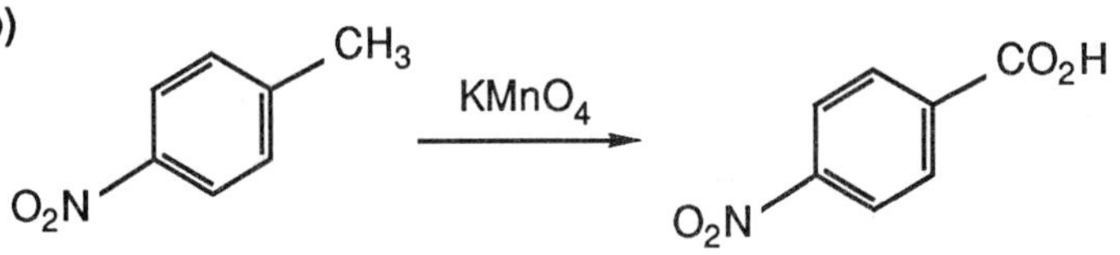

(c)

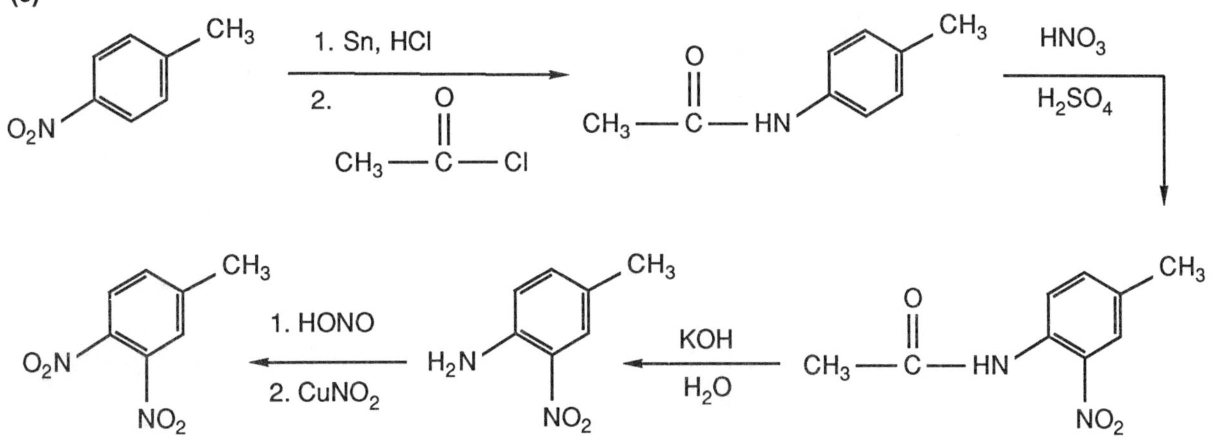

(d)

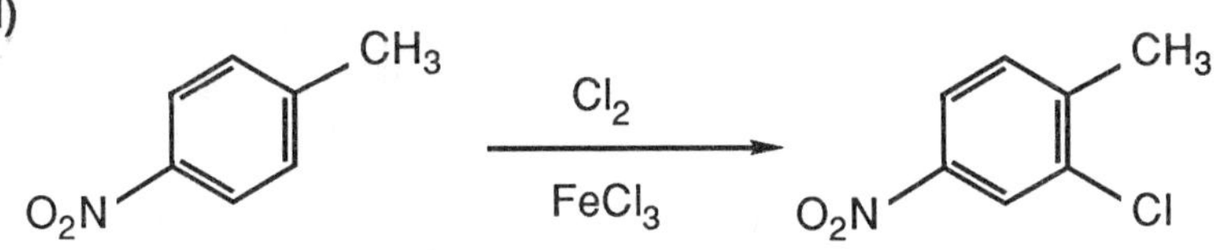

18.23

(a)

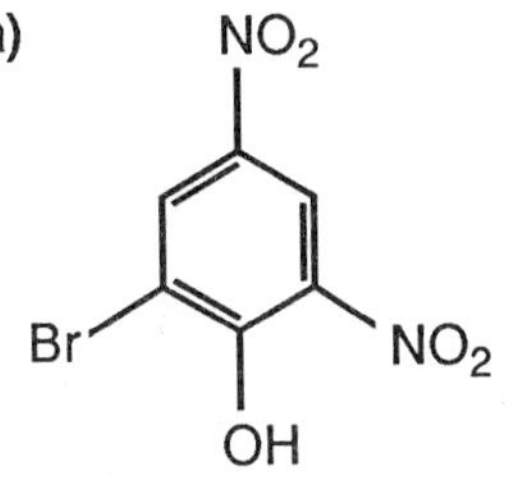

(b)

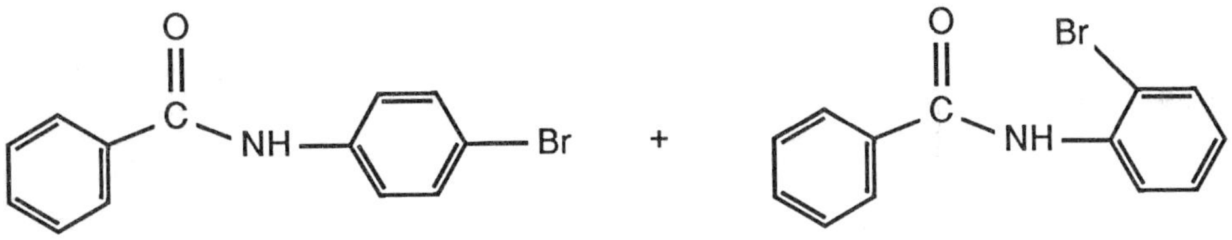

(c)

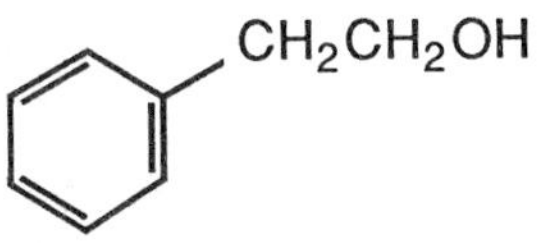

(d)

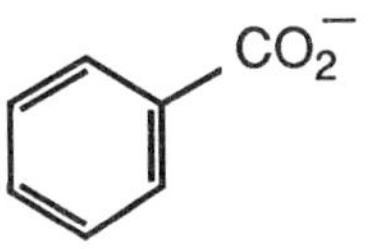

(e)

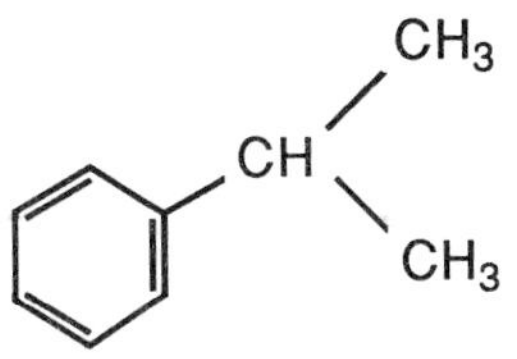

18.24

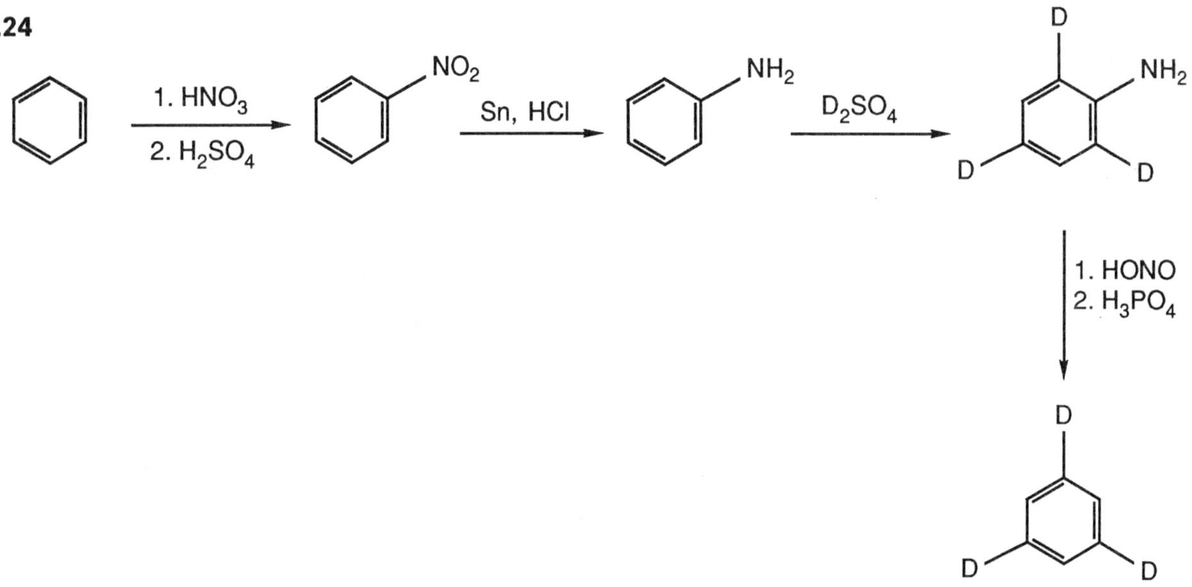

Benzene → (1. HNO_3; 2. H_2SO_4) → NO_2-benzene → (Sn, HCl) → NH_2-benzene → (D_2SO_4) → 2,4,6-trideuterioaniline (NH_2, D, D, D) → (1. HONO; 2. H_3PO_4) → 1,3,5-trideuteriobenzene (D, D, D)

Note that the anilinium ion (which would be *meta*-directing) is unreactive. In contrast, the very small equilibrium amount of free aniline reacts extremely rapidly to give the product of *ortho/para*-substitution.

18.25

(a) Benzene → (Br_2, $FeBr_3$) → Br-benzene → (1. Mg; 2. CO_2; 3. HCl) → CO_2H-benzene → (HNO_3, H_2SO_4) → *m*-NO_2, CO_2H-benzene → (Sn, HCl) → *m*-NH_2, CO_2H-benzene

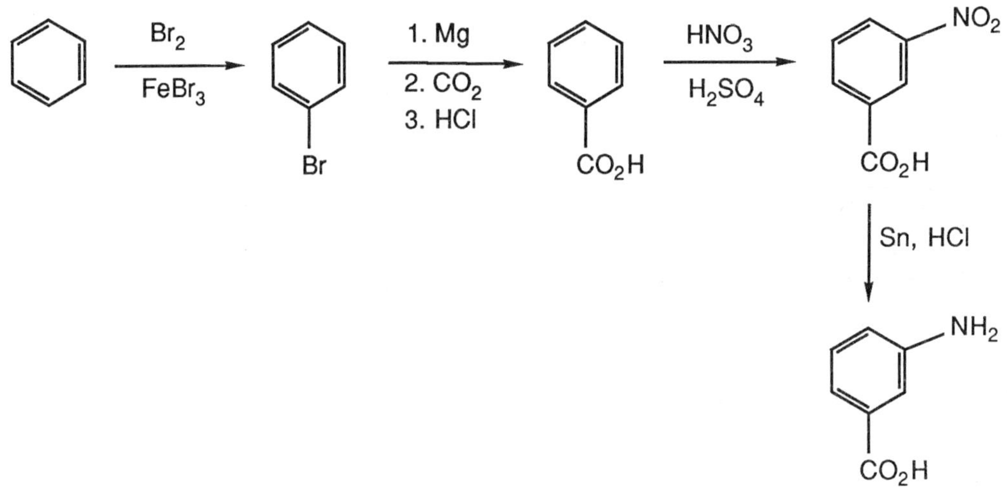

(b)

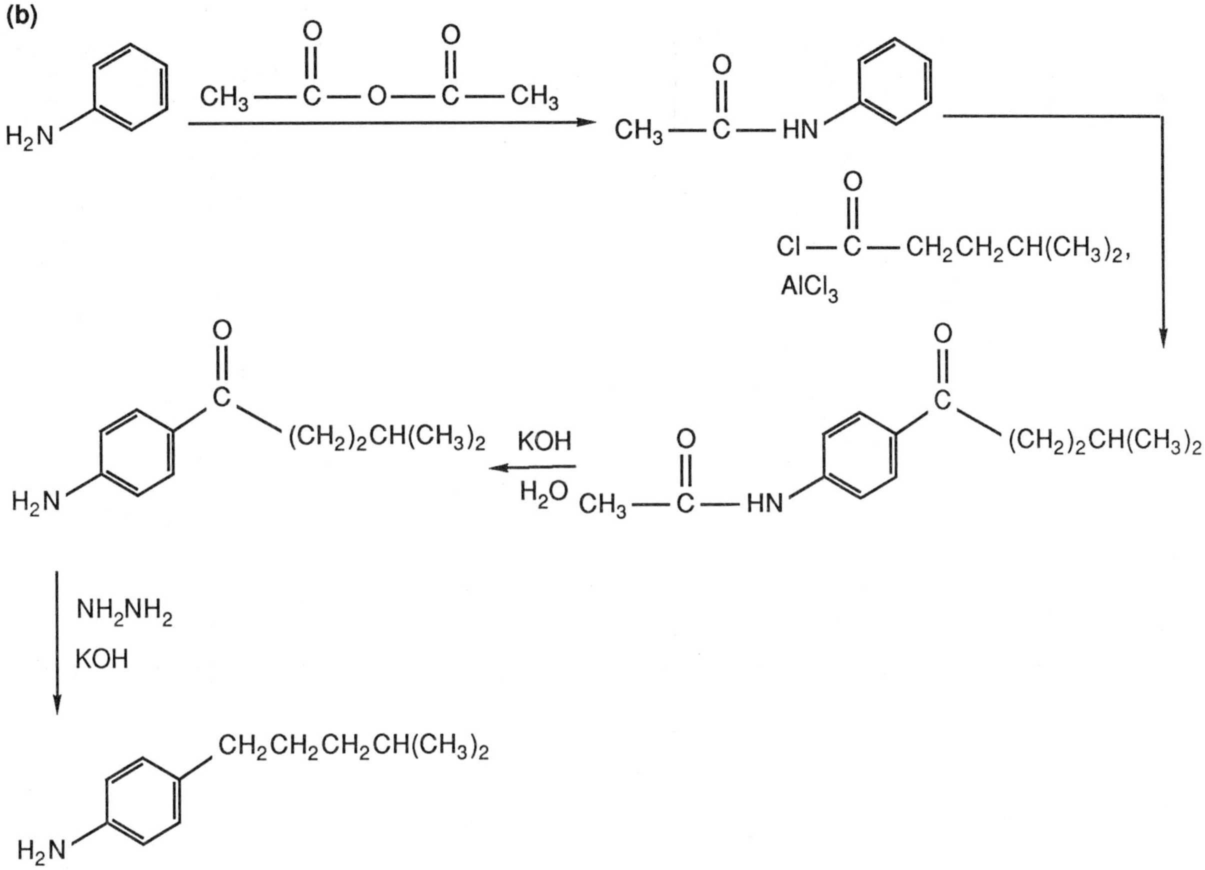

$$H_2N\text{-}C_6H_5 \xrightarrow{CH_3-\overset{O}{\overset{\|}{C}}-O-\overset{O}{\overset{\|}{C}}-CH_3} CH_3-\overset{O}{\overset{\|}{C}}-HN\text{-}C_6H_5$$

$$\xrightarrow[AlCl_3]{Cl-\overset{O}{\overset{\|}{C}}-CH_2CH_2CH(CH_3)_2,} CH_3-\overset{O}{\overset{\|}{C}}-HN\text{-}C_6H_4\text{-}\overset{O}{\overset{\|}{C}}-(CH_2)_2CH(CH_3)_2$$

$$\xrightarrow[H_2O]{KOH} H_2N\text{-}C_6H_4\text{-}\overset{O}{\overset{\|}{C}}-(CH_2)_2CH(CH_3)_2$$

$$\xrightarrow[KOH]{NH_2NH_2} H_2N\text{-}C_6H_4\text{-}CH_2CH_2CH_2CH(CH_3)_2$$

(c)

$$\text{naphthalene} \xrightarrow[FeBr_3]{Br_2} \text{1-bromonaphthalene (Br)}$$

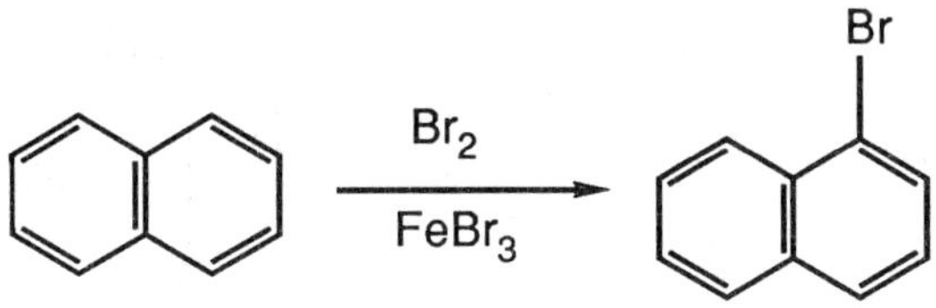

(d)

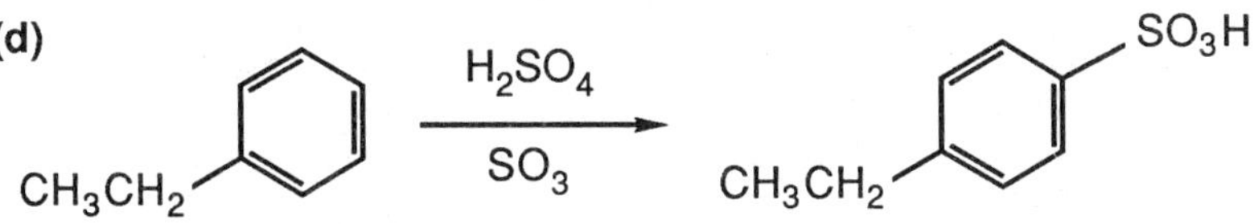

$$CH_3CH_2\text{-}C_6H_5 \xrightarrow[SO_3]{H_2SO_4} CH_3CH_2\text{-}C_6H_4\text{-}SO_3H$$

(e)

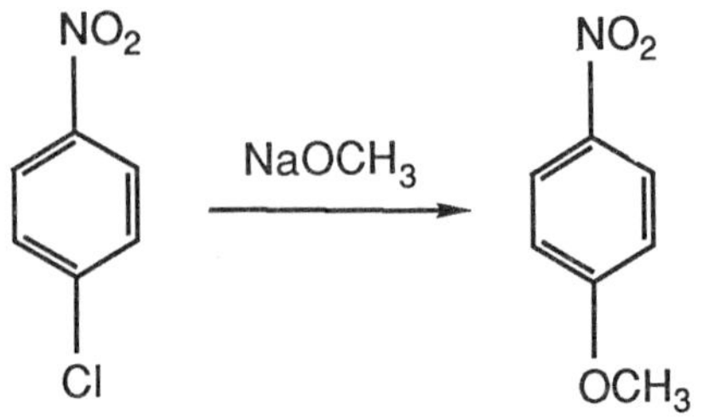

NO_2, Cl (benzene ring) $\xrightarrow{NaOCH_3}$ NO_2, OCH_3 (benzene ring)

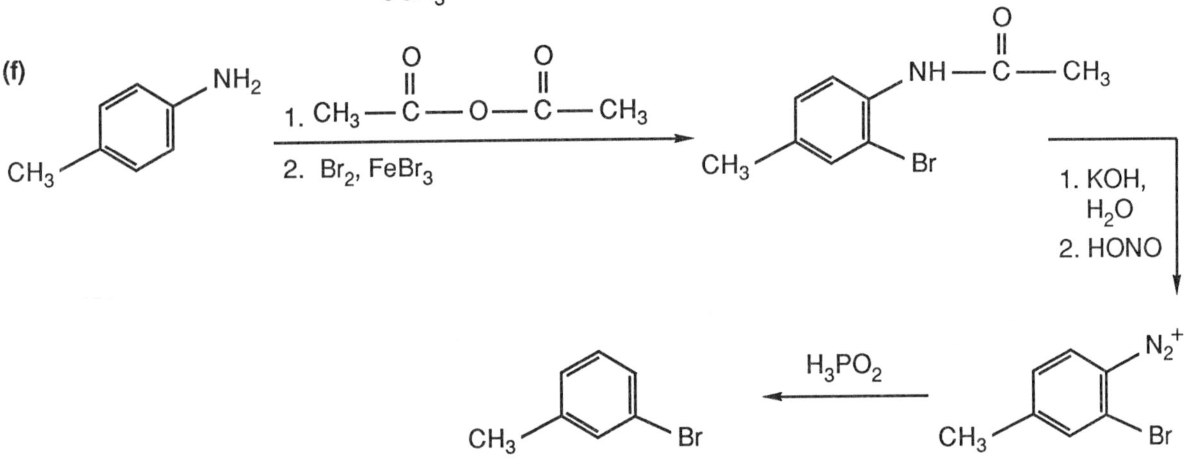

(f)

CH_3, NH_2 (benzene ring) $\xrightarrow[2.\ Br_2,\ FeBr_3]{1.\ CH_3-\overset{O}{\overset{\|}{C}}-O-\overset{O}{\overset{\|}{C}}-CH_3}$ CH_3, $NH-\overset{O}{\overset{\|}{C}}-CH_3$, Br (benzene ring)

$\xrightarrow[2.\ HONO]{1.\ KOH,\ H_2O}$ CH_3, N_2^+, Br (benzene ring) $\xrightarrow{H_3PO_2}$ CH_3, Br (benzene ring)

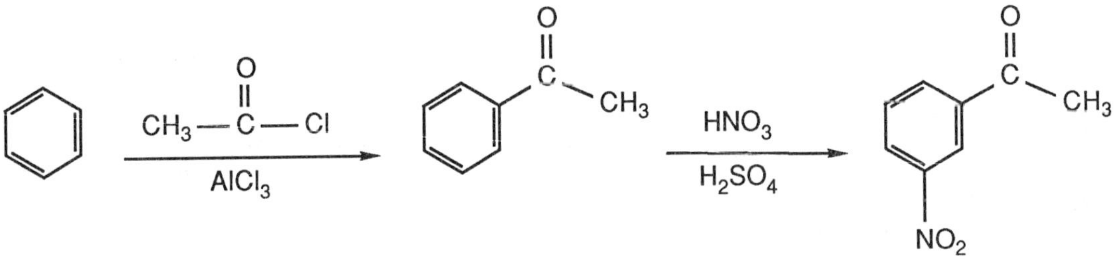

(g)

(benzene) $\xrightarrow[AlCl_3]{CH_3-\overset{O}{\overset{\|}{C}}-Cl}$ $C_6H_5-\overset{O}{\overset{\|}{C}}-CH_3$ $\xrightarrow[H_2SO_4]{HNO_3}$ $\overset{O}{\overset{\|}{C}}-CH_3$, NO_2 (benzene ring)

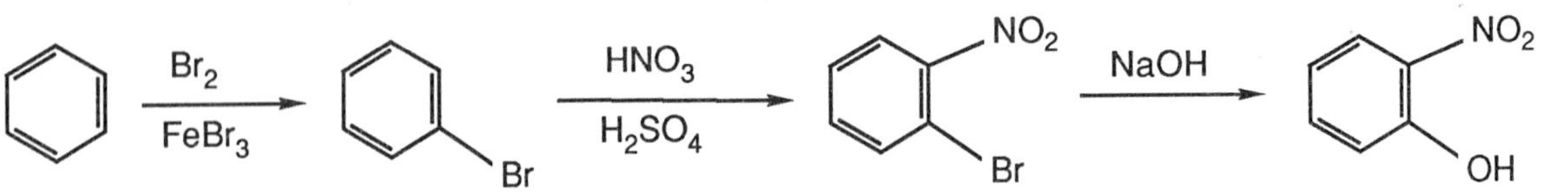

(h)

(benzene) $\xrightarrow[FeBr_3]{Br_2}$ Br (benzene ring) $\xrightarrow[H_2SO_4]{HNO_3}$ NO_2, Br (benzene ring) $\xrightarrow{NaOH}$ NO_2, OH (benzene ring)

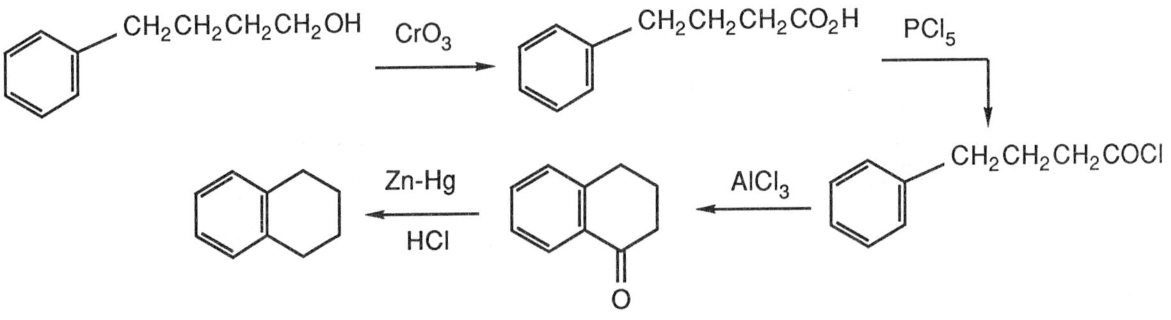

(i)

$C_6H_5-CH_2CH_2CH_2CH_2OH$ $\xrightarrow{CrO_3}$ $C_6H_5-CH_2CH_2CH_2CO_2H$ $\xrightarrow{PCl_5}$ $C_6H_5-CH_2CH_2CH_2COCl$ $\xrightarrow{AlCl_3}$ (tetralone, O) $\xrightarrow[HCl]{Zn\text{-}Hg}$ (tetralin)

18.26

(a)

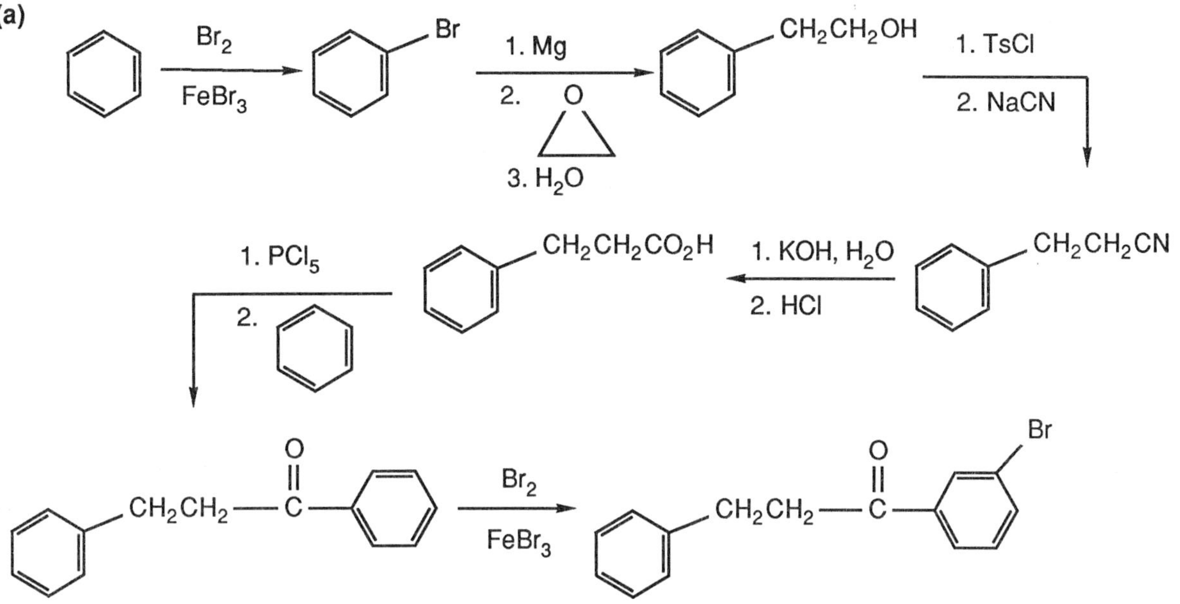

Br$_2$ / FeBr$_3$ — Br

1. Mg
2. O (ethylene oxide)
3. H_2O

CH_2CH_2OH

1. TsCl
2. NaCN

CH_2CH_2CN

1. KOH, H_2O
2. HCl

$CH_2CH_2CO_2H$

1. PCl_5
2. (benzene)

$CH_2CH_2—\overset{O}{\overset{||}{C}}—$

Br$_2$ / FeBr$_3$

$CH_2CH_2—\overset{O}{\overset{||}{C}}—$ Br

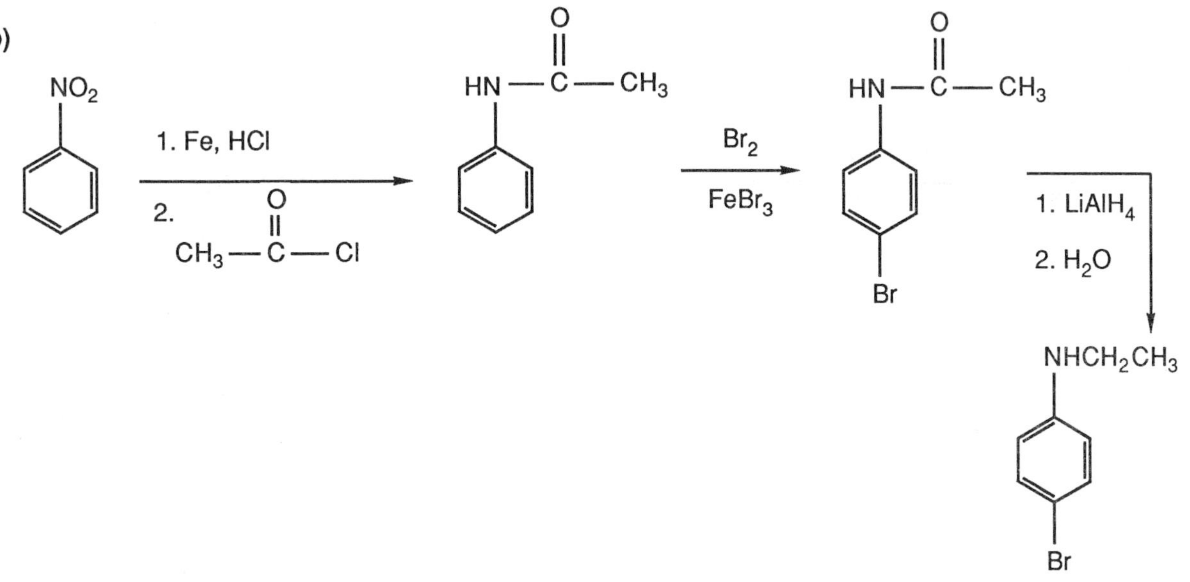

(b)

NO_2

1. Fe, HCl
2. $CH_3—\overset{O}{\overset{||}{C}}—Cl$

$HN—\overset{O}{\overset{||}{C}}—CH_3$

Br$_2$ / FeBr$_3$

$HN—\overset{O}{\overset{||}{C}}—CH_3$ Br

1. $LiAlH_4$
2. H_2O

$NHCH_2CH_3$ Br

(c)

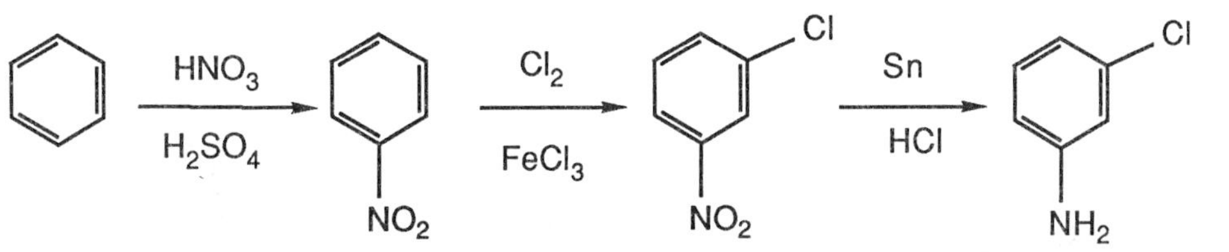

HNO_3 / H_2SO_4 — NO_2

Cl_2 / $FeCl_3$ — Cl, NO_2

Sn / HCl — Cl, NH_2

(d)

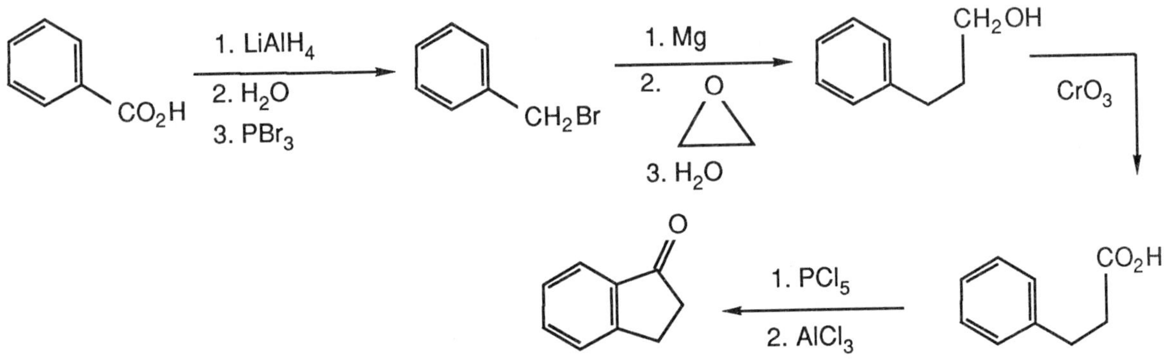

CO_2H — 1. $LiAlH_4$ 2. H_2O 3. PBr_3 → CH_2Br — 1. Mg 2. O 3. H_2O → CH_2OH — CrO_3 → CO_2H — 1. PCl_5 2. $AlCl_3$ → O

(e)

Br_2 / $FeBr_3$ → Br — 1. Mg 2. CO_2 3. HCl → CO_2H — 1. HNO_3, H_2SO_4 2. PCl_5 → O ‖ C — Cl, NO_2 I

H_2N — O ‖ CH_3—C—Cl → O ‖ CH_3—C—HN — O ‖ CH_3—C—Cl / $AlCl_3$ → O ‖ CH_3—C—HN, C—CH_3 ‖ O — NaOH / H_2O → H_2N, C—CH_3 ‖ O II

O ‖ C—Cl, NO_2 I + H_2N, C—CH_3 ‖ O II → O ‖ C—NH, NO_2, C—CH_3 ‖ O

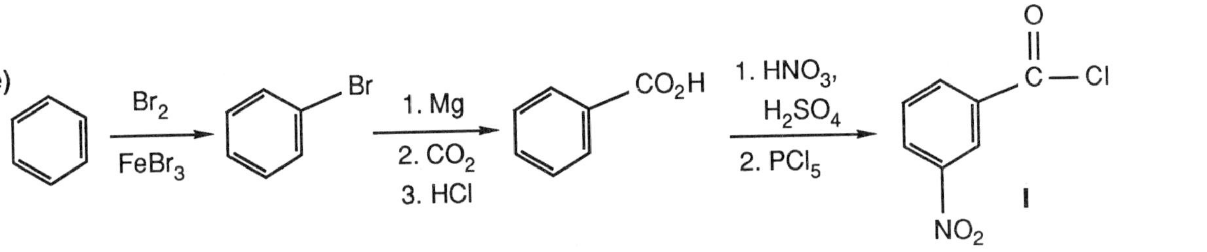

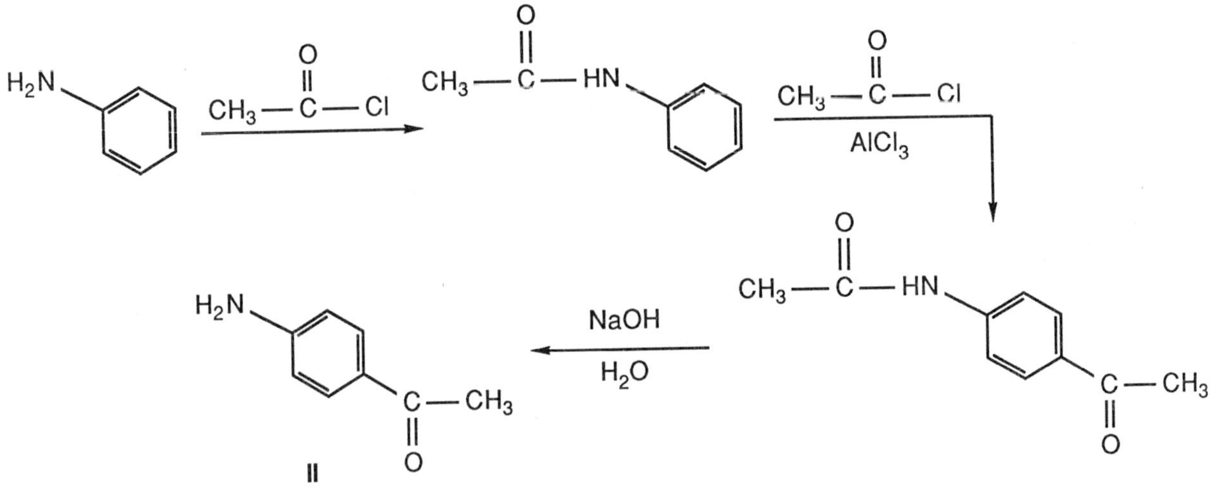

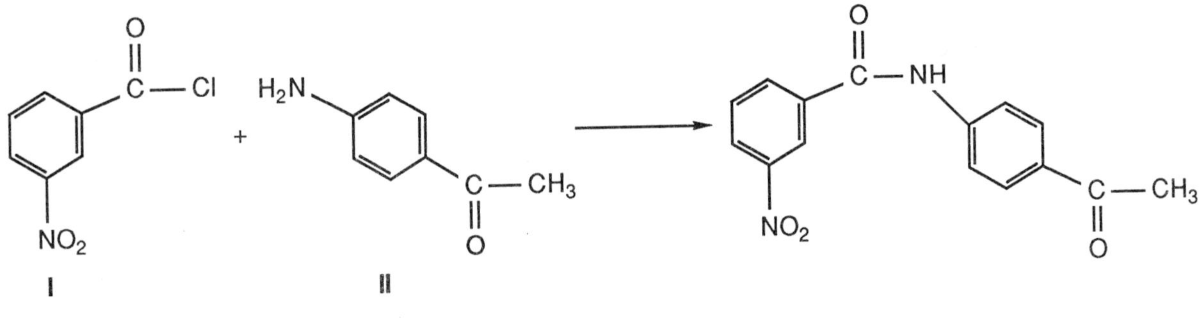

(f)

CH_3 $\xrightarrow[H_2SO_4]{HNO_3}$ O_2N CH_3 $\xrightarrow[2.\ CH_3-C(=O)-Cl]{1.\ Fe,\ HCl}$ $CH_3-C(=O)-NH$ CH_3

HNO_3, H_2SO_4

$CH_3-C(=O)-NH$ CH_3 NO_2

1. KOH, H_2O
2. HONO

$^+N_2$ CH_3 NO_2

1. H_3PO_4
2. Sn, HCl

CH_3 NH_2 $\xrightarrow[2.\ H_2O,\ H_2SO_4]{1.\ HONO}$ CH_3 OH

(g)

$\xrightarrow[FeCl_3]{Cl_2}$ Cl $\xrightarrow[H_2SO_4]{SO_3}$ Cl SO_3H $\xrightarrow{HNO_3}$ NO_2 Cl SO_3H

Sn, HCl

NH_2 Cl SO_3H

1. HONO
2. CuCN

CN Cl SO_3H

1. KOH, H_2O
2. HCl

CO_2H Cl SO_3H

(h)

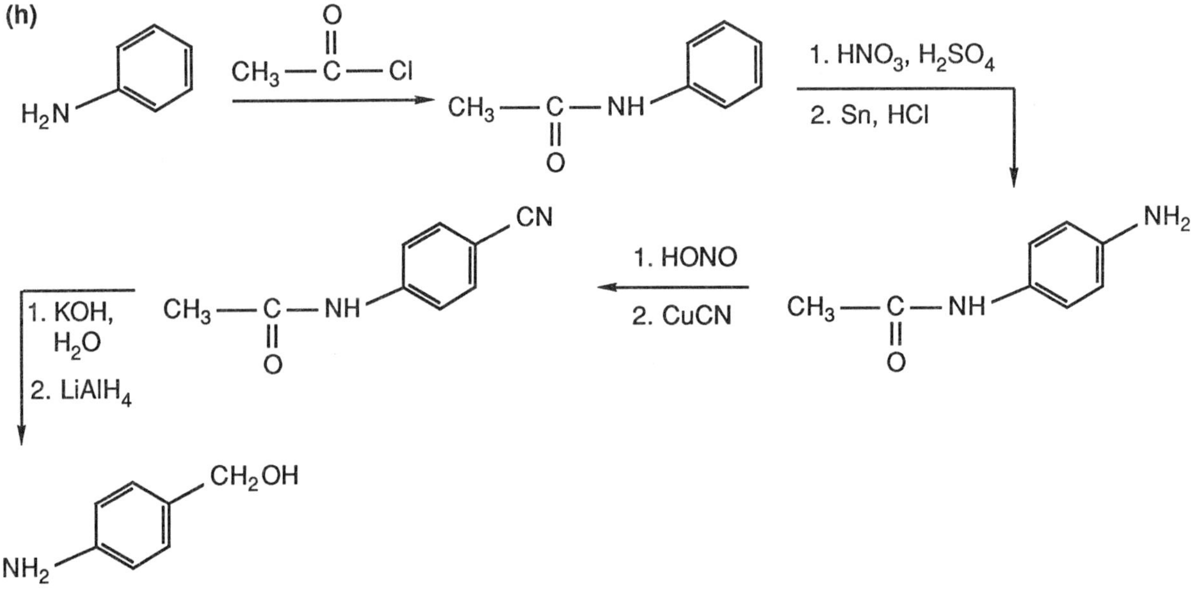

H_2N–C_6H_5 $\xrightarrow{CH_3-\overset{O}{\overset{\|}{C}}-Cl}$ $CH_3-\underset{O}{\underset{\|}{C}}-NH-C_6H_5$ $\xrightarrow[2.\ Sn,\ HCl]{1.\ HNO_3,\ H_2SO_4}$ $CH_3-\underset{O}{\underset{\|}{C}}-NH-C_6H_4-NH_2$ (para)

$\xrightarrow[2.\ CuCN]{1.\ HONO}$ $CH_3-\underset{O}{\underset{\|}{C}}-NH-C_6H_4-CN$ (para) $\xrightarrow[2.\ LiAlH_4]{1.\ KOH,\ H_2O}$ $NH_2-C_6H_4-CH_2OH$ (para)

(i)

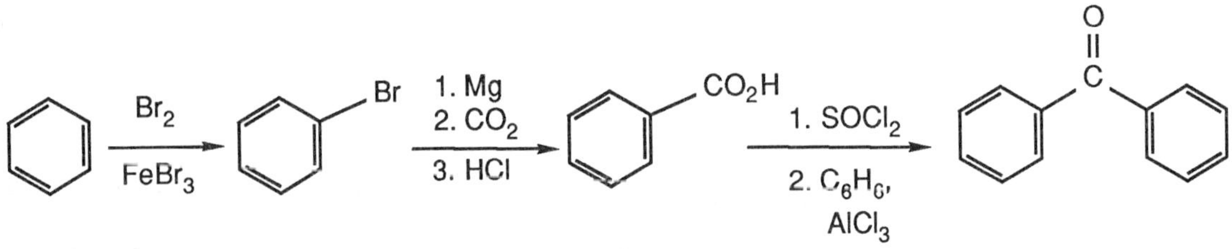

C_6H_6 $\xrightarrow[FeBr_3]{Br_2}$ C_6H_5Br $\xrightarrow[3.\ HCl]{1.\ Mg\quad 2.\ CO_2}$ $C_6H_5CO_2H$ $\xrightarrow[2.\ C_6H_6,\ AlCl_3]{1.\ SOCl_2}$ $C_6H_5-\overset{O}{\overset{\|}{C}}-C_6H_5$

18.27

(a) $CH_3O-C_6H_4-C(CH_3)_2Br$ (para)

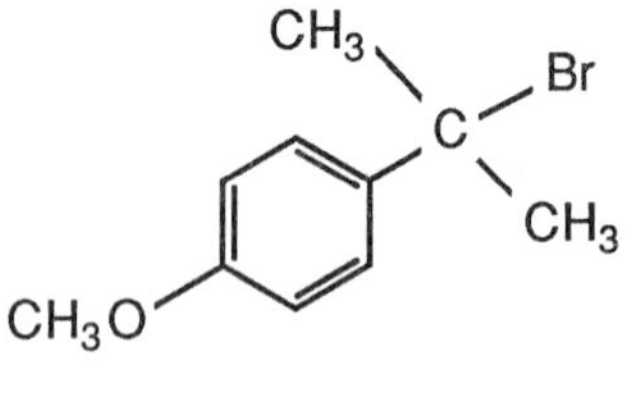

(b) m-chloronitrobenzene (NO_2, Cl)

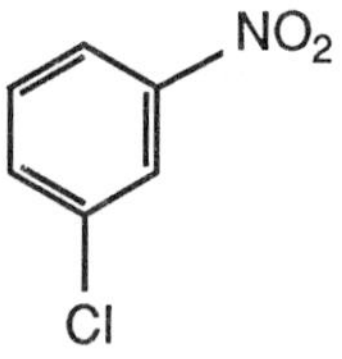

(c) m-nitrobenzoic acid (CO_2H, NO_2)

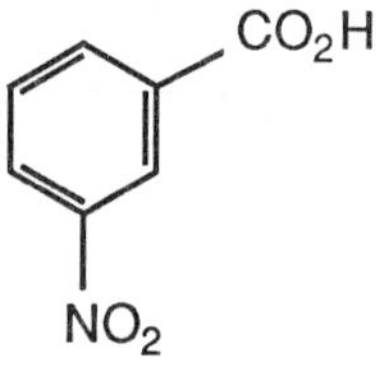

(d) $CH_3-\overset{O}{\overset{\|}{C}}-C_6H_4-OCH_2-C_6H_5$ (para)

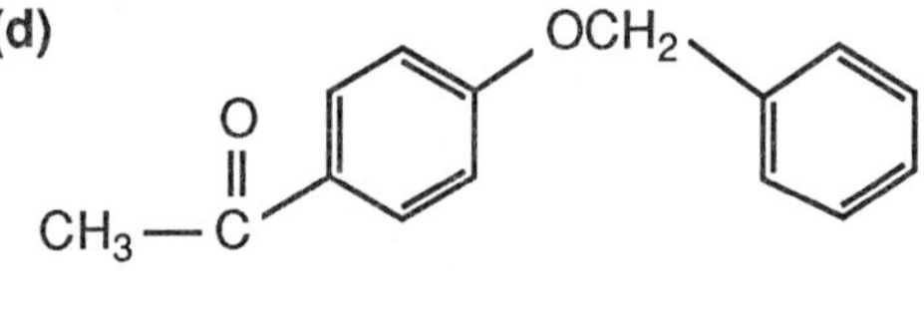

(e) o-$C_6H_4(\overset{O}{\overset{\|}{C}}-OCH_3)(CO_2^-)$

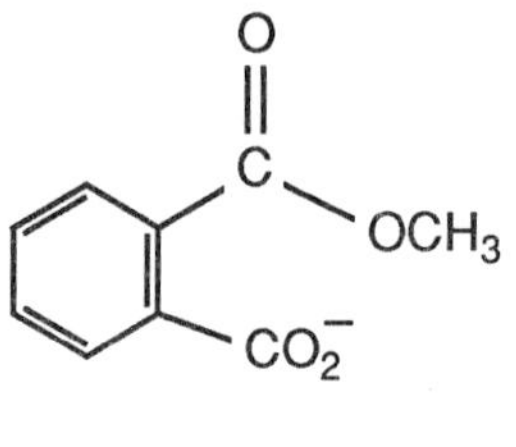

(f) CH_2CH_3, O_2N, NO_2 + CH_2CH_3, NO_2, NO_2

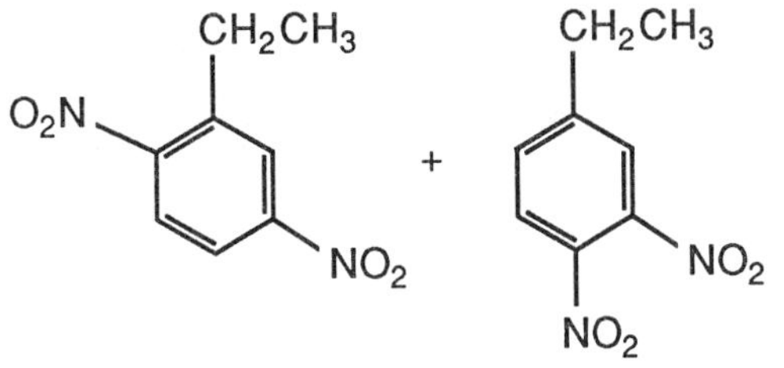

(g) CH_3, $HOSO_3$, CH_3 + CH_3, CH_3, SO_3H

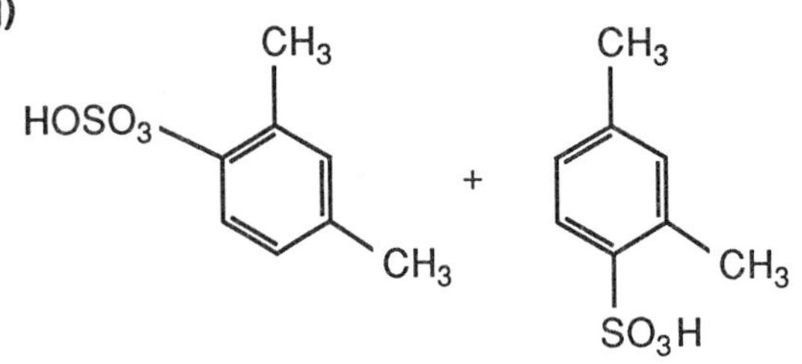

(h) $CH_2CH_2—C(=O)—Cl$

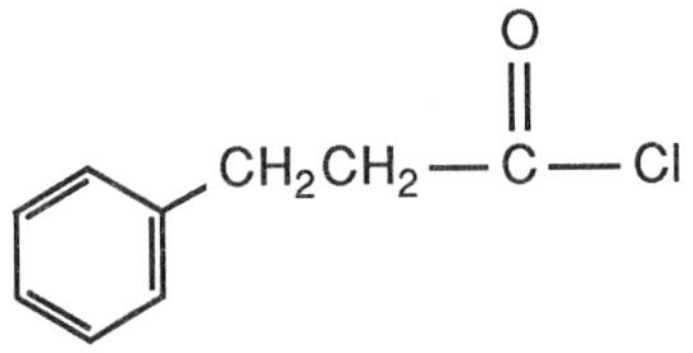

(i) NO_2 + O_2N (O)

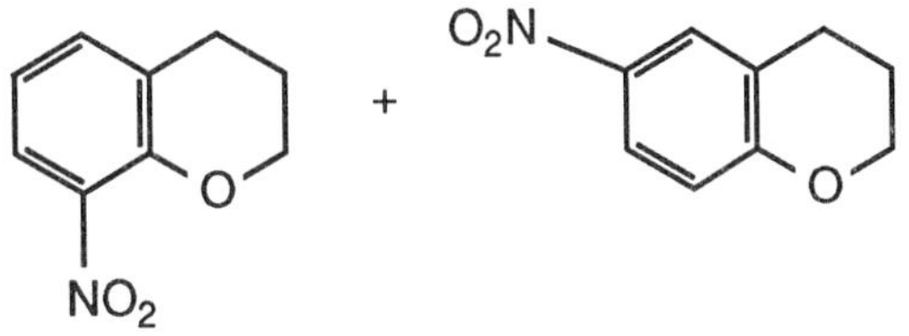

(j) CH_2CH_3, Br

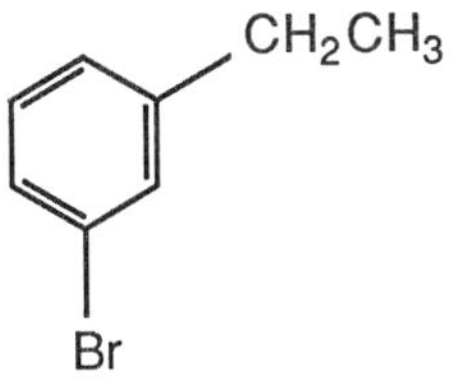

18.28 The molecular formula ($C_8H_8O_3$) of **A** indicates five degrees of unsaturation, four of which correspond to the aromatic ring. The indicated reactions will introduce an acetyl group and then oxidize it (haloform reaction) to a carboxyl group. This means that the methoxy group (along with the second carboxyl group) was already present in **A.** The directive effect of the methoxy group will cause substitution at positions *ortho* or *para* to it, so the *meta* carboxyl was already present in **A**:

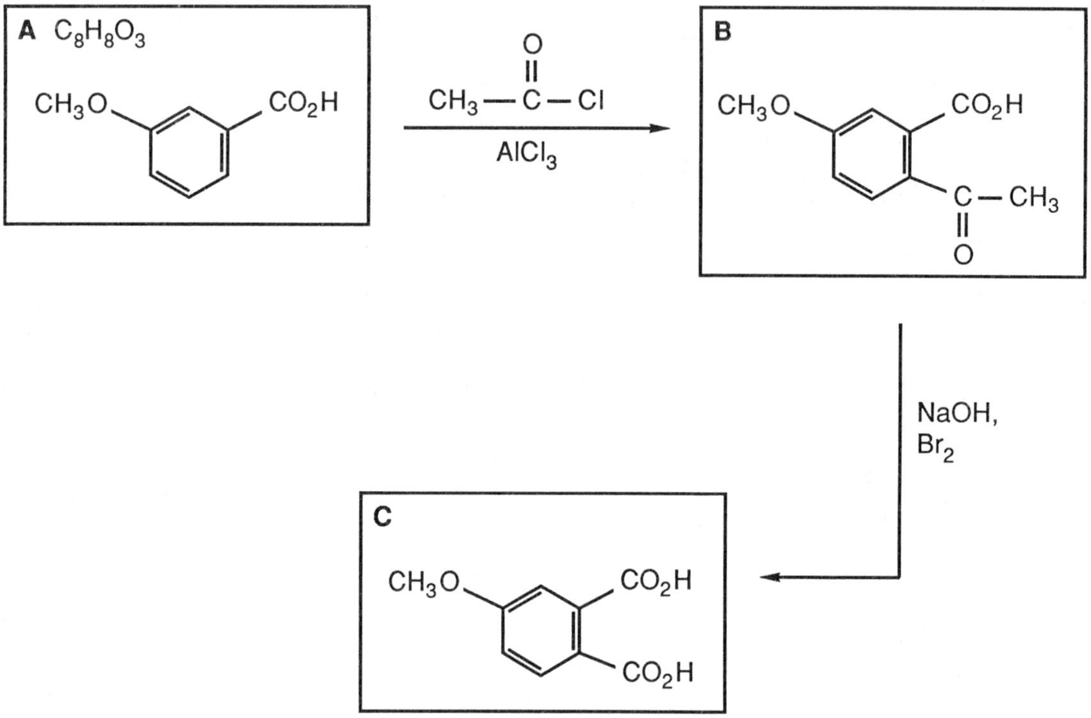

A $C_8H_8O_3$

CH_3O CO_2H

$CH_3—\overset{O}{\overset{\|}{C}}—Cl$, $AlCl_3$ →

B

CH_3O CO_2H $\underset{\underset{O}{\|}}{C}—CH_3$

NaOH, Br_2 ↓

C

CH_3O CO_2H CO_2H

18.29 The structure of **B** tells you that **A** is *para* substituted. The iodoform indicates the presence of a methyl ketone, and the Tollen's test indicates an aldehyde. The simplest compound that would be oxidized readily to **B** by potassium permanganate is 4-formylacetophenone:

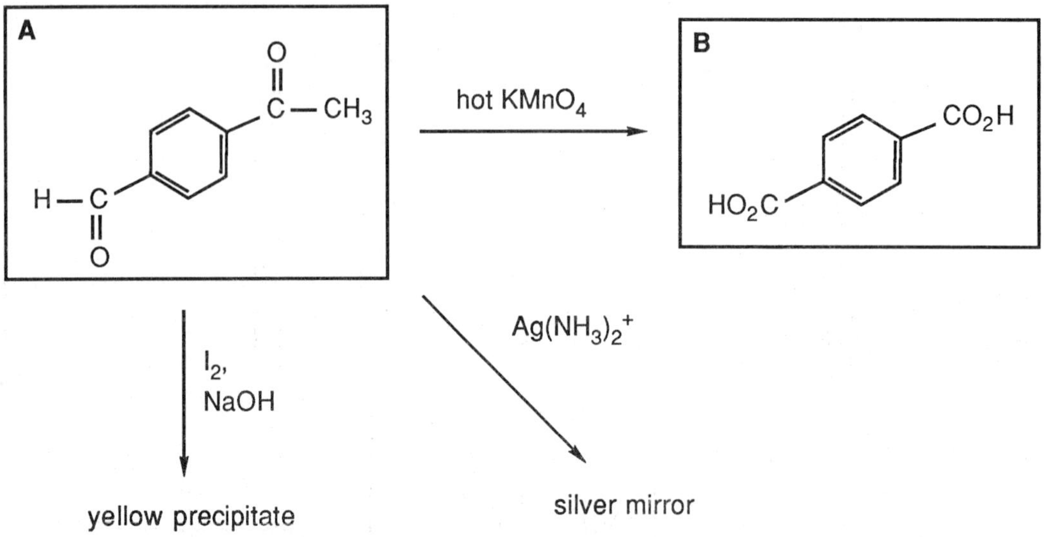

A

$\overset{O}{\overset{\|}{C}}—CH_3$

$H—\underset{\underset{O}{\|}}{C}$

hot $KMnO_4$ →

B

CO_2H

HO_2C

I_2, NaOH ↓

yellow precipitate

$Ag(NH_3)_2^+$ ↘

silver mirror

18.30 (a) $C_6H_5C_2H_5$ $\xrightarrow{Br_2,\ FeBr_3}$ p-$BrC_6H_4C_2H_5$

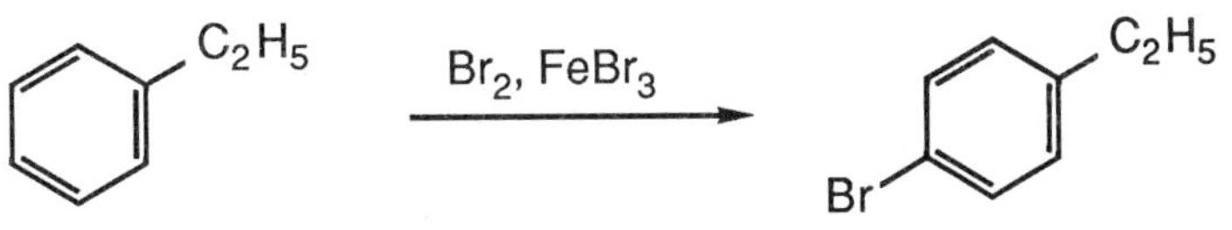

Br, C_2H_5

(b) $C_6H_5C_2H_5$ → $C_6H_5CH{=}CH_2$

1. NBS
2. KOtBu

$CH=CH_2$

(c) $C_6H_5C_2H_5$ → p-$IC_6H_4C_2H_5$

1. HNO_3
2. Fe, HCl
3. HONO
4. CuI

I, C_2H_5

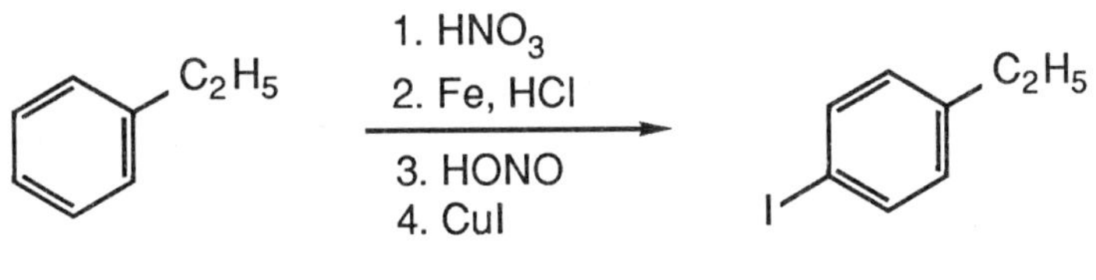

(d) $C_6H_5C_2H_5$ → p-$NCC_6H_4C_2H_5$

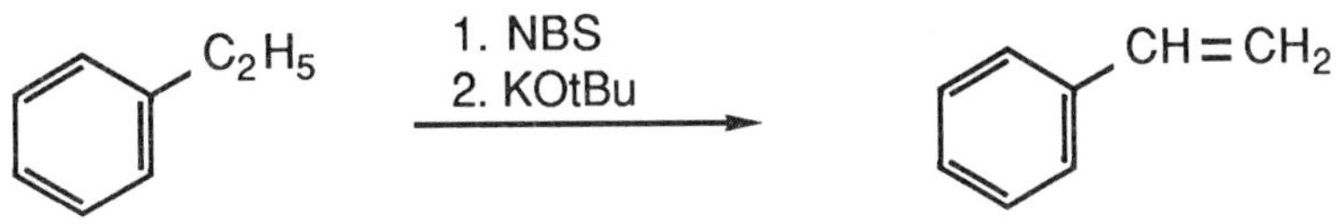

1. HNO_3
2. Fe, HCl
3. HONO
4. CuCN

NC, C_2H_5

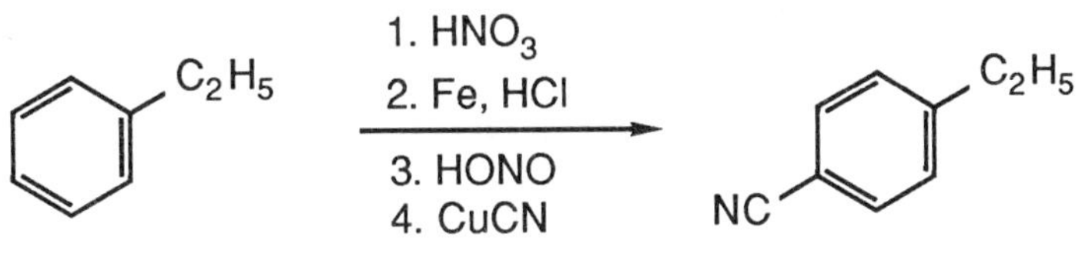

(e) $C_6H_5C_2H_5$ → C_6H_5CN

1. $KMnO_4$
2. PCl_5
3. NH_3
4. P_2O_5

CN

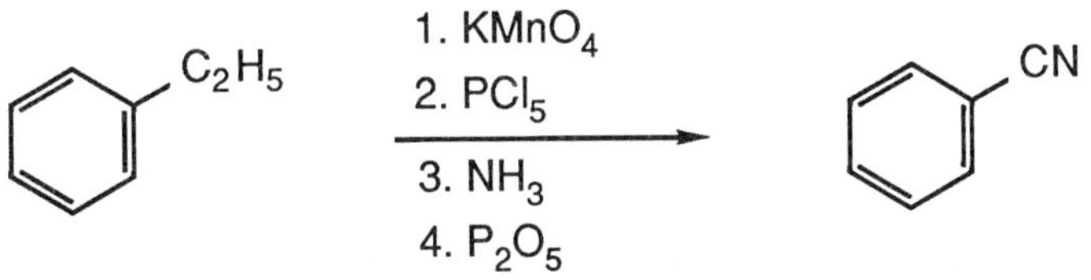

18.31

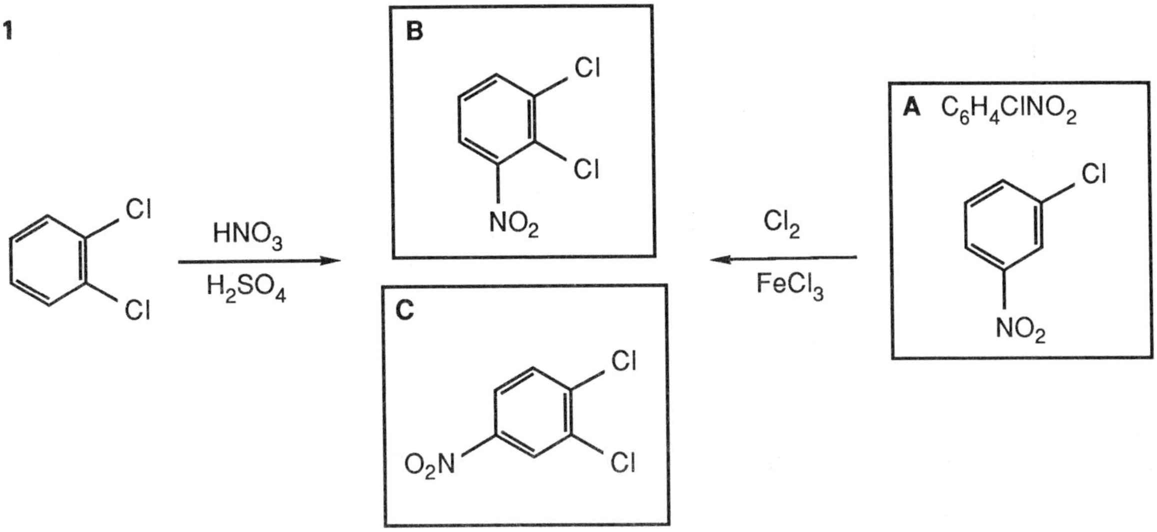

1,2-dichlorobenzene (Cl, Cl) $\xrightarrow[H_2SO_4]{HNO_3}$ **B** + **C**

B (Cl, Cl, NO_2)

C (Cl, Cl, O_2N)

A $C_6H_4ClNO_2$ (Cl, NO_2) $\xrightarrow[FeCl_3]{Cl_2}$ **B**

18.32 As shown in the following equations, protonation (which is a type of electrophilic substitution) should occur at the two *meta* positions. This would afford 2,4,6-[$^2H_3$]-benzoic acid:

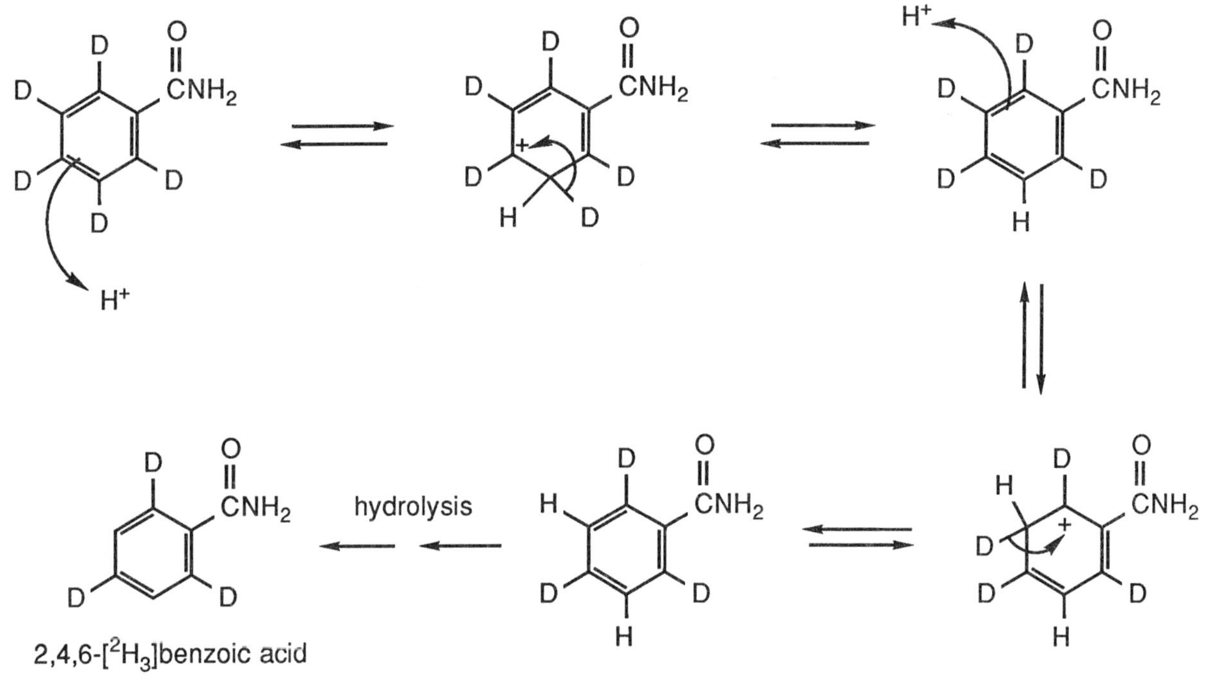

hydrolysis

2,4,6-[$^2H_3$]benzoic acid

18.33

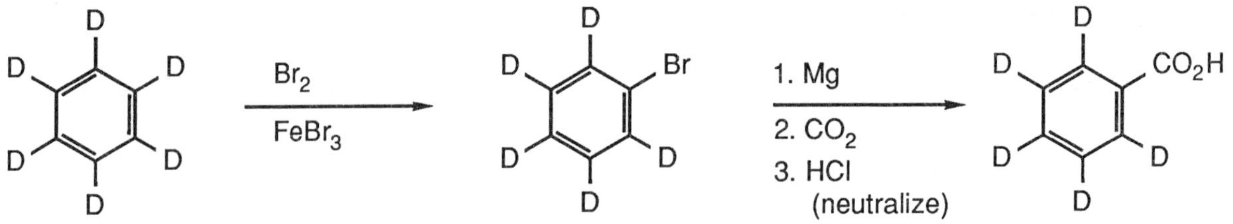

Br_2 / $FeBr_3$

1. Mg
2. CO_2
3. HCl (neutralize)

18.34 By reaction with D_2SO_4:

$$C_6H_6 \underset{}{\overset{D_2SO_4}{\rightleftharpoons}} C_6H_5D \rightleftharpoons \text{etc.} \rightleftharpoons C_6D_6$$

18.35 The Friedel-Crafts alkylation reaction should introduce one or more alkyl substituents onto the aromatic ring. Rearrangement is likely, so either propyl or isopropyl groups can be introduced. Both spectra show a 5H signal in the aromatic region, so it appears that both are monosubstituted products. Compound **A** shows the characteristic 6H methyl doublet and 1H septet for an isopropyl group, while compound **B** exhibits the triplet/sextet/triplet expected for a propyl group. The structures are:

$C_6H_5CH(CH_3)_2$ **A**

$C_6H_5CH_2CH_2CH_3$ **B**

18.36 (a) $p\text{-}CH_3OC_6H_4CH(CH_3)CH_3 \longrightarrow p\text{-}CH_3OC_6H_4C(CH_3)(Br)CH_3$

The major change in the NMR spectrum would be with respect to the isopropyl group of the reactant. The 6H doublet would become a 6H singlet in the product, and the 1H septet of the reactant would be absent.

(b) $C_6H_5NO_2 \longrightarrow m\text{-}ClC_6H_4NO_2$

The most clear cut spectroscopic change for this reaction would be in the aromatic region of the NMR spectrum. The spectrum of the reactant would show the characteristic 3:2 pattern of a polar, monosubstituted benzene, while the product would exhibit a more complex pattern.

(c)

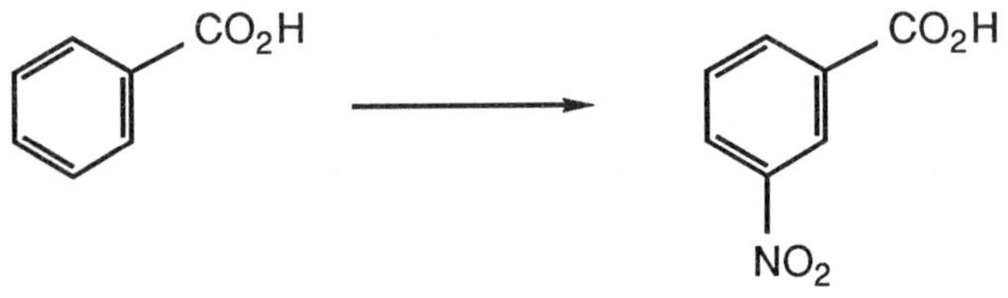

The most clear cut spectroscopic change for this reaction would be in the aromatic region of the NMR spectrum. The spectrum of the reactant would show the characteristic 3:2 pattern of a polar, monosubstituted benzene, while the product would exhibit a more complex pattern.

(d)

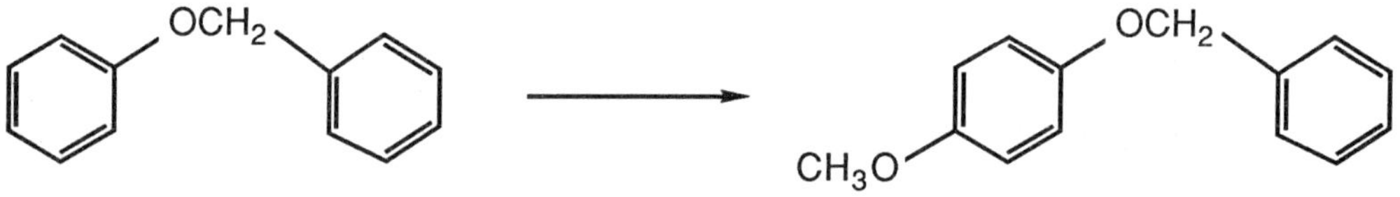

The infrared spectrum of the product would exhibit a characteristic carbonyl absorption in the 1675–1700 cm^{-1} region, and the CH_3 group of the acetyl group would afford a 3H singlet at about 2.4 ppm in the NMR spectrum. The aromatic region of the NMR spectrum would also show some changes. The phenyl of the benzyl group would remain as a 5H "singlet", but the signal for the other aromatic ring would become more symmetrical (showing the typical pattern for a *para*-disubstituted ring).

(e)

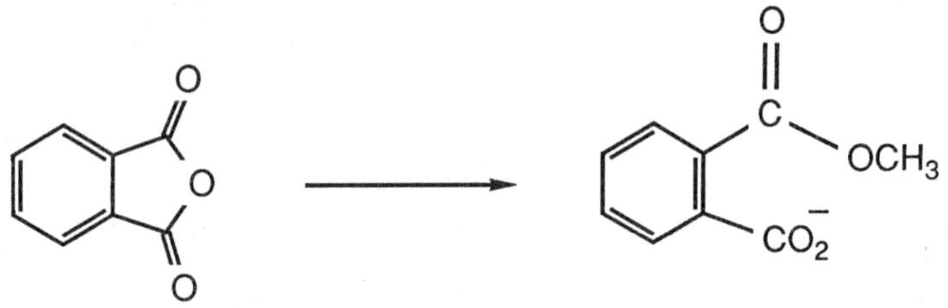

The infrared spectrum of the reactant would exhibit the characteristic peaks for anhydride carbonyls near 1800 cm^{-1}, and these would be shifted to the normal ester/acid region for conjugated carbonyls (1675–1730 cm^{-1}) in the product. The product would also show a less symmetric pattern for the aromatic ring and would exhibit a sharp 3H singlet near 3.7 ppm for the methyl ester.

(f)

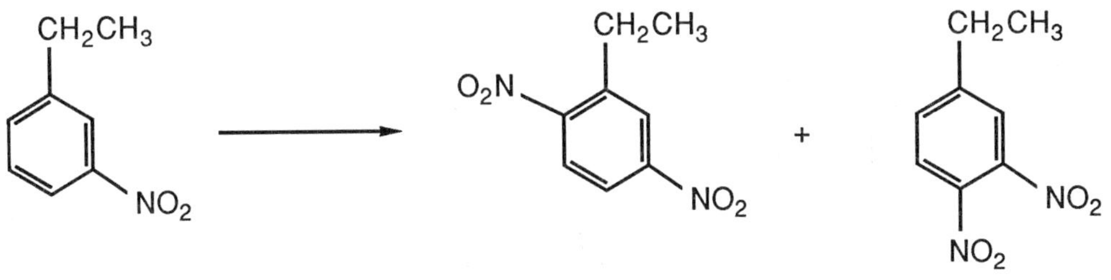

The NMR spectrum of the product would differ from that of the reactant qualitatively, but the integration of the aromatic region relative to the ethyl pattern would show the substitution.

(g)

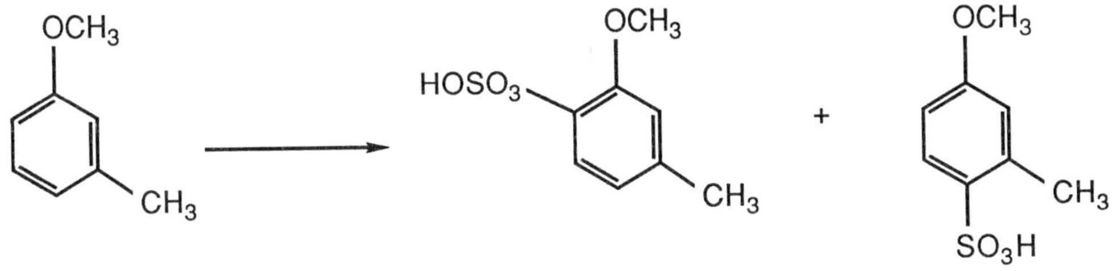

The infrared spectrum of the product would show OH peaks and S=O peaks for the sulfonic acid. The NMR spectrum would also show substantial qualitative changes, and the integration of the aromatic region relative to the methoxy group would demonstrate that substitution had occurred.

(h)

Conversion of the acid to an acid chloride would result in the characteristic change in the C=O peak from near 1800 cm^{-1} to approximately 1700–1725 cm^{-1}. The very broad (2500–3500 cm^{-1}) OH peak would be absent in the product.

(i)

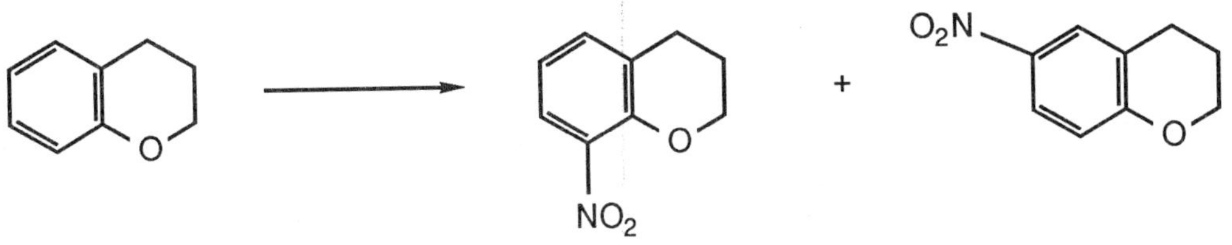

The infrared spectrum of the product would exhibit new peaks corresponding to the nitro group, and the integration (aliphatic/aromatic ratio) would indicate substitution on the aromatic ring.

(j)

$$\text{3-BrC}_6\text{H}_4\text{–C(=O)–CH}_3 \longrightarrow \text{3-BrC}_6\text{H}_4\text{–CH}_2\text{CH}_3$$

Reduction of the acetyl group to an ethyl group would eliminate the C=O peak near 1700 cm^{-1} in the reactant. The 3H singlet near 2.4 ppm in the reactant would be replaced by a characteristic ethyl signal in the product: a 2H quartet near 2.6 ppm and a 3H triplet at about 1.2 ppm.

CHAPTER 19. CARBOHYDRATES

ANSWERS TO EXERCISES

19.1

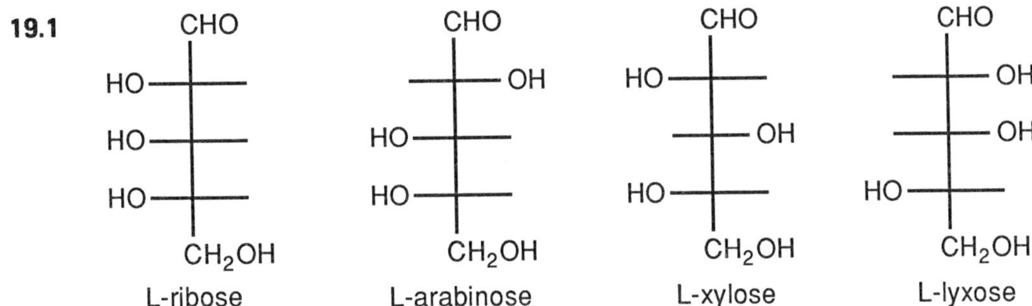

19.2

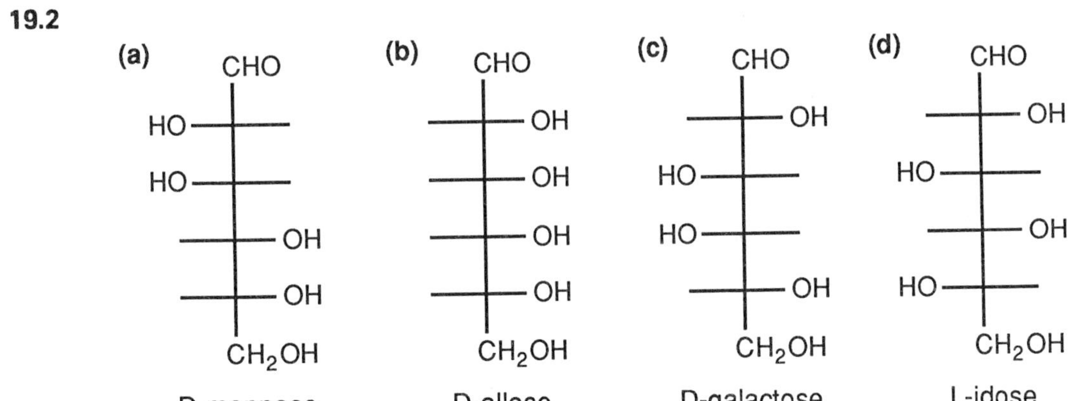

19.3

(a)

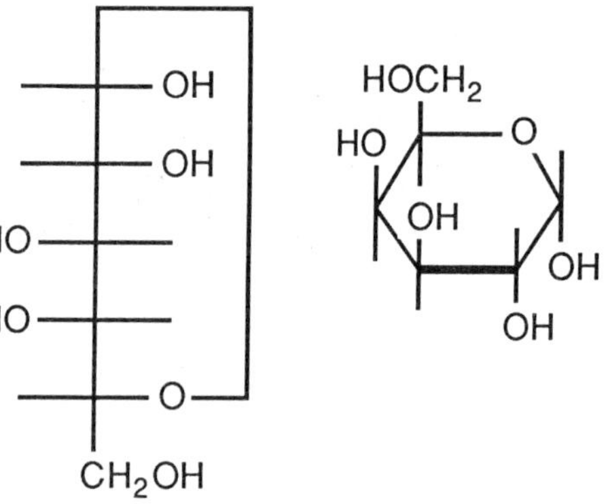

HOCH2, HO, O, OH, OH, HO, HO, O, CH2OH, OH, OH, OH

(b)

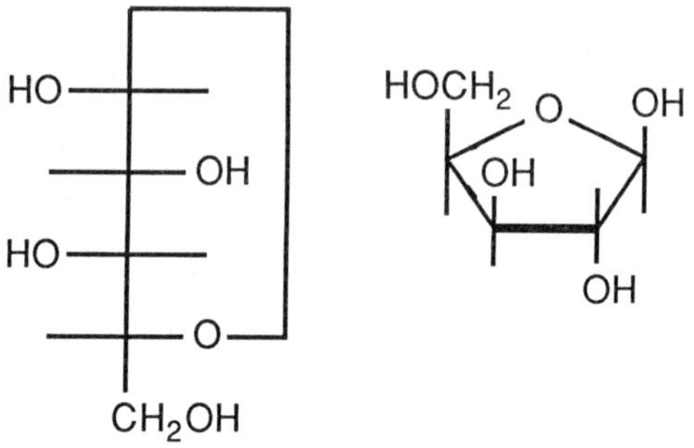

HO, OH, HO, O, CH2OH, HOCH2, O, OH, OH, OH

(c)

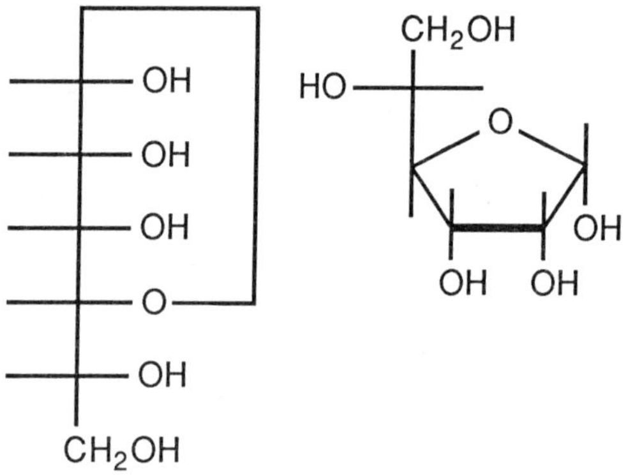

OH, OH, OH, O, OH, CH2OH, CH2OH, HO, O, OH, OH, OH

(d)

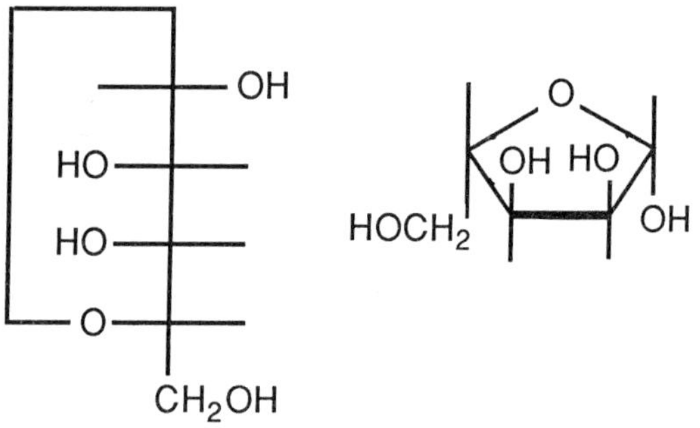

OH, HO, HO, O, CH2OH, O, OH, HO, HOCH2, OH

(e)

HO, CH2OH, HO, OH, O, CH2OH, HOCH2, O, OH, HO, CH2OH, HO

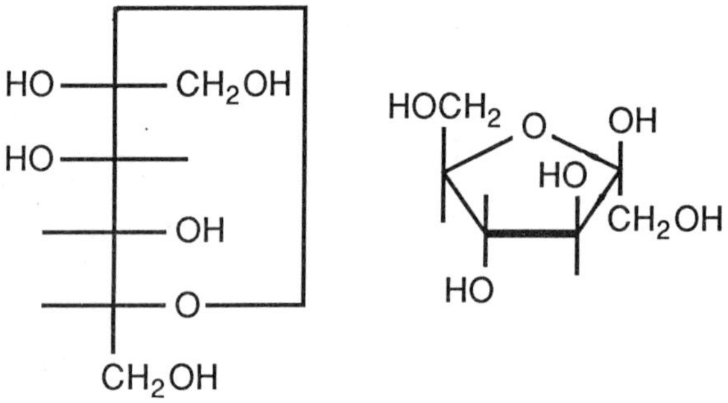

19.4 **(a)** α-*D*-allopyranose **(b)** β-*L*-ribofuranose **(c)** α-*D*-fructofuranose

19.5

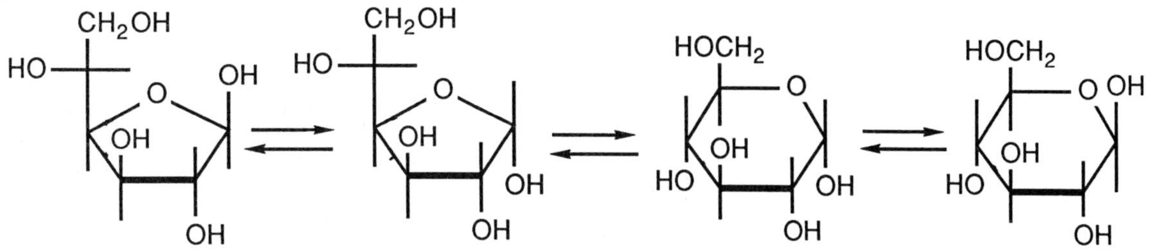

CH_2OH HO OH O OH OH ⇌ CH_2OH HO O OH OH OH ⇌ $HOCH_2$ O OH HO OH OH ⇌ $HOCH_2$ O OH OH HO OH

19.6 (a)

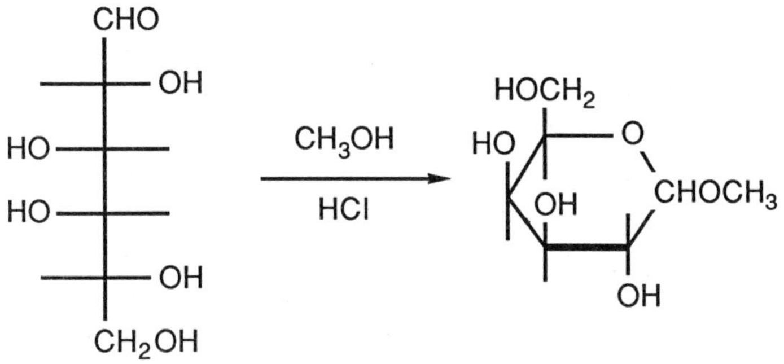

CHO, OH, HO, HO, OH, CH_2OH $\xrightarrow[HCl]{CH_3OH}$ $HOCH_2$ O HO OH $CHOCH_3$ OH

(b)

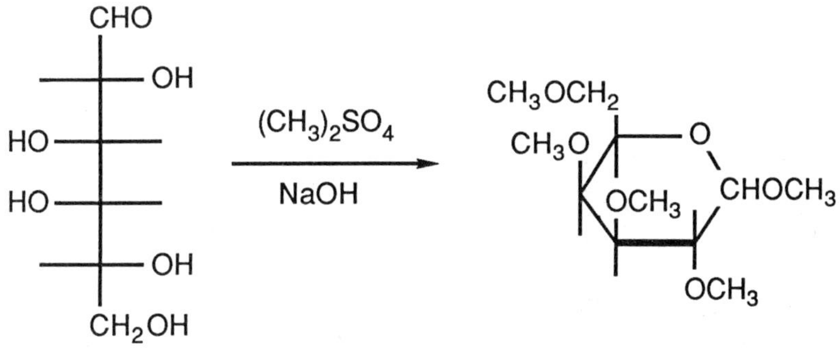

CHO, OH, HO, HO, OH, CH_2OH $\xrightarrow[NaOH]{(CH_3)_2SO_4}$ CH_3OCH_2 O CH_3O OCH_3 $CHOCH_3$ OCH_3

(c)

CHO, OH, HO, HO, OH, CH_2OH $\xrightarrow{Ac_2O}$ $AcOCH_2$ O AcO OAc CHOAc OAc

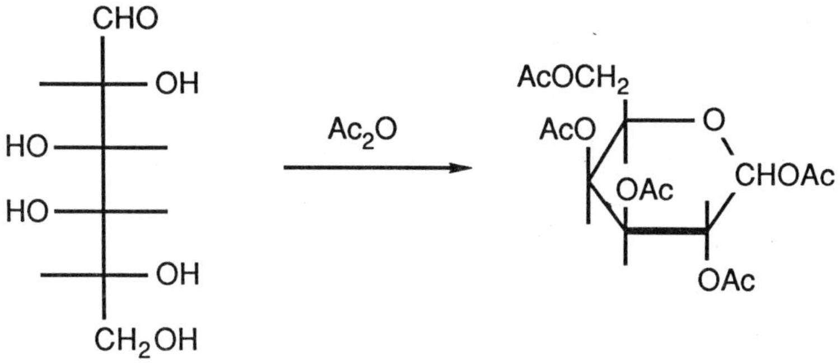

19.7

(a) HOCH₂ furanoside (OMe; HO, OH) $\xrightarrow[Ag_2O]{CH_3I}$ MeOCH₂ furanoside (OMe; MeO, OMe)

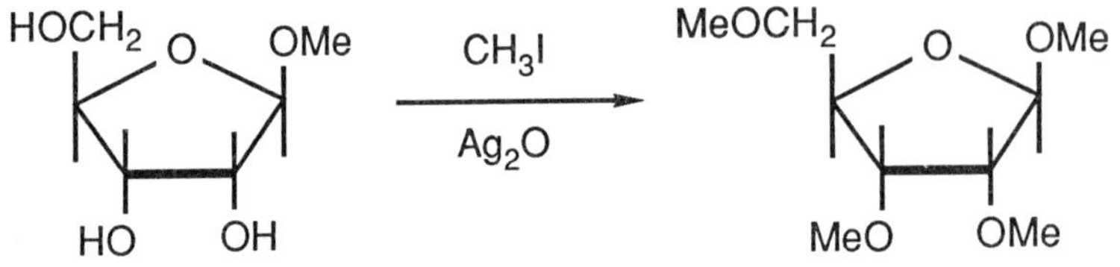

(b) HOCH₂ furanoside (OMe; HO, OH) $\xrightarrow[\text{2. HCl, } H_2O]{\text{1. } CH_3\text{, } Ag_2O}$ MeOCH₂ furanose (} H, OH; MeO, OMe)

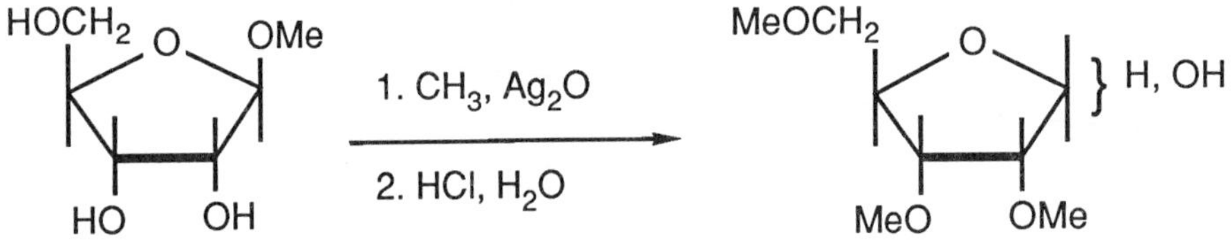

(c) HOCH₂ furanoside (OMe; HO, OH) $\xrightarrow{Ac_2O}$ AcOCH₂ furanoside (OMe; AcO, OAc)

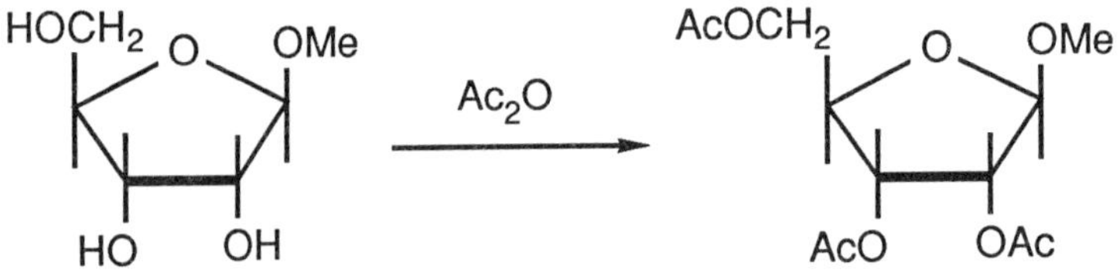

(d) HOCH₂ furanoside (OMe; HO, OH) $\xrightarrow[\text{(1 eqiv.)}]{TsCl}$ TsOCH₂ furanoside (OMe; HO, OH)

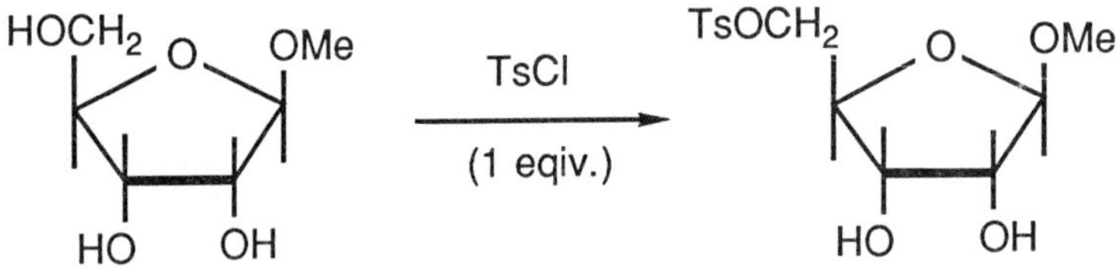

19.8

(a) CHO; HO—; —OH; —OH; CH_2OH ⟶ CH_2OH; HO—; —OH; —OH; CH_2OH

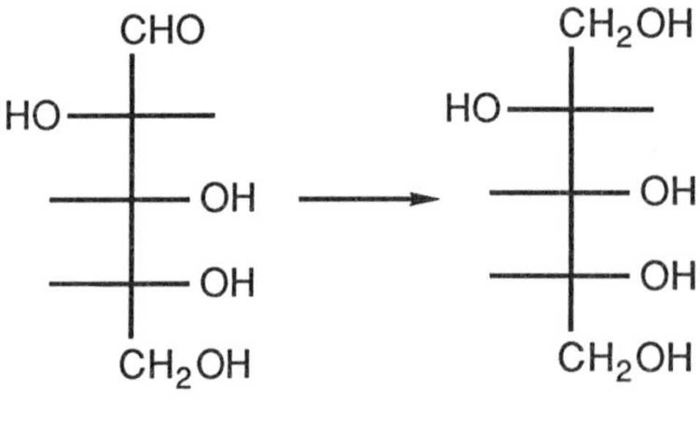

D- arabinitol

(b) CHO; —OH; HO—; —OH; —OH; CH_2OH ⟶ CH_2OH; —OH; HO—; —OH; —OH; CH_2OH

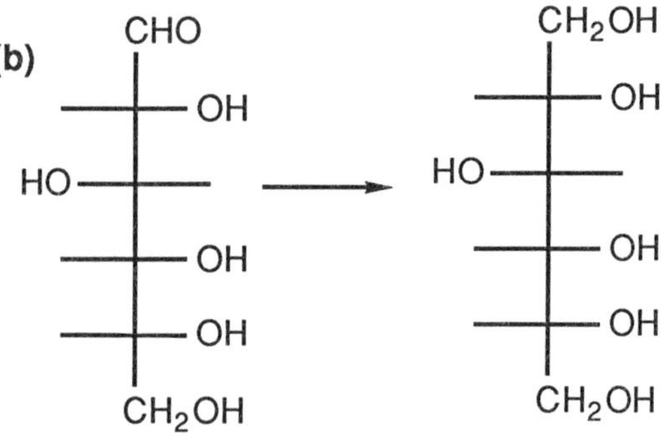

D- glucitol

(c)

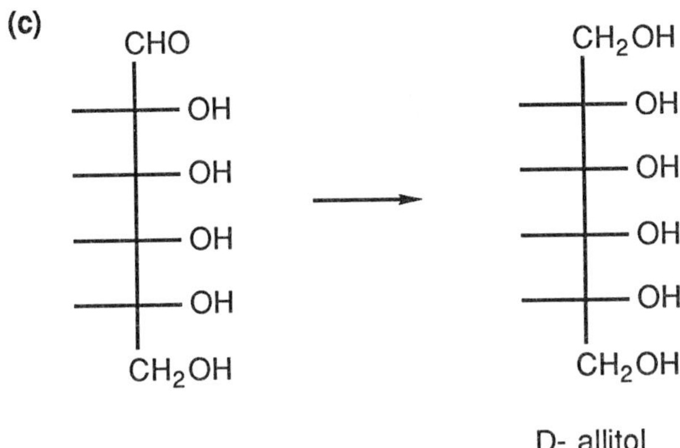

D- allitol

19.9

D-ribose → ribitol ← L-ribose

19.10

(a)

L-sorbose —reduction→ L-iditol + D-glucitol

(b)

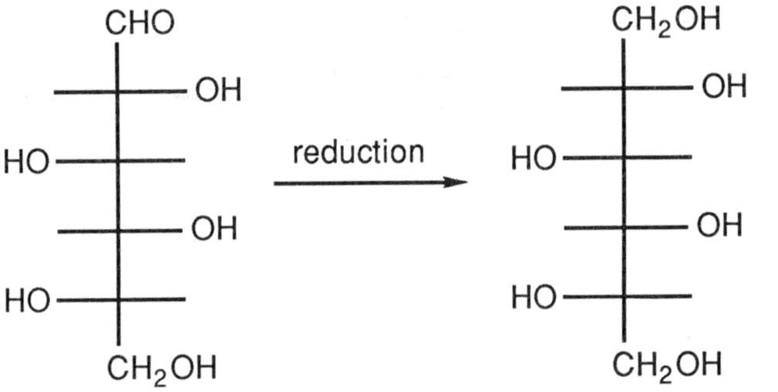

CHO / CH₂OH structures (Fischer projections):

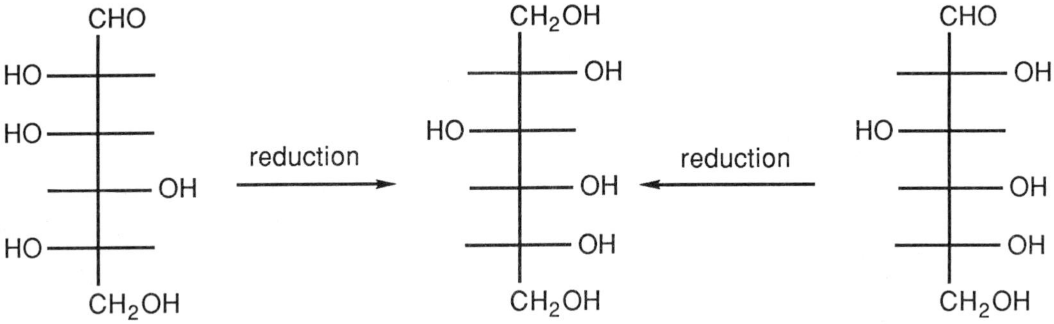

L-idose: CHO; —OH (right); HO— (left); —OH (right); HO— (left); CH_2OH

reduction →

L-iditol: CH_2OH; —OH (right); HO— (left); —OH (right); HO— (left); CH_2OH

L-gulose: CHO; HO—; HO—; —OH; HO—; CH_2OH

reduction →

D-glucitol: CH_2OH; —OH; HO—; —OH; —OH; CH_2OH

← reduction

D-glucose: CHO; —OH; HO—; —OH; —OH; CH_2OH

19.11 (a) CO_2H; —OH; HO—; —OH; —OH; CH_2OH

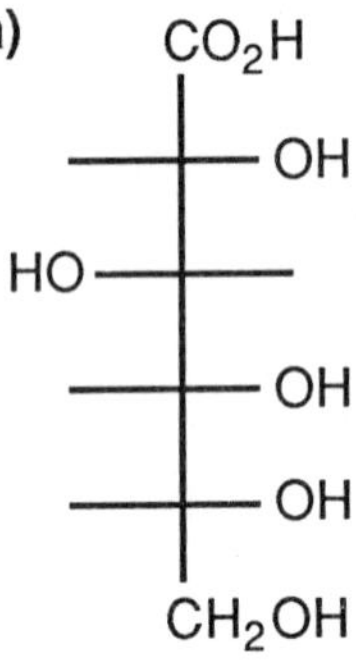

(b) CO_2H; —OH; HO—; —OH; —OH; CO_2H

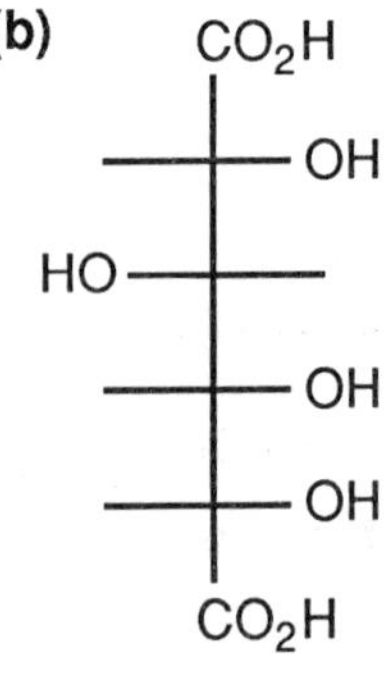

(c) $CH{=}N{-}NHC_6H_5$; $={}N{-}NHC_6H_5$; HO—; —OH; —OH; CO_2H

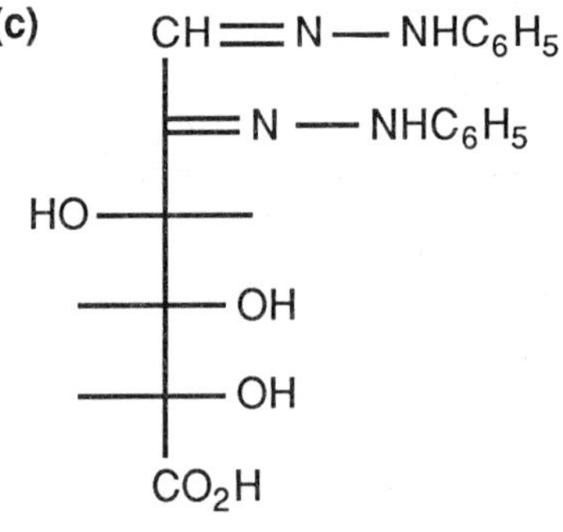

(d) 5 HCO_2H + CH_2O

19.12 Only (c), which is an acetal, would not give a positive test for the aldehyde group. The remaining structures would all exist in solution as equilibrium mixtures of the free aldehyde and hemiacetal.

19.13 (a)

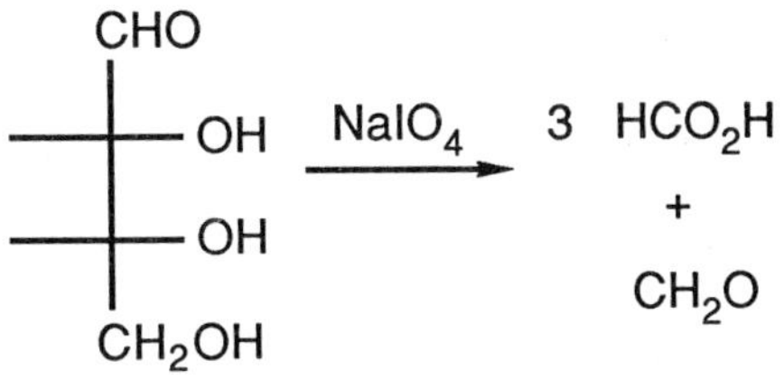

CHO; —OH; —OH; CH_2OH $\xrightarrow{NaIO_4}$ 3 HCO_2H + CH_2O

(b)

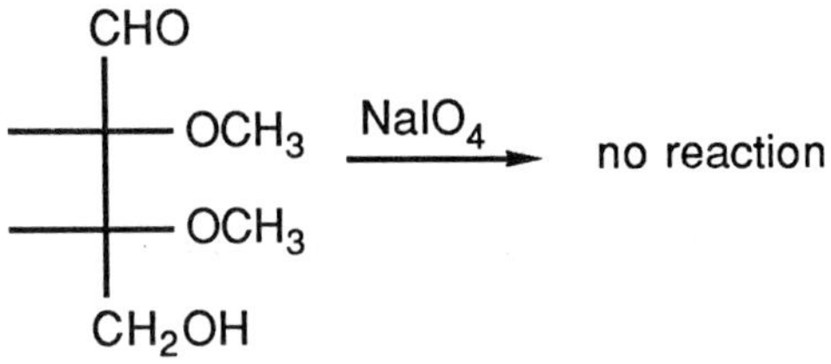

CHO; —OCH_3; —OCH_3; CH_2OH $\xrightarrow{NaIO_4}$ no reaction

(c) $HOCH_2$, O, OH, OCH_3, OH $\xrightarrow{NaIO_4}$ $HOCH_2$, O, OCH_3, CHO CHO

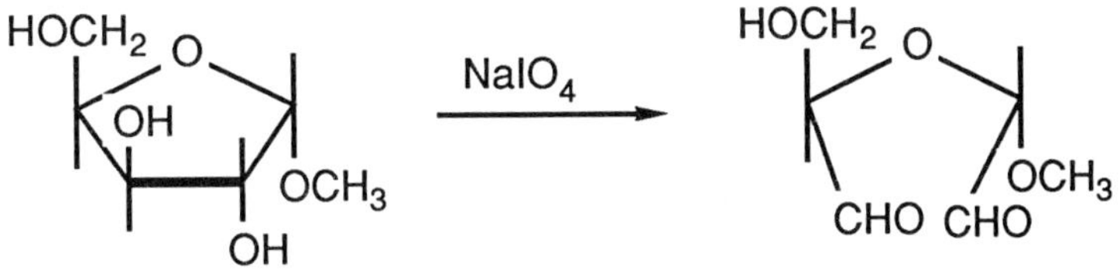

(d) $HOCH_2$, O, OH, HO, HO, OH ⇌ CHO; —OH; —OH; —OH; —OH; CH_2OH $\xrightarrow{NaIO_4}$ 5 HCO_2H + CH_2O

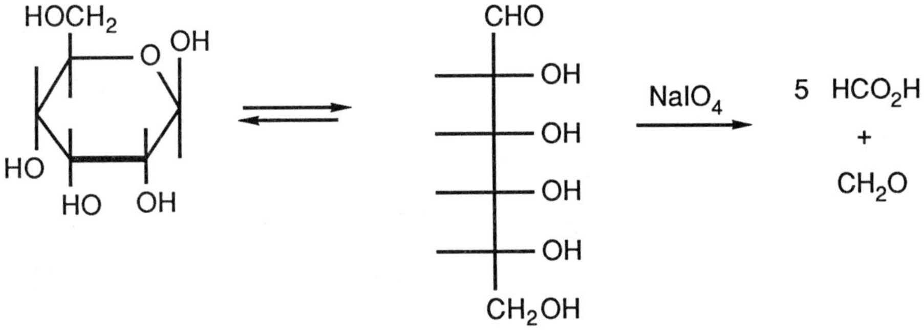

19.14

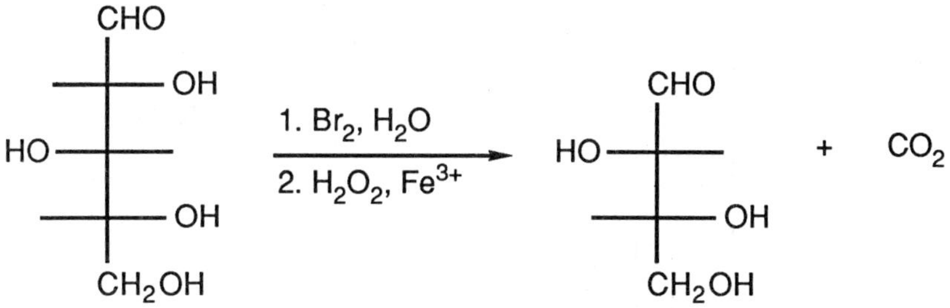

CHO; —OH; HO—; —OH; CH_2OH $\xrightarrow[\text{2. } H_2O_2,\ Fe^{3+}]{\text{1. } Br_2,\ H_2O}$ CHO; HO—; —OH; CH_2OH + CO_2

D-threose

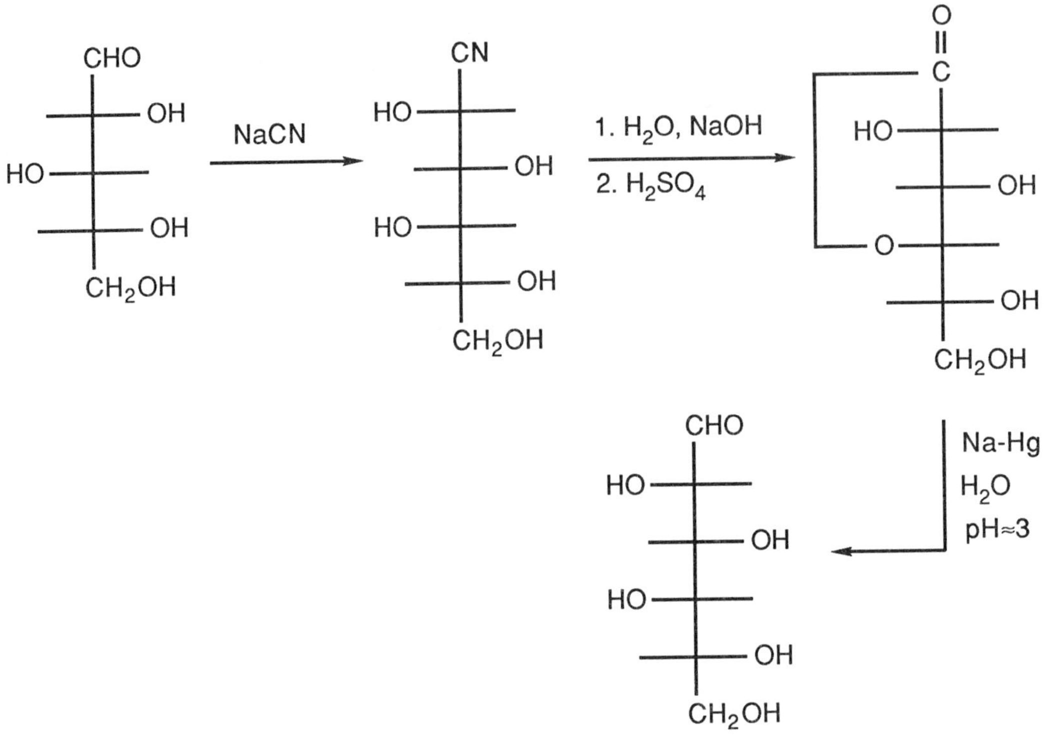

NaCN

1. H_2O, NaOH
2. H_2SO_4

Na-Hg
H_2O
pH≈3

D-idose

19.15

(a)

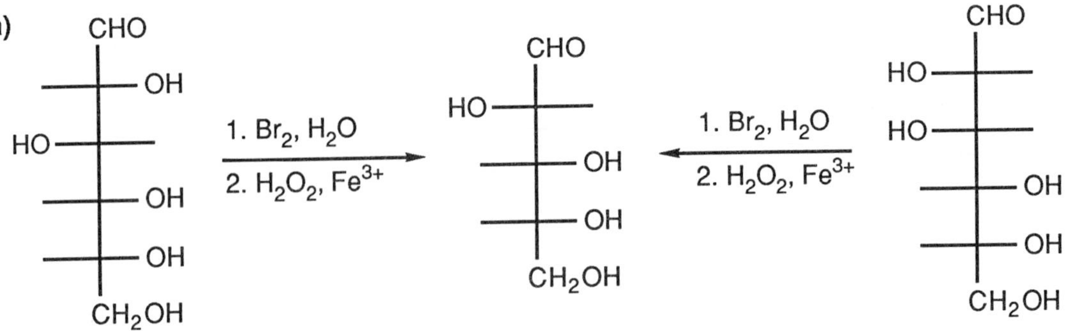

1. Br_2, H_2O
2. H_2O_2, Fe^{3+}

1. Br_2, H_2O
2. H_2O_2, Fe^{3+}

D-glucose D-arabinose D-mannose

(b)

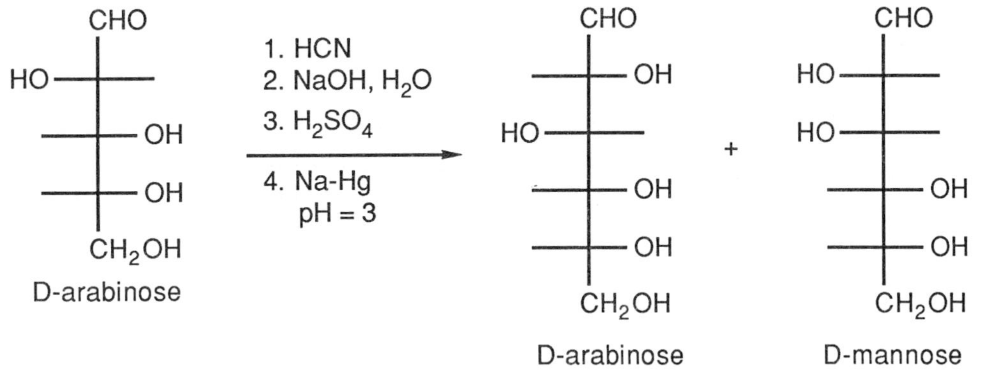

D-arabinose → (1. HCN; 2. NaOH, H_2O; 3. H_2SO_4; 4. Na-Hg, pH = 3) → D-arabinose + D-mannose

19.16

(a)

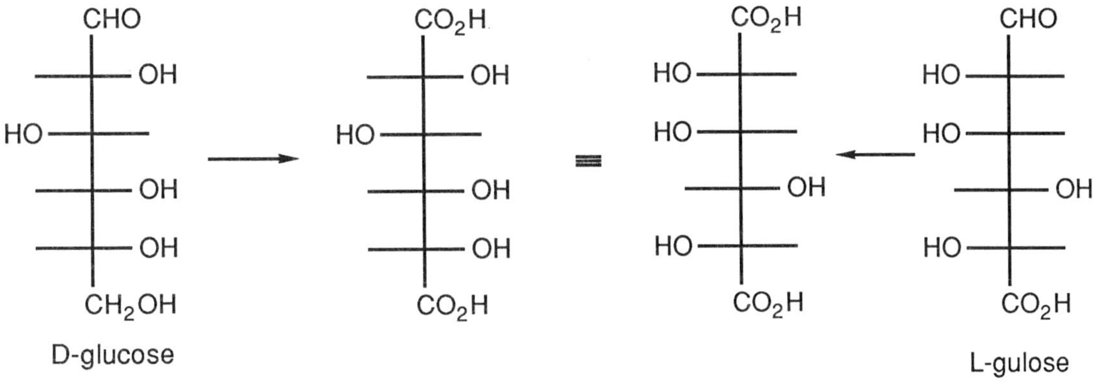

D-glucose → ≡ ← L-gulose

(b)

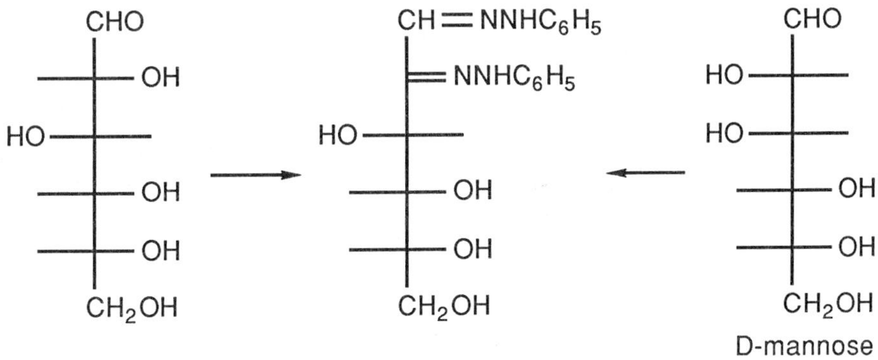

→ $CH{=}NNHC_6H_5$, $=NNHC_6H_5$ ←

D-mannose

19.17 (a)

$HOCH_2$ … ⇌ … $\xrightarrow{NaIO_4}$ 4 HCO_2H + CH_2O

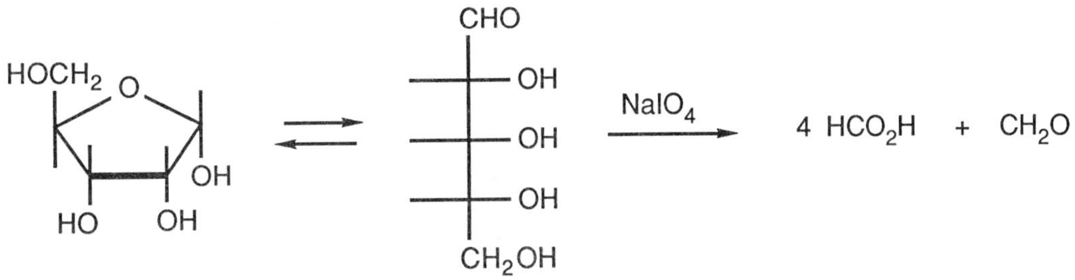

(b) [structure: methyl glycopyranoside] $\xrightarrow{NaIO_4}$ [structure: dialdehyde with $HOCH_2$, O, OCH_3, CHO, CHO] + HCO_2H

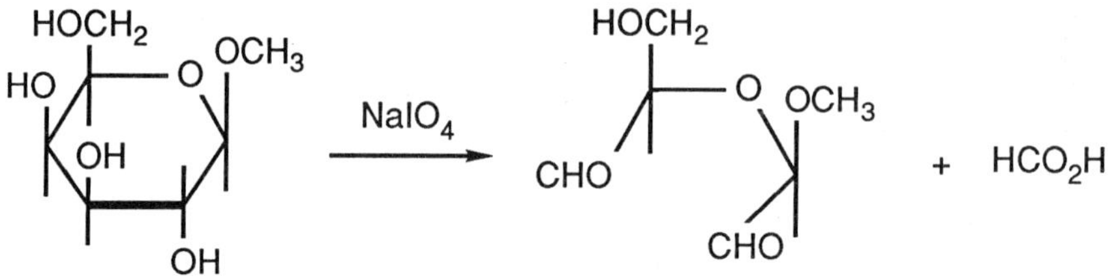

(c) [structure: methyl glycofuranoside with $HOCH_2$, O, OCH_3, HO, CH_2OH, HO] $\xrightarrow{NaIO_4}$ [structure: $HOCH_2$, O, OCH_3, CH_2OH, CHO CHO]

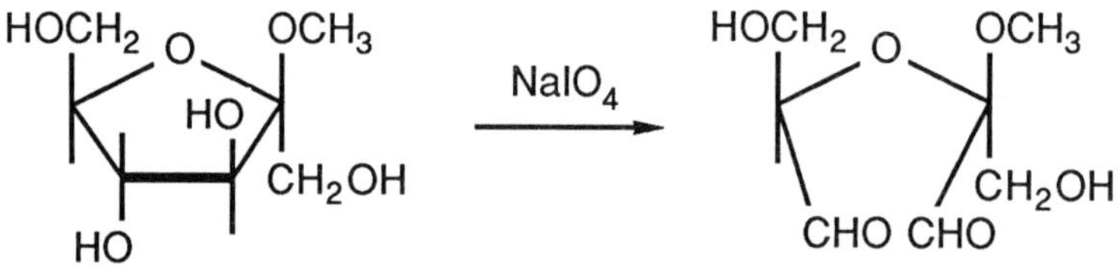

19.18

(a) 4 HCO_2H + CH_3OH

(b) HCO_2H + CH_2OH–CH(OH)–CH_2OH + CH_2OH–CH_2OH (c) 2 CH_2OH–CH(OH)–CH_2OH

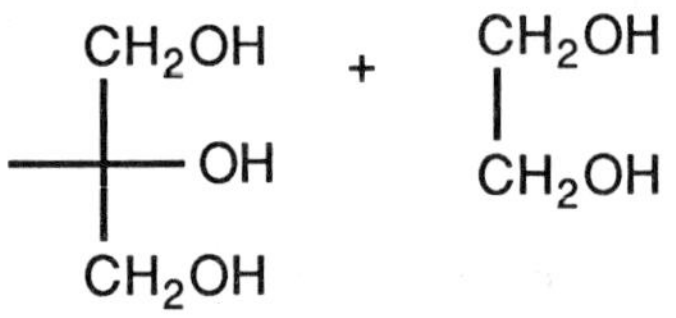

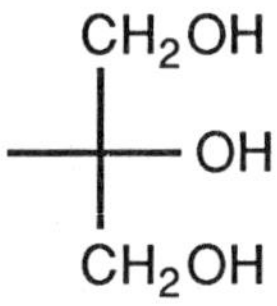

19.19 Both the 5- and 6-membered ring structures would consume two moles of periodate, but the former would yield formaldehyde rather than formic acid. Both the 6- and 7-membered ring structures would yield formic acid, but the latter would consume three moles of periodate:

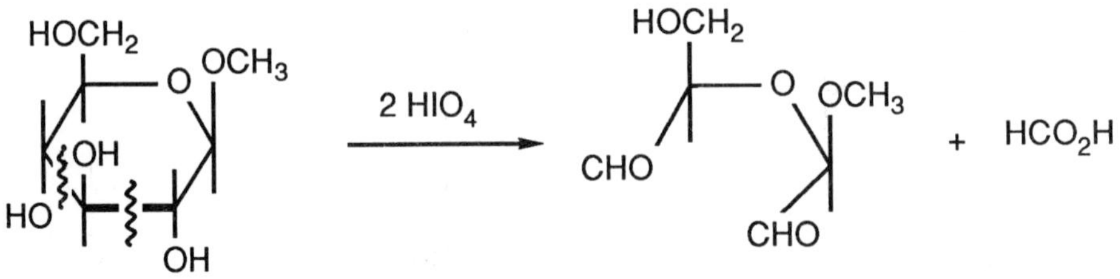

[structure: methyl glycopyranoside with $HOCH_2$, O, OCH_3, OH, HO, OH] $\xrightarrow{2\ HIO_4}$ [structure: $HOCH_2$, O, OCH_3, CHO, CHO] + HCO_2H

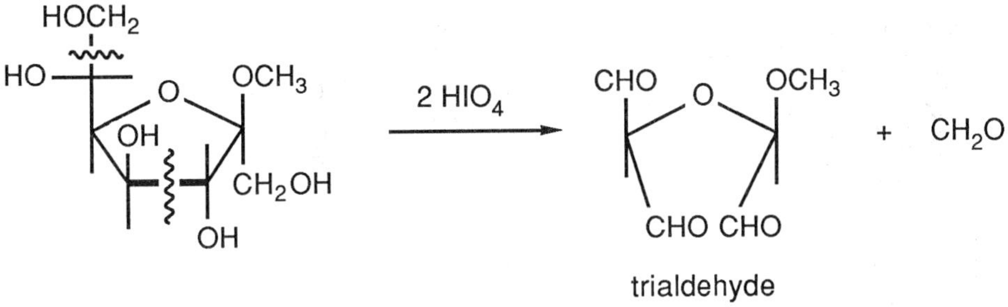

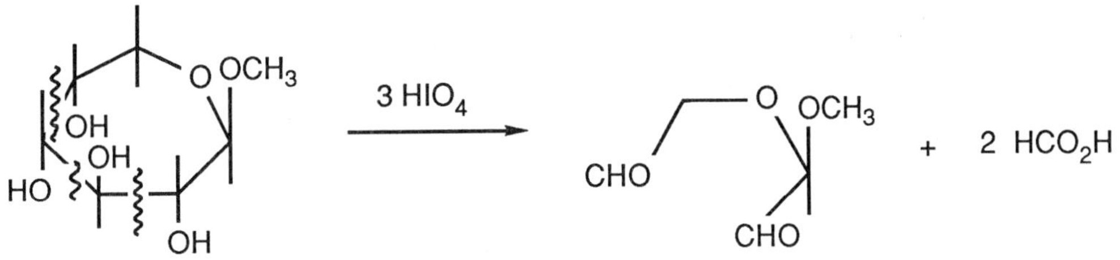

$HOCH_2$ HO O OCH_3 OH CH_2OH OH $\xrightarrow{2\ HIO_4}$ CHO O OCH_3 CHO CHO + CH_2O

trialdehyde

O OCH_3 OH OH HO OH $\xrightarrow{3\ HIO_4}$ O OCH_3 CHO CHO + 2 HCO_2H

ANSWERS TO PROBLEMS

19.1 **(a)** CHO, HO, HO, OH, OH, CH_2OH ⟶ CO_2H, HO, HO, OH, OH, CO_2H

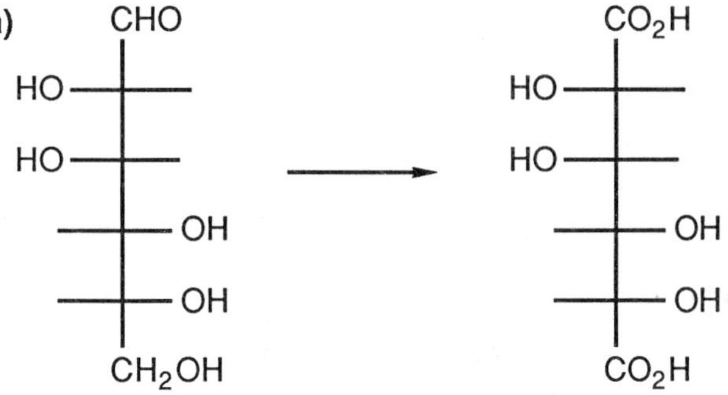

(b) $HOCH_2$ O OH, HO OH ⇌ CHO, OH, OH, OH, CH_2OH ⟶ CO_2H, OH, OH, OH, CO_2H

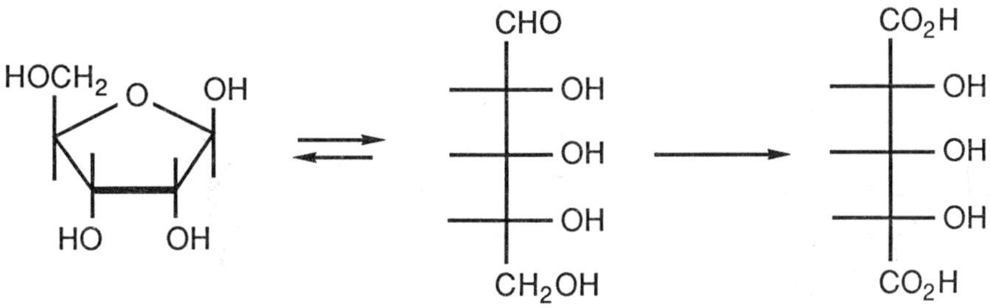

(c)

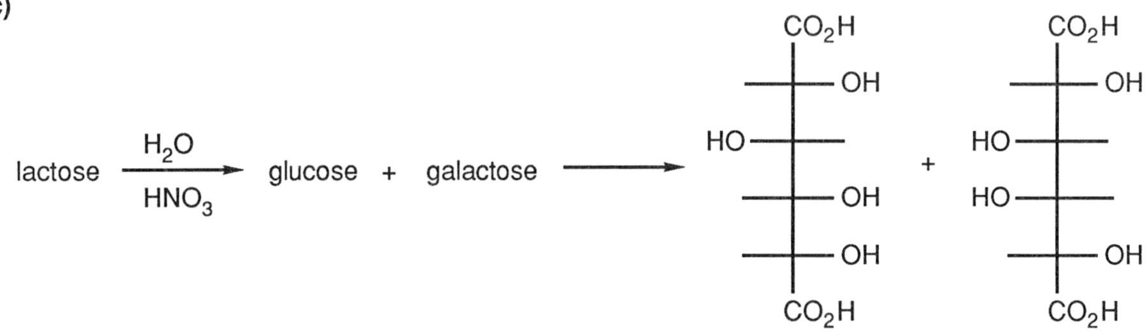

lactose $\xrightarrow[HNO_3]{H_2O}$ glucose + galactose ⟶

CO_2H
—OH
HO—
—OH
—OH
CO_2H

+

CO_2H
—OH
HO—
HO—
—OH
CO_2H

(d)

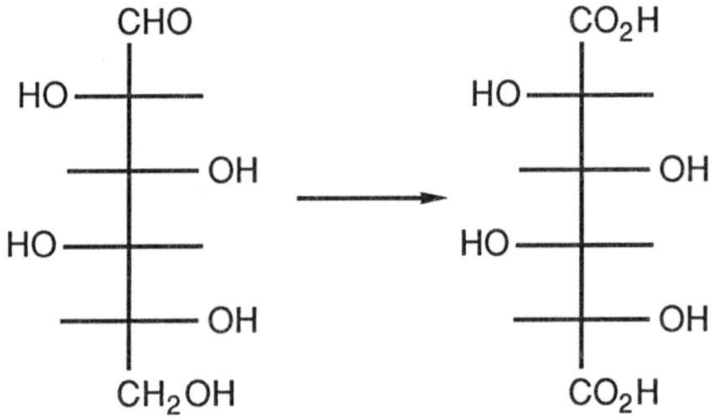

CHO
HO—
—OH
HO—
—OH
CH_2OH

⟶

CO_2H
HO—
—OH
HO—
—OH
CO_2H

(e)

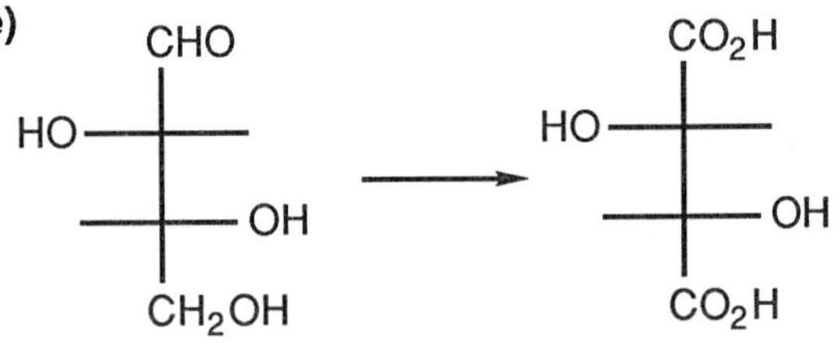

CHO
HO—
—OH
CH_2OH

⟶

CO_2H
HO—
—OH
CO_2H

(f)

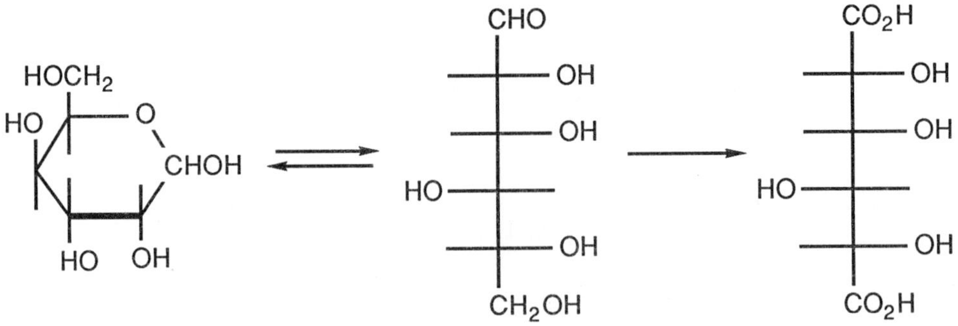

$HOCH_2$
HO
O
CHOH
HO OH

⇌

CHO
—OH
—OH
HO—
—OH
CH_2OH

⟶

CO_2H
—OH
—OH
HO—
—OH
CO_2H

19.2 The products of these reactions are osazones:

(a)
CH=NNHC$_6$H$_5$
=NNHC$_6$H$_5$
HO—
—OH
—OH
CH$_2$OH

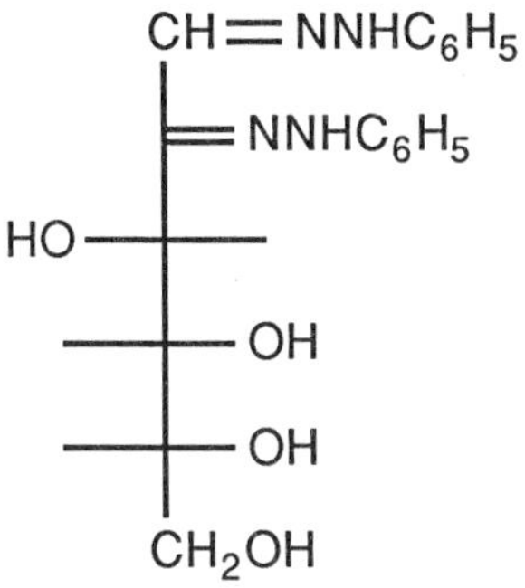

(b)
CH=NNHC$_6$H$_5$
=NNHC$_6$H$_5$
—OH
—OH
CH$_2$OH

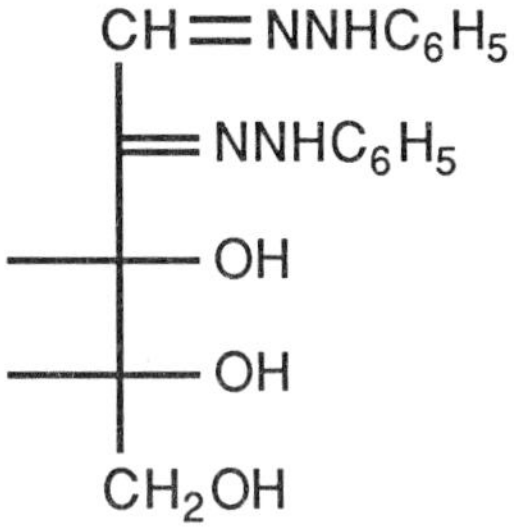

(c)
CH=NNHC$_6$H$_5$
=NNHC$_6$H$_5$
HO—
—O-galactopyranosyl
—OH
CH$_2$OH

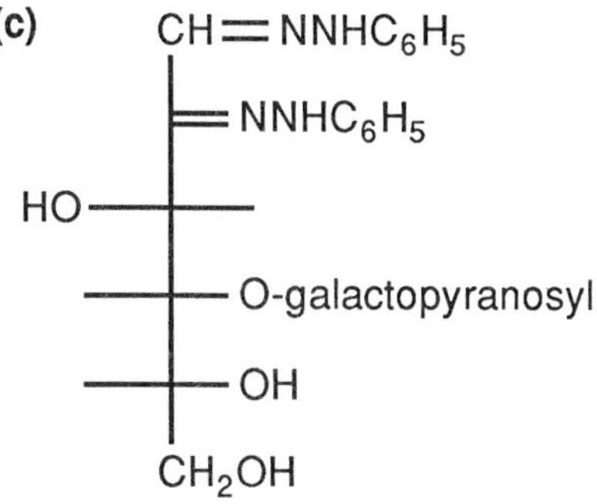

(d)
CH=NNHC$_6$H$_5$
=NNHC$_6$H$_5$
—OH
HO—
—OH
CH$_2$OH

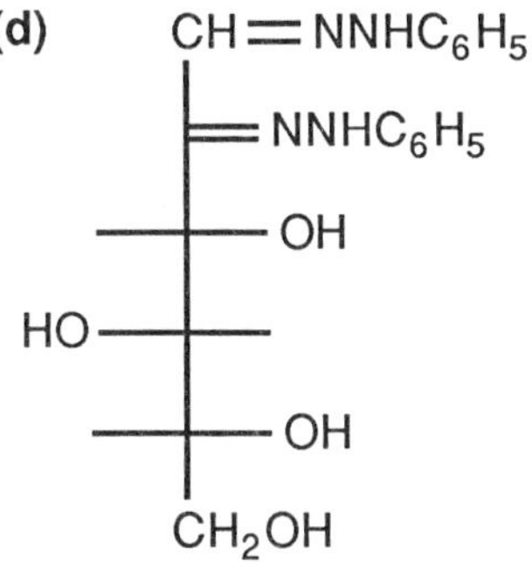

(e)
CH=NNHC$_6$H$_5$
=NNHC$_6$H$_5$
—OH
CH$_2$OH

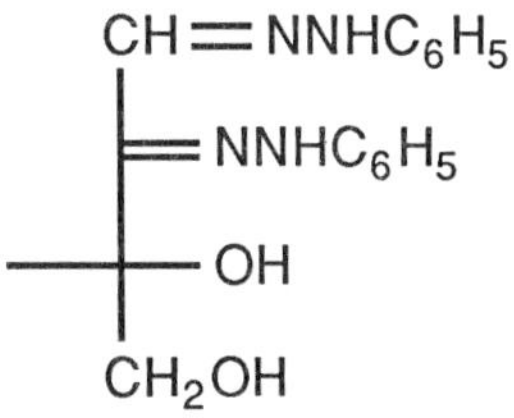

(f)
CH=NNHC$_6$H$_5$
=NNHC$_6$H$_5$
—OH
HO—
—OH
CH$_2$OH

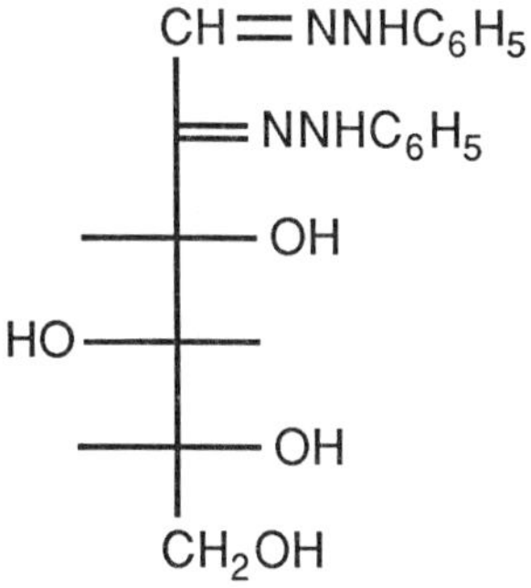

19.3 The products of these reactions are aldonic acids:

(a)
CO$_2$H
HO—
HO—
—OH
—OH
CH$_2$OH

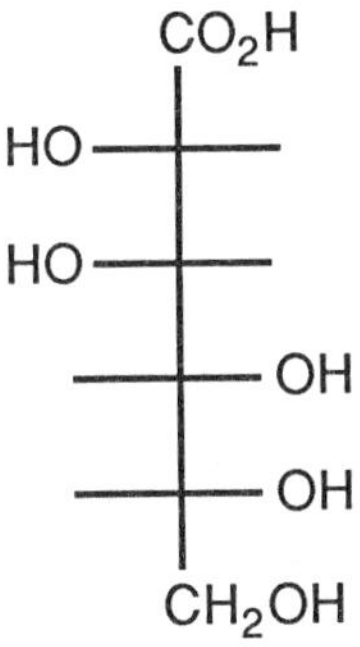

(b)
CO$_2$H
—OH
—OH
—OH
CH$_2$OH

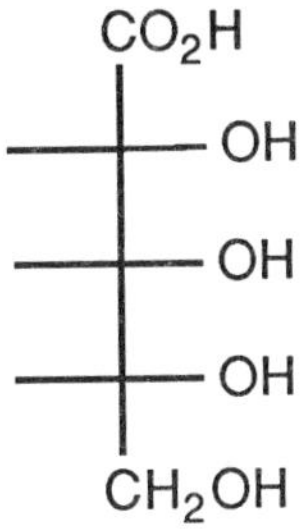

(c)
CO$_2$H
—OH
HO—
—OH
—OH
CH$_2$OH

+

CO$_2$H
—OH
HO—
HO—
—OH
CH$_2$OH

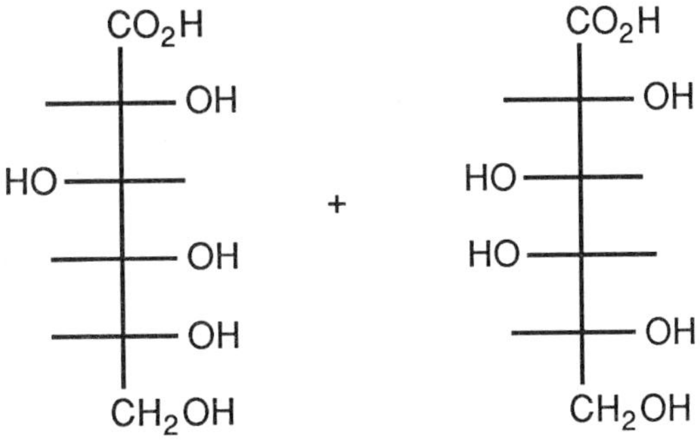

(d)
CO$_2$H
HO—
—OH
HO—
—OH
CH$_2$OH

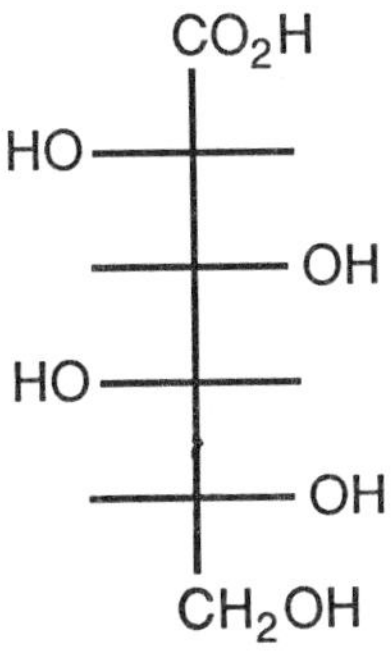

(e)
CO$_2$H
HO—
—OH
CH$_2$OH

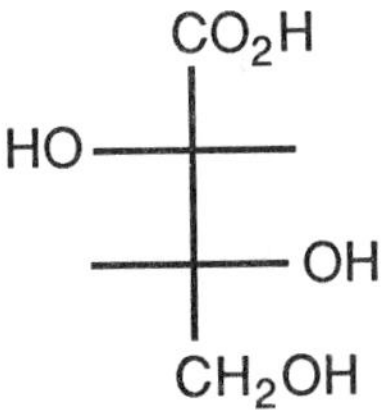

(f)
CO$_2$H
—OH
—OH
HO—
—OH
CH$_2$OH

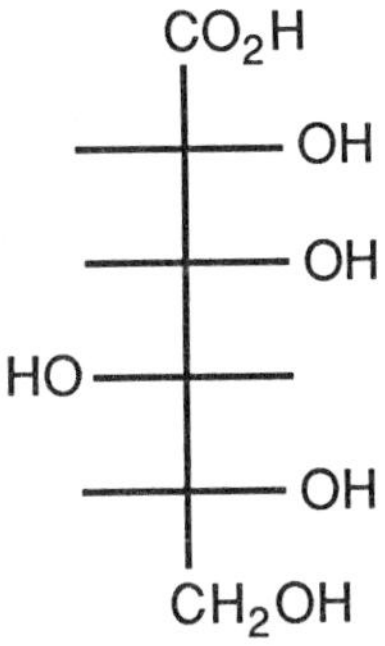

19.4 The products of these reactions are the aldoses with one less carbon:

(a) CO_2 +

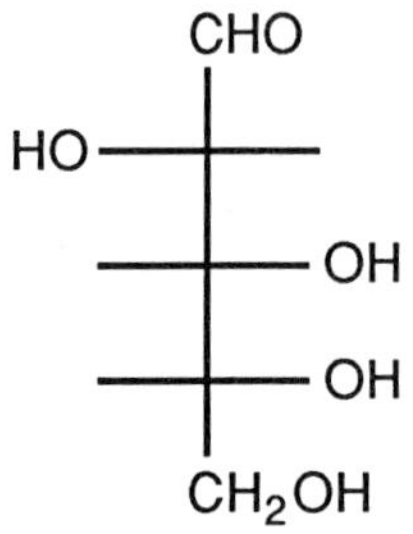

CHO
HO—
—OH
—OH
CH_2OH

(b) CO_2 +

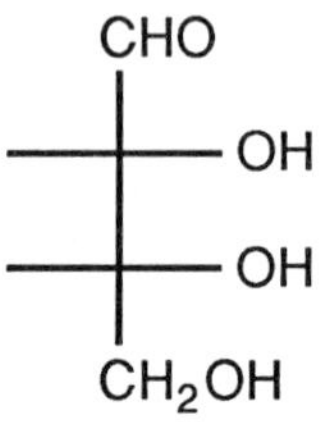

CHO
—OH
—OH
CH_2OH

(c) CO_2 +

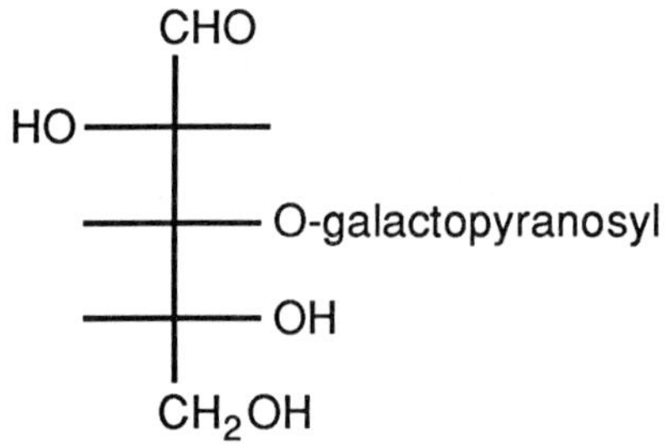

CHO
HO—
—O-galactopyranosyl
—OH
CH_2OH

(d) CO_2 +

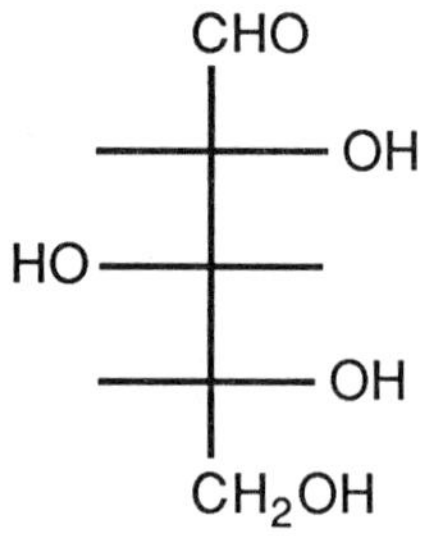

CHO
—OH
HO—
—OH
CH_2OH

(e) CO_2 +

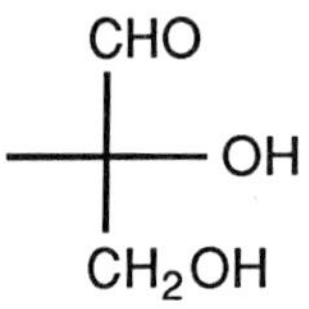

CHO
—OH
CH_2OH

(f) CO_2 +

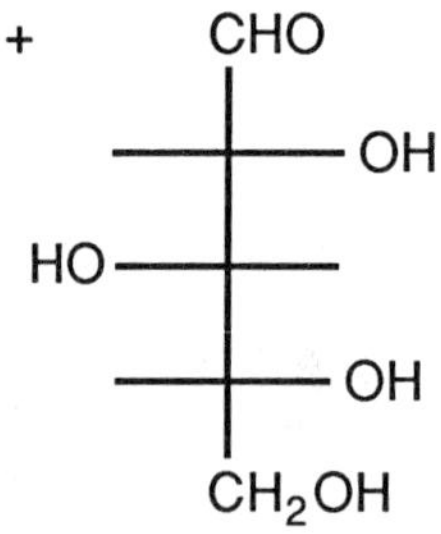

CHO
—OH
HO—
—OH
CH_2OH

19.5

(a)

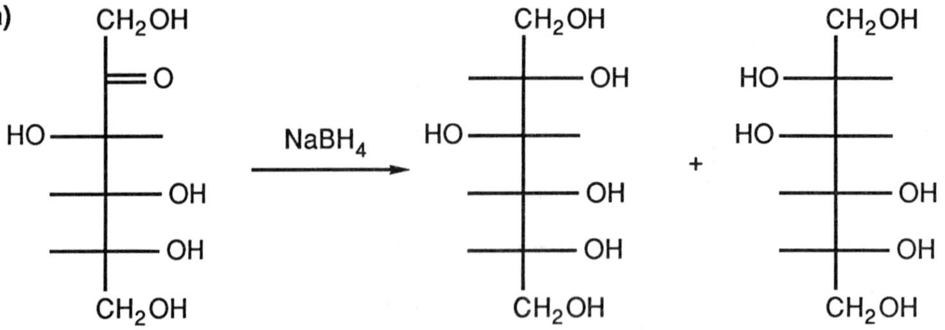

CH_2OH
=O
HO—
—OH
—OH
CH_2OH

$\xrightarrow{NaBH_4}$

CH_2OH
—OH
HO—
—OH
—OH
CH_2OH

+

CH_2OH
HO—
HO—
—OH
—OH
CH_2OH

(b)

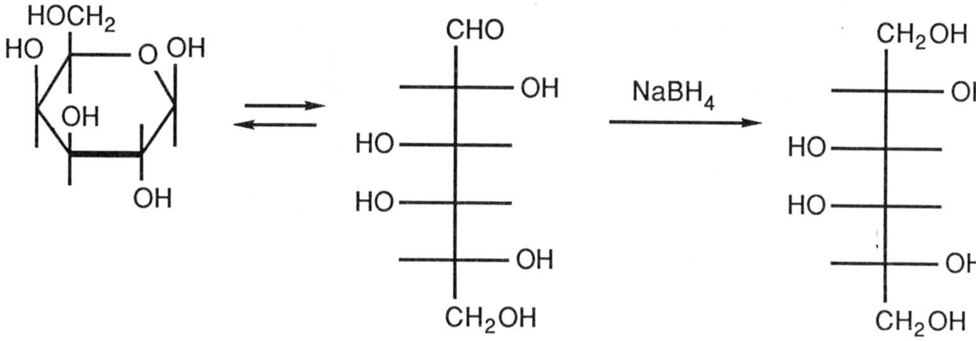

$HOCH_2$
HO O OH
OH
OH

⇌

CHO
OH
HO
HO
OH
CH_2OH

$\xrightarrow{NaBH_4}$

CH_2OH
OH
HO
HO
OH
CH_2OH

(c)

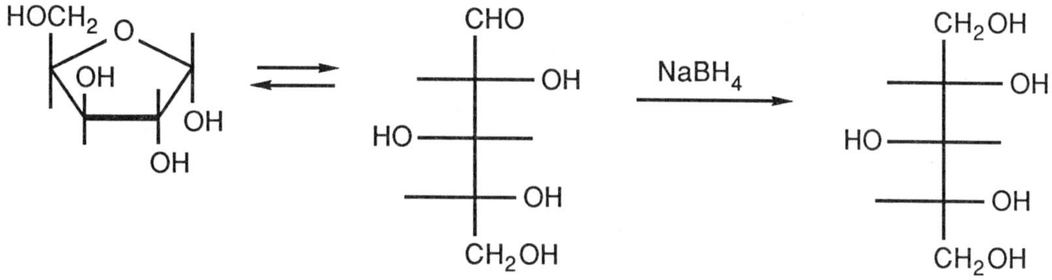

$HOCH_2$ O
OH
OH
OH

⇌

CHO
OH
HO
OH
CH_2OH

$\xrightarrow{NaBH_4}$

CH_2OH
OH
HO
OH
CH_2OH

(d)

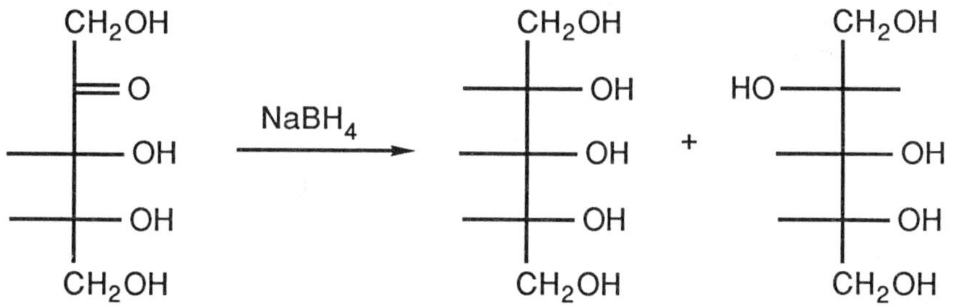

CH_2OH
=O
OH
OH
CH_2OH

$\xrightarrow{NaBH_4}$

CH_2OH
OH
OH
OH
CH_2OH

+

CH_2OH
HO
OH
OH
CH_2OH

(e)

CHO
OH
HO
HO
CH_2OH

$\xrightarrow{NaBH_4}$

CH_2OH
OH
HO
HO
CH_2OH

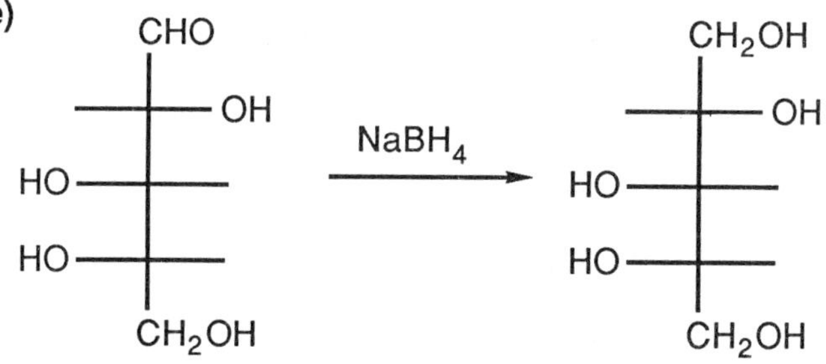

19.6 (a)

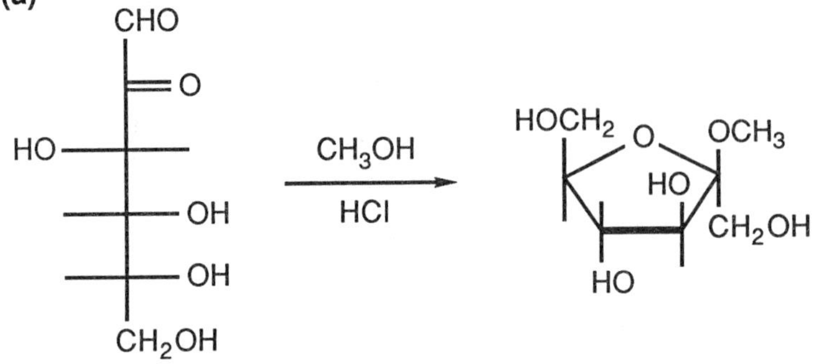

CHO, =O, HO—, —OH, —OH, CH_2OH $\xrightarrow[HCl]{CH_3OH}$ $HOCH_2$, O, OCH_3, HO, CH_2OH, HO

\+ anomer
(+ pyranoside forms)

(b)

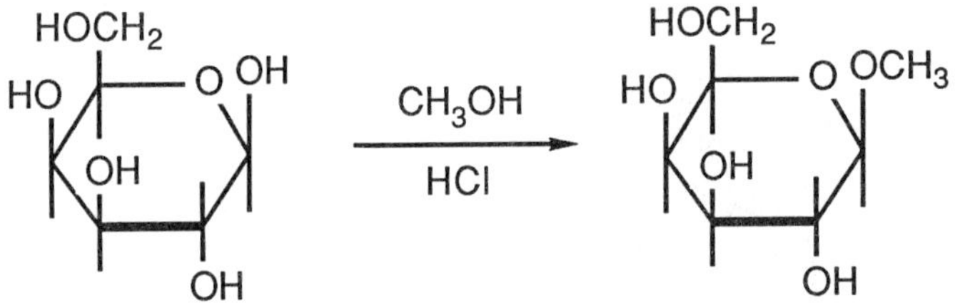

$HOCH_2$, HO, O, OH, OH, OH $\xrightarrow[HCl]{CH_3OH}$ $HOCH_2$, HO, O, OCH_3, OH, OH

\+ anomer
(+ pyranoside forms)

(c)

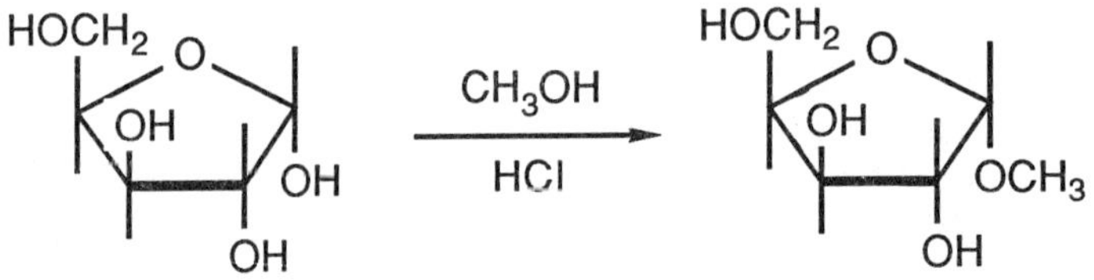

$HOCH_2$, O, OH, OH, OH $\xrightarrow[HCl]{CH_3OH}$ $HOCH_2$, O, OH, OCH_3, OH

\+ anomer
(+ pyranoside forms)

(d)

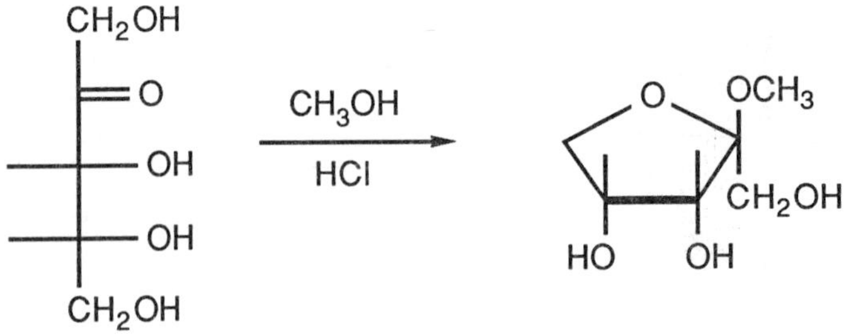

CH_2OH, =O, —OH, —OH, CH_2OH $\xrightarrow[HCl]{CH_3OH}$ O, OCH_3, CH_2OH, HO, OH

\+ anomer

(e)

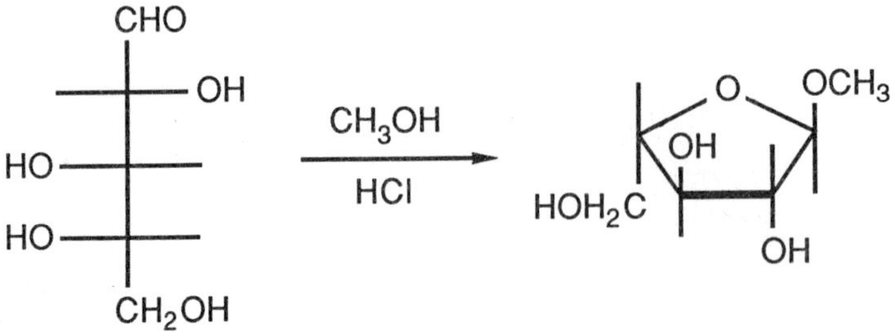

CHO, —OH, HO—, HO—, CH_2OH $\xrightarrow[HCl]{CH_3OH}$ O, OCH_3, OH, HOH_2C, OH

\+ anomer
(+ pyranoside forms)

19.7 The compounds that would give a positive Tollens test are those with a free aldehyde (or hemiacetal) group or an α-hydroxy keto group:

(a) maltose: positive

(b) *D*-lyxose: positive

(c) α-*D*-fructofuranose: positive

(d) methyl β-*D*-ribofuranose: negative

(e) sucrose: negative

(b) *L*-sorbose: positive

19.8

(a) HOCH2, O, OH, HO, OH, O, HOCH2, O, OH, OH, OH ⇌ HOCH2, O, OH, HO, OH, O, HOCH2, OH, OH, CHO, OH

$\xrightarrow{NaIO_4}$ $3\ HCO_2H + CH_2O +$ HOCH2, O, CHO, H, CHO, O, CHO, CHO

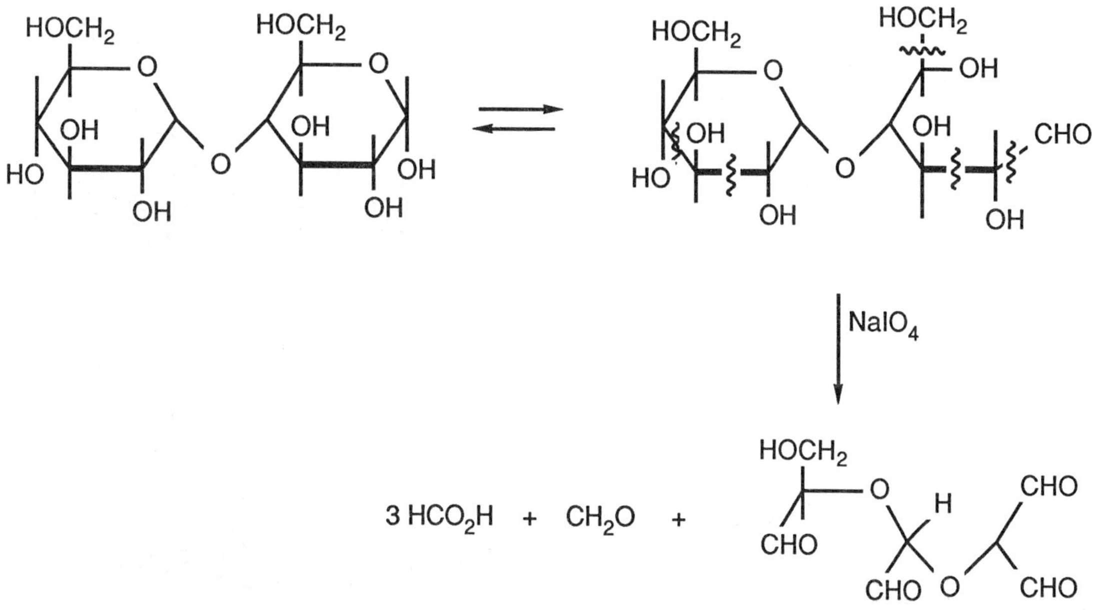

(b) CHO, HO, HO, OH, CH_2OH $\xrightarrow{NaIO_4}$ $4\ HCO_2H + CH_2O$

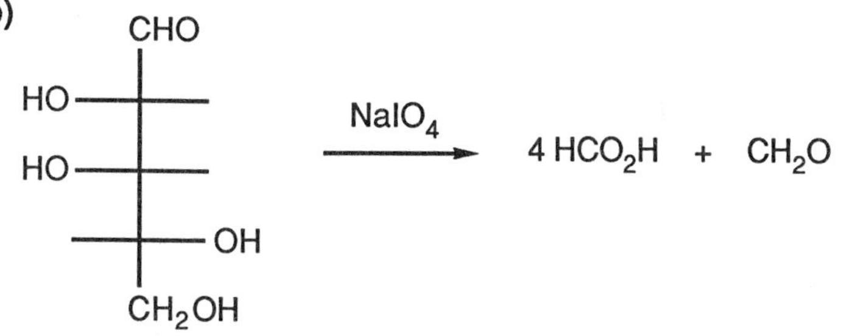

(c) HOCH₂ O CH₂OH / HO / OH / HO ⇌ CHO / =O / HO— / —OH / —OH / CH_2OH $\xrightarrow{NaIO_4}$ $4\ HCO_2H + CH_2O + CO_2$

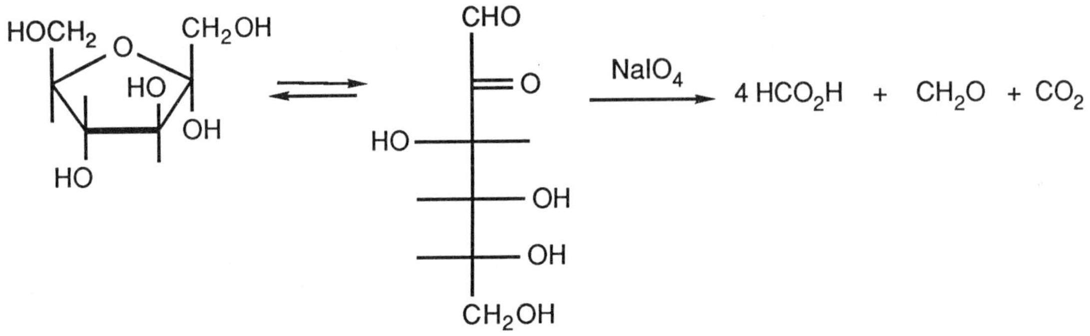

(d) $HOCH_2$ O OCH_3 / HO OH $\xrightarrow{NaIO_4}$ $HOCH_2$ O OCH_3 / CHO CHO

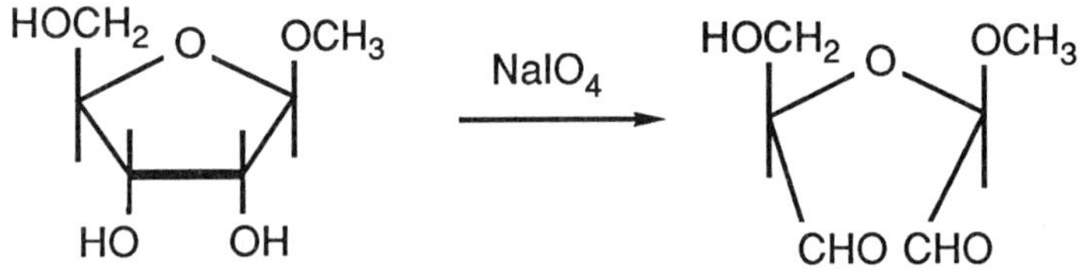

(e) $HOCH_2$ / O / OH / HO / HO / O / $HOCH_2$ O / HO / CH_2OH / HO $\xrightarrow{NaIO_4}$ $HOCH_2$ / O / CHO / CHO / O / $HOCH_2$ O / CH_2OH / CHO CHO + HCOOH

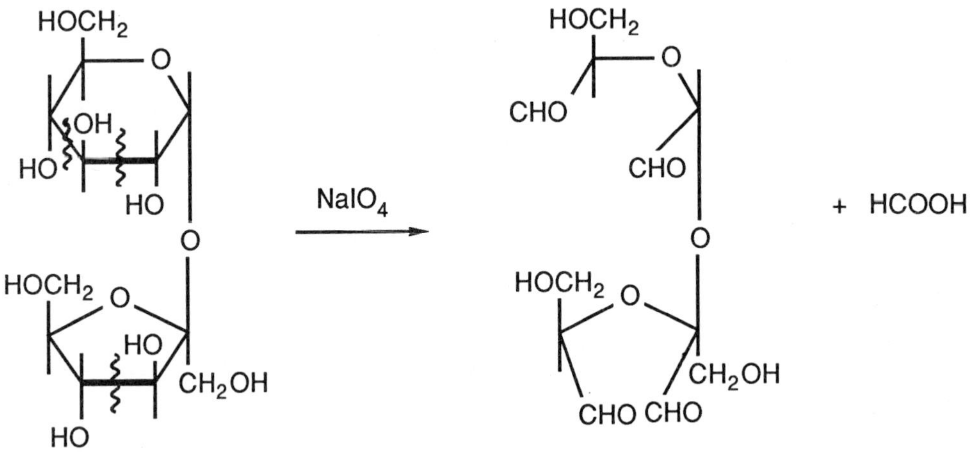

(f) CH_2OH / =O / HO— / —OH / HO— / CH_2OH $\xrightarrow{NaIO_4}$ $3\ HCO_2H + 2\ CH_2O + CO_2$

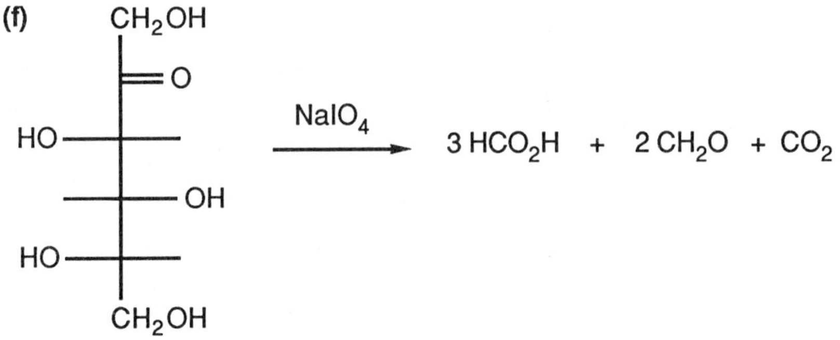

19.9 α-*D*-Glucopyranose can undergo mutarotation to form an equilibrium mixture of α-*D*-glucopyranose, [α] = +111° and β-*D*-glucopyranose, [α] = +19.2°. The β form predominates by a ratio of about 2:1.

19.10

CO_2H / OH / OH / HO / OH / CH_2OH $\xrightarrow{HCl}$ $HOCH_2$ / HO / O / O / HO / OH

D-gulonic acid
negative rotation

D-gulonolactone
positive rotation

19.11

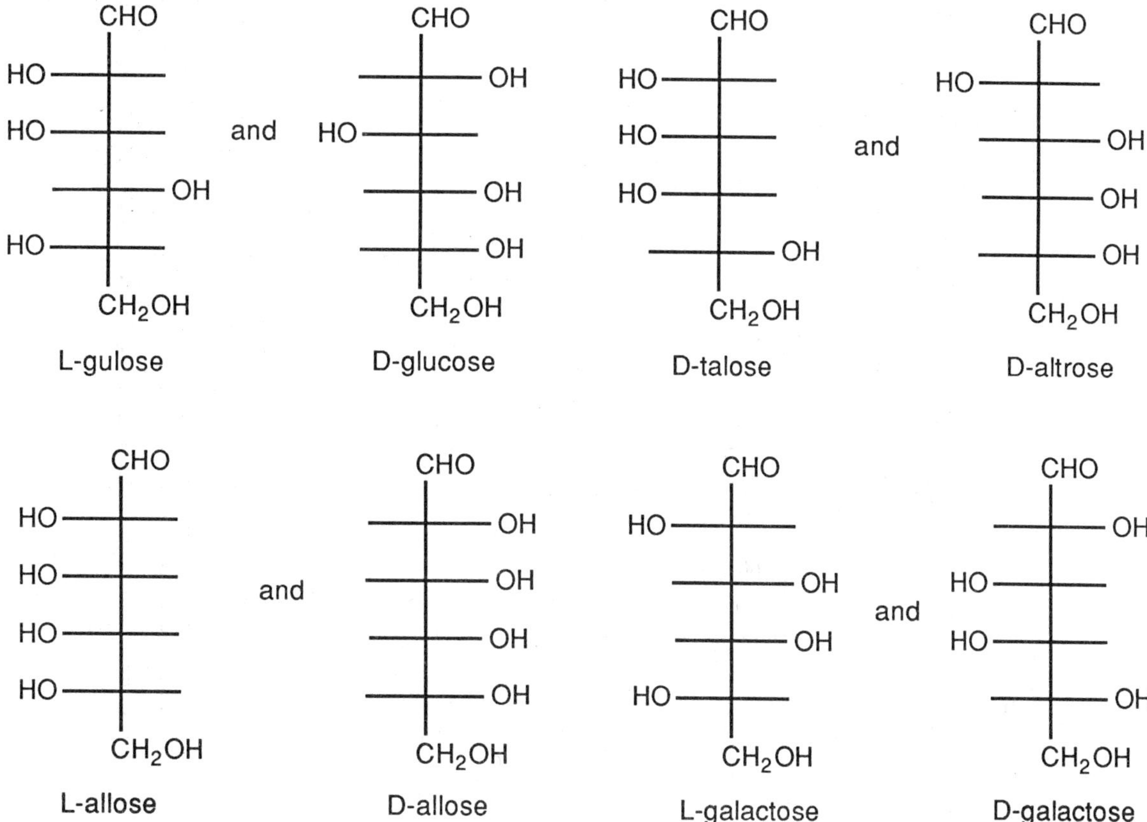

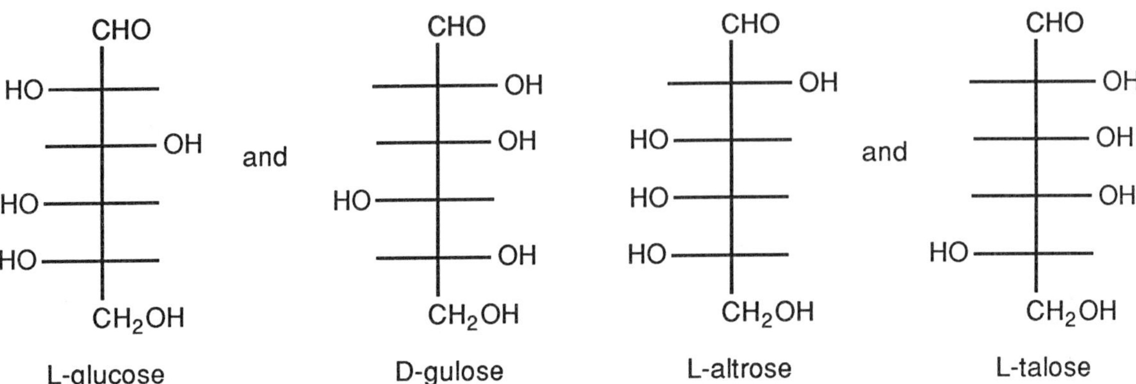

L-glucose D-gulose L-altrose L-talose

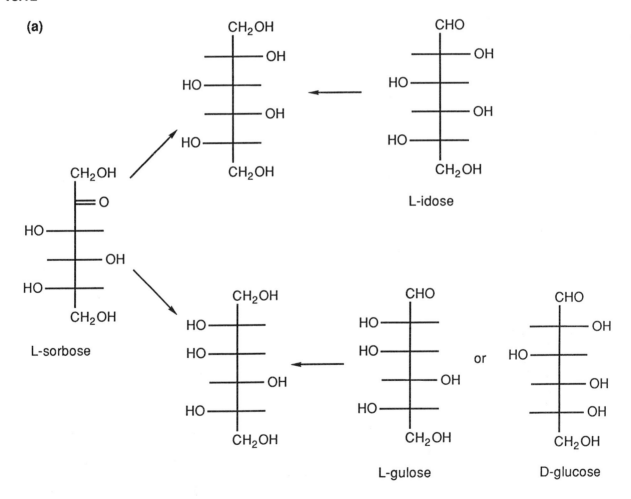

19.12

(a)

CH_2OH OH HO OH HO CH_2OH

CHO OH HO OH HO CH_2OH

L-idose

CH_2OH =O HO OH HO CH_2OH

L-sorbose

CH_2OH HO HO OH HO CH_2OH

CHO HO HO OH HO CH_2OH

or

CHO OH HO OH OH CH_2OH

L-gulose D-glucose

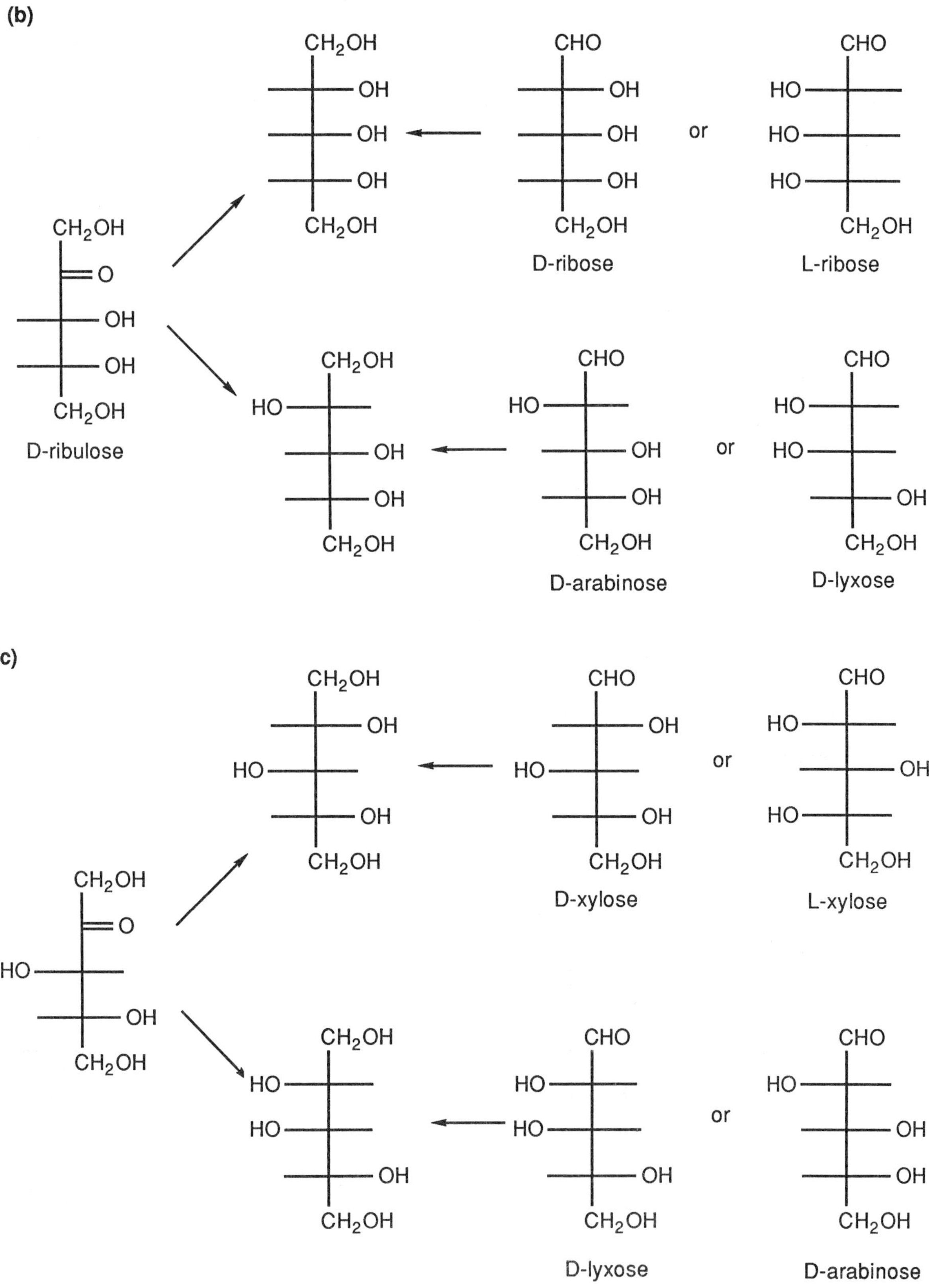
(b)
CH2OH
O
OH
OH
CH2OH
D-ribulose
D-ribose
or
L-ribose
D-arabinose
D-lyxose
(c)
D-xylose
L-xylose
D-lyxose
D-arabinose
CHO
HO

(d)

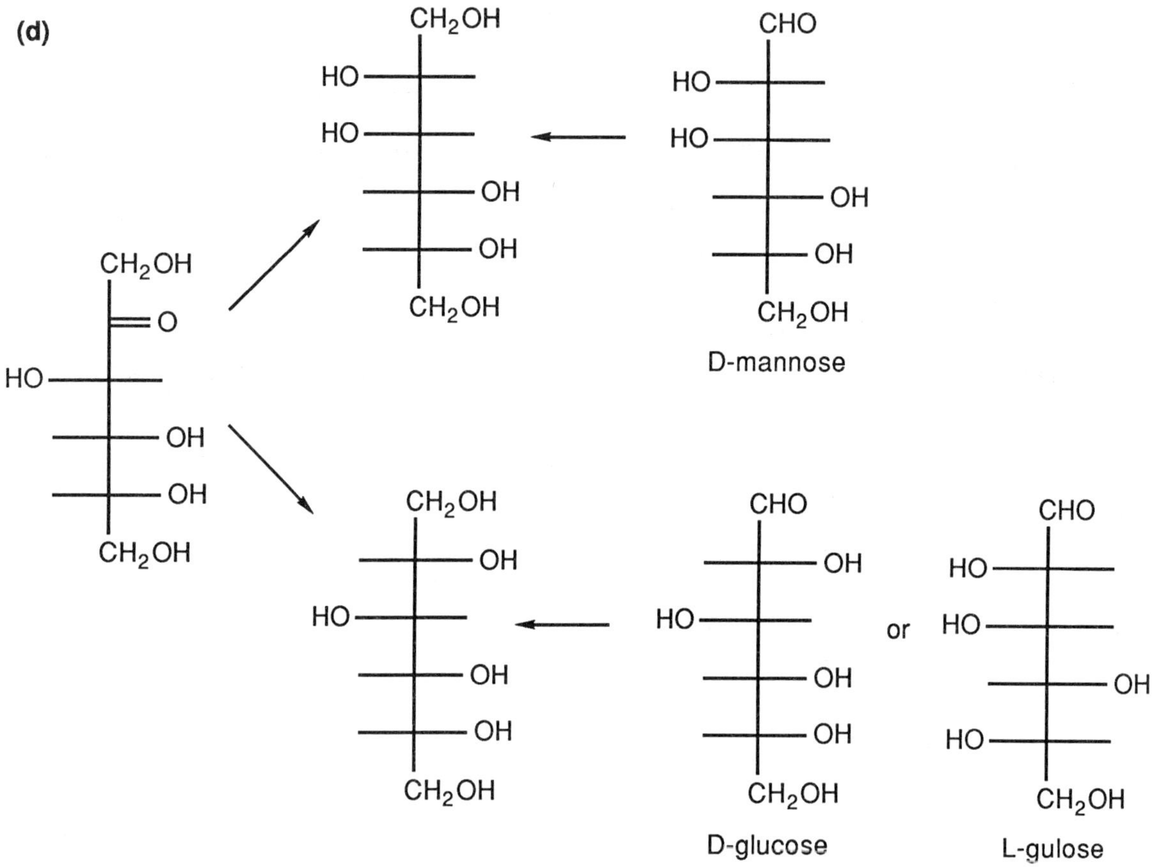

CH_2OH / =O / HO— / —OH / —OH / CH_2OH → CH_2OH / HO— / HO— / —OH / —OH / CH_2OH ← CHO / HO— / HO— / —OH / —OH / CH_2OH

D-mannose

CH_2OH / =O / HO— / —OH / —OH / CH_2OH → CH_2OH / —OH / HO— / —OH / —OH / CH_2OH ← CHO / —OH / HO— / —OH / —OH / CH_2OH or CHO / HO— / HO— / —OH / HO— / CH_2OH

D-glucose L-gulose

19.13

CHO / —OH / —OH / —OH / —OH / CH_2OH → CHO / —OH / —OH / —OH / —OH / —OH / CH_2OH

D-allose

CHO / —OH / HO— / —OH / —OH / CH_2OH → CHO / —OH / —OH / HO— / —OH / —OH / CH_2OH

D-glucose

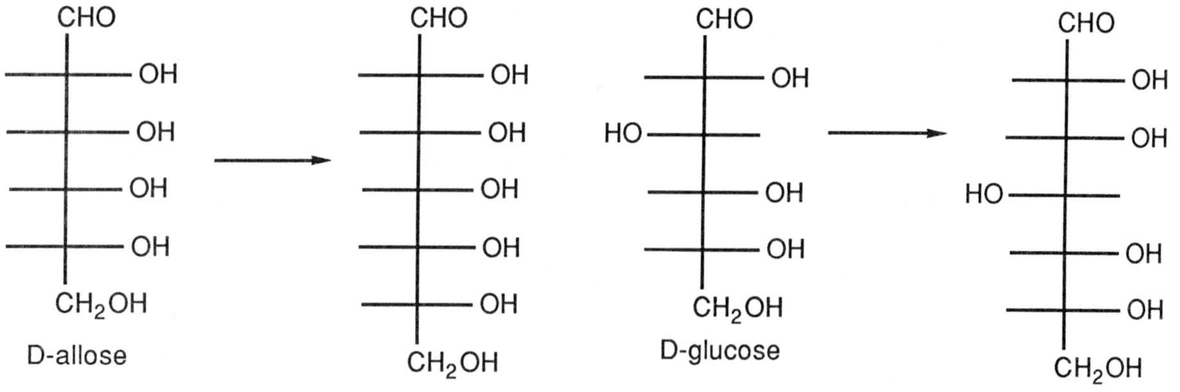

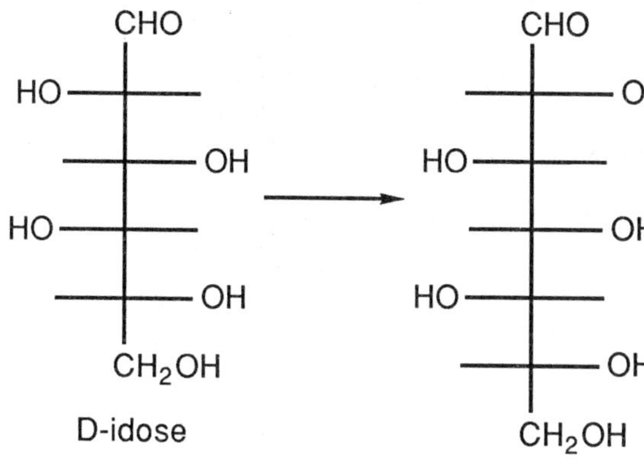
D-idose

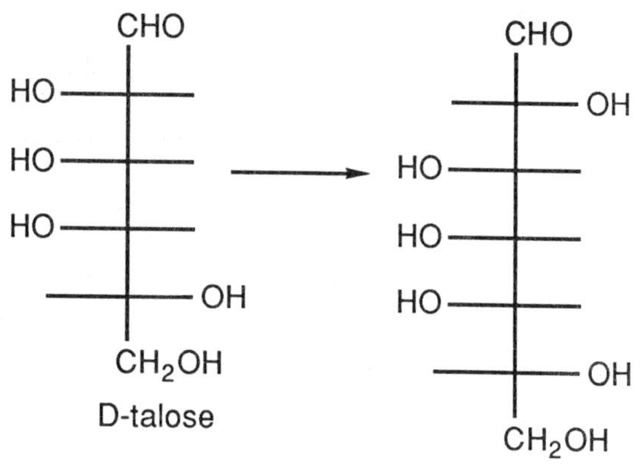
D-talose

19.14

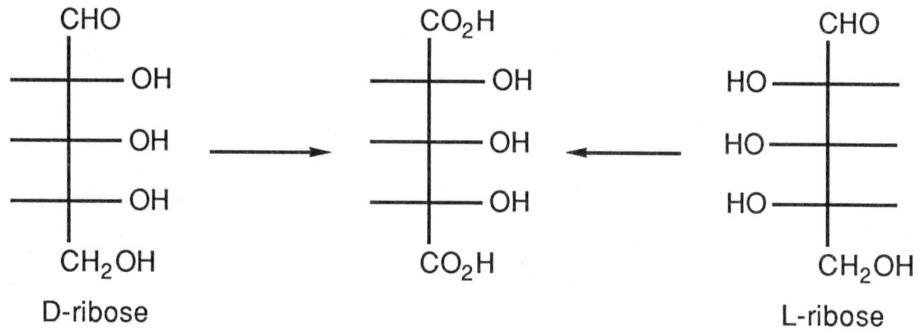
D-ribose
L-ribose

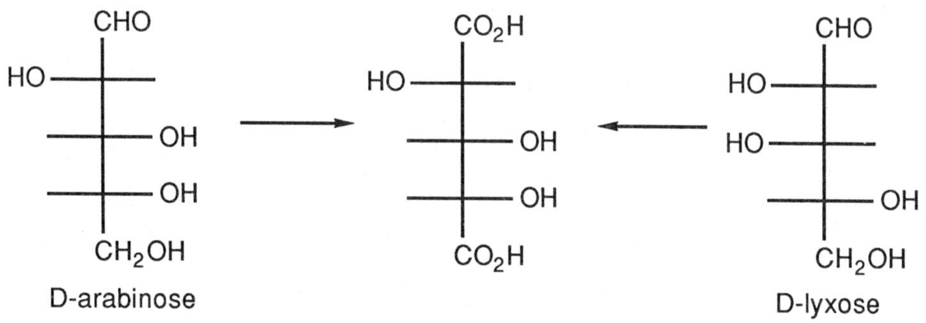
D-arabinose
D-lyxose

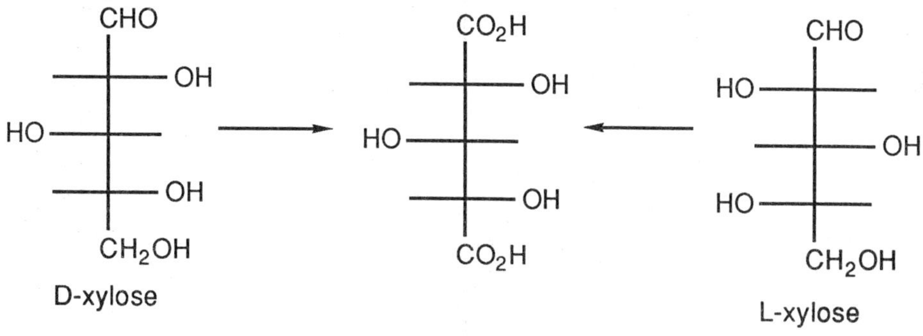
D-xylose
L-xylose

L-arabinose (CHO; OH right; HO left; HO left; CH_2OH) → (CO_2H; OH right; HO left; HO left; CO_2H) ← L-lyxose (CO_2H; OH right; OH right; HO left; CO_2H)

19.15

D-allose (CHO; OH; OH; OH; OH; CH_2OH) → (CO_2H; OH; OH; OH; OH; CO_2H) ← L-allose (CHO; HO; HO; HO; HO; CH_2OH)

D-altrose (CHO; HO left; OH; OH; OH; CH_2OH) → (CO_2H; HO left; OH; OH; OH; CO_2H) ← D-talose (CHO; HO; HO; HO; OH right; CH_2OH)

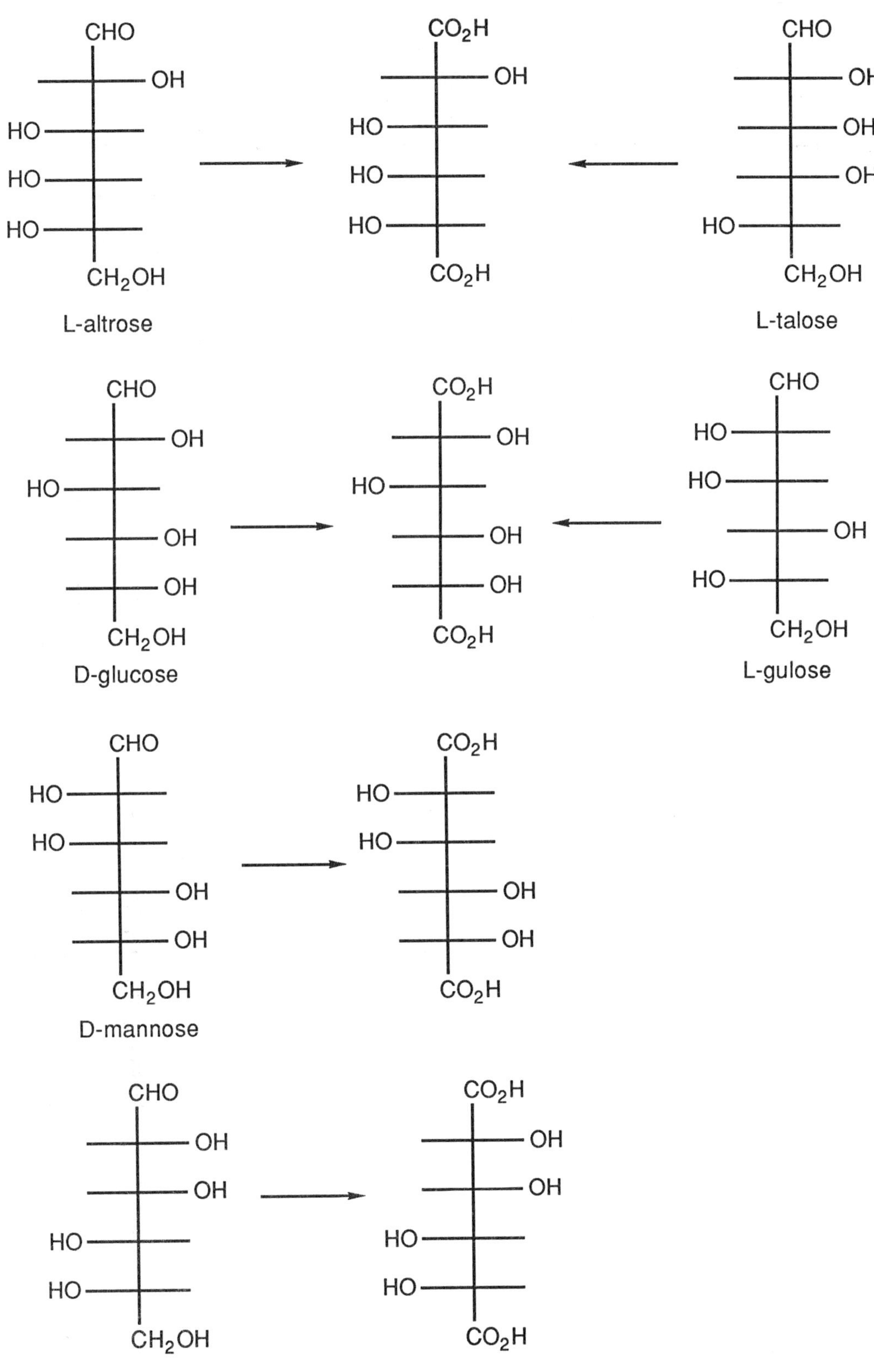
CHO
OH
HO
HO
HO
CH2OH
L-altrose
CO2H
OH
HO
HO
HO
CO2H
CHO
OH
OH
OH
HO
CH2OH
L-talose
CHO
OH
HO
OH
OH
CH2OH
D-glucose
CO2H
OH
HO
OH
OH
CO2H
CHO
HO
HO
OH
HO
CH2OH
L-gulose
CHO
HO
HO
OH
OH
CH2OH
D-mannose
CO2H
HO
HO
OH
OH
CO2H
CHO
OH
OH
HO
HO
CH2OH
L-mannose
CO2H
OH
OH
HO
HO
CO2H

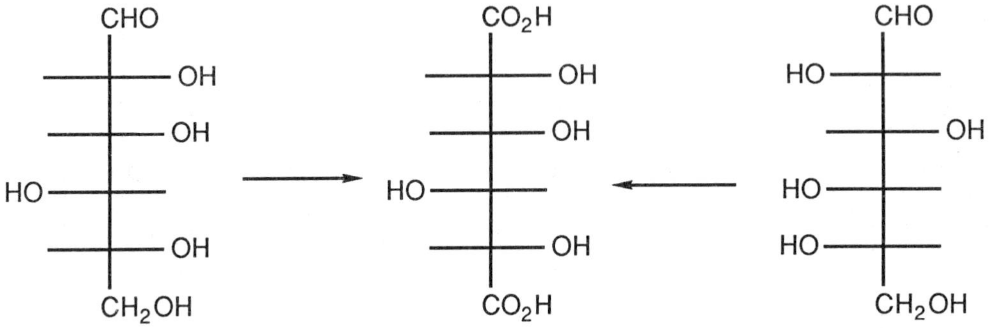

CHO, OH, OH, HO, OH, CH_2OH

D-gulose

CO_2H, OH, OH, HO, OH, CO_2H

CHO, HO, OH, HO, HO, CH_2OH

L-glucose

CHO, HO, OH, HO, OH, CH_2OH

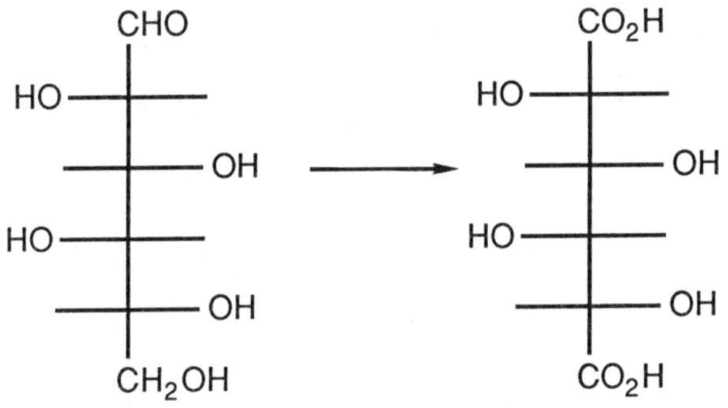

D-idose

CO_2H, HO, OH, HO, OH, CO_2H

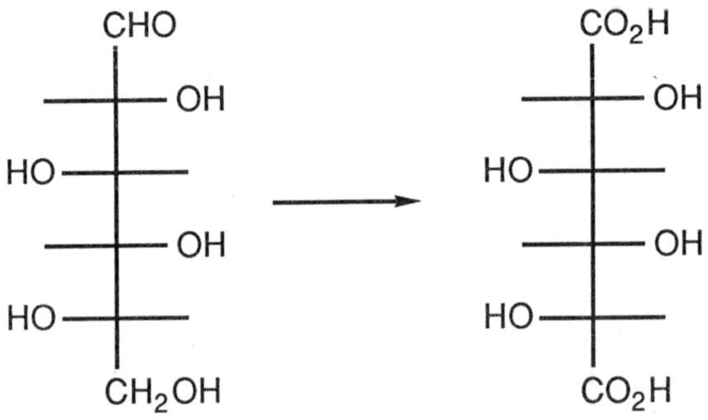

CHO, OH, HO, OH, HO, CH_2OH

L-idose

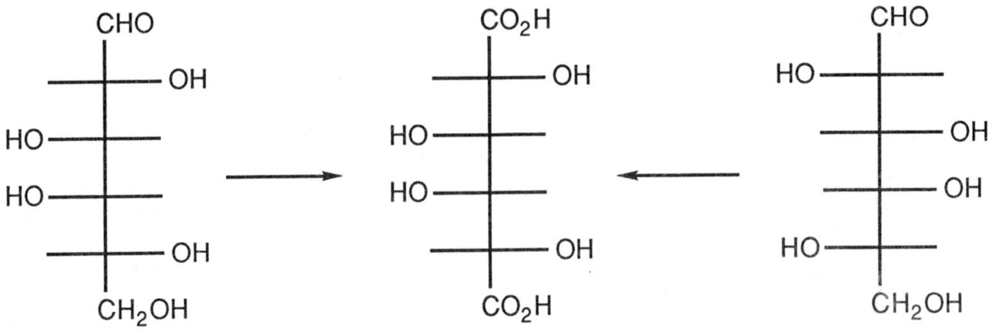

CO_2H, OH, HO, OH, HO, CO_2H

CHO, OH, HO, HO, OH, CH_2OH

D-galactose

CO_2H, OH, HO, HO, OH, CO_2H

CHO, HO, OH, OH, HO, CH_2OH

L-galactose

19.16

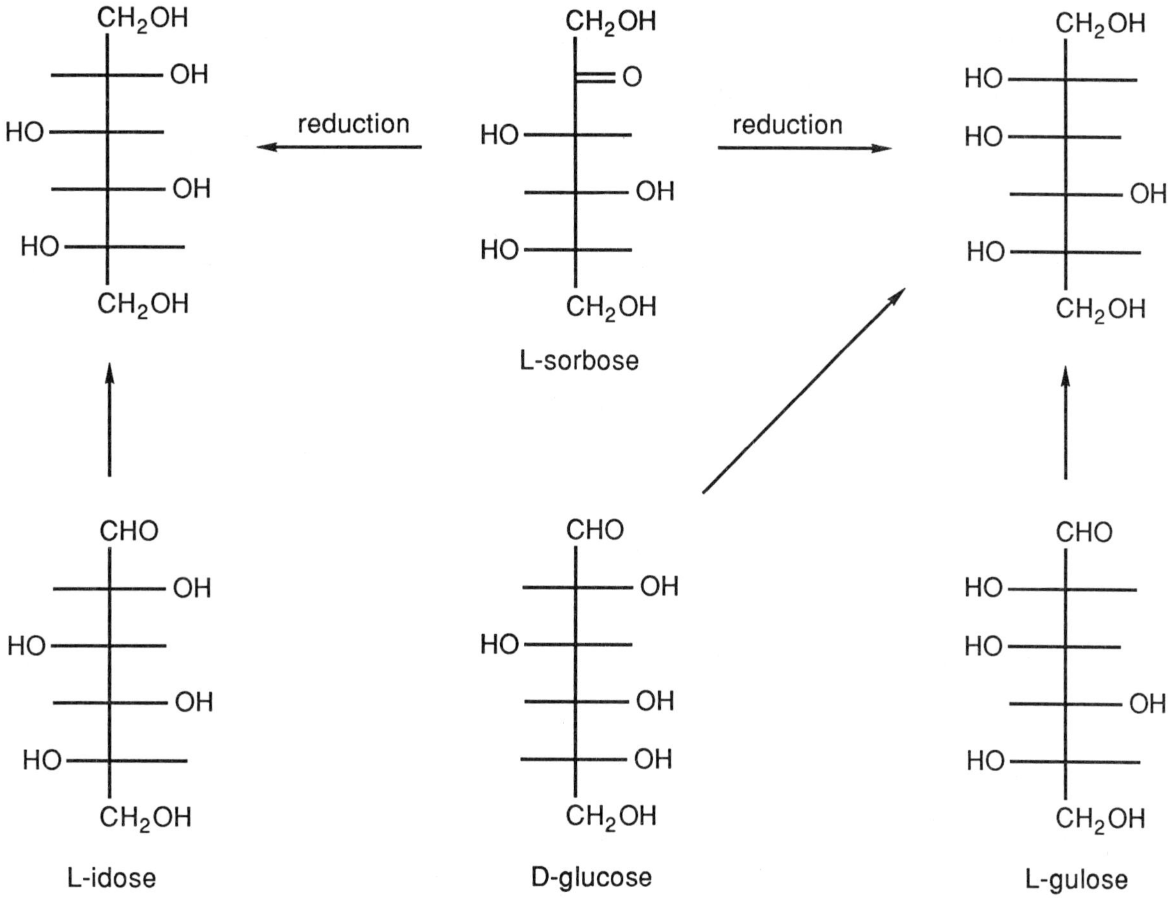
CH2OH
OH
HO
OH
HO
CH2OH
reduction
CH2OH
O
HO
OH
HO
CH2OH
L-sorbose
reduction
CH2OH
HO
HO
OH
HO
CH2OH
CHO
OH
HO
OH
HO
CH2OH
L-idose
CHO
OH
HO
OH
OH
CH2OH
D-glucose
CHO
HO
HO
OH
HO
CH2OH
L-gulose

19.17

CH$_2$OH / —OH / HO— / —OH / —OH / CH$_2$OH ← reduction — D-fructose: CH$_2$OH / =O / HO— / —OH / —OH / CH$_2$OH — reduction → CH$_2$OH / HO— / HO— / —OH / —OH / CH$_2$OH

D-gluose: CHO / —OH / HO— / —OH / —OH / CH$_2$OH

L-gulose: CHO / HO— / HO— / —OH / HO— / CH$_2$OH

D-mannose: CHO / HO— / HO— / —OH / —OH / CH$_2$OH

19.18 Xylose was found to yield an inactive aldaric acid upon oxidation. The diacid must therefore have reflection symmetry, so two possibilities exist:

CO$_2$H / —OH / —OH / —OH / CO$_2$H   CO$_2$H / —OH / HO— / —OH / CO$_2$H

For each of these possibilities there are two alternatives for xylose (one for *D* and one for *L*). The successive conversion of (+)-xylose to (+)-gulose and (+)-glucaric acid establishes a a direct structural relationship with (+)-glucose, which Fischer designated as having the *D* configuration:

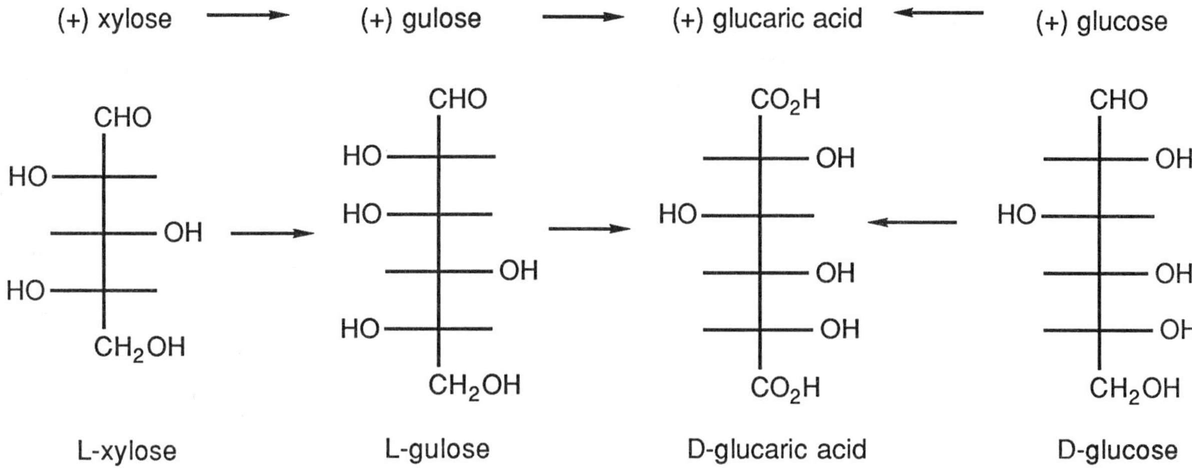

19.19 (+)-Glucose and (+)-mannose yield the same osazone, so they are C-2 epimers. Therefore knowledge of the full structure of (+)-glucose defines both enantiomers of mannose:

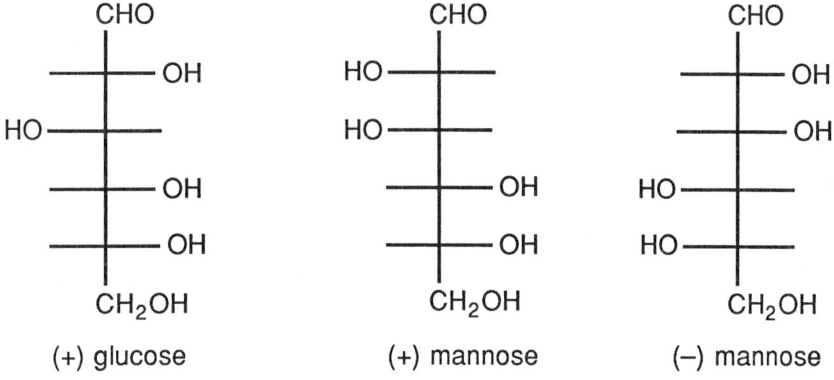

19.20 **(a)** The reaction of sodium in alcohol produces sodium ethoxide, and this can cause some epimerization at C-2 of glucose prior to reduction:

(a)

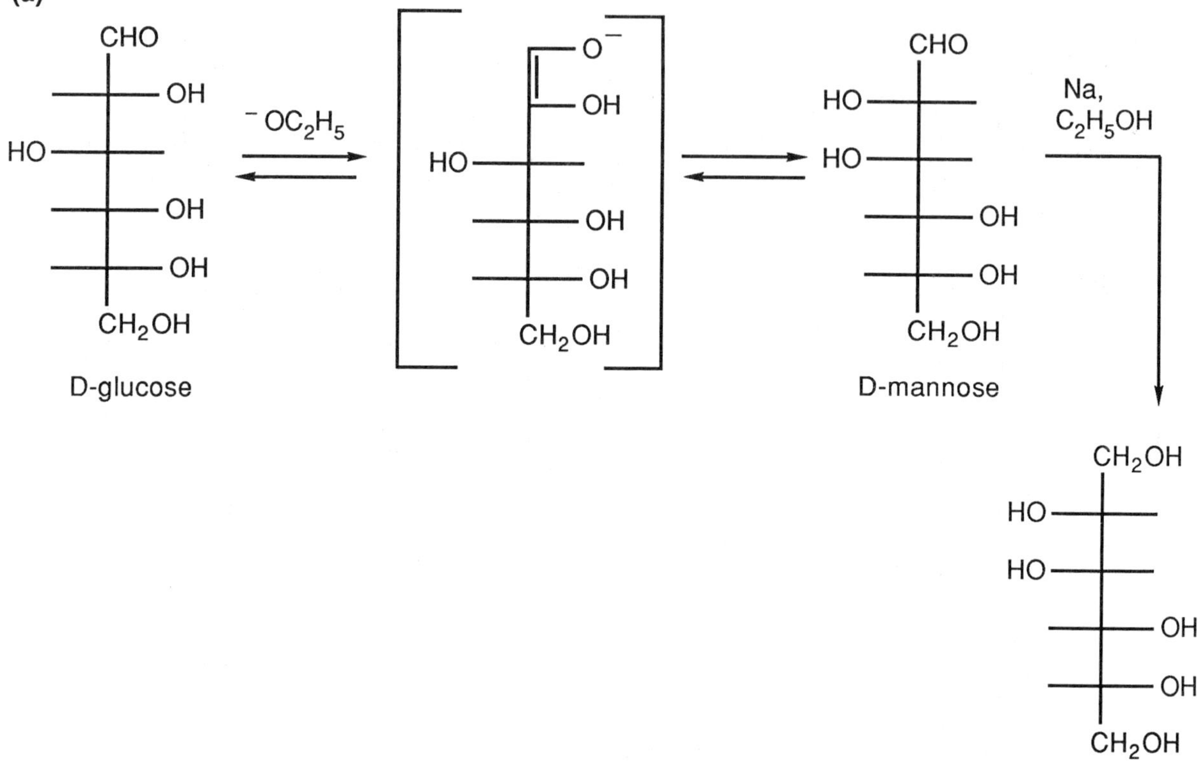

$^{-}OC_2H_5$

D-glucose

D-mannose

Na, C_2H_5OH

(b) The reported rotation of 0.5° is within experimental error of zero.

19.21 Tosyl chloride reacts preferentially with the least hindered, primary alcohol. The negative Tollens test for **C** indicates that this compound is not a free aldehyde, but is instead a cyclic derivative (in which the hemiacetal OH is acylated). For convenience we will assume a β pyranose structure:

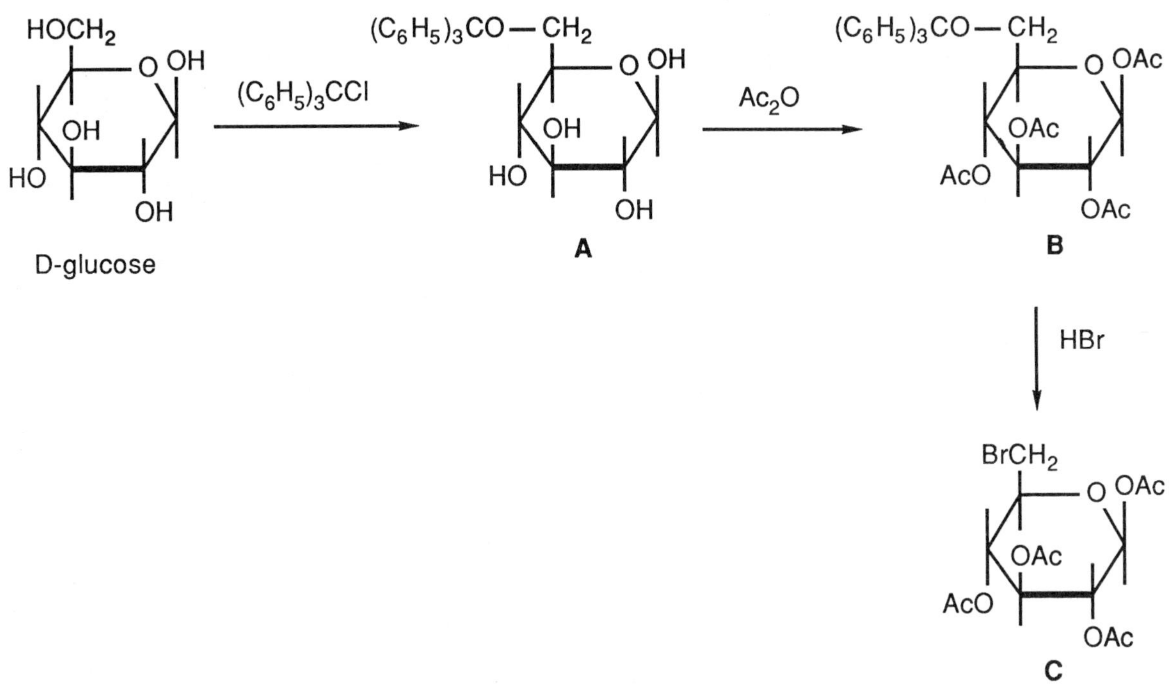

HOCH$_2$ … D-glucose $\xrightarrow{(C_6H_5)_3CCl}$ $(C_6H_5)_3CO-CH_2$ … A $\xrightarrow{Ac_2O}$ $(C_6H_5)_3CO-CH_2$ … B $\xrightarrow{HBr}$ BrCH$_2$ … C

19.22 Cotton is cellulose, a 1,4-polymer of β-*D*-glucopyranose. The reaction product **A** from H_2SO_4/acetic acid is the octaacetate of a C_{12} compound, so it clearly contains two glucose residues. Reaction with sodium methoxide cleaves (via transesterification) the acetates to form 4-*O*-(β-*D*-glucopyranosyl) β-*D*-glucopyranose:

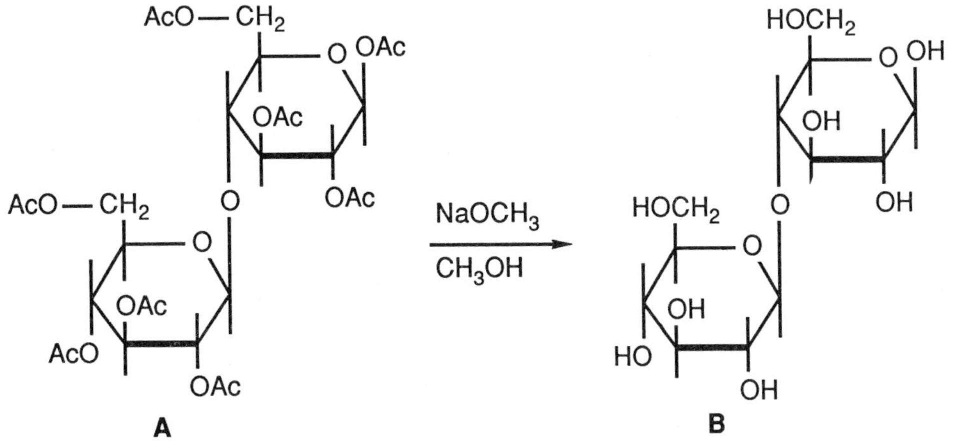

A $\xrightarrow[CH_3OH]{NaOCH_3}$ B

19.23

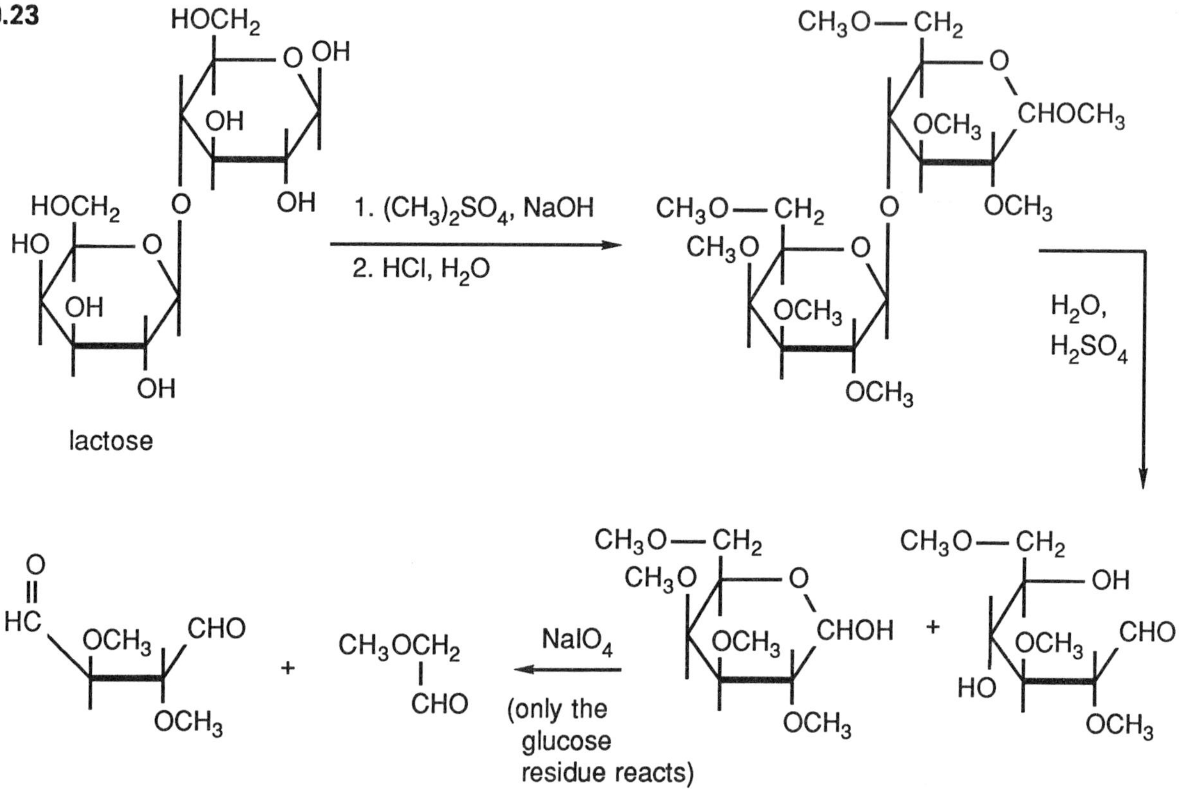

$HOCH_2$ O OH OH OH $HOCH_2$ O HO O OH OH

lactose

1. $(CH_3)_2SO_4$, NaOH

2. HCl, H_2O

$CH_3O—CH_2$ O $CHOCH_3$ OCH_3 OCH_3 $CH_3O—CH_2$ O CH_3O O OCH_3 OCH_3

H_2O, H_2SO_4

$CH_3O—CH_2$ CH_3O O CHOH OCH_3 OCH_3 + $CH_3O—CH_2$ OH OCH_3 CHO HO OCH_3

$NaIO_4$ (only the glucose residue reacts)

O ‖ HC OCH_3 CHO OCH_3 + CH_3OCH_2 | CHO

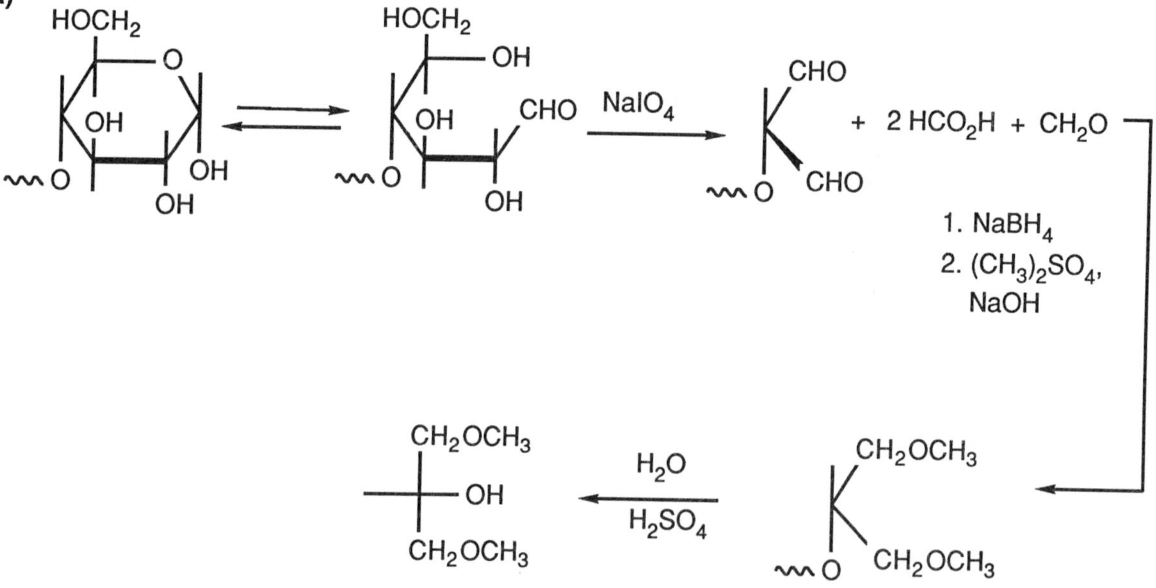

19.24

(a)

$HOCH_2$ O OH OH OH ~O ⇌ $HOCH_2$ OH OH CHO OH ~O

$NaIO_4$ → CHO CHO ~O + 2 HCO_2H + CH_2O

1. $NaBH_4$
2. $(CH_3)_2SO_4$, NaOH

CH_2OCH_3 CH_2OCH_3 ~O

H_2O, H_2SO_4 →

CH_2OCH_3 —OH CH_2OCH_3

(b)

HOCH$_2$ — O — HO — OH — OH — O∿∿∿

$NaIO_4$ →

HOCH$_2$ — O — CH=O — HC=O — O∿∿∿ + HCO_2H

1. $NaBH_4$
2. $(CH_3)_2SO_4$, NaOH

$CH_3O—CH_2$ — O — OCH_3 — OCH_3 — O∿∿

H_2O, H_2SO_4 →

CH_2OCH_3 — OH — CH_2OCH_3 + CHO — CH_2OCH_3

(c)

HOCH$_2$ — O — OH — ∿∿∿O — OH — O∿∿∿

$NaIO_4$ →

HOCH$_2$ — O — ∿∿∿O — CH=O — HC=O — O∿∿∿

1. $NaBH_4$
2. $(CH_3)_2SO_4$, NaOH

$CH_3O—CH_2$ — O — ∿∿∿O — CH_3O — OCH_3 — O∿∿∿

H_2O, H_2SO_4 →

CH_2OCH_3 — OH — OH — CH_2OCH_3 + CHO — CH_2OCH_3

19.25 For a 1:7 (i.e., 2:14) ratio of the glycerol to erythritol derivatives, n must be twelve, and the total (average) number of residues must be sixteen:

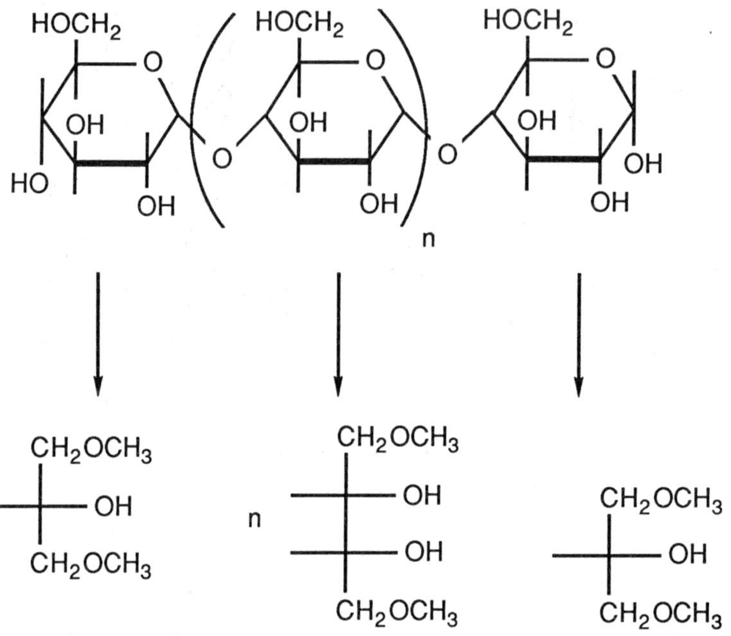

19.26 Amylopectin has 10^5 residues, so the end groups are not significant in this analysis. Only the branch-point residues (i.e., those linked to three glucose subunits) yield *O*-methylerythritol. The other residues all yield 1,4-di-*O*-methylerythritol, so the 26:1 product ratio indicates 26 residues (average) between branch points in this sample:

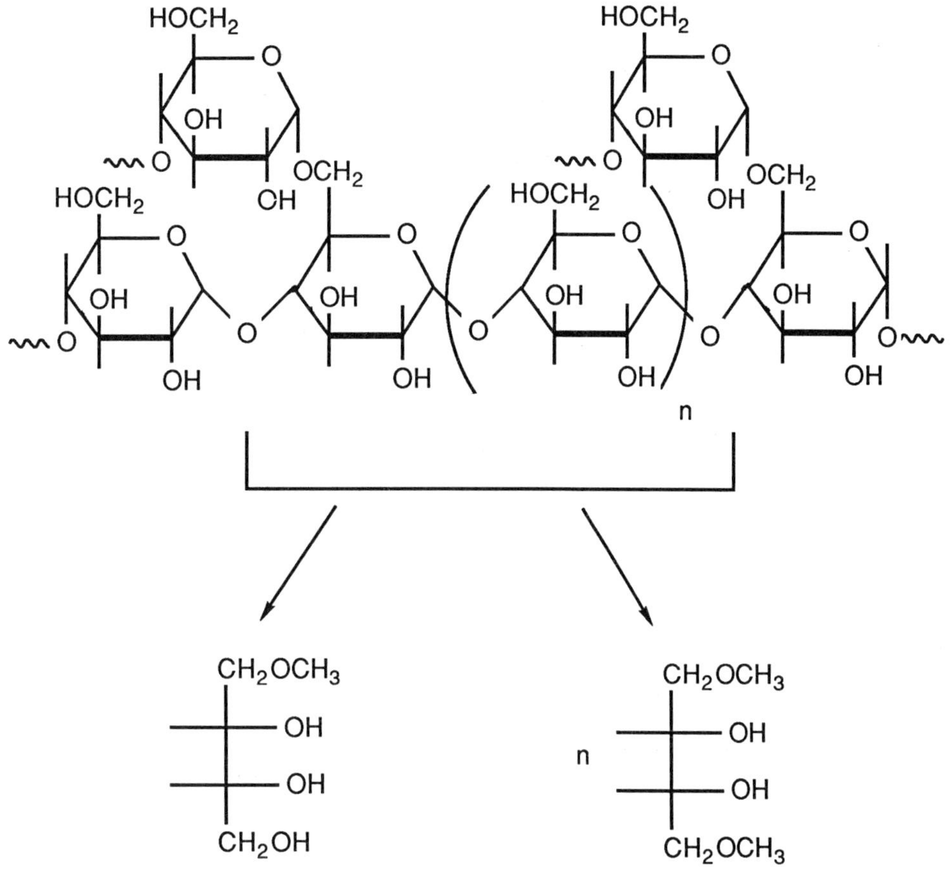

19.27 Key information in this problem comes from the nature of the C_1 fragments. Formic acid arises from CHOH groups between adjacent CHOH or C=O groups. CH_3OH arises only from a structure in which the glycosidic carbon exists as the *hemiacetal* (if C-6 were *also* to have a free hydroxyl group, then C-6 would generate a second CH_3OH). The two-carbon fragment affords direct information on ring substitution, because this can be formed only when C-2 and C-3 both have free hydroxyl groups:

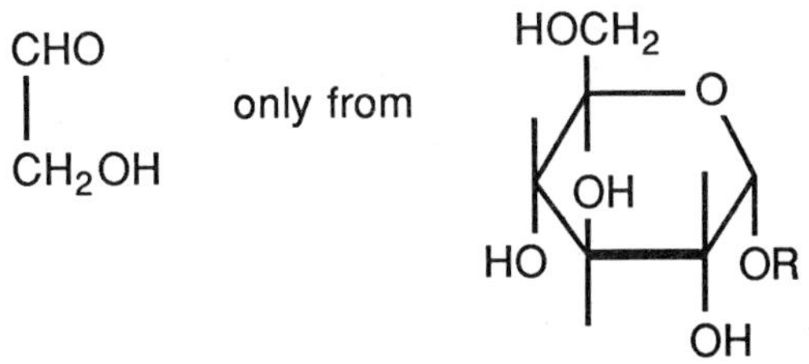

CHO–CH_2OH only from

With no substitution a glucopyranoside would give the following results:

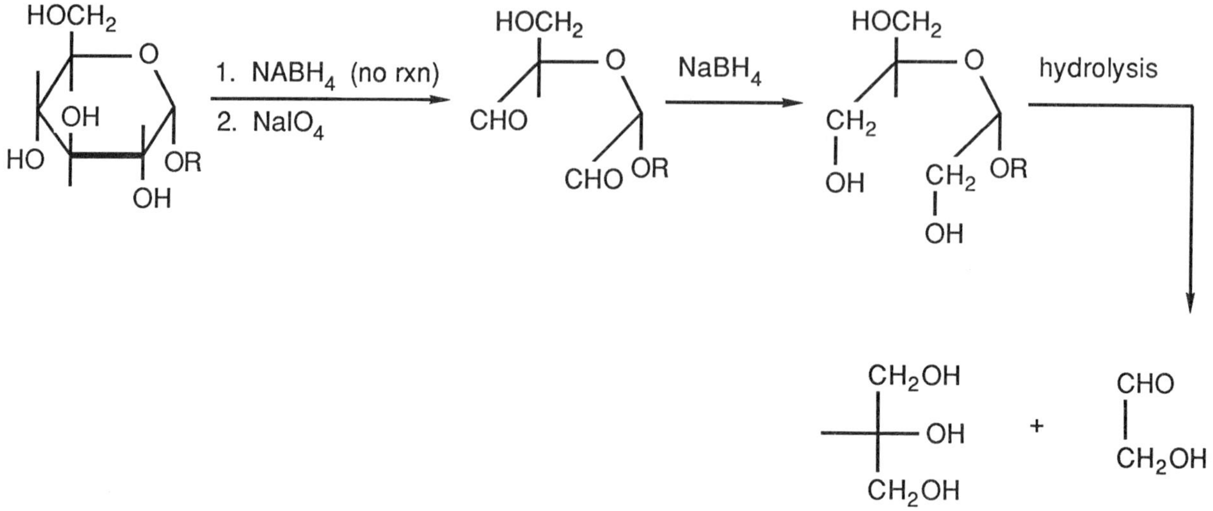
HOCH2
O
OH
HO
OR
OH
1. NABH4 (no rxn)
2. NaIO4
HOCH2
O
CHO
CHO OR
NaBH4
HOCH2
O
CH2
OH
CH2 OR
OH
hydrolysis
CH2OH
OH
CH2OH
+
CHO
CH2OH

The formation of two moles of these products in this problem means that the trisaccharide contains two unsubstituted glucopyranosyl residues.

The erythritol formed in the reaction sequence cannot arise from the aldehyde end of glucose, which would afford threitol instead:

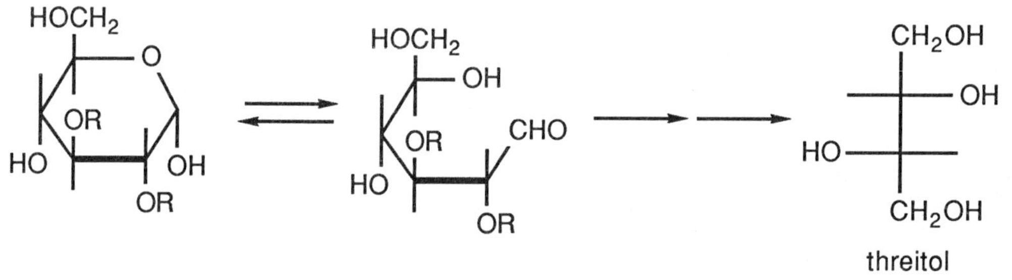
HOCH2
O
OR
HO
OH
OR
HOCH2
OH
OR
CHO
HO
OR
CH2OH
OH
HO
CH2OH
threitol

The erythritol must therefore arise from the other end of the glucose molecule, so the C-2 and C-3 hydroxyls are free but C-4 is blocked:

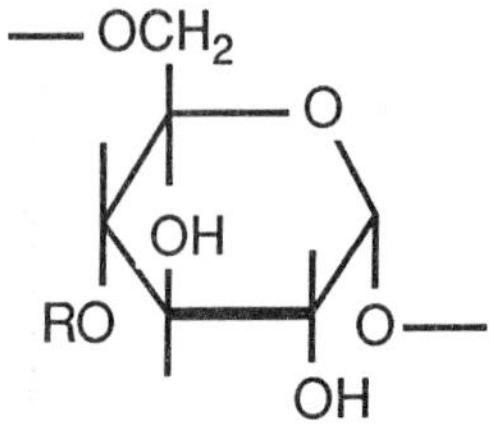
—OCH2
O
OH
RO
O—
OH

The second glucosyl residue must be on either the C-6 or C-1 hydroxyl, but the formation of a single mole of methanol means that it must be on C-6:

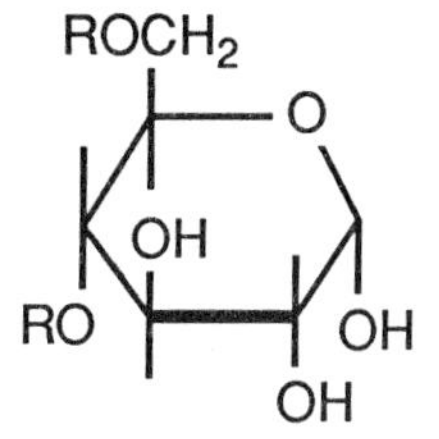

ROCH2, O, OH, RO, OH, OH

The complete structure of the trisaccharide must therefore be:

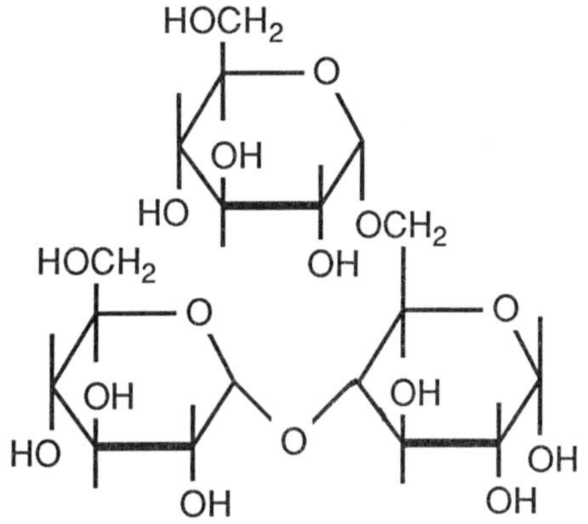

HOCH2, O, OH, HO, OCH2, OH, HOCH2, O, OH, HO, O, OH, O, OH, OH, OH

19.28 (a) maltose

HOCH2, O, OH, HO, OH, O, HOCH2, O, OH, OH, OH

$(CH_3)_2SO_4$, NaOH

$CH_3O—CH_2$, O, OCH_3, CH_3O, OCH_3, O, $CH_3O—CH_2$, O, OCH_3, OCH_3, OCH_3

H_2O, H_2SO_4

CHO, OCH_3, CH_3O, OCH_3, OH, CH_2OCH_3 + CHO, OCH_3, CH_3O, OH, OH, CH_2OCH_3

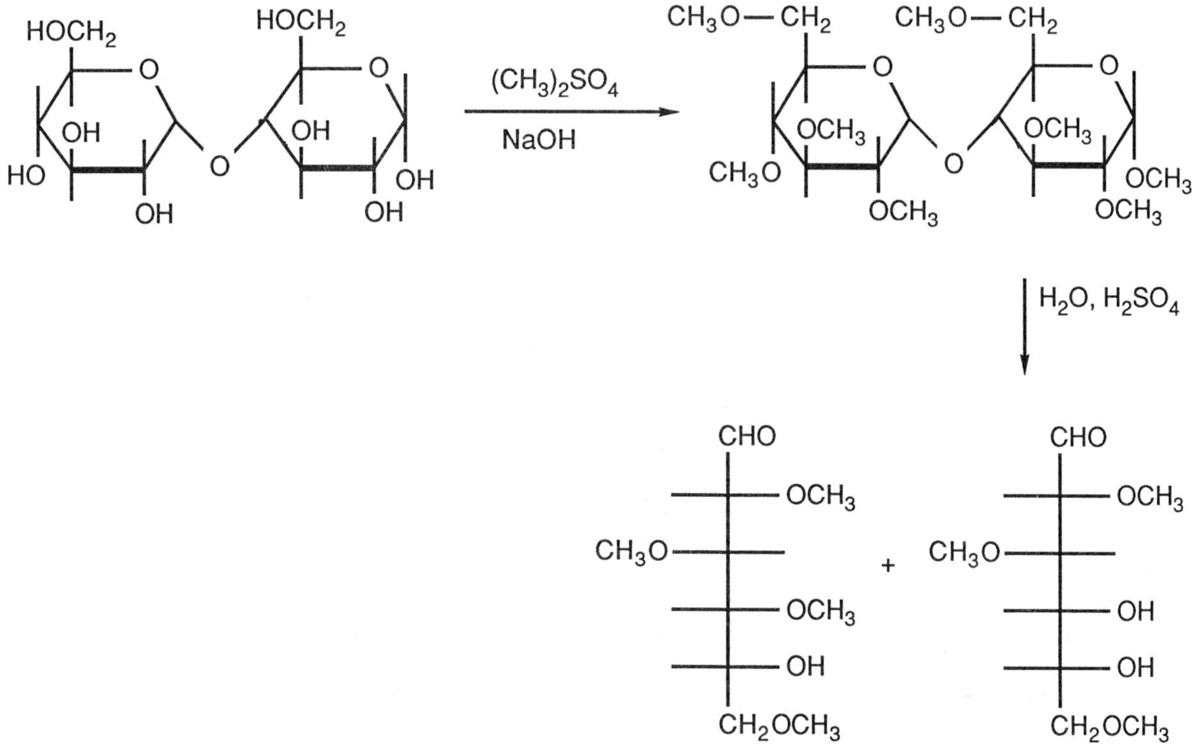

cellobiose

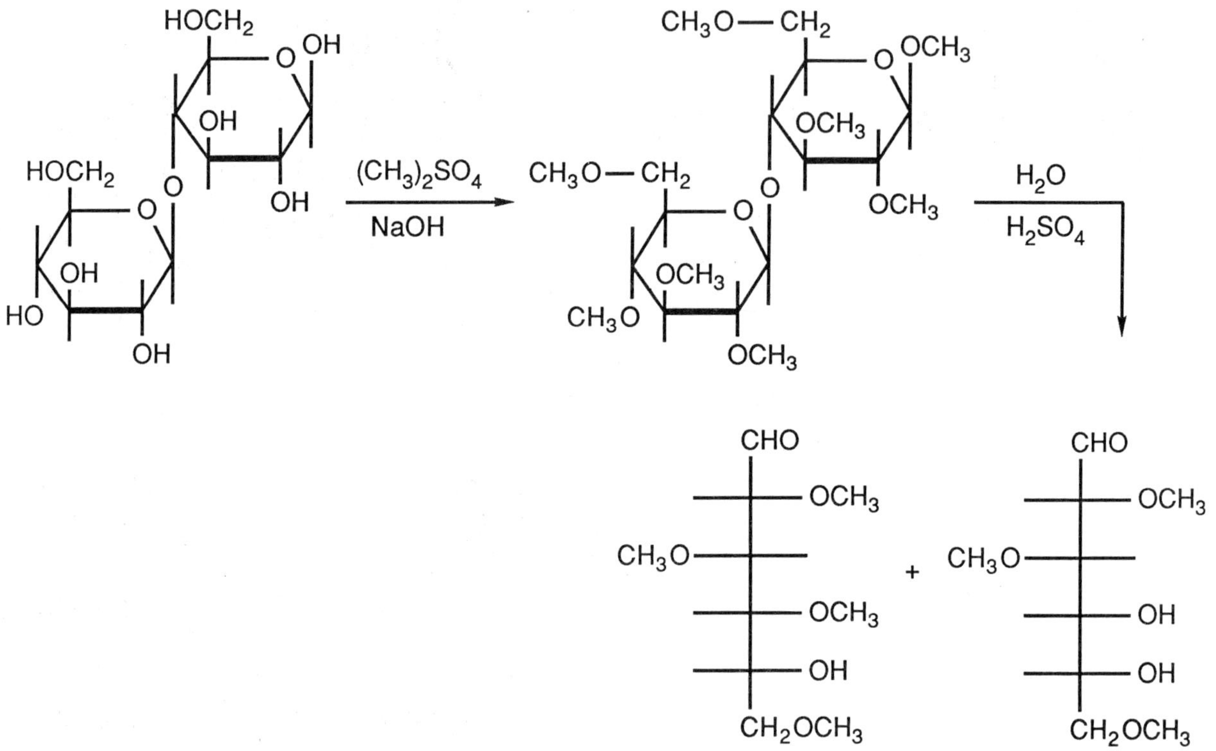

$$\xrightarrow[\text{NaOH}]{(CH_3)_2SO_4}$$

$$\xrightarrow[H_2SO_4]{H_2O}$$

HOCH$_2$, OH, OH, OH, O, HOCH$_2$, O, OH, HO, OH

CH$_3$O—CH$_2$, OCH$_3$, OCH$_3$, OCH$_3$, O, CH$_3$O—CH$_2$, O, OCH$_3$, CH$_3$O, OCH$_3$

CHO, OCH$_3$, CH$_3$O, OCH$_3$, OH, CH$_2$OCH$_3$ + CHO, OCH$_3$, CH$_3$O, OH, OH, CH$_2$OCH$_3$

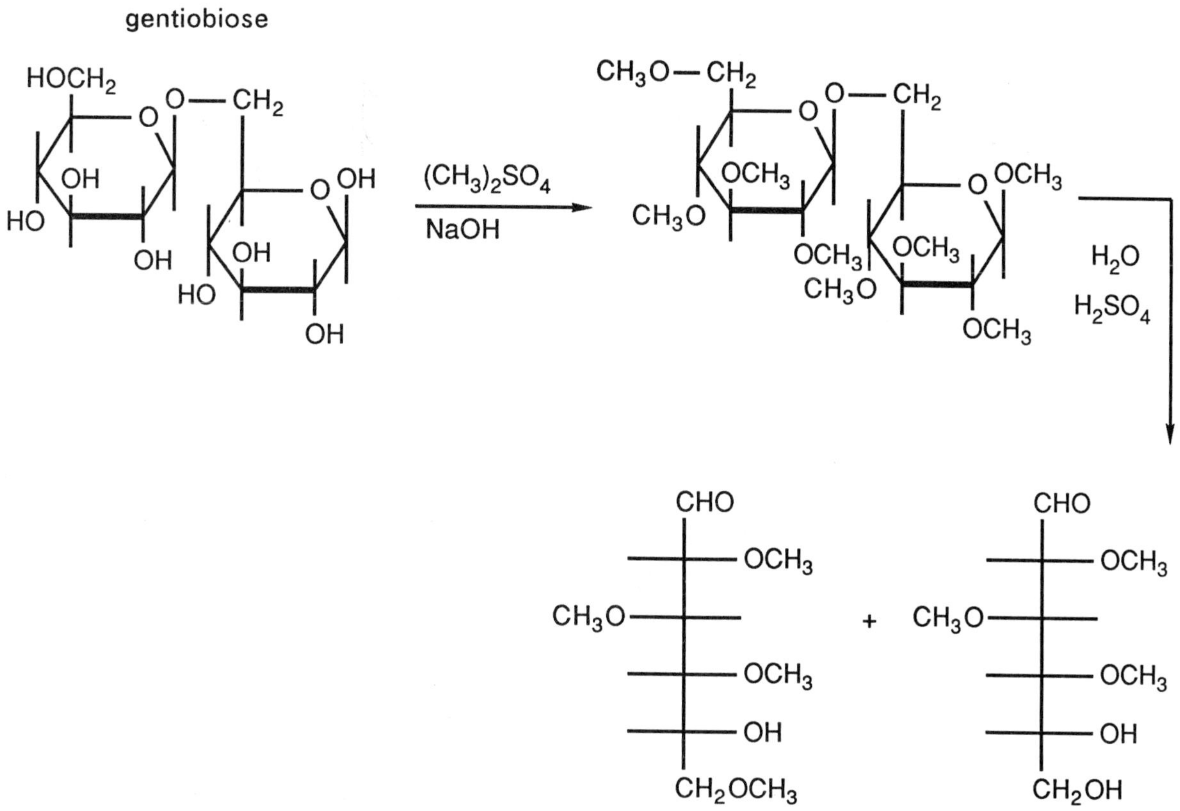

gentiobiose

HOCH2 O—CH2 OH HO OH OH O OH HO OH

$(CH_3)_2SO_4$ / NaOH →

CH_3O—CH_2 O—CH_2 OCH_3 CH_3O OCH_3 OCH_3 OCH_3 CH_3O OCH_3

H_2O / H_2SO_4 ↓

CHO, OCH_3, CH_3O, OCH_3, OH, CH_2OCH_3 + CHO, OCH_3, CH_3O, OCH_3, OH, CH_2OH

(b) In each case two glucose derivatives are formed: one trimethyl and one tetramethyl. The pyranose ring results in a free hydroxyl at C-5 after acid hydrolysis, and the position of the remaining free OH group defines the site of the glycosidic linkage. Only the stereochemistry at C-1 (for each of the rings) is not defined by these experiments. Thus maltose and cellobiose must be epimeric at the glycosidic linkage of 4-*O*-(*D*-glucopyranosyl) *D*-glucopyranose. Similarly, gentiobiose is shown to be 6-*O*-(*D*-glucopyranosyl) *D*-glucopyranose.

19.29

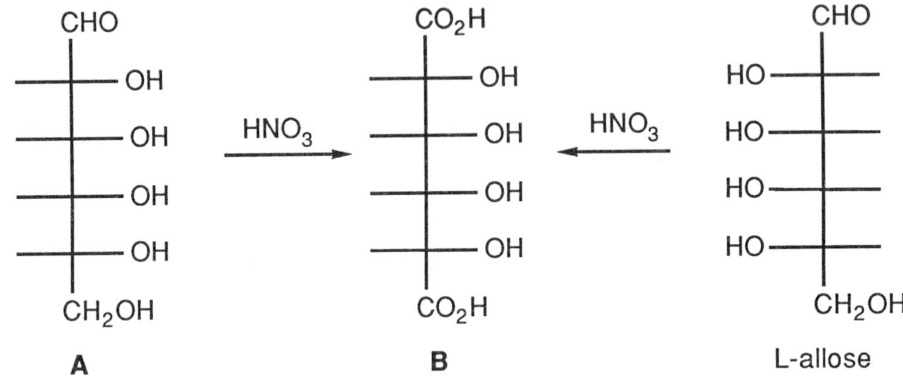

$C_6H_5NHNH_2$

$CH{=}NNHC_6H_5$
$=NNHC_6H_5$
OH
OH
OH
CH_2OH

C

19.30

(a)

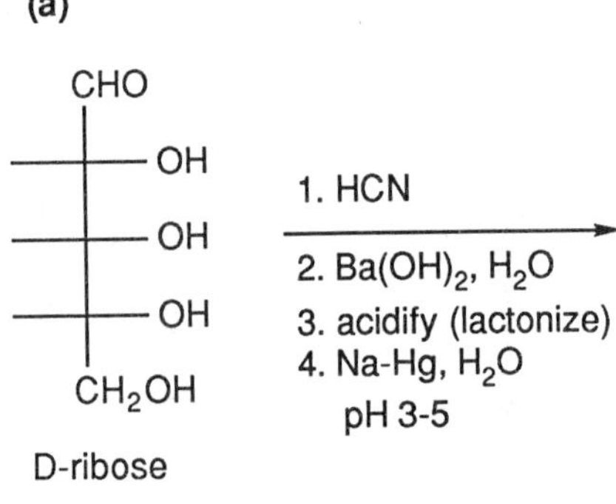

CHO
OH
OH
OH
OH
CH_2OH

+

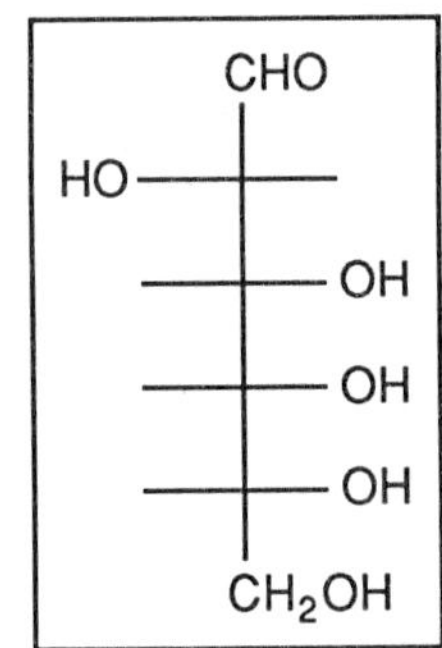

(b)

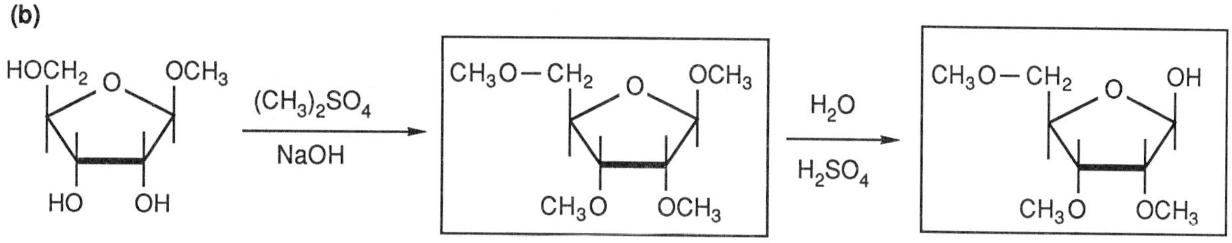

$$HOCH_2\text{-furanoside}(OCH_3)(HO)(OH) \xrightarrow[NaOH]{(CH_3)_2SO_4} \boxed{CH_3O{-}CH_2\text{-furanoside}(OCH_3)(CH_3O)(OCH_3)} \xrightarrow[H_2SO_4]{H_2O} \boxed{CH_3O{-}CH_2\text{-furanose}(OH)(CH_3O)(OCH_3)}$$

(c) two answers are possible:

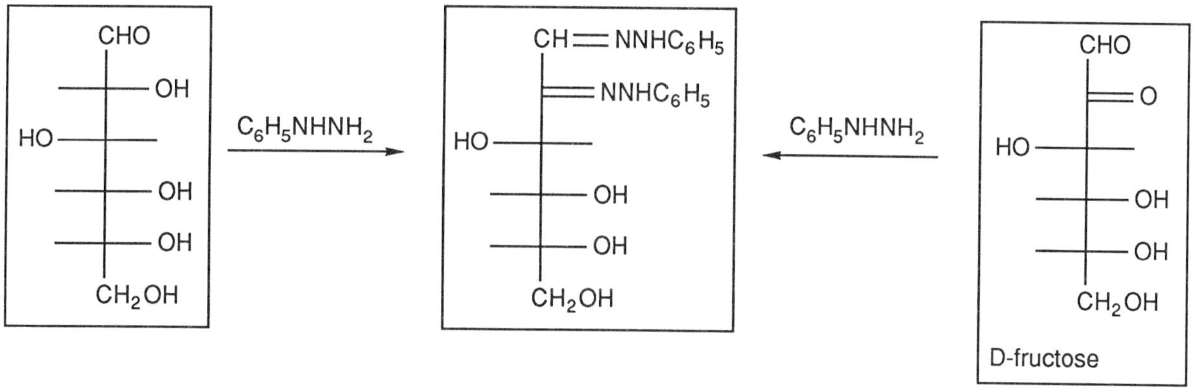

CHO / —OH / HO— / —OH / —OH / CH_2OH $\xrightarrow{C_6H_5NHNH_2}$ $CH{=}NNHC_6H_5$ / $=NNHC_6H_5$ / HO— / —OH / —OH / CH_2OH $\xleftarrow{C_6H_5NHNH_2}$ CHO / =O / HO— / —OH / —OH / CH_2OH

D-fructose

19.31 The key structural information is provided for compound **H**, which is identified as *D*-threose. Chain lengthening will afford two aldopentoses, *D*-xylose and *D*-lyxose, and the latter must be **F** because reduction of xylose would yield an optically inactive alditol (and the problem states that all the compounds are optically active). **E** is formed from two aldopentoses, **D** and **F**, so **D** must be *D*-arabinose. Formation of this compound by chain shortening of **A** means that **A** must be either *D*-glucose or *D*-mannose:

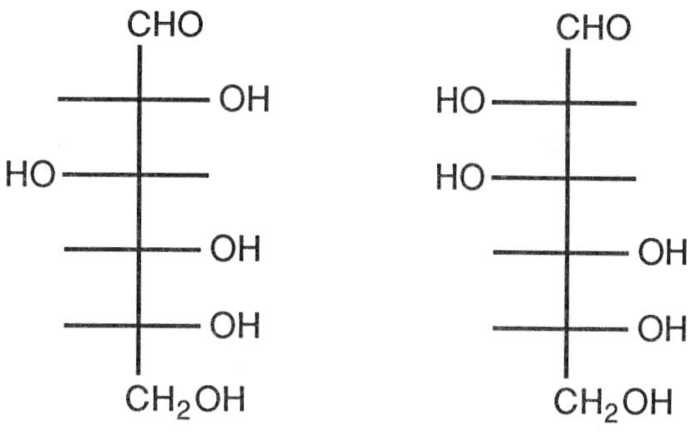

CHO / —OH / HO— / —OH / —OH / CH_2OH

D-glucose

CHO / HO— / HO— / —OH / —OH / CH_2OH

D-mannose

It cannot be mannose because oxidation yields the aldaric acid **B** (which is also formed from **C**), but the aldaric acid formed from mannose cannot be produced from another aldohexose. This logic defines the structures of both **A** and **B**, and you can further conclude that compound **C** is an isomeric aldose, *L*-gulose. The complete answer follows:

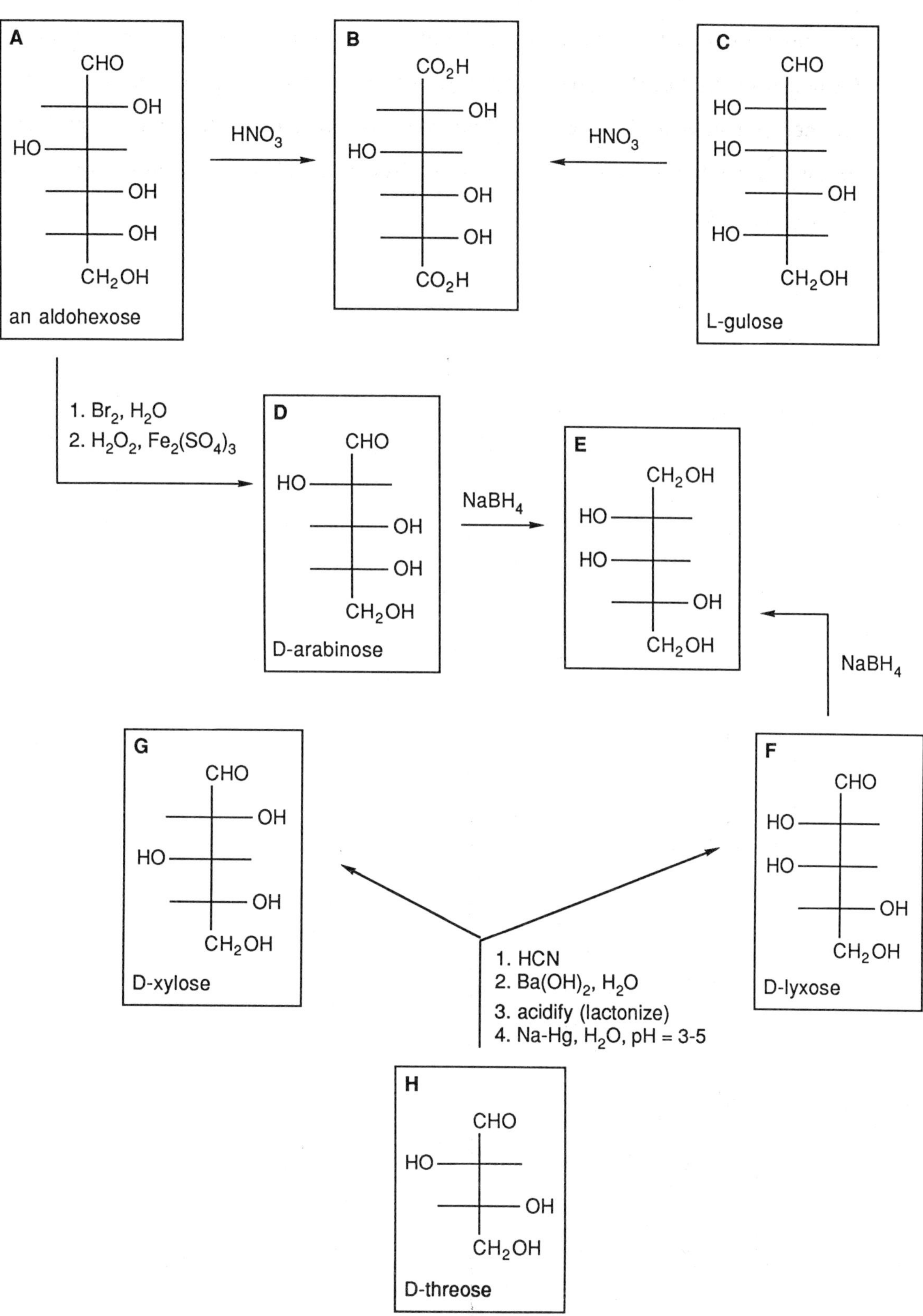
A
CHO
OH
HO
OH
OH
CH2OH
an aldohexose
HNO3
B
CO2H
OH
HO
OH
OH
CO2H
HNO3
C
CHO
HO
HO
OH
HO
CH2OH
L-gulose
1. Br2, H2O
2. H2O2, Fe2(SO4)3
D
CHO
HO
OH
OH
CH2OH
D-arabinose
NaBH4
E
CH2OH
HO
HO
OH
CH2OH
NaBH4
G
CHO
OH
HO
OH
CH2OH
D-xylose
F
CHO
HO
HO
OH
CH2OH
D-lyxose
1. HCN
2. Ba(OH)2, H2O
3. acidify (lactonize)
4. Na-Hg, H2O, pH = 3-5
H
CHO
HO
OH
CH2OH
D-threose

19.32 The key structural information in the problem is the structure of **H** as *D*-erythrose. Chain lengthening of this compound will yield the aldonic acids **D** and **G.** Assignment of the two possible structures is possible from the subsequent reduction of **D** to the aldose **E** and the alditol **F.** Since all compounds in the problem must be optically active, the alditol **F** cannot be symmetrical (as would be the case if the assignments for **D** and **G** were reversed). The degradation to give **D** allows two possible structures for **A**, *D*-mannose and *D*-glucose:

D-mannose	D-glucose
CHO	CHO
HO—	—OH
HO—	HO—
—OH	—OH
—OH	—OH
CH_2OH	CH_2OH

D-mannose D-glucose

Glucose can be ruled out on the basis of the chain lengthening sequence (followed by reduction) to yield the seven-carbon alditols **B** and **C.** One of the compounds derived from glucose in this manner would be optically inactive. Compound **A** is therefore *D*-mannose, and the complete answer follows:

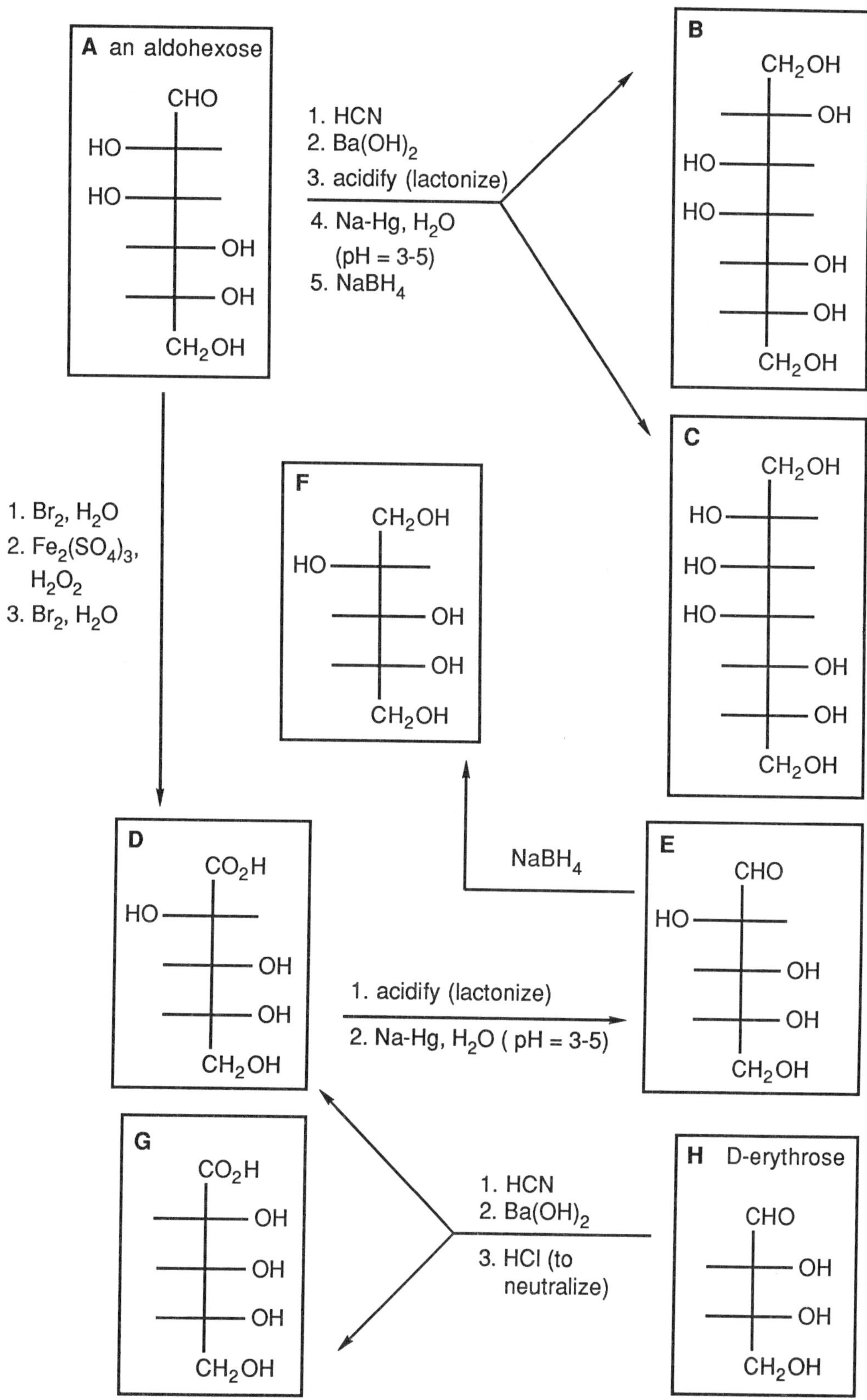
A an aldohexose
CHO
HO
HO
OH
OH
CH2OH
1. HCN
2. Ba(OH)2
3. acidify (lactonize)
4. Na-Hg, H2O (pH = 3-5)
5. NaBH4
B
CH2OH
OH
HO
HO
OH
OH
CH2OH
C
CH2OH
HO
HO
HO
OH
OH
CH2OH
1. Br2, H2O
2. Fe2(SO4)3, H2O2
3. Br2, H2O
F
CH2OH
HO
OH
OH
CH2OH
NaBH4
D
CO2H
HO
OH
OH
CH2OH
1. acidify (lactonize)
2. Na-Hg, H2O (pH = 3-5)
E
CHO
HO
OH
OH
CH2OH
G
CO2H
OH
OH
OH
CH2OH
1. HCN
2. Ba(OH)2
3. HCl (to neutralize)
H D-erythrose
CHO
OH
OH
CH2OH

19.33 The structures of **G** and **H** are known, and these two compounds are C–2 epimers. Therefore, alditols **E** and **F** are also C–2 epimers. Formation of both from **A** requires that **A** must be a ketose:

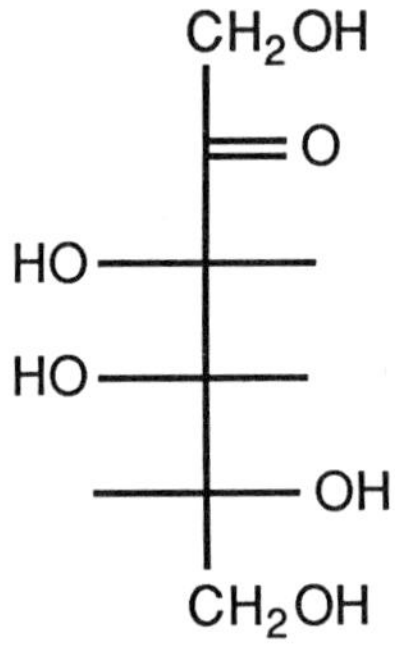

Reaction with methanol/HCl forms the corresponding methyl glycoside:

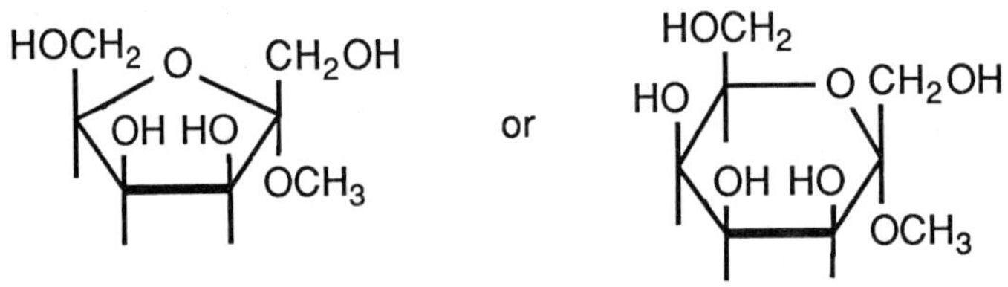

Consumption of only one mole of HIO_4 without loss of any C_1 fragments permits you to rule out the pyranose structure for **B**. Methylation followed by hydrolysis yields a methylated ketose **B** that is not a reducing sugar (i.e., it is no longer an α hydroxy ketone):

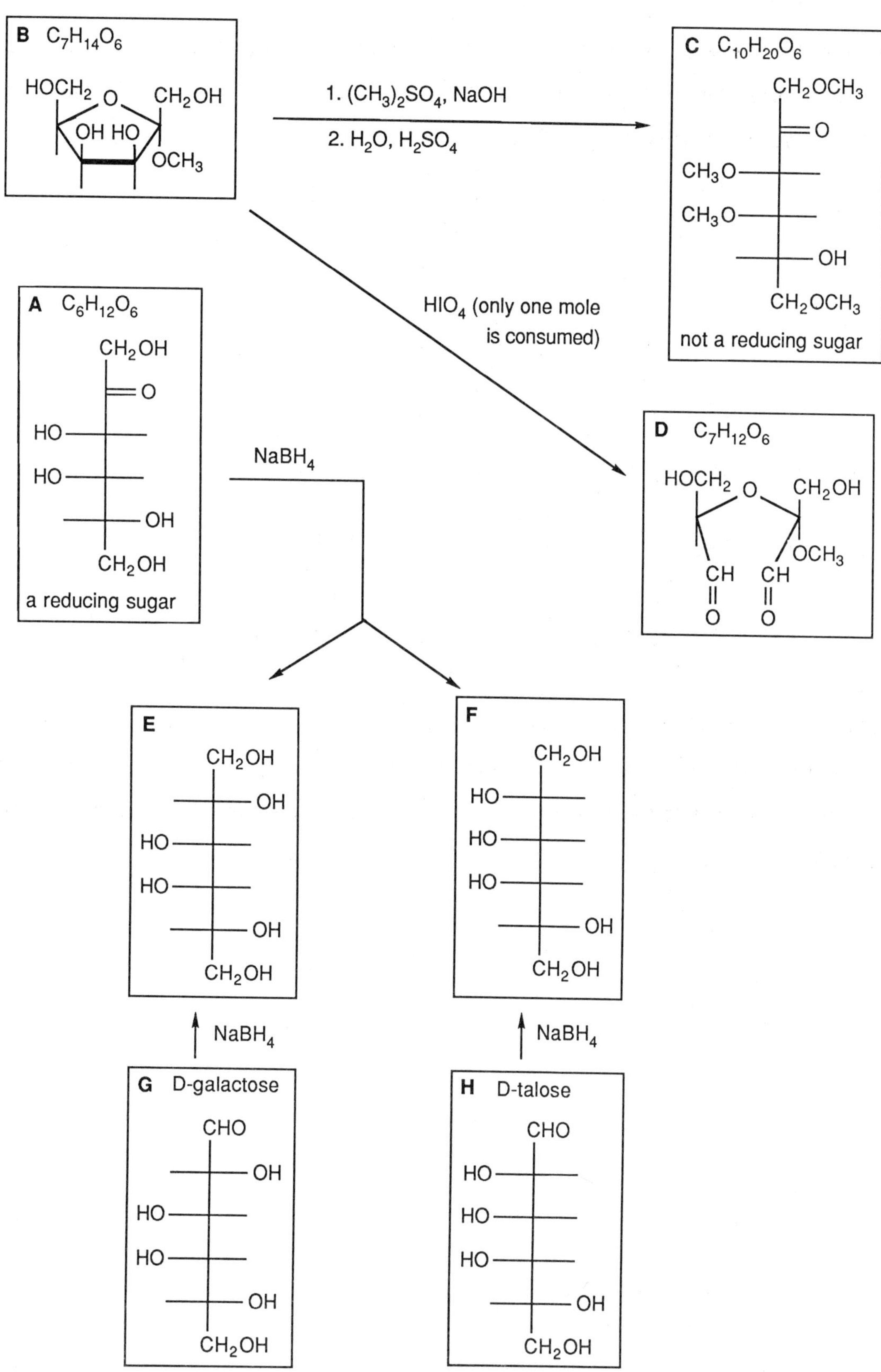
B $C_7H_{14}O_6$
HOCH₂ O CH₂OH
OH HO
OCH₃
1. $(CH_3)_2SO_4$, NaOH
2. H_2O, H_2SO_4
C $C_{10}H_{20}O_6$
CH₂OCH₃
O
CH₃O
CH₃O
OH
CH₂OCH₃
not a reducing sugar
A $C_6H_{12}O_6$
CH₂OH
O
HO
HO
OH
CH₂OH
a reducing sugar
HIO_4 (only one mole is consumed)
D $C_7H_{12}O_6$
HOCH₂ O CH₂OH
OCH₃
CH CH
O O
$NaBH_4$
E
CH₂OH
OH
HO
HO
OH
CH₂OH
F
CH₂OH
HO
HO
HO
OH
CH₂OH
$NaBH_4$
$NaBH_4$
G D-galactose
CHO
OH
HO
HO
OH
CH₂OH
H D-talose
CHO
HO
HO
HO
OH
CH₂OH

19.34 Steps (a)–(c) of the synthesis constitute a chain-end reversal, converting *D*-galactose to *L*-galactonic acid. Ruff degradation in step (d) then generates *L*-lyxose. Osazone formation in (e) followed by hydrolysis in (f) brings about selective oxidation at C-2. Chain lengthening via cyanide addition in step (g) yields the necessary six-carbon chain. Note that chirality remains only at C-4 and C-5 of the final product. After hydrolysis of the cyano group, acid catalyzed lactonization (h) is accompanied by enolization of the C-3 ketone to generate *L*-ascorbic acid.

CHO

OH

HO

HO

OH

CH_2OH

≡

OH

CHO

OH

HO

HO

CH_2OH

D-galactose

(a)

$CH_3-\overset{O}{\overset{\|}{C}}-CH_3$

H_2SO_4

CH_3 CH_3

O O

O

O

O

CH_3

CH_2OH

CH_3

(b)

$KMnO_4$

CH_3 CH_3

O O

O

O

O

CH_3

CO_2H

CH_3

(c-1)

H_2O

H_2SO_4

HO

CHO

OH

HO

HO

CO_2H

(c-2)

$NaBH_4$

HO

CH_2OH

OH

HO

HO

CO_2H

(d)

H_2O_2, Fe^{3+}

HO

CH_2OH

HO

CHO

HO

L-lyxose

(e)

$C_6H_5NHNH_2$

HO

CH_2OH

HO

$CH=NNHC_6H_5$

$NNHC_6H_5$

(f)

H_2O, H_2SO_4

HO

CH_2OH

HO

CHO

O

(g)

HCN

HO

CH_2OH

CN

HO

O

OH

(h)

1. H_2O, NaOH

2. H_2SO_4

H

CH_2OH

HO

C

O

O

HO

OH

CHAPTER 20. AMINO ACIDS AND PEPTIDES

ANSWERS TO EXERCISES

20.1 **(a)** An amine (K_b = 10^{-4}) is more basic than a carboxylate ion (K_b = 10^{-11}):

$$NH_2\text{-}CH(R)\text{-}CO_2^- \rightleftharpoons {}^+NH_3\text{-}CH(R)\text{-}CO_2^-$$

(b) A carboxyl group (K_a = 10^{-5}) is more acidic than an ammonium ion (K_a = 10^{-10}):

$${}^+NH_3\text{-}CH(R)\text{-}CO_2H \rightleftharpoons {}^+NH_3\text{-}CH(R)\text{-}CO_2^-$$

20.2 **(a)** Lysine, arginine, and histidine:

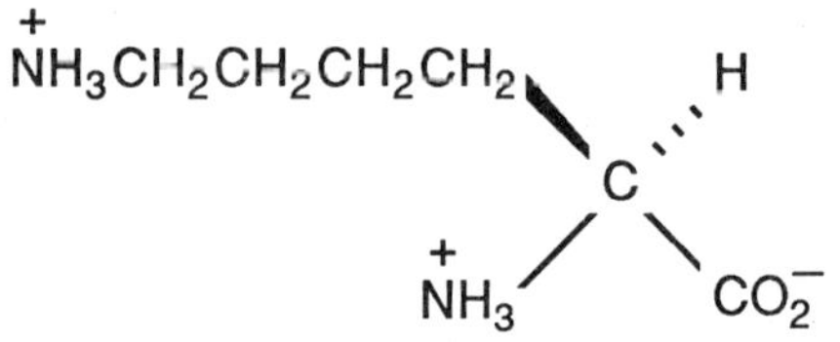

$^+NH_3CH_2CH_2CH_2CH_2$–C(H)($^+NH_3$)(CO_2^-)

Lys

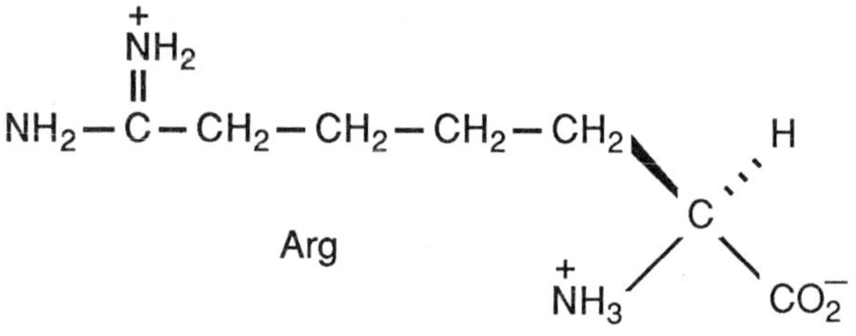

NH_2–C(=$\overset{+}{N}H_2$)–CH_2–CH_2–CH_2–CH_2–C(H)($^+NH_3$)(CO_2^-)

Arg

HN (imidazolium ring, $\overset{+}{N}$–H)–CH_2–C(H)($^+NH_3$)(CO_2^-)

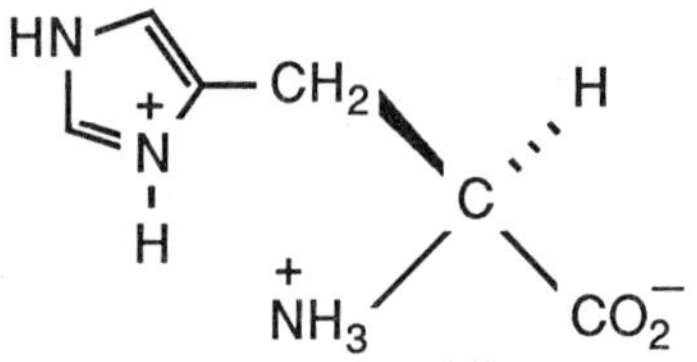

His

(b) Aspartic acid, glutamic acid:

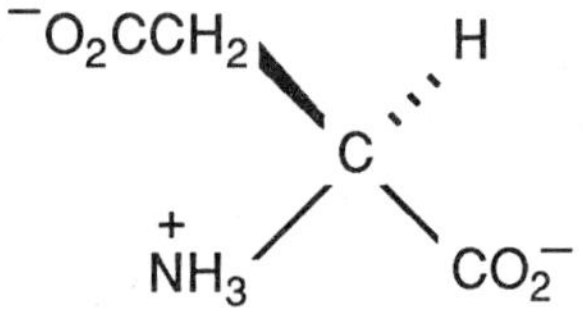

$^-O_2CCH_2$–C(H)($^+NH_3$)(CO_2^-)

Asp

$^-O_2CCH_2CH_2$–C(H)($^+NH_3$)(CO_2^-)

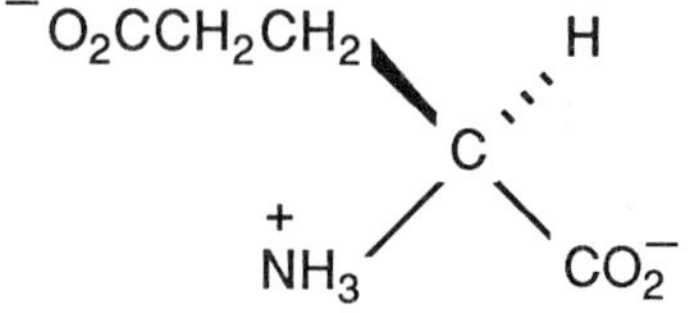

Glu

(c) Valine (isovaleric acid), proline (pyrrolidine), methionine (methyl/thio), phenylalanine (alanine), threonine (threose), and leucine/isoleucine (isomers):

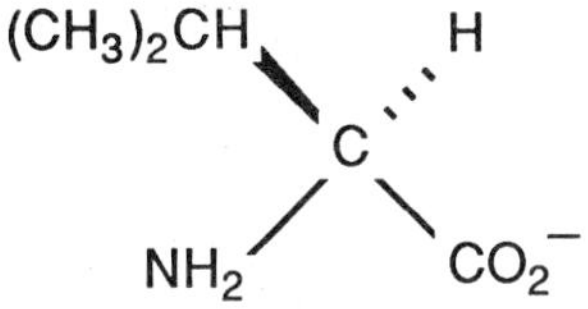

$(CH_3)_2CH$ H C NH_2 CO_2^-

Val

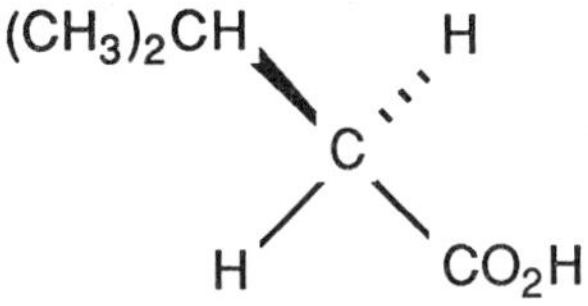

$(CH_3)_2CH$ H C H CO_2H

Isovaleric acid

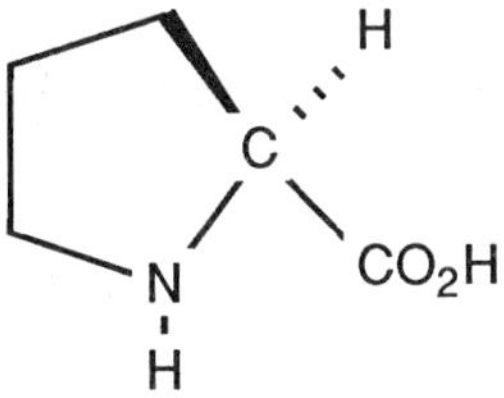

H C N H CO_2H

Pro

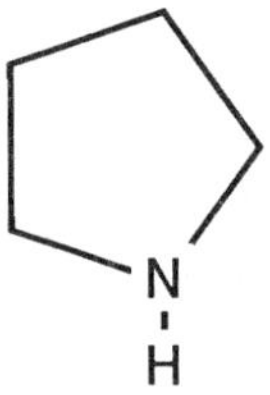

N H

Pyrrolidine

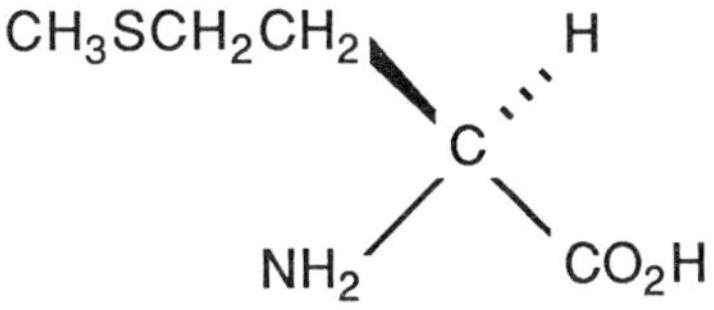

$CH_3SCH_2CH_2$ H C NH_2 CO_2H

Met

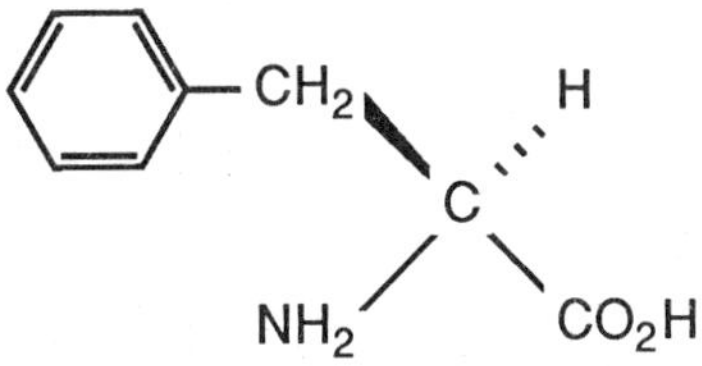

CH_2 H C NH_2 CO_2H

Phe

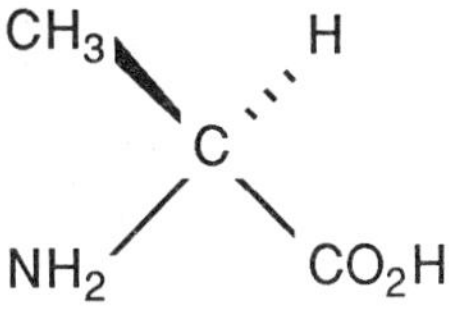

CH_3 H C NH_2 CO_2H

Ala

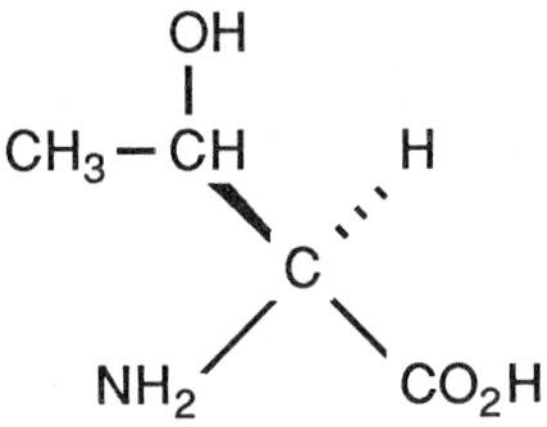

OH CH_3-CH H C NH_2 CO_2H

Thr

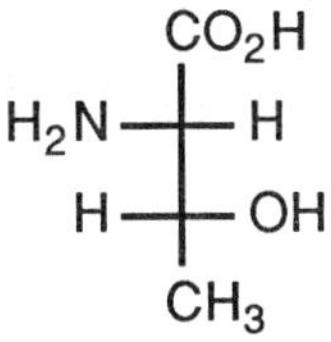

CO_2H H_2N H H OH CH_3

Threonine

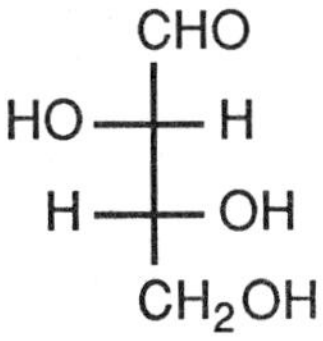

CHO HO H H OH CH_2OH

Threose

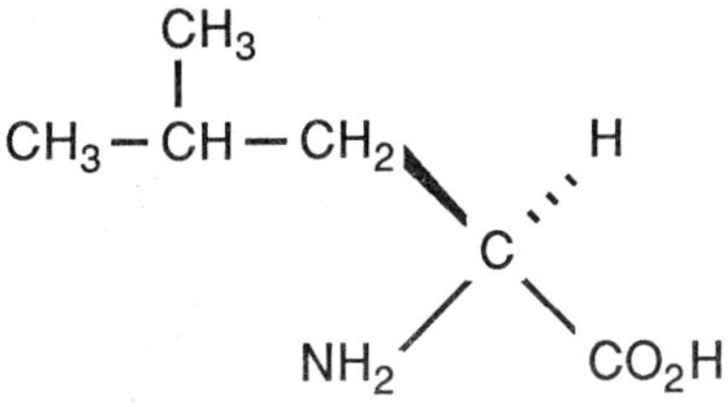

CH_3 $CH_3-CH-CH_2$ H C NH_2 CO_2H

Leu

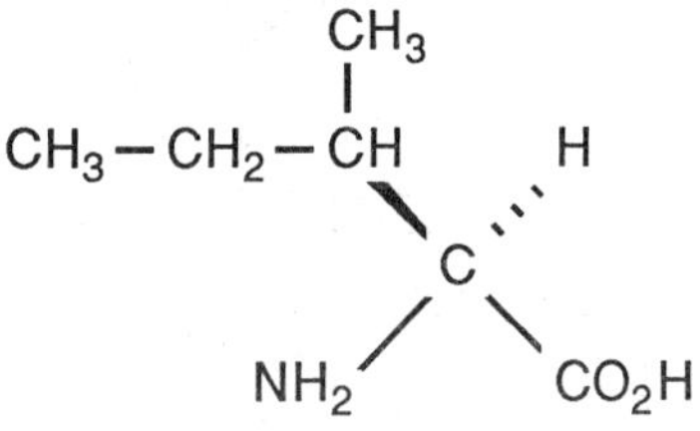

CH_3 CH_3-CH_2-CH H C NH_2 CO_2H

Ile

(d) Aspartic acid (asparagus), glutamic acid (wheat gluten), cysteine (urinary calculi: cysts), serine (*serius*: silk), tyrosine (*tyros*: cheese), and histidine (*histion*: tissue).

20.3 (a)

$$C_6H_5CH_2CH_2CO_2H \xrightarrow[PCl_3]{Br_2} C_6H_5CH_2-\overset{Br}{\overset{|}{C}H}-\overset{O}{\overset{\|}{C}}-OH \xrightarrow{NH_3} C_6H_5CH_2-\overset{NH_2}{\overset{|}{C}H}-\overset{O}{\overset{\|}{C}}-OH$$

(b)

$$C_6H_5CH_2-\overset{O}{\overset{\|}{C}H} \xrightarrow[HCN]{NH_4Cl} C_6H_5CH_2-\overset{NH_2}{\overset{|}{C}H}-CN \xrightarrow[\text{2. neutralize}]{\text{1. HCl, } H_2O} C_6H_5CH_2-\overset{NH_2}{\overset{|}{C}H}-\overset{O}{\overset{\|}{C}}-OH$$

(c)

$$Br-CH(CO_2C_2H_5)_2 \xrightarrow{\text{potassium phthalimide anion}} \text{PhthN}-CH(CO_2C_2H_5)_2 \xrightarrow[\text{2. } C_6H_5CH_2Br]{\text{1. } NaOC_2H_5} \text{PhthN}-C(CO_2C_2H_5)_2-CH_2C_6H_5 \xrightarrow[\text{2. HCl (neut)}]{\text{1. NaOH, } H_2O} C_6H_5CH_2-\overset{NH_2}{\overset{|}{C}H}-\overset{O}{\overset{\|}{C}}-OH$$

20.4

(a)

$$C_6H_5CH_2-\overset{NH_2}{\overset{|}{C}H}-\overset{O}{\overset{\|}{C}}-OH \xrightarrow[\text{2. HCl (neut.)}]{\text{1. TsCl, } NEt_3} C_6H_5CH_2-\overset{NH-SO_2-C_6H_4-CH_3}{\overset{|}{C}H}-CO_2H$$

(b)

$$C_6H_5CH_2-\overset{NH_2}{\overset{|}{C}H}-\overset{O}{\overset{\|}{C}}-OH \xrightarrow[NaHCO_3]{C_6H_5CH_2O\overset{O}{\overset{\|}{C}}-Cl} C_6H_5CH_2-\overset{C_6H_5CH_2O\overset{O}{\overset{\|}{C}}-NH}{\overset{|}{C}H}-\overset{O}{\overset{\|}{C}}-OH$$

(c)
$$C_6H_5CH_2-\underset{}{\overset{NH_2}{\overset{|}{CH}}}-\overset{O}{\overset{\|}{C}}-OH \xrightarrow{(CH_3)_3CO\overset{O}{\overset{\|}{C}}-N_3} C_6H_5CH_2-\overset{(CH_3)_3CO\overset{O}{\overset{\|}{C}}-NH}{\overset{|}{CH}}-\overset{O}{\overset{\|}{C}}-OH$$

20.5

(a)
$$C_6H_5CH_2-\overset{NH_2}{\overset{|}{CH}}-\overset{O}{\overset{\|}{C}}-OH \xrightarrow[HCl]{CH_3OH} C_6H_5CH_2-\overset{NH_2}{\overset{|}{CH}}-\overset{O}{\overset{\|}{C}}-OCH_3$$

(b)
$$C_6H_5CH_2-\overset{NH_2}{\overset{|}{CH}}-\overset{O}{\overset{\|}{C}}-OH \xrightarrow[HCl]{C_6H_5CH_2OH} C_6H_5CH_2-\overset{NH_2}{\overset{|}{CH}}-\overset{O}{\overset{\|}{C}}-OCH_2C_6H_5$$

20.6

(a)
$$HSCH_2-\overset{NH_2}{\overset{|}{CH}}-\overset{O}{\overset{\|}{C}}-OH \xrightarrow[air]{O_2} \begin{array}{l} S-CH_2-\overset{NH_2}{\overset{|}{CH}}-\overset{O}{\overset{\|}{C}}-OH \\ | \\ S-CH_2-\underset{NH_2}{\underset{|}{CH}}-\underset{O}{\underset{\|}{C}}-OH \end{array}$$

(b)
$$HSCH_2-\overset{NH_2}{\overset{|}{CH}}-\overset{O}{\overset{\|}{C}}-OH \xrightarrow[\text{2. HCl (neut)}]{\text{1. } C_6H_5CH_2Cl\text{, NaOH}} C_6H_5CH_2SCH_2-\overset{NH_2}{\overset{|}{CH}}-\overset{O}{\overset{\|}{C}}-OH$$

20.7 removal of amino-protecting groups:

(a)
$$C_6H_5CH_2-\overset{NH-SO_2-C_6H_4-CH_3}{\overset{|}{CH}}-CO_2H \xrightarrow{Na,\ NH_3} C_6H_5CH_2-\overset{NH_2}{\overset{|}{CH}}-\overset{O}{\overset{\|}{C}}-OH$$

(b)
$$C_6H_5CH_2-\overset{C_6H_5CH_2O\overset{O}{\overset{\|}{C}}-NH}{\overset{|}{CH}}-\overset{O}{\overset{\|}{C}}-OH \xrightarrow[\text{or 1. HBr; 2. } NH_3\text{, } H_2O \text{ (neut)}]{H_2\text{, Pd or Na, } NH_3} C_6H_5CH_2-\overset{NH_2}{\overset{|}{CH}}-\overset{O}{\overset{\|}{C}}-OH$$

(c)

$$(CH_3)_3COC(=O){-}NH{-}CH(CH_2C_6H_5){-}C(=O){-}OH \xrightarrow[\text{2. } NH_3,\ H_2O \text{ (neut)}]{\text{1. HCl or } CF_3CO_2H} C_6H_5CH_2{-}CH(NH_2){-}C(=O){-}OH$$

removal of carboxyl-protecting groups:

(a)

$$C_6H_5CH_2{-}CH(NH_2){-}C(=O){-}OCH_3 \xrightarrow[\text{or NaOH, } H_2O]{HCl,\ H_2O} C_6H_5CH_2{-}CH(NH_2){-}C(=O){-}OH$$

(b)

$$C_6H_5CH_2{-}CH(NH_2){-}C(=O){-}OCH_2C_6H_5 \xrightarrow[\text{or HCl, } H_2O]{H_2,\ \text{Pd or Na, } NH_3} C_6H_5CH_2{-}CH(NH_2){-}C(=O){-}OH$$

removal of thiol-protecting groups:

(a)

$$\begin{matrix} S{-}CH_2{-}CH(NH_2){-}C(=O){-}OH \\ | \\ S{-}CH_2{-}CH(NH_2){-}C(=O){-}OH \end{matrix} \xrightarrow[\text{or } HOCH_2CH_2SH]{NaBH_4 \text{ or Na, } NH_3} HSCH_2{-}CH(NH_2){-}C(=O){-}OH$$

(b)

$$C_6H_5CH_2SCH_2{-}CH(NH_2){-}C(=O){-}OH \xrightarrow[\text{or HBr (non aqueous)}]{Na,\ NH_3} HSCH_2{-}CH(NH_2){-}C(=O){-}OH$$

20.8

(a)

$$CbzNHCH_2{-}C(=O){-}OH \xrightarrow{PCl_5} CbzNHCH_2{-}C(=O){-}Cl \xrightarrow{NH_2{-}CH(CH_3){-}C(=O){-}OH} CbzNHCH_2{-}C(=O){-}NH{-}CH(CH_3){-}C(=O){-}OH$$

(b)

$$\mathrm{CbzNHCH_2{-}\overset{O}{\overset{\|}{C}}{-}OH \xrightarrow[NEt_3]{(CH_3)_2CHCH_2\overset{O}{\overset{\|}{C}}{-}Cl} CbzNHCH_2{-}\overset{O}{\overset{\|}{C}}{-}O{-}\overset{O}{\overset{\|}{C}}CH_2CH(CH_3)_2}$$

$$\mathrm{\xrightarrow{NH_2{-}\overset{CH_3}{\overset{|}{C}H}{-}\overset{O}{\overset{\|}{C}}{-}OH} CbzNHCH_2{-}\overset{O}{\overset{\|}{C}}{-}NH{-}\overset{CH_3}{\overset{|}{C}H}{-}\overset{O}{\overset{\|}{C}}{-}OH}$$

(c)

$$\mathrm{CbzNHCH_2{-}\overset{O}{\overset{\|}{C}}{-}OH \xrightarrow[2.\ HO{-}C_6H_4{-}NO_2]{1.\ PCl_5} CbzNHCH_2{-}\overset{O}{\overset{\|}{C}}{-}O{-}C_6H_4{-}NO_2}$$

$$\mathrm{\xrightarrow{NH_2{-}\overset{CH_3}{\overset{|}{C}H}{-}\overset{O}{\overset{\|}{C}}{-}OH} CbzNHCH_2{-}\overset{O}{\overset{\|}{C}}{-}NH{-}\overset{CH_3}{\overset{|}{C}H}{-}\overset{O}{\overset{\|}{C}}{-}OH}$$

(d)

$$\mathrm{CbzNHCH_2{-}\overset{O}{\overset{\|}{C}}{-}OH \xrightarrow[DCC]{NH_2{-}\overset{CH_3}{\overset{|}{C}H}{-}\overset{O}{\overset{\|}{C}}{-}OCH_3} CbzNHCH_2{-}\overset{O}{\overset{\|}{C}}{-}NH{-}\overset{CH_3}{\overset{|}{C}H}{-}\overset{O}{\overset{\|}{C}}{-}OH}$$

20.9

(a)

$$\mathrm{CH_3{-}\overset{Ts-NH}{\overset{|}{C}H}{-}\overset{O}{\overset{\|}{C}}{-}OH \xrightarrow{SOCl_2} CH_3{-}\overset{Ts-NH}{\overset{|}{C}H}{-}\overset{O}{\overset{\|}{C}}{-}Cl \xrightarrow{NH_2CH_2\overset{O}{\overset{\|}{C}}{-}OEt}}$$

$$\mathrm{CH_3{-}\overset{Ts-NH}{\overset{|}{C}H}{-}\overset{O}{\overset{\|}{C}}{-}NH{-}CH_2{-}\overset{O}{\overset{\|}{C}}{-}OEt}$$

(b)

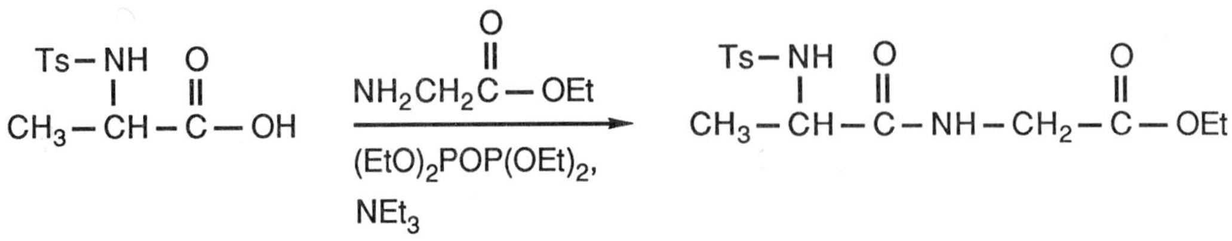

$$CH_3-CH(NH-Ts)-C(=O)-OH \xrightarrow[(EtO)_2POP(OEt)_2,\ NEt_3]{NH_2CH_2C(=O)-OEt} CH_3-CH(NH-Ts)-C(=O)-NH-CH_2-C(=O)-OEt$$

(c)

$$CH_3-CH(NH-Ts)-C(=O)-OH \xrightarrow[2.\ HO-C_6H_4-NO_2]{1.\ PCl_5} CH_3-CH(NH-Ts)-C(=O)-O-C_6H_4-NO_2$$

$$\xrightarrow{NH_2CH_2C(=O)-OEt} CH_3-CH(NH-Ts)-C(=O)-NH-CH_2-C(=O)-OEt$$

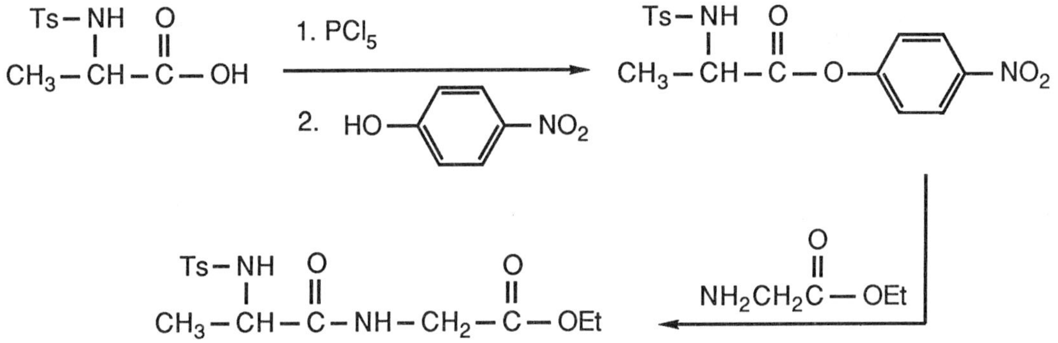

(d)

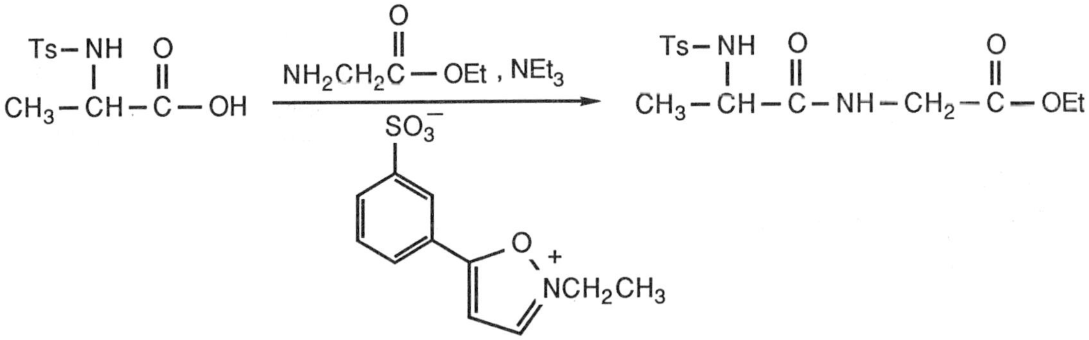

$$CH_3-CH(NH-Ts)-C(=O)-OH \xrightarrow[\text{3-}(SO_3^-)C_6H_4\text{-isoxazolium, } N^+CH_2CH_3]{NH_2CH_2C(=O)-OEt,\ NEt_3} CH_3-CH(NH-Ts)-C(=O)-NH-CH_2-C(=O)-OEt$$

20.10 (a)

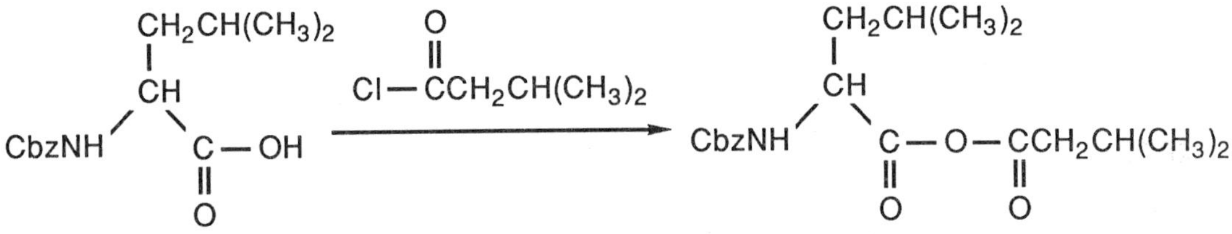

$$CbzNH-CH(CH_2CH(CH_3)_2)-C(=O)-OH \xrightarrow{Cl-C(=O)CH_2CH(CH_3)_2} CbzNH-CH(CH_2CH(CH_3)_2)-C(=O)-O-C(=O)CH_2CH(CH_3)_2$$

(b)

$$\text{N-Cbz-pyrrolidine-2-}CO_2H \xrightarrow{Cl-C(=O)CH_2CH(CH_3)_2} \text{N-Cbz-pyrrolidine-2-}C(=O)-O-C(=O)CH_2CH(CH_3)_2$$

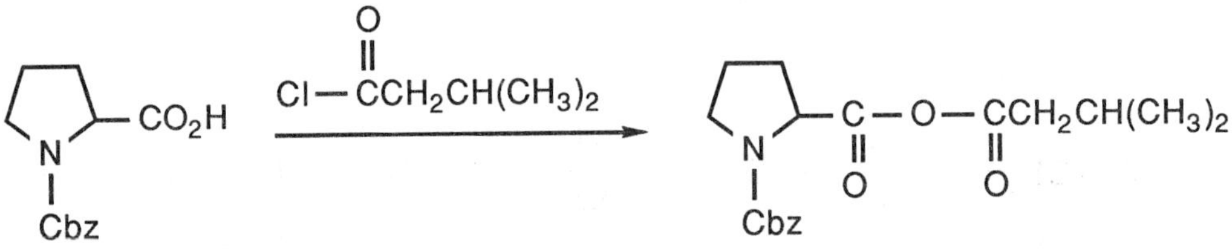

20.11

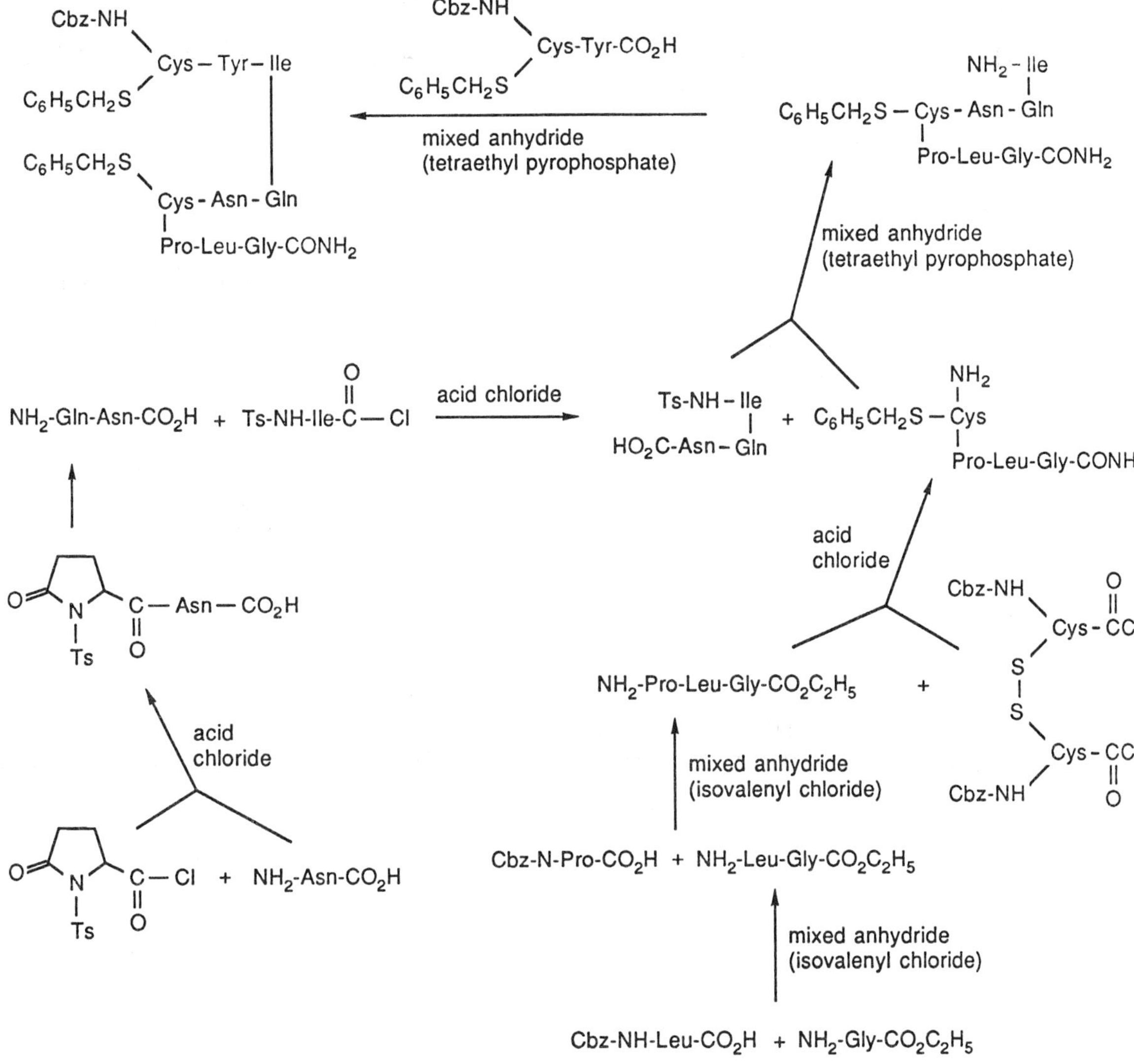

20.12

$$\mathrm{HO_2CCH_2CH_2{-}\underset{}{\overset{NH_2}{\overset{|}{C}H}}{-}CO_2H} \xrightarrow[\mathrm{NEt_3}]{\mathrm{TsCl}} \mathrm{HO_2CCH_2CH_2{-}\overset{Ts-NH}{\overset{|}{C}H}{-}CO_2H}$$

20.13

$$\mathrm{CH_3CH_2{-}\underset{\underset{CH_3}{|}}{C}H{-}\overset{NH_2}{\overset{|}{C}H}{-}CO_2H} \xrightarrow[\mathrm{NEt_3}]{\mathrm{TsCl}} \mathrm{CH_3CH_2{-}\underset{\underset{CH_3}{|}}{C}H{-}\overset{Ts-NH}{\overset{|}{C}H}{-}CO_2H}$$

20.14 **(a)** For a 30-subunit chain 29 coupling reactions must be carried out. With an average yield of 75% the overall yield would be $(0.75)^{29} \times 100 = 2.4 \times 10^{-4} \times 100$ or 0.02%.

(b) With an average yield of 75% the overall yield would be $(0.94)^{29} \times 100 = 0.17 \times 100$ or 17%.

20.15

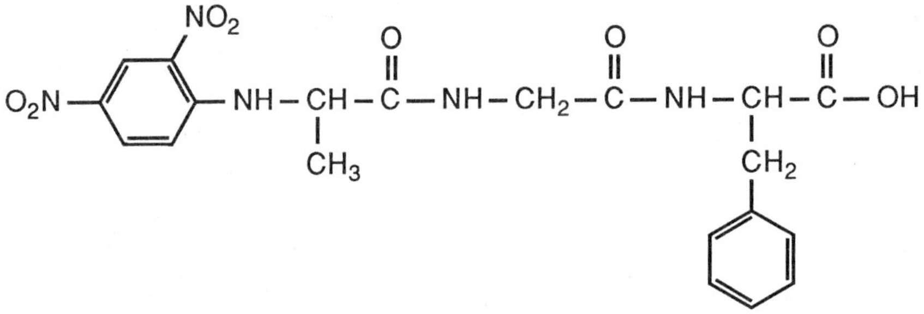

$$2,4\text{-}(O_2N)_2C_6H_3\text{—NH—CH(CH}_3\text{)—C(=O)—NH—CH}_2\text{—C(=O)—NH—CH(CH}_2C_6H_5\text{)—C(=O)—OH}$$

20.16

$$2,4\text{-}(O_2N)_2C_6H_3\text{—NH—CH(CH}_3\text{)—CO}_2\text{H} + NH_2\text{—CH(CH}_2C_6H_5\text{)—C(=O)—OH} + NH_2CH_2CO_2H$$

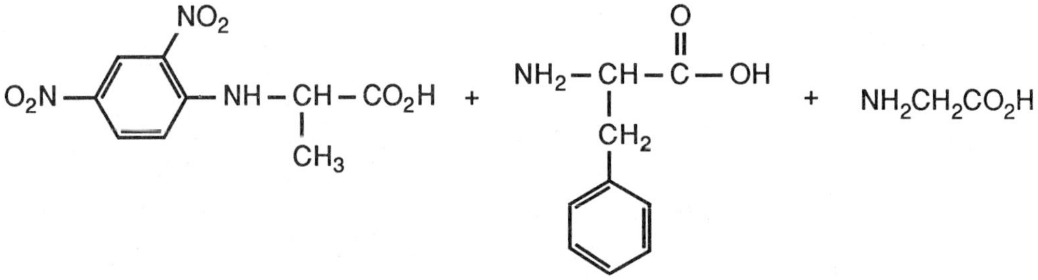

ANSWERS TO PROBLEMS

20.1 **(a)** *L*-alanyl-*L*-seryl-*L*-leucine

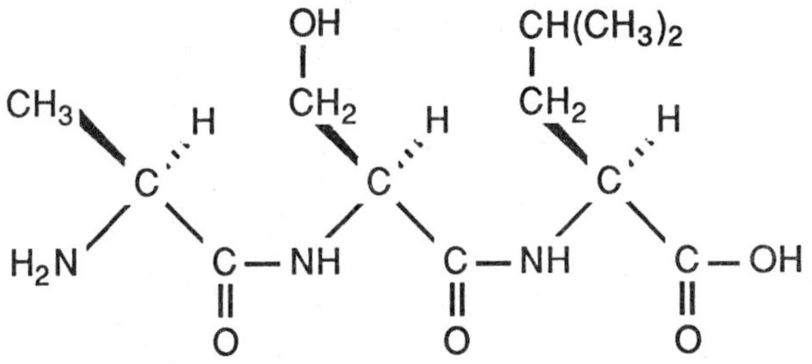

$$H_2N\text{—CH(CH}_3\text{)—C(=O)—NH—CH(CH}_2\text{OH)—C(=O)—NH—CH(CH}_2\text{CH(CH}_3)_2\text{)—C(=O)—OH}$$

L-alanyl-*D*-seryl-*L*-leucine

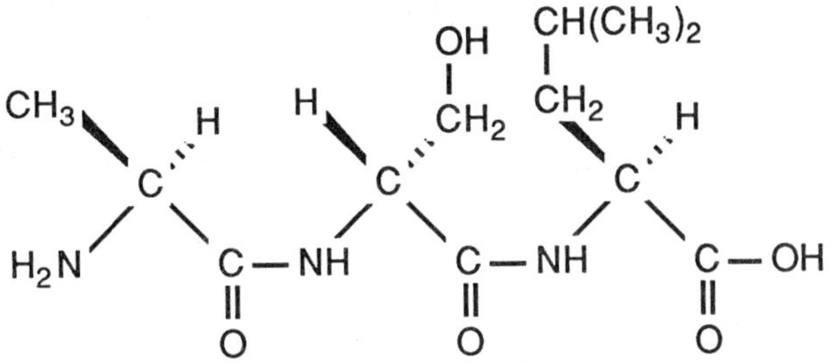

OH CH(CH$_3$)$_2$
CH$_3$ H H CH$_2$ CH$_2$ H
C C C
H$_2$N C—NH C—NH C—OH
O O O

(b) diastereomers

20.2 HOCH$_2$ H C H$_2$N CO$_2$H $\xrightarrow{PBr_3}$ BrCH$_2$ H C H$_2$N CO$_2$H $\xrightarrow{Zn, HCl}$ CH$_3$ H C H$_2$N CO$_2$H

20.3 HOCH$_2$ H C H$_2$N CO$_2$H $\xrightarrow{TsCl}$ TsOCH$_2$ H C TsNH CO$_2$H $\xrightarrow{1.\ Na_2S;\ 2.\ HCl\ (neut.)}$ HSCH$_2$ H C TsNH CO$_2$H $\xrightarrow{Na, NH_3}$ HSCH$_2$ H C NH$_2$ CO$_2$H

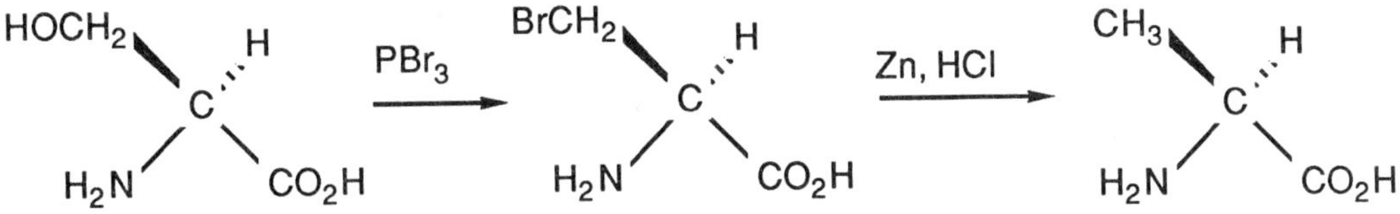

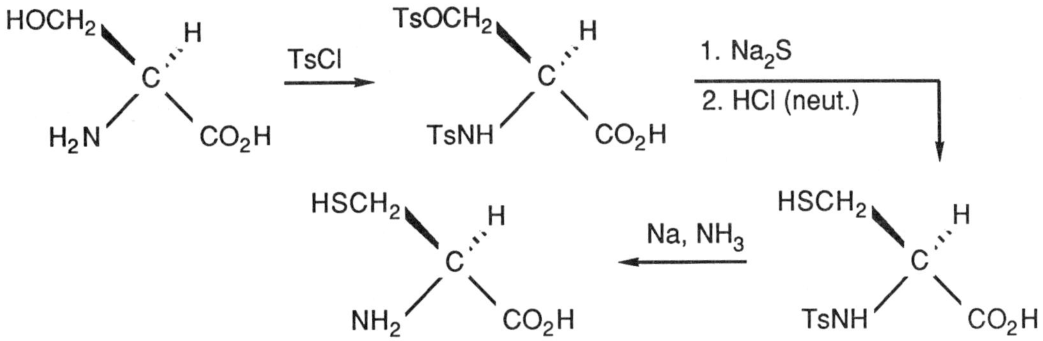

20.4 **(a)** $CH_3-CH(NH_2)-CO_2H$ Ala; $HOCH_2-CH(NH_2)-CO_2H$ Ser; $HSCH_2-CH(NH_2)-CO_2H$ Cys

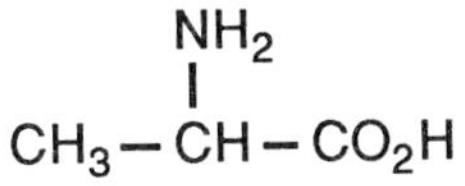

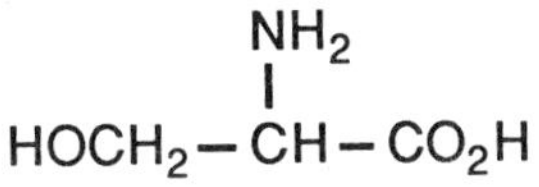

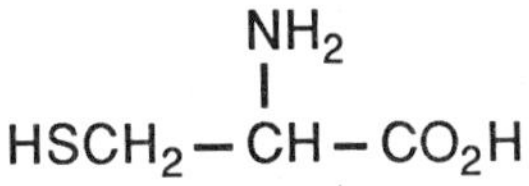

(b) $CH_3SCH_2CH_2-CH(NH_2)-CO_2H$ Met; $CH_3-CH(OH)-CH(NH_2)-CO_2H$ Thr; $HO_2CCH_2-CH(NH_2)-CO_2H$ Asp

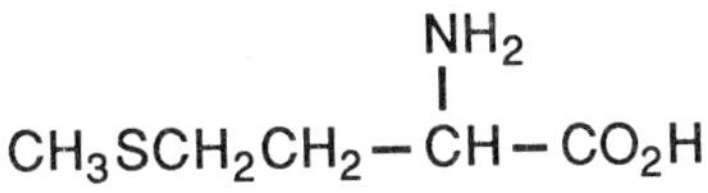

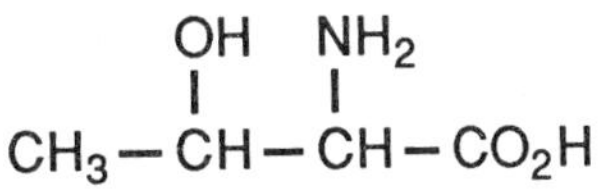

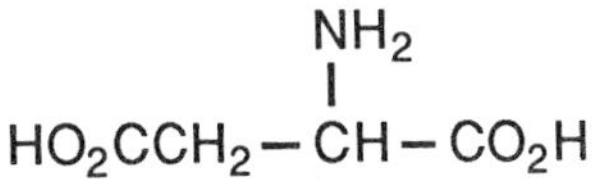

$H_2NC(=O)-CH_2-CH(NH_2)-CO_2H$ Asn

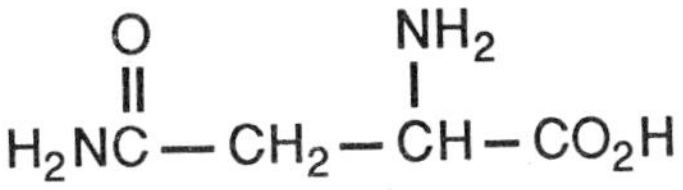

(c)

$$HO_2CCH_2CH_2-\overset{NH_2}{\overset{|}{C}}H-CO_2H$$

Glu

$$H_2N\overset{O}{\overset{\|}{C}}-CH_2-CH_2-\overset{NH_2}{\overset{|}{C}}H-CO_2H$$

Gln

Pro (pyrrolidine ring: NH—CH—CO_2H)

$$H_2N\overset{NH}{\overset{\|}{C}}-NH-CH_2-CH_2-CH_2-\overset{NH_2}{\overset{|}{C}}H-CO_2H$$

Arg

$$\text{(imidazol-4-yl)}-CH_2-\overset{NH_2}{\overset{|}{C}}H-CO_2H$$

His

$$CH_3-\overset{CH_3}{\overset{|}{C}}H-\overset{NH_2}{\overset{|}{C}}H-CO_2H$$

Val

Known biochemical pathways include the following:

(a)

$$HOCH_2-\overset{NH_2}{\overset{|}{C}}H-CO_2H \xrightarrow[\text{substitution}]{\text{nucleophilic}} HSCH_2-\overset{NH_2}{\overset{|}{C}}H-CO_2H$$

Ser → Cys

(b)

$$HO_2CCH_2-\overset{NH_2}{\overset{|}{C}}H-CO_2H \xrightarrow[\text{formation}]{\text{amide}} H_2N\overset{O}{\overset{\|}{C}}-CH_2-\overset{NH_2}{\overset{|}{C}}H-CO_2H$$

Asp → Asn

Asp ↓ reduction

$$HOCH_2CH_2-\overset{NH_2}{\overset{|}{C}}H-CO_2H \xrightarrow[\text{substitution}]{\text{nucleophilic}} CH_3SCH_2CH_2-\overset{NH_2}{\overset{|}{C}}H-CO_2H$$

homoserine → Met

homoserine ↓ elimination of H_2O, then addition of H_2O

$$CH_3-\overset{OH}{\overset{|}{C}}H-\overset{NH_2}{\overset{|}{C}}H-CO_2H$$

Thr

(c)

$$HO_2CCH_2CH_2-\underset{\text{Glu}}{CH(NH_2)}-CO_2H \xrightarrow[\text{formation}]{\text{amide}} H_2N\overset{O}{\overset{\|}{C}}-CH_2-CH_2-CH(NH_2)-CO_2H \text{ (Gln)}$$

Glu $\xrightarrow{\text{reduction}}$

$$H\overset{O}{\overset{\|}{C}}CH_2CH_2-CH(NH_2)-CO_2H \xrightarrow[\text{and reduction}]{\text{cyclization}} \text{Pro (cyclic: } -NH-CH-CO_2H \text{ ring)}$$

Glu $\xrightarrow{\text{reduction and amination}}$

$$H_2NCH_2CH_2-CH(NH_2)-CO_2H \text{ (ornithine)} \xrightarrow[\text{formation}]{\text{amidine}} H_2N\overset{NH}{\overset{\|}{C}}-NH-CH_2-CH_2-CH_2-CH(NH_2)-CO_2H \text{ (Arg)}$$

20.5 Recall that a group will be 50% protonated when the pH is equal to the pK_a of the protonated form. The pK_a values for the peptide carboxyl, the side chain carboxyl, and the amino groups of aspartic acid and glutamic acid are 2, 4, and 10, respectively:

Glu

$$HO_2CCH_2CH_2-CH(\overset{+}{N}H_3)-CO_2H \quad pH \leq 2$$

$$HO_2CCH_2CH_2-CH(\overset{+}{N}H_3)-CO_2^- \quad pH = 2.5 - 3.5$$

$$^-O_2CCH_2CH_2-CH(\overset{+}{N}H_3)-CO_2^- \quad pH = 4.5 - 9.5$$

$$^-O_2CCH_2CH_2-CH(NH_2)-CO_2^- \quad pH > 10$$

Asp

$$HO_2CCH_2-CH(\overset{+}{N}H_3)-CO_2H \quad pH < 2$$

$$HO_2CCH_2-CH(\overset{+}{N}H_3)-CO_2^- \quad pH = 2.5 - 3.5$$

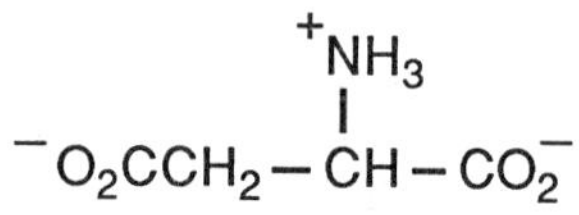

$$^{-}O_2CCH_2-\underset{\overset{|}{^{+}NH_3}}{CH}-CO_2^-$$
pH = 4.5 - 9.5

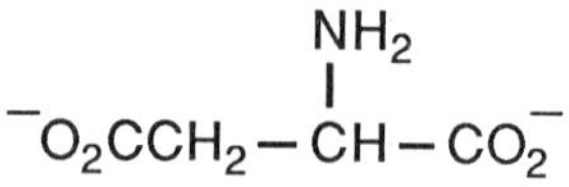

$$^{-}O_2CCH_2-\underset{\overset{|}{NH_2}}{CH}-CO_2^-$$
pH > 10

20.6 The pK_a values for the carboxyl, the basic side chain group, and the α amino groups of lysine 2, 9, and 10.5, respectively, while the corresponding values for arginine are 2, 9, and 12.5:

Lys

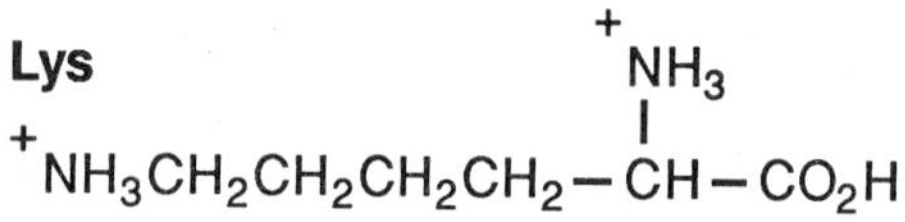

$$^{+}NH_3CH_2CH_2CH_2CH_2-CH(^{+}NH_3)-CO_2H$$
pH < 2

$$^{+}NH_3CH_2CH_2CH_2CH_2-CH(^{+}NH_3)-CO_2^-$$ 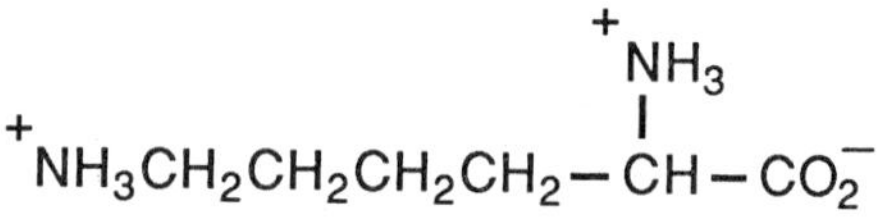
pH = 2.5 - 8.5

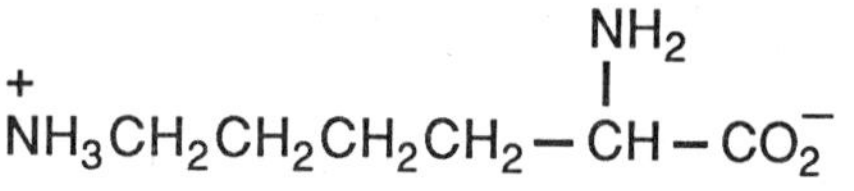

$$^{+}NH_3CH_2CH_2CH_2CH_2-CH(NH_2)-CO_2^-$$
pH = 9.5 - 10

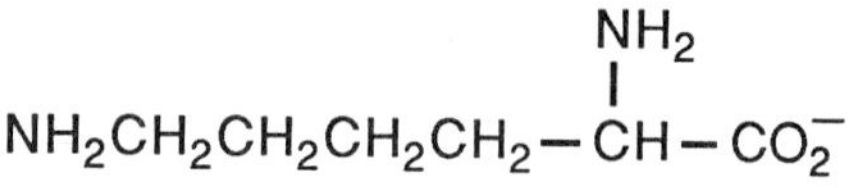

$$NH_2CH_2CH_2CH_2CH_2-CH(NH_2)-CO_2^-$$
pH > 10.5

Arg

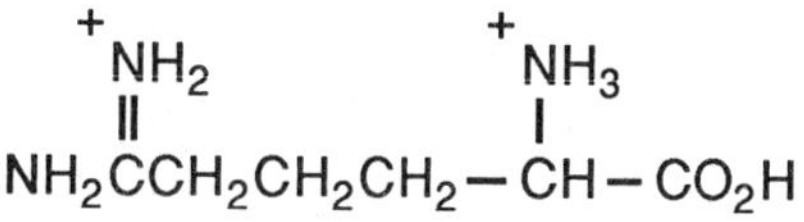

$$NH_2C(={}^{+}NH_2)CH_2CH_2CH_2-CH(^{+}NH_3)-CO_2H$$
pH < 2

$$NH_2C(={}^{+}NH_2)CH_2CH_2CH_2-CH(^{+}NH_3)-CO_2^-$$ 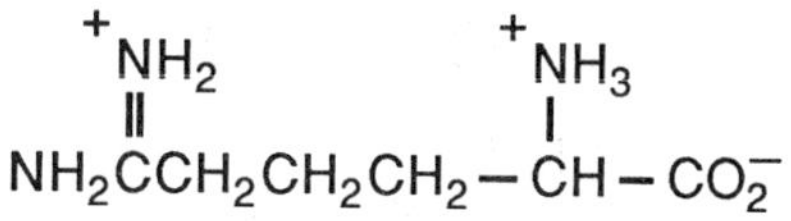
pH = 2.5 - 8.5

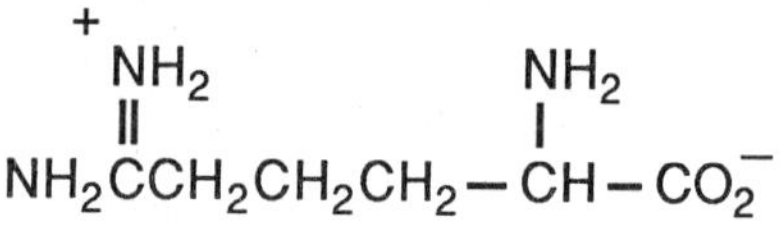

$$NH_2C(={}^{+}NH_2)CH_2CH_2CH_2-CH(NH_2)-CO_2^-$$
pH = 9.5 - 12

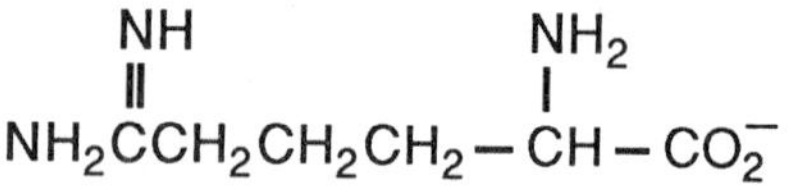

$$NH_2C(=NH)CH_2CH_2CH_2-CH(NH_2)-CO_2^-$$
pH > 12.5

His

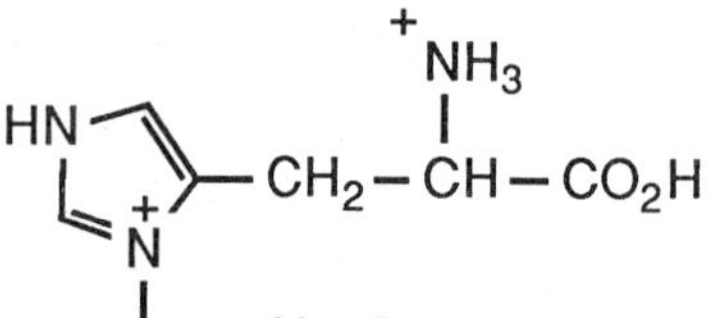

HN(imidazolium ring, $\overset{+}{N}$–H)–$CH_2-CH(^{+}NH_3)-CO_2H$
pH < 2

HN(imidazolium ring, $\overset{+}{N}$–H)–$CH_2-CH(^{+}NH_3)-CO_2^-$ 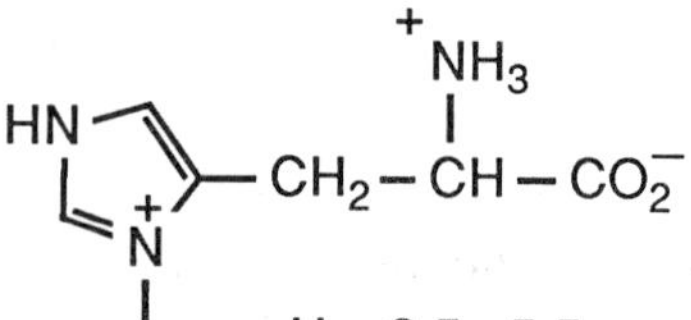
pH = 2.5 - 5.5

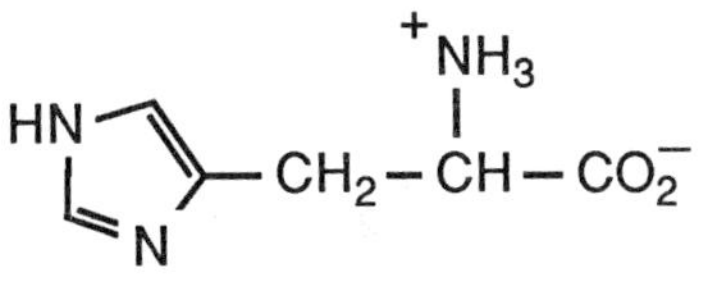

$$HN\text{-imidazolyl}-CH_2-\overset{+NH_3}{CH}-CO_2^-$$

pH = 6.5 - 8.5

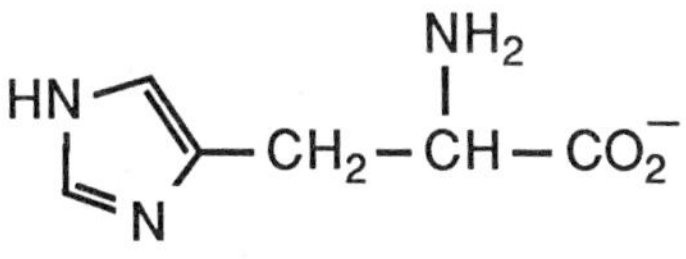

$$HN\text{-imidazolyl}-CH_2-\overset{NH_2}{CH}-CO_2^-$$

pH ≥ 9

20.7 **(a)**

$(CH_3)HC{=}C(H)(CO_2CH_3)$ $\xrightarrow[CH_3OH]{Hg(OAc)_2}$ $CH_3O-CH(CH_3)-CH(HgOAc)-CO_2CH_3$ $\xrightarrow[2.\ Br_2]{1.\ KBr}$ $CH_3O-CH(CH_3)-CH(Br)-CO_2CH_3$ $\xrightarrow[2.\ HCl\ (heat)]{1.\ NaOH,\ H_2O}$ $CH_3O-CH(CH_3)-CH(Br)-CO_2H$ $\xrightarrow{NH_3}$ $CH_3O-CH(CH_3)-CH(NH_2)-CO_2H$ $\xrightarrow[2.\ NH_3\ (heat)]{1.\ conc.\ HBr}$ $HO-CH(CH_3)-CH(NH_2)-CO_2H$

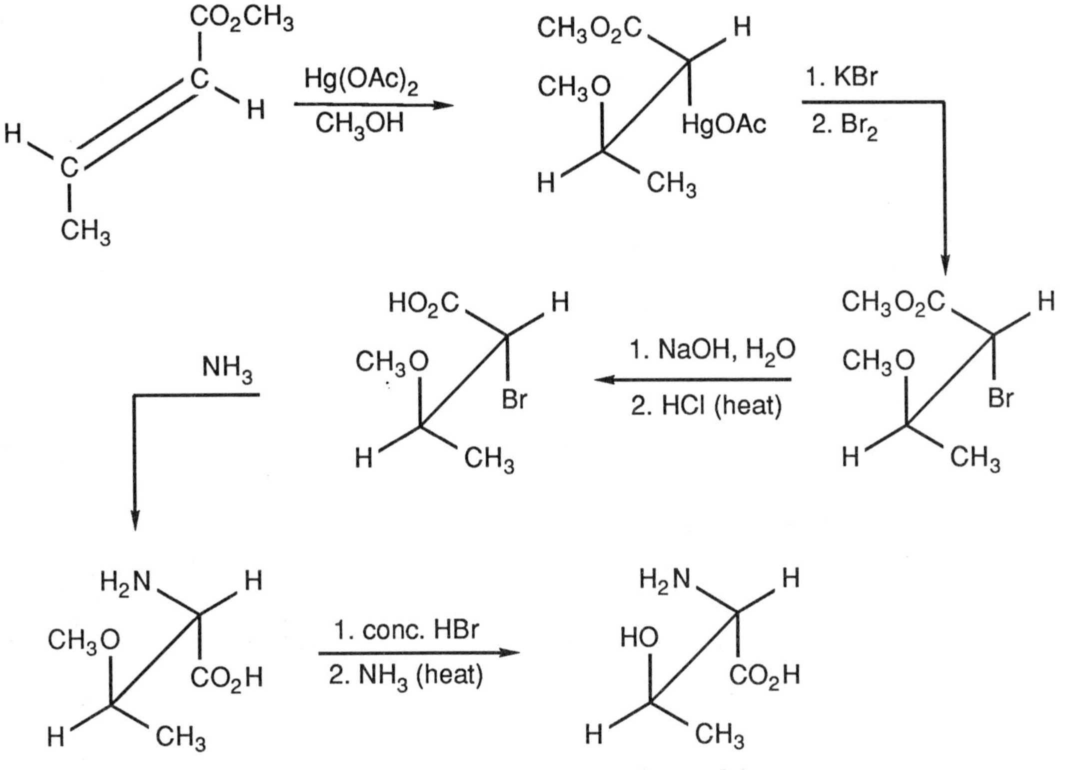

(racemic)

(b) The free hydroxyl group could cyclize to the epoxide:

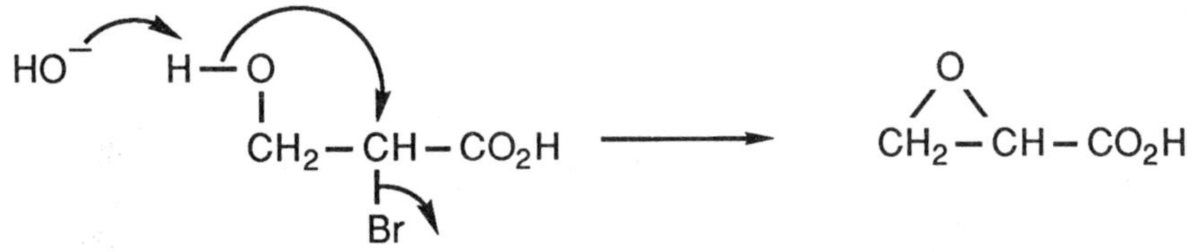

$$HO^- \quad H-O-CH_2-\underset{Br}{CH}-CO_2H \longrightarrow \overset{O}{CH_2-CH}-CO_2H$$

(c) Introduction of oxygen on the more substituted carbon atom would have to proceed via a transition state in which positive charge developed on the same carbon, but this carbon also carries the electron withdrawing carboxyl group (resulting in an unfavorable interaction):

$$CH_2{=}CH{-}CO_2CH_3 \longrightarrow \overset{+}{C}H_2{-}CH(HgOAc){-}CO_2CH_3$$

$$CH_2{=}CH{-}CO_2CH_3 \longrightarrow CH_2(HgOAc){-}\overset{+}{C}H{-}CO_2CH_3$$

20.8 Nucleophilic displacement with ammonia:

$$(CH_3)_2CH{-}CH_2{-}CO_2H \xrightarrow[PCl_3]{Br_2} (CH_3)_2CH{-}CH(Br){-}CO_2H \xrightarrow{NH_3} (CH_3)_2CH{-}CH(NH_2){-}CO_2H$$

Addition of HCN:

$$(CH_3)_2CH{-}CHO \xrightarrow[HCN]{NH_4Cl} (CH_3)_2CH{-}CH(NH_2){-}CN \xrightarrow[\text{2. base (neutralize)}]{\text{1. HCl, } H_2O} (CH_3)_2CH{-}CH(NH_2){-}CO_2H$$

Malonic ester synthesis:

$$Br{-}CH(CO_2Et)_2 \xrightarrow{\text{phthalimide anion } (N^-)} \text{PhthN}{-}CH(CO_2Et)_2 \xrightarrow[\text{2. } (CH_3)_2CHBr]{\text{1. NaOEt}} \text{PhthN}{-}C(CO_2Et)_2{-}CH(CH_3)_2$$

$$\text{PhthN}{-}C(CO_2Et)_2{-}CH(CH_3)_2 \xrightarrow[\text{2. HCl (heat)}]{\text{1. NaOH, } H_2O} (CH_3)_2CH{-}CH(NH_2){-}CO_2H$$

20.9 Nucleophilic displacement with ammonia:

$$CH_3CH_2-CH(CH_3)-CH_2-CO_2H \xrightarrow[PCl_3]{Br_2} CH_3CH_2-CH(CH_3)-CH(Br)-CO_2H \xrightarrow{NH_3} CH_3CH_2-CH(CH_3)-CH(NH_2)-CO_2H$$

Addition of HCN:

$$CH_3CH_2-CH(CH_3)-CHO \xrightarrow[HCN]{NH_4Cl} CH_3CH_2-CH(CH_3)-CH(NH_2)-CN$$

1. HCl, H_2O
2. base (neutralize)

$$\longrightarrow CH_3CH_2-CH(CH_3)-CH(NH_2)-CO_2H$$

Malonic ester synthesis:

$$Br-CH(CO_2Et)_2 \xrightarrow{\text{phthalimide anion } (N^-)} \text{phthalimido-}N-CH(CO_2Et)_2$$

1. NaOEt
2. $CH_3CH_2-CHBr(CH_3)$

$$\longrightarrow \text{phthalimido-}N-C(CO_2Et)_2-CHCH_3-CH_2CH_3$$

1. NaOH, H_2O
2. HCl (heat)

$$\longrightarrow CH_3CH_2-CH(CH_3)-CH(NH_2)-CO_2H$$

20.10 (a)

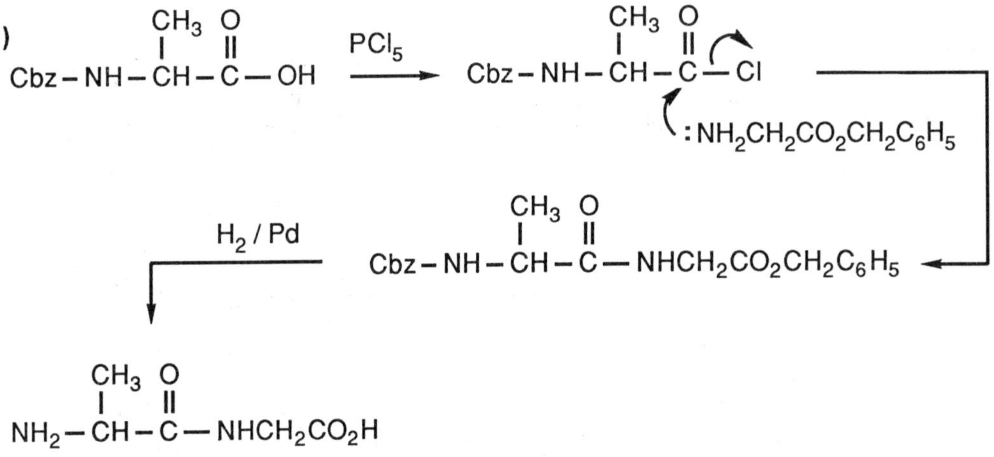

$$\text{Cbz-NH-CH(CH}_3\text{)-C(=O)-OH} \xrightarrow{\text{PCl}_5} \text{Cbz-NH-CH(CH}_3\text{)-C(=O)-Cl}$$

$$:\text{NH}_2\text{CH}_2\text{CO}_2\text{CH}_2\text{C}_6\text{H}_5$$

$$\longrightarrow \text{Cbz-NH-CH(CH}_3\text{)-C(=O)-NHCH}_2\text{CO}_2\text{CH}_2\text{C}_6\text{H}_5 \xrightarrow{\text{H}_2\text{ / Pd}} \text{NH}_2\text{-CH(CH}_3\text{)-C(=O)-NHCH}_2\text{CO}_2\text{H}$$

(b)

$$\text{Cbz-NH-CH(CH}_3\text{)-C(=O)-OH} \underset{}{\overset{\text{NEt}_3}{\rightleftharpoons}} \text{Cbz-NH-CH(CH}_3\text{)-C(=O)-O}^- \quad \text{Cl-C(=O)OC}_2\text{H}_5$$

$$\longrightarrow \text{Cbz-NH-CH(CH}_3\text{)-C(=O)-O-C(=O)OC}_2\text{H}_5$$

$$\xrightarrow{\text{C}_6\text{H}_5\text{CH}_2\text{O}_2\text{CCH}_2\ddot{\text{N}}\text{H}_2} \text{Cbz-NH-CH(CH}_3\text{)-C(=O)-NHCH}_2\text{CO}_2\text{CH}_2\text{C}_6\text{H}_5 \xrightarrow{\text{H}_2\text{ / Pd}} \text{NH}_2\text{-CH(CH}_3\text{)-C(=O)-NHCH}_2\text{CO}_2\text{H}$$

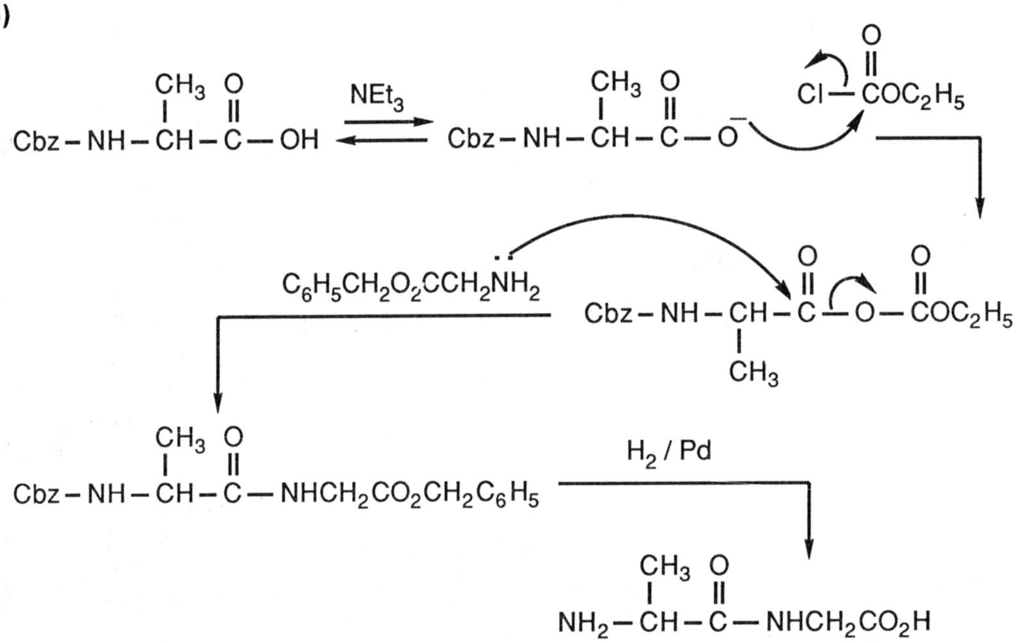

(c)

$$\text{Cbz-NH-CH(CH}_3\text{)-C(=O)-OH} \underset{}{\overset{\text{NEt}_3}{\rightleftharpoons}} \text{Cbz-NH-CH(CH}_3\text{)-C(=O)-O}^- \quad (\text{EtO})_2\text{P-O-P(OEt)}_2$$

$$\downarrow$$

$$\text{Cbz-NH-CH(CH}_3\text{)-C(=O)-OP(OEt)}_2$$

$$\xrightarrow{\text{C}_6\text{H}_5\text{CH}_2\text{O}_2\text{CCH}_2\text{NH}_2}$$

$$\text{Cbz-NH-CH(CH}_3\text{)-C(=O)-NHCH}_2\text{CO}_2\text{CH}_2\text{C}_6\text{H}_5 \xrightarrow{\text{H}_2\,/\,\text{Pd}} \text{NH}_2\text{-CH(CH}_3\text{)-C(=O)-NHCH}_2\text{CO}_2\text{H}$$

(d)

$$\text{Cbz-NH-CH(CH}_3\text{)-C(=O)-OH} \xrightarrow{\text{PCl}_5} \text{Cbz-NH-CH(CH}_3\text{)-C(=O)-Cl}$$

$$\xrightarrow{^-\text{O-C}_6\text{H}_4\text{-NO}_2} \text{Cbz-NH-CH(CH}_3\text{)-C(=O)-O-C}_6\text{H}_4\text{-NO}_2$$

$$\xrightarrow{\text{C}_6\text{H}_5\text{CH}_2\text{O}_2\text{CCH}_2\text{NH}_2}$$

$$\text{Cbz-NH-CH(CH}_3\text{)-C(=O)-NHCH}_2\text{CO}_2\text{CH}_2\text{C}_6\text{H}_5 \xrightarrow{\text{H}_2\,/\,\text{Pd}} \text{NH}_2\text{-CH(CH}_3\text{)-C(=O)-NHCH}_2\text{CO}_2\text{H}$$

(e)

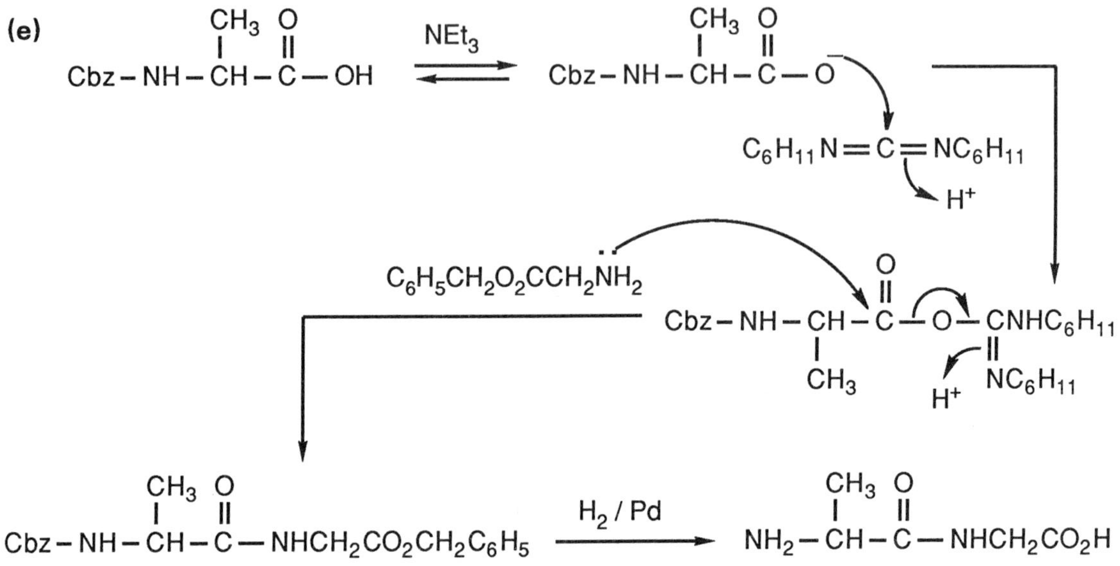

$$\text{Cbz-NH-CH(CH}_3\text{)-C(=O)-OH} \xrightleftharpoons{NEt_3} \text{Cbz-NH-CH(CH}_3\text{)-C(=O)-O}^-$$

$C_6H_{11}N{=}C{=}NC_6H_{11}$; H^+

$C_6H_5CH_2O_2CCH_2NH_2$

$$\text{Cbz-NH-CH(CH}_3\text{)-C(=O)-O-C(=NC}_6\text{H}_{11}\text{)NHC}_6\text{H}_{11}$$; H^+

$$\text{Cbz-NH-CH(CH}_3\text{)-C(=O)-NHCH}_2\text{CO}_2\text{CH}_2\text{C}_6\text{H}_5 \xrightarrow{H_2/Pd} \text{NH}_2\text{-CH(CH}_3\text{)-C(=O)-NHCH}_2\text{CO}_2\text{H}$$

(f)

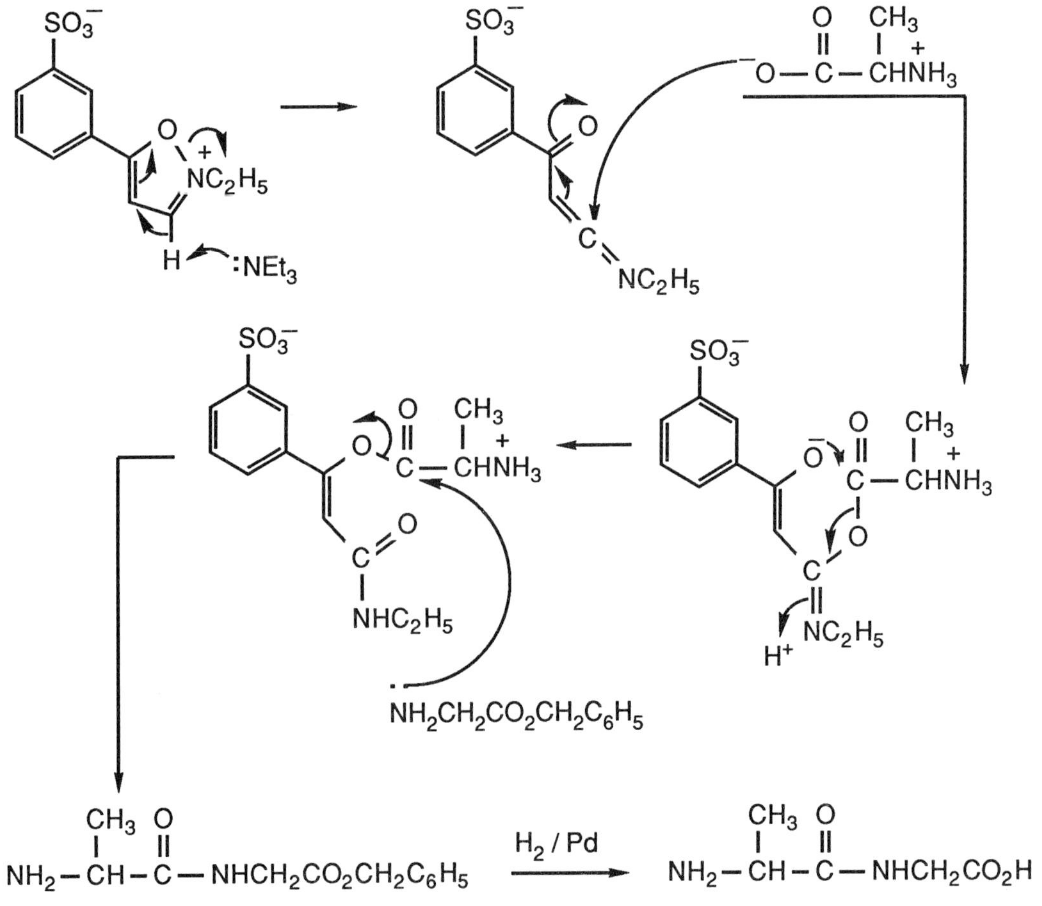

SO_3^- ; NC_2H_5 ; H ; $:NEt_3$ → SO_3^- ; O ; C ; NC_2H_5

$^-O-C(=O)-CH(CH_3)\overset{+}{N}H_3$

SO_3^- ; O^- ; $C-CHNH_3^+$; O ; C ; NC_2H_5 ; H^+

← SO_3^- ; O ; $C(=O)-CH(CH_3)\overset{+}{N}H_3$; C=O ; NHC_2H_5

$NH_2CH_2CO_2CH_2C_6H_5$

$$\text{NH}_2\text{-CH(CH}_3\text{)-C(=O)-NHCH}_2\text{CO}_2\text{CH}_2\text{C}_6\text{H}_5 \xrightarrow{H_2/Pd} \text{NH}_2\text{-CH(CH}_3\text{)-C(=O)-NHCH}_2\text{CO}_2\text{H}$$

20.11 (a)

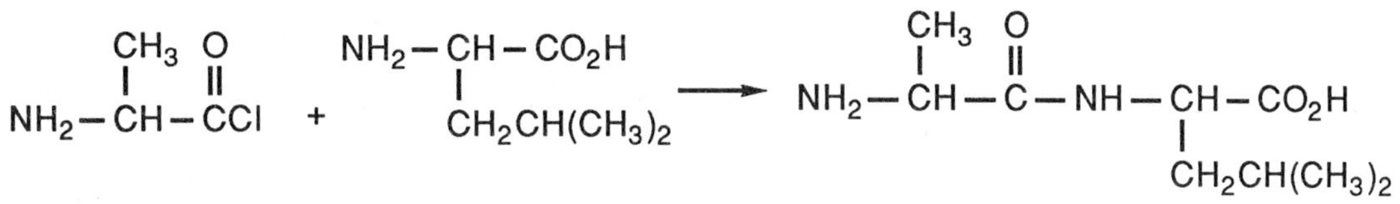

(b)

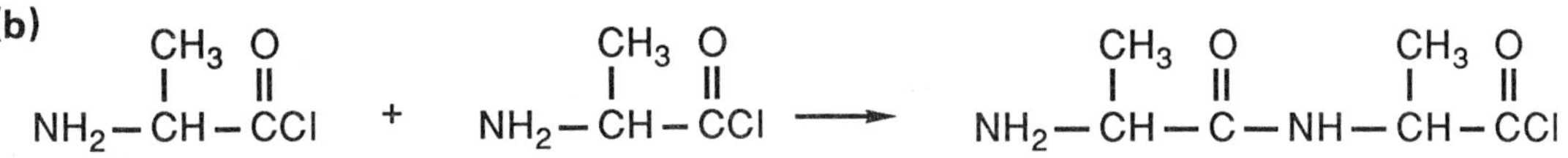

(c)

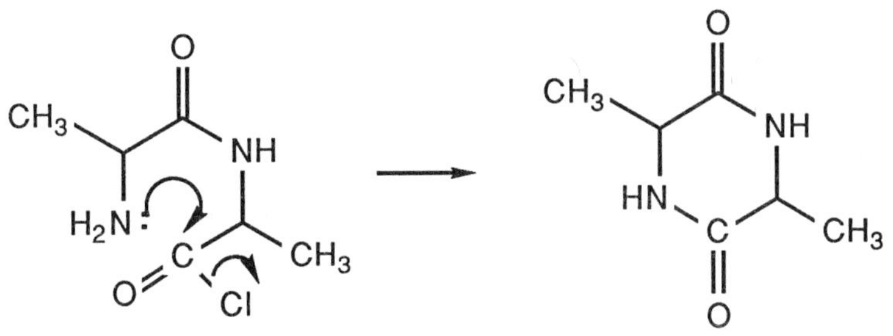

20.12

$$NH_2-\underset{CH_2CH(CH_3)_2}{CH}-\overset{O}{\overset{\|}{C}}-Cl + NH_2-\underset{CH_2OH}{CH}-\overset{O}{\overset{\|}{C}}-OH \longrightarrow \mathbf{A} + \mathbf{B} + \mathbf{C}$$

A: $NH_2-\underset{CH_2CH(CH_3)_2}{CH}-\overset{O}{\overset{\|}{C}}-NH-\underset{CH_2CH(CH_3)_2}{CH}-\overset{O}{\overset{\|}{C}}-Cl$

B: $NH_2-\underset{CH_2CH(CH_3)_2}{CH}-\overset{O}{\overset{\|}{C}}-NH-\underset{CH_2OH}{CH}-\overset{O}{\overset{\|}{C}}-OH$

C: $NH_2-\underset{CH_2CH(CH_3)_2}{CH}-\overset{O}{\overset{\|}{C}}-O-CH_2-\overset{NH_2}{\overset{|}{C}}HCO_2H$

A + Leu—C(=O)—Cl →

$$NH_2-\underset{CH_2CH(CH_3)_2}{CH}-\overset{O}{\overset{\|}{C}}-NH-\underset{CH_2CH(CH_3)_2}{CH}-\overset{O}{\overset{\|}{C}}-NH-\underset{CH_2CH(CH_3)_2}{CH}-\overset{O}{\overset{\|}{C}}-Cl$$

A + Ser → ; B + Leu—C(=O)—Cl →

$$NH_2-\underset{CH_2CH(CH_3)_2}{CH}-\overset{O}{\overset{\|}{C}}-NH-\underset{CH_2CH(CH_3)_2}{CH}-\overset{O}{\overset{\|}{C}}-NH-\underset{CH_2OH}{CH}-\overset{O}{\overset{\|}{C}}-OH$$

A + Ser → ; C + Leu—C(=O)—Cl →

$$NH_2-\underset{CH_2CH(CH_3)_2}{CH}-\overset{O}{\overset{\|}{C}}-NH-\underset{CH_2CH(CH_3)_2}{CH}-\overset{O}{\overset{\|}{C}}-O-CH_2-\overset{NH_2}{\overset{|}{C}}HCO_2H$$

B + Leu—C(=O)—Cl → ; C + Leu—C(=O)—Cl →

$$NH_2-\underset{CH_2CH(CH_3)_2}{CH}-\overset{O}{\overset{\|}{C}}-NH-\underset{CH_2-O-\underset{O}{\underset{\|}{C}}-\underset{CH_2CH(CH_3)_2}{CH}-NH_2}{CH}-\overset{O}{\overset{\|}{C}}-OH$$

20.13

$$NH_2-\underset{CH_3}{CH}-\overset{O}{\overset{\|}{C}}-Cl \;+\; NH_2-\underset{CH_2(CH_2)_3NH_2}{CH}-\overset{O}{\overset{\|}{C}}-OH \longrightarrow \mathbf{A} + \mathbf{B} + \mathbf{C}$$

A: $NH_2-\underset{CH_3}{CH}-\overset{O}{\overset{\|}{C}}-NH-\underset{CH_3}{CH}-\overset{O}{\overset{\|}{C}}-Cl$

B: $NH_2-\underset{CH_3}{CH}-\overset{O}{\overset{\|}{C}}-NH-\underset{CH_2(CH_2)_3NH_2}{CH}-\overset{O}{\overset{\|}{C}}-OH$

C: $NH_2-\underset{CH_3}{CH}-\overset{O}{\overset{\|}{C}}-NH(CH_2)_4-\overset{NH_2}{\overset{|}{C}}HCO_2H$

A + Ala–$\overset{O}{\overset{\|}{C}}$–Cl →

$NH_2-\underset{CH_3}{CH}-\overset{O}{\overset{\|}{C}}-NH-\underset{CH_3}{CH}-\overset{O}{\overset{\|}{C}}-NH-\underset{CH_3}{CH}-\overset{O}{\overset{\|}{C}}-Cl$

A + Lys →; B + Ala–$\overset{O}{\overset{\|}{C}}$–Cl →

$NH_2-\underset{CH_3}{CH}-\overset{O}{\overset{\|}{C}}-NH-\underset{CH_3}{CH}-\overset{O}{\overset{\|}{C}}-NH-\underset{CH_2(CH_2)_3NH_2}{CH}-\overset{O}{\overset{\|}{C}}-OH$

A + Lys →; C + Ala–$\overset{O}{\overset{\|}{C}}$–Cl →

$NH_2-\underset{CH_3}{CH}-\overset{O}{\overset{\|}{C}}-NH-\underset{CH_3}{CH}-\overset{O}{\overset{\|}{C}}-NH(CH_2)_4-\overset{NH_2}{\overset{|}{C}}HCO_2H$

B + Ala–$\overset{O}{\overset{\|}{C}}$–Cl →; C + Ala–$\overset{O}{\overset{\|}{C}}$–Cl →

$NH_2-\underset{CH_3}{CH}-\overset{O}{\overset{\|}{C}}-NH(CH_2)_4-CHCO_2H$, with $-NH-\underset{O}{\underset{\|}{C}}-\underset{NH_2}{CHCH_3}$ on the CH of $CHCO_2H$

20.14

(a)

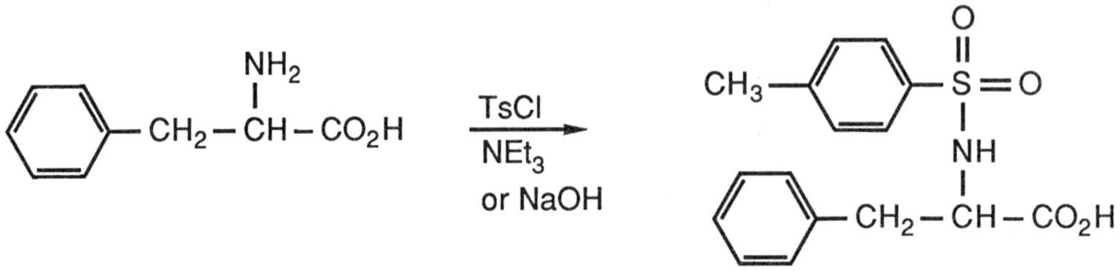

$$C_6H_5-CH_2-CH(NH_2)-CO_2H \xrightarrow[\text{NEt}_3 \text{ or NaOH}]{\text{TsCl}} C_6H_5-CH_2-CH(NH-SO_2-C_6H_4-CH_3)-CO_2H$$

(b)

$$C_6H_5-CH_2-CH(NH_2)-CO_2H \xrightarrow[\text{NaHCO}_3 \text{ or NaOH}]{Cl-C(=O)OCH_2C_6H_5} C_6H_5-CH_2-CH(NH-C(=O)OCH_2C_6H_5)-CO_2H$$

(c)

$$C_6H_5-CH_2-CH(NH_2)-CO_2H \xrightarrow[\text{NEt}_3]{Cl-C(=O)OC(CH_3)_3} C_6H_5-CH_2-CH(NH-C(=O)OC(CH_3)_3)-CO_2H$$

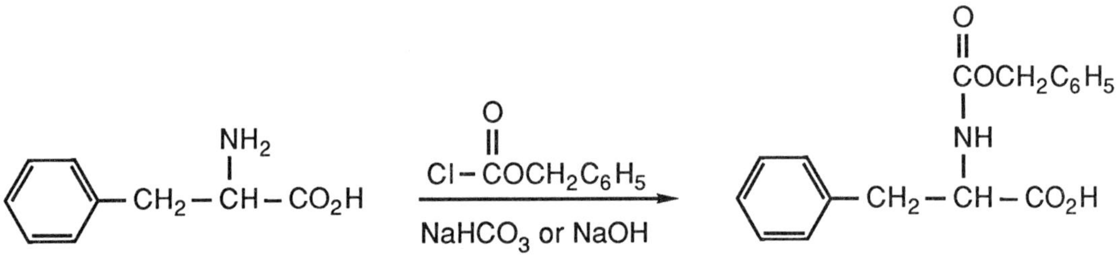

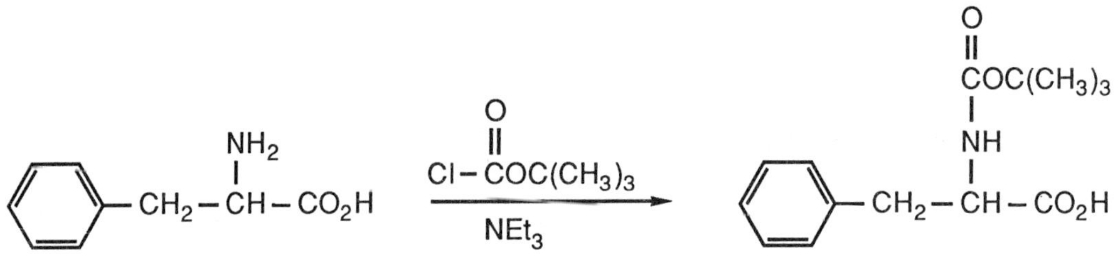

20.15

(a)

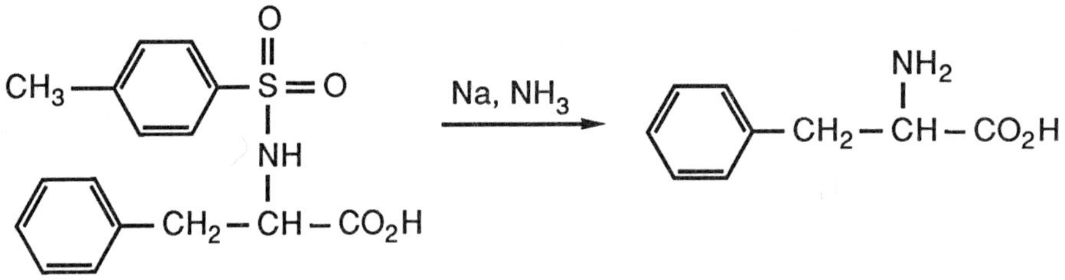

$$C_6H_5-CH_2-CH(NH-SO_2-C_6H_4-CH_3)-CO_2H \xrightarrow{\text{Na, NH}_3} C_6H_5-CH_2-CH(NH_2)-CO_2H$$

(b)

$$C_6H_5-CH_2-CH(NH-C(=O)OCH_2C_6H_5)-CO_2H \xrightarrow[\text{or Na, NH}_3 \text{ or HBr}]{\text{H}_2\text{, Pd}} C_6H_5-CH_2-CH(NH_2)-CO_2H$$

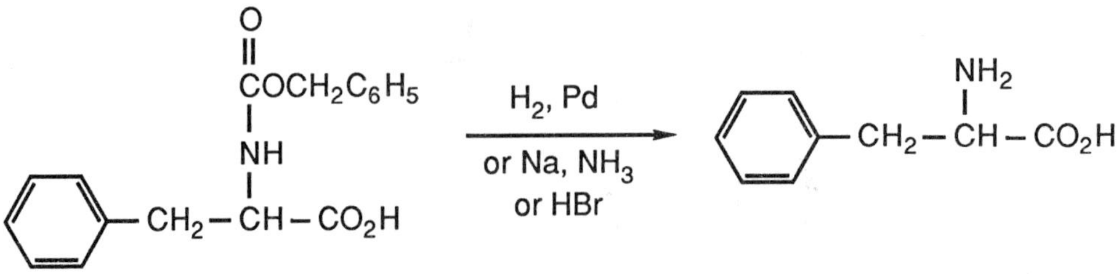

(c)

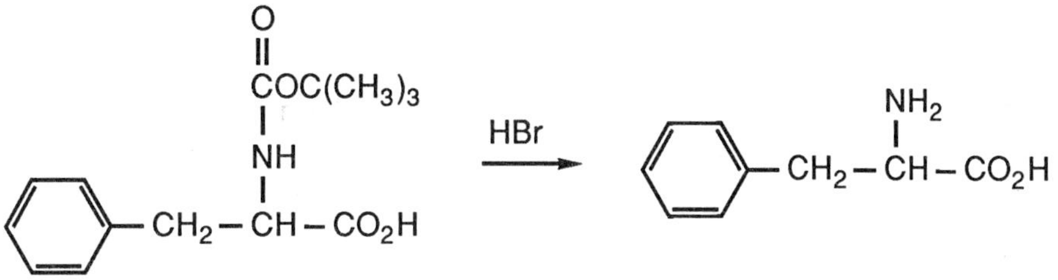

$$C_6H_5-CH_2-CH(NH-COC(CH_3)_3{=}O)-CO_2H \xrightarrow{HBr} C_6H_5-CH_2-CH(NH_2)-CO_2H$$

20.16

(a)

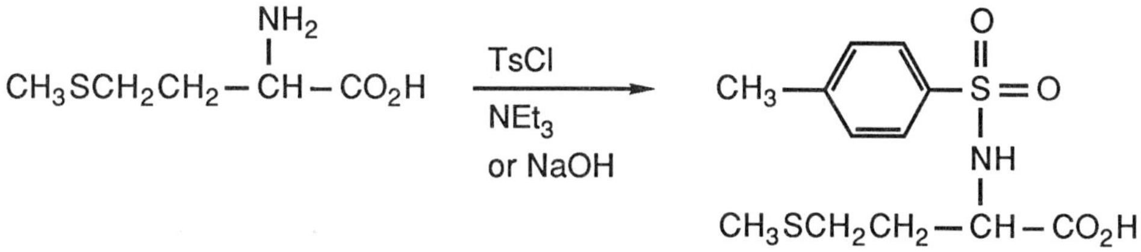

$$CH_3SCH_2CH_2-CH(NH_2)-CO_2H \xrightarrow[\text{NEt}_3 \text{ or NaOH}]{\text{TsCl}} CH_3-C_6H_4-S(=O)_2-NH-CH(CH_2CH_2SCH_3)-CO_2H$$

(b)

$$CH_3SCH_2CH_2-CH(NH_2)-CO_2H \xrightarrow[\text{NaHCO}_3 \text{ or NaOH}]{Cl-C(=O)OCH_2C_6H_5} CH_3SCH_2CH_2-CH(NH-C(=O)OCH_2C_6H_5)-CO_2H$$

(c)

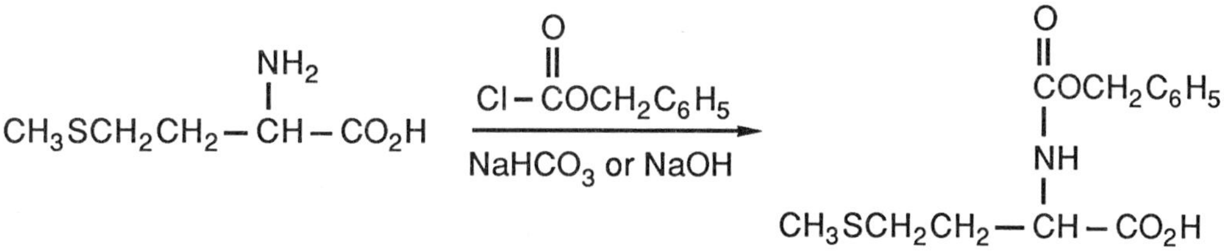

$$CH_3SCH_2CH_2-CH(NH_2)-CO_2H \xrightarrow[\text{NEt}_3]{Cl-C(=O)OC(CH_3)_3} CH_3SCH_2CH_2-CH(NH-C(=O)OC(CH_3)_3)-CO_2H$$

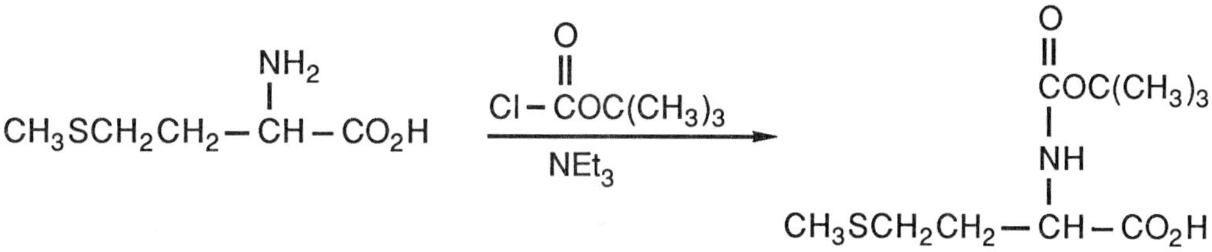

20.17

(a)

$$CH_3-C_6H_4-S(=O)_2-NH-CH(CH_2CH_2SCH_3)-CO_2H \xrightarrow{\text{Na, NH}_3} CH_3SCH_2CH_2-CH(NH_2)-CO_2H$$

Low yield expected.

CH_3-S-CH_2- cleavage probable.

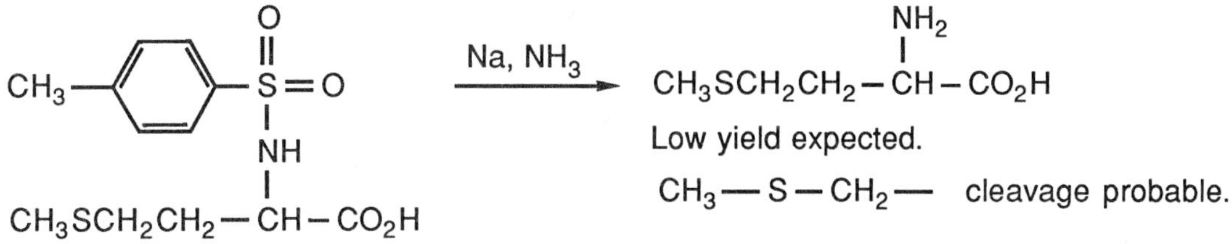

(b)

$$CH_3SCH_2CH_2-CH(NH-COCH_2C_6H_5)-CO_2H \xrightarrow{HBr} CH_3SCH_2CH_2-CH(NH_2)-CO_2H$$

(C=O on the $COCH_2C_6H_5$ group)

Na, NH3 or H2/Pd would cause carbon-sulfur cleavage.

(c)

$$CH_3SCH_2CH_2-CH(NH-COC(CH_3)_3)-CO_2H \xrightarrow{HBr} CH_3SCH_2CH_2-CH(NH_2)-CO_2H$$

(C=O on the $COC(CH_3)_3$ group)

20.18

$$CH_3-CH(NH-COCH_2C_6H_5)-CO_2H \xrightarrow{Na,\ NH_3} CH_3-CH(NH_2)-CO_2H + C_6H_5CH_2OH,\ C_6H_5CH_3, \text{ and products of further reduction}$$

$$\xrightarrow{H_2,\ Pd} CH_3-CH(NH_2)-CO_2H + C_6H_5CH_3$$

$$\xrightarrow{HBr} CH_3-CH(NH_2)-CO_2H + C_6H_5CH_2Br$$

20.19 A fluorine substituent is quite small, so the CF_3 group produces little steric hindrance. Moreover, the CF_3 group is strongly electronegative, which makes the carbonyl group of the trifluoroacetamide more susceptible to nucleophilic attack.

$$CF_3-C(=O)-NHR \leftarrow {}^{-}OH$$

20.20 As with all carboxylic esters the carbonyl carbon is susceptible to nucleophilic attack. In addition the benzylic carbon is readily attacked by nucleophiles:

$$R{-}\overset{\overset{\large O:}{\|}}{C}{-}O{-}CH_2{-}C_6H_5 \;\xrightleftharpoons{H^+}\; R{-}\overset{\overset{\large OH}{\|}}{C}{-}O{-}CH_2{-}C_6H_5 \;\; (Br^-) \longrightarrow RCO_2H + C_6H_5CH_2Br$$

20.21

(a)

$$CH_3CH_2{-}\underset{CH_3}{\underset{|}{CH}}{-}\overset{NH_2}{\overset{|}{CH}}{-}CO_2H \xrightarrow[C_6H_5CH_2OH]{HCl} CH_3CH_2{-}\underset{CH_3}{\underset{|}{CH}}{-}\overset{\overset{+}{N}H_3}{\overset{|}{CH}}{-}\overset{O}{\overset{\|}{C}}OCH_2C_6H_5$$

NaOH (neutralize)

$$CH_3CH_2{-}\underset{CH_3}{\underset{|}{CH}}{-}\overset{NH_2}{\overset{|}{CH}}{-}\overset{O}{\overset{\|}{C}}O^- \xrightarrow{C_6H_5CH_2Cl} CH_3CH_2{-}\underset{CH_3}{\underset{|}{CH}}{-}\overset{NH_2}{\overset{|}{CH}}{-}\overset{O}{\overset{\|}{C}}OCH_2C_6H_5$$

(b)

$$CH_3CH_2{-}\underset{CH_3}{\underset{|}{CH}}{-}\overset{NH_2}{\overset{|}{CH}}{-}CO_2H \xrightarrow{HO{-}C_6H_4{-}NO_2\text{ , HCl}} CH_3CH_2{-}\underset{CH_3}{\underset{|}{CH}}{-}\overset{NH_2}{\overset{|}{CH}}{-}\overset{O}{\overset{\|}{C}}{-}O{-}C_6H_4{-}NO_2$$

20.22

(a)

$$CH_3CH_2{-}\underset{CH_3}{\underset{|}{CH}}{-}\overset{NH_2}{\overset{|}{CH}}{-}\overset{O}{\overset{\|}{C}}OCH_2C_6H_5 \xrightarrow[\text{or Na, } NH_3 \text{; or } H_2\text{, Pd; or HBr}]{H_2O,\ H_2SO_4} CH_3CH_2{-}\underset{CH_3}{\underset{|}{CH}}{-}\overset{NH_2}{\overset{|}{CH}}{-}CO_2H$$

(b)

$$CH_3CH_2-\underset{CH_3}{\underset{|}{CH}}-\overset{NH_2}{\overset{|}{CH}}-\overset{O}{\overset{\|}{C}}-O-C_6H_4-NO_2 \xrightarrow{H_2O,\ NaOH} CH_3CH_2-\underset{CH_3}{\underset{|}{CH}}-\overset{NH_2}{\overset{|}{CH}}-CO_2H$$

20.23

(a)

$$HS-CH_2-\overset{NH_2}{\overset{|}{CH}}-CO_2H \xrightarrow[(air)]{O_2} \begin{array}{l} S-CH_2-\overset{NH_2}{\overset{|}{CH}}-CO_2H \\ | \\ S-CH_2-\underset{NH_2}{\underset{|}{CH}}-CO_2H \end{array}$$

(b)

$$HS-CH_2-\overset{NH_2}{\overset{|}{CH}}-CO_2H \xrightarrow[ClCH_2C_6H_5]{NaOH} C_6H_5CH_2S-CH_2-\overset{NH_2}{\overset{|}{CH}}-CO_2H$$

20.24

(a)

$$\begin{array}{l} S-CH_2-\overset{NH_2}{\overset{|}{CH}}-CO_2H \\ | \\ S-CH_2-\underset{NH_2}{\underset{|}{CH}}-CO_2H \end{array} \xrightarrow[\text{or } HOCH_2CH_2SH]{Na,\ NH_3 \text{ or } NaBH_4} HS-CH_2-\overset{NH_2}{\overset{|}{CH}}-CO_2H$$

(b)

$$C_6H_5CH_2S-CH_2-\overset{NH_2}{\overset{|}{CH}}-CO_2H \xrightarrow[\text{or } HBr]{Na,\ NH_3} HS-CH_2-\overset{NH_2}{\overset{|}{CH}}-CO_2H$$

20.25 **(a)**

$$\text{Cbz-NH-CH}(\text{CH}_2\text{CH}(\text{CH}_3)_2)\text{-}\overset{O}{\overset{\|}{C}}\text{-OH} \xrightarrow{\text{Cl-}\overset{O}{\overset{\|}{C}}\text{CH}_2\text{CH}(\text{CH}_3)_2} \text{Cbz-NH-CH}(\text{CH}_2\text{CH}(\text{CH}_3)_2)\text{-}\overset{O}{\overset{\|}{C}}\text{-O-}\overset{O}{\overset{\|}{C}}\text{CH}_2\text{CH}(\text{CH}_3)_2$$

The electron withdrawing effect of the amino group makes the carbonyl of the amino acid residue more electrophilic. In addition, the isovaleryl group provides some steric hindrance to attack at the other carbonyl.

(b)

$$\text{Cbz-NH-CH}(\text{CH}_2\text{SCH}_2\text{C}_6\text{H}_5)\text{-}\overset{O}{\overset{\|}{C}}\text{-OH} \xrightarrow{\text{Cl-}\overset{O}{\overset{\|}{C}}\text{OCH}_2\text{CH}_2\text{CH}_2\text{CH}_3} \text{Cbz-NH-CH}(\text{CH}_2\text{SCH}_2\text{C}_6\text{H}_5)\text{-}\overset{O}{\overset{\|}{C}}\text{-O-}\overset{O}{\overset{\|}{C}}\text{OCH}_2\text{CH}_2\text{CH}_2\text{CH}_3$$

The $-OCO_2R$ group is a better leaving group than RCO_2^-.

(c)

$$\text{Cbz-NH-CH}(\text{CH}_2\text{CH}(\text{CH}_3)_2)\text{-}\overset{O}{\overset{\|}{C}}\text{-OH} \xrightarrow{(\text{C}_2\text{H}_5\text{O})_2\text{POP}(\text{OC}_2\text{H}_5)_2} \text{Cbz-NH-CH}(\text{CH}_2\text{CH}(\text{CH}_3)_2)\text{-}\overset{O}{\overset{\|}{C}}\text{-O-P}(\text{OC}_2\text{H}_5)_2$$

The C=O group is more susceptible to nucleophilic attack than is a phosphorus atom.

20.26 $NH_2-Leu-CO_2H$ + $NH_2-Ala-CO_2H$ $\longrightarrow$ $NH_2-Leu-Ala-CO_2H$

(a)

$$(CH_3)_2CHCH_2-\overset{NH_2}{\overset{|}{C}}H-\overset{O}{\overset{\|}{C}}OH \xrightarrow[NaHCO_3]{C_6H_5CH_2O\overset{O}{\overset{\|}{C}}-Cl} (CH_3)_2CHCH_2-\overset{Cbz-NH}{\overset{|}{C}}H-\overset{O}{\overset{\|}{C}}OH$$

$$\xrightarrow{PCl_5} (CH_3)_2CHCH_2-\overset{Cbz-NH}{\overset{|}{C}}H-\overset{O}{\overset{\|}{C}}-Cl$$

$$\xrightarrow{NH_2-\overset{CH_3}{\overset{|}{C}}H-CO_2H} (CH_3)_2CHCH_2-\overset{Cbz-NH}{\overset{|}{C}}H-\overset{O}{\overset{\|}{C}}-NH-\overset{CH_3}{\overset{|}{C}}H-CO_2H$$

$$\xrightarrow{H_2,\ Pd} (CH_3)_2CHCH_2-\overset{NH_2}{\overset{|}{C}}H-\overset{O}{\overset{\|}{C}}-NH-\overset{CH_3}{\overset{|}{C}}H-CO_2H$$

(b)

$$NH_2-\overset{CH_3}{\overset{|}{C}}H-CO_2H \xrightarrow[2.\ C_6H_5CH_2Cl]{1.\ NaOH} NH_2-\overset{CH_3}{\overset{|}{C}}H-CO_2CH_2C_6H_5$$

$$\xrightarrow[(CH_3)_2CHCH_2-\overset{Cbz-NH}{\overset{|}{C}}H-\overset{O}{\overset{\|}{C}}OH]{(C_2H_5O)_2POP(OC_2H_5)_2} (CH_3)_2CHCH_2-\overset{Cbz-NH}{\overset{|}{C}}H-\overset{O}{\overset{\|}{C}}-NH-\overset{CH_3}{\overset{|}{C}}H-CO_2CH_2C_6H_5$$

$$\xrightarrow{H_2,\ Pd} (CH_3)_2CHCH_2-\overset{NH_2}{\overset{|}{C}}H-\overset{O}{\overset{\|}{C}}-NH-\overset{CH_3}{\overset{|}{C}}H-CO_2H$$

(c)

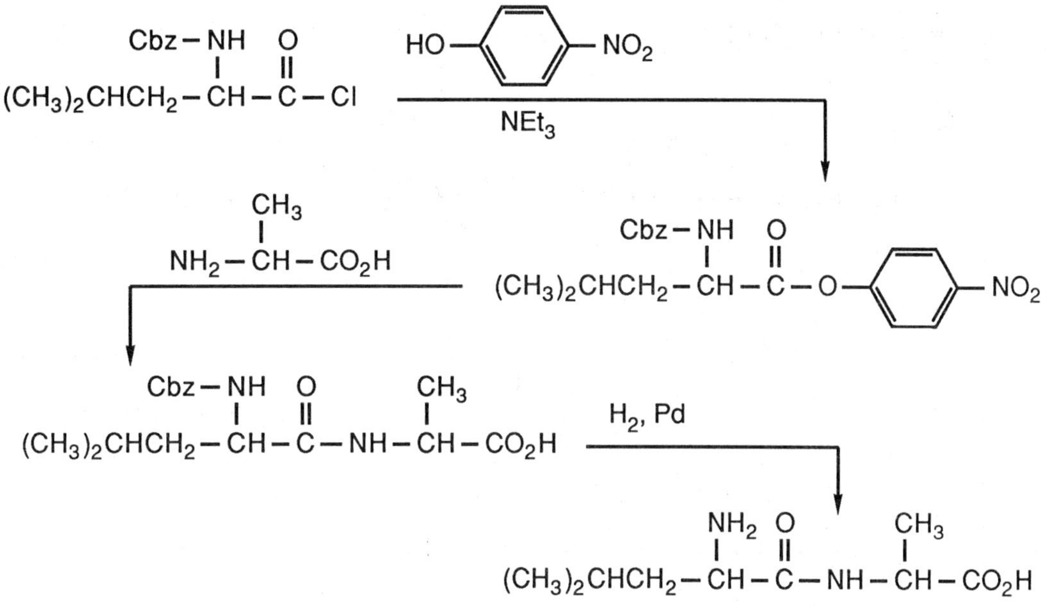

$$(CH_3)_2CHCH_2-CH(NH-Cbz)-C(=O)-Cl \xrightarrow[NEt_3]{HO-C_6H_4-NO_2} (CH_3)_2CHCH_2-CH(NH-Cbz)-C(=O)-O-C_6H_4-NO_2$$

$$\xrightarrow{NH_2-CH(CH_3)-CO_2H} (CH_3)_2CHCH_2-CH(NH-Cbz)-C(=O)-NH-CH(CH_3)-CO_2H$$

$$\xrightarrow{H_2,\ Pd} (CH_3)_2CHCH_2-CH(NH_2)-C(=O)-NH-CH(CH_3)-CO_2H$$

(d)

$$(CH_3)_2CHCH_2-CH(NH-Cbz)-C(=O)-OH \xrightarrow[DCC]{NH_2-CH(CH_3)-CO_2CH_2C_6H_5} (CH_3)_2CHCH_2-CH(NH-Cbz)-C(=O)-NH-CH(CH_3)-CO_2CH_2C_6H_5$$

$$\xrightarrow{H_2,\ Pd} (CH_3)_2CHCH_2-CH(NH_2)-C(=O)-NH-CH(CH_3)-CO_2H$$

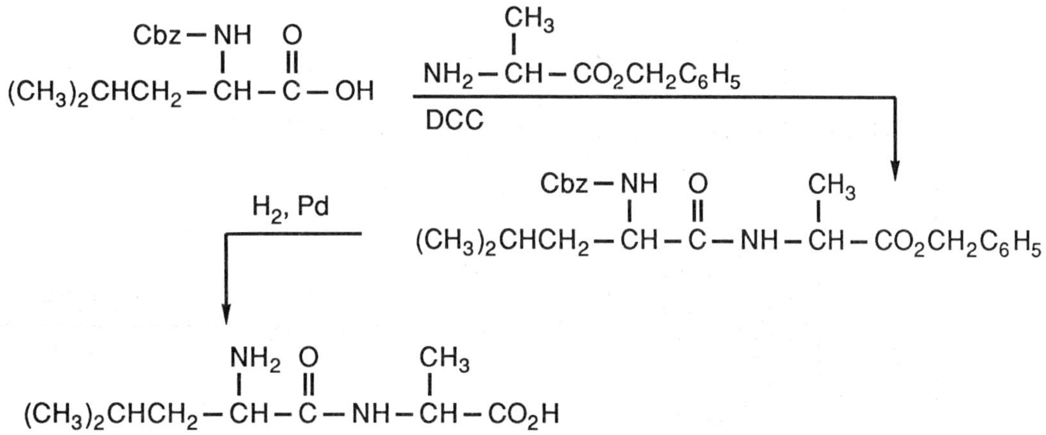

(e)

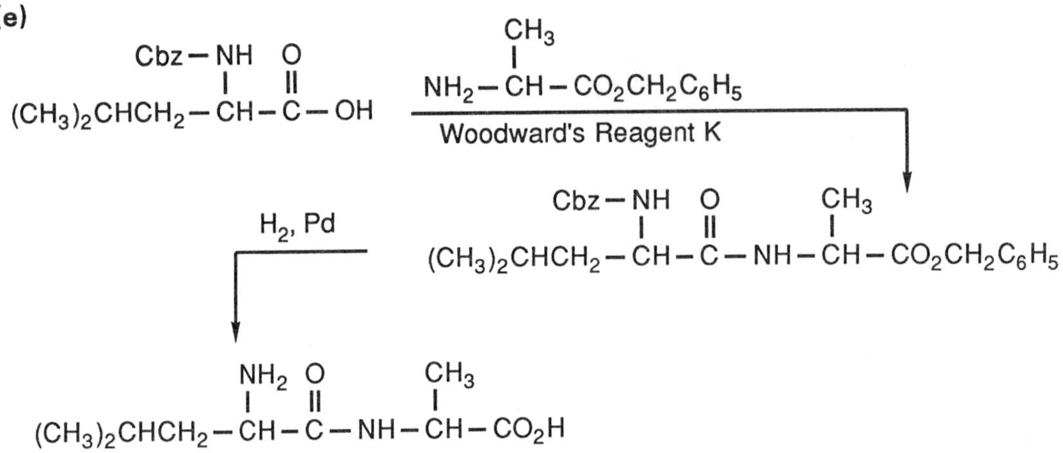

$$(CH_3)_2CHCH_2-\underset{}{CH}(NH-Cbz)-C(=O)-OH \xrightarrow[\text{Woodward's Reagent K}]{NH_2-CH(CH_3)-CO_2CH_2C_6H_5} (CH_3)_2CHCH_2-CH(NH-Cbz)-C(=O)-NH-CH(CH_3)-CO_2CH_2C_6H_5$$

$$\xrightarrow{H_2,\ Pd} (CH_3)_2CHCH_2-CH(NH_2)-C(=O)-NH-CH(CH_3)-CO_2H$$

(f)

$$NH_2-CH(CH_3)-CO_2H \xrightarrow[NaHCO_3]{C_6H_5CH_2OC(=O)-Cl} CbzNH-CH(CH_3)-CO_2H$$

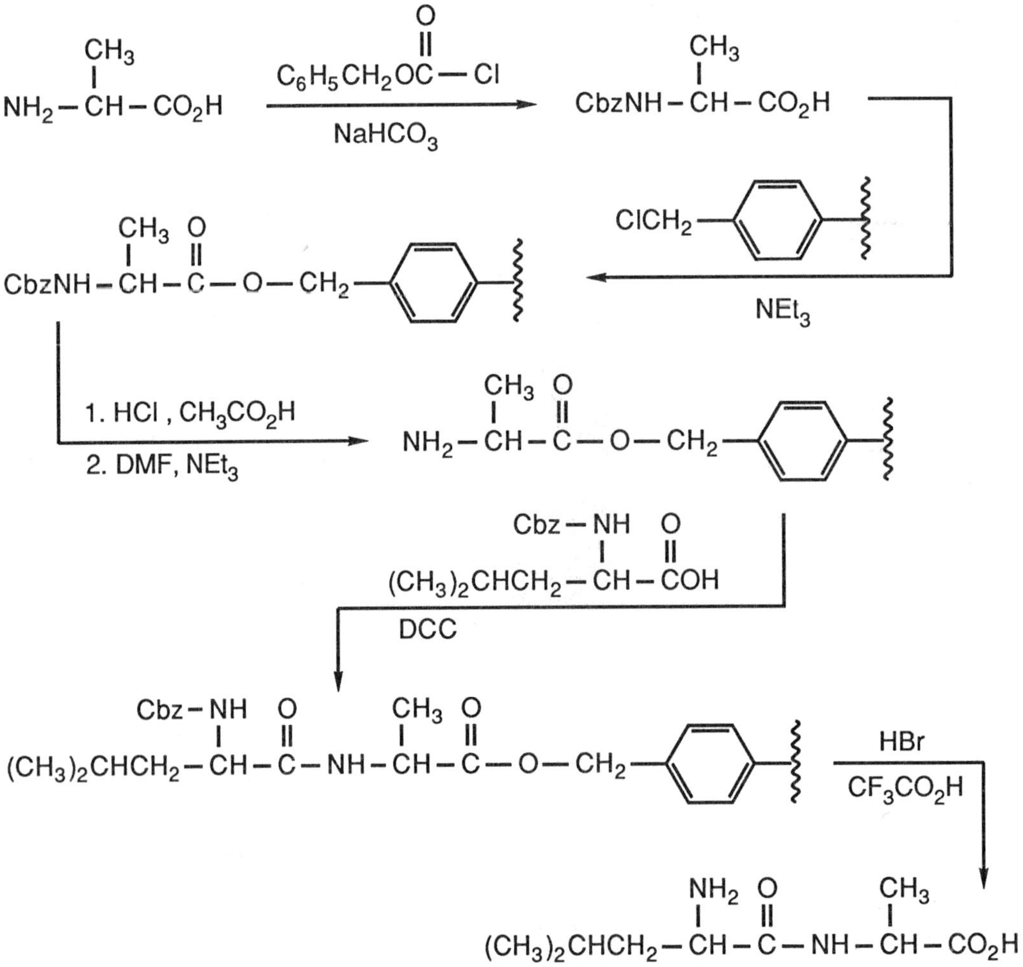

$$\xrightarrow[NEt_3]{ClCH_2-C_6H_4-\text{(resin)}} CbzNH-CH(CH_3)-C(=O)-O-CH_2-C_6H_4-\text{(resin)}$$

$$\xrightarrow[\text{2. DMF, } NEt_3]{\text{1. HCl, } CH_3CO_2H} NH_2-CH(CH_3)-C(=O)-O-CH_2-C_6H_4-\text{(resin)}$$

$$\xrightarrow[DCC]{(CH_3)_2CHCH_2-CH(NH-Cbz)-COH(=O)} (CH_3)_2CHCH_2-CH(NH-Cbz)-C(=O)-NH-CH(CH_3)-C(=O)-O-CH_2-C_6H_4-\text{(resin)}$$

$$\xrightarrow[CF_3CO_2H]{HBr} (CH_3)_2CHCH_2-CH(NH_2)-C(=O)-NH-CH(CH_3)-CO_2H$$

20.27

$NH_2-Gly-CO_2H + NH_2-Cys-CO_2H \longrightarrow NH_2-Gly-Cys-CO_2H$

Protection:

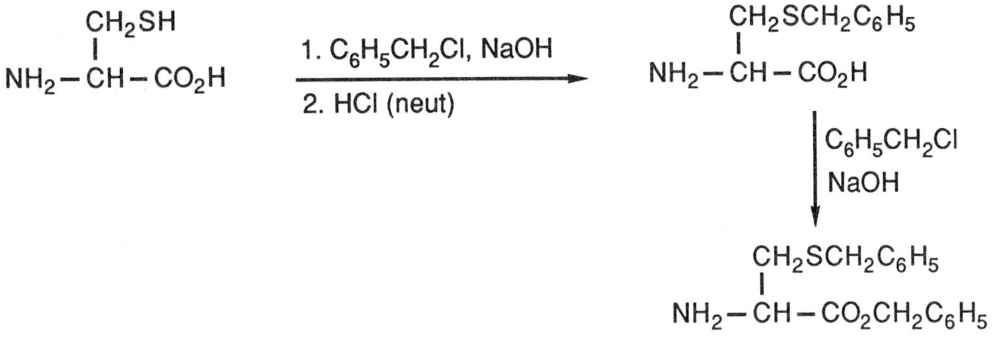

$$NH_2-CH(CH_2SH)-CO_2H \xrightarrow[\text{2. HCl (neut)}]{\text{1. } C_6H_5CH_2Cl,\ NaOH} NH_2-CH(CH_2SCH_2C_6H_5)-CO_2H$$

$$NH_2-CH(CH_2SCH_2C_6H_5)-CO_2H \xrightarrow{C_6H_5CH_2Cl,\ NaOH} NH_2-CH(CH_2SCH_2C_6H_5)-CO_2CH_2C_6H_5$$

(a)

$$NH_2-CH_2-\overset{O}{\overset{\|}{C}}OH \xrightarrow[NaHCO_3]{C_6H_5CH_2O\overset{O}{\overset{\|}{C}}-Cl} CbzNH-CH_2-\overset{O}{\overset{\|}{C}}OH$$

$$CbzNH-CH_2-\overset{O}{\overset{\|}{C}}OH \xrightarrow{PCl_5} CbzNH-CH_2-\overset{O}{\overset{\|}{C}}-Cl$$

$$CbzNH-CH_2-\overset{O}{\overset{\|}{C}}-Cl \xrightarrow{NH_2-CH(CH_2SCH_2C_6H_5)-CO_2CH_2C_6H_5} CbzNH-CH_2-\overset{O}{\overset{\|}{C}}-NH-CH(CH_2SCH_2C_6H_5)-CO_2CH_2C_6H_5$$

$$CbzNH-CH_2-\overset{O}{\overset{\|}{C}}-NH-CH(CH_2SCH_2C_6H_5)-CO_2CH_2C_6H_5 \xrightarrow{Na,\ NH_3} NH_2-CH_2-\overset{O}{\overset{\|}{C}}-NH-CH(CH_2SH)-CO_2H$$

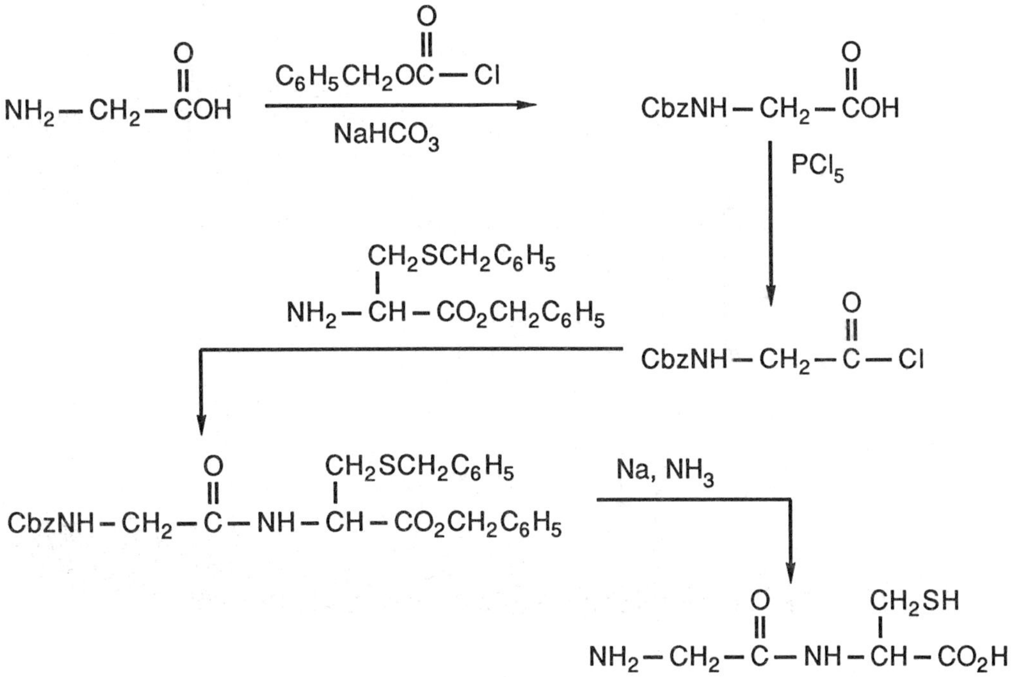

(b)

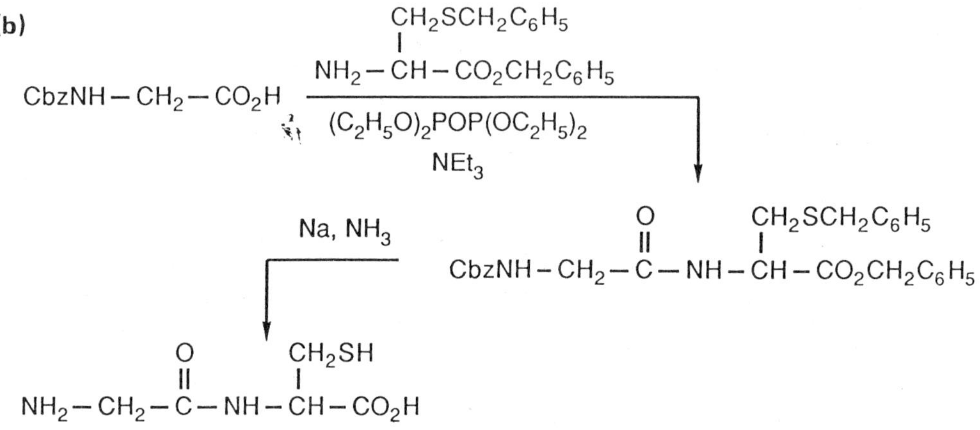

$$CbzNH-CH_2-CO_2H \xrightarrow[\substack{(C_2H_5O)_2POP(OC_2H_5)_2 \\ NEt_3}]{NH_2-CH(CH_2SCH_2C_6H_5)-CO_2CH_2C_6H_5} CbzNH-CH_2-C(=O)-NH-CH(CH_2SCH_2C_6H_5)-CO_2CH_2C_6H_5$$

$$\xrightarrow{Na,\ NH_3} NH_2-CH_2-C(=O)-NH-CH(CH_2SH)-CO_2H$$

(c)

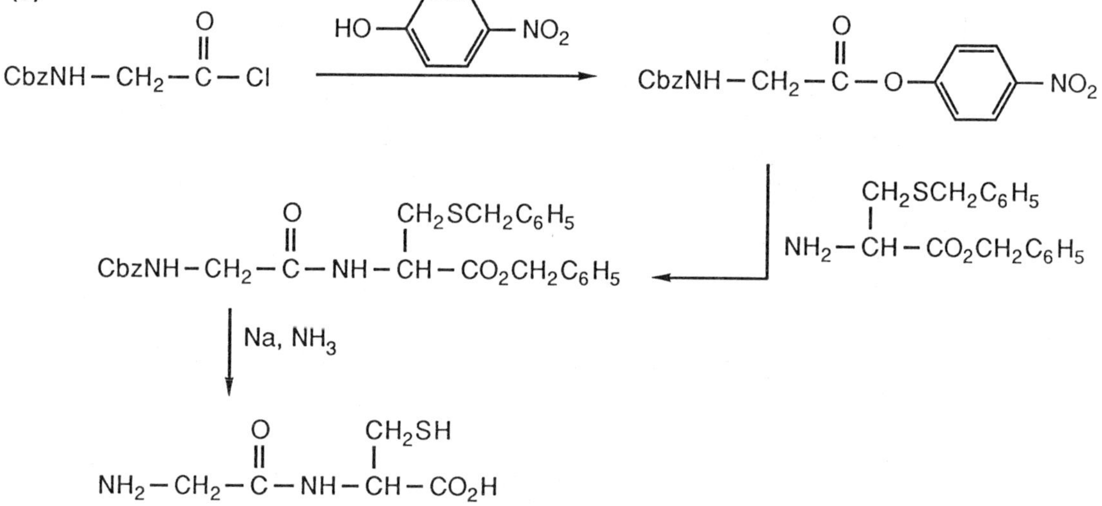

$$CbzNH-CH_2-C(=O)-Cl \xrightarrow{HO-C_6H_4-NO_2} CbzNH-CH_2-C(=O)-O-C_6H_4-NO_2$$

$$\xrightarrow{NH_2-CH(CH_2SCH_2C_6H_5)-CO_2CH_2C_6H_5} CbzNH-CH_2-C(=O)-NH-CH(CH_2SCH_2C_6H_5)-CO_2CH_2C_6H_5$$

$$\xrightarrow{Na,\ NH_3} NH_2-CH_2-C(=O)-NH-CH(CH_2SH)-CO_2H$$

(d)

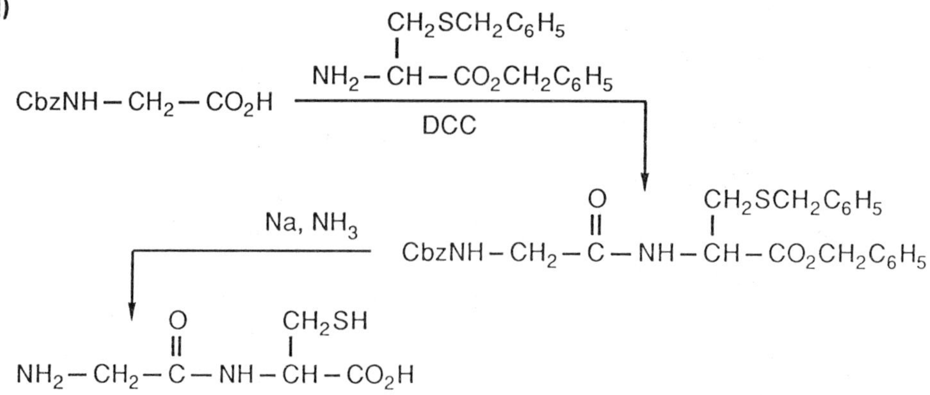

$$CbzNH-CH_2-CO_2H \xrightarrow[DCC]{NH_2-CH(CH_2SCH_2C_6H_5)-CO_2CH_2C_6H_5} CbzNH-CH_2-C(=O)-NH-CH(CH_2SCH_2C_6H_5)-CO_2CH_2C_6H_5$$

$$\xrightarrow{Na,\ NH_3} NH_2-CH_2-C(=O)-NH-CH(CH_2SH)-CO_2H$$

(e)

$$\text{CbzNH}-\text{CH}_2-\text{CO}_2\text{H} \xrightarrow[\text{Woodward's Reagent K}]{\text{NH}_2-\text{CH}(\text{CH}_2\text{SCH}_2\text{C}_6\text{H}_5)-\text{CO}_2\text{CH}_2\text{C}_6\text{H}_5}$$

$$\text{CbzNH}-\text{CH}_2-\text{C}(=\text{O})-\text{NH}-\text{CH}(\text{CH}_2\text{SCH}_2\text{C}_6\text{H}_5)-\text{CO}_2\text{CH}_2\text{C}_6\text{H}_5 \xrightarrow{\text{Na, NH}_3}$$

$$\text{NH}_2-\text{CH}_2-\text{C}(=\text{O})-\text{NH}-\text{CH}(\text{CH}_2\text{SH})-\text{CO}_2\text{H}$$

(f)

$$\text{NH}_2-\text{CH}(\text{CH}_2\text{SCH}_2\text{C}_6\text{H}_5)-\text{CO}_2\text{H} \xrightarrow[\text{NaHCO}_3]{\text{C}_6\text{H}_5\text{CH}_2\text{OC}(=\text{O})-\text{Cl}} \text{CbzNH}-\text{CH}(\text{CH}_2\text{SCH}_2\text{C}_6\text{H}_5)-\text{CO}_2\text{H}$$

$$\xrightarrow{\text{NEt}_3,\ \text{ClCH}_2-\text{C}_6\text{H}_4-\text{(polymer)}}$$

$$\text{CbzNH}-\text{CH}(\text{CH}_2-\text{SCH}_2\text{C}_6\text{H}_5)-\text{C}(=\text{O})-\text{O}-\text{CH}_2-\text{C}_6\text{H}_4-\text{(polymer)} \xrightarrow[\text{2. DMF, NEt}_3]{\text{1. HCl, CH}_3\text{CO}_2\text{H}}$$

$$\text{NH}_2-\text{CH}(\text{CH}_2-\text{SCH}_2\text{C}_6\text{H}_5)-\text{C}(=\text{O})-\text{O}-\text{CH}_2-\text{C}_6\text{H}_4-\text{(polymer)} \xrightarrow[\text{DCC}]{\text{CbzNH}-\text{CH}_2-\text{CO}_2\text{H}}$$

$$\text{CbzNH}-\text{CH}_2-\text{C}(=\text{O})-\text{NH}-\text{CH}(\text{CH}_2-\text{SCH}_2\text{C}_6\text{H}_5)-\text{C}(=\text{O})-\text{O}-\text{CH}_2-\text{C}_6\text{H}_4-\text{(polymer)} \xrightarrow[\text{CF}_3\text{CO}_2\text{H}]{\text{HBr}}$$

$$\text{NH}_2-\text{CH}_2-\text{C}(=\text{O})-\text{NH}-\text{CH}(\text{CH}_2\text{SH})-\text{CO}_2\text{H}$$

20.28 $NH_2-Tyr-CO_2H + NH_2-Val-CO_2H \longrightarrow NH_2-Tyr-Val-CO_2H$

Protection:

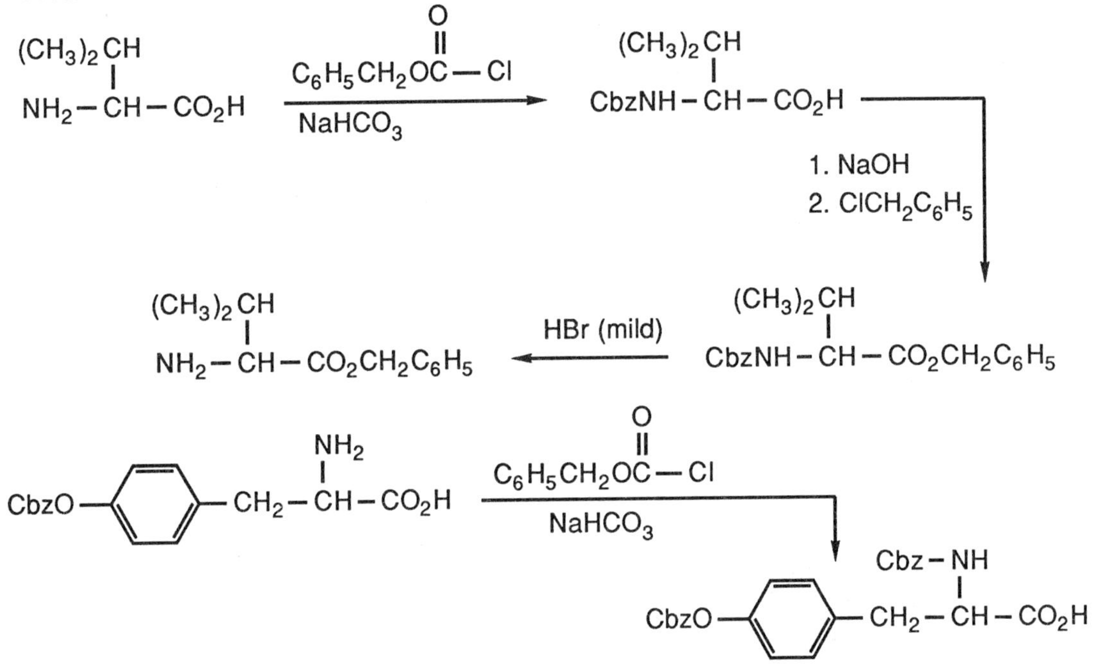

$(CH_3)_2CH-CH(NH_2)-CO_2H \xrightarrow[NaHCO_3]{C_6H_5CH_2OC(=O)-Cl} (CH_3)_2CH-CH(CbzNH)-CO_2H \xrightarrow{1.\ NaOH;\ 2.\ ClCH_2C_6H_5} CbzNH-CH(CH(CH_3)_2)-CO_2CH_2C_6H_5$

$CbzNH-CH(CH(CH_3)_2)-CO_2CH_2C_6H_5 \xrightarrow{HBr\ (mild)} NH_2-CH(CH(CH_3)_2)-CO_2CH_2C_6H_5$

$CbzO-C_6H_4-CH_2-CH(NH_2)-CO_2H \xrightarrow[NaHCO_3]{C_6H_5CH_2OC(=O)-Cl} CbzO-C_6H_4-CH_2-CH(NH-Cbz)-CO_2H$

(a)

$CbzO-C_6H_4-CH_2-CH(NH-Cbz)-CO_2H \xrightarrow{PCl_5} CbzO-C_6H_4-CH_2-CH(NH-Cbz)-C(=O)-Cl$

$\xrightarrow{NH_2-CH(CH(CH_3)_2)-CO_2CH_2C_6H_5} CbzO-C_6H_4-CH_2-CH(NH-Cbz)-C(=O)-NH-CH(CH(CH_3)_2)-CO_2CH_2C_6H_5$

$\xrightarrow{Na,\ NH_3} HO-C_6H_4-CH_2-CH(NH_2)-C(=O)-NH-CH(CH(CH_3)_2)-CO_2H$

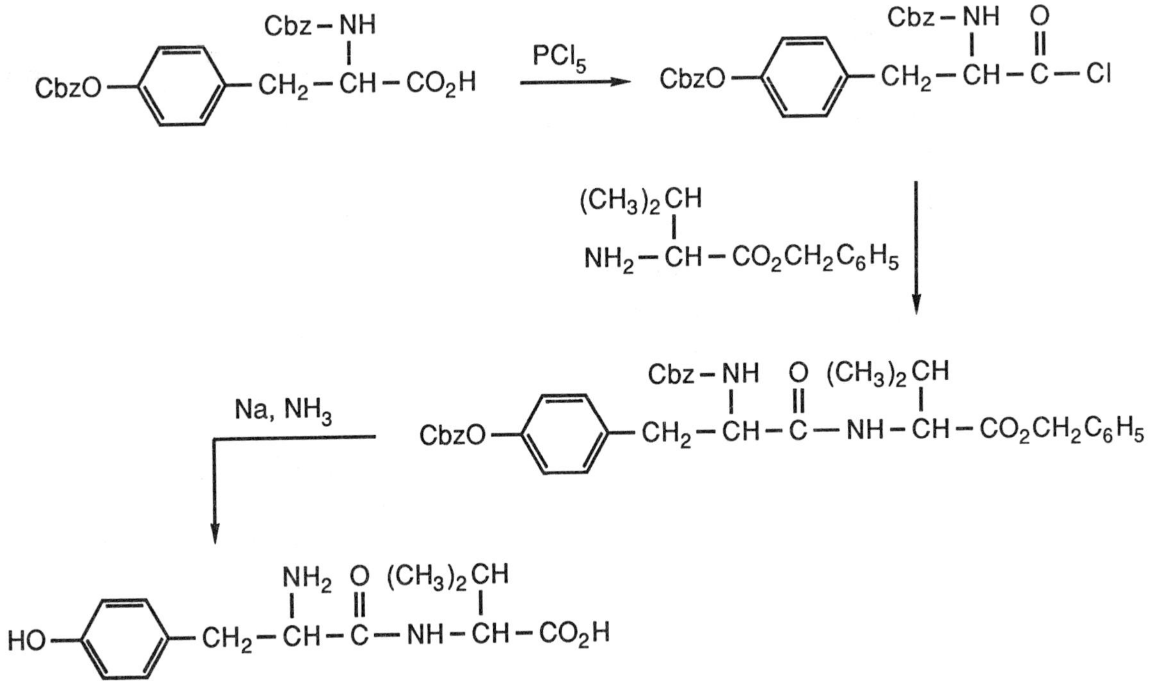

(b)

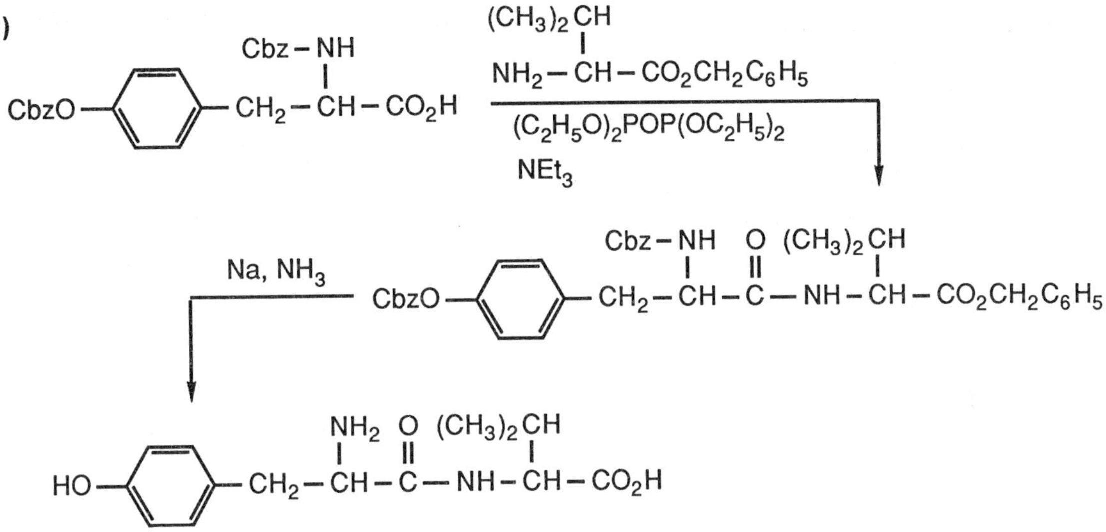

$$\text{CbzO}-C_6H_4-CH_2-CH(NH-\text{Cbz})-CO_2H \xrightarrow[(C_2H_5O)_2POP(OC_2H_5)_2,\ NEt_3]{(CH_3)_2CH-CH(NH_2)-CO_2CH_2C_6H_5}$$

$$\text{CbzO}-C_6H_4-CH_2-CH(NH-\text{Cbz})-C(=O)-NH-CH(CH(CH_3)_2)-CO_2CH_2C_6H_5 \xrightarrow{Na,\ NH_3}$$

$$HO-C_6H_4-CH_2-CH(NH_2)-C(=O)-NH-CH(CH(CH_3)_2)-CO_2H$$

(c)

$$\text{CbzO}-C_6H_4-CH_2-CH(NH-\text{Cbz})-C(=O)-Cl \xrightarrow{HO-C_6H_4-NO_2}$$

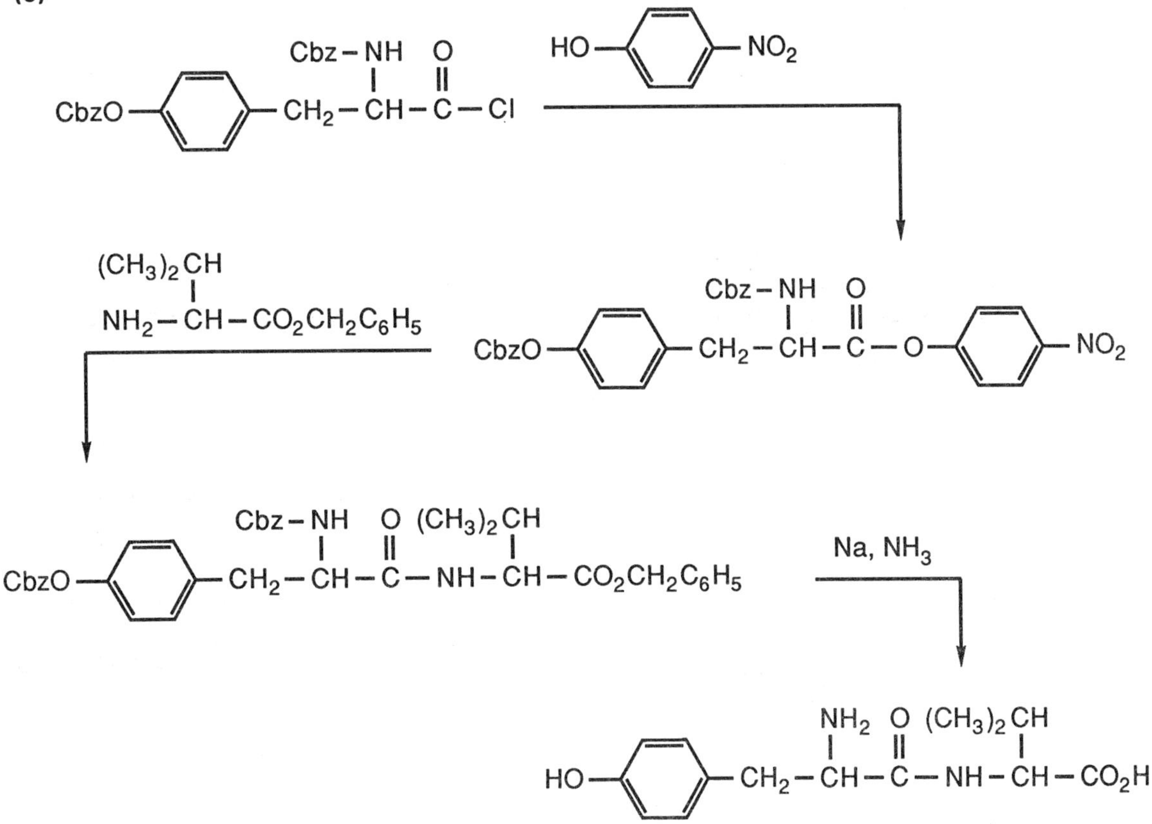

$$\text{CbzO}-C_6H_4-CH_2-CH(NH-\text{Cbz})-C(=O)-O-C_6H_4-NO_2 \xrightarrow{(CH_3)_2CH-CH(NH_2)-CO_2CH_2C_6H_5}$$

$$\text{CbzO}-C_6H_4-CH_2-CH(NH-\text{Cbz})-C(=O)-NH-CH(CH(CH_3)_2)-CO_2CH_2C_6H_5 \xrightarrow{Na,\ NH_3}$$

$$HO-C_6H_4-CH_2-CH(NH_2)-C(=O)-NH-CH(CH(CH_3)_2)-CO_2H$$

(d)

$$\text{CbzO}-\text{C}_6\text{H}_4-\text{CH}_2-\underset{\text{Cbz}-\text{NH}}{\text{CH}}-\text{CO}_2\text{H} \xrightarrow[\text{DCC}]{\text{NH}_2-\text{CH}(\text{CH(CH}_3)_2)-\text{CO}_2\text{CH}_2\text{C}_6\text{H}_5}$$

$$\text{CbzO}-\text{C}_6\text{H}_4-\text{CH}_2-\underset{\text{Cbz}-\text{NH}}{\text{CH}}-\overset{\text{O}}{\overset{\|}{\text{C}}}-\text{NH}-\underset{(\text{CH}_3)_2\text{CH}}{\text{CH}}-\text{CO}_2\text{CH}_2\text{C}_6\text{H}_5 \xrightarrow{\text{Na, NH}_3}$$

$$\text{HO}-\text{C}_6\text{H}_4-\text{CH}_2-\underset{\text{NH}_2}{\text{CH}}-\overset{\text{O}}{\overset{\|}{\text{C}}}-\text{NH}-\underset{(\text{CH}_3)_2\text{CH}}{\text{CH}}-\text{CO}_2\text{H}$$

(e)

$$\text{CbzO}-\text{C}_6\text{H}_4-\text{CH}_2-\underset{\text{Cbz}-\text{NH}}{\text{CH}}-\text{CO}_2\text{H} \xrightarrow[\text{Woodward's Reagent K}]{\text{NH}_2-\text{CH}(\text{CH(CH}_3)_2)-\text{CO}_2\text{CH}_2\text{C}_6\text{H}_5}$$

$$\text{CbzO}-\text{C}_6\text{H}_4-\text{CH}_2-\underset{\text{Cbz}-\text{NH}}{\text{CH}}-\overset{\text{O}}{\overset{\|}{\text{C}}}-\text{NH}-\underset{(\text{CH}_3)_2\text{CH}}{\text{CH}}-\text{CO}_2\text{CH}_2\text{C}_6\text{H}_5 \xrightarrow{\text{Na, NH}_3}$$

$$\text{HO}-\text{C}_6\text{H}_4-\text{CH}_2-\underset{\text{NH}_2}{\text{CH}}-\overset{\text{O}}{\overset{\|}{\text{C}}}-\text{NH}-\underset{(\text{CH}_3)_2\text{CH}}{\text{CH}}-\text{CO}_2\text{H}$$

(f)

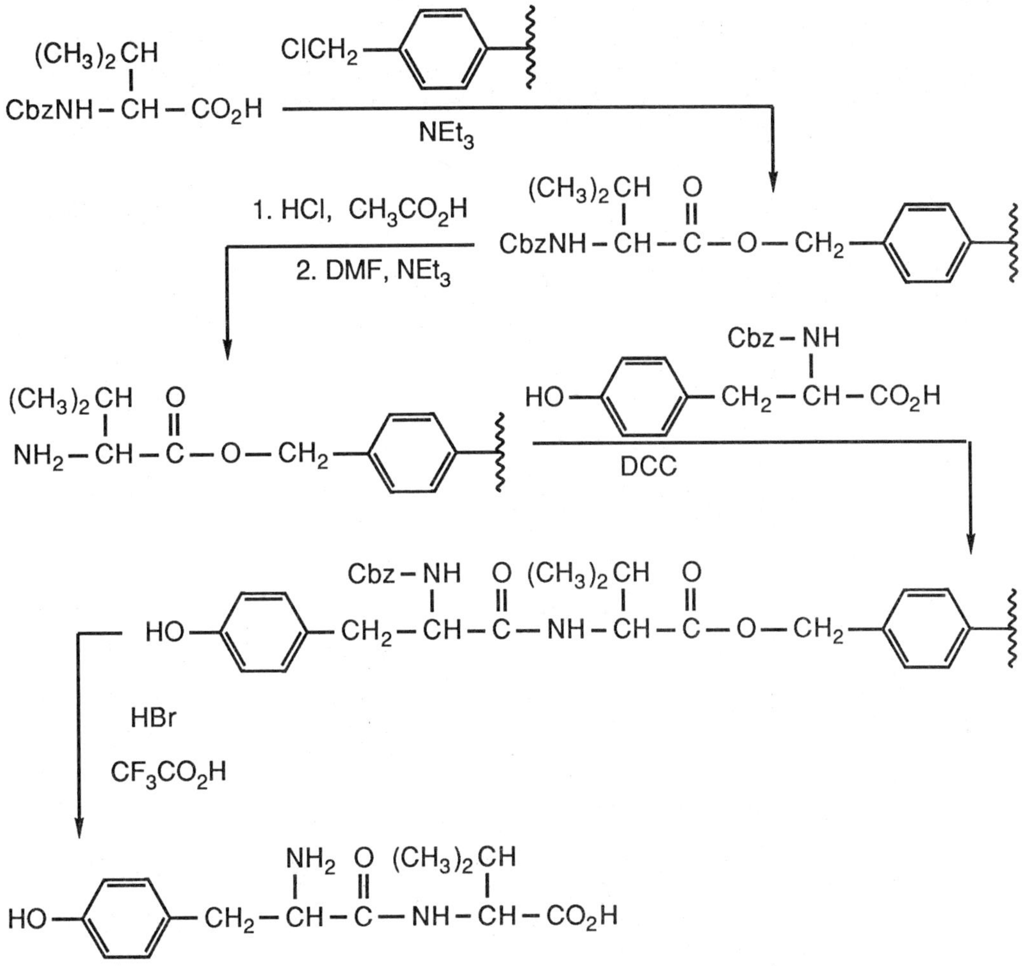

$(CH_3)_2CH$

CbzNH−CH−CO_2H → ($ClCH_2$–C₆H₄–resin, NEt_3)

$(CH_3)_2CH$–CH(NHCbz)–C(=O)–O–CH_2–C₆H₄–resin

1. HCl, CH_3CO_2H
2. DMF, NEt_3

NH_2−CH($(CH_3)_2CH$)−C(=O)−O−CH_2–C₆H₄–resin

+ HO–C₆H₄–CH_2−CH(NH−Cbz)−CO_2H, DCC

HO–C₆H₄–CH_2−CH(NH−Cbz)−C(=O)−NH−CH($(CH_3)_2CH$)−C(=O)−O−CH_2–C₆H₄–resin

HBr

CF_3CO_2H

HO–C₆H₄–CH_2−CH(NH_2)−C(=O)−NH−CH($(CH_3)_2CH$)−CO_2H

20.29

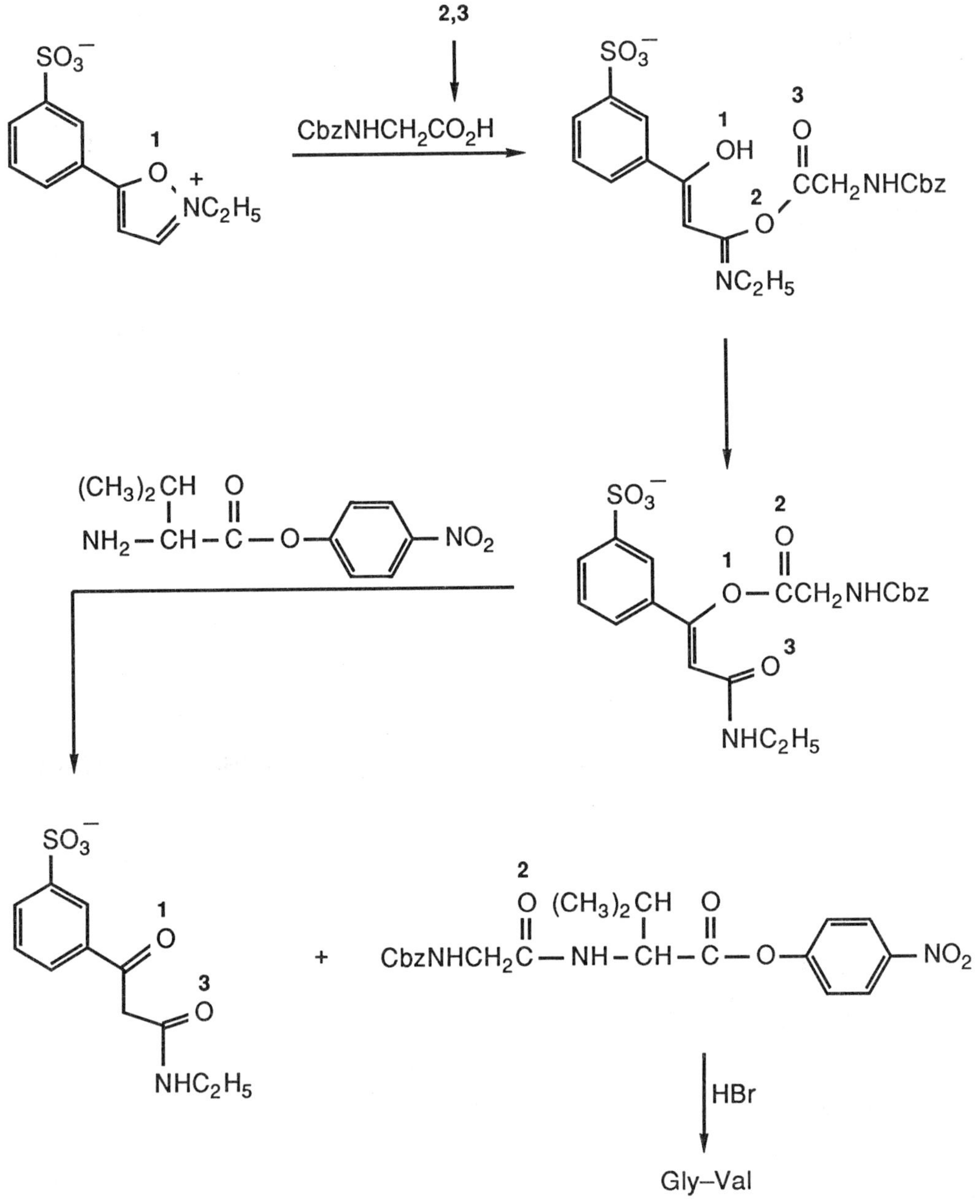

1

SO_3^-

O +

NC_2H_5

2,3

$CbzNHCH_2CO_2H$

SO_3^-

1 OH

3 O

CCH_2NHCbz

2 O

NC_2H_5

$(CH_3)_2CH$ O

$NH_2-CH-C-O-$ $-NO_2$

SO_3^-

2 O

1 $O-CCH_2NHCbz$

3 O

NHC_2H_5

SO_3^-

1 O

3 O

NHC_2H_5

\+

2

O $(CH_3)_2CH$ O

$CbzNHCH_2C-NH-CH-C-O-$ $-NO_2$

HBr

Gly–Val

20.30 One possible sequence follows:

$$HS-CH-(NH_2)-CO_2H \xrightarrow[NaOH]{C_6H_5CH_2Cl} C_6H_5CH_2S-CH-(NH_2)-CO_2H$$

$$\xrightarrow{C_6H_5CH_2OC(=O)-Cl} C_6H_5CH_2S-CH-(CbzNH)-CO_2CH_3$$

$$HO-C_6H_4-CH_2-CH(NH_2)-CO_2H \xrightarrow[HCl]{CH_3OH} HO-C_6H_4-CH_2-CH(NH_2)-CO_2CH_3$$

$$\xrightarrow{DCC} C_6H_5CH_2S-CH_2-CH(CbzNH)-C(=O)-NH-CH(CH_2-C_6H_4-OH)-CO_2CH_3$$

$$\xrightarrow[2.\ HCl\ (neut)]{1.\ NaOH} C_6H_5CH_2S-CH_2-CH(CbzNH)-C(=O)-NH-CH(CH_2-C_6H_4-OH)-CO_2H$$

20.31 **(a)** Reduction of the disulfide linkage was carried out using sodium in ammonia. This reaction would also have reduced the glycine ethyl ester groups if they had not first been hydrolyzed to the free carboxylic acids. (The COOH group is converted to the carboxylate anion, which is unreactive to Na/NH_3.)

(b) As expected, the reaction of an ester occurs in preference to reaction at peptide linkages.

(c) Attempted generation of the acid chloride would have produced a compound containing both acid chloride and free amino groups. Such an intermediate would have undergone self-reaction prior to the addition of ammonia.

20.32 As illustrated by the following reaction in the oxytocin synthesis, the sulfide anion is more nucleophilic than either a carboxylate anion or a free amino group:

$$\underset{^{-}S}{\overset{NH_2}{}}\text{Cys-Pro-Leu-Gly-}CO_2^- \xrightarrow[\text{2. HCl (neut)}]{\text{1. } C_6H_5CH_2Cl} \underset{C_6H_5CH_2S}{\overset{NH_2}{}}\text{Cys-Pro-Leu-Gly-}CO_2H$$

20.33 For a synthesis beginning with 1.0 g (3.3 mmol) of *N*-tosylglutamic acid the theoretical yield of oxytocin is 3.3 g. Using the yields stated in the text and starting from the bottom left portion of Figure 20.4, the overall yield would be 1.3% or 44 mg.

20.34 As illustrated in the following sequence, the incorrect quantity of reagent would result in the formation of a pentapeptide in 33% yield together with a 33% yield of the desired hexapeptide.

Gly-polymer —90%→ Ser-Gly-polymer —90%→ Phe-Ser-Gly-polymer

Phe-Ser-Gly-polymer —50%→ Ile-Phe-Ser-Gly-polymer

Phe-Ser-Gly-polymer —50%, wrong amount of DCC→ Phe-Ser-Gly-polymer

Ile-Phe-Ser-Gly-polymer —90%→ Leu-Ile-Phe-Ser-Gly-polymer —90%→ Ala-Leu-Ile-Phe-Ser-Gly-polymer (33% overall)

Phe-Ser-Gly-polymer —90%→ Leu-Phe-Ser-Gly-polymer —90%→ Ala-Leu-Phe-Ser-Gly-polymer (33% overall)

20.35

$(CH_3)_2CH{-}CH_2$ O ‖ $NH_2{-}CH{-}C{-}NH{-}CH{-}CO_2H$ (CH_2–phenyl) $\xrightarrow{\text{hydrolysis}}$ $(CH_3)_2CH{-}CH_2$ $NH_2{-}CH{-}CO_2H$ + $NH_2{-}CH{-}CO_2H$ (CH_2–phenyl)

↓ F, O_2N, NO_2 (fluorodinitrobenzene)

$(CH_3)_2CH{-}CH_2$ O ‖ $NH{-}CH{-}C{-}NH{-}CH{-}CO_2H$ (O_2N, NO_2 phenyl on NH; CH_2–phenyl) $\xrightarrow{\text{hydrolysis}}$ $(CH_3)_2CH{-}CH_2$ $NH{-}CH{-}CO_2H$ (O_2N, NO_2 phenyl on NH) + $NH_2{-}CH{-}CO_2H$ (CH_2–phenyl)

20.36 The molecular weights of the amino acid residues range from 131 to 184, and a tripeptide containing one residue of each of the three amino acids would have a molecular weight of 445. One additional residue would increase the molecular weight to about 600, so you can conclude that the original material is a tetrapeptide. The reaction with fluorodinitrobenzene demonstrates that phenylalanine is the amino-terminal residue. On the other hand, the isolation of the dipeptides Phe-Leu as well as Leu-Phe requires that the peptide contain two phenylalanine residues. This establishes (from the amino end) the partial structure, NH_2-Phe-Leu-Phe---. Inclusion of the fourth amino residue provides the complete structure, NH_2-Phe-Leu-Phe-Tyr-CO_2H:

peptide $\xrightarrow{\text{hydrolysis}}$ Phe + Leu + Tyr

peptide $\xrightarrow{\text{partial hydrolysis}}$ Phe-Tyr + Phe-Leu + Leu-Phe

peptide $\xrightarrow{\text{1. F, } O_2N\text{, } NO_2 \text{ (fluorodinitrobenzene); 2. hydrolysis}}$ NH-Phe-CO_2H (O_2N, NO_2 phenyl)

CHAPTER 21. β-DICARBONYL COMPOUNDS

ANSWERS TO EXERCISES

21.1 (a) $C_2H_5OC(=O)-CH_3 \longrightarrow C_2H_5OC(=O)-\bar{C}H_2 \longleftrightarrow C_2H_5OC(O^-)=CH_2$

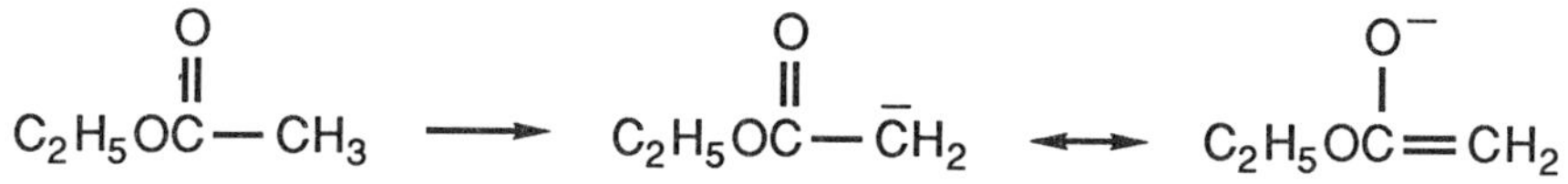

(b) $N\equiv C-CH_3 \longrightarrow N\equiv C-\bar{C}H_2 \longleftrightarrow {}^-N=C=CH_2$

(c) $CH_3-C(=O)-CH_3 \longrightarrow CH_3-C(=O)-\bar{C}H_2 \longleftrightarrow CH_3-C(O^-)=CH_2$

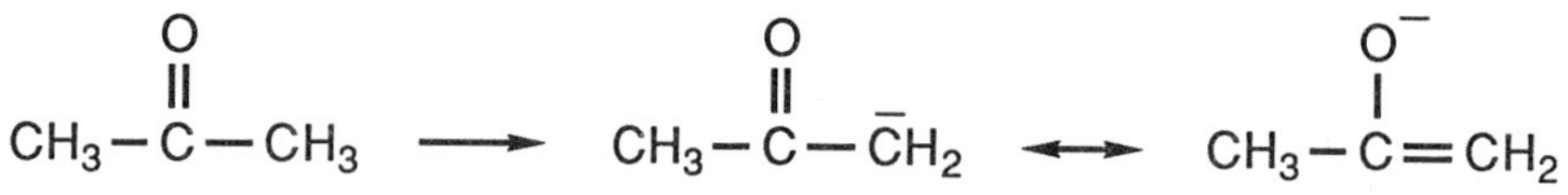

(d) $C_2H_5OC(=O)-CH_2-C(=O)OC_2H_5 \longrightarrow C_2H_5OC(=O)-\bar{C}H-C(=O)OC_2H_5$

$\updownarrow$

$C_2H_5OC(=O)-CH=C(O^-)OC_2H_5 \longleftrightarrow C_2H_5OC(O^-)=CH-C(=O)OC_2H_5$

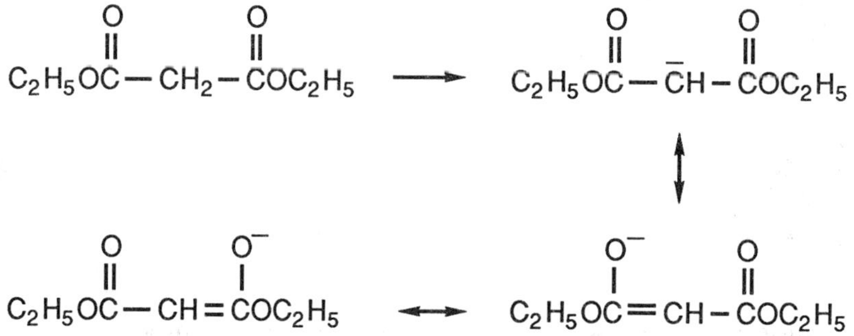

(e) $CH_3C(=O)-CH_2-C(=O)OC_2H_5 \longrightarrow CH_3C(=O)-\bar{C}H-C(=O)OC_2H_5$

$\updownarrow$

$CH_3C(=O)-CH=C(O^-)OC_2H_5 \longleftrightarrow CH_3C(O^-)=CH-C(=O)OC_2H_5$

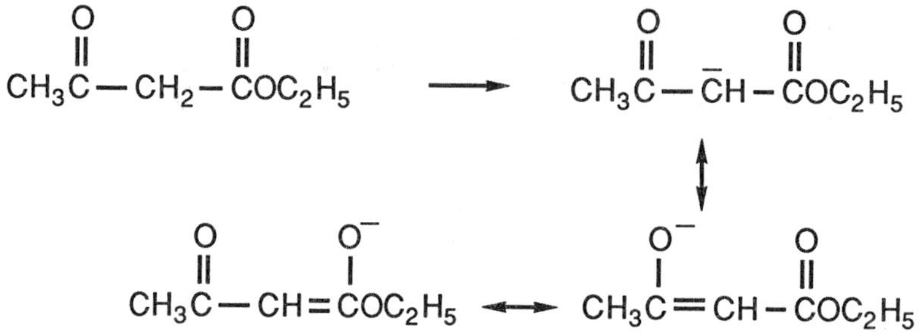

(f) $N\equiv C-CH_2-C\equiv N \longrightarrow N\equiv C-\bar{C}H-C\equiv N$

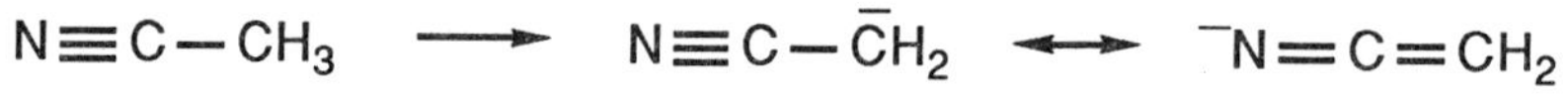

$\updownarrow$

$N\equiv C-CH=C=\bar{N} \longleftrightarrow \bar{N}=C=CH-C\equiv N$

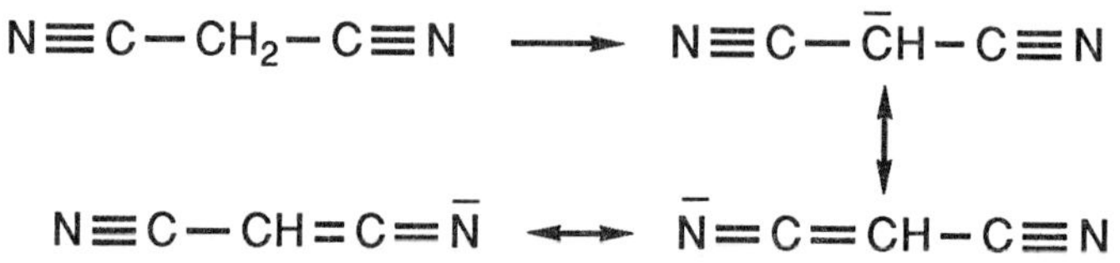

(g) $N{\equiv}C{-}CH_2{-}\overset{O}{\overset{\|}{C}}OC_2H_5 \longrightarrow N{\equiv}C{-}\overline{C}H{-}\overset{O}{\overset{\|}{C}}OC_2H_5$

$\updownarrow$

$N{\equiv}C{-}CH{=}\overset{O^-}{\overset{|}{C}}OC_2H_5 \longleftrightarrow \overline{N}{=}C{=}CH{-}\overset{O}{\overset{\|}{C}}OC_2H_5$

(h) $CH_3\overset{O}{\overset{\|}{C}}{-}CH_2{-}\overset{O}{\overset{\|}{C}}CH_3 \longrightarrow CH_3\overset{O}{\overset{\|}{C}}{-}\overline{C}H{-}\overset{O}{\overset{\|}{C}}CH_3$

$\updownarrow$

$CH_3\overset{O}{\overset{\|}{C}}{-}CH{=}\overset{O^-}{\overset{|}{C}}CH_3 \longleftrightarrow CH_3\overset{O^-}{\overset{|}{C}}{=}CH{-}\overset{O}{\overset{\|}{C}}CH_3$

21.2

(a) 1,3-cyclohexanedione with H's on C-2: H ← more

(b) 3-(CO_2CH_3)cyclohexanone with H's on C-2: H ← more

(c) $CH_2(CO_2H)_2$: both CO_2H ← more

(d) 2-cyanocyclopentanone: H ← more

(e) 2-phenylcyclopentanone (C_6H_5): H ← less

(f) methyl salicylate (CO_2CH_3): OH ← more

21.3 Condensation would be accompanied by transesterification:

$$CH_3{-}\overset{O}{\overset{\|}{C}}{-}OC_2H_5 \xrightarrow[\text{2. HCl}]{\text{1. NaOCH}_3\text{, CH}_3\text{OH}} CH_3\overset{O}{\overset{\|}{C}}{-}CH_2{-}\overset{O}{\overset{\|}{C}}{-}OCH_3$$

21.4

Cyclooctanone $\xrightarrow{\text{NaH}}$ enolate (O^-) + $C_2H_5OCOC_2H_5$ (with C=O) $\longrightarrow$ tetrahedral intermediate with C_2H_5O, O^- and OC_2H_5 on C, attached to the cyclooctanone ring $\longrightarrow$ 2-($CO_2C_2H_5$)cyclooctanone + $^-OC_2H_5$ (removes the H) $\longrightarrow$ enolate (O^-) with $CO_2C_2H_5$ $\xrightarrow{\text{HCl workup}}$ 2-($CO_2C_2H_5$)cyclooctanone

21.5 (a) $CH_3CH_2CH_2C(=O)-CH(CH_2CH_3)-C(=O)OC_2H_5$

(b) $C_6H_5C(=O)-CH_2-C(=O)OC_2H_5$

(c) $CH_3-CH(CH_3)-C(=O)-CH_2-C(=O)OC_2H_5$

Note that the other crossed-condensation product could not form a stable enolate and would be disfavored:

$CH_3-C(=O)-C(CH_3)(CH_3)-C(=O)OC_2H_5$

(d) $C_2H_5OC(=O)-C(=O)-CH(CH_3)-C(=O)OC_2H_5$

21.6 (a) $HO_2C-CH_2-C(CO_2CH_3)=CHCH_2CH_2CH_3$

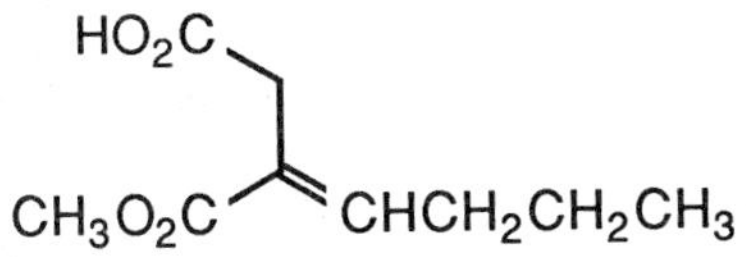

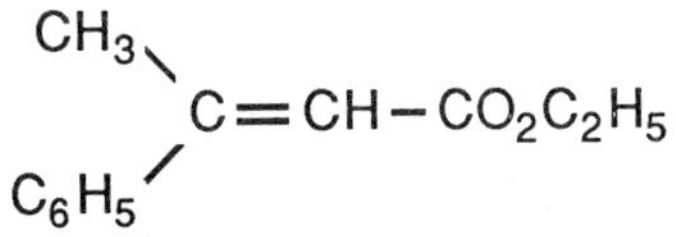

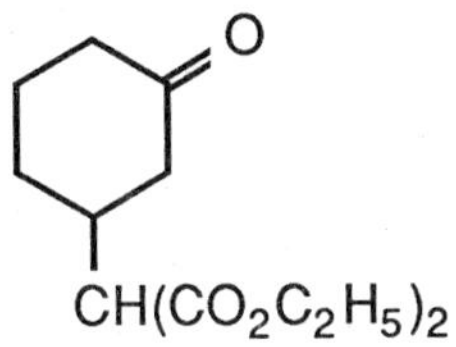

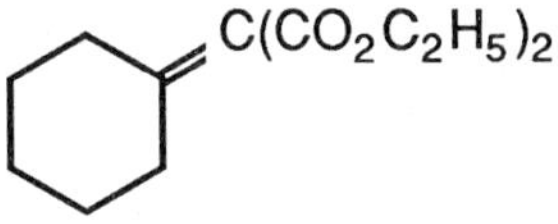

(b) $(CH_3)(C_6H_5)C=CH-CO_2C_2H_5$

(c) $o\text{-}ClC_6H_4-CH=CHCO_2H$

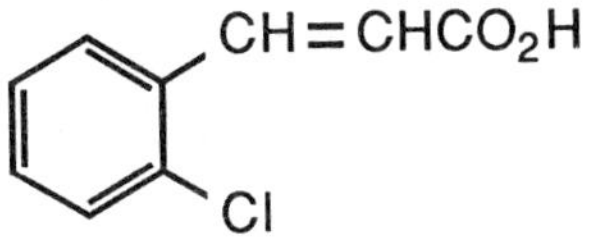

21.7 (a) 3-$CH(CO_2C_2H_5)_2$-cyclohexanone (ring C=O)

(b) cyclohexylidene$=C(CO_2C_2H_5)_2$

(c) cyclohexylidene$=CH-CO_2C_2H_5$

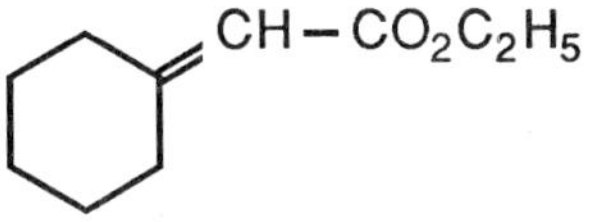

21.8

(a) $$C_6H_5CHO \xrightarrow[\text{NaH}]{(C_2H_5O)_2P(=O)-CH_2CO_2C_2H_5} C_6H_5CH{=}CHCO_2C_2H_5$$

(b) $$C_6H_5CHO \xrightarrow[NH_4^+\ {}^-O_2CCH_3]{NCCH_2CO_2C_2H_5} C_6H_5CH{=}C(CN)CO_2C_2H_5$$

(c) $$C_6H_5CH{=}CHCO_2C_2H_5 \xrightarrow[NaOC_2H_5]{CH_2(CN)_2} C_6H_5CH(CH(CN)_2)-CH_2CO_2C_2H_5$$

21.9

(a) $C_2H_5CH(CO_2C_2H_5)_2$

(b) $CH_3-C(=O)-C(CH_3)_2-CO_2C_2H_5$

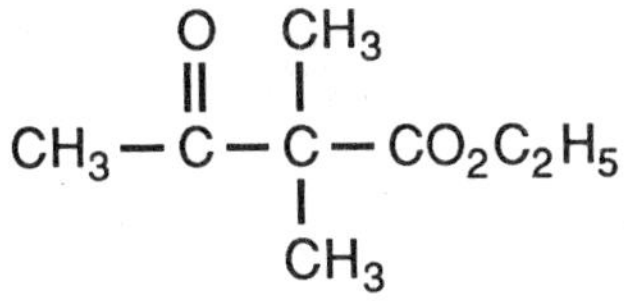

(c) 2-methyl-2-($CH_2-C(CH_3)=CH_2$)cyclohexanone (ring C=O, with CH_3 and $CH_2-C(CH_3)=CH_2$ on the α-carbon)

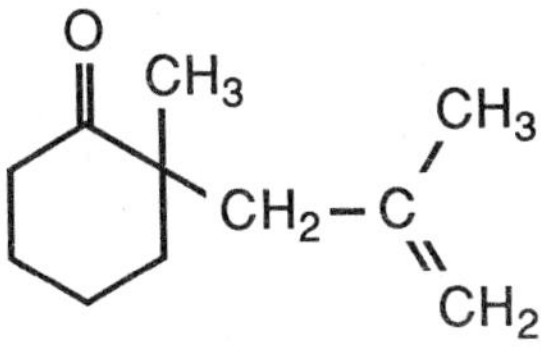

21.10 (a) $CH_3C(=O)CH_2CO_2CH_3$ $\xrightarrow[\text{2. NaOC}_2\text{H}_5\text{, followed by CH}_3\text{CH}_2\text{I}]{\text{1. NaOC}_2\text{H}_5\text{, followed by CH}_3\text{I}}$ $CH_3C(=O)C(CH_3)(CH_2CH_3)CO_2C_2H_5$

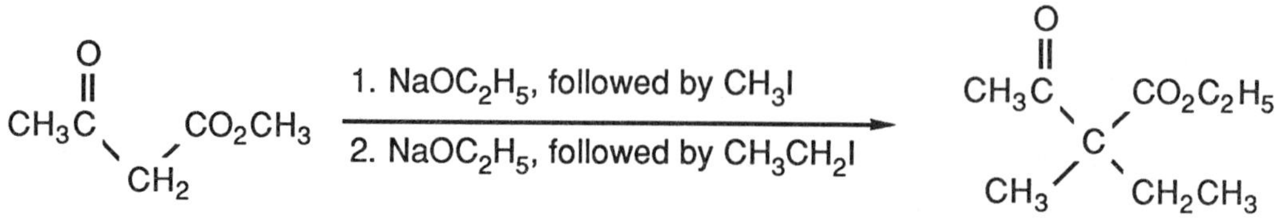

(b) 2-(CO_2CH_3)cyclopentanone $\xrightarrow[\text{2. CH}_3\text{CH}_2\text{CH}_2\text{Br}]{\text{1. NaOCH}_3}$ 2-(CO_2CH_3)-2-($CH_2CH_2CH_3$)cyclopentanone

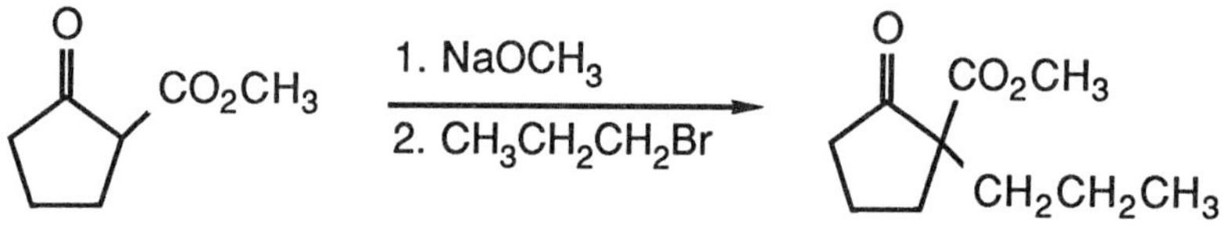

(c) $NC-CH_2-CO_2C_2H_5$ $\xrightarrow[\text{2. CH}_3\text{CH}_2\text{CHBr(CH}_3\text{)}]{\text{1. NaOCH}_3}$ $CH_3CH_2-CH(CH_3)-CH(CN)CO_2C_2H_5$

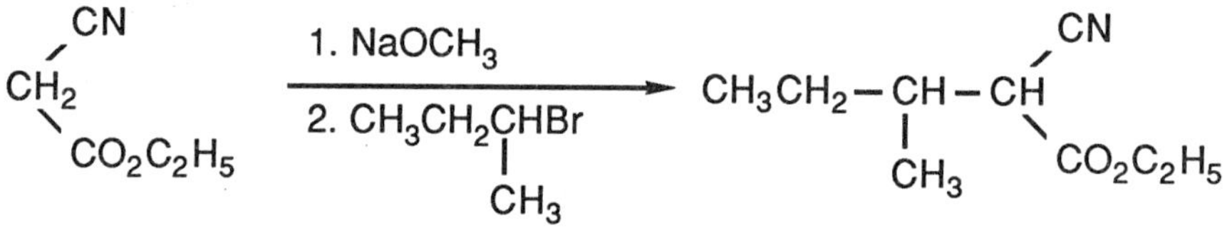

21.11

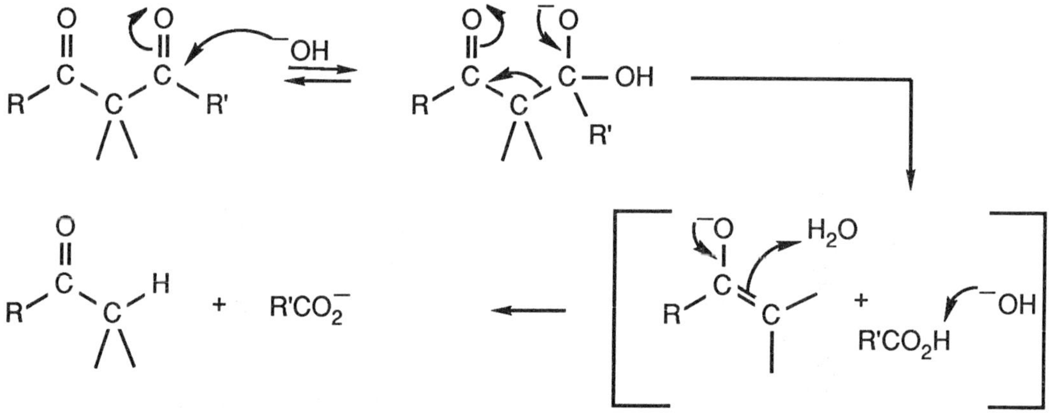

$RC(=O)-C(-)(-)-C(=O)R'$ + ^{-}OH ⇌ $RC(=O)-C(-)(-)-C(O^{-})(OH)R'$ ⟶ [$RC(O^{-})=C<$ + $R'CO_2H$; H_2O, ^{-}OH] ⟶ $RC(=O)-C(-)(-)H$ + $R'CO_2^{-}$

21.12

(a) $CH_3-CH(CO_2C_2H_5)_2$ $\xrightarrow[\text{2. HCl}]{\text{1. KOH, H}_2\text{O}}$ $CH_3-CH(CO_2H)_2$ or $CH_3CH_2CO_2H$ if heated

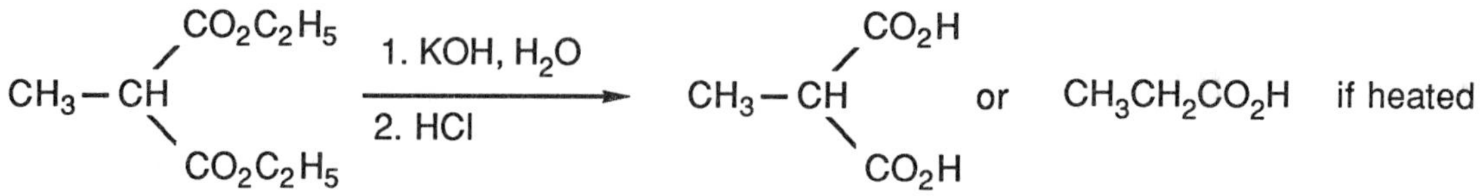

(b)

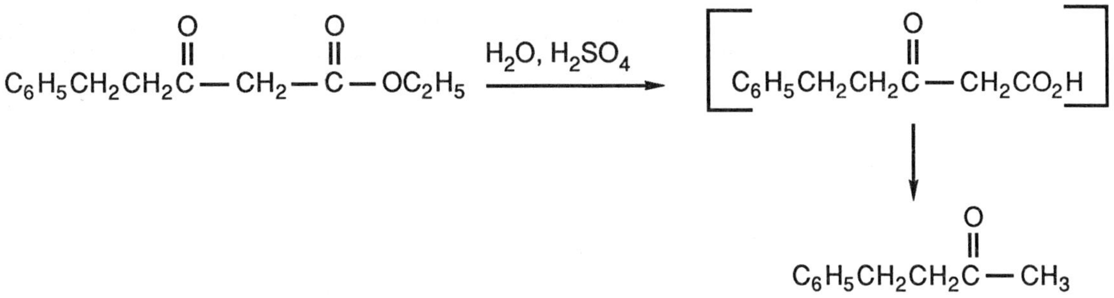

$$C_6H_5CH_2CH_2\overset{\overset{\large O}{\|}}{C}-CH_2-\overset{\overset{\large O}{\|}}{C}-OC_2H_5 \xrightarrow{H_2O,\ H_2SO_4} \left[C_6H_5CH_2CH_2\overset{\overset{\large O}{\|}}{C}-CH_2CO_2H\right] \longrightarrow C_6H_5CH_2CH_2\overset{\overset{\large O}{\|}}{C}-CH_3$$

21.13

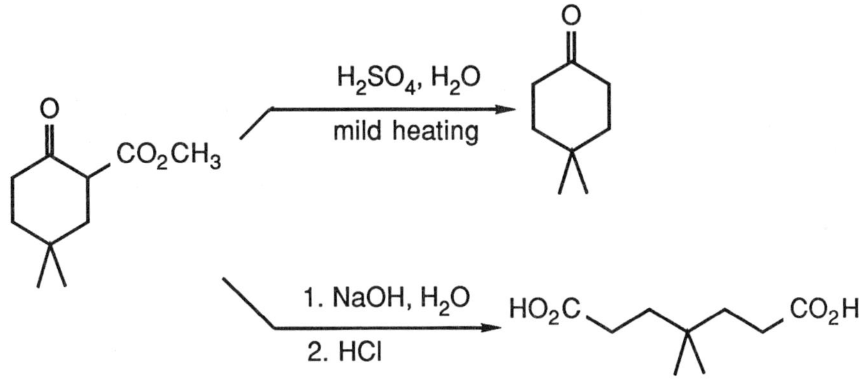

H_2SO_4, H_2O

mild heating

CO_2CH_3

1. NaOH, H_2O

2. HCl

HO_2C CO_2H

21.14 (a)

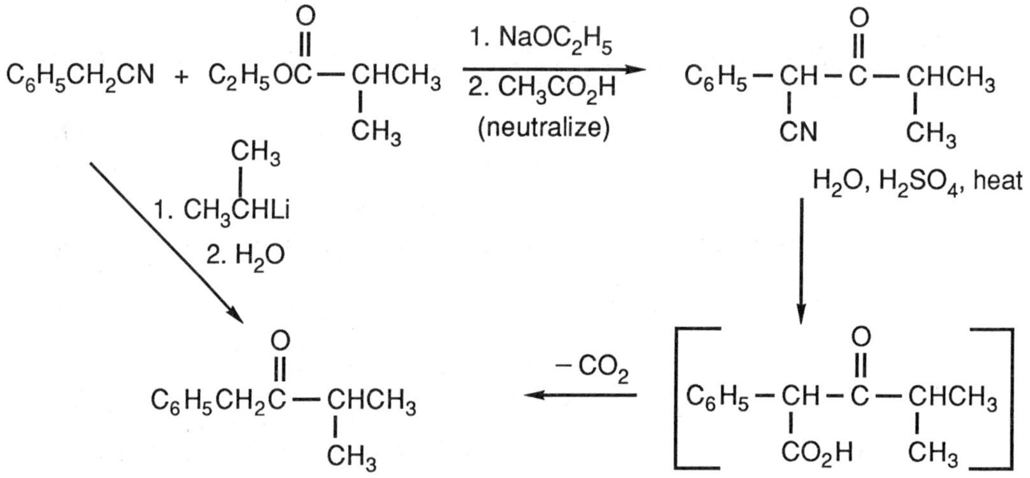

$$C_6H_5CH_2CN + C_2H_5OC(=O)CH(CH_3)CH_3 \xrightarrow[\text{2. } CH_3CO_2H \text{ (neutralize)}]{\text{1. } NaOC_2H_5} C_6H_5-CH(CN)-C(=O)-CH(CH_3)CH_3$$

$$C_6H_5-CH(CN)-C(=O)-CH(CH_3)CH_3 \xrightarrow{H_2O,\ H_2SO_4,\ \text{heat}} \left[C_6H_5-CH(CO_2H)-C(=O)-CH(CH_3)CH_3\right] \xrightarrow{-CO_2} C_6H_5CH_2C(=O)-CH(CH_3)CH_3$$

$$C_6H_5CH_2CN \xrightarrow[\text{2. } H_2O]{\text{1. } CH_3CH(CH_3)Li} C_6H_5CH_2C(=O)-CH(CH_3)CH_3$$

Note that the product could be formed in a single step via the reaction of isopropyllithium.

(b)

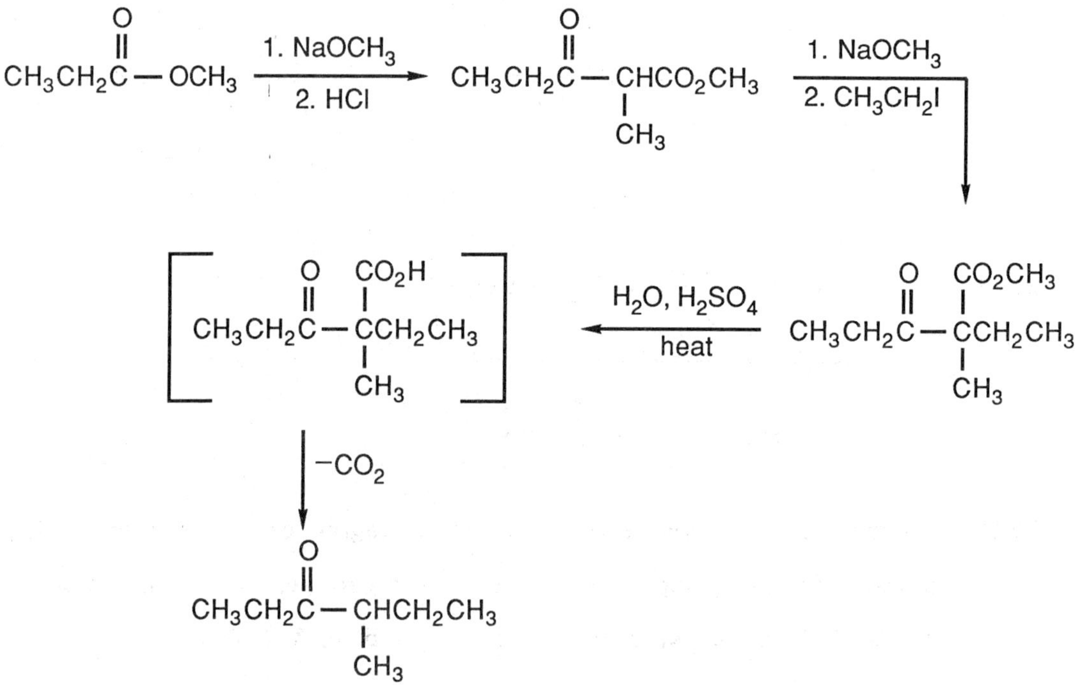

$$CH_3CH_2C(=O)-OCH_3 \xrightarrow[\text{2. HCl}]{\text{1. } NaOCH_3} CH_3CH_2C(=O)-CH(CH_3)CO_2CH_3 \xrightarrow[\text{2. } CH_3CH_2I]{\text{1. } NaOCH_3} CH_3CH_2C(=O)-C(CO_2CH_3)(CH_3)CH_2CH_3$$

$$CH_3CH_2C(=O)-C(CO_2CH_3)(CH_3)CH_2CH_3 \xrightarrow[\text{heat}]{H_2O,\ H_2SO_4} \left[CH_3CH_2C(=O)-C(CO_2H)(CH_3)CH_2CH_3\right] \xrightarrow{-CO_2} CH_3CH_2C(=O)-CH(CH_3)CH_2CH_3$$

(c)

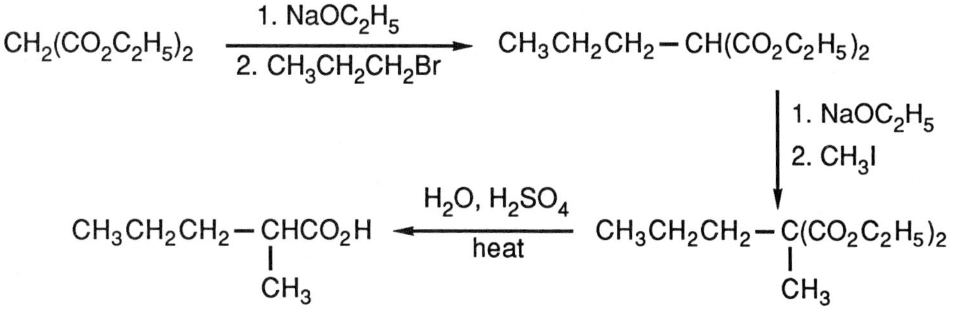

$$CH_2(CO_2C_2H_5)_2 \xrightarrow[\text{2. } CH_3CH_2CH_2Br]{\text{1. } NaOC_2H_5} CH_3CH_2CH_2-CH(CO_2C_2H_5)_2$$

$$\xrightarrow[\text{2. } CH_3I]{\text{1. } NaOC_2H_5} CH_3CH_2CH_2-C(CH_3)(CO_2C_2H_5)_2 \xrightarrow[\text{heat}]{H_2O,\ H_2SO_4} CH_3CH_2CH_2-CH(CH_3)CO_2H$$

(d)

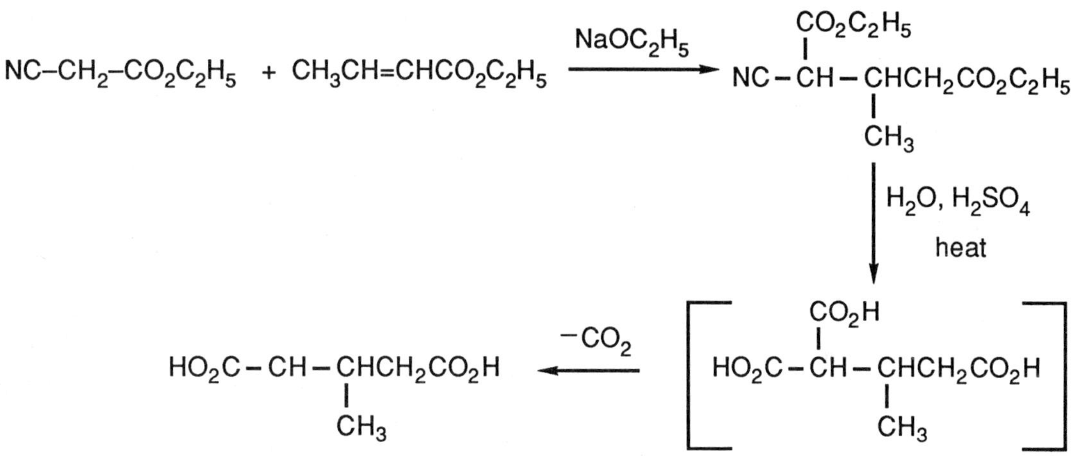

$$NC{-}CH_2{-}CO_2C_2H_5 + CH_3CH{=}CHCO_2C_2H_5 \xrightarrow{NaOC_2H_5} NC-CH(CO_2C_2H_5)-CH(CH_3)CH_2CO_2C_2H_5$$

$$\xrightarrow[\text{heat}]{H_2O,\ H_2SO_4} \left[HO_2C-CH(CO_2H)-CH(CH_3)CH_2CO_2H\right] \xrightarrow{-CO_2} HO_2C-CH_2-CH(CH_3)CH_2CO_2H$$

21.15 $CH_3CH_2-CH_2(CH_2)_{15}CH_2CO_2H$

CH_3CH_2 ← acetyl CoA; $CH_2(CH_2)_{15}CH_2CO_2H$ ← malonyl CoA

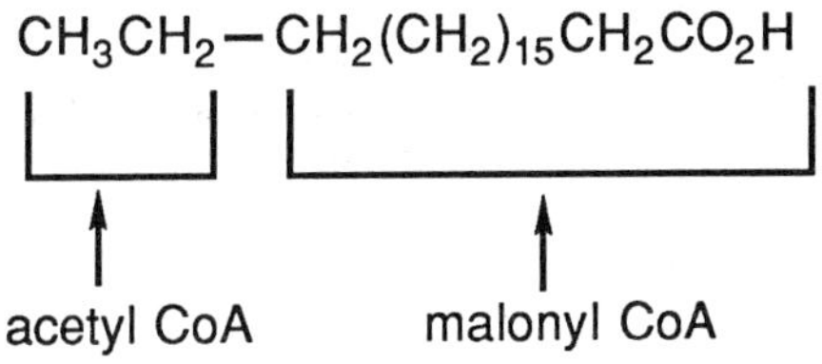

21.16 The molecular formula indicates a single degree of unsaturation, which corresponds to the carbonyl group of the acid. Hydrogen atoms on the carbons marked with asterisks in the following drawing are derived from NADPH:

$$CH_3CH_2CH_2\overset{*}{C}H_2CH_2\overset{*}{C}H_2CH_2\overset{*}{C}H_2CH_2\overset{*}{C}H_2CH_2\overset{*}{C}H_2CH_2CO_2H$$

ANSWERS TO PROBLEMS

21.1

(a) $CH_3CH_2C(=O)-CH_2CO_2H$ (arrow at the C–CH$_2$ bond)

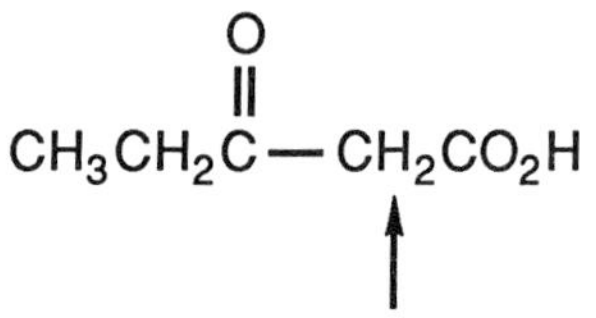

(b) $CH_3CH(CN)C(=O)-CH_2C_6H_5$ (arrow at the CH–C bond)

(c) $C_6H_5C(=O)-C(=O)CH_3$ (arrow at the C–C bond)

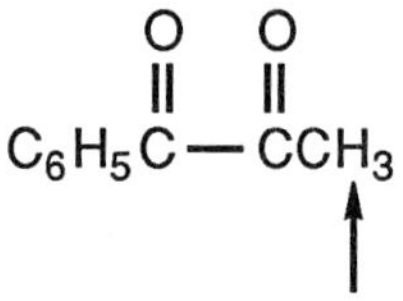

(d) $CH_3CH_2CH_2C(=O)-CH_2CO_2CH_3$ (arrow at the C–CH$_2$ bond)

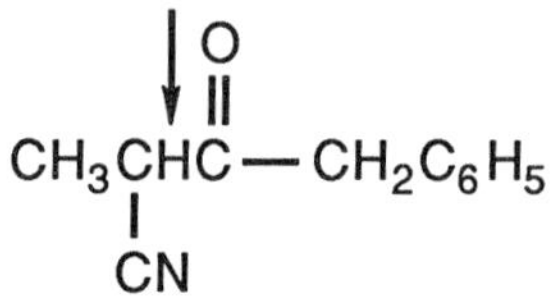

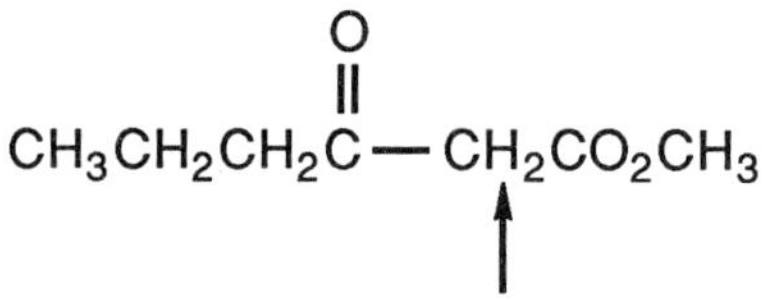

(e) $(C_2H_5O)_2P(=O)-CH_2-C(=O)CH_2CH_2CO_2C_2H_5$ (arrow at CH$_2$)

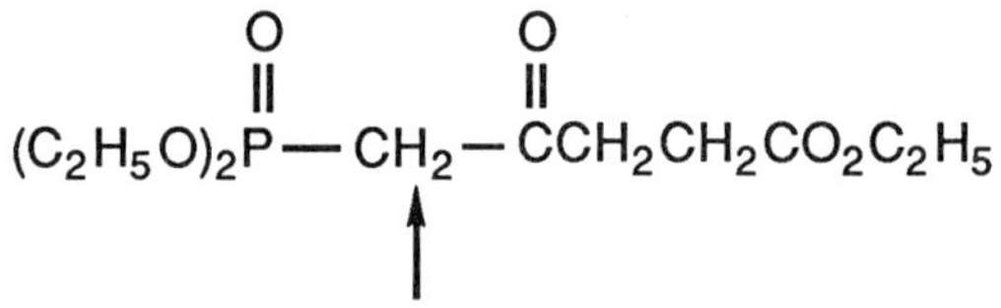

(f) $CH_2(CO_2CH_2CH_2OH)(CO_2CH_3)$ (arrow at CH$_2$)

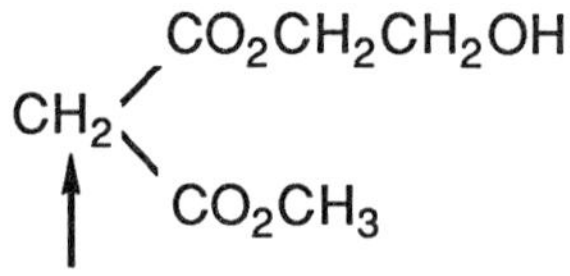

21.2

(a) $CH_3CH_2CH_2CO_2^-\ Na^+$

(b) $CH_3CH_2CH_2C(=O)-C(CH_2CH_3)=C(O^-\ Na^+)OC_2H_5$

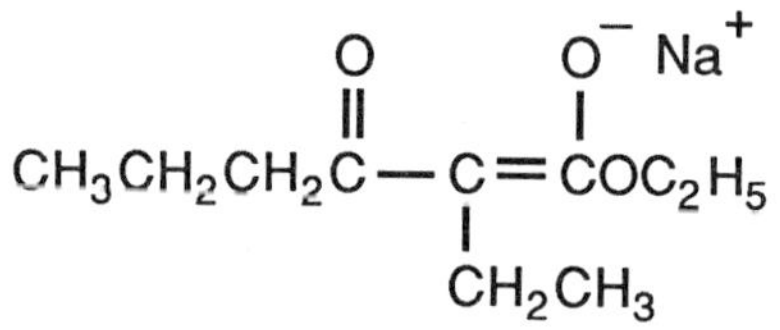

(c) $CH_3CH_2CH_2CO_2H$

(d) $CH_3CH_2CH_2C(=O)-C(CH_2CH_3)=C(O^-\ Na^+)OC_2H_5$

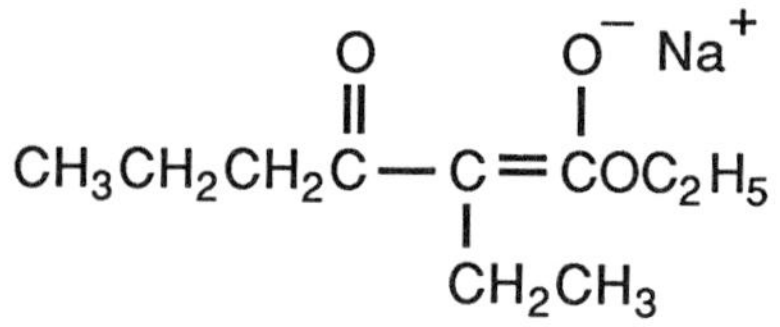

(e) $C_6H_5C(=O)-C(CH_2CH_3)=C(O^-\ Na^+)OC_2H_5$

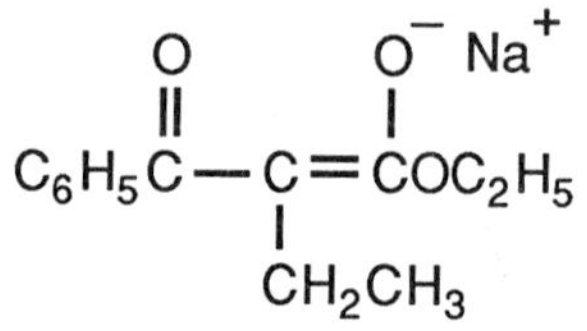

(f) $CH_3-CH(CH_3)-C(=O)-CH(CH_2CH_3)-C(=O)OC_2H_5$

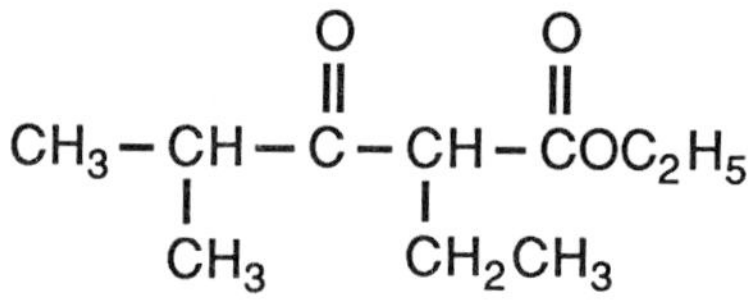

(g) $CH_3CH_2CH_2C(=O)-CH(C_6H_5)CN$

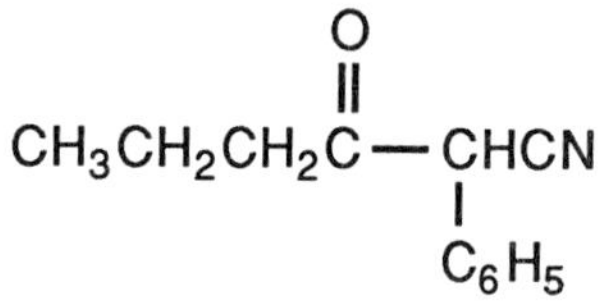

21.3

(a) CH_3; O; $C-COC_2H_5$; O O

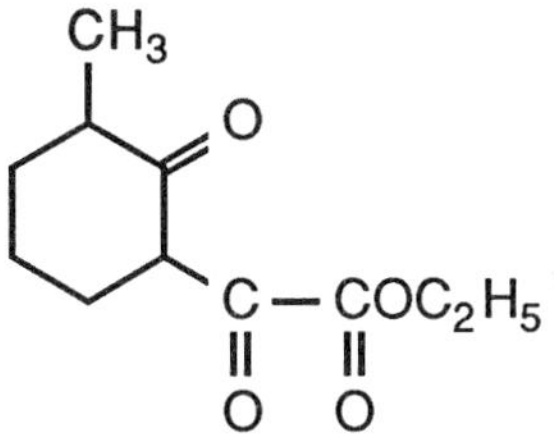

(b) O; O; CH

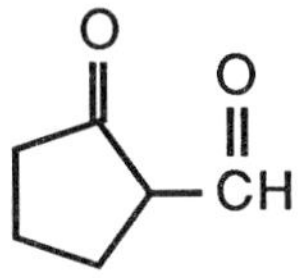

(c) $CH(CO_2C_2H_5)_2$

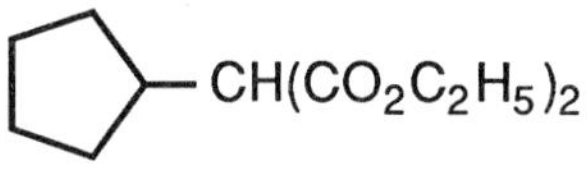

(d) HO_2C; $C_2H_5O_2C$; CHC_6H_5

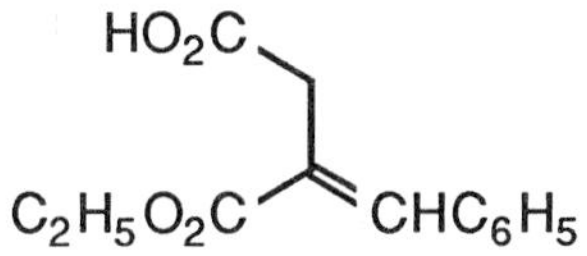

(e) $CHCO_2H$

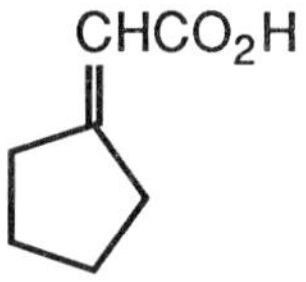

(f) $CH_3CH_2-C(OH)(CH_3)-CH_2CO_2C_2H_5$

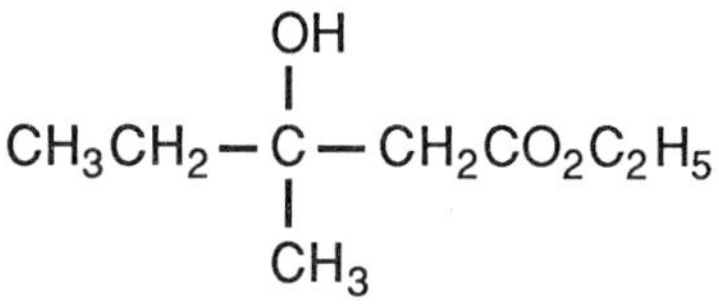

(g) $C_6H_5C(=O)-CH_2-C(=O)C_6H_5$

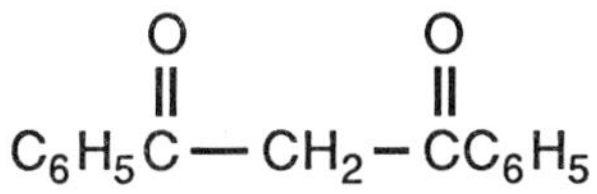

(h) $CH_3CH_2CH_2C(=O)-CH(C_6H_5)CN$

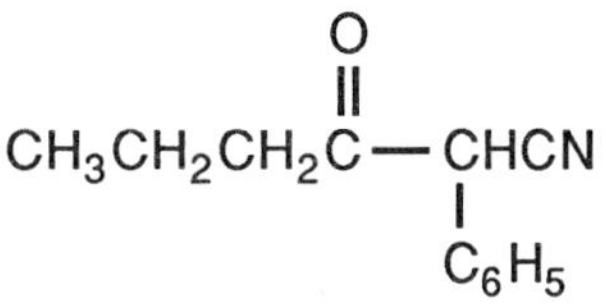

21.4

CHO; Cl

1. $BrCH_2CO_2CH_3$, Zn
2. HCl (neut)

OH; $CHCH_2CO_2CH_3$; Cl

H_2SO_4, H_2O

$CH{=}CHCO_2H$; Cl

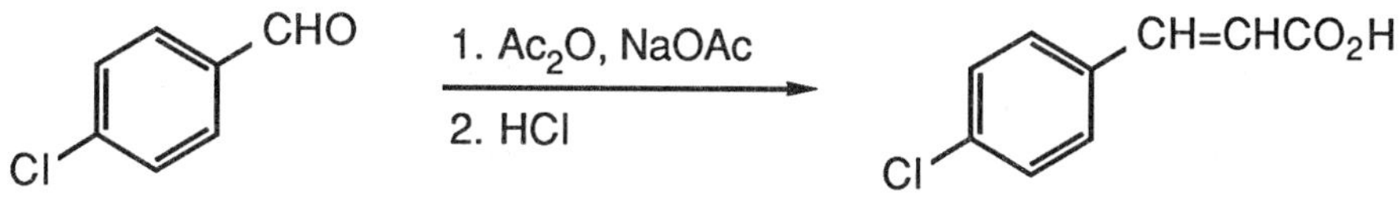

CHO; Cl

1. Ac_2O, NaOAc
2. HCl

$CH{=}CHCO_2H$; Cl

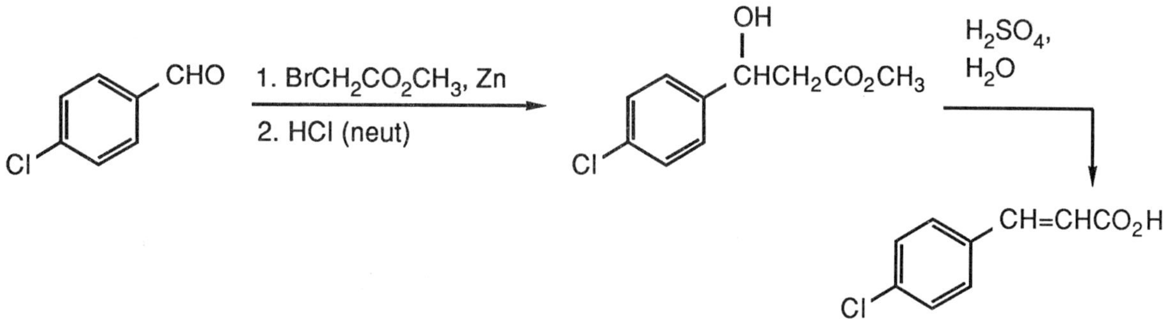

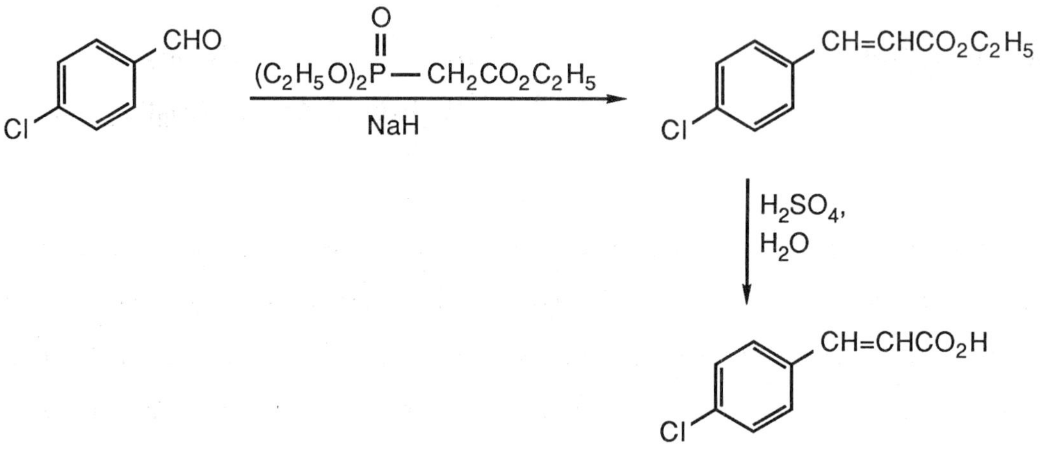

p-ClC_6H_4CHO $\xrightarrow[\text{NaH}]{(C_2H_5O)_2P(=O)-CH_2CO_2C_2H_5}$ p-$ClC_6H_4CH{=}CHCO_2C_2H_5$ $\xrightarrow{H_2SO_4,\ H_2O}$ p-$ClC_6H_4CH{=}CHCO_2H$

21.5

(a) $C_6H_5C(CH_3){=}C(CN)(CO_2C_2H_5)$

(b) $C_6H_5C(CH_3){=}CHCO_2C_2H_5$

(c) $C_6H_5C(OH)(CH_3)-CH_2CO_2C_2H_5$

(d) $C_6H_5C(CH_3){=}C(C_6H_5)CO_2^-\ Na^+$

(e) $C_6H_5C(=O)-CH_2CHO$

21.6

(a) $C_6H_5-CH(CN)CH_2CO_2CH_3$

(b) $C_6H_5CH{=}CHCO_2H$

(c) $\left[C_6H_5-CH(CH_2CO_2CH_3)-CH(CN)CO_2CH_3\right] \longrightarrow C_6H_5-CH(CH_2CO_2H)-CH_2CO_2H$

21.7

(a) $CH_3(CH_2)_6CH(CO_2C_2H_5)_2$

(b) 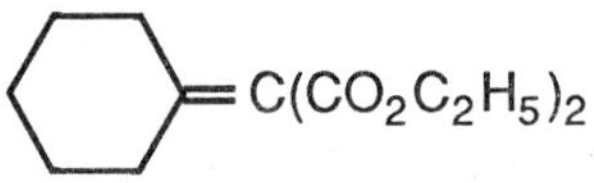 cyclohexylidene$=C(CO_2C_2H_5)_2$

(c) CH_3CO_2H

(d)
$$(C_2H_5O_2C)_2CH-\underset{}{\overset{CH_2CH_3}{|}}{C}HCH_2CO_2CH_3$$

21.8 Since compound **A** is a keto ester, the formation of 1-phenyl-2-butanone must result from hydrolysis of the ester followed by decarboxylation. 1-Phenyl-2-butanone is missing only two carbon atoms of the original **A**, so one of these must be the carboxyl carbon and the other must be a methyl group (for a methyl ester). The reaction of **A** in base to generate phenyl acetic acid must be a retro condensation, so this together with the decarboxylation requires that **A** be a β-keto ester. Two isomeric compounds would undergo the indicated reactions:

$$C_6H_5\underset{CO_2CH_3}{\underset{|}{C}}H\overset{O}{\overset{\|}{C}}-CH_2CH_3 \quad \text{or} \quad C_6H_5CH_2\overset{O}{\overset{\|}{C}}-\underset{CO_2CH_3}{\underset{|}{C}}HCH_3$$

The reactions are shown for just one of the compounds:

$$C_6H_5\underset{CO_2CH_3}{\underset{|}{C}}H\overset{O}{\overset{\|}{C}}-CH_2CH_3 \xrightarrow[\text{2. HCl}]{\text{1. conc. NaOH}} C_6H_5CH_2CO_2H + HO_2CCH_2CH_3$$

$$\xrightarrow[\text{H}_2\text{SO}_4,\ \text{heat}]{\text{H}_2\text{O}} \left[C_6H_5\underset{CO_2H}{\underset{|}{C}}H\overset{O}{\overset{\|}{C}}-CH_2CH_3\right] \longrightarrow C_6H_5CH_2\overset{O}{\overset{\|}{C}}-CH_2CH_3$$

21.9

(a) Cyclohexanone $\xrightarrow{\text{1. NaH; 2. } C_2H_5OCOC_2H_5}$ 2-($CO_2C_2H_5$)cyclohexanone $\xrightarrow{\text{1. } NaOC_2H_5\text{; 2. } CH_3CH_2CH_2Br}$ 2-($CO_2C_2H_5$)-2-($CH_2CH_2CH_3$)cyclohexanone $\xrightarrow{H_2O,\ H_2SO_4,\ \text{heat}}$ 2-($CH_2CH_2CH_3$)cyclohexanone

(b) Cyclopentanone $\xrightarrow{\text{1. } (CH_3CO)_2O,\ NaO_2CCH_3\text{; 2. HCl}}$ cyclopentylidene=$CHCO_2H$ $\xrightarrow{H_2/\ Pt}$ cyclopentyl–CH_2CO_2H

(c) cyclopentyl–CH_2Br + $CH_3\overset{O}{\overset{\|}{C}}-CH_2CO_2C_2H_5$ $\xrightarrow{NaOCH_2CH_3}$ cyclopentyl–$CH_2-\underset{CO_2C_2H_5}{\underset{|}{CH}}-\overset{O}{\overset{\|}{C}}CH_3$ $\xrightarrow{H_2O,\ H_2SO_4,\ \text{heat}}$ cyclopentyl–$CH_2-CH_2-\overset{O}{\overset{\|}{C}}CH_3$

(d) Cyclohexanone + $(C_2H_5O)_2\overset{O}{\overset{\|}{P}}-CH_2CO_2C_2H_5$ $\xrightarrow{NaH}$ cyclohexylidene=$CHCO_2C_2H_5$ $\xrightarrow{\text{1. } (CH_3)_2CuLi\text{; 2. } H_2O}$ 1-CH_3-1-($CH_2CO_2C_2H_5$)cyclohexane $\xrightarrow{H_2O,\ H_2SO_4}$ 1-CH_3-1-(CH_2CO_2H)cyclohexane

(e)

$$C_6H_5-CH_2Br \xrightarrow{\overset{+}{Na}\ \overset{-}{C}H(CO_2C_2H_5)_2} C_6H_5CH_2-CH(CO_2C_2H_5)_2$$

$$C_6H_5CH_2-CH(CO_2C_2H_5)_2 \xrightarrow{\text{1. NBS; 2. KOtBu, tBuOH}} C_6H_5CH{=}C(CO_2C_2H_5)_2$$

$$C_6H_5CH{=}C(CO_2C_2H_5)_2 \xrightarrow{\text{1. } (CH_3)_2CuLi\text{; 2. } H_2O} C_6H_5CH(CH_3)-CH(CO_2C_2H_5)_2$$

$$C_6H_5CH(CH_3)-CH(CO_2C_2H_5)_2 \xrightarrow[\text{heat}]{H_2O,\ H_2SO_4} C_6H_5-CH(CH_3)-CH_2CO_2H$$

(f)

$$C_6H_5CO_2H \xrightarrow[H_2SO_4]{C_2H_5OH} C_6H_5CO_2C_2H_5 \xrightarrow[\text{NaOC}_2\text{H}_5\text{; 2. HCl}]{\text{1. } CH_3CO_2C_2H_5,} C_6H_5C(=O)-CH_2CO_2C_2H_5$$

$$C_6H_5C(=O)-CH_2CO_2C_2H_5 \xrightarrow{NaBH_4} C_6H_5-CH(OH)-CH_2CO_2C_2H_5$$

$$C_6H_5-CH(OH)-CH_2CO_2C_2H_5 \xrightarrow[\text{heat}]{H_2O,\ H_2SO_4} C_6H_5CH{=}CHCO_2H \xrightarrow{H_2/\ Pt} C_6H_5CH_2CH_2CO_2H$$

21.10

(a)

$$CH_2(CO_2C_2H_5)_2 \xrightarrow[\text{2. } C_6H_5CH_2Br]{\text{1. } NaOC_2H_5} C_6H_5CH_2CH(CO_2C_2H_5)_2 \xrightarrow[\text{2. } CH_3I]{\text{1. } NaOC_2H_5} C_6H_5CH_2-C(CH_3)(CO_2C_2H_5)_2$$

$$C_6H_5CH_2-C(CH_3)(CO_2C_2H_5)_2 \xrightarrow[\text{heat}]{H_2O,\ H_2SO_4} C_6H_5CH_2-CH(CH_3)CO_2H$$

(b)

$$CH_3-\overset{\displaystyle CH_3}{\underset{|}{C}}HCHO \xrightarrow[NH_4^{+}\ ^{-}O_2CCH_3]{CH_2(CO_2C_2H_5)_2} CH_3-\overset{\displaystyle CH_3}{\underset{|}{C}}H-CH=C(CO_2C_2H_5)_2 \xrightarrow{NaCN}$$

$$CH_3-\overset{\displaystyle CH_3}{\underset{|}{C}}H-\underset{\displaystyle CN}{CH}-CH(CO_2C_2H_5)_2 \xrightarrow[heat]{H_2O,\ H_2SO_4} CH_3-\overset{\displaystyle CH_3}{\underset{|}{C}}H-\underset{\displaystyle CO_2H}{CH}-CH_2CO_2H$$

(c)

$$CH_3CH_2\overset{\displaystyle O}{\overset{\|}{C}}-CH_3 \xrightarrow[2.\ H_2O]{1.\ BrCH_2CO_2CH_3,\ Zn} CH_3CH_2\overset{\displaystyle OH}{\underset{\displaystyle CH_3}{C}}-CH_2CO_2H$$

(d)

$$C_2H_5O_2CCH_2-\overset{\displaystyle O}{\overset{\|}{C}}CH_3 \xrightarrow[2.\ (CH_3)_2CHCH_2Br]{1.\ NaOC_2H_5} (CH_3)_2CHCH_2-\underset{\displaystyle CO_2C_2H_5}{CH}-\overset{\displaystyle O}{\overset{\|}{C}}CH_3$$

$$\xrightarrow[2.\ BrCH_2CH_2CH_3]{1.\ NaOC_2H_5} (CH_3)_2CHCH_2-\overset{\displaystyle C_2H_5O_2C}{\underset{\displaystyle CH_2CH_2CH_3}{C}}-\overset{\displaystyle O}{\overset{\|}{C}}CH_3 \xrightarrow[heat]{H_2O,\ H_2SO_4} (CH_3)_2CHCH_2-\underset{\displaystyle CH_2CH_2CH_3}{CH}-\overset{\displaystyle O}{\overset{\|}{C}}CH_3$$

21.11 This structure can be formed via Michael addition to the normal product:

$$CH_3CHO \xrightarrow[NH_4^+\ ^-O_2CCH_3]{CH_2(CO_2C_2H_5)_2} CH_3CH{=}C(CO_2C_2H_5)_2 \xleftarrow{} {}^-CH(CO_2C_2H_5)_2$$

$$\longrightarrow CH_3-CH\begin{matrix}{}^-C(CO_2C_2H_5)_2 \\ CH(CO_2C_2H_5)_2\end{matrix} \xrightarrow{H^+} CH_3-CH\begin{matrix}CH(CO_2C_2H_5)_2 \\ CH(CO_2C_2H_5)_2\end{matrix}$$

21.12 The mixed anhydride has two acidic sites, and the phenyl-conjugated enolate is more stable:

$$CH_3\overset{O}{\overset{\|}{C}}-OCH_2C_6H_5 \rightleftharpoons CH_3\overset{O}{\overset{\|}{C}}-O-\overset{O}{\overset{\|}{C}}\bar{C}HC_6H_5 \longrightarrow \mathbf{A}$$

$$CH_3\overset{O}{\overset{\|}{C}}-OCH_2C_6H_5 \rightleftharpoons \bar{C}H_2\overset{O}{\overset{\|}{C}}-O-\overset{O}{\overset{\|}{C}}CH_2C_6H_5 \longrightarrow \mathbf{B}$$

21.13 Attack at a ketone carbonyl occurs more rapidly than attack at an ester carbonyl. For the two ketone carbonyls in this compound, attack at the methyl ketone is less hindered, and this leads to the observed product:

$$C_6H_5\overset{O}{\overset{\|}{C}}-\underset{CO_2C_2H_5}{\underset{|}{CH}}-\overset{O}{\overset{\|}{C}}CH_3 \xleftarrow{} {}^-OH \longrightarrow \longrightarrow C_6H_5\overset{O}{\overset{\|}{C}}-\bar{C}HCO_2C_2H_5$$

observed pathway

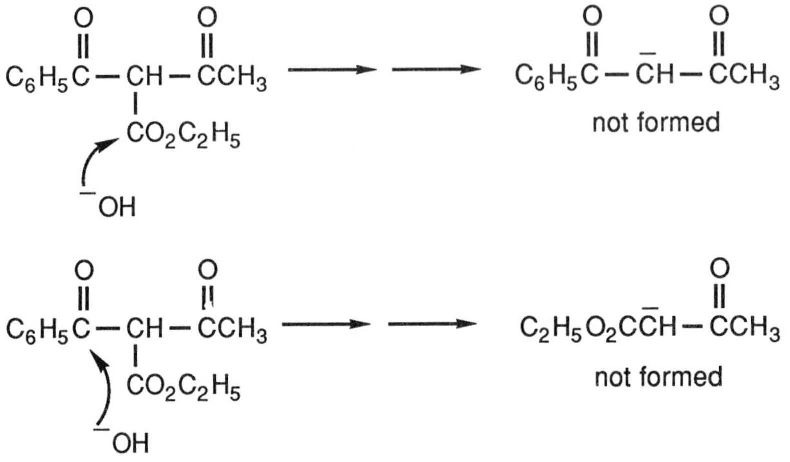

21.14

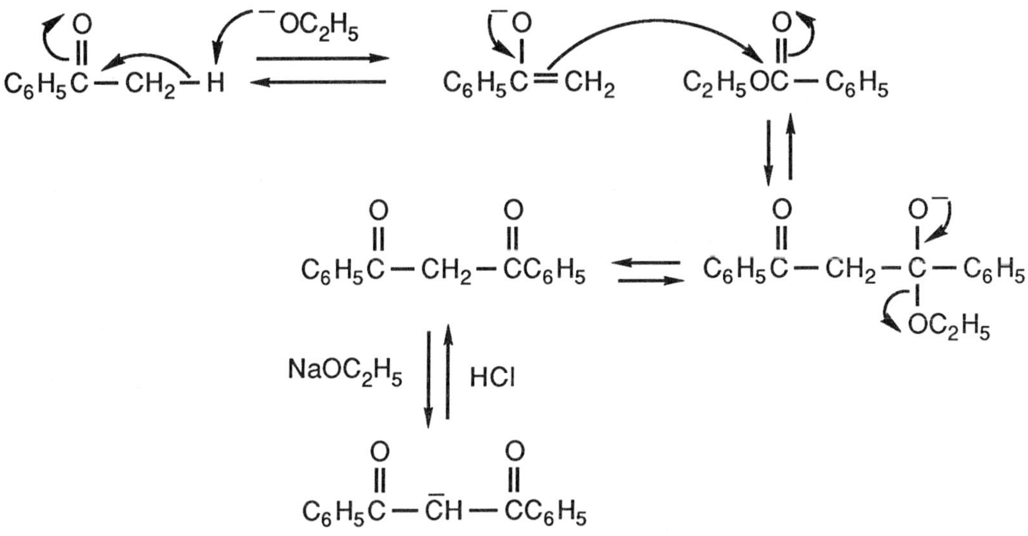

21.15 The enol form of a β-diketone is stabilized by conjugation and by intramolecular hydrogen bonding (Sec. 13.9) . In aprotic solvents the enol can be favored in the equilibrium.

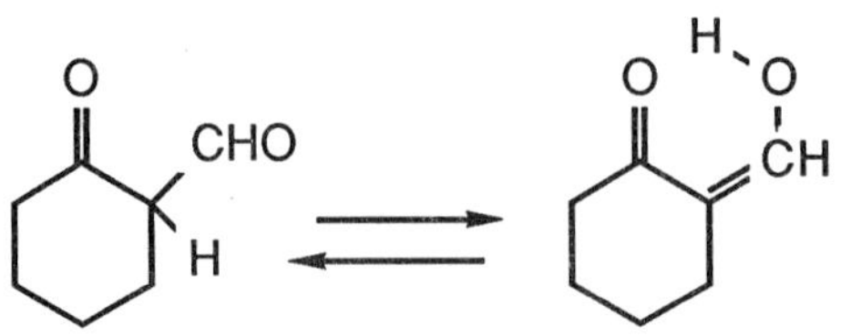

21.16

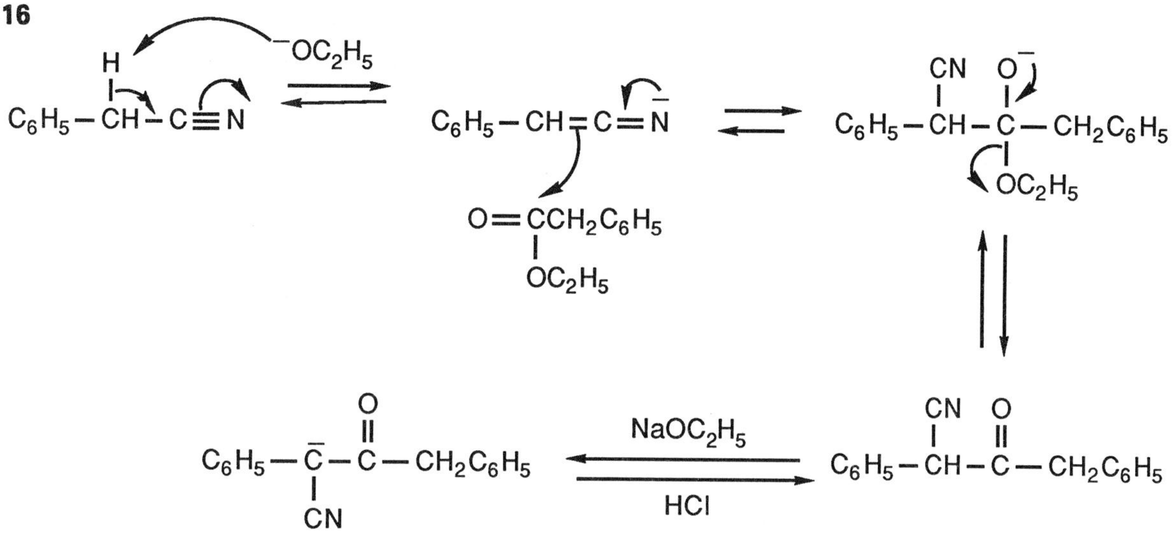

$$C_6H_5-CH_2-C\equiv N \xrightleftharpoons{^-OC_2H_5} C_6H_5-CH=C=\bar{N}$$

$$C_6H_5-CH=C=\bar{N} + O=C(OC_2H_5)CH_2C_6H_5 \rightleftharpoons C_6H_5-CH(CN)-C(O^-)(OC_2H_5)-CH_2C_6H_5$$

$$C_6H_5-CH(CN)-C(O^-)(OC_2H_5)-CH_2C_6H_5 \rightleftharpoons C_6H_5-CH(CN)-C(=O)-CH_2C_6H_5$$

$$C_6H_5-CH(CN)-C(=O)-CH_2C_6H_5 \underset{HCl}{\overset{NaOC_2H_5}{\rightleftharpoons}} C_6H_5-\bar{C}(CN)-C(=O)-CH_2C_6H_5$$

21.17

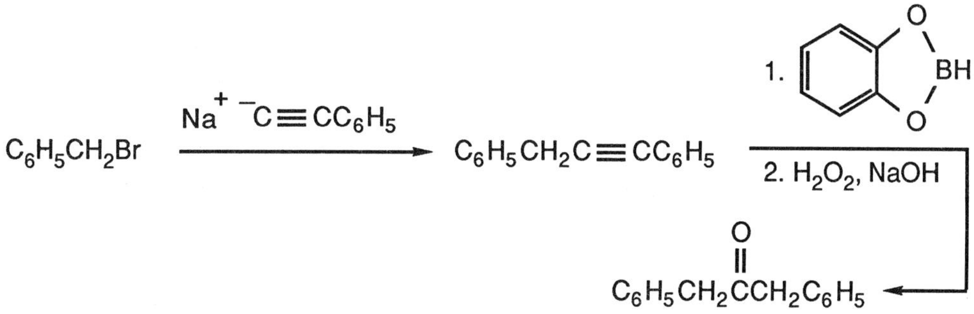

$$C_6H_5CH_2Br \xrightarrow{Na^+\ ^-C\equiv CC_6H_5} C_6H_5CH_2C\equiv CC_6H_5$$

$$C_6H_5CH_2C\equiv CC_6H_5 \xrightarrow[\text{2. } H_2O_2,\ NaOH]{\text{1. catecholborane (benzo-fused } O_2BH)} C_6H_5CH_2\overset{O}{\overset{\|}{C}}CH_2C_6H_5$$

21.18

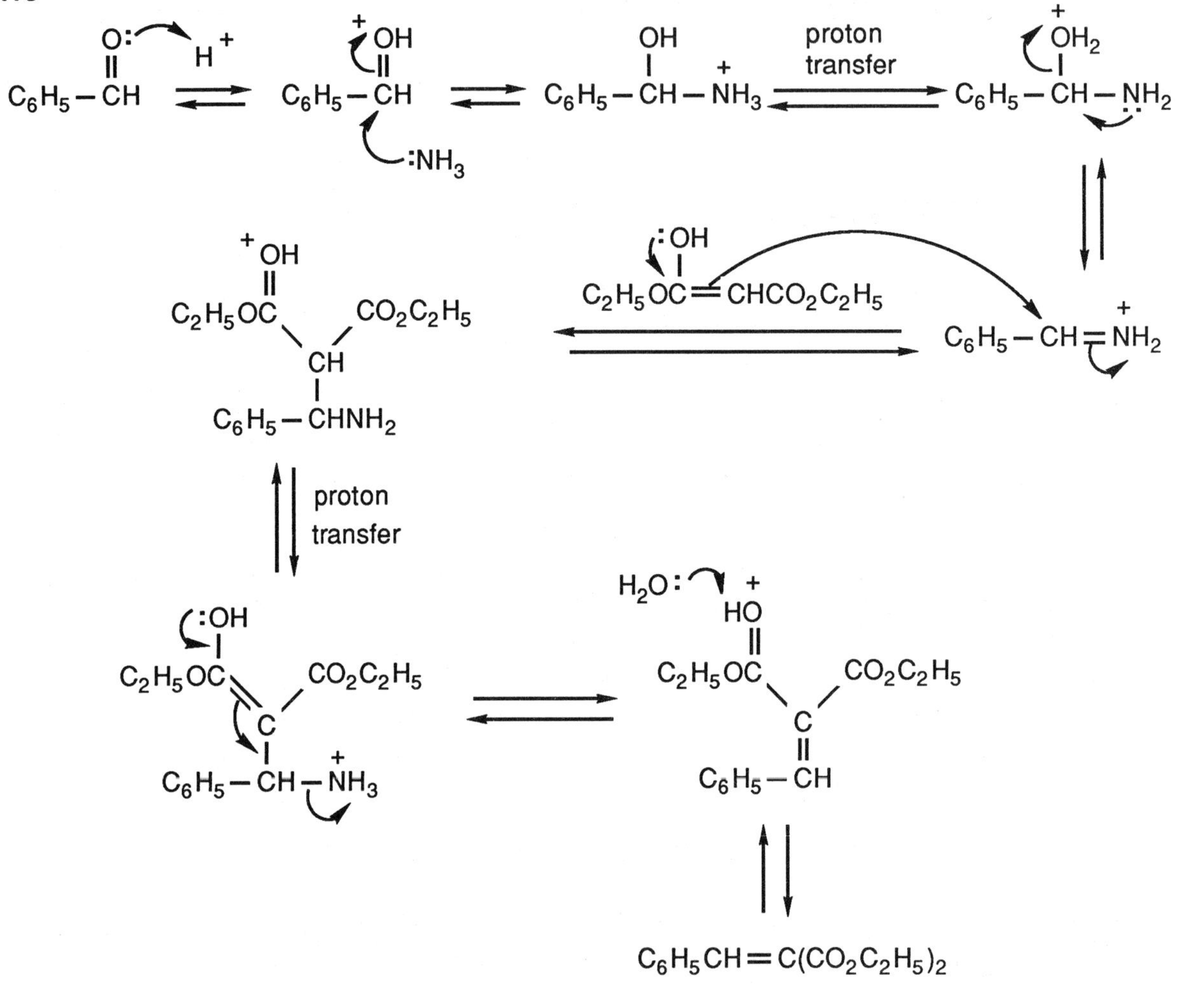

$C_6H_5-CH{=}O$ + H^+ ⇌ $C_6H_5-CH{=}\overset{+}{O}H$ ⇌ (:NH_3) $C_6H_5-CH(OH)-\overset{+}{N}H_3$ ⇌ (proton transfer) $C_6H_5-CH(\overset{+}{O}H_2)-NH_2$ ⇌ $C_6H_5-CH{=}\overset{+}{N}H_2$

$C_6H_5-CH{=}\overset{+}{N}H_2$ + $C_2H_5OC(OH){=}CHCO_2C_2H_5$ ⇌ $(C_2H_5OC(=\overset{+}{O}H))(CO_2C_2H_5)CH-CHNH_2-C_6H_5$

⇌ (proton transfer)

$(C_2H_5OC(OH)=)(CO_2C_2H_5)C-CH(\overset{+}{N}H_3)-C_6H_5$ ⇌ H_2O: + $(C_2H_5OC(=\overset{+}{O}H))(CO_2C_2H_5)C{=}CH-C_6H_5$

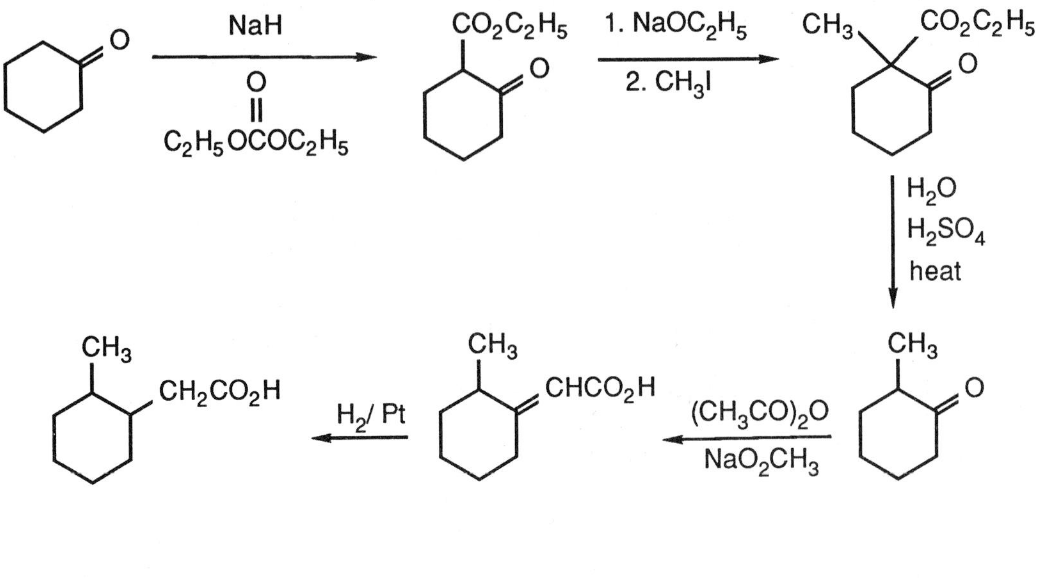

⇌ $C_6H_5CH{=}C(CO_2C_2H_5)_2$

21.19 (a) cyclohexanone $\xrightarrow[C_2H_5OCOC_2H_5 \text{ (C=O)}]{NaH}$ 2-($CO_2C_2H_5$)cyclohexanone $\xrightarrow[2.\ CH_3I]{1.\ NaOC_2H_5}$ 2-CH_3-2-($CO_2C_2H_5$)cyclohexanone $\xrightarrow[\text{heat}]{H_2O,\ H_2SO_4}$ 2-CH_3-cyclohexanone $\xrightarrow[NaO_2CH_3]{(CH_3CO)_2O}$ 2-CH_3-cyclohexylidene $CHCO_2H$ $\xrightarrow{H_2/\ Pt}$ 2-CH_3-cyclohexyl CH_2CO_2H

(b)

$CH_2(CO_2C_2H_5)_2$, $NH_4^+\ {}^-O_2CCH_3$ → $CO_2C_2H_5$, C, $CO_2C_2H_5$ → H_2/ Pt → $CO_2C_2H_5$, CH, $CO_2C_2H_5$

1. $LiAlH_4$ 2. H_2O → CH_2OH, CH, CH_2OH

1. PBr_3 2. $NaCH(CO_2C_2H_5)_2$ → CH, $CH_2CH(CO_2C_2H_5)_2$, $CH_2CH(CO_2C_2H_5)_2$

H_2O, H_2SO_4, heat → CH, $CH_2CH_2CO_2H$, $CH_2CH_2CO_2H$

(c)

O → 1. C_6H_5MgBr 2. H_2O → OH, C_6H_5 → 1. PBr_3 2. CH_3Li → Li, C_6H_5

1. CH_3CN 2. H_2O → O, ‖, CCH_3, C_6H_5

(d)

O → $C_2H_5O_2CCH_2-\overset{O}{\overset{\|}{C}}CH_3$, $NH_4^+\ {}^-O_2CCH_3$ → O, ‖, CCH_3, C, $CO_2C_2H_5$ → H_2/ Pt → O, ‖, CCH_3, CH, $CO_2C_2H_5$

H_2O, H_2SO_4, heat → $CH_2-\overset{O}{\overset{\|}{C}}CH_3$

$NaBH_4$ → $CH_2-\overset{OH}{\overset{|}{C}}HCH_3$

21.20 The structural information in this problem is provided in a series of clues for the different reactions. Before attempting to identify specific structures, it is worthwhile to evaluate the type of compound involved in each of the reactions. As a starting point, the conversion of **F** to **G** is a Wolff-Kishner reduction (of an aldehyde or ketone), and the two compounds must have the same carbon skeleton. Compound **A** has 2 degrees of unsaturation, and its molecular formula (together with its ready alkylation to give **B**) suggests that it is a β-keto ester with the partial structure:

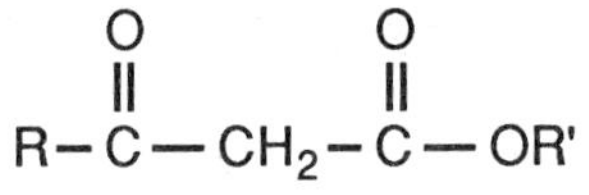

$$R-\overset{\overset{\displaystyle O}{\|}}{C}-CH_2-\overset{\overset{\displaystyle O}{\|}}{C}-OR'$$

A

The two successive alkylation reactions support the suggestion of a CH_2 group between the two carbonyls, and allow partial structures to be drawn for **B** and **C** as well:

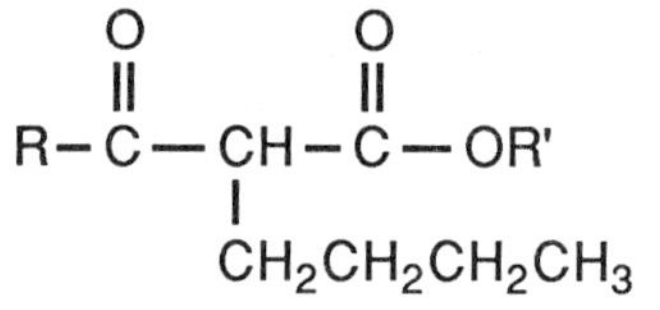

$$R-\overset{\overset{\displaystyle O}{\|}}{C}-\underset{\underset{\displaystyle CH_2CH_2CH_2CH_3}{|}}{CH}-\overset{\overset{\displaystyle O}{\|}}{C}-OR' \qquad R-\overset{\overset{\displaystyle O}{\|}}{C}-\underset{\underset{\displaystyle CH_2CH_2CH_2CH_3}{|}}{\overset{\overset{\displaystyle CO_2R'}{|}}{C}}-CH_2CH_2CH_3$$

B 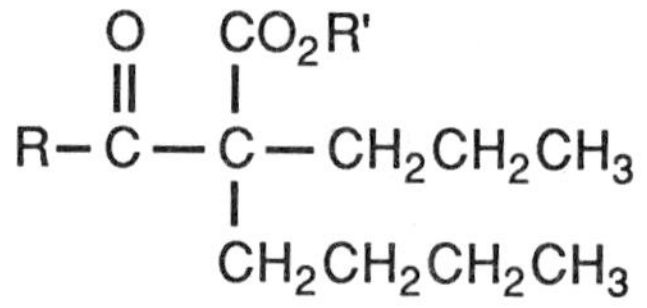 **C**

The alkylation of **B** ($C_{10}H_{18}O_3$) with propyl bromide means that **E** must be a C_{13} compound, and its conversion to **F** involves loss of the $-CO_2R'$ group via hydrolysis and decarboxylation. Since **F** has only 10 carbons, the loss of three carbons means that R′ is an ethyl group. This allows a revision of the structure for **A:**

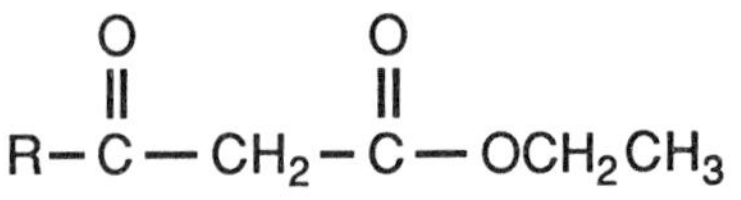

$$R-\overset{\overset{\displaystyle O}{\|}}{C}-CH_2-\overset{\overset{\displaystyle O}{\|}}{C}-OCH_2CH_3$$

A

The molecular formula of **A** (six carbon atoms) permits its full structure to be determined as:

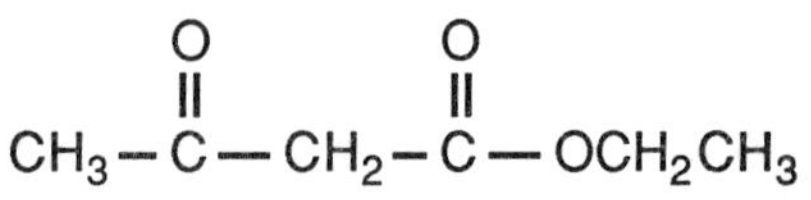

$$CH_3-\overset{\overset{\displaystyle O}{\|}}{C}-CH_2-\overset{\overset{\displaystyle O}{\|}}{C}-OCH_2CH_3$$

This allows deduction of the remaining structures, and the complete answer follows:

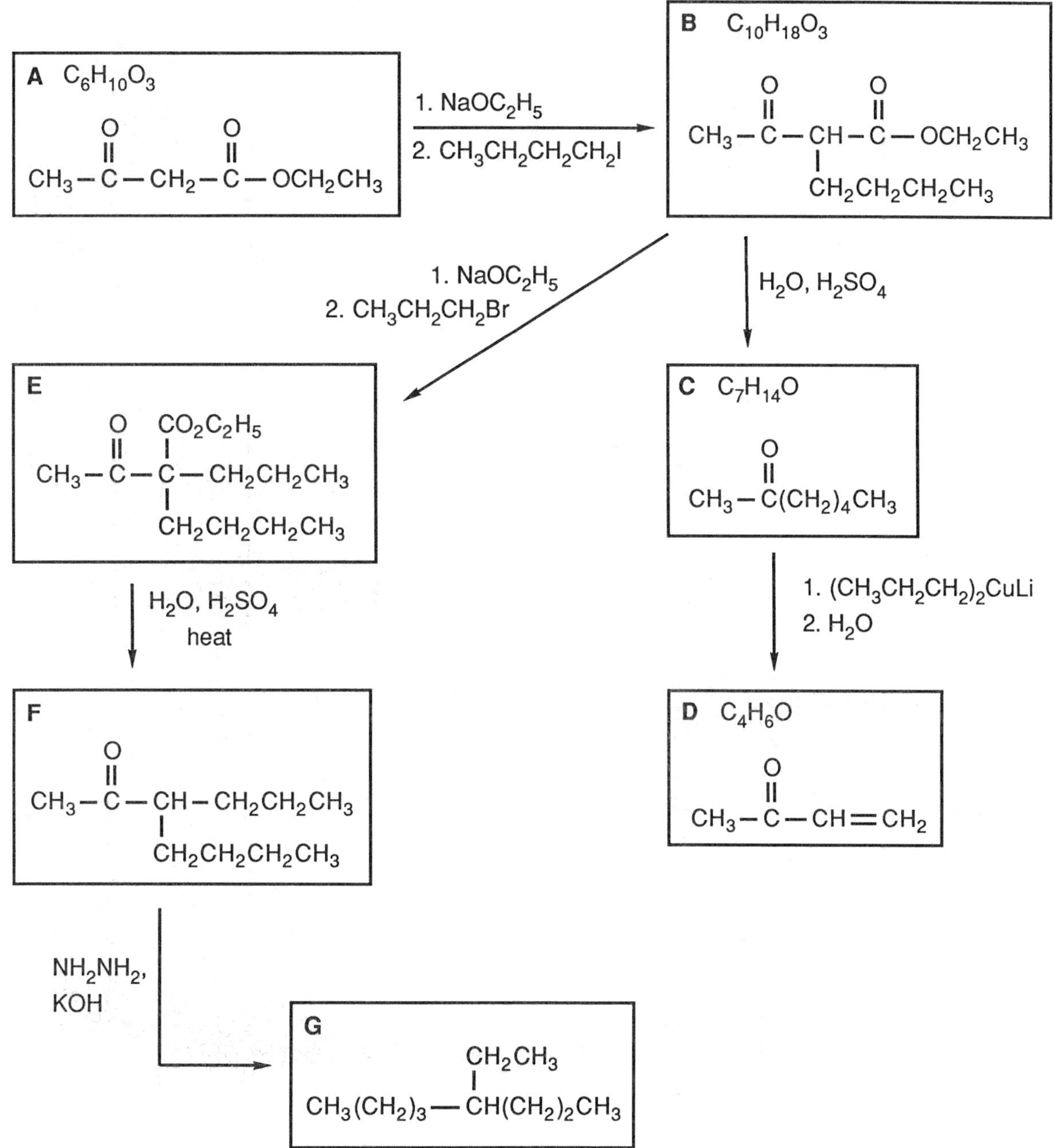
A $C_6H_{10}O_3$
$CH_3-C(=O)-CH_2-C(=O)-OCH_2CH_3$
1. $NaOC_2H_5$
2. $CH_3CH_2CH_2CH_2I$
B $C_{10}H_{18}O_3$
$CH_3-C(=O)-CH(CH_2CH_2CH_2CH_3)-C(=O)-OCH_2CH_3$
1. $NaOC_2H_5$
2. $CH_3CH_2CH_2Br$
H_2O, H_2SO_4
E
$CH_3-C(=O)-C(CO_2C_2H_5)(CH_2CH_2CH_2CH_3)-CH_2CH_2CH_3$
C $C_7H_{14}O$
$CH_3-C(=O)(CH_2)_4CH_3$
H_2O, H_2SO_4
heat
1. $(CH_3CH_2CH_2)_2CuLi$
2. H_2O
F
$CH_3-C(=O)-CH(CH_2CH_2CH_2CH_3)-CH_2CH_2CH_3$
D C_4H_6O
$CH_3-C(=O)-CH=CH_2$
NH_2NH_2,
KOH
G
$CH_3(CH_2)_3-CH(CH_2CH_3)(CH_2)_2CH_3$

21.21 As you would expect, **3** is indeed formed initially, but it rearranges to **2** in the presence of base. A major reason for this is the ability of **2** to form a stable enolate ion under basic conditions (whereas this is not possible for **3**). The mechanism for the reaction follows:

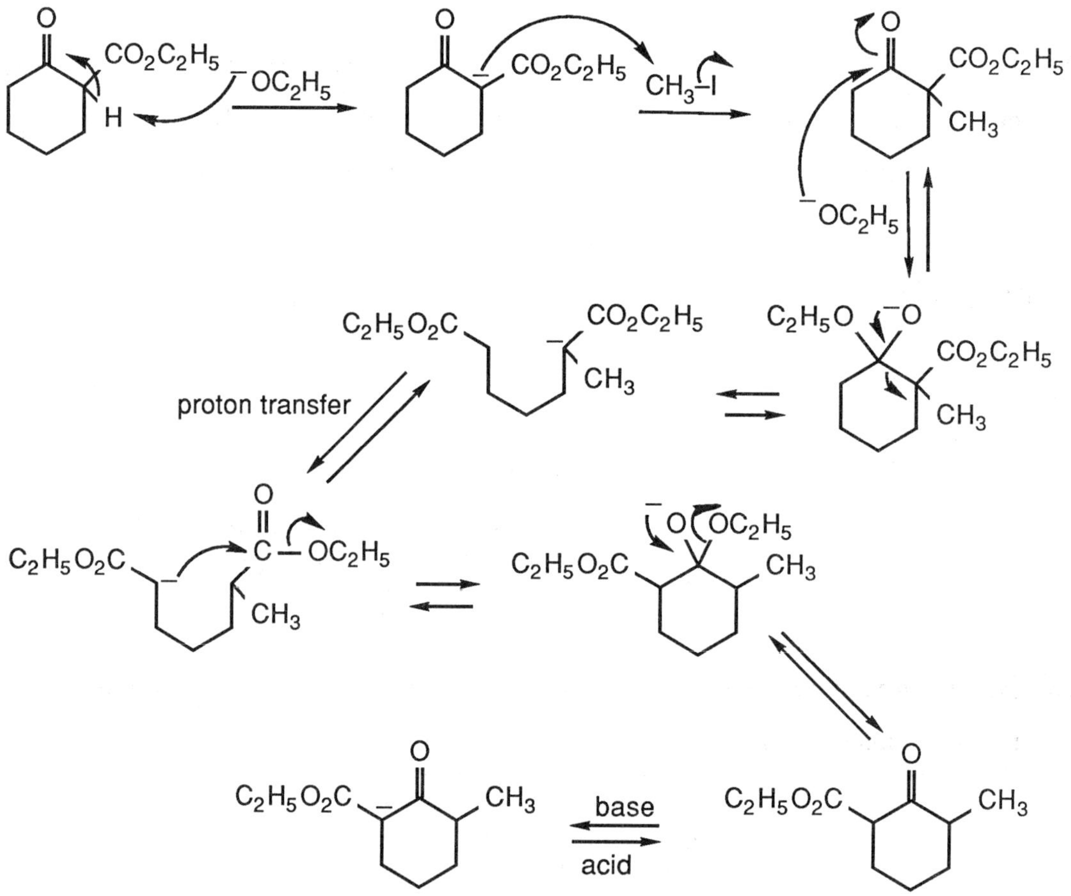

21.22 The 5H aromatic signal is exhibits the typical pattern for a benzoyl group, and the 3H signal at 2.1 ppm is characteristic for an acetyl group. The structure corresponds to the condensation product of acetone with ethyl benzoate.

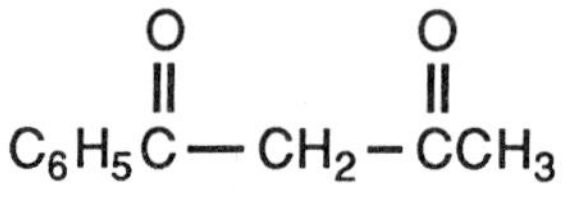

$$C_6H_5\overset{O}{\overset{\|}{C}}-CH_2-\overset{O}{\overset{\|}{C}}CH_3$$

21.23 The keto and enol forms are in equilibrium, although the former predominates.

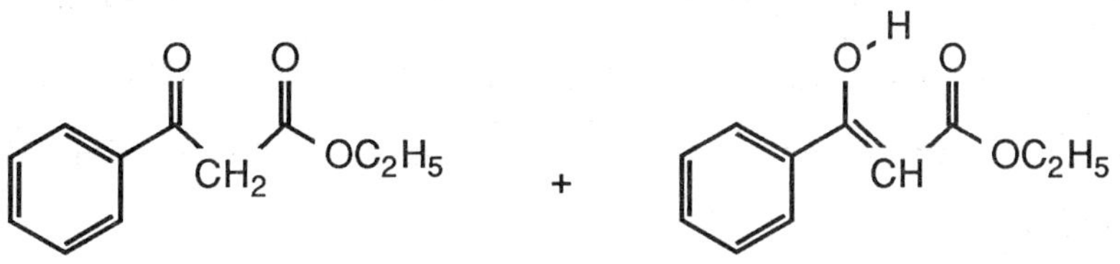

$$\text{PhC(=O)CH}_2\text{C(=O)OC}_2\text{H}_5 \quad + \quad \text{PhC(OH)=CHC(=O)OC}_2\text{H}_5$$

21.24 An isopropyl group is still recognizable in the product, so the enolate derived from ethyl acetate must have attacked the carbonyl of ethyl isobutyrate. The following structure shows the peak assignments for the resulting product:

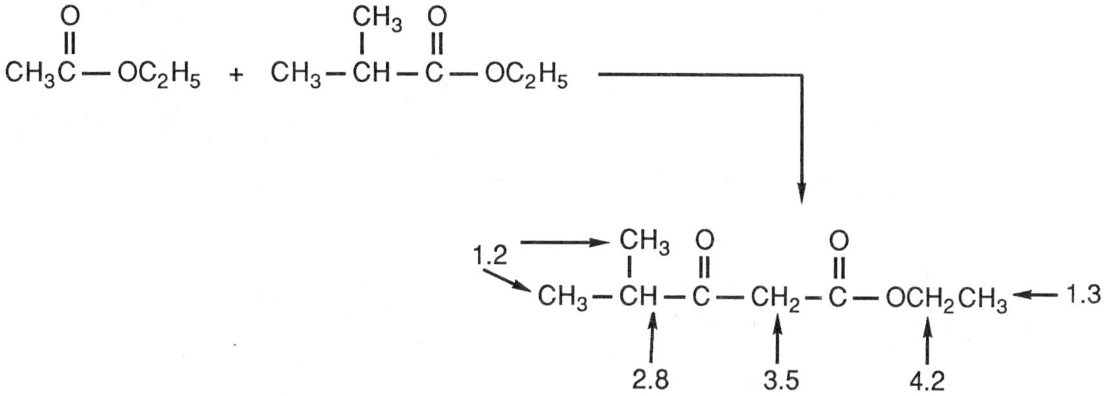

$$CH_3\overset{O}{\overset{\|}{C}}-OC_2H_5 + CH_3-\overset{CH_3}{\overset{|}{CH}}-\overset{O}{\overset{\|}{C}}-OC_2H_5 \longrightarrow CH_3-\overset{CH_3}{\overset{|}{CH}}-\overset{O}{\overset{\|}{C}}-CH_2-\overset{O}{\overset{\|}{C}}-OCH_2CH_3$$

Peak assignments: CH_3 groups of isopropyl, 1.2; CH, 2.8; CH_2, 3.5; OCH_2, 4.2; OCH_2CH_3, 1.3.

21.25 **(a)** A large ester:halide ratio favors the formation of a product that incorporates two molecules of ester for each dihalide:

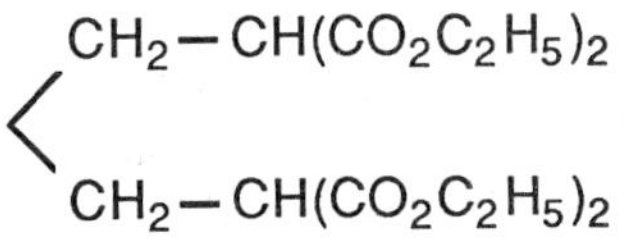

$$CH_2(CH_2-CH(CO_2C_2H_5)_2)_2$$

(b) A low ester:halide ratio favors formation of the product in which the initial adduct reacts in an intramolecular fashion to afford a cyclic product:

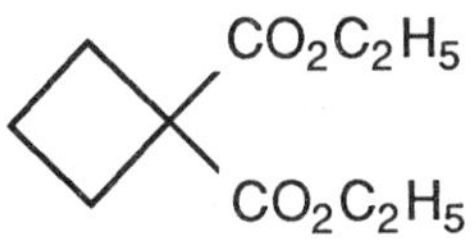

$$\text{cyclobutane-1,1-}(CO_2C_2H_5)_2$$

The spectrum corresponds to the product of (b).

CHAPTER 22. ORGANIC SYNTHESIS USING POLYFUNCTIONAL COMPOUNDS

ANSWERS TO EXERCISES

22.1 **(a)** The anhydride would contain a strained four-membered ring and would not be formed in good yield.

(b) (succinic anhydride structure)

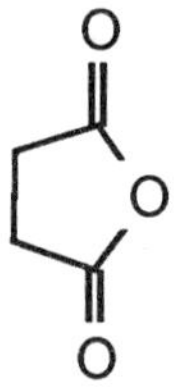

(c) (maleic anhydride structure)

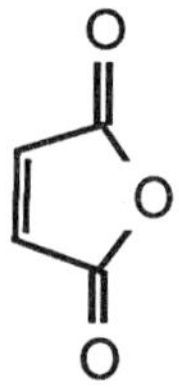

(d) The *E*-double bond would preclude formation of the cyclic anhydride.

22.2 **(a)**

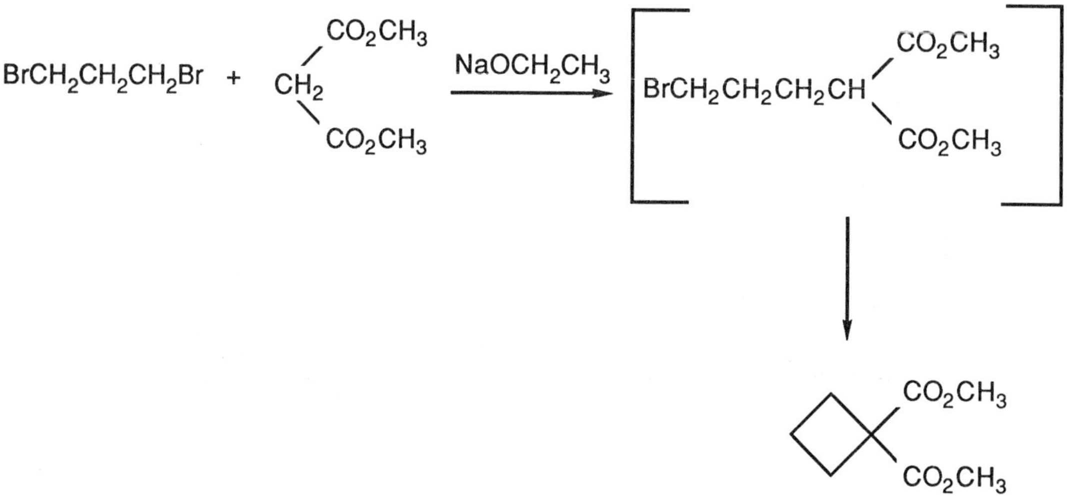

$$BrCH_2CH_2CH_2Br + CH_2(CO_2CH_3)_2 \xrightarrow{NaOCH_2CH_3} [BrCH_2CH_2CH_2CH(CO_2CH_3)_2] \longrightarrow \text{cyclobutane-1,1-}(CO_2CH_3)_2$$

(b)

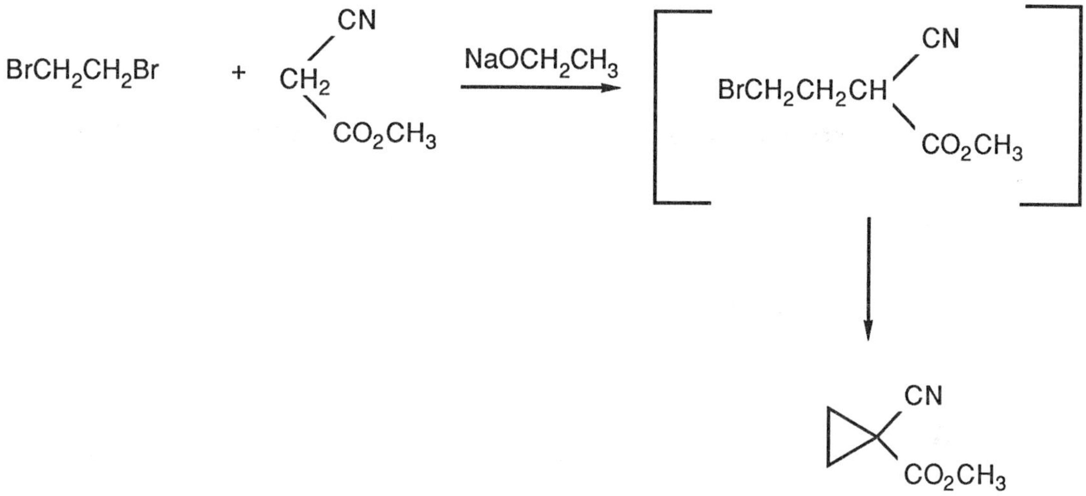

$$BrCH_2CH_2Br + CH_2(CN)(CO_2CH_3) \xrightarrow{NaOCH_2CH_3} [BrCH_2CH_2CH(CN)(CO_2CH_3)] \longrightarrow$$

(c)

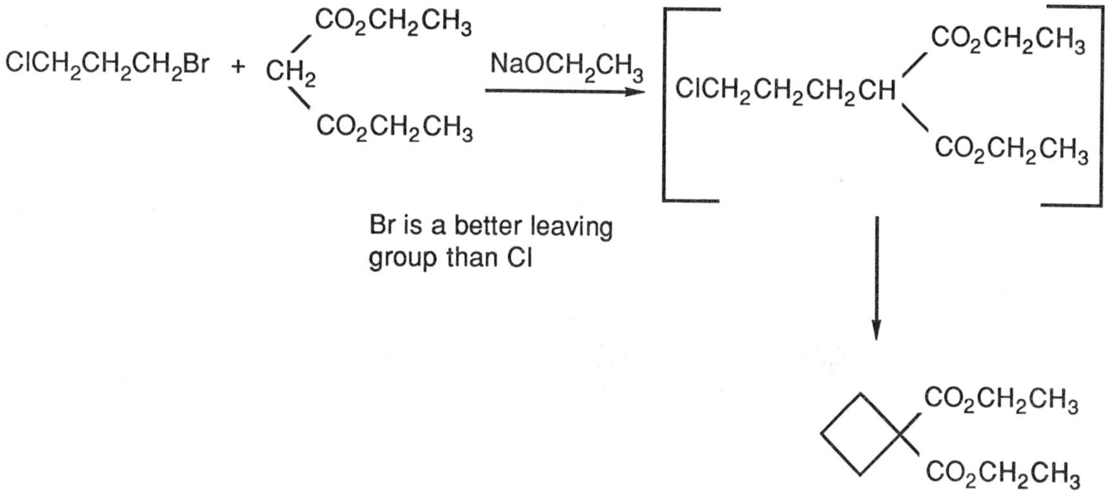

$$ClCH_2CH_2CH_2Br + CH_2(CO_2CH_2CH_3)_2 \xrightarrow{NaOCH_2CH_3} [ClCH_2CH_2CH_2CH(CO_2CH_2CH_3)_2] \longrightarrow$$

Br is a better leaving group than Cl

(d)

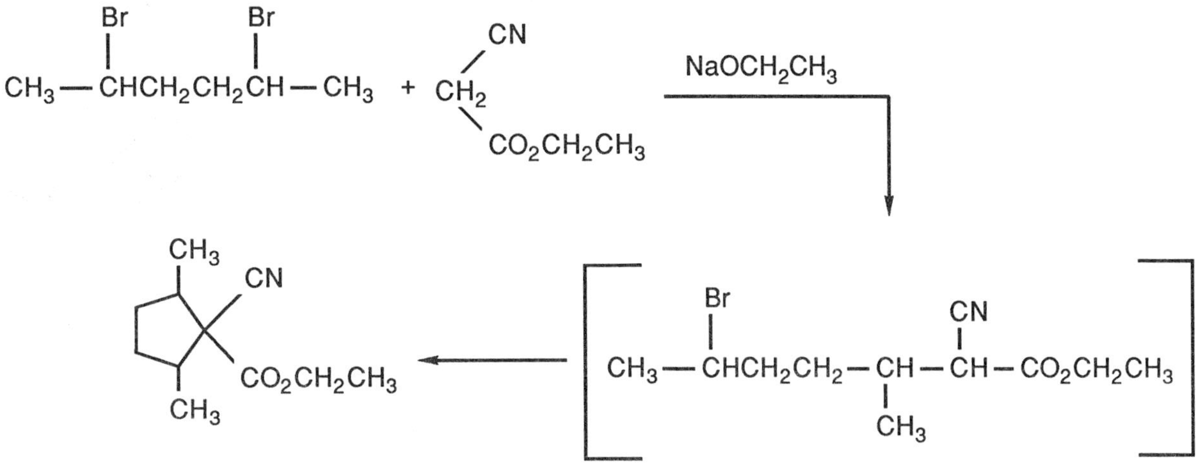

$$CH_3—CH(Br)CH_2CH_2CH(Br)—CH_3 + CH_2(CN)(CO_2CH_2CH_3) \xrightarrow{NaOCH_2CH_3}$$

$$[CH_3—CH(Br)CH_2CH_2—CH(CH_3)—CH(CN)—CO_2CH_2CH_3] \longrightarrow$$

(e)

CH_2OTs ... CH_2OTs + $CH_2(CO_2CH_2CH_3)_2$ $\xrightarrow{NaOCH_2CH_3}$ [$CH_2CH(CO_2CH_2CH_3)_2$, CH_2OTs] ⟶ $CO_2CH_2CH_3$, $CO_2CH_2CH_3$

22.3 (a) CH_3, HC(=O), CH_3, CH_3 $\xrightarrow{NaOCH_2CH_3}$ HO, CH_3, CH_3 ⟶ CH_3, CH_3

(b) CH_2CH_3, O, O $\xrightarrow{KOH}$ HO, CH_3, O ⟶ CH_3, O

(c) CCH_3(=O), CCH_3(=O) $\xrightarrow{KOH}$ HO, CH_3, CCH_3(=O) ⟶ CH_3, CCH_3(=O)

(d) CCH_3(=O), CCH_3(=O) $\xrightarrow{H_2SO_4}$ CCH_3(=O), CH_3, OH ⟶ CCH_3(=O), CH_3

(e)

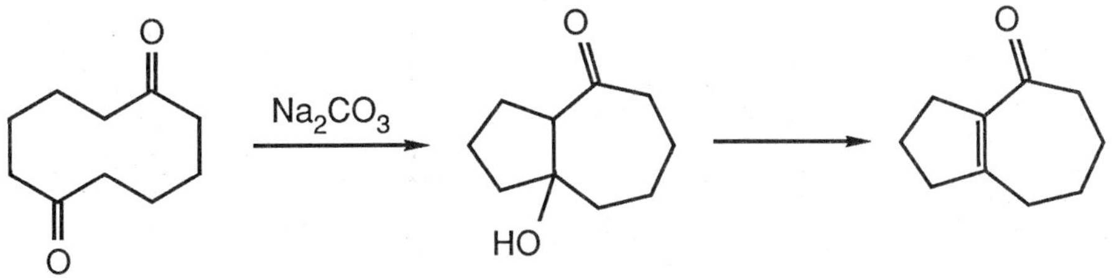

Na_2CO_3

HO

22.4 (a) Reaction via formation of an enolate at either of the indicated positions could only lead to three-membered ring formation.

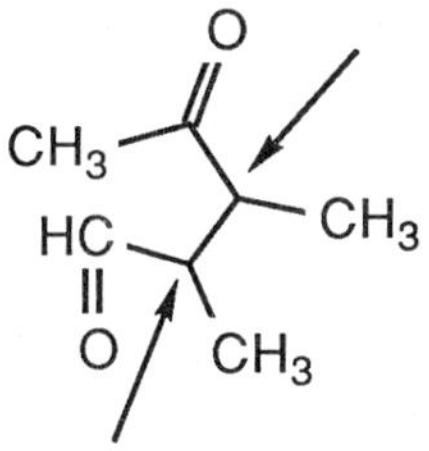

CH_3 HC O CH_3 CH_3

(b) Reaction via formation of an enolate at the indicated position (arrow) would also generate a six-membered ring, but dehydration would require formation of a very strained double bond at the bridgehead of the bicyclic system. Reaction via enolate formation at the other α carbons would produce four-membered rings.

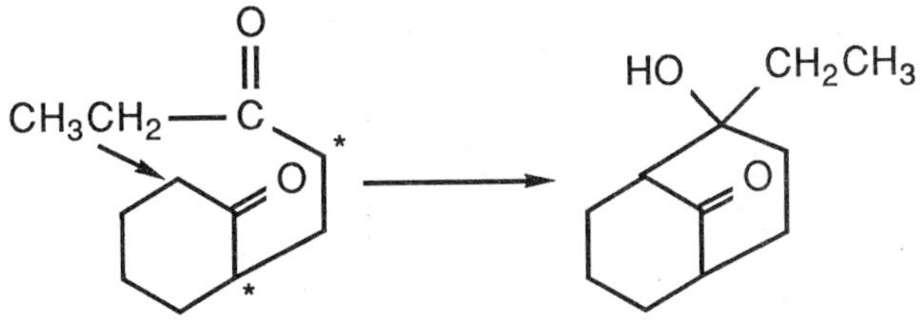

CH_3CH_2—C HO CH_2CH_3

(c) The reactant is symmetrical, so reaction at either α CH_2 group would afford the same product. Reaction via enolate formation at the α methyl positions would be unfavorable, producing a seven-membered ring.

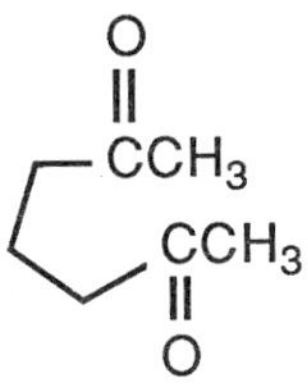

CCH_3 CCH_3

(d) The reactant is symmetrical, so reaction at either α CH_2 group would afford the same product. Reaction via enolate formation at the α methyl positions would be unfavorable, producing an eight-membered ring.

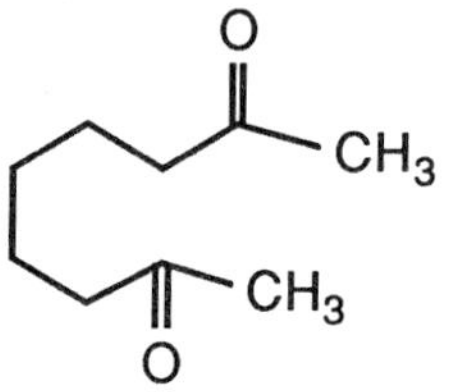

(e) 1,6-Cyclodecanedione is symmetrical, and all four α CH_2 groups are equivalent. No other aldol product could be formed.

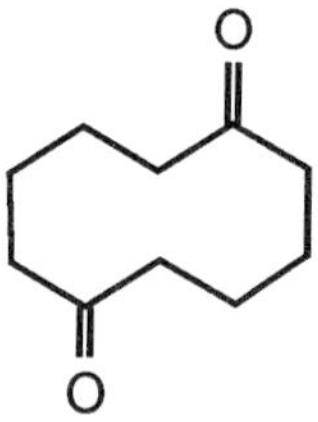

22.5 **(a)**

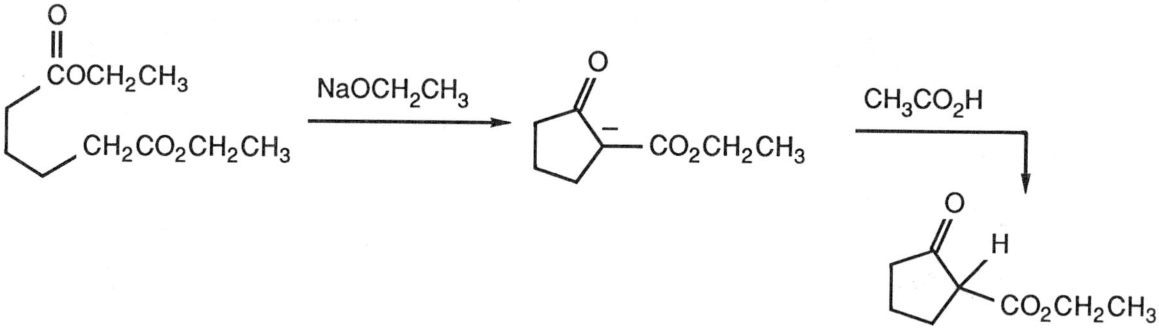

$CH_3CH_2O_2C(CH_2)_3CH_2CO_2CH_2CH_3$ $\xrightarrow{NaOCH_2CH_3}$ 2-oxocyclopentyl carbanion $-CO_2CH_2CH_3$ $\xrightarrow{CH_3CO_2H}$ ethyl 2-oxocyclopentanecarboxylate (H, $CO_2CH_2CH_3$)

(b)

$COCH_3$ / $CH_2CO_2CH_2CH_3$ / CO_2CH_3 $\xrightarrow{NaOCH_3}$ cyclohexanone carbanion with CO_2CH_3, CO_2CH_3 $\longrightarrow$ cyclohexanone with CO_2CH_3, CO_2CH_3

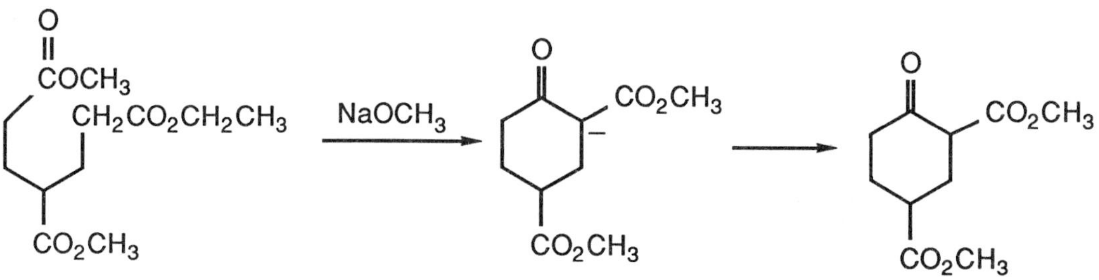

(c)

$CH_2CO_2CH_2CH_3$, $CH_2CO_2CH_2CH_3$ —NaH→ H, H, O, $CO_2CH_2CH_3$ —HCl→ H, H, O, $CO_2CH_2CH_3$

(d)

CO_2CH_3, $CH_2CH_2CO_2CH_3$ —$NaOCH_3$→ O, CO_2CH_3 —HCl→ O, CO_2CH_3

(e)

CO_2CH_3, CO_2CH_3, C_6H_5 —$NaOCH_3$→ CO_2CH_3, O, C_6H_5 —CH_3CO_2H→ CO_2CH_3, O, C_6H_5

22.6 (a)

$CO_2CH_2CH_3$, O + $CH_3-\overset{+}{N}(CH_2CH_3)_2$, CH_3, O —$NaOCH_2CH_3$→ $CO_2CH_2CH_3$, O, O, CH_3 → $CO_2CH_2CH_3$, O

(b)

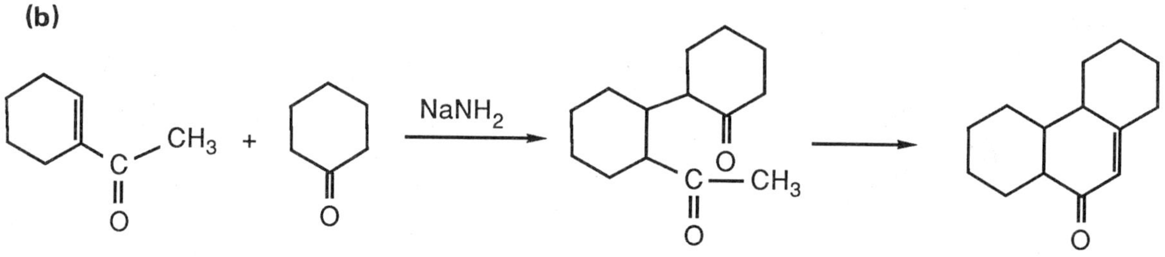

NaNH$_2$

CH$_3$ + ... C—CH$_3$

22.7 (a)

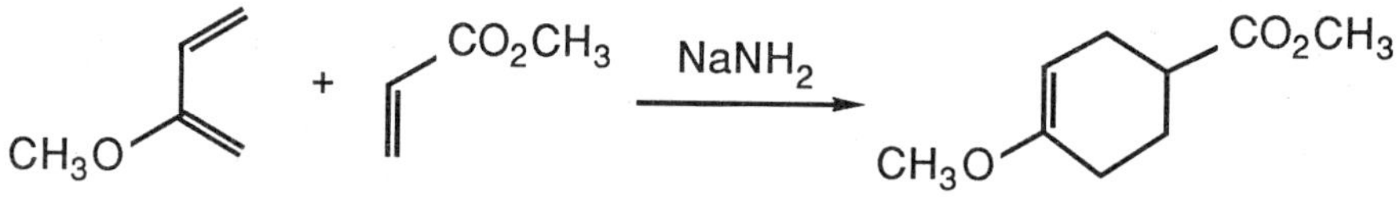

CH_3O + CO_2CH_3 —NaNH$_2$→ CO_2CH_3, CH_3O

(b)

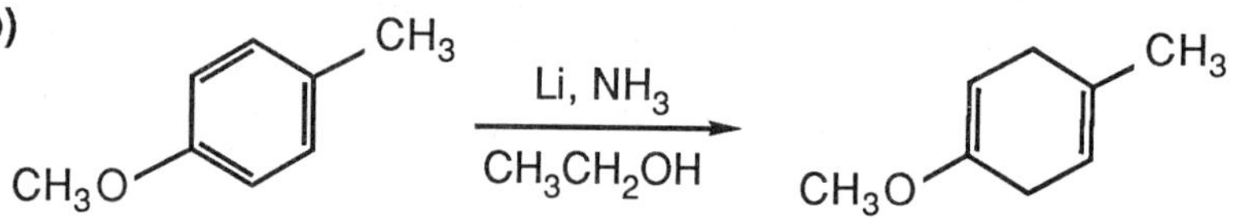

CH_3O, CH_3 —Li, NH$_3$ / CH_3CH_2OH→ CH_3O, CH_3

(c)

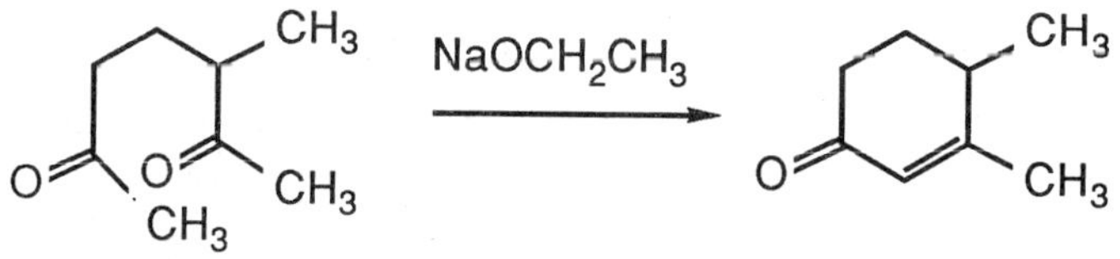

CH_3, CH_3, CH_3 —$NaOCH_2CH_3$→ CH_3, CH_3

(d)

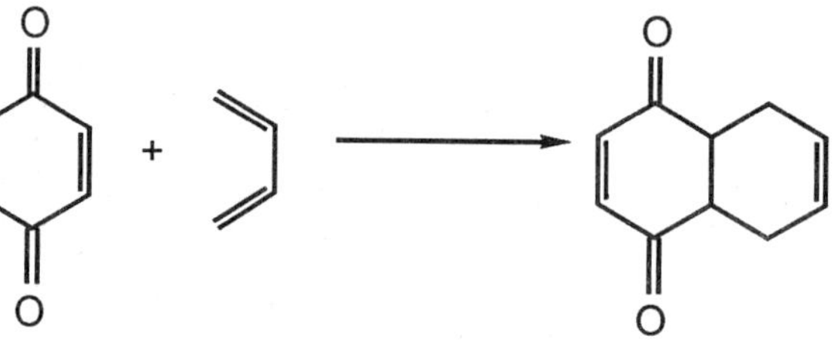

(e)

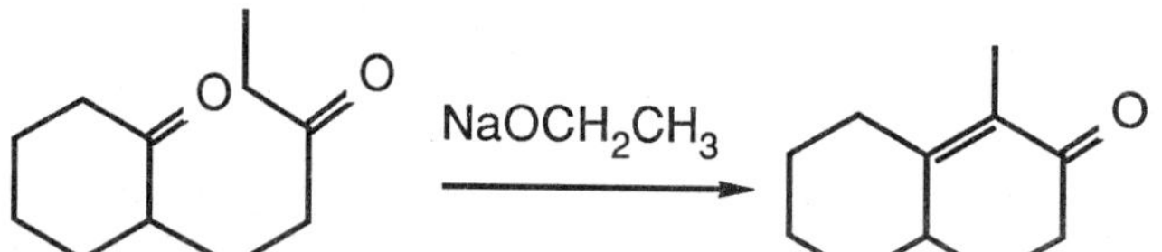

—$NaOCH_2CH_3$→

(f)

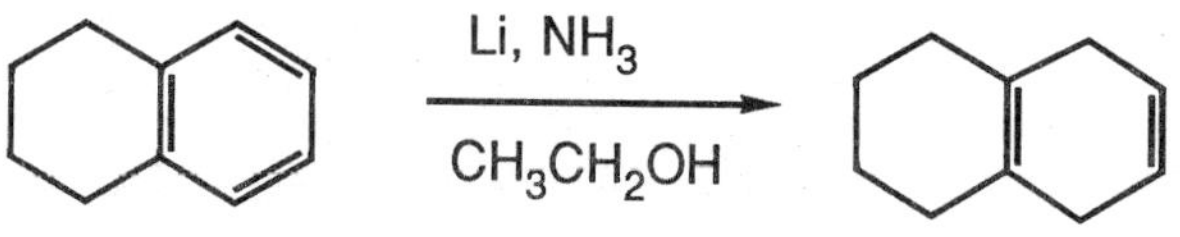

—Li, NH$_3$ / CH_3CH_2OH→

22.8

(a) CO_2CH_3, CO_2CH_3

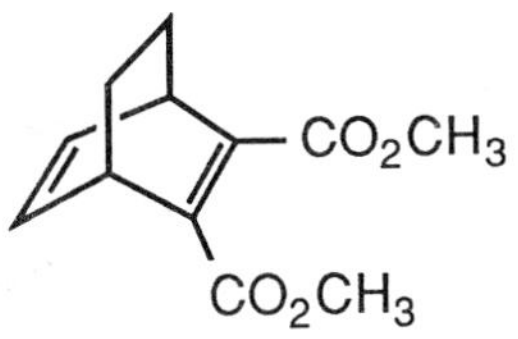

(b) CH_3, CH_3, CHO

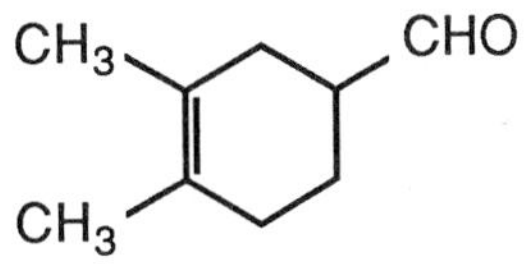

(c)

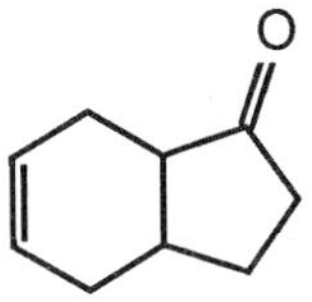

(d) CH_3O

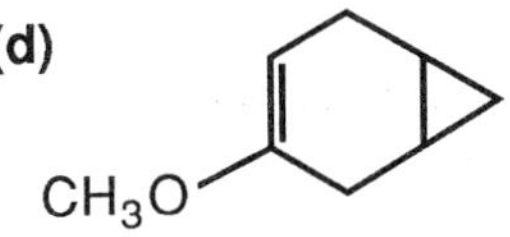

22.9

CH_2OH + $CH_3—\overset{O}{\overset{\|}{O}}—CH_3$ $\xrightarrow{TsOH}$ O, CH_2OH + O O

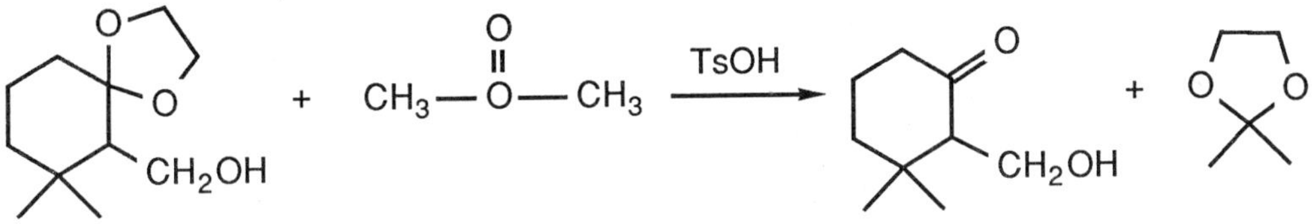

22.10

CH_3CH_2OH from $CH_3—\overset{O}{\overset{\|}{C}}—OR$

22.11

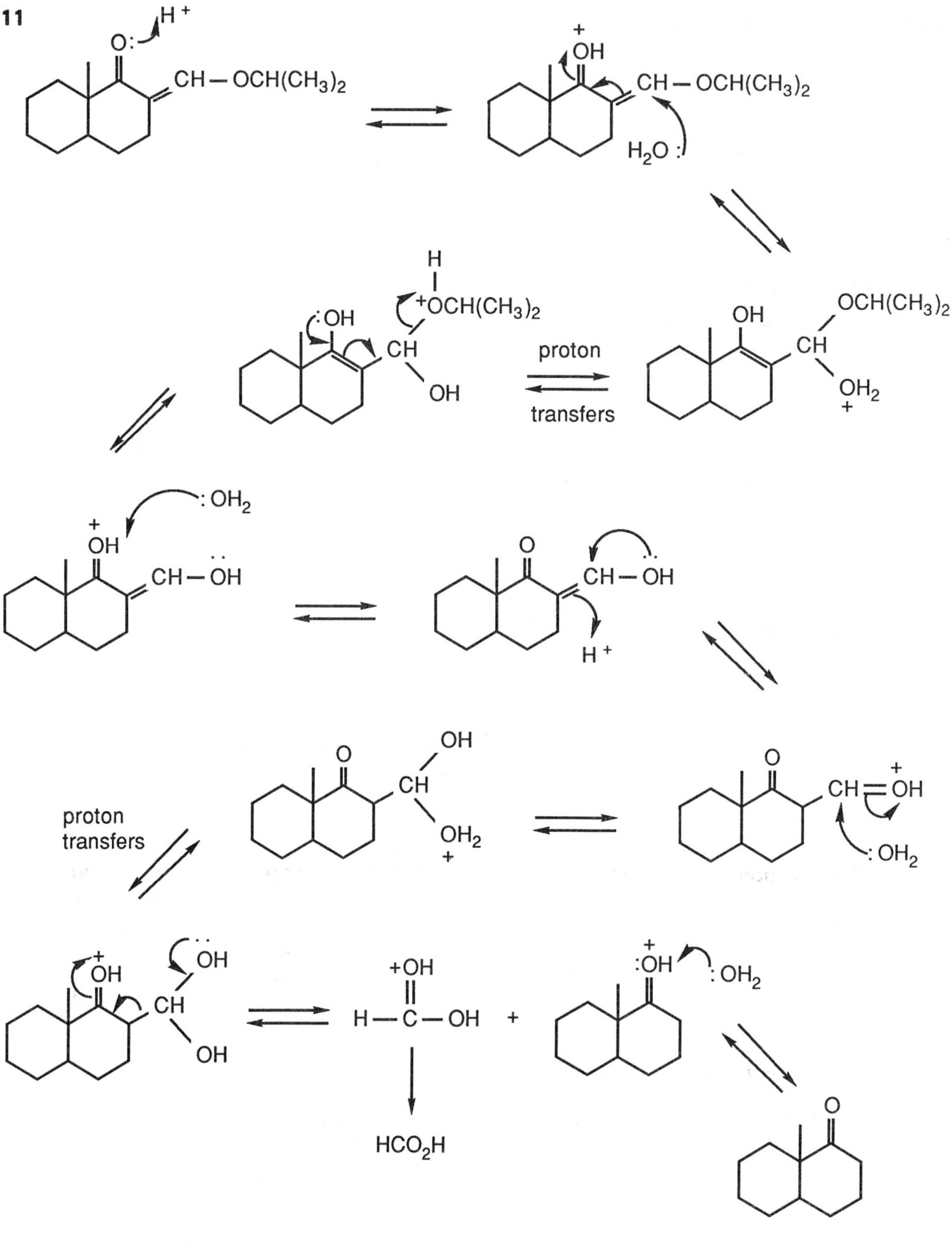

H^+

CH—OCH$(CH_3)_2$ ⇌ $\overset{+}{O}H$ CH—OCH$(CH_3)_2$ H_2O:

⇅

OH OCH$(CH_3)_2$ CH $\overset{+}{O}H_2$ ⇌ proton transfers ⇌ :OH H $\overset{+}{O}$CH$(CH_3)_2$ CH OH

⇅

$\overset{+}{O}H$ CH—$\ddot{O}$H :OH_2 ⇌ O CH—$\ddot{O}$H H^+

⇅

O CH=$\overset{+}{O}H$:OH_2 ⇌ O OH CH $\overset{+}{O}H_2$

proton transfers ⇅

$\overset{+}{O}H$ CH $\ddot{O}$H OH ⇌ +OH H—C—OH → HCO_2H + :$\overset{+}{O}H$:OH_2

⇅

O

ANSWERS TO PROBLEMS

22.1 **(a)** n=1

$$ClCH_2-\overset{O}{\overset{\|}{C}}-CH_2CO_2CH_3 \xrightarrow[\text{2. HCl}]{\text{1. NaOCH}_3}$$ (cyclopropanone with CO_2CH_3)

This reaction should not work: the product would be a strained three-membered, and a retro condensation could occur.

n=4

$$Cl(CH_2)_4-\overset{O}{\overset{\|}{C}}-CH_2CO_2CH_3 \xrightarrow[\text{2. HCl}]{\text{1. NaOCH}_3}$$ (cyclohexanone with CO_2CH_3)

This reaction should work well.

n=7

$$Cl(CH_2)_7-\overset{O}{\overset{\|}{C}}-CH_2CO_2CH_3 \xrightarrow[\text{2. HCl}]{\text{1. NaOCH}_3}$$ (cyclononanone with CO_2CH_3)

This reaction should not work well: neither entropy nor enthalpy favors formation of a nine-membered ring.

(b) n=1

$$CH_3O_2CCH_2CH_2CO_2CH_3 \xrightarrow[\text{2. HCl}]{\text{1. NaOCH}_3}$$ (cyclopropanone with $CHCO_2CH_3$)

The reaction is reversible, and the strained three-membered ring would not be favored.

n=2 $$CH_3O_2CCH_2CH_2CH_2CH_2CO_2CH_3 \xrightarrow[\text{2. HCl}]{\text{1. NaOCH}_3}$$ (cyclopentanone with CO_2CH_3)

This reaction should work well.

n=5 $$CH_3O_2C(CH_2)_5CH_2CO_2CH_3 \xrightarrow[\text{2. HCl}]{\text{1. NaOCH}_3}$$ (cycloheptanone with CO_2CH_3)

Seven-membered ring formation is not ordinarily favorable, so this reaction would not be expected to work well.

(c) n=2

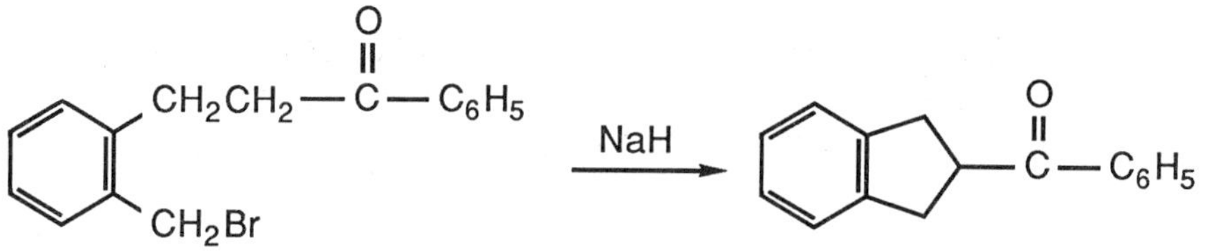

This reaction should work well.

n=3

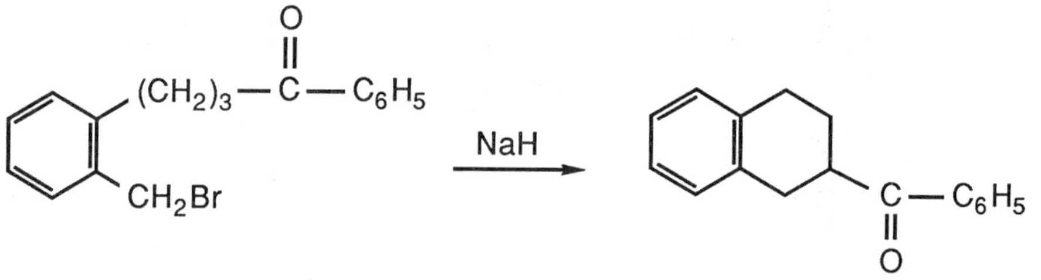

This reaction should work well.

n=5

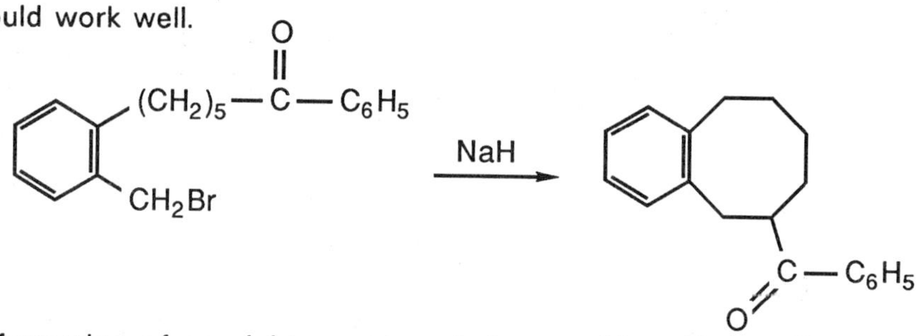

The unfavorable formation of an eight-membered ring would make it unlikely for this reaction to work well.

n=6

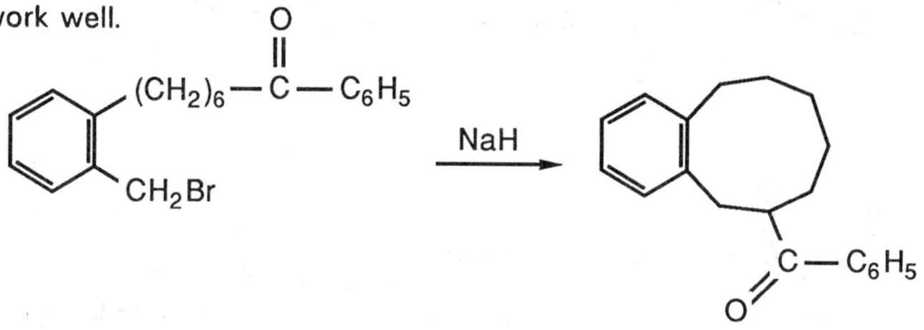

The unfavorable formation of a nine-membered ring would make it unlikely that this reaction would work well.

22.2

(a)

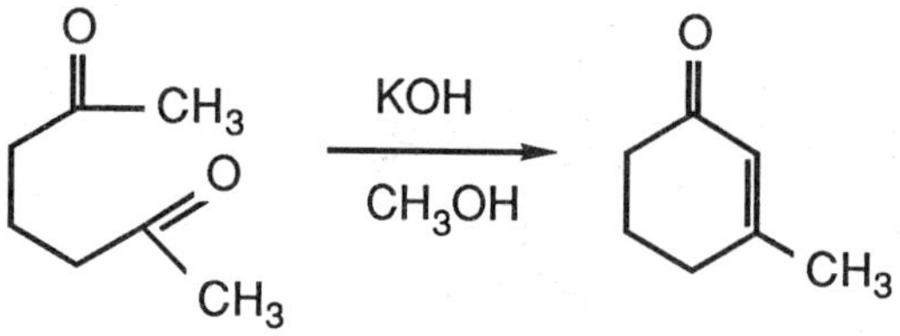

(b)

$CH_3CH(CHO)CH_2CH_2CH_2C(=O)C_6H_5 \xrightarrow[CH_3CH_2OH]{NaOCH_2CH_3}$ 3-methylcyclopentyl phenyl ketone (CH_3, $C(=O)C_6H_5$)

(c)

$C_6H_5CH_2C(=O)CH_2CH_2CH(CH_3)CHO \xrightarrow[CH_3CH_2OH]{NaOCH_2CH_3}$ cyclohexenone with C_6H_5 and CH_3

(d)

$CH_3CH(CHO)CH_2C(=O)CH_2CH_2C_6H_5 \xrightarrow[CH_3CH_2OH]{NaOCH_2CH_3}$ cyclopentenone with CH_3 and $CH_2C_6H_5$

(e)

$CH_3CH(CH_2CH_2C(=O)CH_3)CH_2CH_2C(=O)CH_3 \xrightarrow{KOH}$ cyclohexene with CH_3, CH_3, and $C(=O)CH_3$

22.3

(a)

$CH_3OC(=O)CH_2CH_2CH_2CH_2C(=O)C_6H_5 \xrightarrow[2.\ HCl]{1.\ NaOCH_3}$ 2-benzoylcyclopentanone ($C(=O)C_6H_5$)

(b)

$CH_3OC(=O)CH_2CH_2CH_2C(=O)CH_2C_6H_5 \xrightarrow[2.\ HCl]{1.\ NaOCH_3}$ 2-phenylcyclohexane-1,3-dione (C_6H_5)

(c) $C(=O)OCH_2CH_3$ / $CH_2CH(CH_3)_2$ $\xrightarrow{NaH}$ Na^+ $CH(CH_3)_2$

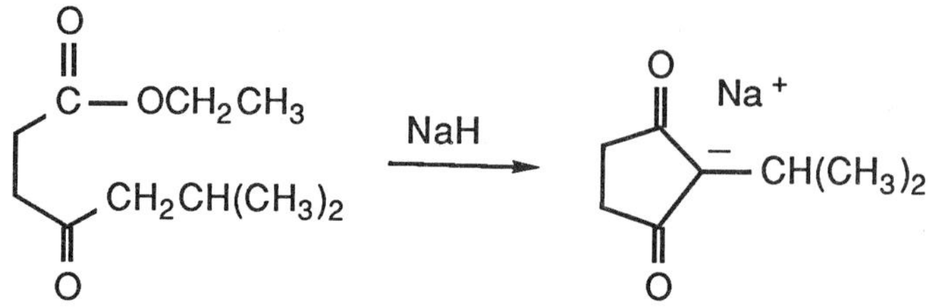

(d) CH_3 $C(=O)OCH_3$ / $CH_2CO_2CH_3$ / CH_3 $\xrightarrow{NaOCH_3}$ CH_3 Na^+ CO_2CH_3 / CH_3

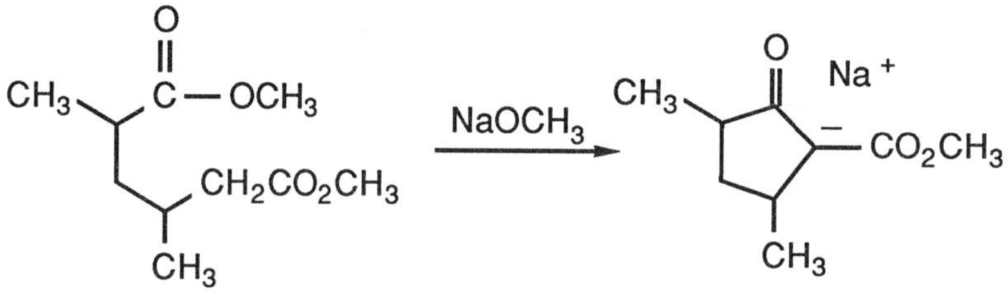

(e) $CH_3CH_2O_2C$ $C(=O)OCH_3$ / $CH_2CO_2CH_2CH_3$ $\xrightarrow[\text{2. HCl}]{\text{1. NaH}}$ $CH_3CH_2O_2C$ CO_2CH_3

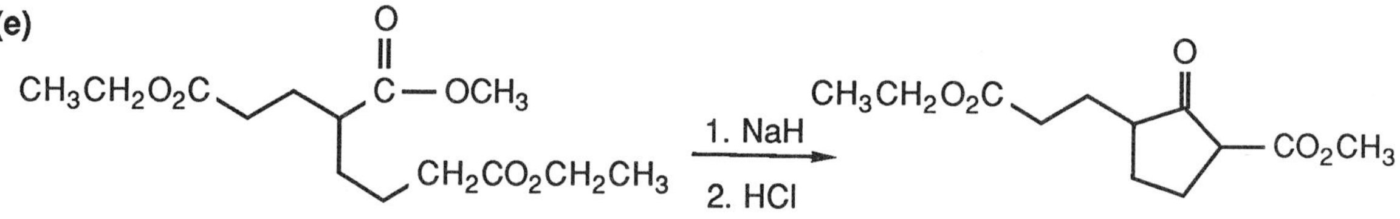

22.4 Cyclizations (NaH followed by acidification) would be the result of the following reactions:

(a) O CH_3 CO_2CH_3 ⟶ O O

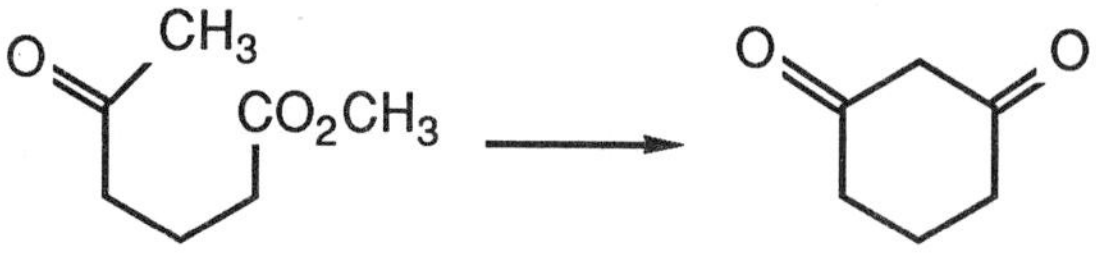

(b) O CH_3 CHO / CH_3 ⟶ O / CH_3

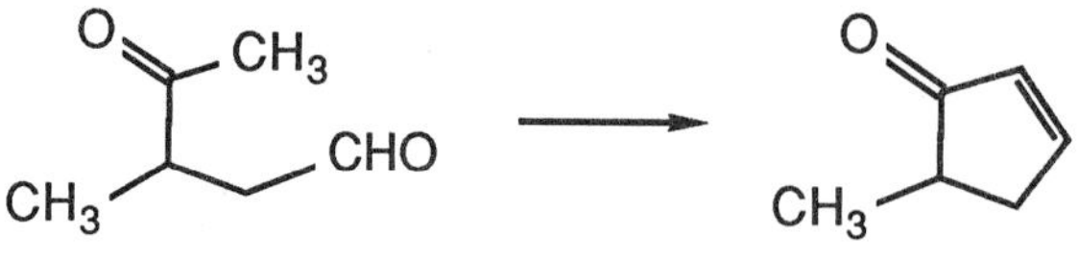

(c) O CH_3 CHO ⟶ O

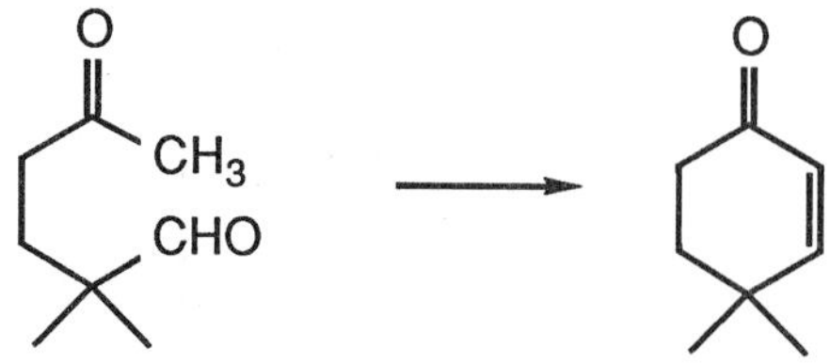

(d) O OCH$_3$ CH$_2$CO$_2$CH$_3$ → O CO$_2$CH$_3$

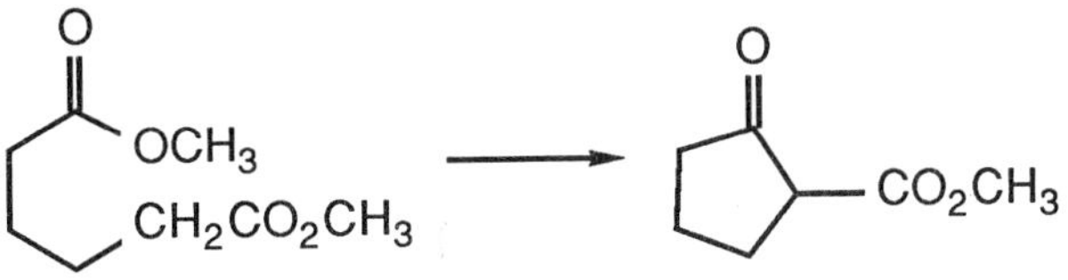

(e) C$_6$H$_5$CH$_2$—C(=O) C$_6$H$_5$ O → C$_6$H$_5$ C$_6$H$_5$CH$_2$ O

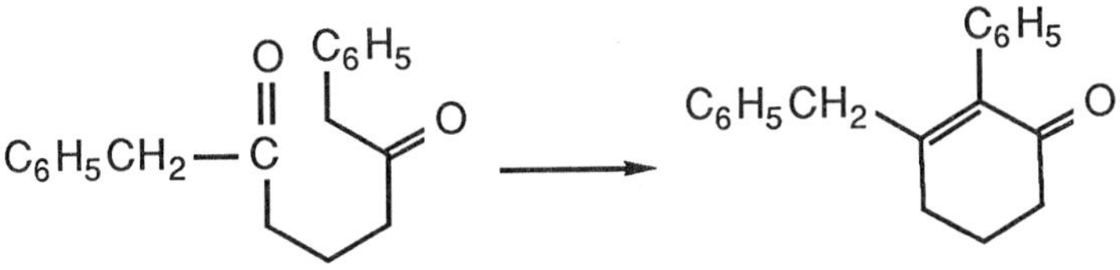

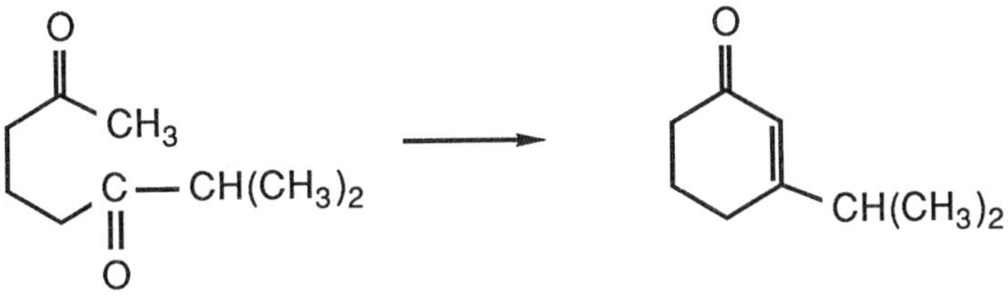

(f) O CH$_3$ C—CH(CH$_3$)$_2$ O → O CH(CH$_3$)$_2$

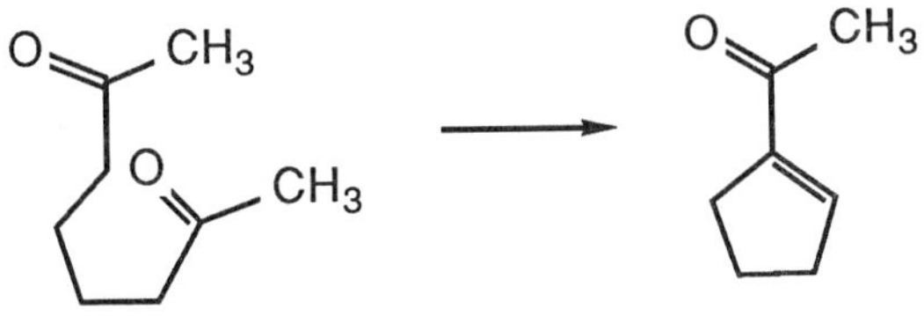

(g) O CH$_3$ O CH$_3$ → O CH$_3$

(h) O O ≡ O O → O

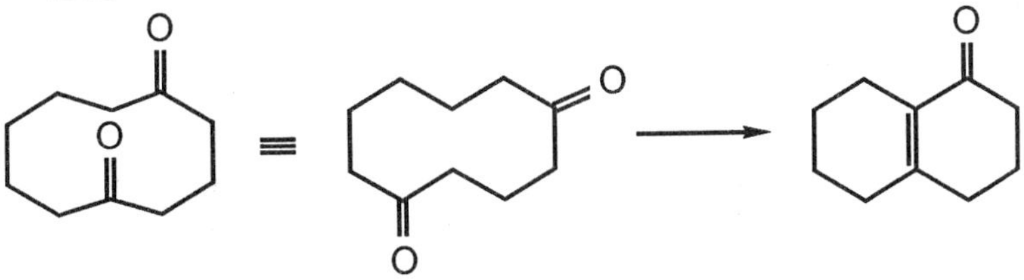

22.5

(a) CH$_3$ O CH$_3$ + O CH$_3$ $\xrightarrow[\text{CH}_3\text{OH}]{\text{NaOCH}_3}$ CH$_3$ O O CH$_3$ CH$_3$ → CH$_3$ O CH$_3$ CH$_3$

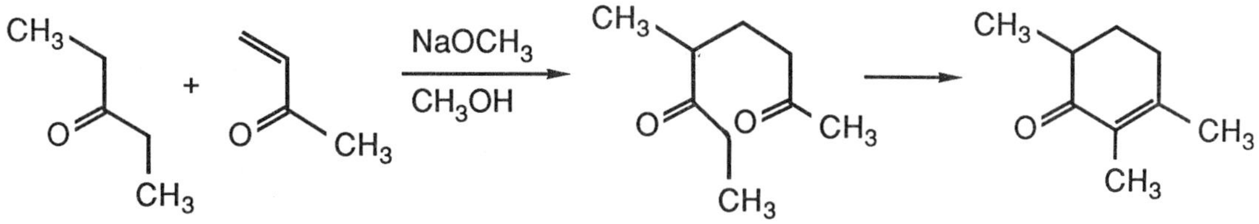

(b)

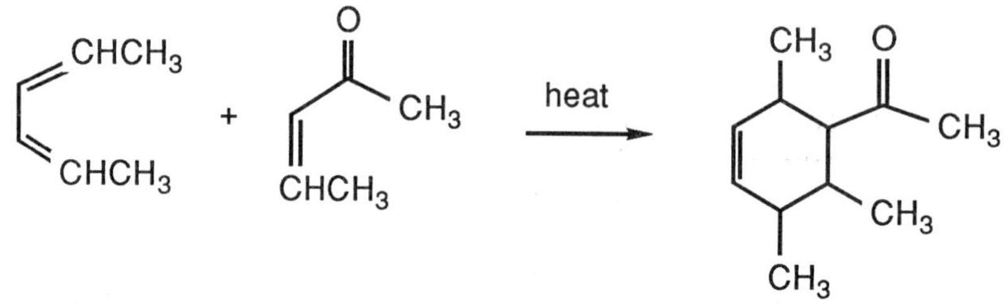
CHCH3
CHCH3
+
O
CH3
CHCH3
heat
CH3
O
CH3
CH3
CH3

(c)

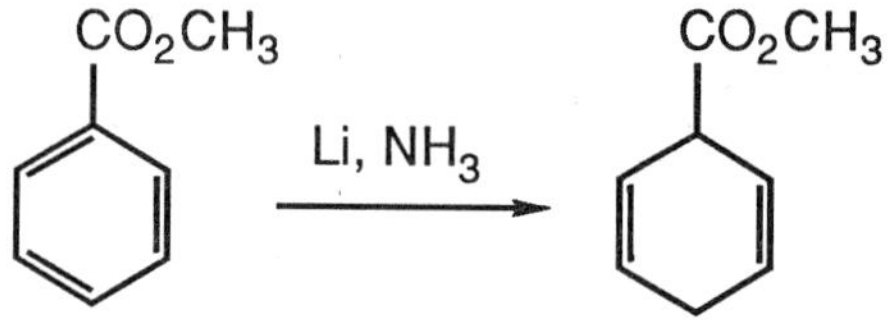
CO2CH3
Li, NH3
CO2CH3

(d)

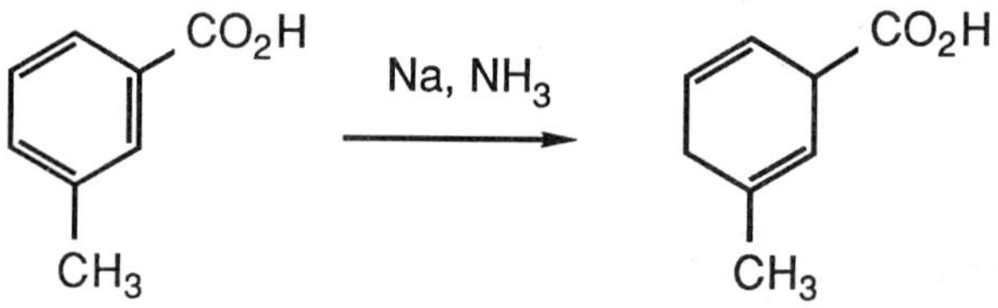
CO2H
CH3
Na, NH3
CO2H
CH3

(e)

+
CHCN
CH2
H
CN

(f)

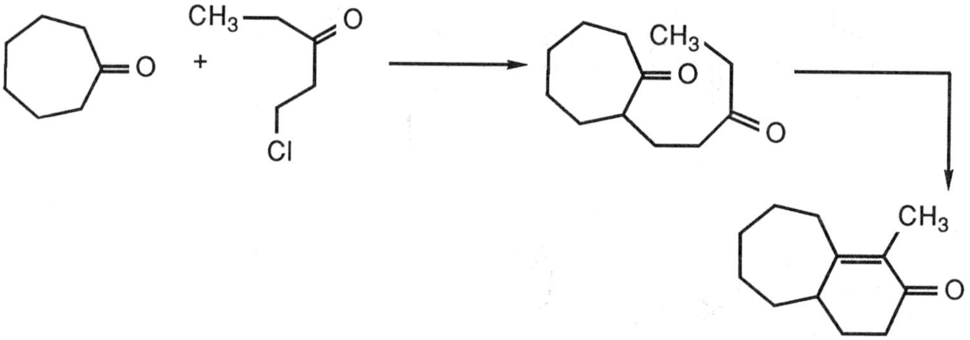
O
+
CH3
O
Cl
CH3
O
O
CH3
O

22.6

(a)

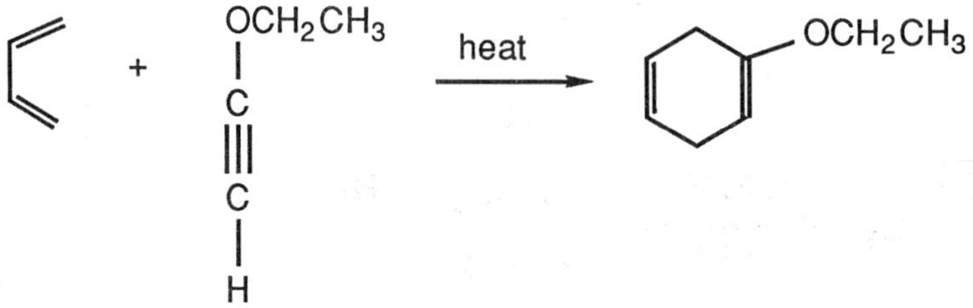
+
OCH2CH3
C
C
H
heat
OCH2CH3

(b)

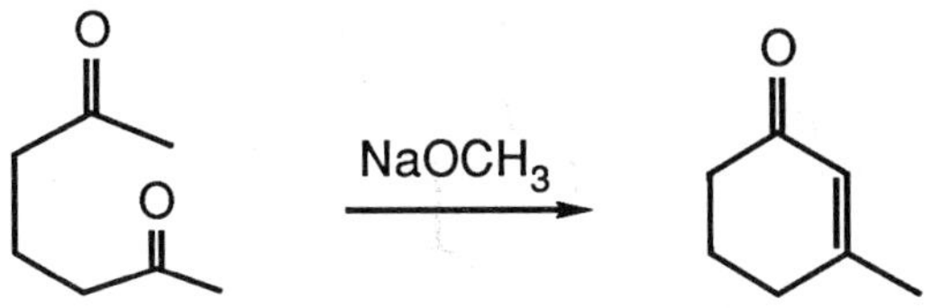

NaOCH$_3$

(c)

CH_3O_2C + heat CH_3O_2C

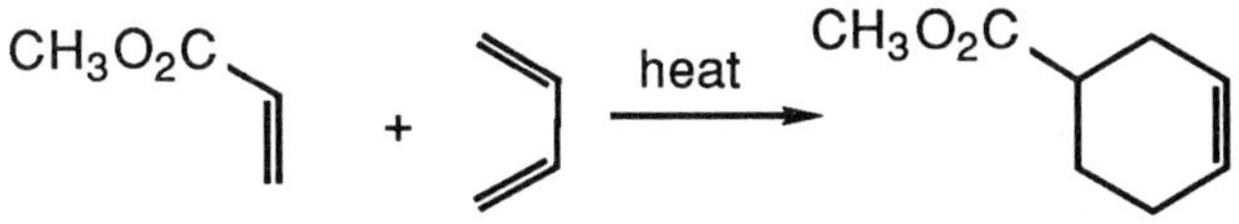

(d)

NaOCH$_3$

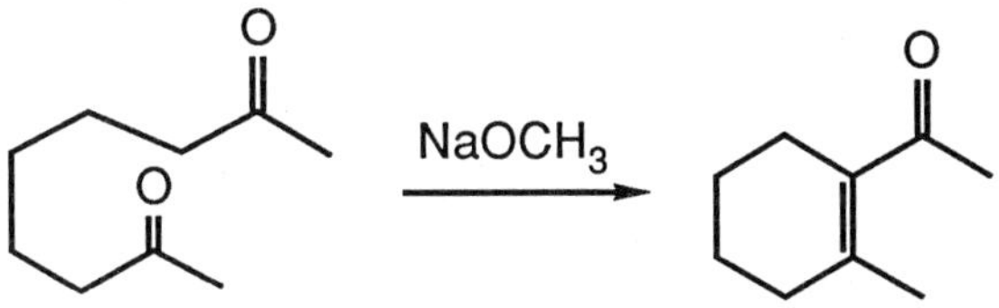

(e)

+ $CH_3-C\equiv C-CH_3$ heat

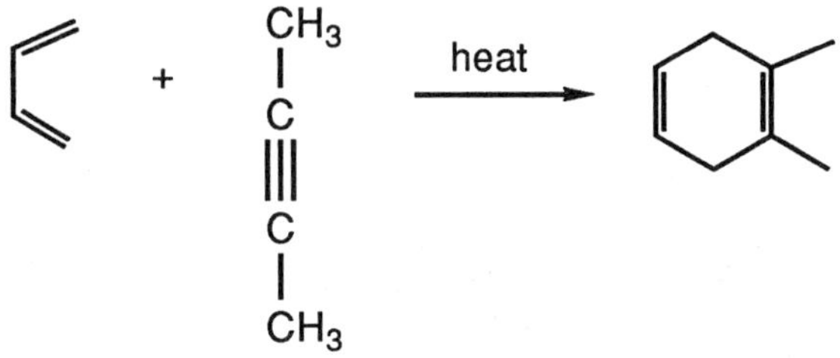

(f)

+ heat

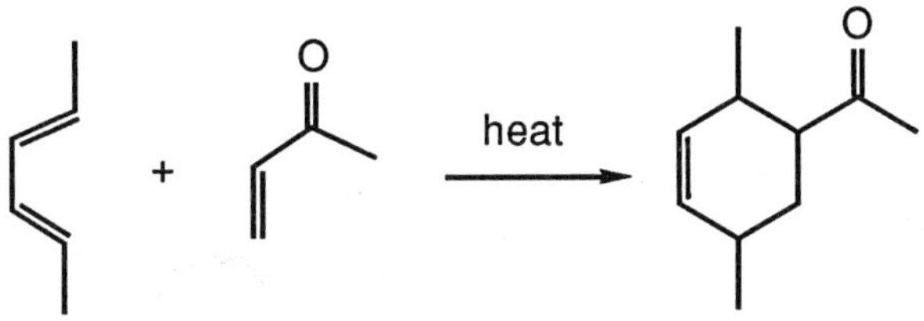

(g)

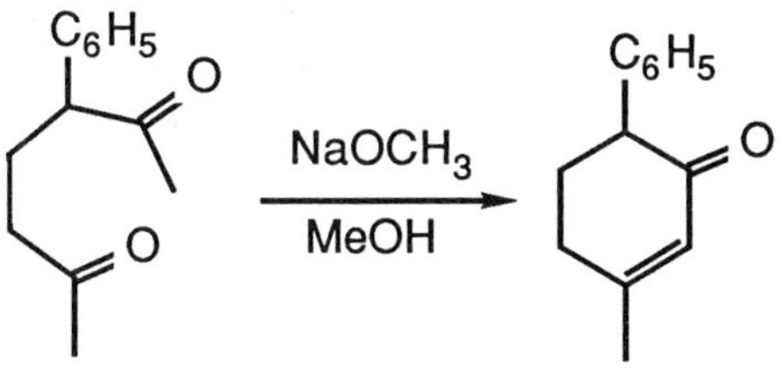

C_6H_5 O O NaOCH$_3$ MeOH C_6H_5 O

(h)

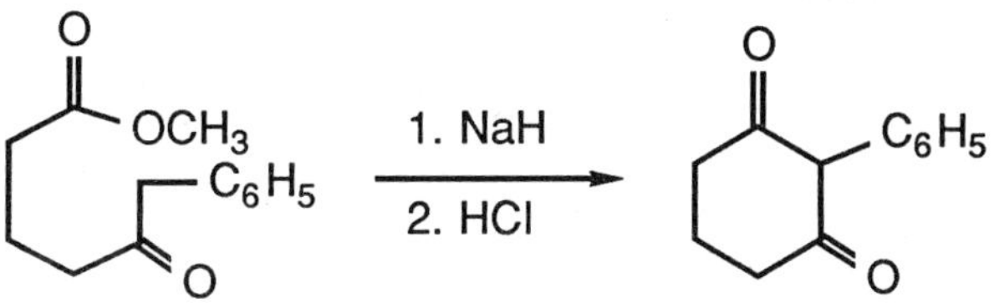

O OCH_3 C_6H_5 O 1. NaH 2. HCl O C_6H_5 O

(i)

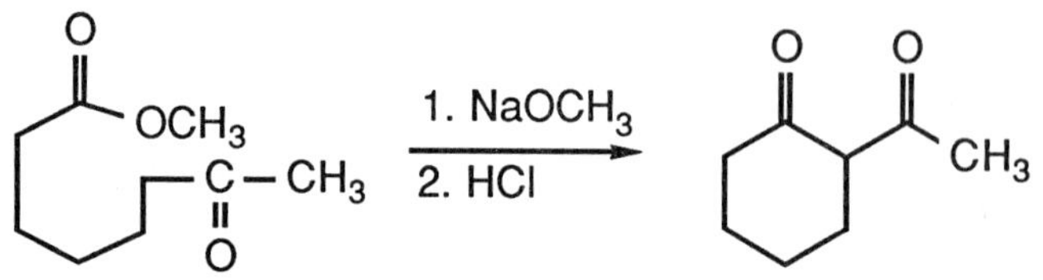

1. $NaOCH_3$
2. HCl

OCH_3 $C-CH_3$ O O O O CH_3

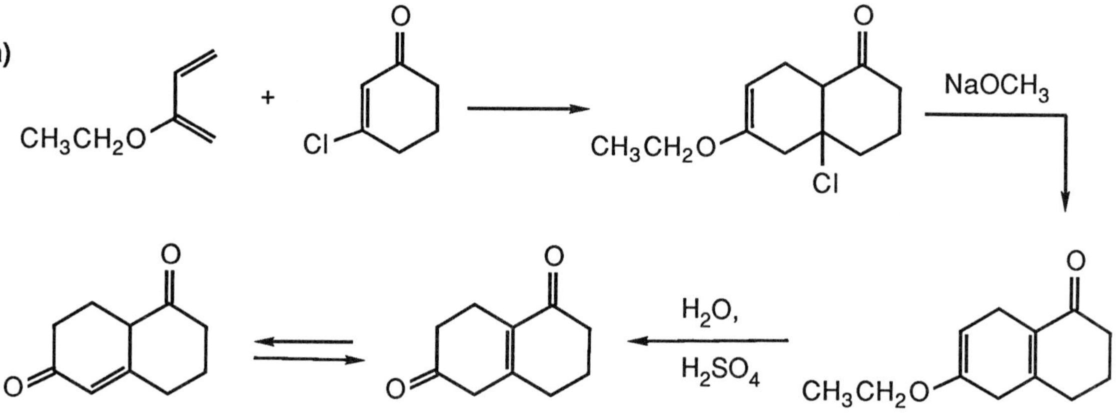

22.7

(a)

CH_3CH_2O + Cl O → CH_3CH_2O Cl O

$NaOCH_3$

H_2O, H_2SO_4

CH_3CH_2O

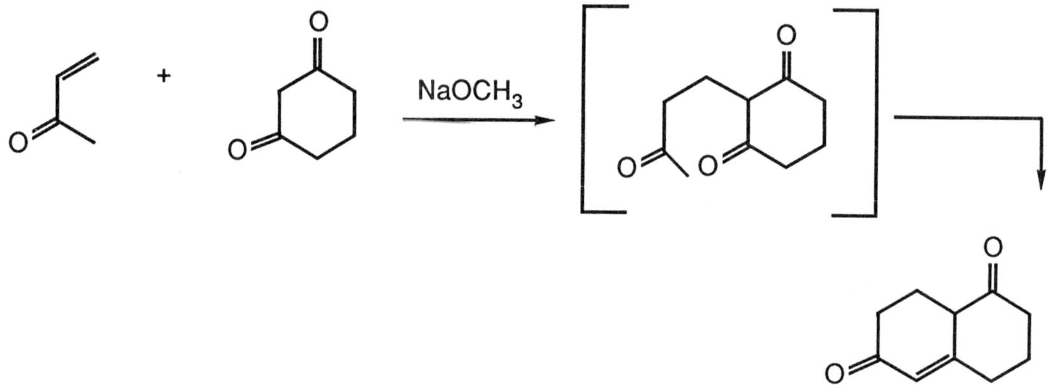

(b)

+ $NaOCH_3$

(c)

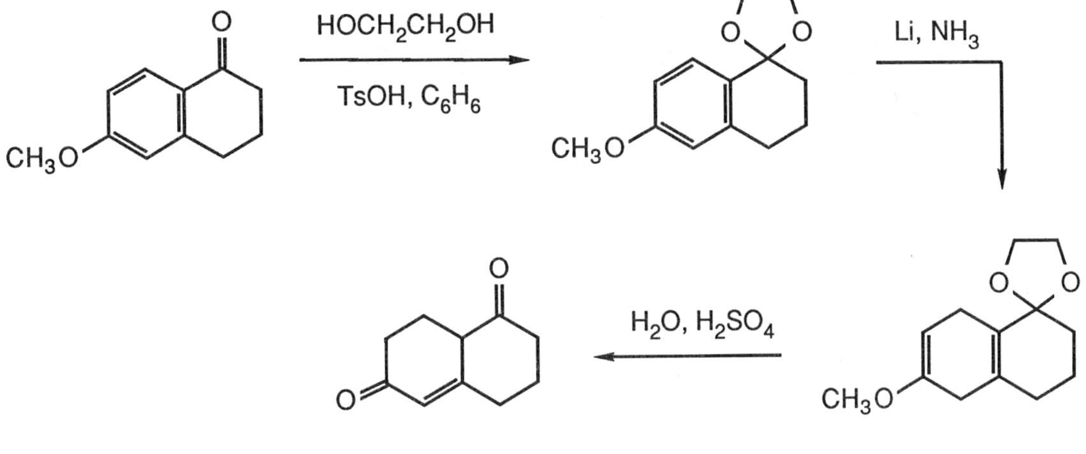

HOCH$_2$CH$_2$OH
TsOH, C$_6$H$_6$
CH$_3$O
O O
Li, NH$_3$
H$_2$O, H$_2$SO$_4$

22.8

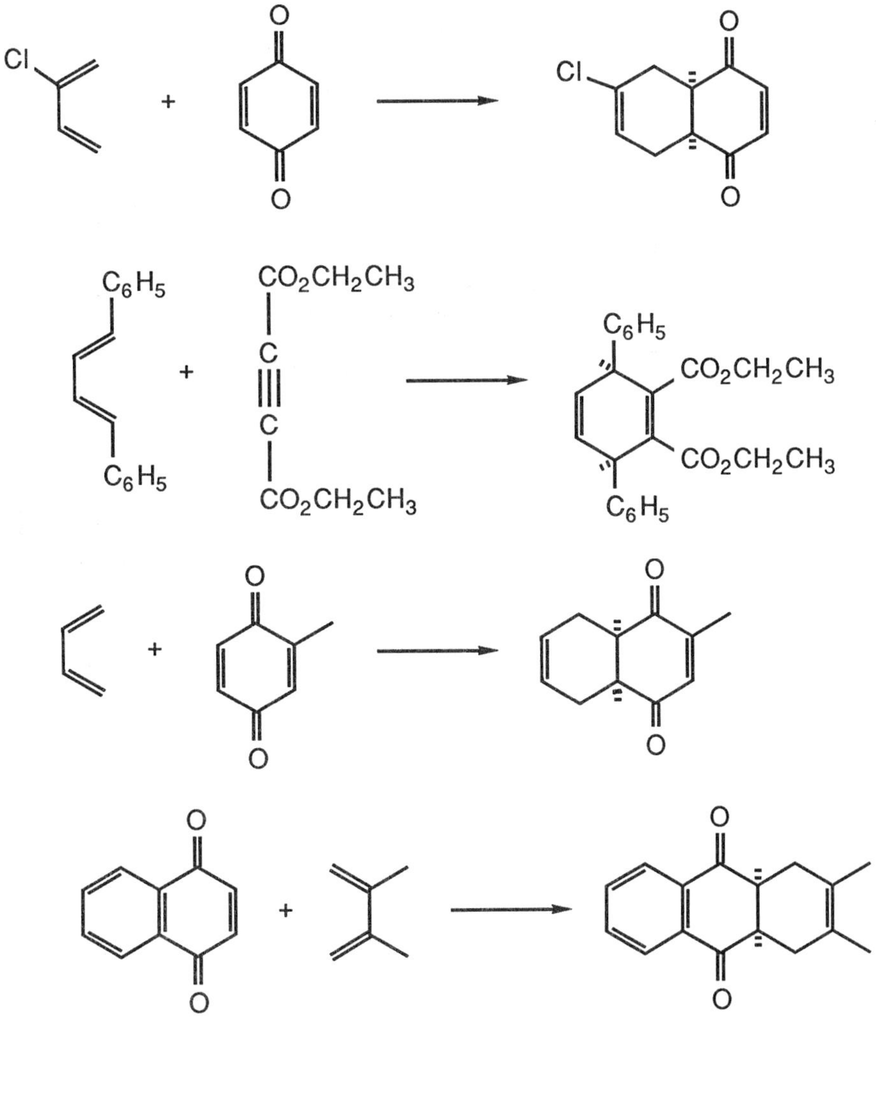

Cl

C_6H_5
$CO_2CH_2CH_3$
C
C
$CO_2CH_2CH_3$

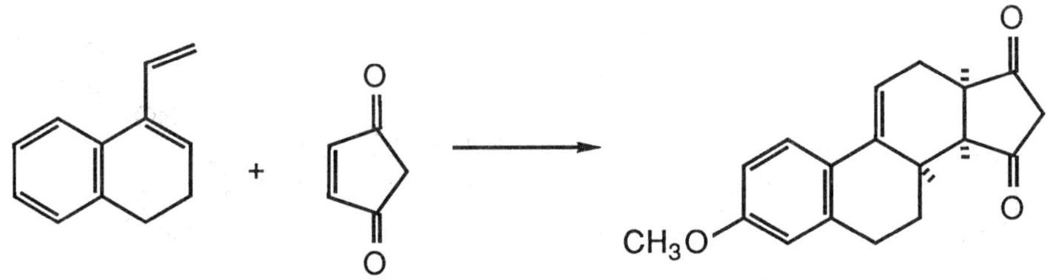

+ ⟶ CH_3O

CH_3O + ⟶ CH_3O H + CH_3O H

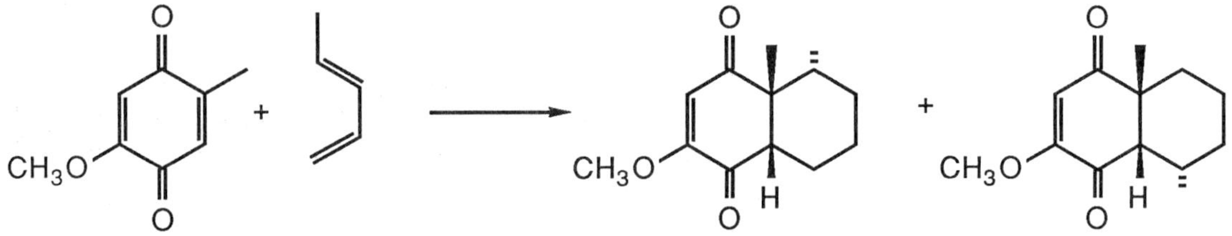

22.9

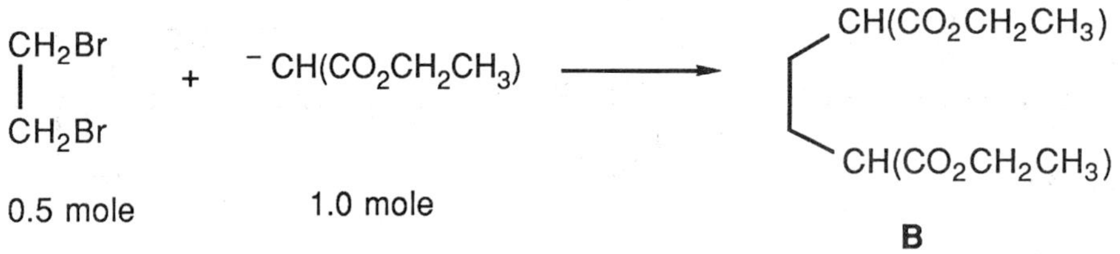

CH_2Br–CH_2Br (0.5 mole) + $^-CH(CO_2CH_2CH_3)$ (1.0 mole) ⟶ $CH(CO_2CH_2CH_3)$ / $CH(CO_2CH_2CH_3)$

B

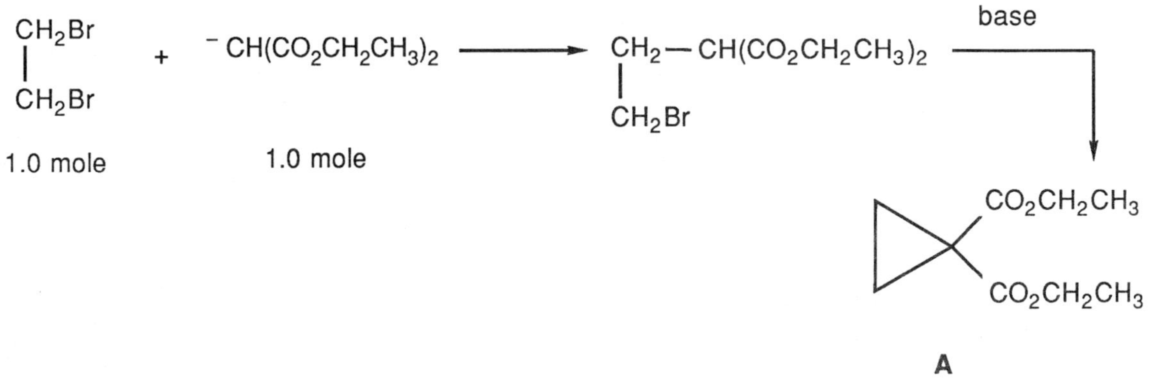

CH_2Br–CH_2Br (1.0 mole) + $^-CH(CO_2CH_2CH_3)_2$ (1.0 mole) ⟶ CH_2—$CH(CO_2CH_2CH_3)_2$ / CH_2Br —base→ $CO_2CH_2CH_3$ / $CO_2CH_2CH_3$

A

22.10

(a)

Benzene + succinic anhydride $\xrightarrow{AlCl_3}$ HO–C(=O)–CH$_2$CH$_2$–C(=O)–Ph $\xrightarrow[2.\ HCl]{1.\ NH_2NH_2,\ KOH}$ HO–C(=O)–CH$_2$CH$_2$CH$_2$–Ph $\xrightarrow{PCl_5}$ Cl–C(=O)–CH$_2$CH$_2$CH$_2$–Ph $\xrightarrow{AlCl_3}$ 1-tetralone

(b)

Methyl vinyl ketone + 2-(H_3CO_2C)cyclopentanone $\xrightarrow{NaOCH_3}$ 2-(CO_2CH_3)-2-(3-oxobutyl)cyclopentanone $\longrightarrow$ bicyclic enone with angular CO_2CH_3

(c)

$CH_3O—\overset{O}{\overset{\|}{C}}—\underset{-}{CH}—\overset{O}{\overset{\|}{C}}—OCH_3$ Na^+ + $BrCH_2$ $BrCH_2$ (–CH$_2$CH$_2$CH$_2$–) $\longrightarrow$ 1,1-cyclohexane with CH_3O_2C and CO_2CH_3

(d)

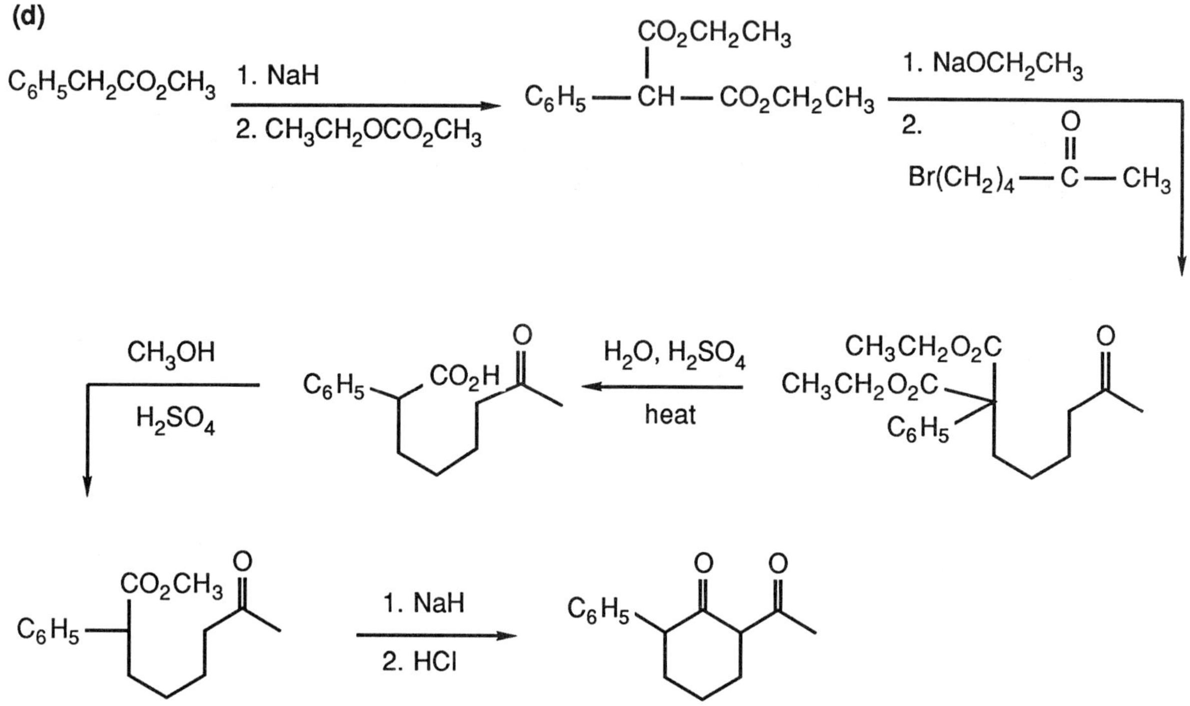

$C_6H_5CH_2CO_2CH_3$ —(1. NaH; 2. $CH_3CH_2OCO_2CH_3$)→ $C_6H_5—CH(CO_2CH_2CH_3)—CO_2CH_2CH_3$ —(1. $NaOCH_2CH_3$; 2. $Br(CH_2)_4—C(=O)—CH_3$)→

$(CH_3CH_2O_2C)_2C(C_6H_5)(CH_2)_4C(=O)CH_3$ —(H_2O, H_2SO_4; heat)→ $C_6H_5CH(CO_2H)(CH_2)_4C(=O)CH_3$ —(CH_3OH; H_2SO_4)→

$C_6H_5CH(CO_2CH_3)(CH_2)_4C(=O)CH_3$ —(1. NaH; 2. HCl)→ 2-phenyl-6-acetylcyclohexanone (C_6H_5, O, O)

(e)

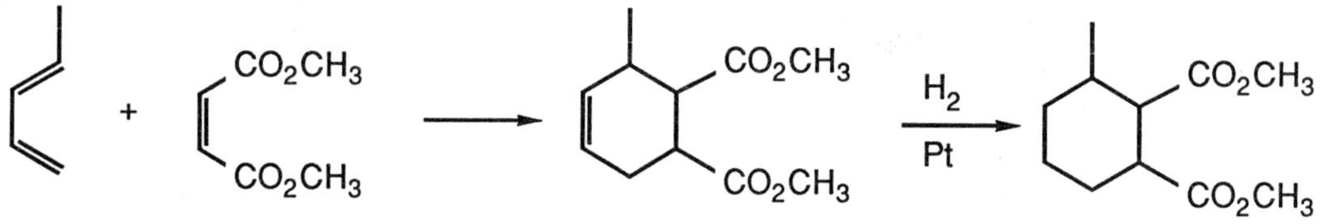

1,3-pentadiene + $CH_3O_2C–CH=CH–CO_2CH_3$ (cis) → methylcyclohexene with CO_2CH_3, CO_2CH_3 —(H_2; Pt)→ methylcyclohexane with CO_2CH_3, CO_2CH_3

22.11

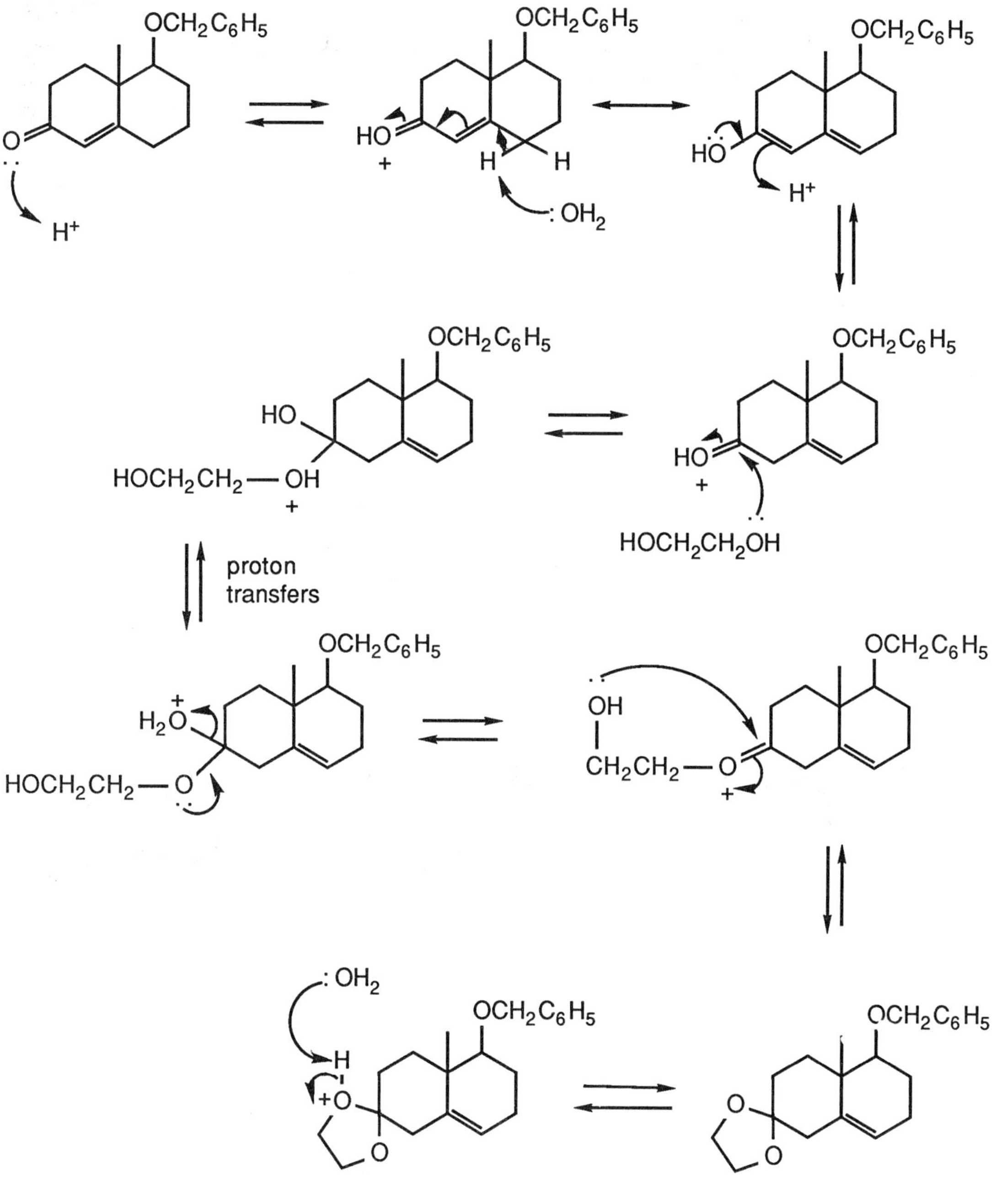

$OCH_2C_6H_5$ H^+ HO^+ H H $:OH_2$ HO H^+

HO $HOCH_2CH_2—\overset{+}{O}H$ HO^+ $HOCH_2CH_2OH$

proton transfers

$H_2\overset{+}{O}$ $HOCH_2CH_2—O$ OH $CH_2CH_2—\overset{+}{O}$

$:OH_2$ H ^+O O O

22.12

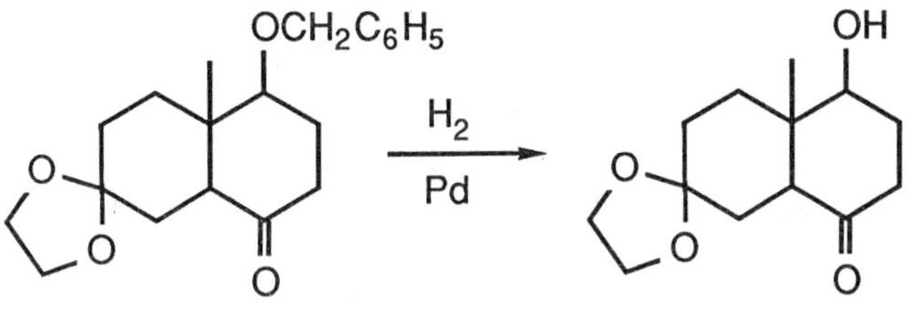

$OCH_2C_6H_5$ O O O $\xrightarrow[Pd]{H_2}$ OH O O O

22.13

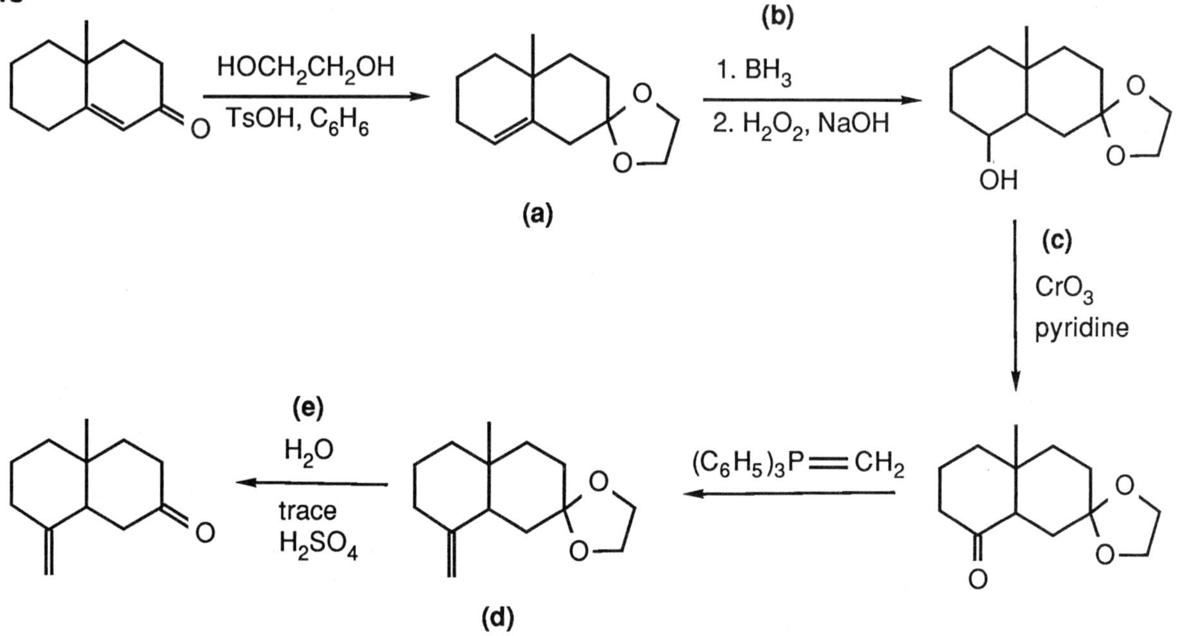

HOCH$_2$CH$_2$OH

TsOH, C$_6$H$_6$

(a)

(b)

1. BH$_3$

2. H$_2$O$_2$, NaOH

O

O

OH

(c)

CrO$_3$

pyridine

(C$_6$H$_5$)$_3$P=CH$_2$

(d)

(e)

H$_2$O

trace

H$_2$SO$_4$

22.14

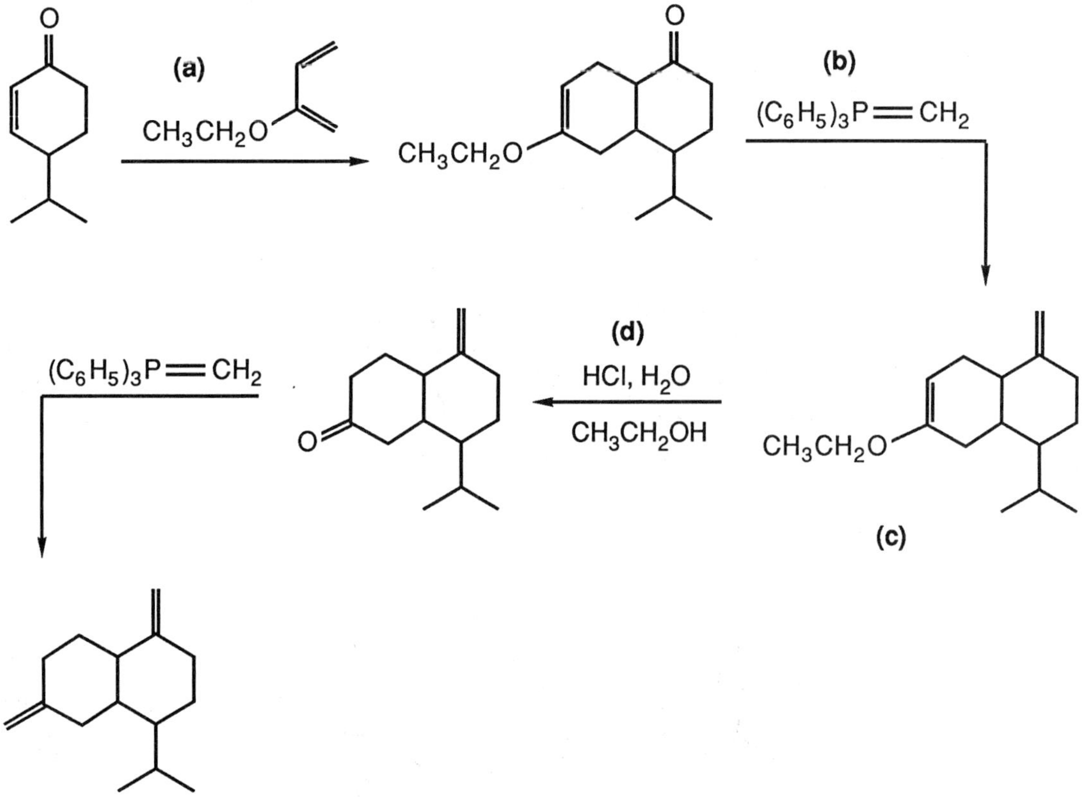

(a)

CH$_3$CH$_2$O

(b)

(C$_6$H$_5$)$_3$P=CH$_2$

(c)

(d)

HCl, H$_2$O

CH$_3$CH$_2$OH

(C$_6$H$_5$)$_3$P=CH$_2$

22.15

OH

(a)

O

TsOH

O

O

O

1. $LiAlH_4$

2. H_2O

O

O

OH

(b)

(c)

TsCl

pyridine

O

O

OTs

(d)

(e)

NaCN

O

O

CN

CH_3OH,

H_2SO_4

HO

CO_2CH_3

22.16

O

O

(a)

$HOCH_2CH_2OH$

TsOH, C_6H_6

O O

O

H_2

Pd

O O

O

(b)

(c)

1. $(CH_3)_2CHLi$

2. H_2O

O O

OH

(d)

H_2O,

H_2SO_4

O

22.17

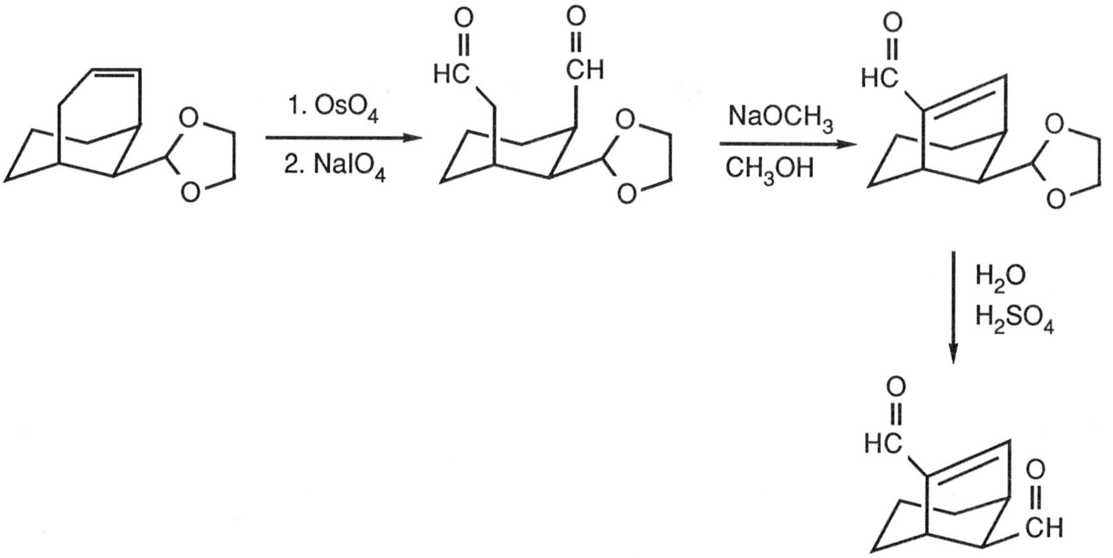

1. OsO_4
2. $NaIO_4$

$NaOCH_3$
CH_3OH

H_2O
H_2SO_4

HC(=O) ... CH(=O)

CHAPTER 23. NATURAL PRODUCTS

ANSWERS TO EXERCISES

23.1 $(CH_3)_2C{=}O$ + $CH_2(CO_2CH_2CH_3)_2$ $\xrightarrow{NH_4^{+}\,^{-}OAc}$ $(CH_3)_2C{=}C(CO_2CH_2CH_3)_2$

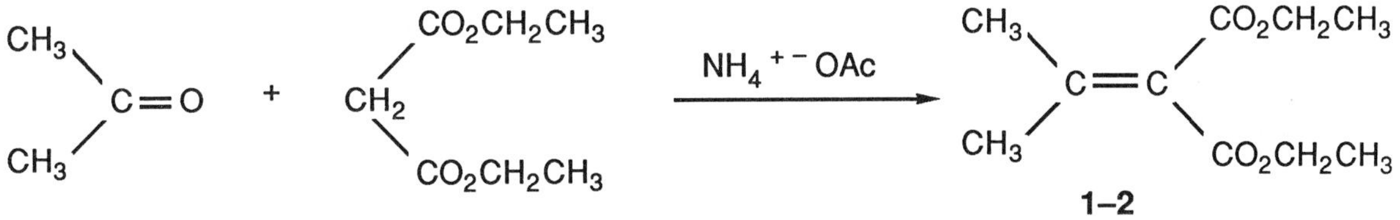

1–2

23.2 Decarboxylation occurs only when the carboxyl group has a carbonyl group in the β position (arrow):

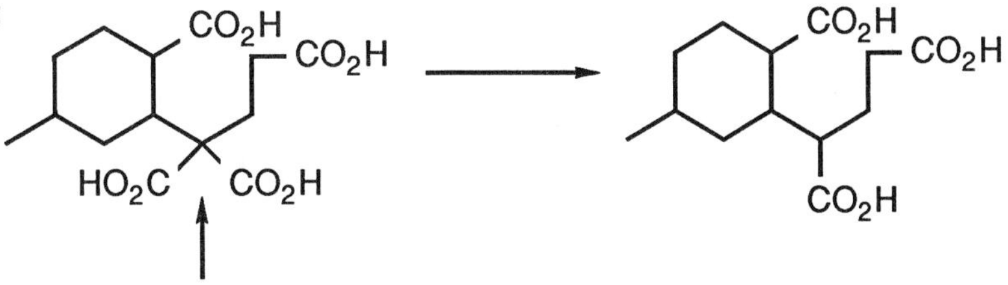

CO_2H, CO_2H, HO_2C, CO_2H → CO_2H, CO_2H, CO_2H

23.3

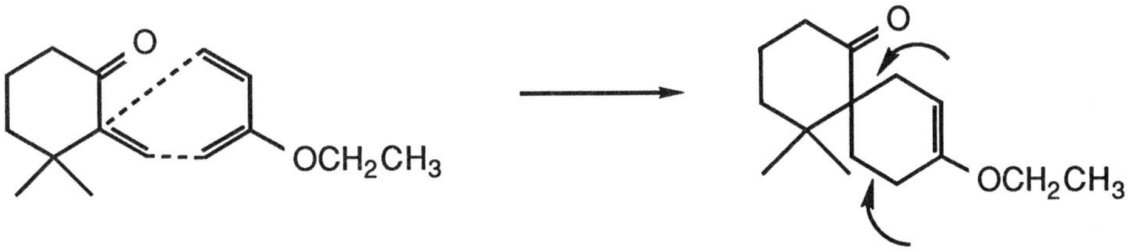

O, OCH_2CH_3 → O, OCH_2CH_3

23.4

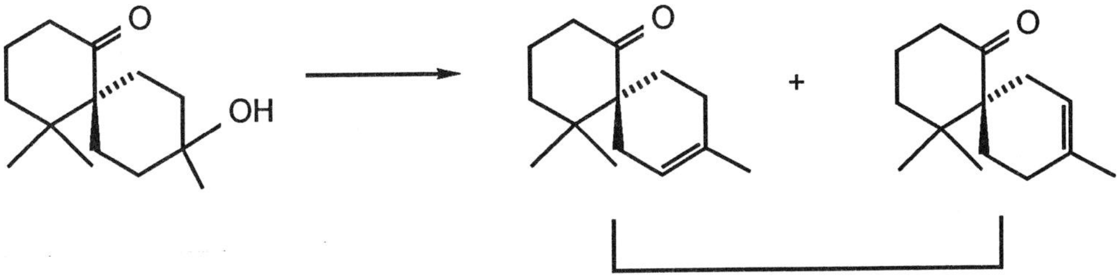

O, OH → O + O

enantiomers

23.5 For the yields specified in the text the overall yield of equilinen starting from **5-2** would be 3-5%. In order to obtain 1 g of equilinen (MW 266) it would be necessary to start with 17 g or 28 g of **5-2** (assuming 5% or 3% overall yield, respectively).

23.6 Assuming 100% yield for several steps where yields were not shown, an overall yield of 0.6% would be obtained for cortisone acetate:

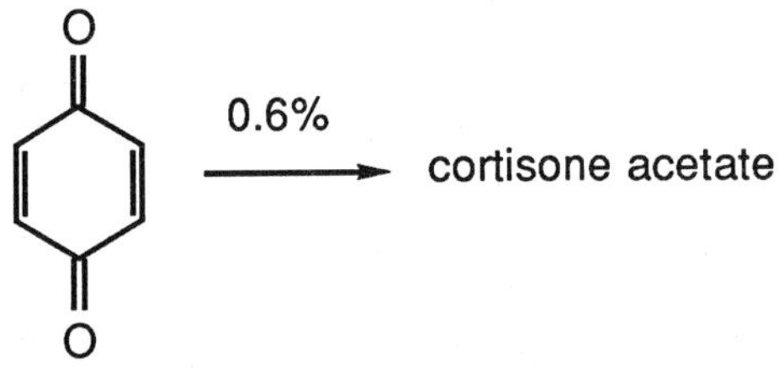

0.6%

cortisone acetate

To obtain 1 g of the final product (MW 402) it would be necessary to start with 45 g of benzoquinone (MW 108).

23.7 The actual oxidizing agent is cyclohexanone:

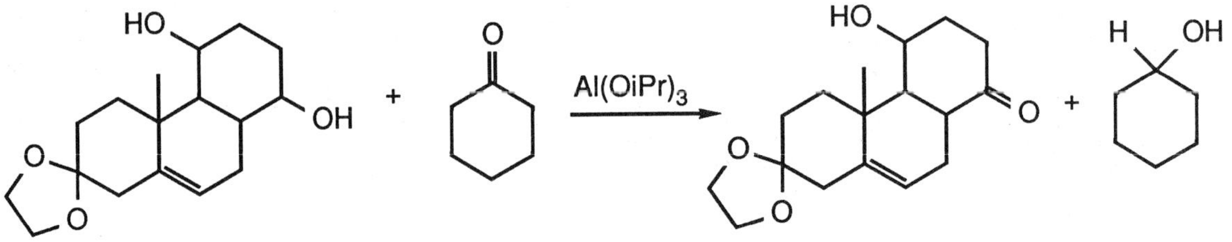

HO OH + O $Al(OiPr)_3$ HO O + H OH

23.8

HO O KOtBu CH_3I HO CH_3 O CH_3 + HO CH_3 CH_3 O

possible C_{19} by-products

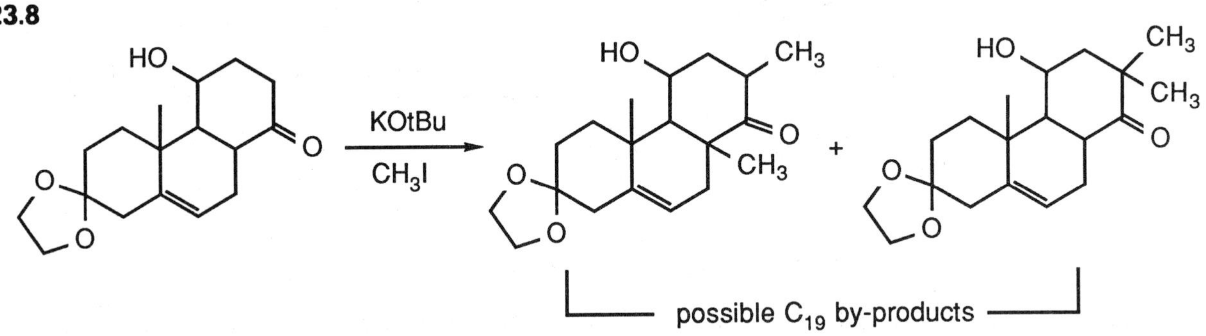

23.9

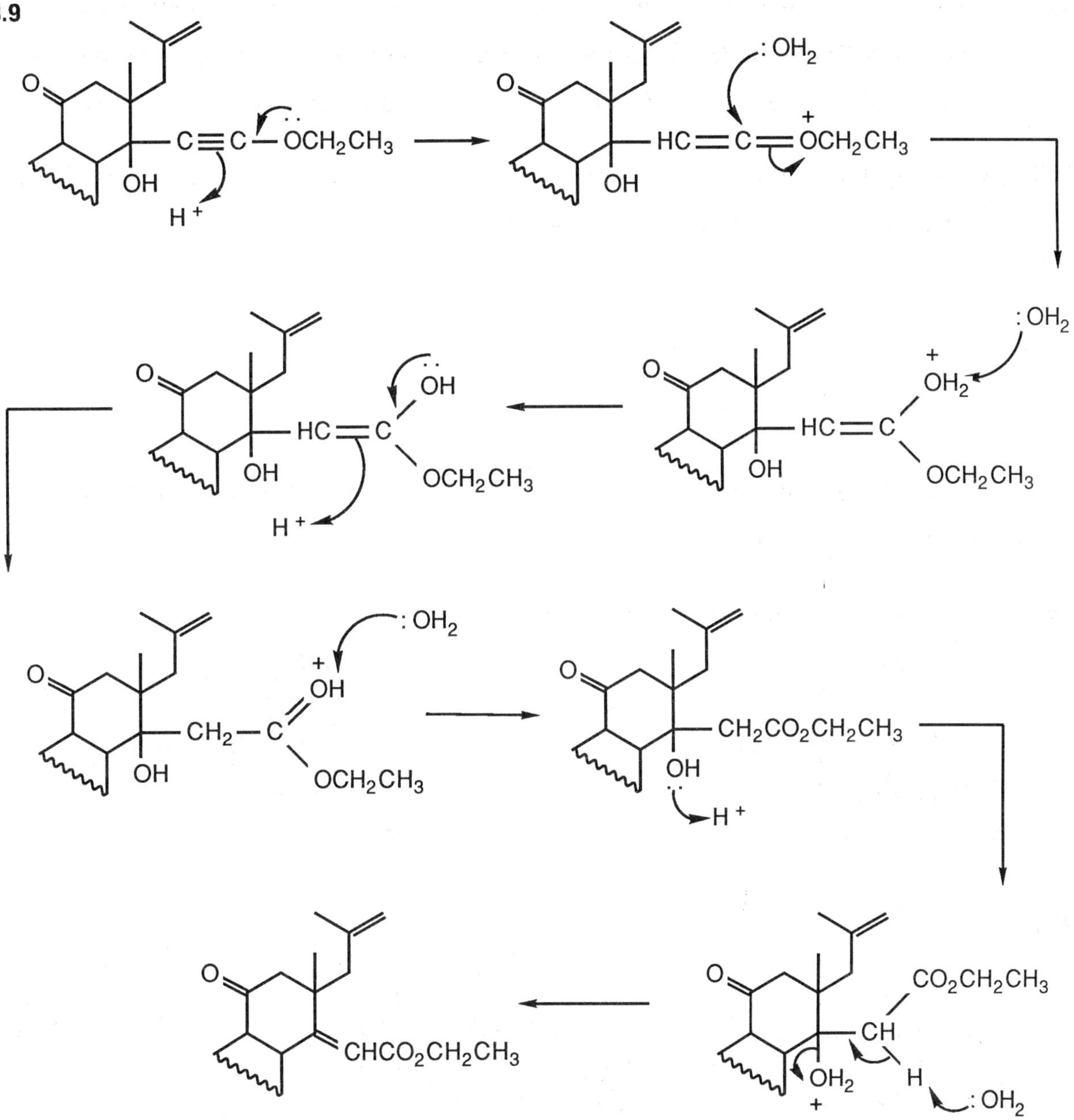

6-15

23.10 Acid hydrolysis would result in cleavage of the ketal to generate the conjugated ketone. Isomerization of the double bond of the ring C side chain (not shown) could also occur.

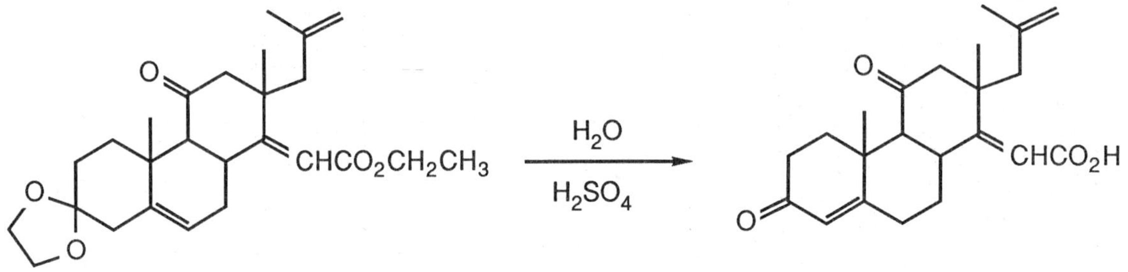

$CHCO_2CH_2CH_3$ → (H_2O, H_2SO_4) → $CHCO_2H$

6-15

23.11 Acid hydrolysis would result in cleavage of the ketal to generate the conjugated ketone. Isomerization of the double bond of the ring C side chain (not shown) could also occur.

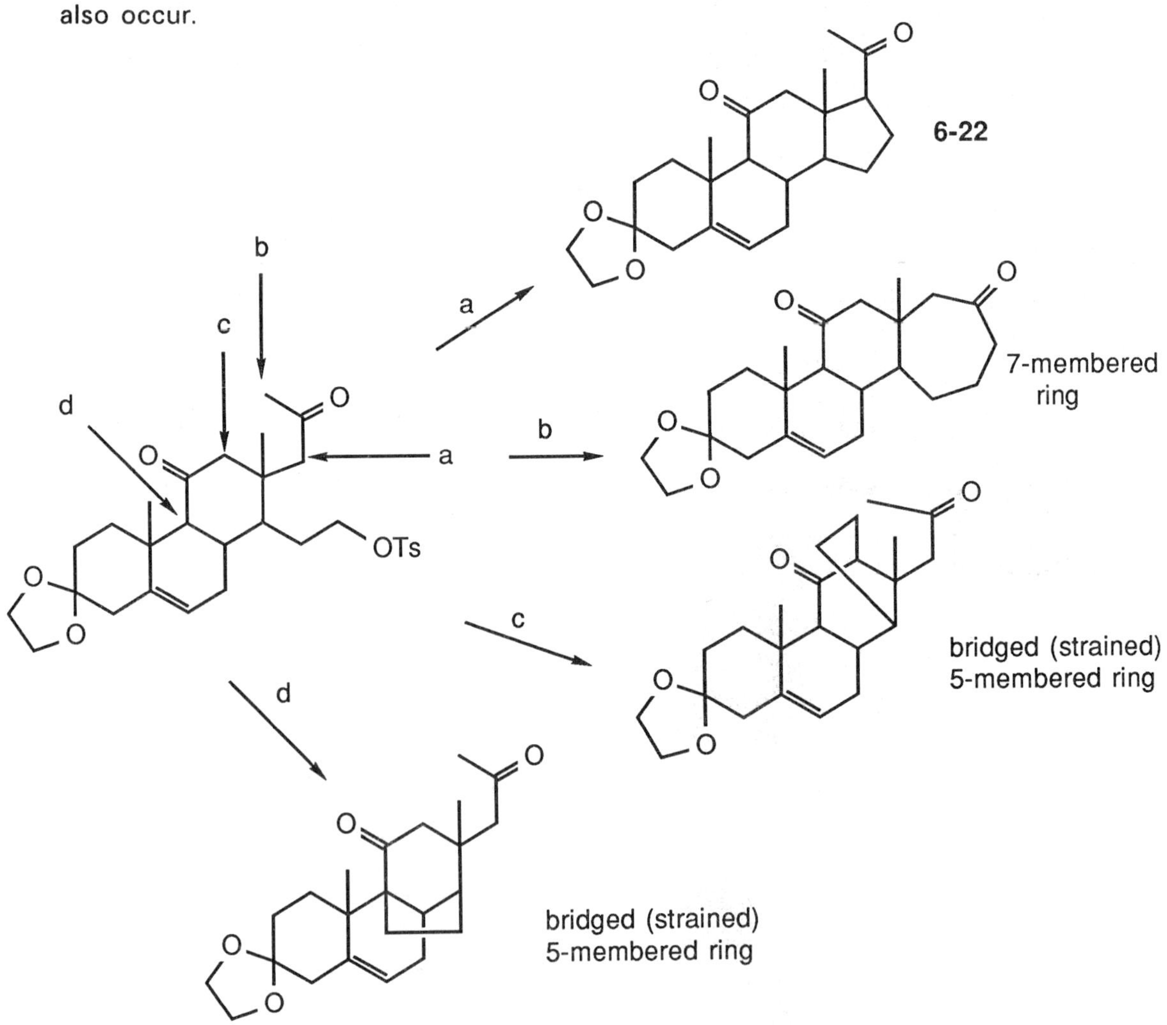

OTs

a → 6-22

b → 7-membered ring

c → bridged (strained) 5-membered ring

d → bridged (strained) 5-membered ring

23.12 (a) The carbonyl group designated (a) is less hindered.

I—CH(C=O–bicyclic)—C(=O)—C(=O)—OCH$_3$ —a→ bicyclic–C(=O)CH$_2$I **6-26**

—b→ bicyclic–C(=O)OCH$_3$

(b) The iodo substituent stabilizes the enolate that acts as a leaving group:

I—CH(C=O–bicyclic)—C(=O)—C(=O)—OCH$_3$ + :OCH$_3^-$ → I—CH$^-$(C=O–bicyclic) + HOCH$_3$ → bicyclic–C(=O)CH$_2$I **6-26**

23.13

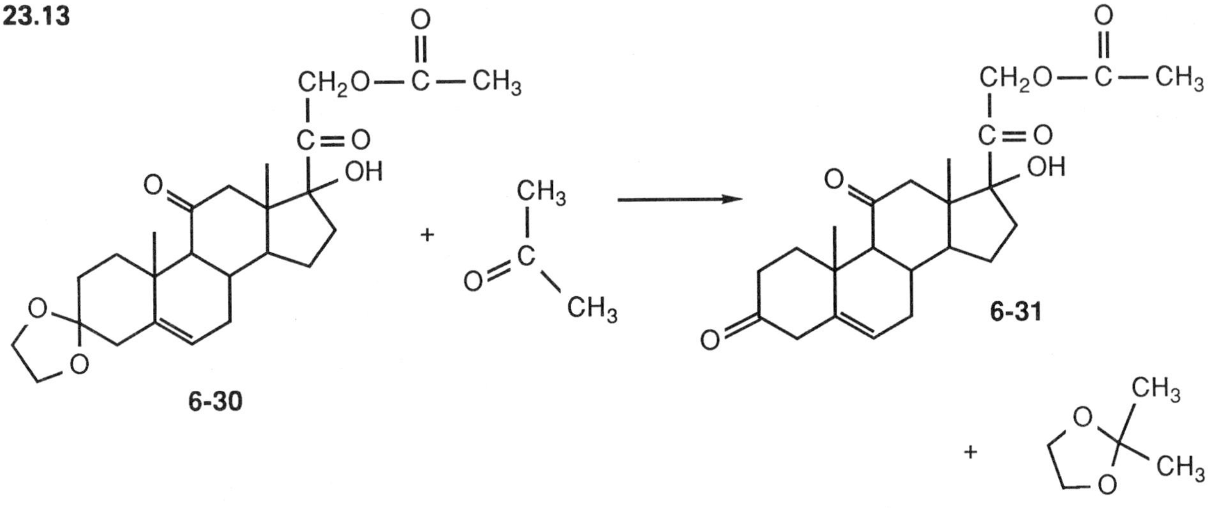

$CH_2O-\overset{O}{\overset{\|}{C}}-CH_3$

$C=O$

OH

O

O

O

6-30

+

CH_3

$O=C$

CH_3

$CH_2O-\overset{O}{\overset{\|}{C}}-CH_3$

$C=O$

OH

O

O

6-31

+

CH_3

O

O

CH_3

23.14 (a)

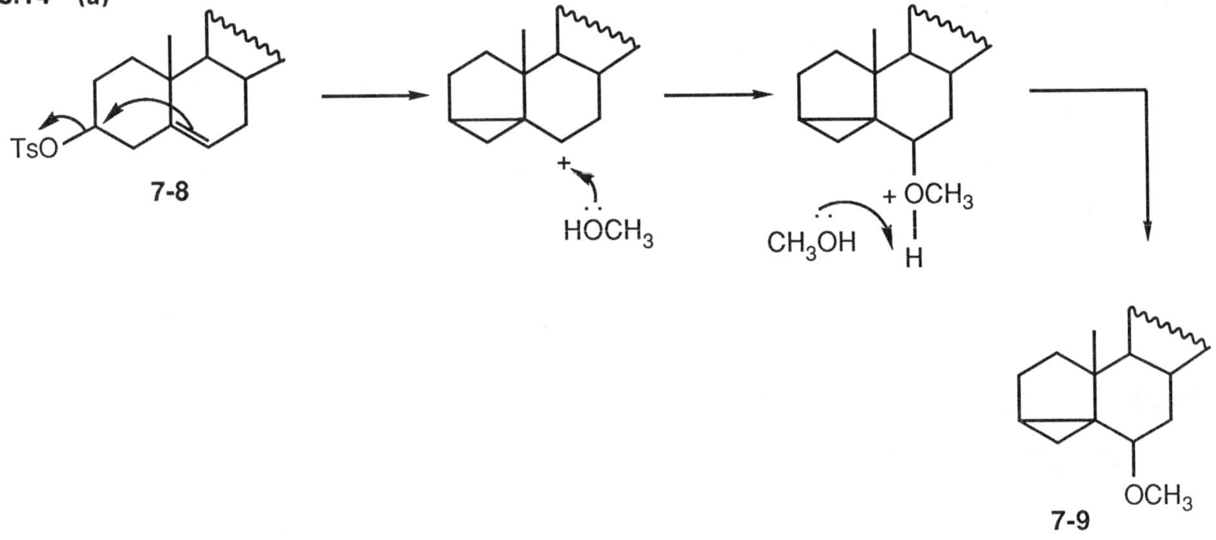

TsO

7-8

+

$H\ddot{O}CH_3$

$CH_3\ddot{O}H$

$+\ OCH_3$

H

OCH_3

7-9

(b)

TsO

CH_3OH

CH_3O

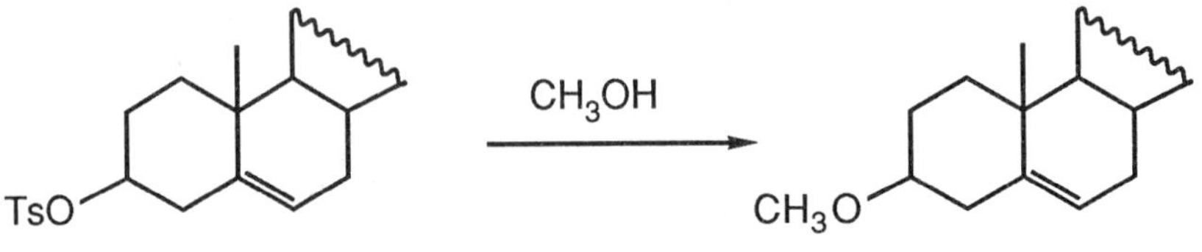

minor product

23.15

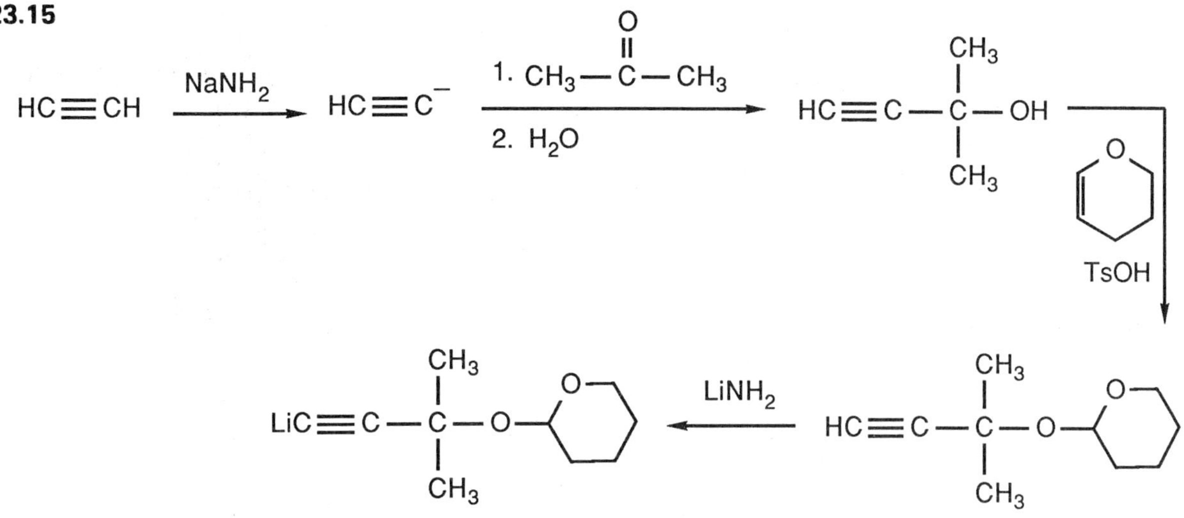

$$HC\equiv CH \xrightarrow{NaNH_2} HC\equiv C^- \xrightarrow[2.\ H_2O]{1.\ CH_3-\overset{O}{\overset{\|}{C}}-CH_3} HC\equiv C-C(CH_3)_2-OH$$

$\xrightarrow{\text{dihydropyran},\ TsOH}$ $HC\equiv C-C(CH_3)_2-O-\text{THP}$ $\xrightarrow{LiNH_2}$ $LiC\equiv C-C(CH_3)_2-O-\text{THP}$

TsOH

LiNH$_2$

23.16

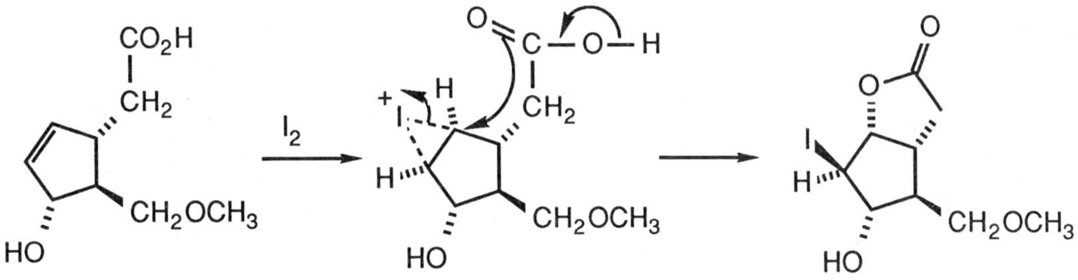

CO_2H, CH_2, CH_2OCH_3, HO $\xrightarrow{I_2}$ H, I$^+$, $C-O-H$, O, CH_2, CH_2OCH_3, HO $\longrightarrow$ O, I, H, CH_2OCH_3, HO

23.17

$$(CH_3O)_2\overset{O}{\overset{\|}{P}}: + BrCH_2-\overset{O}{\overset{\|}{C}}-(CH_2)_4CH_3 \longrightarrow (CH_3O)_2-\overset{O}{\overset{\|}{P}}-CH_2-\overset{O}{\overset{\|}{C}}-(CH_2)_4CH_3$$

NaH

$$(CH_3O)_2-\overset{O}{\overset{\|}{P}}-\overset{-}{C}H-\overset{O}{\overset{\|}{C}}-(CH_2)_4CH_3 \quad Na^+$$

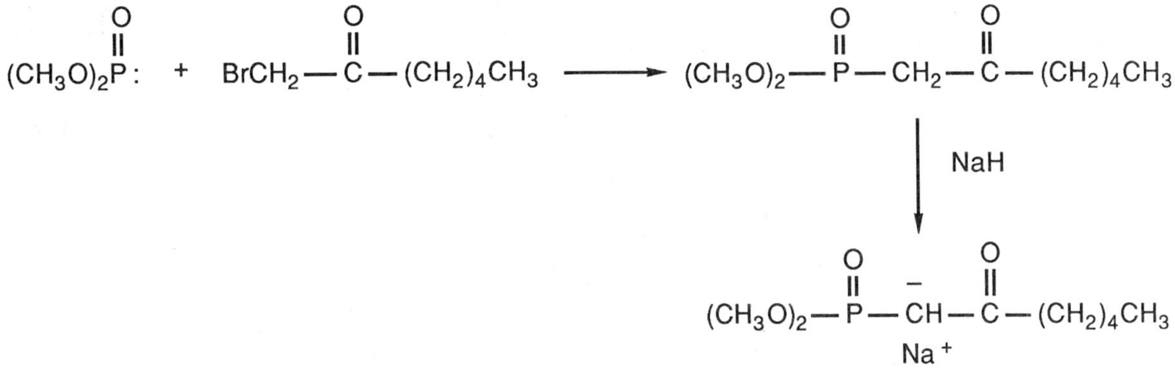

23.18 The minor isomer could be reoxidized to the starting ketone; reduction would then convert most of it to the desired isomer:

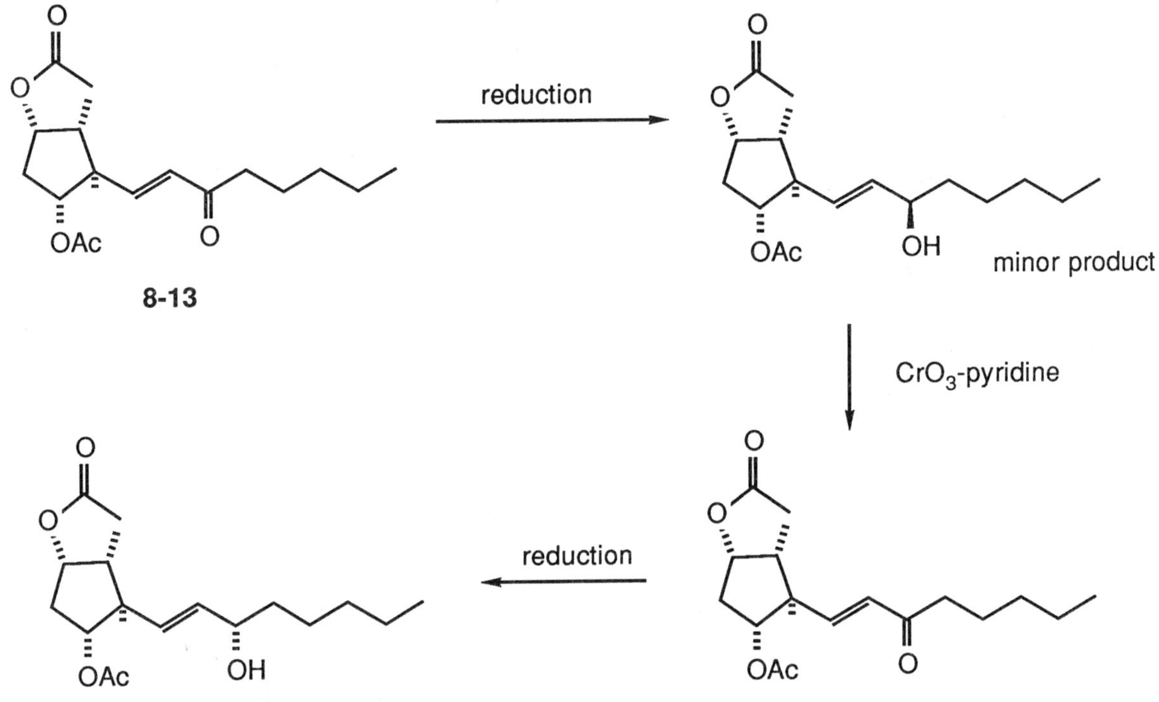

reduction

OAc O

8-13

OAc OH minor product

CrO_3-pyridine

reduction

OAc OH

8-14 major product

OAc O

8-13

23.19 (a) The other product would be methyl acetate:

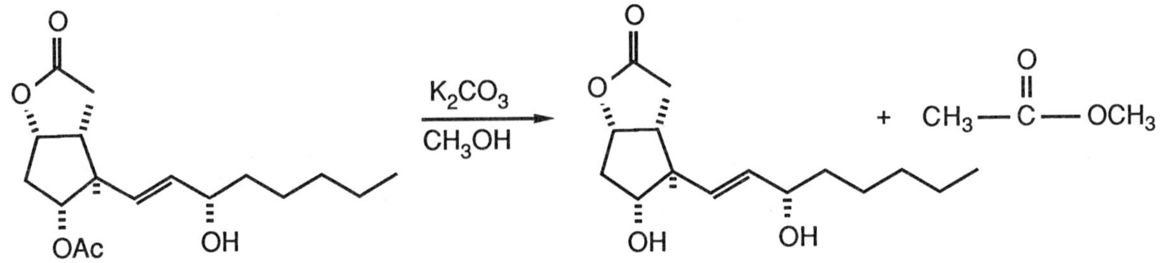

K_2CO_3 / CH_3OH

OAc OH

8-14

OH OH

8-15

+ $CH_3—C(=O)—OCH_3$

(b) If nucleophilic attack also occurred at the lactone carbonyl, ring opening would product an ester side chain:

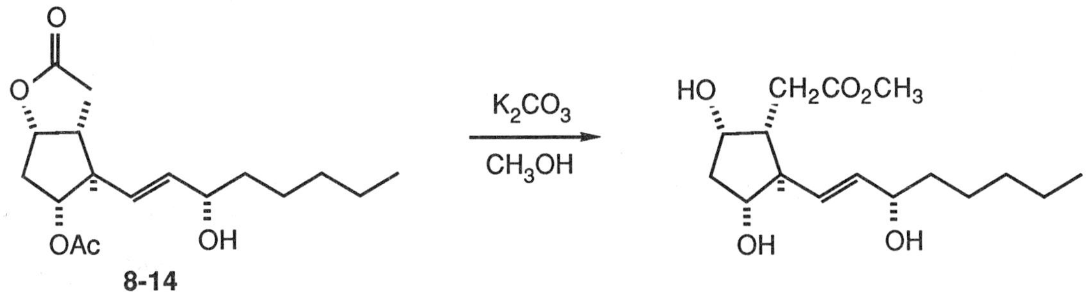

K_2CO_3, CH_3OH

HO $CH_2CO_2CH_3$

OAc OH OH OH

8-14

23.20 **(a)** The other double bond is less susceptible to epoxidation with peracid because it is conjugated with two (electron withdrawing) carbonyl groups.

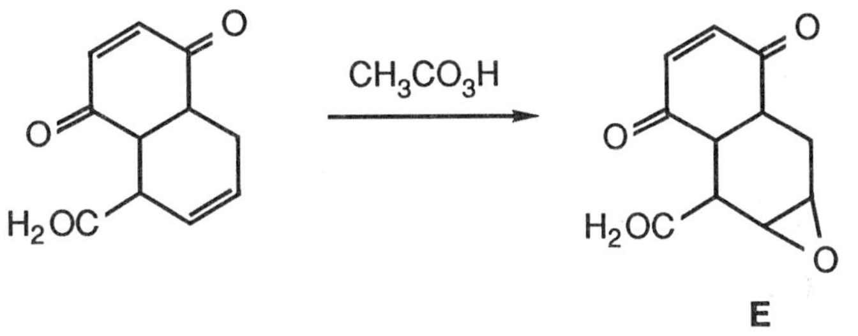

CH_3CO_3H

H_2OC H_2OC

E

(b) Reaction occurs at the less hindered face:

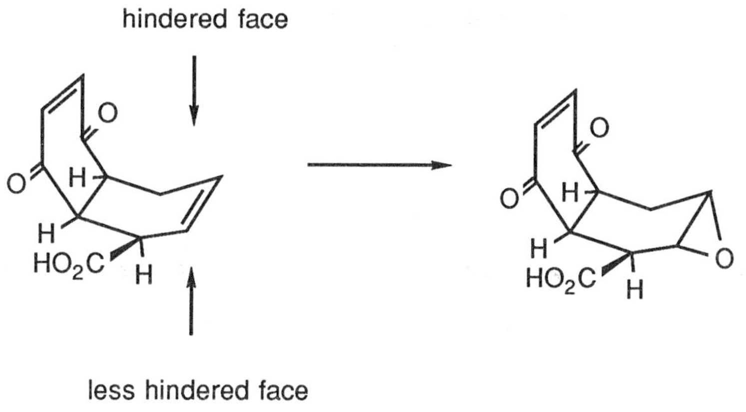

hindered face

H H HO_2C H

less hindered face

23.21 The species that is oxidized (i.e., the actual reducing agent) is 2-propanol.

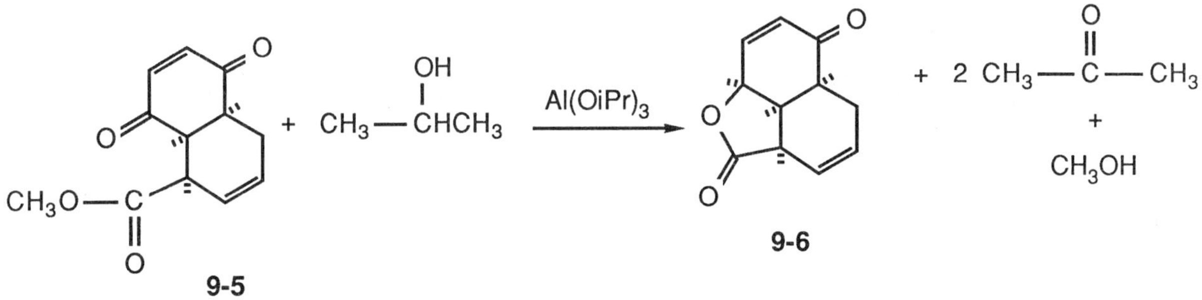

+ CH_3—CHCH$_3$ (OH) → Al(OiPr)$_3$ → + 2 CH_3—C(=O)—CH_3 + CH_3OH

CH_3O—C(=O)

9-5 9-6

23.22 The observed carbonyl peak at 1760 cm^{-1} supports the assignment of **6-1** as the product:

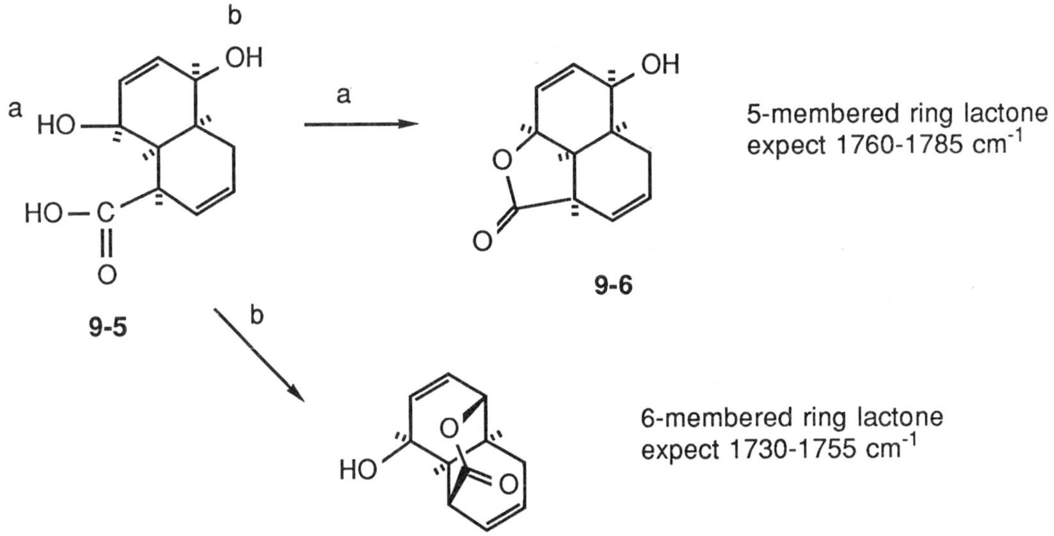

a HO b OH HO—C(=O) 9-5 → a → OH 9-6

5-membered ring lactone expect 1760-1785 cm^{-1}

b → HO

6-membered ring lactone expect 1730-1755 cm^{-1}

23.23 Methoxide ion attacks from the less hindered face. Subsequent protonation of the resulting enolate ion proceeds to give the less strained lactone:

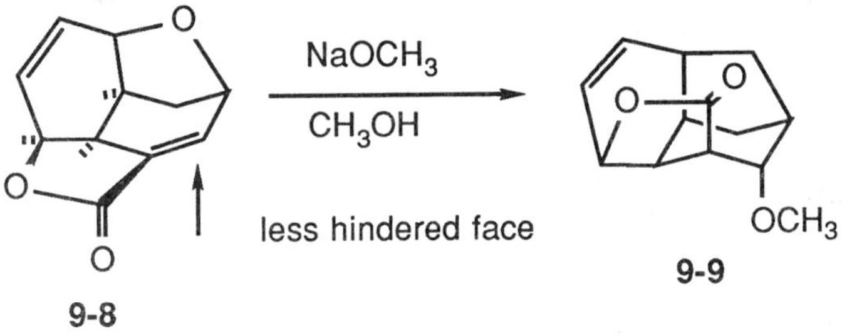

9-8 → $NaOCH_3$ / CH_3OH → 9-9 (OCH_3)

less hindered face

23.24

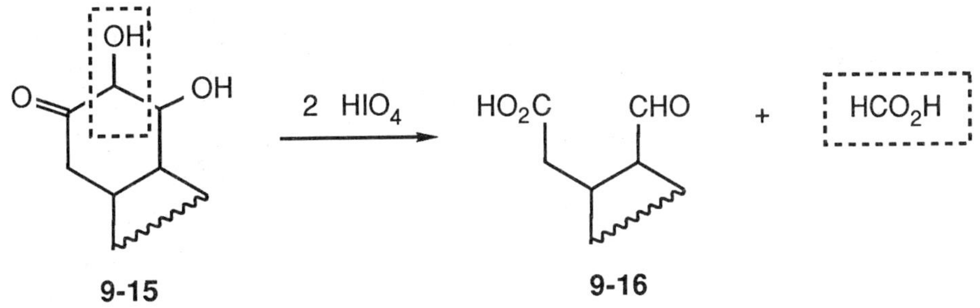

$$\text{9-15} \xrightarrow{2\ HIO_4} \text{9-16} + HCO_2H$$

23.25 **(a)** The amino of the indole group is much less nucleophilic because the "nonbonding" electrons are in fact strongly interacting with the aromatic π system. Recall that pyrrole is a 4n+2 aromatic system:

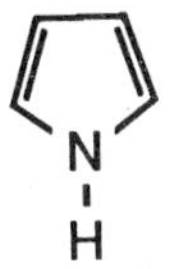

(b)

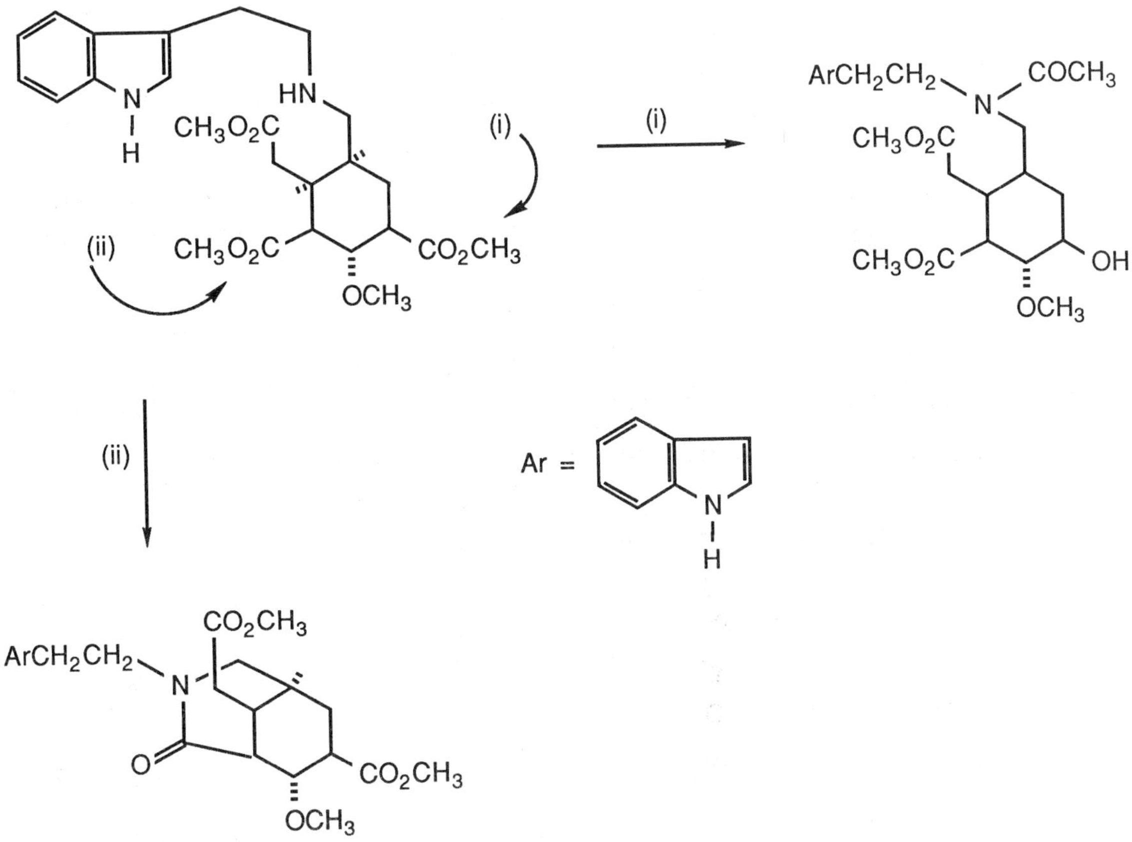

Ar =

(c) Reaction (i) is disfavored because attack by nitrogen at the carbonyl group would require a 7-membered ring transition state. Reaction (ii) is disfavored because the product would be a bridged ring system with unfavorable steric interactions.

23.26 The conformation with fewer axial substituents on the two saturated six membered rings is more stable. (Bonds to nitrogen are not considered here, because nitrogen can undergo inversion of configuration to form the more stable geometry):

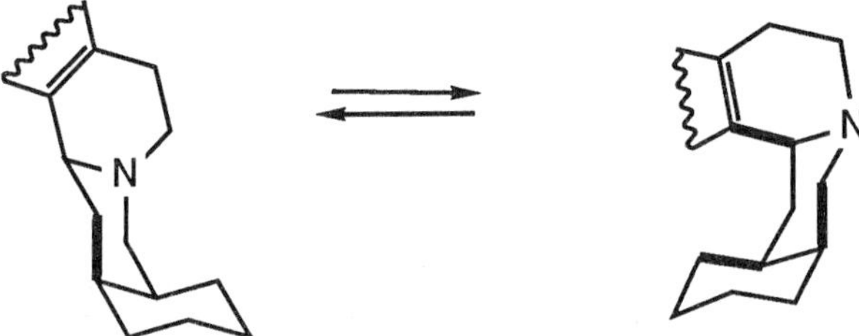

ANSWERS TO PROBLEMS

23.1

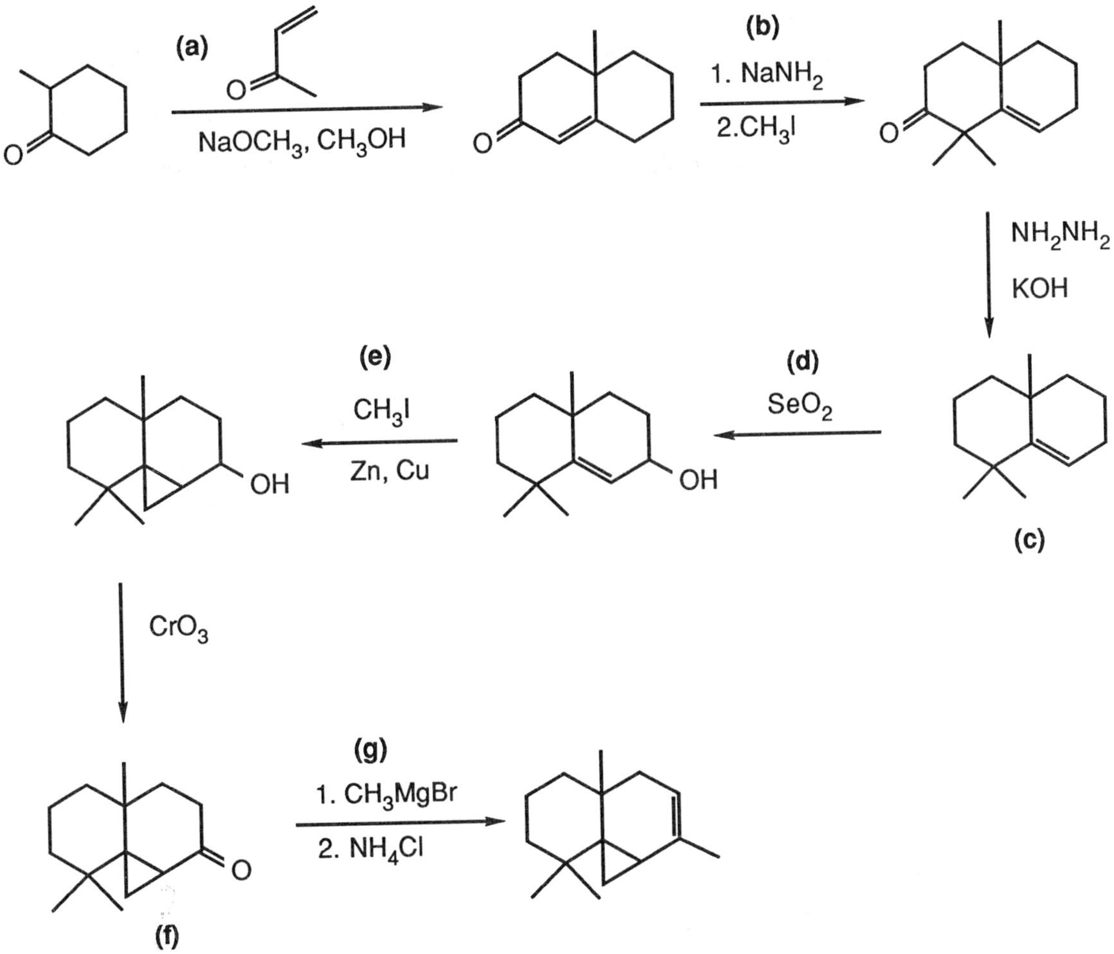

(a) NaOCH$_3$, CH$_3$OH

(b) 1. NaNH$_2$ 2.CH$_3$I

NH$_2$NH$_2$ KOH

(c)

(d) SeO$_2$

OH

(e) CH$_3$I Zn, Cu

OH

CrO$_3$

(f)

O

(g) 1. CH$_3$MgBr 2. NH$_4$Cl

23.2

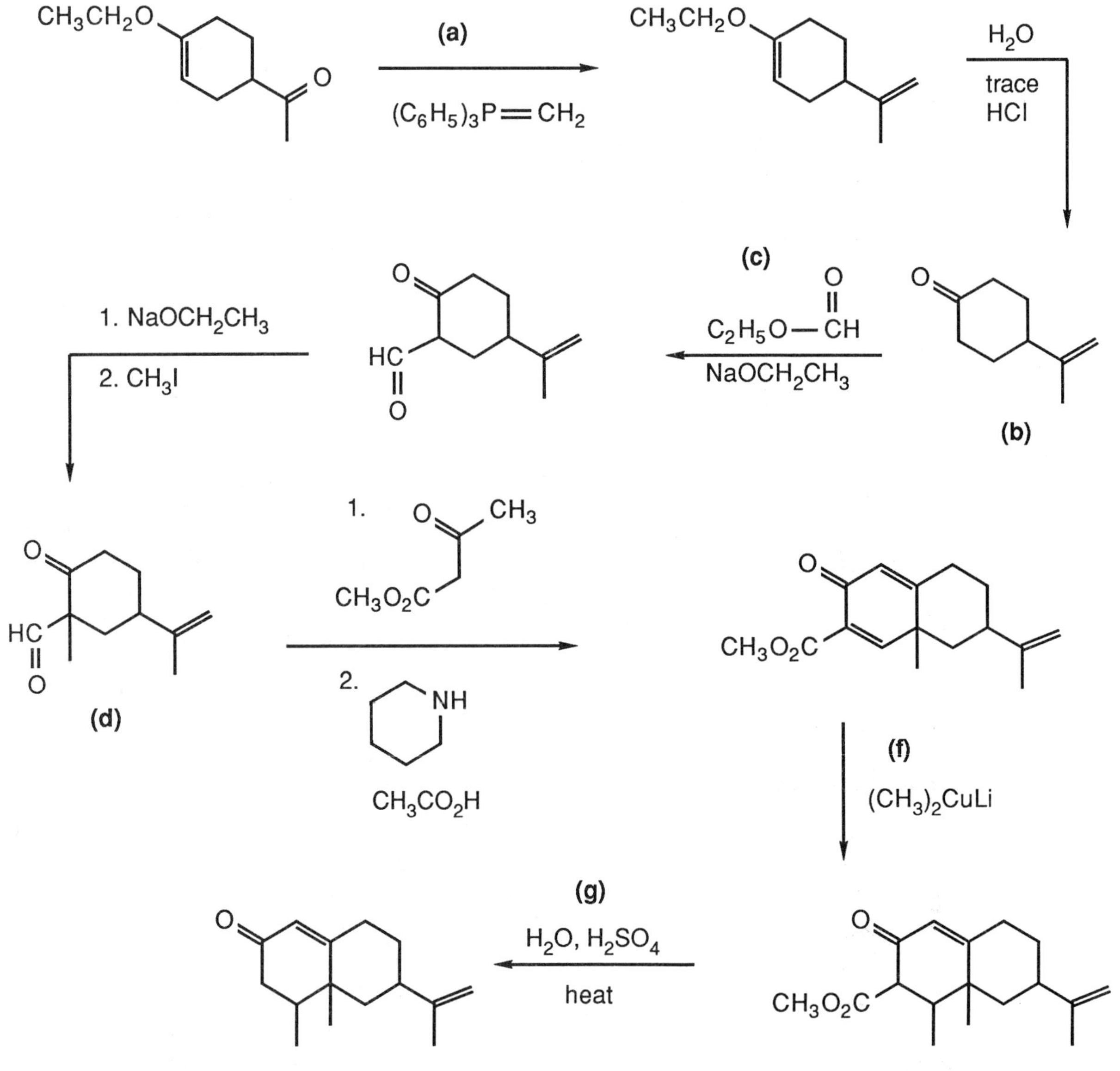

CH_3CH_2O — O

(a)

$(C_6H_5)_3P{=}CH_2$

CH_3CH_2O

H_2O

trace HCl

(c)

O

C_2H_5O-CH

$NaOCH_2CH_3$

O

(b)

1. $NaOCH_2CH_3$

2. CH_3I

O

HC

O

(d)

1. O CH_3

CH_3O_2C

2. NH

CH_3CO_2H

O

CH_3O_2C

(f)

$(CH_3)_2CuLi$

(g)

H_2O, H_2SO_4

heat

O

O

CH_3O_2C

23.3

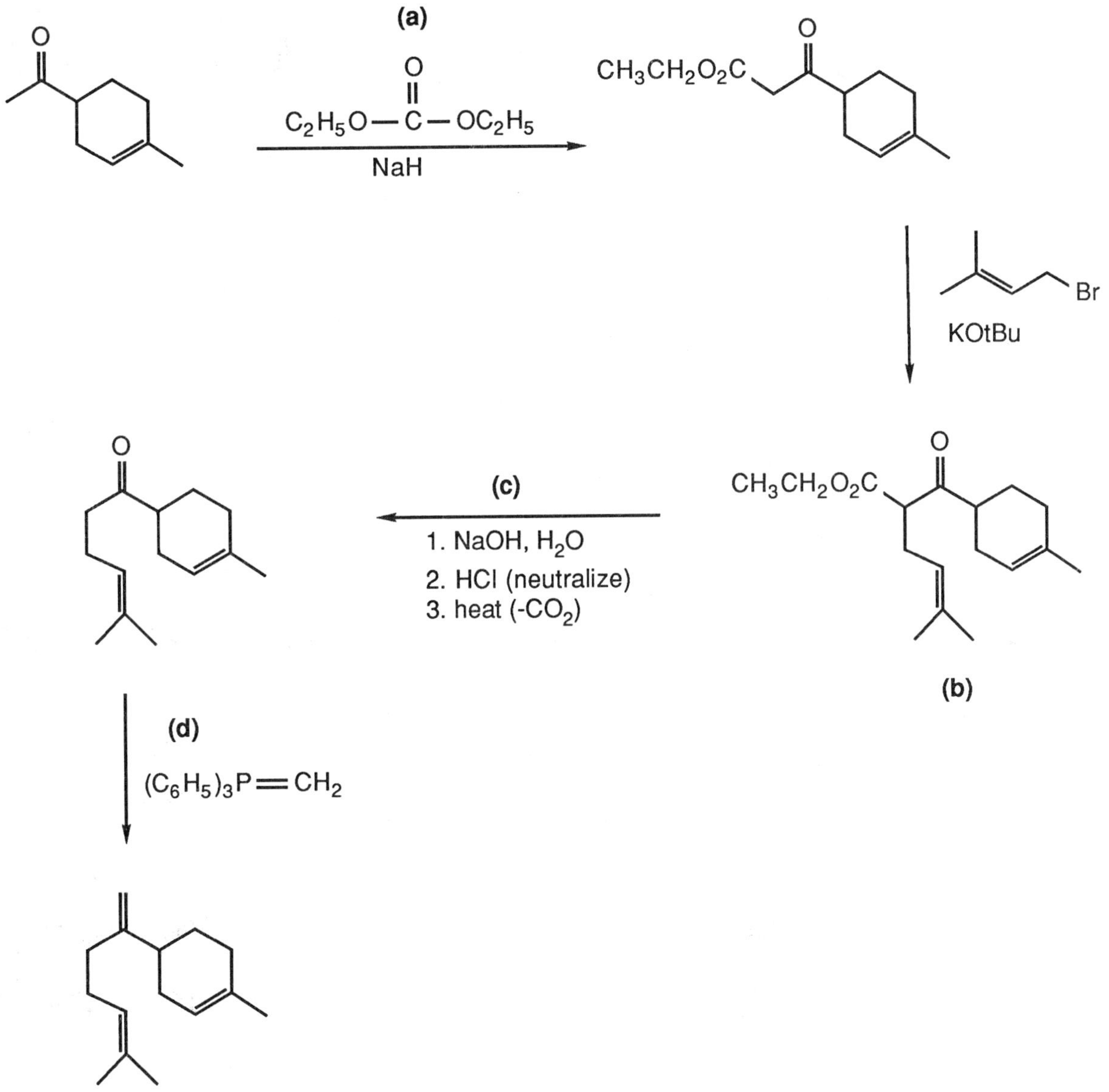

(a)

$C_2H_5O-C(=O)-OC_2H_5$

NaH

$CH_3CH_2O_2C$

Br

KOtBu

$CH_3CH_2O_2C$

(b)

(c)

1. NaOH, H_2O
2. HCl (neutralize)
3. heat ($-CO_2$)

(d)

$(C_6H_5)_3P{=}CH_2$

23.4

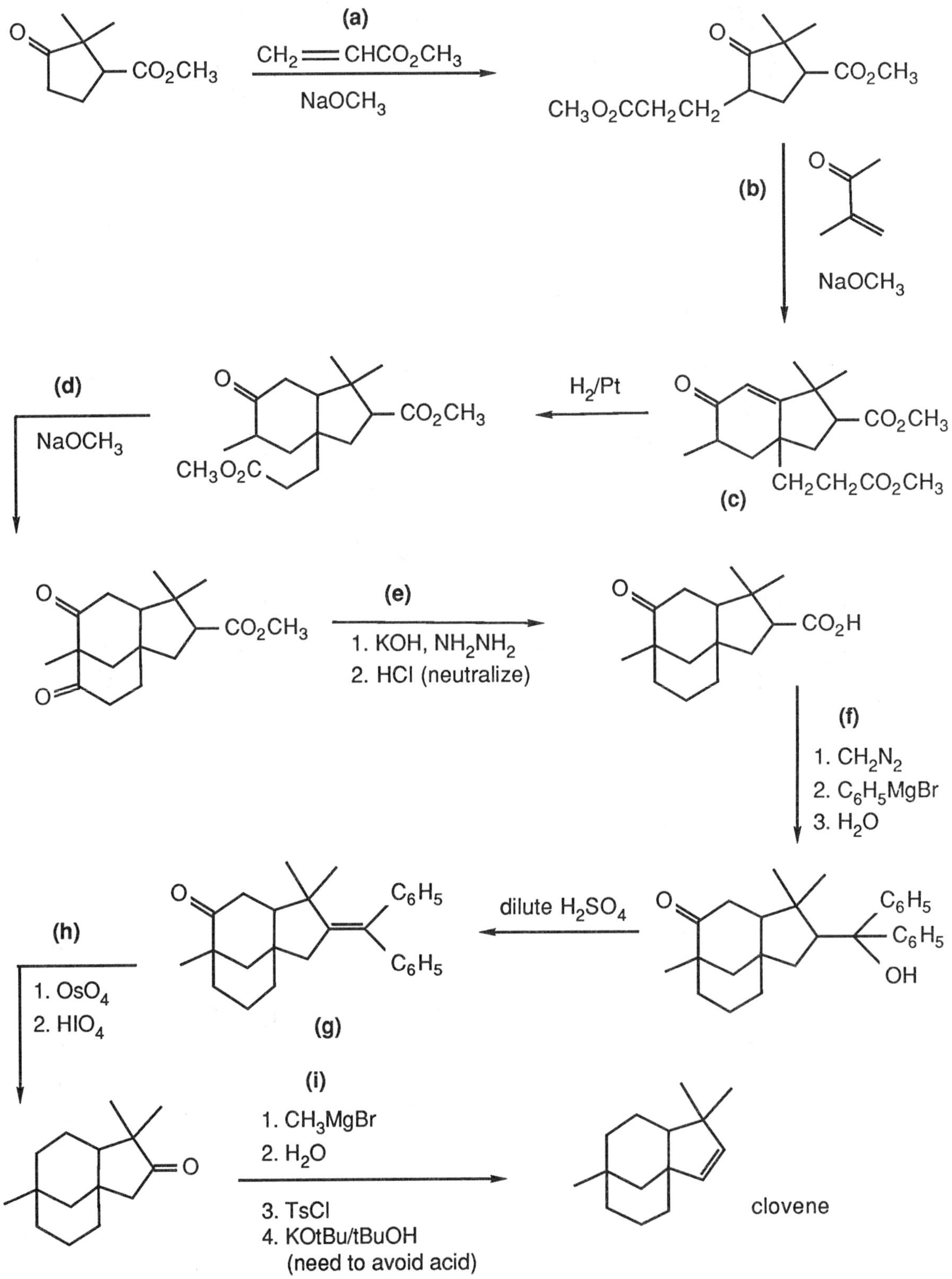

CO_2CH_3

(a)
$CH_2{=}CHCO_2CH_3$
$NaOCH_3$

CO_2CH_3
$CH_3O_2CCH_2CH_2$

(b)
$NaOCH_3$

CO_2CH_3
$CH_2CH_2CO_2CH_3$
(c)

H_2/Pt

CO_2CH_3
CH_3O_2C

(d)
$NaOCH_3$

CO_2CH_3

(e)
1. KOH, NH_2NH_2
2. HCl (neutralize)

CO_2H

(f)
1. CH_2N_2
2. C_6H_5MgBr
3. H_2O

C_6H_5
C_6H_5
OH

dilute H_2SO_4

C_6H_5
C_6H_5
(g)

(h)
1. OsO_4
2. HIO_4

(i)
1. CH_3MgBr
2. H_2O
3. TsCl
4. KOtBu/tBuOH
(need to avoid acid)

clovene

23.5

CHO

HBr

CHO

Br

(a)

(b)

KOtBu
tBuOH

CHO

(c)

$(C_6H_5)_3P{=}CH_2$

(e)

1. $LiAlH_4$, then H_2O
2. CH_3SO_2Cl
3. $LiAlH_4$

CHO

(d)

CHO

cold, alkaline
$KMnO_4$

HO OH

(f)

(g)

$C_6H_5SO_2Cl$

HO $OSO_2C_6H_5$

mild
heating

O

(h)

(i)

$(C_6H_5)_3P{=}CH_2$

aromadendrene

23.6

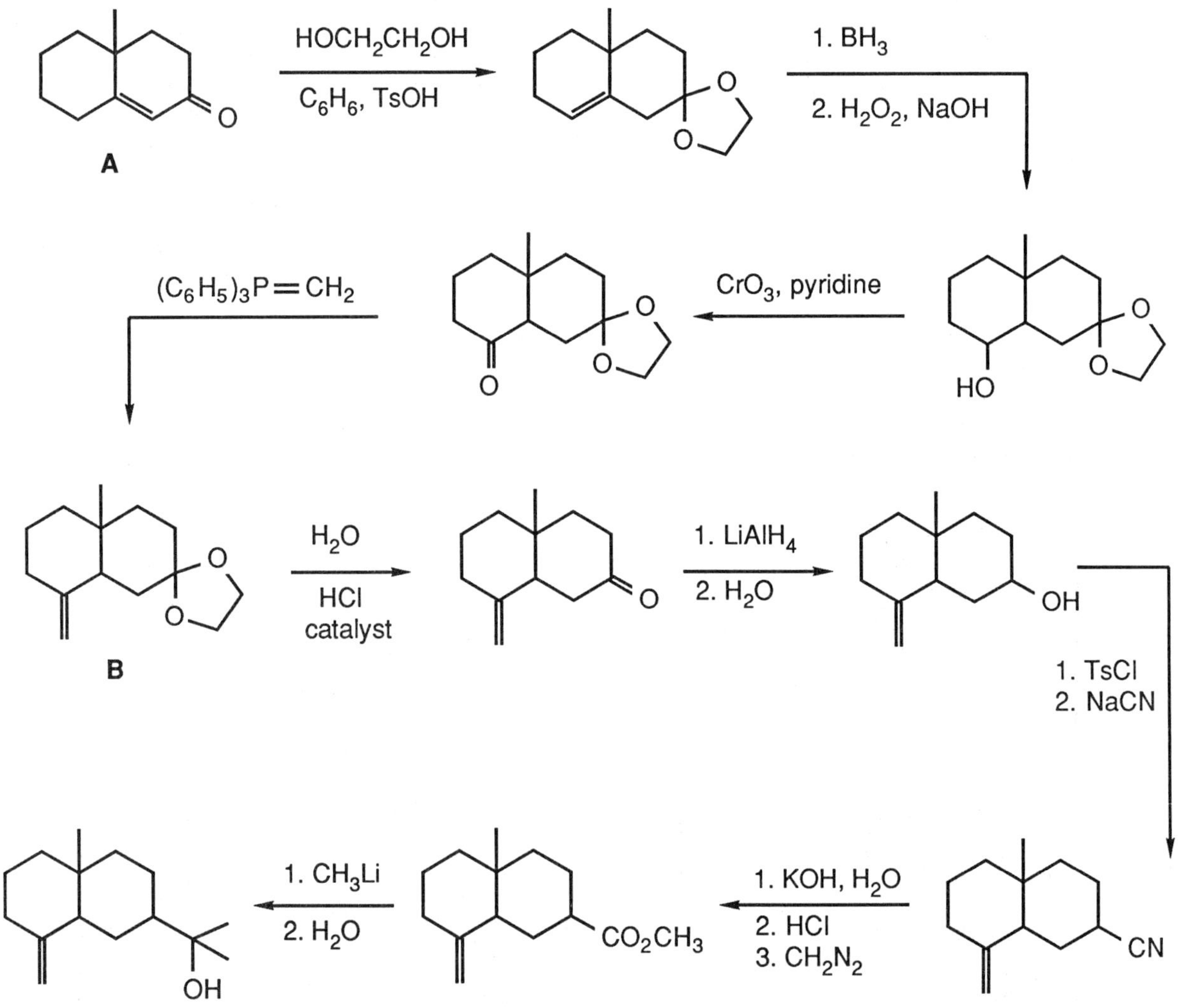

A

$HOCH_2CH_2OH$

C_6H_6, TsOH

1. BH_3

2. H_2O_2, NaOH

$(C_6H_5)_3P{=}CH_2$

CrO_3, pyridine

HO

B

H_2O

HCl catalyst

1. $LiAlH_4$

2. H_2O

OH

1. TsCl

2. NaCN

1. CH_3Li

2. H_2O

OH

CO_2CH_3

1. KOH, H_2O

2. HCl

3. CH_2N_2

CN

23.7

1 → (a) 1. $NaOCH_3$ 2. CH_3I → 2

2 → 1. $CH_2{=}CH{-}C(=O){-}CH_3$, $NaOCH_3/CH_3OH$ 2. HCl → (b) 3

3 → (c) $NaBH_4$ (reduction of non-conjugated ketone) → 4

4 → (d) dihydropyran, TsOH → 5

5 → (e) 1. $NaNH_2$ 2. $CH_3CH_2C(OCH_2CH_2O)CH_2CH_2Br$ → 6

6 → H_2, Pd → 7

7 → (f) dilute HCl, reflux → (g) 8

8 → (h) Li, NH_3 → 9

9 → $HOCH_2CH_2OH$, TsOH, C_6H_6 →

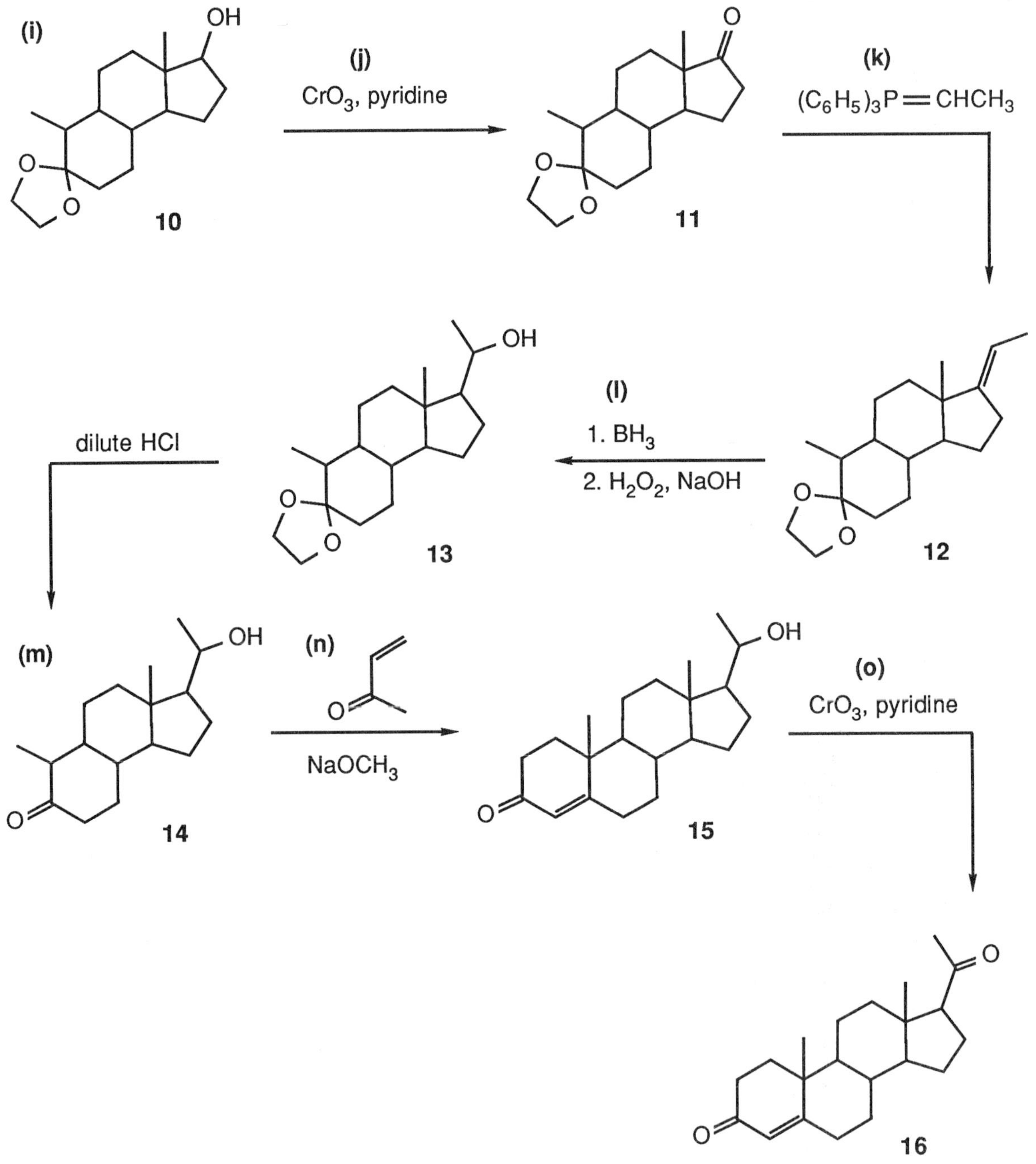

(i) OH 10

(j) CrO_3, pyridine

O 11

(k) $(C_6H_5)_3P{=}CHCH_3$

12

(l) 1. BH_3 2. H_2O_2, NaOH

OH 13

dilute HCl

(m) OH 14

(n) $NaOCH_3$

OH 15

(o) CrO_3, pyridine

O 16

23.8

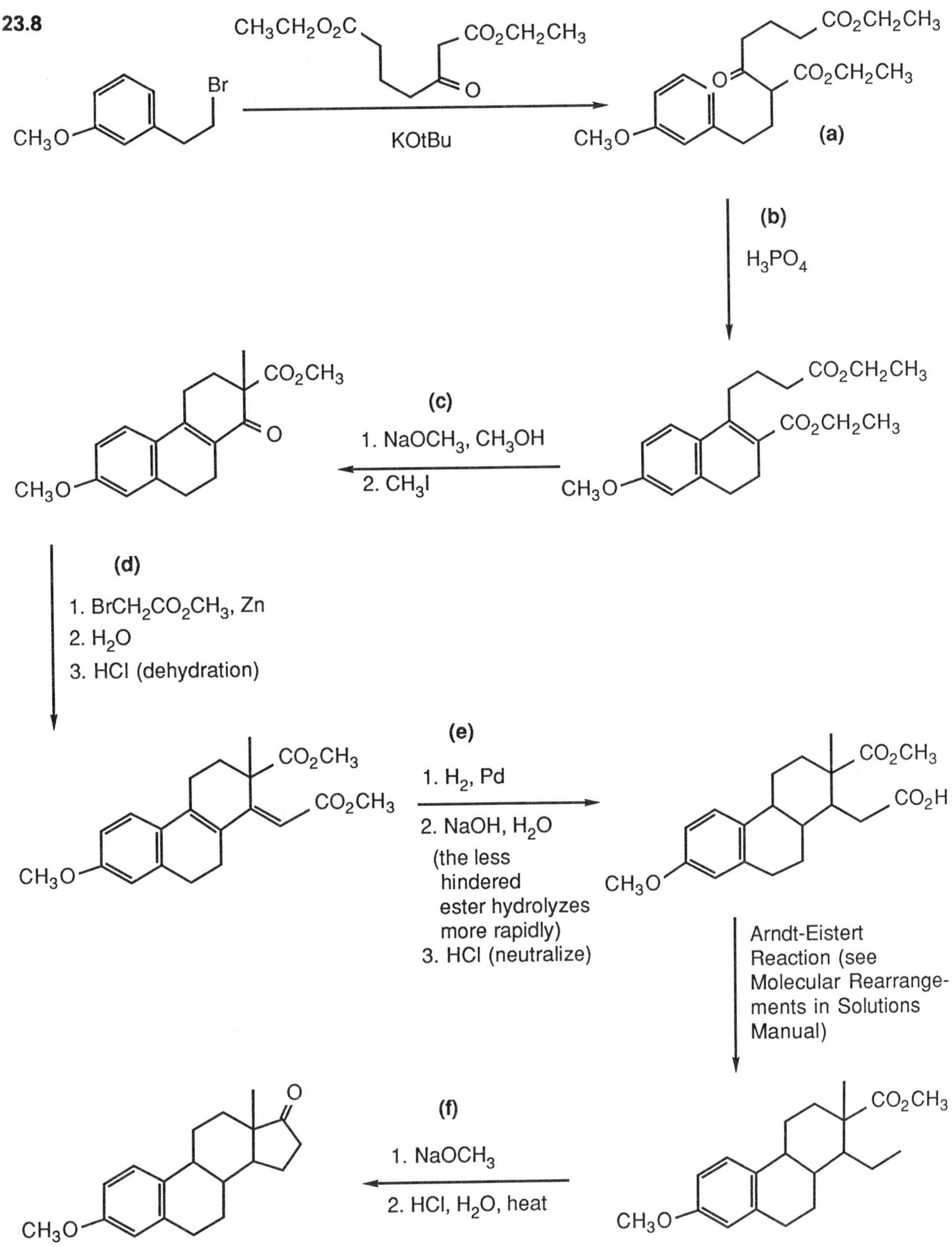

CH_3O–C$_6$H$_4$–CH_2CH_2Br + $CH_3CH_2O_2C$ ... $CO_2CH_2CH_3$ (keto diester, =O) → (KOtBu) → (a)

$CO_2CH_2CH_3$
$CO_2CH_2CH_3$
CH_3O

(b) H_3PO_4

$CO_2CH_2CH_3$
$CO_2CH_2CH_3$
CH_3O

(c)
1. $NaOCH_3$, CH_3OH
2. CH_3I

CO_2CH_3
O
CH_3O

(d)
1. $BrCH_2CO_2CH_3$, Zn
2. H_2O
3. HCl (dehydration)

CO_2CH_3
CO_2CH_3
CH_3O

(e)
1. H_2, Pd
2. NaOH, H_2O
(the less hindered ester hydrolyzes more rapidly)
3. HCl (neutralize)

CO_2CH_3
CO_2H
CH_3O

Arndt-Eistert Reaction (see Molecular Rearrangements in Solutions Manual)

CO_2CH_3
CH_3O

(f)
1. $NaOCH_3$
2. HCl, H_2O, heat

O
CH_3O

23.9

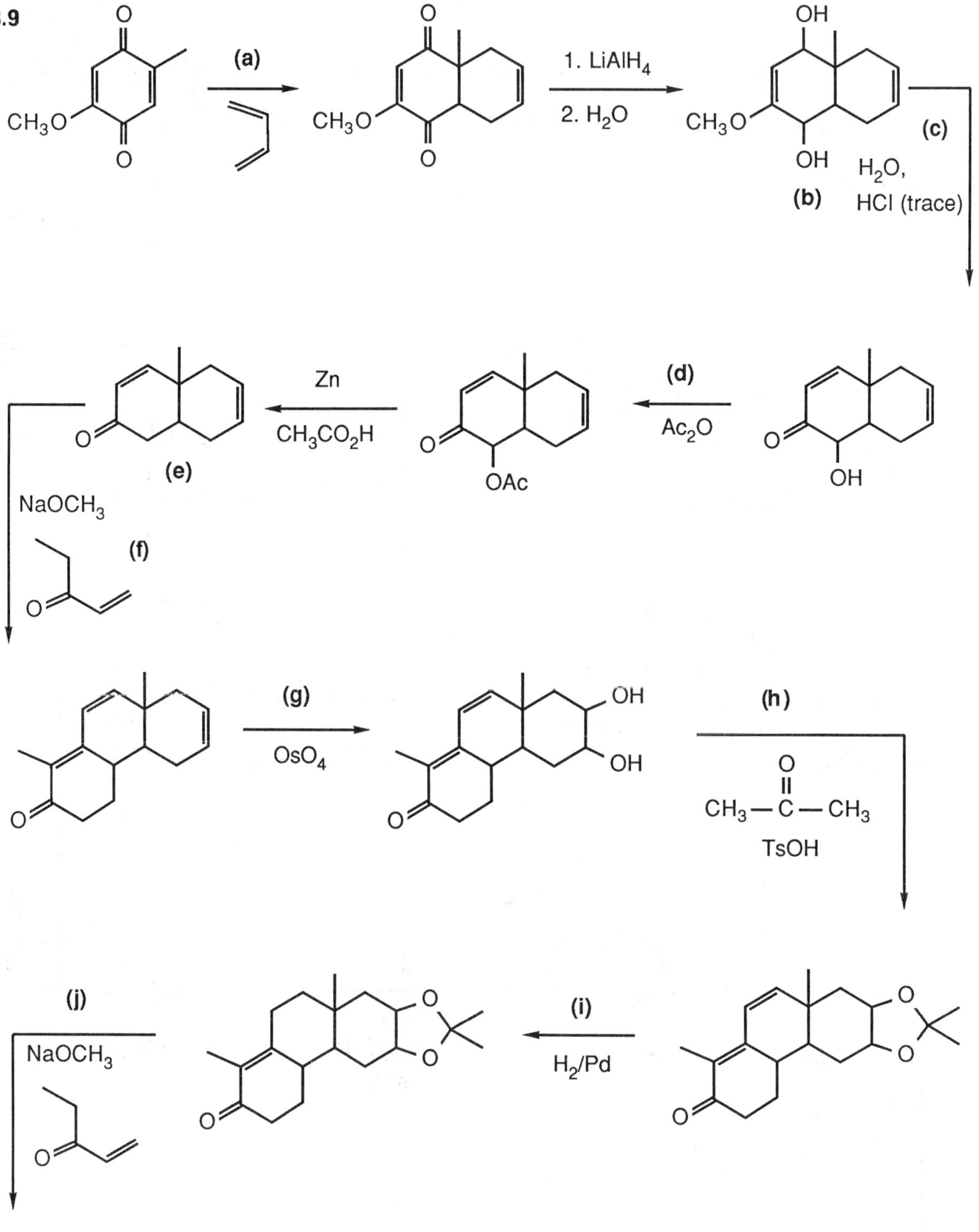

(a)

1. $LiAlH_4$
2. H_2O

CH_3O CH_3O CH_3O

OH OH (b)

(c) H_2O, HCl (trace)

(d) Ac_2O

OH

OAc

Zn CH_3CO_2H

(e)

$NaOCH_3$

(f)

(g) OsO_4

OH OH

(h) $CH_3-\overset{O}{\overset{\|}{C}}-CH_3$ TsOH

(i) H_2/Pd

(j) $NaOCH_3$

1. dilute aqueous acid
2. HIO_4

CHO
CHO

(k)

(l)
$NaOCH_3$

CHO

(m)
1. CrO_3
2. CH_2N_2

CO_2CH_3

(n)
H_2/Pt

CO_2CH_3

1. $NaBH_4$
2. KOH, H_2O

CO_2H

HO

(o)

(p)
Ac_2O

CO_2H

AcO

(q)
1. $SOCl_2$
2. $(CH_3)_2CuLi$, then H_2O

O

AcO

(r)
1. $LiCH_2$
2. H_2O

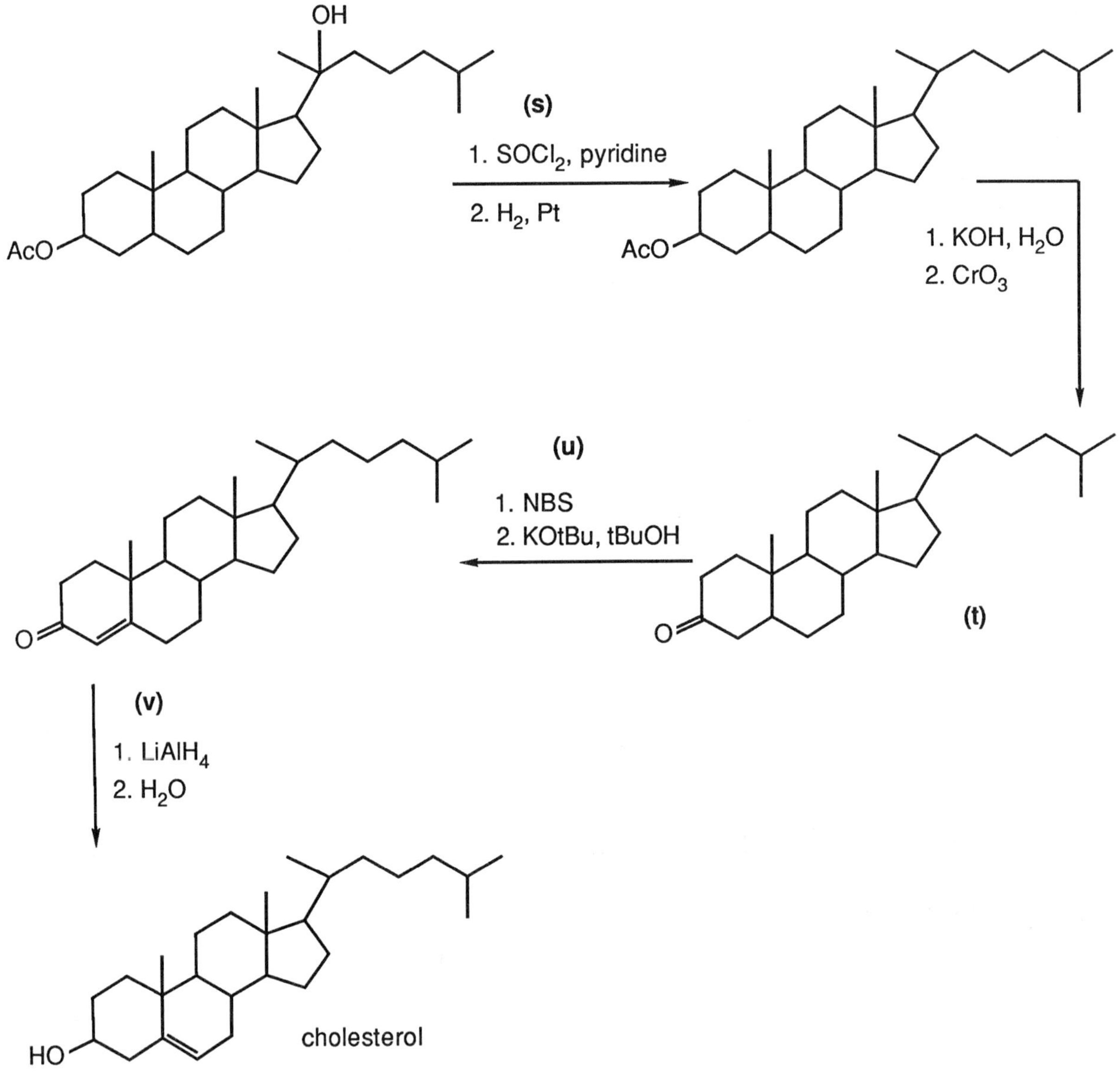

OH

AcO

(s)

1. $SOCl_2$, pyridine

2. H_2, Pt

AcO

1. KOH, H_2O

2. CrO_3

(u)

1. NBS

2. KOtBu, tBuOH

O

O

(t)

(v)

1. $LiAlH_4$

2. H_2O

HO

cholesterol

23.10

(a)

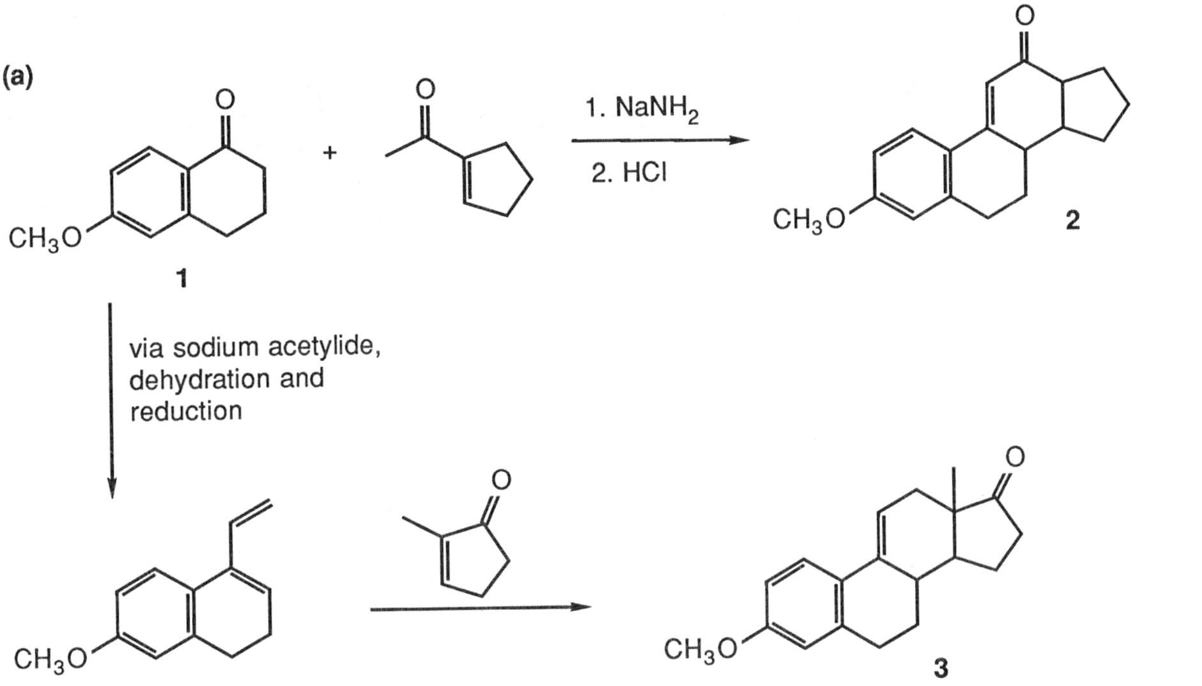
O
CH3O
1
+
O
1. NaNH2
2. HCl
O
CH3O
2
via sodium acetylide,
dehydration and
reduction
CH3O
O
O
CH3O
3

(b)

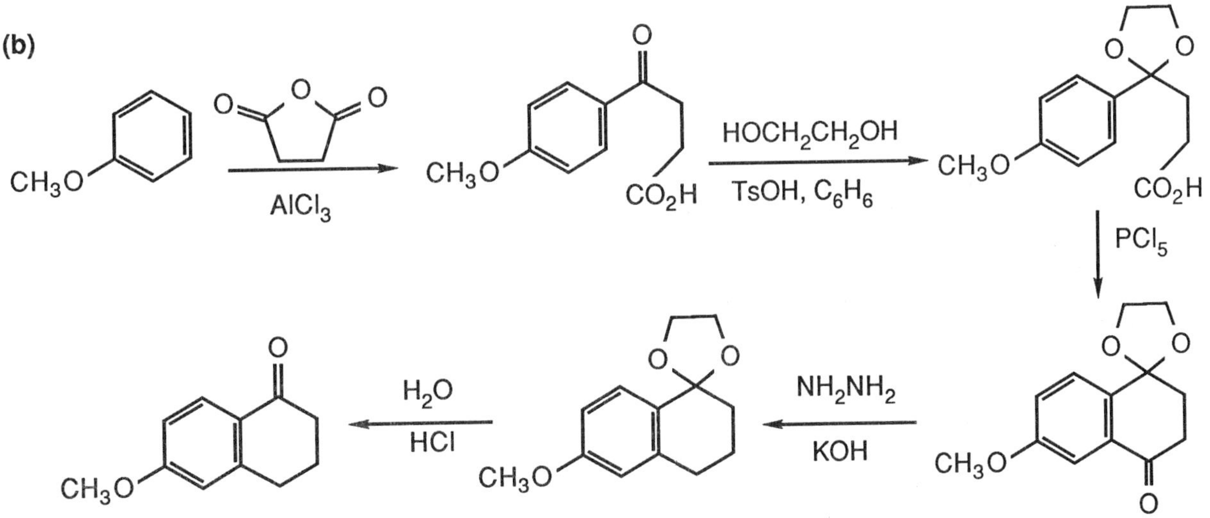
CH3O
O
O
O
AlCl3
O
CH3O
CO2H
HOCH2CH2OH
TsOH, C6H6
O
O
CH3O
CO2H
PCl5
O
O
CH3O
O
NH2NH2
KOH
O
O
CH3O
H2O
HCl
O
CH3O

23.11

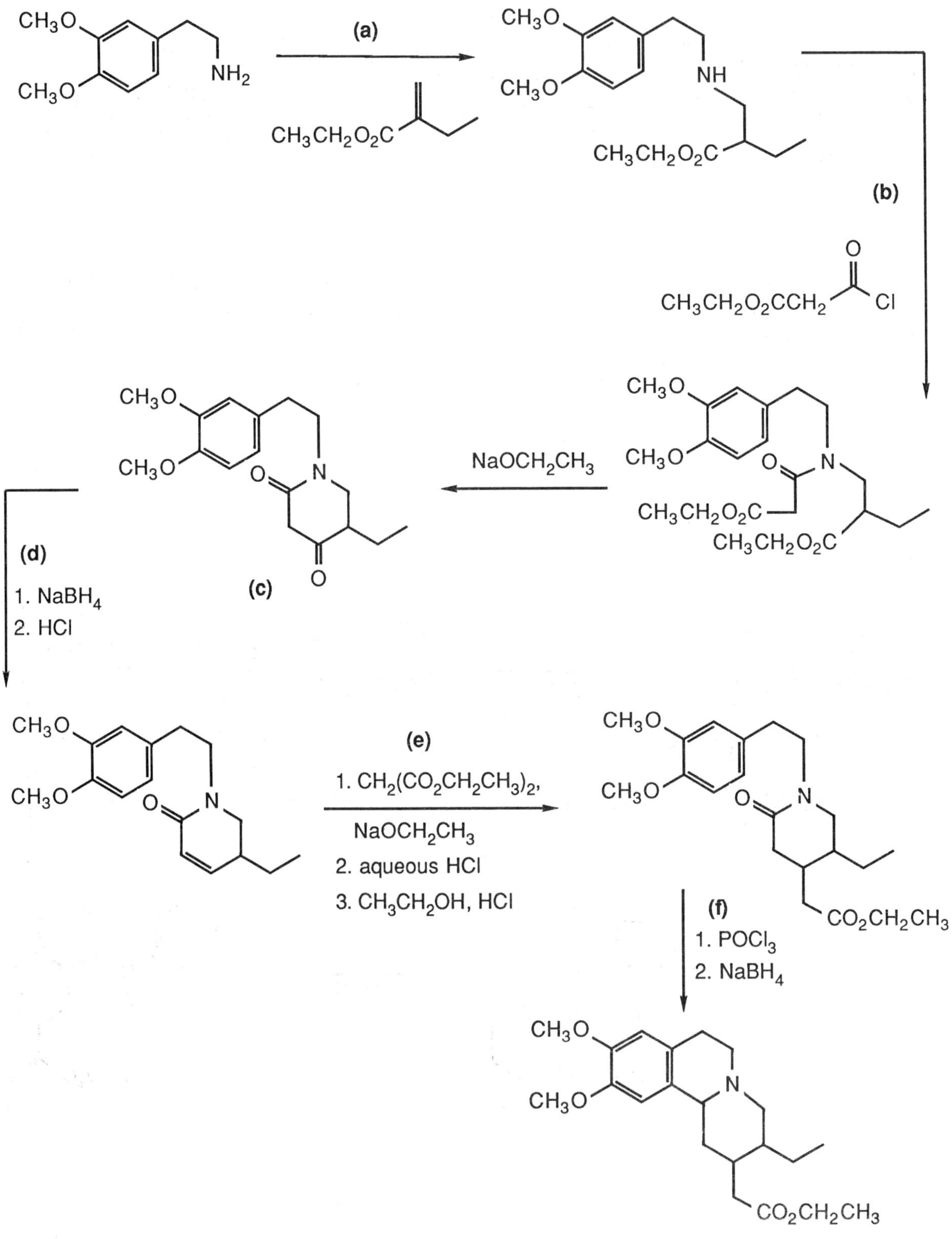
CH3O
CH3O
NH2
(a)
CH3CH2O2C
NH
CH3CH2O2C
(b)
O
CH3CH2O2CCH2
Cl
N
O
CH3CH2O2C
CH3CH2O2C
NaOCH2CH3
O
N
O
(c)
(d)
1. NaBH4
2. HCl
(e)
1. CH2(CO2CH2CH3)2,
NaOCH2CH3
2. aqueous HCl
3. CH3CH2OH, HCl
CO2CH2CH3
(f)
1. POCl3
2. NaBH4
CO2CH2CH3

23.12

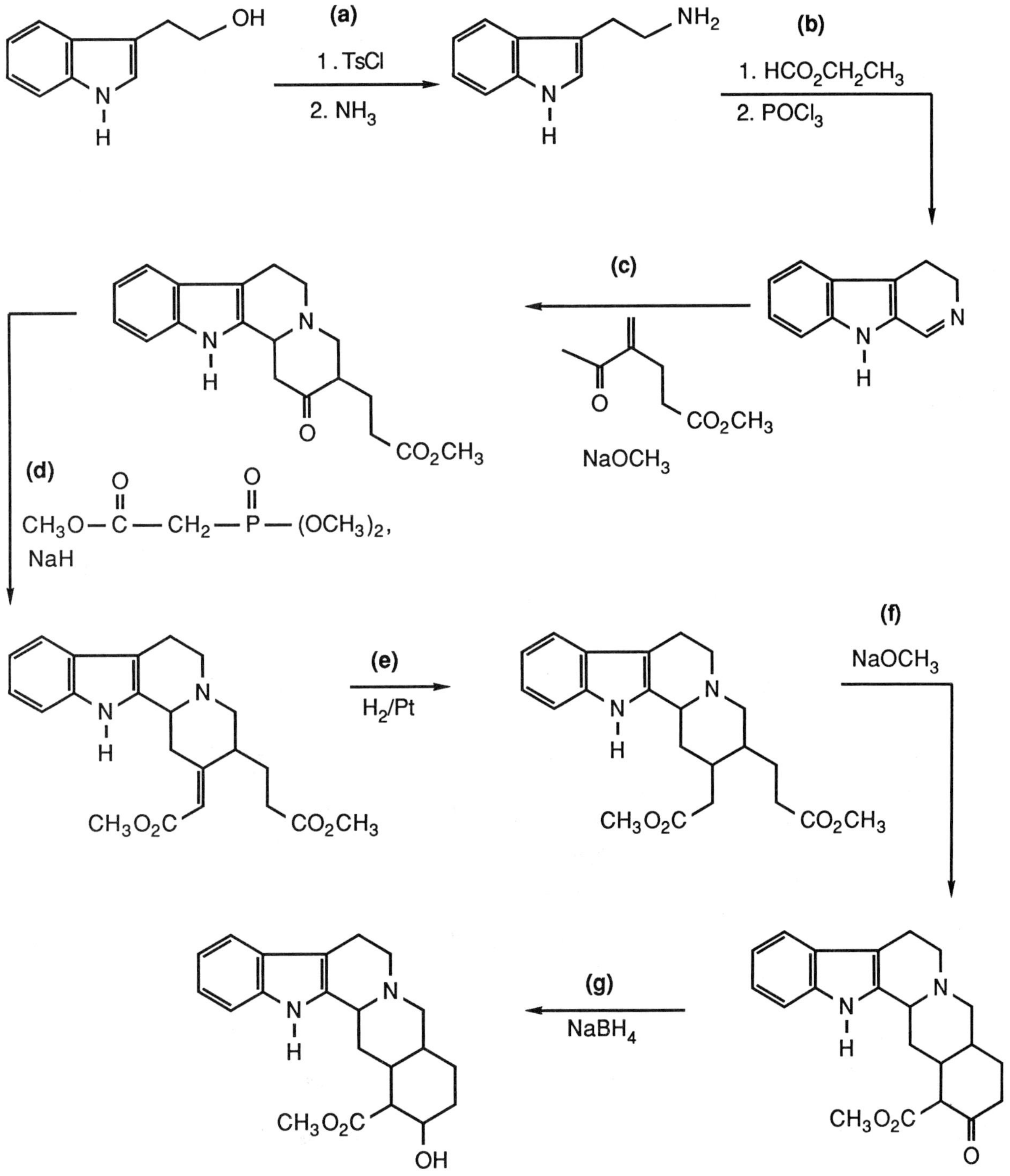

(a) 1. TsCl 2. NH_3

(b) 1. $HCO_2CH_2CH_3$ 2. $POCl_3$

(c) $CH_2=C(COCH_3)CH_2CH_2CO_2CH_3$, $NaOCH_3$

(d) $CH_3O-\overset{O}{\overset{\|}{C}}-CH_2-\overset{O}{\overset{\|}{P}}-(OCH_3)_2$, NaH

(e) H_2/Pt

(f) $NaOCH_3$

(g) $NaBH_4$

CO_2CH_3 CH_3O_2C OH O NH₂ N-H

23.13

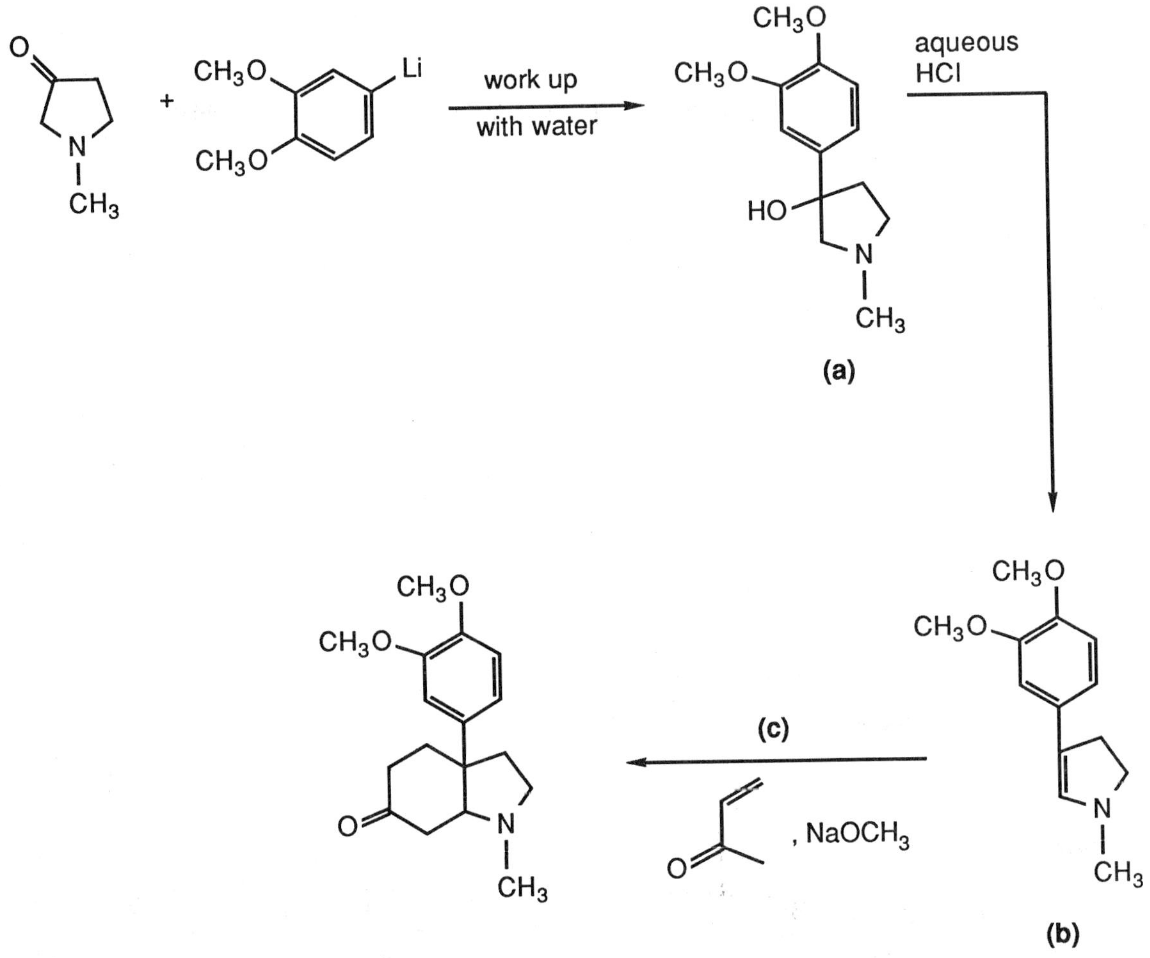

CH_3O | Li | work up with water | CH_3O | CH_3O | HO | N | CH_3 | (a) | aqueous HCl

(b) | CH_3O | CH_3O | N | CH_3

(c) | , $NaOCH_3$ | O | CH_3O | CH_3O | N | CH_3

CHAPTER 24. HETEROCYCLIC COMPOUNDS

ANSWERS TO EXERCISES

24.1

(a) pyridine $\xrightarrow{HCl}$ pyridinium (N^+–H) ^-Cl

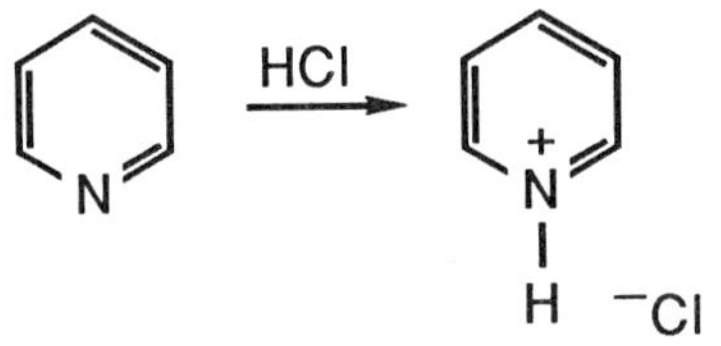

(b) pyridine $\xrightarrow{C_6H_5CH_2Br}$ N-benzylpyridinium (N^+–$CH_2C_6H_5$) ^-Br

(c) pyridine $\xrightarrow{C_6H_5Li}$ 2-C_6H_5-pyridine

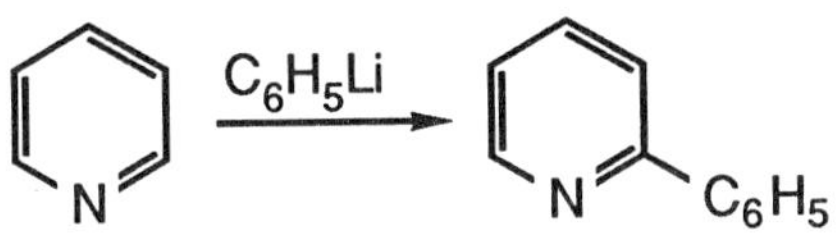

(d) pyridine $\xrightarrow{C_6H_5CO_2H}$ pyridinium (N^+–H) $C_6H_5CO_2^-$

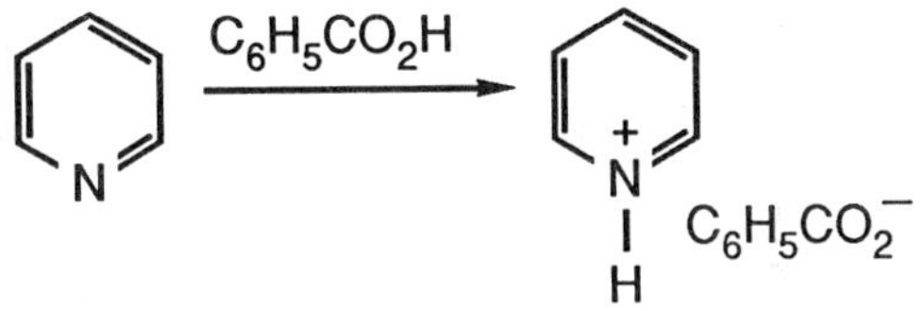

24.2

(a) pyrrole (N–H) $\xrightarrow{Br_2}$ 2-Br-pyrrole (N–H)

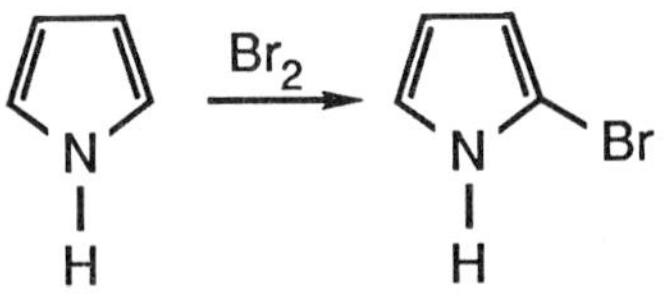

(b) pyrrole (N–H) $\xrightarrow{HNO_3}$ 2-NO_2-pyrrole (N–H)

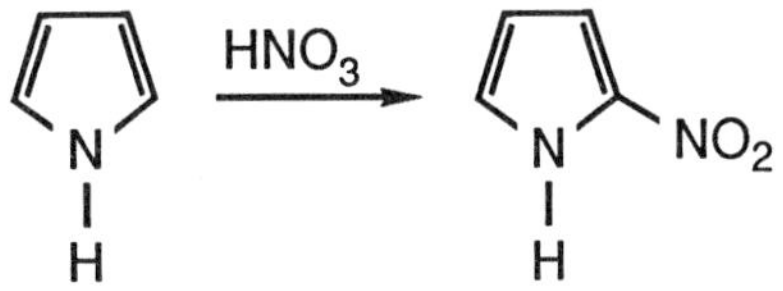

(c) pyrrole (N–H) $\xrightarrow[HCl]{CH_2O}$ 2-CH_2OH-pyrrole (N–H)

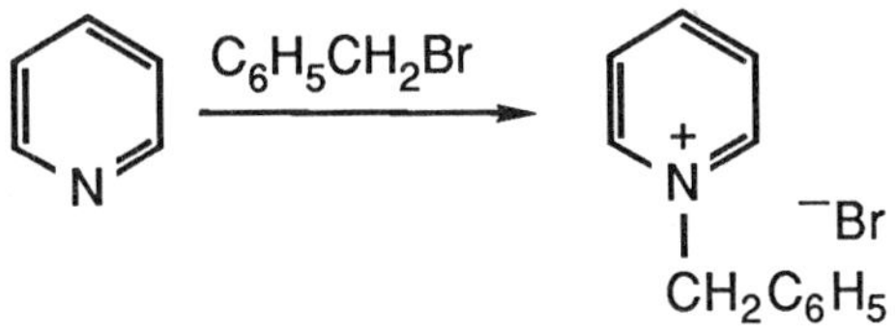

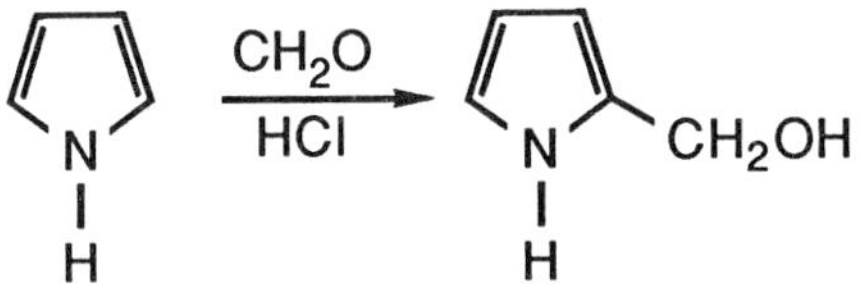

24.3

(a) furan (O) $\xrightarrow{Br_2}$ 2-Br-furan

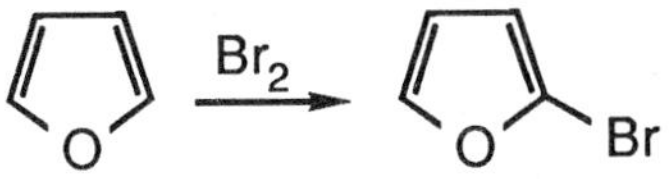

(b) furan (O) $\xrightarrow{HNO_3}$ 2-NO_2-furan

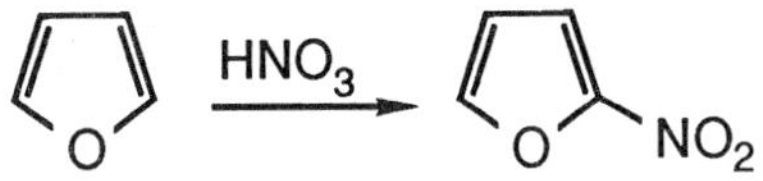

(c) furan (O) $\xrightarrow[HCl]{CH_2O}$ 2-CH_2OH-furan

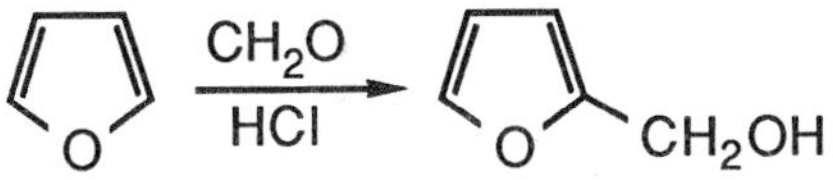

24.4

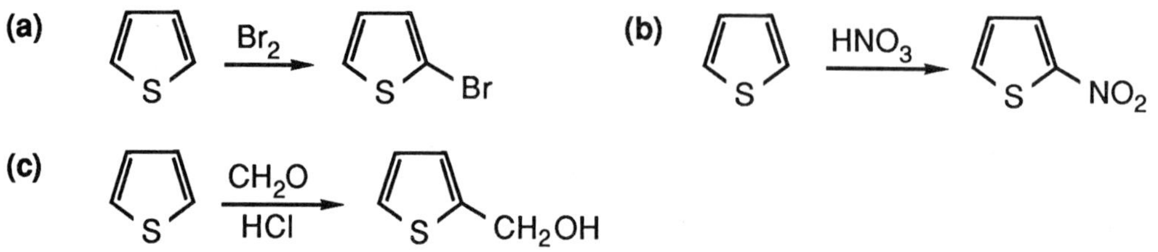

(a) Thiophene $\xrightarrow{Br_2}$ 2-bromothiophene

(b) Thiophene $\xrightarrow{HNO_3}$ 2-nitrothiophene (NO_2)

(c) Thiophene $\xrightarrow[HCl]{CH_2O}$ 2-thiophenemethanol (CH_2OH)

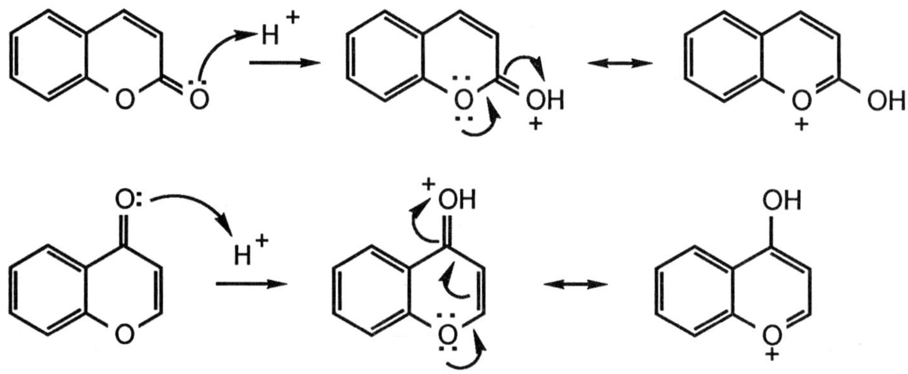

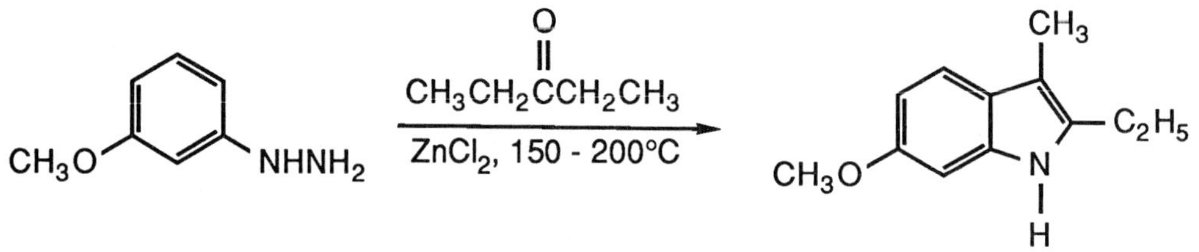

24.5

H^+ ⟶ OH, + ⟷ O, +, OH

O: ⟶ H^+ ⟶ +OH ⟷ OH, O, +

24.6

CH_3O–C_6H_4–$NHNH_2$ $\xrightarrow[ZnCl_2,\ 150 - 200°C]{CH_3CH_2\overset{O}{\overset{\|}{C}}CH_2CH_3}$ CH_3O indole with CH_3, C_2H_5, N–H

24.7 **(a)** Quinoline reacts as follows:

$^{+}NO_2$ H NO_2 H NO_2 H NO_2 N N N + + + N

H NO_2 H NO_2 N + N +

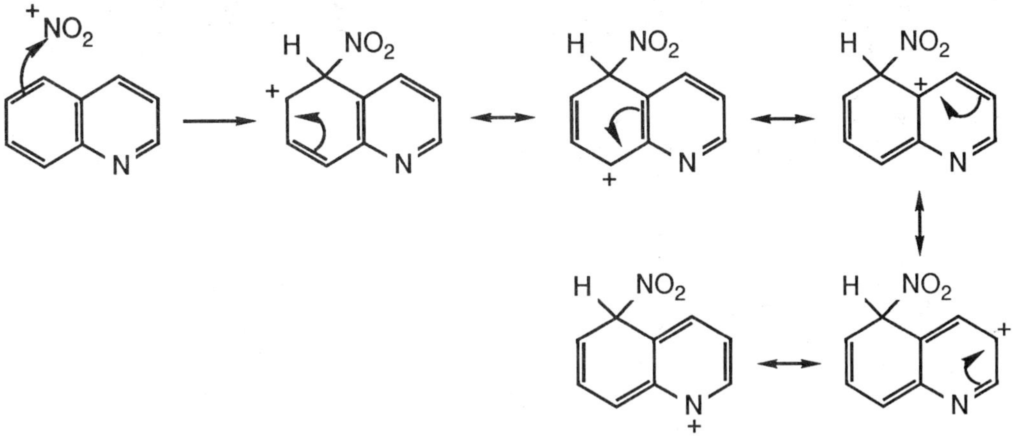

Positive charge is placed on both carbon and nitrogen. Notice that the resonance form with a positively charged nitrogen is not an onium ion but is electron deficient.

Isoquinoline reacts as follows:

$^{+}NO_2$ H NO_2 H NO_2 H NO_2 N N N + + + N

H NO_2 H NO_2 N + N +

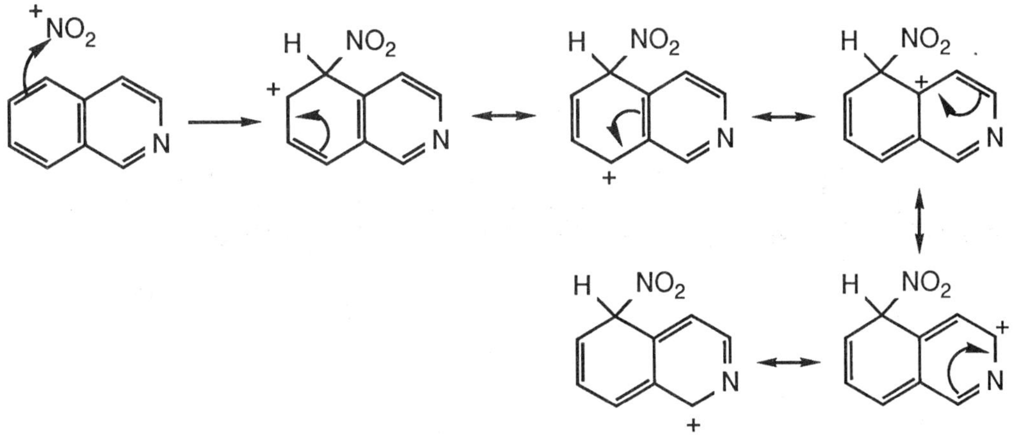

Isoquinoline yields an intermediate with resonance forms analogous to those drawn for reaction at the same site in quinoline. None of these resonance forms have positive charge on nitrogen, so isoquinoline is more reactive than quinoline.

24.8 (a) O $\xrightarrow[\text{H}_2\text{SO}_4 \text{ catalyst}]{\text{CH}_3\text{OH}}$ $CH_3OCH_2CH_2OH$

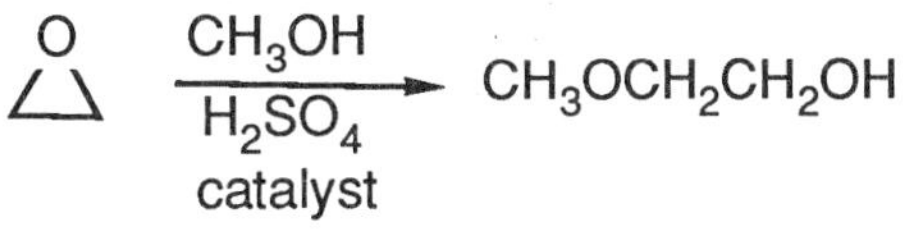

(b) H–N $\xrightarrow[\text{H}_2\text{SO}_4 \text{ catalyst}]{\text{C}_2\text{H}_5\text{OH}}$ $CH_3CH_2OCH_2CH_2NH_2$

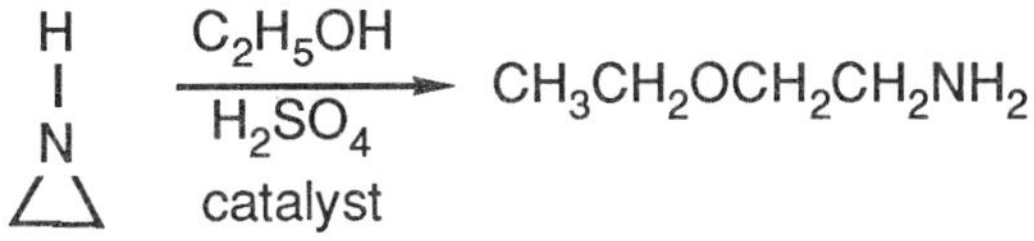

(c)

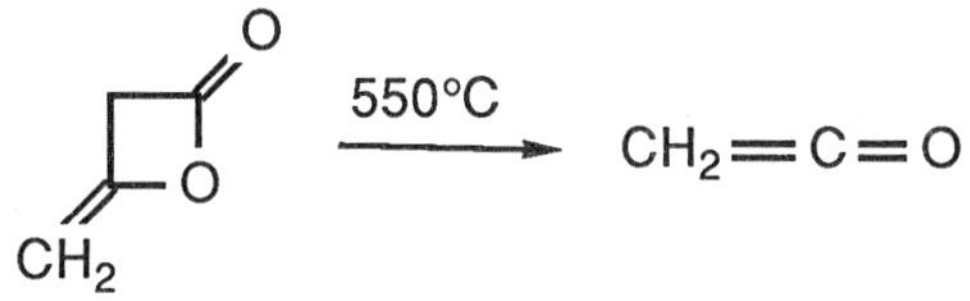

$$\text{(4-methylene-2-oxetanone, } CH_2\text{)} \xrightarrow{550°C} CH_2{=}C{=}O$$

(d)

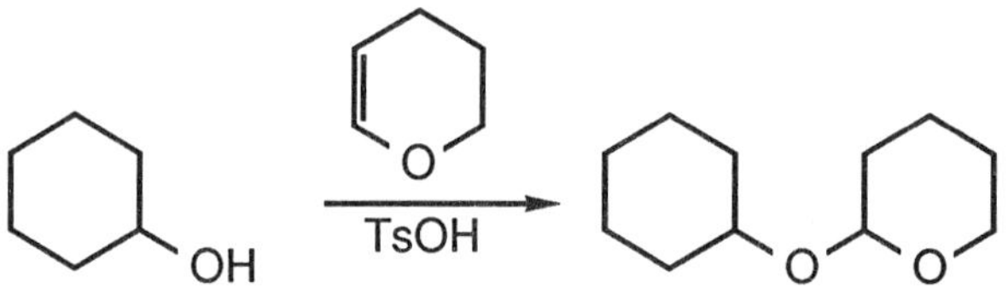

$$\text{cyclohexanol (OH)} \xrightarrow[\text{TsOH}]{\text{dihydropyran}} \text{cyclohexyl–O–tetrahydropyranyl ether}$$

(e)

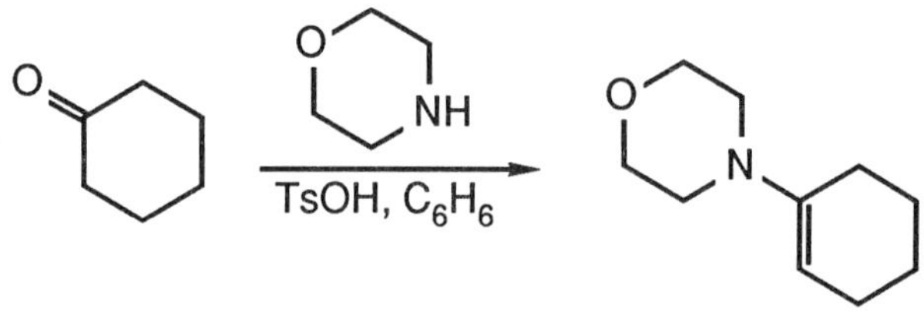

$$\text{cyclohexanone} \xrightarrow[\text{TsOH, } C_6H_6]{\text{morpholine (O, NH)}} \text{N-(1-cyclohexenyl)morpholine}$$

24.9

(a) $C_6H_5CH{=}CH_2 \xrightarrow{Cl{-}C_6H_4{-}CO_3H} C_6H_5{-}CH{-}CH_2$ (epoxide, O bridging CH and CH_2)

(b) $C_6H_5CH{=}CH_2 \xrightarrow{Cl_2} C_6H_5{-}CH(Cl){-}CH_2Cl \xrightarrow{NH_3} C_6H_5{-}CH{-}CH_2$ (aziridine, N–H bridging CH and CH_2)

(c)

$$\text{cyclohexanone} \xrightarrow[\text{TsOH, } C_6H_6]{HOCH_2CH_2OH} \text{cyclohexanone ethylene ketal (spiro-1,3-dioxolane)}$$

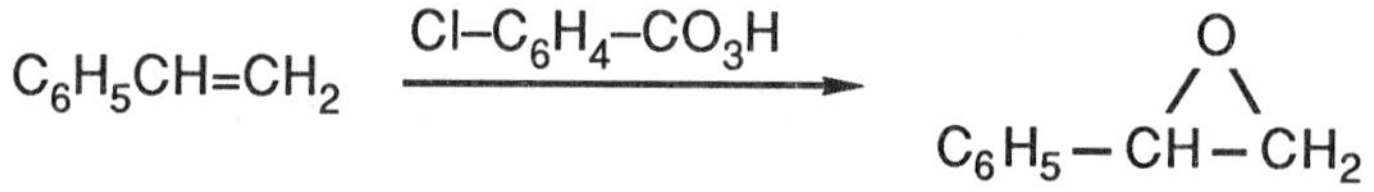

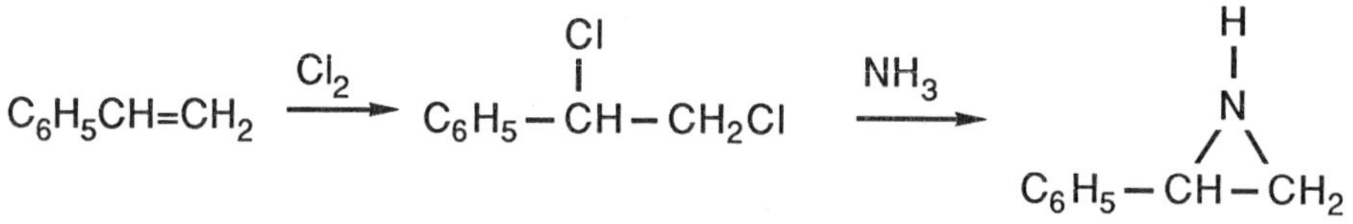

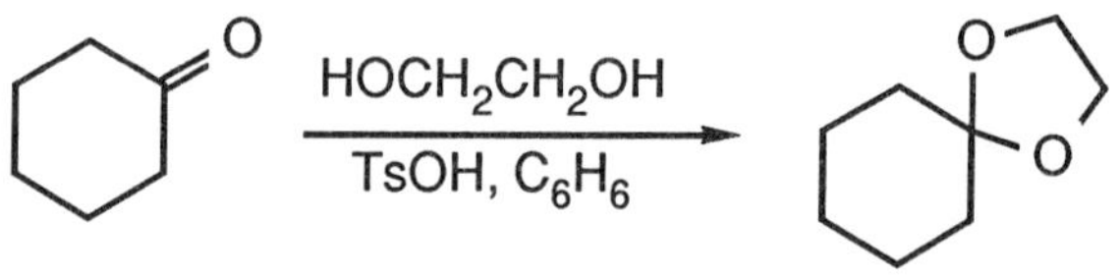

(d) $ClCH_2CH_2{-}CH(CH_2Cl){-}C_6H_5 \xrightarrow{NH_3}$ 3-phenylpyrrolidine (C_6H_5, N–H)

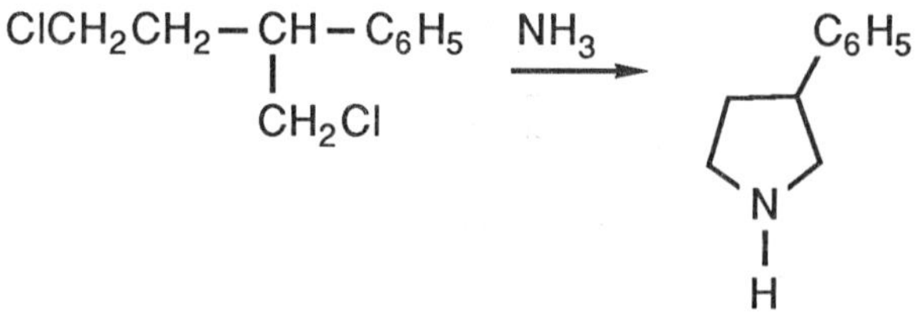

(e) $$C_6H_5CHO \xrightarrow[\text{TsOH, } C_6H_6]{HOCH_2CH_2CH_2OH} \text{2-phenyl-1,3-dioxane}$$

C_6H_5

O O

24.10 **Guanine**

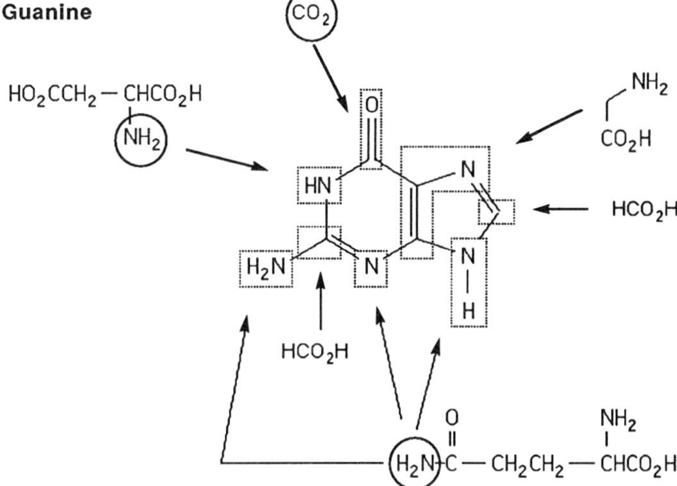

Cytosine

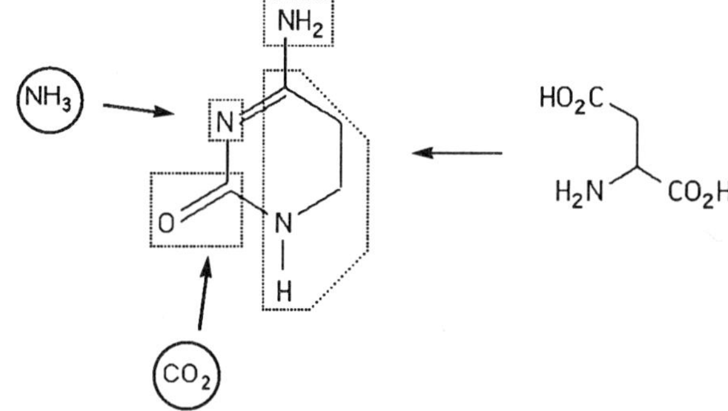

Thymine

methylenetetrahydrofolic acid

O

NH_3 HN CH_3 HO_2C

O N H_2N CO_2H

H

CO_2

24.11 Both have acidic NH groups that are activated by two carbonyls:

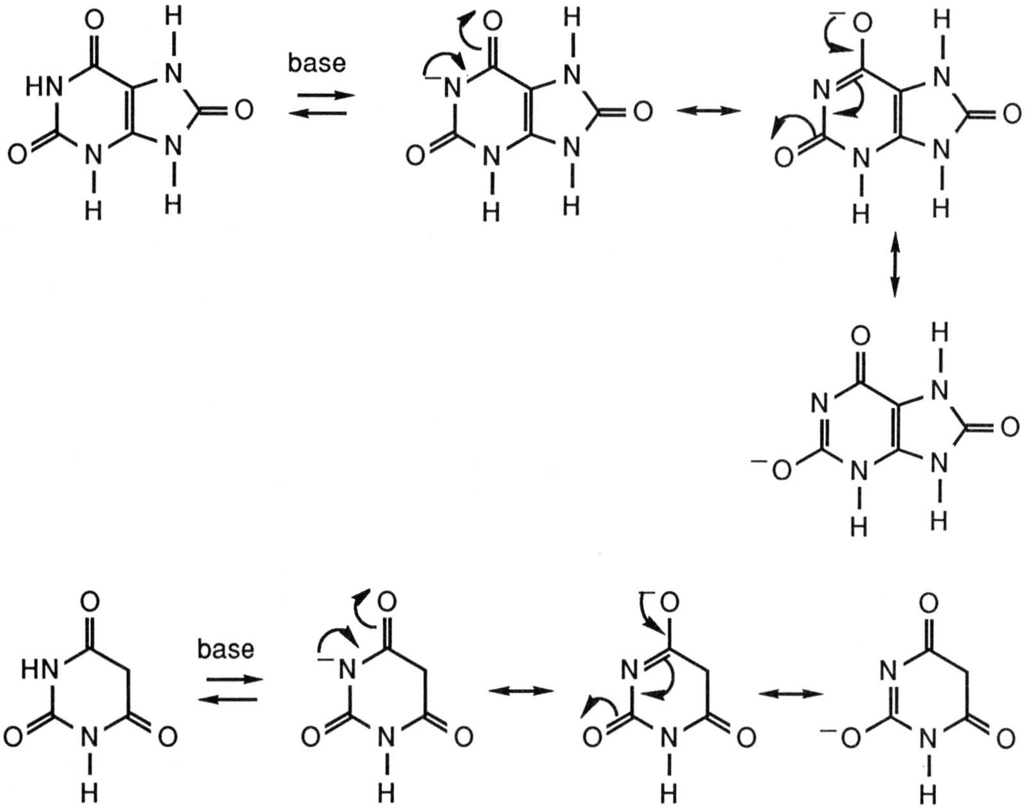

base

24.12

CO_2H / CO_2H —(C_2H_5OH, H_2SO_4)→ $CO_2C_2H_5$ / $CO_2C_2H_5$ —(1. $NaOC_2H_5$ 2. C_2H_5I)→ $C_2H_5CH(CO_2C_2H_5)_2$

—(1. $NaOC_2H_5$ 2. C_2H_5I)→ $(C_2H_5)_2C(CO_2C_2H_5)_2$

—(1. $NaOC_2H_5$, H_2NCNH_2 (O) 2. HCl)→ 5,5-diethylbarbituric acid (C_2H_5, C_2H_5, NH, O)

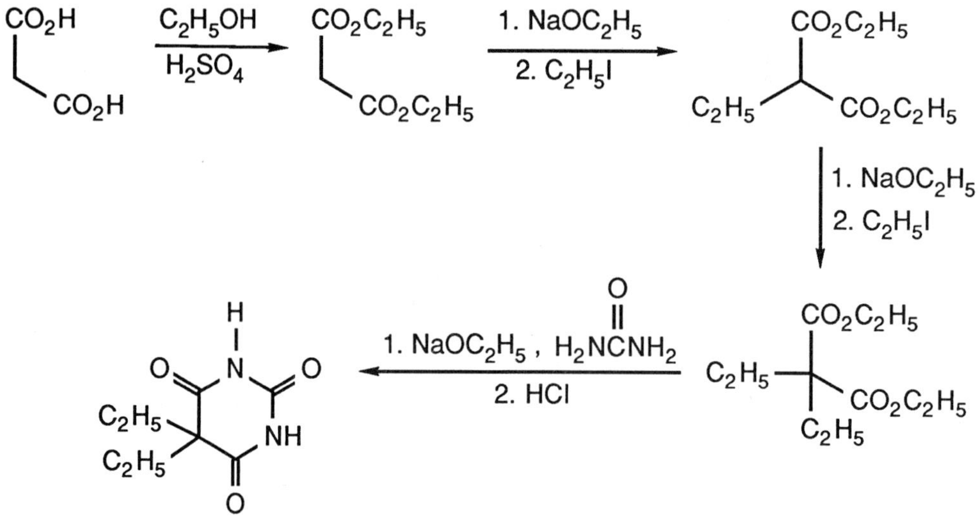

24.13

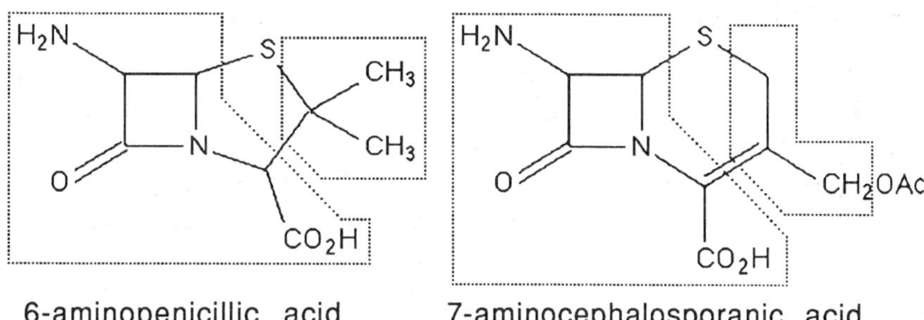

6-aminopenicillic acid 7-aminocephalosporanic acid

24.14 Cephalosporin C contains three linkages that are susceptible to hydrolytic cleavage, two amides and an ester. The most reactive is the β-lactam, and this is followed by the acetate and then by the amide of the side chain:

$HO_2CCH(NH_2)CH_2CH_2CH_2C(=O)—NH$– (β-lactam, S, N, CO_2H, $CH_2O—C(=O)CH_3$)

$\xrightarrow{H_2O\ (acid\ or\ base)}$

$HO_2CCH(NH_2)CH_2CH_2CH_2C(=O)—NH$– ($HO_2C$ HN, S, CO_2H, $CH_2O—C(=O)CH_3$)

$\longrightarrow$

$HO_2CCH(NH_2)CH_2CH_2CH_2C(=O)—NH$– ($HO_2C$ HN, S, CO_2H, CH_2OH)

$\longrightarrow$

H_2N– (HO_2C HN, S, CO_2H, CH_2OH)

ANSWERS TO PROBLEMS

24.1

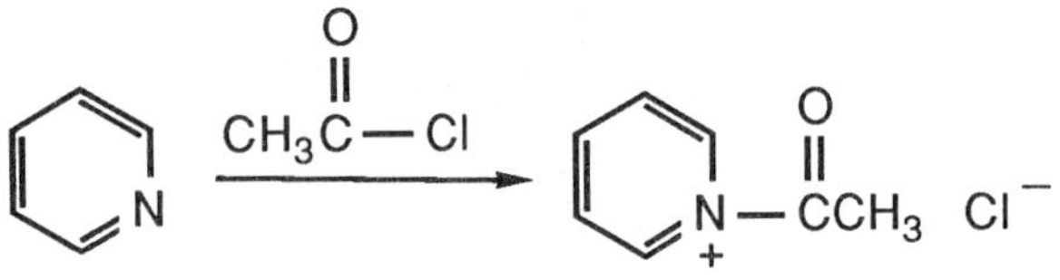

$CH_3\overset{O}{\overset{||}{C}}-Cl$

$N-\overset{O}{\overset{||}{C}}CH_3$ Cl^-

24.2

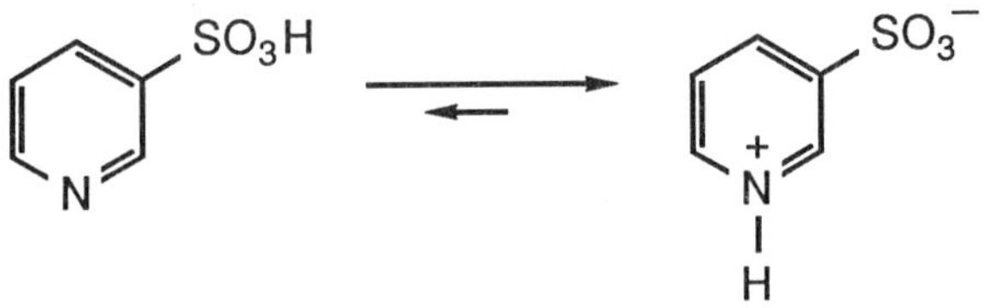

SO_3H SO_3^-

N H

24.3

E^+ E H E H E H E

destabilizing interaction

E^+ H E H E H E

E^+ H E H E H E

destabilizing interaction

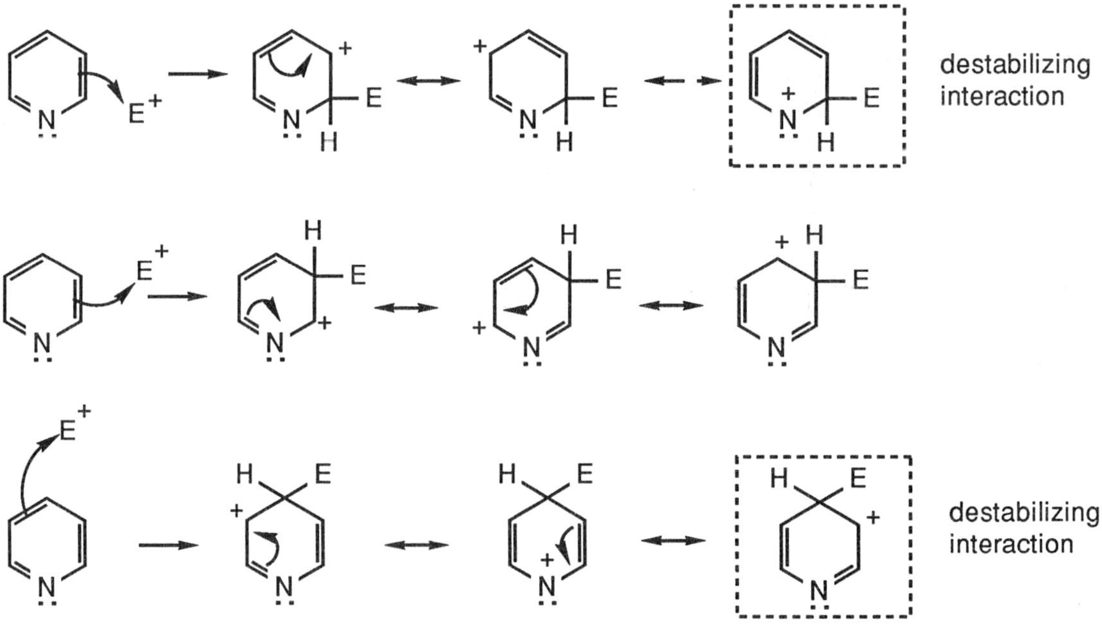

24.4

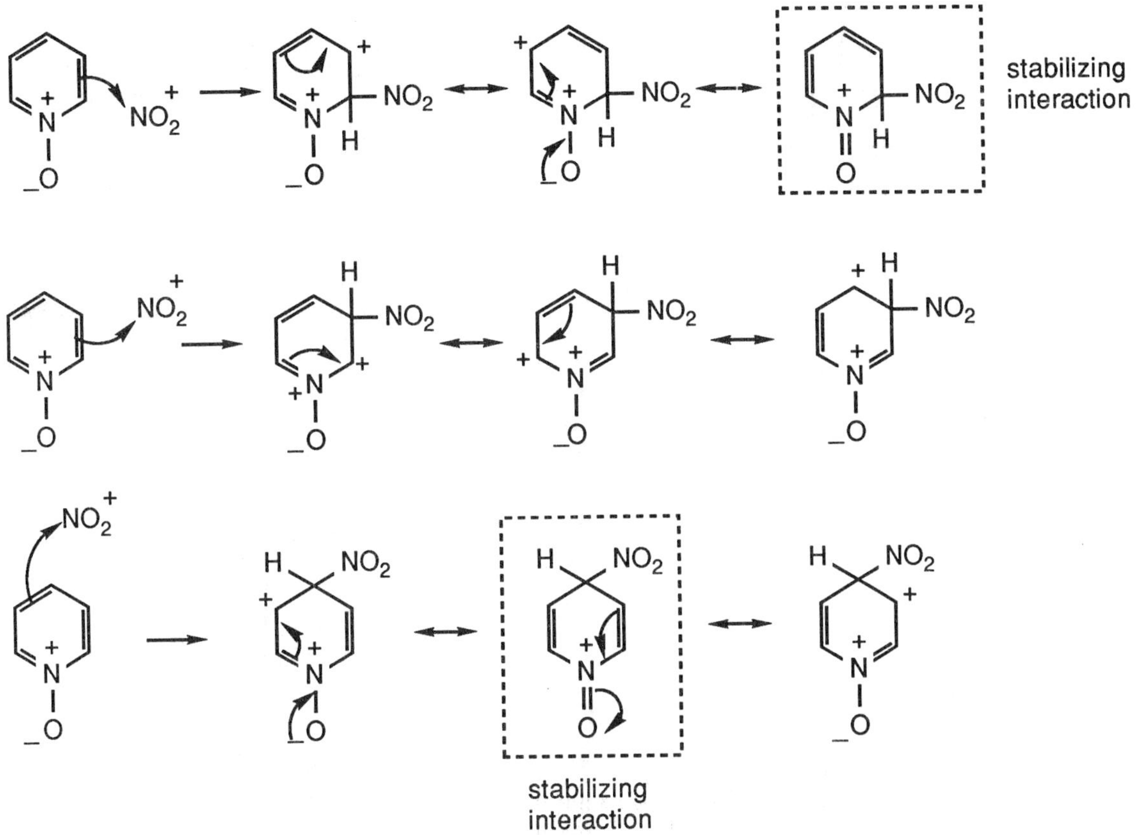

stabilizing interaction

stabilizing interaction

24.5 Nucleophilic attack of chloride ion can only occur at C-2 or at C-6, which is equivalent (i.e., by direct addition to the C=N double bond) or at C-4 (via conjugate addition to the C=N double bond).

or

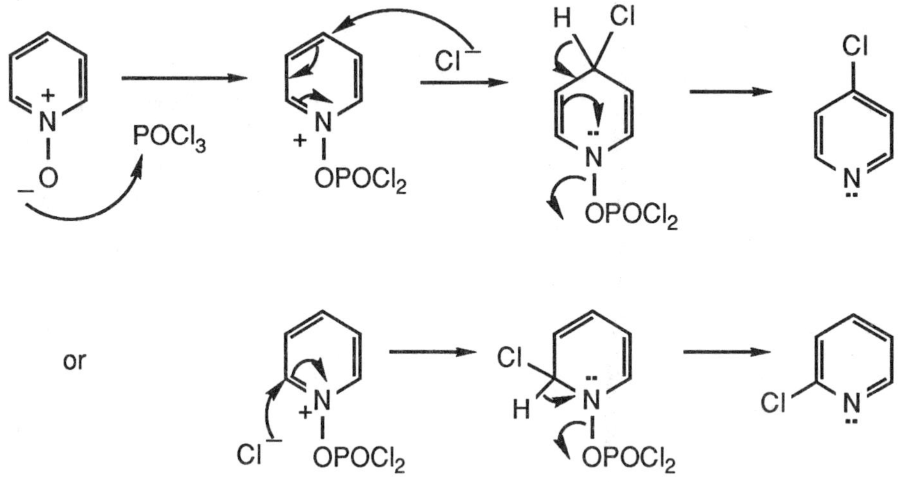

24.6 The *N*-oxide reacts more rapidly because negative charge is not placed on nitrogen in the intermediate (as it is in the reaction of 2-chloropyridine). Localization of the lone pair of electrons on nitrogen actually neutralizes the (formal) positive charge on the nitrogen of the amine oxide.

(a)

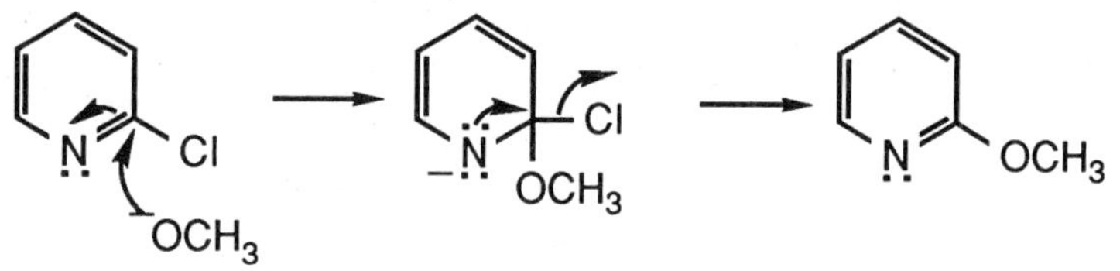

(b)

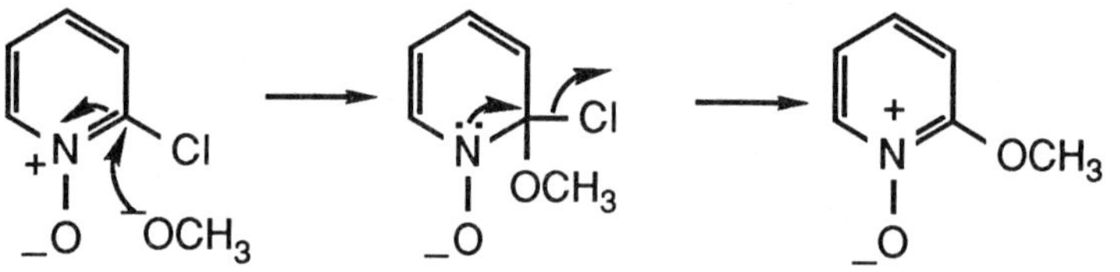

24.7 For the 3-chloro isomer negative charge can only be placed on carbon. This is much less favorable than for either the 2- or 4-chloro isomers, for which the electron pair can be localized on nitrogen as shown in problem 24.6.

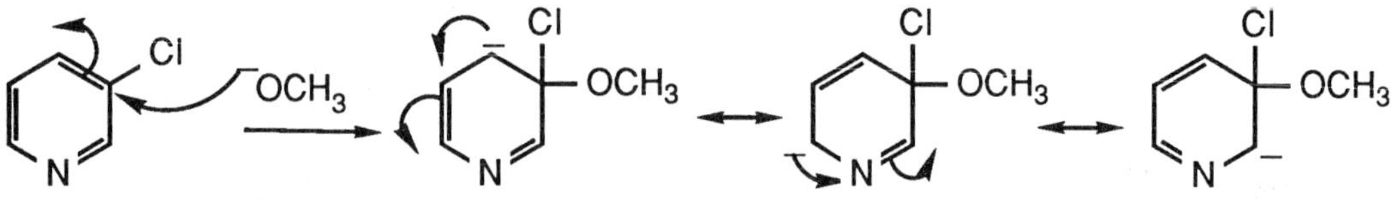

24.8 Attack at C-3 (or C-4):

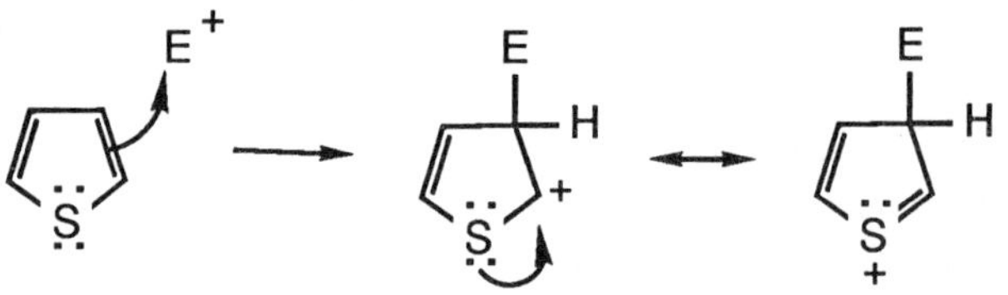

Attack at C-2 (or C-5):

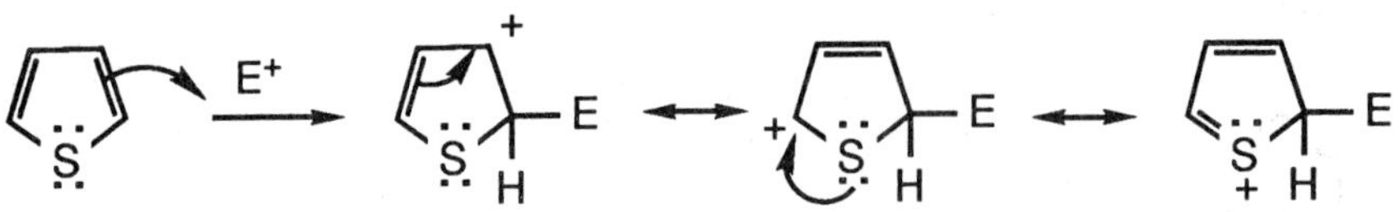

The second pathway (C–2, C–5) is normally preferred as suggested by the third resonance form that can be drawn. The presence of a methoxy group on C–2 allows additional resonance stabilization of the intermediate formed by attack at C–3:

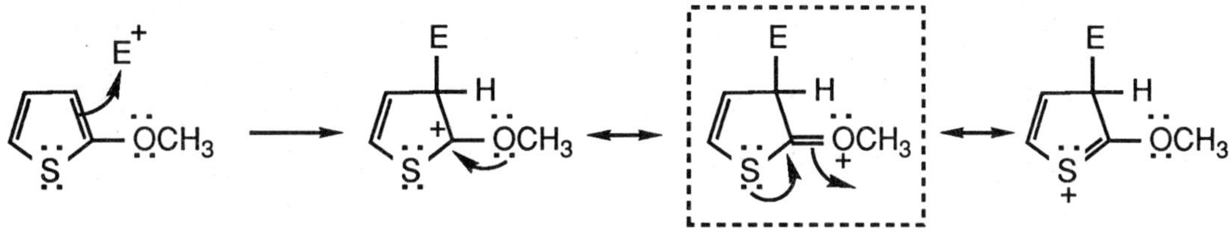

24.9

(a) pyridine $\xrightarrow{HBr}$ pyridinium (N^+–H) Br^-

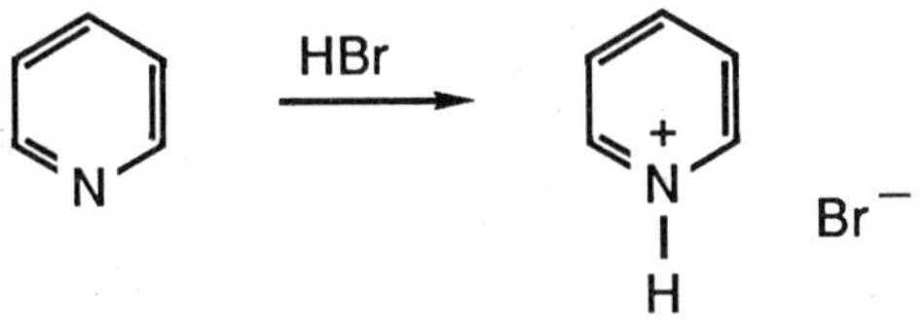

(b) pyridine $\xrightarrow{C_2H_5I}$ N-ethylpyridinium (N^+–C_2H_5) I^-

(c) pyridine $\xrightarrow{C_6H_5Li}$ [N(Li)-2-H-2-C_6H_5 dihydropyridine] → 2-C_6H_5-pyridine

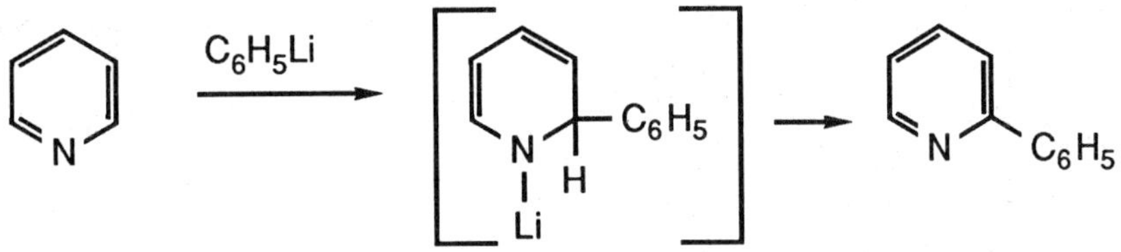

(d) pyridine N-oxide (N^+–O^-) $\xrightarrow{C_6H_5Li}$ 2-C_6H_5-pyridine

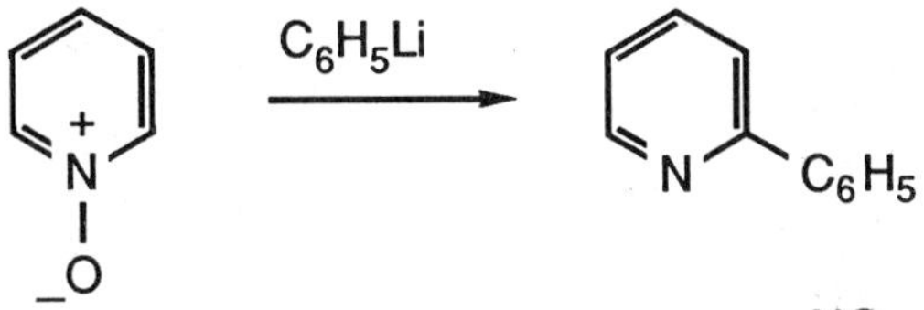

(e) pyridine N-oxide (N^+–O^-) $\xrightarrow{HNO_3,\ H_2SO_4}$ 4-NO_2-pyridine N-oxide (N^+–O^-)

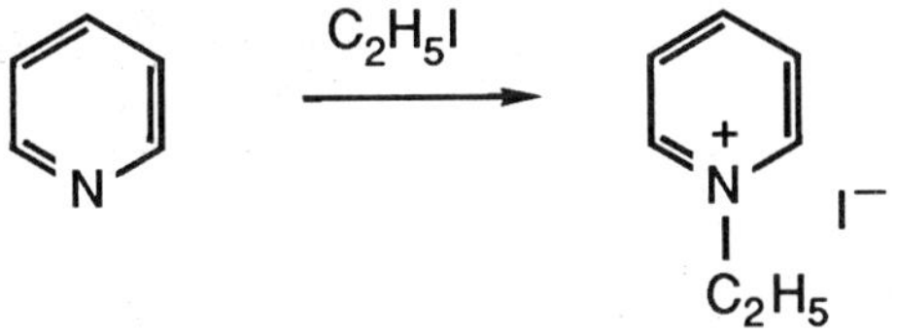

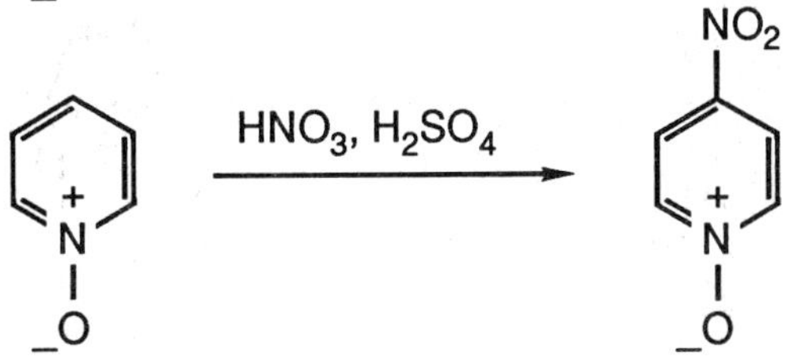

(f) pyridine N-oxide $\xrightarrow{PBr_3}$ 2-bromopyridine + 4-bromopyridine

(g) pyrylium ClO_4^- $\xrightarrow{CH_3Li}$ 2-methyl-2H-pyran (CH_3, H) + 4-methyl-4H-pyran (H, CH_3)

(h) pyrylium ClO_4^- $\xrightarrow{CH_3CH_2NH_2}$ N-ethylpyridinium (C_2H_5) $^-ClO_4$

(i) pyrylium ClO_4^- $\xrightarrow{NaBH_4}$ 2H-pyran (H, H) + 4H-pyran (H, H)

24.10

(a) 2-methylpyrrole (CH_3, N–H) $\xrightarrow{(CH_3)_2NH,\ CH_2O,\ HCl}$ CH_3-pyrrole(N–H)-$CH_2N(CH_3)_2$

(b) 2-methylpyrrole (CH_3, N–H) $\xrightarrow[POCl_3]{CH_3-\overset{O}{\overset{\|}{C}}N(CH_3)_2}$ CH_3-pyrrole(N–H)-$\overset{O}{\overset{\|}{C}}H$

(c) furan $\xrightarrow{CH_3-\overset{O}{\overset{\|}{C}}-O-\overset{O}{\overset{\|}{C}}-CH_3}$ furan-$\overset{O}{\overset{\|}{C}}CH_3$

(d) furan $\xrightarrow{CH_2O,\ HCl}$ 2-furyl–CH_2OH

(e) 2-methylthiophene (CH_3, S) $\xrightarrow{CH_3-\overset{O}{\overset{\|}{C}}Cl,\ ZnCl_2}$ 5-methylthiophene–$\overset{O}{\overset{\|}{C}}CH_3$ (CH_3, S)

(f) 2-methylthiophene (CH_3, S) $\xrightarrow[POCl_3]{CH_3-\overset{O}{\overset{\|}{C}}N(CH_3)_2}$ 5-methylthiophene–$\overset{O}{\overset{\|}{C}}H$ (CH_3, S)

(g) 2-methylthiophene (CH_3, S) $\xrightarrow{Br_2}$ 5-methylthiophene–Br (CH_3, S)

(h) 2-methylthiophene (CH_3, S) $\xrightarrow{HNO_3,\ H_2SO_4}$ 5-methylthiophene–NO_2 (CH_3, S)

24.11

(a) 2-methylpyridine (N, CH_3) $\xrightarrow[2.\ D_2O]{1.\ C_6H_5Li}$ pyridine (N)–CH_2D

(b) 2-methylpyridine (N, CH_3) $\xrightarrow{CH_3CO_3H}$ 2-methylpyridine N-oxide ($\overset{+}{N}$–O^-, CH_3)

(c) $\xrightarrow{CH_3I}$ I^-

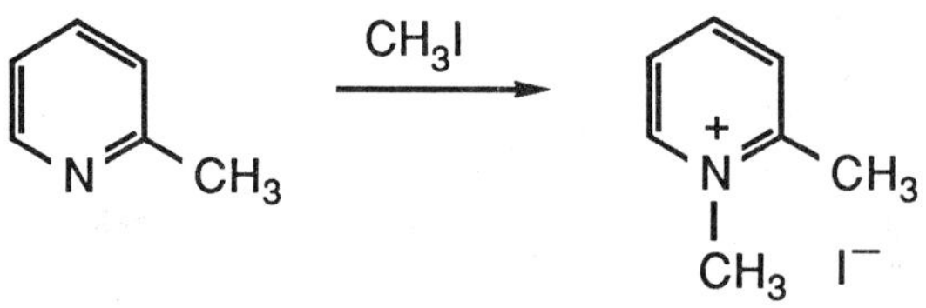

24.12 In both cases the nonbonding electrons on oxygen are able to stabilize the intermediate cation:

(a)

(b)

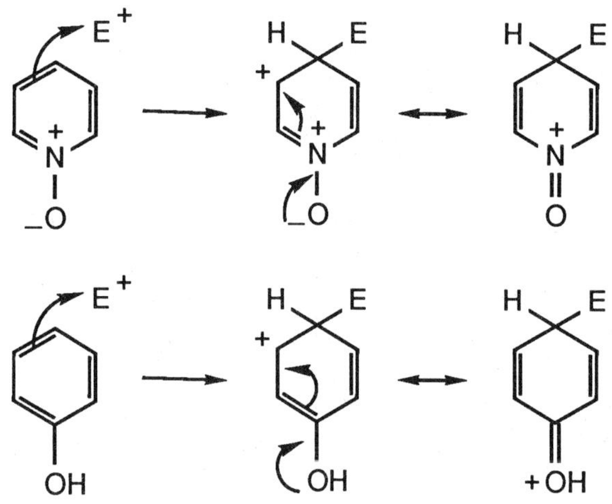

24.13

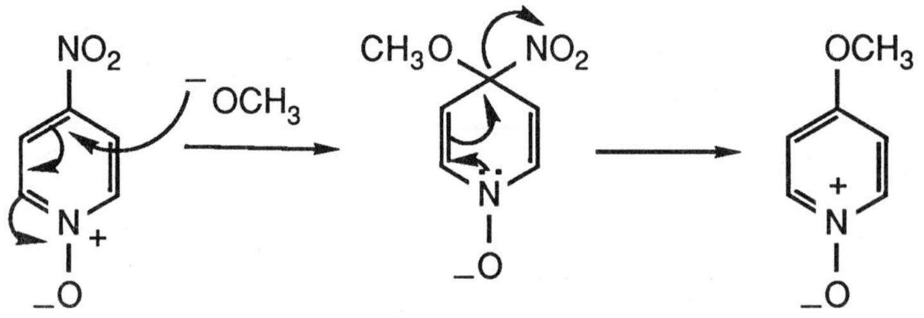

24.14

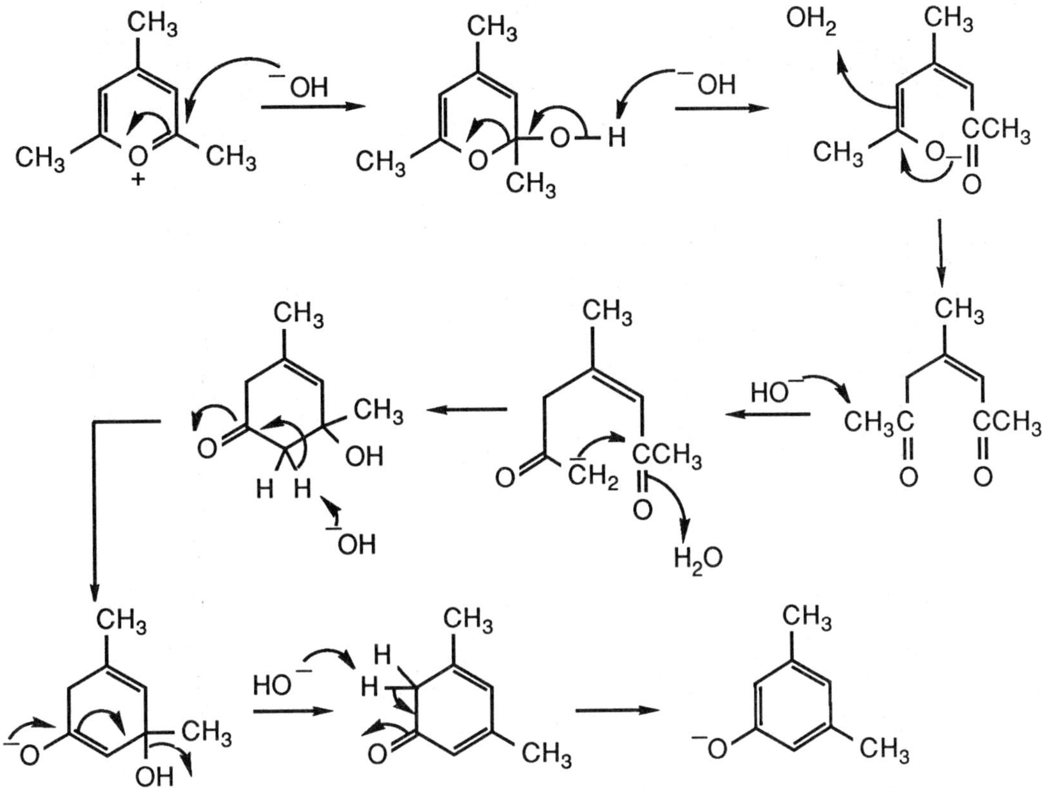

CH_3 ^-OH CH_3 CH_3 O^+ CH_3 → CH_3 CH_3 O O—H CH_3 ^-OH → OH_2 CH_3 CH_3 O^- CCH_3 O

→ CH_3 CH_3C CCH_3 O O HO^- → CH_3 O $\bar{C}H_2$ CCH_3 O H_2O → CH_3 CH_3 O OH H H ^-OH

→ CH_3 ^-O CH_3 OH HO^- → H H CH_3 O CH_3 → CH_3 ^-O CH_3

24.15

CH_3 $:NH_3$ CH_3 O + CH_3 → CH_3 + NH_3 CH_3 O CH_3 proton transfer → NH_4^+ CH_3 $\ddot{N}H_2$ CH_3 O CH_3

CH_3 CH_3 O CCH_3 $_+NH_2$ proton transfer → CH_3 CH_3C CCH_3 HO + :NH → CH_3 H + N CH_3 HO H :NH_3

:NH_3 H CH_3 H H N CH_3 HO → CH_3 CH_3 N CH_3

24.16 Reaction at C-4:

E^+ CH_3 S CO_2CH_3 → E H CH_3 + S CO_2CH_3 ↔ E H CH_3 S + CO_2CH_3

Reaction at C-3:

E^+ CH_3 S CO_2CH_3 → E H CH_3 S + CO_2CH_3 ↔ E H CH_3 S + CO_2CH_3

The intermediate generated by reaction at C-3 is stabilized by the methyl group as shown by the first of the two resonance forms drawn. In contrast, the effect of the methoxycarbonyl group destabilizes the intermediate formed by attack at C-3.

24.17 Reaction at C–2:

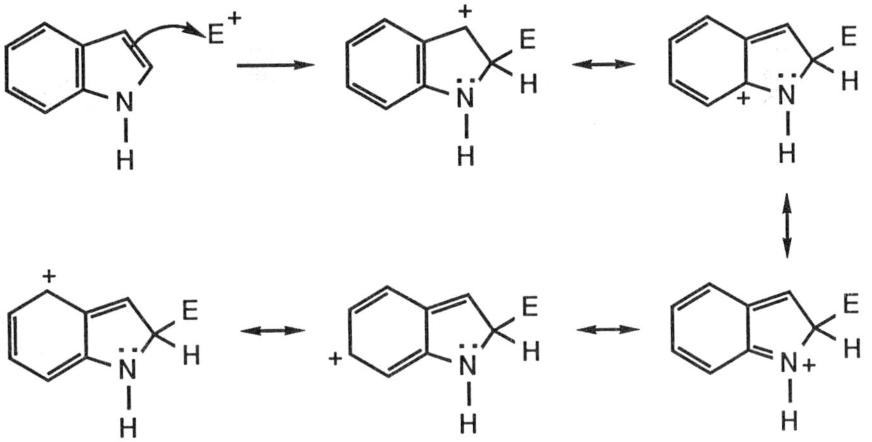

Reaction at C–3:

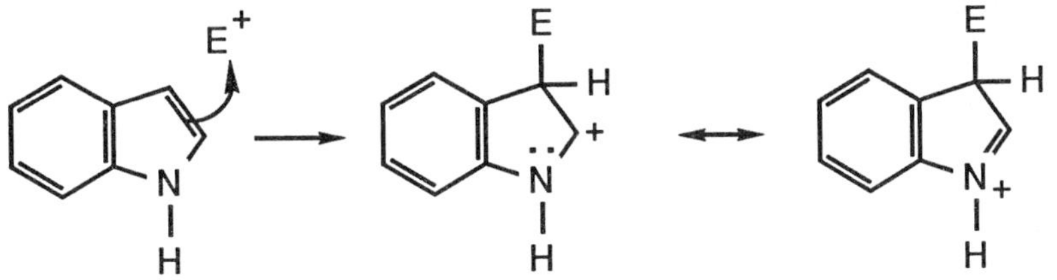

24.18

(a)

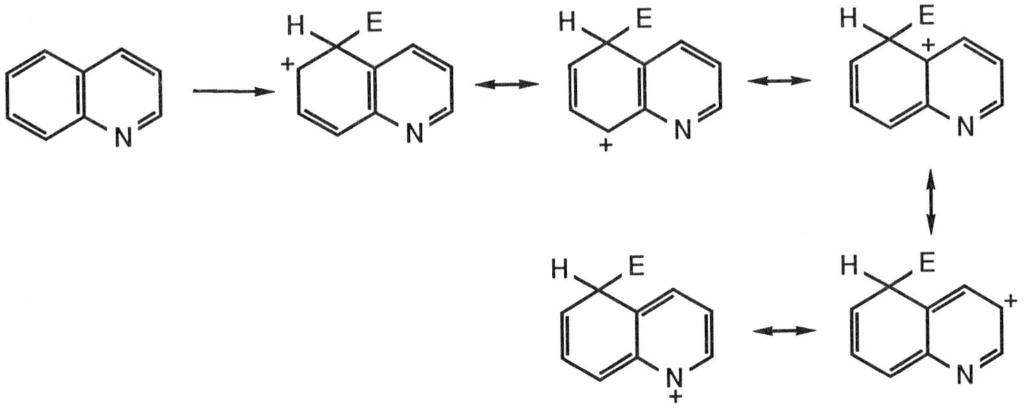

(b)

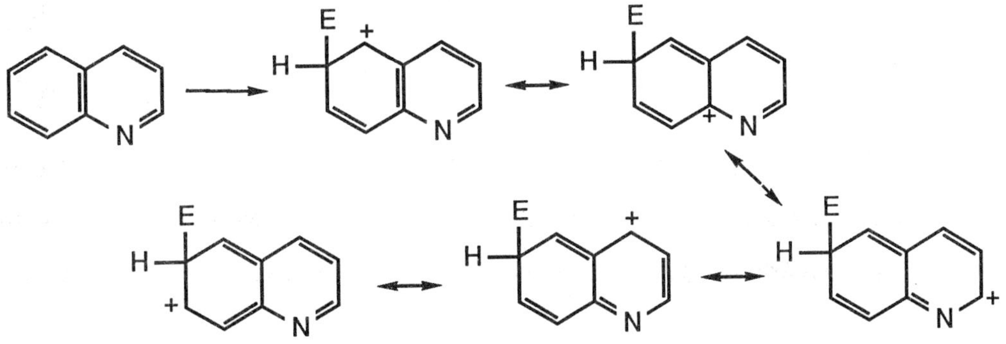

(c)

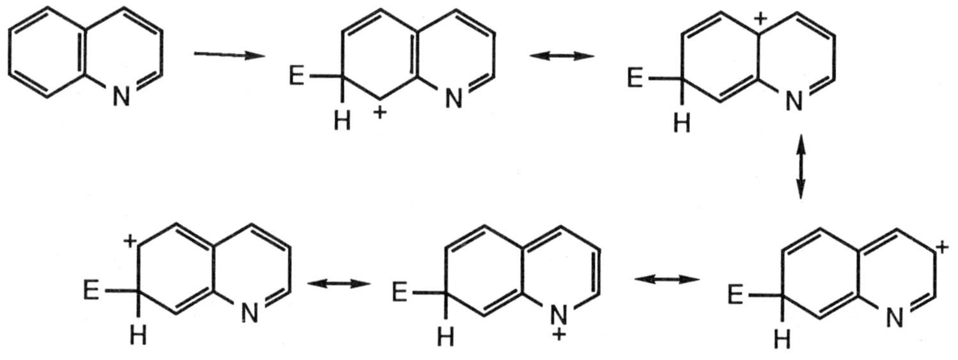

(d)

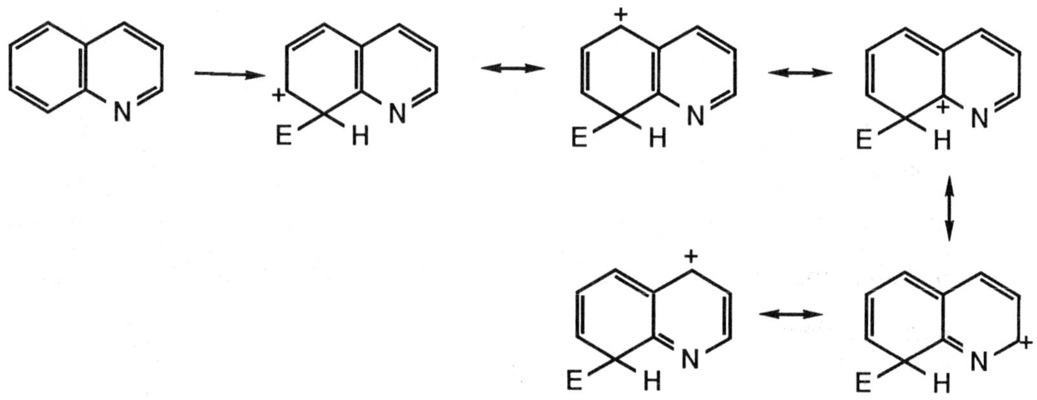

For the unprotonated forms, aniline should be more reactive than quinoline. The nonbonding electrons on the nitrogen of aniline can stabilize a positively charged intermediate, but the nonbonding electrons of quinoline are in the plane of the carbon skeleton and cannot interact with the π system.

24.19 Indole:

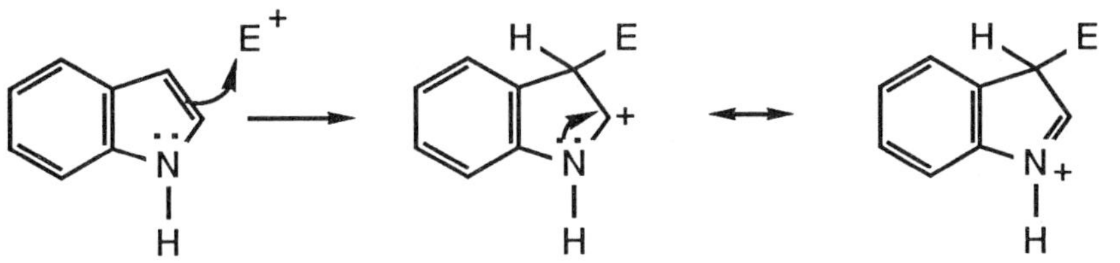

Quinoline:

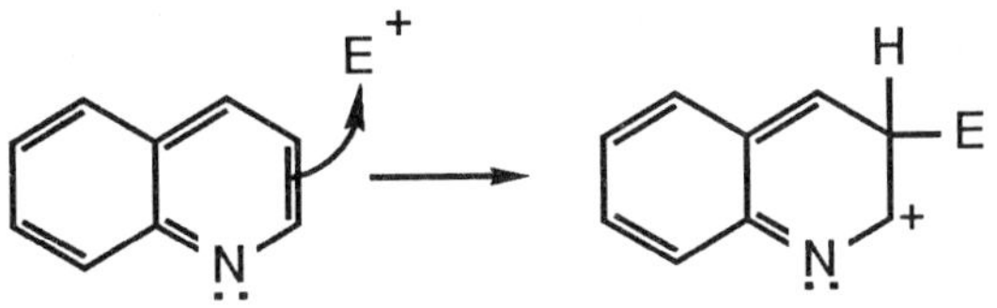

Isoquinoline:

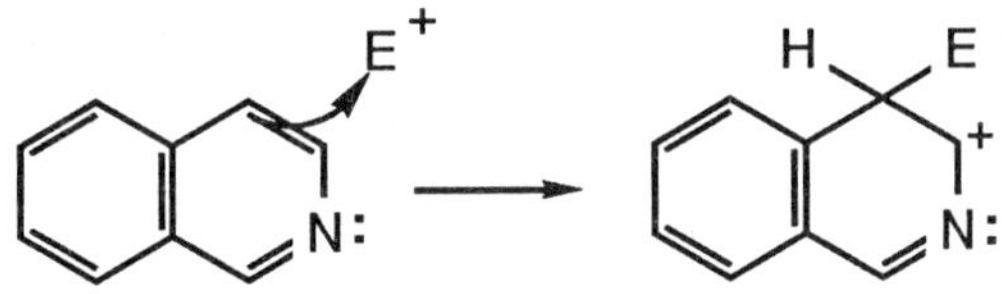

Only with indole can the lone pair of electrons on nitrogen stabilize an intermediate cation. For the other two compounds the electron withdrawing inductive effect of nitrogen makes it preferable for reaction to take place in the other ring (i.e., the benzene ring).

24.20 **(a)** Reaction occurs in such a way that additional positive charge is not placed on the electronegative nitrogen atom:

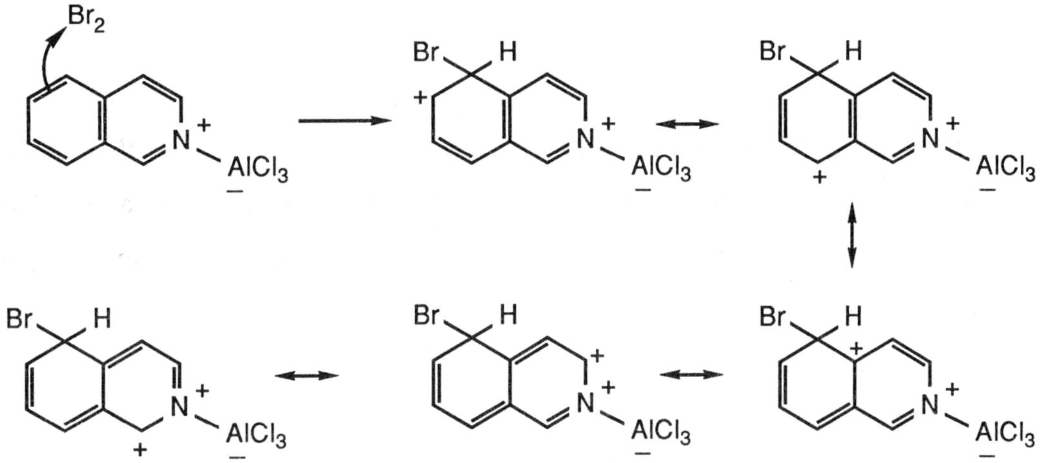

(b) The orientation of the nitration step is governed by the bromine substituent, so no direct information is available regarding the aluminum complex.

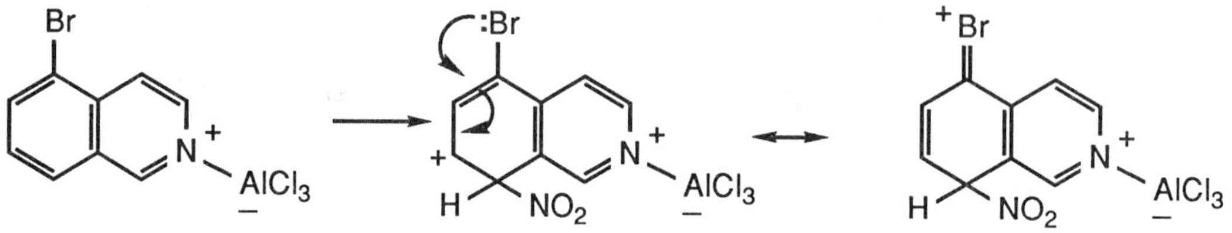

Br, N+, $AlCl_3^-$ → :Br, +, H, NO_2, N+, $AlCl_3^-$ ↔ +Br, H, NO_2, N+, $AlCl_3^-$

On the other hand, the aluminum complex would be extremely unreactive, so it is possible that nitration occurs with the small quantity of uncomplexed (and unprotonated) material that is present at equilibrium.

24.21

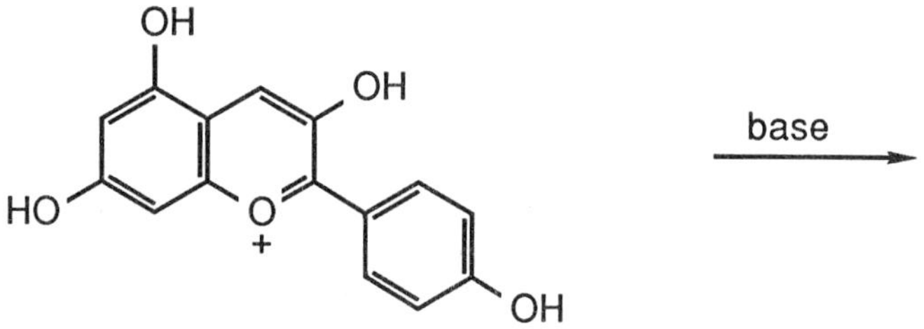

OH, OH, HO, O+, OH —base→

(1) OH, OH, HO, O+, O⁻ ↔ OH, OH, HO, O, O

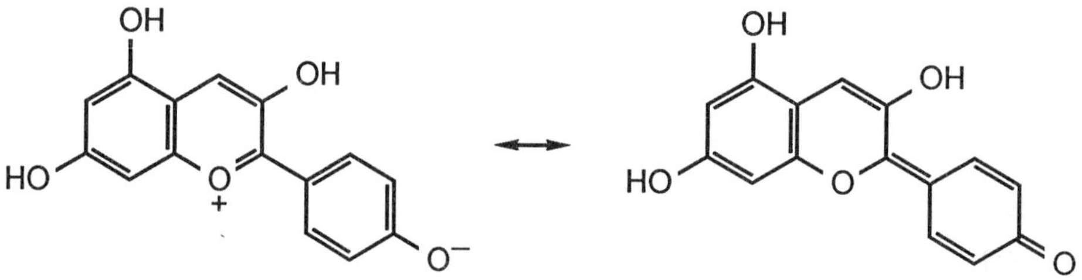

(2) OH, O⁻, HO, O+, OH

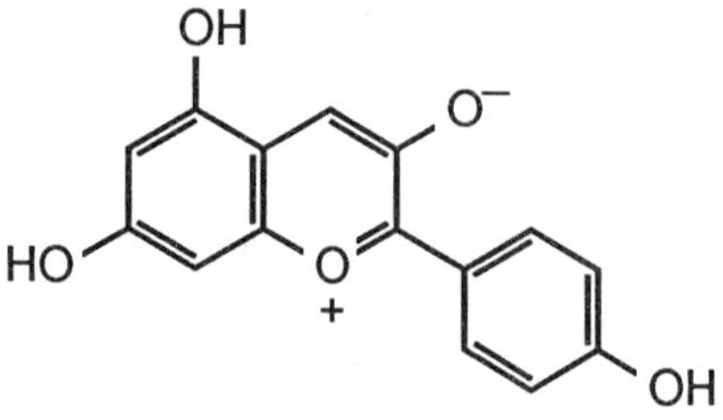

no uncharged resonance forms can be drawn

(3)

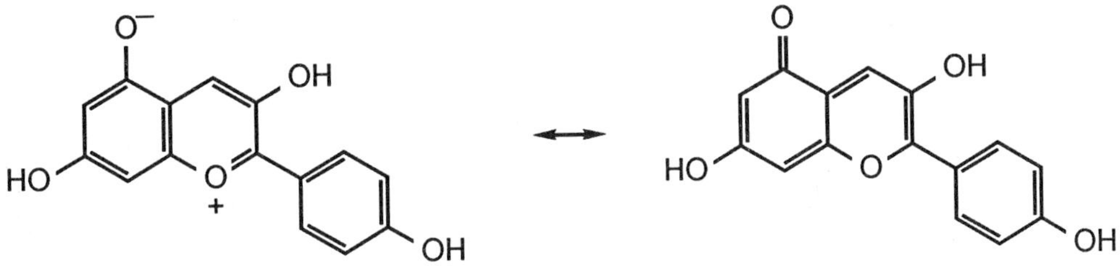

(4)

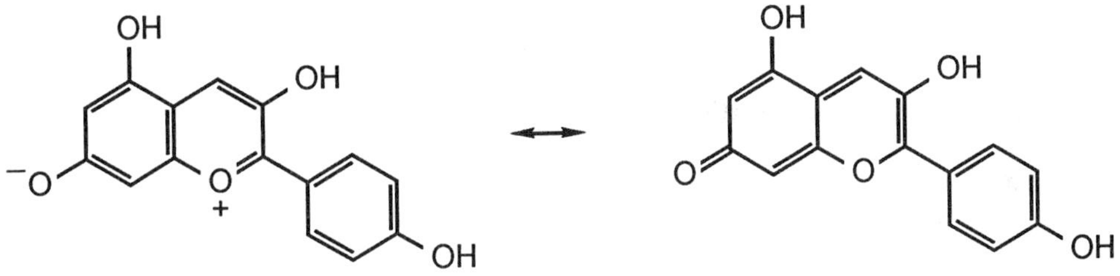

(b)

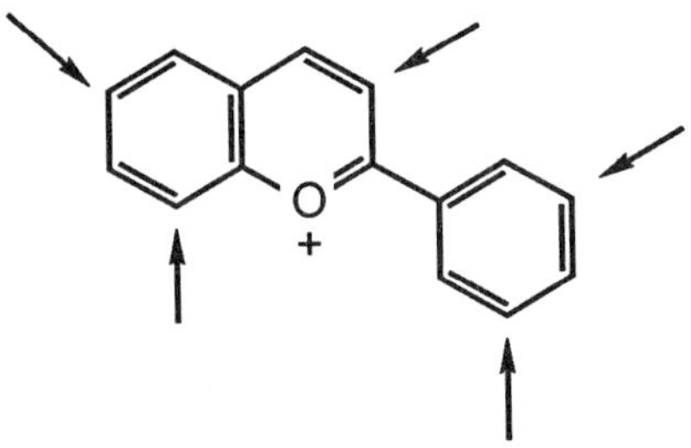

hydroxyls at the indicated positions would not be directly conjugated with the $-\overset{+}{O}{=}C$ group

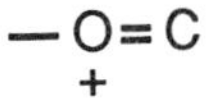

24.22 (a) CH_2O, HCl

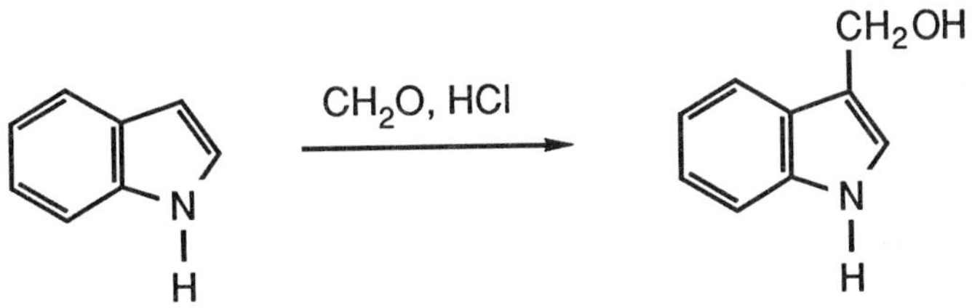

(b) HNO_3, H_2SO_4

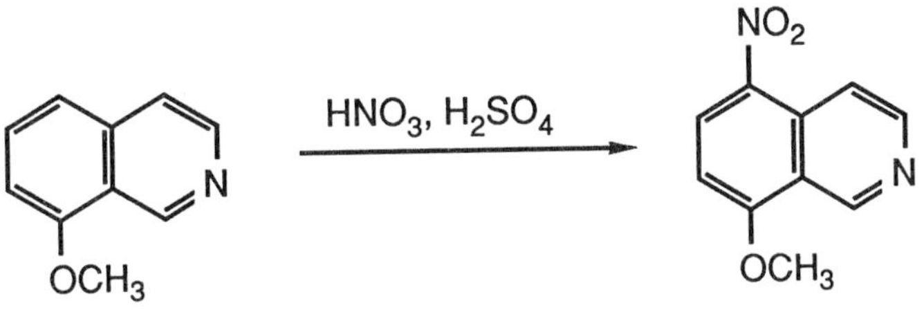

(c)

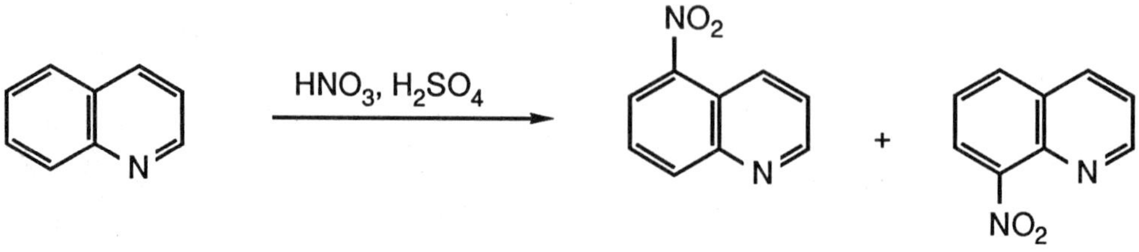

$\xrightarrow{HNO_3,\ H_2SO_4}$ +

N NO$_2$ N NO$_2$

(d)

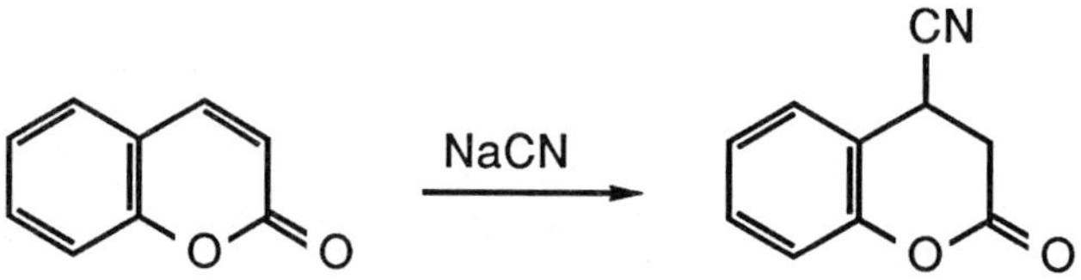

O O $\xrightarrow{NaCN}$ CN O O

(e)

O O $\xrightarrow{NaBH_4}$ OH O

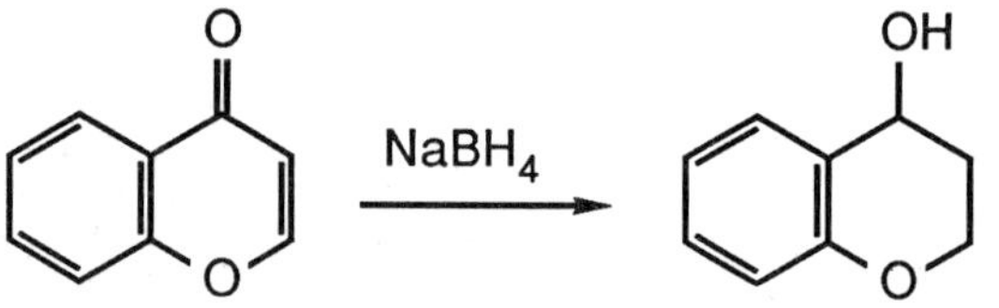

(conjugate addition is followed by carbonyl addition)

(f)

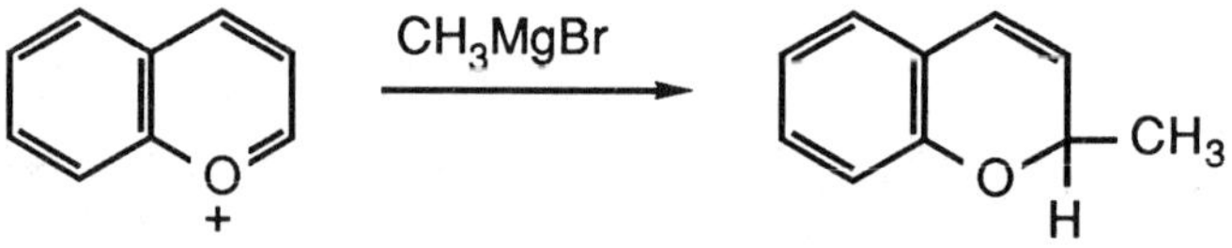

O + $\xrightarrow{CH_3MgBr}$ O H CH$_3$

24.23

(a)

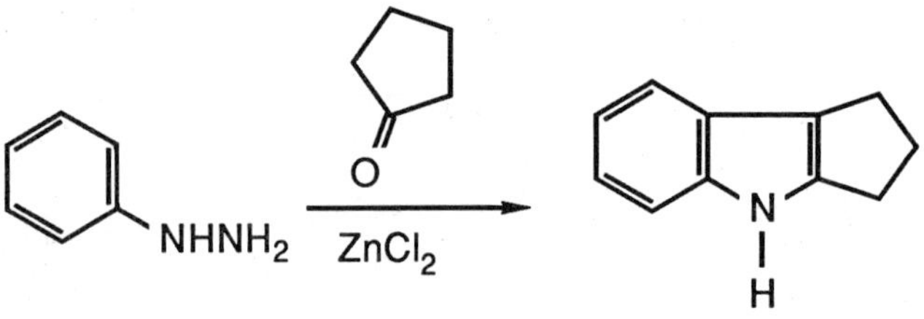

NHNH$_2$ $\xrightarrow[ZnCl_2]{}$ O N H

(b) O$_2$N NH$_2$ $\xrightarrow[H_3PO_4]{CH_2{=}CHCHO}$ O$_2$N N

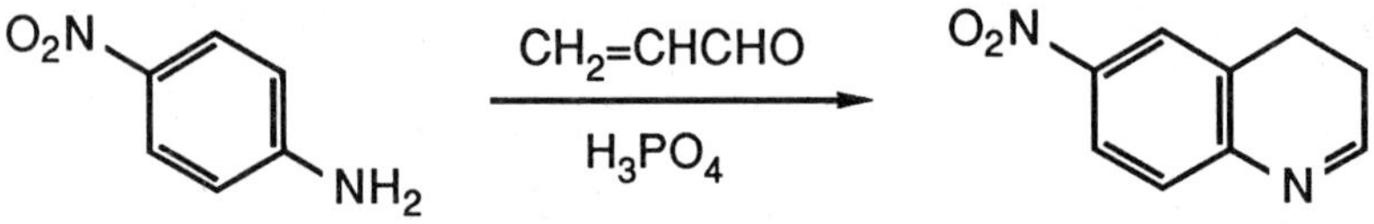

(c)

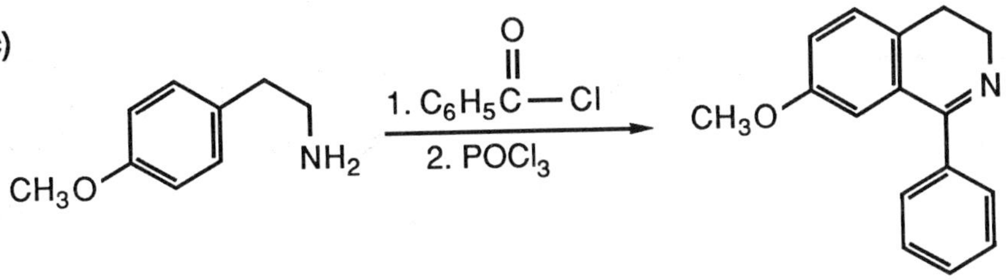
1. $C_6H_5C(O)$—Cl
2. $POCl_3$
CH_3O
NH_2
N

(d)

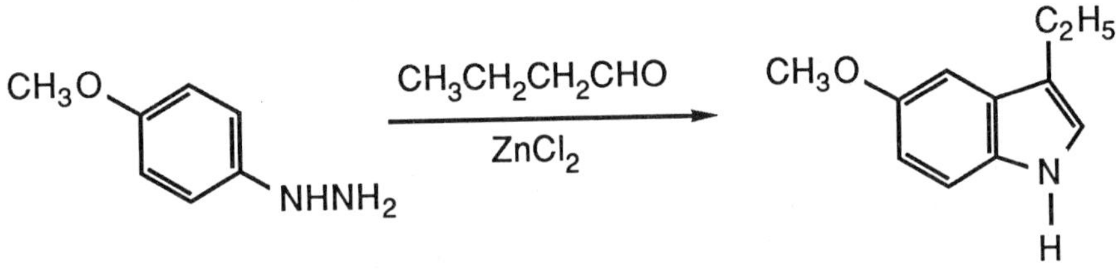
CH_3O
NHNH$_2$
$CH_3CH_2CH_2CHO$
$ZnCl_2$
C_2H_5
N
H

(e)

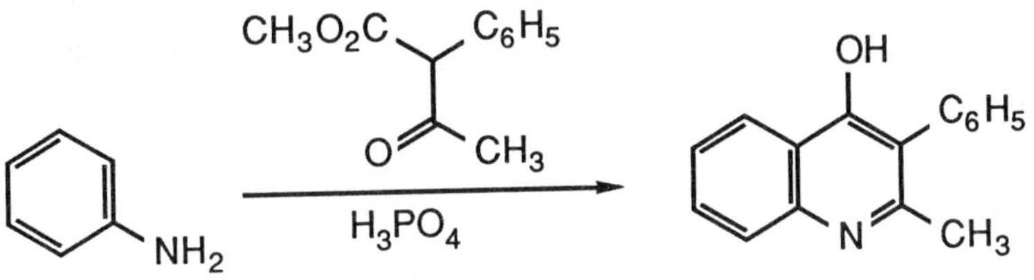
CH_3O_2C
C_6H_5
O
CH_3
NH_2
H_3PO_4
OH
N

24.24 In the five-membered ring heterocycles the positively charged intermediates have resonance forms in which the positive charge resides on the heteroatom as an onium ion:

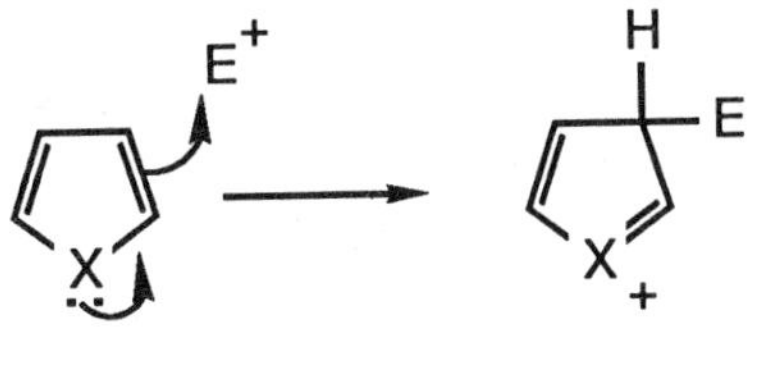
E+
X
H
E
X
+

In contrast the positive charge for intermediates in pyridine derivatives can be localized on the heteroatom only by making it electron deficient:

N
E+
+
N
E
H

24.25 **(a)**

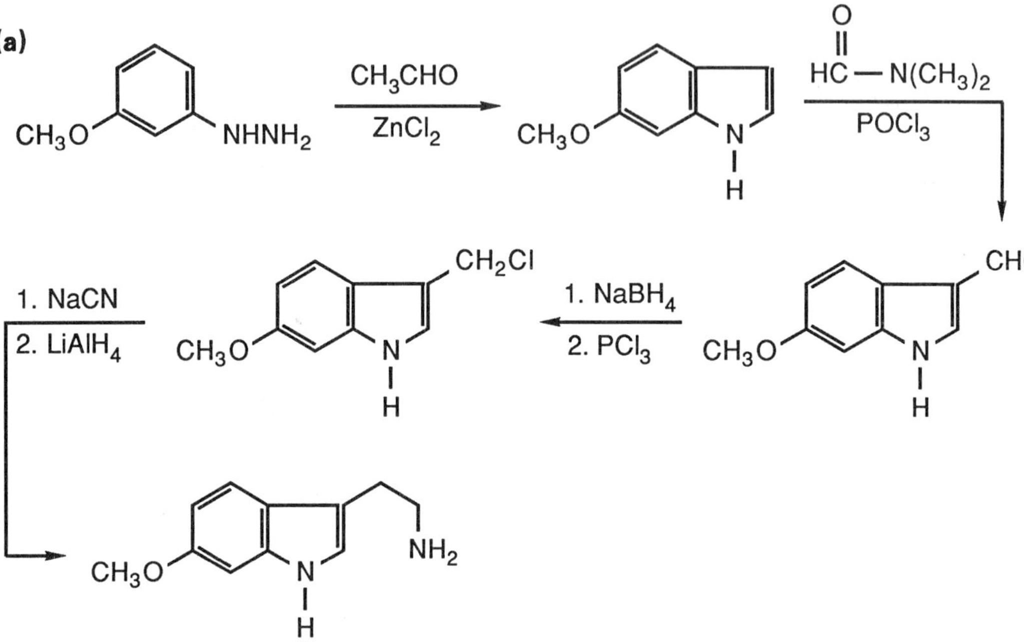

CH_3O–C₆H₄–$NHNH_2$ $\xrightarrow[ZnCl_2]{CH_3CHO}$ 6-methoxyindole $\xrightarrow[POCl_3]{HC(=O)—N(CH_3)_2}$ 3-CHO indole $\xrightarrow[2.\ PCl_3]{1.\ NaBH_4}$ 3-CH_2Cl indole $\xrightarrow[2.\ LiAlH_4]{1.\ NaCN}$ 3-($CH_2CH_2NH_2$) indole

(b) The methoxy group aids in the cyclization reaction:

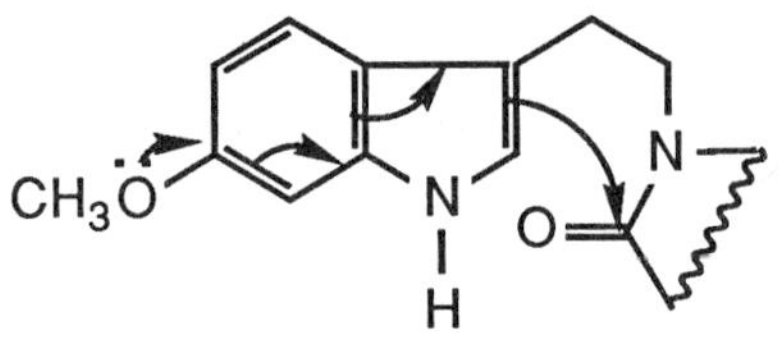

24.26 Attack occurs at the 4-position in each case:

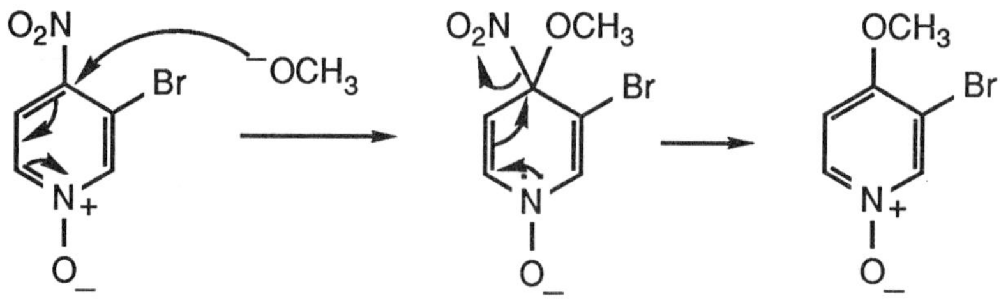

O_2N, Br, $^{-}OCH_3$ → O_2N OCH_3, Br → OCH_3, Br, $N+$, O_-

Br NO₂ ⁻OCH₃ → Br OCH₃ O⁻ ⁺N O ↔ Br OCH₃ O⁻ ⁺N O⁻ +N → OCH₃ NO₂ N+ O⁻

The 3-nitro-4-bromo derivative would react more rapidly because the negative charge of the intermediate is stabilized by interaction with the nitro group as well as with the amine oxide group.

24.27

(a) cyclobutene $\xrightarrow{C_6H_5C(=O)-N_3}$ N—C(=O)C₆H₅ (bicyclic aziridine)

(b) $CH_2-CHCH_2CH_3$ (epoxide) $\xrightarrow{NH_3}$ $NH_2CH_2-CH(OH)CH_2CH_3$ $\xrightarrow{HBr}$ $\overset{+}{N}H_3CH_2-CH(Br)CH_2CH_3$ → $CH_2-CHCH_2CH_3$ (aziridine, N–H)

(c) $CH_2{=}CHCH_2CH_3$ $\xrightarrow{CH_3CO_3H}$ $CH_2-CHCH_2CH_3$ (epoxide)

(d) CH_3CH_2CHO $\xrightarrow{CH_2{=}S(CH_3)_2}$ $CH_2-CHCH_2CH_3$ (epoxide)

24.28

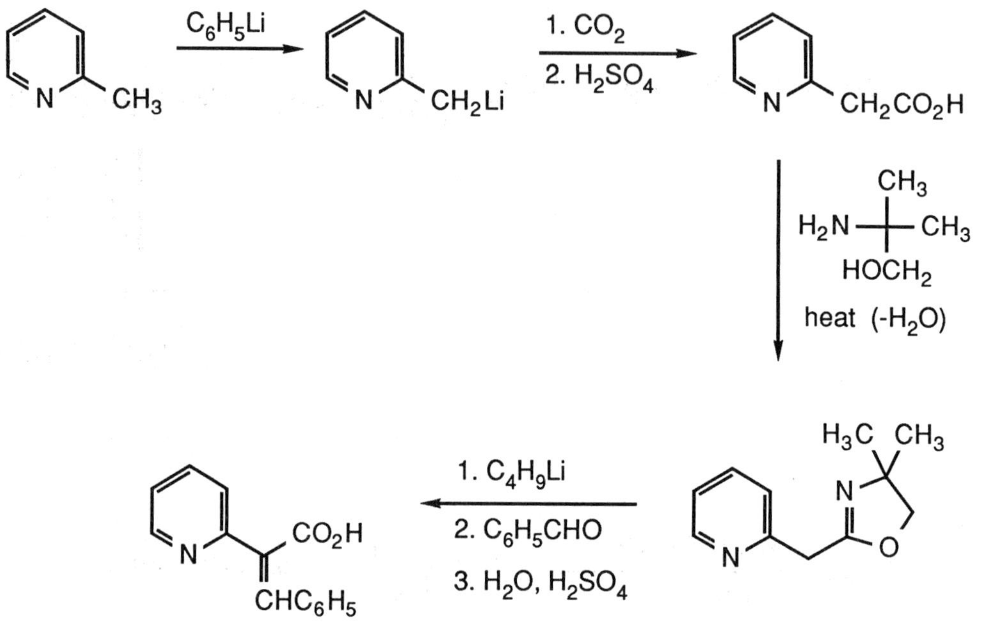

2-methylpyridine (CH_3) → (C_6H_5Li) → 2-pyridyl-CH_2Li → (1. CO_2; 2. H_2SO_4) → 2-pyridyl-CH_2CO_2H

↓ H_2N–C(CH_3)(CH_3)–$HOCH_2$, heat ($-H_2O$)

2-[(4,4-dimethyl-oxazolinyl)methyl]pyridine (H_3C, CH_3, N, O) → (1. C_4H_9Li; 2. C_6H_5CHO; 3. H_2O, H_2SO_4) → 2-pyridyl–C(CO_2H)=CHC_6H_5

24.29

(a) 3-furylmethanol (CH_2OH) → (H_2, Pt) → tetrahydrofuran-3-yl CH_2OH → (CrO_3, pyridine) → tetrahydrofuran-3-yl CHO

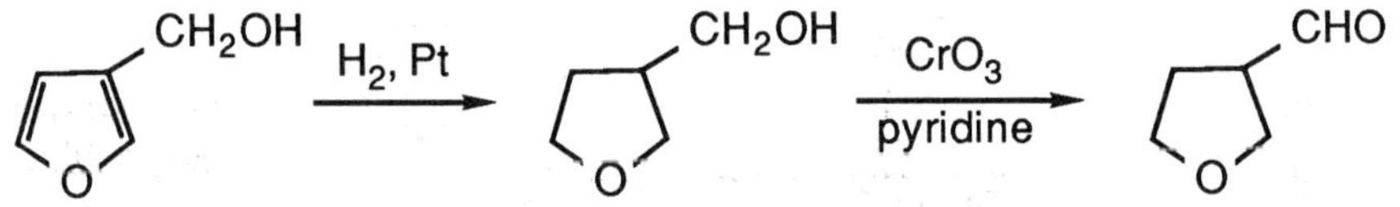

(b) $CH_3CH(OH)CH_2-CH(CH_3)CH_2OH$ → (TsCl (1 equiv.)) → $CH_3CH(OH)CH_2-CH(CH_3)CH_2OTs$ → (NaOH) → 2,4-dimethyltetrahydrofuran (CH_3, CH_3, O)

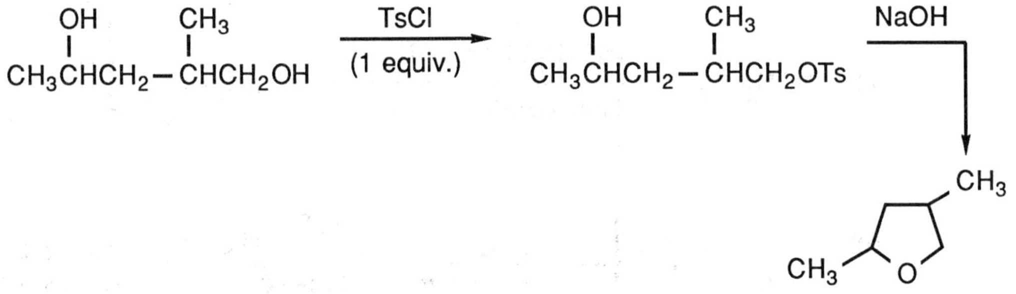

(c) $CH_2{=}CHCH_2-CHBrCH_2CH_3$ → (1. BH_3; 2. H_2O_2, NaOH) → $HOCH_2CH_2CH_2-CHBrCH_2CH_3$ → (1. TsCl; 2. NH_3) → 2-ethylpyrrolidine (CH_2CH_3, N–H)

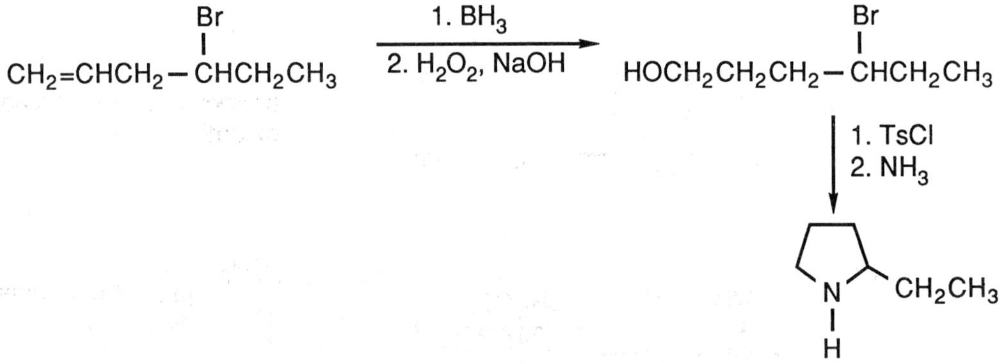

(d)

$$CH_3CH_2CH_2CH_2CH_2CH_2NHCH_3 \xrightarrow[NaOH]{Cl_2} CH_3CH_2CH_2CH_2CH_2CH_2N(Cl)CH_3$$

1. H_2SO_4, heat
2. NaOH (neutralize)

2-ethylpyrrolidine: N–H ring with CH_2CH_3

24.30

(a) $RCHO \xrightarrow[TsOH,\ C_6H_6]{HOCH_2CH_2OH}$ R–CH(1,3-dioxolane) protected against nucleophilic attack

(b) $RC(=O)—CH_2—$ $\xrightarrow[TsOH,\ C_6H_6]{\text{pyrrolidine (N–H)}}$ R–C(N-pyrrolidinyl)=CH— activated as a nucleophile at the α–position

(c) $RCH_2CO_2H \xrightarrow[\text{heat } (-H_2O)]{HOCH_2C(CH_3)_2NH_2}$ RCH_2–(4,4-dimethyloxazoline) protected against nucleophilic attack

(d) RCH_2–(4,4-dimethyloxazoline) $\xrightarrow{C_4H_9Li}$ Li^+ $R\bar{C}H$–(4,4-dimethyloxazoline) activated as a nucleophile at the α–position

(e) ROH $\xrightarrow[BF_3]{\text{dihydropyran}}$ RO–(tetrahydropyran-2-yl) protected against alkylation, acylation, oxidation

(f) —CH(OH)–CH(OH)— $\xrightarrow[TsOH,\ C_6H_6]{CH_3C(=O)CH_3}$ $(CH_3)_2C$ dioxolane (O, O) protected against alkylation, acylation, oxidation

24.31

(a) $C_6H_5N_3$; $-N_2$; NC_6H_5 ; C_6H_5

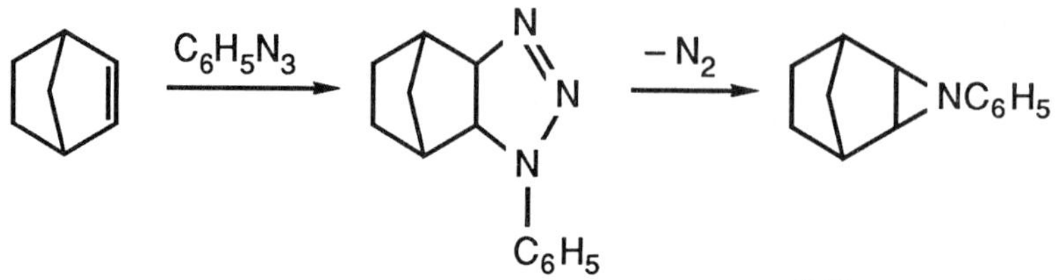

(b) $+ SO_2$; CN ; CN

(c) 550°C ; $CH_2{=}C{=}O$

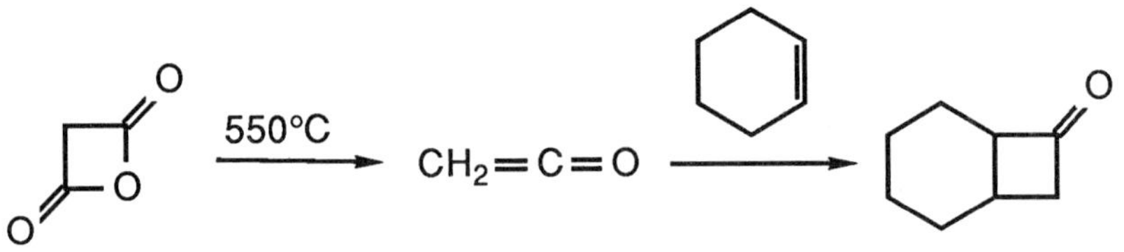

24.32

(a) $CO_2C_2H_5$ / $CO_2C_2H_5$ — 1. $NaOC_2H_5$ 2. $CH_3CH_2CH_2-\underset{Br}{CH}CH_3$ → $CH_3CH_2CH_2-\underset{CH_3}{CH}-CH(CO_2C_2H_5)_2$

1. $NaOC_2H_5$
2. $CH_2{=}CHCH_2Br$

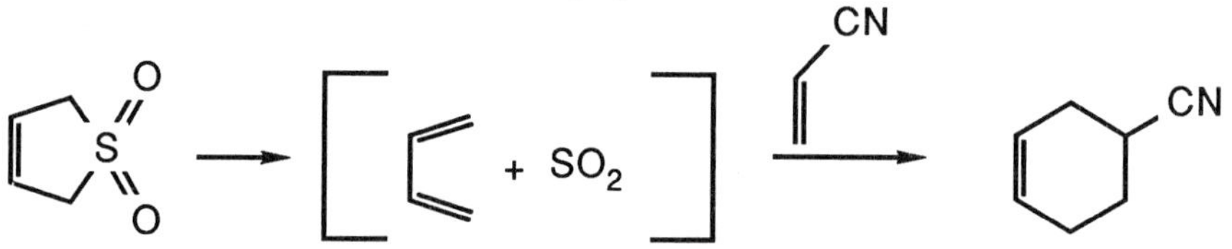

→ $CH_2{=}CHCH_2-C(CO_2C_2H_5)_2-\underset{CH_3}{CH}CH_2CH_2CH_3$

$H_2N\overset{O}{\overset{\|}{C}}NH_2$, $NaOC_2H_5$

→ barbiturate with $CH_2{=}CHCH_2$ and $CH_3CH_2CH_2-\underset{CH_3}{CH}$ substituents (ring: O, N–H, O, N–H, O)

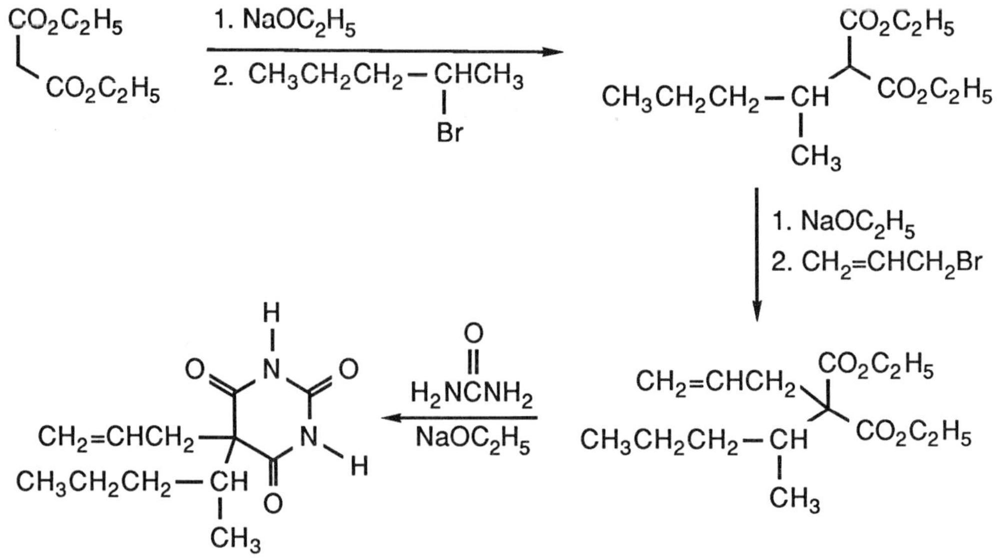

(b) $CO_2C_2H_5$–CH_2–$CO_2C_2H_5$ $\xrightarrow[\text{2. } CH_3CH_2CH_2-CHCH_3 \text{ (Br on CH)}]{\text{1. } NaOC_2H_5}$ $CH_3CH_2CH_2-CH(CH_3)-CH(CO_2C_2H_5)_2$

$\xrightarrow[\text{2. } CH_3CH_2I]{\text{1. } NaOC_2H_5}$ $CH_3CH_2CH_2-CH(CH_3)-C(CH_2CH_3)(CO_2C_2H_5)_2$

$\xrightarrow[NaOC_2H_5]{H_2NCNH_2 \text{ (C=O)}}$ 5-ethyl-5-(1-methylbutyl)barbituric acid (pyrimidine ring with two N–H and three C=O; substituents CH_3CH_2 and $CH_3CH_2CH_2-CH(CH_3)$)

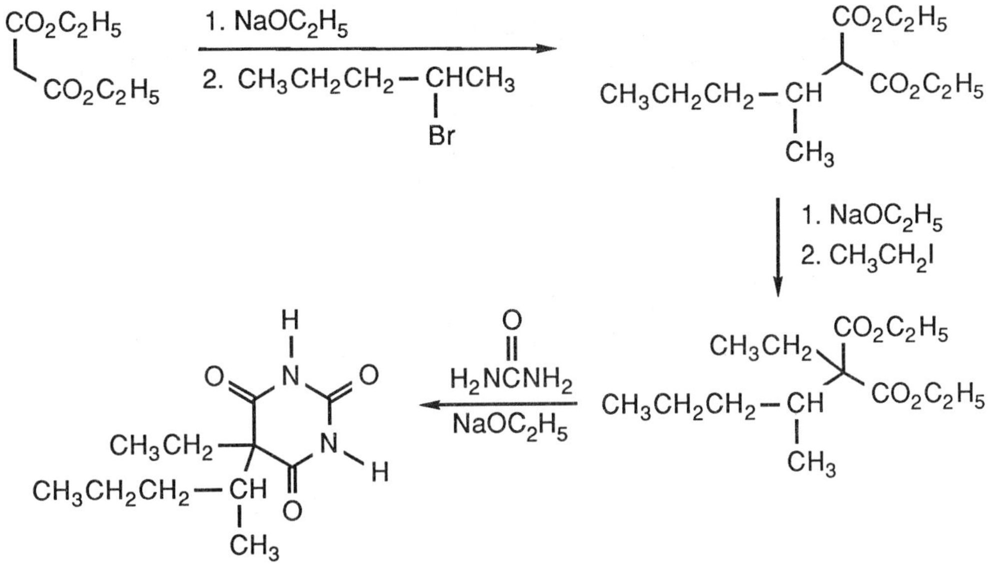

24.33 Preparation of phenobarbital from diethyl malonate would require introduction of the phenyl substituent by "alkylation" with bromobenzene, and such a reaction would not work:

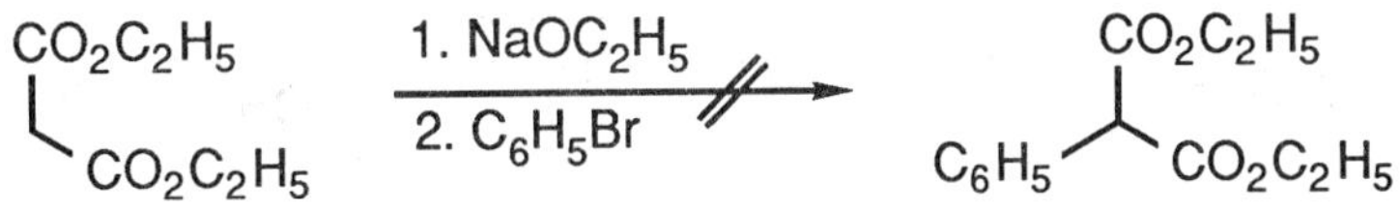

$CO_2C_2H_5$–CH_2–$CO_2C_2H_5$ $\xrightarrow[\text{2. } C_6H_5Br]{\text{1. } NaOC_2H_5}$ (does not occur) $C_6H_5-CH(CO_2C_2H_5)_2$

24.34

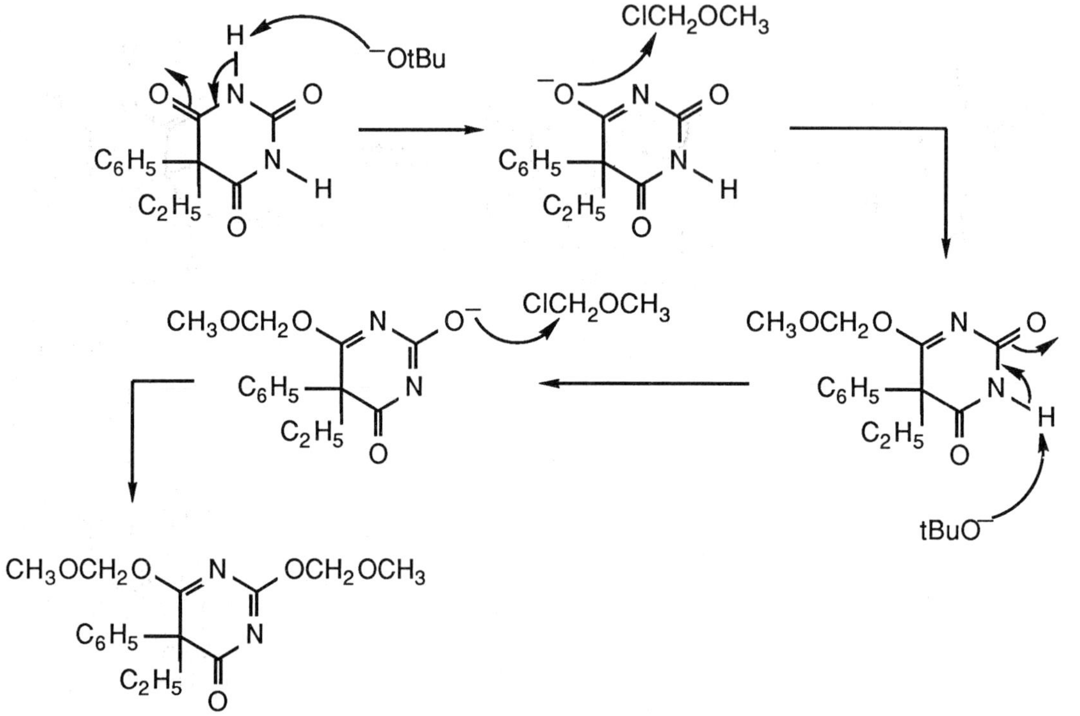

$^{-}$OtBu

$ClCH_2OCH_3$

C_6H_5 C_2H_5

CH_3OCH_2O

$ClCH_2OCH_3$

CH_3OCH_2O

tBuO^{-}

CH_3OCH_2O OCH_2OCH_3

C_6H_5 C_2H_5

24.35 (a)

most acidic

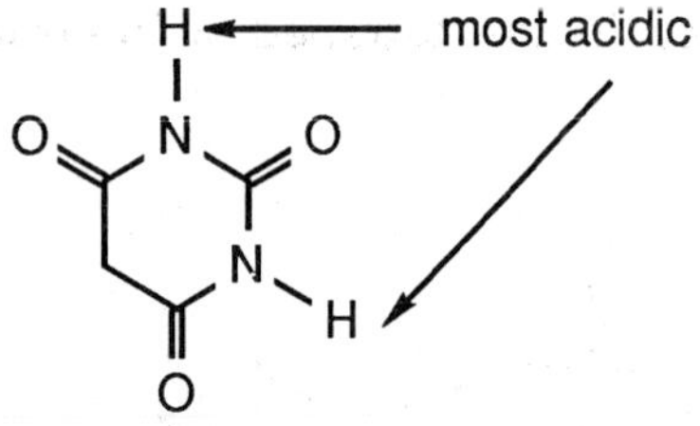

(b) Although the NH hydrogens are most acidic, a complete condensation can occur via enol formation of the active CH_2 group:

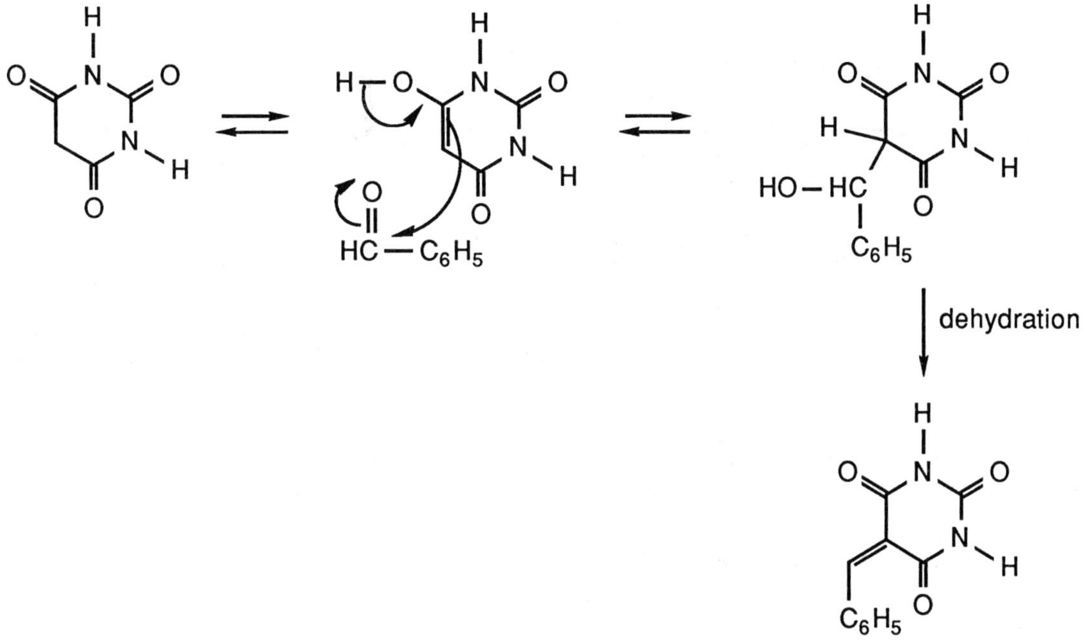

24.36 A bulky R group inhibits approach of the attacking nucleophile (shown here as Nuc^-) and also results in increased steric hindrance in the tetrahedral intermediate that is formed when Nuc^- adds to the C=O group:

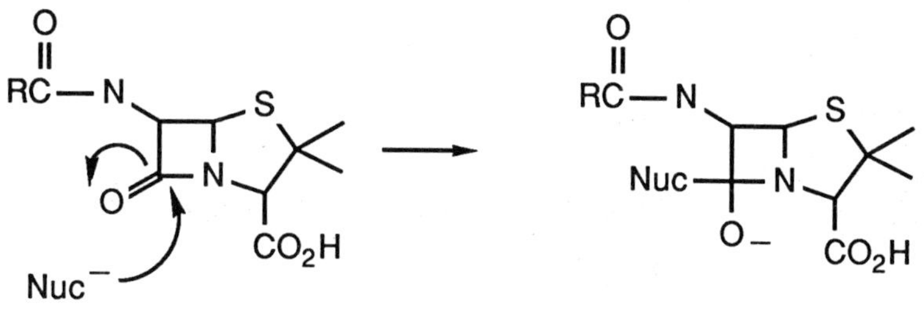

24.37 Nitrosyl chloride is a more selective reagent. The use of aqueous nitrous acid presumably gives the desired 7-aminocephalosporanic acid, which then reacts a second time with nitrous acid:

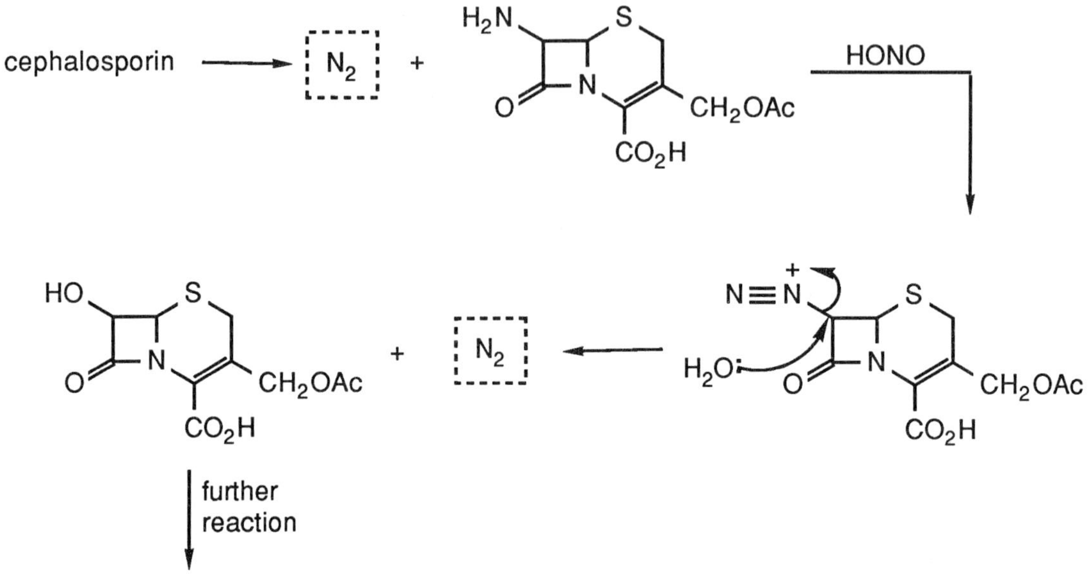

cephalosporin → N_2 + [7-amino cephalosporanic acid: H_2N, S, N, O, CH_2OAc, CO_2H] —HONO→ [diazonium: $N{\equiv}\overset{+}{N}$, H_2O:, S, N, O, CH_2OAc, CO_2H] → N_2 + [HO, S, N, O, CH_2OAc, CO_2H]

further reaction

24.38 Formation of a six-membered ring requires a less strained transition state for intramolecular hydrogen abstraction. (In either case the reaction must proceed via the higher energy boat conformation of the piperidine ring:

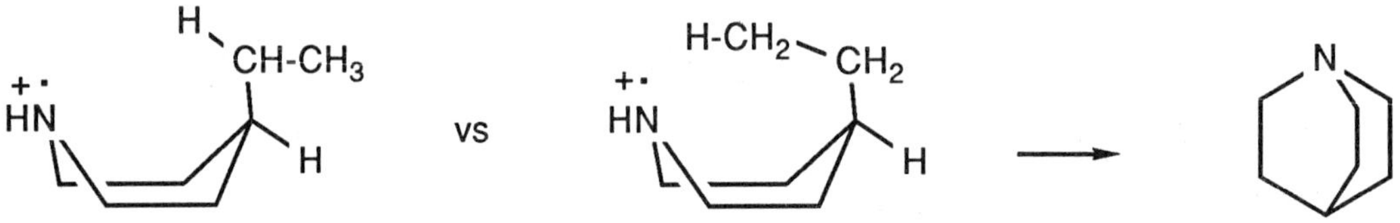

[$\overset{+\bullet}{HN}$ boat, H, CH-CH_3, H] vs [$\overset{+\bullet}{HN}$ boat, H-CH_2-CH_2, H] → [quinuclidine, N]

CHAPTER 25. POLYMERS

ANSWERS TO EXERCISES

25.1 polyethylene $\sim CH_2CH_2—CH_2CH_2—CH_2CH_2—CH_2CH_2—CH_2CH_2\sim$

poly(vinyl chloride)

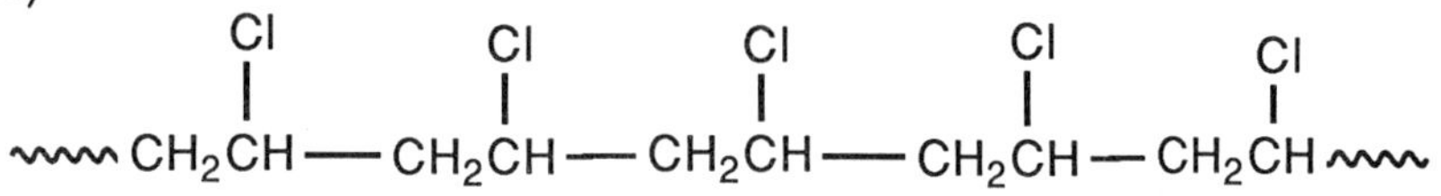

$$\sim CH_2\underset{|}{\overset{Cl}{C}}H—CH_2\overset{Cl}{C}H—CH_2\overset{Cl}{C}H—CH_2\overset{Cl}{C}H—CH_2\overset{Cl}{C}H\sim$$

25.2

$$\sim CH_2—\overset{CH_3}{\underset{CH_3}{C}}—CH_2—\overset{Cl}{CH}-CH_2—\overset{CH_3}{\underset{CH_3}{C}}—CH_2—\overset{Cl}{CH}-CH_2—\overset{CH_3}{\underset{CH_3}{C}}\sim$$

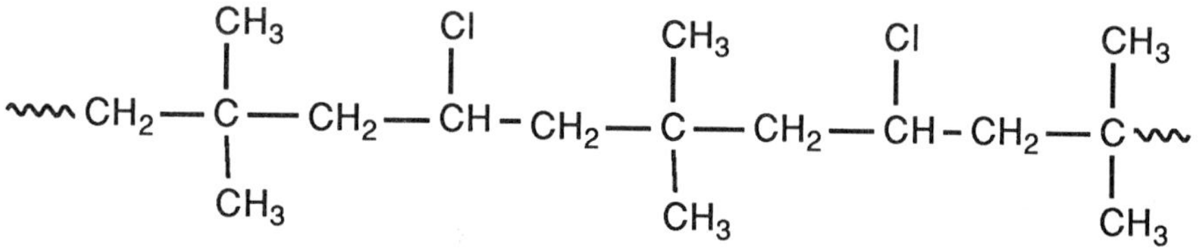

25.3 **(a)** Linear biopolymers: natural rubber, gutta percha, amylose, cellulose, chitin, carrageenin, DNA, RNA, proteins

(b) Branched biopolymers: amylopectin, glycogen

(c) Cross-linked biopolymers: some proteins

25.4 **(a)** Most biopolymers have random molecular weights: natural rubber, gutta percha, polysaccharides.

(b) Structurally homogeneous: proteins, DNA, RNA

25.5 Heteropolymers: polysaccharides, proteins, nucleic acids. (Natural rubber and gutta percha are the only homopolymers discussed in this section).

25.6

(a) $\sim OCH_2CH_2OCH_2CH_2OCH_2CH_2OCH_2CH_2\sim$

(b)

$$\text{ethylene oxide} \overset{H^+}{\rightleftharpoons} \text{protonated epoxide } (^{+}O-H) \xrightarrow{\text{ethylene oxide}} HOCH_2CH_2-\overset{+}{O}(\text{epoxide}) \xrightarrow{\text{ethylene oxide}} \cdots$$

$$HOCH_2CH_2(OCH_2CH_2)_nOCH_2CH_2-\overset{+}{O}(\text{epoxide}) \xrightarrow{H_2O} HOCH_2CH_2(OCH_2CH_2)_nOCH_2CH_2OCH_2CH_2OH$$

(c)

$$CH_3O^- + \text{ethylene oxide} \longrightarrow CH_3OCH_2CH_2O^- \xrightarrow{\text{ethylene oxide}} CH_3OCH_2CH_2OCH_2CH_2O^-$$

$$\xrightarrow{\text{ethylene oxide}} CH_3OCH_2CH_2O(CH_2CH_2O)_nCH_2CH_2O^- \xrightarrow{H_2O} CH_3OCH_2CH_2O(CH_2CH_2O)_nCH_2CH_2OH$$

25.7 (a) linear polymers

(i) $CH_2{=}CH_2 \xrightarrow{BF_3} -\!\!\left(CH_2CH_2\right)_n\!\!-$

(ii) $CH_2{=}CH{-}CH_3 \xrightarrow{BF_3} -\!\!\left(CH_2\underset{CH_3}{\underset{|}{C}H}\right)_n\!\!-$

(iii) $\text{2-pyrrolidinone (cyclic NH, C=O)} \xrightarrow[\text{trace } H_2O]{\text{heat}} -\!\!\left(\overset{O}{\overset{\|}{C}}-CH_2CH_2CH_2NH\right)_n\!\!-$

(iv) $\text{2-piperidinone (cyclic NH, C=O)} \xrightarrow[\text{trace } H_2O]{\text{heat}} -\!\!\left(\overset{O}{\overset{\|}{C}}-CH_2CH_2CH_2CH_2NH\right)_n\!\!-$

(b) linear (random) copolymer

$$CH_2{=}CH_2 + CH_2{=}CH{-}CH_3 \xrightarrow{\text{acid}} \sim\!\!\left(CH_2CH_2\right)_n\left(CH_2\underset{\displaystyle CH_3}{\underset{|}{C}H}\right)_x\left(CH_2CH_2\right)_m\left(CH_2\underset{\displaystyle CH_3}{\underset{|}{C}H}\right)_y\!\!\sim$$

$$\text{2-piperidone (cyclic, NH)} + \text{2-pyrrolidone (cyclic, NH)} \xrightarrow[\text{trace } H_2O]{\text{heat}} \left(\overset{O}{\overset{\|}{C}}{-}(CH_2)_3NH\right)_n\left(\overset{O}{\overset{\|}{C}}{-}(CH_2)_4NH\right)_x\left(\overset{O}{\overset{\|}{C}}{-}(CH_2)_3NH\right)_m\left(\overset{O}{\overset{\|}{C}}{-}(CH_2)_4NH\right)_y\!\!\sim$$

(c) block copolymer

$$HO\left(\overset{O}{\overset{\|}{C}}{-}CH_2CH_2CH_2NH\right)_nH + HO\left(\overset{O}{\overset{\|}{C}}{-}CH_2CH_2CH_2CH_2NH\right)_xH \xrightarrow{\text{dehydration}} HO\left(\overset{O}{\overset{\|}{C}}{-}CH_2CH_2CH_2NH\right)_n\left(\overset{O}{\overset{\|}{C}}{-}CH_2CH_2CH_2CH_2NH\right)_x\!\!\sim$$

ANSWERS TO PROBLEMS

25.1 The end groups of a macromolecule constitute just a tiny portion of the overall structure. Consequently, the chemical and physical properties of the polymer reflect the large number of atoms in the chain rather than the few at the ends.

25.2 **(a)** 7142 subunits

(b) 3361

(c) 1770

(d) 1092

(e) 1235

(f) 2000

(g) 1042

25.3

(a) polyethylene

$$\sim CH_2CH_2{-}CH_2CH_2{-}CH_2CH_2{-}CH_2CH_2{-}CH_2CH_2\sim$$

(b) poly(vinyl chloride)

$$\sim CH_2\overset{\overset{\large Cl}{|}}{C}H{-}CH_2\overset{\overset{\large Cl}{|}}{C}H{-}CH_2\overset{\overset{\large Cl}{|}}{C}H{-}CH_2\overset{\overset{\large Cl}{|}}{C}H{-}CH_2\overset{\overset{\large Cl}{|}}{C}H\sim$$

(c) nylon 6

$$\sim\overset{\overset{\large O}{\|}}{C}{-}(CH_2)_5NH{-}\overset{\overset{\large O}{\|}}{C}{-}(CH_2)_5NH{-}\overset{\overset{\large O}{\|}}{C}{-}(CH_2)_5NH{-}\overset{\overset{\large O}{\|}}{C}{-}(CH_2)_5NH\sim$$

(d) nylon 11

$$\sim\overset{\overset{\large O}{\|}}{C}{-}(CH_2)_{10}NH{-}\overset{\overset{\large O}{\|}}{C}{-}(CH_2)_{10}NH{-}\overset{\overset{\large O}{\|}}{C}{-}(CH_2)_{10}NH{-}\overset{\overset{\large O}{\|}}{C}{-}(CH_2)_{10}NH\sim$$

(e) cellulose

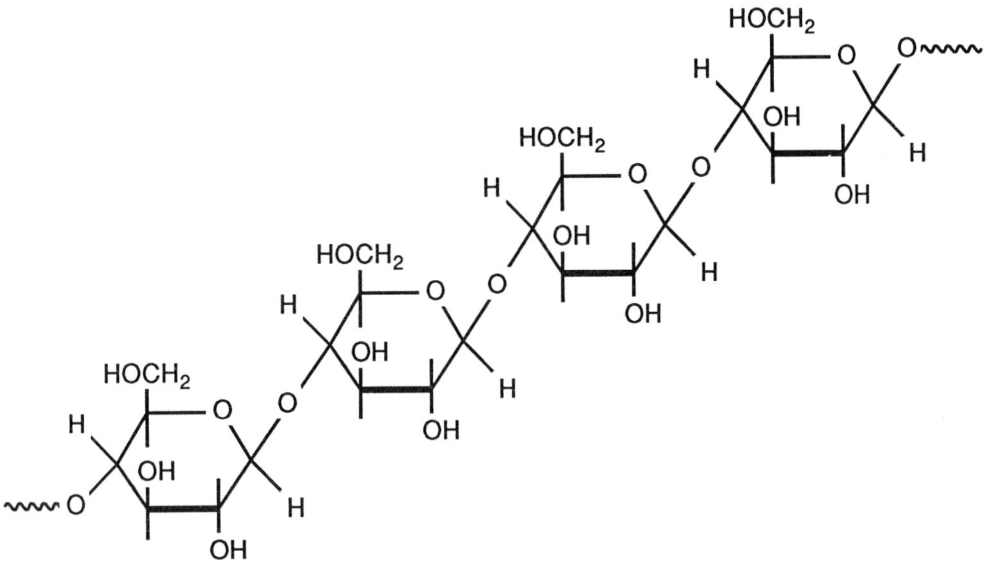

(f) poly(tetrafluoroethylene)

$$\sim CF_2CF_2—CF_2CF_2—CF_2CF_2—CF_2CF_2\sim$$

(g) poly(ethylene terephthalate)

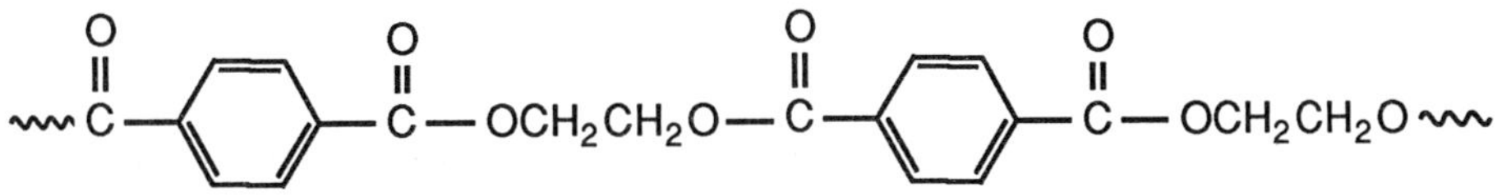

$$\sim \overset{O}{\overset{\|}{C}}-C_6H_4-\overset{O}{\overset{\|}{C}}—OCH_2CH_2O—\overset{O}{\overset{\|}{C}}-C_6H_4-\overset{O}{\overset{\|}{C}}—OCH_2CH_2O\sim$$

25.4

$$R\cdot + CH_2{=}CH_2 \longrightarrow R{-}CH_2CH_2\cdot \longrightarrow \longrightarrow R{-}CH_2CH_2{-}CH_2CH_2\cdot$$

$$\sim CH_2CH(H){-}(CH_2CH_2)_y\sim$$

$$\longrightarrow R{-}CH_2CH_2{-}CH_2CH_3 + \sim CH_2\dot{C}H{-}(CH_2CH_2)_y\sim$$

$$\sim CH_2\dot{C}H{-}(CH_2CH_2)_y\sim \xrightarrow{CH_2=CH_2} \sim CH_2CH(CH_2CH_2\cdot){-}(CH_2CH_2)_y\sim$$

25.5 RNA is more susceptible to hydrolytic cleavage because the 2′-hydroxyl group can attack the phosphate linkage in an intramolecular fashion. This is not possible in the 2′-deoxy system (i.e., in DNA).

RNA **DNA**

25.6

(a)

$$C_6H_5-\overset{\overset{O}{\|}}{C}-O-O-\overset{\overset{O}{\|}}{C}-C_6H_5 \longrightarrow 2\ C_6H_5-\overset{\overset{O}{\|}}{C}-O^\bullet \xrightarrow{-CO_2} C_6H_5^\bullet$$

$$C_6H_5^\bullet \xrightarrow{CH_2{=}CH_2} C_6H_5CH_2-CH_2^\bullet \longrightarrow \longrightarrow C_6H_5-(CH_2CH_2)_n\sim\sim$$

(b)

$$CH_3-\underset{\underset{CN}{|}}{\overset{\overset{CH_3}{|}}{C}}-N{=}N-\underset{\underset{CN}{|}}{\overset{\overset{CH_3}{|}}{C}}-CH_3 \longrightarrow N_2 + CH_3-\underset{\underset{CN}{|}}{\overset{\overset{CH_3}{|}}{C}}{}^\bullet$$

$$\xrightarrow{CH_2{=}CH_2} CH_3-\underset{\underset{CN}{|}}{\overset{\overset{CH_3}{|}}{C}}-CH_2CH_2^\bullet \longrightarrow \longrightarrow CH_3-\underset{\underset{CN}{|}}{\overset{\overset{CH_3}{|}}{C}}-(CH_2CH_2)_n\sim\sim$$

25.7 In order for a plasticizer to function properly, it must not evaporate during the functional lifetime of the polymer or commercial product for which it is used.

25.8

(a)

Ethylene oxide + H^+ ⇌ protonated epoxide ($H-O^+$ ring); attack by a second ethylene oxide gives $HOCH_2CH_2-O^+$ (ring); attack by another ethylene oxide gives $HOCH_2CH_2OCH_2CH_2-O^+$ (ring) $\longrightarrow \longrightarrow$ $HO{+}CH_2CH_2O{+}_n\sim\sim$

(b)

$$CH_3O^- + \text{ethylene oxide} \longrightarrow CH_3OCH_2CH_2O^- \longrightarrow CH_3OCH_2CH_2OCH_2CH_2O^- \longrightarrow CH_3O\text{+}CH_2CH_2O\text{+}_n$$

25.9 The poly(methyl methacrylate) is subjected to transesterification using base (sodium methoxide) and cyclohexanol:

$$\text{+}CH_2\text{—}C(CH_3)(C(=O)OCH_3)\text{+}_n \xrightarrow{^-O\text{-cyclohexyl}} \text{+}CH_2\text{—}C(CH_3)(C(=O)O\text{-cyclohexyl})\text{+}_n$$

25.10 Poly(ethylene terephthalate) could be prepared by a direct esterification or by transesterification using dimethyl terephthalate and ethylene glycol:

$$CH_3O\text{—}C(=O)\text{—}C_6H_4\text{—}C(=O)\text{—}OCH_3 + HOCH_2CH_2OH \xrightarrow{\text{high temp.}} \text{+}C(=O)\text{—}C_6H_4\text{—}C(=O)\text{—}OCH_2CH_2O\text{+}_n$$

25.11

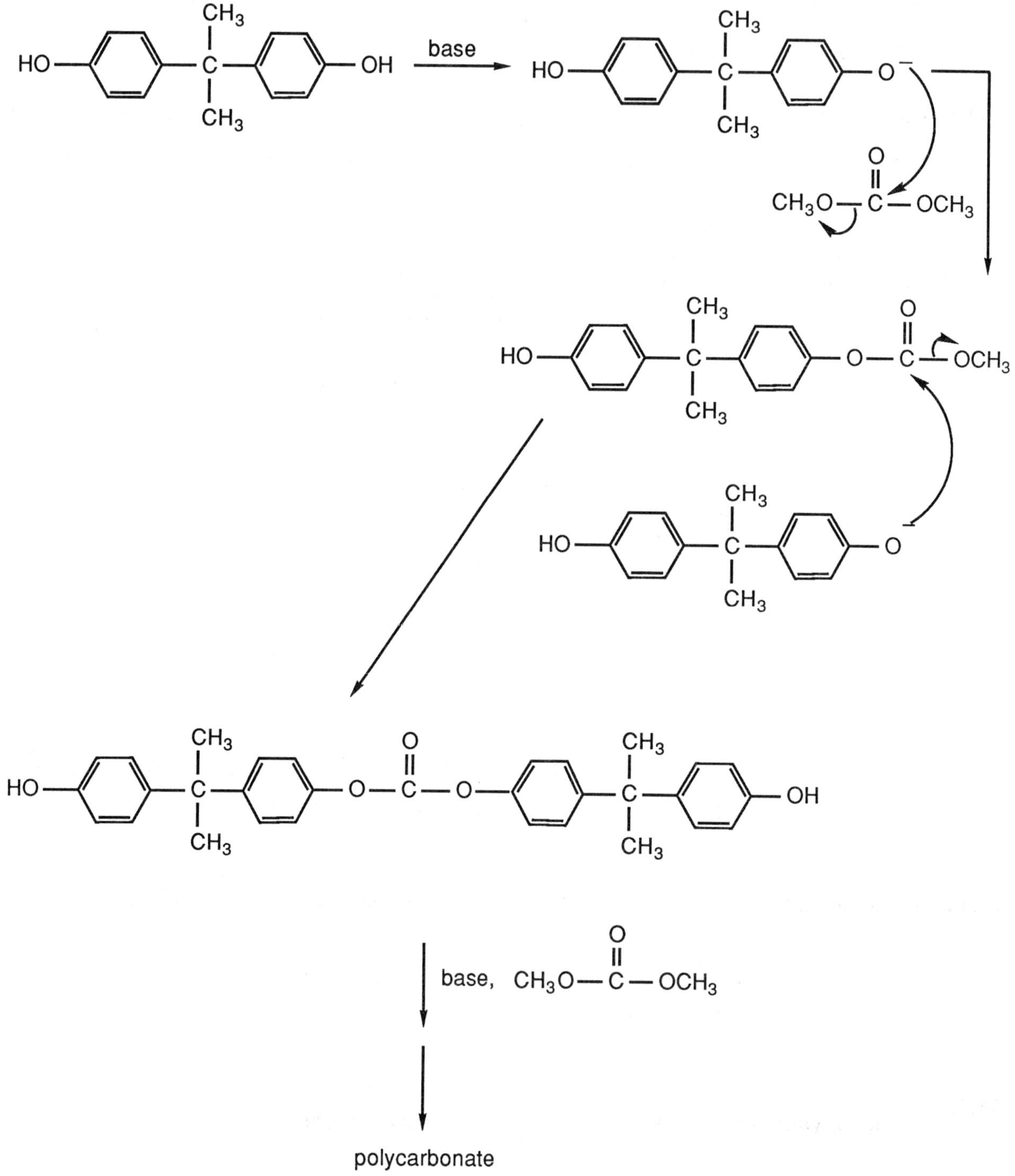

$HO-C_6H_4-C(CH_3)_2-C_6H_4-OH \xrightarrow{\text{base}} HO-C_6H_4-C(CH_3)_2-C_6H_4-O^-$

$CH_3O-\overset{O}{\overset{\|}{C}}-OCH_3$

$HO-C_6H_4-C(CH_3)_2-C_6H_4-O-\overset{O}{\overset{\|}{C}}-OCH_3$

$HO-C_6H_4-C(CH_3)_2-C_6H_4-O^-$

$HO-C_6H_4-C(CH_3)_2-C_6H_4-O-\overset{O}{\overset{\|}{C}}-O-C_6H_4-C(CH_3)_2-C_6H_4-OH$

base, $CH_3O-\overset{O}{\overset{\|}{C}}-OCH_3$

polycarbonate

25.12 Poly(ethylene glycol) is water soluble because it contains one oxygen atom for every two carbons. The repeating $-CH_2CH_2OCH_2CH_2O-$ groups interact favorably with water by hydrogen bonding.

25.13 Nucleophiles such as bisphenol A attack epichlorohydrin at the epoxide group to generate a chloro alcohol, but this undergoes cyclization to form a new epoxide ring:

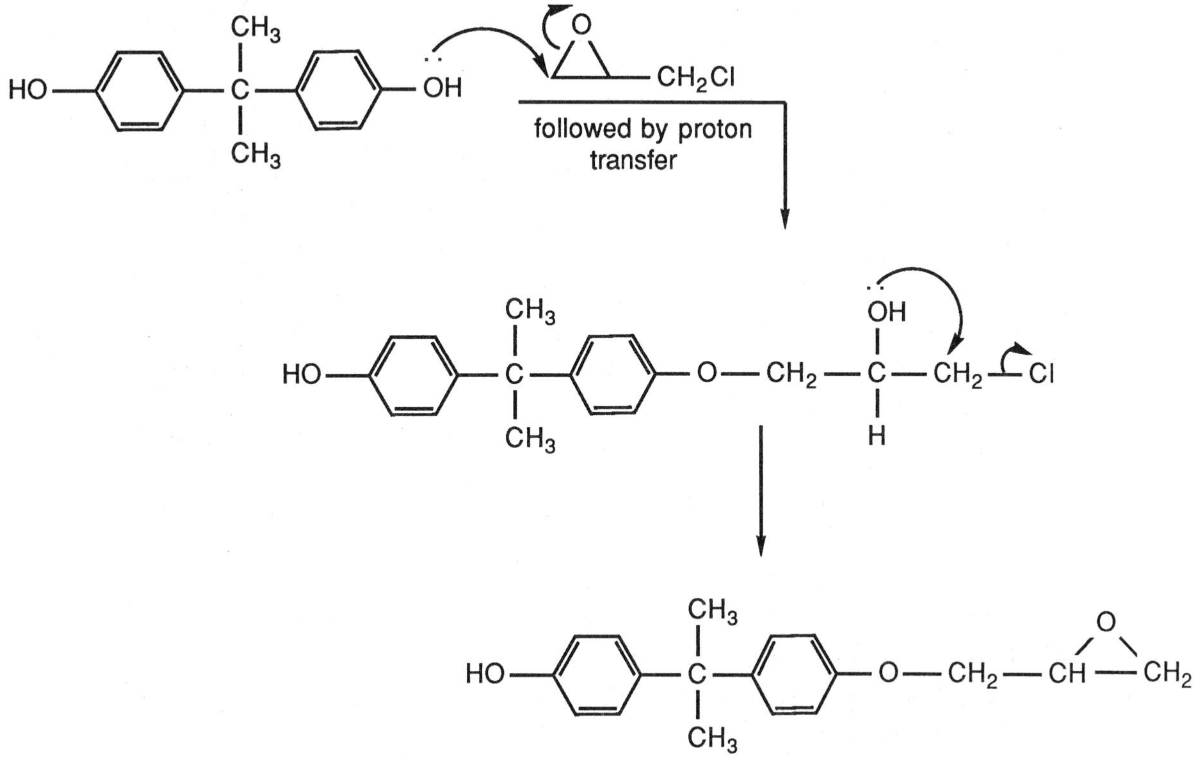

HO–C₆H₄–C(CH₃)₂–C₆H₄–OH + epichlorohydrin (epoxide–CH_2Cl)

followed by proton transfer

$HO-C_6H_4-C(CH_3)_2-C_6H_4-O-CH_2-CH(OH)-CH_2-Cl$

$HO-C_6H_4-C(CH_3)_2-C_6H_4-O-CH_2-\overset{O}{CH-CH_2}$ (epoxide)

25.14 1,4-Butanediol could be produced via acid-catalyzed hydrolysis of tetrahydrofuran:

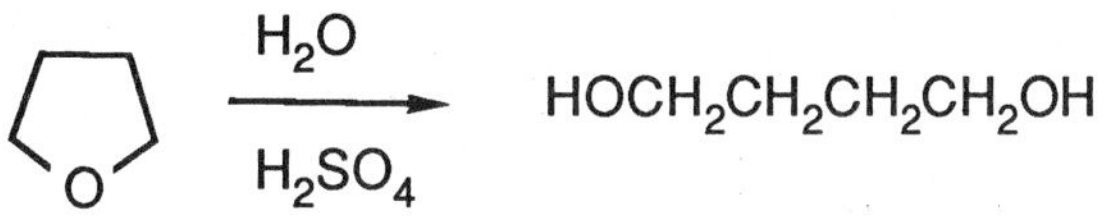

$$\text{THF} \xrightarrow[H_2SO_4]{H_2O} HOCH_2CH_2CH_2CH_2OH$$

25.15

(a) polyethylene $-\!(CH_2CH_2)_n\!-$ no chiral centers

(b) polystyrene $-\!(CH_2CH(C_6H_5))_n\!-$ stereochemical features possible

(c) teflon $-\!\!(\!CF_2CF_2\!)_{n}\!\!-$ no chiral centers

(d) PVC $-\!\!(\!CH_2CH(Cl)\!)_{n}\!\!-$ stereochemical features possible

(e) poly(methyl methacrylate) $-\!\!(\!CH_2-C(CO_2CH_3)(CH_3)\!)_{n}\!\!-$

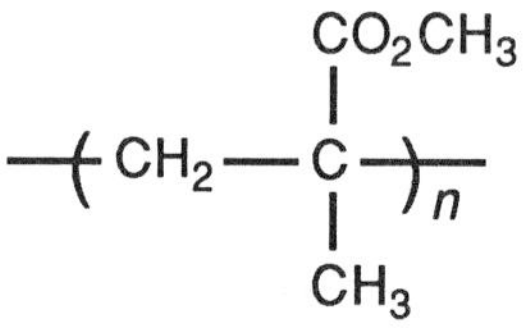

(f) nylon 66 $-\!\![\!C(=O)-CH_2CH_2CH_2CH_2-C(=O)-NH(CH_2)_6NH\!]_{n}\!\!-$ stereochemical features possible

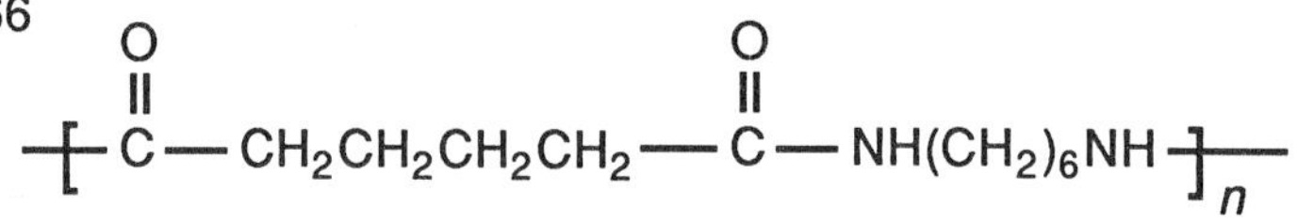

25.16

(b) isotactic polystyrene

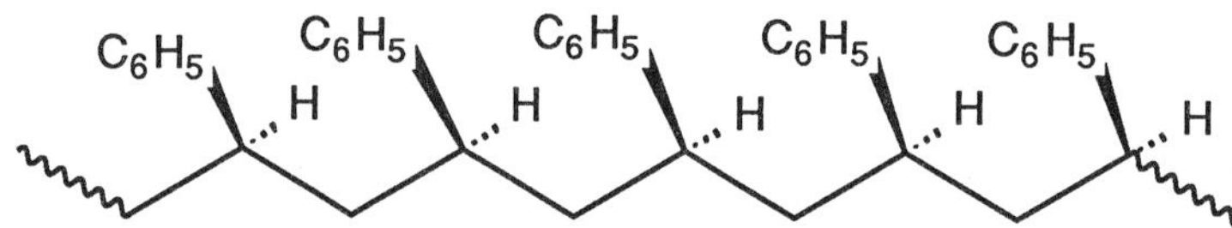

(d) isotactic poly(vinyl chloride)

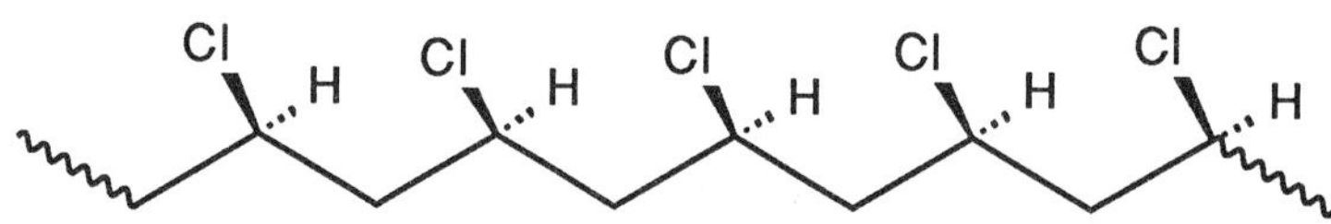

(e) isotactic poly(methyl methacrylate)

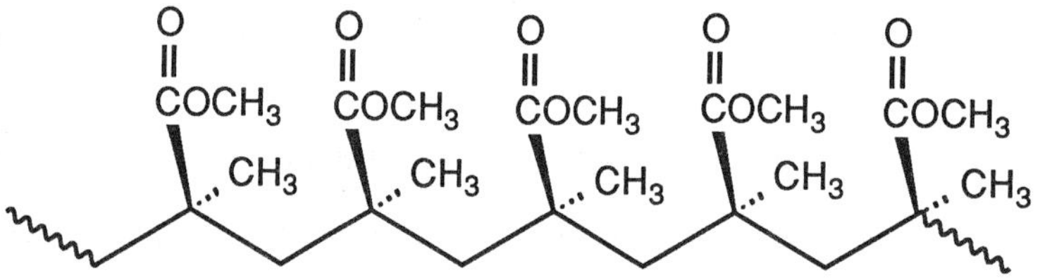

CHAPTER 26. INDUSTRIAL ORGANIC CHEMISTRY

ANSWERS TO EXERCISES

26.1 Polyethylene, polystyrene, poly(vinyl chloride), SBR rubber, poly(ethylene terephthalate) and other polyesters, epoxy resins.

26.2 The major industrial source of acetone at the present time is oxidation of cumene, and the three-carbon skeleton of acetone can be traced back to propylene (i.e., petroleum). As petroleum resources become depleted in the future, fermentation would appear to offer a useful alternative for acetone production. In addition, the development of one-carbon feedstocks may provide still other pathways for acetone production.

26.3 The least toxic (to rats) pesticide in Table 26.1 is methoprene, which requires a dose of 35 g per kg of body weight for 50% mortality:

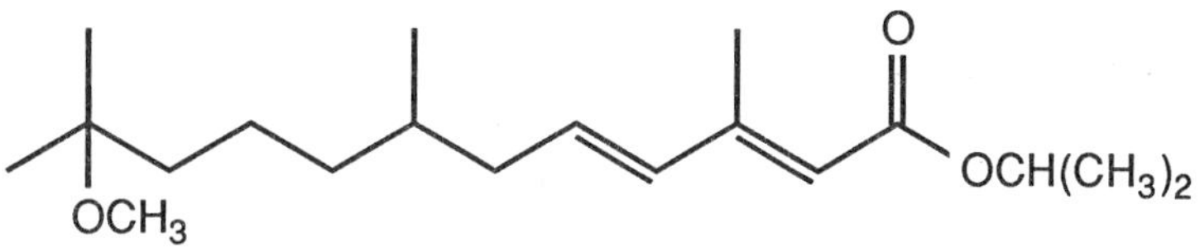

O

$OCH(CH_3)_2$

OCH_3

Methoprene is a highly insect-specific mimic of juvenile hormone. The most toxic pesticide in Table 26.1 is parathion, an organophosphate derivative. It has an LD_{50} of only 3 mg per kg of body weight in rats:

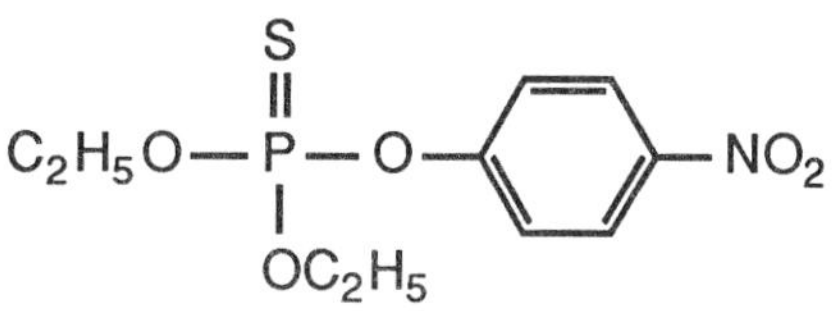

S

$C_2H_5O—P—O—C_6H_4—NO_2$

OC_2H_5

ANSWERS TO PROBLEMS

26.1 Polymers based on cyclic petroleum-derived compounds include polystyrene and BSR rubber, nylon 6 and nylon 66, poly(ethylene terephthalate, polycarbonates, and epoxy resins.

26.2 **(a)** methane: teflon

(b) ethane: poly(chlorotrifluoroethylene)

(c) ethylene: poly(ethylene terephthalate) and other polyesters, poly(ethylene glycol) and other polyethers, polyethylene, poly(vinyl chloride), polystyrene, and SBR rubber

(d) propylene: polypropylene, polyacrylonitrile, polycarbonate, poly(methyl methacrylate)

(e) butadiene: neoprene, SBR rubber

(f) 2-methylbutene: elastomers

(g) benzene: polyamides, polystyrene, SBR rubber, polycarbonate

(h) *p*-xylene: poly(ethylene terephthalate) and other polyesters

26.3 **(a)**

$C_6H_5CH(CH_3)_2 \xrightarrow{O_2\ (air)} C_6H_5C(CH_3)_2OOH \xrightarrow{H_2SO_4} C_6H_5OH + CH_3COCH_3$

$\xrightarrow{H_2SO_4}$ $HO-C_6H_4-C(CH_3)_2-C_6H_4-OH$

(b) Phenol and acetone are generated in equal amounts from cumene hydroperoxide, but twice as much phenol as acetone is needed in the subsequent condensation:

$$2\ C_6H_5OH + CH_3\overset{\overset{\displaystyle O}{\|}}{C}CH_3 \longrightarrow HO-C_6H_4-\underset{\underset{\displaystyle CH_3}{|}}{\overset{\overset{\displaystyle CH_3}{|}}{C}}-C_6H_4-OH$$

Therefore, excess acetone is produced.

(c) The excess acetone could be used in some other process, or else an alternative (secondary) source of phenol could be employed.

26.4

(a)

$$CH_3-\overset{\overset{\displaystyle O}{\|}}{C}-CH_3 \xrightarrow[\text{catalyst}]{\text{base}} CH_3-\underset{\underset{\displaystyle CH_3}{|}}{\overset{\overset{\displaystyle OH}{|}}{C}}-CH_2-\overset{\overset{\displaystyle O}{\|}}{C}-CH_3 \quad \text{(aldol condensation)}$$

diacetone alcohol

(b)

$$CH_3-\underset{\underset{\displaystyle CH_3}{|}}{\overset{\overset{\displaystyle OH}{|}}{C}}-CH_2-\overset{\overset{\displaystyle O}{\|}}{C}-CH_3 \xrightarrow[\text{catalyst}]{\text{acid}} (CH_3)_2C=CH-\overset{\overset{\displaystyle O}{\|}}{C}-CH_3 \quad \text{(dehydration)}$$

mesityl oxide

(c)

$$(CH_3)_2C=CH-\overset{\overset{\displaystyle O}{\|}}{C}-CH_3 \xrightarrow{H_2,\ Ni} CH_3-\overset{\overset{\displaystyle CH_3}{|}}{CH}-CH_2-\overset{\overset{\displaystyle O}{\|}}{C}-CH_3 \quad \text{(catalytic hydrogenation of mesityl oxide)}$$

methyl isobutyl ketone

(d)

$$(CH_3)_2C=CH-\overset{\overset{\displaystyle O}{\|}}{C}-CH_3 \xrightarrow[\text{high temp., pressure}]{H_2,\ Ni} CH_3-\overset{\overset{\displaystyle CH_3}{|}}{CH}-CH_2-\overset{\overset{\displaystyle OH}{|}}{CH}-CH_3 \quad \text{(catalytic hydrogenation of mesityl oxide)}$$

methyl isobutyl carbinol

(e)

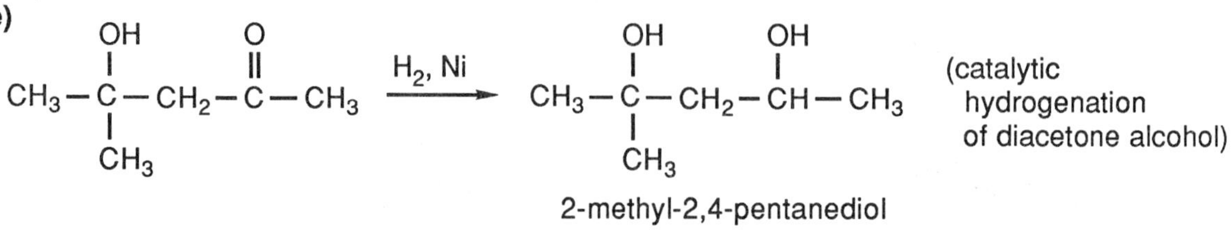

$$CH_3-\underset{CH_3}{\overset{OH}{C}}-CH_2-\overset{O}{\overset{\|}{C}}-CH_3 \xrightarrow{H_2,\ Ni} CH_3-\underset{CH_3}{\overset{OH}{C}}-CH_2-\overset{OH}{CH}-CH_3$$

2-methyl-2,4-pentanediol

(catalytic hydrogenation of diacetone alcohol)

(f)

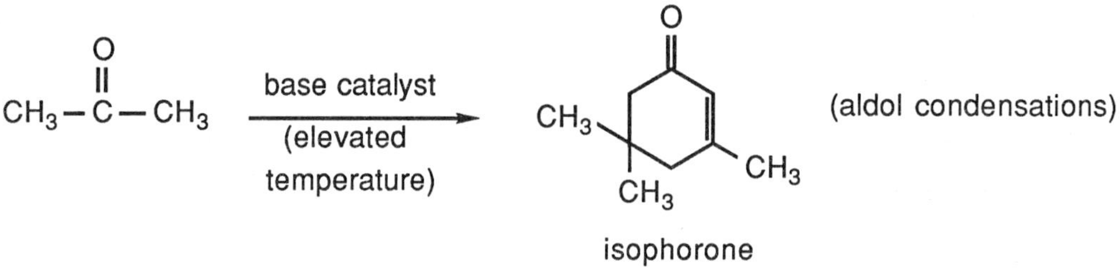

$$CH_3-\overset{O}{\overset{\|}{C}}-CH_3 \xrightarrow[\text{(elevated temperature)}]{\text{base catalyst}}$$ isophorone

(aldol condensations)

26.5 A large portion of the acetone produced industrially is used in the reaction with phenol to make bisphenol A. Hence the formation of both reactants via the oxidation of cumene avoids the necessity for manufacturing the second component in a separate process.

26.6

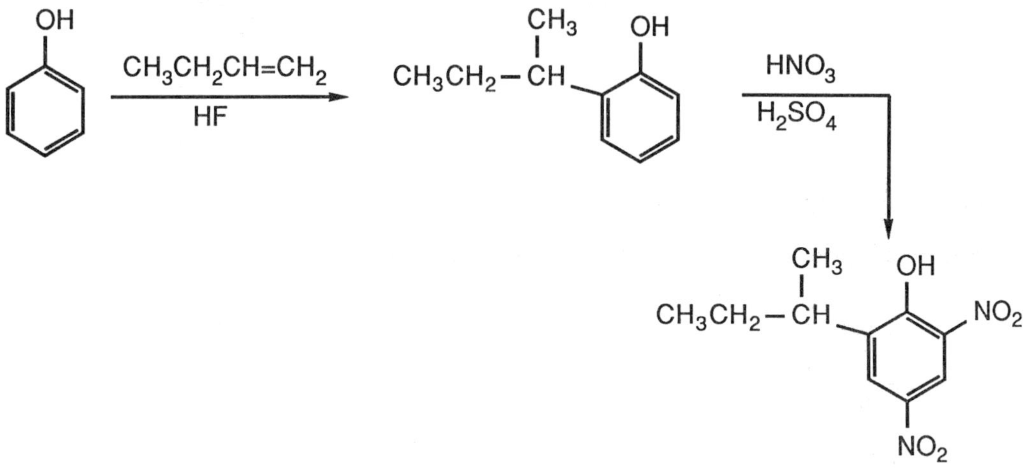

$$C_6H_5OH \xrightarrow[HF]{CH_3CH_2CH=CH_2} \text{2-}(CH_3CH_2-\overset{CH_3}{CH})C_6H_4OH \xrightarrow[H_2SO_4]{HNO_3} \text{2-}(CH_3CH_2-\overset{CH_3}{CH})\text{-4,6-}(NO_2)_2C_6H_2OH$$

26.7 The oxidation of *o*-xylene with air at 400°C with a vanadium oxide catalyst produces phthalic anhydride:

$$C_6H_4(CH_3)_2 + 3\,O_2 \xrightarrow[\text{catalyst}]{O_2} C_6H_4(CO)_2O + 3\,H_2O^{\dagger}$$

26.8 Chlorination of benzene produces chlorobenzene together with di–, tri–, and polychlorinated benzenes:

Cl$_2$, catalyst: benzene → chlorobenzene + 1,2-dichlorobenzene + 1,4-dichlorobenzene

by- products: 1,2,4-trichlorobenzene + 1,2,4,5-tetrachlorobenzene

Reaction of the chlorobenzenes with NaOH at high temperatures would then produce the chlorophenols, and alkylation with chloroacetic acid would yield the herbicides:

NaOH: 1,2,4-trichlorobenzene → 2,4-dichlorophenoxide (O^-); $ClCH_2CO_2H$, then H^+ → OCH_2CO_2H product

2,4-dichlorophenoxyacetic acid

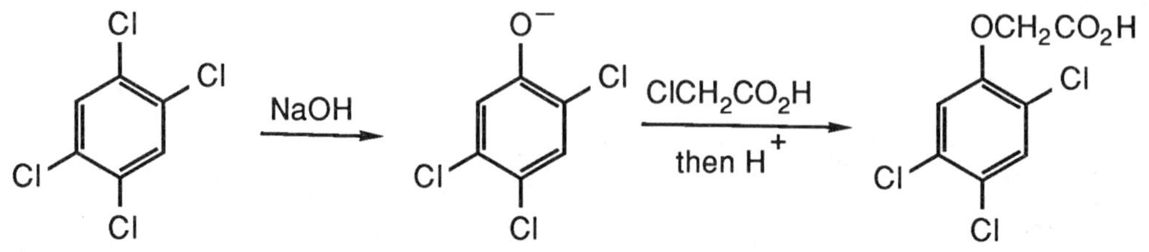

NaOH; $ClCH_2CO_2H$ then H^+

2,4,5-trichlorophenoxyacetic acid

26.9 The mechanisms are very similar, although the intermediate cations are less stable in the DDT reactions:

Bisphenol A

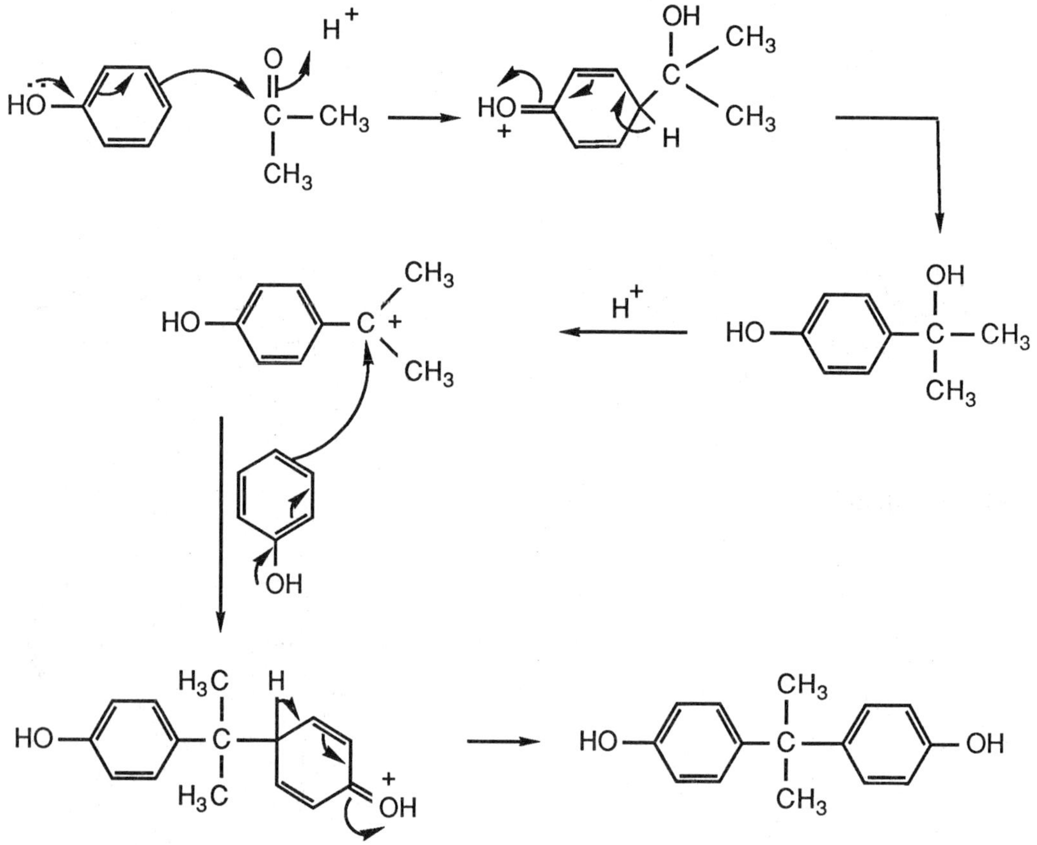

DDT

Cl—C₆H₄ + CH₃CHO (H⁺) → Cl⁺=C₆H₄(H)—CH(OH)—CH₃ → Cl—C₆H₄—CH(OH)—CH₃

Cl—C₆H₄—CH(OH)—CH₃ —(H⁺)→ Cl—C₆H₄—CH⁺—CH₃

Cl—C₆H₄—CH⁺—CH₃ + C₆H₅Cl → Cl—C₆H₄—CH(CH₃)—C₆H₅(H)=Cl⁺ → Cl—C₆H₄—CH(CH₃)—C₆H₄—Cl

26.10 (a) Parathion:

$$(C_2H_5O)_2P(=S)—O—C_6H_4—NO_2 \xrightarrow{H_2O} (C_2H_5O)_2P(=S)—OH + HO—C_6H_4—NO_2$$

$$\downarrow$$

$$C_2H_5OH + H_2S + H_3PO_4$$

(b) Malathion:

$$CH_3O-\overset{\overset{S}{\|}}{\underset{\underset{CH_3O}{|}}{P}}-S-\underset{\underset{CH_2CO_2C_2H_5}{|}}{CHCO_2C_2H_5} \xrightarrow{H_2O} CH_3O-\overset{\overset{S}{\|}}{\underset{\underset{CH_3O}{|}}{P}}-OH + HS-\underset{\underset{CH_2CO_2C_2H_5}{|}}{CHCO_2C_2H_5}$$

$$\downarrow \qquad\qquad \downarrow$$

$$CH_3OH + H_2S + H_3PO_4 \quad + \quad HS-\underset{\underset{CH_2CO_2H}{|}}{CHCO_2H} + C_2H_5OH$$

(c) Fenitrothion:

$$CH_3O-\overset{\overset{S}{\|}}{\underset{\underset{CH_3O}{|}}{P}}-O-C_6H_3(CH_3)-NO_2 \xrightarrow{H_2O} CH_3O-\overset{\overset{S}{\|}}{\underset{\underset{CH_3O}{|}}{P}}-OH + HO-C_6H_3(CH_3)-NO_2$$

$$\downarrow$$

$$CH_3OH + H_2S + H_3PO_4$$

26.11 **(a)** Carbaryl:

$$C_{10}H_7-O-\overset{\overset{O}{\|}}{C}NHCH_3 \xrightarrow{H_2O} C_{10}H_7-OH + CO_2 + CH_3NH_2$$

(b) Carbofuran:

$$\text{(2,2-dimethyl-2,3-dihydrobenzofuran-7-yl)}-O-\underset{\underset{O}{\|}}{C}NHCH_3 \xrightarrow{H_2O} \text{(2,2-dimethyl-2,3-dihydrobenzofuran-7-yl)}-OH + CO_2 + CH_3NH_2$$

26.12 **(a)** Three examples of catalytic hydrogenation were actually presented in this chapter, but only the second and third shown below are processes that yield a specific product:

(1) Catalytic reforming of petroleum:

$$\text{petroleum} \longrightarrow H_2 + C_6H_6 + C_6H_5CH_3 + C_6H_{6-n}(CH_3)_n$$

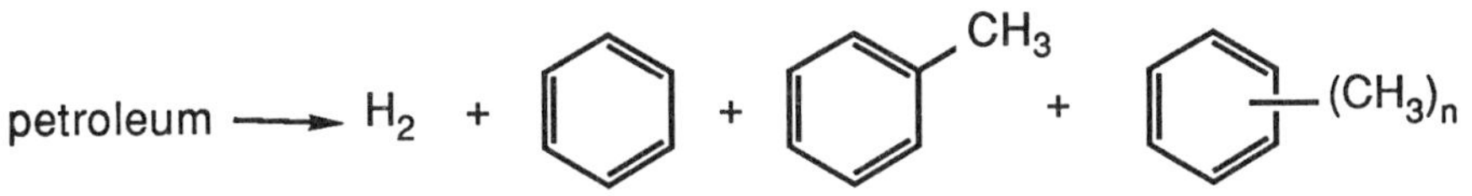

(2) Dehydrogenation of ethylbenzene:

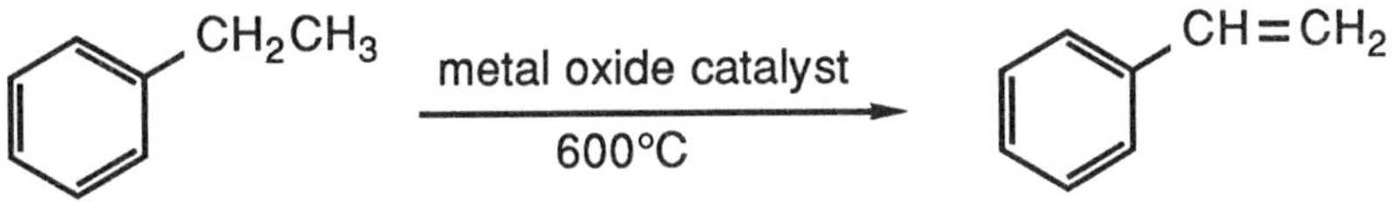

$$C_6H_5CH_2CH_3 \xrightarrow[600^\circ C]{\text{metal oxide catalyst}} C_6H_5CH{=}CH_2$$

(3) Dehydrogenation of 2-methylbutenes:

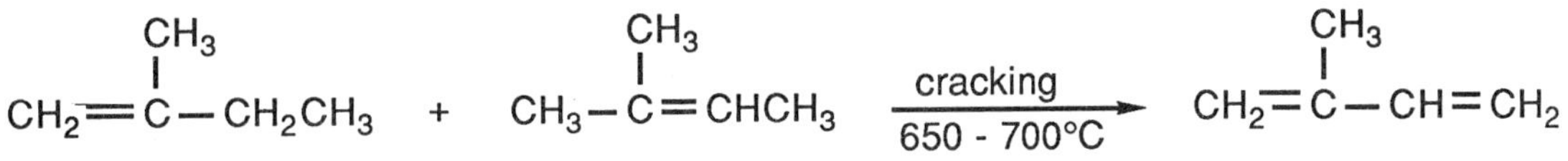

$$CH_2{=}C(CH_3){-}CH_2CH_3 + CH_3{-}C(CH_3){=}CHCH_3 \xrightarrow[650 - 700^\circ C]{\text{cracking}} CH_2{=}C(CH_3){-}CH{=}CH_2$$

(b) The production of benzene, toluene, xylenes, etc. in the catalytic reforming of petroleum reflects the stability of these aromatic systems. In the production of styrene and isoprene the formation of a single product is a consequence of the fact that there is only one way in which two hydrogens can be removed; no other isomeric product can be formed.

CHAPTER 27. MOLECULAR ORBITAL THEORY

ANSWERS TO EXERCISES

27.1 Each second row element contributes four valence-level atomic orbitals and a total of five atomic orbitals when the 1*s* level is also counted. Hydrogen has a single orbital (1*s*), and it is a valence level orbital. For each species the number of contributing valence atomic orbitals is equal to the number of valence level MOs. When 1*s* orbitals are included, the total number of atomic orbitals is equal to the total number of molecular orbitals.

(a)	CO_2	12 valence	15 total
(b)	CH_4	8 valence	9 total
(c)	CF_4	20 valence	25 total
(d)	NH_3	7 valence	8 total
(e)	H_2O	6 valence	7 total
(f)	HCN	9 valence	11 total
(g)	C_6H_6	30 valence	36 total
(h)	$C_{15}H_{30}$	90 valence	105 total

27.2 **(a)** This is not a legitimate molecular orbital. The atomic *p* orbital at the left is symmetric with respect to the plane of the paper, while that on the right is antisymmetric.

(b) This is a legitimate antibonding orbital.

(c) This is a legitimate bonding σ orbital.

(d) This is not a legitimate bonding orbital, because the contribution of the hydrogen atom lacks the π symmetry of the *p* orbital on carbon. If a contribution from the appropriate hydrogen were included, a legitimate π-type orbital would result:

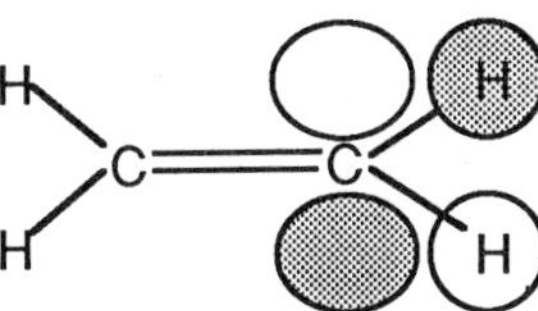

(e) This is a legitimate antibonding orbital.

27.3 **(a)** BeH_2 has a total of four valence level electrons. Both ϕ_1 and ϕ_2 will be doubly occupied, favoring a linear geometry.

(b) BH_2^+ has a total of four valence level electrons. As in the case of BeH_2, a linear geometry is favored.

(c) NH_2^- has a total of eight valence level electrons. ϕ_1–ϕ_3 will all be doubly occupied. The large effect of ϕ_3 favors a nonlinear geometry.

(d) H_2O has a total of eight valence level electrons. As in the case of NH_2^- this favors a nonlinear geometry.

27.4 To evaluate the cleavage of cyclobutane into two ethylenes, you only need to consider the molecular orbitals corresponding to bonds being broken or formed. For cyclobutane these are the two lowest energy orbitals shown in Figure 27.8, $\sigma_1 + \sigma_2$ and $\sigma_1 - \sigma_2$. Both of these orbitals are doubly occupied, and they correlate with $\pi_1 - \pi_2$ and $\pi^*_1 - \pi^*_2$, respectively. The latter is antibonding for both ethylenes, so the concerted fragmentation would have a very high activation energy and is not a favorable process.

27.5 **(a)** The photochemical ring opening of 7-dehydrocholesterol will initially generate an excited state of previtamin D_3. The HOMO of this species is π_4 of the triene system, which has three nodes:

This orbital governs the ring closure (or ring opening of the reverse process), and *conrotatory* motion is necessary to obtain in-phase overlap between the *p* orbitals at the ends of the triene system.

(b) For simplicity the photochemical conrotatory reaction is shown with a partial structure:

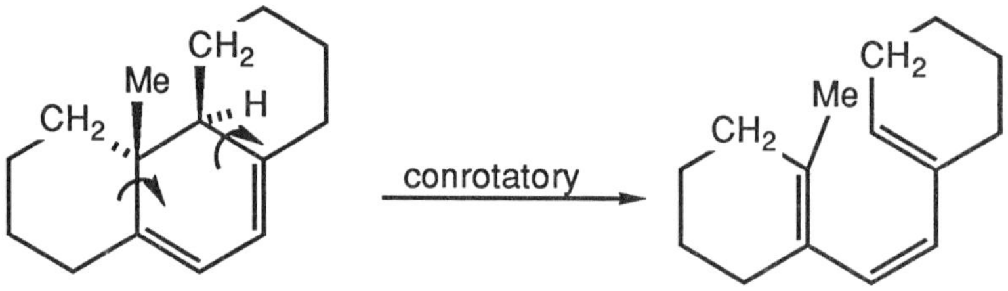

Next, using "R" to designate the CH_2 groups of the adjacent rings, the disrotatory process is shown:

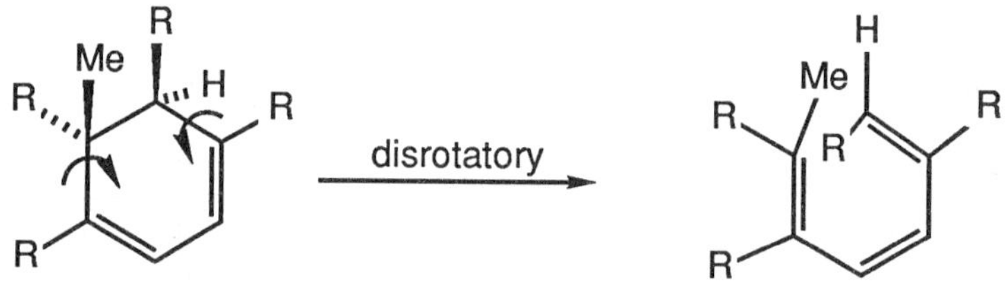

As seen in the preceding drawing, the disrotatory process leads to a product in which the R groups must be *trans* on one of the double bonds. In other words, an *E*-cyclohexene would result, and it is the high energy of such a compound that prevents the thermal ring opening.

27.6 The frontier orbitals that must be considered are those of a hydrogen atom and of a heptatrienyl radical:

H and C=C-C=C-C=C-C

The heptatrienyl radical has seven contributing atomic *p* orbitals and seven electrons, so four of the resulting π MO's will be occupied. The HOMO, which is only singly occupied, has three nodes. As you can see in the following drawing, when there is an odd number of nodes, the contributing atomic p orbitals at the termini must be of opposite phase.

The opposite phase of the p orbitals at the termini of the 7-carbon chain requires that the hydrogen migrate from one face to the other -- i.e., by the antarafacial mode.

27.7 For the answers to this question we will show orbitals that are generated by combining localized group or atomic orbitals. These can interact further when they have the same symmetry, but we will not consider that level of complexity.

(a) To generate the molecular orbitals of formaldehyde:

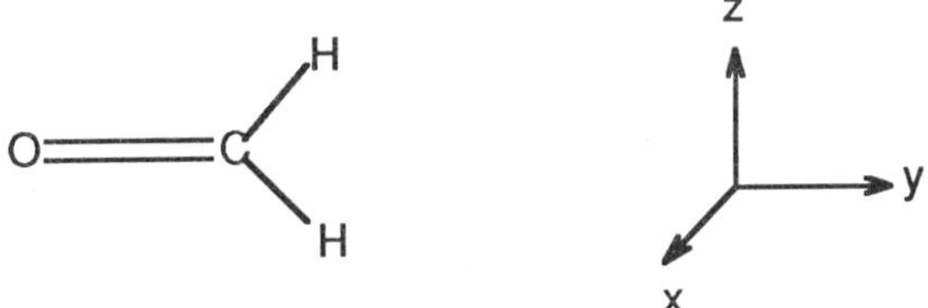

the group orbitals for CH_2 (Figure 27.11) can be used along with the atomic orbitals for the oxygen atom. There are a total of 10 valence atomic orbitals (four each from carbon and oxygen; one from each hydrogen), and these correspond to the four oxygen orbitals plus the six CH_2 group orbitals. These will result in a total of 10 molecular orbitals. There will be a total of 12 valence electrons (one from each hydrogen, four from carbon, and six from oxygen), so six of the MOs will be doubly occupied. To account for bonding between carbon and oxygen, we will consider a hybrid of oxygen s orbitals and the p orbital that lies on the bonding axis (p_y). The hybridized atomic orbitals on oxygen are:

First consider σ bonding. Between carbon and oxygen this results from interaction of the *sp* hybrids on C and O that point toward the other atom:

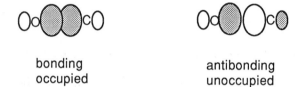

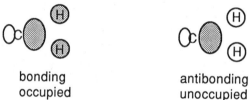

bonding
occupied

antibonding
unoccupied

The carbon–hydrogen σ interactions correspond to the appropriate CH_2 group orbitals:

bonding
occupied

antibonding
unoccupied

Next consider π interactions. Two of the molecular orbitals are the appropriate CH_2 group orbitals (C–H localized):

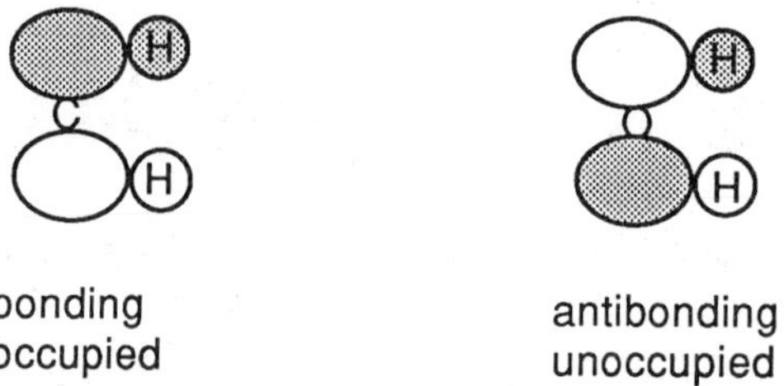

bonding
occupied

antibonding
unoccupied

and two more involve interaction of the p_x atomic orbitals on carbon and oxygen:

bonding
occupied

antibonding
unoccupied

Construction of the preceding orbitals has employed all six of the CH_2 group orbitals, and the remaining two oxygen atomic orbitals represent molecular orbitals that are nonbonding:

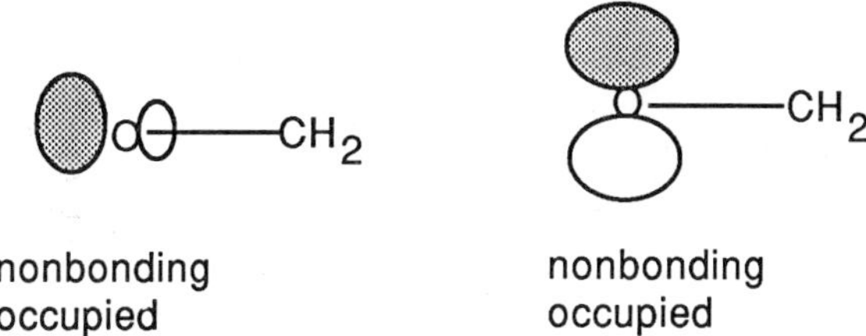

The six doubly occupied MOs will be the four orbitals listed here as bonding plus the two that are designated as bonding.

(b) The orbitals of ethylene can be constructed from the group orbitals for CH_2 (Figure 27.11). The two sets of six CH_2 group orbitals will yield 12 MOs, and the total of 12 electrons will produces six doubly occupied MOs. The C–H σ orbitals result from combination of the equivalent group orbitals:

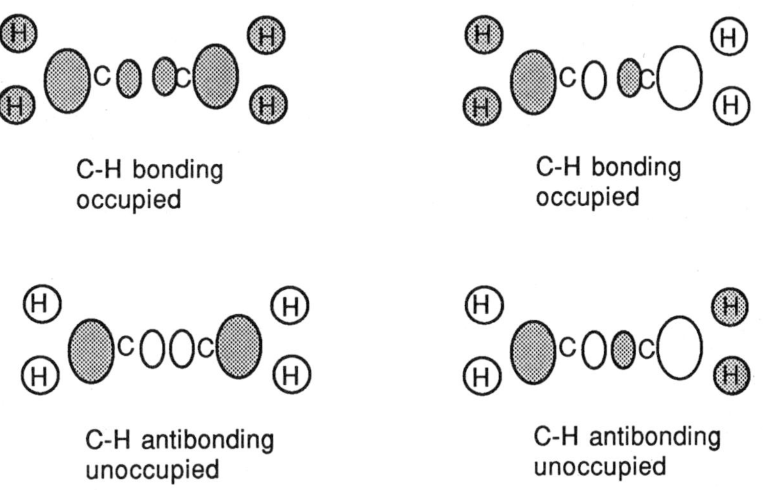

The C–C σ orbitals result from combination of the *sp* hybrid orbital on carbon that points away from the hydrogens:

There are a total of four π-type orbitals that result from combination of the appropriate CH_2 group orbitals. These are all predominantly C–H bonding (or antibonding) in nature, regardless of whether the C–C interaction is in phase or out of phase:

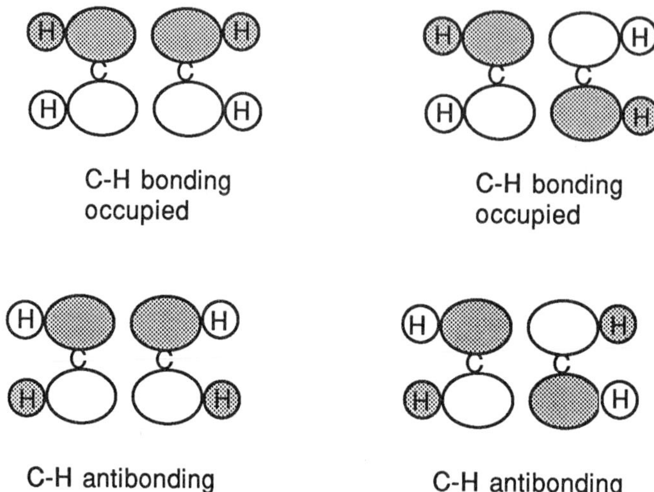

Finally, there are in-phase and out-of-phase combinations of the p_x orbitals that produce the C–C π molecular orbitals:

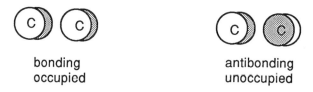

All of the bonding molecular orbitals are doubly occupied, and none of the antibonding orbitals are occupied.

ANSWERS TO PROBLEMS

27.1 Each second row element contributes four valence-level atomic orbitals and a total of five atomic orbitals when the 1*s* level is also counted. Hydrogen has a single orbital (1*s*), and it is a valence level orbital. For each species the number of contributing valence atomic orbitals is listed, and this number is equal to the total number of valence level molecular orbitals.

(a) $CH_2{=}CH_2$

12 valence 6 bonding 6 antibonding

(b) CH_3NH_2

13 valence 6 bonding 6 antibonding 1 nonbonding

(c) $CH_3OH_2^+$

13 valence 6 bonding 6 antibonding 1 nonbonding

(d) BH_3

7 valence 3 bonding 3 antibonding 1 nonbonding (unoccupied)

(e) CH_3^+

7 valence 3 bonding 3 antibonding 1 nonbonding (unoccupied)

(f) $CH_2{=}CH{-}CH{=}CH_2$

22 valence 11 bonding 11 antibonding

(g) $CH_2{=}C{=}CH_2$

16 valence 8 bonding 8 antibonding

27.2 **(a)**

(b)-(d)

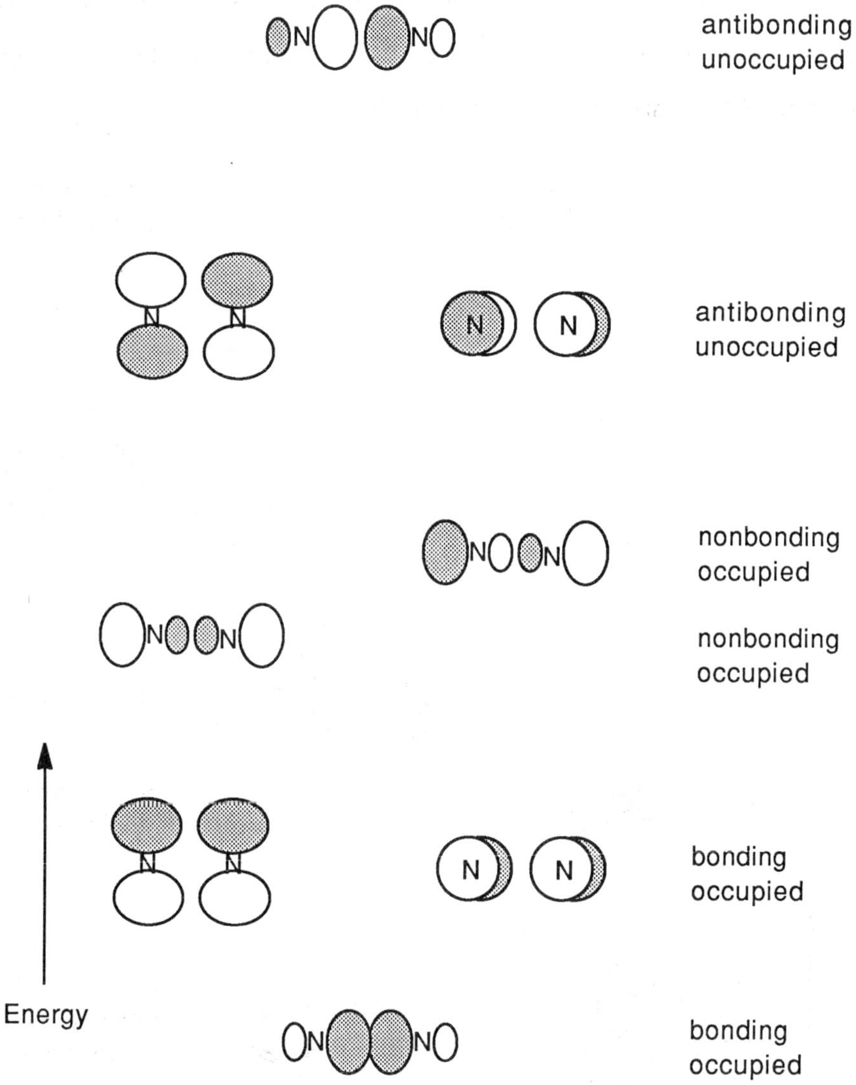

antibonding
unoccupied

antibonding
unoccupied

nonbonding
occupied

nonbonding
occupied

bonding
occupied

Energy

bonding
occupied

The orbitals are shown in order of increasing energy; the three bonding orbitals and both nonbonding orbitals are doubly occupied.

27.3 The atomic and molecular orbitals shown in Problem 27.2 also apply to this question:

This time, however, there are 12 electrons. The five lowest-energy orbitals will each be doubly occupied, but the next two orbitals (both π^*) have the same energy. The energy of the molecule is less if one electron occupies each of these degenerate orbitals than if they are spin-paired in a single orbital:

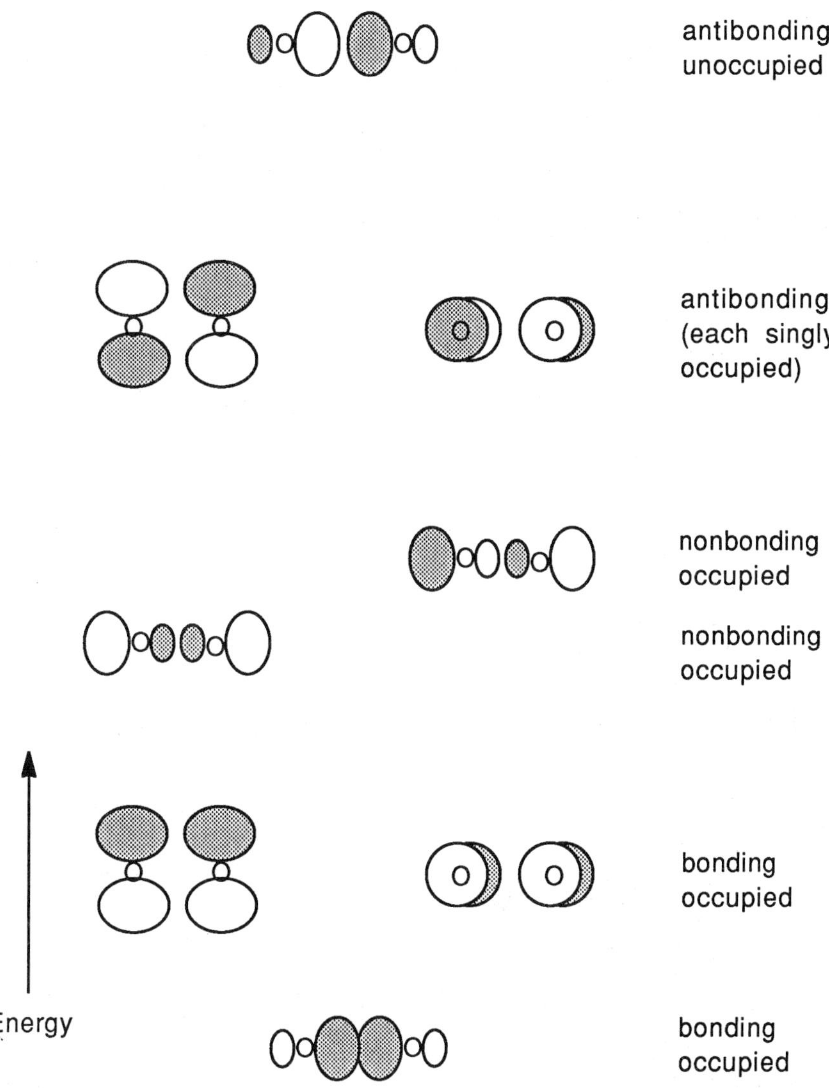

27.4 **(a)** The hybrid orbitals shown in Problem 27.2 for oxygen are those needed for both carbon and nitrogen:

(b)-(d) The major difference between the molecular orbitals for HCN is the interaction of the H atom with the sp_y hybrid orbital. This generates bonding and antibonding orbitals) that are primarily C–H localized, and the corresponding sp hybrid on nitrogen remains as a nonbonding orbital:

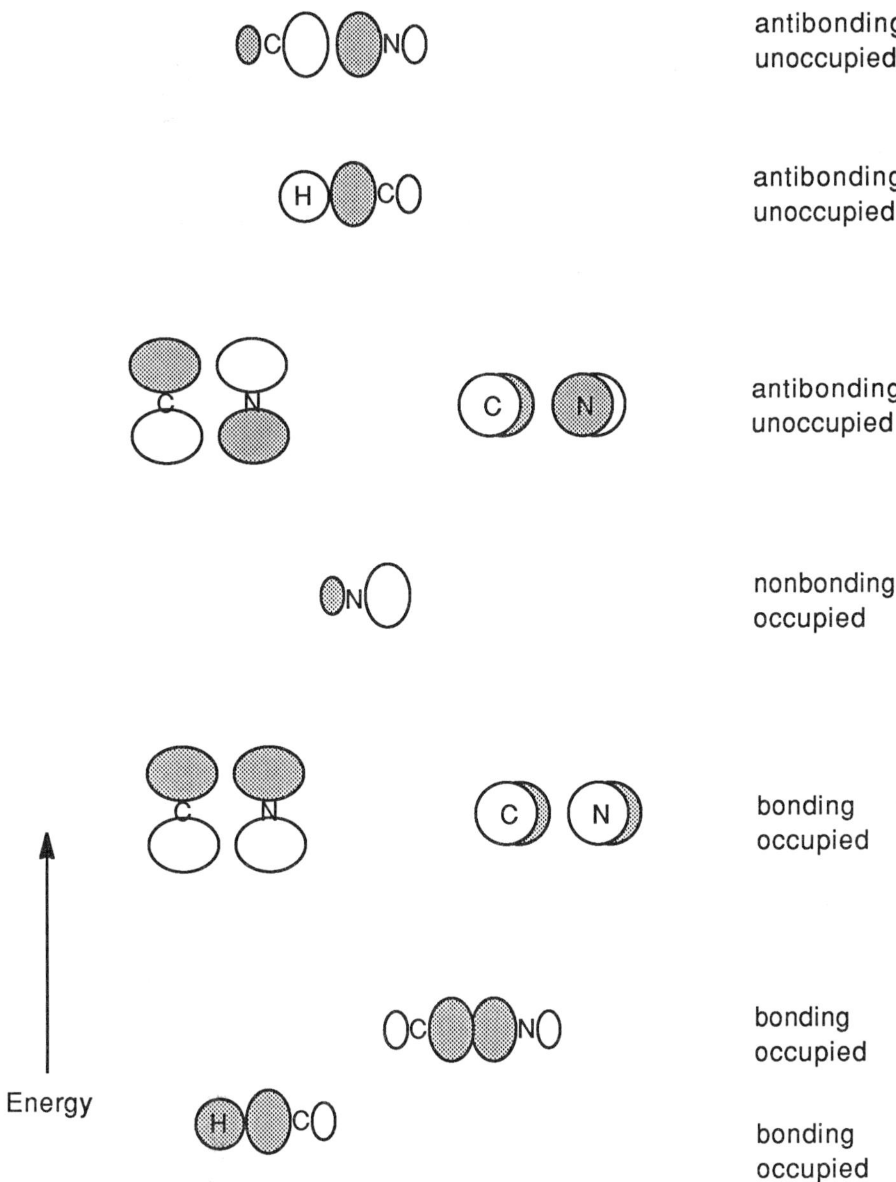

There are 10 valence electrons for HCN, so the four bonding orbitals and the nonbonding orbital (nitrogen) will each be doubly occupied.

27.5 **(a)** Not a legitimate MO: the p orbital on oxygen is symmetric with respect to the vertical plane of symmetry through the oxygen (perpendicular to the page), but the hydrogen orbitals are antisymmetric with respect to this plane.

(b) A legitimate MO: both the p orbital on carbon and the s orbitals on hydrogen are antisymmetric with respect to the vertical symmetry plane that is perpendicular to the page.

(c) A legitimate MO: there are three symmetry planes (including the plane of the paper). For any one of these all three *p* orbitals are either symmetric or antisymmetric.

(d) Not a legitimate MO: the *p* orbital on oxygen is antisymmetric with respect to the vertical plane of symmetry that is perpendicular to the page, but the hydrogen orbitals are symmetric with respect to this plane.

(e) Not a legitimate MO: the *p* orbitals on the carbons are antisymmetric with respect to the horizontal plane of symmetry that is perpendicular to the page, but the hybrid orbital on oxygen is symmetric with respect to this plane.

(f) A legitimate MO: all orbitals are symmetrical with respect to both the plane of the paper and the vertical plane of symmetry that is perpendicular to the page.

(g) Not a legitimate MO: There are three molecular planes of symmetry for cyclobutane. The orbital is symmetric with regard to the plane of the paper and the horizontal plane perpendicular to the page. But with respect to the vertical plane perpendicular to the page the orbital shown is neither symmetric nor antisymmetric -- there is no contribution for the other side of the cyclobutane ring.

(h) Not a legitimate MO: The symmetries of the *p* orbital on oxygen is different from that of the two *s* hydrogen orbitals. For example, the oxygen *p* orbital is antisymmetric with respect to the plane of the paper while the *s* orbitals are symmetric.

27.6 The orbitals of acetylene can be constructed from the group orbitals for CH (Figure 27.10). The two sets of five CH group orbitals will yield 10 MOs, and the total of 10 electrons will produces five doubly occupied MOs. The C–H σ orbitals result from in-phase and out-of-phase combination of the equivalent group orbitals:

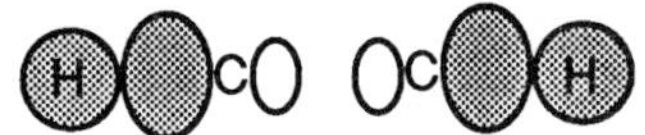

C-H bonding

The C–C σ orbitals result from combination of the *sp* hybrid orbital on carbon that points away from the hydrogens:

Finally, there are in-phase and out-of-phase combinations of the p_x and p_z orbitals that produce the π molecular orbitals:

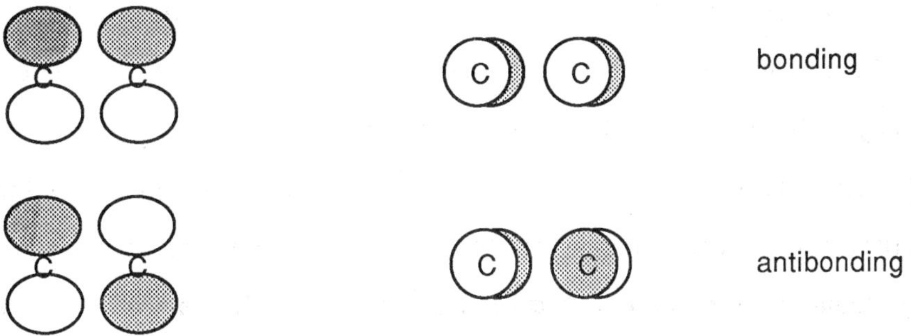

The five occupied molecular orbitals are those designated as bonding; none of the antibonding orbitals will be occupied.

27.7 A linear H–C–H fragment would have three symmetry planes: the plane of the paper, a horizontal plane perpendicular to the page, and a vertical plane horizontal to the page. As shown in the following drawings, the *sp* hybrid orbital directed toward one of the hydrogens is neither symmetric nor antisymmetric with respect to the vertical plane. By itself, however, the *p* orbital is antisymmetric and the *s* orbital is symmetric:

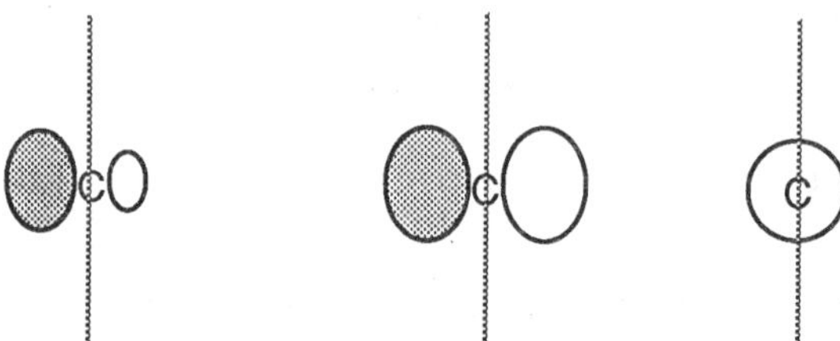

27.8 The occupied orbitals of water in its ground state are ϕ_1–ϕ_4. The two remaining orbitals (both antibonding) are ϕ_5 and ϕ_6. These are the O–H out-of-phase counterparts of ϕ_1 and ϕ_2, and their relative energies for linear to nonlinear geometry distortions are shown below:

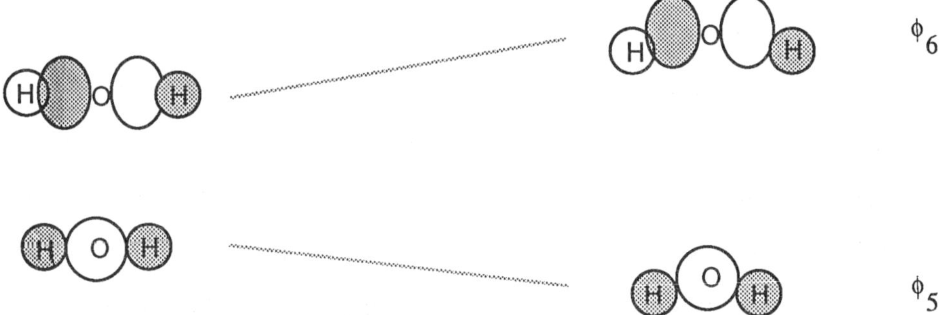

Generation of the lowest excited state of water requires a transition in which an electron moves from ϕ_4 to ϕ_5. This removes an electron from an orbital for which the energy is independent of bending to an orbital for which the energy decreases upon bending. Consequently, the lowest excited state of water should have an H–O–H angle that is smaller than the 105° of water.

27.9 The appropriate orbitals for N_2, O_2, and F_2 were already drawn for oxygen in the answer to Problem 27.3. These molecular orbitals consist of three bonding orbitals, two nonbonding orbitals, and three antibonding orbitals.

For molecular nitrogen there is a total of 10 electrons, so five MOs are doubly occupied: three of these are bonding and the other two are nonbonding. Therefore, a triple bond is normally drawn between the two nitrogens.

In the case of molecular oxygen two additional electrons occupy *antibonding* orbitals. This can be considered to negate the effect of one of the doubly occupied bonding orbitals. This leaves two other bonding MOs that are doubly occupied, so a double bond is normally drawn between the two oxygens.

In the case of molecular fluorine there is a total of 14 electrons, and seven MOs are doubly occupied. Three bonding, two nonbonding, and two antibonding. Considering the two doubly occupied antibonding MOs to cancel the effect of two doubly occupied bonding MOs, only one bonding MO remains (and it is doubly occupied). Consequently, the interaction between the two fluorines is written as a single bond.

27.10 The molecular orbitals for acetylene were drawn in the answer to Problem 27.6, so you should refer to that Problem for an explanation. Here we will only consider the effect of a nonlinear distortion on the energies of the orbitals. Only small changes in energy would be expected for the two lowest energy (N–H) orbitals, and the next MO is primarily N–N so it should be virtually unchanged in energy. For the π orbitals in which the contributing atomic p orbitals are perpendicular to the molecular plane, symmetry restrictions preclude any interaction with the hydrogens so the energies of these orbitals are unaffected by nonlinear distortions. But the situation is quite different for the other π orbitals, both of which should decrease substantially in energy for a nonlinear geometry because favorable N–H interactions develop. Since diimide has a total of 12 electrons, the six lowest orbitals will be doubly occupied. The net energy will be lower for the nonlinear geometry:

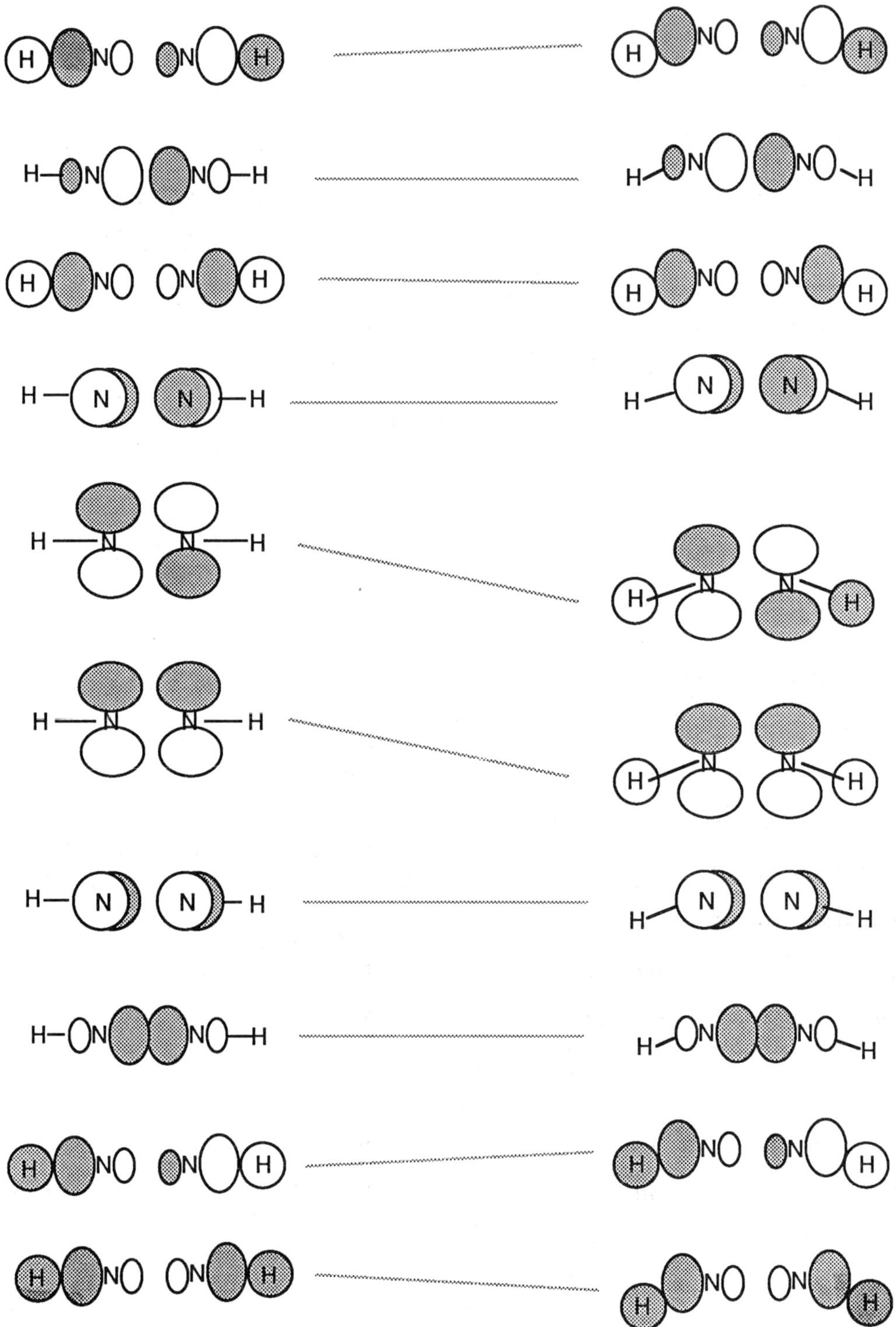
H
N
H
N
H
N
H
N
H
N
H
N
H
N
H
N
H
N
H
N

27.11 **(a)** The H_2F^+ ion has a total of eight valence electrons, so each of the four orbitals depicted in Figure 27.4 is doubly occupied. The major effect of a nonlinear distortion is on ϕ_3, and its energy decreases as the molecule bends. Consequently, a nonlinear geometry is expected.

(b) The $:NH_2^+$ ion has a total of six valence electrons, the same situation that we analyzed in Section 27.2 for methylene ($:CH_2$). The lowest three orbitals of Figure 27.4 are doubly occupied, and the predominant influence again is that of ϕ_3. Consequently, a nonlinear geometry is expected.

27.12 BH_2^+ has a total of four valence level electrons, so only ϕ_1 and ϕ_2 (Figure 27.4) are doubly occupied in the ground state. In the lowest energy triplet state ϕ_2 and ϕ_3 will each have a single electron. The energy of ϕ_2 increases with distortion from linearity, whereas the energy of ϕ_2 decreases. The change will therefore favor a nonlinear geometry for the excited state.

27.13 Photochemical reactions require absorption of light, and this requires occupied and unoccupied orbitals with the appropriate energy differences. Ethylene has a characteristic absorption at 180 nm. This transition corresponds to promotion of an electron from the HOMO to the LUMO (i.e., it is a $\pi \rightarrow \pi^*$ transition of the carbon-carbon double bond). Cyclobutane, on the other hand, has no multiple bonds and does not absorb in the ultraviolet region (the $\sigma \rightarrow \sigma^*$ transition is of much higher energy). Consequently, irradiation with 180 nm light can cause dimerization of two ethylenes, but it cannot induce the reverse process.

27.14 The π system of butadiene has four electrons, so the two lowest π orbitals are occupied. The LUMO is the third π orbital, and it has two out-of-phase interactions. The double bond of ethylene has two electrons, and the occupied orbital (the HOMO) is the in-phase bonding MO:

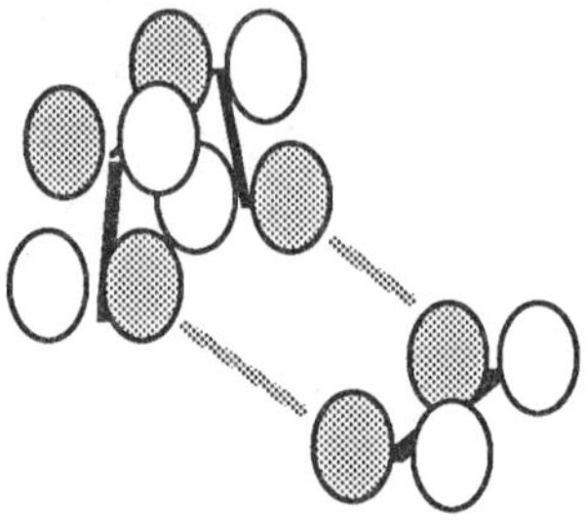

These can interact in a way that would predict a concerted cycloaddition. Both orbitals have the same symmetry (symmetric with respect to the plane of the paper) and interactions between the two (i.e., the dotted lines) are in phase.

27.15 Concerted thermal ring opening of a cyclobutadiene is a conrotatory process:

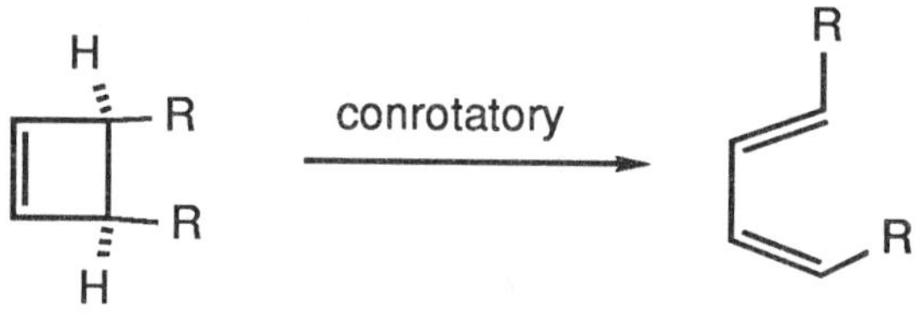

Dewar benzene is resistant to thermal ring opening because the conrotatory process would generate a highly strained cyclohexatriene in which one of the double bonds would have the *E* configuration:

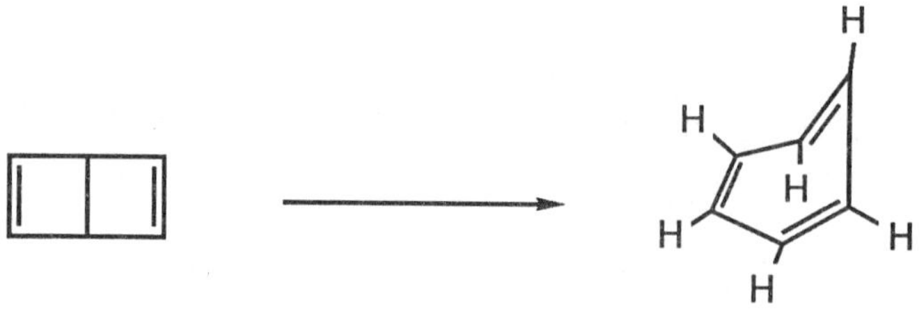

27.16 It is easier to analyze the (hypothetical) reverse reaction, using the π orbitals of the allyl cation; the orbital of interest is the HOMO. Since the allyl cation has only two π electrons, the HOMO is ϕ_1, and ring closure would be *disrotatory*:

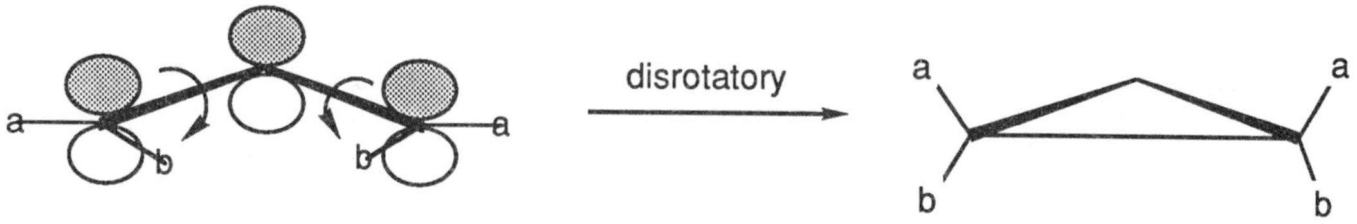

The disrotatory motion could occur in either of two directions, but experiments have shown that the groups *trans* to the leaving group preferentially move away from each other. For the compound in question these substituents are hydrogens, so the methyls would move toward each other:

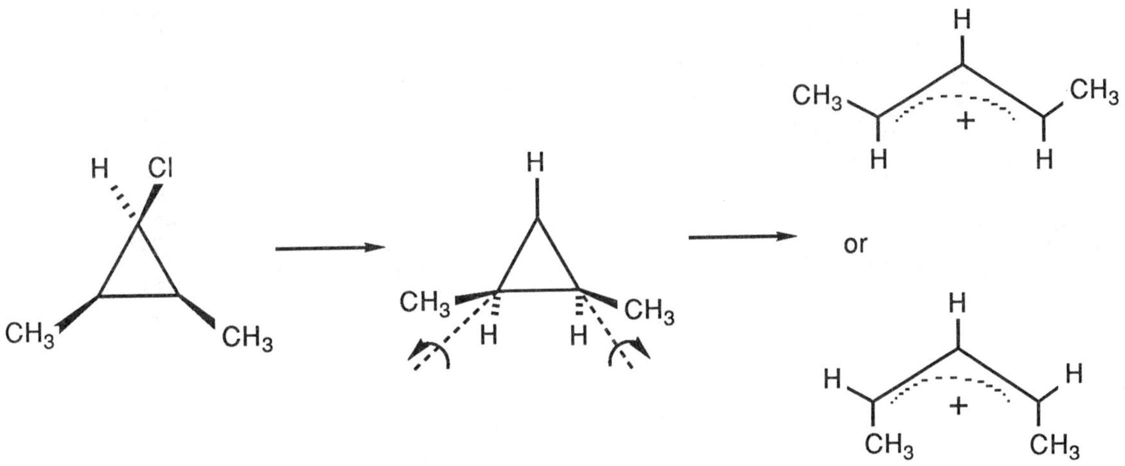

H Cl
CH_3 CH_3
H
CH_3 CH_3
H H
H
CH_3 CH_3
H H
or
H
H H
CH_3 CH_3

27.17 This reaction can be evaluated at the conversion of a cyclohexadiene to a hexatriene, a process that should be *disrotatory*. Two substituents that are *cis* in the cyclohexadiene will either be both *cis* or both *trans* to the carbon chain in the acyclic product:

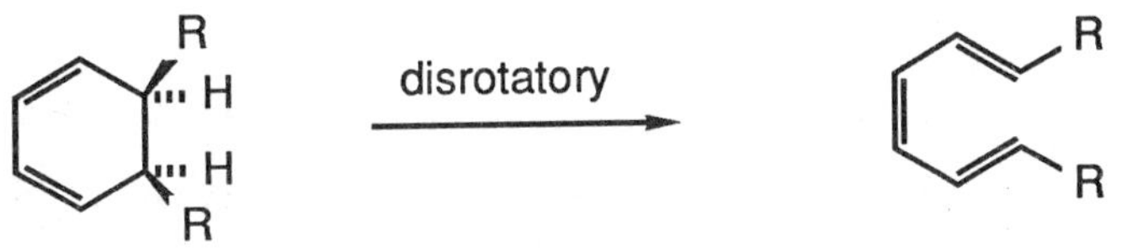

R H H R
disrotatory
R R

Conversely, substituents that are *trans* in the cyclohexadiene will result in one *cis* or and one *trans* group in the product:

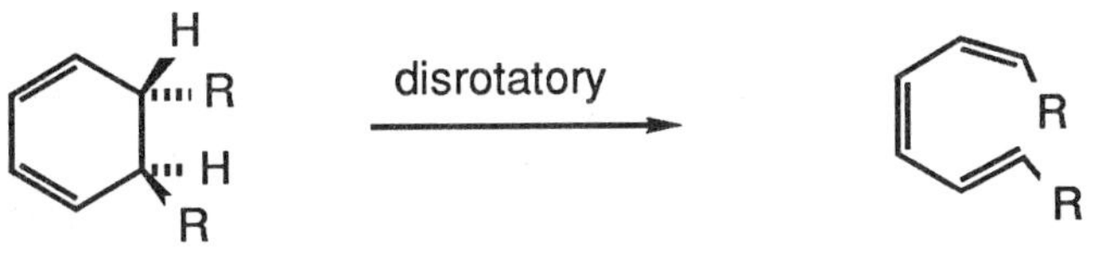

H R H R
disrotatory
R R

For the dihydronaphthalene the *cis* and *trans* compounds would be expected to yield the following isomeric products:

H
H *cis*

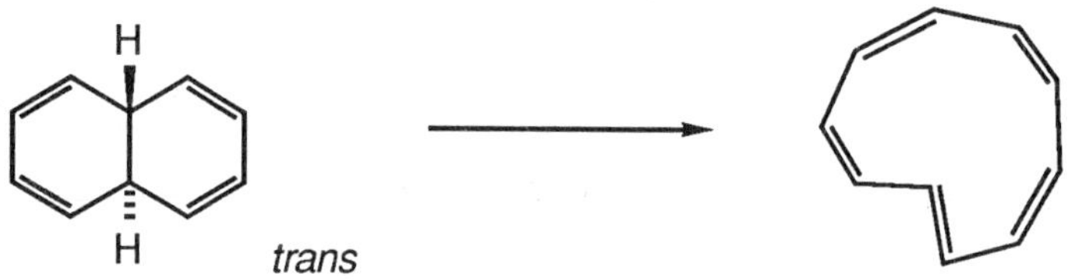

H
H *trans*

27.18 The reaction in question is:

CHCH$_3$
CHCH$_3$

CH$_3$
CH$_3$

To determine the preferred mode of cyclization, you must determine whether the *p* orbital contributions at the terminal carbons of the HOMO of the tetraene are in phase or out of phase. The tetraene system (four double bonds) has eight electrons, so four π orbitals will be occupied. The lowest π MO has no nodes, and these orbitals will be in phase. The next three orbitals will alternate with respect to this property, and the atomic *p* orbital contributions at the termini will be out of phase for the HOMO. This favors a *conrotatory* cyclization:

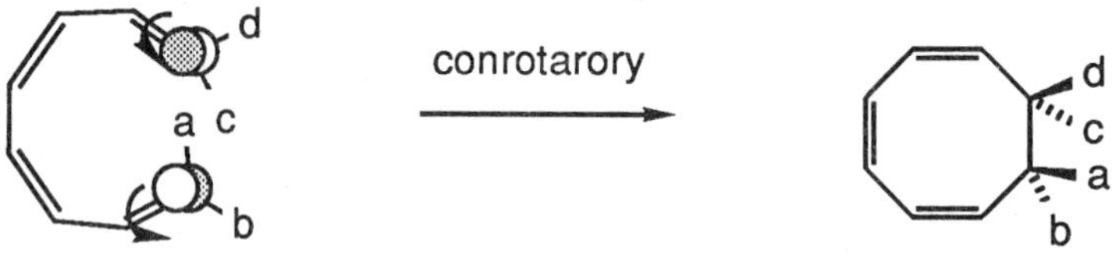

d
a c
b
conrotarory
d
c
a
b

If the terminal double bonds have different configurations (i.e., one *E* and one *Z*), the methyl groups will be *cis* in the product. Alternatively, if the double bonds of the tetraene either both have the *Z* configuration or both have the *E* configuration, as in the following equation, the methyls will be *trans* in the product:

27.19 For the analysis of 1,5-hydrogen shifts we will utilize the orbitals of the acyclic analog, the 1,3-pentadienyl radical. The frontier orbital interactions can most easily be seen in the interaction between the highest occupied molecular orbitals of the hydrogen radical and the pentadienyl radical. The π system of the pentadienyl is occupied by five electrons, so the HOMO is ϕ_3. The *p* orbital contributions on the terminal carbons will be in-phase for ϕ_1, out-of-phase for ϕ_2, and in-phase again for ϕ_3. The following diagram shows that the interaction with ϕ_3 favors a *suprafacial* migration to maintain the in-phase interaction with the 1*s* orbital of hydrogen.

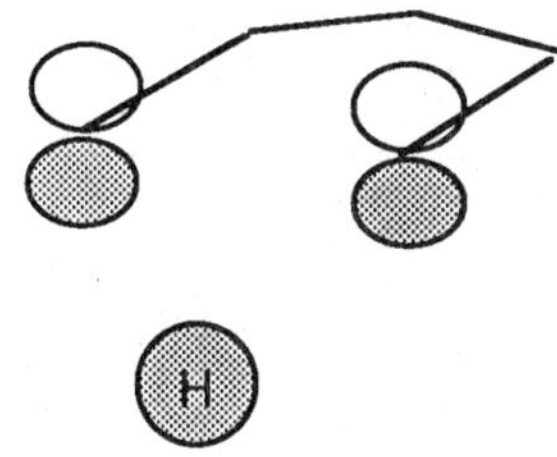

APPENDIX A. MOLECULAR REARRANGEMENTS

In various places in the text we introduced reactions that yield products having carbon skeletons quite different from those of the reactants. Included in this description are carbocation rearrangements (Sections 5.9, 12.7, 12.9, 23.1), oxidation of organoboranes (Section 5.11), Baeyer-Villiger oxidation of ketones (Section 13.11) and the Hofmann degradation of amides (Section 17.7). We will now return to these reactions in order to show the common mechanistic thread that they all share: each of these reactions proceeds by the 1,2-migration of a substituent to an electron deficient atom. This is illustrated by the migration of R in the following equation after departure of the leaving group Y.

$$-\overset{|}{\underset{R}{\overset{|}{C}}}-X-Y \longrightarrow -\overset{|}{\underset{R}{\overset{|}{C}}}-X^{+} \longrightarrow {}^{+}C-\underset{R}{\underset{|}{X}}$$

In some instances molecular rearrangements may be unwanted side reactions that result in reduced yields of desired products, but there are also many situations in which rearrangements can be exploited as valuable and reliable synthetic reactions. A discussion of all known molecular rearrangements is far beyond the scope of this Appendix. Consequently, we will restrict our treatment to a variety of synthetically useful rearrangements. We will begin with migration to electron deficient carbon, nitrogen and oxygen (Sections A.1, A.2, and A.3, respectively) and then conclude with electrocyclic rearrangements analogous to the Cope reaction (Section 27.4).

A.1 REARRANGEMENT TO ELECTRON DEFICIENT CARBON

In our previous discussions of carbocation rearrangements (Sections 5.9, 12.7) we have stressed that rearrangement will ordinarily yield a more stable cation. This has been extensively exploited for synthetic purposes in reactions that yield hydroxy-substituted carbocations, in other words *oxonium ions*, which are much more stable than unsubstituted carbocations.

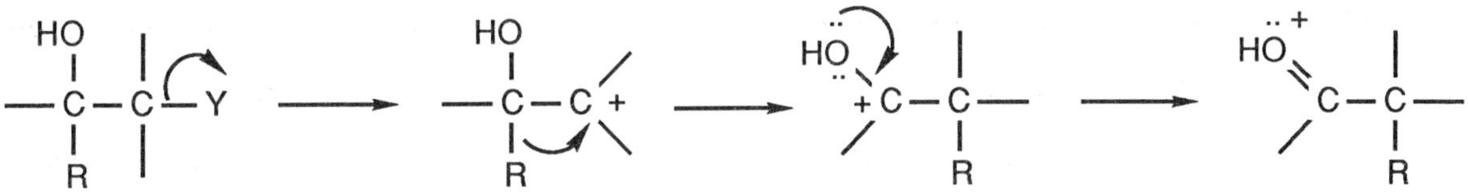

The stability of the rearranged cation tends to result in high yields of the corresponding product in these reactions. In many cases the ionization of leaving group Y may be concerted with the rearrangement, and the nonbonding electrons on oxygen can then assist in the overall reaction as shown in the following equation:

If the initial ionization would produce a cation that is fairly stable (e.g., tertiary or benzylic), the stepwise mechanism may provide a more accurate description of the reaction. But when a less stable intermediate would be formed, the rearrangement step may occur before the cation is fully developed. We will frequently show these reactions as proceeding in a stepwise fashion so that they will be easier to understand.

Pinacol and Related Carbocation Rearrangements

When 2,3-dimethylbutane-2,3-diol (sometimes called *pinacol*) is treated with acid, 3,3-dimethyl-2-butanone is produced.

$$CH_3-\underset{CH_3}{\overset{OH}{C}}-\underset{CH_3}{\overset{OH}{C}}-CH_3 \xrightarrow{H_2SO_4} CH_3-\overset{O}{\overset{\|}{C}}-\underset{CH_3}{\overset{CH_3}{C}}-CH_3 \qquad 65\text{-}72\%$$

This type of acid-catalyzed rearrangement is general for 1,2-diols (*glycols*), and it is often called the *pinacol rearrangement*. Protonation of one of the hydroxyl groups, followed by loss of water, leads to a carbocation:

$$CH_3-\underset{CH_3}{\overset{OH}{C}}-\underset{CH_3}{\overset{OH}{C}}-CH_3 \underset{}{\overset{H^+}{\rightleftharpoons}} CH_3-\underset{CH_3}{\overset{OH}{C}}-\underset{CH_3}{\overset{\overset{+}{O}H_2}{C}}-CH_3 \rightleftharpoons CH_3-\underset{CH_3}{\overset{OH}{C}}-\overset{+}{C}(CH_3)_2$$

A 1,2-alkyl shift then produces an oxonium ion, which yields the rearranged ketone upon transfer of a proton (to water or some other proton acceptor in the reaction mixture).

$$CH_3-\underset{CH_3}{\overset{:\ddot{O}H}{C}}-\overset{+}{C}(CH_3)_2 \longrightarrow CH_3-\overset{\overset{+}{\ddot{O}}H}{\overset{\|}{C}}-\underset{CH_3}{\overset{CH_3}{C}}-CH_3 \underset{}{\overset{-H^+}{\rightleftharpoons}} CH_3-\overset{:O:}{\overset{\|}{C}}-\underset{CH_3}{\overset{CH_3}{C}}-CH_3$$

Other examples of pinacol rearrangements are provided by the following equations:

$$CH_3CH_2-\underset{OH}{\overset{CH_3CH_2}{C}}-\underset{OH}{\overset{C_6H_5}{C}}-C_6H_5 \xrightarrow{H_2SO_4} CH_3CH_2-\overset{O}{\overset{\|}{C}}-\underset{C_6H_5}{\overset{C_6H_5}{C}}-CH_2CH_3 \qquad 78\%$$

$$CH_3-\underset{CH_3}{\overset{OH}{C}}-CH_2OH \xrightarrow{H_2SO_4} CH_3-\underset{CH_3}{CH}-CHO \qquad \sim 100\%$$

As illustrated by the two preceding reactions, pinacol rearrangements proceed via loss of the hydroxyl group that gives the more stable of two possible cations. (Notice also that the actual rearrangement can be a hydride shift rather than an alkyl shift, as seen in the second reaction.)

$$CH_3CH_2-\overset{\displaystyle HO}{\underset{\displaystyle CH_3CH_2}{C}}-\overset{\displaystyle OH}{\underset{\displaystyle C_6H_5}{C}}-C_6H_5 \longrightarrow CH_3CH_2-\overset{\displaystyle H\ddot{O}:}{\underset{\displaystyle CH_3CH_2}{C}}-\overset{+}{C}(C_6H_5)_2 \longrightarrow CH_3CH_2-\overset{\displaystyle H\ddot{O}^+}{\overset{\|}{C}}-\overset{\displaystyle C_6H_5}{\underset{\displaystyle C_6H_5}{C}}-CH_2CH_3 \longrightarrow \text{product}$$

$$CH_3-\overset{\displaystyle \overset{+}{O}H_2}{\underset{\displaystyle CH_3}{C}}-CH_2-OH \longrightarrow CH_3-\underset{\displaystyle CH_3}{\overset{+}{C}}-\overset{\displaystyle H}{\underset{\displaystyle H}{C}}-\ddot{O}H \longrightarrow CH_3-\underset{\displaystyle CH_3}{CH}-\overset{\displaystyle \overset{+}{\ddot{O}}H}{\overset{\|}{CH}} \longrightarrow \text{product}$$

The pinacol and related rearrangements that we will discuss in this section can all be summarized according to the following equation:

Pinacol-Type Rearrangement

$$-\overset{\displaystyle H\ddot{O}:}{\underset{\displaystyle R}{C}}-\overset{+}{\underset{|}{C}}- \longrightarrow -\overset{\displaystyle H\ddot{O}^+}{\overset{\|}{C}}-\overset{|}{\underset{\displaystyle R}{C}}- \xrightarrow{\text{proton transfer}} -\overset{\displaystyle :O:}{\overset{\|}{C}}-\overset{|}{\underset{\displaystyle R}{C}}-$$

One type of closely related rearrangement involves reaction of an epoxide with acid. Opening of the three-membered ring occurs in the direction that affords the more stable cation, the same cation that would be generated from the corresponding diol.

Pinacol-type rearrangements of epoxides are often brought about with Lewis acids such as BF_3, as illustrated by the following example.

$$C_6H_5-CH\overset{O}{-}CH-C_6H_5 \xrightarrow{BF_3} C_6H_5-\underset{C_6H_5}{CH}-\overset{O}{\overset{\|}{CH}} \qquad 74\text{-}82\%$$

Rearrangement in this case involves a 1,2-phenyl shift rather than the alternative hydride shift.

$C_6H_5-CH-CH-C_6H_5$ (epoxide) $\overset{BF_3}{\rightleftharpoons}$ $C_6H_5-CH-CH-C_6H_5$ ($\overset{+}{O}-\overset{-}{B}F_3$)

$\Updownarrow$

$C_6H_5-\overset{+}{C}H-CH(O-\overset{-}{B}F_3)-C_6H_5$ $\longrightarrow$ $C_6H_5-CH(C_6H_5)-\underset{+}{C}H(O-\overset{-}{B}F_3)$

$\updownarrow$

$C_6H_5-CH(C_6H_5)-CH=\overset{+}{O}-\overset{-}{B}F_3$ $\overset{-BF_3}{\rightleftharpoons}$ $C_6H_5-CH(C_6H_5)-CH=O$

When the epoxide is unsymmetrical, ring opening occurs to generate the more stable carbocation, for example the tertiary carbocation in the following reaction.

$\ddot{O}$: $\xrightarrow{BF_3}$ [$+$ $O-\overset{-}{B}F_3$ H H] $\xrightarrow{\text{hydride shift}}$ CH=O, H 91%

The reactions of glycols and epoxides provide useful synthetic procedures for converting alkenes to related aldehydes or ketones. If epoxides are prepared via sulfonium ylides (Section 13.8), rearrangement provides an overall method for converting a ketone ($>$C=O) to the aldehyde with one more carbon ($>$CH-CHO). When more than one group is capable of undergoing a 1,2-shift, the preference is largely governed by the *migratory aptitudes* of the substituents (Section 13.11). These aptitudes are influenced by the steric and electronic effects of other substituents in the molecule, but nearly all the rearrangements we describe in this Appendix are characterized by the following migratory aptitudes:

Migratory Aptitudes

greatest tendency to undergo 1,2-shift	3° alkyl > aryl or hydrogen > 2° alkyl > 1° alkyl > methyl	least tendency to undergo 1,2-shift

Organic chemists tend to refer to all these rearrangements as "alkyl shifts." Nevertheless, you should recognize that *aryl* substituents readily undergo 1,2-migrations as well.

In each of the preceding rearrangements the overall reaction pathway is governed by the initial reaction of a diol or epoxide to produce the more stable of two possible carbocations. Some ingenious procedures have been developed that permit formation of the *other* product. One way is to take advantage of the fact that the more stable carbocation usually corresponds to the more highly substituted carbon of a diol. This means that the hydroxyl group on the other carbon is less hindered, so it will react selectively with an acylating agent.

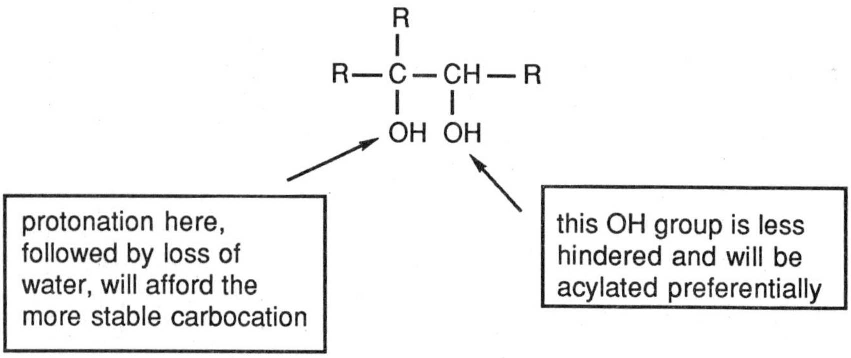

R
R—C—CH—R
OH OH

protonation here, followed by loss of water, will afford the more stable carbocation

this OH group is less hindered and will be acylated preferentially

Formation of a monotosylate from a diol (using a single equivalent of *p*-toluenesulfonyl chloride) converts the less substituted hydroxyl substituent to a good leaving group. The following reaction shows how this strategy has been used to direct a pinacol rearrangement in a specific direction.

HO OH → TsCl (82%) → HO OTs → KOtBu (85%) → O 70%

The potassium *tert*-butoxide served to neutralize the *p*-toluenesulfonic acid generated in the rearrangement and may also have initiated a concerted reaction in the following manner.

HO OTs → KOtBu → ⁻O OTs → O

Another way that pinacol-type reactions can be directed along a specific path is by diazotization (Section 17.4) of a β-amino alcohol. Loss of nitrogen then forces the rearrangement to proceed by migration of a substituent from the hydroxyl-bearing carbon to the carbon that originally had the NH_2 group.

$$-\overset{OH}{\underset{R}{C}}-\overset{|}{\underset{|}{C}}-NH_2 \xrightarrow{HONO} -\overset{OH}{\underset{R}{C}}-\overset{|}{\underset{|}{C}}-\overset{+}{N}\equiv N \xrightarrow{-N_2} -\overset{:\ddot{O}H}{\underset{R}{C}}-C^{+}$$

$$\rightarrow -\overset{+\ddot{O}H}{\overset{\|}{C}}-\underset{R}{C}- \underset{\longleftarrow}{\overset{-H^+}{\longrightarrow}} -\overset{:O:}{\overset{\|}{C}}-\underset{R}{C}-$$

The β-amino alcohols are frequently available from the carbonyl compounds of one less carbon via cyanohydrin formation (Section 13.7) followed by reduction of the cyano group (Sections 16.5, 17.6).

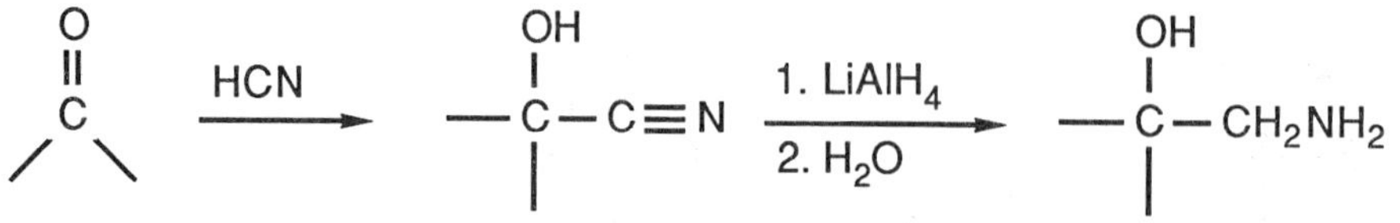

$$C{=}O \xrightarrow{HCN} -\overset{OH}{\underset{|}{\overset{|}{C}}}-C{\equiv}N \xrightarrow[2.\ H_2O]{1.\ LiAlH_4} -\overset{OH}{\underset{|}{\overset{|}{C}}}-CH_2NH_2$$

The β-amino alcohol method has been employed as a procedure for *ring expansion,* as cyclic compounds generally give better yields than do simple amines (Section 17.4):

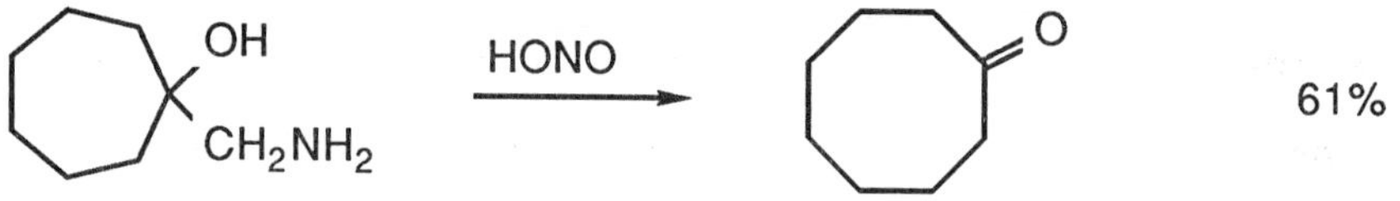

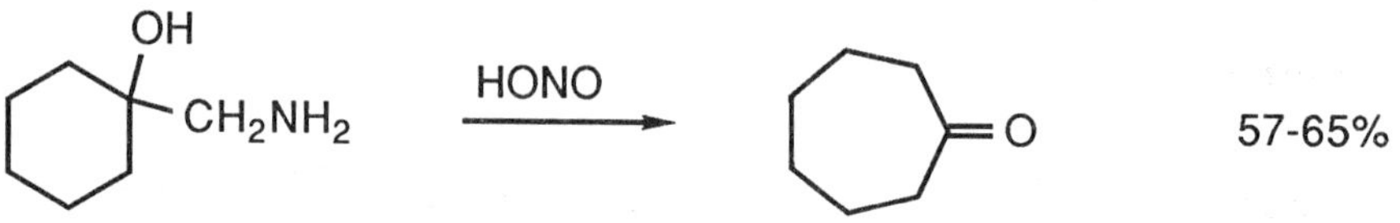

1-(aminomethyl)cycloheptanol (OH, CH_2NH_2) $\xrightarrow{HONO}$ cyclooctanone (=O) 61%

1-(aminomethyl)cyclohexanol (OH, CH_2NH_2) $\xrightarrow{HONO}$ cycloheptanone (=O) 57-65%

A closely related reaction involves the reaction of a ketone with diazomethane:

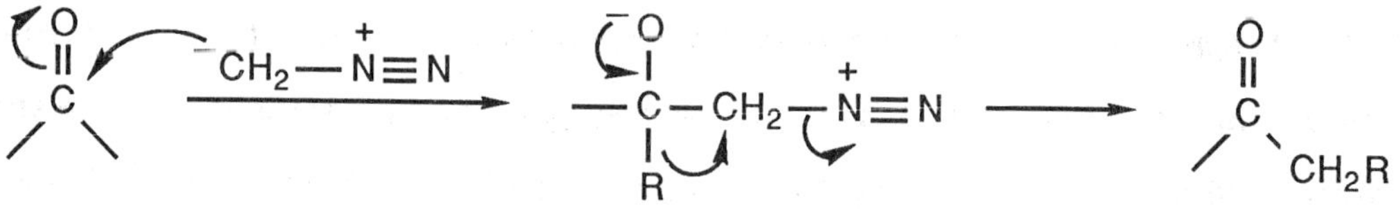

$$C{=}O + {}^{-}CH_2-\overset{+}{N}{\equiv}N \longrightarrow -\overset{{}^{-}O}{\underset{R}{\overset{|}{\underset{|}{C}}}}-CH_2-\overset{+}{N}{\equiv}N \longrightarrow \overset{O}{\overset{\|}{C}}-CH_2R$$

This has also been exploited as a method for ring expansion of ketones. Although the yields are apt to be low, only a single step is required.

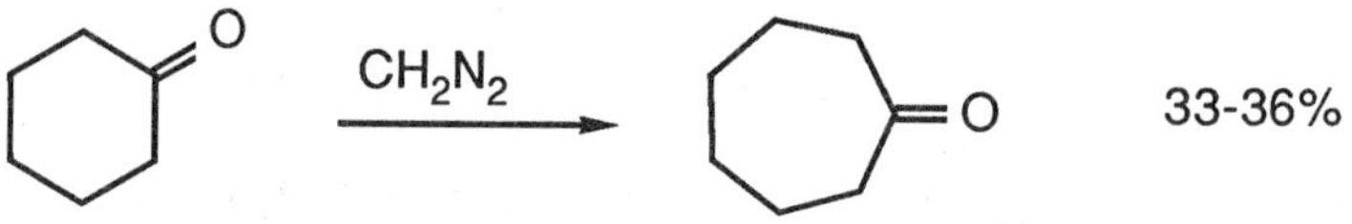

cyclohexanone (=O) $\xrightarrow{CH_2N_2}$ cycloheptanone (=O) 33-36%

In accord with the migratory aptitudes shown above, aldehydes usually react with diazomethane via *hydride* shift to give the corresponding methyl ketone.

$$
\begin{array}{c} O{=}CH \\ AcO{-}\!\!\!\mid\!\!\!{-} \\ {-}\!\!\!\mid\!\!\!{-}OAc \\ {-}\!\!\!\mid\!\!\!{-}OAc \\ CH_2OAc \end{array}
\xrightarrow{CH_2N_2}
\begin{array}{c} CH_3 \\ \mid \\ O{=}C \\ AcO{-}\!\!\!\mid\!\!\!{-} \\ {-}\!\!\!\mid\!\!\!{-}OAc \\ {-}\!\!\!\mid\!\!\!{-}OAc \\ CH_2OAc \end{array}
\qquad 62\%
$$

In most instances there are better ways to convert an aldehyde to the corresponding methyl ketone, so the reaction with aldehydes is not a widely used procedure.

Arndt-Eistert Synthesis

Just as carbocations can undergo rearrangements, so can carbenes, another class of electron-deficient compounds. Carbocations and carbenes both have only six valence-level electrons at the reactive center, so it should not be surprising that they share some patterns of reactivity. We will not discuss the reactions of carbenes in detail but will instead focus on one particular reaction that has found widespread use in synthetic chemistry. The Arndt-Eistert synthesis (developed by the German chemists F. Arndt and B. Eistert in the 1920s) results in *chain lengthening* of a carboxylic acid by a single CH_2 group:

Arndt-Eistert Synthesis

$$
R{-}CO_2H \xrightarrow[\text{2. } CH_2N_2 \quad \text{3. } H_2O,\ Ag_2O \text{ catalyst}]{\text{1. } SOCl_2} R{-}CH_2{-}CO_2H
$$

The carboxylic acid is first converted to its acid chloride by the action of a reagent such as thionyl chloride, and subsequent treatment with diazomethane affords a *diazoketone*.

$$R{-}\overset{O}{\overset{\|}{C}}{-}OH \xrightarrow{SOCl_2} R{-}\overset{O}{\overset{\|}{C}}{-}Cl \xrightarrow{\bar{:}CH_2{-}\overset{+}{N}{\equiv}N} R{-}\overset{O}{\overset{\|}{C}}{-}CH_2{-}\overset{+}{N}{\equiv}N$$

$$\xrightarrow{-H^+} R{-}\overset{O}{\overset{\|}{C}}{-}\overset{\bar{..}}{C}H{-}\overset{+}{N}{\equiv}N \longleftrightarrow R{-}\overset{O}{\overset{\|}{C}}{-}CH{=}\overset{+}{N}{=}\overset{\bar{..}}{N}:$$

Diazoketone

Treatment of the diazoketone with water and a silver (or sometimes copper) catalyst then produces the rearranged acid. Loss of nitrogen first affords a carbene, and the other substituent on the carbonyl group undergoes a 1,2-migration to the electron deficient carbon.

$$R{-}\overset{O}{\overset{\|}{C}}{-}CH{=}\overset{+}{N}{=}\overset{..}{N}: \longleftrightarrow R{-}\overset{O}{\overset{\|}{C}}{-}\overset{\bar{..}}{C}H{-}\overset{+}{N}{\equiv}N \xrightarrow[\text{loss of nitrogen}]{Ag_2O\ \text{catalyzed}} R{-}\overset{O}{\overset{\|}{C}}{-}CH:$$

$$\xrightarrow{\text{1,2-alkyl shift}} O{=}\overset{+}{C}{-}\overset{-}{\underset{..}{C}}H{-}R \longleftrightarrow O{=}C{=}CHR$$

The rearrangement initially produces a *ketene* ($>C{=}C{=}O$), but this quickly adds water under the conditions of the reaction.

$$O{=}C{=}CHR \xrightarrow{:\ddot{O}H_2} \overset{-}{O}{-}C(\overset{+}{\ddot{O}}H_2){=}CHR \xrightarrow{\text{proton transfers}} O{=}C(H_2O){-}CH_2R$$

Alternatively, the reaction can be carried out in an alcoholic solvent, in which case reaction of the ketene will afford an ester.

$$>C{=}C{=}O \xrightarrow{ROH} {-}\overset{|}{C}H{-}C({=}O)OR$$

We showed an example of the Arndt–Eistert synthesis in Example 23.5:

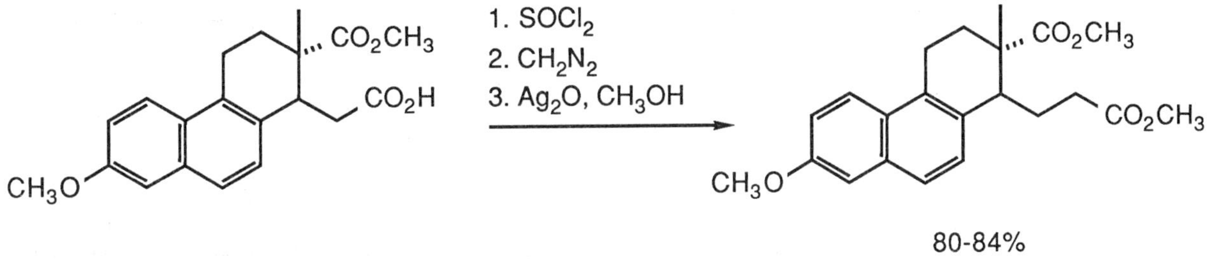
CO2CH3
CO2H
1. SOCl2
2. CH2N2
3. Ag2O, CH3OH
CH3O
CO2CH3
CO2CH3
CH3O
80-84%

The method is further illustrated by the following reactions.

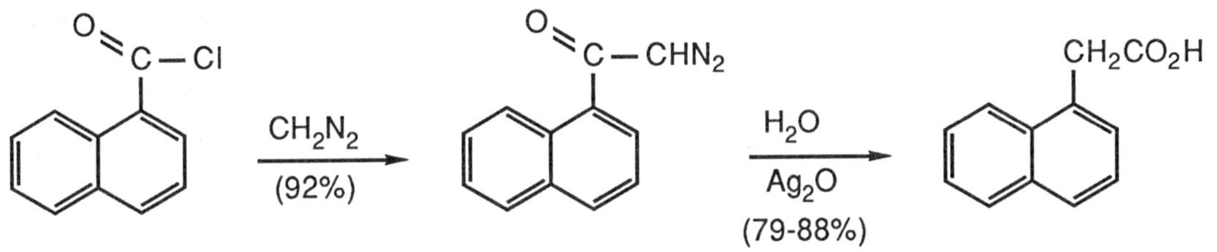
C—Cl
CH2N2
(92%)
C—CHN2
H2O
Ag2O
(79-88%)
CH2CO2H

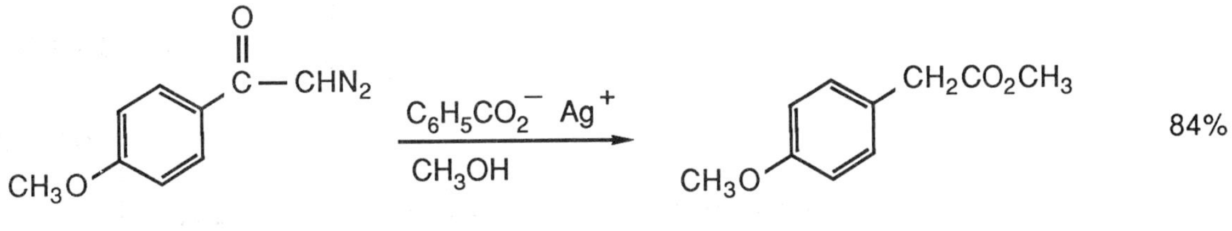
C—CHN2
CH3O
C6H5CO2− Ag+
CH3OH
CH2CO2CH3
CH3O
84%

A.2 REARRANGEMENT TO ELECTRON DEFICIENT NITROGEN

Rearrangement of Carboxylic Acid Derivatives

A variety of reactions are known that involve migration to electron deficient nitrogen, and several of these are closely related to the Hofmann degradation (Section 17.7). In each case a carboxylic acid derivative rearranges to produce the corresponding amine with one less carbon.

$$R\text{-}CO_2H \longrightarrow R\text{-}NH_2$$

The key intermediate in these reactions is formally a *nitrene*, the monovalent nitrogen analog of a carbene. This undergoes a 1,2-shift to generate an isocyanate, which is subsequently hydrolyzed to the free amine.

$$R{-}C({=}O){-}OH \longrightarrow R{-}C({=}O){-}\ddot{N}: \xrightarrow{\text{1,2-alkyl migration}} O{=}C{=}\ddot{N}{-}R \xrightarrow{\text{hydrolysis}} \ddot{N}H_2{-}R$$

Carboxylic acid derivative — Acyl nitrene — Isocyanate

As with rearrangements to electron-deficient carbon, rearrangement in many cases may be concerted with formation of the electron-deficient nitrogen. The free nitrene is therefore not always an actual intermediate, but it is convenient to write all the reactions in the same way. The nitrogen may be protonated in some instances, so that the actual intermediate is a *nitrenium ion* ($R\text{-}CO\text{-}NH^+$) rather than a nitrene. But for convenience we will usually depict these reactions as proceeding via the nitrene.

The Hofmann degradation of primary amides occurs when the amide is treated with bromine (or sometimes chlorine) in base. The first step is *N*-bromination, and base then removes the remaining proton on nitrogen. Ionization of the nitrogen-bromine bond generates the acyl nitrene, which then rearranges.

$$R{-}C({=}O){-}\ddot{N}H_2 \xrightarrow[\text{KOH}]{Br_2} R{-}C({=}O){-}\ddot{N}HBr \xrightarrow{^{-}OH} R{-}C({=}O){-}\ddot{N}^{-}{-}Br \longrightarrow R{-}C({=}O){-}\ddot{N}:$$

The overall reaction can be summarized in the following way:

Hofmann Degradation

$$R{-}C({=}O){-}NH_2 \xrightarrow[\text{KOH}]{Br_2} \left[R{-}C({=}O){-}\ddot{N}:\right] \longrightarrow O{=}\overset{+}{C}{-}\ddot{N}^{-}{-}R \longleftrightarrow O{=}C{=}\ddot{N}{-}R \xrightarrow{H_2O,\ KOH} NH_2R$$

The following reactions illustrate the Hofmann degradation of primary amides (see Section 17.7 for additional examples):

$$m\text{-}BrC_6H_4\text{-}C(=O)\text{-}NH_2 \xrightarrow[\text{KOH}]{Br_2} m\text{-}BrC_6H_4\text{-}NH_2 \quad 87\%$$

$$3,4\text{-}(CH_3O)_2C_6H_3\text{-}C(=O)\text{-}NH_2 \xrightarrow[\text{NaOH}]{Cl_2} 3,4\text{-}(CH_3O)_2C_6H_3\text{-}NH_2 \quad 80\text{-}82\%$$

A closely related reaction is the Curtius rearrangement (named for T. Curtius, the German chemist who discovered it in the late 1800s), which involves decomposition of an acyl azide (R-CO-N_3). The acyl azide is produced either by reaction of the acid chloride with sodium azide,

$$R\text{—}\overset{O}{\overset{\|}{C}}\text{—}Cl \xrightarrow{NaN_3} R\text{—}\overset{O}{\overset{\|}{C}}\text{—}N_3$$

or by diazotization of an acyl hydrazide.

$$R\text{—}\overset{O}{\overset{\|}{C}}\text{—}NHNH_2 \xrightarrow{HONO} R\text{—}\overset{O}{\overset{\|}{C}}\text{—}N_3$$

The acyl azides are unstable compounds that readily lose molecular nitrogen upon heating to temperatures around 80–100°C. (Like other compounds that can liberate N_2, they sometimes decompose explosively.) The resulting isocyanate can then be hydrolyzed to the free amine (or can be isolated if desired). The overall reaction can be summarized as follows.

Curtius Rearrangement

$$R-\overset{O}{\overset{\|}{C}}-\ddot{N}-\overset{+}{N}\equiv N: \longrightarrow \left[R-\overset{O}{\overset{\|}{C}}-\ddot{N} \right] \longrightarrow O=\overset{+}{C}-\overset{-}{\ddot{N}}-R \longleftrightarrow O=C=\ddot{N}-R \xrightarrow{H_2O,\ KOH} NH_2R$$

The following reactions illustrate the Curtius rearrangement.

$$C_6H_5-CH_2-\overset{O}{\overset{\|}{C}}Cl \xrightarrow[\text{2. HCl, } H_2O]{\text{1. } NaN_3} C_6H_5-CH_2\overset{+}{N}H_3\ Cl^- \qquad 94\%$$

$$NH_2NH-\overset{O}{\overset{\|}{C}}-(CH_2)_4-\overset{O}{\overset{\|}{C}}-NHNH_2 \xrightarrow[\text{2. } H_2O\text{, HCl}]{\text{1. HONO}} H_3\overset{+}{N}-(CH_2)_4-\overset{+}{N}H_3 \qquad 73\text{-}77\%$$

The third related rearrangement of carboxylic acid derivatives is the Schmidt reaction, developed by K.F. Schmidt in Germany during the 1920s. When a carboxylic acid is treated with hydrazoic acid (HN_3) in the presence of a strong acid, the amine with one less carbon is produced. Protonation of the carbonyl group of the acid renders it susceptible to nucleophilic attack by hydrazoic acid.

$$R-C(=O)OH \overset{H^+}{\rightleftharpoons} R-C(=\overset{+}{O}H)OH \xrightarrow{:N_3H} \rightleftharpoons R-\underset{OH}{\overset{OH}{\underset{|}{\overset{|}{C}}}}-N_3 \rightleftharpoons R-\overset{O}{\overset{\|}{C}}-N_3$$

The resulting dihydroxy intermediate could undergo loss of N_2 and then rearrange directly, or it might first lose water to give the acyl azide. We have shown it in the second way to emphasize the similarity to the Hofmann and Curtius reactions. Hydrolysis to the amine occurs when the reaction is quenched with water, and the overall reaction can therefore be summarized as follows:

Schmidt Reaction

$$R-\overset{O}{\overset{\|}{C}}-OH \xrightarrow[H_2SO_4]{HN_3} \left[R-\overset{O}{\overset{\|}{C}}-\ddot{N}\right] \longrightarrow O=\overset{+}{C}-\ddot{N}-R \longleftrightarrow O=C=\ddot{N}-R \xrightarrow{H_2O} NH_2R$$

The following reactions show how the Schmidt rearrangement has been used for synthetic purposes:

$$\text{cyclobutane-}CO_2H \xrightarrow[\text{2. NaOH (neutralize)}]{\text{1. } HN_3,\ H_2SO_4} \text{cyclobutane-}NH_2 \qquad 60\text{-}80\%$$

$$CH_3(CH_2)_{16}-\overset{O}{\overset{\|}{C}}OH \xrightarrow[\text{2. } H_2O]{\text{1. } HN_3,\ H_2SO_4} CH_3(CH_2)_{16}NH_3^+\ HSO_4^- \qquad 96\%$$

$$HO_2C\text{(bicyclic, H)} \xrightarrow[\text{2. NaOH}]{\text{1. } HN_3,\ H_2SO_4} H_2N\text{(bicyclic, H)} \qquad 61\%$$

The last of these examples also demonstrates that these rearrangements are highly stereospecific. The rearrangement proceeds with retention of configuration at the carbon which migrates. In fact, such 1,2-migrations to electron deficient centers invariably occur with retention of configuration at the migrating carbon, and the generalization holds for all the rearrangements discussed in this Appendix.

Rearrangement of Aldehyde and Ketone Derivatives

There are two closely related rearrangements of aldehyde and ketone derivatives in which a 1,2-migration to electron-deficient nitrogen produces a carboxylic acid derivative. The first of these is a variation of the Schmidt reaction in which hydrazoic acid reacts with an aldehyde or ketone to produce an *N*-substituted amide. Its synthetic use has been largely restricted to ketones.

Schmidt Reaction of Ketones

$$RC(=O)R' \xrightarrow[2.\ H_2O]{1.\ HN_3,\ H_2SO_4} \left[R-\underset{R'}{\overset{HO}{C}}-\ddot{N}H^{+} \right] \longrightarrow RNH-\overset{O}{\overset{\|}{C}}-R' + R-\overset{O}{\overset{\|}{C}}NHR'$$

As with most other acid catalyzed reactions of ketones the reaction is initiated by protonation of the carbonyl oxygen, which is followed by nucleophilic addition of hydrazoic acid.

$$RC(=O)R \underset{}{\overset{H^+}{\rightleftharpoons}} R_2C{=}\overset{+}{O}H \xrightarrow{:N_3H} \rightleftharpoons R-\underset{R}{\overset{HO}{C}}-\ddot{\underset{..}{N}}-\overset{+}{N}{\equiv}N:$$

The rearrangement occurs as a consequence of loss of nitrogen, and could in principle involve formation of a nitrene as we showed for the Schmidt reaction of carboxylic acids.

$$R-\underset{R}{\overset{HO}{C}}-\ddot{\underset{..}{N}}-\overset{+}{N}{\equiv}N \longrightarrow \left[R-\underset{R}{\overset{HO}{C}}-\ddot{\underset{..}{N}} \right] \longrightarrow \text{rearrangement}$$

However, we will depict the reaction as proceeding via a *nitrenium ion* in order to emphasize the tremendous similarity of this reaction to the pinacol rearrangement. Strongly acidic conditions are employed for the Schmidt reaction, and protonation can occur on the nitrogen that is bonded to carbon. Loss of molecular nitrogen then affords a nitrenium ion ($-C-NH^+$, analogous to a carbocation) rather than a neutral nitrene. Finally, the nitrenium ion rearranges to produce an amide.

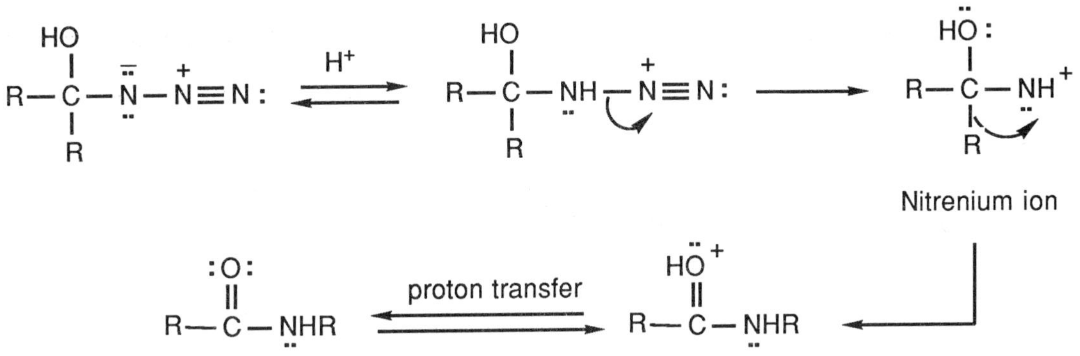

Nitrenium ion

proton transfer

If the two substituents on the carbonyl carbon of the original ketone are not the same, a mixture of two products is expected, although the major product is often that predicted from migratory aptitudes. The following reactions illustrate the Schmidt reaction of ketones.

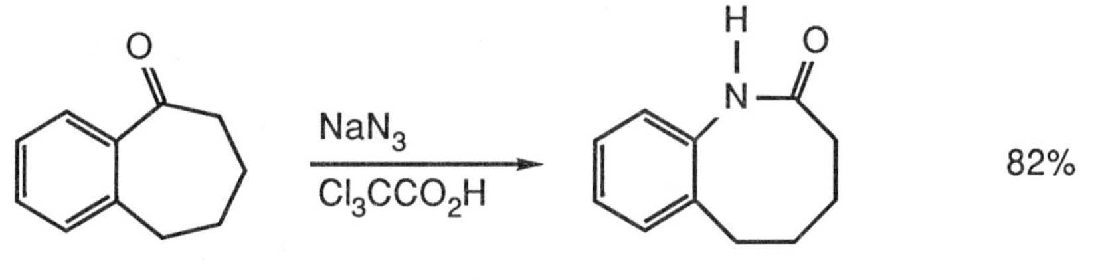

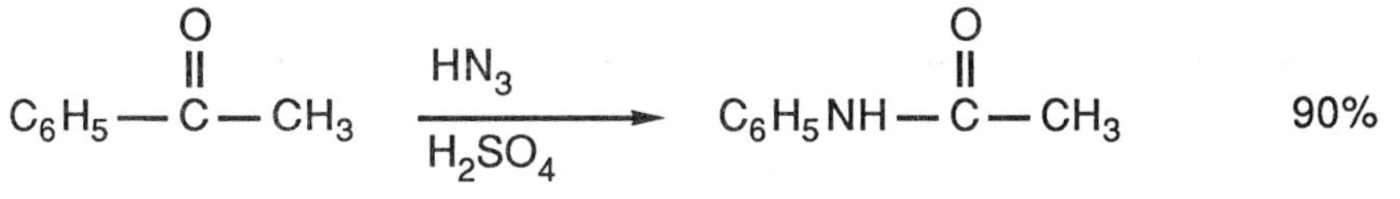

NaN_3 / Cl_3CCO_2H 82%

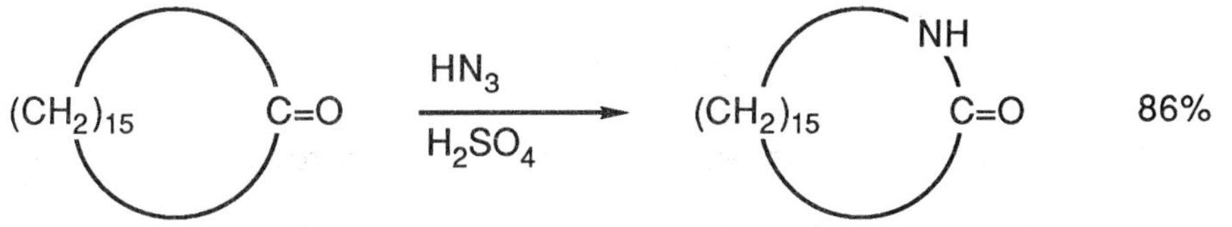

$(CH_2)_{15}$ C=O → HN_3 / H_2SO_4 → $(CH_2)_{15}$ NH C=O 86%

$C_6H_5-C(=O)-CH_3 \xrightarrow[H_2SO_4]{HN_3} C_6H_5NH-C(=O)-CH_3$ 90%

A reaction that has been used even more extensively for synthetic purposes is the Beckmann rearrangement of oximes, a reaction developed in Germany by E.O. Beckmann in the late 1800s. After conversion of an aldehyde or ketone to the corresponding oxime

(C=N-OH, see Section 17.6), treatment with acid causes rearrangement to the corresponding amide.

Beckmann Rearrangement

$$R-\overset{\overset{O}{\|}}{C}-R' \xrightarrow{NH_2OH} R-\overset{\overset{N-OH}{\|}}{C}-R' \xrightarrow{H^+} RNH-\overset{\overset{O}{\|}}{C}-R'$$

The Beckmann rearrangement is stereospecific in two distinct ways. First, there is retention of configuration at the migrating carbon as with other 1,2-migrations. Second, it is specifically the substituent *anti* to the OH group of the oxime which migrates. Aldehydes and unsymmetrical ketones often yield a mixture of stereoisomeric oximes, but each affords a specific rearrangement product.

$$R-\overset{\overset{:N-\ddot{O}H}{\|}}{C}-R' \underset{}{\overset{H^+}{\rightleftharpoons}} R-\overset{\overset{N-\overset{+}{O}H_2}{\|}}{C}-R' \longrightarrow R-\ddot{N}=\overset{+}{C}-R' \xrightarrow{H_2O} R-NH-\overset{\overset{O}{\|}}{C}-R'$$

$$R-\overset{\overset{HO-N:}{\|}}{C}-R' \underset{}{\overset{H^+}{\rightleftharpoons}} R-\overset{\overset{H_2\overset{+}{O}-N:}{\|}}{C}-R' \longrightarrow R-\overset{+}{C}=\ddot{N}-R' \xrightarrow{H_2O} R-\overset{\overset{O}{\|}}{C}-NHR'$$

This stereospecificity indicates that a free nitrenium ion is not involved as an intermediate in the reaction, although we can draw such an ion in order to help explain the rearrangement.

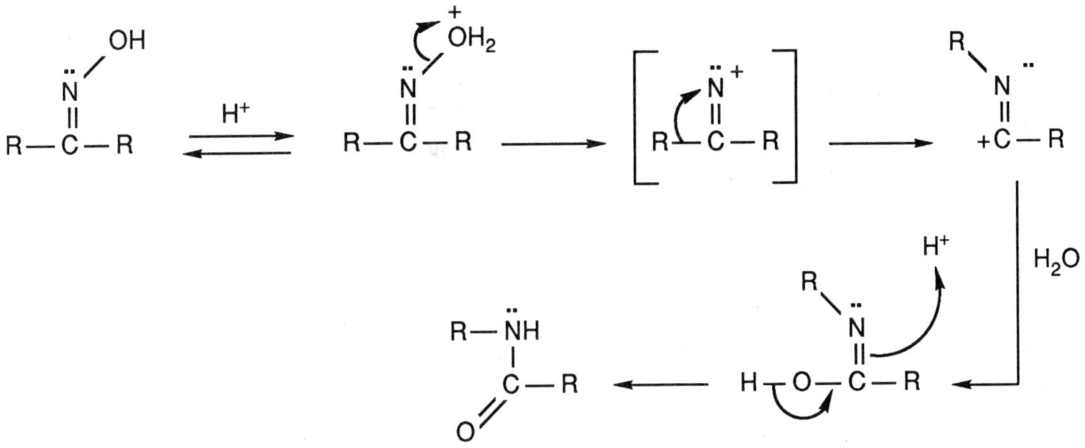

OH N R—C—R $\underset{}{\overset{H^+}{\rightleftharpoons}}$ $\overset{+}{O}H_2$ N R—C—R ⟶ [$\overset{+}{N}$ R—C—R] ⟶ R N +C—R

H_2O

H–O–C—R (with N–R), H^+ ⟶ R—NH, C—R, O

When the oxime is formed from an aldehyde, the stereoisomer with the OH *anti* to hydrogen produces an intermediate that loses a proton to give a nitrile.

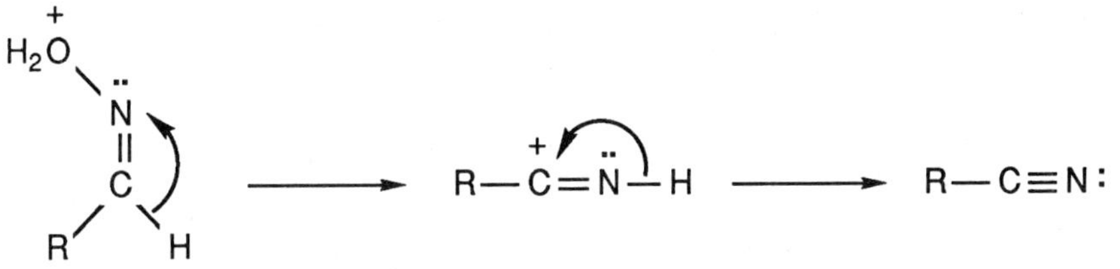

$H_2\overset{+}{O}$ N C R H ⟶ R—$\overset{+}{C}$=N—H ⟶ R—C≡N:

The following reactions illustrate the use of the Beckmann rearrangement for conversion of ketones (via their oximes) to amides:

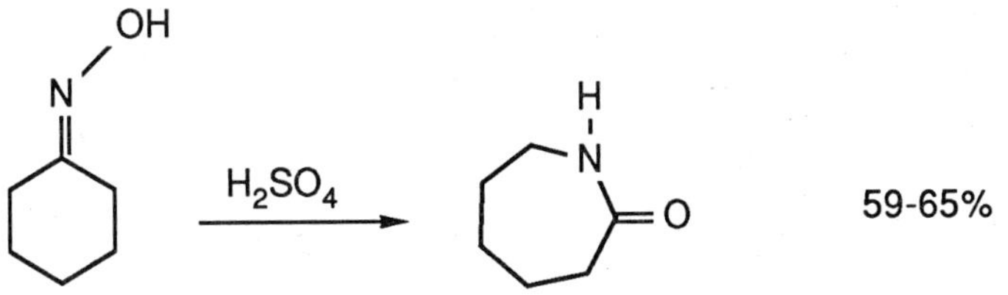

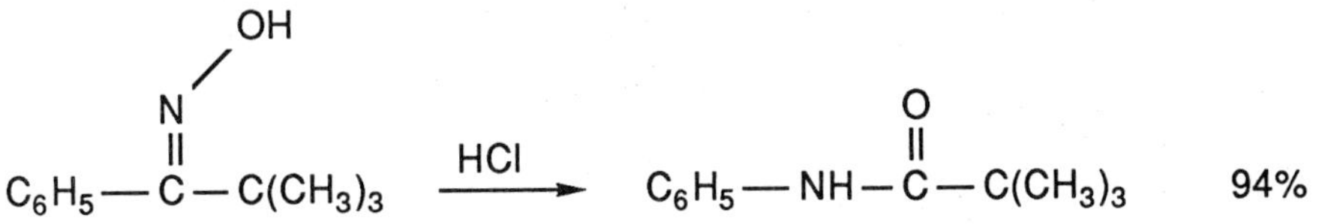

OH N (cyclohexanone oxime) $\xrightarrow{H_2SO_4}$ H N =O (caprolactam) 59-65%

OH N C_6H_5—C—$C(CH_3)_3$ $\xrightarrow{HCl}$ C_6H_5—NH—C(=O)—$C(CH_3)_3$ 94%

A.3 REARRANGEMENT TO ELECTRON DEFICIENT OXYGEN

Rearrangements to electron-deficient oxygen are relatively uncommon, largely because intermediates with electron deficient oxygen are particularly unstable. The only reactions in this category that have general synthetic value are the Baeyer-Villiger reaction and the oxidation of organoboron intermediates. We presented a number of examples of the Baeyer-Villiger reaction in Section 13.11, but our discussion here will emphasize its similarity to the other rearrangements considered in this Appendix. The reaction converts ketones to esters by "insertion" of an oxygen between the carbonyl carbon and one of its substituents, and mechanistically it is closely related to the pinacol rearrangement. Acid-catalyzed addition of a peracid (such as peracetic acid, CH_3CO_3H) to the carbonyl group of a ketone generates a tetrahedral intermediate.

$$R-\overset{O}{\overset{\|}{C}}-R \;+\; CH_3-\overset{O}{\overset{\|}{C}}OOH \;\rightleftharpoons\; R-\underset{R}{\overset{HO}{C}}-O-O-\overset{O}{\overset{\|}{C}}-CH_3$$

Cleavage of the weak oxygen-oxygen bond then affords a carboxylate ion (or the corresponding acid under acidic conditions) and an electron-deficient oxygen cation that undergoes rearrangement to yield an ester.

$$R-\underset{\boxed{R}}{\overset{HO}{C}}-\ddot{O}-O-\overset{O}{\overset{\|}{C}}-CH_3 \longrightarrow {}^{-}O-\overset{O}{\overset{\|}{C}}-CH_3 \;+\; R-\underset{\boxed{R}}{\overset{H\ddot{O}:}{C}}-\ddot{O}^{+}$$

$$R-\overset{O}{\overset{\|}{C}}-O-\boxed{R} \;\underset{\text{proton transfer}}{\rightleftharpoons}\; R-\overset{H\ddot{O}^{+}}{\overset{\|}{C}}-\ddot{O}-\boxed{R}$$

The ionization and rearrangement may actually occur in a concerted manner:

$$\text{R}-\overset{\overset{\text{H}\ddot{\text{O}}:}{|}}{\underset{\underset{\text{R}}{|}}{\text{C}}}-\text{O}-\text{O}-\overset{\overset{\text{O}}{\|}}{\text{C}}-\text{CH}_3 \longrightarrow \text{R}-\overset{\overset{\text{H}\ddot{\text{O}}^+}{\|}}{\text{C}}-\text{OR} + {}^-\text{O}-\overset{\overset{\text{O}}{\|}}{\text{C}}\text{CH}_3$$

Nevertheless, the stepwise mechanism emphasizes its similarity to the other reactions discussed in this Appendix. The overall reaction can therefore be summarized in the following way:

Baeyer-Villiger Reaction

$$\text{R}-\overset{\overset{\text{O}}{\|}}{\text{C}}-\text{R}' \xrightarrow[\text{(e.g., CH}_3\text{CO}_3\text{H)}]{\text{peracid}} \left[\text{R}-\overset{\overset{\text{HO}}{|}}{\underset{\underset{\text{R}}{|}}{\text{C}}}-\ddot{\text{O}}^{+}\right] \longrightarrow \text{R}-\overset{\overset{\text{O}}{\|}}{\text{C}}-\text{OR}' + \text{R}-\text{O}-\overset{\overset{\text{O}}{\|}}{\text{C}}\text{R}'$$

As indicated by the preceding equation, unsymmetrical ketones can yield two isomeric products. Usually one of them will predominate, the major product being determined by the migratory aptitudes of the two substituents. Aldehydes nearly always yield carboxylic acids because of the high migratory aptitude of hydrogen.

$$m\text{-}O_2N\text{-}C_6H_4\text{-CHO} \xrightarrow{CH_3CO_3H} m\text{-}O_2N\text{-}C_6H_4\text{-COOH} \qquad 90\%$$

Since oxidation of aldehydes to carboxylic acids can usually be carried out more easily with other reagents, the Baeyer-Villiger reaction is usually restricted to ketones. The following reactions illustrate its use (see also Section 13.11):

$$p\text{-}O_2N\text{-}C_6H_4\text{-CO-}C_6H_5 \xrightarrow{CH_3CO_3H} p\text{-}O_2N\text{-}C_6H_4\text{-CO-O-}C_6H_5 \qquad 95\%$$

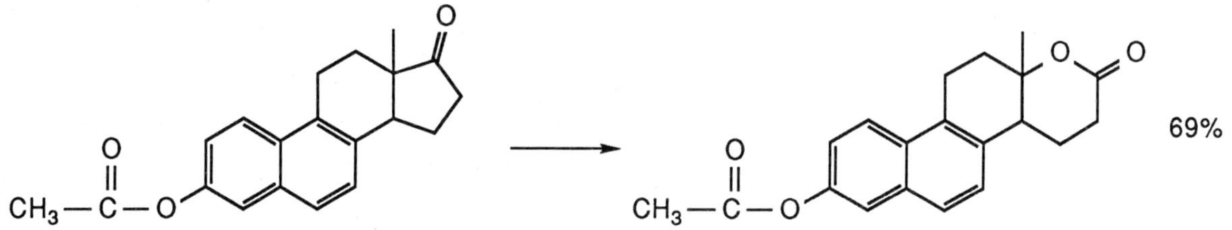

69%

We have shown you numerous examples of hydroboration-oxidation in the text. The reaction is used to add water to a carbon-carbon double bond (Section 5.11) or a carbon-carbon triple bond (Section 7.5). In each case a carbon-boron bond in the organoboron intermediate is oxidized to yield a product having a carbon-oxygen bond instead.

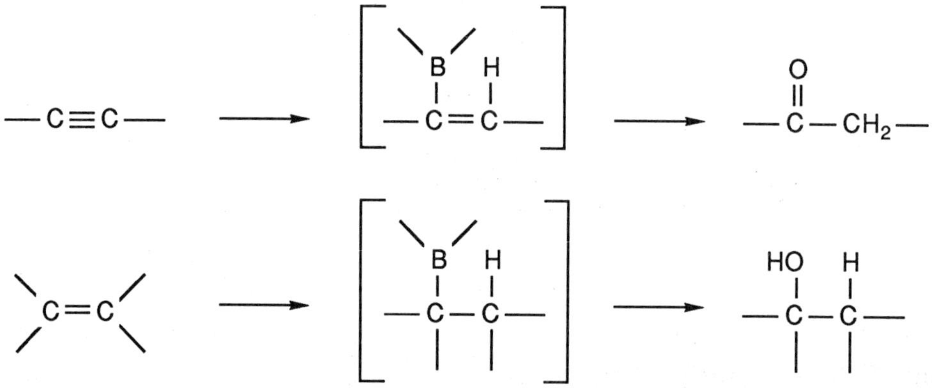

The oxidation occurs by attack of hydroperoxide ion on the electron-deficient boron to generate a tetrahedral intermediate. This undergoes a cleavage of the oxygen-oxygen bond together with a concerted migration of an alkyl (or alkenyl) group to electron-deficient oxygen.

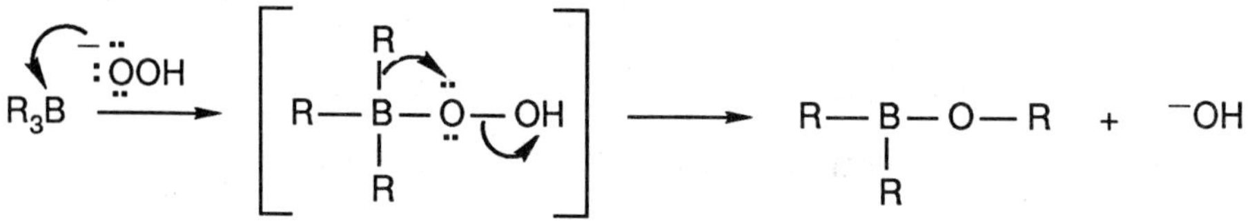

$$R_3B \xrightarrow{\;^{-}:\ddot{O}OH\;} \left[R-\underset{R}{\overset{R}{B^{-}}}-\ddot{O}-OH \right] \longrightarrow R-\underset{R}{B}-O-R \;+\; {}^{-}OH$$

Repetition of this for the remaining two alkyl (or alkenyl) groups generates a trialkyl borate, which then undergoes hydrolysis to the corresponding alcohol. (In the case of an alkenylborane, this produces the enol, which isomerizes to the corresponding aldehyde or ketone.)

$$R{-}\underset{\displaystyle R}{\underset{|}{B}}{-}OR \xrightarrow[NaOH]{H_2O_2} RO{-}\underset{\displaystyle OR}{\underset{|}{B}}{-}OR \xrightarrow[^{-}OH]{H_2O} ROH$$

The oxidation of organoboranes can be summarized in the following way, and we refer you to the text for specific examples.

Oxidation of Organoboranes

$$R_3B \xrightarrow[NaOH]{H_2O_2} ROH$$

$$\left(\;\rangle C{=}\overset{|}{C}\;\right)_3B \xrightarrow[NaOH]{H_2O_2} {-}CH_2{-}\overset{\displaystyle O}{\overset{\|}{C}}{-}$$

A.4 SIGMATROPIC REARRANGEMENTS

Cope Rearrangements

In Section 27.4 we showed that 1,5-hexadienes can undergo *thermal* rearrangement to produce isomeric 1,5-hexadienes. The reaction requires no catalyst, and no intermediates are formed. These rearrangements are reversible, the position of the equilibrium being determined by the relative stabilities of the reactants and products. This process, studied extensively by A.C. Cope at Harvard, is known as the *Cope rearrangement*.

Cope Rearrangement

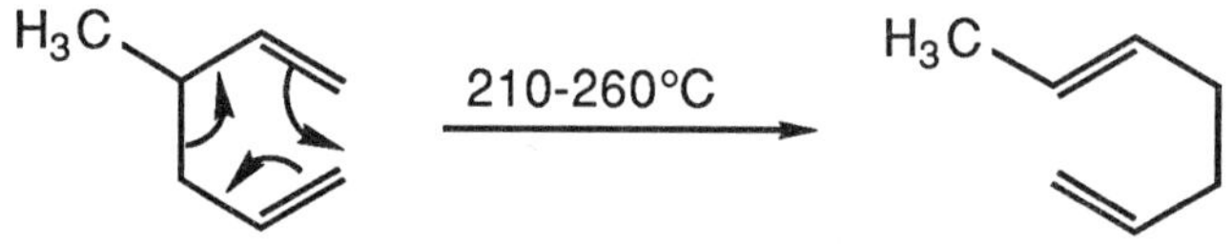

H_3C — 210-260°C → H_3C

Although the example depicted in the preceding equation has a single alkyl substituent, a great variety of substituted 1,5-dienes react in this way. The Cope rearrangement requires moderately high temperatures, but it typically proceeds smoothly at 150–250°C. Temperatures at the low end of this range (or even lower) suffice when the double bonds formed in the rearrangement are conjugated with another unsaturated group, and the reaction is also rapid with compounds for which the double bonds are held in the appropriate geometry. The following examples show how the reaction is favored by these effects:

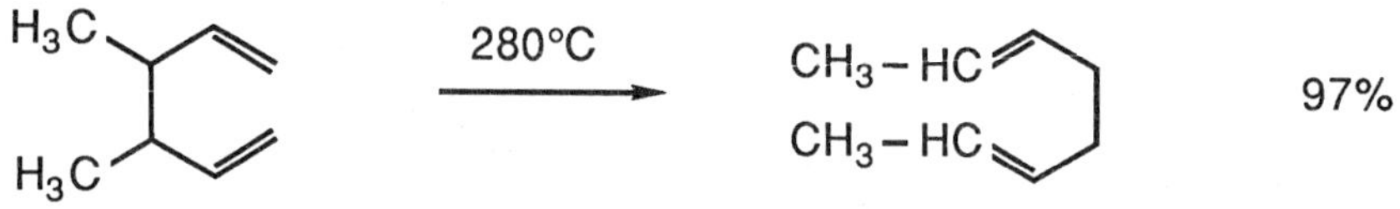

H_3C, H_3C — 280°C → CH_3-HC, CH_3-HC 97%

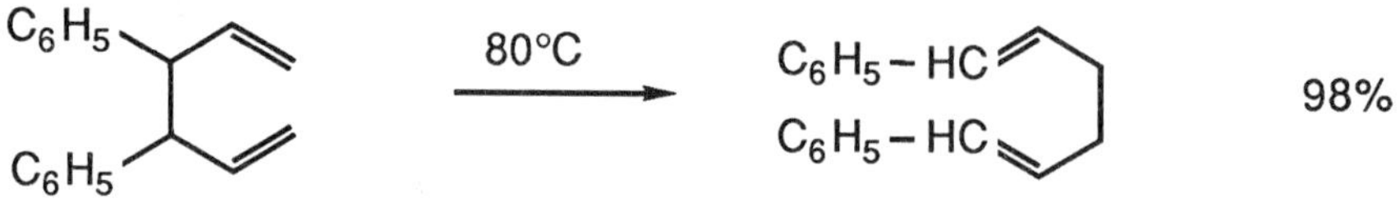

C_6H_5, C_6H_5 — 80°C → C_6H_5-HC, C_6H_5-HC 98%

70°C → ~100%

A particularly useful type of compound for this reaction is one with a hydroxyl group on one of the saturated carbons of the reactant. Initial rearrangement generates an *enol*, and isomerization to the keto form drives the equilibrium toward formation of the final product, an aldehyde or ketone. This is sometimes called the *oxy-Cope* rearrangement.

Oxy-Cope Rearrangement

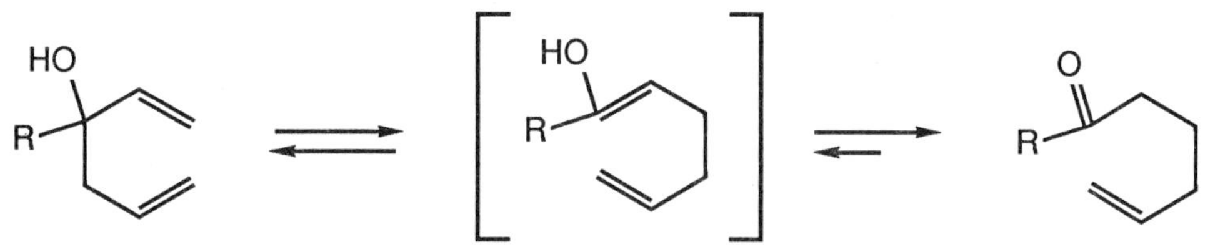

The presence of a carbonyl group in the product makes the oxy-Cope reaction valuable for synthetic purposes, because the product can be transformed into a variety of other compounds. The following examples illustrate the use of this rearrangement.

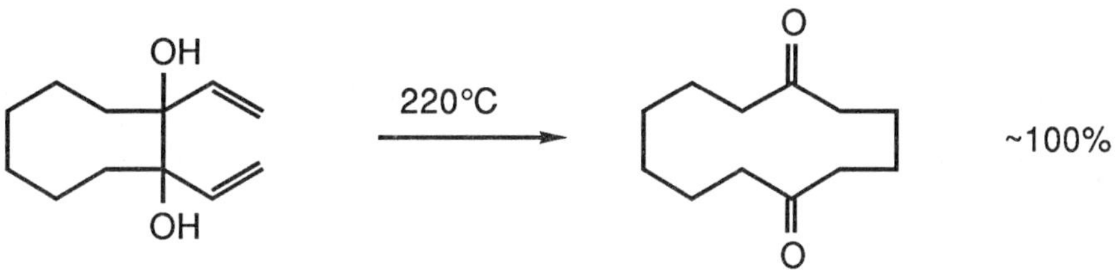

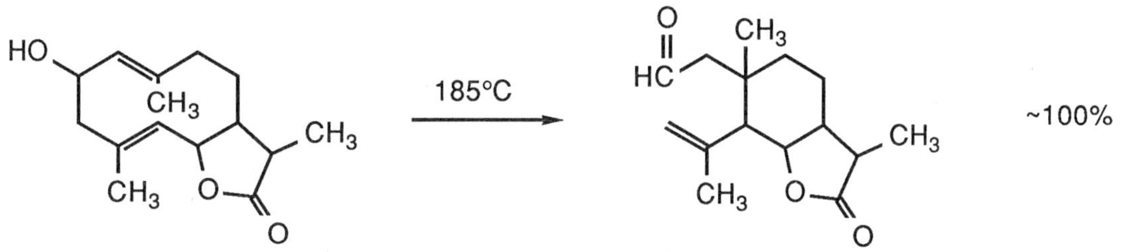

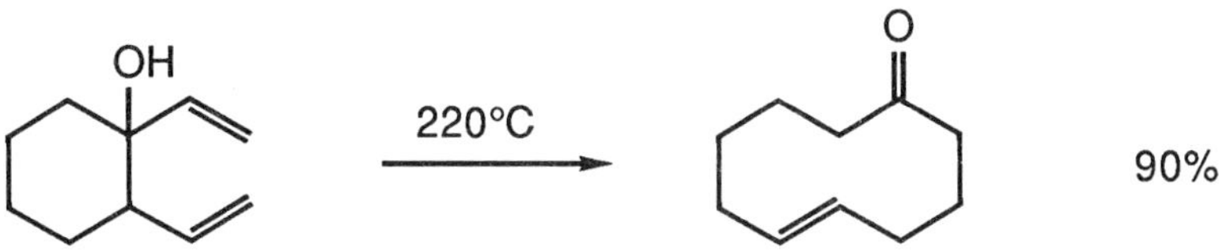

220°C ~100%

185°C ~100%

220°C 90%

Claisen Rearrangements

In 1912 the German chemist L. Claisen heated the allyl ether of the enol of ethyl 3-ketobutyrate to temperatures of 150–200°C during distillation. Surprisingly, this produced an isomeric compound, ethyl 2-allyl-3-ketobutyrate, in which a new carbon-carbon bond had been formed.

$$CH_3-\underset{\displaystyle O-CH_2-CH=CH_2}{C}=CH-CO_2C_2H_5 \xrightarrow{150\text{-}200°C} CH_3-\overset{O}{\overset{\|}{C}}-\underset{\displaystyle CH_2-CH=CH_2}{CH}-CO_2C_2H_5 \quad > 85\%$$

Subsequent investigations demonstrated that this thermal rearrangement is a general reaction for allyl ethers of enols or phenols. When such a compound is heated to temperatures near 200°C the allyl group migrates to the position α to the carbon-oxygen bond. The reaction is called the Claisen rearrangement after its discoverer.

Claisen Rearrangement

$CH_2-CH=CH_2$, O, C, C ~200°C O, C, C, $CH_2-CH=CH_2$

Rearrangement of an allyl aryl ether produces a ketone, which then isomerizes to the enol form, a phenol.

$OCH_2-CH=CH_2$ → O, $CH_2-CH=CH_2$, H → OH, $CH_2-CH=CH_2$ 73%

At first the Claisen rearrangement was thought to be a simple 1,3-migration of the allyl group, but in 1925 it was shown that the allyl group becomes attached at the other end of the three-carbon chain.

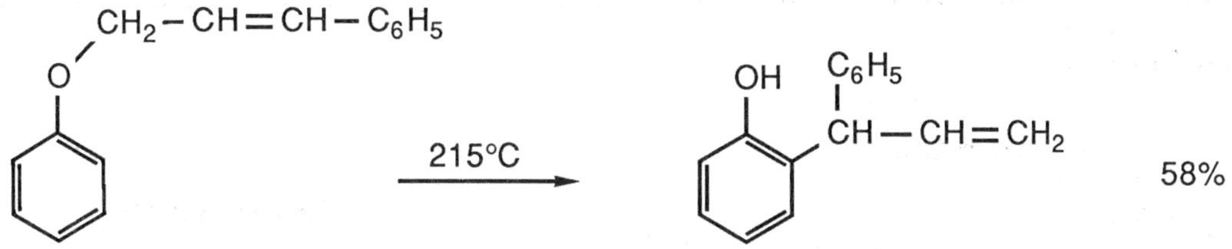

$CH_2-CH=CH-C_6H_5$ (O-phenyl ether) →(215°C) o-($CH(C_6H_5)-CH=CH_2$)phenol (OH) 58%

The complexity of the rearrangement may have been perplexing to the earlier chemists, but you can now recognize it as another example of a sigmatropic rearrangement. Indeed, it is highly similar to the Cope rearrangement except that one of the carbons of the 1,5-diene has been replaced by an oxygen.

Cope and Claisen Rearrangements

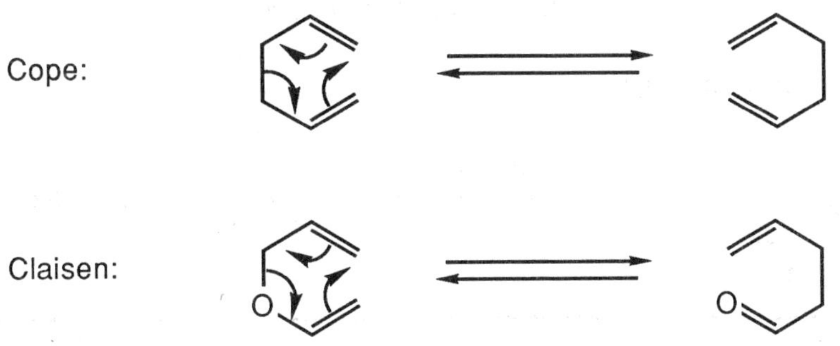

Cope:

Claisen:

As in the case of the oxy-Cope rearrangement, the formation of a carbonyl group (or its isomerization to a phenol) in the Claisen rearrangement favors the rearrangement product in an equilibrium. The cyclic nature of the rearrangement shows you why the allyl group becomes attached at the other end of the three-carbon subunit in the reaction of compounds such as phenyl cinnamyl ether.

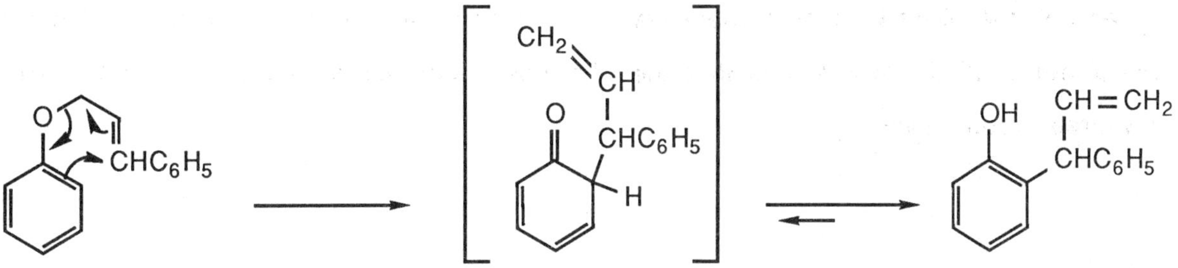

CHC_6H_5 → [CH_2 = CH, O, CHC_6H_5, H] ⇌ OH, $CH=CH_2$, CHC_6H_5

When the *ortho* positions of an aryl allyl ether are substituted, the ketone produced by Claisen rearrangement cannot enolize. In such cases the initial product (which is also a 1,5-diene) usually undergoes a Cope rearrangement. Enolization then produces a *para*-substituted phenol, and the double rearrangement means that the allyl group is attached at the *same* carbon as in the original allyl ether.

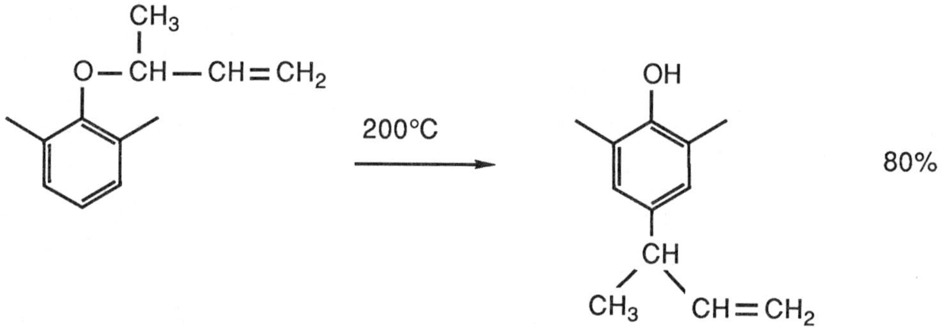

CH_3
O—CH—CH=CH_2

200°C

OH

CH

CH_3 CH=CH_2

80%

The preceding reaction proceeds via the following pathway:

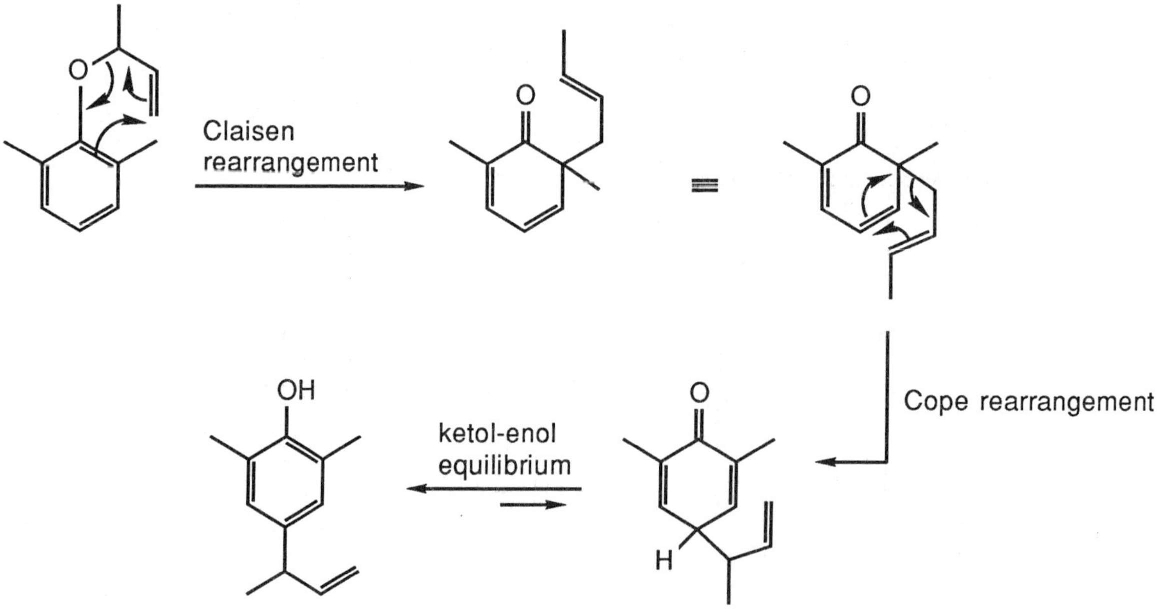

O

Claisen rearrangement

O

O

OH

ketol-enol equilibrium

O

H

Cope rearrangement

Sometimes the subsequent Cope rearrangement competes with enol formation even when the *ortho* positions are not substituted.

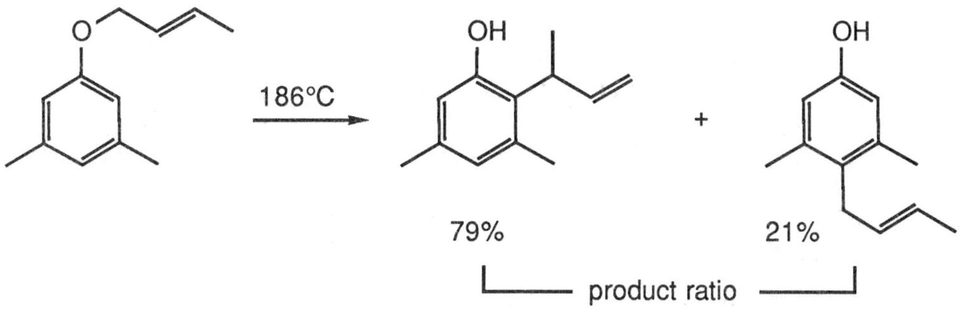

186°C

OH + OH

79% 21%

product ratio

However, the use of a polar solvent (which facilitates the keto-enol equilibrium) usually results in a high preference for simple Claisen rearrangement to produce the *ortho*-allyl phenol.

The Claisen rearrangement has been used extensively for synthetic purposes. It provides a reliable and specific method for introducing a carbon substituent on an aromatic ring, and with aliphatic compounds it permits "alkylation" at the α position of a carbonyl group without the harsh basic conditions of enolate reactions (Section 14.3). The overall processes are summarized in the following Figure.

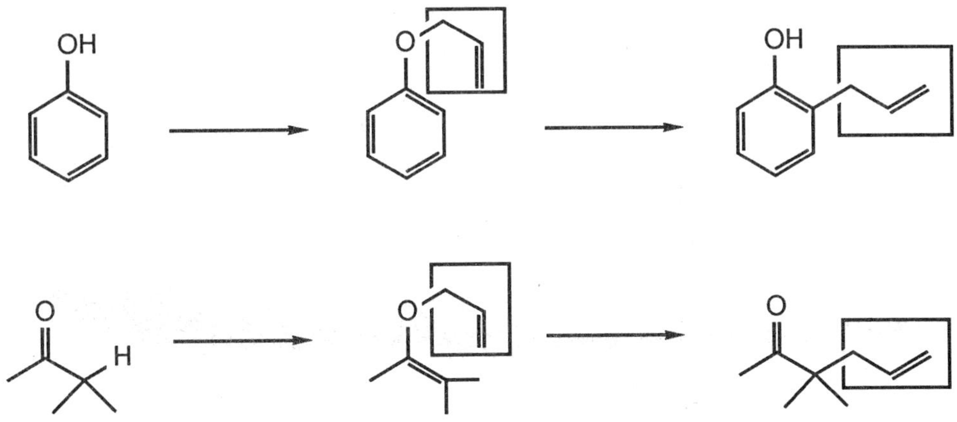

Figure. Introduction of Allyl Substituents by Using the Claisen Rearrangement.

Allyl ethers of phenols are often formed by alkylation of the phenoxide ion with an allyl halide (Section 15.4), but *O*-alkylation of aliphatic carbonyl compounds is usually not practical. In most cases allyl vinyl ethers are prepared under acidic conditions and are in equilibrium with ketals (or acetals).

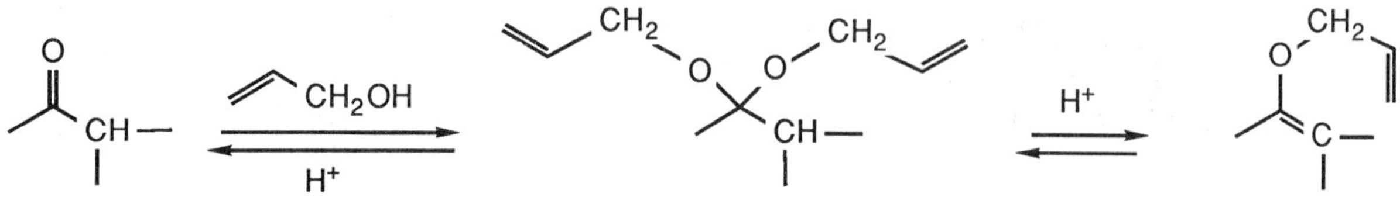

$$\text{(ketone)} \underset{H^+}{\overset{CH_2{=}CHCH_2OH}{\rightleftharpoons}} \text{(diallyl ketal)} \overset{H^+}{\rightleftharpoons} \text{(allyl vinyl ether)}$$

Sometimes the allyl vinyl ether is not isolated but is caused to rearrange directly by forming it at high temperatures. This method has been used with considerable success to introduce an allylic substituent at the α position of a carboxylic acid derivative. When the following diethoxyamino compound (the "ketal" of *N,N*-dimethylacetamide) is heated with an allylic alcohol, two moles of ethanol are lost to produce a vinyl allyl ether, which can undergo a Claisen rearrangement.

$$CH_3-C(OC_2H_5)_2-N(CH_3)_2 \xrightarrow{CH_2{=}CHCH_2OH} CH_2{=}CH-CH_2-O-C(=CH_2)-N(CH_3)_2$$

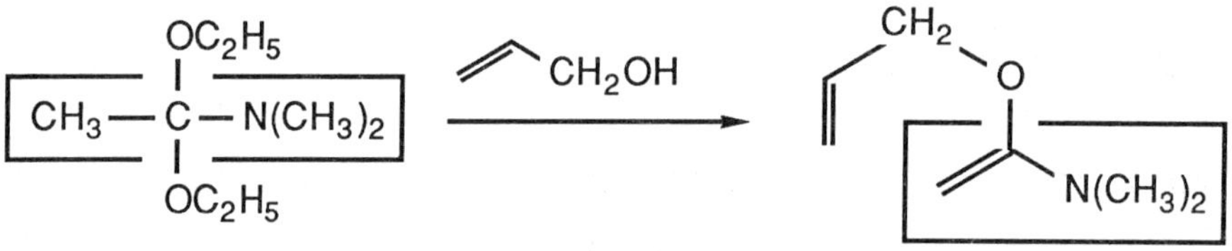

The overall reaction is illustrated by the following example:

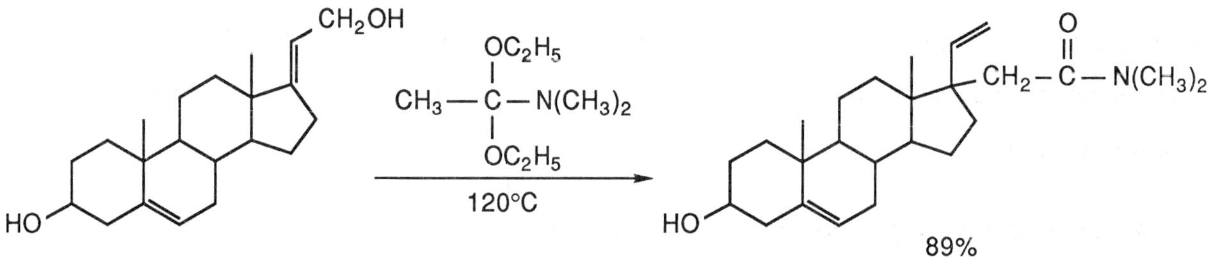

$$\text{HO-steroid}{=}CH-CH_2OH \xrightarrow[120°C]{CH_3-C(OC_2H_5)_2-N(CH_3)_2} \text{HO-steroid}(CH{=}CH_2)-CH_2-C(=O)-N(CH_3)_2 \quad 89\%$$

The following reactions illustrate additional examples of Claisen rearrangements in both aliphatic and aromatic systems.

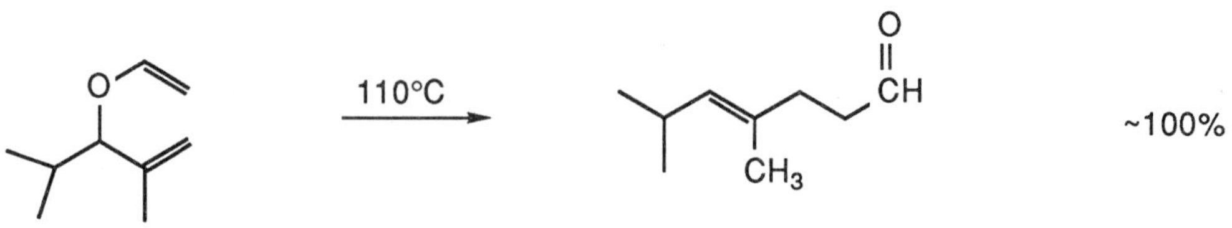

$$\xrightarrow{110°C} \quad \text{CH=O},\ CH_3 \qquad \sim100\%$$

CH_3O — 215°C → OH, CH_3O > 90%

CH_3 — 190-195°C → CH_2CHO, CH_3 93%

A.5 REARRANGEMENTS IN BIOLOGICAL SYSTEMS

Molecular rearrangements are usually thought of as laboratory reactions, but they have also been found in biological systems. In the text we illustrated several biochemical processes that involve carbocation rearrangements (Section 12.9), and we showed you how electrocyclic and sigmatropic rearrangements are both involved in vitamin D biosynthesis. In this section we will show you two additional examples of biologically important rearrangements.

An epoxide rearrangement very similar to the pinacol-type rearrangements discussed in Section A.1 constitutes a key step in the biological hydroxylation of aromatic rings. Enzymatic oxidation of the ring generates an epoxide, and this undergoes ring opening followed by a hydride shift.

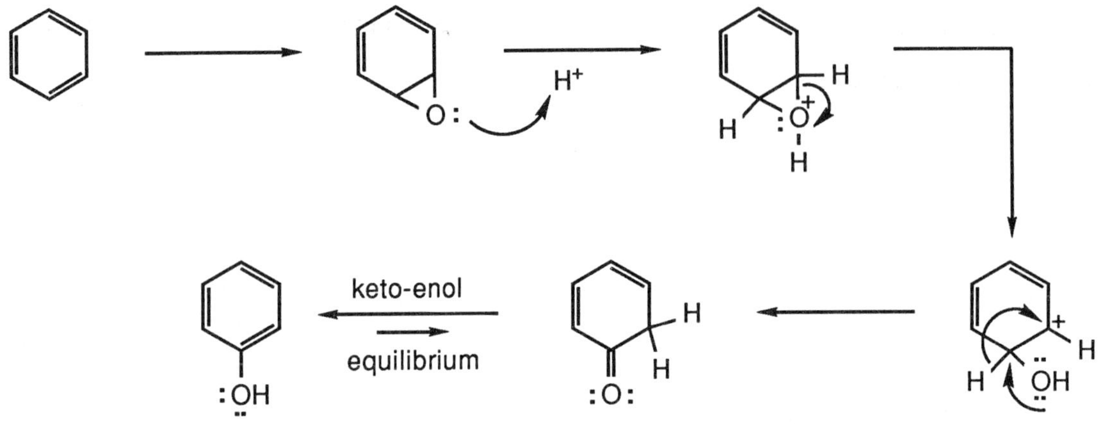

keto-enol equilibrium

This mechanism of aromatic hydroxylation was elucidated by workers at the National Institutes of Health, and the rearrangement has been dubbed the "NIH shift."

A biological example of the Claisen rearrangement is also known. In a sequence of reactions known as the shikimic acid pathway (which leads to many phenolic plant pigments), chorismic acid is converted to prephenic acid. The examples in the preceding section demonstrate that relatively low temperatures are adequate for some sigmatropic reactions, and presumably the enzyme must serve to hold the reactant in an ideal geometry for rearrangement. (Although the carboxyl groups are likely to be ionized at physiological pH, we have shown them in protonated form for convenience).

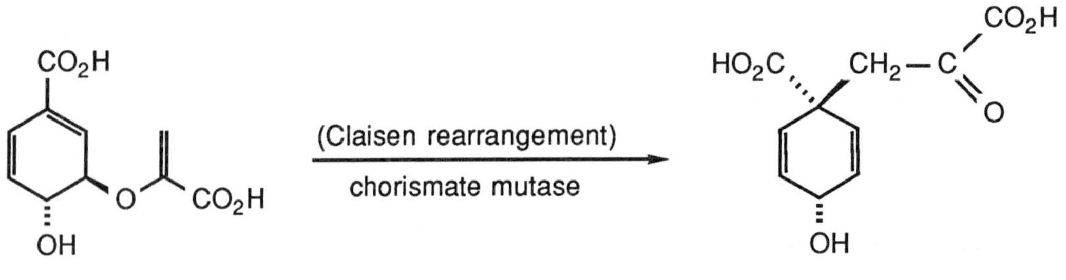

(Claisen rearrangement)
chorismate mutase

Chorismic acid → Prephenic acid

As with other examples shown in the text, these reactions further emphasize the close relationships between organic and biochemistry. The reactions that take place in biological systems are usually very similar to the organic reactions that can be carried out in the laboratory. Enzymes merely direct these reactions to specific products by providing a pathway with a low activation energy.

A.6 SUMMARY OF REACTIONS

The rearrangement reactions discussed in this Appendix are summarized in Table A.1.

Table A.1 Rearrangement Reactions

Reaction	Comments
1. Rearrangement to electron deficient carbon	Appendix A, Section 1
a. Carbocation rearrangement	
$-\underset{R}{\overset{OH}{C}}-\overset{OH}{C}- \xrightarrow{\text{acid}} -\overset{O}{\overset{\Vert}{C}}-\underset{R}{C}-$	Pinacol rearrangement (R=H, alkyl or aryl)
$-\underset{R}{\overset{OH}{C}}-\overset{X}{C}- \longrightarrow -\overset{O}{\overset{\Vert}{C}}-\underset{R}{C}-$	Pinacol-type rearrangement (e.g., $X=NH_2$ with HONO; X=OTs; or $X=N_2^+$ from the reaction of CH_2N_2 with a ketone or aldehyde)
$-\underset{R}{C}\overset{O}{-}C- \xrightarrow{BF_3} -\overset{O}{\overset{\Vert}{C}}-\underset{R}{C}-$ (epoxide)	Epoxide rearrangement (pinacol type)
b. Carbene rearrangement	
$R-\overset{O}{\overset{\Vert}{C}}-OH \xrightarrow[\text{2. } CH_2N_2 \text{; 3. } H_2O,\ Ag_2O]{\text{1. } SOCl_2} R-CH_2CO_2H$	Arndt-Eistert synthesis

2. Rearrangement to electron deficient nitrogen

(Appendix A, Section 2)

a. Carboxylic acid derivatives

$$R{-}\overset{O}{\overset{\|}{C}}{-}NH_2 \xrightarrow[Br_2 \text{ or } Cl_2]{KOH} R{-}NH_2$$

Hofmann degradation (primary amides)

$$R{-}\overset{O}{\overset{\|}{C}}{-}N_3 \xrightarrow[\text{2. } H_2O \text{ acid or base}]{\text{1. heat}} R{-}NH_2$$

Curtius rearrangement (acyl azides)

$$R{-}\overset{O}{\overset{\|}{C}}{-}OH \xrightarrow[\text{2. NaOH}]{\text{1. } HN_3,\ H_2SO_4} R{-}NH_2$$

Schmidt reaction of carboxylic acids

b. Ketones and aldehydes

$$R{-}\overset{O}{\overset{\|}{C}}{-}R \xrightarrow[H_2SO_4]{HN_3} R{-}\overset{O}{\overset{\|}{C}}{-}NHR$$

Schmidt reaction of ketones; rearrangement predicted by migratory aptitudes of R groups.

$$R{-}\overset{N{-}OH}{\overset{\|}{C}}{-}R \longrightarrow R{-}\overset{O}{\overset{\|}{C}}{-}NHR$$

Beckmann rearrangement of oximes migration of R group that is *anti* to the -OH.

3. Rearrangement to electron deficient oxygen

Appendix A, Section 3, Sections 5.11, 7.5, 13.15

$$R{-}\overset{O}{\overset{\|}{C}}{-}R \xrightarrow{R'CO_3H} R{-}\overset{O}{\overset{\|}{C}}{-}OR$$

Baeyer-Villiger oxidation; migration of R group with higher migratory aptitude.

$$R_3B \xrightarrow[NaOH]{H_2O_2} ROH$$

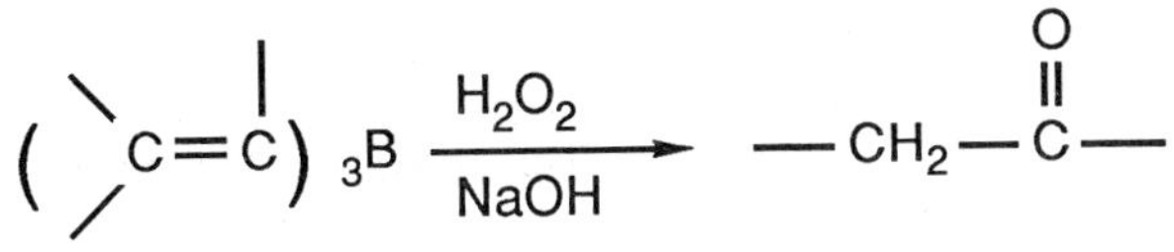

$$\left(\begin{array}{c}\diagdown \\ \diagup\end{array} C{=}C \Big| \right)_3 B \xrightarrow[NaOH]{H_2O_2} -CH_2-\overset{O}{\overset{\|}{C}}-$$

Oxidation of organoboranes. Trialkylboranes yield alcohols; trialkenylboranes yield aldehydes or ketones.

Migratory Aptitudes for Rearrangement to Electron-Deficient Carbon, Nitrogen, or Oxygen		
greatest tendency to undergo 1,2-shift	3° alkyl > aryl or hydrogen > 2° alkyl > 1° alkyl > methyl	least tendency to undergo 1,2-shift

4. Sigmatropic Rearrangements

Appendix A, Section 4

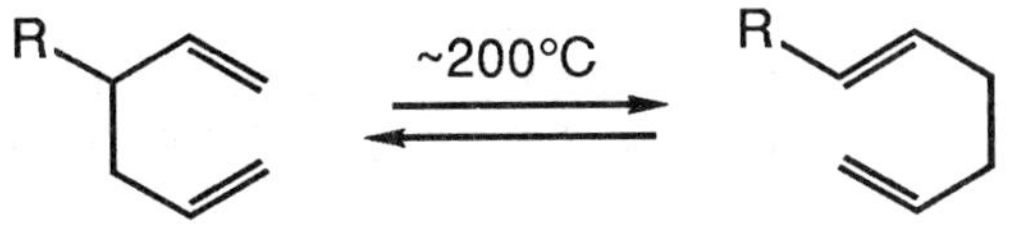

R ⇌ R (~200°C)

Thermal rearrangement of 1,5-dienes equilibrium reaction; more stable diene predominates.

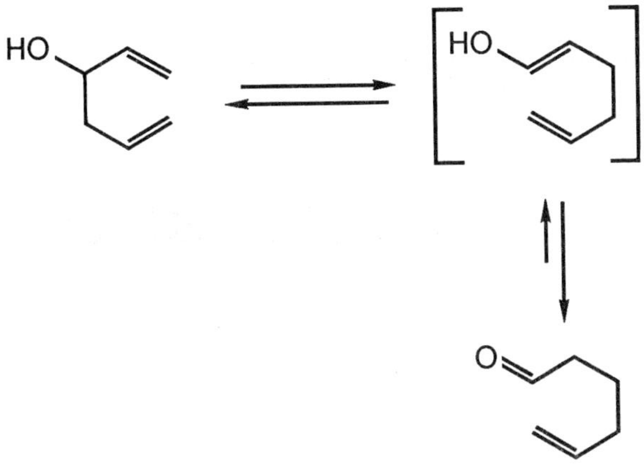

HO ⇌ [HO] ⇌ O

Oxy-Cope reaction. Keto-enol equilibrium favors product formation.

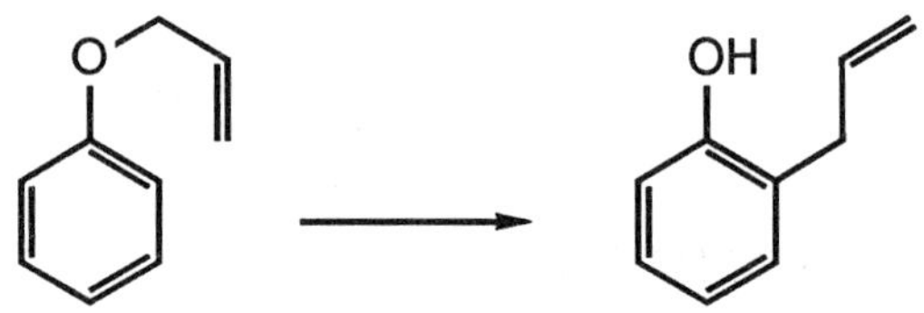

Claisen rearrangement. Allyl aryl ethers *ortho* allyl phenol is produced unless *ortho* positions are substituted)

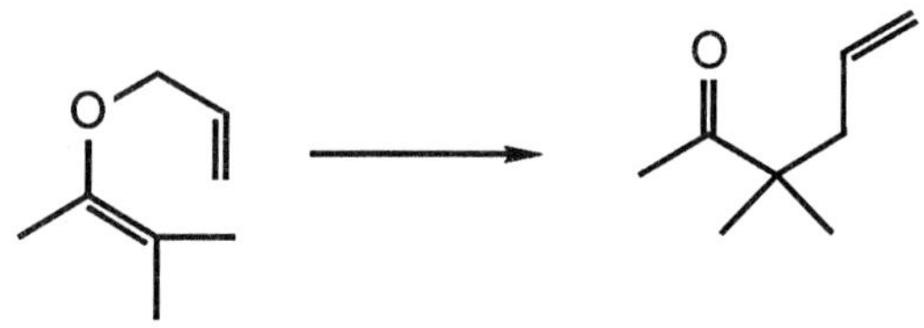

Enol allyl ethers produces γ,δ-unsaturated ketone.

APPENDIX B. MASS SPECTROMETRY

B.1 INTRODUCTION

Every time you turn on your television set you are employing a device that has played an extensive and important role in the development of modern science. At the heart of a TV set is the *cathode ray tube*, commonly known as the picture tube of the television. (When described by its initials, the cathode ray tube, *CRT*, also provides the name that is frequently used for a computer video terminal.) In a highly simplified view a cathode ray tube operates by emitting a beam of electrons inside an evacuated glass tube (Figure B.1). When the electron beam strikes the fluorescent coating at the viewing end of the screen, visible light is emitted.

Why does the CRT afford an image over the entire screen instead of a single spot? Moving electrons are deflected by electromagnetic fields, and the precise direction of the electron beam is controlled by a *deflection yoke*. By rapidly and accurately varying the electromagnetic field produced by the deflection yoke, the beam is directed to the appropriate screen location and a complete image is seen on the viewing screen.

The experiments that led to the discovery of the electron in the late 1800s were centered on the cathode ray tube. Scientists had established that cathode rays (i.e., electron beams) could be deflected by either a magnetic field or an electronic field. The English scientist Joseph J. Thomson, showed that the deflection of a beam of electrons by an *electric* field could be nullified by a counteracting *magnetic* field. The behavior of particles was known to have a different dependence for charge and mass in the two kinds of fields, and this experiment demonstrated that electrons have both *charge* and *mass*. A mathematical analysis of the experiment afforded the *mass to charge ration*, m/e, of the

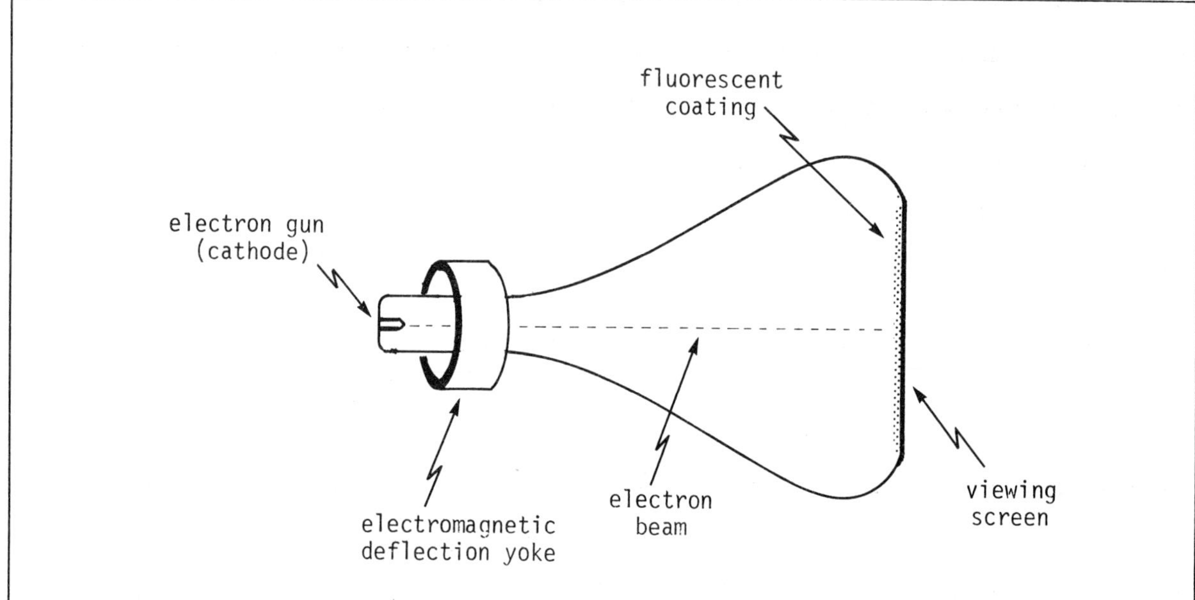

Figure B.1 Simplified view of a cathode ray tube. A beam of electrons emitted at the cathode strikes the fluorescent coating of the screen, and visible light is emitted. The electron beam is deflected to different positions on the screen by electromagnetic fields (deflection yoke), and this produces an image across the entire screen.

electron. The early experiments of Thomson and others showed that the deflection of ions in electromagnetic fields is a function of the mass to charge ratio. This provides the basis of modern *mass spectrometry*: separation of charged particles on the basis of mass.

Mass spectrometry has evolved into a powerful method for analyzing molecular structure, both as a research tool and as a standard analytical technique. But it has also seen some other, less common uses. For example, during the second World War, when Allied scientists with the Manhattan Project were racing to build the first atomic bombs, mass spectrometry was used preparatively to isolate the needed isotope of uranium, ^{235}U. A curious aspect of this episode involved problems in construction of large electromagnets for the mass spectrometers at Oak Ridge, Tennessee. Copper wire was in very short supply, much of the available supply going to munitions production and other war-related production. Another metal with good electrical conductivity was needed, so arrangements were made to borrow fifteen thousand tons of silver from Fort Knox to use as windings for the electromagnets. At today's prices this would be worth about $5,000,000,000 so these were rather valuable instruments.

B.2 HISTORICAL DEVELOPMENT OF THE MASS SPECTROMETER

Toward the end of the 19th century it was found that the production of cathode rays was often accompanied by a second type of emission. These emissions were in the opposite direction from the electron beam, and were described as *positive rays*. These rays were in fact positive ions formed by loss of an electron from the gas molecules that remained in the system. The design of the cathode ray tube was modified to permit study of the positive rays, and this led to some important advances in our understanding of atomic and molecular structure.

In his studies J.J. Thomson found that the positive ions formed in the presence of oxygen had mass to charge ratios of 8, 16 and 32. In accord with common practice for mass spectroscopy the units for m/e are understood to be amu per electronic charge. A singly charged ion therefore has m/e equal to its atomic mass expressed in amu. Thomson concluded that these ions corresponded to O_2^+ (m/e = 32), O^+ (m/e = 16) and the doubly charged species O^{2+} (m/e = 8). Such results provided strong supporting evidence for the existence of oxygen (and subsequently of other elements) as diatomic molecules.

Another very important discovery in Thomson's laboratory involved studies with neon. Although neon was known to have an atomic weight of 20.2 amu, *two different* ions were formed with m/e of 20 and 22. This was the definitive evidence for the existence of *isotopes*, and it also resolved the ambiguity of atomic weights that were nonintegral. Experimental atomic and molecular weights) are the *averages* of the values for the different isotopic species present in the sample. The technique of mass spectrometry (although that name was not used in these early studies) permitted for the first time the direct observation of the different isotopes. While atomic weights are typically nonintegral, the individual isotopes do exhibit atomic masses that are whole numbers.*

The capability of directly measuring the masses of ions (and of the different isotopes present) provided a powerful tool for studying atomic structure in the early 20th century. But mass spectrometry would be of little value for organic analysis if the only information obtained were the molecular weight. Instead a series of decomposition reactions occur to give new ions, and extensive structural information can be obtained by proper interpretation of the resulting *mass spectrum*. The use of mass spectrometry for analysis in organic chemistry was intensively exploited by the petroleum industry starting in the 1940s, and it has now become a standard technique in the field of chemistry. In this appendix we will present a brief overview of how structural information can be obtained from a mass spectrum.

B.3 THE MASS SPECTROMETER

There are many kinds of mass spectrometer, differing in both design and complexity. Costs of such instruments range from only a few thousand dollars up to hundreds of thousands of dollars (or more for some specialized research instruments). A discussion of the different types of mass spectrometer is beyond the scope of this appendix, and we will only illustrate the basic principles by which all these instruments operate.

----

*Actually, integral values of atomic mass are not found with individual isotopes if the measurements are carried out with greater accuracy. For example, ^{12}C (with six protons and six neutrons) has a mass of 12.000 amu, which is not quite six times greater than the 2.014 amu for ^{2}H (one proton and one neutron). The difference of 0.084 amu corresponds to the greater nuclear binding energy for carbon (according to the relationship $E=mc^2$). The pattern is a general one, and relative to $^{12}C=12.000$ amu, atoms with lower atomic numbers exhibit masses that are slightly more than integral values; atoms of the heavier elements show masses that are slightly less than integral values.

A simplified view of a mass spectrometer is shown in Figure B.2, and in many ways it is quite similar to a cathode ray tube. Positive ions from the sample are produced in the region of the spectrometer called the *ion source*,* and these ions emerge through a small slit as an *ion beam*. The ion beam is accelerated in the direction of the ion collector by an electric potential, and the accelerated beam passes through the *mass analyzer*.

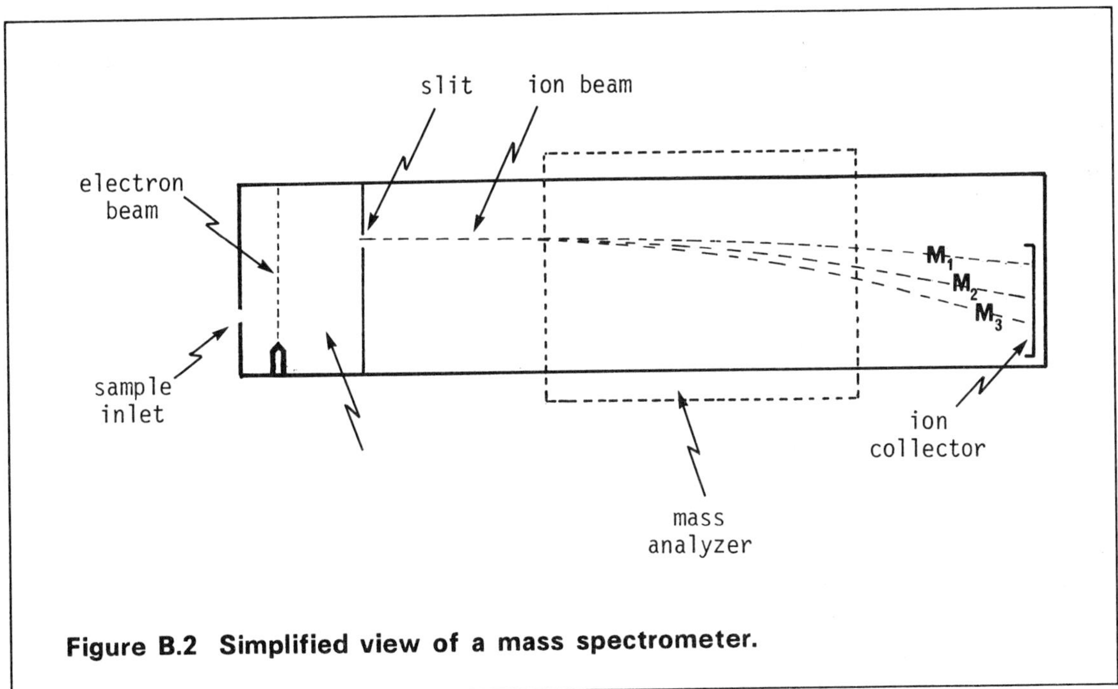

Figure B.2 Simplified view of a mass spectrometer.

There are several types of mass analyzers, but in all cases they rely on the deflection of the ion beam by electric and/or magnetic fields. The magnitude of the deflection of a

---

*The ion source depicted in Figure B.2 is an *electron impact* source, which is the most common type. A beam of high energy electrons passes through the ionization chamber, and when one of these electrons collides with a sample molecule, a positive ion is produced by ejection of a second electron:

$$:\ddot{\underset{\cdot\cdot}{A}}: + e^- \longrightarrow [:\dot{\underset{\cdot\cdot}{A}}:]^+ + 2e^-$$

particular ion varies inversely with its mass to charge ratio, so the heaviest ions are deflected least. When the ions strike the *ion collector*, this registers as an electric current, and it is possible to measure the current corresponding to each different value of m/e. Consequently a record is obtained of each value of m/e for which ions are produced. Moreover, the intensity of the signal at each value of m/e is proportional to the relative number of ions with that mass to charge ratio. Hence the mass spectrum describes not only the mass to charge ratios of the different ions but also the extent to which each contributes to the total ion beam. Under normal conditions of mass spectrometry most of the ions formed have unit positive charge. Therefore we will frequently refer simply to the mass of an ion rather than to its mass to charge ratio.

B.4 THE MASS SPECTRUM

Ejection of an electron from a sample molecule (M) in the ion source of a mass spectrometer produces a positively charged species called the molecular ion. The molecular ion is generated by loss of a single electron from a doubly occupied orbital, so it is actually a *cation radical* ($M^{+\cdot}$).

$$\underset{\text{sample molecule}}{M} \xrightarrow{\text{ionization}} \underset{\text{molecular ion}}{M^{+\cdot}}$$

In most cases the molecular ions decompose into ions with different masses, and the masses of these ions (together with their relative amounts) provide valuable information about the molecular structure of the sample. In this section we will discuss the origin of ions with different mass, and we will show how the resulting information is recorded for subsequent interpretation.

Two general ways are available for recording mass spectral data: graphical display and numerical tabulation. In this appendix we will use only the former method, showing a plot of ion intensity as a function of mass. (Actually it is the mass to charge ratio which is recorded.) The intensity of a peak in the mass spectrum is a measure of the number of ions with that m/e, and it is recorded as the *relative abundance.* For the purpose of simplicity we have omitted some of the smallest peaks in the mass spectra presented in this appendix. Only peaks with intensities at least 1.5% of that for the most intense peak are shown. The presentation of data is illustrated in Mass Spectrum B.1, the mass spectrum for methane.

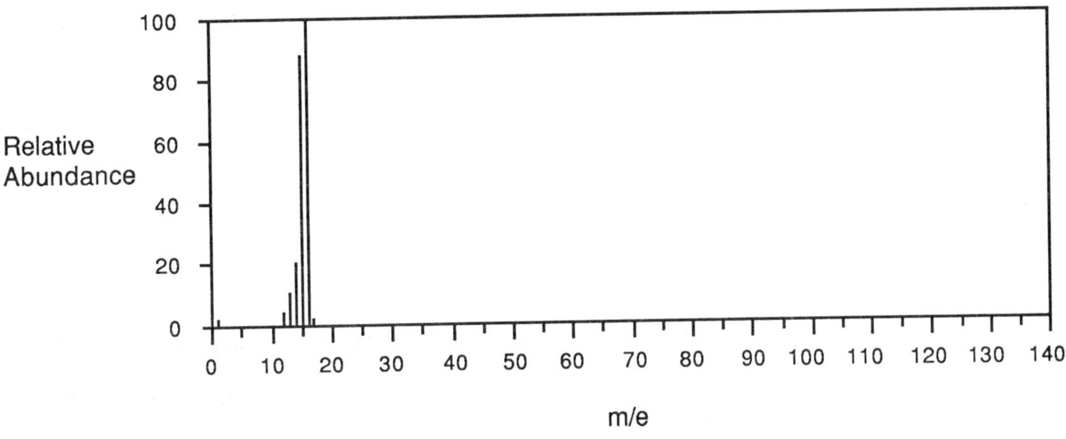

Mass Spectrum B.1 Methane

Notice that *the most intense peak in the spectrum is assigned a relative abundance of 100.* This is standard practice in mass spectrometry, and it allows us to conveniently compare the relative abundance of any ion in the spectrum to that of the most abundant ion. In the case of methane this is the molecular ion, $CH_4^{+\cdot}$, but frequently an ion of different mass will be most abundant. In fact the relative abundance of the molecular ion is sometimes quite low, and this makes interpretation of the mass spectrum much more difficult.

Fragmentation of Ions

At this point we have accounted only for one of the peaks in the mass spectrum of methane, the molecular ion (m/e 16) which is formed by loss of a single electron from CH_4 (MW=16). What are the other peaks in the spectrum? For example, what is the origin of the peak at m/e 15? Starting from the molecular ion, $CH_4^{+\cdot}$, the only way to form a species with a mass of 15 amu is through the loss of one hydrogen. Moreover, since the 15 amu species is positively charged, the 1 amu species that was lost must have been electrically neutral. In other words the following reaction must occur with some of the molecular ions that are formed -- *fragmentation* into a methyl cation and a hydrogen radical:

$$CH_4^{+\cdot} \xrightarrow{\text{fragmentation}} CH_3^{+} + H^{\cdot}$$

The peaks at m/e 14, 13 and 12 must result from further loss of one, two or three hydrogen atoms.

Why do such fragmentations occur? The reactions of ions in a mass spectrometer are quite different from those reactions which occur in solution. When they are first generated by collisions with high energy electrons, the molecular ions usually have considerable potential energy. But ions in a mass spectrometer are formed in the gas phase, and they undergo no collisions with other molecules at the very low pressures that are employed. Consequently, dissipation of the potential energy can only occur through a *unimolecular* reaction such as decomposition into two fragments. As you will see later in this appendix there are certain predictable patterns which fragmentations follow, and these provide information about molecular structure.

There are still two peaks for which we have not accounted in the mass spectrum of methane. These are the peaks at m/e 1 and 17. Clearly, the former can only correspond to H^+, and this must result from an alternative mode of decomposition of one of the ions to generate H^+ and a carbon radical. But the low relative abundance of the m/e 1 peak shows that fragmentation to a hydrogen radical and a carbocation is much more favorable.

Isotopic Contributions

None of the preceding discussion has explained the m/e 17 peak in the mass spectrum of methane. The molecular weight of methane is 16 amu, and you might suspect that m/e 17 corresponds to CH_5^+. But such a species could only result from a *bimolecular* reaction, and this is precluded by the very low pressures inside the mass spectrometer; the concentrations of various species in the gas phase are so low that bimolecular collisions are highly improbable.

To understand the origin of the m/e 17 peak you must recognize a unique aspect of mass spectrometry. Most chemical measurements provide information on the *average* behavior of a large assembly of molecules, 6×10^{23} for one mole of a compound. Mass spectrometry is quite different because the data results from detection of *individual ions* (although a large number of ions is counted during the experiment). Approximately 1% of all carbon atoms are ^{13}C rather than ^{12}C, and in a normal chemical experiment this is reflected by an effective atomic weight of 12.01 amu. But in mass spectrometry we see the results of individual ions and not an average. So $CH_4^{+\cdot}$ ions with ^{12}C will have m/e 16, but those with ^{13}C will have m/e 17. The natural abundance of ^{13}C is 1.1%, and this corresponds precisely to the 100:1 intensity ratio for the peaks at m/e 16 and 17.

B.5 MOLECULAR FORMULAS AND ISOTOPIC CONTRIBUTION

The peak from a molecular ion is conveniently abbreviated as M, and the peak at the next integral value of m/e will be called the M+1 peak. The intensity ratio of these two peaks provides valuable information about the molecular formula of the ions that produce these peaks. In the case of methane the 1.1% natural abundance of ^{13}C results in an M/M+1 ratio of 100:1, but now consider the case of a two-carbon compound, ethane (Mass Spectrum B.2).

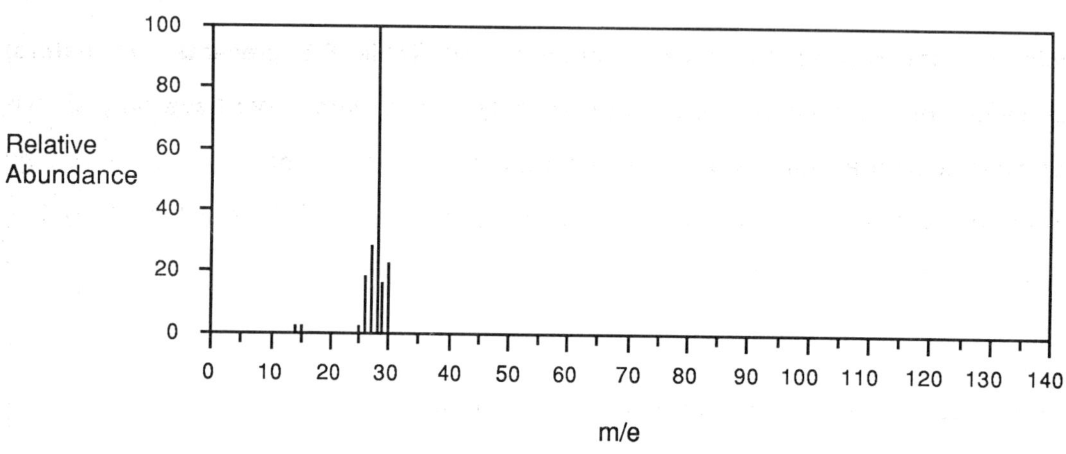

Mass Spectrum B.2 Ethane

The molecular ion for C_2H_6 appears at m/e 30, and the M+1 peak would be found at m/e 31. But now there are *two* carbon atoms, and *each* of them has a 1.1% probability of being ^{13}C rather than ^{12}C. Consequently the probability of finding a ^{13}C in ethane is twice that for methane,* and the M/M+1 ratio is 100:2. (The actual intensities are 26 and 0.5, relative to the most intense peak at m/e 28, and for this reason the weak signal at m/e 31 cannot be seen in Mass Spectrum B.2).

The correlation between molecular formula and the M/M+1 ratio continues for larger molecules, and it can be used to determine the number of carbon atoms in an ion. If the two peaks in question are designated M and M+1, their intensity ratio can be expressed as 100:X. The number X will be a multiple of 1.1 because each carbon in the molecule will have a 1.1% probability of being ^{13}C. Consequently, division of X by 1.1 will yield the number of carbons in the ion M. In fact the analysis can be applied to *any* ion, not just to the molecular ion. It is frequently possible to determine the molecular formulas of many ions in a mass spectrum by carrying out such an analysis. Other elements than carbon can

----

*We have oversimplified this analysis somewhat, because there will also be small contributions to the M+1 peak from the presence of ^{2}H in the sample (Table B.1). In addition the presence of more than one ^{13}C atom in a large molecule can reduce the expected intensity of the M+1 peak. For highly accurate determinations, such factors must be considered explicitly, but for our purposes the analysis presented in the text will be sufficient.

make isotopic contributions to the mass spectrum, and Table B.1 presents the natural isotopic abundances of the elements you are most likely to encounter. We have only shown isotopes that have a natural abundance large enough for a contribution to be observed by mass spectrometry. (For example, the radioactive isotopes ^{3}H and ^{14}C have extremely low natural abundances and are not listed).

Table B.1 Natural Abundances for Isotopes of Common Elements

Element	Isotope	Natural Abundance (%)	Relative Abundance
Hydrogen	^{1}H	99.98	100
	^{2}H	0.02	0.02
Boron	^{10}B	19.7	24.5
	^{11}B	80.3	100
Carbon	^{12}C	98.9	100
	^{13}C	1.1	1.1
Nitrogen	^{14}N	99.6	100
	^{15}N	0.4	0.4
Oxygen	^{16}O	99.8	100
	^{17}O	0.04	0.04
	^{18}O	0.2	0.2
Fluorine	^{19}F	100	100
Silicon	^{28}Si	92.2	100
	^{29}Si	4.7	5.1
	^{30}Si	3.1	3.4
Sulfur	^{32}S	95.0	100
	^{33}S	0.8	0.8
	^{34}S	4.2	4.4

Chlorine	^{35}Cl	75.5	100
	^{37}Cl	24.5	32.
Bromine	^{79}Br	50.5	100
	^{81}Br	49.5	98.
Iodine	^{127}I	100	100

The data in Table B.1 is presented in two ways. In the third column the abundance of each isotope is shown as a percentage of the total. In the fourth column, the same data is presented as relative natural abundance; the most abundant isotope is given as 100, and the values for the other isotopes are scaled accordingly. The relative natural abundances are convenient for comparing peaks in a mass spectrum, where the most intense peak is also assigned a value of 100.

Inspection of Table B.1 shows that many other common elements will make isotopic contributions to the ions observed in mass spectrometry. While this would seem to greatly complicate the interpretation of mass spectra, it actually provides highly useful information in many cases. Several elements have characteristic isotopic abundances that produce easily recognizable patterns in the mass spectrum. Bromine, for example, exists as two isotopes, ^{79}Br and ^{81}Br, in almost equal quantities. Hence, any ion that contains a single bromine atom will give rise to a "doublet" of peaks in the mass spectrum where the individual peaks differ by two mass units. This can be seen in the case of methyl bromide in Mass Spectrum B.3.

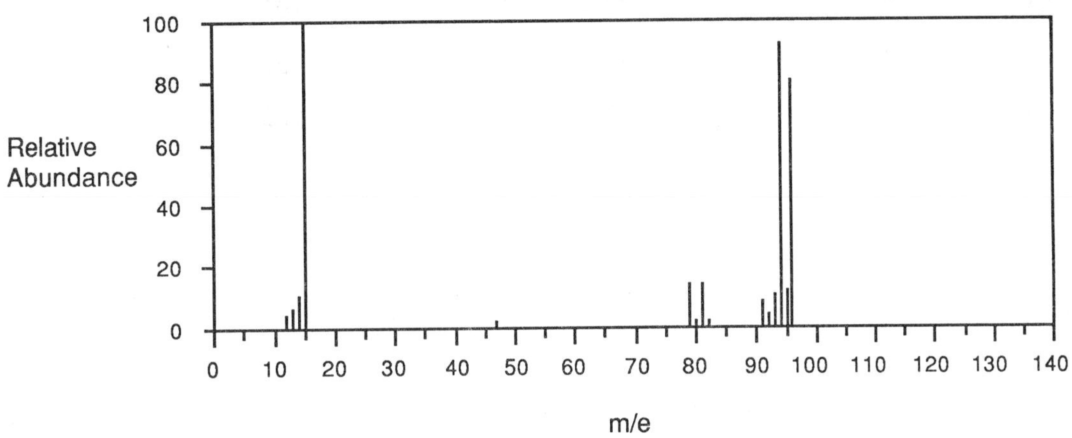

Mass Spectrum B.3 Methyl bromide.

The two peaks of nearly equal intensity at m/e 94 and 96 correspond to the two isotopic forms of the molecular ion, $CH_3{}^{79}Br$ and $CH_3{}^{81}Br$ A second doublet of peaks can also be found at m/e 79 and 81, and these must correspond to the two isotopic forms of Br^+ (formed by fragmentation of the molecular ion into Br^+ and $CH_3^{\cdot}$).

Characteristic patterns are also observed when an ion contains more than one bromine atom. Consider the case of dibromomethane. What should we expect for the molecular ion, $CH_2Br_2{}^{+\cdot}$? There are three possible molecules (if we ignore the minor contributions from ^{13}C): $CH_2{}^{79}Br_2$, $CH_2{}^{79}Br^{81}Br$ and $CH_2{}^{81}Br_2$. The probability that any single bromine atom will be ^{79}Br is 50% or 0.5, so the probability that both bromines of CH_2Br_2 will be ^{79}Br is 0.5 x 0.5 = 0.25 or 25%. The same argument leads to the prediction that 25% of all molecular ions will have two ^{81}Br atoms, and the remaining 50% will have one ^{79}Br together with one ^{81}Br. The net result is that an ion with two bromines will give rise to a 1:2:1 triplet, each peak separated by m/e 2.

These expectations for dibromomethanes are confirmed by Mass Spectrum B.4, as the molecular ion appears as triplet at m/e 172, 174 and 176. Notice also that fragmentation of the molecular ion generates ions which appear at lower m/e as the characteristic doublets for ions containing only a single bromine.

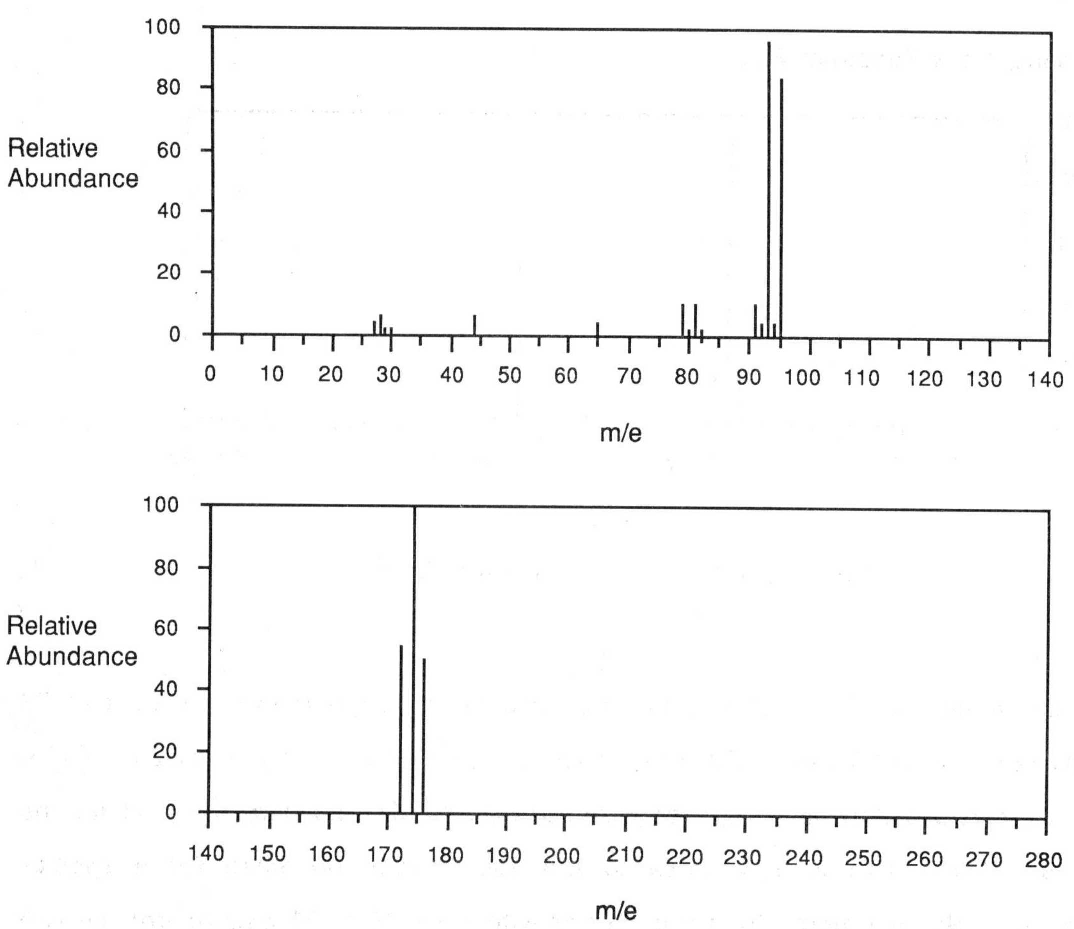

Mass Spectrum B.4 Dibromomethane

The two isotopes of chlorine, ^{35}Cl and ^{37}Cl, lead to similar patterns in the mass spectrum. However, they are not as easily recognized because the natural abundances are 75 and 25%, so the peaks of the doublet for an ion with a single chlorine differ in intensity by a factor of three. The expected pattern for an ion with two chlorine atoms will be a triplet in a ratio of 100:65:11.* Both of these patterns are seen in the mass spectrum of

---

Consider the mass of a chlorine molecule. The probability that a Cl atom will have mass 35 is 75.5% or 0.755. The probability that both will have mass 35 is $(0.755)^2$ or 0.570. Similarly, the probability that both chlorines will have mass 37 is $(0.245)^2$ = 0.060. Therefore 57% of all chlorine molecules will have mass 70 and 6% will have mass 74. The remainder (37%) will have one atom of each isotope with a total mass of 72. Hence the three species will be present in a ratio of 57:37:6 (i.e., 100:65:11).

dichloromethane, Mass Spectrum B.5.

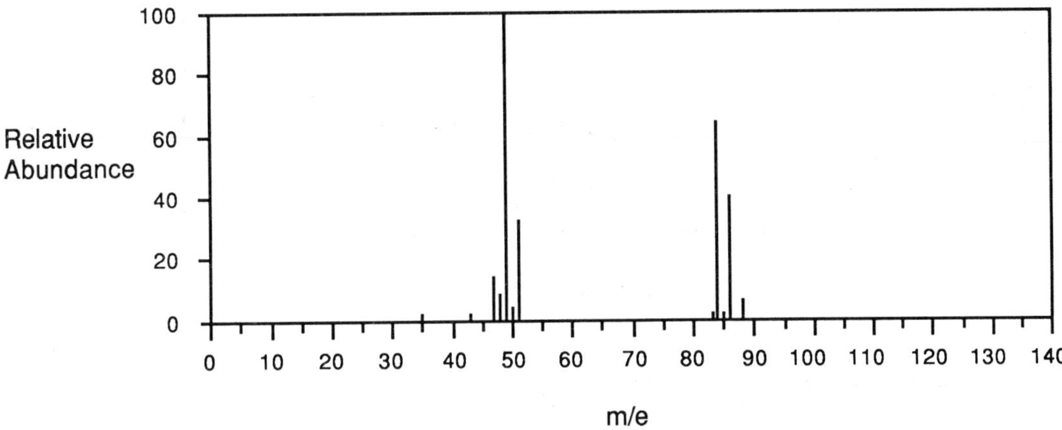

Mass Spectrum B.5 Dichloromethane

The molecular ion, $CH_2Cl_2^{+\cdot}$, affords the expected three peaks at m/e 84, 86 and 88 with intensity ratios of 86:55:9 (i.e., 100:64:10 which are quite close to the predicted relative intensities). A second pattern is easily recognized at m/e 49, 51, and here the ratio for the two peaks is 100:31. This is very close to the 100:32 ratio predicted for a species containing a single chlorine atom. In fact a species with mass 49 or 51 *cannot* contain two chlorines, and subtraction of the mass of a single chlorine (^{35}Cl or ^{37}Cl) leaves 14 which is the mass of a CH_2 group. Hence the peaks near m/e 50 must correspond to CH_2Cl^+ (formed by loss of a neutral chlorine atom from the molecular ion).

Accuracy and Reliability of Peak Intensities

How reliable are peak intensities for identification of the molecular formula? The answer depends to some extent on the type of instrument used. Some mass spectrometers are inherently more accurate than others, but it is never possible to completely remove experimental error. The mass spectra in this appendix have "experimental error" at the level of about 1 unit in relative abundance, because we have drawn each peak in increments of 2 units (i.e., 0, 2, 4, 6, ...100). Significant deviations from the predicted intensity ratios can also be caused by small contributions of other isotopes. When we compare the intensity ratio of two peaks with the isotopic abundances in Table B.1, we are making the assumption that the peaks arise from a single ion (i.e., with only isotopic differences). Consider the molecular ion formed from CH_3Br (Mass Spectrum B.3). The peaks at m/e 94 and 96 exhibit a ratio that is larger than the expected 100:98. The discrepancy might result in part from experimental error, but fragmentation of the molecular ion could also contribute. Recall from the mass spectrum of methane (Mass Spectrum B.1) that loss of hydrogen atoms can yield ions of lower mass units. If the molecular ion from CH_3Br lost one or more hydrogen atoms, the following ions would be formed:

$CH_2{}^{79}Br^+$	m/e 93	$CH_2{}^{81}Br^+$	m/e 95
$CH^{79}Br^+$	92	$CH^{81}Br^+$	94
$C^{79}Br^+$	91	$C^{81}Br^+$	93

The observation of peaks at m/e 91–93 make it quite certain that such fragmentations are occurring. Since we have identified a second ion with m/e 93, it should no longer be surprising that the ratio of peaks at m/e 94 and 96 is slightly too large. Nevertheless, the peaks are sufficiently close in intensity that they can be used to identify the presence of a bromine atom in the ion.

A complete analysis of all isotopic contributions to the peaks in a mass spectrum is rather complicated. Ions that differ by only a few mass units may have different numbers of hydrogens in their molecular formulas or may only differ in isotopic contributions. In addition to the isotopes we have discussed in detail, inspection of Table B.1 will demonstrate that small contributions to peak intensities will also result from isotopes of the other common elements. In this introductory discussion of mass spectrometry we cannot consider all these factors in detail. Even an expert in the field of mass spectrometry would not ordinarily attempt a full analysis by hand, because such an exercise can be done much more effectively by a computer. Indeed, mass spectrometry is an area which has benefited tremendously from the recent advances in computer technology, and it is unrealistic to think that you could fully interpret most mass spectra using only a pencil and paper.

Returning to the original question, can you use peak intensity ratios to determine molecular formulas of ions? The answer is yes, if you work within two limitations. First, you must be certain that the two ions you are comparing have the same molecular formula, i.e., that they differ only in isotopic composition but both arise from the same fragmentation. Second, you should only expect the peak intensity ratios predicted from Table B.1 to be *approximately* the same as those observed experimentally. These ideas are illustrated by Mass Spectrum B.6 (1,4-dibromobenzene), and you should examine this spectrum carefully, and use it to practice spectral interpretation. Use the highest mass peaks to determine the number of bromines in the molecular formula. Calculate the number of carbons in the molecular formula from the ratio of intensities for the peaks at m/e 238 and 239 (the exact intensities are 48.8 and 3.3). Notice that the mass of the molecular ion (with only isotopes ^{13}C and ^{79}Br) allows you to calculate the exact molecular formula. This would then allow you to calculate the degrees of unsaturation, even if you did not know the identity of the compound.

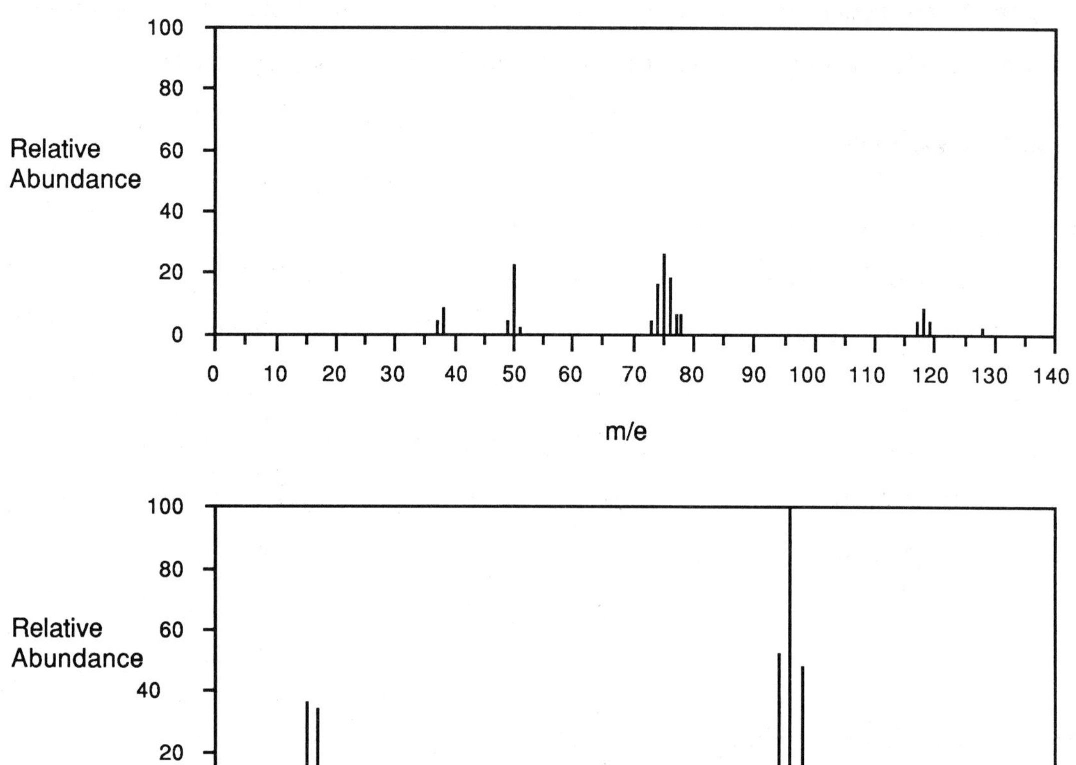

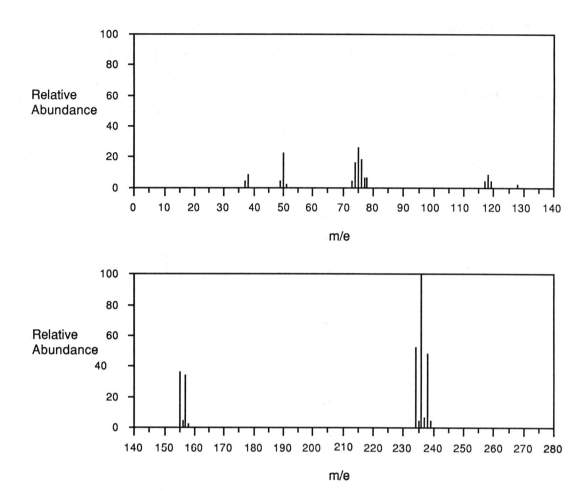

Mass Spectrum B.6 1,4-Dibromobenzene

B.6 MOLECULAR IONS AND ELECTRONIC STRUCTURES

So far we have discussed ways in which the mass spectrum affords information about the *molecular formula* of a compound. Now we will consider the information it provides about *molecular structure*. Most mass spectra contain a large number of peaks at different values of m/e, where these peaks result from fragmentation of the molecular ion into ions of lower mass. The pathway by which an ion undergoes fragmentation is highly dependent upon its molecular structure. For this reason an understanding of fragmentation pathways is the first step in obtaining structural information from a mass spectrum.

Molecular Ion Fragmentation

Before we consider the detailed pathways for ion fragmentation, you should recognize that there are a variety of different processes that can occur. First there is the primary fragmentation of the molecular ion. Since this is a radical cation, the unpaired electron and the positive charge must both appear in the fragments. But only *one* of the two fragments can be positively charged, and only *one* of the two fragments can have an unpaired electron. One major fragmentation pathway involves decomposition of the molecular ion into a cation plus a free radical.

$$M^{+\cdot} \longrightarrow A^{+} + B^{\cdot}$$

In the other major pathway a neutral molecule is lost with the formation of a new radical cation (a process that usually involves a cyclic rearrangement).

$$M^{+\cdot} \longrightarrow C^{+\cdot} + D$$

Only positively charged species can be detected in the mass spectrometer, so no direct information is available about radical $B^{\cdot}$ or molecule D in the preceding equations. But we would know their masses *indirectly* from the difference in m/e between $M^{+\cdot}$ and A^{+} or between $M^{+\cdot}$ and $C^{+\cdot}$. Consequently the *masses* of each of the individual fragments could be deduced.

Secondary Fragmentation

Secondary fragmentations of the ions formed by initial decomposition are also important. In the case of a radical cation, $C^{+\cdot}$, the available decomposition pathways are the same as the two just presented for $M^{+\cdot}$. But a cation such as A^{+} can undergo further fragmentation in only one way, loss of a neutral molecule to form a new cation.

$$A^{+} \longrightarrow E^{+} + F$$

Each of these fragmentation pathways can make an important contribution to the mass spectrum of any particular compound.

Electronic Structure of the Molecular Ion

The molecular ion formed in the ion source of a mass spectrometer is symbolized by $M^{+\cdot}$, but this symbol provides no information about which electron was lost in the ionization.

$$M + e^{-} \longrightarrow M^{+\cdot} + 2e^{-}$$

The formation of the molecular ion must occur by loss of an electron from one of the *occupied molecular orbitals* of the neutral molecule. We can then ask the question, which occupied orbital loses an electron? The answer is quite reasonable. The most probable ionization pathway is that which requires the least energy. In other words the most likely orbital from which an electron could be removed is the *highest occupied molecular orbital*. This may not be the *only* ionization pathway, but it is certainly the *major* pathway in most cases.

Loss of an electron from the HOMO has important consequences, because it allows us to make correlations between the electronic structure of the molecular ion and the reaction pathways for fragmentation. To do this we must first identify the HOMO in typical organic molecules. In a molecule that contains a heteroatom the highest occupied orbital should be that associated with a *nonbonding* pair of electrons. The following equations illustrate formation of molecular ions from such molecules, and the electron deficiency can be drawn as localized on the heteroatom.

halides $$R-\ddot{\underset{\cdot\cdot}{Cl}}: \longrightarrow R-\underset{\cdot\cdot}{\ddot{Cl}}{}^{+\cdot} + e^-$$

alcohols $$R-\underset{\cdot\cdot}{\ddot{O}}-H \longrightarrow R-\underset{\cdot\cdot}{\overset{+\cdot}{O}}-H + e^-$$

carbonyl compounds $$R-C(=\ddot{O}:)-X \longrightarrow R-C(=\ddot{O}{}^{+\cdot})-X + e^-$$

amines $$R-\ddot{N}H_2 \longrightarrow R-\overset{+\cdot}{N}H_2 + e^-$$

In an unsaturated hydrocarbon the highest occupied molecular orbital should be a bonding *π-orbital*, and the electron deficiency of the molecular ion will be delocalized over at least two carbon atoms. The π-orbital will still be occupied by a single electron, so some of the multiple bond character must remain. This makes it difficult to draw an appropriate structural formula. However, it is frequently useful to employ resonance forms in which one carbon of the original π-system has a positive charge and another has a single electron. This is illustrated by the following equation.

$$>C=C< \longrightarrow \left[>\overset{+}{C}-\overset{\cdot}{C}< \longrightarrow >\overset{\cdot}{C}-\overset{+}{C}< \right] + e^-$$

For compounds with several multiple bonds or for aromatic compounds it is even more difficult to draw structures that adequately represent the molecular ion formed by loss of an electron. For example, reference to Figures 8.5 and 10.8 of the text will show you that the highest occupied molecular orbitals extend over 4 and 6 carbon atoms for butadiene and benzene, respectively. Nevertheless, it is often useful to depict these molecular ions with a structural formula having an unpaired electron and a positive charge on adjacent carbons, as shown in the following equation.

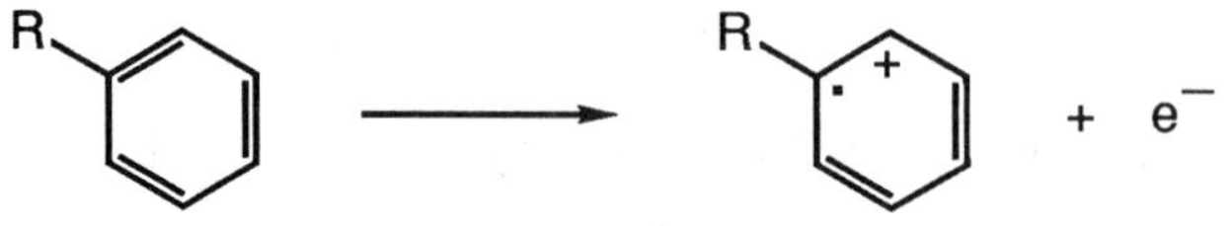

A variety of other resonance forms could be drawn for the molecular ion in the preceding equation, but we merely wanted to show an example of one such structure.

The molecular ion formed from a saturated hydrocarbon cannot be drawn adequately using normal structural formulas. In most cases the highest occupied molecular orbital extends over much or all of the carbon framework, so it serves no purpose to draw the electron deficiency at any single location.

B.7 FRAGMENTATION AND MOLECULAR STRUCTURE

In this section we will consider the mass spectral behavior of compounds containing different functional groups. We will begin with the molecular ions formed from compounds containing a heteroatom and then molecular ions formed from hydrocarbons. Then we will move on to further reactions of cations formed by fragmentation of the molecular ion.

Molecules with a Heteroatom

We begin with fragmentation pathways of the molecular ion in compounds containing a heteroatom (designated X) to emphasize the similarities of reactions in the mass spectrometer to those you already know. This analysis applies to halides (X=F, Cl, Br, I), to alcohols (X=OH), to ethers (X=OR) and to amines (X=NH_2, NHR, NR_2). With minor modification the following equations can also be extended to carbonyl compounds.

Common Fragmentation Pathways for Molecules with a Heteroatom

—C—C—$X^{\cdot+}$	⟶	—C· + C=X^{+}	α-cleavage
—C—C—$X^{\cdot+}$	⟶	—C—C^{+} + ·X:	charged-site cleavage
H—R—$X^{\cdot+}$	⟶	$[R]^{+\cdot}$ + HX	elimination

The first type of fragmentation is called *α-cleavage* because the bond cleavage occurs between the carbon atoms that are α and β to the charged site. This produces a rather stable cation in which the positive charge is localized on a heteroatom which has eight electrons in its valence shell. For example, an alcohol (or ether) would generate an *oxonium ion* by this fragmentation pathway. The other fragment would be a carbon free radical.

—C—C—$\ddot{O}H^{\cdot+}$ ⟶ —C· + C=$\ddot{O}H^{+}$

In the second fragmentation mode a bond to the electron deficient heteroatom is broken. We will call such fragmentation *charged-site cleavage.* This produces a carbocation and a heteroradical as illustrated again for an alcohol. (Both electrons from the C-O bond remain with the oxygen to produce a neutral oxygen fragment -- but one that still has an unpaired electron. The other fragment is simply a carbocation.

$$-\overset{|}{\underset{|}{C}}-\overset{|}{\underset{|}{C}}-\ddot{O}H^{\cdot+} \longrightarrow -\overset{|}{\underset{|}{C}}-C^{+} + :\dot{O}H$$

The third fragmentation mode involves elimination of a neutral molecule, and this generates a new radical cation as the other fragment. The elimination typically involves loss of a remote hydrogen (via a 5- or 6-membered ring transition state). The product ion is electron deficient at not just one but *two* positions, and this leads to greater complexity in subsequent fragmentations. For example, the radical cation from an alcohol could lose water in the following way:

$$\longrightarrow \longrightarrow + H_2O$$

The various fragmentation modes we just discussed are illustrated in Mass Spectra B.7–B.9. For example, the spectrum of 1-butylamine (Mass Spectrum B.7) exhibits an intense peak at m/e 30 (relative abundance=100) that corresponds to α-cleavage.

$$CH_3CH_2CH_2-CH_2-\dot{N}H_2^{+} \xrightarrow{\alpha\text{-cleavage}} CH_3CH_2\dot{C}H_2 + CH_2{=}\overset{+}{N}H_2$$

m/e 30

Charged-site cleavage to produce the primary 1-butyl cation (m/e 57) is not highly favored, and this peak has a relative abundance of only 1. Elimination of ammonia to produce a radical ion with m/e 56 also generates a small peak with a relative intensity of 1, but neither of these peaks is visible in Mass Spectrum B.7.

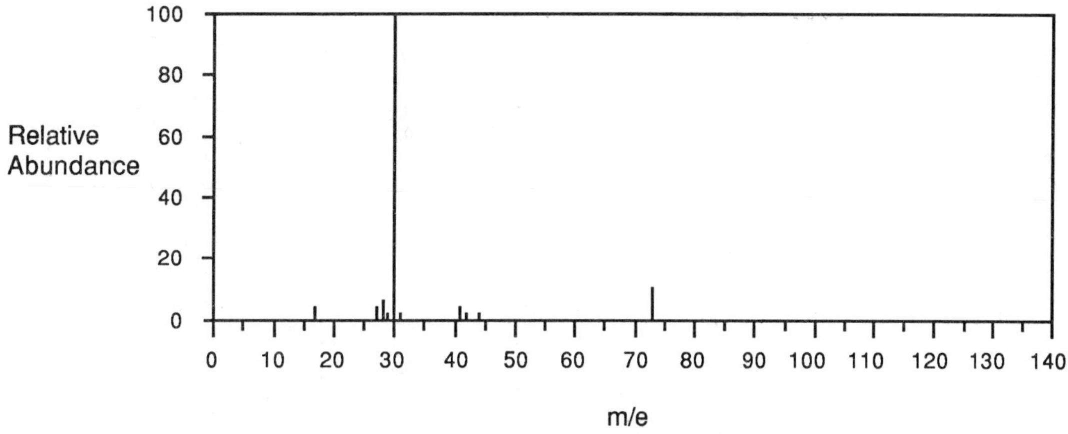

Mass Spectrum B.7 Butylamine

The molecular ion from diisobutyl ether (Mass Spectrum B.8) decomposes predominately via charged-site cleavage to form a butyl cation (m/e 57), so the peak at m/e 87 corresponding to α-cleavage has a low relative intensity of only 8.

$$(CH_3)_2CHCH_2-\ddot{O}^{\cdot +}-CH_2CH(CH_3)_2 \xrightarrow[\text{cleavage}]{\text{charged-site}} (CH_3)_2CHCH_2-\ddot{O}^{\cdot} + \overset{+}{C}H_2CH(CH_3)_2$$

m/e 57

$$(CH_3)_2CH-CH_2-\ddot{O}^{\cdot +}-CH_2CH(CH_3)_2 \xrightarrow{\alpha\text{-cleavage}} (CH_3)_2CH^{\cdot} + CH_2=\overset{+}{\ddot{O}}-CH_2CH(CH_3)_2$$

m/e 87

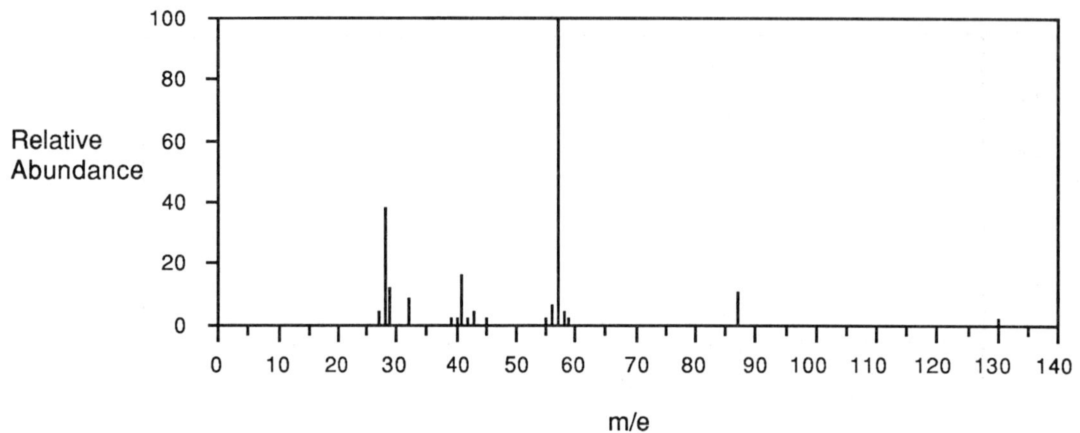

Mass Spectrum B.8 Diisobutyl ether

The molecular ion formed from 1-chloropropane (Mass Spectrum B.9) undergoes initial fragmentation by all three pathways: α-cleavage, charged-site cleavage and elimination (in this case by a 4-membered ring transition state).

$$CH_3CH_2-CH_2-\ddot{Cl}^{\cdot +}: \xrightarrow{\alpha\text{-cleavage}} CH_3CH_2^{\cdot} + CH_2=\overset{+}{Cl}$$

m/e 49

$$CH_3CH_2CH_2-\ddot{Cl}^{\cdot +}: \xrightarrow[\text{cleavage}]{\text{charged-site}} CH_3CH_2CH_2^{+} + :\dot{Cl}:$$

$$CH_3CH_2CH_2-\ddot{Cl}^{\cdot +}: \xrightarrow{\text{elimination}} HCl + [CH_3-CH=CH_2]^{+\cdot}$$

m/e 42

The last of these pathways leads to the most abundant ion (m/e 42) in the spectrum.

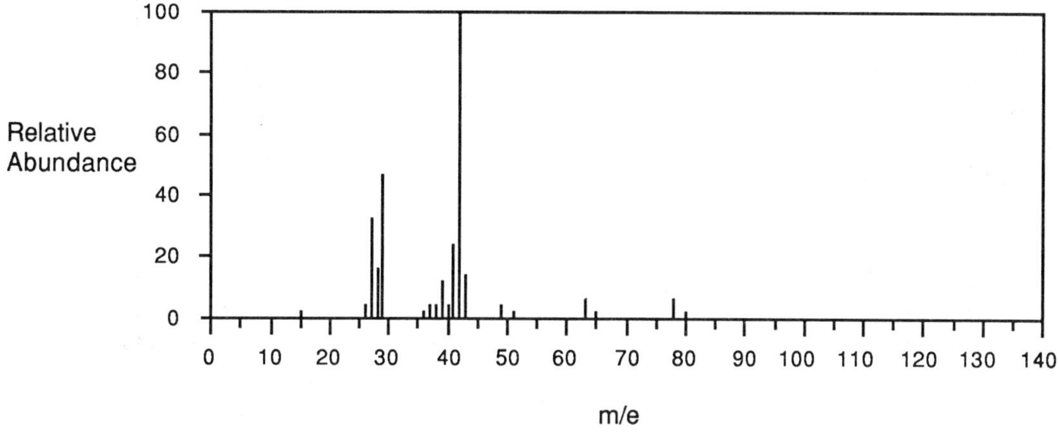

Mass Spectrum B.9 1-Chloropropane

Fragmentation pathways for a molecular ion are highly dependent on structure, but such structure-reactivity patterns are beyond the scope of this appendix. From the information we present you may not know which fragmentation modes should predominate, but you should be able to predict those pathways that are *possible*. This will allow you to evaluate whether or not any particular fragmentation is reasonable for a given structure.

Carbonyl Compounds

A specific type of cyclic elimination has been found to be important with carbonyl compounds. When a hydrogen is appropriately situated so that reaction can occur in a 6-membered cyclic process, a γ-hydrogen is transferred to the electron deficient carbonyl oxygen. This is followed by cleavage of the bond between the carbons that are α,β to the carbonyl group, and the fragmentation produces a neutral alkene plus the radical cation of an enol. This mode of fragmentation is known as the *McLafferty rearrangement*, after F.W. McLafferty of Cornell University who has studied it extensively.

McLafferty Rearrangement

Although experimental evidence indicates a stepwise process, we have depicted the McLafferty rearrangement in a way that emphasizes its similarity to ester pyrolysis (Section 6.9) and electrocyclic processes (Section 27.3).

The importance of the McLafferty rearrangement can be seen from a comparison of Mass Spectra B.10 and B.11, the spectra for 3-pentanone and 2-hexanone, respectively.

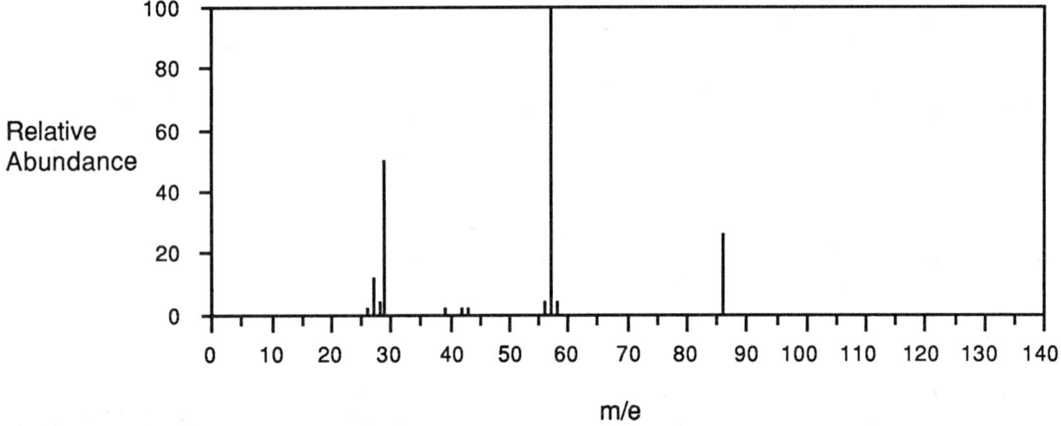

Mass Spectrum B.10 3-Pentanone

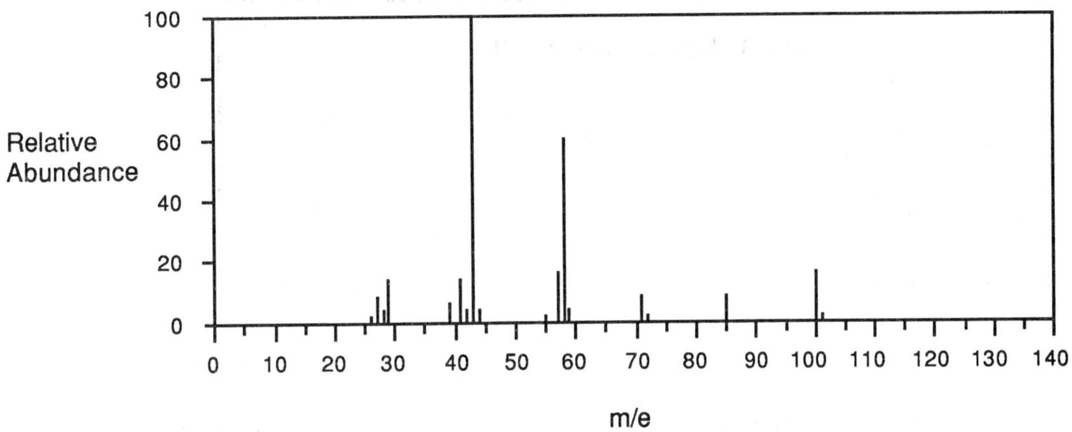

Mass Spectrum B.11 2-Hexanone

3-Pentanone has only ethyl substituents, so there are no γ-hydrogens and McLafferty rearrangements are not possible. Instead the two main fragmentation pathways involve α-cleavage and charged-site cleavage. Notice that the same bond is broken in each of the following equations, but both electrons remain with the carbonyl fragment in the second pathway.

$$CH_3CH_2-\overset{\overset{\cdot +}{O:}}{\overset{\|}{C}}-CH_2CH_3 \xrightarrow{\alpha\text{-cleavage}} CH_3\dot{C}H_2 + \overset{+}{\underset{..}{O}}\equiv C-CH_2CH_3$$

m/e 86 → m/e 57

$$CH_3CH_2-\overset{\overset{\cdot +}{O:}}{\overset{\|}{C}}-CH_2CH_3 \xrightarrow[\text{cleavage}]{\text{charged-site}} CH_3\overset{+}{C}H_2 + \underset{..}{\overset{..}{O}}\equiv \underset{\cdot}{C}-CH_2CH_3$$

m/e 29

The same modes of fragmentation are seen in the case of 2-hexanone.

$$CH_3-\overset{\overset{+\cdot}{:O}}{\overset{\|}{C}}-CH_2CH_2CH_2CH_3 \xrightarrow{\alpha\text{-cleavage}} CH_3-C\equiv\overset{+}{\underset{..}{O}} + \cdot CH_2CH_2CH_2CH_3$$

m/e 43

$$CH_3-\overset{\overset{+\cdot}{:O:}}{\overset{\|}{C}}-CH_2CH_2CH_2CH_3 \xrightarrow[\text{cleavage}]{\text{charged-site}} CH_3-\dot{C}\equiv\ddot{O} \quad + \quad {}^{+}CH_2CH_2CH_2CH_3$$

m/e 57

But another peak for 2-hexanone, at m/e 58, results from a McLafferty rearrangement.

$$CH_3-\overset{\overset{\cdot+}{:O}}{\overset{\|}{C}}-CH_2-CH_2-CH(H)CH_3 \longrightarrow CH_3-\overset{\overset{+\cdot}{:OH}}{\overset{|}{C}}=CH_2 \quad + \quad CH_2=CH-CH_3$$

m/e 58

Hydrocarbons: Alkanes

As we discussed in the previous section, the HOMO of a hydrocarbon is not localized on a single atom, and this makes it somewhat more difficult to predict fragmentation modes. Nevertheless, decomposition of the molecular ion normally proceeds in the manner which gives the most stable cation and radical products. For example, there is a pronounced tendency for a saturated hydrocarbon to cleave at a branching point, because this affords a more substituted cation (or radical).

$$\left[-CH_2-\underset{R}{\overset{R}{\overset{|}{\underset{|}{C}}}}-CH_2CH_2-\right]^{+\cdot} \longrightarrow -CH_2-C^{+}R_2 \quad + \quad {}^{\cdot}CH_2-CH_2-$$

$$\left[-CH_2-\underset{R}{\overset{R}{\overset{|}{\underset{|}{C}}}}-CH_2CH_2-\right]^{+\cdot} \longrightarrow -CH_2-C^{\cdot}R_2 \quad + \quad {}^{+}CH_2-CH_2-$$

Usually the former pathway predominates, and this indicates that *carbocation* stability is the major factor governing the fragmentation. Notice also that the preceding equation shows only one of three possibilities for cleavage at the branched carbon atom. Fragmentation via cleavage of the other two bonds would be expected as well.

The fragmentation of the molecular ion of an alkane is illustrated by Mass Spectrum

The fragmentation of the molecular ion of an alkane is illustrated by Mass Spectrum B.12, the spectrum for 2,2-dimethylpentane.

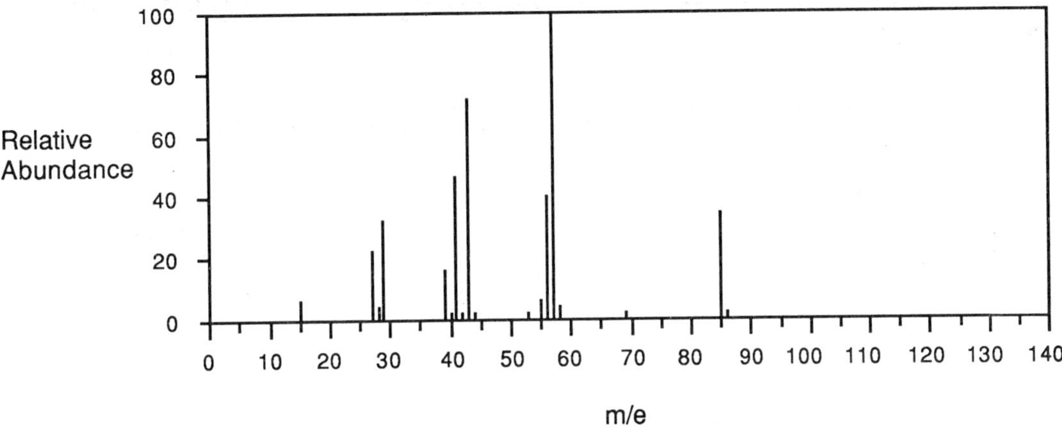

Mass Spectrum B.12 2,2-Dimethylpentane

Cleavage at the quaternary center to form a tertiary cation can involve loss of either a methyl or a propyl radical. The latter radical is more stable, and that pathway gives rise to the most intense peak in the spectrum at m/e 57.

$$\left[CH_3{-}C(CH_3)_2{-}CH_2CH_2CH_3\right]^{+\cdot} \text{ (m/e 100)} \longrightarrow CH_3{-}\overset{+}{C}(CH_3){-}CH_2CH_2CH_3 \text{ (m/e 85)} + CH_3^{\cdot}$$

$$\left[CH_3{-}C(CH_3)_2{-}CH_2CH_2CH_3\right]^{+\cdot} \text{ (m/e 100)} \longrightarrow CH_3{-}C^{+}(CH_3)_2 \text{ (m/e 57)} + {}^{\cdot}CH_2CH_2CH_3$$

Fragmentation can also occur (to a lesser extent) to give a *t*-butyl radical and an 1-propyl cation (m/e 43). The peak at m/e 56 apparently results from *elimination* of propane to form an unsaturated radical cation.

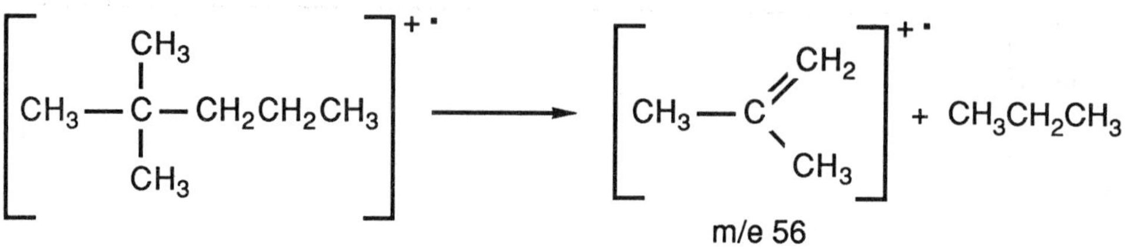

$$\left[CH_3-\overset{\overset{CH_3}{|}}{\underset{\underset{CH_3}{|}}{C}}-CH_2CH_2CH_3 \right]^{+\cdot} \longrightarrow \left[CH_3-C\begin{matrix} \nearrow CH_2 \\ \searrow CH_3 \end{matrix} \right]^{+\cdot} + CH_3CH_2CH_3$$

m/e 56

Hydrocarbons: Alkenes

The cleavage patterns for the molecular ions of *alkenes* are more predictable, because the electron deficiency is associated with a specific group -- the carbon-carbon double bond. An important mode of fragmentation involves cleavage to produce an allylic cation.

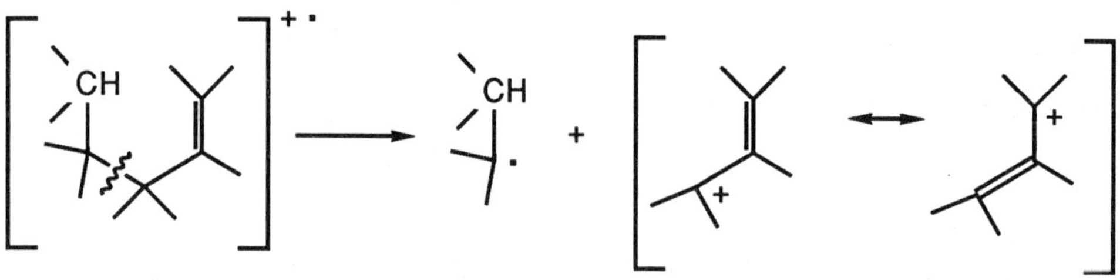

CH CH + +

If a hydrogen is appropriately situated, a cyclic elimination could also occur (shown here for convenience as a one-step process).

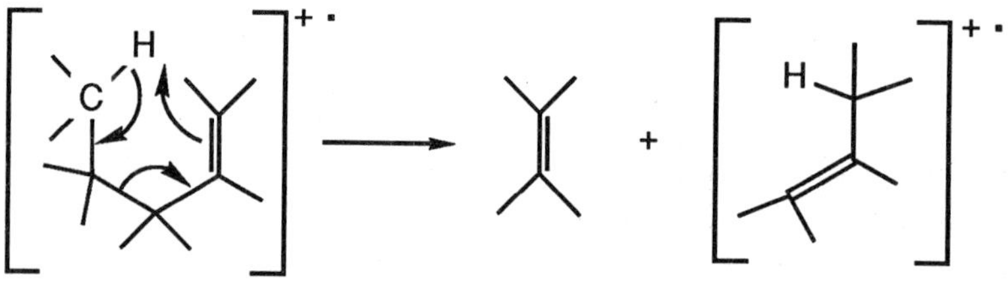

H C + H

The same carbon-carbon bond is cleaved in both cases, but in the second pathway the positively charged species will appear at the next higher m/e.

The behavior of alkenes is illustrated by Mass Spectrum B.13 for 5-methyl-2-hexene.

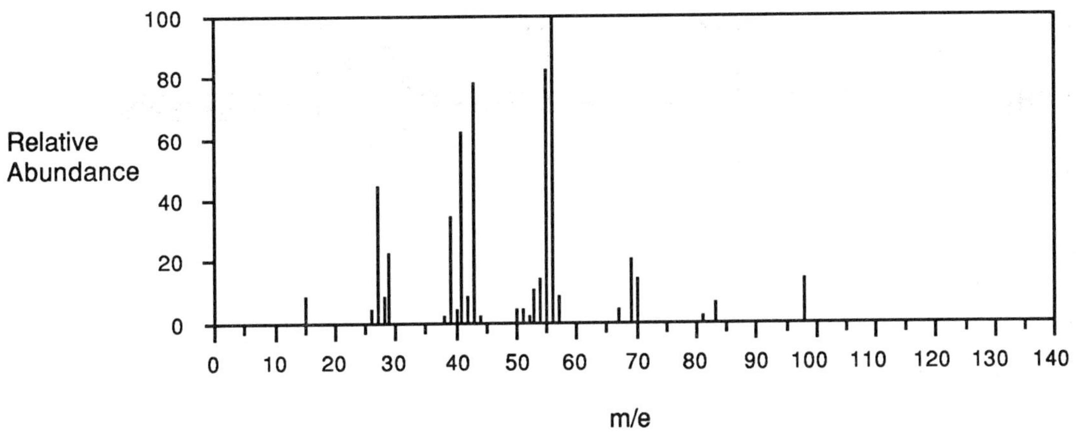

Mass Spectrum B.13 5-Methyl-2-hexene

The two most abundant ions in the spectrum are at m/e 56 and 55, which correspond to allylic cleavage and elimination, respectively.

CH_2^+

m/e 56

H

m/e 57

Hydrocarbons: Aromatics

Hydrocarbons that contain an aromatic ring behave similarly to alkenes, and cleavage to form a *benzylic* cation is a major decomposition pathway of the molecular ion. Cyclic elimination via a 6-membered cyclic process also occurs when possible. As with alkenes these two fragmentation pathways yield ions that differ by only a single mass unit.

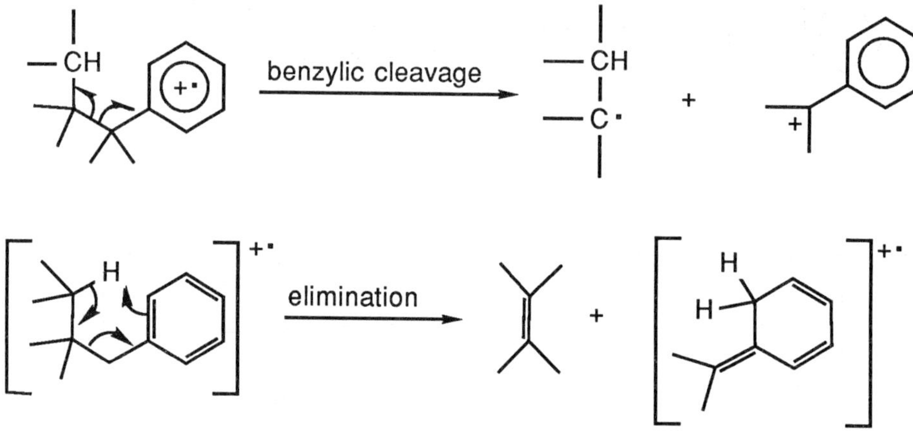

benzylic cleavage

elimination

You must keep track of the electrons in these reactions. In the first of the preceding equations the product could be drawn as a benzyl radical which is still electron deficient in the π-system of the ring:

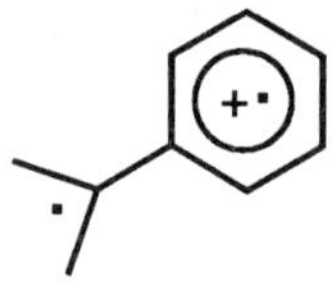

However, a cation *diradical* is certainly not a good way to depict the species that is formed. If we arbitrarily select a particular resonance form of the molecular ion, it becomes clear how a benzylic cation is formed.

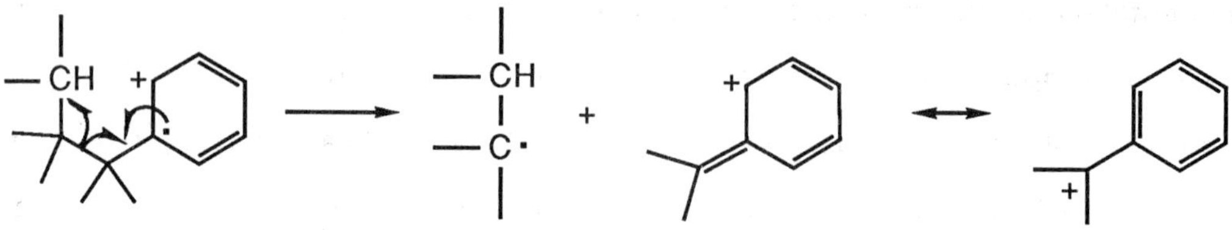

The mass spectrum of butylbenzene (Mass Spectrum B.14) shows peaks corresponding to both cleavage and elimination. The peak at m/e 91 is that for the benzyl cation,

$$[C_6H_5{-}CH_2CH_2CH_2CH_3]^{+\bullet} \longrightarrow C_6H_5CH_2^{+} + {}^{\bullet}CH_2CH_2CH_3$$

m/e 134 m/e 91

and the peak at m/e 92 is the cation produced by elimination of propene.

m/e 92

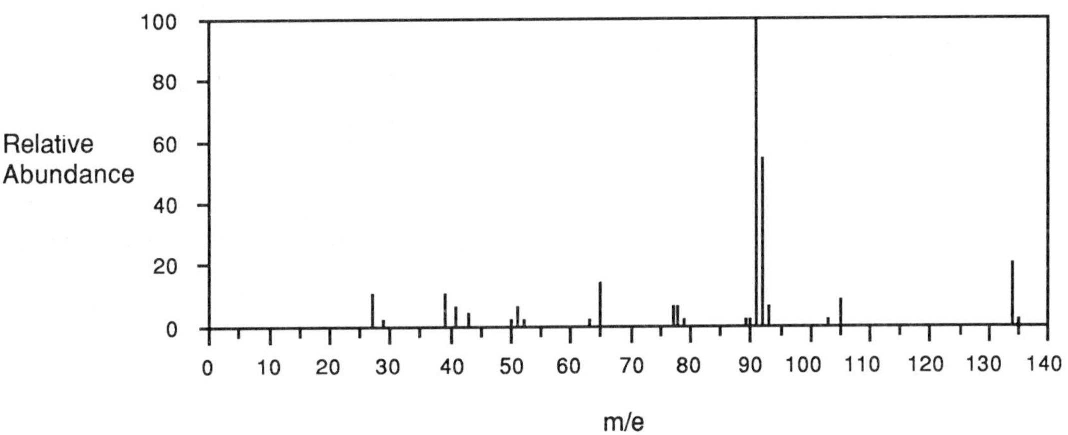

Mass Spectrum B.14 Butylbenzene

Secondary Decomposition of Cations

Fragmentation of the molecular ion produces a new cation, and we will now consider the further decomposition of these daughter cations. If the daughter ion is also a radical cation, subsequent decomposition will proceed in the same ways expected for a molecular ion. But in many instances fragmentation of the molecular ion generates a free radical and a carbocation. How does such a carbocation (with all its electrons paired) behave?

Some of the most important pathways are very similar to reactions that occur in solution: loss of a stable neutral molecule to generate a new carbocation. The following equations show both primary and secondary decomposition steps in typical fragmentation pathways.

$$-\overset{|}{\underset{|}{C}}-\overset{|}{\underset{|}{C}}-\ddot{O}R^{\cdot+} \xrightarrow{\alpha\text{-cleavage}} -\dot{C}< + >C=\overset{+}{\ddot{O}}-R \xrightarrow[\text{molecule}]{\text{loss of neutral}} >C=O + R^+$$

$$X^{+\cdot}-\overset{|}{\underset{|}{C}}-\overset{|}{\underset{|}{C}}-R \xrightarrow[\text{cleavage}]{\text{charged-site}} \ddot{X}\cdot + \overset{+}{C}-\overset{|}{\underset{|}{C}}-R \xrightarrow[\text{molecule}]{\text{loss of neutral}} >C=C< + R^+$$

Each of the preceding pathways for secondary decomposition can be recognized in examples that we have presented in this section. For example, the first of the preceding reactions would yield the following fragmentation for diisobutyl ether (Mass Spectrum B.8).

$$(CH_3)_2CH-CH_2-\ddot{O}^{\cdot+}-CH_2CH(CH_3)_2 \xrightarrow{\alpha\text{-cleavage}} (CH_3)_2CH\cdot + CH_2=\overset{+}{\ddot{O}}-CH_2CH(CH_3)_2$$

$$\xrightarrow{\text{loss of } CH_2O} CH_2=O + {}^+CH_2CH(CH_3)_2 \quad (m/e\ 57)$$

The ion with m/e 57 is also formed by charged-site cleavage, and the extent to which each of the pathways contributes is not obvious. Loss of carbon monoxide from the ion with m/e 57 would also contribute to the peak at m/e 29 in the mass spectrum of 3-pentanone (Mass Spectrum B.10).

$$\overset{+}{\ddot{O}}\equiv C-CH_2CH_3 \xrightarrow{\text{loss of CO}} CO + {}^+CH_2CH_3 \quad (m/e\ 29)$$

Loss of an alkene fragment to form a new carbocation is a common pathway for decomposition of alkyl cations, and this is the source of the ethyl cations produced from 2-hexanone (Mass Spectrum B.11).

$$^{+}CH_2—CH_2—CH_2—CH_3 \xrightarrow[CH_2{=}CH_2]{\text{loss of}} CH_2{=}CH_2 + \underset{m/e\ 29}{^{+}CH_2CH_3}$$

Cyclic elimination can also be important for further fragmentation of cations. This is particularly true for alcohols, where elimination via a 4-membered cyclic transition state is important for both molecular ion decomposition and for secondary decomposition of the oxonium ion formed by α-cleavage. This can be seen in the mass spectrum of 3-octanol (Mass Spectrum B.15). A predominant mode of decomposition for secondary and tertiary alcohols is α-cleavage, and loss of either a propyl radical or a pentyl radical can occur. (Notice that the intensity of the molecular ion is extremely low, a typical result for an alcohol.)

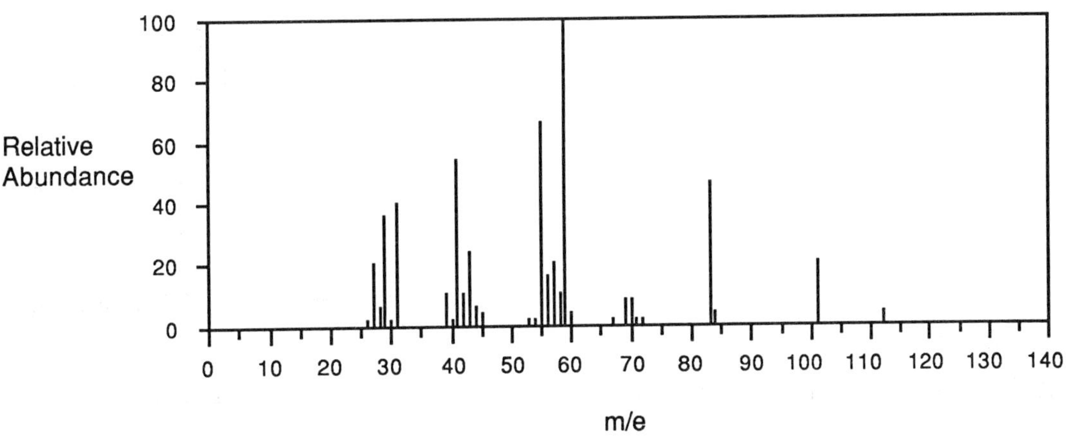

Mass Spectrum B.15 3-Octanol

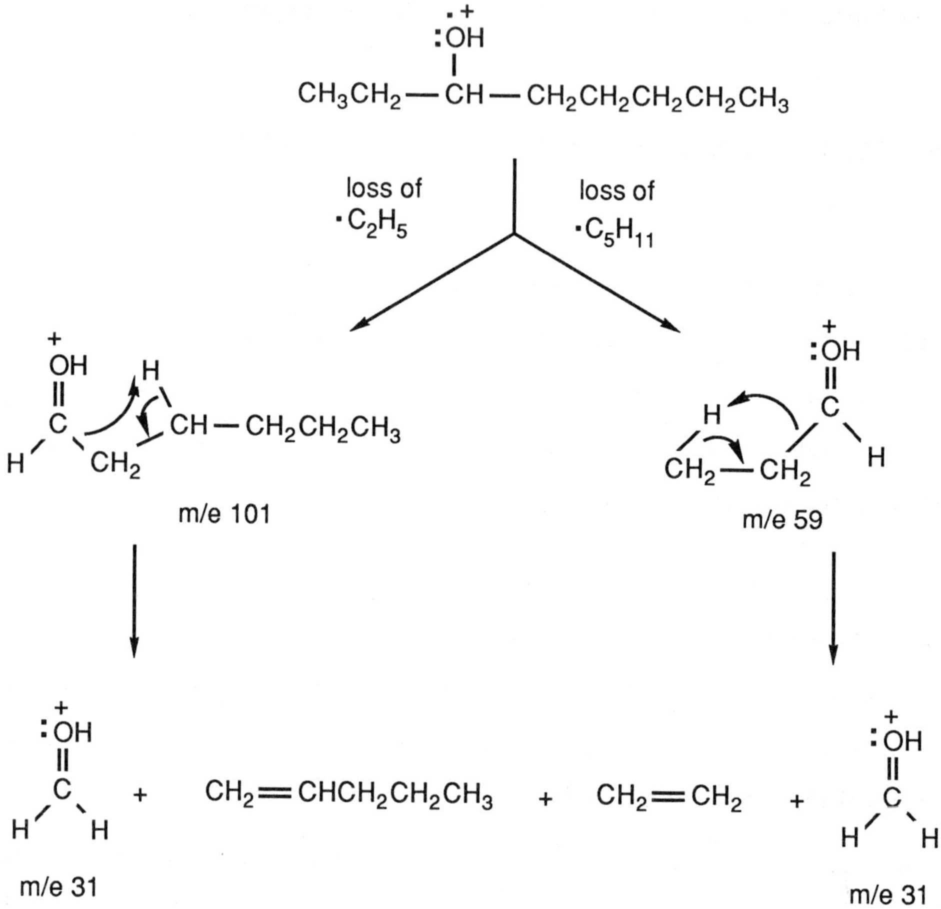

$CH_3CH_2-CH(\overset{\cdot+}{:OH})-CH_2CH_2CH_2CH_2CH_3$

loss of $\cdot C_2H_5$

loss of $\cdot C_5H_{11}$

m/e 101

m/e 59

$H_2C{=}\overset{+}{O}H$ + $CH_2{=}CHCH_2CH_2CH_3$ + $CH_2{=}CH_2$ + $H_2C{=}\overset{+}{O}H$

m/e 31

m/e 31

Each of the oxonium ions shown in the preceding equation can lose a neutral alkene to form protonated formaldehyde, and a small peak is observed at the appropriate value (m/e 31).

Rearrangements

The energetic ions produced in a mass spectrometer frequently can undergo a variety of rearrangements. Many of these rearrangements generate ions that are structurally quite different from the original ion, and this makes it difficult to correlate the observed fragment ions with the molecular structure. Once again the complexity of the topic is beyond the scope of this appendix, and we will not attempt to assign all peaks in a mass spectrum.

Instead we will attempt only to identify the peaks that we expect for a particular structure.

B.8 STRUCTURE DETERMINATION WITH MASS SPECTROMETRY

After an introductory discussion such as this you cannot be expected to deduce the structures of complex molecules from only their mass spectra. However, when combined with chemical or other spectroscopic information, a mass spectrum will frequently enable you to decide between structural alternatives. In this section we will show you how this can be done by working through several examples.

Example B.1 In an effort to synthesize 3-methyl-2-pentanone, 2-pentene was oxidized with perbenzoic acid. The resulting epoxide was treated with methyl lithium to yield an alcohol that was oxidized with CrO_3. The mass spectrum of the final product is shown in Mass Spectrum B.16. Was the synthesis successful?

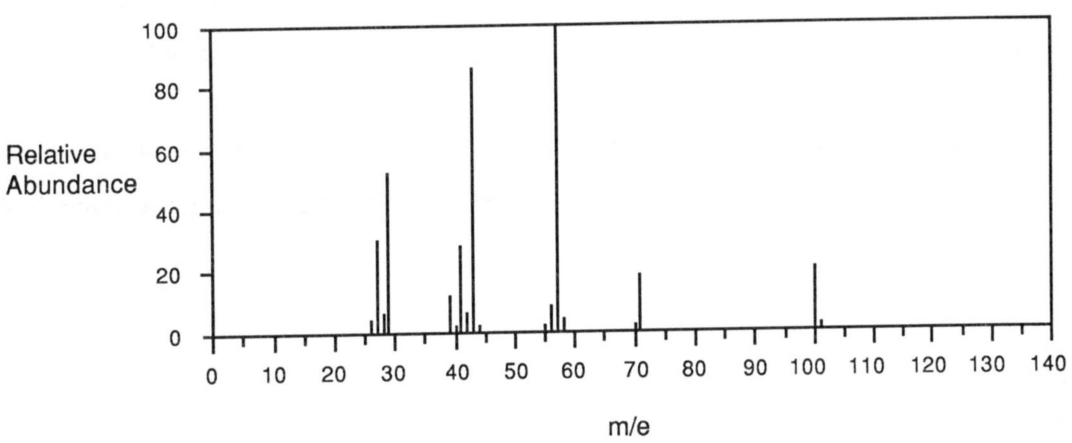

Mass Spectrum B.16 Mass spectrum of an unknown ketone

Solution: The products of the synthesis are reasonably certain except for reaction of the epoxide which could yield two isomeric alcohols. These would afford the desired 3-methyl-2-pentanone (**1**) and 2-methyl-3-pentanone (**2**) respectively.

$$CH_3CH_2CH{=}CHCH_3 \xrightarrow{C_6H_5CO_3H} CH_3CH_2\overset{O}{CH{-}CH}CH_3 \xrightarrow[2.\ H_2O]{1.\ CH_3Li}$$

$$CH_3CH_2{-}CH(CH_3){-}C(=O){-}CH_3 \ (\mathbf{1}) \xleftarrow{CrO_3} CH_3CH_2{-}CH(CH_3){-}CH(OH){-}CH_3$$

$$+$$

$$CH_3CH_2{-}C(=O){-}CH(CH_3){-}CH_3 \ (\mathbf{2}) \xleftarrow{CrO_3} CH_3CH_2{-}CH(OH){-}CH(CH_3){-}CH_3$$

Both ketones would afford a molecular ion at m/e 100, and this is consistent with the observed mass spectrum. The differences should arise in the fragment ions. Ketone **1** should show loss of $CH_3^{\cdot}$ (15 mass units) and $C_4H_9^{\cdot}$ (57 mass units), resulting in peaks at m/e 85 and 43. Formation of the corresponding carbocations would yield peaks at m/e 15 and 57. Of the four peaks expected only that at m/e 43 is observed.

Ketone **2** on the other hand, should show loss of $C_2H_5^{\cdot}$ (29 mass units) and $C_3H_7^{\cdot}$ (43 mass units) to give peaks at m/e 71 and 57. In addition, peaks at m/e 29 and 43 would be expected for the ethyl and isopropyl cations, respectively. Relatively intense peaks are found at all four values of m/e in Mass Spectrum B.16, so it is clear that the spectrum is that of ketone **2**. Hence the synthesis afforded not the desired 3-methyl-2-pentanone (**1**) but the isomeric compound 2-methyl-3-pentanone (**2**).

Example B.2 A Friedel-Crafts alkylation of benzene was carried out using 1-bromo-2-methylpropane. Several products were formed and one of them exhibited the mass spectrum shown in Mass Spectrum B.17. Deduce the structure of this compound.

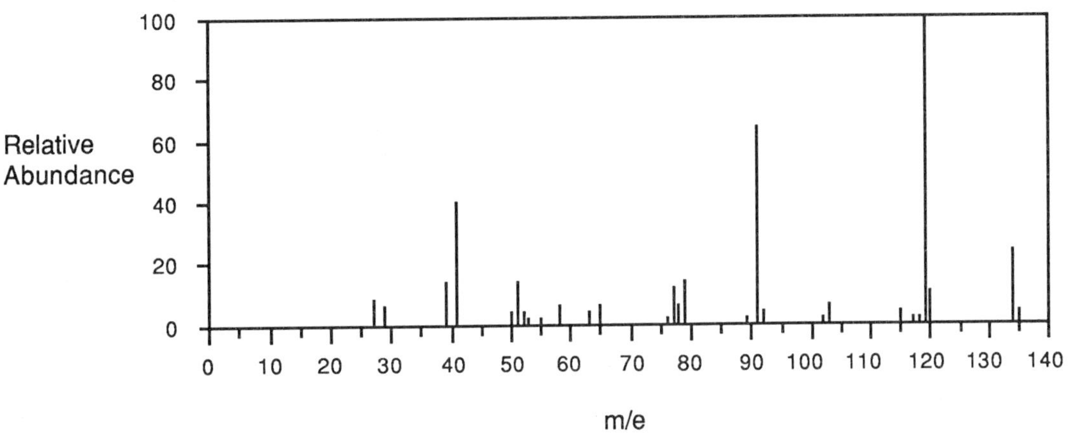

Mass Spectrum B.17 Mass spectrum of alkylation product of benzene

Solution: There are no peaks above m/e 140, suggesting that the ion with m/e 134 is the molecular ion. This corresponds to *monoalkylation*:

$$C_6H_6 + C_4H_9Br \xrightarrow{AlBr_3} \underset{\text{MW 134}}{C_6H_5\text{—}C_4H_9} + HBr$$

There remains a question regarding the structure of the butyl group, however, because carbocation rearrangements frequently accompany Friedel-Crafts alkylations (Section 18.3). Rearrangement in this case could easily produce a *tert*-butyl cation.

$$CH_3\text{—}\underset{H}{\overset{CH_3}{C}}\text{—}CH_2\text{—}\ddot{Br} \xrightarrow{AlBr_3} CH_3\text{—}\underset{+}{\overset{CH_3}{C}}\text{—}CH_3$$

Hence the two most reasonable structures for the product are isobutylbenzene (**1**) with an unrearranged isobutyl group, and *t*-butylbenzene (**2**).

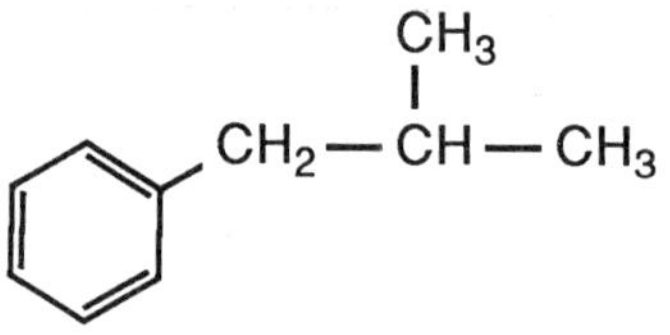

1

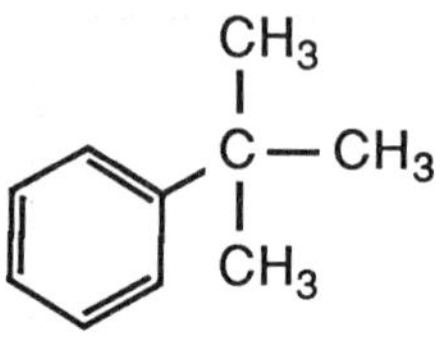

2

The predominant mode of fragmentation for the molecular ions of such alkylbenzenes is α-cleavage to form a benzylic cation.

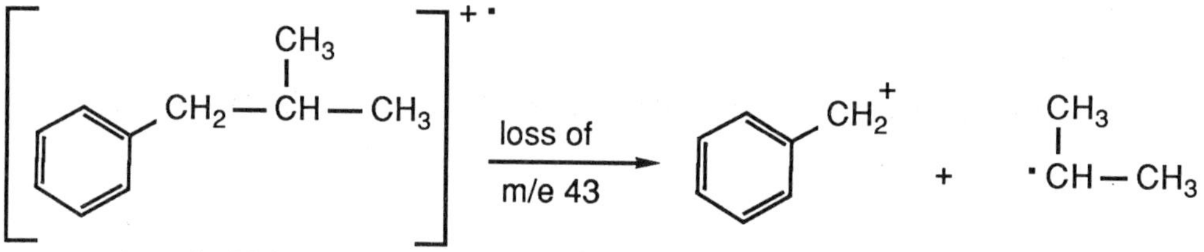

$$[C_6H_5CH_2-CH(CH_3)-CH_3]^{+\cdot} \xrightarrow[\text{m/e 43}]{\text{loss of}} C_6H_5\overset{+}{C}H_2 + \cdot CH(CH_3)-CH_3$$

m/e 134 m/e 91

$$[C_6H_5C(CH_3)_2-CH_3]^{+\cdot} \xrightarrow[\text{m/e 15}]{\text{loss of}} C_6H_5\overset{+}{C}(CH_3)-CH_3 + \cdot CH_3$$

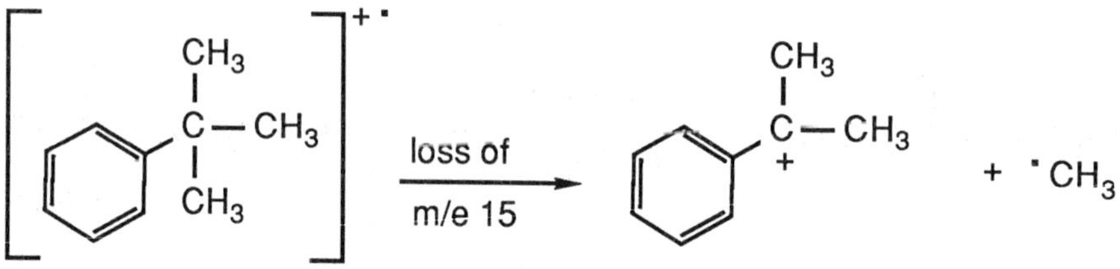

m/e 134 m/e 119

Although there is a modest peak at m/e 91 in Mass Spectrum B.17, the strong peak at m/e 129 makes it clear that the compound is *t*-butylbenzene (**2**).

Example B.3 A sample of 3-hexanone was allowed to react with methylmagnesium bromide, and the reaction mixture was heated with sulfuric acid. An alkene was isolated from this reaction and its mass spectrum is shown in Mass Spectrum B.18. Deduce the structure of the alkene (but ignore stereochemistry).

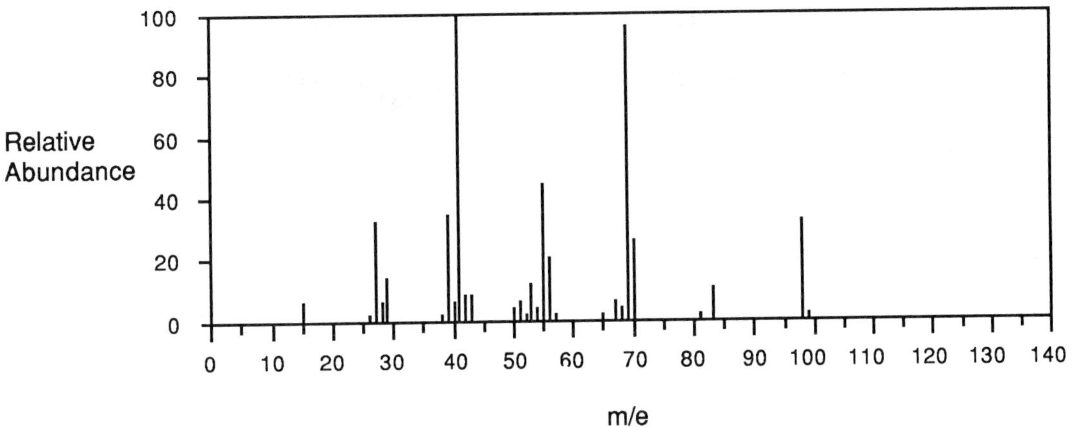

Mass Spectrum B.18 Mass spectrum of unknown alkene

Solution: The reaction with methylmagnesium bromide would lead to 3-methyl-3-hexanol, but two isomeric alkenes (**1** and **2**) could readily be formed by dehydration.

$$CH_3CH_2CH_2{-}\overset{\overset{\displaystyle O}{\|}}{C}{-}CH_2CH_3 \xrightarrow[\text{2. } H_2SO_4]{\text{1. } CH_3MgBr} \left[CH_3CH_2CH_2{-}\underset{\underset{\displaystyle CH_3}{|}}{\overset{\overset{\displaystyle OH}{|}}{C}}{-}CH_2CH_3 \right]$$

$$\underset{\mathbf{1}}{(CH_3CH_2CH_2)(CH_3)C{=}CHCH_3} \quad + \quad \underset{\mathbf{2}}{CH_3CH_2{-}CH{=}C(CH_2CH_3)(CH_3)}$$

The key to this problem is found in the expected fragmentation of the molecular ion to form an allylic cation, and the following alternatives are possible:

$$\mathbf{1} \longrightarrow \underset{m/e\ 98}{\left[(CH_3CH_2CH_2)(CH_3)C{=}CHCH_3 \right]^{+\cdot}} \xrightarrow[m/e\ 29]{\text{loss of}} \underset{m/e\ 69}{({}^{+}CH_2)(CH_3)C{=}CHCH_3} + CH_3CH_2^{\cdot}$$

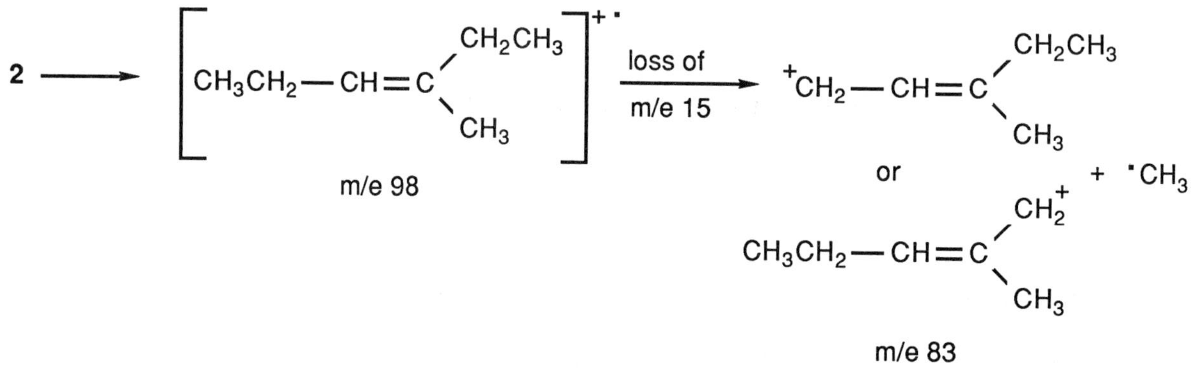

$$\mathbf{2} \longrightarrow \left[CH_3CH_2{-}CH{=}C\begin{matrix} CH_2CH_3 \\ CH_3 \end{matrix} \right]^{+\cdot} \xrightarrow[\text{m/e 15}]{\text{loss of}} {}^{+}CH_2{-}CH{=}C\begin{matrix} CH_2CH_3 \\ CH_3 \end{matrix} \text{ or } CH_3CH_2{-}CH{=}C\begin{matrix} \overset{+}{C}H_2 \\ CH_3 \end{matrix} + {}^{\cdot}CH_3$$

m/e 98

m/e 83

Inspection of the actual spectrum shows that there is a strong peak at m/e 69 but only a weak peak at m/e 83. Consequently, the correct structure of the alkene must be **1**, 3-methyl-2-hexene.

APPENDIX C. TERMS AND DEFINITIONS

Absorption spectrum. A graphical description of the absorption of radiation as a function of wavelength (or energy) of the radiation.

Acetal. A compound containing a saturated carbon atom with two ether linkages, one bond to hydrogen, and the remaining bond to carbon.

Achiral. Posessing reflection symmetry. Describes an object or molecule that is identical to its mirror image.

Acid halide. An acyl halide; a carboxylic acid derivative in which the two substituents on the carbonyl group are alkyl (or aryl) and halo.

Activation energy. The amount of energy that the reactants must have in order for a reaction to take place; the energy difference (usually abbreviated as E_a) between the reactants and the transition state for the reaction.

Acyclic. Non-cyclic; containing no rings.

Acyl group. The RCO group (*i.e.*, C=O with an alkyl or aryl group as one of its substituents).

Acyl halide. An acid halide; a carboxylic acid derivative in which the two substituents on the carbonyl group are alkyl (or aryl) and halo.

Alcohol. A member of the class of organic compounds with the general structure, R-OH, in which an alkyl group is covalently bonded to the oxygen of a hydroxyl group.

Aldaric acid. A carboxylic acid corresponding to oxidation of both termini of an aldose.

Aldehyde. A carbonyl compound in which one of the two substituents on the carbonyl group is a hydrogen and the other is an alkyl (or aryl) group.

Alditol. A polyalcohol corresponding to reduction of the carbonyl group of an aldose or ketose.

Aldonic acid. A monocarboxylic acid corresponding to oxidation of an aldehyde group of an aldose.

Aldose. A monosaccharide having the oxidation state of an aldehyde at C−1 and that of an alcohol at all other carbons.

Aliphatic. A term describing those organic compounds that are not aromatic.

Alkaloid. Any of a variety of naturally occuring, physiologically active amines; usually obtained from plants.

Alkane. A hydrocarbon with the general formula C_nH_{2n+2}, in which each carbon atom is bonded to four other atoms (either carbon or hydrogen).

Alkene. A hydrocarbon with the general formula, C_nH_{2n}, characterized by the presence of a carbon−carbon double bond.

Alkyl group. A hydrocarbon portion of a molecule corresponding to an alkane from which a hydrogen atom has been removed.

Alkyne. A hydrocarbon with the general formula, C_nH_{2n-2}, characterized by the presence of a carbon−carbon triple bond.

Allene. A hydrocarbon with the general formula, C_nH_{2n-2}, characterized by the presence of two carbon−carbon double bonds between three contiguous carbon atoms (*i.e.*, C=C=C).

Amide. A carboxylic acid derivative characterized by an acyl group bonded to an NH_2 (or NHR or NR_2) group.

Ammonium ion. A cation containing a tetrasubstituted, positively charged nitrogen atom (*i.e.*, an alkyl- or aryl-substituted derivative of NH_4^+).

Anhydride. A carboxylic acid derivative characterized by two acyl groups bonded to a single oxygen atom.

Anomeric carbon. The carbonyl carbon of an aldose or ketose, which becomes a chiral center in the cyclic (acetal or ketal) forms.

Anomers. Stereoisomeric sugars that differ only in configuration at the anomeric carbon (a specific type of epimer).

Anti. A term describing the orientation of two substituents (on adjacent carbons) in which they are at a dihedral angle of 180°; for alkanes this is a stable arrangement corresponding to an energy minimum.

Antibonding. A description of a molecular orbital produced by the out-of-phase combination of two atomic orbitals. When occupied by electrons this corresponds to a repulsive interaction between the two atoms.

Aromatic. Belonging to a class of conjugated unsaturated compounds with characteristic features of structure and reactivity, usually related to those of benzene.

Atomic orbital. An allowed energy state for an electron in an atom; the orbital is described mathematically by a wave function. The three-dimensional distribution of electron density is described by the square of the wave function.

Benzyl group. The group, $C_6H_5CH_2$, where C_6H_5 is a phenyl group and the C of the CH_2 is the benzylic carbon.

Benzylic. A term describing the carbon atom of a benzyl (or a substituted benzyl) group that is bonded to the aromatic ring; also used to describe a substituent on a benzylic carbon or a reactive intermediate in which the reactive center is a benzylic carbon.

Bimolecular. A term describing a reaction step in which two molecules collide.

Blocking group. A group that is introduced *temporarily* during a synthesis to modify one reactive center and allow selective reaction at another site; also known as a *masking group* or *protecting group*.

Boat. A conformation for the cyclohexane ring system. It is almost always much higher in energy than the chair conformations, and only an extremely small fraction of the total molecules will exist in the boat form at any one time.

Boltzmann distribution. The statistical distribution of energy for a large number of molecules; the average kinetic energy varies with the temperature, but most individual molecules have either more or less energy than this average.

Bond dissociation energy. The energy required to cleave a bond and form two free radicals.

Bonding. The attractive interaction (*i.e.*, a bond) between two atoms that that results when electrons occupy a molecular orbital formed by the in-phase interaction between two atomic orbitals.

Bonding axis. The imaginary line connecting two nuclei that are bonded to each other.

Branched-chain. A term describing a molecule in which all the carbon atoms are not connected sequentially; *i.e.*, in which it is not possible to encounter all atoms of the carbon chain by proceeding consecutively from one to the next without backtracking.

Bronsted-Lowry classification. Classification of acids as H^+ donors and bases as H^+ acceptors.

Carbohydrate. A general term for compounds with the general formula, $C_x{\cdot}(H_2O)_y$ (but also used to describe derivatives containing other atoms). The terms *sugar* and *saccharide* are synonyms.

Carbon-carbon bond formation. A class of reactions in which a new carbon-carbon bond is formed. Such reactions are the key synthetic steps in building up a new carbon skeleton.

Carbonyl group. The C=O group; the functional group corresponding to a carbon-oxygen double bond.

Carboxyl group. The CO_2H group, which characterizes carboxylic acids; this consists of a carbonyl group for which one substituent is a hydroxyl group.

Carboxylic acid. An organic compound having a carboxyl group; the general structure is $R-CO_2H$. The K_a is usually about 10^{-5}.

Catalyst. A material that acts to facilitate a reaction but which is not consumed in the reaction. A catalyst causes the reaction to proceed more rapidly or to occur at a lower temperature.

Chain reaction. A reaction that operates in a cyclic manner because the reactive intermediate is regenerated. Chain reactions are characterized by three stages: *initiation*, in which a reactive intermediate is formed; *propagation*, in which the intermediate reacts to form product with regeneration of the intermediate; and *termination*, in which the reactive intermediate is consumed.

Chair. A conformation of the cyclohexane ring system, which is almost always the optimum structure. Two chair conformations are always possible, differing by interchange of axial and equatorial positions of the substituents.

Chemical shift. The variation in radiofrequency energy absorbed by nonequivalent nuclei in an NMR experiment as a function of their chemical environments.

Chemoselectivity. Selective reaction at one of several different reactive centers in a molecule.

Chiral. Lacking reflection symmetry. Describes an object or molecule which is different from its mirror image.

Chiral center. A site in a molecule that is characterized by the absence of reflection symmetry. For organic compounds this is usually a tetrahedral carbon atom which has four different substituents. A molecule with a single chiral center will always be a chiral molecule.

Cis. A term describing a relationship between two substituents on a ring or double bond. The substituents (which must be on different carbon atoms) are on the same side of the ring or on the same side of the double bond.

Common name. A name that is based only partially (or not at all) on the molecular structure of a compound.

Concerted. A term describing a process in which the reactants proceed through a single transition state to give the product directly, without the formation of any intermediate.

Configuration. The fixed spatial arrangement of substituents on an atom or at a double bond. A chiral center has two possible configurations, as does a double bond if neither carbon has two equivalent substituents. (If two substituents on a tetrahedral carbon atom are equivalent, only a single configuration is possible for that carbon). Configuration at a chiral center is specified by *R/S* nomenclature (Section 6.3), and configuration at a double bond is specified by *E/Z* nomenclature (Section 5.5).

α,β-**Configuration.** A term describing the configuration of the anomeric carbon of a sugar in its cyclic form. When drawn in a Haworth representation a *D*-sugar has the α-configuration if the oxygen substituent lies below the "plane" of the ring and the β-configuration if it lies above. (The meaning of α and β is reversed for *L*-sugars.)

D,L **Configuration.** A term describing the configuration of the highest numbered chiral carbon of a monosaccharide. A simple *D*-sugar has the *R* configuration at this chiral center. *D*-Glyceraldehyde is the reference compound.

Conformation. The three-dimensional arrangement of a molecule that differs from other arrangements only by rotation about single bonds. Under normal conditions the individual conformations of a molecule cannot be isolated; therefore they are not considered isomers.

Conjugate acid. The species produced by reaction of a base with H^+.

Conjugate addition. Addition to an α,β-unsaturated carbonyl system that is initiated by nucleophilic attack at the β position.

Conjugate base. The species produced by loss of an H^+ from an acid.

Conjugated. Describes two multiple bonds on four contiguous carbon atoms (*i.e.*, C=C-C=C for conjugated double bonds), which form a continuous π-system over the four carbons. Conjugation can also extend over a larger number of carbon atoms.

Connectivity. The specific sequence in which atoms in a molecule are joined (connected) by bonds.

Constitution. The features of a molecular structure that result from the connectivity of the individual atoms. The connectivity sequence by which atoms in a molecule are bonded to each other. The constitution can be defined either by drawing a structural formula or by stating the IUPAC name.

Constitutional equivalence. The relationship between two atoms or groups that cannot be differentiated according to their connectivity.

Constitutional isomers. Two molecules (compounds) with the same molecular formula that have different connectivities (*i.e.*, different constitutions).

Core electrons. The electrons on an atom that are not part of the valence level; *i.e.*, the 1s electrons for carbon and other second-row elements.

Coupling constant. The magnitude of the NMR coupling interaction between two nuclei, usually denoted by the letter J.

Covalent bonding. Bonding between two atoms that results from sharing of electrons by the two atoms in a bonding molecular orbital. (This contrasts with ionic bonding, which is an electrostatic attraction between two ions of opposite charge).

Cyclic. Having atoms that are connected to form a ring.

Cycloalkene. A cyclic alkene in which the double bond is specifically contained within the ring.

Degree of unsaturation. The number of "missing" pairs of hydrogen atoms in a molecule that corresponds to multiple bonds or rings. Each ring or double bond corresponds to one degree of unsaturation, each triple bond corresponds to two degrees of unsaturation.

Derivative. A compound that is formed or derived from another compound, either in an actual reaction or in a hypothetical sense by changing a structural formula on paper.

Diastereomers. Stereoisomers that are not mirror images of each other. Usually two diastereomers differ either in the configuration at a carbon–carbon double bond or in the configurations at some but not all chiral centers.

Diastereotopic. A term describing two atoms or groups in a molecule that are different in the sense that replacement by another substituent would lead to diastereomeric products, depending on which of the two were replaced.

Diene. A hydrocarbon that contains two carbon–carbon double bonds.

Dihedral angle. The angle between two substituents on adjacent carbons that is perceived by viewing along the C–C bond. The dihedral angle, X–C–C–Y, is equal to the angle between the C–X and C–Y bonds when the structure is viewed along the C–C bond (*i.e.*, with a Newman projection).

Dipolar attraction. The attractive interaction between the oppositely charged ends of two polarized bonds; one of the main types of intermolecular forces.

Double bond. The combination of a π bond and a σ bond between two atoms; the carbon–carbon double bond is the characteristic functional group of an alkene.

Downfield. A term describing transitions of lower energy in NMR spectroscopy; it indicates peaks to the left of some point of reference.

E. Entgegen (the German word for opposite); defines the configuration of a double bond in which the two substituents of highest priority are on opposite sides of the double bond.

Eclipsed. A term describing a three-dimensional arrangement in which the dihedral angle between substituents on adjacent carbons is 0°. When viewed as a Newman projection, the substituent on the rear carbon is directly behind (*i.e.*, is eclipsed by) the substituent on the front carbon.

Electron deficient. A term describing an atom lacking the 8 electrons of the inert gas configuration in its valence shell (*e.g.*, the reactive center in a free radical or carbocation).

Electron-withdrawing. The inductive or resonance effect of an electronegative substituent; electron density is shifted to the substituent from the atom to which it is bonded.

Electronic effects. The combination of inductive and resonance effects.

Electrophilic. Electron seeking; the interaction of an alkene (or some other electron donor) with a Lewis acid is described as electrophilic attack on the double bond.

Emission spectrum. A graphical description of the emission of radiation as a function of wavelength (or energy) of the radiation.

Enamine. A compound containing an amine that is directly bonded to a carbon-carbon double bond; a vinyl amine.

Enantiomers. Two stereoisomers that are nonsuperimposable mirror images. Each enantiomer is necessarily chiral.

Endothermic. Description of a process that absorbs heat (*i.e.*, $\Delta H > 0$). In contrast, an **exothermic** process occurs with the liberation of heat (*i.e.*, $\Delta H < 0$).

Energy. The energy of a system is its ability to liberate heat and to do work. Experimental measurement gives an energy change, and this is the sum of the heat evolved and the work done by the system.

Energy barrier. The activation energy for a reaction or conformational change. Molecules with less than this amount of energy cannot undergo the change.

Energy minimum. A position on a reaction coordinate, for which any change results in an increase in the energy. This corresponds to the optimum geometry (or at least one of the stable conformations) of the species.

Enol. A compound containing a hydroxyl group bonded directly to a carbon–carbon double bond; ordinarily undergoes rapid rearrangement to an isomeric carbonyl compound.

Enol ether. An ether in which one of the two groups bonded to oxygen is an alkenyl group; also called a vinyl ether.

Enthalpy. A thermodynamic quantity for which the change, ΔH, is the heat evolved or absorbed in a process occurring at constant pressure.

Entropy. A thermodynamic quantity related to probability; experimental measurements provide the change in entropy (ΔS) for a process, which is related to the enthalpy and free energy changes by $\Delta G = \Delta H - T\Delta S$

Epimers. Stereoisomers which differ in configuration at only one of several chiral centers. All other chiral centers have the same configuration.

Epoxide. A 3-membered ring with two sp^3 carbon atoms and an oxygen atom.

Equivalent. The same. A term describing two or more groups that are indistinguishable on the basis of symmetry or connectivity. *Constitutional equivalence* of groups in a molecule requires either rotational symmetry (the groups are identical) or reflection symmetry (the groups are mirror images).

Ester. A carbonyl compound in which one of the substituents on the carbonyl group is alkoxy (or aryloxy) and the other substituent is alkyl, aryl, or hydrogen.

Ether. A compound containing two R groups bonded to an oxygen, where R can be alkyl, aryl or alkenyl.

Excited state. A state of a species that has absorbed radiation; the energy can be liberated by emission of a photon or by transfer to the surroundings in the form of heat.

Exothermic. Description of a process that liberates heat (*i.e.*, $\Delta H < 0$). In contrast, an **endothermic** process occurs with the absorption of heat (*i.e.*, $\Delta H > 0$).

Fischer projection. A three-dimensional representation used to show the orientation of substituents on a carbon atom, in which the carbon is denoted by the intersection of a horizontal and a vertical line. The substituents on the vertical line are further away from you than the chiral center, and the substituents on the horizontal line are closer to you.

Formal charge. The charge that is assigned to an atomic center according to the formalism that electrons in covalent bonds are shared equally by the bonded atoms.

Free energy. A thermodynamic quantity for which the difference (ΔG) describes the combined enthalpy and entropy differences between two species. The free energy difference between reactants and products of a reaction is related to the equilibrium constant (K) by: $\Delta G = -2.3\ RT \log K$

Free radical. An electron-deficient species having an odd number of electrons (*i.e.*, having at least one unpaired electron).

Free rotation. The situation in which interconversion of conformations occurs rapidly (*i.e.*, freely) at room temperature. In other words the energy barrier for rotation about single bonds is small.

Functional class. A class of compounds characterized by a particular type of reactivity; the members of a functional class all contain a specific functional group.

Functional group. A structural feature, such as a carbon–carbon double bond, that imparts a uniform type of reactivity to molecules of a particular functional class.

Functional group modification. A class of reactions in which the carbon skeleton is unchanged and only the substituents on the carbon skeleton are affected.

Furanose. The cyclic form of a sugar (hemiacetal or hemiketal) that has a 5-membered ring. Corresponding acetals or ketals are called *furanosides*; the entire residue, excluding the anomeric oxygen substituent, is called a *furanosyl* group.

Fused ring system. A type of cyclic structure in which two adjacent atoms of one ring are also part of a second ring.

Gauche. The orientation of two substituents (on adjacent carbons) in which they are at a dihedral angle of about 60°. This corresponds to a staggered conformation of the molecule, but the two substituents may have unfavorable steric interactions.

Geminal. Having two identical substituents on a single carbon atom, *e.g.*, a geminal diol (or *gem*-diol).

Glycoside. A general term for a cyclic acetal or ketal of a sugar (*i.e.*, a furanoside or pyranoside).

Ground state. The lowest energy state of a species; for electronic configurations this is the energy state normally occupied at room temperature.

Haloalkane. A derivative of an alkane in which a hydrogen atom has been replaced by a halogen atom; also known as an alkyl halide.

Halohydrin. A compound with adjacent halo and hydroxy substituents.

Hammond postulate. A general rule about a reaction profile: the transition state for a reaction tends to resemble more closely whichever is higher in energy, the reactant or the product.

Hemiacetal. The product of the addition of an alcohol to the carbonyl group of an aldehyde, *i.e.*, a compound having a carbon on which the four substituents are an alkyl or aryl group, a hydrogen, an alkoxy and a hydroxyl group.

Hemiketal. The product of the addition of an alcohol to the carbonyl group of a ketone, *i.e.*, a compound having a carbon for which two of the substituents are alkyl or aryl groups and the remaining two are an alkoxy group and a hydroxyl group.

Heteroatom. An atom other than carbon in a chain or ring; *e.g.*, nitrogen, oxygen, sulfur, phosphorus.

Heterocycle. A cyclic molecule in which one or more of the atoms of the ring is not carbon.

HOMO. The highest occupied molecular orbital of a species.

Homolog. A molecule that differs from a reference compound by a single CH_2 group.

Hückel aromaticity. A criterion for aromaticity of fully conjugated cyclic molecules with $(4n+2)$ π-electrons.

Hybrid orbital. Atomic orbital for which the wave function has been generated by mathematical combination of other wave functions (usually for s and p atomic orbitals) on a single atom.

Hydrocarbon. A compound composed exclusively of carbon and hydrogen atoms.

Hydrogen bonding. The favorable interaction between the hydrogen of an N–H or O–H group and the nonbonding electrons of another electronegative atom (usually nitrogen or oxygen).

Hydroxylic. A term describing a solvent that contains OH (*i.e.*, hydroxyl) groups.

Imide. A cyclic derivative of a dicarboxylic acid in which the two acyl groups are bonded to a single NH (or NR) group.

Imine. A compound containing a C=N double bond; the nitrogen analog of a ketone.

Iminium ion. An ion with a $C{=}N^+$ group.

In phase. Term describing the interaction between atomic orbitals of two different atoms that leads to a bonding molecular orbital. This results from overlap of lobes for which the wave function has the same sign (phase).

Inductive effect. The influence on structure or reactivity produced by a substituent as a result of polarization of individual bonds.

Integral. A second tracing on an NMR spectrum that indicates the relative area under each peak.

Integration. In NMR spectroscopy, measurement of the area under the peaks of a spectrum.

Intermediate. A reactive intermediate; also a compound that is obtained at an intermediate stage in a synthesis.

Internal alkyne. An alkyne in which the triple bond is located in the interior part of the carbon chain (*i.e.*, C–C≡C–C).

Inversion of configuration. The result of a substitution reaction at a chiral center in which the new group does not occupy the same stereochemical site as the group that it replaces. (If the new and original groups were the same, reaction via inversion of configuration would produce the enantiomer of the reactant).

Ion pair. A cation and anion that remain in close proximity as a result of electrostatic attraction between the two.

Isolated multiple bond. A multiple bond that is not in conjugation with another multiple bond; all carbons attached directly to the multiple bond are saturated (sp^3) carbons.

Isomers. Different compounds (molecules) having the same molecular formula but differing in their molecular structures.

IUPAC name. The name of a compound that is systematically derived from its molecular structure according to the rules set down by the International Union of Pure and Applied Chemistry.

Ketal. A compound containing a saturated carbon atom with two ether linkages and two bonds to carbon atoms.

Ketone. A carbonyl compound in which both substituents on the carbonyl group are alkyl (or aryl groups); the general formula is $R_2C=O$.

Ketose. A monosaccharide having one carbon (usually C–2) at the oxidation state of a ketone and the remaining carbons at the oxidation state of an alcohol.

Kinetic control. Describes a reaction for which the amount of each product obtained is governed by the rate at which it is produced.

Kinetic energy. The energy of a molecule that results from motion; it equals $1/2\ mv^2$, where m is the mass of the molecule and v is its velocity.

Lactam. A cyclic amide that is derived from an amino–substituted carboxylic acid; the ring includes the carbonyl carbon and the nitrogen.

Lactone. A cyclic ester that is derived from a hydroxy–substituted carboxylic acid; the ring includes the carbonyl carbon and the ester oxygen.

Leaving group. The substituent that is replaced in a nucleophilic substitution reaction. It leaves with its bonding pair of electrons.

Lewis acid. A species that can act as an electron pair acceptor; it must have an atomic center with a low–energy unoccupied orbital.

Lewis base. A species that can act as an electron pair donor; it can also act as an H^+ acceptor.

Lobe. A region of an orbital for which the wavefunction describing the orbital has a single phase.

Lone pair. An unshared pair of valence level electrons on an atomic center; a pair of nonbonding electrons.

LUMO. The lowest unoccupied molecular orbital of a species.

Masking group. A group that is introduced *temporarily* during a synthesis to modify one reactive center and allow selective reaction at another site; also known as a *blocking group* or *protecting group*.

Mass spectrometry. An instrumental technique for studying molecular structure, which is based on the separation of charged particles according to mass.

Meso compound. A molecule that is achiral despite having more than one chiral center. It has reflection symmetry, and it contains an even number of constitutionally equivalent chiral centers.

Meta. A term describing the relative positions of two substituents having a 1,3-relationship on a benzene ring.

Migratory aptitude. The relative tendency of a substituent to undergo a 1,2-shift to an electron deficient atom, *e.g.* to migrate in the Baeyer-Villiger oxidation.

Mirror image. The image that would be seen if an object or molecule were viewed in a mirror.

Molecular formula. The chemical formula denoting how many atoms of each element are present in a single molecule.

Molecular ion. The cation radical that is produced in a mass spectrometer by ejection of an electron from a neutral molecule.

Molecular orbital. An orbital derived from interaction of two or more atomic orbitals. The molecular orbital can be either bonding or antibonding, but normally only the bonding molecular orbitals are occupied by electrons.

π-Molecular orbital. A molecular orbital formed by the interaction of two parallel atomic p-orbitals.

σ-Molecular orbital. A molecular orbital formed by the interaction of two atomic s orbitals or two hybrid orbitals (s and p) that are directed toward each other.

Multiplicity. The number of lines in an NMR signal that are observed for a particular nucleus as a result of spin coupling to other nuclei.

Mutarotation. A change in optical rotation of a sample over a period of time that results from a chemical reaction. Once equilibrium is reached, the rotation no longer changes.

Newman projection. A type of drawing used to convey three-dimensional information about two adjacent carbon atoms in an organic structure.

Nitrile. A carboxylic acid derivative characterized the the presence of a cyano group (*i.e.*, R-C≡N).

Nitrilium ion. An ion with a $-C{\equiv}NR^+$ ion.

Node. A region (usually a plane or a spherical surface) of an atomic (or molecular) orbital that separates lobes of opposite phase. At the nodal surface the electron density for that orbital is zero.

Nonbonded interaction. The interaction between the electron clouds of atoms that are not bonded to each other (or to the same central atom); when two atoms or groups approach each other closely, the interaction is repulsive.

Nonbonding electrons. Valence-level electrons that are not involved in bonding; lone pair electrons; an unshared pair of electrons.

Nonpolar. A term describing a covalent bond in which the electrons are shared equally; a bond between two atoms with the same electronegativity. The term nonpolar is also used to describe a compound containing no polar substituents.

Nuclear magnetic resonance (NMR) spectroscopy. A technique for studying molecular structure based on transitions between energetically different spin states of nuclei in a strong magnetic field.

Nucleophile. A species with nonbonding electrons that can attack a carbon atom bearing full or partial positive charge.

Optical activity. The rotation of the plane of polarized light by chiral molecules. One enantiomer must be present in excess of the other for optical activity to be observed.

Organometallic compound. A reactive organic compound characterized by a carbon–metal bond. Organometallic reagents can be symbolized by the formula R–M, where M is a metal atom.

Orientation. The position of a molecule or drawing relative to its surroundings. Reorienting a molecule does not affect the constitution, configuration, or conformation. Only its position relative to its environment is affected.

Ortho. A term describing the relative positions of two substituents having a 1,2-relationship on a benzene ring.

Osazone. A bishydrazone (usually a bisphenylhydrazone) corresponding to an α-ketoaldose.

Out of phase. A term describing the interaction between atomic orbitals of two different atoms that leads to an antibonding molecular orbital. This results from overlap of lobes for which the wavefunction has the opposite sign (phase).

Overlap. The interaction (either favorable or unfavorable) between orbitals of two or more atoms that results from their occupying the same region in space.

Oxidation. A reaction that causes the oxidation state of a compound to increase. For organic compounds this usually involves an increase in the number of bonds between carbon and oxygen or other electronegative atoms, or a decrease in the number of bonds between carbon and hydrogen.

Oxonium ion. An ion characterized by an oxygen that has three bonds (and therefore a formal positive charge).

Para. A term describing the relative positions of two substituents having a 1,4-relationship on a benzene ring.

Phase. The sign (positive or negative) of the wavefunction that describes an orbital. We have used shaded and unshaded drawings to distinguish between orbitals or lobes of opposite sign in this text.

Phenyl group. The C_6H_5 group; the group that would be produced by removal of a hydrogen from a benzene molecule.

Pheromone. A chemical messenger, secreted by an animal, that elicits a specific response in another animal (often of the opposite sex) of the same species.

Polarized bond. A bond between two atoms with different electronegativities; this forms a dipole, placing partial positive charge on one atom and partial negative charge on the other.

Polyene. A hydrocarbon that contains multiple carbon-carbon double bonds.

Polyfunctional. A molecule that contains two (or more) independent functional groups.

Polymer. A molecule that consists of repeating subunits; the number of subunits can be very large.

Potential energy. The energy a molecule posesses that results from the relative positions of its constituent atoms; potential energy reflects the chemical energy that might be released or absorbed in a reaction.

Precursor. A compound that precedes another in a synthetic sequence; the compound that will afford the target molecule when subjected to the appropriate sequence of reactions.

Primary. A term describing a carbon atom that is bonded to only one other carbon atom. When used to describe an alkyl group, it refers to the carbon from which a hydrogen has been removed to generate the group (*i.e.*, the carbon that is the point of attachment for the group).

Principal chain. The longest carbon chain in a molecule that contains the functional group used to name the compound; for an alkene this is the longest carbon chain that contains the double bond.

Prochiral. Capable of becoming chiral by a simple reaction: by *adding* a fourth substituent to an sp^2 carbon or by *modifying* one of two equivalent substituents on an sp^3 carbon.

Protecting group. A group that is introduced *temporarily* during a synthesis to modify one reactive center and allow selective reaction at another site; also known as a *blocking group* or *masking group*.

Proton affinity. The energy released when a base combines with an H^+ ion in the gas phase.

Pyranose. The cyclic form of a sugar (hemiacetal or hemiketal) that has a 6-membered ring. Corresponding acetals or ketals are called *pyranosides*; the entire residue, excluding the anomeric oxygen substituent, is called a *pyranosyl* group.

Quaternary. A term describing a carbon atom that is bonded to four other carbon atoms.

Quaternary ammonium ion. An ammonium ion in which all four substituents on the positively charged nitrogen are alkyl (or aryl) groups.

Racemic. Consisting of exactly equal amounts of two enantiomers. A racemic substance is optically inactive and behaves as a pure compound since the two enantiomers cannot be separated by physical methods such as distillation.

Rate constant. A constant related to the activation energy of a reaction; the rate of a reaction is equal to the product of the rate constant and the appropriate concentration terms for the reactants.

Rate-limiting step. The slow step in a reaction; the step with the greatest activation energy.

Rate of reaction. The rate at which product is formed in a reaction (or the rate at which reactant is consumed); It is expressed as a change in concentration per unit time, typically moles per liter per second M sec^{-1}).

Reaction coordinate. The horizontal axis of a reaction profile; describes the progress of the reaction (for example, the distance X---Y between two atoms, X and Y).

Reaction mechanism. A detailed step-by-step description of the events of a reaction at the molecular level.

Reaction profile. A plot of the potential energy of the molecules involved in a reaction during the course of the reaction (as specified by the reaction coordinate).

Reaction rate. The rate at which reactants are converted into products.

Reactive center. The atom (or group of atoms) in a molecule at which reaction occurs.

Reactive intermediate. A high-energy species that is a minimum-energy structure, but which cannot be isolated (*i.e.*, a free radical, carbocation, carbene or carbanion).

Reducing sugar. A sugar (hemiacetal or hemiketal) that is in equilibrium with the free aldehyde or α-hydroxyketone and readily reacts with mild oxidizing agents such as silver ion and cupric ion.

Reduction. A reaction that causes the oxidation state of a compound to decrease. For organic compounds this usually involves an decrease in the number of bonds between carbon and oxygen or other electronegative atoms, or an increase in the number of bonds between carbon and hydrogen.

Reflection symmetry. A property of an achiral object or molecule, wherein a structure is symmetric about its midpoint, and the two "halves" are mirror images of each other. The most common types of reflection symmetry are a plane of symmetry and a center of symmetry.

Regiospecific. Describes a reaction such as addition to a double bond for which there is a specific correlation between the structure of the reactant and the orientation of the addition.

Relative configuration. Specification of the configuration of a chiral center (or of an entire molecule) in relation to the chiral center of another compound. If one compound can be converted to the other, the relative configuration can be discussed even if the absolute configuration is not known for either.

Resolution. The process by which a mixture of two enantiomers is separated (resolved) into the two optically active, enantiomeric forms. Since the enantiomers have identical physical properties this is accomplished by reaction with a chiral (optically active) reagent.

Resonance contributor. One of the individual structures that is drawn when a single conventional drawing does not adequately represent a molecule or ion; also called a resonance form.

Resonance energy. The energy by which a conjugated molecule is stabilized through resonance effects.

Resonance form. One of the individual structures that is drawn when a single conventional drawing does not adequately represent a molecule or ion; also called a resonance contributor. The individual resonance forms differ only in the positions of multiple bonds and formal charges; there can be no difference in constitution.

Resonance hybrid. The overall, "average" structure corresponding to a group of individual resonance forms.

Resonance stabilization. The stabilization of a molecule or ion for which several resonance forms can be drawn; the stabilization is associated with electron delocalization.

Restricted rotation. The absence of free rotation about a bond; restricted rotation results from a structural feature that makes rotation about the bond difficult or impossible.

Rotational energy. The kinetic energy of a molecule that results from rotational motion of all or part of the molecule relative to its surroundings.

Rotational symmetry. A property of a molecule such that rotation about an axis passing through the center of the molecule (by some angle of less than 360°) generates a structure that is indistinguishable from the original.

Saccharide. Sugar, carbohydrate. The term *saccharide* is usually used with a prefix indicating the number of subunits joined by glycosidic linkages; a *monosaccharide* has only a single subunit, a *disaccharide* has two, *oligosaccharides* have up to 10, and *polysaccharides* are complex carbohydrates with more than 10 subunits.

Saturated. Describes an sp^3 carbon, which is bonded to the maximum number of possible substituents; all four substituents are connected via single bonds.

Saturation. The result in NMR spectroscopy of irradiation with a strong RF signal. The populations become nearly equal for the two spin states of the irradiated nucleus, and there is no net absorption or emission. Coupling interactions with other nuclei are thereby removed.

Sawhorse representation. A type of drawing used to convey three-dimensional information about two adjacent carbon atoms in an organic structure.

Secondary. A term describing a carbon atom that is bonded to exactly two other carbon atoms. When used to describe an alkyl group, it refers to the carbon from which a hydrogen has been removed to generate the group (*i.e.*, the carbon that is the point of attachment for the group).

Selectivity. Preferential reaction at a particular site in a molecule.

Sigma bond. A single bond, formed when electrons occupy a bonding σ orbital.

Single bond. A sigma bond; a normal two-electron bond between two atoms.

Solvolysis. A nucleophilic substitution reaction in which the solvent acts as the nucleophile.

Spectroscopy. The study of absorption or emission of different wavelengths of electromagnetic radiation in order to gain information about molecular structure.

Spectrum. The different colors of visible light that can be seen when a beam of light passes through a prism; more generally, the different wavelengths of electromagnetic radiation. The term is also used to describe a graphical representation of the absorption (or emission) of radiation by a sample as a function of wavelength.

Spin decoupling. Elimination of spin coupling between nuclei by double irradiation of one of them to produce saturation.

Spin-spin coupling. The interaction between nuclei in different spin states; it can result in multiple peaks for a particular nucleus in an NMR spectrum. The term is often shortened to *spin coupling*.

Stability. Used to describe relative energy; the more stable a species, the lower is its energy. Also used to denote reactivity; the more stable a species, the lower is its reactivity.

Staggered. The orientation of the substituents on adjacent carbons in a stable conformation of an alkane; the dihedral angles between such substituents are either 60°or 180°.

Stereochemistry. The study of three-dimensional aspects of molecular structures and chemical reactions; also a term used to describe the 3-dimensional structure of a molecule.

Stereoisomers. Isomers that have the same constitution but differ in some aspect of their three-dimensional structure. Stereoisomers are either enantiomers or diastereomers.

Stereospecific. A term describing a reaction in which there is a specific relationship between the configuration of the reactants and the configuration of the products.

Steric effect. A chemical effect on reactivity resulting from nonbonded interactions.

Steroids. A biochemically important group of compounds with a ring skeleton having three 6-membered rings and a 5-membered ring. Examples are cholesterol, cortisone, and other hormones.

Straight-chain. A term describing molecule for which all the carbon atoms are connected sequentially; *i.e.*, in which it is possible to encounter all atoms of the carbon chain by proceeding consecutively from one to the next without backtracking.

Structural formula. A representation of a molecule that denotes the constitution (and sometimes the three-dimensional structure as well).

Substituent. Any atom or group that is bonded to a particular (carbon) atom is a substituent of that (carbon) atom.

Substituent effect. The change in structure or reactivity that results when one substituent is replaced by another.

Substrate. A molecule that undergoes reaction, *e.g.*, the compound R–X that undergoes nucleophilic substitution to yield a product in which a nucleophile has replaced the leaving group, X.

Sugar. Carbohydrate, saccharide. In nonchemical usage the term sugar denotes table sugar (sucrose).

Symmetry axis. An axis passing through the center of a molecule such that rotation about the axis (by some angle of less than 360°) generates a structure which is indistinguishable from the original.

Synthetic intermediate. A compound, produced at an intermediate stage of an organic synthesis, that is converted to the final product in subsequent reactions.

Target molecule. The compound that is the intermediate goal in a synthetic sequence.

Terminal alkyne. An alkyne in which the triple bond is located at the end of the carbon chain (*i.e.*, C–C≡C–H).

Terpenes. A class of hydrocarbons found in plants that have structures composed of C_5H_8 subunits.

Tertiary. A term describing a carbon atom that is bonded to exactly three other carbon atoms. When used to describe an alkyl group, it describes the carbon from which a hydrogen has been removed to generate the group (*i.e.*, the carbon that is the point of attachment for the group).

Tetrahedral. A term describing the (approximate) geometry of a saturated carbon atom. In methane the four hydrogen substituents on the carbon atom of methane define the corners of a tetrahedron, and the H–C–H angle is 109.5° (called the tetrahedral angle).

Thermodynamic control. Reaction conditions such that the amount of each product obtained results from establishment of an equilibrium between them. The equilibrium constant is related to the change in the thermodynamic quantity, free energy, by the relationship: $\Delta G = -2.3\ RT \log K$

Trans. A term describing a relationship between two substituents on a ring or on a double bond in which the substituents (which must be on different carbon atoms) are on opposite sides of the ring or double bond.

Transition state. An energy-maximum on a reaction profile; the energy difference between the transition state and the reactant defines the activation energy, E_a.

Trigonal. The geometry of a trisubstituted atom (such as a doubly bonded carbon) in which the three substituents and the central atom all lie in a plane; the three substituents are separated by angles of approximately 120°.

Unimolecular. A term describing a reaction step that involves only a single molecule of reactant.

Unsaturated. Describes a molecule that contains atoms with less than the maximum number of substituents; for example, an alkene (for which the doubly bonded carbon atoms have only three substituents).

Unshared pair. A pair of valence level electrons that is not involved in bonding; nonbonding electrons; a lone pair of electrons.

Upfield. A term describing transitions of higher energy in NMR spectroscopy; it indicates peaks to the right of some point of reference.

Valence atomic orbital. An atomic orbital of the level that is occupied by electrons involved in bonding (*i.e.*, 1s for hydrogen and 2s, 2p for carbon and other second-row elements).

Van der Waals forces. Weak interactions between molecules (or between atoms in the same molecule); the net result of attractions and repulsions for all electrons and nuclei.

Vibrational energy. The kinetic energy of a molecule that results from vibrational motion, *i.e.*, oscillating motion of one part of the molecule relative to the remainder.

Vicinal. A term describing the relationship between two substituents that are bonded to adjacent carbon atoms.

Vinyl ether. An ether in which one of the two groups bonded to oxygen is an alkenyl group; also called an *enol ether*.

Visible light. That region of the electromagnetic spectrum that can be detected by the human eye; the region with wavelengths that are shorter than those of the infrared region but longer than those of the ultraviolet region.

Yield. The quantity of material actually obtained from a reaction; usually expressed as a percentage of the theoretical yield (which is the quantity predicted for complete reaction according to the stoichiometry of the balanced equation).

Ylide. A molecule having a negative charge on a carbon atom and a positive charge on an adjacent heteroatom such as phosphorus or sulfur; a double bond between the carbon and the heteroatom can be drawn only if more than 8 electrons are considered to occupy the valence shell of the heteroatom.

Z. Zusammen (the German word for together); defines the configuration of a double bond in which the two substituents of highest priority are on the same side of the double bond.

TABLE 1.4 Reactions of Alkanes

Reaction	Comments
1. Combustion (oxidation) $C_nH_{2n+2} \xrightarrow{O_2} CO_2 + H_2O$	Section 1.10. Used as an energy source; not applicable to synthesis.
2. Halogenation $C_nH_{2n+2} \xrightarrow[\text{heat or light}]{X_2} C_nH_{2n+1}X$ isomers	Section 1.10. X may be Cl or Br. Products typically consist of the various possible constitutional isomers together with polyhalo derivatives. In most cases not a useful reaction for laboratory synthesis.
3. Catalytic cracking, reforming (a) Hydrocracking $C_nH_{2n+2} \xrightarrow[\text{catalyst}]{H_2}$ smaller alkanes (b) Reforming $C_nH_{2n+2} \xrightarrow[\text{catalyst}]{} C_nH_{2n+2}$ isomers	Section 1.10. High-temperature reactions used in petroleum refining; not applicable to laboratory synthesis.

TABLE 1.5 Preparation of Alkanes

Reaction	Comments
1. Catalytic cracking, reforming	Section 1.10. Industrial processes (see Reactions of Alkanes)
2. Reduction of alkyl halides $R{-}X \xrightarrow{Zn/HCl} R{-}H$	Section 1.11. X may be Cl, Br, or I. A useful procedure for replacement of halogen by hydrogen. (Other metals can sometimes be used.)
3. Coupling of alkyl halides	Section 1.11.
(a) Wurtz reaction $R{-}X \xrightarrow{Na} R{-}R$	Affords alkanes derived from two identical alkyl groups only. Not very useful for organic synthesis.
(b) With organocopper reagents $R{-}X \xrightarrow{Li} R{-}Li \xrightarrow{CuI} R_2CuLi \xrightarrow{R'CH_2X} R{-}CH_2{-}R'$	Can be used to prepare a wide variety of alkanes. Must use primary alkyl halide in the reaction with the organocopper reagent. A good synthetic reaction.
4. Reduction of alkenes	Section 5.5

TABLE 5.1 Reactions of Alkenes

Reaction	Comments
1. Reduction: catalytic hydrogenation $>C{=}C< \xrightarrow[\text{cat.}]{H_2} H{-}\overset{\vert}{\underset{\vert}{C}}{-}\overset{\vert}{\underset{\vert}{C}}{-}H$	Section 5.6. Either Pt or Pd/C is a satisfactory catalyst.
2. Addition of carbenes $>C{=}C< \xrightarrow{CH_2I_2,\ ZnCu}$ cyclopropane (CH_2 bridging the two C)	Section 5.7. Simmons-Smith reaction.
$>C{=}C< \xrightarrow{CHX_3,\ \text{base}}$ dihalocyclopropane (CX_2 bridging the two C)	X is usually Cl or Br; KOH and KOtBu are common bases.
3. Polar addition reactions (a) Acid-mediated additions $>C{=}C< \xrightarrow[H^+]{HX} H{-}\overset{\vert}{\underset{\vert}{C}}{-}\overset{\vert}{\underset{\vert}{C}}{-}X$	Section 5.8. Carbocation rearrangements may occur. No H^+ catalyst needed if HX itself a strong acid.
(b) Halogenation: inert solvent $>C{=}C< \xrightarrow{X_2} X{-}\overset{\vert}{\underset{\vert}{C}}{-}\overset{\vert}{\underset{\vert}{C}}{-}X$	Carbocation rearrangements may occur.
(c) Halogenation: nucleophilic solvent $>C{=}C< \xrightarrow[SOH]{X_2} X{-}\overset{\vert}{\underset{\vert}{C}}{-}\overset{\vert}{\underset{\vert}{C}}{-}OS$	Carbocation rearrangements may occur. NBS/H_2O is a good synthetic procedure for adding Br, OH.
4. Other addition reactions	
(a) Hydration via hydroboration $>C{=}C< \xrightarrow[2.\ NaOH,\ H_2O_2]{1.\ BH_3} H{-}\overset{\vert}{\underset{\vert}{C}}{-}\overset{\vert}{\underset{\vert}{C}}{-}OH$	Section 5.11. Yields the less substituted alcohol.
(b) Hydration via oxymercuration $>C{=}C< \xrightarrow[2.\ NaBH_4]{1.\ Hg(O\overset{O}{\overset{\Vert}{C}}CH_3)_2,\ H_2O} H{-}\overset{\vert}{\underset{\vert}{C}}{-}\overset{\vert}{\underset{\vert}{C}}{-}OH$	Section 5.11. Yields the more substituted alcohol. Use of alcohols (ROH) as solvent in place of water allows synthesis of ether by addition of H, OR.
(c) Free-radical addition of HBr $>C{=}C< \xrightarrow[\text{peroxides}]{HBr} H{-}\overset{\vert}{\underset{\vert}{C}}{-}\overset{\vert}{\underset{\vert}{C}}{-}Br$	Section 5.10. Affords the less substituted bromide. Not applicable to other hydrogen halides.

TABLE 5.1 Reactions of Alkenes (continued)

Reaction	Comments
5. Oxidation (a) Addition of OH groups $>C=C< \xrightarrow[\text{or alkaline } KMnO_4]{OsO_4,\ NaClO_3} HO-\overset{\vert}{\underset{\vert}{C}}-\overset{\vert}{\underset{\vert}{C}}-OH$	Section 5.12. Formation of glycols. The high toxicity of osmium compounds makes potassium permanganate the reagent of choice.
(b) Epoxidation $>C=C< \xrightarrow{CH_3-\overset{O}{\overset{\Vert}{C}}-OOH}$ epoxide $\left(>C\overset{O}{—}C<\right)$	Hydrolysis of the epoxide with aqueous acid provides an alternative preparation of glycols.
(c) Ozonolysis $>C=C< \xrightarrow[2.\ Zn]{1.\ O_3} >C=O + O=C<$	Oxidative cleavage yields aldehydes and ketones with carbon skeletons corresponding to the two "halves" of the alkene. Good for structure proof.
(d) Allylic oxidation $>C=C<\!-\!CH< \xrightarrow{NBS} >C=C<\!-\!C(Br)<$	Section 8.4.

TABLE 5.2 Preparation of Alkenes

Reaction	Comments
Functional Group Modification: No Change in Carbon Skeleton	
1. Base-mediated elimination $H-\overset{\vert}{\underset{\vert}{C}}-\overset{\vert}{\underset{\vert}{C}}-X \xrightarrow{\text{base}} >C=C<$	Section 5.13. X must be a good leaving group, i.e., HX must be a strong acid. Common bases are sodium and potassium salts of water and of alcohols.
2. Dehydration of alcohols $H-\overset{\vert}{\underset{\vert}{C}}-\overset{\vert}{\underset{\vert}{C}}-OH \xrightarrow{\text{acid}} >C=C<$	Section 5.13. Normally use conc. sulfuric or phosphoric acid. Rearrangements should be expected.

TABLE 5.2 Preparation of Alkenes (continued)

Reaction	Comments
Functional Group Modification: No Change in Carbon Skeleton	
3. Dehalogenation $\mathrm{Br{-}\overset{\vert}{\underset{\vert}{C}}{-}\overset{\vert}{\underset{\vert}{C}}{-}Br} \xrightarrow{Zn} \mathrm{>C{=}C<}$	Section 5.13. The dibromide is prepared from the alkene; the two-step sequence can be useful in purification.
4. Pyrolysis of esters $\mathrm{H{-}\overset{\vert}{\underset{\vert}{C}}{-}\overset{\vert}{\underset{\vert}{C}}{-}O\overset{O}{\overset{\Vert}{C}}R} \xrightarrow{\text{pyrolysis}} \mathrm{>C{=}C<}$	Section 6.9.
5. Pyrolysis of amine oxides $\mathrm{H{-}\overset{\vert}{\underset{\vert}{C}}{-}\overset{\vert}{\underset{\vert}{C}}{-}\overset{O^-}{\overset{\vert}{\underset{\underset{R}{\vert}}{N^+}}}{-}R} \xrightarrow{\text{pyrolysis}} \mathrm{>C{=}C<}$	Sections 6.9, 17.5.
6. Reduction of alkynes $\mathrm{{-}C{\equiv}C{-}} \longrightarrow \mathrm{{-}CH{=}CH{-}}$	Section 7.6.
Reaction	**Comments**
Reactions Yielding Alkenes via Carbon-Carbon Bond Formation	
7. Reactions of organocuprates	Section 5.14 (Cf. Section 1.11)
(a) With a *remote* double bond $\mathrm{R'{-}X} \xrightarrow{R_2CuLi} \mathrm{R'{-}R}$	Either R or R′ can contain a remote double bond, i.e., neither Cu nor X is attached directly to unsaturated carbon. See Section 1.11 for other limitations.
(b) With an *alkenyl* derivative $\mathrm{>C{=}C<^{X}} \xrightarrow{R_2CuLi} \mathrm{>C{=}C<^{R}}$ or $\mathrm{R{-}X} \xrightarrow{(>C=C<)_2CuLi} \mathrm{>C{=}C<^{R}}$	The reaction proceeds with retention of stereochemistry of the double bond.
8. Wittig reaction $\mathrm{>C{=}O} \xrightarrow{(C_6H_5)_3P=C(R)(R')} \mathrm{>C{=}C(R)(R')}$	Section 13.9.

TABLE 7.2 Reactions of Alkynes

Reactions	Comments
1. Salt formation	Section 7.4. Terminal alkynes only.
(a) Alkali metal salts	
$R{-}C{\equiv}C{-}H \xrightarrow{NaNH_2} R{-}C{\equiv}C^{-}Na^{+}$ $R{-}C{\equiv}C{-}H \xrightarrow{LiNH_2} R{-}C{\equiv}C^{-}Li^{+}$	Used for alkylation to form new carbon-carbon bonds.
(b) Copper and silver salts	
$R{-}C{\equiv}C{-}H \xrightarrow[\text{aqueous } NH_3]{CuCl} R{-}C{\equiv}C{-}Cu$ $R{-}C{\equiv}C{-}H \xrightarrow[\text{alcohol}]{AgNO_3} R{-}C{\equiv}C{-}Ag$	Used to identify terminal alkynes. Precipitate indicates a terminal alkyne; internal alkynes give no reaction.
2. Hydration	Section 7.5.
(a) Direct hydration	
$-C{\equiv}C- \xrightarrow[H_2SO_4,\ HgSO_4]{H_2O} -\overset{O}{\overset{\Vert}{C}}-CH_2-$	Terminal alkynes yield ketones; unsymmetrical internal alkynes yield mixtures of the two possible ketones.
(b) Via hydroboration	
$-C{\equiv}C- \xrightarrow[2.\ H_2O_2,\ NaOH]{1.\ (CH_3-\underset{}{\overset{CH_3}{\overset{\vert}{C}H}}-\overset{CH_3}{\overset{\vert}{C}H}\text{)}_2BH} -\overset{O}{\overset{\Vert}{C}}-CH_2-$ $-C{\equiv}C- \xrightarrow[2.\ H_2O_2,\ NaOH]{1.\ \text{catecholborane (benzo-fused } O_2BH)} -\overset{O}{\overset{\Vert}{C}}-CH_2-$	Terminal alkynes yield aldehydes; internal alkynes yield a predominance of the ketone with oxygen on the less hindered carbon.
3. Reduction	Section 7.6.
(a) Catalytic hydrogenation	
$-C{\equiv}C- \xrightarrow[\text{catalyst}]{H_2} -CH_2-CH_2-$	Reduction proceeds to the alkane unless a deactivated catalyst is used.
$-C{\equiv}C- \xrightarrow[Pd-BaSO_4]{H_2} \overset{H}{\diagdown}C{=}C\overset{H}{\diagup}$	Reaction stops at the alkene with a deactivated catalyst. *Syn* addition.

TABLE 7.2 Reactions of Alkynes (continued)

Reactions	Comments
(b) Via hydroboration	
$-C{\equiv}C- \xrightarrow[\text{2. } CH_3CO_2H]{\text{1. } (CH_3-CH(CH_3)-CH(CH_3)-)_2BH}$ (H)C=C(H)	A good synthetic method for laboratory use. *Syn* addition.
$-C{\equiv}C- \xrightarrow[\text{2. } CH_3CO_2H]{\text{1. catecholborane (BH)}}$ (H)C=C(H)	
(c) Metal reduction	
$-C{\equiv}C- \xrightarrow[NH_3]{\text{Na or Li}}$ (H)C=C(H)	A good synthetic method for laboratory use. *Anti* addition.
4. Addition of hydrogen halides	Section 7.7.
(a) Direct addition	
$-C{\equiv}C- \xrightarrow{HX}$ (H)C=C(X) $\longrightarrow$ $-C(H)(H)-C(X)(X)-$	Reaction proceeds via the more stable cation. (Addition of HBr proceeds by a radical mechanism if peroxides are present.) Difficult to stop at the alkenyl halide.
(b) Via hydroboration with catecholborane	
$R-C{\equiv}C-H \xrightarrow[\text{2. } H_2O;\ \text{3. } I_2,\ NaOH]{\text{1 catecholborane (BH)}}$ (H)(R)C=C(I)(H)	Terminal alkynes only. Overall *syn* addition of HI.
$-C{\equiv}C- \xrightarrow[\text{2. } Br_2;\ \text{3. } NaOCH_3,\ CH_3OH]{\text{1. catecholborane (BH)}}$ (H)C=C(Br)	Terminal alkynes yield 1-bromo-1-alkenes. Internal alkynes yield predominantly the product with bromine on the less hindered carbon atom. Overall *anti* addition of HBr.

TABLE 7.3 Preparation of Alkynes

Reaction	**Comments**
Functional Group Modification: No Change in Carbon Skeleton	
1. Elimination of HX	Section 7.8. Many bases will work: $NaNH_2/NH_3$, KOH/alcohol, KOtBu/DMSO, etc.
(a) From 1,2-dihalides	
$-\overset{H}{\underset{X}{C}}-\overset{H}{\underset{X}{C}}- \xrightarrow{\text{base}} -C{\equiv}C-$	The halides are available from addition of X_2 to the alkene.
(b) From 1,1-dichlorides	
$-\overset{H}{\underset{H}{C}}-\overset{Cl}{\underset{Cl}{C}}- \xrightarrow{\text{base}} -C{\equiv}C-$	The dichlorides are available from treatment of the aldehyde or ketone with PCl_5.
Reactions Yielding Alkynes via Carbon-Carbon Bond Formation	
2. Alkylation of sodium acetylides	Section 7.10.
$-C{\equiv}C-H \xrightarrow{NaNH_2} -C{\equiv}C^-Na^+ \xrightarrow{RX} -C{\equiv}C-R$	RX must be a primary alkyl halide.

TABLE 11.2 Reactions of Alcohols

Reaction	Comments
1. Acid-base reactions	Section 11.1
$R{-}OH \overset{H^+}{\rightleftharpoons} R{-}OH_2^+$ $R{-}OH \overset{base}{\rightleftharpoons} R{-}O^-$	Analogous to the reactions of water
2. Reaction with active metals	
$R{-}OH \xrightarrow{Na} R{-}O^-\ Na^+ + H_2$	Analogous to the reactions of water
3. Alkylation	Section 11.8
$R{-}\ddot{\underset{..}{O}}{-}H \xrightarrow{CH_3{-}X} R{-}\overset{+}{\underset{..}{O}}(CH_3){-}H$	Discussed more fully in Chapter 12
$R{-}O^- \xrightarrow{CH_3{-}X} R{-}O{-}CH_3$	

TABLE 11.2 Reactions of Alcohols (continued)

Reaction	Comments
4. Acylation	Section 11.8
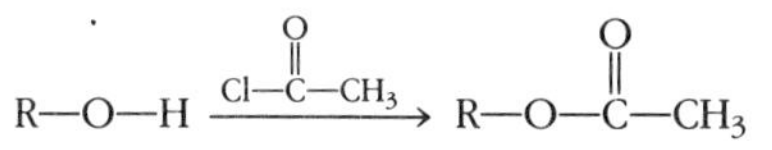 $R{-}O{-}H \xrightarrow{Cl-C(=O)-CH_3} R{-}O{-}C(=O){-}CH_3$	Discussed more fully in Chapter 16
5. Formation of sulfonate esters	Section 11.9
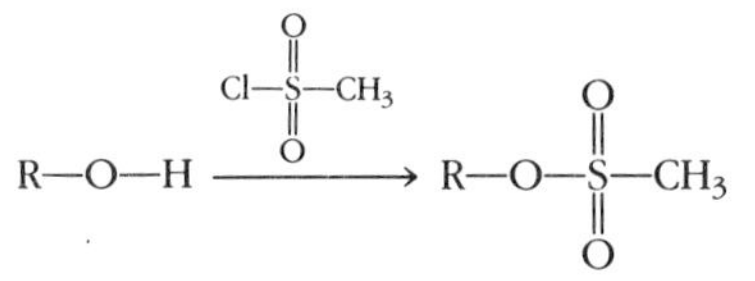 $R{-}O{-}H \xrightarrow{Cl-S(=O)_2-CH_3} R{-}O{-}S(=O)_2{-}CH_3$	The reaction is usually carried out in pyridine.
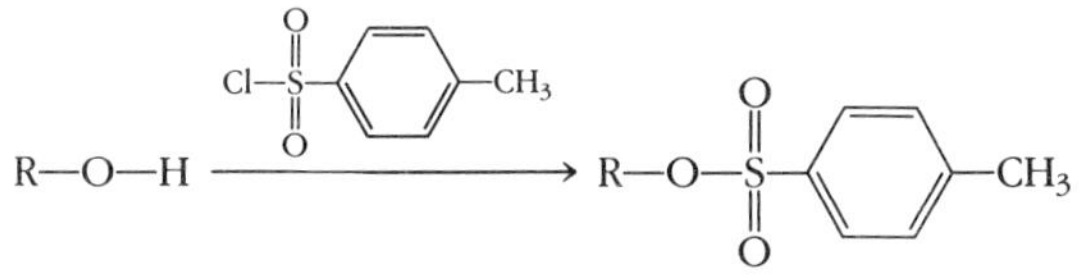 $R{-}O{-}H \xrightarrow{Cl-S(=O)_2-C_6H_4-CH_3} R{-}O{-}S(=O)_2{-}C_6H_4{-}CH_3$	
6. Oxidation of alcohols	Section 11.3
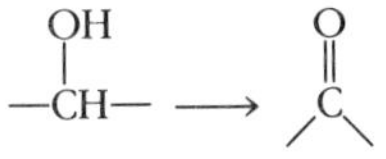 $-CH(OH)- \longrightarrow >C{=}O$	Industrial oxidations may use O_2 and a metal catalyst. Typical laboratory oxidizing agents are $KMnO_4$ and Cr(VI) reagents such as CrO_3, $Na_2Cr_2O_7$, H_2CrO_4, and CrO_3-pyridine.
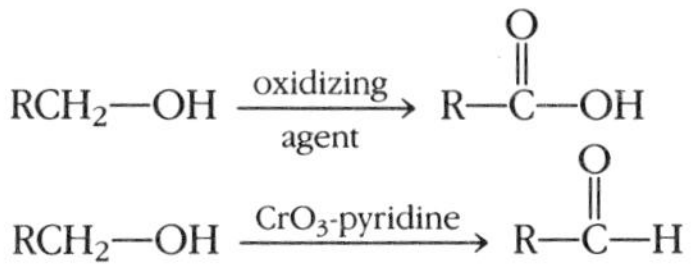 $RCH_2{-}OH \xrightarrow[\text{agent}]{\text{oxidizing}} R{-}C(=O){-}OH$ $RCH_2{-}OH \xrightarrow{CrO_3\text{-pyridine}} R{-}C(=O){-}H$	Primary alcohols usually yield carboxylic acids. If the CrO_3-pyridine reagent is used, the oxidation stops at the aldehyde stage.
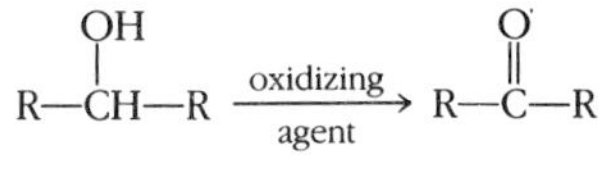 $R{-}CH(OH){-}R \xrightarrow[\text{agent}]{\text{oxidizing}} R{-}C(=O){-}R$	Secondary alcohols yield ketones (including oxidation with CrO_3-pyridine).
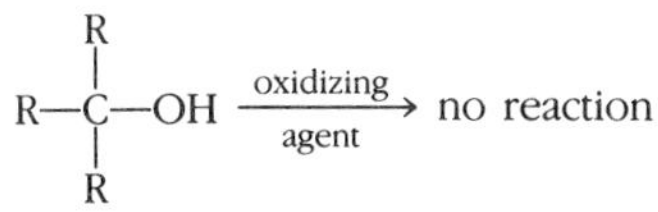 $R{-}C(R)_2{-}OH \xrightarrow[\text{agent}]{\text{oxidizing}}$ no reaction	Tertiary alcohols do not react under conditions used to oxidize primary and secondary alcohols.

TABLE 11.2 Reactions of Alcohols (continued)

Reaction	Comments
7. Conversion to halides (X = Cl, Br)	Section 11.9
R—OH ⟶ R—X	Sometimes conc. HX can be used, but rearrangements may result. Superior laboratory reagents are PCl_3, PCl_5, $POCl_3$, $SOCl_2$, and PBr_3.
8. Dehydration	Section 5.13
$-CH-C-OH \xrightarrow{\text{conc. } H_2SO_4} >C=C<$	Acid-catalyzed rearrangements may result.
$-CH-C-OH \longrightarrow -CH-C-Cl \xrightarrow{\text{base}} >C=C<$ $-CH-C-OH \longrightarrow -CH-C-OTs \xrightarrow{\text{base}} >C=C<$	Alternatively, the alcohol can first be converted to a halide or a sulfonate ester. Elimination can then be carried out under basic conditions (Section 11.9).

TABLE 11.3 Preparation of Alcohols

Reaction	Comments
Functional Group Modification, No Change in Carbon Skeleton	
1. Hydration of alkenes	
(a) Acid-catalyzed addition of H_2O	Section 5.8
$>C=C< \xrightarrow[H^+]{H_2O} -CH-C-OH$	May involve acid-catalyzed rearrangements
(b) Hydroboration-oxidation	Section 5.11
$>C=C< \xrightarrow[\text{2. } H_2O_2,\text{ NaOH}]{\text{1. } BH_3} -CH-C-OH$	No rearrangements. The OH group is introduced onto the less substituted carbon atom. A *syn* addition.

TABLE 11.3 Preparation of Alcohols (continued)

Reaction	Comments
Functional Group Modification, No Change in Carbon Skeleton	
(c) Oxymercuration-demercuration	Section 5.11
$>C{=}C< \xrightarrow[\text{2. NaBH}_4]{\text{1. Hg(OCCH}_3)_2 \text{ (C=O)}} -CH-C-OH$	No rearrangements. The OH group is introduced onto the more substituted carbon atom.
2. Oxidation of alkenes: glycol formation	Section 5.12
$>C{=}C< \xrightarrow[\text{or alkaline KMnO}_4]{\text{OsO}_4\text{, NaClO}_3} HO-C-C-OH$	A *syn* addition
3. Allylic oxidation	Section 8.4
$>C{=}C<(CH-) \xrightarrow{\text{SeO}_2} >C{=}C<(C-OH)$	
4. Reduction of carbonyl compounds	Section 11.4
(a) Catalytic hydrogenation	
$>C{=}O \xrightarrow[\text{catalyst}]{\text{H}_2} -CH(OH)-$	Requires high pressures. Other multiple bonds may be reduced preferentially.
(b) Dissolving metal reactions	
$>C{=}O \xrightarrow[\text{CH}_3\text{CH}_2\text{OH}]{\text{Na}} -CH(OH)-$	Most commonly done with sodium-ethanol but other metals and other proton sources can also be used.
$-C(=O)OR \xrightarrow[\text{CH}_3\text{CH}_2\text{OH}]{\text{Na}} -CH_2OH$	Aldehydes, ketones, and esters are all reduced to alcohols.

TABLE 11.3 Preparation of Alcohols (continued)

Reaction	Comments
Functional Group Modification, No Change in Carbon Skeleton	
(c) Metal hydride reductions	
$>C=O \xrightarrow[CH_3CH_2OH]{NaBH_4} -\overset{OH}{\underset{\vert}{C}}-H$	Sodium borohydride reduces aldehydes and ketones to alcohols. Carboxylic acids and esters are not reduced.
$>C=O \xrightarrow[2.\ H_2O]{1.\ LiAlH_4} -\overset{OH}{\underset{\vert}{C}}H$	Aldehydes, ketones, carboxylic acids, and esters are all reduced to alcohols with $LiAlH_4$.
$-\overset{O}{\overset{\Vert}{C}}-OR \xrightarrow[2.\ H_2O]{1.\ LiAlH_4} -CH_2OH$	
5. Reduction of epoxides	Section 11.6
$-\underset{\vert}{C}\overset{O}{-}\underset{\vert}{C}- \xrightarrow[2.\ H_2]{1.\ LiAlH_4} H-\underset{\vert}{\overset{\vert}{C}}-\underset{\vert}{\overset{\vert}{C}}-OH$	
6. Hydrolysis of epoxides: glycol formation	Section 5.12
$-\underset{\vert}{C}\overset{O}{-}\underset{\vert}{C}- \xrightarrow[H^+]{H_2O} HO-\underset{\vert}{\overset{\vert}{C}}-\underset{\vert}{\overset{\vert}{C}}-OH$	
7. Hydrolysis of alkyl halides or tosylates (X = Cl, Br, OTs)	Discussed in Chapter 12
$R-X \xrightarrow{H_2O} R-OH$	
8. Hydrolysis of esters	Discussed in Section 16.6
$R-\overset{O}{\overset{\Vert}{C}}-OR' \xrightarrow[\text{acid or base}]{H_2O} R-\overset{O}{\overset{\Vert}{C}}OH + R'OH$	
9. Hydrolysis of ethers	Discussed in Section 15.3
$R-O-R' \xrightarrow[acid]{H_2O} R-OH + R'-OH$	

TABLE 11.3 Preparation of Alcohols (continued)

Reaction	Comments
Reactions Yielding Alcohols via Carbon-Carbon Bond Formation	
10. Reaction of organometallic reagents with carbonyl compounds	Section 11.5
(a) Aldehydes and ketones	
$>C{=}O \xrightarrow[2.\ H_2O]{1.\ RMgX\ \text{or}\ RLi} -\overset{OH}{\underset{\vert}{\overset{\vert}{C}}}-R$	The halogen (X) of the Grignard reagent can be Cl, Br, I.
$>C{=}O \xrightarrow[2.\ H_2O]{1.\ NaC{\equiv}CR\ \text{or}\ LiC{\equiv}Cr} -\overset{OH}{\underset{\vert}{\overset{\vert}{C}}}-C{\equiv}CR$	
(b) Esters	
$-\overset{O}{\overset{\Vert}{C}}-OCH_3 \xrightarrow[2.\ H_2O]{1.\ RMgX\ \text{or}\ RLi} -\overset{OH}{\underset{R}{\overset{\vert}{\underset{\vert}{C}}}}-R$	Two identical alkyl groups are introduced in the reaction of an ester.
11. Reaction of organometallic reagents with epoxides	Section 11.6
$-\overset{\vert}{C}\overset{O}{-}\overset{\vert}{C}- \xrightarrow[2.\ H_2O]{1.\ RMgX} -\overset{OH}{\overset{\vert}{C}}-\overset{\vert}{C}-R$	The alkyl group is introduced at the less substituted carbon atom of the epoxide. The halogen, X, of the Grignard reagent can be Cl, Br, I.
$-\overset{\vert}{C}\overset{O}{-}\overset{\vert}{C}- \xrightarrow[2.\ H_2O]{1.\ RLi\ \text{or}\ R_2CuLi} -\overset{OH}{\overset{\vert}{C}}-\overset{\vert}{C}-R$ $-\overset{\vert}{C}\overset{O}{-}\overset{\vert}{C}- \xrightarrow[2.\ H_2O]{1.\ LiC{\equiv}C{-}R\ \text{or}\ NaC{\equiv}CR} -\overset{OH}{\overset{\vert}{C}}-\overset{\vert}{C}-C{\equiv}CR$	Magnesium salts sometimes cause rearrangements with substituted epoxides. Lithium dialkylcuprates are superior reagents.
12. Condensation reactions of carbonyl compounds	Chapter 14
$>C{=}O + -\overset{\vert}{C}H-\overset{O}{\overset{\Vert}{C}}- \xrightarrow[\text{or base}]{\text{acid}} -\overset{OH}{\overset{\vert}{C}}-\overset{\vert}{C}-\overset{O}{\overset{\Vert}{C}}-$	

TABLE 13.5 Reactions of Aldehydes and Ketones

Reaction	Comments
1. Oxidation reactions	Section 13.11
(a) Oxidation of aldehydes $R\text{—}\overset{O}{\overset{\|}{C}}H \xrightarrow[\text{agent}]{\text{oxidizing}} R\text{—}\overset{O}{\overset{\|}{C}}OH$	A wide variety of oxidizing agents can be used: Cr(VI) and Mn(VII) are common. Reaction with Ag^+ to form Ag° is used as a test for aldehydes (Section 13.12).
(b) Oxidative cleavage $\text{—}CH_2\text{—}C\text{=}O \xrightarrow[\text{KMnO}_4]{\text{CrO}_3\text{ or}} \text{—}CO_2H \;\; \text{—}CO_2H$	Infrequently used as a laboratory method. Cyclic ketones yield dicarboxylic acids on vigorous oxidation.
(c) Baeyer-Villiger oxidation of ketones $R\text{—}\overset{O}{\overset{\|}{C}}\text{—}R' \xrightarrow{\text{peroxy acid}} R\text{—}\overset{O}{\overset{\|}{C}}\text{—}OR'$	A variety of peroxy acids can be used. The preference of alkyl group migration (R vs R′) is $3° > \text{aryl} > 2° > 1° > CH_3$. Aldehydes are usually oxidized to the corresponding carboxylic acids.
(d) Epoxidation of α,β-unsaturated carbonyl compounds $\text{C=C—}\overset{O}{\overset{\|}{C}}\text{—} \xrightarrow[\text{NaOH}]{H_2O_2} \text{—}\underset{O}{C\text{—}C}\text{—}\overset{O}{\overset{\|}{C}}\text{—}$	Isolated carbonyl groups and double bonds are unaffected by alkaline H_2O_2.
(e) Selenium dioxide oxidation $\text{—}CH_2\text{—}\overset{O}{\overset{\|}{C}}\text{—} \xrightarrow[\text{NaOH}]{SeO_2} \text{—}\overset{O}{\overset{\|}{C}}\text{—}\overset{O}{\overset{\|}{C}}\text{—}$	Formation of α-dicarbonyl compounds.
2. Reduction of aldehydes and ketones	Sections 13.11, 11.4
(a) Hydride reduction $\overset{O}{\overset{\|}{C}} \longrightarrow \text{—}\overset{OH}{\overset{\|}{C}}\text{—H}$	Both $LiAlH_4$ and $NaBH_4$ work well. (Note that alkali metal hydrides such as LiH and NaH function as bases rather than as reducing agents.)
(b) Catalytic hydrogenation $\overset{O}{\overset{\|}{C}} \xrightarrow[\text{catalyst}]{H_2} \text{—}\overset{OH}{\overset{\|}{C}}\text{—H}$	Carbon-carbon double bonds are reduced in preference to carbonyl groups.
(c) Dissolving metal reduction $\overset{O}{\overset{\|}{C}} \longrightarrow \text{—}\overset{OH}{\overset{\|}{C}}\text{—H}$	Common reagents are sodium or lithium in ammonia. (Sodium-alcohol is sometimes used with saturated ketones or aldehydes.)

TABLE 13.5 Reactions of Aldehydes and Ketones (continued)

Reaction	Comments
$>C{=}C{-}C({=}O){-} \longrightarrow >CH{-}CH{-}C({=}O){-}$	α,β-Unsaturated aldehydes and ketones can be partially reduced to the saturated carbonyl compound (Section 13.13).
(d) Deoxygenation of aldehydes and ketones	
$>C{=}O \xrightarrow[\text{KOH}]{NH_2NH_2} {-}CH_2{-}$	The Wolff-Kishner reduction is generally superior to the Clemmensen reduction. Both procedures can yield other products in the reaction with α,β-unsaturated aldehydes or ketones.
$>C{=}O \xrightarrow[\text{HCl}]{\text{Zn—Hg}} {-}CH_2{-}$	
$>C{=}O \xrightarrow[\text{2. } H_2\text{, Ni}]{\text{1. HS—}CH_2CH_2\text{—SH}} {-}CH_2{-}$	Desulfurization (Section 15.5).
3. Reaction with carbon nucleophiles	Carbon-carbon bond-forming reactions.
(a) Organometallic reagents	Sections 13.5 and 11.5 Carbonyl addition is favored with conjugated aldehydes and ketones.
$>C{=}O \xrightarrow[\text{2. } H_2O]{\text{1. RMgX or RLi}} {-}C(OH)(R){-}$	
(b) Lithium cuprates	Section 13.10. Saturated aldehydes and ketones are unreactive; α,β-unsaturated compounds undergo conjugate addition.
$>C{=}C{-}C({=}O){-} \xrightarrow[\text{2. } H_2O]{\text{1. } R_2CuLi} {-}C(R){-}CH{-}C({=}O){-}$	
(c) Phosphorus ylides	Section 13.8. Attack occurs preferentially at carbonyl carbon, even with α,β-unsaturated systems.
$>C{=}O \xrightarrow{(C_6H_5)_3P{=}C<} >C{=}C<$	
(d) Sulfur ylides	Section 13.8. Attack occurs preferentially at carbonyl carbon, even with α,β-unsaturated systems.
$>C{=}O \xrightarrow{(CH_3)_2S{=}C<} {-}C{-}C{-}$ (epoxide, O bridging both C)	

TABLE 13.5 Reactions of Aldehydes and Ketones (continued)

Reaction	Comments
(e) Hydrogen cyanide $>C{=}O \underset{\text{acid or base}}{\overset{HCN}{\rightleftharpoons}} -\overset{OH}{\underset{\vert}{C}}-CN$	Section 13.7 The reaction is reversible, and conjugate addition is favored with α,β-unsaturated aldehydes and ketones.
4. Reaction with heteroatom nucleophiles	Section 13.7. Reversible reactions catalyzed by either acid or base.
(a) Water $>C{=}O \overset{H_2O}{\rightleftharpoons} -\overset{OH}{\underset{\vert}{C}}-OH$	Equilibrium usually favors the free carbonyl compound.
(b) Alcohols $>C{=}O \overset{ROH}{\rightleftharpoons} -\overset{OH}{\underset{\vert}{C}}-OR$	Hemiacetal or hemiketal formation. Equilibrium usually favors the free carbonyl unless a ring is formed.
$>C{=}O \underset{H^+}{\overset{ROH}{\rightleftharpoons}} -\overset{OR}{\underset{\vert}{C}}-OR$	Ketal formation (Section 15.5).
(c) Hydrazine and primary amine derivatives $>C{=}O \overset{R-NH_2}{\rightleftharpoons} >C{=}N-R$	The products may be intermediates in other reactions (such as the Wolff-Kishner reduction, Section 13.11). Hydrazones are useful for characterization of aldehydes and ketones (Section 13.12).
(d) Secondary amines $>\overset{O}{\overset{\Vert}{C}}-\underset{\vert}{CH}- \overset{R_2NH}{\rightleftharpoons} >\overset{NR_2}{\overset{\vert}{C}}{=}\underset{\vert}{C}-$	Enamine formation (Section 17.8).

TABLE 13.5 Reactions of Aldehydes and Ketones (continued)

Reaction	Comments
(e) Thiols $\text{>C=O} \xrightleftharpoons{\text{RSH}} \text{—C(SR)—SR}$	Preparation of thioketals (Section 15.5).
5. Halogenation	Section 13.9.
(a) Acid-catalyzed $\text{>C(=O)—CH<} \xrightarrow[\text{H}^+]{\text{X}_2} \text{>C(=O)—C(X)<}$	Reaction via enol form. Can be used for *mono*halogenation.
(b) Base-catalyzed $\text{>C(=O)—CH<} \xrightarrow[\text{base}]{\text{X}_2} \text{>C(=O)—C(X)<}$	Reaction via enolate ion. Typically more than one halogen is introduced.
$\text{—C(=O)—CH}_3 \xrightarrow[^-\text{OH}]{\text{X}_2} \text{—C(=O)—O}^- + \text{HCX}_3$	Methyl ketones are cleaved (haloform reaction).
6. Alkylation of aldehydes and ketones $\text{>C(=O)—CH<} \xrightarrow[\text{2. RX}]{\text{1. base}} \text{>C(=O)—C(R)<}$	Section 14.3. Reaction via enolate ion. Carbon-carbon bond formation.
7. Condensation reactions $\text{>C(=O)—CH}_2\text{—} \xrightarrow[\text{base or acid}]{\text{>C=O}} \text{>C(=O)—C=C<}$	Sections 14.1, 14.2 Reaction via the enol or enolate ion. Carbon-carbon bond formation.

TABLE 13.6 Preparation of Aldehydes and Ketones

Reaction	Comments
1. Oxidation of alcohols	Sections 13.3, 11.3.
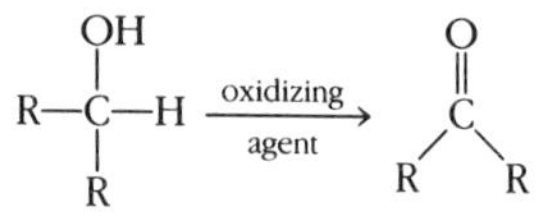 $R-C(OH)(R)-H \xrightarrow[\text{agent}]{\text{oxidizing}} R-C(=O)-R$	A variety of oxidizing agents can be used to convert 2° alcohols to ketones: CrO_3, $KMnO_4$, $K_2Cr_2O_7$, etc.
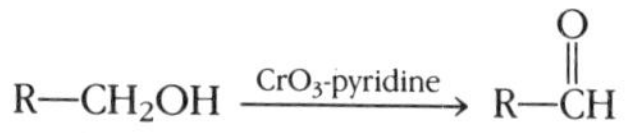 $R-CH_2OH \xrightarrow{CrO_3\text{-pyridine}} R-C(=O)H$	Oxidation of 1° alcohols can continue past the aldehyde state to give carboxylic acids with other, more reactive oxidizing agents.
2. Hydration of alkynes	Section 7.5.
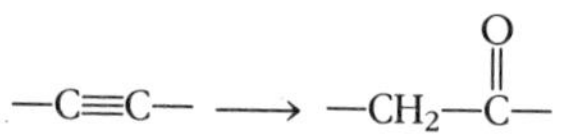 $-C\equiv C- \longrightarrow -CH_2-C(=O)-$	
3. Ozonolysis of alkenes	Sections 13.3 and 5.12.
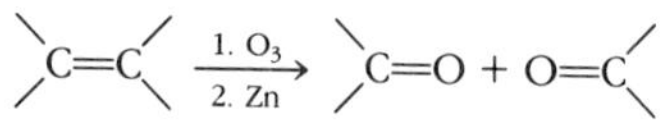 $>C=C< \xrightarrow[\text{2. Zn}]{\text{1. } O_3} >C=O + O=C<$	The intermediate ozonide can also be reduced with $(CH_3)_2S$.
4. Periodate cleavage of 1,2-diols	Section 13.3.
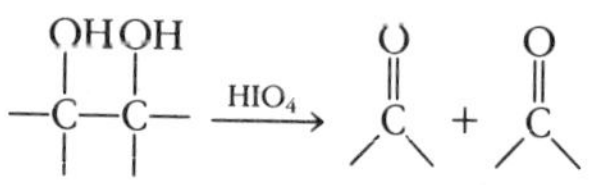 $-C(OH)-C(OH)- \xrightarrow{HIO_4} >C=O + O=C<$	
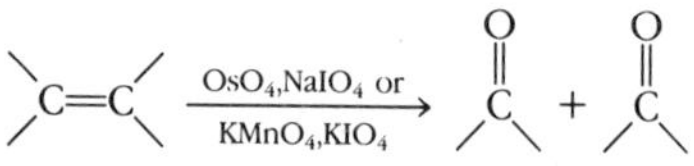 $>C=C< \xrightarrow[KMnO_4, KIO_4]{OsO_4, NaIO_4 \text{ or}} >C=O + O=C<$	Direct cleavage of an alkene via the diol without isolation.
5. Reduction of acid chlorides	Section 13.3, aldehyde synthesis.
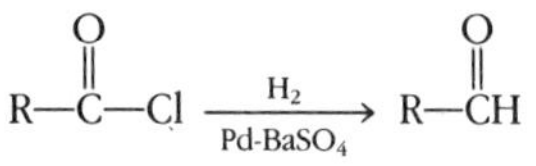 $R-C(=O)-Cl \xrightarrow[\text{Pd-BaSO}_4]{H_2} R-C(=O)H$	A deactivated catalyst is needed.
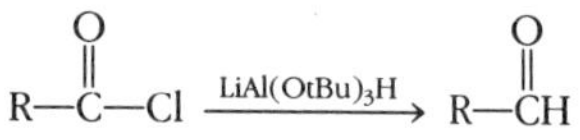 $R-C(=O)-Cl \xrightarrow{LiAl(OtBu)_3H} R-C(=O)H$	With other hydride reducing agents such as $LiAlH_4$ or $NaBH_4$ the reduction proceeds further to give the 1° alcohol.

TABLE 13.6 Preparation of Aldehydes and Ketones (continued)

Reaction	Comments
6. Hydrolysis of blocked carbonyl groups	
$\mathrm{X(Y)C{<}} \underset{}{\overset{H_2O}{\rightleftharpoons}} \mathrm{C{=}O}$	Reversal of carbonyl addition reactions (Section 13.7). X, Y = OR, OH, CN, etc.
$\mathrm{HO(X)C{<}} \xrightarrow{H_2O} \mathrm{C{=}O}$	Hydrolysis of acetals and ketals (Section 15.5).
$\mathrm{R{-}N{=}C(R){<}} \overset{H_2O}{\rightleftharpoons} \mathrm{C{=}O}$	Section 13.7.
7. Reaction of acid chlorides with organometallic reagents	Section 13.4, ketone synthesis.
$-\mathrm{C(=O)}-\mathrm{Cl} \xrightarrow{R_2Cd} -\mathrm{C(=O)}-\mathrm{R}$ $-\mathrm{C(=O)}-\mathrm{Cl} \xrightarrow{R_2CuLi} -\mathrm{C(=O)}-\mathrm{R}$	Organocadmium reagents have been largely replaced by organocuprates in modern organic chemistry. Other organometallic reagents react further to yield 3° alcohols.
8. Reaction of carboxylic acids with alkyllithium reagents	Section 13.4, ketone synthesis.
$-\mathrm{C(=O)}-\mathrm{OH} \xrightarrow{R-Li} -\mathrm{C(=O)}-\mathrm{R}$	Limited to alkyllithium reagents (primarily CH_3Li).
9. Reaction of nitriles with organometallic reagents	Section 13.4, ketone synthesis.
$\mathrm{R'}-\mathrm{C{\equiv}N} \xrightarrow[2.\ H_2O]{1.\ R-M} \mathrm{R'}-\mathrm{C(=O)}-\mathrm{R}$	Alkyllithium reagents are preferable.
10. Friedel-Crafts acylation	Section 18.3.
$\mathrm{Ar}-\mathrm{H} \xrightarrow[AlCl_3]{RC(=O)Cl} \mathrm{Ar}-\mathrm{C(=O)}-\mathrm{R}$	Synthesis of aromatic ketones.

TABLE 13.6 Preparation of Aldehydes and Ketones (continued)

Reaction	Comments
11. Acylation of enolate ions Synthesis of β-dicarbonyl compounds	Section 21.2.
enolate carbanion (C=O attached to C^-) + R—C(=O)—X ⟶ β-dicarbonyl (C(=O)—C—C(=O)—R)	Reaction via enol or enolate ion. Carbon-carbon bond formation.
12. Alkylation of aldehydes and ketones	Sections 14.3, 14.4.
C(=O)—CH— ⟶ (1. base; 2. RX) ⟶ C(=O)—C—R	Reaction via enolate ion. Carbon-carbon bond formation.
13. Condensation reactions	Sections 14.1, 14.2.
C(=O)—CH_2^- + C=O ⟶ (base or acid) ⟶ C(=O)—C=C—	Reaction via enol or enolate ion. Carbon-carbon bond formation.

TABLE 14.1 Reactions of Aldehydes and Ketones: Carbon-Carbon Bond Forming Reactions of Enols and Enolate Ions

Reaction	Comments
1. Aldol condensation	Sections 14.1, 14.2
C(=O)—CH— + C=O ⇌ (acid or base) C(=O)—C—C(OH)—	Base catalysis is more common. Typical bases are NaOH, $NaOCH_3$, and other salts of alcohols. The reaction is reversible.

TABLE 14.1 Reactions of Aldehydes and Ketones: Carbon-Carbon Bond Forming Reactions of Enols and Enolate Ions (continued)

Reaction	Comments
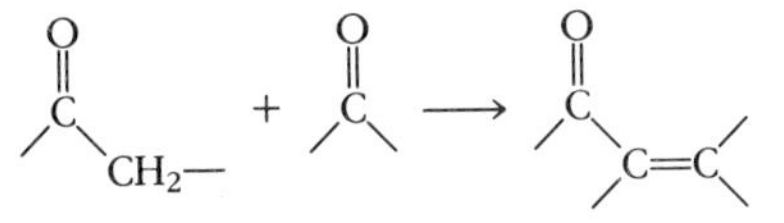 $\text{—C(=O)—CH}_2\text{—} + \text{>C=O} \longrightarrow \text{—C(=O)—C=C<}$	Dehydration often occurs during the reaction if there is a second α hydrogen.
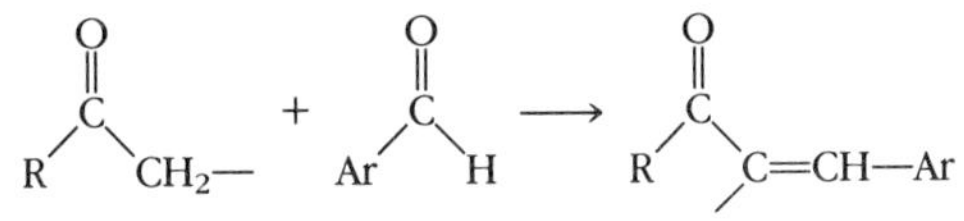 $\text{RC(=O)CH}_2\text{—} + \text{ArC(=O)H} \longrightarrow \text{RC(=O)C(—)=CH—Ar}$	Mixed aldol condensations work well when one compound is an aldehyde, especially an aromatic aldehyde. (The unhindered aldehyde carbonyl is highly susceptible to nucleophilic attack, and an aromatic aldehyde cannot form an enolate ion.)
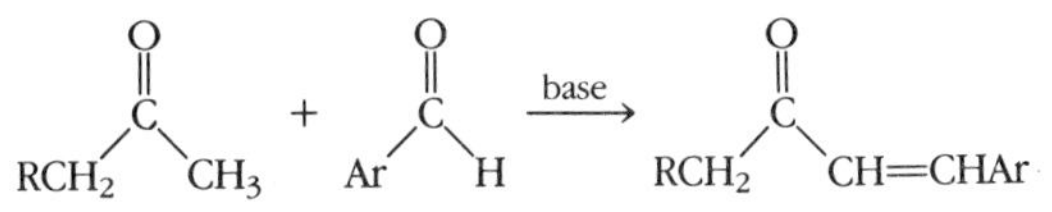 $\text{RCH}_2\text{C(=O)CH}_3 + \text{ArC(=O)H} \xrightarrow{\text{base}} \text{RCH}_2\text{C(=O)CH=CHAr}$	Under basic conditions an unsymmetrical ketone usually reacts at the less substituted α carbon. (Reaction at the more substituted α carbon is sometimes found with acid catalysis.)
2. Alkylation of enolate ions: **saturated carbonyl compounds**	Sections 14.3, 14.4
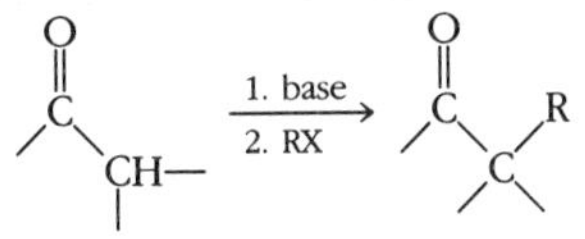 $\text{—C(=O)—CH<} \xrightarrow[\text{2. RX}]{\text{1. base}} \text{—C(=O)—C(R)<}$	A full equivalent of base is required. Side reactions (polyalkylation, aldol condensations, elimination) may occur. The alkyl halide (or sulfonate) must be methyl or primary. The base must be nonnucleophilic (KOtBu) or strong enough to cause complete conversion to the enolate ion [$NaNH_2$, LiH, NaH, $(C_6H_5)_3CK$, $LiN(iPr)_2$]. Lithium enolates lead to fewer side reactions.

TABLE 14.1 Reactions of Aldehydes and Ketones: Carbon-Carbon Bond Forming Reactions of Enols and Enolate Ions (continued)

Reaction	Comments
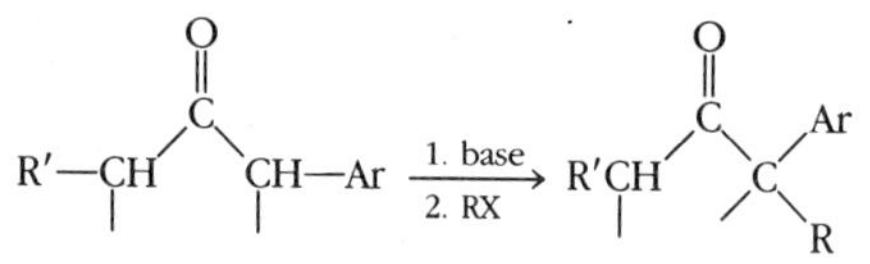 R′—CH(—)—C(=O)—CH(—)—Ar $\xrightarrow[\text{2. RX}]{\text{1. base}}$ R′CH(—)—C(=O)—C(Ar)(R)(—)	Formation of an enolate ion conjugated with an aryl substituent is favored.
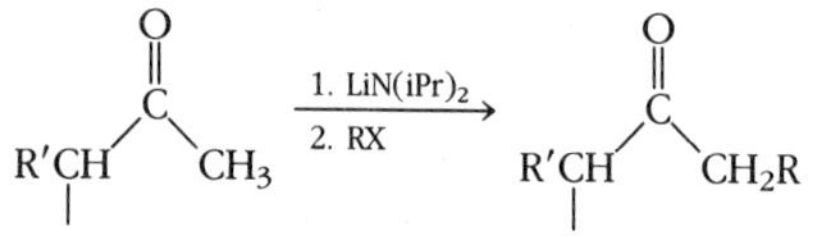 R′CH(—)—C(=O)—CH_3 $\xrightarrow[\text{2. RX}]{\text{1. LiN(iPr)}_2}$ R′CH(—)—C(=O)—CH_2R	Lithium diisopropylamide usually affords the less substituted enolate ion.
unsaturated carbonyl compounds	
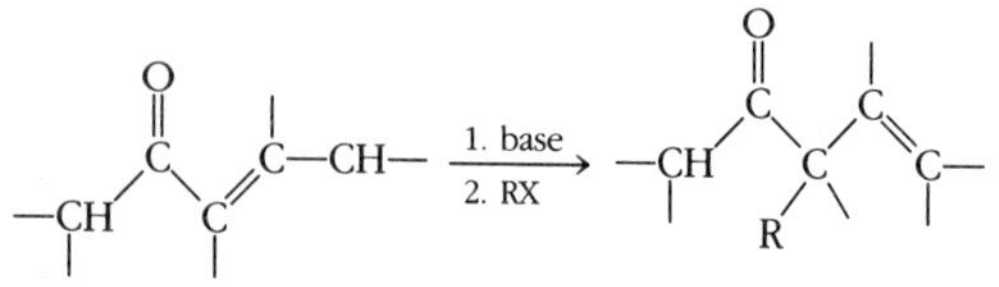 —CH(—)—C(=O)—C(—)=C(—)—CH— $\xrightarrow[\text{2. RX}]{\text{1. base}}$ —CH(—)—C(=O)—C(R)(—)—C(—)=C(—)—	In reversible reactions α,β-unsaturated carbonyl compounds afford the more stable enolate ion (by removal of a γ hydrogen), but this alkylates at the α position.
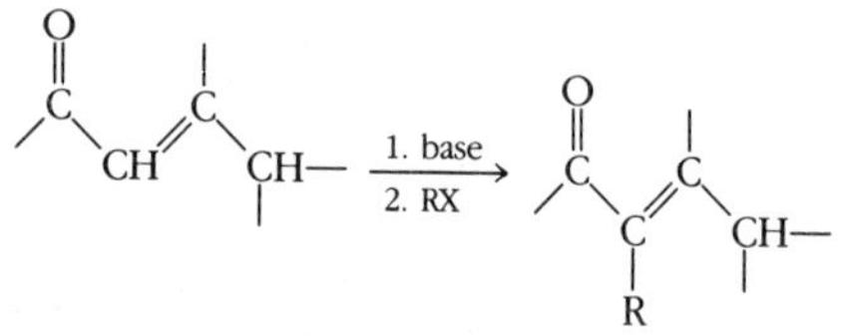 —C(=O)—CH=C(—)—CH(—)— $\xrightarrow[\text{2. RX}]{\text{1. base}}$ —C(=O)—C(R)=C(—)—CH(—)—	If the α carbon also has a hydrogen, the initial product usually rearranges to an α,β-unsaturated compound.
3. Reductive alkylation of an α,β-unsaturated carbonyl compound	Section 14.4
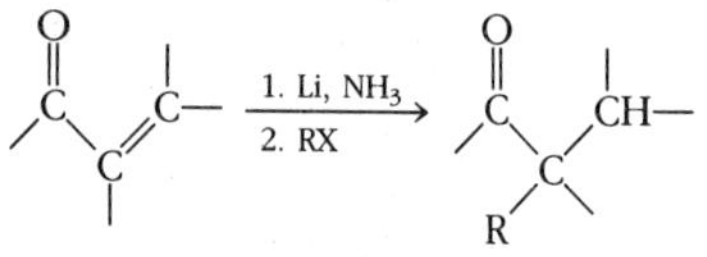 —C(=O)—C(—)=C(—)— $\xrightarrow[\text{2. RX}]{\text{1. Li, NH}_3}$ —C(=O)—C(R)(—)—CH(—)—	Reduction generates the enolate ion, which can then be alkylated. RX must be methyl or primary.

TABLE 14.1 Reactions of Aldehydes and Ketones: Carbon-Carbon Bond Forming Reactions of Enols and Enolate Ions (continued)

Reaction	Comments
4. Michael reaction	Section 14.5
α,β-unsaturated carbonyl (C=C–C=O) + –CH–C(=O)– $\overset{\text{base}}{\rightleftharpoons}$ –C(=O)–CH–C–C–C(=O)–	The reaction is reversible, but a catalytic quantity of base favors product formation. Typical bases are $NaOCH_3$, $NaOC_2H_5$. (Acid catalysis has been used on occasion.)
α,β-unsaturated carbonyl + CH_3–C(=O)–CH_2R $\xrightarrow{\text{base}}$ –C(=O)–CH–C–CH_2–C(=O)–CH_2R	Reaction via the less substituted enolate ion is typical.
α,β-unsaturated carbonyl + $ArCH_2$–C(=O)– $\longrightarrow$ –C(=O)–CH–C–CH(Ar)–C(=O)–	But aryl substitution favors enolate formation at that α carbon.

TABLE 14.2 Some Commonly Used Bases

Base	Conjugate Acid	K_a	Uses, Comments
NaOH, KOH, $R_4N^+OH^-$	H_2O	10^{-16}	Used to neutralize acidic reaction mixtures; catalysts in aldol condensations. Unsatisfactory if anhydrous or aprotic conditions are needed.
$NaOCH_3$, $NaOC_2H_5$	CH_3OH, C_2H_5OH	10^{-16}–10^{-18}	Used as catalysts in aldol condensations, Michael reactions. The alkoxide ions are weaker bases than enolate ions of aldehydes and ketones; only a small fraction of the carbonyl compound is converted to enolate at any instant.

TABLE 14.2 Some Commonly Used Bases (continued)

Base	Conjugate Acid	K_a	Uses, Comments
KOtBu	tBuOH	$\sim 10^{-18}$	Less nucleophilic than other alkoxide bases; can therefore be used in alkylation reactions of carbonyl compounds.
$(C_6H_5)_3CK$	$(C_6H_5)_3CH$	10^{-32}	More basic than enolate ions; produces essentially complete conversion of an aldehyde or ketone to the enolate ion.
$CH_3{-}S(=O){-}CH_2^-Na^+$	$CH_3{-}S(=O){-}CH_3$	10^{-35}	Prepared by reaction of DMSO with NaH. Often used to form ylides from the onium salts.
$NaNH_2$, $LiNH_2$	NH_3	10^{-35}	Often used with liquid ammonia as solvent. Converts terminal alkynes to the corresponding salts; nearly complete conversion of a ketone to the enolate.
$LiN(iPr)_2$	$HN(iPr)_2$	$\sim 10^{-36}$	Forms lithium enolate at the less substituted α carbon of a ketone. Sterically hindered, poor nucleophile. Excellent reagent for alkylation of a ketone.
LiH, NaH	H_2	(very small)	Powerful bases; evolution of H_2 allows complete conversion of ketone to enolate ion.
$LiAlH_4$	H_2	(very small)	Powerful base and powerful nucleophile. Converts OH groups to salts but acts as nucleophile at carbon. Rarely used as a base. (Note that $NaBH_4$ is a much weaker base, and it can be used in aqueous solution.)
C_6H_5Li, *tert*-BuLi, $CH_3CH_2CH_2CH_2Li$	Benzene, isobutane, butane	(very small)	The organolithium reagents are powerful bases and powerful nucleophiles. Rarely used to generate enolates but are used to form ylides from onium salts.
RMgBr, etc.	R—H	(very small)	Grignard reagents are strongly basic and strongly nucleophilic. Almost always react as nucleophiles with carbonyl compounds but have been used as bases to convert terminal alkynes and NH groups to the corresponding salts.

TABLE 15.2 Reactions of Ethers (Section 15.3)

Reaction	Comments
1. Acid-catalyzed cleavage	Ethers are unreactive under basic conditions.
(a) Dialkyl ethers	
$R{-}O{-}R \xrightarrow{HX} RX$	Reaction requires conc. HBr or HI at high temperatures.
$R{-}O{-}tBu \xrightarrow{HX} ROH + tBu{-}X$	Epoxides and *tert*-butyl ethers react under less stringent conditions.
(b) Aryl alkyl ethers	
$Ar{-}O{-}R \xrightarrow{HX} ArOH + R{-}X$	Aryl alkyl ethers yield the phenol.
(c) Acetals and ketals	
$>C(OR)_2 \xrightarrow[H^+]{H_2O} >C{=}O + 2\ ROH$	Hydrolysis occurs under very mild acidic conditions.
(d) Thioacetals and thioketals	
$R_2C(SCH_2CH_2S) \xrightarrow{H_2,\ Ni} R_2CH_2$	Hydrogenolysis with Raney nickel yields the hydrocarbon.
2. Base-catalyzed cleavage	
$R{-}CH{-}CH_2$ (epoxide, O bridging) $\xrightarrow[H_2O]{NaOH} R{-}CH(OH){-}CH_2OH$	Epoxides are an exception; other ethers do not react under basic conditions. Other nucleophiles will also open an epoxide ring by attack at the less substituted carbon (Sections 11.6, 12.8).

TABLE 15.3 Preparation of Ethers

Reaction	Comments
1. Acid-catalyzed condensation of alcohols	Section 15.4.
$2\,ROH \xrightarrow{H^+} ROR$	Produces symmetrical ethers. Many side reactions; yields variable.
2. Solvolysis of alkyl halides or sulfonates	
$R{-}X + R'OH \longrightarrow ROR'$	Best results if RX can ionize to stable carbocation. Alcohol is typically methanol or ethanol.
3. Alkylation of alkoxide ions	
$RO^- + R'X \longrightarrow ROR'$	Limited to 1° and 2° halides and sulfonates (Williamson synthesis).
4. Addition to alkenes	
(a) Acid-catalyzed	
$R{-}OH + {>}C{=}C{<} \xrightarrow{H^+} RO{-}\overset{\mid}{\underset{\mid}{C}}{-}\overset{\mid}{\underset{\mid}{C}}{-}H$	Acid conditions; rearrangements.
(b) Oxymercuration	
${>}C{=}C{<} \xrightarrow[\text{2. } NaBH_4]{\text{1. } Hg(O\overset{O}{\overset{\Vert}{C}}CH_3)_2,\ ROH} {-}\overset{OR}{\overset{\mid}{\underset{\mid}{C}}}{-}\overset{\mid}{\underset{\mid}{C}}H$	Nucleophilic attack by ROH occurs on the *more* substituted carbon atom. No rearrangements.
5. Epoxides	
(a) Epoxidation of alkenes	
${>}C{=}C{<} + R{-}CO_3H \longrightarrow {-}\underset{\mid}{C}{-}\underset{\mid}{C}{-}$ (bridged by O)	Most general method for forming epoxides.
(b) Via sulfur ylides	
${>}{=}O + (CH_3)_2S{=}CH_2 \longrightarrow$ (epoxide: C–O–CH_2 ring)	One additional carbon is introduced.
(c) Cyclization of halo alcohols	
${>}C{=}C{<} \longrightarrow {-}\overset{HO}{\overset{\mid}{\underset{\mid}{C}}}{-}\overset{\mid}{\underset{\underset{X}{\mid}}{C}}{-} \xrightarrow{\text{base}} {-}\underset{\mid}{C}{-}\underset{\mid}{C}{-}$ (bridged by O)	

TABLE 15.3 Preparation of Ethers (continued)

Reaction	Comments
6. Acetals and ketals	
(a) Protection of aldehydes and ketones	
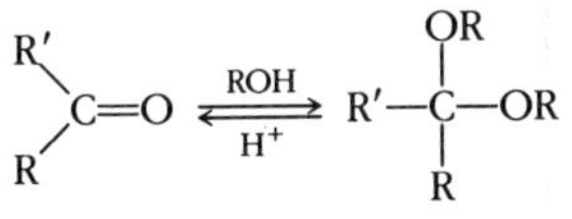 $R'R C{=}O \underset{H^+}{\overset{ROH}{\rightleftharpoons}} R'{-}C(OR)_2{-}R$ R = alkyl, R′ = alkyl or H	Reactions are reversible. Mild acid hydrolysis liberates aldehyde or ketone.
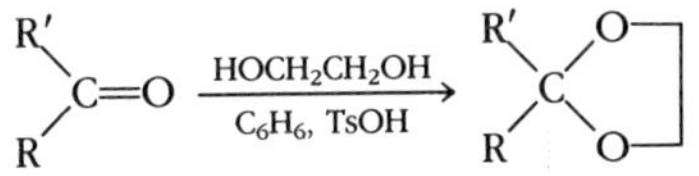 $R'R C{=}O \xrightarrow[C_6H_6,\ TsOH]{HOCH_2CH_2OH}$ cyclic acetal/ketal ($R'RC(OCH_2CH_2O)$)	Ethylene glycol yields cyclic acetal or ketal.
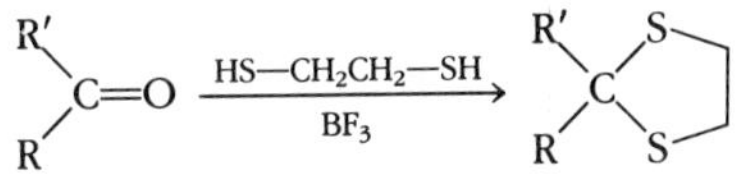 $R'R C{=}O \xrightarrow[BF_3]{HS{-}CH_2CH_2{-}SH}$ cyclic thioacetal/thioketal ($R'RC(SCH_2CH_2S)$)	Ethanedithiol yields cyclic thioacetal or thioketal. (Hydrogenolysis with Raney Ni/H_2 yields hydrocarbon).
(b) Protection of alcohols	
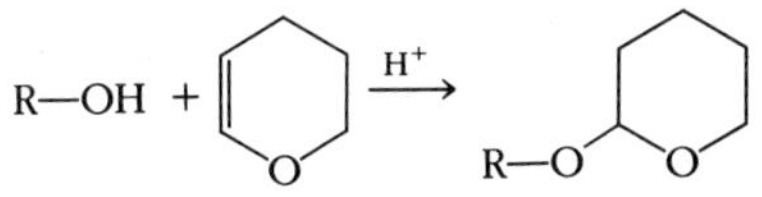 R—OH + dihydropyran $\xrightarrow{H^+}$ R—O–tetrahydropyranyl	Tetrahydropyranyl ether: mild acid hydrolysis regenerates alcohol.

TABLE 16.8 Reactions of Carboxylic Acids and Carboxylic Acid Derivatives

Reaction	Comments
1. Carboxylic acids	
(a) Reaction as bases	Section 16.3
$R-C(=O)OH \underset{}{\overset{H^+}{\rightleftharpoons}} R-C(=\overset{+}{O}H)OH$	Weaker bases than H_2O.
(b) Reaction as acids	Section 16.3
$RCO_2H \rightleftharpoons RCO_2^- + H^+$	Much stronger acids than H_2O. Anion is unreactive toward nucleophilic attack. Acidification regenerates carboxylic acid.
(c) Reduction	Section 16.4
$R-\overset{O}{\overset{\Vert}{C}}-OH \xrightarrow[2.\ H_2O]{1.\ LiAlH_4} RCH_2OH$	The most commonly used reducing agent is $LiAlH_4$.
(d) Oxidation	Section 16.4
$R-CO_2H \xrightarrow{Pb(OAc)_4} R-OAc + CO_2$	Unreactive (already in highest oxidation state), but oxidative decarboxylation can occur with certain transition metal salts.

TABLE 16.8 Reactions of Carboxylic Acids and Carboxylic Acid Derivatives (continued)

Reaction	Comments
(e) Reaction with organometallic compounds	Sections 13.5, 16.5
$R'{-}C(=O){-}OH \xrightarrow{RLi} R'{-}C(=O){-}R$	Only with alkyllithium reagent. Other organometallics just form salt of acid.
(f) Conversion to carboxylic acids	Not applicable.
(g) Conversion to acid chlorides	Section 16.6
$R{-}C(=O){-}OH \xrightarrow{PCl_5} R{-}C(=O){-}Cl$	$SOCl_2$ and PCl_3 also work well.
(h) Conversion to anhydrides	Section 16.6
$R{-}C(=O){-}OH + Ac_2O \longrightarrow R{-}C(=O){-}O{-}C(=O){-}R$	Anhydrides are infrequently used unless commercially available. Can sometimes be prepared using acetic anhydride.
(i) Conversion to esters	Section 16.6
$R'{-}CO_2H \xrightleftharpoons{ROH,\ H^+} R'CO_2{-}R$	Also via acid chloride (see entry 2i of this table). Equilibrium reaction.
$R{-}CO_2H \xrightarrow{CH_2N_2} RCO_2CH_3$	Section 11.8. Methyl esters with diazomethane; *O*-alkylation.
$R{-}CO_2H \xrightarrow[2.\ Et_3O^+BF_4^-]{1.\ (iPr)_2NEt} R{-}CO_2Et$	Section 11.8. Ethyl (and methyl) esters with oxonium salts; *O*-alkylation.
(j) Conversion to amides	Via acid chlorides, anhydrides, or esters. See entries 2j, 3j, and 4j of this table.
(k) Conversion to nitriles	Via amides; See entry 5k of this table.
2. Acid chlorides	
(a) Reaction as bases	Section 16.3
$R{-}C(=O){-}Cl \xrightleftharpoons{H^+} R{-}C(={}^+OH){-}Cl$	Weaker bases than H_2O.

TABLE 16.8 Reactions of Carboxylic Acids and Carboxylic Acid Derivatives (continued)

Reaction	Comments
(b) Reaction as acids	Section 16.3
$-\overset{\vert}{\underset{\vert}{C}}(H)-C(=O)Cl \underset{}{\overset{base}{\rightleftharpoons}} >C=C(O^-)Cl$	Enolate ion formation; comparable with ketones in acidity, but most bases will attack carbonyl.
(c) Reduction	Section 13.4
$R-\overset{O}{\overset{\Vert}{C}}-Cl \xrightarrow[\text{or } H_2,\ Pd\text{–}BaSO_4]{LiAl(OtBu)_3H} R\overset{O}{\overset{\Vert}{C}}H$	Selective reduction to aldehyde.
$R-\overset{O}{\overset{\Vert}{C}}-Cl \xrightarrow[2.\ H_2O]{1.\ LiAlH_4} R-CH_2OH$	Section 16.5 Other reducing agents will also yield primary alcohol.
(d) Oxidation	See entry 1d of this table.
(e) Reaction with organometallic reagents	Section 16.5
$R-\overset{O}{\overset{\Vert}{C}}-Cl \xrightarrow[2.\ H_2O]{1.\ R'-M} R-\overset{OH}{\underset{R'}{C}}-R'$	Highly reactive.
$R-\overset{O}{\overset{\Vert}{C}}-Cl \xrightarrow{R'_2CuLi} R-\overset{O}{\overset{\Vert}{C}}-R'$	Organocopper reagents are more selective than other organometallics.
(f) Conversion to carboxylic acids	Section 16.6
$R-\overset{O}{\overset{\Vert}{C}}-Cl \xrightarrow{H_2O} R-CO_2H + HCl$	Unintentional hydrolysis may occur if moisture is not excluded.
(g) Conversion to acid chlorides	Not applicable.
(h) Conversion to anhydrides	Section 16.6
$R-\overset{O}{\overset{\Vert}{C}}-Cl \xrightarrow{R'-\overset{O}{\overset{\Vert}{C}}-OH} R-\overset{O}{\overset{\Vert}{C}}-O-\overset{O}{\overset{\Vert}{C}}-R'$	Mixed anhydride formation.
(i) Conversion to esters	Sections 11.8, 16.6
$R-\overset{O}{\overset{\Vert}{C}}-Cl \xrightarrow{R'OH} R-\overset{O}{\overset{\Vert}{C}}-OR'$	Acylation of alcohol.

TABLE 16.8 Reactions of Carboxylic Acids and Carboxylic Acid Derivatives (continued)

Reaction	Comments
(j) Conversion to amides	Section 16.6
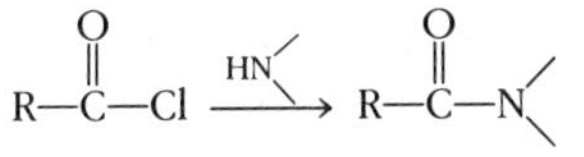 $R{-}\overset{O}{\overset{\parallel}{C}}{-}Cl \xrightarrow{HN<} R{-}\overset{O}{\overset{\parallel}{C}}{-}N<$	Amine must have at least one replaceable hydrogen.
(k) Conversion to nitriles	Via amides. See entry 5k in this table.
(l) α-Halogenation	Section 16.8
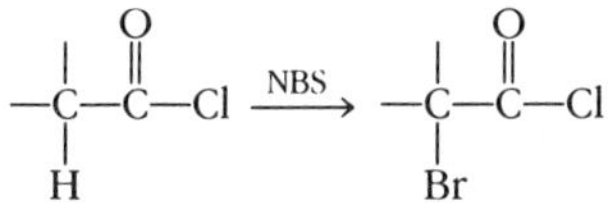 $-\underset{H}{\overset{\vert}{C}}{-}\overset{O}{\overset{\parallel}{C}}{-}Cl \xrightarrow{NBS} -\underset{Br}{\overset{\vert}{C}}{-}\overset{O}{\overset{\parallel}{C}}{-}Cl$	Hydrolysis during workup affords the α-halo acid.
3. Acid anhydrides	
(a) Reaction as bases	Section 16.3
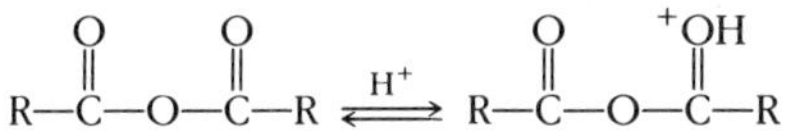 $R{-}\overset{O}{\overset{\parallel}{C}}{-}O{-}\overset{O}{\overset{\parallel}{C}}{-}R \overset{H^+}{\rightleftharpoons} R{-}\overset{O}{\overset{\parallel}{C}}{-}O{-}\overset{^+OH}{\overset{\parallel}{C}}{-}R$	Weaker bases than H_2O.
(b) Reaction as acids	Section 16.3
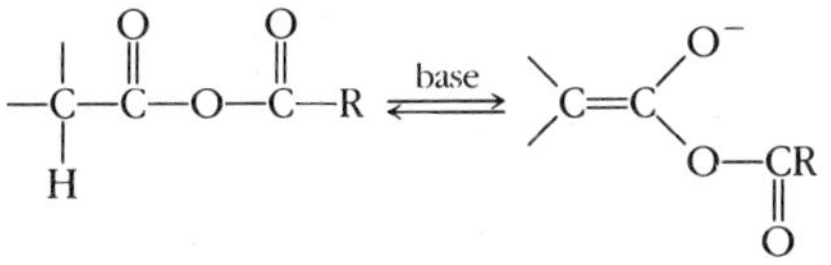 $-\underset{H}{\overset{\vert}{C}}{-}\overset{O}{\overset{\parallel}{C}}{-}O{-}\overset{O}{\overset{\parallel}{C}}{-}R \overset{base}{\rightleftharpoons} {>}C{=}C(O^-)(O{-}\overset{\parallel}{\underset{O}{C}}R)$	Enolate ion formation; comparable with ketones in acidity. Most bases will attack carbonyl.
(c) Reduction	Section 16.4
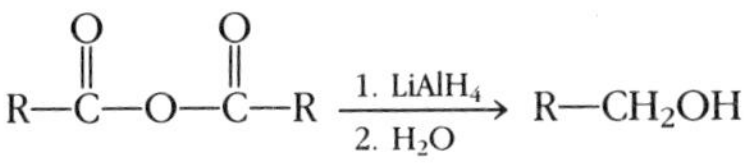 $R{-}\overset{O}{\overset{\parallel}{C}}{-}O{-}\overset{O}{\overset{\parallel}{C}}{-}R \xrightarrow[2.\ H_2O]{1.\ LiAlH_4} R{-}CH_2OH$	Not used preparatively.
(d) Oxidation	See entry 1d of this table.
(e) Reaction with organometallic reagents	Section 16.5
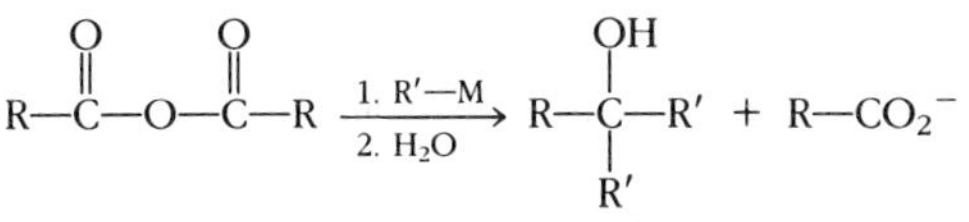 $R{-}\overset{O}{\overset{\parallel}{C}}{-}O{-}\overset{O}{\overset{\parallel}{C}}{-}R \xrightarrow[2.\ H_2O]{1.\ R'{-}M} R{-}\underset{R'}{\overset{OH}{C}}{-}R' + R{-}CO_2^-$	Highly reactive, not normally used preparatively. The two "halves" react differently when the R groups are not the same.

TABLE 16.8 Reactions of Carboxylic Acids and Carboxylic Acid Derivatives (continued)

Reaction	Comments
(f) Conversion to carboxylic acids	Section 16.6
$R{-}\overset{O}{\overset{\|}{C}}{-}O{-}\overset{O}{\overset{\|}{C}}{-}R \xrightarrow{H_2O} RCO_2H$	Unintentional hydrolysis may occur if moisture is not excluded.
(g) Conversion to acid chlorides	Not applicable.
(h) Conversion to anhydrides	Not applicable.
(i) Conversion to esters	Section 16.6
$R{-}\overset{O}{\overset{\|}{C}}{-}O{-}\overset{O}{\overset{\|}{C}}{-}R \xrightarrow{R'OH} R{-}\overset{O}{\overset{\|}{C}}{-}OR' + RCO_2H$	Only one of the two acyl groups is converted to the ester.
(j) Conversion to amides	Section 16.6
$R{-}\overset{O}{\overset{\|}{C}}{-}O{-}\overset{O}{\overset{\|}{C}}{-}R \xrightarrow{HN<} R{-}\overset{O}{\overset{\|}{C}}{-}N< + RCO_2^-$	Only one of the two acyl groups is converted to the amide.
(k) Conversion to nitriles	Via amides; See entry 5k of this table.
4. Esters	
(a) Reaction as bases	Section 16.3
$R{-}\overset{O}{\overset{\|}{C}}{-}OR' \overset{H^+}{\rightleftharpoons} R{-}\overset{^+OH}{\overset{\|}{C}}{-}OR'$	Weaker bases than H_2O.
(b) Reaction as acids	Section 16.3
$-\underset{H}{\underset{\|}{\overset{\|}{C}}}{-}CO_2R \overset{\text{base}}{\rightleftharpoons} {>}C{=}C{<}^{O^-}_{OR}$	Enolate formation; comparable with ketones in acidity.
(c) Reduction	Section 16.4
$R{-}\overset{O}{\overset{\|}{C}}{-}OR' \xrightarrow[2.\ H_2O]{1.\ LiAlH_4} RCH_2OH + R'OH$	Two alcohols are formed
(d) Oxidation	See entry 1d of this table.

TABLE 16.8 Reactions of Carboxylic Acids and Carboxylic Acid Derivatives (continued)

Reaction	Comments
(e) Reaction with organometallic reagents	Section 13.6
$R{-}\overset{O}{\overset{\|}{C}}{-}OR' \xrightarrow[2.\ H_2O]{1.\ R''{-}M} R{-}\underset{R''}{\overset{OH}{C}}{-}R'' + R'OH$	A tertiary alcohol is formed. with alkyllithium and Grignard reagents. A second alcohol is produced from the —OR′ portion of the ester.
(f) Conversion to carboxylic acids	Section 16.6. Hydrolysis
$R{-}\overset{O}{\overset{\|}{C}}{-}OR' \xrightleftharpoons{H_2O,\ H^+} R{-}\overset{O}{\overset{\|}{C}}{-}OH + R'OH$	An equilibrium reaction under acidic conditions.
$R{-}\overset{O}{\overset{\|}{C}}{-}OR' \xrightarrow{H_2O,\ ^-OH} R{-}\overset{O}{\overset{\|}{C}}{-}O^- + R'OH$	Alkaline conditions (saponification); the salt of the acid is formed.
(g) Conversion to acid chlorides	Not applicable.
(h) Conversion to anhydrides	Not applicable.
(i) Conversion to esters	Section 16.6
$R{-}\overset{O}{\overset{\|}{C}}{-}OR' \xrightleftharpoons{R''OH} R{-}\overset{O}{\overset{\|}{C}}{-}OR'' + R'OH$	Transesterification occurs under either acidic or basic conditions.
(j) Conversion to amides	Section 16.6
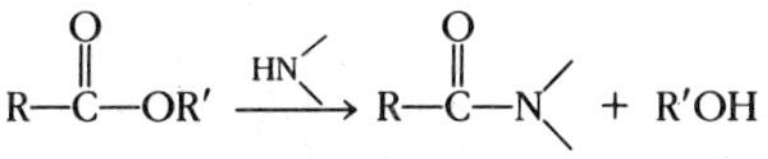 $R{-}\overset{O}{\overset{\|}{C}}{-}OR' \xrightarrow{HN<} R{-}\overset{O}{\overset{\|}{C}}{-}N< + R'OH$	
(k) Conversion to nitriles	Via amides; see entry 5k of this table.
(l) Ester pyrolysis	Section 6.9
5. Amides	
(a) Reaction as bases	Section 16.3
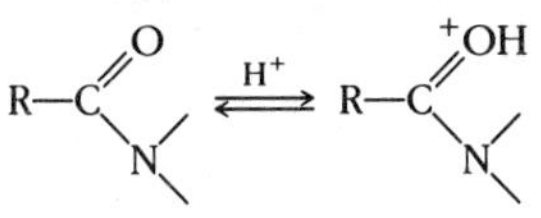 $R{-}C({=}O){-}N< \xrightleftharpoons{H^+} R{-}C({=}\overset{+}{O}H){-}N<$	Comparable with H_2O as bases. Protonation can also occur on nitrogen.
(b) Reaction as acids	Section 16.3
$R{-}\overset{O}{\overset{\|}{C}}{-}NHR \xrightleftharpoons{base} R{-}\overset{O}{\overset{\|}{C}}{-}\overset{-}{N}{-}R$	Comparable with H_2O as acids.

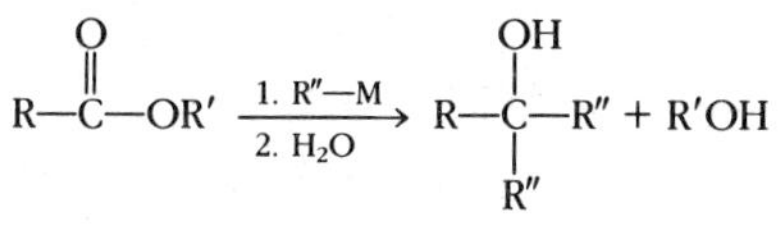

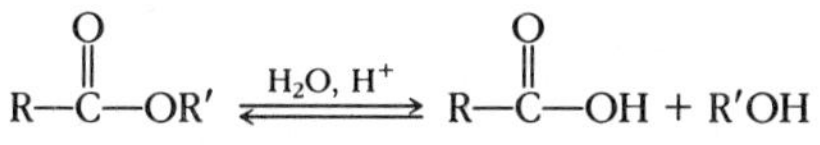

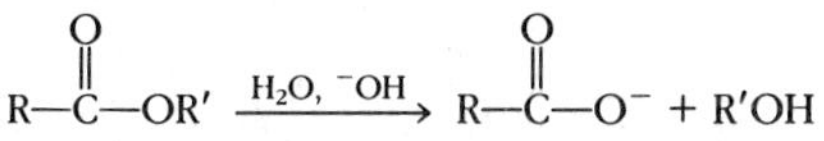

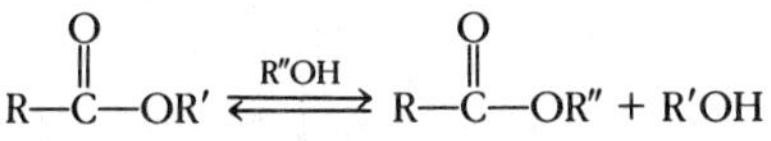

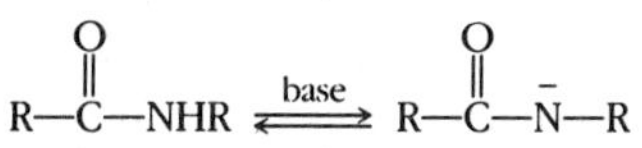

TABLE 16.8 Reactions of Carboxylic Acids and Carboxylic Acid Derivatives (continued)

Reaction	Comments
$-\overset{\vert}{\underset{\vert}{C}}(H)-\overset{O}{\overset{\Vert}{C}}-NR_2 \overset{\text{base}}{\rightleftharpoons} \;>C{=}C(O^-)(NR_2)$	Tertiary amides can only form enolate ions; comparable with ketones in acidity.
(c) Reduction	Section 16.4
$R-\overset{O}{\overset{\Vert}{C}}-N(R')(R'') \xrightarrow[\text{2. } H_2O]{\text{1. } LiAlH_4} R-CH_2-N(R')(R'')$	R′ and R″ may be alkyl, aryl, or hydrogen.
(d) Oxidation	See entry 1d of this table.
(e) Reaction with organometallic reagents	Section 16.5
$R-\overset{O}{\overset{\Vert}{C}}-NH_2 \xrightarrow{R-M} R-\overset{O}{\overset{\Vert}{C}}-\bar{N}H$ $R-\overset{O}{\overset{\Vert}{C}}-NH-R' \xrightarrow{R-M} R-\overset{O}{\overset{\Vert}{C}}-\bar{N}R'$	Not used preparatively. Primary and secondary amides undergo acid-base reaction. Tertiary amides can undergo nucleophilic attack at carbonyl group.
(f) Conversion to carboxylic acids	Section 16.6
$R-\overset{O}{\overset{\Vert}{C}}-N< \xrightarrow[H^+]{H_2O} R-CO_2H + \;>\overset{+}{N}H_2$	Hydrolysis. Acidic or basic conditions can be employed.
$R-\overset{O}{\overset{\Vert}{C}}-N< \xrightarrow[^-OH]{H_2O} R-CO_2^- + \;>NH$	Carboxylate salt is produced in base.
(g) Conversion to acid chlorides	Not applicable.
(h) Conversion to anhydrides	Not applicable.
(i) Conversion to esters	Not applicable.
(j) Conversion to amides	Not applicable.
(k) Conversion to nitriles	Section 16.6
$R-\overset{O}{\overset{\Vert}{C}}-NH_2 \xrightarrow{P_2O_5} R-C{\equiv}N$	Dehydration. Primary amides only. $SOCl_2$, $POCl_3$ and PCl_5 can also be used.

TABLE 16.8 Reactions of Carboxylic Acids and Carboxylic Acid Derivatives (continued)

Reaction	Comments
6. Nitriles	
(a) Reaction as bases	Section 16.3
$R{-}C{\equiv}N \overset{H^+}{\rightleftharpoons} R{-}C{\equiv}\overset{+}{N}H$	Much weaker bases than H_2O.
(b) Reaction as acids	Section 16.3
$-\underset{H}{\underset{\vert}{\overset{\vert}{C}}}-C{\equiv}N \overset{base}{\rightleftharpoons} {>}C{=}C{=}N^-$	Enolate ion formation; comparable with ketones in acidity.
(c) Reduction	Section 16.4
$R{-}C{\equiv}N \xrightarrow[2.\ H_2O]{1.\ LiAlH_4} R{-}CH_2NH_2$	
(d) Oxidation	Not usually oxidized.
(e) Reaction with organometallic reagents	Section 16.5
$R{-}C{\equiv}N \xrightarrow[2.\ H_2O]{1.\ R'M} R{-}\overset{NH}{\overset{\Vert}{C}}{-}R' \longrightarrow R{-}\overset{O}{\overset{\Vert}{C}}{-}R'$	Grignard and alkyllithium reagents yield ketones.
(f) Conversion to carboxylic acids	Section 16.6
$R{-}C{\equiv}N \xrightarrow{H_2O,\ H^+} R{-}CO_2H$ $R{-}C{\equiv}N \xrightarrow{H_2O,\ ^-OH} R{-}CO_2^-$	Hydrolysis; acidic or basic conditions may be employed.
(g) Conversion to acid chlorides	Not applicable.
(h) Conversion to anhydrides	Not applicable.
(i) Conversion to esters	Not applicable.
(j) Conversion to amides	Section 16.6
$R{-}C{\equiv}N \xrightarrow[H^+\ or\ ^-OH]{H_2O} R{-}\overset{O}{\overset{\Vert}{C}}{-}NH_2$	Partial hydrolysis can be carried out in either acid or base under carefully controlled conditions.
(k) Conversion to nitriles	Not applicable.

TABLE 16.9 Preparation of Carboxylic Acids and Carboxylic Acid Derivatives

Reaction	Comments
1. Carboxylic acids	
(a) Oxidation of alcohols and aldehydes	Sections 11.3, 13.11, 16.7
$R{-}\overset{O}{\overset{\|}{C}}H \xrightarrow{\text{oxidation}} RCO_2H$	
$R{-}CH_2OH \xrightarrow{\text{oxidation}} RCO_2H$	
(b) Oxidation of alkenes	Section 16.7
$R{-}CH{=}CH{-}R' \xrightarrow{KMnO_4} RCO_2H + R'CO_2H$	Further oxidation may occur.
$R{-}CH{=}CH{-}R' \xrightarrow[\text{2. } H_2O_2,\ NaOH]{\text{1. } O_3} RCO_2H + R'CO_2H$	Oxidative workup of ozonide.
(c) Oxidation of methyl ketones	Section 13.9
$R{-}\overset{O}{\overset{\|}{C}}{-}CH_3 \xrightarrow[\text{2. HCl}]{\text{1. NaOCl}} R{-}CO_2H + CHCl_3$	Haloform reaction (Cl_2, Br_2 or I_2 in aqueous base)
(d) Oxidation of alkyl substituents on aromatic rings	Section 16.7
$Ar{-}R \xrightarrow[Na_2Cr_2O_7]{KMnO_4 \text{ or}} Ar{-}CO_2H$	
(e) Carboxylation of organometallic reagents	Section 16.7
$R{-}M \xrightarrow[\text{2. } H^+]{\text{1. } CO_2} R{-}CO_2H$	
(f) Koch-Haaf carboxylation of carbocations	Section 16.7
$R^+ \xrightarrow[H_2SO_4]{CO} R{-}CO_2H$	Must be moderately stable carbocation. Carbocation is generated by reactions of alcohol, alkene, or alkyl halide in H_2SO_4.
2. Esters	
(a) Oxidation of ketones	Sections 13.11, 16.7
$R{-}\overset{O}{\overset{\|}{C}}{-}R' \xrightarrow{CH_3CO_3H} R{-}\overset{O}{\overset{\|}{C}}{-}OR'$	Baeyer-Villiger oxidation.

TABLE 16.9 Preparation of Carboxylic Acids and Carboxylic Acid Derivatives (continued)

Reaction	Comments
(b) Condensation reactions of esters	Chapter 21
$-\overset{\vert}{\underset{\vert}{C}}H-CO_2R \xrightarrow[\text{base}]{R'-\overset{O}{\overset{\Vert}{C}}-X} R'-\overset{O}{\overset{\Vert}{C}}-\overset{\vert}{\underset{\vert}{C}}-CO_2R$	
(c) Alkylation of activated esters	Chapter 21
$-\overset{O}{\overset{\Vert}{C}}-CH_2-CO_2R \xrightarrow[\text{2. R'X}]{\text{1. base}} -\overset{O}{\overset{\Vert}{C}}-\underset{R'}{\underset{\vert}{C}}H-CO_2R$	
3. Nitriles	
(a) Cyanide displacement	Sections 12.8, 16.7
$R-X \xrightarrow{\text{NaCN}} R-CN$	RX must be primary or secondary halide or sulfonate.
(b) Alkylation of cyano esters	Chapter 21
$RO_2C-CH_2-CN \xrightarrow[\text{2. R'X}]{\text{1. base}} RO_2C-\underset{R'}{\underset{\vert}{C}}H-CN$	

TABLE 17.2 Reactions of Amines

Reaction	Comments
1. Reaction as base	Section 17.4
$R{-}\ddot{N}< + H^+ \rightleftharpoons R{-}\overset{+}{N}{-}H$	K_a of ammonium ion is approximately 10^{-5}
2. Reaction as acid	Section 17.4
$R{-}\ddot{N}{-}H \rightleftharpoons H^+ + R{-}\ddot{N}:^-$	$K_a \sim 10^{-35}$. Amide ion is a strong base.
3. Acylation	Sections 16.6, 17.4
$R{-}\ddot{N}{-}H \xrightarrow{R'CCl(=O)} R{-}\ddot{N}{-}C(=O){-}R'$	Formation of amides. Only primary and secondary amines react.
$R{-}\ddot{N}{-}H \xrightarrow{ArSO_2Cl} R{-}\ddot{N}{-}SO_2Ar$	Formation of sulfonamides.
4. Alkylation	Section 17.4
$R{-}\ddot{N}{-} \xrightarrow{R'X} R{-}\overset{+}{N}{-}R'$	Reaction with alkyl halide or sulfonate. Polyalkylation and elimination are important side reactions.
5. Reaction with nitrous acid	Section 17.4
$R{-}NH_2 \xrightarrow{HONO} R{-}\overset{+}{N}{\equiv}N \longrightarrow$ substitution and elimination	Primary aliphatic amines form unstable diazonium ions; secondary amines yield *N*-nitroso compounds; tertiary amines yield complex mixtures.
$Ar{-}NH_2 \xrightarrow{HONO} Ar{-}\overset{+}{N}{\equiv}N$	Primary aromatic amines form diazonium salts that are stable in solution (see Section 18.5).

TABLE 17.2 Reactions of Amines (continued)

Reaction	Comments
6. Hofmann elimination 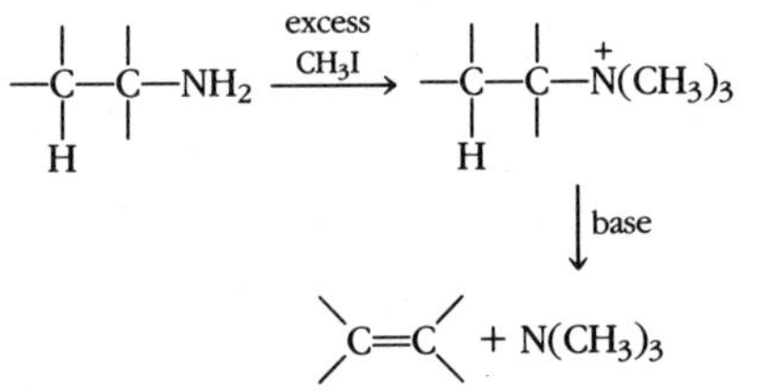 $-\underset{H}{\overset{\vert}{C}}-\overset{\vert}{\underset{\vert}{C}}-NH_2 \xrightarrow{\text{excess } CH_3I} -\underset{H}{\overset{\vert}{C}}-\overset{\vert}{\underset{\vert}{C}}-\overset{+}{N}(CH_3)_3$ $\downarrow$ base $>C=C< + N(CH_3)_3$	Section 17.5 Decomposition of quaternary ammonium salts. Less substituted alkene predominates. *Anti* elimination.
7. Amine oxide pyrolysis 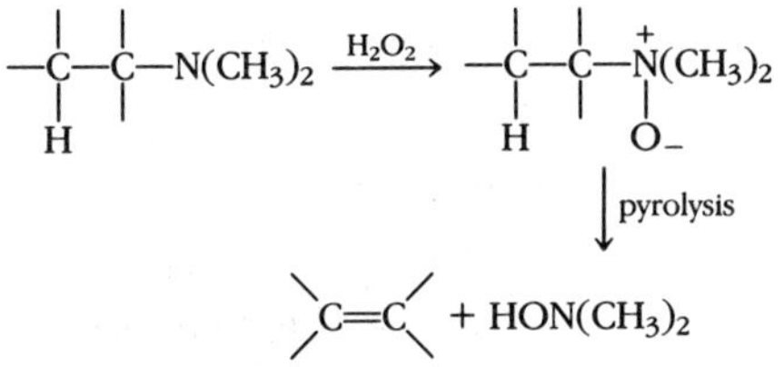 $-\underset{H}{\overset{\vert}{C}}-\overset{\vert}{\underset{\vert}{C}}-N(CH_3)_2 \xrightarrow{H_2O_2} -\underset{H}{\overset{\vert}{C}}-\overset{\vert}{\underset{\vert}{C}}-\underset{O^-}{\overset{+}{N}}(CH_3)_2$ $\downarrow$ pyrolysis $>C=C< + HON(CH_3)_2$	Section 17.5 *Syn* elimination.
8. Enamine formation 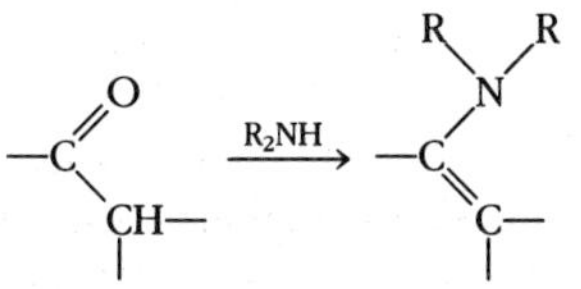 $-C(=O)-\underset{\vert}{CH}- \xrightarrow{R_2NH} -C(NR_2)=\underset{\vert}{C}-$	Section 17.8 Enamines are useful for carrying out reaction at the α carbon of a ketone.

TABLE 17.3 Preparation of Amines

Reaction	Comments
1. Reduction of nitro compounds $R{-}NO_2 \xrightarrow{H_2,\ Pt} R{-}NH_2$ (or $Ar{-}NO_2$)	Section 17.6 Other reducing agents can be used: Zn, HCl; Sn, HCl; Fe, HCl; H_2S; $LiAlH_4$.
2. Reduction of amides 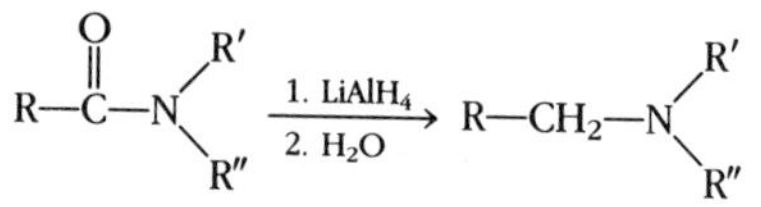 $R{-}\overset{O}{\overset{\Vert}{C}}{-}N{<}^{R'}_{R''} \xrightarrow[2.\ H_2O]{1.\ LiAlH_4} R{-}CH_2{-}N{<}^{R'}_{R''}$	Sections 16.4, 17.6 R′ and R″ can be hydrogen, alkyl, or aryl.
3. Reduction of nitriles $R{-}C{\equiv}N \xrightarrow[2.\ H_2O]{1.\ LiAlH_4} RCH_2NH_2$	Section 16.4, 17.6 Yields primary amines.
4. Reduction of oximes 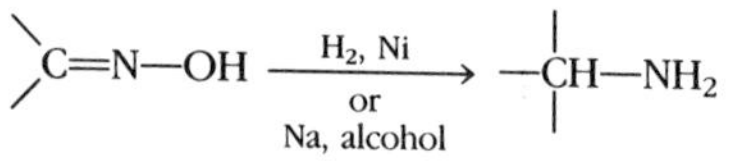 $>C{=}N{-}OH \xrightarrow[\text{or Na, alcohol}]{H_2,\ Ni} {-}\overset{\vert}{\underset{\vert}{C}}H{-}NH_2$	Section 17.6 Yields primary amines. $LiAlH_4$ can also be used for reduction.
5. Reductive amination $>C{=}O \overset{HN{<}^{R}_{R'},\ H^+}{\rightleftharpoons} >C{=}\overset{+}{N}{<}^{R}_{R'} \xrightarrow{NaBH_3CN} {-}\overset{\vert}{\underset{\vert}{C}}H{-}N{<}^{R}_{R'}$ $R{-}NH_2 \xrightarrow{CH_2O,\ HCO_2H} R{-}N{<}^{CH_3}_{CH_3}$	Sections 13.7, 17.7 R and R′ can be hydrogen, alkyl, or aryl. Catalytic hydrogenation is sometimes used for reduction. Special case of reductive amination with excess formaldehyde. Hydrogens are replaced by methyl groups to yield tertiary amines.
6. Alkylation of ammonia and amines 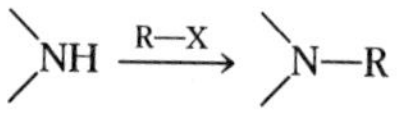 $>NH \xrightarrow{R{-}X} >N{-}R$	Section 17.7 Polyalkylation and elimination may occur.
7. Ritter reaction 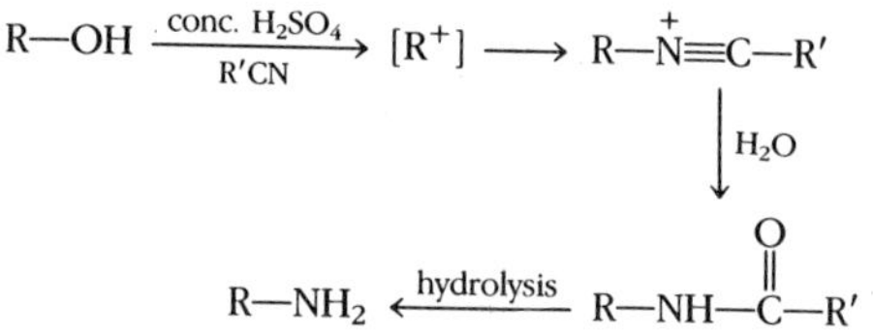 $R{-}OH \xrightarrow[R'CN]{conc.\ H_2SO_4} [R^+] \longrightarrow R{-}\overset{+}{N}{\equiv}C{-}R'$ $\downarrow H_2O$ $R{-}NH_2 \xleftarrow{hydrolysis} R{-}NH{-}\overset{O}{\overset{\Vert}{C}}{-}R'$	Section 17.7 Alkylation of carbocations with nitriles. Nitrilium ion reacts with water to form amide. R^+ must be tertiary (or resonance-stabilized) cation.

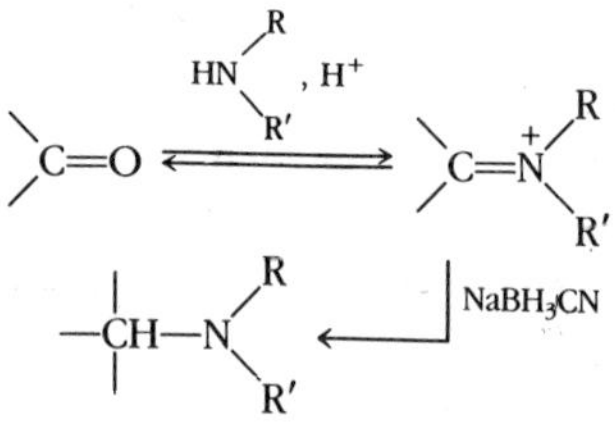

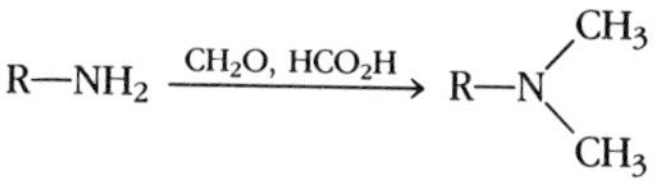

TABLE 17.3 Preparation of Amines (continued)

Reaction	Comments
8. Hydroboration	Section 17.7
$\mathrm{>C{=}C<} \xrightarrow{BH_3} \mathrm{-\overset{H}{\underset{\mid}{\overset{\mid}{C}}}-\overset{\mid}{\underset{\mid}{\overset{B-}{C}}}-} \xrightarrow{H_2N-Cl \text{ or } H_2N-OSO_3H} \mathrm{-\overset{H}{\underset{\mid}{\overset{\mid}{C}}}-\overset{NH_2}{\underset{\mid}{\overset{\mid}{C}}}-}$	*Syn* addition.
9. Hofmann degradation of primary amides	Section 17.7
$R-\overset{O}{\overset{\|}{C}}-NH_2 \xrightarrow{X_2,\ NaOH} R-NH_2$	X can be Cl, Br, or I.

TABLE 18.2 Substitution Reactions of Benzene and Benzene Derivatives

Reaction	Comments
1. Hydrogen exchange	Section 10.7
$Ar{-}H \xrightarrow{*H^+} Ar{-}H^*$	This reaction can only be observed by using isotopic labels.
2. Chlorination, bromination	Section 10.8
$Ar{-}H \xrightarrow[FeCl_3]{Cl_2} Ar{-}Cl$ $Ar{-}H \xrightarrow[FeBr_3]{Br_2} Ar{-}Br$	The ferric halide catalyst may not be necessary with highly reactive aromatic compounds.
3. Nitration	Section 10.8
$Ar{-}H \xrightarrow[H_2SO_4]{HNO_3} Ar{-}NO_2$ $Ar{-}H \xrightarrow{NO_2^+BF_4^-} Ar{-}NO_2$	
4. Sulfonation	Section 10.8, 18.2
$Ar{-}H \xrightarrow[H_2SO_4]{SO_3} Ar{-}SO_3H$	Sulfonation is reversible. Equilibrium mixtures of *ortho, meta,* and *para* isomers may be formed. Hydrolysis in strong acid replaces $-SO_3H$ by $-H$.
5. Friedel-Crafts alkylation	Section 18.3
$Ar{-}H \xrightarrow[AlCl_3]{R{-}X} Ar{-}R$	Carbocation rearrangements of R—X and of Ar—R are common. Polyalkylation is a major problem. Important for industrial synthesis; seldom used in laboratory synthesis. Deactivated compounds (Figure 18.1) do not react.
6. Friedel-Crafts acylation	Section 18.3
$Ar{-}H \xrightarrow[AlCl_3]{R{-}C(=O)Cl} Ar{-}C(=O){-}R$	A versatile carbon-carbon bond-forming reaction for laboratory as well as industrial synthesis. Deactivated compounds (Figure 18.1) do not react.
$Ar{-}H \xrightarrow[AlCl_3]{R{-}C(=O){-}O{-}C(=O){-}R} Ar{-}C(=O){-}R$	Reduction of the acyl group provides a method for introducing an alkyl side chain.

TABLE 18.3 Functional Group Interconversions in Aromatic Compounds

Reaction	Comments
1. Sulfonic acids	Section 18.4
(a) In acid	
$Ar{-}SO_3H \xrightarrow[H_2SO_4]{H_2O} Ar{-}H$	Reverse of sulfonation: $-SO_3$ is replaced by H.
(b) In base	
$Ar{-}SO_3H \xrightarrow[250\text{–}300°C]{NaOH} Ar{-}OH$	Very stringent conditions; not generally useful for laboratory synthesis.
2. Halides (X=F, Cl, Br, I)	Section 18.7
(a) Reaction with strong base	
$Ar{-}X \xrightarrow[200\text{–}350°C]{NaOH} Ar{-}OH$ $Ar{-}X \xrightarrow[NH_3]{NaNH_2} Ar{-}NH_2$	Elimination-addition via benzyne intermediate. Requires stringent conditions; not generally useful for laboratory synthesis.
(b) Nucleophilic substitution of activated aromatic halides	
o-nitrophenyl-X (NO_2, X) $\xrightarrow{Nuc^-}$ o-nitrophenyl-Nuc (NO_2, Nuc)	Limited to reaction of halides with an *o*- or *p*-NO_2 (or carbonyl) substituent. Nucleophile can be alkoxide ion or amino group.
(c) Formation of organometallic reagent	Sections 11.5, 13.4
$Ar{-}X \xrightarrow{M} Ar{-}M$	Useful intermediates for forming new carbon-carbon bonds. Other functional groups on the aromatic ring (e.g., OH, nitro, carbonyl) will interfere.

TABLE 18.3 (continued)

Reaction	Comments
3. Alkyl derivatives	
(a) Bromination	Sections 8.4, 18.4
$Ar{-}CH_2{-}R \xrightarrow{NBS} Ar{-}\overset{\overset{\Large Br}{\vert}}{C}H{-}R$	Benzylic bromination with NBS
(b) Oxidation	Sections 16.7, 18.4
$Ar{-}R \xrightarrow[Na_2Cr_2O_7]{KMnO_4 \text{ or}} Ar{-}CO_2H$	Other substituents may be oxidized also. (Halo and nitro groups are unaffected.)
4. Phenols	Sections 11.7, 11.8
(a) Alkylation	
$Ar{-}OH \xrightarrow[base]{RX} Ar{-}OR$	
(b) Acylation	
$Ar{-}OH \xrightarrow{R{-}\overset{O}{\overset{\Vert}{C}}Cl} Ar{-}O\overset{O}{\overset{\Vert}{C}}R$	
5. Amino derivatives	
(a) Acylation	Sections 16.6, 17.4, 18.3
$Ar{-}NH_2 \xrightarrow[\text{or } Ac_2O]{CH_3\overset{O}{\overset{\Vert}{C}}Cl} Ar{-}NH\overset{O}{\overset{\Vert}{C}}CH_3$	Useful as blocking group. Electrophilic substitution of amide is usually preferable to reaction of free amine. Secondary amines react similarly.
(b) Diazotization	Section 18.5
$Ar{-}NH_2 \xrightarrow{HONO} Ar{-}\overset{+}{N}{\equiv}N{:}$	Primary aromatic amines only. The diazonium salts are versatile synthetic intermediates.

TABLE 18.3 (continued)

Reaction	Comments
6. Diazonium salts	Section 18.5
(a) Replacement by —Cl, —Br, —I	
$Ar{-}N_2^+ \xrightarrow{CuCl} Ar{-}Cl$ $Ar{-}N_2^+ \xrightarrow{CuBr} Ar{-}Br$ $Ar{-}N_2^+ \xrightarrow{CuI} Ar{-}I$	Sandmeyer reaction. Copper powder plus a halide salt can be used in place of cuprous halide.
(b) Replacement by —F	
$Ar{-}N_2^+ \xrightarrow[\text{2. mild heating}]{\text{1. } HBF_4} Ar{-}F$	
(c) Replacement by —CN	
$Ar{-}N_2^+ \xrightarrow{CuCN} Ar{-}CN$	Sandmeyer reaction. Alternative method of carboxylic acid synthesis (cf. Section 16.7).
(d) Replacement by $-NO_2$	
$Ar{-}N_2^+ \xrightarrow{CuNO_2} Ar{-}NO_2$	Sandmeyer reaction. Permits sequence nitro → amino → nitro.
(e) Replacement by H	
$Ar{-}N_2^+ \xrightarrow[NaBH_4]{H_3PO_2 \text{ or}} Ar{-}H$	Reductive deamination: ($-NH_2 \rightarrow -N_2^+ \rightarrow H$). Permits use of nitro and amino groups as blocking or activating groups.
(f) Replacement by OH	
$Ar{-}N_2^+ \xrightarrow[H_2SO_4]{H_2O} Ar{-}OH$	Useful synthesis of phenols.
(g) Coupling with ArX ($X{=}NH_2$, OH)	
$Ar{-}N_2^+ \xrightarrow{C_6H_5{-}X} Ar{-}N{=}N{-}C_6H_4{-}X$	Preparation of azo dyes.

TABLE 18.3 (continued)

Reaction	Comments
7. Reduction of aromatic ring	
(a) Catalytic hydrogenation	Section 8.6
benzene $\xrightarrow[\text{cat.}]{H_2}$ cyclohexane	High pressures sometimes required.
(b) Birch reduction	Section 22.4
benzene $\xrightarrow[NH_3]{Li}$ 1,4-cyclohexadiene	Hydrogens not usually added to positions with electron-donating groups. (Na or K can also be used.)

TABLE 19.1 Reactions of Carbohydrates

Reaction	Comments
1. Intramolecular cyclization	Yields five- or six-membered rings preferentially.
(a) Hemiacetal-hemiketal formation	Section 19.2
C=O … CHOH ⇌ cyclic C(OH)–O–CH	Rapid equilibrium favoring cyclic form.
(b) Acetal-ketal formation	Section 19.3
C=O … CH—OH $\xrightarrow[H_2SO_4]{CH_3OH}$ cyclic C(OCH_3)–O–CH	Formation of methyl glycosides is most common. The glycoside is not a reducing sugar.
(c) Lactone formation	Section 19.5
CO_2H … CH—OH $\overset{acid}{\rightleftharpoons}$ cyclic C(=O)–O–CH	Illustrated for an aldonic acid. Aldaric acids can yield products with two lactone rings. Ring opens to the salt of the hydroxy acid in base.
(d) Intramolecular *O*-alkylation	Section 19.3
CHOH … TsO—CH $\xrightarrow{base}$ cyclic CH–O–CH	Three-membered rings (epoxides) can also be formed in this reaction.

TABLE 19.1 (continued)

Reaction	Comments
2. Protection of OH groups as acetals or ketals	Section 19.3
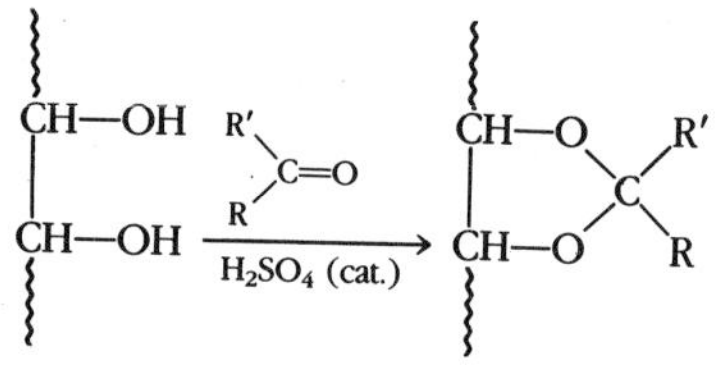 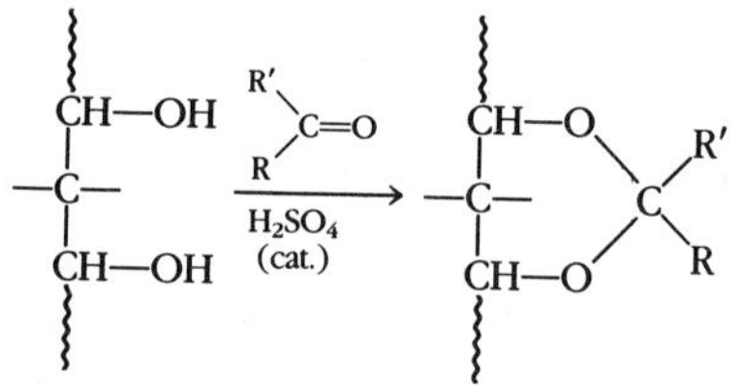 $\wr$CH—OH, CH—OH$\wr$ + R′(R)C=O $\xrightarrow{H_2SO_4\ (cat.)}$ $\wr$CH—O, CH—O$\wr$ >C(R′)(R)	Five- or six-membered rings are formed. Common reagents include acetone (R = R′ = CH_3), acetaldehyde (R = CH_3, R′ = H), benzaldehyde (R = C_6H_5, R′ = H).
$\wr$CH—OH, —C—, CH—OH$\wr$ + R′(R)C=O $\xrightarrow{H_2SO_4\ (cat.)}$ $\wr$CH—O, —C—, CH—O$\wr$ >C(R′)(R)	
3. Methylation of hydroxyl groups	Section 19.3
—C—OH $\xrightarrow[\text{or } CH_3I,\ Ag_2O]{(CH_3)_2SO_4,\ NaOH}$ —C—OCH_3	All free OH groups are alkylated. Silver salts cannot be used with reducing sugars.
4. Acylation of hydroxyl groups	Section 19.3
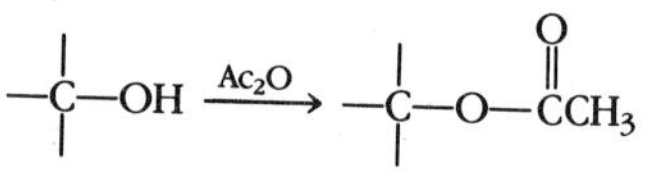 —C—OH $\xrightarrow{Ac_2O}$ —C—O—C(=O)CH_3	All free OH groups can be acylated. Acetic anhydride is a common reagent.
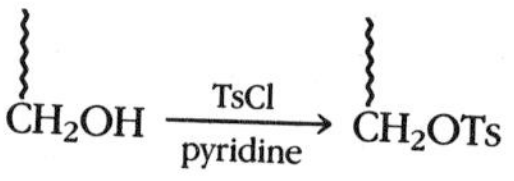 $\wr CH_2OH \xrightarrow[\text{pyridine}]{TsCl} \wr CH_2OTs$	Primary hydroxyl groups react preferentially.

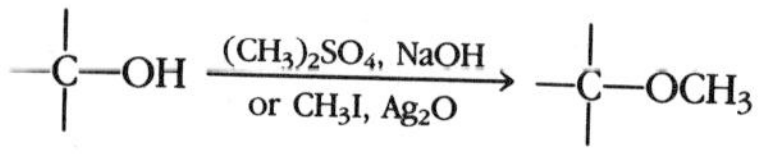

TABLE 19.1 (continued)

Reaction	Comments
5. Reduction	Section 19.4
(a) Alditol formation	
CHO–CHOH~ or CH_2OH–C=O~ $\xrightarrow{H_2/\text{cat. or Na/alcohol or } NaBH_4}$ CH_2OH–CHOH~	The product from an aldose or ketose is an alditol.
(b) Lactone reduction	
(lactone: C(=O)–O–CH ring) $\xrightarrow[H_2O,\ pH\ 3]{Na\text{—}Hg}$ (CH(OH)–O–CH ring)	With sodium amalgam the product is an aldose (sodium in alcohol would reduce the lactone to an alditol). Used in Fischer-Kiliani chain lengthening process.
6. Oxidation	Section 19.5
(a) Bromine water oxidation	
CHO~ $\xrightarrow{Br_2,\ H_2O}$ CO_2H~	An aldose yields an aldonic acid.
(b) Oxidation with nitric acid	
CHO~CH_2OH $\xrightarrow{HNO_3}$ CO_2H~CO_2H	An aldose yields an aldaric acid.

TABLE 19.1 (continued)

Reaction	Comments
(c) Oxidation with silver ion	
CHO or CH_2OH–C=O $\xrightarrow{Ag(NH_3)_2{}^+OH^-}$ Ag^0 (Precipitate)	Used for identification of free aldehyde or α-hydroxy ketone (Tollens test).
(d) Oxidation with cupric ion	
CHO or CH_2OH–C=O $\xrightarrow{CuSO_4}$ Cu_2O (Precipitate)	Used for identification (or quantitative analysis) of free aldehyde or α-hydroxy ketone. (Reagent is called Fehling solution.)
(e) Oxidation with phenylhydrazine	
CHO–CHOH or CH_2OH–C=O $\xrightarrow{NH_2NHC_6H_5}$ C=N—NHC_6H_5 / C=N—NHC_6H_5	Osazone formation.
(f) Periodate oxidation	
CH=O / C=O / CHOH / CH_2OH $\xrightarrow{NaIO_4}$ HCO_2H / CO_2 / HCO_2H / CH_2O	Cleaves bonds between carbons bearing —OH or =O substituents. C—H bonds remain intact. No cleavage if OH groups are protected as ethers.
7. Chain lengthening	Section 19.6
CHO $\xrightarrow{HCN}$ CN–CH—OH $\xrightarrow[2.\ neutralize]{1.\ H_2O,\ ^-OH}$ CO_2H–CH—OH	Product is acid (or lactone) with one more carbon (two epimers are formed). Reduction of lactone with sodium amalgam completes the Fischer-Kiliani synthesis.
8. Chain shortening	Section 19.6
CO_2H–CHOH $\xrightarrow[Fe^{3+}]{H_2O_2}$ CHO + CO_2	Ruff degradation. The aldonic acid is prepared by bromine water oxidation of the aldose.

TABLE 20.2 Blocking Groups for Peptide Synthesis

Blocking Group (Abbreviation)	Reagents for Introduction	Reagents for Removal, Comments
Protection of $>N{-}H$		
p-Toluenesulfonyl (Ts) 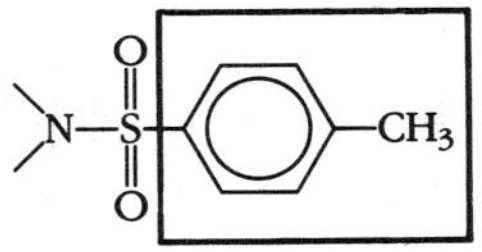 $>N{-}S(=O)_2{-}C_6H_4{-}CH_3$	TsCl and base (NaOH or 3° amine)	Na, NH_3—strongly reducing conditions; will also cleave carbon-sulfur bonds, sulfur-sulfur bonds and benzyl groups; will reduce esters
Benzyloxycarbonyl (Cbz) 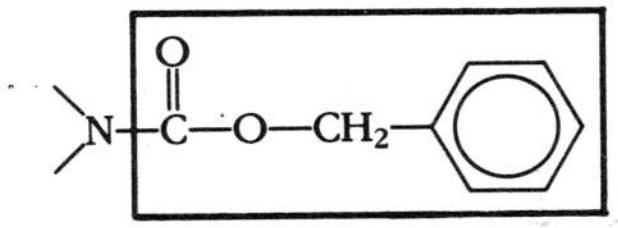 $>N{-}C(=O){-}O{-}CH_2{-}C_6H_5$	Benzyl chloroformate ($C_6H_5CH_2{-}O{-}C(=O){-}Cl$) and base (NaOH or $NaHCO_3$)	H_2, Pd (not applicable for sulfur-containing peptides); HBr (in nonaqueous solvent); Na, NH_3
tert-Butoxycarbonyl (tBOC)co 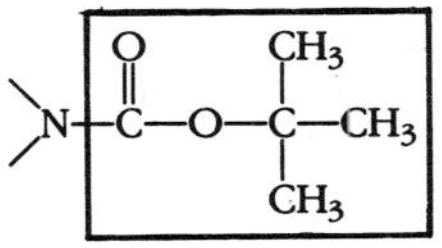 $>N{-}C(=O){-}O{-}C(CH_3)_3$	*tert*-Butyl azidoformate ($t\text{-}C_4H_9O{-}C(=O){-}N_3$) and a 3° amine, e.g., $N(C_2H_5)_3$	HBr (in nonaqueous solvent)
Trifluoroacetyl (TFA) 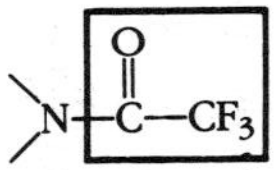 $>N{-}C(=O){-}CF_3$	Trifluoroacetic anhydride $CF_3{-}C(=O){-}O{-}C(=O){-}CF_3$	aqueous NH_3; aqueous NaOH (dilute)

TABLE 20.2 Blocking Groups for Peptide Synthesis

Blocking Group (Abbreviation)	Reagents for Introduction	Reagents for Removal, Comments
Protection of $-\overset{O}{\overset{\parallel}{C}}-O-H$		
Methyl or ethyl esters $-\overset{O}{\overset{\parallel}{C}}-O-\boxed{CH_3,}$ $-\overset{O}{\overset{\parallel}{C}}-O-\boxed{C_2H_5}$	CH_3OH, C_2H_5OH, acid (cf. Section 15.7)	Hydrolysis in acid or base (some hydrolysis of peptide bonds may occur also)
Benzyl esters $-\overset{O}{\overset{\parallel}{C}}-O-\boxed{CH_2-C_6H_5}$	$C_6H_5CH_2OH$, acid (cf., Section 15.7) or $C_6H_5CH_2Cl$, base (can also be used to form benzyl ethers for protection of OH groups)	H_2, Pd (not applicable to sulfur-containing peptides; will remove *N*-Cbz groups); Na, NH_3 (see comments under *N*-Ts); HBr in nonaqueous solvent; hydrolysis (acid or base catalysis—see comments under methyl or ethyl esters)
Protection of —S—H Groups		
Benzyl thioethers $-S-\boxed{CH_2-C_6H_5}$	$C_6H_5CH_2Cl$, NaOH	Na, NH_3 (see comments under *N*-Ts); HBr in nonaqueous solvent (will also cleave *O*-benzyl groups)
Disulfide bond formation —S—S—	Mild oxidation (e.g., O_2)	Na, NH_3; milder reducing agents such as $NaBH_4$ or $HOCH_2CH_2SH$ (mercapto-ethanol) also cleave the disulfide bond.

TABLE 21.2 Preparation of β-Dicarbonyl Compounds

Reaction	Comments
1. Ester condensation	Section 21.2 Produces β-ketoesters. Strong bases such as NaH can be used. The reactions are reversible.
(a) Self-condensation $RCH_2\overset{O}{\overset{\parallel}{C}}OC_2H_5 \xrightarrow[\text{2. HCl (neutralize)}]{\text{1. NaOC}_2\text{H}_5} RCH_2\overset{O}{\overset{\parallel}{C}}—\underset{R}{\underset{\mid}{C}}HCO_2C_2H_5$	
(b) Mixed condensation $Ar—\overset{O}{\overset{\parallel}{C}}—OC_2H_5 + RCH_2CO_2C_2H_5 \xrightarrow[\text{2. HCl (neutralize)}]{\text{1. NaOC}_2\text{H}_5} Ar—\overset{O}{\overset{\parallel}{C}}—\underset{R}{\underset{\mid}{C}}HCO_2C_2H_5$	Best results when one ester has no α hydrogens.
2. Acylation of ketones or nitriles	Section 21.2
$R—CO_2C_2H_5 + R'CH_2CN \xrightarrow[\text{2. HCl (neutralize)}]{\text{1. NaOC}_2\text{H}_5} R\overset{O}{\overset{\parallel}{C}}—\underset{R'}{\underset{\mid}{C}}HCN$	Strong bases such as NaH are often used. Mixed condensation; preferential formation of just one enolate is needed. Yields β-keto nitriles or β-diketones.
$CH_3\overset{O}{\overset{\parallel}{C}}CH_2— + R\overset{O}{\overset{\parallel}{C}}OC_2H_5 \xrightarrow[\text{2. HCl (neutralize)}]{\text{1. NaOC}_2\text{H}_5} CH_3\overset{O}{\overset{\parallel}{C}}—\underset{\mid}{C}H—\overset{O}{\overset{\parallel}{C}}R$	
$CH_3\overset{O}{\overset{\parallel}{C}}CH_2— + C_2H_5O\overset{O}{\overset{\parallel}{C}}OC_2H_5 \xrightarrow[\text{2. HCl (neutralize)}]{\text{1. NaH}} CH_3\overset{O}{\overset{\parallel}{C}}—\underset{\mid}{C}H—CO_2C_2H_5$	Acylation of a ketone with diethyl carbonate yields a β-keto ester.

TABLE 21.3 Aldol-Type Condensations of Carboxylic Acid Derivatives

Reaction	Comments
1. Reformatsky reaction	Section 21.3
$R_2C{=}O + BrCH(R')-CO_2C_2H_5 \xrightarrow{\text{1. Zn; 2. HCl (neutralize)}} R_2C(OH)-CH(R')-CO_2C_2H_5$	Yields β-hydroxy esters via zinc enolate.
2. Stobbe condensation	Section 21.3
$C_2H_5O_2C-CH_2-CH_2-CO_2C_2H_5 + O{=}CR_2 \xrightarrow{\text{1. NaH; 2. HCl (neutralize)}} R_2C{=}C(CH_2CO_2H)-CO_2C_2H_5$	Condensation between succinic esters and ketones to give the unsaturated ester-acid. (Alkoxide salts can also be used as bases.)
3. Perkin condensation	Section 21.3
$ArCHO + (RCH_2\overset{O}{\overset{\|}{C}})_2O \longrightarrow ArCH{=}C(R)-\overset{O}{\overset{\|}{C}}-OH$	Condensation between an anhydride and an aromatic aldehyde. (The corresponding carboxylate salt serves as the base.)
4. Knoevenagel condensation	Section 21.4
$R_2C{=}O + CH_2(X)(Y) \xrightarrow{NH_3} R_2C{=}C(X)(Y)$	Condensation of a β-dicarbonyl compound catalyzed by amine or ammonium salt. X and Y can be CN, COR, CO_2R, CO_2H.

TABLE 21.3 Aldol-Type Condensations of Carboxylic Acid Derivatives (continued)

Reaction	Comments
5. Modified Wittig reaction	Section 21.4
$R{-}C(=O){-}R + (C_2H_5O)_2P(=O){-}CH_2{-}X \xrightarrow{NaH} R(R)C{=}CH{-}X$	Effectively a condensation of a β-dicarbonyl compound. The phosphonyl group is lost during the reaction. X can be CN, COR, CO_2R.
6. Michael addition	Section 21.4
$CH_3COCH{=}CH_2 + R{-}CH(X)(Y) \xrightarrow{base} CH_3COCH_2CH_2{-}C(R)(X)(Y)$	Conjugate addition of enolate ion from a β-dicarbonyl compound. (See also Section 14.5). X and Y can be CN, COR, CO_2R.

TABLE 21.4 Alkylation and Decarboxylation of β-Dicarbonyl Compounds

Reaction	Comments
1. Alkylation	
$-CH(X)(Y) \xrightarrow[2.\ RX]{1.\ NaOC_2H_5} R-C(X)(Y)$	RX is an alkyl halide or tosylate. R cannot be tertiary, and best results are obtained when R is primary or methyl. X and Y can be CN, COR, CO_2R.
2. Decarboxylation	Section 21.6
$CH_3C(=O)-C-CO_2H \longrightarrow CH_3C(=O)-C-H$	Removal of activating group after alkylation or condensation reaction. The carboxyl group must be β to the carbonyl group.
$CH_3C(=O)-C-CN \xrightarrow[H_2O]{H_2SO_4} CH_3C(=O)-C-CO_2H \xrightarrow{-CO_2} CH_3C(=O)-C-H$	Nitriles and esters must first be hydrolyzed to $-CO_2H$ (acid hydrolysis is preferred in order to avoid retrocondensation).

TABLE 22.4 Some Common Protecting Groups

Protecting Group	Reagents for Introduction	Reagents for Removal; Comments
	1. Protection of Aldehydes and Ketones	
(a) Acetals, ketals (acyclic)		Section 15.7
$>C{=}O \xrightarrow{ROH,\ H^+} >C(OR)_2$	Alcohol, acid	Dilute aqueous acid (typically, HCl or H_2SO_4)
(b) Acetals, ketals (cyclic)		Section 15.7
$>C{=}O \xrightarrow[TsOH,\ C_6H_6]{HOCH_2CH_2OH}$ 1,3-dioxolane	Ethylene glycol, TsOH, benzene	Dilute aqueous acid (typically, HCl or H_2SO_4)
(c) Vinyl ethers	(Usually not prepared directly from aldehyde or ketone)	Section 15.7 Dilute aqueous acid (typically, HCl or H_2SO_4)
(d) Thioketals, thioacetals		Section 15.7
$>C{=}O \xrightarrow[BF_3]{HSCH_2CH_2SH}$ 1,3-dithiolane	Ethanedithiol, BF_3	Hydrolysis is slow; catalysis by silver or mercury salts is usually needed.
1,3-dithiolane $\xrightarrow[H_2]{Ni} >CH_2$		Raney nickel cleaves the sulfurs by hydrogenolysis (to generate $—CH_2—$).

TABLE 22.4 (continued)

Protecting Group	Reagents for Introduction	Reagents for Removal; Comments
	2. Protection of Carboxyl Groups	
(a) Methyl, ethyl esters		Section 16.6
	Methanol, ethanol (acid catalysis)	Hydrolysis in acid or base
(b) Benzyl esters		Section 16.6
	Benzyl alcohol (acid catalysis) or benzyl chloride, base	Hydrogenolysis (Pd catalyst), HBr in nonaqueous solvent, or sodium, ammonia
(c) β-Bromo esters		cf. Section 22.4
$RCO_2H \xrightarrow[H^+]{HOCH_2CH_2Br} RCO_2CH_2CH_2Br$ $RCO_2CH_2CH_2Br \xrightarrow{Zn} RCO_2H$	β-Bromoethanol, acid (i.e., standard esterification methods)	Zinc, acetic acid (reductive elimination); trichloroethyl esters can also be used.

TABLE 22.4 (continued)

Protecting Group	Reagents for Introduction	Reagents for Removal; Comments
3. Protection of Amino Groups		
(a) Acetyl group		Section 16.6
$>N{-}H \xrightarrow[NEt_3]{Ac_2O \text{ or } AcCl} >N{-}Ac$ $>N{-}Ac \xrightarrow[^{-}OH \text{ or } H^+]{H_2O} >NH$	Acetic anhydride or acetyl chloride and a tertiary amine	Vigorous hydrolysis in acid or base
(b) Trifluoroacetyl group		Section 20.3
	Trifluoroacetic anhydride and a tertiary amine	Mild alkaline hydrolysis using dilute NaOH or aqueous ammonia
(c) *p*-Toluenesulfonyl group		Section 20.3
	p-Toluenesulfonyl chloride and a tertiary amine	Sodium in ammonia; esters, ketones, etc. are reduced under these conditions.
(d) Benzyloxycarbonyl group		Section 20.3
	Benzyl chloroformate and base (NaOH or $NaHCO_3$)	HBr in nonaqueous solvent, sodium in ammonia, or H_2/Pd
(e) *tert*-Butoxycarbonyl group (tBOC)		Section 20.3
	tert-Butyl azidoformate and a tertiary amine	HBr in nonaqueous solvent

TABLE 22.4 (continued)

Protecting Group	Reagents for Introduction	Reagents for Removal; Comments
	4. Protection of the Hydroxyl Group	
(a) Esters, e.g., acetates		Sections 11.8, 16.6
$—O—\overset{\overset{O}{\|\|}}{C}—R$	Acetic anhydride or acetyl chloride and a tertiary amine	Hydrolysis (potassium carbonate); methanolic ammonia
(b) Benzyl ether		Sections 8.6, 11.8, 15.4
	Sodium amide or sodium hydride, then benzyl bromide	Hydrogenolysis with Pd catalyst
(c) Tetrahydropyranyl ether		Section 15.5
	Dihydropyran, *p*-toluenesulfonic acid	Mild acid hydrolysis (or alcoholysis), e.g., HOAc, H_2O, THF
(d) Silyl ethers		
	$(CH_3)_3SiCl$ and a tertiary amine	$Bu_4N^+ F^-$ (Primary hydroxyl groups are also regenerated by mild acid hydrolysis; secondary and tertiary derivatives react more slowly)
(e) Protection of diols: acetonide		Section 15.5
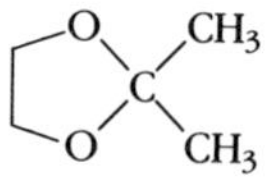 (cyclic O–C(CH$_3$)$_2$–O acetonide ring: O, C, CH$_3$, CH$_3$)	2,2-Dimethoxypropane and TsOH (ketal exchange, Section 22.4)	Dilute aqueous acid

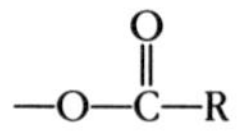